Handbook of
Plant
Nutrition

SECOND EDITION

Handbook of
Plant
Nutrition

SECOND EDITION

Edited by
Allen V. Barker
David J. Pilbeam

CRC Press
Taylor & Francis Group
Boca Raton London New York

CRC Press is an imprint of the
Taylor & Francis Group, an **informa** business

CRC Press
Taylor & Francis Group
6000 Broken Sound Parkway NW, Suite 300
Boca Raton, FL 33487-2742

First issued in hardback 2019
First issued in paperbak 2021

ISBN 13: 978-1-03-209863-0 (pbk)
ISBN 13: 978-1-4398-8197-2 (hbk)

Visit the Taylor & Francis Web site at
http://www.taylorandfrancis.com

and the CRC Press Web site at
http://www.crcpress.com

Contents

SECTION I Introduction

SECTION II Essential Elements: Macronutrients

SECTION III Essential Elements: Micronutrients

SECTION IV   Beneficial Elements

SECTION V   Conclusion

Preface

In 2007, we edited *Handbook of Plant Nutrition*, a compendium of knowledge at that time on the mineral nutrition of plants. This handbook was inspired by Homer D. Chapman's 1965 book, *Diagnostic Criteria for Plants and Soils*, and had contributions from eminent plant and soil scientists from around the world. Its purpose was to provide a current source of information on the nutritional requirements of world crops, and it covered the uptake and assimilation of elements essential or beneficial for plant growth, the availability of these elements to plants, fertilizers used to enhance the supply of the elements, and diagnostic testing of plants and soils to determine when they are in short supply. Each element considered was given its own chapter, with macronutrients being covered in more detail than micronutrients and beneficial elements.

Since the publication of the first edition, there have been advances in many aspects of plant nutrition, so a second edition is now timely. The second edition seeks to outline recent advances but also to put these advances into the context of our historical understanding of the importance of nutrient and beneficial elements to plants. This action means that inevitably some repetition of ideas that appeared in the first edition will occur, but repetition has been minimized in a variety of ways.

First, most of the chapters are written by different experts, or a group of experts, from those who contributed to the first edition. This change does not reflect on the outstanding contributions made by those earlier authors, some of whom have contributed chapters on different elements in this second edition, but merely gives a stimulus to cover each element in a different manner. Different workers inevitably have different interests in their research and expressions of their knowledge, and these interests and expressions are reflected in the way in which they have written about a particular element.

Second, the chapter sections have been modified to reflect current interests within plant nutrition. For example, the chapters in the second edition have more extensive coverage of the relationship between plant genetics and the accumulation and use of nutrients by plants, but most chapters contain less on the discovery of essentiality or beneficial action, as in most instances this information is covered well in the first edition. The first edition had extensive tables showing elemental composition of different plants and different plant parts. To avoid repetition, such tables are not nearly as extensive in the second edition and concentrate on the summarization of values reported from the period of time between the publications of the two editions. A final difference is that more knowledge about the importance of lanthanides in plant nutrition has become available since the first edition, so these elements are featured in a new chapter in the second edition.

Chapters on the different mineral elements follow the general pattern of a description of the determination of essentiality (or beneficial effects of the element), uptake and assimilation, physiological responses of plants to the element, genetics of its acquisition by plants, concentrations of the element and its derivatives and metabolites in plants, interaction of the element with uptake of other elements, diagnosis of concentrations of the element in plants, forms and concentrations of the element in soils and its availability to plants, and soil tests and fertilizers used to supply the element. These vary slightly depending on whether elements are assimilated, or function unchanged in plants; whether they are macronutrients, micronutrients, or beneficial elements; whether or not we use fertilizers to improve their supply; and whether we know about genetic differences in their plant requirements. There is a color insert of some of the images of plants showing deficiencies or toxicities of different nutrients that appear in black and white in the chapters, some images of subcellular structures, and also some of the diagrams. In addition, there is an introduction in which the editors discuss world population growth, trends in the use of inorganic fertilizers, developments in improving the efficiency of fertilizer use, the ionic composition of plants and its manipulation, and techniques used in plant nutrition research. The conclusion notes key points discussed in the chapters. This new handbook describes interactions between the elements in plants and highlights areas of rapid change in the study of plant nutrition and in the application of our knowledge.

The editors are grateful to the authors of the chapters in both editions for their detailed and informative coverage of topics in plant nutrition. With the world population, and the number of mouths to feed, increasing rapidly, with pollution from agricultural wastes still an environmental problem in many parts of the world, and with great interest in lowering the energy requirements of agriculture (and the greenhouse gas emissions associated with this), we hope that the second edition will provide a useful stimulus to people working to overcome these problems.

It is with sadness that we note the death of one of the contributors, Dr. Nand Kumar Fageria, during the final stages of preparation of this book. Dr. Fageria, a former senior soil scientist at Empresa Brasileira de Pesquisa Agropecuária (EMBRAPA) in Brazil, was a leading contributor to work on plant nutrition, with more than 200 publications to his name, including some well-known books on crop nutrition. His chapter on potassium in this book is a detailed and informative account of current ideas on the role of this element in plant nutrition, although in truth he could have also contributed outstanding chapters on many other elements required by plants. We also note the passing of Professor Volker Römheld, who jointly contributed the chapter on iron, and Professor Konrad Mengel, who contributed the chapter on potassium to the first edition. Professor Römheld was professor of plant nutrition at Hohenheim University, and published extensively on acquisition, uptake, and physiology of micronutrients in plants. Professor Mengel was professor of plant nutrition at the Justus Liebig University, Giessen, and not only was an expert on the potassium nutrition of plants but also had a broad interest across the whole of plant nutrition. This interest was put to good use in *Principles of Plant Nutrition* (K. Mengel and E.A. Kirkby), which ran to five editions and was a stimulating introduction to the subject for many students worldwide.

Allen V. Barker
Amherst, Massachusetts

David J. Pilbeam
Leeds, United Kingdom

Editors

Allen V. Barker is a professor at the University of Massachusetts, Amherst, where he has taught organic farming, soil fertility, and plant nutrition for 50 years. His research has addressed nitrogen nutrition of crops with emphasis on ammonium nutrition and on the interactions of nitrogen with other elements in affecting crop growth and nutrient accumulation. He is a member of editorial boards of several journals that publish articles on plant nutrition. He wrote *Science and Technology of Organic Farming*, which is also published by CRC Press, and edited the first edition of *Handbook of Plant Nutrition* with David Pilbeam.

David J. Pilbeam has over 30 years of experience in research and teaching on plant nutrition and physiology at the University of Leeds, United Kingdom. He has published particularly on the physiology of uptake and assimilation of inorganic nitrogen by plants, but also on the accumulation of other elements. Aside from research on the physiological aspects of plant nutrition, Dr. Pilbeam has published on more agronomic aspects of plant growth and nutrition, including work on intercropping, novel crops, and agroforestry. Together with Allen Barker, Dr. Pilbeam edited the first edition of *Handbook of Plant Nutrition* in 2007. He is currently a member of the editorial board of the *Journal of Plant Nutrition*.

Contributors

Allen V. Barker
Stockbridge School of Agriculture
University of Massachusetts
Amherst, Massachusetts

F. Pax C. Blamey
School of Agriculture and Food Sciences
The University of Queensland
Brisbane, Queensland, Australia

María De Los Ángeles Bustamante Muñoz
Group of Applied Research on Agrochemistry
    and Environment
Polytechnic High School of Orihuela
Universidad Miguel Hernández de Elche
Orihuela, Spain

Khaled Drihem
School of Biology
University of Leeds
Leeds, United Kingdom

Touria E. Eaton
Cooperative Extension and Research
Lincoln University of Missouri
Jefferson City, Missouri

Nand Kumar Fageria (Deceased)
National Rice and Bean Research Center
Brazilian Corporation of Agricultural Research
Santo Antônio de Goiás, Brazil

Sabine Goldberg
U.S. Salinity Laboratory
Riverside, California

Cynthia Grant
Agriculture and Agri-Food Canada
Brandon, Manitoba, Canada

Henry M.R. Greathead
School of Biology
University of Leeds
Leeds, United Kingdom

Witold Grzebisz
Department of Agricultural Chemistry
    and Environmental Biogeochemistry
Poznan University of Life Science
Poznań, Poland

Umesh C. Gupta
Agriculture and Agri-Food Canada
Charlottetown, Prince Edward Island, Canada

Russell L. Hamlin
Cottonwood Ag Management
Naperville, Illinois

Silvia H. Haneklaus
Institute for Crop and Soil Science
Federal Research Institute for Cultivated Plants
Braunschweig, Germany

Malcolm J. Hawkesford
Rothamsted Research
Harpenden, United Kingdom

Bryan G. Hopkins
College of Life Sciences
Brigham Young University
Provo, Utah

Zhengyi Hu
College of Resources and Environment
University of Chinese Academy of Sciences
Beijing, People's Republic of China

Peter M. Kopittke
School of Agriculture and Food Sciences
The University of Queensland
Brisbane, Queensland, Australia

David E. Kopsell
Department of Agriculture
Illinois State University
Normal, Illinois

Dean A. Kopsell
Department of Plant Sciences
The University of Tennessee
Knoxville, Tennessee

Bernd G. Lottermoser
Environment and Sustainability Institute
University of Exeter
Penryn, United Kingdom

Jian Feng Ma
Institute of Plant Science and Resources
Okayama University
Kurashiki, Japan

Neal W. Menzies
School of Agriculture and Food Sciences
The University of Queensland
Brisbane, Queensland, Australia

Raúl Moral Herrero
Group of Applied Research on Agrochemistry
    and Environment
Polytechnic High School of Orihuela
Universidad Miguel Hernández de Elche
Orihuela, Spain

Concepción Paredes Gil
Group of Applied Research on Agrochemistry
    and Environment
Polytechnic High School of Orihuela
Universidad Miguel Hernández de Elche
Orihuela, Spain

Aurelia Pérez-Espinosa
Group of Applied Research on Agrochemistry
    and Environment
Polytechnic High School of Orihuela
Universidad Miguel Hernández de Elche
Orihuela, Spain

María Dolores Pérez-Murcia
Group of Applied Research on Agrochemistry
    and Environment
Polytechnic High School of Orihuela
Universidad Miguel Hernández de Elche
Orihuela, Spain

David J. Pilbeam
School of Biology
University of Leeds
Leeds, United Kingdom

Ewald Schnug
Institute for Crop and Soil Science
Federal Research Institute for Cultivated Plants
Braunschweig, Germany

Sven Schubert
Institute of Plant Nutrition
Interdisciplinary Research Center for
    Environmental Research (IFZ)
Justus Liebig University
Giessen, Germany

Margie L. Stratton
Stockbridge School of Agriculture
University of Massachusetts
Amherst, Massachusetts

J. Bernhard Wehr
School of Agriculture and Food Sciences
The University of Queensland
Brisbane, Queensland, Australia

Philip J. White
The James Hutton Institute
Dundee, United Kingdom

Monika A. Wimmer
Institute of Crop Science and Resource
    Conservation—Plant Nutrition
University of Bonn
Bonn, Germany

Bruce W. Wood
Southeastern Fruit and Tree Nut Research
    Laboratory
Agricultural Research Service
United States Department of Agriculture
Byron, Georgia

Inmaculada Yruela
Aula Dei Experimental Station
Spanish National Research Council (CSIC)
Zaragoza, Spain

Section I

Introduction

Section 1

Introduction

1 Introduction

Allen V. Barker and David J. Pilbeam

CONTENTS

1.1 NEED FOR EFFICIENT CROP PRODUCTION

During the past 30 years, both the size of the world population and the production of crops to feed these people have increased considerably. Based on data of the Food and Agriculture Organization (FAO) of the United Nations, production of the common crop groups increased by 47% between 1985 and 2005, although if all of the 174 crops covered in the FAO reports are considered this increase is approximately 28% (Foley et al., 2011). In this time period, the area of cropland only increased by 2.4% (although taking into account multiple cropping, the decreasing proportion of land left fallow, and decreasing incidence of crop failures, the area of cropland that was harvested increased by 7%). This means that average crop yields per unit land area increased by 20% between 1985 and 2005 (Foley et al., 2011). These yield increases were brought about by advances in crop production techniques, including in the use and application of fertilizers.

Between 1987 and 2012, the world population increased from 5 billion to about 7 billion, and the rate of population growth is such that it is estimated that by 2030 the number of people requiring food will exceed 8 billion (UN Department of Economic and Social Affairs, Population Division, 2013). Very little wilderness can be converted to croplands, and, furthermore, wild land remaining often is poorly suited to agriculture, meaning that extending agriculture across the globe would have only a small impact on overall food production. Although agriculture is responsible for emissions of greenhouse gases (GHGs), through nitrous oxide (N_2O) released from the use of nitrogenous fertilizer, methane (CH_4) from livestock nutrition, CH_4 and N_2O from manure management, CH_4 from rice growing, CH_4 and N_2O from burning of residues, energy consumption in the manufacture of fertilizers and pesticides and their application, and energy consumption in other farm operations, conversion of wildlands would give a greater emission of GHGs than would agricultural intensification (Burney et al., 2010).

Consequently, the increasing population will have to be fed through improved productivity of current cropland. It has been suggested that over the next few decades production of food will have to double, yet at the same time emissions of GHGs from the farming sector will have to fall by at least 80%, the loss of biodiversity and habitats will have to be reduced, the amount of water used in crop production will have to be lowered, and pollution by agrochemicals will have to be reduced (Foley et al., 2011). These requirements certainly produce big challenges, and also big opportunities, in plant nutrition.

1.2 USE OF FERTILIZERS

As global cereal yields increased between 1960 and 2000, matching the increase in use of N and other fertilizers, the nitrogen efficiency of cereal production (the yield of cereals/amount of fertilizer N used) declined (Tilman et al., 2002). This result matches what would be expected from the law of diminishing returns, where supplying one extra unit of nutrient at a current low rate of supply gives a higher increment of extra crop yield than supplying one extra unit of nutrient onto an already high rate (Tilman et al., 2002). This shows us that although we ought to be able to increase the efficiency of fertilizer use in productive areas, there should be big opportunities to increase crop production in areas of the world where productivity is currently low, and increased use of fertilizers will have a big role to play here. World demand for total fertilizer nutrients has been calculated to grow at an annual rate of 1.9% between 2012 and 2016, with the biggest rates of increase being in sub-Saharan Africa, South Asia, Eastern Europe, and Central Asia (FAO, 2012).

Currently, there is a big discrepancy over how much fertilizer is used in different places, and areas where yields are low typically have lower use of agrochemicals in general. In a study of a corn-based system in western Kenya in 2004–2005, inputs of N and P averaged 7 and 8 kg ha^{-1} year^{-1}, respectively, and with total outputs of 59 and 7 kg ha^{-1} year^{-1} it can be seen that not only were there low rates of production of nitrogen- and phosphorus-containing metabolites in crops but that the nitrogen balance of the fields was negative (Vitousek et al., 2009). Data for 2009 indicate that consumption of nitrogen, phosphate, and potassium sources (excluding manures) was as low as 1.1 kg ha^{-1} of agricultural land in Angola and only 0.5 kg ha^{-1} in the Democratic Republic of Congo (World Bank, 2013), so increased use of fertilizers here should have a big impact on crop yields. Some crops have better prospects for having their yields increased by more application of fertilizers than others. Whereas global variation in yields of sorghum, millet, and groundnut seem to be explained largely by differences in climate, variability in yields of barley, sugar beet, and palm seem to relate to differences in crop management, in particular application of fertilizers and irrigation (Mueller et al., 2012). Increasing yields of corn in West Africa to 75% of the attainable yield could be achieved by removing current nutrient limitation, whereas increasing corn yields in sub-Saharan Africa to 75% of the achievable yields would require both removing nutrient deficiencies and applying irrigation (although increasing these yields to only 50% of the achievable yields mostly would require only increased application of nutrients) (Mueller et al., 2012).

Indeed, there are some areas of the world where fertilizer application rates are higher than required for maximum crop yields. One area where fertilizer overuse occurs is China (Vitousek et al., 2009; Foley et al., 2011; Mueller et al., 2012), because of the intensive nature of Chinese agriculture and government subsidies for fertilizers (Yang and Zhang, 2006). In the cereal production area of the North China Plain, annual nitrogen fertilizer applications of over 500 kg N ha^{-1} are used, and water leaching at a depth of 1.4 m in the soil profile was found to contain 12–39 mg nitrate-N L^{-1} after maize harvest (Ju et al., 2006). Similarly, in a survey of 916 orchards in northern China the average annual rates of input of nitrogen and phosphorus were 588 and 157 kg ha^{-1}, respectively, two to three times higher than fruit demand (Lu et al., 2012). However, despite the risk of groundwater pollution, and the waste of valuable nitrogen that is not being taken up by the crop plants, it should be borne in mind that the North China Plain is the area of highest agricultural productivity in China, with cereal production commonly being based on growing wheat and corn crops on the same parcel of land in a rotation that can be completed in one calendar year (Ju et al., 2006; He et al., 2010). The high use of fertilizers on both crops gives rise to waste of valuable resources, although the potential leaching of nitrate from this cereal rotation is no higher than the increase in nitrate leaching of 36 kg N ha^{-1} year^{-1} that was estimated to have occurred in the United Kingdom between the early 1940s and the 1990s as annual fertilizer nitrogen use increased from 20 to 190 kg ha^{-1} and wheat yields increased from 2.35 to 5.92 t ha^{-1} (Davies and Sylvester-Bradley, 1995). Although some of the nitrogen and phosphorus applied to crops may be wasted, China has managed to feed a population that has grown by 9 million per year between 1991 and 2010 (UN Department of Economic and Social Affairs, Population Division, 2013).

In other areas, the use of artificial fertilizers is comparatively efficient. In the United Kingdom, the annual use of fertilizer nitrogen on winter wheat between 2007 and 2011 was 188 kg ha^{-1}, approximately the same as the rate applied in the 1990s, yet wheat yields over this period averaged 7.76 t ha^{-1} (National Statistics, 2011). More than 150 years have elapsed since Liebig published his classic work on plant nutrition (van der Ploeg et al., 1999), and in many countries numerous studies since then on the relationship between crop yields and nutrient supply have enabled agronomists to accurately predict the optimum rate of nutrient supply for different crops in different fields and under different climatic conditions. It is not surprising that only small increases in demand for total fertilizer nutrients are projected for North America and Western Europe between 2012 and 2016 (FAO, 2012).

Over recent years, there has been a trend to use less fertilizer on some farms in North America and Western Europe through adopting extensive agricultural systems as a means of growing crops in a more sustainable and less polluting manner. Saving the energy costs, and preventing pollution at the same time, should give more sustainability to agriculture and horticulture. However, this can have consequences of lowered productivity, and with a world population of 7 billion rising by 80 million per year, that strategy could be risky. Also, although there are decreased GHG emissions from the lower use of fertilizers, life cycle analysis has shown that the efficient use of fertilizers to stimulate crop growth can give higher yields of crops per unit of GHG emission resulting from the manufacture and use of fertilizers and the crop production procedures than in less intensively grown crops (Brentrup and Pallière, 2008; Brentrup and Lammel, 2011). The implication here is that in terms of feeding the increasing population and lowering GHG emissions from the agricultural sector, more intensification of agriculture will be required. However, there will still be a need to lower GHG emissions and pollution arising from agriculture even further.

Another risk associated with increased use of fertilizers is that it may become increasingly difficult to maintain their supplies. Although nitrogen fertilizers may become increasingly expensive as world energy supplies become more expensive, ultimately they can continue to be synthesized indefinitely. The Haber–Bosch process for the synthesis of ammonia from atmospheric N_2 is carried out under high temperature and pressure and so has a very high energy requirement. The recent identification of an iron complex that efficiently catalyzes the conversion under the milder conditions of −78°C at 1 atm of N_2 by scientists at the California Institute of Technology in Pasadena offers possibilities of lowering energy use in the production of N fertilizers (Anderson et al., 2013; editorial, *Nature*, September 5, 2013). This should ensure a reliable supply of nitrogen fertilizers in years to come. However, production of phosphorus and potassium fertilizers relies on mined ores, and in recent years it has been suggested that phosphate supplies may be running out.

The total world reserves of phosphates are currently estimated to be in excess of 300 billion tons (U.S. Geological Survey, 2013), and with worldwide consumption of P_2O_5 in fertilizers expected to be 45.3 million tons in 2016 (U.S. Geological Survey, 2013) and these reserves are sufficient for many years of use in agriculture. Nevertheless, as reserves are utilized the reserves that remain tend to be more difficult to extract, and possibly less pure, making phosphorus fertilizers more expensive. Furthermore, many of the reserves are in parts of the world where political instability could put supplies at risk. A further hazard of using phosphorus fertilizers is that rock phosphate contains the heavy metal cadmium at concentrations between 1 and 200 mg Cd (kg P_2O_5)$^{-1}$, and this cadmium ends up in the commercial fertilizers (Smolders, 2013). Cd^{2+} is sufficiently similar to Zn^{2+} to be taken up by plants and, if ingested in crop products, has the potential to be hazardous to humans. Durum wheat grain (*Triticum durum* Desf., syn. *Triticum turgidum* L. ssp. *durum*), for example, can accumulate cadmium supplied in phosphorus fertilizers (Grant et al., 2013). In European fertilizers, cadmium can be present at up to 120 mg Cd (kg P_2O_5)$^{-1}$, although it is usually at a much lower level, so there have been proposals to limit the amount of cadmium allowed in these fertilizers (Smolders, 2013).

Global reserves of potassium are currently estimated to be 250 billion tons, with mine production of the element running at 34 million tons in 2012 and consumption expected to increase annually by 3% until 2016 (U.S. Geological Survey, 2013). Forty new potash mines are expected to be opened worldwide by 2017 (U.S. Geological Survey, 2013), so potassium supplies appear to be at low risk of diminishing.

One strategy to cope with environmental pollution from the use of fertilizers, to minimize emissions of GHGs arising from their manufacture, and to cope with decreasing availability of essential precursors is to use fertilizers more efficiently, and the wheat–corn system of the North China Plain could have its yield sustained with as little as only half of the current N fertilizer supply (Vitousek et al., 2009). Fertilizer use can be lowered through physical methods such as better forecasting of when (and where) fertilizer application would give worthwhile yield responses, better application methods, and better formulation of fertilizers to release the nutrients at the optimum time for plant growth and biological methods, such as the breeding of plants with a better nutrient-use efficiency.

Improvements in nutrient-use efficiency through better forecasting of when and where fertilizers are required can be achieved in those areas where currently fertilizer recommendations are not well developed. These areas include the productive agricultural systems of China, where work is in progress to formalize recommendations for different crops. For example, Chuan et al. (2013) used data on nutrient responses of wheat in the field between 2000 and 2011 to evaluate a new *Nutrient Expert for wheat* fertilizer decision support system that is site specific. Such systems should enable high yields to be maintained with lower inputs of fertilizers, thereby lowering some pollution problems and lessening the GHG emissions from unnecessary synthesis and application of mineral fertilizers. If there were more energy-efficient production of nitrogen fertilizers, and these were used more efficiently in the field, China could save emissions of between 102 and 357 Tg CO_2 equivalents per year, approximately equal to the emissions reduction targets for 2020 from the entire economies of some of the largest European nations (Zhang et al., 2013).

If efficiencies in fertilizer use can be gained by developing recommendation systems in productive agriculture where traditionally such systems have not existed, the opposite problem is seen in the productive agriculture of Europe. The continent currently has at least 10 soil-P tests currently in use, giving more than threefold differences in fertilizer-P recommendations for similar soil–crop situations (Jordan-Meille et al., 2012). The efficiency of P fertilizer use could be improved on a continent-wide basis through the development of a better model of soil-P availability (Jordan-Meille et al., 2012). The third fertilizer use scenario, subsistence agriculture with low levels of inputs and low yields, would benefit from even small increases in the use of fertilizers. The law of diminishing returns shows that in subsistence agriculture supply of even a small additional increment of fertilizer would give a substantial increase in crop yield.

1.3 IMPROVING CROP QUALITY THROUGH PLANT NUTRITION

As well as trying to produce heavier yields of crops, plant nutritionists are also interested in the production of crops of higher quality. Considerable interest exists currently in the production of crops that are rich in micronutrients that are in low supply in the modern human diet. For example, selenium, zinc (and possibly iron), and other micronutrients seem to be in low supply in the modern Western diet, and there is interest in using plant nutrition to increase concentrations of these elements in crop products. Worldwide, the most prevalent nutrient deficiencies in the human diet are due to shortage of iron, zinc, and iodine, with calcium and selenium also being important and magnesium and copper deficiencies giving problems in some areas and multiple micronutrient deficiencies also occurring (Ramakrishnan, 2002; White and Broadley, 2009; Stein, 2010).

It is possible to give supplements to populations under threat of these deficiencies, but increasing the concentrations of deficient elements in the plant and animal foodstuffs that they consume is a better strategy. One way this can be done is by improving the diet of people and livestock, to increase the consumption of species naturally high in elements otherwise deficient in the diet.

For example, wheat grains are typically low in selenium, whereas *Brassica* species tend to concentrate this element (Watson et al., 2012), so more consumption of brassicas in the Western diet would be of benefit.

The fact that concentrations of different elements vary between plant species has been known about for many years, but more recently research on the extent to which the elemental composition of a species is under genetic control has increased. The mineral elemental composition of plants and other organisms now has its own name, the *ionome* (Lahner et al., 2003; Salt, 2004; Salt et al., 2008). In an analysis of 21 species from 7 plant families grown in 6 plots receiving different fertilizer treatments since 1856, it was possible to distinguish between plants from different families based on their shoot mineral element concentration, irrespective of what fertilizer treatment had been applied to the individuals (White et al., 2012). The elements analyzed, in order of their importance in distinguishing between plant families, were Ca > Mg > Ni > S > Na > Zn > K > Cu > Fe > Mn > P.

This knowledge that there are genetic differences between plants in their accumulation of mineral elements has given a scientific basis to breed increased micronutrient concentration into crop species, thereby generating genetic biofortification of crop species (White and Brown, 2010). For example, Chatzav et al. (2010) investigated the concentrations of micronutrients in grains of ancestral wheat plants and found they contain up to two times the concentrations of zinc and iron of modern bread wheat. This finding suggests the possibility of breeding new wheat cultivars with higher contents of these essential elements. The Consultative Group on International Agricultural Research (CGIAR) runs the HarvestPlus program to breed and distribute developing world staple crop species rich in iron and zinc, along with a high provitamin A content (Welch and Graham, 2005; Nestel et al., 2006; Bouis and Welch, 2010). Breeding crop species with increased concentrations of essential metabolites, including trace elements, is a key policy for improving human health in many countries.

Another plant-breeding approach may be to breed antinutritional compounds out of crop plants. Phytic acid, which is not digestible by humans or monogastric farm animals, chelates calcium, zinc, and iron and is the major store of phosphate in plants. Attempts are being made to breed crops with low concentrations of phytic acid in their grains, an action that will give increased availability of essential nutrients in the diet of humans and livestock and will lower the excretion of phosphate by the farm animals (Raboy, 2009).

Although different plant species may accumulate different nutrients to different extents, there are also effects of environment on their uptake and relative proportions in plants. In a study of one natural clone and five hybrid willow (*Salix* sp.) varieties, concentrations of N, P, S, Mn, and Cu were significantly higher in a treatment with irrigation and multinutrient fertilizer than in control plants, whereas concentrations of Fe, B, Zn, and Al were significantly lower (Ågren and Weih, 2012). K, Ca, Mg, and Na did not show any differences. Despite the current attention to the ionome, in this experiment, environmental conditions (water and nutrient supply) had a bigger effect on elemental composition of the plants than genotype. There are many examples of the influence of environmental conditions on the uptake of individual elements in the following chapters in this book.

As nutrient accumulation in plants is influenced by the environment, it is obvious that low concentrations of essential elements in crop plants can also be tackled by changes in agronomic practices, in particular by changes to plant nutrition that increase the availability of scarce elements. This action has been termed agronomic biofortification (White and Brown, 2010). This practice can involve more growth of species mixtures, where the presence of one species helps make a particular micronutrient more available to a second species or provides a different range of micronutrient concentrations to grazing herbivores, more use of crop residues, farmyard manures, and composts to increase micronutrient levels in the soil, and more analysis of farmgate micronutrient balances so that particular deficiencies can be identified and rectified by the supply of micronutrient fertilizers (Watson et al., 2012).

Work on evaluating micronutrient concentrations in crop plants and then increasing amounts of essential micronutrients in the crop products by supply of micronutrient fertilizers has become increasingly important. For example, many soils of Scandinavia have low bioavailability of selenium, so crops grown there tend to be low in this element, as are animal products arising from these crops. From 1985, the government of Finland arranged for amendment of NPK fertilizers with sodium selenate to help correct a selenium deficiency in the Finnish diet, and daily Se intake of the population increased considerably (Hartikainen, 2005). Another example is the study of iodine concentrations; iodine is not an essential or beneficial element for plants, but it is essential for humans, so studies have investigated the uptake and accumulation of iodine by plants in order to improve human nutrition (e.g., Smoleń and Sady, 2012).

Another reason for low availability of micronutrients is because soil stocks have become depleted as yields of crops removed from the land have increased following the increased use of NPK fertilizers. This depletion of micronutrients in the soil is often referred to as nutrient mining, and it poses a threat not only to the nutritional quality of the crop products harvested but also to the overall yields of crops (Moran, 2011). Nutrient mining is likely to be more of a problem in old weathered soils (e.g., in Australia and Southeast Asia) than in soils formed more recently (e.g., those formed from volcanic activity or glaciation in Northern Europe and North America) (Jones et al., 2013). Indeed, in a survey of 132 soils in the United Kingdom, there appeared to be no biologically significant decreases in Zn, Cu, and Mn since the National Soil Inventory was drawn up 30 years ago (McGrath et al., 2013). In contrast, although cereal yields have increased markedly in India since 1960, many soils there are now deficient in potassium or sulfur, and also at least one of boron, iron, copper, or manganese (Jones et al., 2013). Where soil concentrations of micronutrients are below a critical threshold, the threat to crop production arises because, as described in Liebig's law of the minimum, a shortage of one nutrient limits the growth of crops even if the other nutrients are available at sufficiency levels (Moran, 2011).

The Law of the Minimum, as originally postulated by Sprengel and promoted by Liebig, indicates that plant growth should be limited by whichever nutrient resource is in most limiting supply for plant growth. However, more recently, the multiple limitation hypothesis (Bloom et al., 1985; Gleeson and Tilman, 1992) has postulated that plant growth is constrained by many limitations simultaneously, so that plants trade off one resource to acquire another resource that is in limited supply. For example, a photosynthesizing plant that grows in nutrient-poor soil is able to use some of its assimilated carbon to support mycorrhizal fungi that make phosphorus available to the plant.

Indeed, Ågren and Weih (2012) pointed out that although there tends to be a linear relationship between internal concentrations of nitrogen or phosphorus (and also sulfur and manganese in their experiments) and relative growth rate (RGR) across the range of concentrations of these elements that plants accumulate, for other essential and beneficial elements plants frequently accumulate amounts above the concentration at which there ceases to be any relationship with RGR. This does not fit the law of the minimum. Further modeling of the relationship between supply of nitrogen and phosphorus to plants and their rate of growth indicates that if both of the elements are limiting the supply of an additional increment of only one of them will increase RGR; resources are presumably used to make the other element more available, and the multiple limitation hypothesis seems to hold (Ågren et al., 2012). According to the model, this situation seems to occur only at concentrations of nitrogen or phosphorus close to the optimum for plant growth, and at other concentrations the Law of the Minimum is more realistic.

Even if some elements can accumulate in plants without stimulating additional plant growth, they can still be beneficial for growth processes to the extent that growth would be limited if they are not available at all. We have known about the essentiality of the most well-known nutrients for some time, and we have a reasonably good idea about why they are essential. This volume delineates the current state of our knowledge in this respect. However, as analytical techniques become more sensitive, and it becomes possible to increase the purity of water for growing control plants in experiments, more elements are found to be essential. In the first edition we included nickel, which had been found to be essential for plant development not long before publication. Various elements had recently been classified as being beneficial, and this relatively new category of plant nutrient (aluminum, cobalt, selenium, silicon, sodium, and vanadium) was also included. Although these elements are generally regarded as having physiological functions (Pilon-Smits et al., 2009), some authors doubt whether beneficial elements really do give clear benefits for plant growth. Such doubt is clearly discussed for aluminum in Chapter 16. Since the first edition, there has been convincing research showing the beneficial nature of lanthanides (rare earth elements), and these are covered in this second edition.

1.4 METHODS OF RESEARCH IN PLANT NUTRITION

Research in plant nutrition takes advantage of new techniques to move in directions that were not possible in earlier times and to check the accuracy of earlier findings. Detailed discussion of techniques currently used to investigate concentrations of different ions in plants is given by Conn and Gilliham (2010). Many of the techniques used are also listed in the chapters of this book.

Recent methods for following the uptake of ions by plants involve the use of planar optode technology. This is a noninvasive technology that uses reversible changes in fluorescence of fluorophores specific for particular analytes, so by capturing an image of part of a root, a whole root, or a root system and then from measuring change in fluorescence density or fluorescence decay time, a quantitative picture of changes in the amount of an analyte round the root can be built up (Blossfeld, 2013). The technology so far mostly has been used for looking at changes in concentration of rhizospheric oxygen, ammonium, and pH and can give confirmation or rejection of previous studies using microelectrodes (Blossfeld, 2013). This procedure is a quantitative equivalent of the dye techniques used to visualize the uptake of different nutrients and pH changes round roots of plants growing in gel in the laboratory of Horst Marschner in Hohenheim in the 1980s. Another visualization technique, used to measure cytosolic concentrations of micronutrient ions, is that of Förster resonance energy transfer (FRET). A zinc-binding domain flanked by two fluorescent proteins that overlap spectrally was genetically expressed in the root cells of arabidopsis (*Arabidopsis thaliana* Heynh.), and where there was no zinc present the proteins interacted by FRET to give fluorescence, whereas binding to zinc altered their position relative to each other so that the fluorescence was reduced in proportion to zinc concentration (Lanquar et al., 2014).

Earlier studies looked at the uptake of isotopically labeled forms of plant nutrients, and work has inevitably followed to make images of such uptake in real time. Beta (β) particles have high energy and travel some distance through biological samples, so it is possible to monitor the movement of β-emitters such as ^{35}S, ^{45}Ca, ^{55}Fe, ^{32}P, and ^{33}P in whole plant/soil systems in a quantitative manner (Kanno et al., 2012). Other workers have used stable isotopes. For example, Metzner et al. (2011) fed $^{26}MgCl_2$, ^{41}KCl, and $^{44}CaCl_2$ to cut stems of common bean (*Phaseolus vulgaris* L.) and then followed the movement of the stable Mg^{2+}, K^+, and Ca^{2+} ions from the xylem vessels and into the surrounding cells by secondary ion mass spectrometry (SIMS) after freezing the plant material and checking the integrity of the frozen samples by scanning electron microscopy.

A less direct method of observing the uptake of an element is to accurately visualize the microenvironment of the rhizosphere and then to estimate the uptake of an element based on the overall soil concentration of that element and known parameters of its uptake into isolated roots. Accurate pictures of the relationship between root hairs and soil particles could theoretically be made by x-ray microscopy. However, unless x-rays with sufficient energy can be generated, the pictures taken of the changing concentration must have a long exposure and a resolution that is lower than allows for growing root hairs to be seen clearly. This need has led to the use of synchrotrons. A synchrotron source produces radiation that is more intense (it has a greater brilliance) than conventional x-ray tubes. The rays tend to be more parallel (it has high collimation). It has energy ranges from the infrared through high energy x-rays. It is highly polarized and is pulsed (enabling it to be used in studies carried out on a time-resolved basis) (Lombi and Susini, 2009). A synchrotron can generate very high energy x-rays, allowing images to be captured with as short an exposure time of 5 min and giving a resolution of about 1 µm, which is suitable for investigating uptake into growing root hairs.

Keyes et al. (2013) used synchrotron-produced x-rays and a tomographic microscope to visualize the arrangement of soil particles in one plane around root hairs of wheat growing in soil. From this, they were able to model the flux of P across the soil particles and its uptake by the root hairs, based on the P adsorption and desorption characteristics of the soil and uptake kinetic parameters from the literature. Uses of a range of synchrotron techniques in the study of soils, plants, and their interaction in the rhizosphere are discussed by Lombi and Susini (2009).

Observation of uptake of amino acids has been carried out by the use of quantum dots. Different nanoparticle chromophores can be attached to amino or carboxyl groups of organic compounds, so if quantum dots are bound to amino acids and these are taken up, they can then be detected inside the plant. Whiteside et al. (2009) fed arbuscular mycorrhizal annual bluegrass (or annual meadow grass, *Poa annua* L.) grown in sand–vermiculite, and also in a minirhizotron, with glycine labeled with quantum dots on its amino groups, and saw uptake into the fungi and plants. This technique obviously is specialized for investigating the uptake of organic nitrogen into plants, and it is not without problems as Al-Salim et al. (2011) did not see the uptake of quantum-dot-labeled glycine into ryegrass (*Lolium perenne* L.), onion (*Allium cepa* L.), or arabidopsis grown in solution. None of these plants had formed mycorrhizal associations, but arabidopsis never forms mycorrhizal associations yet amino acid uptake has been seen in this species (see Chapter 2).

Not only has it been difficult up to now to quantify the uptake of different nutrients into different parts of root systems growing in solid media, it has sometimes proved problematic to merely visualize morphological changes in root systems in response to nutrient supply. Studies of root responses to deficiencies of specific nutrients have often been carried out on plants grown in agar gel, yet the gelling agent used can contribute nutrients to the plants that make observations on root responses to deficiency of dubious reliability. Recent research has identified suitable agar and agarose sources for use as solid media to study the effects of deficiencies of nine different elements on root characteristics in arabidopsis (Gruber et al., 2013).

One extremely valuable resource in plant nutrition research is the availability of long-term field sites. New techniques can be developed at regular intervals, but the ability to be able to sample plants from plots that have particular nutrient applications over long periods of time is priceless regardless of the techniques used for analysis (Rasmussen et al., 1998; Peterson et al., 2012). Examples include the Park Grass and Broadbalk wheat experiments at Rothamsted, United Kingdom (e.g., Jenkinson et al., 2008; White et al., 2012); the Morrow Plots in Illinois, United States (e.g., Nafziger and Dunker, 2011); and the Rengen Grassland Experiment in Germany (e.g., Hejcman et al., 2010). Long-term experiments are always difficult to finance, as even if they are established by governments or private donors there has to be an ongoing commitment to financial support on an annual basis. The ideal way to manage these experiments is to include storage facilities for plant and soil samples that are taken from the experiment, so that future generations can go back and reanalyze samples in the light of advances in analytical techniques. At the very least, efforts should be made to preserve the actual plots of long-term experiments and to find means of financing their ongoing maintenance, as these are resources that cannot be replaced in a short time once they are lost.

Research in plant nutrition also changes as developments in other disciplines influence the interests of researchers. One of the major areas of research in plant nutrition currently is the genetics of nutrient acquisition and assimilation. The principle of genetic biofortification has been discussed earlier, but as well as increasing concentrations of essential micronutrients in crop products it may be possible to increase the responses of plants to nutrient supply in terms of their overall yields. This possibility has led to considerable interest in understanding the genetic control of expression of genes coding for different transporters and different assimilatory enzymes in plants. As well as following changes in expression of genes for different transporter proteins as a plant is exposed to varying concentrations of a particular nutrient, another approach is to identify quantitative trait loci (QTLs) for broader characteristics, such as nutrient-use efficiency, and then, through searching databases for similar DNA sequences, to identify what the genes in these loci are (e.g., Chardon et al., 2012).

These genes can then be overexpressed or deleted in the target species to further investigate their importance in plant nutrition. These modifications lead to the possibility of breeding similar changes to the gene into crop species in order to improve their agronomic performance. However, many of the traits that could be improved by breeding have multigene control, which makes the problem challenging for the breeders.

Many of the characteristics required to improve the conversion of nutrients into crop yield may not be obvious, so care has to be taken in screening plants for suitable attributes and then breeding them into crops. Research programs on the efficiency of use of individual nutrients by different genotypes of plants, particularly under conditions of low nutrient supply, are currently common and should help refine our views of what genes in plants need to be manipulated so that we can achieve bigger crop yields with lower rates of application of fertilizers.

REFERENCES

Ågren, G.I. and M. Weih. 2012. Plant stoichiometry at different scales: Element concentration patterns reflect environment more than genotype. *New Phytol.* 194:944–952.

Ågren, G.I., J.Å.M. Wetterstedt, and M.F.K. Billberger. 2012. Nutrient limitation on terrestrial plant growth—Modeling the interaction between nitrogen and phosphorus. *New Phytol.* 194:953–960.

Al-Salim, N., E. Barraclough, E. Burgess et al. 2011. Quantum dot transport in soil, plants, and insects. *Sci. Total Environ.* 409:3237–3248.

Anderson, J.S., J. Rittle, and J.C. Peters. 2013. Catalytic conversion of nitrogen to ammonia by an iron model complex. *Nature* 501:84–87.

Bloom, A.J., F.S. Chapin III, and H.A. Mooney. 1985. Resource limitation in plants—An economic analogy. *Annu. Rev. Ecol. Sys.* 16:363–392.

Blossfeld, S. 2013. Light for the dark side of plant life: Planar optodes visualizing rhizosphere processes. *Plant Soil* 369:29–32.

Bouis, H.E. and R.M. Welch. 2010. Biofortification—A sustainable agricultural strategy for reducing micronutrient malnutrition in the global south. *Crop Sci.* 50:S-20–S-32.

Brentrup, F. and J. Lammel. 2011. LCA to assess the environmental impact of different fertilisers and agricultural systems. *Proceedings of the International Fertiliser Society 687*, Colchester, U.K.

Brentrup, F. and C. Pallière. 2008. GHG emissions and energy efficiency in European nitrogen fertiliser production and use. *Proceedings of the International Fertiliser Society 639*, Colchester, U.K.

Burney, J.A., S.J. Davis, and D.B. Lobell. 2010. Greenhouse gas mitigation by agricultural intensification. *Proc. Natl. Acad. Sci. USA* 197:12052–12057.

Chardon, F., V. Noël, and C. Masclaux-Daubresse. 2012. Exploring NUE in crops and in *Arabidopsis* ideotypes to improve yield and seed quality. *J. Exp. Bot.* 63:3401–3412.

Chatzav, M., Z. Peleg, L. Ozturk et al. 2010. Genetic diversity for grain nutrients in wild emmer wheat: Potential for wheat improvement. *Ann. Bot.* 105:1211–1220.

Chuan, L., P. He, M.F. Pampolino et al. 2013. Establishing a scientific basis for fertilizer recommendations for wheat in China: Yield response and agronomic efficiency. *Field Crop. Res.* 140:1–8.

Conn, S. and M. Gilliham. 2010. Comparative physiology of elemental distributions in plants. *Ann. Bot.* 105:1081–1102.

Davies, D.B. and R. Sylvester-Bradley. 1995. The contribution of fertiliser nitrogen to leachable nitrogen in the UK: A review. *J. Sci. Food Agric.* 68:399–406.

FAO. 2012. Current world fertilizer trends and outlook to 2016. Food and Agriculture Organization of the United Nations, Rome, Italy.

Foley, J.A., N. Ramankutty, K.A. Brauman et al. 2011. Solutions for a cultivated planet. *Nature* 478:337–342.

Gleeson, S.K. and D. Tilman. 1992. Plant allocation and the nutrient limitation hypothesis. *Am. Nat.* 139:1322–1343.

Grant, C., D. Flaten, M. Tenuta, S. Malhi, and W. Akinremi. 2013. The effect of rate and Cd concentration of repeated phosphate fertilizer applications on seed Cd concentration varies with crop type and environment. *Plant Soil* 372:221–233.

Gruber, B.D., R.F.H. Giehl, S. Friedel, and N. von Wirén. 2013. Plasticity of the *Arabidopsis* root system under nutrient deficiencies. *Plant Physiol.* 163:161–179.

Hartikainen, H. 2005. Biogeochemistry of selenium and its impact on food chain quality and human health. *J. Trace Elem. Med. Biol.* 18:309–318.

He, C.-E., X. Wang, X. Liu, A. Fangmeier, P. Christie, and F. Zhang. 2010. Nitrogen deposition and its contribution to nutrient inputs to intensively managed agricultural ecosystems. *Ecol. Appl.* 20:80–90.

Hejcman, M., J. Szaková, J. Schellberg, and P. Tlustoš. 2010. The Rengen Grassland Experiment: Relationship between soil and biomass chemical properties, amount of elements applied, and their uptake. *Plant Soil* 333:163–179.

Jenkinson, D.S., P.R. Poulton, and C. Bryant. 2008. The turnover of organic carbon in subsoils. Part 1. Natural and bomb radiocarbon in soil profiles from the Rothamsted long-term field experiments. *Eur. J. Soil Sci.* 59:391–399.

Jones, D.L., P. Cross, P.J.A. Withers et al. 2013. Nutrient stripping: The global disparity between food security and soil nutrient stocks. *J. Appl. Ecol.* 50:851–862.

Jordan-Meille, L., G.H. Rubæk, P.A.I. Ehlert et al. 2012. An overview of fertilizer-P recommendations in Europe: Soil testing, calibration and fertilizer recommendations. *Soil Use Manag.* 28:419–435.

Ju, X.T., C.L. Kou, F.S. Zhang, and P. Christie. 2006. Nitrogen balance and groundwater nitrate contamination: Comparison among three intensive cropping systems on the North China Plain. *Environ. Pollut.* 143:117–125.

Kanno, S., M. Yamawaki, H. Ishibashi et al. 2012. Development of real-time radioactive isotope systems for plant nutrient uptake studies. *Phil. Trans. R. Soc. B* 367:1501–1508.

Keyes, S.D., K.R. Daly, N.J. Gostling et al. 2013. High resolution synchrotron imaging of wheat root hairs growing in soil and image based modelling of phosphate uptake. *New Phytol.* 198:1023–1029.

Lahner, B., J. Gong, M. Mahmoudian et al. 2003. Genomic scale profiling of nutrient and trace elements in *Arabidopsis thaliana*. *Nat. Biotechnol.* 21:1215–1221.

Lanquar, V., G. Grossmann, J.L. Vinkenborg, M. Merkx, S. Thomine, and W.B. Frommer. 2014. Dynamic imaging of cytosolic zinc in *Arabidopsis* roots combing FRET sensors and RootChip technology. *New Phytol.* 202:198–208.

Lombi, E. and J. Susini. 2009. Synchrotron-based techniques for plant and soil science: Opportunities, challenges and future perspectives. *Plant Soil* 320:1–35.

Lu, S., Z. Yan, Q. Chen, and F. Zhang. 2012. Evaluation of conventional nitrogen and phosphorus fertilization and potential environmental risk in intensive orchards of North China. *J. Plant Nutr.* 35:1509–1525.

McGrath, S.P., R. Stobart, M.M. Blake-Kalff, and F.J. Zhao. 2013. Current status of soils and responsiveness of wheat to micronutrient applications. Project Report 518, Home Grown Cereals Authority, Kenilworth, U.K.

Metzner, R., H.U. Schneider, U. Breuer, M.R. Thorpe, U. Schurr, and W.H. Schroeder. 2011. Tracing cationic nutrients from xylem into stem tissue of French bean by stable isotope tracers and cryo-secondary ion mass spectrometry. *Plant Physiol.* 152:1030–1043.

Moran, K. 2011. Role of micronutrients in maximising yields and in bio-fortification of food crops. *Proceedings International Fertiliser Society* 702, Colchester, U.K.

Mueller, N.D., J.S. Gerber, M. Johnston, D.K. Ray, N. Ramankutty, and J.A. Foley. 2012. Closing yield gaps through nutrient and water management. *Nature* 490:254–257.

Nafziger, E.D. and R.E. Dunker. 2011. Soil organic carbon trends over 100 years in the Morrow plots. *Agron. J.* 103:261–267.

National Statistics. 2011. Department of Food, Rural Affairs and Agriculture, U.K. http://www.defra.gov.uk/statistics, accessed January 23, 2013.

Nestel, P., H.E. Bouis, J.V. Meenakshi, and W. Pfeiffer. 2006. Biofortification of staple food crops. *J. Nutr.* 136:1064–1067.

Peterson, G.A., D.J. Lyon, and C.R. Fenster. 2012. Valuing long-term field experiments: Quantifying the scientific contribution of a long-term tillage experiment. *Soil Sci. Soc. Am. J.* 76:757–765.

Pilon-Smits, E.A.H., C.F. Quinn, W. Tapken, M. Malagoli, and M. Schiavon. 2009. Physiological functions of beneficial elements. *Curr. Opin. Plant Biol.* 12:267–274.

Raboy, V. 2009. Approaches and challenges to engineering seed phytate and total phosphorus. *Plant Sci.* 177:281–296.

Ramakrishnan, U. 2002. Prevalence of micronutrient malnutrition worldwide. *Nutr. Rev.* 60(5):S46–S52.

Rasmussen, P.E., K.W.T. Goulding, J.R. Brown, P.R. Grace, H.H. Janzen, and M. Körschens. 1998. Long-term agroecosystem experiments: Assessing agricultural sustainability and global change. *Science* 282:893–896.

Salt, D.E. 2004. Update on plant ionomics. *Plant Physiol.* 136:2451–2456.

Salt, D.E., I. Baxter, and B. Lahner. 2008. Ionomics and the study of the plant ionome. *Annu. Rev. Plant Biol.* 59:709–733.

Smolders, E. 2013. Revisiting and updating the effect of phosphorus fertilisers on cadmium accumulation in European agricultural soils. *Proceedings International Fertiliser Society* 724, Colchester, U.K.

Smoleń, S. and W. Sady. 2012. Influence of iodine form and application method on the effectiveness of iodine biofortification, nitrogen metabolism as well as the content of mineral nutrients and heavy metals in spinach plants (*Spinacia oleracea* L.). *Sci. Hortic.* 143:176–183.

Stein, A.J. 2010. Global impacts of human mineral malnutrition. *Plant Soil* 335:133–154.

Tilman, D., K.G. Cassman, P.A. Matson, R. Naylor, and S. Polasky. 2002. Agricultural sustainability and intensive production practices. *Nature* 418:671–677.

UN Department of Economic and Social Affairs, Population Division. 2013. World population prospects: The 2012 revision. http://esa.un.org/unpd/wpp/Excel-Data/population.htm, accessed January 23, 2013.

U.S. Geological Survey. 2013. Mineral commodity summaries, United States Geological Survey. http://minerals.usgs.gov/minerals, accessed February 4, 2013.

van der Ploeg, R.R., W. Böhm, and M.B. Kirkham. 1999. On the origin of the theory of mineral nutrition of plants and the law of the minimum. *Soil Sci. Soc. Am. J.* 63:1055–1062.

Vitousek, P.M., R. Naylor, T. Crews et al. 2009. Nutrient imbalances in agricultural development. *Science* 324:1519–1520.

Watson, C.A., I. Öborn, A.C. Edwards et al. 2012. Using soil and plant properties and farm management practices to improve the micronutrient composition of food and feed. *J. Geochem. Explor.* 121:15–24.

Welch, R.M. and R.D. Graham. 2005. Agriculture: The real nexus for enhancing bioavailable micronutrients in food crops. *J. Trace Elem. Med. Biol.* 18:299–307.

White, P.J. and M.R. Broadley. 2009. Biofortification of crops with seven mineral elements often lacking in human diets—Iron, zinc, copper, calcium, magnesium, selenium and iodine. *New Phytol.* 182:49–84.

White, P.J., M.R. Broadley, J.A. Thompson et al. 2012. Testing the distinctiveness of shoot ionomes of angiosperm families using the Rothamsted Park Grassland Continuous Hay Experiment. *New Phytol.* 196:101–109.

White, P.J. and P.H. Brown. 2010. Plant nutrition for sustainable development and global health. *Ann. Bot.* 105:1073–1080.

Whiteside, M.D., K.K. Treseder, and P.R. Atsatt. 2009. The brighter side of soils: Quantum dots track organic nitrogen through fungi and plants. *Ecology* 90:100–108.

World Bank. 2013. Indicators. Agriculture & rural development. Fertilizer consumption (kilograms per hectare of arable land). http://data.worldbank.org/indicator/AG.CON.FERT.zs, accessed January 22, 2013.

Yang, J. and J. Zhang. 2006. Grain filling of cereals under soil drying. *New Phytol.* 169:223–236.

Zhang, W.-F., Z.-X. Dou, P. He et al. 2013. New technologies reduce greenhouse gas emissions from nitrogenous fertilizer in China. *Proc. Natl. Acad. Sci. USA* 110:8375–8380.

Section II

Essential Elements: Macronutrients

2 Nitrogen

David J. Pilbeam

CONTENTS

2.1 INTRODUCTION

Crop plants typically are comprised of C, H, O, N, and then other elements, in that order, so it is obvious that nitrogen is required by plants in large amounts. The fact that nitrogen is taken up by plant roots was shown by de Saussure in 1804, who demonstrated that an inorganic nitrogen source such as nitrate is essential for plant growth (Hewitt, 1966). From these facts, it follows that as plants have a high nitrogen requirement there are many situations in which its supply could limit crop growth, and amending soils and growing media with additional nitrogen can increase yields.

The earliest sources of nitrogen to increase crop growth were animal excreta and plant remains, and analysis of $\delta^{15}N$ in cereal grains and pulse seeds from Neolithic sites across Europe showed that farmers have used manures for at least 8000 years (Bogaard et al., 2013). The Roman writer Cato the Elder, writing a farming manual *circa* 160 BC, gave advice on how to use manure, including which crops needed to be sown in well-manured or naturally strong soil (turnips, kohlrabi, and radish). He also was aware that lupins, beans, and vetch were useful for enhancing the yields of cereals (Hooper and Ash, 1934). We had to wait until the nineteenth century when the German chemists Carl Sprengel and Justus von Liebig realized that the nitrogen in animal manures can be supplied to plants as ammonium ions in inorganic fertilizers (van der Ploeg et al., 1999) and until later that century when Hellriegel and Wilfarth showed that legume root nodules are able to carry out the conversion of atmospheric dinitrogen to ammonium (Nutman, 1987).

2.2   UPTAKE OF NITROGEN BY PLANTS

Despite the abundance of dinitrogen (N_2) gas in the atmosphere, only plants in the Fabaceae and a few other families are able to utilize this as their N source. For most plant species, the major N forms available to them are nitrate (NO_3^-) and ammonium (NH_4^+).

In experiments where NO_3^- or NH_4^+ is supplied as the sole N source, many plants tend to grow better with the NO_3^-. The exception is plants adapted to acid soils, soils that contain a higher concentration of NH_4^+ than NO_3^- and where NH_4^+ uptake predominates. As acid soils are infrequent in agriculture, it might seem that the NO_3^- ion is the major N source for most crop plants (except rice).

More detailed investigation of this assumption showed that with low N supply to wheat (*Triticum aestivum* L.), the rate of growth was approximately similar, irrespective of whether NO_3^- or NH_4^+ was supplied, or even slightly faster with NH_4^+ than with NO_3^- (Cox and Reisenauer, 1973). However, yields became saturated at concentrations of NH_4–N at which NO_3–N was still giving increased growth, and at even higher NH_4–N supply growth was inhibited. This response usually is referred to as *ammonium toxicity*. The growth rate eventually reached a plateau at higher NO_3^- concentrations, but if the high-NO_3^- nutrient solution was then supplemented with low rates of NH_4^+ there was further stimulation of plant growth (Cox and Reisenauer, 1973). It can be concluded that at low rates of supply either NH_4–N or NO_3–N gives equal amounts of wheat growth, or slightly better growth with NH_4–N; at higher rates of N supply NO_3^- gives noticeably better growth; but the best growth comes from a high rate of supply of NO_3^- supplemented with a small amount of NH_4^+.

These two ions interact for uptake, and the presence of NH_4^+ in the rooting medium slows the uptake of NO_3^- (e.g., Taylor and Bloom, 1998), although the presence of NO_3^- does not affect NH_4^+ uptake. This interaction is affected by external pH, and at acid pH, NO_3^- uptake tends to be much higher than uptake of NH_4^+, whereas at more neutral pH the rates of uptake of both ions are similar (Michael et al., 1965). There is also a temperature effect, and uptake of NO_3^- is depressed more by low temperature than is the uptake of NH_4^+ (Clarkson and Warner, 1979). Uptake of NO_3^- follows a diurnal pattern, with the fastest rate occurring during the light period then declining in the dark (Pearson and Steer, 1977; Matt et al., 2001).

2.2.1   NITRATE

Movement of NO_3^- into roots occurs through different systems that work at different concentrations of the ion in the external medium. Siddiqi et al. (1990), working on barley (*Hordeum vulgare* L.) seedlings, demonstrated that at external concentrations of NO_3^- between 0 and 0.5 mol m^{-3} there were two different uptake systems showing Michaelis–Menten kinetics, and therefore dependent on an energy source to pump NO_3^- into roots against its concentration gradient. In contrast, when the seedlings were supplied with NO_3^- between 1 and 50 mol m^{-3} its rate of uptake showed a linear relationship with external concentration, indicating uptake by means of passive diffusion.

There are transmembrane transporter proteins in plants that act as NO_3^-/H^+ symporters and move NO_3^- ions into and between cells, and even within one plant species there are different forms of these transporters depending on whether they are in the plasmalemma or in the tonoplast of roots or shoots and also on the developmental stage of the plant. The transporters are divided into three families, NRT1, NRT2, and NRT3.

In arabidopsis (*Arabidopsis thaliana* Heynh.), there are two high-affinity nitrate transport systems (HATS) that work at low external NO_3^- concentrations (1 mmol m^{-3} to 1 mol m^{-3}) (Miller et al., 2007). These two systems are a constitutive form (cHATS) that is present all the time, even when the plant is not exposed to NO_3^-, and a form that is induced when the plant is exposed to the ion (iHATS). There is one low-affinity nitrate uptake system (LATS), which is constitutive and works on external NO_3^- concentrations above 1 mol m^{-3}. Soil concentrations of NO_3^- are typically above this value, at least in agricultural soils.

Generally, LATS is composed of NRT1 transporters, although one of its constituents in arabidopsis, NRT1.1 (a member of the NRT1 family), is a dual-affinity transporter that is active at both low and high external NO_3^- concentrations. This dual activity is regulated by changes to a threonine residue. In a dephosphorylated state, the protein is a low-affinity transporter, but with low nitrate supply the threonine is phosphorylated, and the protein becomes a high-affinity transporter (Liu and Tsay, 2003). NRT2 transporters, such as NRT2.1 in arabidopsis, are high-affinity transporters. The *AtNRT2.1* gene is induced by NO_3^- and is repressed by metabolites of N assimilation, so AtNRT2.1 seems to be involved in iHATS. However, it also appears to have a role in cHATS (Miller et al., 2007). Expression of the *NRT2.1* gene is repressed by NH_4^+ if NO_3^- supply is also high, which may account for the inhibitory effect of NH_4^+ on NO_3^- uptake (Krouk et al., 2006).

The activity of NRT2.1 depends on another protein, NAR2.1 (NRT3.1). Together, they form a complex in the plasmalemma, probably a tetramer comprising two molecules of each (Yong et al., 2010). At least three other transporters are involved in HATS in arabidopsis (NRT1.1, NRT2.2, and NRT2.4), but NRT2.1 seems to make the predominant contribution as mutants with no *NRT2.1* gene expression have as little as 25% of the normal high-affinity uptake and cannot grow when NO_3^- is below 1 mol m^{-3} and is the sole N source (Laugier et al., 2012). Like NRT1.1, this protein is regulated by posttranscriptional control as well as by transcription and represents a point at which the N demand of the whole plant can regulate the rate of NO_3^- uptake (Laugier et al., 2012). Not only can the amount of NRT2.1 transporter protein be regulated by external or internal N concentrations affecting the expression of the *NRT2.1* gene, but activity of existing molecules of the transporter can be regulated very quickly.

These changes to the transporters are important in regulating plant responses to changes in NO_3^- supply. In N-starved barley seedlings, supply of NO_3^- gave immediate uptake (due to constitutive mechanisms), and then there was an induction of uptake that was linked to synthesis of proteins (Siebrecht et al., 1995). Tomato root *NRT1.2* and *NRT2.1* are both upregulated quickly on resupply of NO_3^- to N-deficient plants, an upregulation that lasts for 24 h (Wang et al., 2001). With normal supply to arabidopsis, there is usually high expression of *NRT2.1* in the roots and low expression of the gene for another high-affinity transporter, NRT2.4 (Kiba et al., 2012). Removal of the nitrate source gave a short-term increase in *NRT2.1* expression and then a reversal to the original level by 3 days. However, there was a very large increase in expression of *NRT2.4* in the roots, and it also became detectable in the shoot (Kiba et al., 2012). The NRT2.4 transporter seems to become more important in times of N starvation.

Different parts of the root system express different nitrate transporter genes. The *NRT1.1* gene is expressed in the epidermis at the root tip, allowing synthesis of the NRT1.1 transporter to take up NO_3^- at that point, but is also expressed in the cortex and endodermis further back in the root (Huang et al., 1996). The *NRT1.2* gene in arabidopsis is expressed in the root hairs and epidermis (Huang et al., 1999), allowing low-affinity uptake to occur from the root tip back to the root hair zone. The expression of *NRT2.4* in N-starved plants occurs in the lateral roots, in the epidermal cells, and in young parts of the primary root, whereas NRT2.1 seems to occur in the old parts of the primary root irrespective of whether nitrate supply is abundant or deficient (Kiba et al., 2012). In these more mature parts of the root, the *AtNRT2.1* gene is expressed in the epidermis and cortex (Nazoa et al., 2003). The uptake rate in barley is higher in middle and old zones of the roots than in the tips, an action that could be linked to the fact that the old root cells are vacuolated and have a site to store the NO_3^- taken up (Siebrecht et al., 1995). In corn (*Zea mays* L.), likewise, the uptake rate of NO_3^- is higher in old parts of the root than at the tip (Taylor and Bloom, 1998), although in all parts of the root uptake seems to exceed local demand (Bloom et al., 2012). Up to 60% of the NO_3^- taken up by corn seems to enter through the secondary roots, with much of the remaining uptake occurring into the basal regions of the primary root; the NO_3^- entering the secondary roots is translocated quickly to the shoot (Lazlof et al., 1992).

In the experiments of Siebrecht et al. (1995), although much of the NO_3^- taken up in the root tips was assimilated locally, some was exported back to the old parts of the roots and was able to induce nitrate uptake mechanisms there. This action gives a role to the root tip as a sensor of NO_3^- in the rooting environment. Further evidence for this response was found in N-depleted maize roots, where rapid induction of iHATS occurred if NO_3^- was supplied, particularly in the tip region, although uptake reverted to the background level after approximately 24 h (Sorgonà et al., 2011). This action seemed to involve induction of the NRT2.1 transporter.

Once NO_3^- is taken up into root cells, some is assimilated there, some is moved across the tonoplast into the vacuoles (in the old parts of the roots, where vacuoles occur), and much of it is moved to adjacent cells and ultimately into the xylem and up to the shoot. The accumulation of NO_3^- in vacuoles (in root and leaf cells) is facilitated through CLCa anion channels in the tonoplast (De Angeli et al., 2006). Although the CLC family were originally named as being chloride transporters, CLCa gives much greater permeability of the tonoplast to NO_3^- than to Cl^- (De Angeli et al., 2006).

The low-affinity NRT1.5 transporter present in the plasmalemma of the pericycle cells around the protoxylem loads NO_3^- into the xylem, although this action is not the only mechanism for nitrate loading (Lin et al., 2008; Dechorgnat et al., 2011). In barley and corn, other xylem-loading systems have been identified, including the general anion channels xylem parenchyma quickly activating anion conductance (X-QUAC) and xylem parenchyma inwardly rectifying conductance (X-IRAC); of these, X-QUAC seems to be the most important in nitrate loading (Köhler and Raschke, 2000; Köhler et al., 2002; Gilliham and Tester, 2005). As well as being expressed in epidermal and cortical cells of roots, the *AtNRT2.1* gene is expressed in the endodermis (Nazoa et al., 2003).

Some crop plants with a high N requirement increase the hydraulic conductivity of their root systems in response to increase in NO_3^- concentration in the root environment (Clarkson et al., 2000). Nitrate is water soluble and moves toward roots in the bulk flow of water through the soil driven by the pull of transpiration, so there could be a link between uptake of NO_3^- and water. The change in hydraulic conductivity is quick, it seems to work on cell membrane hydraulic properties, and it involves aquaporins (Gloser et al., 2007; Górska et al., 2008a). Aquaporin genes were upregulated in roots of N-deficient tomato (*Solanum lycopersicum* L. syn. *Lycopersicon esculentum* Mill.) plants resupplied with NO_3^- (Wang et al., 2001), and under such conditions, the activity of existing aquaporins may increase (Górska et al., 2008b). It is also possible that an increase in hydraulic conductivity arising from a high-nitrate zone in the soil may allow NO_3^- to move between parallel xylem vessels due to differences in water potential breaking down sectoriality and allowing some lateral movement across the root before movement up to the shoot (Thorn and Orians, 2011).

In many plants, hydraulic conductivity of the roots is much higher with NO_3–N than with NH_4–N, and expression of aquaporins seems to be higher in nitrate-supplied plants (Guo et al., 2007). However, in rice hydraulic conductivity is higher with NH_4^+ than with NO_3^- (Gao et al., 2010). It may be that in plants in which nitrate is the main N source, response to localized NO_3^- supply is also a response to availability of water. Increase in root hydraulic conductivity in response to increased NO_3^- supply seems to be higher in fast-growing species, and the ability to use rapid changes in root hydraulic conductivity to make the most of NO_3^- available by bulk flow may be important for plants with a high-nitrate requirement (Górska et al., 2010).

There were shown to be peaks of flux of NO_3^- into roots of dwarf maize (*Zea mays*) from 50 to 250 mmol m^{-3} concentration (within the HATS range), one during early vegetative growth and one just before flowering (Garnett et al., 2013). The rates of flux matched total plant uptake, indicating that HATS can meet requirements of nitrate for growth throughout the life of a plant without involvement of LATS. Of all the *NRT* genes, *ZmNRT2.1* and *ZmNRT2.2* were expressed the most in the roots, and changes in the levels of their transcripts matched the NO_3^- uptake patterns (Garnett et al., 2013).

2.2.2 AMMONIUM

The pK_a value for the dissociation of ammonia (NH_3) dissolved in water into ammonium (NH_4^+) and hydroxyl (OH^-) ions is 9.25, so at normal soil and cellular pH a plant is exposed predominantly to the NH_4^+ ion. Although NH_3 is able to permeate through membranes, NH_4^+ is taken up by means of specific transporters, the ammonium transporter (AMT) family of proteins. These are split into the AMT1, AMT2, AMT3, and AMT4 subfamilies (Koegel et al., 2013).

In lowland rice (*Oryza sativa* L.), a crop plant that is adapted to anaerobic soils where NH_4^+ is the predominant N species, there are three members of each of the AMT1, AMT2, and AMT3 subfamilies and one member of AMT4 (Gaur et al., 2012). Members of the AMT1 subfamily comprise a high-affinity uptake system (HATS), and as several AMT1 genes are expressed at any one time it seems that NH_4^+ uptake occurs through several transporters (Mota et al., 2011). AMT1.1 gives approximately 30% of the NH_4^+ uptake in arabidopsis roots (Camañes et al., 2012) and works best at low external NH_4^+ concentration. In fact, a period of N starvation gives increased expression of *AtAMT1.1* in the roots and decreased expression of the gene in the shoots (Engineer and Kranz, 2007). Expression of the *AMT1.1* gene also is upregulated by N starvation in rice roots, as are *OsAMT1.2*, *OsAMT3.2*, and *OsAMT3.3* (Li et al., 2012). Higher external ammonium supply leads to the phosphorylation of a key threonine residue in the AMT1.1 protein, giving a slower rate of transport. This phosphorylation seems to be controlled directly by the NH_4^+ itself, so AMT1.1 works as a transceptor (a transporter and a receptor) that detects high soil concentrations of NH_4^+ and lowers its rate of uptake (Lanquar et al., 2009). This act may prevent accumulation of toxic NH_4^+ ions in the roots.

Expression of the *AMT1.1* gene in rice roots is downregulated on the supply of nitrate to N-starved plants, as is *AMT1.2* (Li et al., 2012). In arabidopsis, expression of the gene increases the rate of NO_3^- uptake, and conversely expression of the *NRT2.1* gene increases the rate of NH_4^+ uptake (Camañes et al., 2012). Arabidopsis may grow better with a balanced N supply. AMT1.3 appears to function as a transceptor, and a component of HATS, in rice (Gaur et al., 2012). Its expression is downregulated by N starvation and is upregulated by resupply of NH_4^+ (Li et al., 2012).

There is a low-affinity system that works best at high external pH, and this action could be due to undissociated NH_3 crossing the membranes (and may explain the higher uptake of ammonia/ammonium relative to nitrate at high pH). However, *AMT2* genes seem to be involved in a LATS (Gaur et al., 2012). Ammonium influx occurs at a relatively uniform rate along a corn root from tip to the base of the zone of root growth and is then higher at more basal regions (Taylor and Bloom, 1998), but growth of the tissues at the root tip seems to also rely on NH_4^+ or its precursors taken up further back (Bloom et al., 2012). If NH_4NO_3 is supplied to corn, the NH_4^+ taken up at the root tip is used preferentially there, and the NO_3^- taken up remains unassimilated (Bloom et al., 2012). The *AMT2* gene is upregulated quickly on the resupply of NO_3^- to N-deficient tomato roots, and this upregulation remains for 24 h (Wang et al., 2001).

2.2.3 ORGANIC NITROGEN FORMS

It is likely that plants may be able to take up quaternary ammonium compounds as intact molecules (Warren, 2013b). They can also take up urea, in pathways similar to those involved in uptake of NH_4^+ (Mérigout et al., 2008; Witte, 2011). In arabidopsis, there is an active system that transports urea into roots at external concentrations up to 50 mmol m^{-3}, but at concentrations 10 times higher the rate of uptake is related linearly to external concentration and therefore occurs by passive diffusion (Wang et al., 2012b). Urea also is taken up through leaves, and the N derived from its application to leaves of potato (*Solanum tuberosum* L.) was found in the tubers within 48 h (Witte et al., 2002). Genes expressing high- and low-affinity urea transporters have been identified, and detection of ^{15}N-labeled urea in arabidopsis roots after supplying ^{15}N-urea for 5 min seems to confirm that it is taken up as an intact molecule (Mérigout et al., 2008).

One transporter involved is DUR3, a high-affinity transporter with an affinity constant for urea of 3–4 mmol m^{-3} (Kojima et al., 2007). It is expressed in the root epidermal cells, including in root hairs, and also near root xylem and in the shoots, indicating a role in uptake and in transport of urea within a plant (Kojima et al., 2007). Its activity increases under N deficiency (Arkoun et al., 2013) and decreases after resupply of NO$_3^-$ or NH$_4^+$ (Kojima et al., 2007), and as DUR3 is a high-affinity transporter it seems that its presence is an adaptation to enable plants to use the urea present in soils otherwise deficient in nitrogen but not in soils high in mineral N (Kojima et al., 2007). Supply of NO$_3^-$ with urea seems to increase the rate of uptake of the urea (Garnica et al., 2009). In rice, seedlings grown on urea have less shoot growth than seedlings grown on NH$_4$–N or even NO$_3$–N alone (Wang et al., 2013a).

Plants take up amino acids, with several systems being shown to be important in arabidopsis. The lysine histidine transporter 1 (LHT1) is responsible for the uptake of neutral and acidic amino acids, and amino acid permease 1 (AAP1) absorbs the acidic amino acid glutamate, the basic amino acid histidine, and neutral amino acids (Lee et al., 2007), and AAP5 is responsible for the uptake of the cationic amino acids L-lysine and L-arginine (Näsholm et al., 2009; Svennerstam et al., 2011). LHT1 from arabidopsis has K_m values for proline, glutamate, and histidine of 10, 13.6, and 362 mmol m^{-3}, respectively (Hirner et al., 2006). There are other amino acid transporters in plants, although many are doubtless involved with transport within the plants, with 67, 134, and 96 genes involved in amino acid transport having been identified in arabidopsis, black cottonwood (*Populus trichocarpa* Torr. and A. Gray), and rice, respectively (Näsholm et al., 2009). LHT1 is present in the epidermis of emerging roots and lateral roots of arabidopsis, but not in the main root (Hirner et al., 2006). Its expression is induced if plants are grown with amino N rather than inorganic N (Hirner et al., 2006). AAP1 is present in the plasmalemma of the arabidopsis root epidermis and the outer layer of the root cap (Lee et al., 2007).

Arabidopsis plants take up L-amino acids preferentially over D-amino acids, with L-arginine being taken up from a mixture of 10 amino acids and NH$_4^+$ at two-thirds of the rate of NH$_4^+$ and with glycine, L-alanine, L-serine, L-valine, and L-isoleucine being taken up at approximately one-third of that rate (Forsum et al., 2008). Uptake of L-glutamine and L-asparagine occurred at a slightly slower rate, but as they contain two NH$_2$ groups they represented a good source of N for the plants. Indeed, these two amino acids gave the best growth of the plants if the 11 N sources were supplied individually, and they also gave a significantly greater biomass if supplied with NO$_3^-$ than plants grown with NO$_3^-$ on its own (as did L-aspartate) (Forsum et al., 2008).

Many amino acid uptake studies have been carried out at concentrations much higher than those occurring in soil, and there has been little quantification until recently of the extent to which amino acids could meet plant N requirements (Näsholm et al., 2009). This assessment is difficult, as where ^{15}N amino acids are supplied to soils the ^{15}N could be released as inorganic N during mineralization and then taken up in that form. Supply of dual (^{13}C, ^{15}N)-labeled amino acids gives some idea if individual molecules are taken up intact and conjugating glycine with fluorescent quantum dots has enabled the movement of the intact conjugate into hyphae and roots and shoots of annual bluegrass (*Poa annua* L.) to be measured (Whiteside et al., 2009), but it is still difficult to evaluate the potential contribution of all the amino acids in a soil. It was shown that supplying N to perennial ryegrass (*Lolium perenne* L.) in hydroponic culture as an equimolar mixture of NH$_4^+$, NO$_3^-$, and glycine gave an increased proportion of the N being taken up as glycine compared with what would have been predicted from supplying the three N forms individually. The rates of uptake of NH$_4^+$ and NO$_3^-$ were noticeably lower in the mixture, whereas the glycine uptake was only slightly lower (Thornton and Robinson, 2005). Five plant species in a seminatural temperate grassland were shown to take up ^{15}N-labeled ammonium nitrate more than ^{15}N-labeled amino acids (Harrison et al., 2007).

Some plants can take up intact peptides, and wheat has been demonstrated to take up L-trialanine only marginally slower than NH$_4^+$, although D-tripeptides are taken up more slowly (Hill et al., 2011). In an annual, arabidopsis, and a perennial, angled lobelia (*Lobelia anceps* L. f.), peptides of four amino acid residues in length are taken up (Soper et al., 2011). In arabidopsi*s*, the PTR1 transporter (peptide transporter 1) expressed in the roots seems to carry out the uptake of dipeptides (Komarova et al., 2008). The supply of small peptides in the soil solution also may be increased by plants themselves, as leek (*Allium porrum* L.) seedlings were shown to release into the nutrient solution proteases that degrade proteins to low-molecular-weight peptides (Adamczyk et al., 2009). Some plants seem to use proteins more directly. Arabidopsis, a species from nutrient-rich environments, and mulloway needlebush (*Hakea actites*), a species that grows in nutrient-poor heathland and forms cluster roots, were shown to have proteolytic activity on the root surface and possibly in the root apoplast. They also seemed to have the ability to take up proteins into their root hairs and into cortical cells of roots with root hairs (possibly by endocytosis) (Paungfoo-Lonhienne et al., 2008). However, the plants did not grow nearly as well with protein as the sole N source as with ammonium nitrate, but protein seemed to be a valuable supplemental source of nitrogen to plants grown in mixed protein and inorganic N sources.

The uptake of amino acids by plants may involve mycorrhizas, and several amino acid transporters have been identified in mycorrhizal fungi (Näsholm et al., 2009). In ectomycorrhizal symbioses, the amino acids and peptides have been known for some time to be taken up by the fungus with the N being passed to the plant (Melin and Nilsson, 1953). However, ectomycorrhizal fungi also increase the uptake of NH_4^+ and NO_3^- by their host plant (Plassard et al., 1991, 1994). In a study of 10 ectomycorrhizal fungal species, NH_4^+ was generally a more suitable N source than NO_3^- (Finlay et al., 1992). Some ectomycorrhizal fungal species can live with an N source of either peptides or proteins (Abuzinadah and Read, 1986), and the production of extracellular proteinases has been demonstrated in ectomycorrhizal fungi (Maijala et al., 1991). In ericoid mycorrhizas (symbioses between fungi and plants in Ericaceae in northern hemisphere tundra and heaths and in the understory of boreal forests), the fungi seem to be at least partly responsible for the uptake of not only amino acids into the plants but also NH_4^+ (Rains and Bledsoe, 2007) and NO_3^- (Grelet et al., 2009). Ericoid mycorrhizal fungi also produce extracellular proteinases (Leake and Read, 1990).

Most crop species form arbuscular mycorrhizas (AMs), and being mycorrhizal is their normal state (Smith and Smith, 2011). AM fungal species can take up NO_3^-, NH_4^+, and glycine and pass the N onto plants (Hawkins et al., 2000). However, unlike ectomycorrhizas and ericoid mycorrhizas, AMs are thought to be unable to release the N in complex organic molecules (Smith and Smith, 2011), although in the experiments of Whiteside et al. (2009) annual bluegrass (*Poa annua* L.) plants received glycine from uptake via AM fungi. Ammonium seemed to be the N form taken up most readily into the AM hyphae in the experiments of Hawkins et al. (2000), and in experiments on AM pygmy cypress (*Cupressus pigmaea* Sarg.) trees NH_4^+ was taken up more readily than glycine (Rains and Bledsoe, 2007). In experiments with carrot (*Daucus carota* L.) roots inoculated with AM fungi, the extraradical mycelium was able to take up NO_3^- and NH_4^+ ions, with both being converted in the fungus to glutamine and then to other amino acids and particularly arginine. The arginine moved to the intraradical mycelium, where it broke down with the release of NH_4^+ that was passed into the root cells (Govindarajulu et al., 2005). The *AMT3.1* and *AMT4* genes are expressed in cortical cells of roots of mycorrhizal, but not in nonmycorrhizal, sorghum (*Sorghum bicolor* Moench) plants, and the AMT3.1 protein is localized around the mature arbuscules at the periarbuscular membrane (Koegel et al., 2013). These two transporters therefore seem to be important in the transfer of NH_4^+ from fungus to plant. Another six *AMT* genes were identified in the roots of the nonmycorrhizal plants and in other organs (Koegel et al., 2013), so there appear to be different AMTs involved in the uptake of NH_4^+ directly from the soil.

The extraradical mycelium of AM fungi seems to have both a LATS and a HATS for NH_4^+ (Pérez-Tienda et al., 2012). It is thought that forest trees with ectomycorrhizal associations take up amino acid N more efficiently than inorganic N, whereas trees with AMs take up inorganic N more efficiently than amino acid N (McFarland et al., 2010). In an experiment on durum wheat (*Triticum durum* Desf.), total growth and uptake of N were enhanced by the plants having AM associations, but although the presence of fungi increased soil N mineralization it actually decreased the ability of the plants to acquire N from organic N (Saia et al., 2014). Recent research has indicated that in

N-limited soils of boreal forests the ectomycorrhizal fungi use carbon from the trees but hang on to the N that they acquire, thus exacerbating the effects of N deficiency (Näsholm et al., 2013).

2.2.4 DINITROGEN GAS

Those few plant species that can utilize dinitrogen (N_2) gas as their N source depend on symbiotic microorganisms to fix it. The most commonly studied of these relationships is between gram-negative members of the alpha subgroup of the proteobacteria (rhizobia) that live in root nodules of plants in the family Fabaceae and the nonlegume Parasponia (*Parasponia andersonii*) in the family Cannabaceae (Santi et al., 2013). Other associations within root nodules involve filamentous bacteria (*Frankia* spp., gram-positive actinomycetes), with mostly woody plants in eight different families (Santi et al., 2013).

The nitrogen made available to leguminous plants by the plant–*Rhizobium* symbiosis is also available to other plant species. In pastures, N fixed in the nodules of legumes is passed onto grass species by mycorrhizal fungi, although it is also possible for N in the grasses to be passed to the legume species (Pirhofer-Walzl et al., 2012). Similarly, N has been demonstrated to pass from N-fixing soybean to corn (Bethlenfalvay et al., 1991). It seems likely that in low-input agricultural systems, N fixed by leguminous crops can be passed to nonleguminous weeds by mycorrhizal hyphae connecting the plants, enabling the weeds to grow vigorously (Moyer-Henry et al., 2006). The amount of N passed from N-fixing species to nonfixing species can be up to 80% of the total N in the combined biomass of both species, although in most examples studied it is considerably less, and usually with a small amount of N moving in the reverse direction (He et al., 2009).

A few plant species form symbiotic associations with N-fixing cyanobacteria, either an intercellular association (plants in the Gunneraceae) or an extracellular association (liverworts, hornworts, the aquatic fern azolla (*Azolla* spp. Lam.), and gymnosperms of the Cycadaceae) (Santi et al., 2013) and sphagnum moss (*Sphagnum* spp. L.) (Berg et al., 2013). Although these plants do not include any crop species, such symbioses are potentially possible (Ahmed et al., 2010). Loose associations between N-fixing bacteria (diazotrophic endophytes) and crop plants may certainly be possible, and the Trenton cultivar of wheat inoculated with a strain of klebsiella (*Klebsiella pneumoniae* Trevisan) was shown to receive fixed N from the bacteria present in intercellular spaces of the root cortex and to have improved growth as a consequence (Iniguez et al., 2004). Such associations seem to be common in the family Poaceae, to which most cereal crops belong, and involve a range of alpha- and beta-proteobacteria collectively known as plant growth–promoting rhizobacteria (PGPR) (Santi et al., 2013). The presence of the diazotrophs may enhance crop yields, and yield increases of 13%–22% in rice and 5.9%–33% in corn have been seen in different studies comparing inoculated with uninoculated plants (Santi et al., 2013, and references therein). However, the major stimulation to plant yield may come about not through enhanced N availability but through the PGPRs producing phytohormones such as IAA and cytokinins that modify root architecture.

Understanding uptake of N compounds is a flourishing area of current research, and new discoveries are being made almost on a daily basis. The reader is referred to the reviews of Tegeder and Rensch (2010), Dechorgnat et al. (2011), Kraiser et al. (2011), and Wang et al. (2012c) for further details.

2.3 ASSIMILATION OF NITROGEN IN PLANTS

Nitrate taken up by a plant is reduced to nitrite (NO_2^-) and then to NH_4^+ in reactions that require energy. Ammonium that arises from nitrate, or by uptake from the soil, is added to organic acids to make amino acids. The amino acids are then assembled into proteins or are the starting point for the synthesis of other N-containing molecules such as nucleic acids, chlorophyll, alkaloids, cyanogenic glycosides, and glucosinolates.

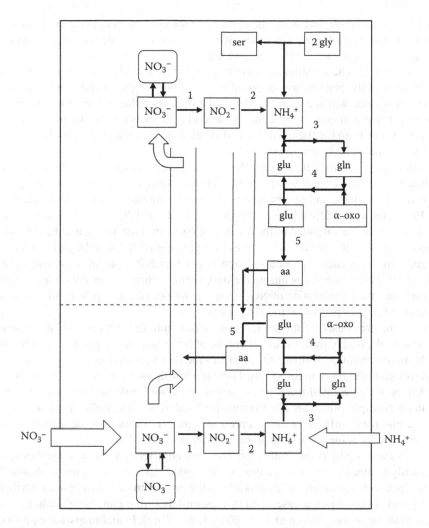

FIGURE 2.1 Assimilation of nitrate and ammonium by plants. *Abbreviations*: α-oxo, α-oxoglutarate; aa, amino acids; glu, glutamate; gln, glutamine; gly, glycine; ser, serine; NO_3^-, nitrate; NO_2^-, nitrite; NH_4^+, ammonium. Enzymes: 1 nitrate reductase, 2 nitrite reductase, 3 GS, 4 GOGAT, 5 transaminases.

Where NO_3^- is the form of nitrogen taken up by plants, the first step of assimilation is reduction to NO_2^- (Figure 2.1). This reaction is catalyzed by the enzyme nitrate reductase (NR), located in the cytoplasm (cytosol). The enzyme requires energy to carry out the reduction, with one form of the enzyme using NADH from respiratory reactions in the mitochondria and another using NADPH from photosystem 1. Nitrate reductase uses molybdenum as a cofactor in the reduction of nitrate (Campbell, 1999).

Nitrite produced from NO_3^- moves into the chloroplasts (or the plastids in the roots), where the enzyme nitrite reductase catalyzes its reduction to NH_4^+. This process requires electrons from reduced ferredoxin, which is a product of photosystem 1. Since ferredoxin does not move around the plants, there must be some other reductant in roots for the reduction of NO_2^- arising from nitrate that is reduced there.

The NH_4^+ produced in the chloroplast (or root plastid) joins with a molecule of the amino acid glutamate to make another amino acid, glutamine, in a reaction catalyzed by glutamine synthetase (GS). This enzyme uses ATP as its energy source, and it has a very high affinity for NH_4^+. The glutamine then passes the amino (NH_2) group acquired from the NH_4^+ to the organic acid α-oxoglutarate, reverting to glutamate itself and generating another molecule of glutamate in the process. This reaction is catalyzed by glutamate synthase (GOGAT). The first molecule of glutamate can then become an acceptor for another NH_4^+ ion, and the second molecule is converted to other amino acids by transaminases. These amino acids are then exported from the site of synthesis to form proteins in other parts of the plant or are converted into N-containing secondary metabolites. GOGAT exists in two forms, one which uses reduced ferredoxin as its reductant and one that uses NADH. The first form is present in chloroplasts, and the second is in roots (especially in root nodules). The α-oxoglutarate comes from the tricarboxylic acid (Krebs) cycle, which is replenished with substrate to prevent the organic acids becoming depleted.

Nitrate reductase is substrate inducible and is subject to feedback inhibition by amino acids (particularly glutamine) (Campbell, 1999). Its activity changes over a 24 h cycle, increasing in the light and decreasing in the dark, and this pattern is general for nitrate assimilation in plants (Pearson and Steer, 1977; Matt et al., 2001). In fact, the pattern closely matches the expression of the *NRT1.5* gene, which is responsible for much of the loading of NO_3^- into the xylem and hence root to shoot nitrate transport (Lin et al., 2008). Light causes the synthesis of NR mRNA and synthesis of more of the enzyme, and dark causes phosphorylation of existing NR molecules, slowing their activity (Campbell, 1999). These then decay during the dark period. Nitrate assimilation in the light occurs at a faster rate than the translocation of NO_3^- from the roots and uses up NO_3^- that is stored in the leaf pools during the dark period (Matt et al., 2001).

The NH_4^+ in the previously mentioned reactions arises from the reduction of NO_3^- (mostly in the leaves, but also in the roots), from the direct uptake of NH_4^+ or in a few plant species from the conversion of N_2 in root nodules. Unlike NO_3^-, NH_4^+ in the root system is nearly all assimilated there, and little moves up to the shoot in the xylem. There is a fourth source of NH_4^+ in plants that is exclusive to the leaves, NH_4^+ released from the formation of one molecule of serine from two molecules of glycine in photorespiration. This NH_4^+ is reassimilated by leaf cytosolic GS (Hirel et al., 2007), and its release and reassimilation seem to represent a greater N flux than the primary assimilation of nitrogen that occurs at the same time (Tcherkez and Hodges, 2008).

Amino acids taken up by plants, rather than being used directly in protein synthesis, are transaminated readily to other amino acids, either in the roots or after transport to the shoots (Näsholm et al., 2009). Urea seems to be assimilated mostly in the roots (through conversion to carbamate and then NH_4^+ by nickel-containing ureases, and then assimilation into glutamine), although some may be translocated to the shoots (Mérigout et al., 2008; Witte, 2011). In arabidopsis and potato, the urease gene is transcribed in all tissues, so urea is certainly likely to be assimilated in shoots in these species (Witte et al., 2002; Witte, 2011). In fact, the concentration of urea in shoots is higher than would be predicted from the rates of potential urease activity, and it is probably stored in vacuoles, whereas the urease is located in the cytoplasm (Witte, 2011).

Regardless of the N form taken up, the ultimate end products of its assimilation are the amino acid constituents of proteins. These amino acids accumulate in developing leaves, roots, and other physiological sinks of a growing plant. A high proportion of plant N is present in leaf proteins, many of which form the enzymes involved in photosynthesis. The enzyme ribulose bisphosphate carboxylase/oxygenase is the major leaf protein, particularly in C3 plants, and it can account for up to 30% of the N in leaves (Lawlor et al., 1989). In the flag leaf of wheat, the concentrations of soluble proteins and chlorophyll reach their maximum at the time of full expansion and then decline to nearly zero by senescence (Lawlor et al., 1989). The amino N remobilized from senescing leaves is partitioned to other processes in the plant, primarily reproductive growth, and is transported mainly to the developing storage organs in the phloem as asparagine and glutamine (Masclaux-Daubresse et al., 2010). In the later stages of growth, typically after flowering in an annual plant, the degradation of proteins in the leaves exceeds the synthesis of new proteins.

In some annual plant species (e.g., oilseed rape, *Brassica napus* L.), N uptake almost ceases once flowering occurs with nitrogen in the developing storage organs coming almost exclusively from senescing leaves (Ulas et al., 2012), but in other species (e.g., corn), a sizeable proportion of N in the developing grains comes from uptake from the soil, as well as from mobilization (Masclaux-Daubresse et al., 2010). In corn, nitrogen taken up at the later stages of growth is partitioned more to the developing ear than to vegetative structures (Ta and Wieland, 1992), although it may be allocated to the vegetative tissues of the plant first and from there to the kernels as turnover of leaf proteins releases amino acids (Gallais et al., 2006). From 45% to 65% of the N in corn grains comes from mobilization from the leaves, with the rest coming from uptake after silking (Hirel et al., 2007). It has been estimated that 60%–85% of the N in a corn plant at anthesis eventually is located in the cob (Ta and Wieland, 1992).

About 70% of the maximum N content of spring and winter wheat plants has been absorbed by ear emergence, but uptake still occurs after this (Watson et al., 1963). Kichey et al. (2007) found that an average of 71% of winter wheat grain N comes from mobilization. Bogard et al. (2010), following field experiments on cultivars of wheat grown in seven locations in northern France and in a range of environmental conditions, estimated that 84% of winter wheat grain N came from mobilization from the leaves and stems and the remaining 16% from soil uptake. The efficiency of mobilization can be seen from the fact that they calculated that 78% of the N absorbed before anthesis ended up in the grains. Of the N in winter wheat crops grown at recommended fertilizer N rates in the United Kingdom and New Zealand, 14% was structural N, 43% was photosynthetic N, and 43% was reserve N (mainly located in the true stem) at anthesis (Pask et al., 2012). Although most of the grain N came from the leaves, the reserve N appeared to be a major source of N for the grain initially, so that green leaf area was maintained for some time.

Modern *stay-green* cultivars of cereals keep taking N up from the soil for translocation to the developing grains at a time when uptake is slowing in the traditional cultivars, as a longer duration of shoot activity maintains root activity (Borrell et al., 2001). Although much of the N in corn grains already is present in the leaves before silking, most of the total grain dry matter comes from photosynthesis post-silking (Ning et al., 2013). In stay-green cultivars with a larger leaf area and longer leaf area duration than traditional cultivars, total yields were higher than in older cultivars, but a high proportion of the N in the grains still came from mobilization from the leaves, and the extra N taken up after silking was used to maintain leaf area (Ning et al., 2013).

In perennial plants, those that are deciduous tend to scavenge some of the nitrogen present in the leaves before they are shed. In aspen (*Populus tremula* L.) trees, it has been calculated that up to 80% of the N in the leaves is moved to the remaining parts of the tree before leaf fall (Keskitalo et al., 2005).

Other N-containing compounds in plants are formed from amino acids. All plants contain purine and pyrimidine bases (in nucleic acids), chlorophyll, and the indole hormones, and these have amino acid precursors. The distribution of N-containing, nonprotein amino acids, alkaloids, cyanogenic glycosides, and glucosinolates varies between plant species, with some being in only a few species, genera, or families. For example, glucosinolates are restricted largely to the order Brassicales. If amino acids accumulate, they can cause slowing of nitrate uptake as well as nitrate reduction. For example, glutamine represses transcription of the *AtNRT2.1* gene (Nazoa et al., 2003).

2.4   PHYSIOLOGICAL RESPONSES OF PLANTS TO NITROGEN SUPPLY

Plants subjected to nitrogen deficiency tend to be smaller and to have leaves that are paler green than normal (Figures 2.2 and 2.3). This color effect is due to lower chlorophyll concentration.

2.4.1   EFFECTS OF NITROGEN SUPPLY ON SHOOTS AND LEAVES

Not only are plants grown under N deficiency small, but they have a lower shoot:root ratio, as ongoing shortage of nitrogen leads to lowered shoot growth before root growth is slowed (Brouwer, 1962). As well as giving a large shoot and increased plant size overall, higher N supply also gives a larger leaf area. For example, flag leaves of winter and spring wheat grown in sand had a larger area

(a)   (b)

FIGURE 2.2 Nitrogen deficiency in soybean (*Glycine max* Merr.). (a) N-deficient plant (note low number of leaves, small area of individual leaves, and paler coloration) and (b) N-sufficient plant.

FIGURE 2.3 **(See color insert.)** Nitrogen-deficient young pepper (*Capsicum* sp. L.) plants showing chlorosis (paler coloration) of oldest leaves. (Photograph by A.V. Barker.)

and a higher fresh and dry mass with high N supply than with low N supply (Lawlor et al., 1989). The high-N plants had more cells in their flag leaf, and furthermore these cells had a larger volume.

Removal of N supply results in existing leaves undergoing cell expansion slowly, and leaves that form later have a slower rate of cell division (Roggatz et al., 1999; Trápani et al., 1999; Broadley et al., 2001). Plants grown with NH_4^+ have a smaller leaf area than plants grown with NO_3^- (Walch-Liu et al., 2000), and as NO_3^- accumulates in the leaves during the light period (Matt et al., 2001) and is an osmoticum, it could be assumed that the NO_3^- ion itself may drive leaf expansion by its effect on leaf water potential. However, direct measurement of turgor pressure in epidermal cells of expanding sunflower (*Helianthus annuus* L.) leaves failed to show any effect of withdrawal of NO_3^- (Palmer et al., 1996). Nitrate makes only a small contribution to changes in the water potential of leaf cells, and organic solutes probably make a much larger contribution to driving leaf expansion (Fricke and Flowers, 1998). An increase in leaf growth rate can be seen upon the supply of NO_3^- to plants previously supplied

NH_4^+, and since this result is apparent with NO_3^- supplied at as low a concentration as 10 mmol m^{-3}, it cannot be an osmotic effect (Rahayu et al., 2005). There is an increased concentration of cytokinins in the xylem and leaves of plants within 4 h of supply of NO_3^-, and that increase appears to control leaf expansion, possibly through affecting cell wall extensibility (Rahayu et al., 2005).

Increased leaf area has to be matched by more translocation of water and nutrients to the shoots, and in rooted cuttings of poplar (*Populus trichocarpa* Torr. and Gray x *deltoides* Bartr. ex Marsh) exposed to high ammonium nitrate, expression of seven aquaporin genes was upregulated in the secondary xylem of the stem. This action may have enabled the inflow of water into areas of xylem cell growth, as the high N supply gave increased xylem cross-sectional area and specific conductivity that matched the increased leaf area seen in these plants (Hacke et al., 2010).

Another effect that nitrogen has on leaves is to maintain leaf area duration, and if plants are subjected to N deficiency leaves tend to senesce early (Spiertz and de Vos, 1983; Lawlor et al., 1989). Nitrogen supply also affects total leaf area by affecting the number of leaves, and under extreme N deficiency the rate of wheat leaf primordia initiation relative to accumulated thermal time is slower (Longnecker et al., 1993). This deficiency also is reflected in the formation of tillers by cereals, with plants receiving high rates of N having more tillers than plants in low N (Tanaka and Garcia, 1965; Spiertz and de Vos, 1983; Wada et al., 1986; Longnecker et al., 1993; Oscarson, 2000; Xue et al., 2013).

Within a leaf, nitrogen is allocated to enable its functions (mainly photosynthesis) to proceed at an optimum rate. The higher the leaf N concentration, the higher the concentration of chlorophyll and proteins (Evans, 1983). In experiments on wheat supplied N fertilizer at 200 kg N ha^{-1}, concentrations of soluble proteins were nearly three times higher, and concentrations of ribulose bisphosphate carboxylase/oxygenase and chlorophyll were more than double those in unfertilized plants (Lawlor et al., 1989). The rate of photosynthesis per unit leaf area was higher in the fertilized plants. However, the relationship between N concentration in the leaf and rate of carbon assimilation (at light saturation) is asymptotic (Evans, 1983). At very high internal N concentrations, less photosynthesis occurs per unit of N present.

Within the canopy, N is allocated differently to the leaves at the top than to those toward the bottom (Kull, 2002; Hirose, 2005). The intensity of photosynthetically active radiation (PAR) declines downward (Gallagher and Biscoe, 1978), so toward the bottom of a dense canopy the amount of energy available is much less than at the top. The lower leaves develop as *shade* leaves, with a higher specific leaf area (the ratio of leaf area to leaf mass), so the N concentration (amount per unit leaf area) is lower as the tissues of the leaf are distributed over a bigger area (e.g., Laisk et al., 2005). This distribution of tissues, and the N in them, optimizes the ability of the plant to benefit from the high irradiance at the top of the canopy. However, this distribution does not match directly the distribution of PAR down the profile of the canopy, with N being at higher concentrations than would be required to give maximum rate of photosynthesis (Kull, 2002; Hirose, 2005). At any one time, photosynthesis usually is downregulated, so although it would be expected that the availability of N in a leaf would affect the ability of the leaf to carry out photosynthesis, it is likely that the rate of photosynthesis determines the concentration of N (Kull, 2002; Laisk et al., 2005).

If N deficiency arises, different crop species react in different ways, with there being a spectrum from maintaining leaf area and losing leaf N content (and the ability to utilize the PAR intercepted efficiently) through to maintaining the assimilatory capacity of existing leaves at the expense of leaf area (Lemaire et al., 2008).

In wheat, the effect of N on leaf initiation and growth gives a bigger leaf area index (and a longer leaf area duration) with N fertilization than in unfertilized crops (Watson et al., 1963; Fischer, 1993). As in other crops, a linear relationship occurs between rate of crop growth and interception of PAR (Monteith, 1977; Gallagher and Biscoe, 1978), so the increased leaf area results in a higher yield. Wheat grain yield is correlated strongly with grain number per unit land area, rather than individual grain size, and was shown to be related linearly to a photothermal quotient (the ratio of PAR intercepted: (mean temperature −4.5°C) over a 30-day period spanning anthesis) (Abbate et al., 1995).

2.4.2 Effects of Nitrogen Supply on Developing Seeds

Grain number is dependent on the number of flower spikes formed, the number of spikelets in each spike, and the number of viable florets in each spikelet. One reason for the increased grain number with N fertilization is that the increased tiller number gives more spikes, and another reason is that the number of spikelets per spike is increased (Longnecker et al., 1993; Ewert and Honermeier, 1999; Oscarson, 2000). However, the increase in number of grains per plant seems to come about more due to the increased number of tillers (and therefore spikes) than through there being more grains per main stem spike (Oscarson, 2000). Within the spikelets, N supply may influence also the development of floret primordia, but Ferrante et al. (2010) found that availability of N does not influence floret initiation in durum wheat. However, many florets that are initiated can degrade, and high N supply prevents degeneration of some of the florets from occurring and gives a larger number of grains than in plants with low N supply (Ferrante et al., 2010). N appears to have an indirect effect, with improved vegetative growth giving more assimilates to lower the extent of floret degeneration that would otherwise occur (Ferrante et al., 2013).

Similar effects are seen in other cereals. In corn, kernel number is also the major determinant of crop yield and is increased by N fertilization. This action occurs because of increased fertilization of ovules or decreased kernel abortion with an adequate N supply, rather than differences in numbers of ovules per cob or cobs per plant, and it arises from the indirect effect of the nitrogen in giving more assimilates for grain filling (Uhart and Andrade, 1995). Some increase in mean kernel weight occurs with increased N supply, although to a smaller extent than kernel number (Uhart and Andrade, 1995; Ciampitti and Vyn, 2011). Rice develops fewer panicles per unit land area and fewer spikelets per panicle, having a smaller leaf area index and lower leaf area duration under nitrogen deficiency (Wada et al., 1986; Xue et al., 2013).

2.4.3 Effects of Nitrogen Supply on Root Systems

As well as affecting the development of leaves, stems, and seeds, N supply also affects the development of the root system. Nitrogen deficiency in corn crops gives rise to increased angles from the horizontal of crown and base roots, particularly in genotypes that normally have shallow angles. This change in morphology gives a deeper distribution of the roots in the soil profile (Trachsel et al., 2013). In addition, N availability affects the proportions of different parts of the root system. It was seen that lateral roots of barley grow into patches of high concentration of NO_3^- and NH_4^+ (Drew, 1975). A similar response occurs in arabidopsis, where lateral roots grow into patches of high NO_3^- when the overall N supply is low, through having a higher density and a faster elongation rate, although lateral root elongation is inhibited with a high, homogeneous supply of NO_3^- (Linkohr et al., 2002). A high C:N ratio in the rooting medium suppresses the initiation of lateral roots, but supply of L-glutamic acid inhibits primary root growth and stimulates the growth of lateral roots (Zhang et al., 2007). It is well known that plants grown in hydroponic culture have smaller roots in NH_4–N than in NO_3–N, and supply of NH_4^+ to arabidopsis inhibits the elongation of primary and lateral roots although the number of lateral roots per unit length of primary root is increased (Li et al., 2010). Contact of arabidopsis shoots with NH_4^+ inhibits the formation of lateral roots (Li et al., 2013).

It seems that uptake of NO_3^- by the NRT1.1 transporter in arabidopsis signals to the plant (through interaction with the *ANR1* gene that helps regulate lateral root growth) that its primary root is growing through a high-nitrate patch (Zhang et al., 2007). The suppression of lateral root initiation by high C:N ratio seems to involve the NRT2.1 transporter, possibly acting as a nitrate sensor. The inhibition of the growth of lateral roots by high nitrate supply overall seems to be signalled by NO_3^- accumulating in the plant, possibly in the shoot, which may inhibit the synthesis or translocation of auxins such as IAA and may cause the accumulation of abscisic acid. The inhibition of formation of lateral roots by NH_4^+ supplied to the shoots is mediated by the formation of ethylene in the shoots, an action that may work by limiting the transport of auxins from the shoot to the roots

(Li et al., 2013). The inhibition of lateral root growth by NH_4^+ supplied to the roots appears to be independent of auxins (Li et al., 2010).

Infection of arabidopsis with the PGPR strain *Phyllobacterium brassicacearum* STM196 antagonizes the inhibition of lateral root development by high extracellular NO_3^- (Mantelin et al., 2006). This effect is thought to be brought about by auxins and cytokinins produced by the PGPR. Both the suppression of lateral root growth and the stimulation of plant growth by the rhizobacteria require increased expression of *NRT2.5* and *NRT2.6* genes in the shoots (Mantelin et al., 2006; Kechid et al., 2013). However, nitrate influx and expression of the *AtNRT1.1* and *AtNRT2.1* genes were decreased by 8 days after inoculation, but the expression of *AMT* genes was not affected (Mantelin et al., 2006). It appears that these plants still are able to absorb the NH_4^+ made available by the rhizobacteria, but they lose the capacity to limit their lateral roots to the high-nitrate patches and expand root growth into other zones and obtain other nutrients that may be in short supply.

Another root system change that occurs with N deficiency is that density and length of root hairs are increased (Foehse and Jungk, 1983). In arabidopsi*s*, N starvation for 2 days gave longer root hairs, and by 5 days root hair number also was increased (Engineer and Kranz, 2007). A similar pattern of increase in length and number of lateral roots occurred. Within the 2 days, the expression of the *AtAMT1.1* gene occurred in the distal parts of the root system, particularly in the new and existing root hairs and in the tips of newly emerging lateral roots and in the junctions of primary and lateral roots. By 5 days of N starvation, this expression was throughout the root system (Engineer and Kranz, 2007).

Growing phosphorus-deficient plants of mulloway needlebush with no nitrogen gave a much higher mass of cluster roots than plants grown with limiting supplies of either ammonium nitrate or glycine (Paungfoo-Lonhienne et al., 2009). No cluster roots were formed if the N supply was nonlimiting, regardless of whether it was ammonium nitrate, glycine, or protein. Although cluster roots are thought of as being involved primarily with the acquisition of phosphate by plants, their formation is regulated partly by the availability of nitrogen, and they have a proteolytic activity that helps make protein N available to plants (see Section 2.2.3).

2.4.4 EFFECTS OF NITROGEN SUPPLY ON WHOLE PLANT GROWTH

Within a plant, nitrogen is partitioned to optimize the growth potential, so that proportions of shoot to root vary according to N availability, the proportions of the different parts of the root system vary to optimize the uptake of N from a heterogeneous soil environment, and N is distributed within the canopy to maximize the interception of PAR. As these processes serve to make conditions optimal for the assimilation of carbon, the main contributor to plant dry mass, it can be seen that there must be a critical relationship between N in a plant and its rate of growth. Relative growth rate (RGR) is directly related to the internal concentration of nutrients such as nitrogen, so that

$$\mathrm{RGR} = P_n \times (C - C_{\min})$$

where
  P_n is the nutrient productivity (in this case nitrogen productivity, the amount of biomass produced per amount of N it contains per unit time)
  C is the concentration of N in the plant
  $C_{\min}$ is the minimum N concentration below which it has no effect on plant growth (Ingestad and Ågren, 1992)

As $C_{\min}$ is small, P_n is in effect the slope of a linear relationship between RGR and internal N concentration at values well above $C_{\min}$. At the early stages of vegetative growth, the internal N concentration remains relatively constant, giving a constant value of RGR (and an exponential increase in mass of the plants). However, as plants age, the proportion of strengthening material (cellulose and lignin) increases, particularly as the stem develops, and as these compounds do not contain N the concentration of N in the whole plant decreases (Ingestad and Ågren, 1992). RGR decreases with this decrease.

If N supply to a plant is interrupted, the plant N concentration decreases, and growth rate slows. However, initially at least, the decline in growth rate is not as great as would be predicted from the relationship between internal N concentration and RGR. In nitrate-grown plants, it seems that the free NO_3^- accumulating in the vacuoles buffers the plants until it is used up, and then RGR decreases dramatically (Walker et al., 2001). Although this effect of free NO_3^- in maintaining RGR can last a few days, it is largely an artificial position brought about by the removal of N supply from plants grown in hydroponic culture. These plants go from a considerable oversupply of N to withdrawal of N very quickly, which would not happen in a field soil. In a mature horticultural crop of tomato grown on rock wool, the size of pools of free NO_3^- may be sufficient to buffer plants against a sudden withdrawal of nitrate for time periods of 2–3 weeks (Le Bot et al., 2001).

The total N content of a crop is related to the mass of the crop according to the equation

$$N_{content} = aW^b$$

where
$N_{content}$ is the mass of N in the crop on a given land area (kg ha^{-1})
a and b are constants
W is the mass of crop on the same land area

Parameter a is the N content where $W = 1$ t ha^{-1}, and b is the ratio of accumulation of nitrogen to biomass (and has a value lower than 1) (Greenwood et al., 1990; Gastal and Lemaire, 2002). It therefore follows that N concentration is given by the equation

$$N\% = a'W^{1-b}$$

where
$N\%$ is the concentration of N in the crop (as a percentage of dry matter)
a' differs from a to take into account the difference in units between $N_{content}$ and W (Greenwood et al., 1990; Gastal and Lemaire, 2002)

This makes it clear that there is a decline in $N\%$ with increase in W (i.e., a decline in N concentration with crop growth stage) (Figure 2.4), and it follows a similar pattern across a range of crops, with the main difference between groups of crops being that the $N\%$ value is lower at a given W value for C4 plants than for C3 plants; b is relatively constant between the two groups, and a is lower in C4 plants than in C3 plants (Greenwood et al., 1990). However, within these broad groupings, there are differences between individual crop species, although the relationship between $N\%$ and W for any crop seems to be constant across different environmental conditions (Lemaire et al., 2007). The relationship only holds for values of $W = 1$ t ha^{-1} and above. Growth rate under N sufficiency is exponential up to this value of W, but above it becomes linear (Greenwood et al., 1990).

The relationship between $N\%$ and W gives the idea of a critical N concentration, the minimum concentration of N in a crop that generates maximum possible rate of growth of the crop at the value of W at that time (Greenwood et al., 1990). As with accumulation of unassimilated NO_3^- in storage pools, this leads to the concept of luxury consumption of N, with plants accumulating N at higher concentration than is required for maximum growth rate. $N\%$ in excess of the critical value has been demonstrated in a range of crops under different environmental conditions, at least at higher values of W (Lemaire et al., 2007). $N\%_{critical}$ is constant up to 1 t ha^{-1} and then decreases with further

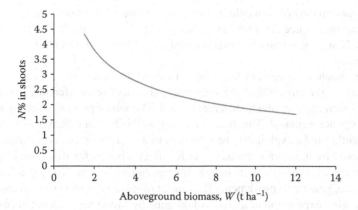

FIGURE 2.4 Change in N concentration in shoot biomass during crop growth. Based on nominal values, $N\% = aW^{1-b}$, $a = 5.2$, and $b = 0.55$, so that $N\% = 5.2W^{-0.45}$. Values for different crops are given in Table 2.1.

TABLE 2.1

Critical Nitrogen Concentrations of Some Crops in Relation to Shoot Biomass (W)

Crop	$N\%_{critical}$	Reference
Cereals		
Corn (*Zea mays* L.)	$3.40W^{-0.37}$	Plénet and Lemaire (2000)
Spring wheat (*Triticum aestivum* L.)	$38.5W^{-0.57}$	Ziadi et al. (2010)
Winter wheat (*Triticum aestivum* L.)	$5.35W^{-0.442}$	Justes et al. (1994)
Rice (*Oryza sativa* L.)	$5.18W^{-0.52}$	Sheehy et al. (1998)
Other grasses		
Annual ryegrass (*Lolium multiflorum* Lam.)	$40.73W^{-0.379}$	Marino et al. (2004)
Other crops		
Cabbage (*Brassica oleracea* var. *capitata* L.)	$5.11W^{-0.33}$	Ekbladh and Witter (2010)
Cotton (*Gossypium* spp.)	$4.969W^{-0.131}$	Xiaoping et al. (2007)
	$4.296W^{-0.131}$	
Linseed (*Linum usitatissimum* L.)	$4.69W^{-0.53}$	Flénet et al. (2006)
Winter oilseed rape (*Brassica napus* L.)	$4.48W^{-0.25}$	Colnenne et al. (1998)
Sunflower (*Helianthus annuus* L.)	$4.53W^{-0.42}$	Debaeke et al. (2012)

Note: Values (in g N per 100 g shoot dry mass) for different crops of a and $1 - b$ relative to shoot biomass (W) in the equation $N\% = aW^{1-b}$. The values presented give the relationship shown in Figure 2.4 for each crop species to maintain maximum growth rate at any stage of crop growth once the biomass yield has reached 1 t ha⁻¹.

---

growth, and any value below the line in Figure 2.4 shows where plants are below this critical value and growth is likely to become limited by shortage of nitrogen. Values for the relationship between $N\%$ and W for a range of crop species are shown in Table 2.1.

2.5 GENETICS OF ACQUISITION OF NITROGEN BY PLANTS

One of the big differences between plant species that could affect N uptake is whether they have tap roots or a fibrous root system, and this difference is controlled genetically. Less substantial differences in root architecture also may have an effect on N acquisition, and with oilseed rape it has been shown that an N-efficient cultivar has a higher root length density and more fine roots than an older cultivar. Although the efficient cultivar has a slower rate of N uptake during vegetative growth, it continues taking up N once seed formation has begun (Ulas et al., 2012). It outyields the older cultivar with low N supply, so this observation forms the basis of a potential breeding strategy for low-N conditions.

As has already been seen, each of the different transporters for NO_3^-, NH_4^+, amino acids, and urea, and also each of the enzymes of N metabolism, is coded by a different gene, and these genes are expressed at different times and in different tissues. The whole process of acquisition of nitrogen by plants is under genetic control. This fact gives the possibility that differences between species in the genes responsible can be exploited to breed plants that utilize N more efficiently. This advantage also can be achieved by improving the use of N fertilizers, but as fertilizer-use efficiency is influenced strongly by the weather, which in many countries cannot be forecast very reliably more than a few days in advance, genetic improvement offers more reliable prospects (Barraclough et al., 2010).

However, changing expression of an individual gene by breeding does not necessarily improve the agronomic performance of the plant, even if the gene has been chosen carefully. Suppose that you identify an important carrier for NO_3^-; its expression is linked to expression and repression of many other genes, so increasing expression of one does not necessarily increase nitrate uptake. If it does, will the additional ions taken up be assimilated into amino acids? Perhaps a better approach would be to breed for increased expression of nitrate reductase, but as this enzyme is substrate inducible its activity is linked already to how much NO_3^- previously has been taken up by the plant. However, the activity of individual enzymes has been manipulated, and overexpression of NADH-glutamate synthase (NADH-GOGAT) in tobacco (*Nicotiana tabacum* L.) gave bigger shoots at flowering (Chichkova et al., 2001). NADH–GOGAT has an important role in assimilation of NH_4^+ from uptake and mobilization (Quraishi et al., 2011), and increased expression of the enzyme in wheat and barley to improve grain filling is now the subject of a patent (EP2534250 A2). Nevertheless, picking out individual transporters and enzymes such as this approach is problematic, as is discussed clearly by Chardon et al. (2012).

As these authors explain, many of the characters that could be manipulated to improve the acquisition of nitrogen by plants are controlled by more than one gene. A useful approach is to identify quantitative trait loci (QTLs), areas of the DNA that contain genes that give rise to the trait under consideration. Then, by searching databases for similar DNA sequences that already have been identified, it can be seen what those genes are. For example, in considering corn grain yield it was found that the QTLs involved linked to genes for GS (Chardon et al., 2012), showing that this enzyme is very important for grain filling. In winter wheat, GS activity has been found to be correlated strongly with the amount of N mobilized from foliage to grains and to grain yield (Kichey et al., 2007), although other workers found QTLs for GS to colocate with QTLs for grain N content, rather than overall yield (Habash et al., 2007).

In breeding for improved N nutrition, it might be possible to select for nitrogen-use efficiency (NUE), and many studies have identified QTLs for NUE. However, although NUE commonly is defined as the yield per unit of N available to the crop (Hirel et al., 2007), it also can be defined as the amount of N removed in a crop as a proportion of the N supplied in fertilizer or even as the reciprocal of N concentration in the tissues (i.e., biomass/mass of N in the tissues). NUE needs to be defined clearly, and QTLs for the trait as defined can then be investigated under conditions of high or low N supply, to give crops that use N more efficiently in intensive or extensive agriculture, respectively.

Selection of traits to improve needs to be imaginative. It is known that tropical grasses in the genus *Brachiaria* (signalgrass) exude compounds from their roots that inhibit nitrification in the soil, so it has been suggested that this capacity could be bred into cereals in the tribe Triticeae, also in the Poaceae family (Subbarao et al., 2009). This potentially would give barley, rye, and wheat crops that would lower the oxidation of NH_4^+ to NO_3^- in the soil, thereby minimizing the losses of NO_3^- by leaching and making more N available to the plants. As discussed earlier, the abortion of embryos under N deficiency gives rise to lower yields of corn through causing fewer kernels to develop. Consequently, a good breeding strategy could be to breed corn plants with lower rates of embryo abortion under such conditions so that higher yields can be obtained with low N inputs (Gallais and Coque, 2005). Photosynthesis tends to be downregulated in plants, and as there can be considerable loss of leaf mass without loss of yield, the rate at which crop photosynthesis occurs probably is controlled by sink strength (Greenwood et al., 1986), so breeding for less seed abortion could be a good strategy in all cereals.

The big increases in wheat yields in the second half of the twentieth century came about following the introduction of stem-shortening genes into the germplasm. Less investment of assimilates into the stem automatically gave more investment into other plant parts, including the developing grains, thus giving a higher harvest index and higher yields. An added bonus was that because the stems were shorter, the plants were less prone to lodging, which was particularly common where N fertilizers had been used as the ears were heavy under those circumstances. Wheat varieties with short stems could therefore be grown with high-N application without risk of lodging, so overall yields increased (Barraclough et al., 2010). A breeding modification that has no apparent connection to N metabolism enabled higher yields to occur through more use of N fertilizers.

Perhaps not surprisingly, QTLs for wheat grain yield, grain protein yield, and N harvest index (the amount of N in the grain/total N in the plant) under contrasting conditions of high or low N supply have been linked to the expression of the short allele of the *Rht* dwarfing gene (Laperche et al., 2007; Quraishi et al., 2011). A QTL for NUE has been shown to colocate with two alleles of the *VRN1* (vernalization) gene, *Vrn-A1* and *Vrn-D1* (Quraishi et al., 2011). QTLs for yield traits under contrasting conditions of high and low N supply have been shown to colocalize with alleles of the *PPD1* (photoperiod) gene, in particular *Ppd-B1* and *Ppd-D1* (Laperche et al., 2007). In winter wheat, flowering occurs only in the longer days of spring, after a period of cold weather in the short days of winter. In the comparatively long days of autumn, the *VRN2* gene is expressed and downregulates *VRN3*. *VRN1* is expressed at low level. After a period of cold during the short days of winter, *VRN1* is expressed more, and this action downregulates the expression of *VRN2*. That action in turn allows for the upregulation of *VRN3* and gives more expression of *VRN1* and flowering. The *VRN3* gene also is controlled by *PPD1*, so that in winter wheat, *VRN3* is expressed only after day length has increased (Distelfeld et al., 2009). *Vrn-A1* and *Vrn-D1* are dominant alleles of the *VRN1* gene located in the A and D components of the wheat genome that do away with the requirement for a period of cold (vernalization) in order for flowering to occur (Cockram et al., 2007). *Ppd-B1* and *Ppd-D1* are semidominant alleles of *PPD1* located in the B and D components of the genome involved with the sensitivity of wheat to photoperiod, and the *Ppd-D1a* form gives early flowering in short days or long days (Cockram et al., 2007; Distelfeld et al., 2009). It can be seen that, as with stem shortening, genes controlling developmental pattern rather than aspects of N nutrition can be directly responsible for improving the efficiency with which plants utilize N.

Identification of QTLs for traits associated with efficient uptake and utilization of N should highlight to plant breeders some of the genes that do not otherwise appear to be associated with N nutrition, but the breeders have to be aware of two problems. First, the traits themselves have to be identified precisely. Are we trying to breed crops with the highest possible yields under conditions of abundant N supply, the highest yields with low N supply, highest seed/grain N concentration, highest nitrogen harvest index, or some other characteristic? Second, identification of genes that colocalize with QTLs depends on the environmental conditions where the work is carried out. Manipulation of *Ppd* genes in wheat may appear to offer a chance of improving NUE, but early-flowering *PPD1* genotypes may give higher yields in southeastern Europe where summers are hot and dry, whereas their yields can be lower than later-flowering genotypes in northwestern Europe (Cockram et al., 2007). The *stay-green* trait, which extends the period over which the leaves provide energy to allow uptake of N, may have the potential to increase cereal yields but only in environments where water supply is adequate to maintain a longer period of crop growth.

2.6　CONCENTRATIONS OF NITROGENOUS COMPOUNDS IN PLANTS

Most of the N within a plant is in the proteins, with the highest concentrations being in the seeds, and as protein is important in human and livestock diets there is considerable interest in growing crops with a high protein content. However, as shown earlier, grain filling with carbohydrates from photosynthesis and N from mobilization of foliar amino acids gives an imbalance that leads to grain N concentration (and protein content) varying with environmental conditions and cultivar.

Use of plant N in making grain proteins comes at the expense of further vegetative growth, so there is usually a negative correlation between grain yield and grain protein concentration in wheat crops grown in the field (Bly and Woodard, 2003; Barraclough et al., 2010; Bogard et al., 2010). New stay-green cultivars of corn produce a higher grain yield than older cultivars, as photosynthesis carries on for longer, but the grain N concentration is lower (Ning et al., 2013). In recent times, increases in crop yields have been matched by decreased seed or grain protein concentrations in a variety of crops (Triboi and Triboi-Blondel, 2002), but the negative relationship between wheat grain yield and grain N concentration was even noted by Lawes and Gilbert as early as 1857 (Barraclough et al., 2010). Because of this relationship, there is a negative correlation between concentrations of proteins and the other major component of a seed, so that protein accumulation in cereal grains comes at the expense of starch accumulation, and protein accumulation in oilseed crops such as oilseed rape and sunflower comes at the expense of oil accumulation (Triboi and Triboi-Blondel, 2002).

Where a high seed protein content is required, breeding has given us the potential to select for genotypes that typically have low yields and high seed protein concentration. Wheat cultivars used for bread making have a higher grain protein content and lower grain yields than other wheat cultivars (Barraclough et al., 2010), and cultivar is a more important determinant of milling and baking properties than either level of N fertilization or its timing (Otteson et al., 2008). The major protein present in mature wheat grains is gluten, itself comprising gliadins (protein subunits present as monomers) and glutenins (protein subunits assembled into polymers stabilized by interchain disulfide bonds) in approximately equal amounts (Wan et al., 2013). Gluten forms a viscoelastic matrix in the dough that prevents carbon dioxide from the yeast escaping, with gliadins being the major determinants of dough viscosity and extensibility and glutenins determining dough strength (elasticity) (Wan et al., 2013).

Even allowing for the negative relationship between seed yield and protein concentration, there is some potential to increase seed N content by soil management. In winter wheat, the N harvest index (the proportion of plant N present in the harvested grains) is increased by the application of N at anthesis (Wuest and Cassman, 1992b), and the actual grain N concentration also is increased (Wuest and Cassman, 1992a). In fact, where individual wheat crops varied from a strong negative correlation between grain yield and grain protein concentration, those crops that had a higher protein concentration than was predicted from the yield mostly had a higher uptake of N post-anthesis (Bogard et al., 2010). This feature could mean that stay-green cultivars should have higher grain protein concentration than older cultivars, but in winter wheat, it seems to be less consistently linked to delayed leaf senescence than is grain yield (Bogard et al., 2011).

As N supplied late to wheat plants is partitioned mostly to the developing grain, it is theoretically possible to supply N early in vegetative growth to give maximum yield and then to supply N late to ensure a high grain protein content. Indeed, Bly and Woodard (2003) found that postpollination application of N to bread wheats gave a higher grain protein concentration, particularly in those crops that yielded sufficiently well to be above a target yield. The practical difficulties here include that fact that driving over a high-yielding crop to supply late nitrogen can cause damage to the plants. Also, gliadins are increased more than glutenins, with various γ-gliadin genes being upregulated, so baking quality can be affected adversely (Wan et al., 2013). The N fertilization policy has to match the product that an individual crop is grown for. Bread wheat must have a high grain protein content, so overall yield can be sacrificed, whereas in cereals grown for livestock overall yield may be more important, although protein contents should be as high as possible. Oilseed rape is grown for its oil content, so seed protein content is less important; corn for human consumption ideally has a reasonable protein content, but if corn is grown for bioethanol, high grain protein is not required (Chardon et al., 2012). Barley used for brewing beer should have low protein and a correspondingly high starch content to optimize the production of monosaccharides to be fermented, so the farmer has to ensure that little N is taken up after anthesis. In order to ensure low soil N at anthesis, fertilizers can be applied only early and at a low rate.

Nitrogen in amino acids, particularly asparagine, can react with reducing sugars in the Maillard reaction, giving rise to the carcinogen acrylamide (Mottram et al., 2002; Halford et al., 2012). This reaction is favored by heat, and as plants contain both precursors plant-based foods may contain acrylamide, particularly when cooked for a long time. Tareke et al. (2002) found concentrations of 150–1000 μg kg^{-1}, or even higher, in carbohydrate-rich foods of plant origin, such as potatoes. This potential for acrylamide formation has led the European Food Safety Authority to monitor the concentrations of acrylamide in foods, with the as yet unsuccessful aim of lowering its concentrations, although epidemiological studies of possible links between acrylamide and cancers have mostly given negative results (Sanderson, 2012). Asparagine can accumulate in wheat or potatoes, often when protein synthesis is low, yet N supply is high, so there is interest in breeding low-asparagine cultivars to help minimize the formation of acrylamide in foodstuffs (Halford et al., 2012). In potatoes, reducing sugars accumulate in the tubers with N deficiency, so maintaining N fertilizer levels not only helps achieve high yields but should also give a product with lower tendency to form acrylamide on cooking (Halford et al., 2012).

Aromatic rice was shown to be more flavored in high-N soil and contained a higher concentration of the imino acid L-proline, which is a precursor of the odiferous compound 2-acetyl-1-pyrroline (Yang et al., 2012). It certainly seems that good availability of N to plants can give high concentrations of nitrogenous secondary metabolites. For example, high N supply can give increased concentrations of alkaloids (Nowacki et al., 1976), and in tobacco a higher rate of N supply gives higher concentrations of the alkaloid nicotine (Matt et al., 2002). However, there is a trade-off with the use of N in primary metabolism, and tobacco genotypes that naturally have higher nicotine contents tend to have lower biomass yields (Matzinger et al., 1972). In the salad vegetable rocket (*Eruca sativa* Mill.), there is a link between the concentrations of different glucosinolates and N supply. Increasing N supply from a low rate gave higher glucosinolate content up to a critical level (Omirou et al., 2012). Increased N supply gave increased concentrations of some volatile aromatic compounds responsible for flavor in tomato (although excess N gave decreased concentrations of some of them) and also gave higher concentrations of soluble sugars and soluble solids that help give taste, along with higher titratable acidity (Wang et al., 2007).

Unassimilated mineral N can accumulate in plants to varying degrees (Table 2.2). While NH_4^+ concentrations are always low in healthy plants, NO_3^- accumulates in the vacuoles. Accumulation of NO_3^- in leaves occurs particularly under conditions of low photon flux density, when there are lower concentrations of soluble photosynthetic products and plants maintain a balance in osmotic potential in the leaves from NO_3^- and soluble organic molecules (Burns et al., 2011).

The accumulation of NO_3^- in plant foodstuffs may give a risk of methemoglobinemia in infants. This malady occurs from the oxidation of ferric (Fe^{2+}) oxyhemoglobin brought about by NO_2^-, which could arise from the NO_3^-, although there is no firm evidence that nitrate alone in the diet would produce such an effect (Hord et al., 2009). In the past, accumulation of NO_3^- in leafy vegetables was thought to give a risk of stomach cancer to consumers, but evidence for such an effect has also not been forthcoming (Santamaria, 2006; Hord et al., 2009). Despite this lack of evidence, nitrate concentrations in leafy vegetables have to be kept low according to legislation enacted in the EU. This regulation leads to attempts by growers to limit nitrate accumulation in salad crops, for example, by increasing NH_4^+ supply to limit NO_3^- uptake, but it can also limit the commercial opportunities to grow leafy salad crops in winter in more northerly latitudes, where photon flux density is low and photoperiod is short, even if the temperature can be kept warm by growing them under glass.

TABLE 2.2

Concentrations of Nitrogenous Compounds in Plants

Compound	Plant	Concentration	Reference
Nitrate	Cabbage crops (*Brassica oleracea* var. *capitata* L.)	500–1000 mg kg^{-1} FM	Santamaria (2006)
	Endive crops (*Cichorium endivia* L.)	1000–2500 mg kg^{-1} FM	Santamaria (2006)
	Leek crops (*Allium porrum* L., syn. *A. ampeloprasum* var. *porrum* (L.) J. Gay)	1000–2500 mg kg^{-1} FM	Santamaria (2006)
	Lettuce (*Lactuca sativa* L.)	Mean 3266 (2291–4833) mg kg^{-1} FM, nutrient film technology Mean 1190 (772–1907) mg kg^{-1} FM, soil grown	Burns et al. (2012)
	Rocket crops (*Eruca sativa* Mill.)	>2500 mg kg^{-1} FM	Santamaria (2006)
	Spinach crops (*Spinacia oleracea* L.)	>2500 mg kg^{-1} FM	Santamaria (2006)
	Tobacco roots (*Nicotiana tabacum* L.)	1500–1750 μmol g^{-1} DM	Camacho-Cristóbal and González-Fontes (2007)
	Corn roots (*Zea mays* L.)	Up to 70 mol m^{-3} NO$_3^-$ in tissues	Bloom et al. (2012)
Ammonium	Tobacco leaf	30–40 μmol g^{-1} DM	Camacho-Cristóbal and González-Fontes (2007)
	Tobacco roots	11–17 μmol g^{-1} DM	Camacho-Cristóbal and González-Fontes (2007)
	Corn roots	Up to 25 mol m^{-3} NH$_4^+$ (per water volume in tissues)	Bloom et al. (2012)
	Rice shoots (*Oryza sativa* L.)	13.0–75.7 μmol g^{-1} FM	Balkos et al. (2010)
	Rice roots	12.4–37.6 μmol g^{-1} FM	Balkos et al. (2010)
	Rice shoots	0.4–3.0 μmol g^{-1} FM	Wang et al. (2013a)
	Rice roots	0.1–2.5 μmol g^{-1} FM	Wang et al. (2013a)
Urea	Rice shoots	0.1–0.4 μmol g^{-1} FM	Wang et al. (2013a)
	Rice roots	0.04–0.2 μmol g^{-1} FM	Wang et al. (2013a)
Amino acids	Cucumber leaf (*Cucumis sativus* L.)	Free glu 2.653, gln 0.442, arg 0.153 nmol mg^{-1} FM	Borlotti et al. (2012)
	Cucumber root	Free glu 0.904, gln 0.495, arg 0.721 nmol mg^{-1} FM	Borlotti et al. (2012)
	Tobacco leaf	175–200 μmol g^{-1} DM total free amino acids	Camacho-Cristóbal and González-Fontes (2007)
	Tobacco roots	110–130 μmol g^{-1} DM total free amino acids	Camacho-Cristóbal and González-Fontes (2007)
Proteins	Winter wheat flag leaf (*Triticum aestivum* L.)	Soluble proteins, up to 14 g m^{-2} Rubisco, up to 7 g m^{-2}	Lawlor et al. (1989)
	Tobacco leaf	180–200 mg g^{-1} DM	Camacho-Cristóbal and González-Fontes (2007)
	Tobacco roots	175–200 mg g^{-1} DM	Camacho-Cristóbal and González-Fontes (2007)
	Rice shoots	11–19 mg g^{-1} FM	Balkos et al. (2010)
	Rice roots	6–8 mg g^{-1} FM	Balkos et al. (2010)
	Winter wheat grain	6.2%–15.9% DM[a]	Barraclough et al. (2010)
	Corn grain	6.9%–10.7% DM[a]	Ning et al. (2013)
	Rapeseed (*Brassica napus* L.)	16.8%–23.2%	Triboi and Triboi-Blondel (2002)
	Sunflower seed (*Helianthus annuus* L.)	17.1%–21.1%	Triboi and Triboi-Blondel (2002)

(Continued)

TABLE 2.2 (*Continued*)

Concentrations of Nitrogenous Compounds in Plants

Compound	Plant	Concentration	Reference
Chlorophyll	Winter wheat leaves	Up to 0.9 g m^{-2}	Lawlor et al. (1989)
	Cucumber leaf	14 mg g^{-1} FM	Borlotti et al. (2012)
	Aspen leaves (*Populus tremula* L.)	Up to 6 mg g^{-1} DM	Keskitalo et al. (2005)

[a] Reported by authors as N concentration, converted to protein concentration using conversion factor from $N\%$ of 5.7. DM, dry mass; FM, fresh mass.

2.7 INTERACTIONS OF NITROGEN WITH OTHER ELEMENTS

It is known that there are reasonably constant ratios of tissue concentrations of phosphorus, potassium, calcium, and magnesium relative to nitrogen across a big range of plant species. Adherence to these proportions gives growth that occurs at an optimum rate, although deviation from optimum proportions can be found (Knecht and Göransson, 2004). Local deficiencies of these and other elements can have effects on the internal ratios, but other factors are also important.

As N is the root-acquired element taken up in the largest amounts by plants, whether it is taken up as the NO_3^- anion or the NH_4^+ cation has a big effect on the uptake of other anions and cations. Arnon (1939) showed that barley grown in NH_4^+ contained lower concentrations of the cations Ca^{2+}, Mg^{2+}, and K^+ than plants grown in NO_3^-. Not only is there a general effect in which extensive uptake of one or other of these ions makes it energetically more difficult for a plant to acquire other anions or cations, but it is likely that similarly sized ions may compete with the N forms for binding to the transporter proteins.

This action can be seen in experiments on common bean (*Phaseolus vulgaris* L.), where the rate of uptake of K^+ was much higher with NO_3^- as the N source than with NH_4^+ (Guo et al., 2007). In rice, which has NH_4^+ as its major N source, fresh weight was greatest with seedlings grown in 10 mol NH_4^+ m^{-3} and 5 mol K^+ m^{-3}, but seedlings grown in 10 mol NH_4^+ m^{-3} and 0.02 mol K^+ m^{-3} had lower mass than seedlings grown in lower concentrations of NH_4^+ and different concentrations of K^+ or seedlings grown in 10 mol NO_3^- m^{-3} (Balkos et al., 2010). NH_4^+ was shown to be taken up by K^+ transporters and channels in barley and arabidopsis, and there is direct competition between the two ions for uptake (ten Hoopen et al., 2010).

Cations other than K^+ are less similar in size and properties to NH_4^+, yet there are still interactions between them and nitrogen nutrition because of their importance in nitrogenous compounds. Magnesium is a constituent of chlorophyll, and so could affect photosynthesis. Supplying high concentrations of Mg^{2+} in the hydroponic solution gave sunflower plants supplied NH_4^+ similar shoot dry weight and leaf area to plants supplied NO_3^-, which at low Mg^{2+} supply had much lower shoot dry weight and leaf area (Lasa et al., 2000). The high Mg^{2+} supply increased the rate of photosynthesis per unit leaf area to a similar value to that found in the NO_3-supplied plants, but this result was not an effect due to the role of Mg^{2+} in chlorophyll synthesis as chlorophyll concentrations were not affected by Mg^{2+} supply (Lasa et al., 2000).

Rapid growth of temperate zone pasture crops at temperatures above 14°C following freezing or near freezing temperatures, and after supply of N fertilizers, gives a risk of grass tetany in livestock caused by low Mg^{2+} concentrations in the forage (Robinson et al., 1989). Nitrogen fertilization can increase K^+ concentrations in the forage and also produces more young plant tissues that have a high K^+:(Ca^{2+} + Mg^{2+}) ratio. However, NO_3^- can encourage Mg^{2+} uptake, and a temperature above 14° should encourage oxidation of urea or NH_4^+ to NO_3^-, so the grass tetany risk may come from the increased concentrations of crude protein, fatty acids, and organic acids and decreased concentrations of water-soluble carbohydrates in the forage, all of which decrease the availability of Mg^{2+} to the animals (Robinson et al., 1989). In multispecies pastures, the supply of N fertilizers encourages the growth of grasses more than forbs and legumes, and as grasses have lower Mg^{2+} concentrations the forage obtained presents a bigger risk to livestock than unfertilized pastures (Robinson et al., 1989).

Supply of N as NH_4^+ rather than NO_3^- also depresses the uptake of Ca^{2+}. Although the calcium-deficiency disorders of horticultural crops, such as blossom-end rot in tomatoes and peppers, are caused by rapid fruit growth outstripping the potential for transpiration to supply Ca^{2+} ions in adequate quantities, supply of NH_4^+ increases the risk of these diseases occurring by lowering Ca^{2+} uptake (Wilcox et al., 1973).

Nitrogen supply affects the uptake of iron (Fe) by plants, principally because uptake of NO_3^- leads to alkalinization of the rhizosphere due to the cotransport of H^+ with the NO_3^- ions, and this alkalinization makes Fe^{3+} less available. Genes for nicotianamine synthase are induced by NO_3^-, thus facilitating the synthesis of nicotianamine, which is involved in uptake and homeostasis of Fe (Wang et al., 2003). However, Fe also affects nitrogen nutrition. For example, Fe-deficient cucumber (*Cucumis sativus* L.) was found to have lower nitrate reductase activity (and a lower level of NR transcript) in the leaves than control plants and slightly lower nitrate reductase activity in the roots. GS and GOGAT activities were increased by Fe deficiency in both organs (Borlotti et al., 2012).

Another divalent cation, Zn^{2+}, had higher concentrations in wheat with N supplied as NO_3^- than with mixed NH_4^+/NO_3^- (Wang and Below, 1998; Drihem and Pilbeam, 2002) and also than with NH_4^+ alone (Wang and Below, 1998). Zinc uptake is enhanced by good N supply, and the rate of uptake and the rate of translocation from roots to shoots of ^{65}Zn were higher in durum wheat seedlings with adequate supply of NO_3^- than in seedlings with low NO_3^- supply, particularly in Zn-deficient seedlings (Erenoglu et al., 2011). There is considerable interest in increasing Zn concentrations in cereal grains, and more accumulates in grains if plant senescence is delayed after anthesis. One way in which senescence can be delayed is by maintaining high-N availability, and use of N fertilizers at anthesis has been shown to increase grain Zn^{2+} accumulation (Kutman et al., 2012).

Nickel (Ni^{2+}) is also taken up more when NO_3^- is the N source. Arabidopsis plants fed NO_3^- had greater expression of the *IRT1* gene (which codes for a transporter of divalent cations including Fe^{2+} and Ni^{2+}) in the roots and contained higher concentrations of Ni^{2+} in roots and shoots than plants supplied NH_4–N (Hu et al., 2013). These high Ni^{2+} concentrations gave the nitrate-fed plants more signs of nickel toxicity. Although Ni^{2+} can be toxic, it is a plant micronutrient as it is a component of ureases (Dixon et al., 1975), and Ni^{2+} deficiency depresses urea uptake (Arkoun et al., 2013).

In the same way that NH_4^+ uptake affects cation uptake, it could be expected that supply of N as NO_3^- might interfere with uptake of another major anion required by plants, SO_4^{2-}. However, interaction at the point of uptake appears to occur more between NO_3^- and Cl^-. Influx of Cl^- in barley is inhibited by NO_3^- in the rooting medium (Glass and Siddiqi, 1985), and there was a negative effect on influx of NO_3^- into barley and carrot root cells from Cl^- ions accumulated in the vacuoles (Cram, 1973). As noted earlier, the CLCa nitrate transporter in the tonoplast has a higher selectivity for NO_3^- than for Cl^- (De Angeli et al., 2006). In *Citrus* species, HATS for NO_3^- uptake is competitively inhibited by external Cl^-, although the LATS that operates at higher external NO_3^- concentrations is not (Cerezo et al., 1997). Within plants, NO_3^- moves more readily through the X-QUAC system (and hence into the xylem in the roots) than Cl^-, which itself moves more readily than SO_4^{2-} (Gilliham and Tester, 2005). There are further interactions between NO_3^- and Cl^- as nitrate reductase is inhibited by Cl^-, which acts on its molybdenum component (Barber et al., 1989). All of these interactions are often masked by the fact that when plants are exposed to excessive Cl^- ions, they may be experiencing salinity, and many of the processes of uptake and assimilation of nitrogen also are affected by the low water potential that occurs.

The effect of Cl^- on molybdenum in nitrate reductase highlights the fact that activity of the enzyme is also dependent on the supply of molybdate (MoO_4^{2-}) to a plant, and its internal concentration (Shaked and Bar-Akiva, 1967). Another anion whose uptake is affected by the N form supplied is phosphate (Pi), which has been known for a long time to be present at higher concentrations in plants supplied NH_4^+ than in plants supplied NO_3^- (Arnon, 1939). Rice grown with NH_4^+ takes up more Pi than rice grown with NO_3^-, possibly because the low rhizosphere pH generated by NH_4^+ nutrition stimulates the H^+-ATPase that generates the energy for Pi uptake (Zeng et al., 2012).

However, as already discussed, the ratios of N:P in plants are relatively constant at values that give optimum growth (Knecht and Göransson, 2004). Optimum N:P ratios of 11.83:1 have been found in growth-related tissues of many crop species, and average values of 12.65:1 and 13.64:1 occur in terrestrial and freshwater autotrophs (Greenwood et al., 2008). This ratio is not found under all circumstances, and in timothy (*Phleum pratense* L.), higher concentrations of P occurred in the tissues when N supplies were limiting growth (Bélanger and Richards, 1999). In tomato supplied ratios of NO_3–N and P ranging from 18:1 to 2:1, the plants took up N and P at rates that gave internal concentrations closer to 14:1 than to the supply ratios, and where internal homeostasis was unable to give internal values close to 14:1, the plants grew less well (Abduelghader et al., 2011). This occurrence reinforces the idea of there being an optimum N:P ratio for plant growth, although the N:P ratio and the RGR of whole plants decline with age, due to the decline in proportion of tissues involved in growth (Ågren, 2004; Greenwood et al., 2008).

Work on arabidopsis has shown antagonism between uptake of nitrate and phosphate, so that high NO_3^- suppresses phosphate accumulation and high phosphate suppresses N accumulation, although to a lesser extent (Kant et al., 2011). It seems that in plants supplied NO_3^-, low N supply gives rise to responses controlled by different genes, including the *Nitrogen Limitation Adaptation* (*NLA*) gene, which is involved in the regulation of phosphate uptake (Kant et al., 2011). Nitrogen deprivation gave rise to accumulation of inorganic phosphate in corn leaves and down-regulation of genes involved in phosphate homeostasis (Schlüter et al., 2012). In experiments on arbuscular mycorrhizal barrel medic (*Medicago truncatula* Gaertn.) in association with the fungus *Rhizophagus irregularis* C. Walker & A. Schüßler (syn. *Glomus irregulare* Blaszk., Wubet, Renker & Buscot; *Glomus intraradices* G. N. Schenck & G. S. Sm.), plants that were subjected to simultaneous N and P deprivation had more mycorrhizal colonization than plants subjected to low P alone, despite having higher internal P concentrations (Bonneau et al., 2013). More genes were induced in the plants subjected to low N and P together, and N deficiency induced several genes for phosphate transporters.

Sulfur (S) and N nutrition interact at many levels, as the uptake and assimilation of NO_3^- and SO_4^{2-} have much in common, and there are many common products of N and S metabolism (Hesse et al., 2004). Nitrate induces gene expression for sulfate transporters (Vidmar et al., 1999; Wang et al., 2003) and for sulfate assimilatory enzymes (Koprivova et al., 2000; Wang et al., 2003). Ammonium supply to N-deficient plants also increases the expression of a sulfate transporter gene (Vidmar et al., 1999). Conversely, low S depresses the uptake of NO_3^- and NH_4^+ (Clarkson et al., 1989), and there is a negative effect on NR gene expression with S deficiency in tobacco (Migge et al., 2000). This gene repression was caused by an accumulation of glutamine or asparagine following the withdrawal of S. Sulfur-deficient plants accumulate arginine and asparagine in particular, but have lower levels of the sulfur-containing amino acids cysteine and methionine (Hesse et al., 2004). Guinea grass (*Panicum maximum* Jacq.) accumulates asparagine with low S supply, the leguminous pasture crop stylo (*Stylosanthes guianensis* Sw.) accumulates arginine, and both species accumulate free NO_3^- (Schmidt et al., 2013). With low S supply, the N:S ratio can be above 60:1 in these species, whereas for optimal growth a more normal ratio of 20:1 is required (Schmidt et al., 2013). Accumulation of asparagine in wheat grains under S deficiency gives an increased risk of formation of acrylamide when the flour products are cooked (Halford et al., 2012).

Some of the interaction between N and S metabolism comes from *O*-acetylserine, the immediate precursor of cysteine that does not itself contain S. For assimilation of sulfate to occur, plants must contain adequate levels of this precursor, and as it is an amino acid its concentration is dependent on

N nutrition (Hesse et al., 2004). As well as affecting concentrations of S-containing amino acids, S supply affects concentrations of other metabolites that contain S and N. This effect includes alliins (cysteine sulfoxides), concentrations of which were increased considerably in bulbs and leaves of onion and garlic by increased S fertilization but not much by increased N fertilization (Bloem et al., 2005). Concentrations of glucosinolates were high in the heads of broccoli (*Brassica oleracea* L. var. *italica* Plenck.) grown with low rates of N supply, but were low if enough N was supplied to give good yields when S supply was low (Schonhof et al., 2007). Heads with tissue N:S ratios above 10:1 had low glucosinolate concentrations.

Imposition of boron deficiency on tobacco plants led to decreased rates of nitrate uptake within 2 days, and expression of *NRT2* genes decreased (Camacho-Cristóbal and González-Fontes, 2007). This result gave noticeably lower concentrations of NO_3^- in leaves and roots, and the accumulation of carbohydrates in boron-deficient plants could arise from interference with N nutrition.

Soil aluminum is often deleterious to crop production, although there are some plants in which it is a beneficial element. In a comparison of 15 accessions of subspecies *indica* rice, it was found that they preferentially took up NO_3^- as their N source, and they tended to be sensitive to aluminum, whereas 15 subspecies *japonica* accessions were found to take up NH_4^+ preferentially and were aluminum tolerant and even had their growth stimulated by aluminum (Zhao et al., 2013). Although many rice genotypes preferentially acquire their nitrogen as NH_4^+, most other cereal species preferentially take up NO_3^- and are sensitive to aluminum (Zhao et al., 2013).

2.8 DIAGNOSIS OF NITROGEN STATUS IN PLANTS

Plants deficient in N have yellow leaves, due to lowered synthesis of protein and chlorophyll, and traditionally farmers have known when their crops require additional nitrogen by this observation. The logical progression for such observations was the measurement of chlorophyll concentration in crops with small, handheld meters that measure transmissions in the red (R) and near-infrared (NIR) wavelengths (Olfs et al., 2005). Strong negative relationships occurred between the NIR:R reflectance ratio at the prepanicle initiation or the panicle differentiation growth stages of rice and the extra yield that could be obtained by topdressing with nitrogen (Turner and Jund, 1991). Measurement of the reflectance ratio in the uppermost fully expanded leaves of wheat, barley, oats, and rye crops relative to the values with no leaf present gave accurate predictions of the leaf N concentrations and allowed the calculation of critical chlorophyll meter values for fertilizer N requirement (Peltonen et al., 1995). More recently, chlorophyll meters have been replaced by leaf color charts for poor rice farmers (Balasubramanian et al., 1999; Yang et al., 2003).

These methods give a more rapid diagnosis of crop N requirement than laboratory tests. Another quick, field test is to measure NO_3^- concentration in the sap with test strips (Scaife and Stevens, 1983). This test is based on the idea that if NO_3^- concentrations in leaves are low, the potential capacity for the formation of amino acids and proteins will not be realized, so additional N fertilizer should be supplied. It also identifies crops that have accumulated NO_3^- above limits for human consumption. The corn stalk nitrate test can be used to measure N concentration in the stalks of corn at harvest and gives an indication as to whether the crop is N deficient or contains excessive N, and serves as a tool to adjust the N fertilizer supply in following seasons (e.g., Sawyer, 2010).

Initially, chlorophyll (and NO_3^- concentration) measurements were made on individual leaves from representative plants, and it was established if crop yield would be increased by the application of a N fertilizer. Modeling of the relationship between sensor readings and plant yield subsequently became more complex. For example, models were established for wheat to enable farmers to apply N at a 1 m^2 scale in the field to give the maximum grain yield possible at harvest, based on midseason predictions of what the yield would otherwise have been in each square without N application (Raun et al., 2002).

The ratio of reflectance in the NIR wavelengths to the reflectance in the red wavelengths gives a ratio-based vegetation index (VI). From this determination, people developed the normalized difference vegetation index (NDVI), the ratio of the difference between the NIR and red reflectances and their sum (Jackson and Huete, 1991):

$$\frac{\left(R_{NIR} - R_R\right)}{\left(R_{NIR} + R_R\right)}$$

where R_{NIR} and R_R are the reflectances in the NIR and red wavelengths, respectively. This index is used for remote sensing of chlorophyll concentration by satellites to estimate plant biomass across areas of land. Other VIs include orthogonal VIs, which have a correction component to account for the background reflectance from the soil, and atmospheric corrected indices, to take into account the spectral properties of the atmosphere (Daughtry et al., 2000; Zhao et al., 2005; Dorigo et al., 2007). Some researchers developed VIs based on derivatives of the reflectance values measured.

Commonly used derivatives include the red-edge effect. Here, the fact that in moving from 680 to 800 nm you pass from the wavelength of maximum absorbance (and minimum reflectance) of chlorophyll (between 660 and 680 nm) to almost zero absorbance (and nearly maximum reflectance) gives the opportunity to plot the gradient of the line between reflectances at the wavelengths chosen in the red and NIR bands to give the chlorophyll concentration (Daughtry et al., 2000; Zhao et al., 2005; Dorigo et al., 2007). Furthermore, if the relationship between reflectance and wavelength is plotted continuously between contiguous wavelengths, it is possible to see at which wavelength the gradient reaches its maximum value, and this value can be used as an indication of the density of vegetation (Daughtry et al., 2000; Dorigo et al., 2007). Determination of the red-edge inflection point is more sensitive than NDVI when biomass is dense, but is less sensitive when biomass is scarce, so when reflectance is used to evaluate N requirements early in wheat crop growth, the extra expense and lower sensitivity of red-edge position measurements make NDVI a more useful technique (Kanke et al., 2012).

Work has been carried out to optimize the choice of NIR and R wavelengths for such models and to evaluate how wide the bandwidth can be for accurate and repeatable measurements. In drawing up an NDVI for the evaluation of N stress in cotton, it was found that although the red channels of most commercial sensors typically operate at 640–660 nm, the best combination of wavelengths to use is between 680 and 730 nm in the red band and between 750 and 850 nm in the NIR (Zhao et al., 2005). However, relationships such as this based on two wavelengths cause problems when an entire canopy is being analyzed, as a dense canopy causes saturation of the NDVI (Wang et al., 2012a). For this reason, indices have been developed based on three different wavelengths (Dorigo et al., 2007). A three-band NDVI has been evaluated on rice and wheat, based on the relationship between

$$\frac{\left(R_{924} - R_{703} + 2 \times R_{423}\right)}{\left(R_{924} + R_{703} - 2 \times R_{423}\right)}$$

and leaf N concentration (Wang et al., 2012a). The regression of leaf N concentration on the result of the three-wavelength term was robust and repeatable for both species, even if relatively broad bandwidths (36 nm for 924 nm, 15 nm for 703 nm, and 21 nm for the 423 nm measurement) were used. This determination facilitated the development of an inexpensive remote sensor for accurate estimation of plant N concentration. In a study on wheat and barley, reflectance in the red, red-edge, and NIR bands (670 nm with a bandwidth of 25 nm, 720 nm with a bandwidth of 10 nm, and 790 nm with a bandwidth of 25 nm, respectively) was measured by satellite, from the air, or by motorbike, and after processing the data through VIs, plant N concentration was plotted at a field scale (Perry et al., 2012).

Other wavelengths can be used, as in studies on the optimal N fertilizer rate in maize (Solari et al., 2008; Sripada et al., 2008; Barker and Sawyer, 2010). The sensors used by these authors measure in the green wavelengths and in the NIR, and Sripada et al. (2008) found that on testing a range of different VIs based on sensor measurements at the V6 stage of maize growth, the relative green normalized difference vegetation index ($RGNDVI_R$) was the best predictor of the economic optimum N rate (EONR):

$$RGNDVI_R = \frac{GNDVI_{plot}}{GNDVI_{reference\ plot}}$$

$$GNDVI = \frac{(NIR - G)}{(NIR + G)}$$

where

NIR is the reflectance at 880 nm

G is the reflectance at 590 nm

$GNDVI_{plot}$ is the value of the green normalized difference vegetation index (GNDVI) in the trial plot

$GNDVI_{reference\ plot}$ is the value of the GNDVI in a high-N strip (Sripada et al., 2008)

Early sensors were passive, and their accuracy was affected by light conditions in the field. That problem restricted the hours in which sensing could be carried out. With the advent of cheap light-emitting diodes, active sensors have been developed that measure the reflectance of light emitted on to the canopy, and many recent studies have used them. Active sensors allow for the measurement of chlorophyll fluorescence rather than reflectance. They facilitate the measurement of the chlorophyll/flavonoid ratio in plants, which should be more sensitive to the N status as flavonoids increase with N deficiency while chlorophyll decreases (Samborski et al., 2009). In a study on the turfgrasses seashore paspalum (*Paspalum vaginatum* Sw.) and manila grass (*Zoysia matrella* Merr.), the nitrogen balance index (NBI_1) discriminated between six different levels of applied N:

$$NBI_1 = \frac{FRF_G FRF_{UV}}{FRF_R^2}$$

where FRF_G, FRF_{UV}, and FRF_R are the values of fluorescence excited by green, ultraviolet, and red wavelengths, respectively (Agati et al., 2013).

An alternative approach that is cheaper than using specialist sensors is to use digital photography. Measurements of green channel minus red channel values in a rice crop image have been shown to be related to N content (Wang et al., 2013b). A comparison of commercial sensors used for plant N assessment is given by Samborski et al. (2009).

2.9 FORMS AND CONCENTRATIONS OF NITROGEN IN SOILS AND THEIR AVAILABILITY TO PLANTS

In soils that do not receive fertilizers, N occurs mainly in humus, in the proteins of microorganisms, in plant roots, and in soil-living animals. As these organisms die, proteins are hydrolyzed to amino acids, which are then converted to NH_4^+, NO_2^-, and NO_3^- by soil microorganisms. However, there is some input of inorganic N in rainwater as NO_3^- and NO_2^- from the fixation of N by lightning and combustion of fossil fuels, and as N_2 gas that exchanges with N_2 in the atmosphere. This N_2 can be reduced to NH_4^+ by free-living and symbiotic microorganisms.

Microorganisms outcompete plants for mineral N in a natural soil, but if microbial biomass remains constant some of their N content is released as they respire CO_2 from carbon in the organic matter that they feed on, making N available for plants (Kuzyakov and Xu, 2013). Even where there has been supply of inorganic N in fertilizers, the natural nitrogen cycle operates and N forms other than those applied can arise. Much of the ^{15}N in $^{15}NH_4^{15}NO_3$, or ^{15}N urea was immobilized in soil microbes within a month of supplying it as fertilizer (Recous et al., 1988). However, after application of N fertilizer there may be more N present than the soil microbes can utilize, increasing the risk of it being lost (Kuzyakov and Xu, 2013).

The microbial conversion of NH_4^+ to NO_2^- and then NO_3^- depends on there being oxygen available, and in an aerobic soil occurs to the extent that NO_3^- is the major N form present. Under anaerobic conditions, such as waterlogging, these two microbial steps are inhibited, and furthermore, existing NO_3^- is reduced by denitrifying soil microbes and is lost as nitrogen gases (N_2 or N_2O), so NH_4^+ tends to be the major inorganic N form present. Even if it does not give rise to anaerobic soil conditions, heavy rainfall can remove some of the NO_3^- ions by leaching, although NH_4^+ ions are retained in the soil by cation exchange. Soil acidity also inhibits the reactions of nitrification without significantly affecting proteolysis, again giving rise to higher proportions of the soil inorganic N occurring as NH_4^+ and also giving high concentrations of amino acids (Näsholm et al., 2009). Microbial reactions are temperature dependent, and cold conditions such as those in the taiga and tundra tend to slow down many of the steps of the N cycle.

In the seasonally cold soils of the taiga, proteases function well, and amino acids are abundant (Kielland et al., 2007). Amino acid uptake theoretically could account for between 10% and 90% of N requirement in some plant species, although there is adsorption of amino acids to clay minerals and other soil components and there is competition with soil microbes for their uptake (Lipson and Näsholm, 2001; Kuzyakov and Xu, 2013). In recent times, amino acids have come to be regarded as being an important N source for plants and trees in many soils. Measurement of diffusive flux of amino acids to roots of plants in 15 boreal forest soils from northern Sweden showed that in those soils, amino acids supply 74%–89% of the total N flux, with NH_4^+ contributing 5%–15% and NO_3^- 5%–11% (Inselbacher and Näsholm, 2012). This sample included soils that had received mineral N fertilizer. However, the actual concentrations of N forms in the soils differed considerably, with NH_4^+ contributing 79% of the total soil solution N, amino acids 11% and NO_3^- 10% (Table 2.3).

Differences between N flux and concentrations of various N forms can arise because of imbalances between the different steps of mineralization and also because the different N forms are taken up by plants and trees at differing rates. Amino acids may be present mostly in bound form, and in a study of three sites in Sweden with four different land uses, bound amino acids were typically in the concentration range of 20–30 mmol m^{-3}, whereas free amino acids occurred in the range of 0–29 mmol m^{-3} for individual compounds (Jämtgård et al., 2010). These concentrations are within the range of K_m values seen for amino acid transporters. Serine, glycine, and alanine were the major amino acids present. Although bound amino acids may not be instantaneously available to plants, they represent a pool from which free amino acids may arise.

As well as amino acids, soil solution also contains amino sugars (Mulvaney et al., 2001) and quaternary ammonium compounds such as betaine, carnitine, acetyl carnitine, choline, and ergothioneine (Warren, 2013a). In a study of soils from 18 arable sites in Illinois (USA), the concentration of amino sugar N ranged from 31% to 98% as high as amino acid N (Mulvaney et al., 2001). In a subalpine Australian grassland, the quaternary ammonium pool in the soil water was up to 25% the size of the amino acid pool (Warren, 2013a).

Currently, it is thought that in soils of low fertility amino acids are the main N form available to plants, that in soils of high fertility NO_3^- is the predominant form, and that in soils of intermediate fertility it is NH_4^+. The most important N form taken up by plants in each of these different soils matches the form that predominates (Rothstein, 2009). In a study of five forest sites of different levels of fertility in Michigan, USA, the concentration of free amino acids was highest at low fertility and decreased with increase in fertility. The concentration of NO_3^- showed the reverse trend, and the concentration of NH_4^+ was highest in the sites with intermediate fertility. Amino acid concentrations were particularly high in the spring and so could represent an important N

TABLE 2.3
Some Representative Soil N Contents

Amino Acids	Ammonium	Nitrate	Notes	References
3–24 mg N kg^{-1} soil	2–8 mg N kg^{-1} soil	<1.0 mg N kg^{-1} soil	Five forest taiga ecosystems	Kielland et al. (2007)
0.6 mg N kg^{-1} soil (free)	2.8 mg N kg^{-1} soil (free)	0.35 mg N kg^{-1} soil (free)	Unfertilized boreal soil in northern Sweden	Inselbacher and Näsholm (2012)
6 mg N kg^{-1} soil (free + exchangeable)	5 mg N kg^{-1} soil (free + exchangeable)	0.4 mg N kg^{-1} soil (free + exchangeable)		
30 to <5 mg N m^{-2} (September)	80–40 mg N m^{-2} (May)	70–10 mg N m^{-1} (April)	Temperate forest, United States	Rothstein (2009)
150 to <5 mg N m^{-2} (May)	310–70 mg N m^{-2} (September)	400–30 mg N m^{-2} (September)	Soluble peptides 20–150 mg N m^{-2}	
<5 μmol L^{-1} soil solution (free), up to 30 μmol L^{-1} (bound)	Up to 10 μmol L^{-1} soil solution	<5 μmol L^{-1} soil solution	Thinned birch forest, mid Sweden	Jamtgård et al. (2010)
<5 μmol L^{-1} soil solution (free), up to 30 μmol L^{-1} (bound)	Up to 7 μmol L^{-1} soil solution	<5 μmol L^{-1} soil solution	Old grassland, mid Sweden	Jamtgård et al. (2010)
Up to 5 μmol L^{-1} soil solution (free), 30 μmol L^{-1} (bound)	Up to 6 μmol L^{-1} soil solution	5 μmol L^{-1} soil solution	Organic ley, mid Sweden	Jamtgård et al. (2010)
Up to 10 μmol L^{-1} soil solution (free), up to 70 μmol L^{-1} (bound)	Up to 10 μmol L^{-1} soil solution	Up to 2200 μmol L^{-1} soil solution	Organic lettuce, mid Sweden	Jamtgård et al. (2010)
nd	2.0 (range 0–11) mg kg^{-1} soil 0.05–0.8 mol m^{-3} in soil solution	28.7 (range 0–97) mg kg^{-1} soil 0–7 mol m^{-3} in soil solution	256 sites in 4 winter cereal fields, United Kingdom	Lark et al. (2004) and Miller et al. (2007)

Amino Acids	Ammonium	Amino Sugars		
70–908 mg N kg^{-1} soil (in hydrolysate)	182–604 mg N kg^{-1} soil (in hydrolysate)	116–511 mg N kg^{-1} soil (in hydrolysate)	Maize fields preplanting, Illinois, USA	Mulvaney et al. (2001)

Amino Acids	Quaternary Ammonium Compounds	Peptides		
4.8 μmol L^{-1} soil solution	1.2 μmol L^{-1} soil solution	0.4 μmol L^{-1} soil solution	Subalpine grassland, Australia	Warren (2013a)

source for the trees, particularly as the trees in the low fertility sites tended to have ectomycorrhizal associations, whereas those in the high fertility sites were more likely to have AM associations (Rothstein, 2009).

In agricultural soils that receive fertilizers and manures, the concentrations of inorganic N forms are higher than in unfertilized soils, and the predominant ion is NO_3^-. However, concentrations vary considerably among soils, among different fields on one soil type, and even within fields. In a study of soil samples taken in October from 4 m intervals along a transect across four fields in a winter cereal rotation in the United Kingdom, concentrations of NO_3^- ranged from below 1 mol m^{-3} to above 7 mol m^{-3} and of NH_4^+ from below 0.1 mol m^{-3} to up to 0.8 mol m^{-3} (Lark et al., 2004; Miller et al., 2007). Although concentrations of both N forms varied considerably between consecutive transect positions, most of the samples contained NO_3^- at concentrations above 1 mol m^{-3}, and so in the range where uptake into plants would occur by LATS rather than HATS. Most of the samples contained NH_4^+ at between 0.1 and 0.2 mol m^{-3}. The significance of urea uptake in relation to crop yields is probably small other than when it is applied to foliage, as the presence of free and microorganism-bound urease means that the urea in a field soil is probably hydrolyzed before much of it is taken up (Engels and Marschner, 1990).

2.10   TESTING SOILS FOR NITROGEN CONTENT

A key requirement for predicting fertilizer N requirements in many North American and European countries is to know the soil mineral nitrogen (SMN) content (the concentrations of NH_4^+ and NO_3^- ions). Laboratory analysis is carried out on representative soil samples (15–20 ha^{-1}) taken within the rooting depth of the crop being grown or about to be grown, typically in 2–3 different soil layers (Olfs et al., 2005). The soil is extracted with a mild extractant such as 1.0 kmol m^{-3} KCl or 0.0125 kmol m^{-3} CaCl$_2$, and the extract is tested for NO_3^- and NH_3 (Olfs et al., 2005). Tests for SMN are typically carried out before planting a crop, so they can be used to predict its N requirement, but can also be carried out during vegetative growth to improve the accuracy of these predictions. An example here is the pre-sidedress nitrate test recommended for corn growing in several U.S. states, where soil cores are collected by the farmer in spring just before rapid crop growth starts and are then sent to a soil-testing laboratory (e.g., Iowa State University, 1997). The pre-sidedress nitrate test has been shown to be more reliable for fertilizer recommendations in corn crops in Argentina than presowing soil nitrate tests (Sainz Rosas et al., 2008). NO_3^- concentration in the soil also can be measured quickly in the field with nitrate test strips (Schmidhalter, 2005).

   In addition to measurement of SMN, there can be a requirement to measure the potential for N mineralization in a soil. It is possible to measure the total soil organic N concentration by difference between total N and inorganic N concentrations, but this does not distinguish between organic N that may be mineralized and recalcitrant organic N. The standard method of estimating mineraliz-able N is to incubate soil for 210 days in an aerobic environment and to measure mineral N at the end of that time (Schomberg et al., 2009). However, if mineralization potential is being measured to predict N fertilizer requirement, results are required quickly. One means of achieving this action is to measure NH_4–N concentration after anaerobic incubation of a soil sample for 7 or 14 days (Schomberg et al., 2009).

   Another way of evaluating the N fertilizer requirement from how much mineralizable N there is in a soil is based on the fact that in well-fertilized corn fields amino sugar-N concentrations are high, whereas in poorly fertilized fields they are low (Mulvaney et al., 2001). This measurement is the Illinois Soil Nitrogen Test, where amino sugar-N concentration is measured in soil at 0–15 cm depth before planting corn and the N fertilizer requirement is calculated based on the values obtained. The method involves making hydrolysates by heating soil under reflux in hydrochloric acid and measur-ing total hydrolyzable N by Kjeldahl analysis, then hydrolyzable NH_4–N, NH_4–N + amino sugar-N, and amino acid-N concentrations in subsequent steps (Mulvaney and Khan, 2001). The test has been used since 2001, but gives a poor level of accuracy in some areas (Laboski et al., 2008).

2.11   NITROGEN FERTILIZERS

Since the amount of crop biomass formed is proportional to PAR intercepted, providing crops with sufficient N to develop and maintain a large leaf area enables maximum interception to occur. Fertilizer policy needs to give sufficient N for RGR to remain constant at its maximum possible value (i.e., growth is exponential) for as long as possible. As N deficiency causes a proportionally larger decline in dry matter accumulation when W (mass of crop per given land area) is small and growth rate under N sufficiency is proportional to W, it is more important to ensure N supplies are sufficient early on in crop growth than when W is larger, and growth rate is approximately linear (Greenwood et al., 1986). Eventually, RGR will decline anyway (due to increased structural material and self-shading, but also as environmental cues switch on reproductive growth). Prolonging a fast growth rate is advantageous, but it is vital to ensure growth goes as fast as possible to start with.

The relationship between W and $N\%$ is constant for each crop across a range of environmental conditions (Greenwood et al., 1990; Lemaire et al., 2007). Therefore, it is possible to calculate easily how much N already must have been taken up per unit of land area in a crop of a given biomass and to know how much more has to be available to the plants to achieve any particular target yield. This concept is the basis of simple fertilizer recommendations, where subtraction of the value of available N in the soil from the total required gives the value of N that needs to be supplied. As crop growth is reduced in proportion to the ratio of N concentration actually in a crop to the critical N concentration at that growth stage, this ratio can also be used as a nitrogen nutrition index to evaluate the extent of any N deficiency that occurs (Gastal and Lemaire, 2002), and N can be supplied at an appropriate rate.

This N supply can be provided in fertilizers, and the main N fertilizers produced globally are shown in Table 2.4. In commercial horticulture, plants often are grown in nutrient culture (either in true hydroponics, aeroponics, or trickle supply of nutrient solution over an inert support such as rock wool). Here, N is supplied predominantly as NO_3^-, with some NH_4^+ (e.g., in a 23:1 ratio in the work of Le Bot et al., 2001) and at slightly acid pH, ideal for uptake of the NO_3^- ion, thus minimizing competition between the uptake of NH_4^+ and other cations such as K^+ and Ca^{2+}.

In some farming systems, N is provided to farmland in manures and composts. Pig slurry supplies approximately 4.4 kg N per tonne at 6% dry matter (approximately 70% as NH_4–N and 30% as organic N), cattle slurry 2.6 kg N per tonne at 6% dry matter (40% NH_4–N and 60% organic N), fresh farmyard manure about 6.0 kg N per tonne (20% NH_4–N and 80% organic N), and broiler litter 30 kg N per tonne (25% NH_4–N, 65% organic N, and 10% uric acid) (Defra, 2010). Farmland also has some additional input of inorganic N through deposition of N forms that originally entered the atmosphere as pollutants. Although the NOx gases have a harmful effect on the environment, causing acid rain and raising the fertility of natural ecosystems that need to be of low fertility to flourish, they do represent a source of soil mineral N that can be taken up by crop plants. Indeed, total depositions of 117 kg N ha^{-1} $year^{-1}$ have been measured in a wheat–corn cropping system in the North China Plain (He et al., 2010).

Too much N fertilizer application can lead to pollution through leaching of NO_3^-, especially if it occurs at times before crops are growing fast and taking up NO_3^- quickly. Furthermore, excessive use of N on cereals gives lodging and delayed senescence, causing many of the grains to fail to develop completely (Yang and Zhang, 2006). It also gives a risk of foliar diseases developing and silage crops not fermenting properly (Defra, 2010). The ideal is to supply the EONR (N_{opt}, the amount of N supplied that gives the optimum economic return). This optimum fertilization

---

TABLE 2.4
Annual World Production of Major N Fertilizers

Fertilizer	Production in 2012 (tonnes of N)
Anhydrous ammonia	136,455,000
Urea	74,395,000
Ammonium nitrate	16,030,000
Diammonium phosphate	6,291,000
Ammonium sulfate	4,724,000
Calcium ammonium nitrate	3,783,000
Monoammonium phosphate	2,610,000

Source: Figures from International Fertilizer Industry Association (IFA), www.fertilizer.org/ifa, accessed March 18, 2014. With permission.

varies considerably among fields, because of differences in organic matter content and soil chemical properties and also between years, because of climatic variability (Lory and Scharf, 2003; Olfs et al., 2005; Torres and Link, 2010).

2.11.1 Fertilizer Recommendation Techniques

One way of arriving at N fertilizer recommendations is through an index method, such as the soil nitrogen supply (SNS) index used in the United Kingdom (Defra, 2010). SNS is the SMN + the N already in the crop + the N that will have been made available by mineralization by harvest time. Farmers work out the SNS index value, either by the field assessment method or by measurement of SMN, and use the index value to work out how much N fertilizer to supply (Defra, 2010). The field assessment method requires them to arrive at an SNS value from knowledge of usual rainfall, soil type, and previous crop. The measurement method requires them to take representative soil samples to 90 cm depth for measurement of SMN, to estimate the N already in the crop (from density of cereal shoots or green area index of oilseed rape), and to allow for future mineralization of N in soils with high organic matter content (Defra, 2010).

Measurement of SMN is important in the N_{min} method developed for use on cereal crops in Germany (Wehrmann and Scharpf, 1979). Here a target yield was established, from which the total N requirement of the crop could be deduced, and the SMN measured at the start of the growth period + estimated mineral N that would arise from mineralization was subtracted from this target N requirement to give the amount of fertilizer N required (Olfs et al., 2005).

Another forecasting technique is the balance sheet method (also called a forecast balance sheet), commonly used in the United States, France, and other countries for a variety of arable crops. N_{opt} is estimated as being the total N content of the harvested part of the crop minus all the N coming from sources other than fertilizer and adjusted for the efficiency of the crop in recovering N from the soil (Lory and Scharf, 2003). It is calculated for a selected target yield, so that for a grain crop

$$N_f = \frac{\left(N_g - N_{gs}\right)}{\text{FNUE}}$$

where
   N_f is the estimated N_{opt} for a selected yield goal
   N_g is the N content of the harvested grain
   N_{gs} is the N in the grain that came from the soil rather than the fertilizer
   The fertilizer nitrogen-use efficiency (FNUE) is the proportion of fertilizer N applied recovered
      in the grain (Lory and Scharf, 2003)

It has to take into account the price of the fertilizer:value of grain ratio. It requires information on the availability of N in the soil, a value that can come from an index based on soil properties or previous crop or can come from soil testing.

The problem with models based on testing soil N or NO_3^- content at sowing, or early in the growth period, is that the soil mineralization potential has to be estimated, and as this potential is very weather dependent (Torres and Link, 2010) inaccuracy is introduced into the fertilizer recommendation. Measurement of mineralization potential of the soil by 7-day anaerobic incubation improved the accuracy of fertilizer N recommendations based on the pre-sidedress soil nitrate test in corn (Sainz Rosas et al., 2008) and on soil NO_3^- content at sowing in spring wheat in the important Pampa grain-producing region of Argentina (Reussi Calvo et al., 2013).

Fertilizer recommendations based on soil or plant analysis may be too expensive for small farmers, so recommendation systems that are robust and simple have been produced for Asia and Africa (Chuan et al., 2013). The Nutrient Expert decision support system for cereals uses site-specific, nutrient management to make field-specific recommendations for N and also for P and K (Pampolino et al., 2012). It uses information given by the farmer to estimate how much N is likely to be available, namely soil color, texture, organic matter content, crop sequence, residue management, water supply, fertilizer inputs, and current yield. The software estimates the natural N supply and what potential yield could be achieved, and makes recommendations of how much fertilizer N should be given (and at which growth stages) to achieve the potential yield. The output is based on experiments in which crops grown in different fields were evaluated for nutrient uptake by comparing a zero-N treatment with a treatment of N supply as in best practice, and yield–response curves were generated (Pampolino et al., 2012). The Nutrient Expert system has given good results for wheat in China (Chuan et al., 2013).

2.11.2 TIMING AND AMOUNTS OF NITROGEN FERTILIZER APPLICATION

Because of the between-season variability in N availability due to weather conditions, monitoring of soil and plants can remove some of the uncertainty in making N fertilizer recommendations. Monitoring ensures that N is applied only when the plants are responsive to it. That action has given rise to the use of split applications, which shortens the time between N becoming available and being taken up by a crop and minimizes losses due to leaching or volatilization (Torres and Link, 2010). In an experiment on three split applications of N to winter wheat in Germany, the authors found an N-use efficiency (N uptake relative to N supply) of 83% compared with 62% for average yields across the 27 EU member states (Torres and Link, 2010). The normal practice for N fertilization of winter wheat in northern Europe is now for three applications, one early to promote tillering ($50–80$ kg N ha^{-1}), one at the beginning of stem elongation (50 kg N ha^{-1}), and one at the second node stage ($40–50$ kg N ha^{-1}) (Hirel et al., 2007). A preplanting N application, with the risks of nitrate leaching during the winter rains, is not required as there is usually sufficient N already in the soil for early growth of the crop. Application of N after the main period of vegetative growth would be wasted for increasing grain yield but could improve crop quality. For example, postpollination foliar application of urea ammonium nitrate to hard red winter wheat and hard red spring wheat crops in South Dakota, USA, increased grain protein concentration (Bly and Woodard, 2003).

Split applications are not universally beneficial, and in winter cereals in Mediterranean climates, the usual practice of a presowing application of N (with P and K) plus a further application at tillering does not seem to give a yield advantage over just one application at tillering (Torres and Link, 2010). Split applications did not give a yield advantage in experiments on barley, wheat, or oilseed rape in western Canada, although there may have been some yield advantages in wetter regions (Grant et al., 2012). Furthermore, while it is possible to drive over cereal crops a few times without causing too much damage, for row crops such as sugar beet and potato you cannot drive over after row closure without yield loss (Olfs et al., 2005).

The normal practice for N fertilization of maize in the United States is an application presowing, with a further application (typically after a pre-sidedress test for soil NO$_3$$^{-}$) during vegetative growth. A split application of 90 kg N ha^{-1} presowing and a further 90 kg ha^{-1} at the V6 or V10 growth stages gives the best results (Walsh et al., 2012). In Colombia, three applications (at sowing and at V6 and V10) have been shown to produce better yields for the same N rate than a double split (Torres and Link, 2011). In China there may be merely one N application, before sowing, but high-yielding farms use a basal application and subsequent applications at V8, V12, and VT growth stages (up to 450 kg N ha^{-1}) (Peng et al., 2013). For silage corn, N often is supplied preplanting as manure, as farmers producing silage are usually doing so for their own livestock and have manure to dispose of.

For lowland (paddy) rice in the tropics, fertilizers are used to supplement biological N fixation. Fixation can be from indigenous organisms, with cyanobacteria supplying on average 30 kg N ha^{-1} per crop and photosynthetic bacteria supplying an average of 7 kg N ha^{-1} (Ladha and Reddy, 2003). Alternatively, it can be supplied either from inoculating azolla as a companion crop or growing leguminous plants for feed and fodder or as cover crops between subsequent rice crops; azolla can release 20–30 kg N ha^{-1}, with a high proportion of that N coming from atmospheric N_2 (Ladha and Reddy, 2003). Rice in southeast Asia typically is given N fertilizer at sowing (40–100 kg N ha^{-1}), with a topdressing between the formation of the panicle primordia and the late stage of spikelet initiation (15–45 kg N ha^{-1}) (Hirel et al., 2007). The level of supply for the second split can be determined by use of chlorophyll meters or leaf color charts. Because of the interaction between K^+ and NH_4^+, the main N form taken up by rice, potassium should be supplied at the same time (Balkos et al., 2010).

Oilseed rape requires more N than cereals because the oil in the seeds is more expensive for the plant to produce than carbohydrates, and as little N is taken up during seed filling N fertilizer is supplied to winter crops in two splits between sowing and spring (at an optimum level of 180–200 kg N ha^{-1}) (Hirel et al., 2007). Some N also may be supplied presowing (Defra, 2010). Much of the N remains in the crop residue at harvest, so it becomes available for subsequent cereal crops in a typical northern European arable rotation (Hirel et al., 2007).

Split applications have been used for a winter wheat–summer corn rotation in the North China Plain based on a refinement of the N_{min} method. This system receives excessive fertilizer N, with consequent leaching of NO_3^- into watercourses. The improved N_{min} method is based on the changing requirements for N as crops grow, so fertilization is based on the difference between a target value of N that should be present in the shoots of the wheat or corn at three different growth stages to achieve a target yield and the concentrations of SMN measured at 0–30, 0–60, and 0–90 cm depths in the soil at early, middle, and later growth stages, respectively (Chen et al., 2006; Zhao et al., 2006). In a study of the corn part of the rotation across a large number of farms, no fertilizer N was required at planting due to the residual N from the wheat crop, but then supplying N according to the previously mentioned protocol at the 3-leaf stage (V3) and the 10-leaf stage (V10) gave a grain yield of 8.9 t ha^{-1} (compared with 8.5 t ha^{-1} for treatment based on normal farmer practice) yet with a noticeably lower rate of fertilizer supply (157 kg N ha^{-1} compared with 263 kg N ha^{-1}) (Cui et al., 2008). Other workers have suggested a standard soil N_{min} value that should be maintained for the entire growth of maize crops in China or three more accurate values that should be maintained from sowing to the V8 growth stage, from V8 to VT and from VT to R6 (Peng et al., 2013).

2.11.3 Developments in Nitrogen Fertilizer Use

Simple ways of lowering pollution caused by leaching of NO_3^- can be achieved by N fertilizer formulation, with nitrification inhibitors being added to ammonium fertilizers and urea to slow the oxidation of soil-immobile NH_4^+ ions to water-soluble NO_3^- ions. Nitrification inhibitors currently used in this way include nitrapyrin, dicyandiamide (DCD), and 3,4-dimethylpyrazole phosphate (DMPP), but such use is not common as it is frequently not cost-effective (Subbarao et al., 2009).

Further ways of minimizing losses through formulation come from coating urea fertilizers with protectant coverings (polymer-coated urea). Urea applied to paddy fields is lost rapidly from the system by hydrolysis to volatile ammonia (NH_3) gas, but urea coated with polyolefin was shown to have a lower proportion of the N being lost by volatilization than uncoated urea in the productive rice-growing regions of southern China (Xu et al., 2013). There was a greater accumulation of N in the shoots, due to the slow release of available N matching the plant N demand better. Polymer-coated urea possibly may be used to increase wheat grain protein content as more N should be available for plant uptake later in the growing season than if uncoated urea is supplied, but the slow release may give a shortage of N early in the growing season with a consequent negative effect on overall grain yield (Farmaha and Sims, 2013).

Urea also can be formulated with urease inhibitor, such as phenylphosphorodiamidate (PPD) (Arkoun et al., 2013) or N-(n-butyl)thiophosphoric triamide (Agrotain®) (Suter et al., 2013). Field experiments on maize grown on poorly drained claypan soil in Missouri, USA, showed that there were yield advantages over urea supplied preplanting from urea + urease inhibitor (Agrotain), urea + nitrification inhibitor (nitrapyrin), and most of all with polymer-coated urea (Motavalli et al., 2012).

The application of variable amounts of N fertilizer within and between fields that has been made possible by remote sensing of plant N, coupled with advances in fertilizer formulation, has improved the efficiency of N fertilizer use. However, the biggest gains will come in those countries where models for crop N requirement have not been well developed in the past. This gain could lead to increased crop yields over big areas or to the maintenance of current yields with application of less fertilizer. For example, models of crop N requirement have not been well developed historically in China, and the fact that N fertilizers have been subsidized by the government (Yang and Zhang, 2006) has given rise to excessive N use. This overuse leads to waste of fertilizers (with consequent unnecessary emissions of greenhouse gases in their manufacture) and large-scale nitrate pollution of water courses (Ju et al., 2006). The problem has been made worse by the large amounts of N reaching some agroecosystems through atmospheric deposition. There are highly productive systems in the country that are supported strongly by the use of synthetic N fertilizers, and as techniques for modeling crop yield are perfected even higher yields may be obtained, yet the total amount of N supplied per unit land area will decrease.

REFERENCES

Abbate, P.E., F.E. Andrade, and J.P. Culot. 1995. The effects of radiation and nitrogen on number of grains in wheat. *J. Agric. Sci. (Cambridge)* 124:351–360.

Abduelghader, A.A., F.E. Sanders, and D.J. Pilbeam. 2011. Growth and biomass partitioning in tomato in relation to ratio of nitrogen: Phosphorus supply. *J. Plant Nutr.* 34:2018–2038.

Abuzinadah, R.A. and D.J. Read. 1986. The role of proteins in the nitrogen nutrition of ectomycorrhizal plants. I. Utilization of peptides and proteins by ectomycorrhizal fungi. *New Phytol.* 103:481–493.

Adamczyk, B., M. Godlewski, A. Smolander, and V. Kitunen. 2009. Degradation of proteins by enzymes exuded by *Allium porrum* roots—A potentially important strategy for acquiring organic nitrogen by plants. *Plant Physiol. Biochem.* 47:919–925.

Agati, G., L. Foschi, N. Grossi, L. Guglielminetti, Z.G. Cerovic, and M. Volterrani. 2013. Fluorescence-based versus reflectance proximal sensing of nitrogen content in *Paspalum vaginatum* and *Zoysia matrella* turfgrasses. *Eur. J. Agron.* 45:39–51.

Ågren, G. 2004. The C:N:P stoichiometry of autotrophs—Theory and observations. *Ecol. Lett.* 7:185–191.

Ahmed, M., L.J. Stal, and S. Hasnain. 2010. Association of non-heterocystous cyanobacteria with crop plants. *Plant Soil* 336:363–375.

Arkoun, M., L. Jannin, P. Laîné et al. 2013. A physiological and molecular study of the effects of nickel deficiency and phenylphosphorodiamidate (PPD) application on urea metabolism in oilseed rape (*Brassica napus* L.). *Plant Soil* 362:79–92.

Arnon, D.I. 1939. Effect of ammonium and nitrate nitrogen on the mineral composition and sap characteristics of barley. *Soil Sci.* 48:295–307.

Balasubramanian, V., A.C. Morales, R.T. Cruz, and S. Abdulrachman. 1999. On-farm adaptation of knowledge-intensive nitrogen management technologies for rice systems. *Nutr. Cycling Agroecosyst.* 53:59–69.

Balkos, K.D., D.T. Britto, and H.J. Kronzucker. 2010. Optimization of ammonium acquisition and metabolism by potassium in rice (*Oryza sativa* L. cv. IR-72). *Plant Cell Environ.* 33:23–34.

Barber, M.J., B.A. Notton, C.J. Kay, and L.P. Solomonson. 1989. Chloride inhibition of spinach nitrate reductase. *Plant Physiol.* 90:70–94.

Barker, D.W. and J.E. Sawyer. 2010. Using active canopy sensors to quantify corn nitrogen stress and nitrogen application rate. *Agron. J.* 102:964–971.

Barraclough, P.B., J.R. Howarth, J. Jones et al. 2010. Nitrogen efficiency of wheat: Genotypic and environmental variation and prospects for improvement. *Eur. J. Agron.* 33:1–11.

Bélanger, G. and J.E. Richards. 1999. Relationship between P and N concentrations in timothy. *Can. J. Plant Sci.* 79:65–70.

Berg, A., S. Danielsson, and B.H. Svensson. 2013. Transfer of fixed-N from N_2-fixing cyanobacteria associated with the moss *Sphagnum riparium* results in enhanced growth of the moss. *Plant Soil* 362:271–278.

Bethlenfalvay, G.J., M.G. Reyes-Solis, S.B. Camel, and R. Ferrera-Cerrato. 1991. Nutrient transfer between the root zones of soybean and maize plants connected by a common mycorrhizal mycelium. *Physiol. Plant.* 82:423–432.

Bloem, E., S. Haneklaus, and E. Schnug. 2005. Influence of nitrogen and sulfur fertilization on the alliin content of onions and garlic. *J. Plant Nutr.* 27:1827–1839.

Bloom, A.J., L. Randall, A.R. Taylor, and W.K. Silk. 2012. Deposition of ammonium and nitrate in the roots of maize seedlings supplied with different nitrogen salts. *J. Exp. Bot.* 63:1997–2006.

Bly, A.G. and H.J. Woodard. 2003. Foliar nitrogen application timing influence on grain yield and protein concentration of hard red winter and spring wheat. *Agron. J.* 95:335–338.

Bogaard, A., R. Fraser, T.H.E. Heaton et al. 2013. Crop manuring and intensive land management by Europe's first farmers. *Proc. Natl. Acad. Sci. USA* 110:12589–12594.

Bogard, M., V. Allard, M. Brancourt-Hulmel et al. 2010. Deviation from the grain protein concentration— Grain yield negative relationship is highly correlated to post-anthesis N uptake in winter wheat. *J. Exp. Bot.* 61:4303–4312.

Bogard, M., M. Jourdan, V. Allard et al. 2011. Anthesis date mainly explained correlations between post-anthesis leaf senescence, grain yield, and grain protein concentration in a winter wheat population segregating for flowering time QTLs. *J. Exp. Bot.* 62:3621–3636.

Bonneau, L., S. Huguet, D. Wipf, N. Pauly, and H.-N. Truong. 2013. Combined phosphate and nitrogen limitation generates a nutrient stress transcriptome favorable for arbuscular mycorrhizal symbioses in *Medicago truncatula*. *New Phytol.* 199:188–202.

Borlotti, A., G. Vigani, and G. Zocchi. 2012. Iron deficiency affects nitrogen metabolism in cucumber (*Cucumis sativus* L.) plants. *BMC Plant Biol.* 12:189.

Borrell, A., G. Hammer, and E. van Oosterom. 2001. Stay-green: A consequence of the balance between supply and demand for nitrogen during grain filling? *Ann. Appl. Biol.* 138:91–95.

Broadley, M.R., A.J. Escobar-Gutiérrez, A. Burns, and I.G. Burns. 2001. Nitrogen-limited growth of lettuce is associated with lower stomatal conductance. *New Phytol.* 152:97–106.

Brouwer, R. 1962. Nutritive influences on the distribution of dry matter in the plant. *Neth. J. Agric. Sci.* 10:399–408.

Burns, I.G., J. Durnford, J. Lynn, S. McClement, P. Hand, and D. Pink. 2012. The influence of genetic variation and nitrogen source on nitrate accumulation and iso-osmotic regulation by lettuce. *Plant Soil* 352:321–339.

Burns, I.G., K.-F. Zhang, M.K. Turner, and R. Edmondson. 2011. Iso-osmotic regulation of nitrate accumulation in lettuce. *J. Plant Nutr.* 34:283–313.

Camacho-Cristóbal, J.J. and A. González-Fontes. 2007. Boron deficiency decreases plasmalemma H^+-ATPase expression and nitrate uptake, and promotes ammonium assimilation into asparagine in tobacco roots. *Planta* 226:443–451.

Camañes, G., E. Bellmunt, J. García-Andrade, P. García-Agustín, and M. Cerezo. 2012. Reciprocal regulation between *AtNRT2.1* and *AtAMT1.1* expression and the kinetics of NH_4^+ and NO_3^- influxes. *J. Plant Physiol.* 169:268–274.

Campbell, W.H. 1999. Nitrate reductase structure, function and regulation: Bridging the gap between biochemistry and physiology. *Annu. Rev. Plant Physiol. Plant Mol. Biol.* 50:277–303.

Cerezo, M., P. García-Agustín, M.D. Serna, and E. Primo-Millo. 1997. Kinetics of nitrate uptake by *Citrus* seedlings and inhibitory effects of salinity. *Plant Sci.* 126:105–112.

Chardon, F., V. Noël, and C. Masclaux-Daubresse. 2012. Exploring NUE in crops and in *Arabidopsis* ideotypes to improve yield and seed quality. *J. Exp. Bot.* 63:3401–3412.

Chen, X.P., F. Zhang, V. Römheld et al. 2006. Synchronizing N supply from soil and fertilizer and N demand of winter wheat by an improved N_{min} method. *Nutr. Cycling Agroecosyst.* 74:91–98.

Chichkova, S., J. Arellano, C.P. Vance, and G. Hernández. 2001. Transgenic tobacco plants that overexpress alfalfa NADH-glutamate synthase have higher carbon and nitrogen content. *J. Exp. Bot.* 52:2079–2087.

Chuan, L., P. He, M.F. Pampolino, A.M. Johnston et al. 2013. Establishing a scientific basis for fertilizer recommendations for wheat in China: Yield response and agronomic efficiency. *Field Crops Res.* 140:1–8.

Ciampitti, I.A. and T.J. Vyn. 2011. A comprehensive study of plant density consequences on nitrogen uptake dynamics of maize plants from vegetative to reproductive stages. *Field Crops Res.* 121:2–18.

Clarkson, D.T., M. Carvajal, T. Henzler et al. 2000. Root hydraulic conductance: Diurnal aquaporin expression and the effects of nutrient stress. *J. Exp. Bot.* 51:61–70.

Clarkson, D.T., L.R. Saker, and J.V. Purves. 1989. Depression of nitrate and ammonium transport in barley plants with diminished sulphate status. Evidence of co-regulation of nitrogen and sulphate intake. *J. Exp. Bot.* 40:953–963.

Clarkson, D.T. and A.J. Warner. 1979. Relationships between root temperature and the transport of ammonium and nitrate ions by Italian and perennial ryegrass (*Lolium multiflorum* and *Lolium perenne*). *Plant Physiol.* 64:557–561.

Cockram, J., H. Jones, F.J. Leigh et al. 2007. Control of flowering time in temperate cereals: Genes, domestication, and sustainable productivity. *J. Exp. Bot.* 58:1231–1244.

Colnenne, C., J.M. Meynard, R. Reau et al. 1998. Determination of a critical nitrogen dilution curve for winter oilseed rape. *Ann. Bot.* 81:311–317.

Cox, W.J. and H.M. Reisenauer. 1973. Growth and ion uptake by wheat supplied nitrogen as nitrate, or ammonium, or both. *Plant Soil* 38:363–380.

Cram, W.J. 1973. Internal factors regulating nitrate and chloride influx in plant cells. *J. Exp. Bot.* 24:328–341.

Cui, Z., X. Chen, Y. Miao et al. 2008. On-farm evaluation of the improved soil N_{min}-based nitrogen management for summer maize in North China Plain. *Agron. J.* 100:517–525.

Daughtry, C.S.T., C.L. Walthall, M.S. Kim, E. Brown de Colstoun, and J.E. McMurtrey III. 2000. Estimating corn leaf chlorophyll concentration from leaf and canopy reflectance. *Rem. Sens. Environ.* 74:229–239.

De Angeli, A., D. Monachello, G. Ephritikhine et al. 2006. The nitrate/proton antiporter AtCLCa mediates nitrate accumulation in plant vacuoles. *Nature* 442:939–942.

Debaeke, P., E.J. van Oosterom, E. Justes et al. 2012. A species-specific critical nitrogen dilution curve for sunflower (*Helianthus annuus* L.). *Field Crops Res.* 136:76–84.

Dechorgnat, J., C.T. Nguyen, P. Armengaud et al. 2011. From the soil to the seeds: The long journey of nitrate in plants. *J. Exp. Bot.* 62:1349–1359.

Defra. 2010. Department of Environment, Food and Rural Affairs. *Fertiliser Manual.* 8th edn. The Stationery Office, Norwich, U.K.

Distelfeld, A., C. Li, and J. Dubcovsky. 2009. Regulation of flowering in temperate cereals. *Curr. Opin. Plant Biol.* 12:178–184.

Dixon, N.E., C. Gazzola, R.L. Blakeley, and B. Zerner. 1975. Jack bean urease (EC 3.5.1.5). A metalloenzyme. A simple biological role for nickel? *J. Am. Chem. Soc.* 97:4131–4133.

Dorigo, W.A., R. Zurita-Milla, A.J.W. de Wit, J. Brazile, R. Singh, and M.E. Schaepman. 2007. A review on reflective remote sensing and data assimilation techniques for enhanced agroecosystem modeling. *Int. J. Appl. Earth Obs. Geoinfo.* 9:165–193.

Drew, M.C. 1975. Comparison of the effects of a localized supply of phosphate, nitrate, ammonium and potassium on the growth of the seminal root system, and the shoot, in barley. *New Phytol.* 75:479–490.

Drihem, K. and D.J. Pilbeam. 2002. Effects of salinity on accumulation of mineral nutrients in wheat grown with nitrate-nitrogen or mixed ammonium: Nitrate-nitrogen. *J. Plant Nutr.* 25:2091–2113.

Ekbladh, G. and E. Witter. 2010. Determination of the critical nitrogen concentration of white cabbage. *Eur. J. Agron.* 33:276–284.

Engels, C. and H. Marschner. 1990. Plant uptake and utilization of nitrogen. In *Nitrogen Fertilization in the Environment*, ed. P.E. Bacon, pp. 41–81. Marcel Dekker Inc., New York.

Engineer, C.B. and R.G. Kranz. 2007. Reciprocal leaf and root expression of *AtAmt1.1* and root architectural changes in response to nitrogen starvation. *Plant Physiol.* 143:236–250.

EP2534250 A2. 2012. European patent, Improvement of the grain filling of a plant through the modulation of NADH-glutamate synthase activity, published December 19, 2012.

Erenoglu, E.B., U.B. Kutman, Y. Ceylan, B. Yildiz, and I. Cakmak. 2011. Improved nitrogen nutrition enhances root uptake, root-to-shoot translocation and remobilization of zinc (^{65}Zn) in wheat. *New Phytol.* 189:438–448.

Evans, J.R. 1983. Nitrogen and photosynthesis in the flag leaf of wheat (*Triticum aestivum* L.). *Plant Physiol.* 72:297–302.

Ewert, F. and B. Honermeier. 1999. Spikelet initiation of winter triticale and winter wheat in response to nitrogen fertilization. *Eur. J. Agron.* 11:107–113.

Farmaha, B.S. and A.L. Sims. 2013. Yield and protein response of wheat cultivars to polymer-coated urea and urea. *Agron. J.* 105:229–236.

Ferrante, A., R. Savin, and G.A. Slafer. 2010. Floret development of durum wheat in response to nitrogen availability. *J. Exp. Bot.* 61:4351–4359.

Ferrante, A., R. Savin, and G.A. Slafer. 2013. Floret development and grain setting differences between modern durum wheats under contrasting nitrogen availability. *J. Exp. Bot.* 64:169–184.

Finlay, R.D., A. Frostegård, and A.-M. Sonnerfeldt. 1992. Utilization of organic and inorganic nitrogen sources by ectomycorrhizal fungi in pure culture and in symbiosis with *Pinus contorta* Dougl. ex Loud. *New Phytol.* 120:105–115.

Fischer, R.A. 1993. Irrigated spring wheat and timing and amount of nitrogen fertilizer. II. Physiology of grain yield. *Field Crops Res.* 33:57–80.

Flénet, F., M. Guérif, J. Boiffin, D. Dorvillez, and L. Champolivier. 2006. The critical N dilution curve for linseed (*Linum usitatissimum* L.) is different from other C3 species. *Eur. J. Agron.* 24:367–373.

Foehse, D. and A. Jungk. 1983. Influence of phosphate and nitrate supply on root hair formation of rape, spinach and tomato plants. *Plant Soil* 74:359–368.

Forsum, O., H. Svennerstam, U. Ganeteg, and T. Näsholm. 2008. Capacities and constraints of amino acid utilization in *Arabidopsis*. *New Phytol.* 179:1058–1069.

Fricke, W. and T.J. Flowers. 1998. Control of leaf cell elongation in barley. Generation rates of osmotic pressure and turgor, and growth-associated water potential gradients. *Planta* 206:53–65.

Gallagher, J.N. and P.V. Biscoe. 1978. Radiation absorption, growth and yield of cereals. *J. Agric. Sci. (Camb.)* 91:47–60.

Gallais, A. and M. Coque. 2005. Genetic variation and selection for nitrogen use efficiency in maize: A synthesis. *Maydica* 50:531–547.

Gallais, A., M. Coque, I. Quilléré, J.-L. Prioul, and B. Hirel. 2006. Modelling postsilking nitrogen fluxes in maize (*Zea mays*) using ^{15}N-labelling field experiments. *New Phytol.* 172:696–707.

Gao, Y., Y. Li, X. Yang, H. Li, Q. Shen, and S. Guo. 2010. Ammonium nutrition increases water absorption in rice seedlings (*Oryza sativa* L.) under water stress. *Plant Soil* 331:193–201.

Garnett, T., V. Conn, D. Plett et al. 2013. The response of the maize nitrate transport system to nitrogen demand and supply across the lifecycle. *New Phytol.* 198:82–94.

Garnica, M., F. Houdusse, J.C. Yvin, and J.M. Garcia-Mina. 2009. Nitrate modifies urea root uptake and assimilation in wheat seedlings. *J. Sci. Food Agric.* 89:55–62.

Gastal, F. and G. Lemaire. 2002. N uptake and distribution in crops: An agronomical and ecophysiological perspective. *J. Exp. Bot.* 53:789–799.

Gaur, V.S., U.S. Sing, A.K. Gupta, and A. Kumar. 2012. Understanding the differential nitrogen sensing mechanism in rice genotypes through expression analysis of high and low affinity ammonium transporter genes. *Mol. Biol. Rep.* 39:2233–2241.

Gilliham, M. and M. Tester. 2005. The regulation of anion loading to the maize root xylem. *Plant Physiol.* 137:819–828.

Glass, A.D.M. and M.Y. Siddiqi. 1985. Nitrate inhibition of chloride influx in barley: Implications for a proposed chloride homeostat. *J. Exp. Bot.* 165:556–566.

Gloser, V., M.A. Zwieniecki, C.M. Orians, and N.M. Holbrook. 2007. Dynamic changes in root hydraulic properties in response to nitrate availability. *J. Exp. Bot.* 58:2409–2415.

Górska, A., J.W. Lazor, A.K. Zwieniecka, C. Benway, and M.A. Zwieniecki. 2010. The capacity for nitrate regulation of root hydraulic properties correlates with species' nitrate uptake rates. *Plant Soil* 337:447–455.

Górska, A., Q. Ye, N.M. Holbrook, and M.A. Zwieniecki. 2008a. Nitrate control of root hydraulic properties in plants: Translating local information to whole plant response. *Plant Physiol.* 148:1159–1167.

Górska, A., A. Zwieniecka, N.M. Holbrook, and M.A. Zwieniecki. 2008b. Nitrate induction of root hydraulic conductivity in maize is not correlated with aquaporin expression. *Planta* 228:989–998.

Govindarajulu, M., P.E. Pfeffer, H. Jin et al. 2005. Nitrogen transfer in the arbuscular mycorrhizal symbiosis. *Nature* 435:819–823.

Grant, C.A., R. Wu, F. Selles et al. 2012. Crop yield and nitrogen concentration with controlled release urea and split applications of nitrogen as compared to non-coated urea applied at seeding. *Field Crops Res.* 127:170–180.

Greenwood, D.J., T.V. Karpinets, K. Zhang, A. Bosh-Serra, A. Boldrini, and L. Karawulova. 2008. A unifying concept for the dependence of whole-crop N:P ratio on biomass: Theory and experiment. *Ann. Bot.* 102:967–977.

Greenwood, D.J., G. Lemaire, G. Gosse, P. Cruz, A. Draycott, and J.J. Neeteson. 1990. Decline in percentage N of C3 and C4 crops with increasing plant mass. *Ann. Bot.* 66:425–436.

Greenwood, D.J., J.J. Neeteson, and A. Draycott. 1986. Quantitative relationships for the dependence of growth rate of arable crops on their nitrogen content, dry weight and aerial environment. *Plant Soil* 91:281–301.

Grelet, G.-A., A.A. Meharg, E.I. Duff, I.C. Anderson, and I.J. Alexander. 2009. Small genetic differences between ericoid mycorrhizal fungi affect nitrogen uptake by *Vaccinium*. *New Phytol.* 181:708–718.

Guo, S., R. Kaldenhoff, N. Uehlein, B. Sattelmacher, and H. Brueck. 2007. Relationship between water and nitrogen uptake in nitrate- and ammonium-supplied *Phaseolus vulgaris* L. plants. *J. Plant Nutr. Soil Sci.* 170:73–80.

Habash, D.Z., S. Bernard, J. Schondelmaier, J. Weyen, and S.A. Quarrie. 2007. The genetics of nitrogen use in hexaploid wheat: N utilisation, development and yield. *Theoret. Appl. Genet.* 114:403–419.

Hacke, U.G., L. Plavcová, A. Almeida-Rodriguez, S. King-Jones, W. Zhou, and J.E.K. Cooke. 2010. Influence of nitrogen fertilization on xylem traits and aquaporin expression in stems of hybrid poplar. *Tree Physiol.* 30:1016–1025.

Halford, N.G., T.Y. Curtis, N. Muttucumaru, J. Postles, J.S. Elmore, and D.S. Mottram. 2012. The acrylamide problem: A plant and agronomic science issue. *J. Exp. Bot.* 63:2841–2851.

Harrison, K.A., R. Bol, and R.D. Bardgett. 2007. Preferences for different nitrogen forms by coexisting plant species and soil microbes. *Ecology* 88:989–999.

Hawkins, H.-J., A. Johansen, and E. George. 2000. Uptake and transport of organic and inorganic nitrogen by arbuscular mycorrhizal fungi. *Plant Soil* 226:275–285.

He, C.-E., X. Wang, X. Liu, A. Fangmeier, P. Christie, and F. Zhang. 2010. Nitrogen deposition and its contribution to nutrient inputs to intensively managed agricultural ecosystems. *Ecol. Appl.* 20:80–90.

He, X., M. Xu, G.Y. Qiu, and J. Zhou. 2009. Use of ^{15}N stable isotope to quantify nitrogen transfer between mycorrhizal plants. *J. Plant Ecol.* 2:107–118.

Hesse, H., V. Nikiforova, B. Gakière, and R. Hoefgen. 2004. Molecular analysis and control of cysteine biosynthesis: Integration of nitrogen and sulphur metabolism. *J. Exp. Bot.* 55:1283–1292.

Hewitt, E.J. 1966. *Sand and Water Culture Methods Used in the Study of Plant Nutrition*, 2nd edn., CAB, Buckinghamshire, U.K.

Hill, P.W., R.S. Quilliam, T.H. DeLuca et al. 2011. Acquisition and assimilation of nitrogen as peptide-bound and *D*-enantiomers of amino acids by wheat. *PLoS ONE* 6:e19220.

Hirel, B., J. Le Gouis, B. Ney, and A. Gallais. 2007. The challenge of improving nitrogen use efficiency in crop plants: Towards a more central role for genetic variability and quantitative genetics within integrated approaches. *J. Exp. Bot.* 58:2369–2387.

Hirner, A., F. Ladwig, H, Stransky et al. 2006. *Arabidopsis* LHT1 is a high-affinity transporter for cellular amino acid uptake in both root epidermis and leaf mesophyll. *Plant Cell* 18:1931–1946.

Hirose, T. 2005. Development of the Monsi-Saeki theory on canopy structure and function. *Ann. Bot.* 95:483–494.

Hooper, W.D. and H.B. Ash (translators). 1934. *Cato and Varro on Agriculture*. Loeb Classical Library, vol. 283. Harvard University Press, Cambridge, MA.

Hord, N.G., Y. Tang, and N.S. Bryan. 2009. Food sources of nitrates and nitrites: The physiologic context for potential health benefits. *Am. J. Clin. Nutr.* 90:1–10.

Hu, Y., N.S. Wang, X.J. Hu, X.Y. Lin, Y. Feng, and C.W. Jin. 2013. Nitrate nutrition enhances nickel accumulation and toxicity in *Arabidopsis* plants. *Plant Soil* 371:105–115.

Huang, N.-C., C.-S. Chiang, N.M. Crawford, and Y.-F. Tsay. 1996. *CHL1* encodes a component of the low-affinity nitrate uptake system in *Arabidopsis* and shows cell type-specific expression in roots. *Plant Cell* 8:2183–2191.

Huang, N.-C., K.-H. Liu, H.-J. Lo, and Y.-F. Tsay. 1999. Cloning and functional characterization of an *Arabidopsis* nitrate transporter gene that encodes a constitutive component of low-affinity uptake. *Plant Cell* 11:1381–1392.

Ingestad, T. and G.I. Ågren. 1992. Theories and methods on plant nutrition and growth. *Physiol. Plant.* 84:177–184.

Iniguez, A.L., Y. Dong, and E.W. Triplett. 2004. Nitrogen fixation in wheat provided by *Klebsiella pneumoniae* 342. *Mol. Plant-Microbe Interact.* 17:1078–1085.

Inselbacher, E. and T. Näsholm. 2012. The below-ground perspective of forest plants: Soil provides mainly organic nitrogen for plants and mycorrhizal fungi. *New Phytol.* 195:329–334.

International Fertilizer Industry Association. 2014. IFA data. www.fertilizer.org/ifa (accessed March 18, 2014).

Iowa State University. 1997. *Nitrogen Fertilizer Recommendations for Corn in Iowa*. Iowa State University, University Extension, Ames, IA.

Jackson, R.D. and A.R. Huete. 1991. Interpreting vegetation indices. *Prev. Vet. Med.* 11:185–200.

Jämtgård, S., T. Näsholm, and K. Huss-Danell. 2010. Nitrogen compounds in soil solutions of agricultural lands. *Soil Biol. Biochem.* 42:2325–2330.

Ju, X.T., C.L. Kou, F.S. Zhang, and P. Christie. 2006. Nitrogen balance and groundwater nitrate contamination: Comparison among three intensive cropping systems on the North China Plain. *Environ. Pollut.* 143:117–125.

Justes, E., B. Mary, J.-M. Meynard, J.-M. Machet, and L. Thelier-Huche. 1994. Determination of a critical nitrogen dilution curve for winter wheat crops. *Ann. Bot.* 74:397–407.

Kanke, Y., W. Raun, J. Solie, M. Stone, and R. Taylor. 2012. Red edge as a potential index for detecting differences in plant nitrogen status in winter wheat. *J. Plant Nutr.* 35:1526–1541.

Kant, S., M. Peng, and S.J. Rothstein. 2011. Genetic regulation by *NLA* and microRNA827 for maintaining nitrate-dependent phosphate homeostasis in *Arabidopsis*. *PLoS Genetics* 7:e1002021.

Kechid, M., G. Desbrosses, W. Rokhsi, F. Varoquaux, A. Djekoun, and B. Touraine. 2013. The *NRT2.5* and *NRT2.6* genes are involved in growth promotion of *Arabidopsis* by the plant growth-promoting rhizobacterium (PGPR) strain *Phyllobacterium brassicacearum* STM196. *New Phytol.* 198:514–524.

Keskitalo, J., G. Bergquist, P. Gardeström, and S. Jansson. 2005. A cellular timetable of autumn senescence. *Plant Physiol.* 139:1635–1648.

Kiba, T., A.-B. Feria-Bourrellier, F. Lafouge et al. 2012. The *Arabidopsis* nitrate transporter NRT2.4 plays a double role in roots and shoots of nitrogen-starved plants. *Plant Cell* 24:245–258.

Kichey, T., B. Hirel, E. Heumez, F. Dubois, and J. Le Gouis. 2007. In winter wheat (*Triticum aestivum* L.), post-anthesis nitrogen uptake and remobilisation to the grain correlates with agronomic traits and nitrogen physiological markers. *Field Crops Res.* 102:22–32.

Kielland, K., J.W. McFarland, R.W. Ruess, and K. Olson. 2007. Rapid cycling of organic nitrogen in taiga forest ecosystems. *Ecosystems* 10:360–368.

Knecht, M.F. and A. Göransson. 2004. Terrestrial plants require nutrients in similar proportions. *Tree Physiol.* 24:447–460.

Koegel, S., N.A. Lahmidi, C. Arnould et al. 2013. The family of ammonium transporters (AMT) in *Sorghum bicolour*: Two AMT members are induced locally, but not systemically in roots colonized by arbuscular mycorrhizal fungi. *New Phytol.* 198:853–865.

Köhler, B. and K. Raschke. 2000. The delivery of salts to the xylem. Three types of anion conductance in the plasmalemma of the xylem parenchyma of roots of barley. *Plant Physiol.* 122:243–254.

Köhler, B., L.H. Wegner, V. Osipov, and K. Raschke. 2002. Loading of nitrate into the xylem: Apoplastic nitrate controls the voltage dependence of X-QUAC, the main anion conductance in the xylem parenchyma cells of barley roots. *Plant J.* 30:133–142.

Kojima, S., A. Bohner, B. Gassert, L. Yuan, and N. von Wirén. 2007. AtDUR3 represents the major transporter for high-affinity urea transport across the plasma membrane of nitrogen-deficient *Arabidopsis* roots. *Plant J.* 52:30–40.

Komarova, N.Y., K. Thor, A. Gubler et al. 2008. AtPTR1 and AtPTR5 transport dipeptides *in Planta*. *Plant Physiol.* 148:856–869.

Koprivova, A., M. Suter, R. Op den Camp, C. Brunold, and S. Kopriva. 2000. Regulation of sulfate assimilation by nitrogen in *Arabidopsis*. *Plant Physiol.* 122:737–746.

Kraiser, T., D.E. Gras, A.G. Gutiérrez, B. González, and R.A. Gutiérrez. 2011. A holistic view of nitrogen acquisition in plants. *J. Exp. Bot.* 62:1455–1466.

Krouk, G., P. Tillard, and A. Gojon. 2006. Regulation of the high-affinity NO_3^- uptake system by NRT1.1-mediated NO_3^- demand signaling in *Arabidopsis*. *Plant Physiol.* 142:1075–1086.

Kull, O. 2002. Acclimation of photosynthesis in canopies: Models and limitations. *Oecologia* 133:267–279.

Kutman, U.B., B.Y. Kutman, Y. Ceylan, E.A. Ova, and I. Cakmak. 2012. Contributions of root uptake and remobilization to grain zinc accumulation in wheat depending on post-anthesis zinc availability and nitrogen nutrition. *Plant Soil* 361:177–187.

Kuzyakov, Y. and X. Xu. 2013. Competition between roots and microorganisms for nitrogen: Mechanisms and ecological relevance. *New Phytol.* 198:656–669.

Laboski, C.A.M., J.E. Sawyer, D.T. Walters et al. 2008. Evaluation of the Illinois Soil Nitrogen Test in the North Central Region of the United States. *Agron. J.* 100:1070–1076.

Ladha, J.K. and P.M. Reddy. 2003. Nitrogen fixation in rice systems: State of knowledge and future prospects. *Plant Soil* 252:151–167.

Laisk, A., H. Eichelmann, V. Oja et al. 2005. Adjustment of leaf photosynthesis to shade in a natural canopy: Rate parameters. *Plant Cell Environ.* 28:375–388.

Lanquar, V., D. Loqué, F. Hörmann et al. 2009. Feedback inhibition of ammonium uptake by a phospho-dependent allosteric mechanism in *Arabidopsis*. *Plant Cell* 21:3610–3622.

Laperche, A., M. Brancourt-Hulmel, E. Heumez et al. 2007. Using genotype x nitrogen interaction variables to evaluate the QTL involved in wheat tolerance to nitrogen constraints. *Theoret. Appl. Genet.* 115:399–415.

Lark, R.M., A.E. Milne, T.M. Addiscott, K.W.T. Goulding, C.P. Webster, and S. O'Flaherty. 2004. Scale- and location-dependent correlation of nitrous oxide emissions with soil properties: An analysis using wavelets. *Eur. J. Soil Sci.* 55:611–627.

Lasa, B., S. Frechilla, M. Aleu et al. 2000. Effects of low and high levels of magnesium on the response of sunflower plants grown with ammonium and nitrate. *Plant Soil* 225:167–174.

Laugier, E., E. Bouguyon, A. Mauriès, P. Tillard, A. Gojon, and L. Lejay. 2012. Regulation of high-affinity nitrate uptake in roots of *Arabidopsis* depends predominantly on posttranscriptional control of the NRT2.1/NAR2.1 transport system. *Plant Physiol.* 158:1067–1078.

Lawlor, D.W., M. Kontturi, and A.T. Young. 1989. Photosynthesis by flag leaves of wheat in relation to protein, ribulose *bis*phosphate carboxylase activity and nitrogen supply. *J. Exp. Bot.* 40:43–52.

Lazlof, D.B., T.W. Rufty Jr., and M.G. Redinbaugh. 1992. Localization of nitrate absorption and translocation within morphological regions of the corn root. *Plant Physiol.* 100:1251–1258.

Leake, J.R. and D.J. Read. 1990. Proteinase activity in mycorrhizal fungi. I. The effect of extracellular pH on the production and activity of proteinase by ericoid endophytes from soils of contrasted pH. *New Phytol.* 115:243–250.

Le Bot, J., B. Jeannequin, and R. Fabre. 2001. Growth and nitrogen status of soilless tomato plants following nitrate withdrawal from the nutrient solution. *Ann. Bot.* 88:361–370.

Lee, Y.-H., J. Foster, J. Chen, L.M. Voll, A.P.M. Weber, and M. Tegeder. 2007. AAP1 transports uncharged amino acids into roots of *Arabidopsis*. *Plant J.* 50:305–319.

Lemaire, G., E. van Oosterom, M.-H. Jeuffroy, F. Gastal, and A. Massignam. 2008. Crop species present different qualitative types of response to N deficiency during their vegetative growth. *Field Crops Res.* 105:253–265.

Lemaire, G., E. van Oosterom, J. Sheehy, M.H. Jeuffroy, A. Massignam, and L. Rossato. 2007. Is crop N demand more closely related to dry matter accumulation or leaf area expansion during vegetative growth? *Field Crops Res.* 100:91–106.

Li, G., B. Li, G. Dong, X. Feng, H.J. Kronzucker, and W. Shi. 2013. Ammonium-induced shoot ethylene production is associated with the inhibition of lateral root formation in *Arabidopsis*. *J. Exp. Bot.* 64:1413–1425.

Li, Q., B.-H. Li, H.J. Kronzucker, and W.-M. Shi. 2010. Root growth inhibition by NH_4^+ in *Arabidopsis* is mediated by the root tip and is linked to NH_4^+ efflux and GMPase activity. *Plant Cell Environ.* 33:1529–1542.

Li, S.-M., B.-Z. Li, and W.-M. Shi. 2012. Expression patterns of nine ammonium transporters in rice in response to N status. *Pedosphere* 22:860–869.

Lin, S.-H., H.-F. Kuo, G. Canivenc et al. 2008. Mutation of the *Arabidopsis NRT1.5* nitrate transporter causes defective root-to-shoot nitrate transport. *Plant Cell* 20:2514–2528.

Linkohr, B.I., L.C. Williamson, A.H. Fitter, and H.M.O. Leyser. 2002. Nitrate and phosphate availability and distribution have different effects on root system architecture of *Arabidopsis*. *Plant J.* 29:751–760.

Lipson, D. and T. Näsholm. 2001. The unexpected versatility of plants: Organic nitrogen use and availability in terrestrial ecosystems. *Oecologia* 128:305–316.

Liu, K.-H. and Y.-F. Tsay. 2003. Switching between the two action modes of the dual-affinity nitrate transporter CHL1 by phosphorylation. *EMBO J.* 22:1005–1013.

Longnecker, N., E.J.M. Kirby, and A. Robson. 1993. Leaf emergence, tiller growth, and apical development of nitrogen-deficient spring wheat. *Crop Sci.* 33:154–160.

Lory, J.A. and P.C. Scharf. 2003. Yield goal versus delta yield for predicting fertilizer nitrogen need in corn. *Agron. J.* 95:994–999.

Maijala, P., K.V. Fagerstedt, and M. Raudaskoski. 1991. Detection of extracellular cellulolytic and proteolytic activity in ectomycorrhizal fungi and *Heterobasidion annosum* (Fr.) Bref. *New Phytol.* 117:643–648.

Mantelin, S., G. Desbrosses, M. Larcher, T.J. Tranbarger, J.-C. Cleyet-Marel, and B. Touraine. 2006. Nitrate-dependent control of root architecture and N nutrition are altered by a plant growth-promoting *Phyllobacterium* sp. *Planta* 223:591–603.

Marino, M.A., A. Mazzanti, S.G. Assuero, F. Gastal, H.E. Echeverría, and F. Andrade. 2004. Nitrogen dilution curves and nitrogen use efficiency during winter-spring growth of annual ryegrass. *Agron. J.* 96:601–607.

Masclaux-Daubresse, C., F. Daniel-Vedele, J. Dechorgnat, F. Chardon, L. Gaufichon, and A. Suzuki. 2010. Nitrogen uptake, assimilation and remobilization in plants: Challenges for sustainable and productive agriculture. *Ann. Bot.* 105:1141–1157.

Matt, P., M. Geiger, P. Walch-Liu, C. Engels, A. Krapp, and M. Stitt. 2001. The immediate cause of the diurnal changes of nitrogen metabolism in leaves of nitrate-replete tobacco: A major imbalance between the rate of nitrate reduction and the rates of nitrate uptake and ammonium metabolism during the first part of the light period. *Plant Cell Environ.* 24:177–190.

Matt, P., A. Krapp, V. Haake, H.-P. Mock, and M. Stitt. 2002. Decreased Rubisco activity leads to dramatic changes in nitrate metabolism, amino acid metabolism and the levels of phenylpropanoids and nicotine in tobacco antisense *RBCS* transformants. *Plant J.* 30:663–677.

Matzinger, D.F., E.A. Wernsman, and C.C. Cockerham. 1972. Recurrent family selection and correlated response in *Nicotiana tabacum* L. I. 'Dixie Bright 244' x 'Coker 139'. *Crop Sci.* 12:40–43.

McFarland, J.W., R.W. Ruess, K. Kielland, K. Pregitzer, R. Hendrik, and M. Allen. 2010. Cross-ecosystem comparisons of in situ plant uptake of amino acid-N and NH_4^+. *Ecosystems* 13:177–193.

Melin, E. and H. Nilsson. 1953. Transfer of labelled nitrogen from glutamic acid to pine seedlings through the mycelium of *Boletus variegatus* (Sw.) Fr. *Nature* 171:134.

Mérigout, P., M. Lelandais, F. Bitton et al. 2008. Physiological and transcriptomic aspects of urea uptake and assimilation in *Arabidopsis* plants. *Plant Physiol.* 147:1225–1238.

Michael, G., H. Schumacher, and H. Marschner. 1965. Aufnahme von Ammonium- und Nitratstickstoff aus markiertem Ammoniumnitrat und deren Verteilung in der Pflanze. *Z. Pflanzenern. Bodenk.* 110:225–238.

Migge, A., C. Bork, R. Hell, and T.W. Becker. 2000. Negative regulation of nitrate reductase gene expression by glutamine or asparagine accumulating in leaves of sulfur-deprived tobacco. *Planta* 211:587–595.

Miller, A.J., X. Fan, M. Orsel, S.J. Smith, and D.M. Wells. 2007. Nitrate transport and signalling. *J. Exp. Bot.* 58:2297–2306.

Monteith, J.L. 1977. Climate and the efficiency of crop production in Britain. *Phil. Trans. Roy. Soc. London B* 281:277–294.

Mota, M., C.B. Neto, A.A. Monteiro, and C.M. Oliveira. 2011. Preferential ammonium uptake during growth cycle and identification of ammonium transporter genes in young pear trees. *J. Plant Nutr.* 34:798–814.

Motavalli, P.P., K.A. Nelson, and S. Bardhan. 2012. Development of a variable-source N fertilizer management strategy using enhanced-efficiency N fertilizers. *Soil Sci.* 177:708–718.

Mottram, D.S., B.L. Wedzicha, and A.T. Dodson. 2002. Acrylamide is formed in the Maillard reaction. *Nature* 419:448–449.

Moyer-Henry, K.A., J.W. Burton, D.W. Israel, and T.W. Rufty. 2006. Nitrogen transfer between plants: A ^{15}N natural abundance study with crop and weed species. *Plant Soil* 282:7–20.

Mulvaney, R.L. and S.A. Khan. 2001. Diffusion methods to determine different forms of nitrogen in soil hydrolysates. *Soil Sci. Soc. Am. J.* 65:1284–1292.

Mulvaney, R.L., S.A. Khan, R.G. Hoeft, and H.M. Brown. 2001. A soil organic nitrogen fraction that reduces the need for nitrogen fertilization. *Soil Sci. Soc. Am. J.* 65:1164–1172.

Näsholm, T., P. Högberg, O. Franklin et al. 2013. Are ectomycorrhizal fungi alleviating or aggravating nitrogen limitation of tree growth in boreal forests? *New Phytol.* 198:214–221.

Näsholm, T., K. Kielland, and U. Ganeteg. 2009. Uptake of organic nitrogen by plants. *New Phytol.* 182:31–48.

Nazoa, P., J.J. Vidmar, T.J. Tranberger et al. 2003. Regulation of the nitrate transporter gene *AtNRT2.1* in *Arabidopsis thaliana*: Responses to nitrate, amino acids and developmental stage. *Plant Mol. Biol.* 52:689–703.

Ning, P., S. Li, P. Yu, Y. Zhang, and C. Li. 2013. Post-silking accumulation and partitioning of dry matter, nitrogen, phosphorus and potassium in maize varieties differing in leaf longevity. *Field Crops Res.* 144:19–27.

Nowacki, E., M. Jurzysta, P. Gorski et al. 1976. Effect of nitrogen nutrition on alkaloid metabolism in plants. *Biochem. Physiol. Pflanzen* 169:231–240.

Nutman, P.S. 1987. Centenary lecture. *Phil. Trans. Roy. Soc. London B* 317:69–106.

Olfs, H.-W., K. Blankenau, F. Brentrup, J. Jasper, A. Link, and J. Lammel. 2005. Soil- and plant-based nitrogen-fertilizer recommendations in arable farming. *J. Plant Nutr. Soil Sci.* 168:414–431.

Omirou, M., C. Papastefanou, D. Katsarou et al. 2012. Relationships between nitrogen, dry matter accumulation and glucosinolates in *Eruca sativa* Mills. The applicability of the critical NO_3-N levels approach. *Plant Soil* 354:347–358.

Oscarson, P. 2000. The strategy of the wheat plant in acclimating growth and grain production to nitrogen availability. *J. Exp. Bot.* 51:1921–1929.

Otteson, B.N., M. Mergoum, and J.K. Ransom. 2008. Seeding rate and nitrogen management on milling and baking quality of hard red spring wheat genotypes. *Crop Sci.* 48:749–755.

Palmer, S.J., D.M. Berridge, A.J.S. McDonald, and W.J. Davies. 1996. Control of leaf expansion in sunflower (*Helianthus annuus* L.) by nitrogen nutrition. *J. Exp. Bot.* 47:359–368.

Pampolino, M.F., C. Witt, J.M. Pasuquin, A. Johnston, and M.J. Fisher. 2012. Development approach and evaluation of the Nutrient Expert software for nutrient management in cereal crops. *Comput. Electron. Agric.* 88:103–111.

Pask, A.J.D., R. Sylvester-Bradley, P.D. Jamieson, and M.J. Foulkes. 2012. Quantifying how winter wheat crops accumulate and use nitrogen reserves during growth. *Field Crops Res.* 126:104–118.

Paungfoo-Lonhienne, C., T.G.A. Lonhienne, D. Rentsch et al. 2008. Plants can use protein as a nitrogen source without assistance from other organisms. *Proc. Natl. Acad. Sci.* 105:4524–4529.

Paungfoo-Lonhienne, C., P.M. Schenk, T.G.A. Lonhienne et al. 2009. Nitrogen affects cluster root formation and expression of putative peptide transporters. *J. Exp. Bot.* 60:2665–2676.

Pearson, C.J. and B.T. Steer. 1977. Daily changes in nitrate uptake and metabolism in *Capsicum annuum*. *Planta* 137:107–112.

Peltonen, J., A. Virtanen, and E. Haggrèn. 1995. Using a chlorophyll meter to optimize nitrogen fertilizer application for intensively-managed small-grain cereals. *J. Agron. Crop Sci.* 174:309–318.

Peng, Y., P. Yu, X. Li, and C. Li. 2013. Determination of the critical soil mineral nitrogen concentration for maximizing maize grain yield. *Plant Soil* 372:41–51.

Pérez-Tienda, J., A. Valderas, G. Camañes, P. García-Agustín, and N. Ferrol. 2012. Kinetics of NH_4^+ uptake by the arbuscular mycorrhizal fungus *Rhizophagus irregularis*. *Mycorrhiza* 22:485–491.

Perry, E.M., G.J. Fitzgerald, J.G. Nuttall, G.J. O'Leary, U. Schulthess, and A. Whitlock. 2012. Rapid estimation of canopy nitrogen of cereal crops at paddock scale using a Canopy Chlorophyll Content Index. *Field Crops Res.* 134:158–164.

Pirhofer-Walzl, K., J. Rasmussen, H. Høgh-Jensen, J. Eriksen, K. Søegaard, and J. Rasmussen. 2012. Nitrogen transfer from forage legumes to nine neighbouring plants in a multi-species grassland. *Plant Soil* 350:71–84.

Plassard, C., D. Barry, L. Eltrop, and D. Mousain. 1994. Nitrate uptake in maritime pine (*Pinus pinaster*) and the ectomycorrhizal fungus *Hebeloma cylindrosporum*: Effect of ectomycorrhizal symbiosis. *Can. J. Bot.* 72:189–197.

Plassard, C., P. Scheromm, D. Mousain, and L. Salsac. 1991. Assimilation of mineral nitrogen and ion balance in the two partners of ectomycorrhizal symbiosis: Data and hypothesis. *Experientia* 47:340–349.

Plénet, D. and G. Lemaire. 2000. Relationships between dynamics of nitrogen uptake and dry matter accumulation in maize crops. Determination of critical N concentration. *Plant Soil* 216:65–82.

Quraishi, U.M., M. Abrouk, F. Murat et al. 2011. Cross-genome map based dissection of a nitrogen use efficiency ortho-metaQTL in bread wheat unravels concerted cereal genome evolution. *Plant J.* 65:745–756.

Rahayu, R.S., P. Walch-Liu, G. Neumann, V. Römheld, N. won Wirén, and F. Bangerth. 2005. Root-derived cytokinins as long-distance signals for NO_3^-—Induced stimulation of leaf growth. *J. Exp. Bot.* 56:1143–1152.

Rains, K.C. and C.S. Bledsoe. 2007. Rapid uptake of ^{15}N-ammonium and glycine-^{13}C, ^{15}N by arbuscular and ericoid mycorrhizal plants native to a Northern California coastal pygmy forest. *Soil Biol. Biochem.* 39:1078–1086.

Raun, W.R., J.B. Solie, G.V. Johnson et al. 2002. Improving nitrogen use efficiency in cereal grain production with optical sensing and variable rate application. *Agron. J.* 94:815–820.

Recous, S., C. Fresneau, G. Faurie, and B. Mary. 1988. The fate of labelled ^{15}N urea and ammonium nitrate applied to a winter wheat crop. I. Transformations of nitrogen in the soil. *Plant Soil* 112:205–214.

Reussi Calvo, N.I., H. Sainz Rosas, H. Echevarría, and A. Berardo. 2013. Contribution of anaerobically incubated nitrogen to the diagnosis of nitrogen status in spring wheat. *Agron. J.* 105:321–328.

Robinson, D.L., L.C. Kappel, and J.A. Boling. 1989. Management practices to overcome the incidence of grass tetany. *J. Anim. Sci.* 67:3470–3484.

Roggatz, U., A.J.S. McDonald, I. Stadenberg, and U. Schurr. 1999. Effects of nitrogen deprivation on cell division and expansion in leaves of *Ricinus communis* L. *Plant Cell Environ.* 22:81–89.

Rothstein, D.E. 2009. Soil amino-acid availability across a temperate-forest fertility gradient. *Biogeochemistry* 92:201–215.

Saia, S., E. Benítez, J.M. García-Garrido, L. Settanni, G. Amato, and D. Giambalvo. 2014. The effect of arbuscular mycorrhizal fungi on total plant nitrogen uptake and nitrogen recovery from soil organic material. *J. Agric. Sci. Camb.* 152:370–378.

Sainz Rosas, H., P.A. Calviño, H.E. Echevarría, P.A. Barbieri, and M. Redolatti. 2008. Contribution of anaerobically mineralized nitrogen to the reliability of planting or presidedress soil nitrogen test in maize. *Agron. J.* 100:1020–1025.

Samborski, S.M., N. Tremblay, and E. Fallon. 2009. Strategies to make use of plant sensor-based diagnostic information for nitrogen recommendations. *Agron. J.* 101:800–816.

Sanderson, K. 2012. Bid to curb fried-food chemical goes cold. *Nature* 491:22–23.

Santamaria, P. 2006. Nitrate in vegetables: Toxicity, content, intake and EC regulation. *J. Sci. Food Agric.* 86:10–17.

Santi, C., D. Bogusz, and C. Franche. 2013. Biological nitrogen fixation in non-legume plants. *Ann. Bot.* 111:743–767.

Sawyer, J. 2010. Corn stalk nitrate interpretation. Integrated Crop Management News, Extension and Outreach, Iowa State University, Ames, IA. http://www.extension.iastate.edu/CropNews/2010/0914sawyer.htm (accessed July 24, 2013).

Scaife, A. and K.L. Stevens. 1983. Monitoring sap nitrate in vegetable crops: Comparison of test strips with electrode methods, and effects of time of day and leaf position. *Commun. Soil Sci. Plant Anal.* 14:761–771.

Schlüter, U., M. Mascher, C. Colmsee et al. 2012. Maize source leaf adaptation to nitrogen deficiency affects not only nitrogen and carbon metabolism but also control of phosphate homeostasis. *Plant Physiol.* 160:1384–1406.

Schmidhalter, U. 2005. Development of a quick on-farm test to determine nitrate levels in soil. *J. Plant Nutr. Soil Sci.* 168:432–438.

Schmidt, F., F.D. De Bona, and F.A. Monteiro. 2013. Sulfur limitation increases nitrate and amino acid pools in tropical forages. *Crop Pasture Sci.* 64:51–60.

Schomberg, H.H., S. Wietholter, T.S. Griffin et al. 2009. Assessing indices for predicting potential nitrogen mineralization in soils under different management systems. *Soil Sci. Soc. Am. J.* 73:1575–1586.

Schonhof, I., D. Blankenburg, S. Müller, and A. Krumbein. 2007. Sulfur and nitrogen supply influence growth, product appearance, and glucosinolate concentration of broccoli. *J. Plant Nutr. Soil Sci.* 170:65–72.

Shaked, A. and A. Bar-Akiva. 1967. Nitrate reductase activity as an indication of molybdenum level and requirement of *Citrus* plants. *Phytochemistry* 6:347–350.

Sheehy, J.E., M.J.A. Dionora, P.L. Mitchell et al. 1998. Critical nitrogen concentrations: Implications for high-yielding rice (*Oryza sativa* L.) cultivars in the tropics. *Field Crops Res.* 59:31–41.

Siddiqi, M.Y., A.D.M. Glass, T.J. Ruth, and T.W. Rufty Jr. 1990. Studies on the uptake of nitrate in barley. I. Kinetics of $^{13}NO_3^-$ influx. *Plant Physiol.* 93:1426–1432.

Siebrecht, S., G. Mäck, and R. Tischner. 1995. Function and contribution of the root tip in the induction of NO_3^- uptake along the barley root axis. *J. Exp. Bot.* 46:1669–1676.

Smith, S.E. and F.A. Smith. 2011. Roles of arbuscular mycorrhizas in plant nutrition and growth: New paradigms from cellular to ecosystem scales. *Annu. Rev. Plant Biol.* 62:227–250.

Solari, F., J. Shanahan, R. Ferguson, J. Schepers, and A. Gitelson. 2008. Active sensor reflectance measurements of corn nitrogen status and yield potential. *Agron. J.* 100:571–579.

Soper, F.M., C. Paungfoo-Lonhienne, R. Brackin, D. Rentsch, S. Schmidt, and N. Robinson. 2011. *Arabidopsis* and *Lobelia anceps* access small peptides as a nitrogen source for growth. *Funct. Plant Biol.* 38:788–796.

Sorgonà, A., A. Lupini, F. Mercati, L. Di Dio, F. Sunseri, and M.R. Abenavoli. 2011. Nitrate uptake along the maize primary root: An integrated physiological and molecular approach. *Plant Cell Environ.* 34:1127–1140.

Spiertz, J.H.J. and N.M. de Vos. 1983. Agronomical and physiological aspects of the role of nitrogen in yield formation of cereals. *Plant Soil* 75:379–391.

Sripada, R.P., J.P. Schmidt, A.E. Dellinger, and D.B. Beegle. 2008. Evaluating multiple indices from a canopy reflectance sensor to estimate corn N requirements. *Agron. J.* 100:1553–1561.

Subbarao, G.V., M. Kishii, K. Kakahara et al. 2009. Biological nitrification inhibition (BNI)—Is there potential for genetic interventions in the Triticeae? *Breed. Sci.* 59:529–545.

Suter, H., H. Sultana, D. Turner, R. Davies, C. Walker, and D. Chen. 2013. Influence of urea fertiliser formulation, urease inhibitor and season on ammonia loss from ryegrass. *Nutr. Cycling Agroecosyst.* 95:175–185.

Svennerstam, H., S. Jämtgård, I. Ahmad, K. Huss-Danell, T. Näsholm, and L. Ganeteg. 2011. Transporters in *Arabidopsis* roots mediating uptake of amino acids at naturally occurring concentrations. *New Phytol.* 191:459–467.

Ta, C.T. and R.T. Wieland. 1992. Nitrogen partitioning in maize during ear development. *Crop Sci.* 32:443–451.

Tanaka, A. and C.V. Garcia. 1965. Studies of the relationship between tillering and nitrogen uptake of the rice plant. 2. Relation between tillering and nitrogen metabolism of the plant. *Soil Sci. Plant Nutr.* 11:129–135.

Tareke, E., P. Rydberg, P. Karlsson, S. Eriksson, and M. Törnqvist. 2002. Analysis of acrylamide, a carcinogen formed in heated foodstuffs. *J. Agric. Food Chem.* 50:4998–5006.

Taylor, A.R. and A.J. Bloom. 1998. Ammonium, nitrate, and proton fluxes along the maize root. *Plant Cell Environ.* 21:1255–1263.

Tcherkez, G. and M. Hodges. 2008. How stable isotopes may help to elucidate primary nitrogen metabolism and its interaction with (photo)respiration in C_3 leaves. *J. Exp. Bot.* 59:1685–1693.

Tegeder, M. and D. Rensch. 2010. Uptake and partitioning of amino acids and peptides. *Mol. Plant* 3:997–1011.

ten Hoopen, F., T.A. Cuin, P. Pedas et al. 2010. Competition between uptake of ammonium and potassium in barley and *Arabidopsis* roots: Molecular mechanisms and physiological consequences. *J. Exp. Bot.* 61:2303–2315.

Thorn, A.M. and C.M. Orians. 2011. Patchy nitrate promotes inter-sector flow and ^{15}N allocation in *Ocimum basilicum*: A model and an experiment. *Funct. Plant Biol.* 38:879–887.

Thornton, B. and D. Robinson. 2005. Uptake and assimilation of nitrogen from solutions containing multiple N sources. *Plant Cell Environ.* 28:813–821.

Torres, L.O. and A. Link. 2010. Best management principles and techniques to optimise nutrient use efficiency. *Proceedings International Fertiliser Society* 683, York, U.K.

Trachsel, S., S.M. Kaeppler, K.M. Brown, and J.P. Lynch. 2013. Maize root growth angles become steeper under low N conditions. *Field Crops Res.* 140:18–31.

Trápani, N., A.J. Hall, and M. Weber. 1999. Effects of constant and variable nitrogen supply on sunflower (*Helianthus annuus* L.) leaf cell number and size. *Ann. Bot.* 84:599–606.

Triboi, E. and A.-M. Triboi-Blondel. 2002. Productivity and grain or seed composition: A new approach to an old problem—Invited paper. *Eur. J. Agron.* 16:163–186.

Turner, F.T. and M.F. Jund. 1991. Chlorophyll meter to predict nitrogen topdress requirement for semidwarf rice. *Agron. J.* 83:926–928.

Uhart, S.A. and F.H. Andrade. 1995. Nitrogen deficiency in maize: II. Carbon-nitrogen interaction effects on kernel number and grain yield. *Crop Sci.* 35:1384–1389.

Ulas, A., G. Schulte auf'm Erley, M. Kamh, F. Wiesler, and W.J. Horst. 2012. Root-growth characteristics contributing to genotypic variation in nitrogen efficiency of oilseed rape. *J. Plant Nutr. Soil Sci.* 175:489–498.

van der Ploeg, R.R., W. Böhm, and M.B. Kirkham. 1999. On the origin of the theory of mineral nutrition of plants and the Law of the Minimum. *Soil Sci. Soc. Am. J.* 63:1055–1062.

Vidmar, J.J., J.K. Schoerring, B. Touraine, and A.D.M. Glass. 1999. Regulation of the *hvst1* gene encoding a high-affinity sulfate transporter from *Hordeum vulgare*. *Plant Molec. Biol.* 40:883–892.

Wada, G., S. Shoji, and T. Mae. 1986. Relationship between nitrogen absorption and growth and yield of rice plants. *Jpn. Agric. Res. Q.* 20:135–145.

Walch-Liu, P., G. Neumann, F. Bangerth, and C. Engels. 2000. Rapid effects of nitrogen form on leaf morphogenesis in tobacco. *J. Exp. Bot.* 51:227–237.

Walker, R.L., I.G. Burns, and J. Moorby. 2001. Responses of plant growth rate to nitrogen supply: A comparison of relative addition and N interruption treatments. *J. Exp. Bot.* 52:309–317.

Walsh, O., W. Raun, A. Klatt, and J. Solie. 2012. Effects of delayed nitrogen fertilization on maize (*Zea mays* L.) grain yields and nitrogen use efficiency. *J. Plant Nutr.* 35:538–555.

Wan, Y., P.R. Shewry, and M.J. Hawkesford. 2013. A novel family of γ-gliadin genes are highly regulated by nitrogen supply in developing wheat grain. *J. Exp. Bot.* 64:161–168.

Wang, R., M. Okamoto, X. Xing, and N.M. Crawford. 2003. Microarray analysis of the nitrate response in *Arabidopsis* roots and shoots reveals over 1,000 rapidly responding genes and new linkages to glucose, trehalose-6-phosphate, iron, and sulfate metabolism. *Plant Physiol.* 132:556–567.

Wang, W., X. Yao, X.-F. Yao et al. 2012a. Estimating leaf nitrogen concentration with three-band vegetation indices in rice and wheat. *Field Crops Res.* 129:90–98.

Wang, W.-H., B. Köhler, F.-Q. Cao et al. 2012b. Rice DUR3 mediates high-affinity urea transport and plays an effective role in improvement of urea acquisition and utilization when expressed in *Arabidopsis*. *New Phytol.* 193:432–444.

Wang, W.-H., G.-W. Liu, F.-Q. Cao, X.-Y. Cheng, B.-W. Liu, and L.-H. Liu. 2013a. Inadequate root uptake may represent a major component limiting rice to use urea as a nitrogen source for growth. *Plant Soil* 363:191–200.

Wang, W.-T., S.-W. Huang, R.-L. Liu, and J.-Y. Jin. 2007. Effects of nitrogen application on flavor compounds of cherry tomato fruits. *J. Plant Nutr. Soil Sci.* 170:461–468.

Wang, X. and F.E. Below. 1998. Accumulation and partitioning of mineral nutrients in wheat as influenced by nitrogen form. *J. Plant Nutr.* 21:49–61.

Wang, Y., D. Wang, G. Zhang, and J. Wang. 2013b. Estimating nitrogen status of rice using the image segmentation of a G-R thresholding method. *Field Crops Res.* 149:33–39.

Wang, Y.-H., D.F. Garvin, and L.V. Kochian. 2001. Nitrate-induced genes in tomato roots. Array analysis reveals novel genes that may play a role in nitrogen nutrition. *Plant Physiol.* 127:345–359.

Wang, Y.-Y., P.-K. Hsu, and Y.-F. Tsay. 2012c. Uptake, allocation and signaling of nitrate. *Trends Plant Sci.* 17:458–467.

Warren, C.R. 2013a. High diversity of small organic N observed in soil water. *Soil Biol. Biochem.* 57:444–450.

Warren, C.R. 2013b. Quaternary ammonium compounds can be abundant in some soils and are taken up as intact molecules by plants. *New Phytol.* 198:476–485.

Watson, D.J., G.N. Thorne, and S.A.W. French. 1963. Analysis of growth and yield of winter and spring wheats. *Ann. Bot.* 27:1–22.

Wehrmann, J. and H.C. Scharpf. 1979. Der Mineralstickstoffgehalt des Bodens als Maßstab für den Stickstoffdüngerbedarf (N_{min}—Methode). *Plant Soil* 52:109–126.

Whiteside, M.D., K.K. Treseder, and P.R. Atsatt. 2009. The brighter side of soils: Quantum dots track organic nitrogen through fungi and plants. *Ecology* 90:100–108.

Wilcox, G.E., J.E. Hoff, and C.M. Jones. 1973. Ammonium reduction of calcium and magnesium content of tomato and sweet corn leaf tissue and influence on incidence of blossom end rot of tomato fruit. *J. Am. Soc. Hortic. Sci.* 98:86–89.

Witte, C.-P. 2011. Urea metabolism in plants. *Plant Sci.* 180:431–438.

Witte, C.-P., S.A. Tiller, M.A. Taylor, and H.V. Davies. 2002. Leaf urea metabolism in potato. Urease activity profile and patterns of recovery and distribution of ^{15}N after foliar urea application in wild-type and urea-antisense transgenics. *Plant Physiol.* 128:1129–1136.

Wuest, S.B. and K.G. Cassman. 1992a. Fertilizer-N use efficiency of irrigated wheat: I. Uptake efficiency of preplant versus late-season application. *Agron. J.* 84:682–688.

Wuest, S.B. and K.G. Cassman. 1992b. Fertilizer-N use efficiency of irrigated wheat: I. Partitioning efficiency of preplant versus late-season application. *Agron. J.* 84:689–694.

Xiaoping, X., W. Jianguo, W. Zhiwei et al. 2007. Determination of a critical dilution curve for nitrogen concentration in cotton. *J. Plant Nutr. Soil Sci.* 170:811–817.

Xu, M., D. Li, J. Li et al. 2013. Polyolefin-coated urea decreases ammonia volatilization in a double rice system of southern China. *Agron. J.* 105:277–284.

Xue, Y., H. Duan, L. Liu, Z. Wang, J. Yang, and J. Zhang. 2013. An improved crop management increases grain yield and nitrogen and water use efficiency in rice. *Crop Sci.* 53:271–284.

Yang, J. and J. Zhang. 2006. Grain filling of cereals under soil drying. *New Phytol.* 169:223–236.

Yang, S., Y. Zou, Y. Liang et al. 2012. Role of soil total nitrogen in aroma synthesis of traditional regional aromatic rice in China. *Field Crops Res.* 125:151–160.

Yang, W.-H., S. Peng, J. Huang, A.L. Sanico, R.J. Buresh, and C. Witt. 2003. Using leaf color charts to estimate leaf nitrogen status of rice. *Agron. J.* 95:212–217.

Yong, Z., Z. Kotur, and A.D.M. Glass. 2010. Characterization of an intact two-component high-affinity nitrate transporter from *Arabidopsis* roots. *Plant J.* 63:739–748.

Zeng, H., G. Liu, T. Kinoshita et al. 2012. Stimulation of phosphorus uptake by ammonium nutrition involves plasma membrane H+ ATPase in rice roots. *Plant Soil* 357:205–214.

Zhang, H., H. Rong, and D. Pilbeam. 2007. Signalling mechanisms underlying the morphological responses of the root system to nitrogen in *Arabidopsis thaliana*. *J. Exp. Bot.* 58:2329–2338.

Zhao, D.H., J.L. Li, and J.G. Qi. 2005. Identification of red and NIR spectral regions and vegetative indices for discrimination of cotton nitrogen stress and growth stage. *Comput. Electron. Agric.* 48:155–169.

Zhao, R.-F., X.-P. Chen, F.-S. Zhang, H. Zhang, J. Schroder, and V. Römheld. 2006. Fertilization and nitrogen balance in a wheat-maize rotation system in North China. *Agron. J.* 98:938–945.

Zhao, X.Q., S.W. Guo, F. Shinmachi et al. 2013. Aluminium tolerance in rice is antagonistic with nitrate preference and synergistic with ammonium preference. *Ann. Bot.* 111:69–77.

Ziadi, N., G. Bélanger, A. Claessens et al. 2010. Determination of a critical nitrogen dilution curve for spring wheat. *Agron. J.* 102:241–250.

3 Phosphorus

Bryan G. Hopkins

CONTENTS

3.1   HISTORICAL BACKGROUND

John Emsley (2000) states "… phosphorus was greeted with great acclaim, and yet it was damned from the moment it was born." Emsley artfully and accurately tells the history of the 13th element to be discovered—ranging from its toxicities and dangers in weaponry and industry to its nutritional role.

Elemental phosphorus (P), discovered about 1669, exists in white or red mineral forms and is a multivalent, pnictogen (nitrogen family), nonmetallic element with the atomic number 15. Due to its high reactivity, it never is found naturally as a free element on Earth, but only in its maximally oxidized and hydrated state as inorganic orthophosphate (PO_4^{3-}), or more typically associated with one or two protons as HPO_4^{2-} or $H_2PO_4^-$. The degree of protonation is a function of pH. Phosphate is a trivalent resonating tetraoxyanion that acts as a linkage or binding site and is typically resistant to polarization and nucleophilic reaction, except in metal–enzyme complexes (Clarkson and Hanson, 1980).

Presently, the vast majority of commercially produced P compounds are consumed as plant fertilizers.

Rocks and soil minerals, which contain plant nutrients in their makeup, break down very slowly, but this process is generally not sufficient to raise crops continuously. The art of fertilizing with ash, manure, salts, plant residues, etc., began shortly after the first hunter-gatherers converted their lifestyles to cultivate crops. The science of fertilization began with the discovery of various chemical elements and observations that maintaining or improving crop yields required various chemical inputs.

Emsley (2000) discusses record keeping at Oxford University during the fourteenth century, documenting the decline in yields with the constant cropping necessitated by the small land mass and the large population of England. This loss in yield is attributed to declining soil fertility. Although soils tend to be able to sustain slow to modest plant growth in native ecosystems, lost mineral nutrients must be replenished under the more intensive systems required for sustaining a population consuming agricultural commodities. Phosphorus is one of the minerals most commonly depleted to a point of significantly impacting plant growth.

Although farmers commonly observed plant response to manures and other fertilizer materials and soil amendments, the formal science of soil fertility and plant nutrition was meager prior to the late Renaissance, after which developments in biology, chemistry, and physics resulted in a series of important discoveries that eventually led to the Green Revolution of the Twentieth Century. In the early nineteenth century, Arthur Young described possibly the first formal fertilization experiments with poultry manure, gunpowder, charcoal, ashes, and salts in the *Annals of Agriculture*. In 1799, Erasmus Darwin, grandfather of Charles Darwin, wrote in *The Philosophy of Agriculture and Gardening* that nitrogen (N) and P are plant nutrients taken up by roots and that compost, bone ash, and manures should be applied as fertilizers. He even suggested exploration of P-bearing minerals. According to Emsley (2000), he was largely ignored until decades later.

About the same time, Theodore de Saussure (1767–1845) built upon discoveries by Antoine Lavoisier (1767–1845) in chemistry—confirming that plants absorb specific mineral elements from the soil. Georges Ville (1824–1897) was possibly the first to state that plants take up P in the oxidized form as an essential nutrient. Justus von Liebig (1803–1873) maintained that the "other" mineral nutrients, especially P, were as important as N, carbon (C), hydrogen (H), and oxygen (O). He popularized the *Liebig's Law of the Minimum*, which actually originated with Carl Sprengel's (1786–1859) "theory of minimum"—stating that any nutrient in deficient supply becomes the limiting factor for growth, even if the others are supplied in abundance. Liebig also built upon the work of the French chemist Jean-Baptiste Boussingault (1802–1887), who established the first agricultural experiment station in 1836—making significant advances including promotion of simultaneous application of both N and P. Hall (1909) provides an excellent review of the discoveries of these and other early scientists regarding the essentiality of plant nutrients and the developments in the fertilizer production industry.

Although the essentiality of P as a nutrient was known, the plant availability of P from limited sources other than organic wastes, such as manure, was problematic. It was learned that every living thing contained relatively high amounts of P and that the concentration in bones was especially high. However, application of pulverized bones to soil was not efficient due to the poor solubility of the calcium phosphate compound found therein. In 1840 in Germany, Liebig applied sulfuric acid (H_2SO_4) to powdered bones, showing that the resulting product was more successfully taken up by growing plants than untreated bone powder. This fertilizer material was termed single superphosphate (SSP; 16% to 20% P_2O_5), also known as normal or ordinary superphosphate.

Shortly afterward, John Bennet Lawes (1814–1900) began the testing and commercial production of mineral P in England. Although not as well documented or known, James Murray of Ireland purportedly had developed his own version of an effective P fertilizer prior to Liebig or Lawes. He and Lawes filed for patents about the same time, but Lawes eventually purchased Murray's patent. In collaboration with others, Lawes conducted extensive research efforts at his estate, the now renowned Rothamsted Research Experiment Station, established in 1843 and the longest continuously run station. Several phosphate-containing materials (including animal manures and acidified bones and minerals) were evaluated as fertilizers at Rothamsted. These efforts launched the widespread production and use of effective phosphate fertilizers. These early P fertilizer materials frequently were mixed with N-rich bat guano and/or potassium (K) sources. Manufacturers of fertilizer sprouted up around the world in the latter half of the nineteenth century and the early twentieth century.

Interestingly, although credited with the formal discovery of acidifying bones to create a P fertilizer, Liebig lambasted England for robbing European battlefields, such as Waterloo, and catacombs in various countries of skeletons to feed the ever-increasing demand for increased agricultural productivity (Hall, 1909)—actions that provide some context for the great increase in the demand for P fertilizer during that time. Because of eventual scarcity and the associated costs of bone recovery, and with advances in the processing of apatite or rock phosphate, the practice of applying acidified bone powder largely was abandoned.

Rock phosphate was first used about 1850 to make P fertilizer, and following the introduction of the electric submerged-arc furnace in 1890, elemental P production switched from bone-ash heating to production from mineral P sources. After the depletion of world guano sources about the same time, mineral P became the major source of phosphate fertilizer production. Eventually, H_2SO_4 was replaced with phosphoric acid (H_3PO_4) to create fertilizer with a higher concentration of P, known as triple superphosphate (TSP; 0-45-0, $N-P_2O_5-K_2O$) or treble, double, or concentrated superphosphate. In the 1960s, ammonia (NH_3) was reacted with H_3PO_4 to form monoammonium phosphate (MAP; 11-52-0), diammonium phosphate (DAP; 18-46-0), and ammonium polyphosphate (APP; 10-34-0 or 11-37-0) used commonly today (Mortvedt et al., 1999; Wagganman, 1969). Phosphate rock production greatly increased in the mid-twentieth century, and it remains the primary global source of P (Figure 3.1). Phosphate mines are on every continent, with the United States, China, Morocco, and Russia recognized as the top producers.

Crop responses to P fertilization were widespread at experiment stations around the world as well as readily observed by farmers. Lawes and others at Rothamsted began what was to become a global use of the scientific method to address the need for increased crop production. Lawes and scientists at other locations conducted P fertilizer rate studies with the goal to determine the correct amount to apply. In general, these resulted in curvilinear yield responses with increasing P fertilizer rate up until yield plateaued or even declined at excessively high rates. Although rate studies were vital in guiding farmers to apply sufficient, but not extreme amounts of fertilizer, Hall (1909) discusses that it was not possible at that time to find the correct rate for all soils and circumstances. Early scientists collectively found a limit to fertilizer response by plants. The massive responses to P fertilizer reached a plateau as a result of continued P fertilization and buildup in the soil.

FIGURE 3.1   The phosphate-rich minerals of the park formation at the Simplot Vernal Mine near Vernal, Utah, United States.

Efforts were then concentrated upon methods of determining which plants in which soils would be responsive to fertilizers as well as the economics of fertilization.

Farmers and scientists understood that customized P fertilizer recommendations were needed. Initial efforts were focused on attempts to correlate plant response to P fertilizer with the total elemental concentration in soils, but it was soon discovered that the relationship was not predictive. Hall (1905) proposed plant analysis as a tool to predict fertilizer need. Macy (1936) established the notion of a critical concentration range for each species. Good correlations between plant tissue and yield were achieved for many elements and species. This tool continues to be developed and used. However, P deficiencies, especially for annual plants, often occur very early in the season before tissue can be used to predict P response. And there is convenience of being able to apply fertilizer prior to planting, thus creating a need for a predictive soil test. Therefore, initial failures in soil testing were reexamined.

Scientists studying soil physical chemistry discovered the various mineral and organic P compounds in the soil and their variability in solubility and plant availability. Several soil tests were developed with the intent to provide a reasonable correlation between yield parameters and a diagnostic determination of a P concentration related to plant P availability (Bray and Kurtz, 1945; Dyer, 1894; Hanlon and Johnson, 1984; Morgan, 1941; Olsen et al., 1954; Truog, 1930). Early reviews of plant nutrition and soil–plant relationships are made by Kitchen (1948) and Russell (1961).

The Green Revolution was a function of a wide variety of societal advancements, including advances in pest management, irrigation, crop breeding, mechanization, communications, transportation, and, especially, the advent of soil testing and the widespread availability of inexpensive fertilizers and scientific knowledge related to their use. Unfortunately, the widespread use of fertilizer did not come without its problems. In recent years, the effect of nutrient enrichment of surface water bodies has been attributed mostly to use of manures and fertilizers. There are concerns as well about the depletion of P mineral reserves. These issues have resulted in farmers, industry professionals, and scientists working toward the more efficient use of P in order to provide the food, fuel, and fiber for seven billion plus people on Earth while minimizing the impact to the environment and loss of natural resources. The remainder of this chapter is focused upon an understanding of P as an essential nutrient and its efficient use as a fertilizer.

3.2 UPTAKE OF PHOSPHORUS BY PLANTS

3.2.1 SOIL–PLANT INTERACTIONS IMPACTING UPTAKE

Plant P uptake is relatively less efficient compared with most other nutrients due to its poor solubility in soil. Soil chemistry, discussed in more detail in a later section, has a large impact on P uptake. Plants have to "drink their nutrients" and, as such, these elements must be dissolved into the soil solution for uptake to occur. Once in solution, nutrients are taken up by plants as a function of a combination of root interception, diffusion, and mass flow.

Mass flow is the simple process of dissolved nutrients being carried in the stream of water moving through the soil and to plant roots. For nutrients that are poorly soluble in soil, such as P, mass flow is not a major contributor to uptake because of the very low concentration in the soil solution. The concentration of P in solution, even in fertile soils, rarely exceeds 1 mg kg^{-1} and commonly is less than 0.05 mg kg^{-1}. This small amount of dissolved P does move via mass flow into the root apoplast. However, it is well documented that this minuscule amount of P is not adequate for intensive crop production, as plants need large quantities of this primary macronutrient.

Mass flow is a passive process, with nutrients being carried into the plant as a function of water relations. Plants, however, need a higher amount of nutrients than is supplied via this mechanism. Therefore, plants actively select and take up many of the nutrients they need—including P. This process creates a zone of nutrient salt depletion in the 0.2–1 mm rhizosphere zone near plant roots. The gradient of high salt concentration in the bulk soil compared to the lower levels in the rhizosphere is not chemically stable, and as a result, dissolved ions will move from the area of high salt concentration toward the area of low concentration in order to achieve equilibrium, thus hastening the movement of nutrients toward plant roots. This process of diffusion is relatively more important for P because mass flow contributions are minimal. Barber (1980, 1995) calculates that diffusion provides ~92.5% of P compared with only about 2.5% for mass flow. However, this percentage is impacted by crop species and by many soil factors, including temperature, water, P buffering capacity, and pathway tortuosity (Barber, 1977).

The remaining 5% of P uptake is attributed to root interception. However, although diffusion of P is important, continual root expansion is vital for nutrient uptake (Kissel et al., 1985; Lindsay, 2001; Marschner, 2012; Sposito, 2008). Phosphorus can diffuse only about 0.5 mm and the efficiency of diffusion in any one locale decreases as the supply of easily solubilized P is exhausted quickly. Although Barber (1980, 1995) claims that root interception is a very small contributor to P uptake, the reality is that root interception is a vital partner with diffusion and separating these two mechanisms gives a flawed view of nutrient uptake.

As roots encounter new areas of soil, they exert a considerable impact on the rhizosphere, which then affects the availability of P and other nutrients. One of the main impacts is on soil pH, with roots exuding protons and organic acids that lower pH in the rhizosphere. This action can be especially important for releasing P bound by calcium (Ca) and magnesium (Mg), especially in alkaline and calcareous soils. Acidifying fertilizers and soil amendments can have a similar impact in the soil immediately surrounding these compounds. These impacts on soil pH are localized, with the bulk soil pH not generally changing significantly in any one growing season. Organic acids, such as citrate and oxalate, exuded from roots (and microorganisms) also can displace P from soil minerals via ligand exchange, making it available for plant uptake. Again, all of these mechanisms are partners with mass flow, diffusion, and root interception in aiding in P absorption.

Hopkins et al. (2014) review the impacts on plant P availability as a function of modifications for the rhizosphere by plants and by management impacts, including pH. In the case of strongly acidic soils, it is helpful to raise soil pH to near neutral (pH 7) to maximize plant P availability and uptake. Growing at an optimum pH has other health benefits for plants, except for species that require acidic soil conditions, such as blueberry (*Vaccinium* spp. L.) and azalea (*Rhododendron* spp. L.).

Similarly, most strongly alkaline pH soils are known to have limited P solubility compared with neutral to slightly acidic soils, with the exception of Chernozems, which are high in humus.

Adjusting acidic soil pH is well studied and commonly performed. Adjusting the pH of alkaline soil is also possible with the addition of elemental S, H_2SO_4, and other strong acids and acid-forming materials (Horneck et al., 2007). However, it is not practical to attempt to modify the pH of alkaline soils in most circumstances, especially when there is a superabundance of carbonate (CO_3^{2-}) as in calcareous soils and in most irrigation water. In these cases, the rates of acidifying materials required are exorbitantly high and the CO_3^{2-} buffers the pH against change. The exudation of protons by plants and microbes, fertilization (especially with ammonium (NH_4^+)-based materials), and the replacement of bases (Ca, Mg, Na, and K) with H from natural precipitation and pure sources of irrigation (snow melt and other low EC waters) all work constantly to acidify soil, but the reality is that in most cases in alkaline soils, the balance lies in favor of CO_3^{2-} accumulating in soil and preventing pH change from occurring. Therefore, the management strategy for alkaline soils is to add relatively higher rates of P fertilizer [along with other nutrients that are similarly impacted by high pH, such as zinc (Zn)] and to select species adapted to these soils rather than attempting to lower the bulk soil pH.

Although lowering the pH of alkaline soil is not practical in most circumstances, another approach to enhance P solubility is pH modification of soil microsites in the rhizosphere. Fertilization in a concentrated band with strongly acid P fertilizers, such as H_3PO_4, temporarily lowers the microsite soil pH in the band and can result in short-term increased P-use efficiency (PUE). However, the soil pH rebounds after a few wetting–drying cycles (Thien, 1976), and the long-term precipitates that eventually form may be even less soluble than those formed after traditional sources of fertilizer are applied (Lindsay, 2001). Furthermore, the lowering of the pH of alkaline soil using strong acids can have a negative impact on P solubility if the pH is swung too far in the other direction and an acidic soil condition occurs.

The rate of many of these chemical, and all biological, reactions is impacted by temperature. As such, the occurrence of P deficiencies is often a function of time of season. Deficiency is relatively more common early in the growing season when soils are cool and root growth is minimal. As the soil warms, the rates of most reactions in soil, roots, and microbes increase and quicken the release of P from soil minerals and organic matter and, simultaneously, roots explore new areas of soil (Gardner, 1984; Lingle and Davis, 1959; Locascio et al., 1960; Lorenz et al., 1964). This temperature effect can relieve early-season P deficiency due to these multifaceted impacts, but often the damage is done and not recoverable. Phosphorus deficiencies can also become relatively common in the late season as pathogen infection of root and vascular tissues occurs and/or root growth slows due to a shift in plant resources being applied to reproductive tissue growth.

3.2.2 ABSORPTION OF PHOSPHORUS BY ROOTS

In general, nutrient uptake is greatest in the region just behind the calyptra (hardened root cap) of actively growing roots where root hairs exist (White, 2012a). Root hairs serve as extensions of the epidermal cells and effectively increase the absorptive surface area of the roots. Ernst et al. (1989) found that P uptake in soil-grown corn (*Zea mays* L.) was reduced as distance from the tip increased. However, Clarkson et al. (1987) found that this trend could be reversed in barley (*Hordeum vulgare* L.) when grown under extreme P deficiency. In addition, the surface area can be further increased in many plant species through symbiotic association with mycorrhizal fungi (Hopkins et al., 2014; Robinson, 1986). These fungi penetrate root cells and their long-stranded bodies extend into the soil several centimeters past the root rhizosphere. Mycorrhizae facilitate water and nutrient uptake and are especially beneficial in the uptake of P. The presence of root hairs and mycorrhizae extends the diameter of the nutrient uptake cylinder around each root; however, it is noteworthy that the vast majority (~99%) of soil volume remains unexplored, even by species with high root length density (Fixen and Bruulsema, 2014).

Once P encounters the exterior root surface, it can follow three pathways in its journey to the root vascular tissue for upward transport, namely, the apoplastic, symplastic, and transmembrane routes. The apoplast is the free diffusional space outside the plasma membrane, consisting of gaps between cells and the porous lattice network of cell walls, which serve as a filter to prevent soil and other large particles from entering the roots. The apoplastic route includes movement through these spaces, which does not include the crossing of any cellular membranes. In contrast, the symplastic route begins with P crossing a cell membrane into the cytoplasm where it moves cell to cell via connective channels between cells (plasmodesmata). Once a phosphate ion is inside the cell, it is possible to be transported between connected cells without crossing any additional membranes. However, the transmembrane route involves membrane transport between some cell and organelles, including across the vacuole membranes. This route results in the greatest control over which atoms and molecules are transported.

Although convenient to think of these as distinct and separate mechanisms of transport, these routes are connected and a phosphate molecule can be transported via any combination, even changing pathways at any time before reaching the endodermis. The half-time rate of exchange ($t_{\{1/2\}}$) between external ions and the cytoplasm is between 23 and 115 min for phosphate, which is slower than for other ions (e.g., NH_4^+ is 7–14 min), but is orders of magnitude lower than exchange rates across the vacuole membrane (White, 2012a).

Water and low-molecular-weight solutes are typically the only compounds transported across membranes via the symplastic or transmembrane routes. Phosphorus exists in nature almost exclusively in its most highly oxidized form as phosphate. This form is taken up and utilized by plants. Phosphite (PO_3^{3-}), a more reduced oxide of P, also can be taken up by plants, but it can be detrimental to plants that are already deficient in phosphate as it is an analog that inhibits phosphate uptake (Ratjen and Gerendás, 2009). The mono- and diprotonated phosphate ions (HPO_4^{2-} and $H_2PO_4^-$) are the only significant forms of phosphate in a small enough molecular form to be transported across membranes, although plants can absorb certain soluble organic P forms, such as nucleic acids, in some cases.

However, slightly larger phosphate compounds can begin the journey into the plant by entering the apoplasm where it is theorized that conditions allow a protected environment for chemical transformations. For example, a phosphate ion cleaved from a phosphate ester has a high probability of forming a precipitate, such as calcium phosphate, if it dissociates in the unprotected bulk soil solution. If this reaction occurs, the precipitated P will not enter the plant unless it is resolubilized. Although the phosphate ester may be too large to cross cell membranes, it will enter the apoplasm. If it does so and then dissociates in this protected environment, it has a much higher probability of remaining soluble, due to the low pH and high organic acid concentration of the apoplasm. Once in the apoplasm, the orthophosphate has a much higher probability of being utilized by the plant.

Some phosphate ion molecules entering the root outer epidermal and cortical cells are utilized by these cells, but the vast majority is transported symplastically to and through the endodermis. The endodermis is an inner layer of cells in the root cortex surrounding the stele. Apoplastic movement is halted by the lipophilic Casparian strips of the cylinder of the endodermis, with connecting walls entrenched with suberin that minimizes apoplastic solute and solvent movement into the interior of the root. Atoms and molecules from the apoplasm must pass through the plasma membranes and protoplasts of the endodermal cells to reach the stele. It should be noted, however, that this endodermal seal is not perfect, as it is not fully developed at the root apex or when lateral root branches develop and temporarily sever the membrane system.

In nonsaline soils, the mineral ion salt concentration in the soil water is much lower than in the plant, and as such, an expenditure of energy is required for salt accumulation against a gradient in root cells. For example, although P concentration in soil solution is likely less than 0.050 mg L^{-1}, the concentration in plant cells is thousandfold greater (at least 50–500 mg L^{-1}). Active transport (requiring energy) across the root membranes is required to enable plants to accumulate some nutrients (and exclude other elements) at levels much higher than are in the soil. To facilitate active transport, the plasma membranes of epidermal, cortical, and endodermal cells contain various proton pumps that transport specific ions, including phosphate, against large concentration gradients. Some of these transporter proteins are selective for phosphate. Researchers are currently working on ways to increase the uptake of P, by stimulating these nutrient transport proteins in the root.

Once past the endodermis, phosphate ions are further transported across stele cells (through the same mechanisms previously described) to the vascular system, where they encounter pits that allow entrance into the xylem vessels or tracheids or are loaded actively into the xylem from the xylem parenchyma. At this point, they are able to be transported throughout the plant for essential uses. This soil–root system of P transportation is the dominant form of uptake in plants.

3.2.3 ABSORPTION OF PHOSPHORUS BY SHOOTS

Fertilizer is sometimes applied as a dilute foliar spray directly to leaves and stems or injected into irrigation water (fertigation). Unless the amount of irrigation water applied is very small, a majority of the fertigated P is washed into the soil for root uptake. However, a small amount of P deposited onto plant shoots can be absorbed internally. Foliar transport is similar to root transport, complete with the various transport mechanisms and phosphate transporters previously described.

Phosphorus application to leaves does result in P accumulation in the plant. However, it is important to realize that there are fundamental disadvantages for foliar absorption of P. First, P is washed easily or blown off leaves. Second, while roots have evolved to take in water, the water in a plant leaf is in the process of exiting via the transpirational stream and, therefore, foliar uptake of P requires it to diffuse against the transpirational stream or to pass through the cuticle of leaves. These actions occur, but are not as efficient as with root absorption. Also, there is a limit to the amount of salt that can be in contact with sensitive leaf tissues. Therefore, foliar P applications are not more efficient in supplying adequate P needs than soil applications, but P absorption can occur.

3.3 PHYSIOLOGICAL RESPONSES OF PLANTS TO PHOSPHORUS

Phosphorus is one of the essential elements required by all plants to complete their life cycles and without any substitute for its functions. The primary macronutrients, N, P, and K, are designated as such since they are most commonly deficient and not because of their concentration in plants. Although N and K almost always have the highest mineral nutrient concentrations, the secondary macronutrients [sulfur (S), Mg, and, especially, Ca] are often at higher or equivalent concentrations in plants as P.

Phosphorus is involved in every growth phase in every living cell. In agronomy and horticulture, P is vital in nutrient management for achieving maximum crop yields (Bennett, 1993; Bundy et al., 2005; Grant et al., 2001; Hopkins et al., 2008, 2010a,b,c, 2014; Marschner, 2012; Ozanne, 1980; Stark et al., 2004; Westermann, 2005; Young et al., 1985).

The central role of P in plants is in bioenergetics, as it is a component of the adenosine phosphates (ADP and ATP) used in photosynthesis to convert light energy to chemical energy and in respiration reactions. Therefore, all energy-requiring reactions in living organisms require P. In addition, P modifies enzyme activity in phosphorylation, activates proteins, regulates metabolic processes, and is involved in cell signaling and division (Dubetz and Bole, 1975). Furthermore, P is a structural component of nucleic acids, nucleotides, phospholipids, coenzymes, and phosphoproteins. When in monoester form, P is an essential ligand in enzymatic catalysis. Phytic acid, the hexaphosphate ester of myoinositol phosphate, is the primary P storage in seeds. Phosphates, inorganic or organic, also serve as cellular pH buffers. From a cellular perspective, P has widespread involvement in virtually every physiological process in plants.

The visual response when P is deficient is very different than for other nutrients—which are generally expressed in terms of decreased chlorophyll production (chlorosis). Rather, the development of dark green or purpling of leaves and stems is reported for P-deficient plants (Barben et al., 2010a,b,c, 2011; Bennett, 1993; Hecht-Buchholz, 1967; Hill et al., 2014a,b; Nichols et al., 2012; Summerhays et al., 2014). This darkening or purpling is due to the accumulation of photosynthates

FIGURE 3.2 **(See color insert.)** Phosphorus-deficient corn (*Zea mays* L.) showing discoloration, chlorosis, and necrosis of leaves. (Photograph by A.V. Barker.)

and anthocyanin, which are being inefficiently utilized due to reduced supply of chemical energy in the plant. Reduction in amount of chlorophyll, common to most other nutrient deficiencies, does not generally occur until advanced stages of deficiency and concentration, in fact, can increase (Rao and Terry, 1989) with the combination of slow growth and accumulation of these other compounds overcoming any development of chlorosis (Hecht-Buchholz, 1967).

It would be convenient if obvious visual symptoms were always apparent with plant P deficiency, but visual symptoms are the exception rather than the rule. The shoot purpling is frequently cited, especially for corn (Figure 3.2), as evidence of P deficiency; however, it is far more common to have a yield response to P fertilizer for crops without any purpling or other obvious visual symptoms. Most species are more likely to show a dark green color of leaves or shoots rather than purple or red (Figure 3.3). But these leaf coloration differences are rare, and the main visual symptom is less overall shoot growth with symptoms varying with plant species (Bingham, 1966; Hambridge, 1941; McMurtrey, 1948; Wallace, 1961). Although not generally easily discerned from a visual perspective,

FIGURE 3.3 **(See color insert.)** Phosphorus-deficient cucumber (*Cucumis sativus* L.) showing early necrosis of leaves. (Photograph by A.V. Barker.)

P-deficient plants show suppressions in leaf expansion (Fredeen et al., 1989) and number (Lynch et al., 1991). The zone of cell division is reduced in corn (Assuero et al., 2004). Reproductive tissues also are impacted by P deficiency due to delays in flower initiation, flower number, and seed formation (Barry and Miller, 1989; Bould and Parfitt, 1973; Rossiter, 1978). Also, root hydraulic conductivity is decreased due to a decrease in genes encoding aquaporins (Clarkson et al., 2000).

Root growth also can be restricted with P deficiency, but less so than shoots, resulting in an increase in the shoot/root ratio (Fredeen et al., 1989). Sucrose tends to accumulate in the roots of P-deficient plants in an apparent signaling of P deficiency. Additionally, the elongation rate of root cells may actually increase (Anuradha and Narayanan, 1991). In fact, Smith et al. (1990) found that in the legume and forage crop Caribbean stylo (*Stylosanthes hamata* Taub.), shoot growth decreased, but root growth continued due to P translocation to roots under P-deficient conditions. In some species, root clusters are common for plants growing on the most P-impoverished soils and enable these plants to mine the soil more effectively of P (Hawkesford et al., 2012).

Because plants are dependent upon root interception of P, deficiencies are relatively more common in the early part of the growing season when soils are cold or water logged and have small, slow-growing root systems unable to expand into the soil effectively. Ironically, the situation can be magnified with P deficiency if the situation is extreme enough to cause limited root growth or susceptibility to root and vascular tissue–damaging pathogens (Lambert et al., 2005; Westermann, 2005; Westermann and Kleinkopf, 1985), thus exacerbating the situation.

Potato (*Solanum tuberosum* L.) is arguably the crop species with the greatest susceptibility to P deficiency (Fixen and Bruulsema, 2014; Hopkins et al., 2014; Rosen et al., 2014). Hopkins et al. (2014) reviewed the impacts of P deficiency on potato roots and shoots (leaf size and growth of all plant parts), as well as tuber yield and quality (set, number, size, specific gravity, starch synthesis, maturity). Dyson and Watson (1971) reported that adequate P impacted leaf area index during the first 8 weeks after emergence resulting in a 17% increase in leaf area duration. Even though potato is very susceptible to P deficiency, like other plants, it rarely shows visual symptoms. In fact, it is much more common to have restricted yields with no readily apparent visual indications in the canopy other than slight stunting in some cases (Hopkins et al., 2014). When visual deficiency symptoms do occur (Figure 3.2), they appear as purpling in extreme circumstances (Barben et al., 2010a,b,c, 2011) but more commonly as a dark greening of the leaf tissue (Bennett, 1993; Marschner, 2012; Stark et al., 2003; Stark and Westermann, 2008). However, it is most common to have no evident color differences and only shortened internodes and, therefore, stunted shoots (Barben et al., 2010a,b,c, 2011; Bennett, 1993; Marschner, 2012; Nichols et al., 2012).

This situation, with extreme deficiency resulting in easily observable visual symptoms and the more common scenario with mild deficiency resulting in *hidden hunger*, is also true for other crops less sensitive to P deficiency than potato.

3.4 GENETICS OF PHOSPHORUS NEED AND ACQUISITION BY PLANTS

3.4.1 GENETIC CONTROL

A plant passes the necessary information for need and acquisition of P to its progeny in its DNA. The various needs and uptake properties of plants are controlled genetically, and there are variations of these inherited characteristics that can impact greatly plant P requirement and the ability to obtain the element. For example, P uptake capacity increases after P is withheld from plants and is correlated with an increase in the transcription of genes encoding proton-coupled P transporters (White and Hammond, 2008). This response is controlled by biochemical signals derived from the interplay between root and shoot P status (White, 2012a,b; White and Hammond, 2008). It is theorized that low-P root status initiates a complex regulatory cascade through a transcription factor and that increased transport of sucrose and microRNA in the phloem acts as a signal. It is known that there is a general response with high concentration of sucrose in phloem, an action that results in greater root biomass and upregulates the expression of genes encoding transporters for nitrate (NO_3^-), PO_4^{3-}, sulfate (SO_4^{2-}), NH_4^+, K^+, and iron (Fe^{2+}, Fe^{3+}), but there are fine controls specific to P and controlled by shoot P concentration rather than in the roots (Drew and Saker, 1984; Drew et al., 1984; White, 2012a,b).

Drew and Saker (1984) and Drew et al. (1984) proposed a regulatory mechanism of excess P in shoots being transported back to roots to regulate P uptake. White (2012b) also states that production of specific microRNA compounds in shoots and their translocation to roots regulates the turning on or off of P deficiency response mechanisms. Resupply of adequate P shuts these mechanisms off, but not immediately, which is an important consideration for plants growing in nutrient solutions where P supply can change dramatically, resulting in P toxicity (Cogliatti and Clarkson, 1983). Furthermore, P deficiency decreases root hydraulic conductivity due to a decrease in genes encoding aquaporins (Clarkson et al., 2000).

White and Hammond (2008) stated that plant P status influences plant shoot/root biomass ratio, root morphology, P metabolism, and release of protons, phosphatases, and organic acids. All of these responses facilitate enhanced P uptake but are expressed differentially across and within species. Hawkesford et al. (2012) discuss the wide range of adaptive responses of plants to P deficiency (Lambers et al., 2006) triggered by P-starvation signaling pathways (Rolland et al., 2006).

Genetic differences for P need and acquisition across and within species are significant. Differences can be tied generally to microbial association efficiency, root exudates, and morphology and architecture of root systems (Pearson and Rengel, 1997), as well as P concentrations in plant tissues and total plant biomass. Not only are these differences scientifically interesting, but this knowledge can be exploited in the breeding and genetic alteration of crop plants to increase production and conserve mineral P fertilizer reserves. These differences are briefly discussed in the following, with examples from a few key species. Crop species are the primary focus because of their economic value and the use of fertilizer on them, and the degree of study is higher for these compared with most native species.

3.4.2 Differences across Species

Cereal grain crops provide by far the most calories worldwide, with wheat being the leading crop for direct human consumption in developed countries and rice (*Oryza sativa* L.) in developing countries. Corn is the dominant species grown in the United States, with it and soybean (*Glycine max* Merr.) largely used for animal feedstuffs. Alfalfa (*Medicago sativa* L.) is also used widely as animal feed, especially in developed countries. Soybean is the leading oil crop in developed countries, followed by rapeseed (*Brassica napus* L.) and sunflower (*Helianthus annuus* L.). Soybean and palm (*Elaeus* spp. and *Attalea maripa* Mart.) are the leading oil crops in developing countries. Sugarcane (*Saccharum officinarum* L.) and sugar beet (*Beta vulgaris* L.) are the dominant sources of refined sugars. Potato is the predominant high-starch root and tuber crop in developed countries, with cassava (*Manihot esculenta* L.) followed closely by potato and sweet potato (*Ipomoea batatas* Lam.) as the main sources in developing countries. Although urban landscapes use fertilizers for some food production, the predominant land use is for functionality and aesthetics. Use of fertilizers in urban landscapes is significant, with turfgrass (combined across all species) being grown in all cities and towns and being the number one irrigated crop in the United States.

The vast majority of P fertilizer is applied on these species and, as such, most of the research with regard to P nutrition is focused on these species as well. There is a large focus on corn, which is moderately responsive to P fertilization, due to the sheer abundance of research data available on this key species. Many of the other crops behave similarly to corn, but some known differences are discussed below. Potato will also be a point of special emphasis because of its importance as an important world food crop and its very unique P nutrition needs. At the other end of the spectrum of P fertilization need are the turfgrasses, which rarely provide a response to P fertilization despite their intense cultivation.

The reason that there are large differences across and within species with regard to P need and acquisition efficiency is a complex interaction between genetic and phenotypic differences in tissue concentration, total biomass production, microbial interactions, and, especially, root morphology and architecture.

In general, plants that grow slowly are less likely to become P deficient than rapidly growing plants. Natively grown species in wildland areas, such as Cascade fir (*Abies amabilis* Douglas ex J. Forbes), typically are not fertilized and, as such, are adapted to surviving with the low quantities of naturally mineralized P that become available as rock minerals break down very slowly over time. In contrast, most crop species have been bred to grow rapidly to meet the tremendous demand for food, fuel, and fiber for more than seven billion people on this planet. As such, these plants have an enormous need for supplemental P to achieve high yields required by society.

Although crops generally have relatively high P uptake needs compared with natively grown plants, there are large differences across crop species as well. The difference for potato and corn has already been pointed out. Another example is with slow-growing perennials, such as a dwarf apple (*Malus domestica* Borkh.), having a relatively low P demand despite having a large canopy of leaves that have to be regrown each year. Other deciduous fruit species respond to P fertilizer infrequently even when soil P is low (Childers, 1966). This lack of response is partially due to the ability of these perennials to store P and translocate it to newly forming tissues. Also, the per day growth rate for these fruit trees, while faster than most noncrop native trees, is much slower than corn, for example.

Plants can be classified in terms of their PUE, which is the percentage of fertilizer or bioavailable soil P taken up by plants compared with the total applied and/or available in the soil. Plants with high PUE have a high P influx and/or a high root/shoot ratio (Föhse et al., 1988). Corn is an example of a modestly high P-use-efficient plant. Schenk et al. (1979) reported that corn employs a stress response by increasing the root/shoot ratio when P is limited, with some genotypes doubling the ratio when grown in low versus high P conditions. Baker et al. (1970) found that rooting depth and, to a lesser degree, P influx rate were the reason for corn hybrids having relatively high P efficiency. It seems that the most likely success for breeding for increased P efficiency in corn ought to be focused on increased root expansion in the soil rather than increased P uptake per unit of root length, although the latter may also be effective.

3.4.3 Potato Is an Inefficient Responder

In contrast to corn, potato is considered to be an *inefficient responder* when it comes to P fertilization (Miyasaka and Habte, 2001). Potato has a shallow, poorly effective root system, especially with regard to P uptake (Asfary et al., 1983; Lesczynski and Tanner, 1976; Love et al., 2003; Munoz et al., 2005; Opena and Porter, 1999; Pack et al., 2006; Pan et al., 1998; Peralta and Stockle, 2002; Pursglove and Sanders, 1981; Sattelmacher et al., 1990; Tanner et al., 1982; Yamaguchi and Tanaka 1990). Weaver (1926) studied the root architecture and density of several major crops and found potato roots to be less dense, less branched, and shallower than all of the others. Tanner et al. (1982) found that a majority of potato roots reside in the top 60 cm of soil, with 90% of root length in the top 25 cm, whereas most other crops root more deeply. Lesczynski and Tanner (1976), Yamaguchi and Tanaka (1990), and Iwama (2008) had similar findings, with potato having the lowest root density by a substantial amount than the other five major crop species compared. For example, the root density for wheat was fourfold greater than potato.

In addition, potato has fewer root hairs than most other crops. Dechassa et al. (2003) found that potato and carrot (*Daucus carota* L.) yielded only 16% and 4%, respectively, of maximum when grown under very low P condition, in contrast to cabbage (*Brassica oleracea* var *capitata* L.) at 80%. Potato and carrot had a very low root to shoot ratio and, more importantly, low P influx rates due to a low number of root hairs. When soil P supply is low, the expanded reach of root hairs contributes up to 90% of total P uptake (Föhse et al., 1991). Potato has a relatively high total root length density, about the same as more P-efficient cotton (*Gossypium hirsutum* L.), sugar beet, and many vegetables. However, it has 50% less total root length density than winter wheat and oilseed rape (Stalham and Allen, 2001). However, the main reason for its poor P efficiency is attributed to root hairs comprising only about 21% of the total root mass, compared with 30%–60% for most other crop species (Yamaguchi, 2002).

Furthermore, the root system for potato tends to decline in the late season when P demand is at its highest, a response that is in contrast to many other species that accumulate P relatively earlier in the growing season (Fixen and Bruulsema, 2014). However, Jacob et al. (1949) observed that potato takes up a greater proportion of P later in the growing season compared with other crops. Furthermore, potato P continues to be taken up later in the growing cycle than is either N or K (Carpenter, 1963; Kleinkopf et al., 1981; Lorenz, 1947; Roberts et al., 1991; Soltanpour, 1969). Phosphorus uptake progresses steadily throughout the season, whereas N shows little if any additional uptake after about 80 days after emergence (Kelling et al., 1998). This situation is a significant disadvantage for potato, especially in light of its susceptibility to pathogens that may degrade the root and vascular systems, negatively impacting the uptake and translocation of P. Thornton et al. (2014) suggest that it is likely that PUE in potato could be improved by breeding for more extensive root systems and increased root hair production. They cite Deguchi et al. (2011) already using this approach to breed potato cultivars with improved drought resistance due to better water uptake efficiency, which also likely improves P uptake.

As a result of these and possibly other genetic differences, fertilizer recommendations and soil test cutoff levels for potato are globally much higher than for other crops (Fixen and Bruulsema, 2014; Hopkins et al., 2014; Kelling and Speth, 1997; Lang et al., 1999; Moorhead et al., 1998; Rosen et al., 2014). Fertilizer rates can be higher than 400 kg P_2O_5 ha^{-1} in long-season, high-yielding environments. A survey of state university research–based fertilizer recommendations shows that most other crop species require about half the amount of P that potato requires. For example, in the United States, a large majority of potato production occurs in the Pacific Northwest states. In these states, the maximum recommended fertilizer rate is 134 kg P_2O_5 ha^{-1} for corn (Brown et al., 2010), whereas the maximum rate for potato ranges from 252 to 493 kg P_2O_5 ha^{-1} (Lang et al., 1999; Stark et al., 2004). Similarly, the University of Wisconsin recommends optimum levels for soil test P (Bray P1) at, depending upon soil texture and so forth, between 16 and 50 mg kg^{-1} for various crops except potato, which has optimum test levels of 61–200 mg kg^{-1} (Laboski and Peters, 2012). Truly, potato has unique genetic differences from most other species that need to be examined.

3.4.4 Unique Differences for Phosphorus Uptake across Species

Fast-growing, short-season vegetable crops are often similar to potato in their relatively high P need, although not generally as extreme (Alt, 1987; Greenwood et al., 1980; Itoh and Barber, 1983; Nishomoto et al., 1977; Sanchez, 1990).

A contrasting species with regard to P nutrition is sugar beet. Its root architecture and morphology are vastly different to most other species, sending a taproot largely downward for the first several weeks of growth in an effort to ensure adequate water supply. The negative by-product of this root growth strategy is that the nutrient-rich topsoil is left largely unexplored for several weeks. As a result, yield deficiencies due to early-season P deficiencies caused by the development of fewer and thinner cambial rings are common.

Alfalfa is also a taprooted species, but it has more balanced growth early in its establishment and is very efficient at surface soil feeding and thus is less impacted by P deficiencies at establishment. However, this species is grown as a perennial and is unique in that large amounts of P are removed through the harvested hay over time. As such, alfalfa is very responsive to P fertilizers when grown in soil with low to moderate soil test P levels—especially after years of P removal through crop harvest. The challenge for this and many other perennials is that P can be surface applied only after initial establishment. Fortunately, its surface-feeding rooting efficiency enables this approach when other species are less adept at P recovery from broadcast applications of fertilizer not incorporated into the soil.

Soybean is another major world crop, but it is somewhat unique in terms of P need. Soybean has much less total biomass production than corn and, thus, less total P uptake. Its root system is relatively shallow and less extensive than corn and most other crops, but it is efficient at P recovery in soil due to a majority of its roots being in the P-rich topsoil. In the United States, it is common to grow soybean in rotation with corn, with growers often not applying any supplemental P fertilizer to the soybean crop directly, allowing the plant to feed off of the remnants of the corn crop. Wheat and rice and most other grains approach corn in terms of their P fertilization needs but take up less total P as a function of relatively lower biomass yield.

Although similar to corn in biomass production and canopy architecture, sorghum (*Sorghum bicolor* Moench) thrives in relatively less fertile soils with less fertilizer application, due to its extremely fibrous and extensive root system. Sorghum roots effectively explore a larger volume of soil than most other crop plants due to this fibrous root system. Cassava is similar to sorghum in terms of its ability to thrive in less fertile soil, which enables it to be successfully grown in regions where fertilizers are not readily available or affordable to indigenous populations.

Like sorghum, turfgrass species also have very fibrous and efficient root systems. However, they differ in terms of much lower biomass production. Additionally, turfgrasses tend to root only in the top few cm of the P-rich topsoil. These factors result in very little need for P fertilization. In fact, turfgrass species grown in soil with modestly low residual P levels often have a competitive edge over invasive weeds. Although not typically managed for any type of biomass production, turfgrass species are among the most intensively cultivated plants, often with removal of mowing clippings multiple times in a week, and yet they can often go without P fertilization for many years without suppression in growth. As long as bioavailable P is not extremely low, established turf can often go for several years without supplementary P fertilization (especially if clippings are returned to the soil during mowing). The exception would be newly seeded/sodded fields and sports fields where frequent seeding and sodding are performed. These situations have plants without extensive root systems, and the P need during establishment is relatively higher than when maturity is reached and their root systems are well established.

3.4.5 DIFFERENCES WITHIN SPECIES

Differences in P efficiency are not only observed across species, but also within. Schenk et al. (1979) evaluated five corn genotypes, and although all increased the root/shoot ratio under P-deficient conditions, the magnitude of difference was significant. Buso and Bliss (71) showed differences across lettuce (*Lactuca sativa* L.) varieties, although others found little difference (Nagata et al., 1992; Sanchez and El-Hout, 1995).

Varietal differences for potato yield response to P have been measured (Freeman et al., 1998; Jenkins and Ali, 1999; Moorhead et al., 1998; Murphy et al., 1967; Sanderson et al., 2002, 2003; Thorton et al., 2008). Thornton et al. (2014) report that the Shepody cultivar has greater root density and earlier maturity, which both contribute to its being less responsive to P than the widely grown Russet Burbank cultivar. Thornton et al. (2014) report differences among potato varieties in terms of PUE. The recently released Alturas cultivar has an even greater efficiency with much lower fertilizer P requirement than other cultivated varieties due to its more extensive and efficient root system. McCollum (1978a,b) reported similar findings with "Ranger Russet" and "Premier Russet" reaching maximum yield at a lower level of P fertilization than "Shepody" or "Russet Burbank."

Potato cultivars differ in total root length density and depth of soil penetration due mostly to differences in the time of active root growth and development (Stalham and Allen, 2001). For example, the duration of root growth was almost half (Olsen et al., 1954) for "Cara" compared to "Atlantic" (Wagganman, 1969). The longer the time of active root growth, the deeper and more soil explored.

Iwama et al. (1981) found a wide difference in 268 unselected clones from 1.3 g root dry weight plant^{-1} to 2.8 g for early versus late maturing clones, respectively.

Another difference common within species is related to pathogen susceptibility, which can have a large impact on P status in plants (Lambert et al., 2005). Pathogens that attack and degrade root tissue limit the ability of the plant to encounter P, and pathogens that infect vascular tissues result in inability of plants to move P from roots to shoots and vice versa. As such, varietal differences in disease susceptibility can have a major impact on P nutrition.

For example, potato is highly susceptible to root and vascular system pathogens, a trait that is partly the explanation of why this species is unique with respect to its P need (Rosen et al., 2014; Thornton et al., 2014). Phosphorus deficiency can increase the severity of several important potato diseases, including common scab (*Streptomyces scabies*), Verticillium wilt (*Verticillium dahliae* and *Verticillium albo-atrum*), and late blight (*Phytophthora infestans*). Davis et al. (1994) showed a close relationship between P fertilizer rate, wilt symptoms due to *V. dahliae*, and pathogen colonization of stem tissue. As optimal or higher levels of P can speed tuber maturity and increase skin thickness, Herlihy and Carroll (1969) and Herlihy (1970) found that P reduced tuber infection with late blight. As previously mentioned, P uptake occurs relatively late for potato and yet root system development ceases 60–90 days after planting and, in fact, the roots actually begin to deteriorate at this time. Furthermore, these developments all coincide with when disease development hastens rapidly while tubers are bulking rapidly and nutrient requirements are still high (Pan et al., 1998; Thornton et al., 2014). Although these principles are generally true for potato, there is a very wide difference in disease susceptibility by cultivar. The most commonly grown cultivar is "Russet Burbank," but it is also very susceptible to disease and has a very high P fertilization requirement compared with improved varieties.

Although these interactions between nutrition and pathology are more relevant for the highly susceptible potato compared with most other species, it is important to realize that P deficiency may make a plant more susceptible to pathogen infection, and diseases that degrade root or vascular tissue can impair P uptake even when soil P levels are high.

3.4.6 GENETIC MODIFICATION POTENTIAL

As fertilization and other agricultural improvements led to the Green Revolution, the next revolution is advanced genetic modification potential. Thornton et al. (2014) state that both quantitative trait locus (QTL) mapping and marker-assisted selection have been identified as useful tools to facilitate breeding for complex traits such as PUE. The QTL associated with enhanced P uptake has been identified in several species, such as pearl millet (*Pennisetum glaucum* R. Br.), corn, and rice (Hash et al., 2002), but little work has been done in species such as potato. However, Thornton et al. (2014) state that recent research aimed at improving drought resistance has identified QTLs associated with root length and root dry weight on chromosome 5 in potato (Iwasa et al., 2011). This same approach could be used to improve identify root traits or other characteristics associated with improved PUE.

Another avenue is genetic modification by impacting direct P acquisition or other cellular mechanisms. Miyasaka and Habte (2001) identified the genes for high-affinity P transporters in arabidopsis (*Arabidopsis thaliana* Heynh.), shown to increase P uptake at low P concentrations by almost three-fold. Advances in breeding disease resistance also have large potential to have a secondary impact on P nutrition. Breeders, pathologists, and agronomists will need to work together to provide a better understanding of how disease resistance and control practices impact root health as a way to improve P uptake efficiency throughout the growing season (Thornton et al., 2014). There is significant opportunity for the genetic improvement of PUE in crop plants (Lynch, 1998).

Lambers et al. (2011) suggest that the genetics of native species growth on severely P-deficient soils be explored as an avenue to enhance PUE. One approach that has been suggested by Thornton et al. (2014) to improve PUE in potato is to take advantage of native germplasm from South America. These native plants have evolved under conditions of low soil nutrient availability and may, therefore, have more efficient and extensive root systems. However, Sattlemacher et al. (1990) evaluated 27 native clones and 9 advanced cultivars under conditions of low and high soil nutrient availability and concluded that the advanced group had higher yield potential under both scenarios, and showed no evidence of having reduced nutrient-use efficiency compared with the native group. Nevertheless, the native germplasm should be explored for potential crosses, which may result in improved varieties.

3.5 CONCENTRATIONS OF PHOSPHORUS IN PLANTS

Mineral nutrients generally make up less than 10% of the dry weight of a plant, with N and K generally at ~2%–5% each followed by similar concentrations of P, Ca, S, Mg, and chloride (Cl⁻) at 0.1%–1% each. The micronutrients, other than Cl⁻, are at concentrations several orders of magnitude lower than the macronutrients. Mills and Jones (1996) have published typical P concentrations for a wide variety of aboveground plant parts ranging from 0.08% to 1.3%, although it is possible to have slightly lower levels with roots sometimes down to levels of 0.04% but in some reports as high as 4% (Hawkesford et al., 2012). Typical P levels in plant tissue are shown in Table 3.1, ranging from the very-slow-growing, native Cascade fir with very low P levels in its tissues to the high levels of P in coleus (*Coleus* spp. Lour. now largely *Plectranthus* spp. L'Hér).

Lambers et al. (2010) stated that the P requirement for optimal growth is 0.3%–0.5% but that some plants evolving on P-limiting soils may contain an order of magnitude less. The chance of P toxicity increases at levels above 1%, although this is rare because plants downregulate their P transporters when P levels are high (Dong et al., 1999). However, toxicities do occur in unique circumstances (Hawkesford et al., 2012; Shane et al., 2004). Hawkesford et al. (2012) point out the wide variety of tolerance to levels of P, with toxicity identified in pigeon pea (*Cajanus cajan* Millsp.) and black gram (*Vigna mungo* Hepper) at levels as low as 0.3% and 0.6% shoot P, respectively, while the very-fast-growing green mulla mulla (*Ptilotus polystachyus* F. Muell.) shows no signs of toxicity at 4% shoot P.

TABLE 3.1
Average Phosphorus Plant Tissue Concentrations

Species	Plant Part	Timing	P, %
Alfalfa (*M. sativa* L.)	Whole tops	Prior to flowering	0.26–0.70
Apple (*M. domestica* Borkh.)	Mature new leaves	Summer	0.09–0.40
Cascade fir (*A. amabilis* Doug. ex J. Forbes)	Terminal cuttings	Summer	0.09–0.16
Coleus (*Plectranthus* spp.)	Mature new leaves	Mature plants	1.1–1.3
Cotton (*G. hirsutum* L.)	Petioles	First squares to initial bloom	0.30–0.50
	Petioles	Full bloom	0.25–0.45
Corn (*Z. mays* L.)	Whole tops	<30 cm tall	0.30–0.50
	Leaves below whorl	Prior to tasseling	0.25–0.45
	Ear leaves	Initial silk	0.25–0.50
Rapeseed (*B. napus* L.)	Leaves, 5th from top	Rosette to pod	0.28–0.69
Rice (*O. sativa* L.)	Mature new leaves	Maximum tillering	0.10–0.18
Sorghum (*Sorghum vulgare* Moench)	Whole tops	23–39 days after planting	0.30–0.60
	Mature new leaves	37–56 days after planting	0.13–0.25
	Third leaf below head	Bloom	0.23–0.35
	Third leaf below head	Grain in dough stage	0.15–0.25
Soybean (*G. max* Merr.)	Mature new leaves	Prior to pod set	0.25–0.50
Sugar beet (*B. vulgaris* L.)	Mature new leaves	80 days after planting	0.45–1.1

Source: Based upon Mills, H.A. and Jones, J.B., *Plant Analysis Handbook II*, MicroMacro Publishing Inc., Athens, GA, 1996.

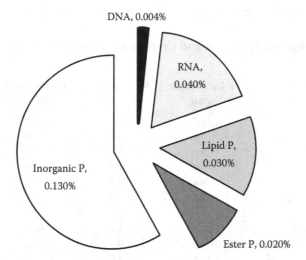

FIGURE 3.4   Fractions of phosphorus occurring in plants.

Bieleski (1973) states that a typical plant contains the fractions as shown in Figure 3.4. There is not a similar comprehensive listing of P concentrations in root tissues. Separating roots from soil is difficult. As a result, P concentrations in roots generally are not measured commercially and rarely are reported in the scientific literature. In general, P concentrations measured across a wide variety of species at the Brigham Young University Environmental Analytical Lab shows that P concentrations in shoots are ~25%–200% higher than in root tissue. In addition, it is difficult to quantitatively gather all of the roots from a plant growing in soil. Therefore, total root uptake of P is rarely measured or reported. However, we know that root biomass is generally much less than shoot biomass, with ratios ranging from 0.10 to 0.3. The combination of lower biomass and lower P concentration results in only a small amount of the total P being in roots.

Succulent new tissues (either roots or shoots) tend to be higher in P than in lignified stems and other older tissues. Even in nonwoody crop tissues, the P concentration will drop dramatically in just a few weeks, as shown for cotton, corn, and sorghum in Table 3.1 and for potato in Figure 3.5. Phosphorus is highly mobile within the plant, with much of it being cannibalized from senescing tissue to newer growth and reproductive organs as the season progresses. Plants mobilize P and

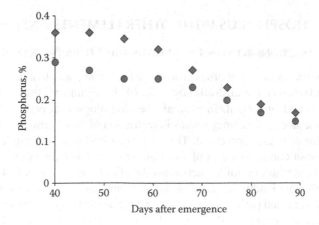

FIGURE 3.5   Average seasonal potato (*S. tuberosum* L.) petiole phosphorus concentrations for unfertilized ● and fertilized ◆ (150 kg P_2O_5 ha^{-1}) plots with Olsen bicarbonate P concentration of 23 mg kg^{-1} prior to fertilization.

TABLE 3.2

Uptake into Aboveground Biomass and Crop Removal for Phosphorus at Average Yields

Crop	P Uptake Rate, kg Mg^{-1}	Removal Rate, kg Mg^{-1}	Average Yield, Mg ha^{-1}	Total P Removal, kg ha^{-1}
Alfalfa (*M. sativa* L.)	5.7	6.0	7.2	43
Corn (*Z. mays* L.), grain	9.6	6.3	10.0	63
Silage	9.6	1.6	42.0	67
Potato (*S. tuberosum* L.)	2.1	1.5	47.0	70
Rice (*O. sativa* L.)	8.4	6.7	8.6	58
Sorghum (*S. vulgare* Moench), grain	13.0	7.8	3.8	29
Soybean (*G. max* Merr.)	18.0	12.0	2.9	35
Sugar beet (*B. vulgaris* L.)	1.4	1.1	64.0	70
Wheat (*Triticum aestivum* L.) spring	13.0	9.5	3.2	30
Winter	11.0	8.0	3.2	25

Source: Based on Mills, H.A. and Jones, J.B. Jr., *Plant Analysis Handbook II*, Athens, GA: MicroMacro Publishing Inc., 1996; USDA, Crop production 2013 summary. *National Agricultural Statistics Service*, 2014. http:// www.usda.gov/nass/PUBS/TODAYRPT/crop1014.pdf. Accessed July 21, 2014.

some other nutrients to seeds, tubers, and other reproductive tissues to provide ample nutrition for the next cycle of growth. When these tissues are harvested, a significant quantity of P is removed from the soil (Table 3.2).

It is noteworthy that the values in Table 3.2 are only averages and, in some cases, the removal rates are substantially higher. For example, irrigated wheat yields in the Pacific Northwest have approximately threefold higher yields than the average shown, and potato yields are double the national average in the Columbia Basin in Oregon and Washington due to a large number of growing degree days and otherwise optimal conditions. Removal rates for nutrients would be proportionally higher based on actual yield removal. Understanding and managing depletion rates are important factors when developing a sustainable plan for crop production, although basing fertilizer recommendations solely on these values is not recommended as in many cases much higher rates are needed or, in other cases, it being appropriate to harvest excess P from soils with excessive levels.

3.6 RATIOS OF PHOSPHORUS WITH OTHER ELEMENTS AND INTERACTIONS

3.6.1 GENERAL INTERACTIONS BETWEEN PHOSPHORUS AND OTHER NUTRIENTS

Interactions among nutrients in soil or other growth media can impact plant health, yield, and nutrient concentrations and ratios (Fageria, 2001; Foy et al., 1978; Reichman, 2002). Nutrient interactions with P are widespread and important. In general, the physiological effects of P on plants impact other nutrients. For instance, P deficiency results in restriction of shoot growth, whereas root growth often remains the same or is even increased. This result can lead to an accumulation of nutrients in shoots due to greater root contact per unit of shoot growth. However, under extreme P deficiency, a lack of energy supply can result in limited active uptake of nutrients, such as K. Other general interactions include overall plant health degradation with any nutrient deficiency or toxicity resulting in reduced root growth, increased pathogen infection, and reduced ability to actively take up nutrients.

A beneficial interaction occurs if P is applied in conjunction with NH_4^+–N. This interaction appears to enhance the plant uptake of both nutrients due to increased P solubility, increased shoot and root growth, and alteration of plant metabolism (Bundy et al., 2005; Engelstad and Teramn, 1980; Leikam et al., 1983; Murphy et al., 1978). These observed benefits are more likely to occur for plants growing in soil with minimal bioavailable P (Engelstad and Teramn, 1980). However, excess N availability can result in decreased shoot-to-root ratios more so than any other nutrient excess, causing P supply issues.

Chloride in close proximity with phosphate also has been shown to restrict P uptake (Berger et al., 1961; Hang, 1993; Kalifa et al., 2000; Zhong, 1993). Reducing the Cl^- content of band-applied K helps avoid this problem (Berger et al., 1961), although Panique et al. (1997) saw no decrease in P uptake even where 448 kg K_2O as KCl ha^{-1} was banded near the row. James et al. (1970) also noted no effect of Cl^- on P uptake. There is also speculation of competitive antagonism with other anions, such as NO_3^-, SO_4^{2-}, borate (BO_3^{3-}), and molybdate (MoO_4^{2-}), although the evidence for competitive antagonism is sparse and conflicting, and field data do not seem to support it.

However, P has a well-documented antagonistic interaction with Ca and Mg in soil due to chemical bonding with P, for which precipitates are not very soluble, especially in alkaline soils. These precipitates can also occur within roots and other plant tissues, but the acidic biological environment allows for solubilization more readily than with soil. Similarly, aluminum (Al), manganese (Mn), Fe, Zn, and copper (Cu) can precipitate with P in soil. These precipitates can form under alkaline conditions, but Al, Mn, and Fe phosphates are known to precipitate under strongly acidic conditions. In the case of Al, this element solubilizes at about pH 5.5 and below. Once in soil solution, Al can precipitate with P. A similar reaction occurs with Mn and Fe. The interaction between P and the various metals, which occurs in plants and soils, is one of the more well-known and studied interactions.

3.6.2 PHOSPHORUS–MICRONUTRIENT METAL INTERACTIONS

The P–Zn interaction is the most documented and well known. Boawn et al. (1963) reported Zn deficiency as a result of high P fertilization levels. The P–Zn antagonism was documented in subsequent works (Boawn and Leggett, 1964; Jackson and Carter, 1976; Soltanpour, 1969). Broadley et al. (2012) discuss the interaction of P–Zn and offer several possible explanations for this antagonistic interaction when high rates of P fertilizer are applied to plants, including (1) decrease of Zn solubility in soil, (2) reduced root growth, (3) reduced arbuscular mycorrhizal colonization, (4) dilution effect due to higher yields, (5) reduction of Zn solubility and mobility in plant tissues, and (6) P toxicity. Barben et al. (2011) found that increasing available Zn generally reduced shoot P and increased root P in potato. This effect of solution Zn was exacerbated by optimal and excessive solution P, an occurrence that is similar to observations in other studies (Barben et al., 2010a; Boawn and Leggett, 1964; Chatterjee and Khurana, 2007). This action was likely due to a P–Zn binding in roots, as was suggested in other research (Leece, 1978; Singh et al., 1988; Terman et al., 1972).

Barben et al. (2011) thoroughly reviewed the complex interactions between P, Zn, Mn, Fe, and Cu. Their data show a strong three-way interaction between P, Zn, and Mn levels that also can impact concentrations of Fe and Cu. Their data mostly support other findings of an antagonistic interaction between Zn and P that has been observed commonly and studied and that can result in excessive P uptake under Zn deficiency (Barben et al., 2010a,b; Bingham, 1963; Boawn and Leggett, 1964; Loneragan et al., 1979; Webb and Loneragan, 1988) or in a P-induced Zn deficiency (Christensen, 1972; Christensen and Jackson, 1981; Soltanpour, 1969). Apparent in the Barben et al. (2011) study, but inconsistent with some previous studies (Barben et al., 2010a,b; Bingham and Leggett, 1963; Boawn and Leggett, 1964; Cakmak and Marschner, 1987), shoot Zn was reduced at excessive solution P relative to deficient or optimal solution P, regardless of Mn level. These findings support reduced shoot Zn with high available P in potato as suggested by others (Christensen, 1972; Christensen and Jackson, 1981; Soltanpour, 1969).

Although the P–Zn interaction is studied most commonly, other cationic micronutrients such as Mn, Fe, and Cu interact with P as well (Barben et al., 2011; Beer et al., 1972; Brown and Tiffin, 1962; James et al., 1995; Safaya, 1976). Phosphorus and Mn interactions have been reported in several species (Barben et al., 2010a,b; Ducic and Polle, 2007; Gunes et al., 1998; Le Mare, 1977;

Marsh et al., 1989; Neilsen et al., 1992; Nogueira et al., 2004; Rhue et al., 1981; Sarkar et al., 2004; Sharma and Arora, 1987; Zhu et al., 2002). Reductions in plant P with increasing Mn were observed in tomato (*Solanum lycopersicum* L.; Gunes et al., 1998) and potato (Sarkar et al., 2004), whereas a rise in P was seen in shoots and roots of sorghum with increasing Mn (Galvez et al., 1989). Barben et al. (2010c) found that when Zn availability was held constant as Mn varied, plant P consistently was depressed, especially in shoots, at optimal solution Mn compared to either deficient or excessive available solution Mn.

Although Zn has a larger impact on plant Cu, P influences depressed Cu levels as well (Forsee and Allison, 1944; Halder and Mandal, 1981). Safaya (1976) reported that a strong P x Zn interaction also affects total Cu uptake. Other studies show that the effects of Zn on Fe uptake and transport in plants do not result only from increasing available Zn, but are also strongly influenced by P x Zn (Hamblin et al., 2003). While most studies agree that reduced shoot Fe results from Mn interference with Fe translocation, root interactions also influence reductions in Fe with some studies suggesting Mn-induced P–Fe binding within roots (Alvarez-Tinault et al., 1980; Cumbus et al., 1977).

Precipitation in roots of micronutrient-bound PO_4^{3-} is indicated by several studies (Leece, 1978; Singh et al., 1988; Terman et al., 1972) and likely explains the results of Barben et al. (2011) with high available P. At low available P, however, little explanation has been given. Reduced nutrient transport from root to shoot due to unavailable compounds involved in P metabolism (e.g., ATP, ATPases, alkaline phosphatase, and phosphoenolpyruvate carboxylase), which are required to provide the necessary energy for nutrient mobilization via proton pumps and other mechanisms in transmembrane transport (Clemens et al., 2002; Fox and Guerinot, 1998; Grusak et al., 1999), and as such, is a reasonable explanation.

3.6.3 MANAGEMENT OF INTERACTIONS

It is apparent that nutrient interactions occur in the soil and in plant tissues. The next logical step may seem to manage nutrients based on ratios of P with other nutrients. However, the empirical evidence raises a flag of caution for this approach. It is not uncommon for soils and crops to be analyzed for management purposes, with many laboratories printing various ratios on these reports. Whether intended or not, these ratios imply that there is an ideal ratio to achieve maximum production efficiency. This approach, while appearing to be logical, is not supported generally by field research. There are exceptions of course, such as with the sodium absorption ratio (SAR) for soil and irrigation water analysis, which have been shown by research and field application to be useful values in managing sodium (Na)-affected conditions (Hopkins et al., 2007b). However, there is no such conclusively proven ratio for P with any other element.

Table 3.3 shows the approximate ratio of all of the essential elements, as compared with P, in plant tissues. Although this information is interesting, it should not be used as a management tool in an attempt to fertilize to achieve a certain ratio of P with any other nutrient. This approach might be especially tempting for the P and Zn interaction outlined earlier, which has been well documented and studied. For example, according to the sufficiency range suggested by Mills and Jones (1996) for corn at the initial silk stage, the concentration ranges are 2500–5000 and 20–60 mg kg^{-1} for P and Zn, respectively. Assuming the widest ratios within these ranges would be acceptable, the ideal P:Zn ratio could be assumed to be from 50 to 250 times as much P as Zn. However, an informal survey of dozens of unresponsive corn P and Zn trials (Hopkins, unpublished data) shows above-average yields in many of these trials where this supposed ideal ratio is violated. For example, in one trial, the extremely high P fertilizer rates resulted in leaf concentrations of 21 and 7200 mg kg^{-1} for Zn and P, respectively, a ratio of 348 times more P than Zn. The yields of this treatment were 17.5 Mg ha^{-1}, which is nearly double the national average and similar to the others with less P fertilizer and resulting in more normal P concentrations. Obviously, the wide ratio did not impact yield. One could argue that this event is an anomaly, but many such instances occur in growers' fields and in field research. More importantly, no strong research evidence shows a good correlation between

TABLE 3.3

Average Ratio on a Mass Basis of Nutrients Relative to Phosphorus

	Nonmineral		Plant Shoots
	Carbon	C	210
	Oxygen	O	210
	Hydrogen	H	25
	Mineral		
Primary macronutrients	Nitrogen	N	15
	Potassium	K	10
Secondary macronutrients	Calcium	Ca	4.5
	Sulfur	S	0.75
	Magnesium	Mg	0.75
Micronutrients	Chloride	Cl	0.5
	Iron	Fe	0.25
	Manganese	Mn	0.025
	Zinc	Zn	0.015
	Copper	Cu	0.0025
	Boron	B	0.0025
	Molybdenum	Mo	0.0001
	Nickel	Ni	0.00025

Note: Based on compilation of a wide variety of mostly crop plants from data at the Brigham Young University Environmental Analytical Lab and Hopkins research data sets.

tissue P:Zn ratios and yield. Despite this fact, it is somewhat commonplace for field managers and tissue-testing services to recommend fertilizer based on these ratios.

Even more common are fertilizer recommendations made based on some ideal soil test ratios. Soils vary much more widely in terms of total and bioavailable ratios as compared with plant tissue concentrations. Elements that typically have much higher total and bioavailable elemental concentrations in soil than P include C, O, H, N, K, Ca, Mg, and Na. Aluminum, silicon (Si), and Fe typically have higher total concentrations than P in soil, but generally have lower bioavailable levels. Nutrients other than those listed earlier are typically equal to or, for most, much lower than total or bioavailable P, although there are many exceptions (such as the S in a soil with high concentrations of gypsum). Although these general trends exist, the ratios of P to other elements in soil vary widely.

Those who promote ideal soil test ratios forget that plants largely self-regulate uptake and exclusion of many elements. Just one example is Ca and P. Calcium typically is found at concentrations much higher than P in soil, and yet plants actively take up P and exclude some Ca so that levels in most plant species are close to equivalent for these nutrients. When comparing ratios, Ca is many-fold higher than P in high base saturation and calcareous soils, giving a very high Ca to P ratio. Although it is true that Ca levels are higher in plants in these circumstances, they do not follow the same proportions as what is found in the soil. Managing by some supposed ideal ratio would not be practical. Rather, P should be managed by bioavailable P concentration, and in some cases, the recommendation needs to be modified based on concentration of Ca minerals (Westermann, 1992), but not on some fictitious ideal ratio.

Another example of a ratio that would possibly be important would be P:Zn, but no strong evidence can be found to support such a claim. In fact, an informal survey from the files of a large crop-consulting corporation in the Midwestern United States (Servi-Tech Inc., Dodge City, Kansas, United States) shows that in fields with above-average yields, these yields were not correlated to the P:Zn soil test ratio. The only facts that have been shown in the P–Zn research cited earlier is that very high rates of P fertilizer can induce micronutrient deficiencies and vice versa. These studies were all based on added P fertilizer and not on existing high soil test levels. Therefore, P and micronutrients need to be managed according to their individual soil test values and experimentally derived information on crop need, with appropriate levels of fertilizer added without excess. In other words, P fertilizer should not be added to a soil just because the soil test Zn, Mn, Fe, or Cu levels are exceptionally high or vice versa. These nutrients are managed on their individual concentration levels and not according to ratios of one to another.

3.6.4 DIFFERENCES ACROSS AND WITHIN SPECIES

Table 3.3 shows a ratio of average plant tissue concentrations. It is vital to understand that these ratios vary widely across species. For example, typical concentrations for P are similar for corn and alfalfa at about 0.25%–0.6%, but Ca is drastically different, with corn having Ca concentrations in the same range as P, but alfalfa typically having Ca at 1.8%–3.0% in its shoots (Mills and Jones, 1996). This difference in P:Ca ratio of about 1:1 for corn and 1:5 for alfalfa shows that ratios can vary widely across species.

Within species, differences are also significant. The temporal nutrient concentration differences that occur through the course of a growing season and differences across varieties/cultivars/hybrids were discussed previously. These differences are not surprising, but it is also important to note that there can be significant phenotypic differences even within the same genotype based on soil and environmental conditions. For example, Table 3.4 shows the results of selected sugar beet trials in Idaho, where the unfertilized treatments had similar sugar yields and quality compared with those fertilized with P. All of the fields had the same variety, and the fields selected had good fertility and management, with the result of above-average yields. These data show that high yields are obtained, and yet the petiole P concentrations vary widely (note that all of these values are above the established critical level). The K concentrations for these fields also are shown, along with the ratio of P to K, showing a very wide range to illustrate that it is unlikely there is some ideal P:K ratio to be achieved in order to obtain high yields. This example is just one, but a survey of farm field results and research publications shows a wide variety in the ratios of essential nutrients, with no strong correlation between some supposed ideal optimum ratio and yield parameters.

TABLE 3.4
Concentration of Phosphorus and Potassium and K:P Ratios in Sugar Beet (*B. vulgaris* L.) Leaves from 11 Field Samples

Field	P, %	K, %	K:P
1	0.45	3.6	8.0
2	0.55	4.3	7.8
3	0.67	6.0	8.9
4	0.56	2.1	3.7
5	0.87	4.1	4.7
6	0.54	5.2	9.7
7	0.98	4.7	4.8
8	1.07	2.5	2.3
9	0.56	5.7	10.0
10	0.46	4.3	9.3
11	0.49	3.2	6.4

3.7   DIAGNOSIS OF PHOSPHORUS STATUS IN PLANTS

Determining whether there is sufficient P available to plants is possible through tissue analysis as a function of critical values and sufficiency ranges (Macy, 1936). These acceptable ranges are typically determined by plotting yields relative to plants having adequate P (yield = 100%) against P concentration in plant tissue (Figures 3.6 and 3.7). Often, the critical level is set at 90% or 95% of maximum for most crops and 98% or 100% for higher-value crops (based on the principle of maximum economic yield).

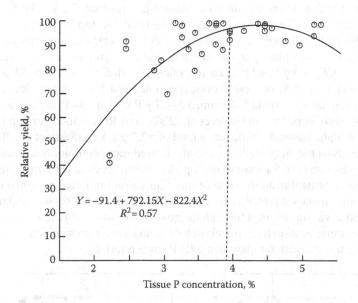

FIGURE 3.6 Critical concentration using a curvilinear model of phosphorus in midribs of endive (*Cichorium endivia* L.) at the eight-leaf stage. (From Sanchez, C.A., Phosphorus, in *Handbook of Plant Nutrition*, Barker, A.V. and Pilbeam, D.J. (eds.), CRC Press, Boca Raton, FL, 2007, pp. 51–90.)

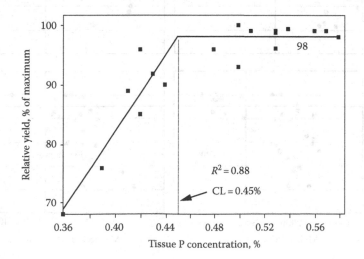

FIGURE 3.7 Critical concentration using a linear and plateau model of phosphorus in radish (*Raphanus sativus* L.) leaves. (From Sanchez, C.A., Phosphorus, in *Handbook of Plant Nutrition*, A.V. Barker and D.J. Pilbeam, (eds.), CRC Press, Boca Raton, FL, 2007, pp. 51–90.)

As P is mobile in plants, sampling and analysis of new growth usually are recommended to ascertain the current status of P uptake. Phosphorus is incorporated into the structure of many cellular compounds, so it is possible to have a relatively high P concentration in the overall tissue, while also having a deficient level of soluble phosphate that is needed for the chemical transactions that require it. Therefore, in many instances, researchers have found that the best tissue to sample is that which contains the pipeline of vascular tissue. Cotton, sugar beet, and potato are examples of plants that commonly are managed for in-season nutrient status by sampling and analyzing petiole tissue connecting leaves to stems.

Westermann and Kleinkopf (1985) found that potato yield is related to the number of days from the time of tuber initiation where the shoots contained greater than 2.2 g P kg^{-1} dry mass, which correlated to 1 g soluble P kg^{-1} in the fourth petiole from the top of the plant. Walworth and Muniz (1993) published a compendium of potato nutrient concentrations in which they defined the "sufficient" level for various plant parts at different stages of growth, with midseason leaf-blade sufficiency at $\geq$2.6–4.7 g P kg^{-1} and midseason petiole P sufficiency of $\geq$1.5–3.1 g P kg^{-1}. Recent work by Freeman et al. (1998) suggested critical petiole total P levels for "Russet Burbank" of 4.5–5.7 g P kg^{-1} when tuber length is 5–10 mm, 3.5–4.7 g P kg^{-1} at 35–45 mm, and 2.1–2.6 g P kg^{-1} at 75–85 mm, although work by Sanderson et al. (2003) and Rosen and Bierman (2008) was less definitive. Work in Idaho showed a sufficiency level of >2.2 g P kg^{-1} (Stark et al., 2003).

Potato tissue analysis has been studied more and is used more commonly as a management tool in commercial production than for most of the top global crop species, with most growers sampling the crop on a weekly basis during the time of tuber initiation until canopy senescence. However, other species are also managed actively with tissue analysis. Figure 3.8 shows yield response curves for lettuce sampled at various timings through the growing season. It is noteworthy that, in the case of lettuce, soluble acetic acid–extractable phosphate is analyzed rather than total P. Other species are managed using this extraction rather than total P concentration.

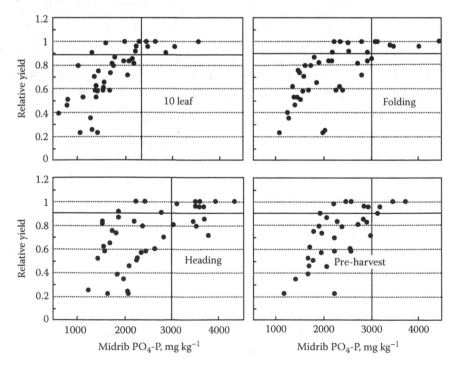

FIGURE 3.8 Critical acetic acid–extractable phosphorus at four growth stages of lettuce (*L. sativa* L.). (From Sanchez, C.A., Phosphorus, in *Handbook of Plant Nutrition*, Barker, A.V. and Pilbeam, D.J. (eds.), CRC Press, Boca Raton, FL, 2007, pp. 51–90.)

Similar information is available for a wide variety of other species for P, such as for corn, soybean, pecan (*Carya illinoinensis* K. Koch), tomato, and lettuce (Sanchez, 2007). Sanchez (2007) and Mills and Jones (1996) provide a listing of plant tissue concentrations for hundreds of species identified by sufficiency ranges as determined by research or, if insufficient information is available, by survey range or average based on nutrient concentrations found through routine analysis in their laboratories. The key is to sample each species uniquely by following established sampling protocols and taking tissue from parts that are highly correlated to P status and relate to fertilizer decisions.

3.8 FORMS, CONCENTRATIONS, AND BIOAVAILABILITY OF PHOSPHORUS IN SOILS

3.8.1 THREE POOLS OF SOIL PHOSPHORUS

Phosphorus exists in solid form as part of the structure of a wide variety of soil minerals, such as rock phosphate, present as fluorapatite $[Ca_5(PO_4)_3F]$ or hydroxyapatite $[Ca_5(PO_4)_3OH]$. It is also present as precipitated forms on other soil and rock particles. Inorganic mineral forms in soil are not found in any typical ratios, but tend to be combinations of sorbed/precipitated P on amorphous Fe and Al oxides and hydrous oxides and $CaCO_3^{2-}$ (Cole et al., 1953; Griffin and Jurinak, 1973; Holford and Mattingly, 1975).

The organic fraction is also an important pool of solid P in soils, existing predominately as the easily degraded phospholipids (~1%) and nucleic acids (5%–10%) and their degradation products, along with the more stable inositol polyphosphates (up to 60%), which are part of the humus fraction of soils (Anderson, 1967; Halstead and McKercher, 1975; Ko and Hora, 1970; Omotoso and Wild, 1970; Steward and Tate, 1971). There are many other organic P compounds, consisting of components of living organisms and their degradation products, with many of them unidentified complex compounds. Soil microbes degrade other organisms and release organic P, including by enzymatic cleavage via phosphatases (Alexander, 1977; Anderson, 1975; Cosgrove, 1977; Feder, 1973).

Plants absorb nutrients from the solution phase and not directly from the solid phase (Figure 3.9). Therefore, solid forms must be converted to liquid and chemically converted to mono- or diprotonated phosphate (HPO_4^{2-} or $H_2PO_4^-$) before plants can obtain P. Unfortunately, total P concentrations are much higher in the solid phase than in soil solution (Young et al., 1985). Bioavailable dissolved P concentration in the soil solution is typically very low; the median is 0.05 mg P kg^{-1} compared with the 200 to >1000 mg kg^{-1} total P typical in soils (Young et al., 1985). Although solution P levels are usually low, there is variability across soils as dictated primarily by soil pH and the presence of various cations, minerals, and organic compounds (Lindsay, 2001; Sposito, 2008).

Some soils have unusually high solution P levels, typically those that have been very highly amended with manure or similar organically complexed P materials (Alva, 1992; Bradley and Sieling, 1953; Davenport et al., 2005; Holford and Mattingly, 1975; Kissel et al., 1985; Lindsay 2001; Nagarajah et al., 1970; Sharpley et al., 2003; Sposito, 2008). Erich et al. (2002) found that soils amended with compost and manure developed higher plant-available and desorbable P. Fixen and Bruulsema (2014) state that some studies have shown that soluble organic compounds can inhibit P sorption in certain soils (Iyamuremye and Dick, 1996) but can increase P-sorption capacity in various tropical soils (Guppy et al., 2005). Malik et al. (2012) suggested that organic P sources can stimulate the formation of slow-release organic P forms. However, use of living cover crops to keep P in a more soluble plant-available form has mixed results (Little et al., 2004).

Much of the P taken up by plants is provided by the mineralization of organic materials, but release from minerals is also important. Minerals break down very slowly over time and can release structural P, but this amount represents very little of the P that plants take up in any one growing season. The majority of P supplied to plants comes from desorption of precipitated mineral deposits.

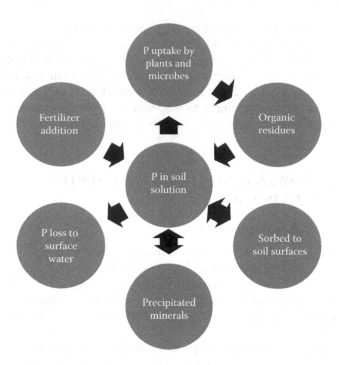

FIGURE 3.9   Phosphorus soil cycle.

However, this process is generally very inefficient because the phosphate molecules are poorly soluble and, as such, can quickly precipitate back out of soil solution as P cycling occurs.

In summary, soil P exists in three pools of availability, although the actual system is much more complex. The bioavailable P is the soil solution P (often termed the *intensity* portion). The labile P is the P in readily soluble form (termed the *quantity* factor). This labile fraction typically includes poorly sorbed mineral P, soluble P minerals, and easily mineralizable organic P. The other pool is the nonlabile P, which is strongly sorbed P, insoluble P minerals, and organic forms resistant to mineralization and cleavage of P.

3.8.2   Phosphorus Equilibrium in Soil

Precipitation/adsorption of P occurs rapidly due to equilibrium chemistry. Just as there is a limit to the amount of table salt (NaCl) that can be dissolved in water, the soil solution will allow only a finite amount of dissolved P (with a notable difference being that the equilibrium concentration for highly soluble NaCl is very high, but it is very low for the poorly soluble P compounds). The reverse reaction (solubilization of precipitates and desorption of labile P) also occurs, with P coming back into soil solution. The rate of this reaction accelerates when nonequilibrium conditions exist due to P being removed from soil solution. As plants take up P, a state of nonequilibrium is created, and dissolution of solid-phase P will occur until equilibrium conditions are satisfied once again. Although these sorption–desorption reactions occur and help maintain P available for plant use, the rate of the dissolution reactions may be too slow to fully meet the P demand for plants growing in some soils.

Soil water levels will have an impact on the equilibrium levels and plant availability of P, with optimum levels at about field capacity (Watanabe et al., 1960). Saturation above this optimum, although resulting in enhanced solubility of Fe-bound phosphates (Bacon and Davey, 1982; Holford and Patrick, 1979; Ponnamperuma, 1972), results in reduced P uptake for crops not adapted to saturated conditions due to poor root health under anaerobic conditions. Low soil water negatively impacts P uptake due to reduced volume of soil from which P can diffuse; it has a more tortuous pathway with higher concentration of other salts interfering in the path (Barber, 1980). Additionally, dry conditions often result in roots drawing water from lower portions of the soil profile where P is in lower concentrations as compared with the topsoil (Hanway and Olson, 1980). Soil temperature also impacts P solubility and availability as a function of increased chemical reaction kinetics and microbial activity (Gardner and Jones, 1973; Sutton, 1969).

The P concentration of the soil solution at equilibrium and, thus, plant P availability are highest at the slightly acidic to neutral pH range and are reduced considerably in strongly acidic or alkaline soil conditions. Solubility is further decreased in the presence of excessive lime ($CaCO_3$) (Kissel et al., 1985; Westermann, 1992). Excessive $CaCO_3$ in the soil increases P sorption on these surfaces and increases the precipitation of soil solution P as Ca–P minerals (Sharpley et al., 1989). Phosphorus combines with Ca and Mg, which are typically at high concentrations in alkaline soils. These Ca–Mg phosphates are poorly soluble at high pH. A similar reaction occurs in acidic soils, but with Al, Fe, and Mn, with the formation of poorly soluble phosphate minerals under low pH conditions. However, George et al. (2012) offer a somewhat different view, stating that inorganic P reactions are not as important as are changes in P stability as a function of pH in most soils with levels of organic matter that are greater than about 1%.

Because of these physical–chemical issues, plants utilize only a small portion (from near 0% to ~35%) of fertilizer P in the first year after application (Jacob and Dean, 1950; Randall et al., 1985; Syers et al., 2008), with the lower end of this range more common. Jacob and Dean (1950) found that two varieties of potato took up just 5% of the fertilizer P applied. Such is the case with a typical broadcast application of traditional P fertilizers.

The remaining P not utilized by plants and precipitated out of solution is often referred to as *fixed P*. Fixation is an unfortunate and confusing term. As such, it is no longer used in reference to P in soil science publications but continues to be used commonly in production agriculture. Fixation refers to very different processes for N, P, and K. Unlike the fixation of K^+ and NH_4^+ into the lattice structure of clay minerals (which removes K and NH_4^+ from the bioavailable pool) and N fixation of atmospheric N_2 gas by N-fixing microbes (which adds N to the bioavailable pool), P fixation is defined as the formation of solid-phase P compounds formed by precipitation or strong adsorption to soil minerals. Despite what is implied with the term fixation, these precipitates and adsorbed forms of P are not all permanently lost to plant uptake, although it may take many years before plants are able to utilize most of the P from a fertilization event (Randall et al., 1985; Syers et al., 2008). This phenomenon is related to the very poor solubility and mobility of P in soil and the fact that roots tend to explore less than 1% of soil at any one time. Since a large portion of fertilizer P is not utilized by plants in the first year after application, the remaining P is available for uptake in future years. Solubility of these P minerals declines slowly with time as the P compounds formed become increasingly less amorphous and more crystalline by geologic processes via P fixation reactions (Kissel et al., 1985; Lindsay, 2001; Sposito, 2008).

3.8.3 FATE OF FERTILIZER PHOSPHORUS IN SOIL

Fixen and Bruulsema (2014) give an excellent explanation of the physical chemistry principles of P availability from fertilizer. They state that reactions of P applied to soils can be described very broadly as sorption, precipitation, or organic interactions (Kissel et al., 1985; McLaughlin et al., 2011; Sposito, 2008). Sorption refers to the adsorption of P to the surface of soil particles, such as Fe and Al oxides or $CaCO_3^{2-}$, often replacing water or hydroxyl (OH^-) ions through ligand exchange. These reversible reactions occur rapidly. Sorption also includes the slow penetration of P below the surfaces of these solid particles. Australian research has shown that the P-sorption capacity of soils is an important factor in determining the amount of P fertilizer required to achieve maximum yields (Hegney et al., 2000).

Fixen and Bruulsema (2014) further stated that precipitation occurs when the addition of a P fertilizer causes the concentration (activity) of P in solution to exceed the solubility product of mineral phases (Bell and Black 1970; Lindsay et al., 1962). Below these concentrations, the sorption reactions described earlier are the dominant reactions. The classic solubility diagrams of Lindsay (1979) are still the generally accepted chemistry of the thermodynamic relationships believed to occur. When soluble fertilizers are added to soil, the most soluble minerals precipitate first because the reactions are faster. There are more ions in solution and a greater chance that collisions with nucleation sites will take place and cause the mineral to grow. With time, the P in soil solution declines due to incorporation into the more soluble minerals, and those same minerals then begin dissolving. The P contained in them precipitates as the less soluble mineral phases grow. So, as time passes, there is a cascade of P from soluble to less soluble minerals forming.

Lindsay (1979) states that Al and Fe phosphates are the most stable (least soluble) forms in acidic soils and Ca phosphates are most stable in alkaline conditions. However, Fixen and Bruulsema (2014) state that recent studies using direct spectroscopic techniques show that soil P is not always behaving as stated by Lindsay (1979), likely due to factors that influence reaction kinetics and the numerous potential reactants in soil solutions that could create somewhat messy mineral phases. For example, significant calcium phosphate minerals have been identified in moderately acidic soils, while Fe oxides have been found to be active sorbers of P in alkaline soils.

Fixen and Bruulsema (2014) summarize the sequence of events following fertilizer P (as MAP) addition to soil as follows:

1. Upon addition to soil, capillary flow of water into the granule begins—bringing Al, Fe, and Ca along with it. This process occurs at the same time P is diffusing out of the granule. This action is much less important with fluid NH_4^+ phosphate forms, such as APP. The flow into the granule can impact the diffusion of P from the granule. Spectroscopic evidence exists that in some soils the flow into the granule results in precipitation of the mineral crandallite, a Ca–Al–OH phosphate (Lombi et al., 2004). Most of the P leaves the granule in a matter of days, but crandallite residues can remain for months.

2. As P leaves the granule, much of it quickly precipitates as several different forms of NH_4^+ phosphate, since NH_4^+ is also diffusing from the granule. These new NH_4^+ phosphates are still very soluble, but less soluble than MAP. Changes in pH and other solution factors are the driving forces behind this precipitation.

3. As the NH_4^+ diffuses away from the reaction site or nitrifies, the NH_4^+ phosphates dissolve and, in nearly all temperate soils, P precipitates as dicalcium phosphate dihydrate ($Ca_2HPO_4 \cdot 2H_2O$), which is a somewhat soluble P form, although much less soluble than the original fertilizer.

4. The soil pH then largely dictates subsequent reactions: In moderately *acidic* soils, P minerals containing Fe, Al, or Ca precipitate are formed after about 35 days. Common examples are strengite ($FePO_4 \cdot 2H_2O$), vivianite ($Fe_3(PO_4)_2 \cdot 8H_2O$), variscite ($AlPO_4 \cdot 2H_2O$), and apatite analogs (various calcium phosphates). After about 180 days, these minerals will have dissolved largely, and much of the P is sorbed by hydrous oxides of Fe and Al or they remain as apatite analogs. In *alkaline* soils, P minerals containing Ca are dominant at 35 days with octacalcium phosphate (OCP; $Ca_8H_2(PO_4)_6 \cdot 5H_2O$) being one of the most common forms, but with apatite analogs occurring as well. Eventually, the more soluble Ca–P minerals, such as OCP, dissolve, and P is sorbed by $CaCO_3$ or Fe oxides. Apatite analogs often are found, persisting for extended periods. The importance of Fe oxides in these alkaline soils, verified by direct spectroscopic evidence, has been surprising considering the solubility relationships described by Lindsay (1979).

Fixen and Bruulsema (2014) stated that a critical component of this cascade of reactions is the massive decline in P solubility, which occurs while these reactions between fertilizer and soil are taking place. The solubility of the original P fertilizer is very high, but the solubility of the final products is several orders of magnitude less. Cultural practices, as well as soil and environmental conditions, greatly impact the timing of the formation of these P minerals in soil. They further stated that crop P uptake in a given growing season is the sum of the P taken up by roots intercepting fertilized zones and the P absorbed from the bulk soil, with the vast majority of the P coming from the bulk soil. For example, Australian research employing P isotopes has indicated that 70% or more of crop P uptake is derived from the bulk soil (McBeath et al., 2011). We apply fertilizer P to fill the gap between what the bulk soil can provide crop roots and the P required for the attainable crop yield.

3.9 SOIL TESTING

3.9.1 BENEFITS OF SOIL TESTING

Although plant tissue analysis is a good tool for predicting yields and in-season fertilizer needs, it needs to be coupled with soil testing. Tissue analysis is especially helpful for highly soluble and mobile nutrients, such as N. Unfortunately, P does not fit into this category of soil mobile nutrients, making soil testing an even more important management tool. Hopkins et al. (2010a,b) found that fertilizer P applied and incorporated into the soil prior to planting resulted in better yield response than when applied via fertigation. This action is logical given the fact that P applied to the soil surface will generally only move a few mm in soil once it converts from liquid to solid phase. Thus, this P is unavailable to all but surface-feeding roots, which are relatively more subject to interferences from soil heat and water limitations.

Therefore, it is vital for managers to be able to predict P fertilizer needs prior to planting through soil testing and field records so that this fertilizer can be incorporated into soil prior to, at, or shortly after planting. Typically, there is a curvilinear relationship between soil test P and yield response (Figure 3.10). The likelihood of response diminishes as the soil test P level approaches the critical level, which is defined as the concentration above which no response to added P is reasonably expected (Kissel et al., 1985; Lindsay, 2001; Marschner, 2012; Sposito, 2008; Young et al., 1985). Responses to P fertilizer are large at very low soil test P levels, with decreasing amounts needed in order to achieve full yield potential as soil test P approaches the critical level. Figure 3.10 shows the correlation between yield responses and soil test P levels for celery at 18 sites in Florida with Histosol soils, illustrating the development of a critical level unique to the crop, soil, and environmental conditions. Figure 3.11 shows the result of compiled fertilizer studies resulting in a P fertilizer recommendation rate chart as a function of soil test for celery, lettuce, sweet corn, and snap beans.

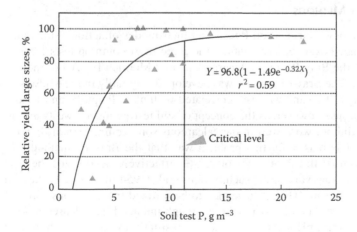

FIGURE 3.10 Critical soil-test phosphorus levels (by water extraction) for production of large, harvest-size celery (*Apium graveolens* Pers.) on Florida Histosols. (From Sanchez, C.A., Phosphorus, in *Handbook of Plant Nutrition*, Barker, A.V. and Pilbeam, D.J. (eds.), CRC Press, Boca Raton, FL, 2007, pp. 51–90.)

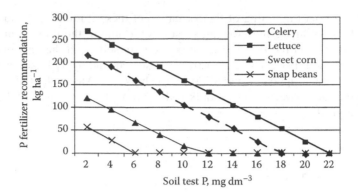

FIGURE 3.11 Recommendations for phosphorus fertilization for selected crops on Everglades Histosols. (From Sanchez, C.A., Phosphorus, in *Handbook of Plant Nutrition*, Barker, A.V. and Pilbeam, D.J. (eds.), CRC Press, Boca Raton, FL, 2007, pp. 51–90.)

Soil testing is not a perfect tool, but several methods are available and are moderately correlated to yield for various soil and crop combinations. Phosphorus soil tests are estimates of bioavailability and are index values. This testing is in contrast to other tests that are quantitative. For example, the NO_3–N test is designed to extract all of this form of N from soil for analytical determination and interpretation. In contrast, the amount of P extracted from a soil includes the very small quantity of solution P plus a variety of labile solid soil P forms, none of which are fully plant available. For example, a Timpanogos loam extracted at the BYU Environmental Analytical Lab had 1645 mg P kg^{-1} of total P (including all mineral and organic components of the soil). Of this quantity, only 0.04 mg P kg^{-1} was soluble and available for immediate plant uptake. Neither of these extractions are valuable for predicting the P status of a soil in terms of plant availability. As previously mentioned, the first attempts at soil testing included analysis of total P and had poor correlation to actual P availability to plants, with many soils having high total P concentration but very low bioavailable concentrations and vice versa. As also previously mentioned, the amount in soil solution at any one time is minute in comparison to what the plant takes up during the season. For this reason, other soil-testing methods were developed to extract the most readily soluble forms of P that can be correlated to plant uptake.

3.9.2 Soil Test Methods

Sanchez (2007) reviews various soil P tests and their modes of action. Although several methods have been introduced over the last century, the three most common tests for estimating P in the United States are the Bray P1, Olsen bicarbonate, and Mehlich III. All of these tests are indexes for bioavailability. The Bray P1 method was developed originally in the Midwestern United States by Bray and Kurtz (1945) and was well correlated with plant P uptake and yield response in corn and many other crops. However, as the concept of soil testing moved westward, it was discovered that this method did not work well in the calcareous soils common in arid and semiarid regions (Mehlich, 1978). The reason for this problem was that the Bray P1 method relies primarily on a weak acid to extract the P, but the soil CO_3^{2-} effectively neutralizes the acid and alters the results, giving false low values. As such, Olsen et al. (1954) developed a neutral salt extraction method using bicarbonate (HCO_3^-) designed to be utilized in calcareous soils. This method correlated well with plant P uptake and yield. Although intended for calcareous soils, it also worked well in acid and neutral pH soils. Later, the soil-testing movement sought to reduce the cost for labor and chemicals, and as such, universal extractants were developed, which were designed to extract a majority of the essential nutrients from soil with one procedure for a wide variety of soils (Hanlon et al., 1984). Mehlich (1984) developed an extractant from a buffered acid solution.

This test correlated well with P nutrition in plants under most soil conditions. Many other soil-P extractants have been developed, but although many are well correlated to P nutrition, they lack widespread adoption. There are other tests available (such as P-sorption isotherms, P fractionation, and isotopic dilution), but they tend to lack calibration data with yield or are too labor intensive to be used for routine soil testing.

It is important to realize that the results of these bioavailability tests must be converted to usable P fertilizer recommendations. Also, the analytical values cannot be related directly to an amount of P that is precisely available for plant uptake. In the case of NO_3–N, it is possible to convert the concentration to a quantity of N available to plants on a kg ha^{-1} basis by multiplying the concentration in the soil by the depth of the soil sample and converting to the proper units. This process is done legitimately because all of NO_3–N is extracted quantitatively from the soil and, assuming it is not lost via leaching or by other processes, it is all plant available. However, many growers and agronomists erroneously employ the same mathematics for P and other nutrients that are extracted using a bioavailability index method. This process should not be done since the commonly used P tests are not quantitative. When a soil P extractant is added to the soil and then shaken for a set period of time, the chemicals dissolve a portion of the labile P, but not all.

For example, the Timpanogos soil cited in the preceding example also was extracted with all three of the common methods described earlier, as well as a variety of fractionation methods to determine organic P, Ca-bound P, Fe-bound P, Al-bound P, etc. All of these extractions resulted in different P concentrations extracted from the same soil as a function of the types and strengths of chemicals added, time of shaking, heat of the solution, etc. Obviously, none of these extracts represents precisely the amount that is plant available, but rather, in the case of the bioavailable extractants, an index of P availability that requires field calibration. The results of the bioavailability tests were 10, 19, and 21 mg P kg^{-1} for the Olsen bicarbonate, Bray P1, and Mehlich III extractions, respectively. The mistake that is made commonly by growers and their agronomists is to convert these concentrations to what they assume to be the equivalent of fertilizer P. In this case, the results would be 20, 38, and 42 kg P ha^{-1} or, in terms of fertilizer amounts, would be 45, 86, and 95 kg P_2O_5 ha^{-1} for the Olsen bicarbonate, Bray P1, and Mehlich III extractions, respectively. They then look up the amount of P taken up or removed by a crop, such as what is shown in the last column in Table 3.2, and then subtract the amount from their calculation to determine the fertilizer rate to be applied. One can see, however, that each of the methods gives a very different result and may lead one to ask which method is correct. The problem is that this is not an appropriate way to determine fertilizer rate.

Rather, the values from these extracts should be interpreted based on fertilizer trials used to correlate fertilizer response to soil test level, such as is shown in Figure 3.10. What is the likelihood of a fertilizer response at 6 mg P kg^{-1}? The answer is dependent on which test was employed. If this value was the result of a total soil P analysis, this would be an impossibly low number and meaningless due to a lack of correlation for fertilizer recommendation. If this is a water-extractable value, this value is impossibly high and, again, meaningless. If it was from the same extraction used to develop the yield response curve in Figure 3.10, then we can legitimately interpret the data to show a high probability of response. In this case, a 6 kg P ha^{-1} value would mean that we would probably only achieve about 65% of maximum yield without fertilizer.

It is important to understand that each soil extractant will result in its own unique response curve with varying results. For example, a value of 20 kg P ha^{-1} for the Olsen bicarbonate test would be *high* for most crops and no fertilizer would be recommended, but the Bray P1 and Mehlich III tests tend to be about twice as high, and therefore, a value of 20 kg P ha^{-1} may trigger a recommendation to apply fertilizer P based on field calibration showing responsiveness at this level. The main point here is that data from each extractant have to be interpreted based on its own independently developed data set calibrating it to crop response to P fertilization.

Some researchers have attempted to make direct comparisons between the commonly used tests (Bishop et al., 1967; Fixen and Groove, 1990; Hooker et al., 1980; Kamprath and Watson, 1980;

Kuo et al., 1996; Maier et al., 1989b; Mallarino, 1997; Smith and Sheard, 1957; Thomas and Peaslee, 1973). Maier et al. (1989b) compared eight extractants at 33 locations and determined that the potato critical values were 17 and 26 mg kg^{-1} for the Bray P1 and Olsen bicarbonate extractants, respectively. The other extractants evaluated did not correlate well under the conditions of their study. This is just one example, but other comparisons exist. It is advisable to use a soil test method for which there are calibrated data for valid comparisons.

Ideally, a soil extractant has been vetted fully with fertilizer response research trials for each crop and soil combination possible, with the results showing at least an adequate relationship between extractable P and crop yield parameters of interest with a significant response to P fertilization when the extracted amount of P is *low*. There are plenty of examples of such research results, but the reality is that there are large gaps in the data due to time and budget constraints of the scientific and agricultural communities. Not every crop has been tested under every soil and environmental condition, and untested cultivars may have unique requirements. The response curve is likely to change even within a species with differences between cultivars or with differences across soils and environments. For example, in potato, there is a wide range of critical levels varying by time (Johnston et al., 1986), soil type and texture (Birch et al., 1967; Bishop et al., 1967; Boyd and Dermott, 1967; Giroux et al., 1984; Kalkafi et al., 1978; Kelling and Speth, 1997; Maier et al., 1989b; Redulla et al., 2002), and cultivar (Freeman et al., 1998; Maier et al., 1989a; Murphy et al., 1967; Sanderson et al., 2003). In the work of Johnston et al. (1986), these researchers determined the critical level for potato with the Olsen bicarbonate test was at 25 mg kg^{-1}, but this value varied between 10 and 54 mg kg^{-1} between years. Similar findings have been discovered for other crops as well. These varying results might leave one doubting the value of soil testing.

3.9.3 CUSTOMIZATION OF SOIL TESTING

However, soil testing is a proven method, even if not perfect, and it is certainly better than not having any assessment of soil P status. Additional research is needed to fine-tune P fertilizer recommendations based on soil testing. For example, it was discovered that adjusting the P fertilizer recommendations based on the concentration of free lime in the soil significantly improved accurate prediction of P response (Lang et al., 1999; Stark et al., 2004; Westermann, 1992). Other researchers have begun to use Al analysis to adjust P recommendations (Khiari et al., 2000). Additional research is needed to examine these and other adjustments that may improve soil test correlations with yield response.

Furthermore, soil test data should be combined with field records of yield performance along with records of tissue analysis and long-term soil P trends to fine-tune future fertilizer recommendations in order to customize fertilizer recommendations for individual fields. For example, an alfalfa field in Idaho had a *medium–high* soil test of 15 kg P ha^{-1} (Olsen bicarbonate). However, analysis of the harvested hay from this field revealed that the P content was unusually low, especially in the cuttings later in the season. Field records showed that other crops grown in this field previously also trended low during the later part of the season. The farmer and his agronomist conducted a fertilizer strip trial in this field, showing significant yield increases with modest amounts of P fertilizer despite the soil test interpretation indicating that none was needed. The soil testing did not fail, but rather this particular soil was below average in terms of its ability to desorb P to provide for late-season plant needs. The soil test interpretations were then adjusted for this soil type in making future P recommendations.

Another important factor to consider when soil testing is the spatial variability of soils and other conditions. Most fields contain a variety of conditions and soil types. It is not uncommon to have areas of a field that are deficient in P when the field average is showing adequate or even excessive P availability. This occurrence is especially true in fields with eroded hilltops with exposed subsoil having very low bioavailable P levels. Subsoil is typically low in organic matter and plant-available P and high in pH and antagonistic (for P and many micronutrients) calcium carbonate ($CaCO_3$), magnesium carbonate ($MgCO_3$), gypsum ($CaSO_4$), or sodium bicarbonate ($NaHCO_3$). Calcium in particular is a known antagonist for P availability to plants. Spatial differences also can develop in the opposite situation in areas of fields that are especially productive and result in greater nutrient removal rates as a function of continual above-average yields.

If only one representative sample is taken per field, it is likely that portions of the field will have suppressed yields in cases where fertility levels are below the average. Over the years, this situation has resulted in growers having their faith in soil testing shaken when, in fact, the problem lies with poor sampling and interpretation. Various methods to account for spatial variability within fields can be employed, but at a minimum, samples should be taken from unique areas likely to be less fertile than the field average. Technology exists to variably apply fertilizer to the areas that are in need with none applied where the P levels are high.

Another approach to soil testing is the use of ion exchange resins (Jones et al., 2013). The concept here is that a porous bag containing the resin lies in the soil and undergoes conditions similar to those plant roots will experience. The soil solution then interacts with the resin, with nutrient absorption taking place. Later, the resin capsule is removed from soil and the nutrients are extracted and analyzed. Although the claim that this system is more true to the mechanisms of nutrient uptake (as compared with extracting a soil by adding chemicals, shaking, and filtering) is partially true, the resins can be located only in one place in the soil, whereas roots expand into the fertile topsoil along with deeper horizons. And, although diffusion toward the resin can occur, mass flow and root interception mechanisms are not operating. Furthermore, there is a lack of rhizosphere interaction with proton and organic acid and other root exudates. Research results are mixed with regard to whether the resins are better or less satisfactory predictors of P availability compared with conventional soil tests.

3.10 PHOSPHORUS FERTILIZER MANAGEMENT

3.10.1 PHOSPHORUS FERTILIZER IMPORTANCE AND CONSERVATION

Successful civilizations are built largely on a foundation of agricultural productivity. A populace with plentiful and inexpensive food is then free to pursue advances in education, communication, engineering, transportation, etc., because the majority of their time and efforts are not focused on providing their next meal. Such is the case for many of the impoverished nations of the world. Furthermore, many civilizations have failed due to the loss of soil or soil productivity. Such losses contributed to the demise of the Fertile Crescent, the Roman Empire, and the Incan Empire, as examples (Pointing, 2007). George Santayana (1863–1952) coined the saying that "Those who cannot remember the past are condemned to repeat it." If humankind hopes for continued prosperity in developed nations, the good earth that currently provides food, fuel, and fiber for more than seven billion people must not be ignored. Achieving prosperity in developing nations hinges upon a high level of self-sufficiency with regard to crop production. Crops contain nutrients, and their removal through harvest and transport to population centers depletes soil fertility. The law of conservation of mass applies insofar as matter, including plant nutrients, can be neither created nor destroyed. In other words, when minerals are removed from the soil, they must eventually be replenished through the colossally slow process of rock minerals breaking down naturally or through the addition of fertilizers.

The manufacture of fertilizer, however, is not without cost of resources, some of which may be eventually exhausted. The majority of P fertilizer used is in mineral form. The resources used include mined P and S minerals (elemental S used to create H_2SO_4, which is then used to create H_3PO_4 from rock phosphate) and fossil fuels for heating during the manufacturing process, as well as natural gas used in the manufacture of N fertilizers. Some of these resources are finite, and all are worthy of conservation. There is much speculation and study on when these resources might be depleted, with some arguing that we will run out of rock phosphate resources and fossil fuels within decades (Cordell et al., 2009; Marschner, 2012). However, these estimates are far too extreme (Van Kauwenbergh et al., 2013). Current sources, along with known undeveloped resources and improvement in mining and recovery technology, mean that the Earth's vast resources will provide phosphate rock for fertilization for many centuries. Regardless, we owe our best efforts of conservation to future generations. We do not want to duplicate the actions of the infamous Easter Island inhabitants who, beginning many centuries prior to their failure, squandered their natural resources to the point of collapse.

There are at least two avenues of conservation. Philosophically, society should be recycling back into soils human and other animal wastes, as well as other wastes from products generated via fertilization of crops. Loss to water bodies should be avoided as much as possible, as P recovery from lake and ocean sediments is not realistic with current technologies. Ideally, these wastes should be applied to the land from which they came, but this act is not practical under current policies and conditions. Typically, crops are transported to centers of animal production and human populations and the wastes are largely applied to land in that locale. This movement results in depletion in the areas of crop production and accumulation near the animal production centers. Transporting animal manures back to the point of origin of the crops from which they came comes at a cost that society is currently not willing to pay. Even if society was willing to transport these wastes, there is an argument again because of the use of fossil fuels that would be required, as well as the pollution derived from the exhaust. Alternatively, many propose that crops should be locally produced, with waste products recycled back into soils, but this would result in increases in food production costs under current conditions.

Even if the ideal of directly applying nutrients from waste back to the soil from which they came was possible, there would not be enough to meet the demand because the system is leaky and solubility of P decreases over time. Recycling P is thwarted partially due to soil reactions described previously, with the cascade of increasing less soluble P minerals forming over time. In other words, recovery of fertilizer P is never 100% efficient. The other inefficiency is due to the leaky nature of the P cycle. Some of the P in the waste stream inevitably enters the surface water system and eventually ends up as sediment in streams, lakes, and oceans. This process occurs through point source (such as from sewage treatment facilities) and nonpoint source (such as soil erosion and P transport via precipitation runoff) avenues of loss. It is not reasonable to stop the loss of P to water completely, but this action should be the goal as much as it is reasonably possible to do so.

Therefore, mineral fertilizers continue to be an important resource to replenish nutrients lost through crop harvest and will remain so indefinitely. In an effort to lead in the wise use of resources, the fertilizer industry has adopted the *4 Rs of fertilizer stewardship*, which are choosing the (1) right rate, applying the correct rate based on experimentation under various environmental conditions; (2) right timing, ensuring peak nutrient availability at the time of greatest uptake; (3) right placement, concentrating nutrients in the soil at locations where there is a large volume of roots; and (4) right product, using sources that are soluble and have release patterns conducive for plant uptake. In addition, factors that enhance root–soil contact will increase PUE (Hopkins et al., 2014).

3.10.2 Phosphorus Sources

Before discussing specific P sources and considerations, it is important to understand fertilizer labeling. The laws in many countries require that any product sold as a fertilizer must contain a guaranteed analysis of the primary macronutrients in the order of $N–P_2O_5–K_2O$. Due to misconceptions regarding the chemical compositions of fertilizer materials at the time of the creation of the fertilizer laws, P is expressed on a container of fertilizer as P_2O_5 rather than its elemental composition, or as it is typically found (as phosphate) in most P fertilizer. For example, the sellers of 0-45-0 are claiming a minimum of 45% P_2O_5, which is 45 kg P_2O_5 for a 100 kg unit. The oxidized state of P is actually PO_4^{3-} and not P_2O_5, but the laws and tradition have institutionalized this labeling. It is not likely to change in the future. This labeling is unfortunate because, although P is expressed consistently as P_2O_5 for fertilizer, there is a lack of consistency in how P is expressed for concentrations and quantities in plants, soils, and waste materials, sometimes being expressed in the oxide form (either P_2O_5 or PO_4) and sometimes in the elemental form (P). This variability frequently causes confusion, and therefore it is important to understand which form is being expressed when dealing with P nutrition data. Converting from elemental P to fertilizer rate is accomplished by multiplying P_2O_5 by 43.7%. For the 45 kg of P_2O_5 in the preceding example, there would be 20 kg of elemental P in this fertilizer.

There are several considerations when choosing a fertilizer source. The first is effectiveness of the material. In the case of P, the material needs to be water soluble or at least should have an eventual slow or controlled-release rate that is predictable. Unless a slow or controlled-release pattern is desirable, P fertilizer should be at least 60% water soluble and slow or controlled-release materials need to be similarly water soluble within the course of a growing season. Most traditional sources, such as MAP and DAP, are greater than 90% soluble. If compared fertilizers have a similar solubility, the choice of which source to use becomes one of price, availability, convenience of application, and accompanying nutrients.

Phosphorus fertilizer materials generally are blended with other nutrients. A PO_4^{3-} anion has to be combined with cations (such as NH_4^+, H^+, K^+, Ca^{2+}, Mg^{2+}) to maintain electrical neutrality. The vast majority of P fertilizers used globally are ammoniated phosphates, such MAP, DAP, and APP. Orthophosphate (phosphate ion) is found in MAP and DAP. APP contains polymerized phosphate ions created through dehydration prior to ammoniating. Generally, N also is needed by plants, and the blend is effective in supplying P and a portion of the N need of crops. Historically, TSP was popular, but the recognition of the synergy between NH_4^+–N and P, perceived problems with P availability in certain soils, and fertilizer industry nuances have resulted in a decline in its use. There are other mineral fertilizers available as well, such as monopotassium phosphate (MKP; KH_2PO_4; 0-52-34), but these are utilized in small quantities.

There are also a wide variety of organically complexed P products, such as raw manure, treated biosolid waste, composted manures, and other wastes, recycled crop residues, blood meal, and fish meal. Immediate water solubility of these materials is not relevant as the release of P is typically dependent upon the mineralization of the organic materials, a process that eventually breaks down the complexed P molecules into plant-available phosphate. Plant availability of P from manure is estimated at between approximately half to nearly the same as compared with mineral P fertilizers, with release occurring mostly in the first year after application but extending into subsequent years as well (Abbott and Tucker, 1973; Curless et al., 2005; Elias-Azar et al., 1980; Gale et al., 2000; Gracey, 1984; Laboski and Lamb, 2003; Meek et al., 1979; Motavalli et al., 1989; Powell et al., 2001).

Much of the P in manure is present as orthophosphate and, as such, is immediately plant available. But a large portion is organically complexed and released slowly. This slow release can be an advantage, as long as the P is supplied in a timely fashion that coincides with plant need, which is sometimes difficult to predict and control. As previously mentioned, availability of P early in the season is critical, and yet, these materials depend upon temperature-driven microbial decay, possibly resulting in delayed release and early-season P deficiency unless adequate amounts of immediately water-soluble P are present as well. Although manure or biosolid material can serve as a very good source of nutrients, there are many potential downsides, including presence of weed seeds, nutrient imbalances, odor, presence of toxins, cost of transportation, and compaction of soil due to heavy axle loads from application. Nevertheless, proper utilization of these by-products is an effective way to recycle P and other nutrients.

Manufactured slow- and controlled-release fertilizers are engineered to release P over time (Hanafi et al., 2002; McLean and Logan, 1970; Yanai et al., 1997). In some cases, release of P from these products also is temperature controlled, having the same problem as organically complexed P early in the growing season. However, others have a time-release mechanism that is not temperature dependent. Unlike soluble nutrients, such as N, problems with fertilizer P efficiency depend more on complexation in the soil than on leaching or gaseous loss from the system (Hanafi et al., 2002; Kissel et al., 1985; McLean and Logan, 1970). Adding immediately soluble P fertilizer results in a temporary increase in soil solution P concentration at levels that exceed chemical equilibrium constants, forcing precipitation of phosphate minerals. A slow or controlled-release P fertilizer may minimize the formation of these phosphate compounds as the soil solution P concentration does not spike at as high a level and the P is released gradually over time as a function of temperature and moisture, with increased PUE as the potential result.

Organically certified P fertilizer is another avenue of fertilization that is popular with many home gardeners and required for organically certified produce (Hopkins and Hirnyck, 2007). The term *organic* here should not be confused with *organically complexed* P referred to previously. The latter is an organic compound, which is defined as a molecule containing C covalently bonded to other atoms (not including salt-forming ionic compounds containing C, such as urea fertilizer $[CO(NH_2)_2]$, CO_2, and CO_3^{2-}). In contrast, Hopkins and Hirnyck (2007a) describe organic production as an ecological production management system and as a labeling term to denote products approved under the authority of the Organic Foods Production Act (in the United States). Traditional P fertilizers and many organically complexed fertilizers, such as raw manure, are not automatically certified organically. Composted manure and other products can become certified if their manufacturer meets the conditions of certification. These products can be effective P sources.

One example of an inefficient organically certified P fertilizer is untreated rock phosphate. This material is very insoluble and, as such, is a poor source of plant-available P in anything but highly acidic soil. Although this and similar materials, such as some untreated bone meals, are advocated commonly for use, applying it as a source of fertilizer is a poor choice from the practical view of P solubility and also from a philosophical view since this represents a wasteful use of the finite rock phosphate reserves.

Another point to be made regarding organically certified fertilizers is that any P atoms finding their way into a plant are chemically indistinguishable from P from any other source. It is not safer, does not make food taste any better, and is the same in every respect. There are other differences possible between organically certified and traditional fertilizers. For example, the reason that DAP, MAP, and APP are not eligible for organic certification has to do with the use of fossil fuels in the manufacture of the NH_4^+, but there is no evidence that these sources of P result in unhealthy conditions for plants or for animals consuming them. There are other important considerations when comparing these sources, such as the possible presence of weed seeds, nematodes, insects, and pathogens found in manures and biosolids unless they are treated to kill these organisms. Modern consumers are, as a whole, misinformed by much of the terminology surrounding organic and related terms, such as *natural* as used in foodstuffs.

Engelstad and Teramn (1980) reviewed the effectiveness of P fertilizers. There are differences across sources, but most data show that equal rates of soluble P result in approximately the same plant response regardless of sources used. Again, an atom of P is the same regardless of whether it came from bones, rock, or manure, as long as it enters the soil solution and then the root. However, there are other differences that may be important as affecting uptake probability.

One difference between fertilizers is the reaction pH values (MAP = 3.5, UAP and DAP = 8.0, TSP = 1.5, and APP = 6.2; Young et al., 1985). However, the pH in the microsite around the fertilizer does not remain acidic for long, and the uptake of P is not impacted much because of the difference in reaction pH among sources (Young et al., 1985). Other fertilizers in the band, particularly N, can also have an effect on pH, most commonly an acidifying effect when nitrification occurs. Rhue et al. (1981) found that APP resulted in reduced P uptake and potato tuber yield and quality compared with DAP on soils with pH of about five. The effect was likely due to the further reduction in pH in the banded application with the APP, which has an acidic reaction pH, as opposed to DAP with an alkaline reaction pH. Young et al. (1985) stated that, in general, there are no major differences across traditional fertilizer P sources in terms of impact on crops, although some studies show minor differences. They speculated that these differences likely were due to pH or micronutrient interactions.

Another difference that can be important is that DAP can result in greater volatilization of NH_3 or accumulation of nitrite (NO_2^-) than MAP due to this short-lived reaction pH and, thus, may be more damaging to seeds and seedlings when in direct contact (Armstrong, 1999). High applications (≥ 270 kg P_2O_5 ha^{-1}) of DAP or UAP placed near (≤ 5 cm) or in contact with potato seed piece delayed emergence, reduced stand, and negatively affected yields (Chu et al., 1984; Fixen et al., 1979–1981; Meisinger et al., 1978). When growing vegetables on very alkaline soil and with high rates of P, DAP is avoided to prevent toxicity to these small-seeded species. The popularity of MAP and DAP is due to a high analysis of water-soluble P, which results in low per unit transportation costs. In contrast, a composted manure with 1% P has transportation costs more than 50 times greater per unit of P than MAP. This cost becomes commercially unmanageable if transporting more than a few miles.

Liquid fertilizers also tend to have higher transportation costs due to lower analysis from extra water weight. The most popular liquid fertilizer is APP because its P analysis is relatively high compared with most other liquids, although lower than MAP and DAP. It is a common myth that the polymerized phosphate molecules in APP are less plant available as the phosphate chains quickly turn to orthophosphate once hydrolyzed in the soil and are available immediately for plant uptake. Legitimate reasons cited for using liquids over solid P fertilizers in some circumstances are homogeneity of blends resulting in uniform application, sequestering of micronutrients to aid in their uptake, ease for combination applications with liquid pesticides, ease of fertigation, no caking of solids, and advantages for the fertilizer dealer in terms of safety and equipment logistics. But if the same amount of P is applied at the same distance from a plant root, it will not likely distinguish between a phosphate ion from MAP, DAP, or APP as the P is chemically identical once entering the soil. This is true for all water-soluble forms of P.

Comparing liquid with various forms of solid P fertilizer in potato, sources have been shown to result in differing yield results. Stark and Ojala (1989) reported yields 9%–15% higher with band-applied APP than with acid–urea phosphate in potato grown in a calcareous Idaho soil. Locasio and Rhue (1990) reported yields 20%–40% higher with APP or TSP than with DAP on a slightly acidic sand soil in Florida. However, DAP resulted in higher yields compared with APP (Rhue et al., 1981), as well as being better than MAP (Sanderson et al., 2003) and TSP (Giroux et al., 1984). However, MAP or TSP had higher yields than APP in Michigan (Christenson and Doll, 1968). Given these convoluted results and other researchers reporting no source differences, it is recommended to choose a source based on pricing and convenience factors (MacLean, 1983; Rosen and Bierman, 2008).

Another source issue is the common practice of the mixing of several nutrients to make up a fertilizer product. Phosphorus fertilizer is mixed commonly with K and/or the secondary macronutrients and micronutrients to make a more complete mixture. These mixtures can be problematic due to segregation of fertilizer prills of differing size, density, and shape and can result in uneven applications. This problem is especially true when using broadcast spreaders throwing the material a long distance. Again, using a well-mixed liquid blend eliminates this problem. Also, this problem can be eliminated for solid fertilizers by liquefying and mixing the nutrients and then solidifying into an even blended material.

A potential drawback of these blends is that a nutrient that is not needed based on soil-testing results may be applied wastefully. This waste is especially common in the urban fertilizer market. Most fertilizer materials available in retail outlets are solid mixtures or blends, such as a 20-10-15 (20% N, 10% P_2O_5, and 15% K_2O). However, many urban landscapes have been overfertilized grossly, and especially, turfgrasses common to urban landscapes are particularly efficient at P uptake. Greater than 90% of soil samples processed at the BYU Environmental Analytical Lab for homeowners show that no additional P is needed, and yet most homeowners and urban landscape managers wastefully apply P due to the convenience of purchasing a fertilizer mixture. Turfgrass is the number one irrigated crop in the United States, and therefore, the amount of wasted P application is substantial. As a result of this and environmental issues surrounding P fertilizer use, many communities have instituted fertilizer bans (Hopkins et al., 2013). This approach is shortsighted. Instead, urban landscapes need to be fertilized similarly to how farmers do business by basing fertilizer recommendations on soil testing and applying custom blends rather than a "one-size-fits-all" approach. More often than not, soil and tissue analyses reveal that all of the fertilizer needs for a typical urban landscape can be met with a combination of urea and ammonium sulfate fertilizer, with no P needed in the foreseeable future.

Fertilizer prill size is another important source consideration, especially for turfgrasses and carrot and other plants that have a very narrow diameter of soil exploration by roots for each plant. For most agronomic crops, this is not an issue because of the wide circle of soil exploration by each plant. But plants with small root systems require small prill sizes for more uniform coverage of nutrient over the landscape. A large prill applied to turfgrass may result in a few plants right around the prill receiving a large dose, but those a few cm away accessing little if any of the nutrient. This result is particularly true for the poorly mobile P fertilizers, and as such, the turfgrass and specialty crop fertilizer industries have products with very small prill size that are available, although the cost is relatively high.

Other aspects related to fertilizer sources are additives and enhanced efficiency fertilizers. One very widely sold, but controversial, product is AVAIL® (Specialty Fertilizer Products, Leawood, Kansas). Hopkins (2013) reviews the proposed mode of action for AVAIL, a high charge density polymer that sequesters interfering cations. He also reviews the work performed on AVAIL and its impact on P soil chemistry. Chien et al. (2014) rebut the effectiveness of AVAIL. However, positive responses were observed and published for rice (Dunn and Stevens, 2008) and two potato studies showed mixed responses (Stark and Hopkins, 2014). Hopkins (2013) reviews other informally reported studies on a variety of crops. In some cases, yield increases have been reported, whereas in other instances, yields were not impacted or results were mixed. One conclusion from this review was that many of the nonresponsive reports cited by Chien et al. (2014) occurred on soils with medium to high soil test P where the probability of P response would be unlikely. For example, McGrath and Binford (2012) found no response to AVAIL in corn, but all of their sites had moderately high to very high soil test P and responded to starter P at only 2 of 8 sites. Karamanos and Puurveen (2011) also observed no response at two field sites with wheat grown in slightly acidic soil, one site with low and the other with high soil test P. Although Chien et al. (2014) conducted a meta-analysis on many of these studies, the fact that most were done on high P testing soils makes the analysis suspect. A similar analysis for MAP trials on high P testing soils would lead to a conclusion that MAP is not needed, when in fact it is an effective product when used on soils in need. Hopkins (2013) states that AVAIL was effective only when the rate of P was reduced. In other words, if a plant already has adequate P due to high rates of P fertilizer or residual P in the soil, AVAIL or any other P fertilizer enhancements will not likely provide any benefit.

Another approach to enhancing P efficiency is to increase solubility. Phosphorus bound to the organic acids in products such as manure, compost, biosolids, or other waste materials has been shown to increase P solubility and plant uptake dramatically when high rates are used (Bradley and Sieling, 1953; Holford and Mattingly, 1975; Nagarajah et al., 1970). This effect can last for decades and is observed commonly in soils that have a history of heavy manure or biosolid applications (Sharpley et al., 2003). There have been many efforts to harness the increase in P solubility when applied in combination with organic acids, but without having to apply the massive quantities of manures or other biosolids. This effect is accomplished potentially by adding concentrated humic, fulvic, or other organic acid additives directly with P fertilizers. Andrade et al. (2007) stated that this practice may improve PUE through a prolonged increase in P solubility. Doing so theoretically promotes the bioavailability of P without the drawbacks listed earlier. However, the sale of humic substances, unlike fertilizer sales, largely is unregulated, and products may not be reliable. Thus, buyers should work with products that are from reliable companies who can provide independent research confirmation. Plants might not benefit from additional application of organic substances in soils that are naturally high in humic substances.

Plants deficient in P have been shown to upregulate root exudation of organic acids into the soil (Grierson, 1992; Zhang et al., 1997), and it is well documented that various organic acids help to mobilize poorly soluble mineral nutrients with citrate, malate, and oxalate, the most common and effective at mobilizing P (Hoffland, 1992; Oburger et al., 2009). Their ability to reduce P precipitation (Grossl and Inskeep, 1991) and even to improve solubility of poorly soluble phosphates (Singh and Amberger, 1998) is potentially valuable in meeting plant P demands. Tan (2003) and Hill et al. (2014a,b) reviewed the potential impact of organic acids on P nutrition.

In the case of potato grown in calcareous soil, Hopkins and Stark (2003) reported that humic acid use increased plant P uptake, resulting in increased tuber quality and yield. More recent developments with use of organic acids combined with P fertilizer have been reported with a unique P fertilizer, Carbond® P (Land View Inc., Rupert, Idaho) (Hill et al., 2014a,b; Summerhays et al., 2014) for (Hopkins unpublished data). The P in this product is bonded chemically with organic acids, which is in contrast to the simple mixing of P fertilizer with an organic acid product prior to application as was reported earlier for the work of Hopkins and Stark (2003), and as is a relatively common practice in the western United States. This work is focused on low-OM soils, but recent research (Hopkins unpublished data) found similar results for moderate-OM soil, but an apparent diminished effect when soil OM was high and also found that the effect of Carbond P is not likely due to plant physiological impacts, but suggested that it is more likely related to impacts of the organic acids on soil P chemistry.

3.10.3 PHOSPHORUS RATE

Choosing the correct rate of fertilizer P to apply has already been eluded to in the soil-testing section. Getting the right rate is a difficult proposition though. Thousands of rate studies conducted on many crops grown in a variety of soils show that the optimum rate is only somewhat predictable. There are many parameters that are integrated by the plant–soil system.

Residual soil P is an important factor impacting rate, as has been discussed previously. However, another factor is yield potential. Some soil and environmental conditions drastically limit yields. Assuming that this yield limitation is not related to P uptake, such as poor soil fertility or root or vascular system diseases, it is likely that the optimum P rate is relatively low. Evaluation of the soil system and the history of the field, including yield history, can be used to help predict rate adjustments. For example, Stark et al. (2004) provided a base P recommendation with an adjustment upward in rate of fertilizer for each increment in yield potential. A similar approach is taken in other circumstances.

Environmental conditions and many pest-related impacts on yield cannot be predicted easily. In the case of N, in-season adjustments can be made easily to cut back or add to the forecasted amount needed for the whole season. This adjustment is not as efficiently done for P due to the lack of mobility previously discussed. However, Hopkins et al. (2010a,b) found that for potato, an in-season adjustment can be made by applying P through the irrigation system if petiole tissue sampling indicates a need. Of course, a similar approach could be done by applying dry fertilizer via airplane or field spreader for nonirrigated fields, but the cost and efficiency of uptake are problematic. Furthermore, preplant-applied fertilizer obviously cannot be removed in situations where the yields are being limited due to some unforeseen problem. Although potato is somewhat efficient for in-season fertilizer use due to a prolific amount of surface-feeding roots once the canopy closes (Westermann, 2005), other crops are not as efficient. Alfalfa seems to be responsive to P fertilization on an established crop, but corn is not as efficient in its in-season uptake. Every effort should be employed to apply the correct rate of fertilizer P to plants preplant and incorporated into the soil based on soil test. Additional in-season applications should be applied if tissue analysis indicates a need and if the crop has been shown to be responsive. However, it is common that the costs for these in-season applications are higher than soil-incorporated applications and the uptake efficiency is less.

Although soil testing is a valuable tool, it is not a perfect predictor of P fertilizer need. This case is particularly problematic with potato and other inefficient P responders (Bishop et al., 1967; Freeman et al., 1998; Johnston et al., 1986). Some researchers have found soil testing to be correlated highly to plant response (Bishop et al., 1967; Giroux et al., 1984; Stark et al., 2004; Westermann, 2005), but in other studies, the results were less conclusive for tuber yield and quality responses at high soil test levels (Bishop et al., 1967; Kelling et al., 1992; Liegel et al., 1981; Nelson and Hawkins, 1947; Sanderson et al., 2003). Regardless, potato responds to fertilizer P at soil test levels higher than what is sufficient for most agronomic crops (Bundy et al., 2005). Rosen et al. (2014) stated that, due to these results and the high value of the potato crop, a method recommended by many states and provinces is to apply 50 to 100% of the predicted P removal rate (Table 3.2); even then, soil test P levels are high. This approach reduces the risk of yield loss without significantly depleting the bank of soil P and also not building it to even higher levels. However, in some cases, environmental regulations or guidelines prohibit this approach. It should be noted that fertilizing based on removal rate alone is not advisable because the rate needed is generally greater than removal rate when soil test levels are low. This approach can be used for other crops that are known to be more efficient in P response, although responses are less likely and the economics of this approach may not be justified for crops with lower value than potato.

Rosen et al. (2014) provided an excellent review of the confounding information available for P rates and stated that economical P rates for potato are clearly well above those required for most other crops. Long-term studies with corn and soybean in Iowa and Minnesota showed that applying P at crop removal rates when soil test P were in the medium range (16–20 mg kg^{-1} Bray P1) achieved maximum economic yields (Mallarino et al., 1991; Randall et al., 1997; Webb et al., 1992). These rates are one-sixth to one-half those required by potato and other P-inefficient crops. Most other agronomic crops are similar to corn and soybean, with typical rates near removal amounts when soil test levels are moderate.

3.10.4 Phosphorus Fertilizer Timing and Placement

Several timing and placement choices exist, including preplant broadcast either left on the surface (no or minimum tillage system) or incorporated into soil, concentrated bands applied with or near the seed, concentrated bands either applied during the season to the surface or injected between rows, in-season broadcast or applied with irrigation water, and small liquid volume foliar sprays. Each of these has pros and cons discussed below in this section.

With N fertilization, it is a good management practice to apply P in-season through slow- or controlled-release sources, with irrigation water, or as foliar or dry broadcast soil applications (Hopkins et al., 2008, 2014; Stark et al., 2004). Loss mechanisms for N result in leaching of NO_3^- and gaseous losses of NH_3 and nitrous oxide (N_2O) via volatilization and denitrification. However, the chemistry of P is very different from N, having none of these loss mechanisms. This principle is not well understood and, as such, it is common lore for growers to assume that the same constraints that hold for N will not hold for P.

Also, there is a common thought process that the cascading loss of P solubility over time is a reason to encourage in-season applications of P or recommend application directly to leaves to avoid soil interactions. Although these seem logical, they are flawed in practice due to the lack of mobility of P through soil and the inefficiency of foliar applications. Although the loss of P solubility with time is a real effect, the reality is that there is not much of a difference when comparing availability from a preplant with an in-season application a few weeks later (Lindsay, 2001; Sposito, 2008). Even when comparing a fall versus spring application, the difference in P availability is not tremendous. For instance, Stark et al. (2014) reported no significant difference between fall- and spring-applied P fertilizers for "Russet Burbank" potato. In the case of lettuce, waiting to apply P in-season resulted in crop losses compared with applying ample P preplant and without interruptions in supply (Burns, 1987; Sanchez et al., 1990). Similar findings were made for muskmelon (*Cucumis melo* L.) and sugar beet (Grunes et al., 1958; Lingle and Wright, 1964). Young et al. (1985) stated that timing of P application is not a critical issue in P management, as long as adequate P is available throughout the season.

As discussed previously, an in-season application of P is inherently inefficient because, unlike N and many other nutrients, P is not mobile in the soil (Lindsay, 2001; Sposito, 2008), and therefore applied P may remain in the surface layer where it is poorly available to plant roots. Broadcast and irrigated in-season applications may result in P deposition in the top few mm of soil where root biomass may be low and soil is dry. High concentration of P in surface soil is also an environmental concern because the primary P loss mechanism from soil is erosional transport into surface water (Sharpley et al., 2003), and the nearer to the surface that P is fixed, the greater the chance of erosion. The surface deposition problem possibly could be overcome with in-season P applications applied as a band knifed into the soil, but the damage from root pruning could offset the benefit of applying P fertilizer to growing plants as compared with a preplant fertilization if the knife application enters the root zone (Fallah, 1979; Stark et al., 2003; Stark and Westermann, 2008).

Despite the fact that in-season application of P is less efficient than when P is incorporated into the soil prior to or soon after planting, this practice is not completely ineffective and is sometimes necessary (MacKay et al., 1988; Westermann, 1984; Westermann and Kleinkopf, 1985). Rosen and Bierman (2008) showed a significant increase in potato P uptake and yield for split application as compared with an untreated control when the in-season application was incorporated into the soil through the cultivation and hilling process immediately prior to emergence. A later application via this method would likely have resulted in root damage, but root development into the affected area was not significant early in the season. Kelling and Speth (1997) found that in-season application of P was generally as effective as preseason application if also incorporated into the soil and if root pruning was not significant.

However, in-season applications that do not include placement into the soil are relatively inefficient (Lucas and Vittum, 1976; Randall and Hoeft, 1988). It makes theoretical sense to apply all of the anticipated P fertilizer required prior to planting in the rooting zone since timing is not a major factor. Although some species, especially perennials such as alfalfa and grasses, are adept at P uptake through roots close to the soil surface, many other species are very poor at P uptake from surface soil when P fertilizer applications concentrate it in this zone. In the case of potato, it has been shown that midseason P applications can be effective (Stark et al., 2003; Stark and Westermann, 2008; Westermann, 2005), likely due to an upright canopy architecture (high percentage of water and P and other solutes follow the stems to be deposited at the base of the plant) and an abundance of surface-feeding roots after the canopy closes and completely covers the soil. However, these in-season P applications are not as effective as when preplant P is mixed in the soil and in better contact with plant roots (Anghinoni et al., 1980a,b; Hopkins et al., 2010a,b; Sleight et al., 1984).

The rate of P fertilization was found to be 50% for banded versus broadcast applications to vegetable crops (Sanchez et al., 1990, 1991). The efficiency of banding versus broadcast is much greater at low versus high soil P test results, with about a threefold increase at low soil P, but approaching equivalent status at high. Similar findings were made for corn, winter wheat, and other agronomic crops (Barber, 1958; Peterson et al., 1981; Welch et al., 1966).

In-season P application probably should be viewed as a means of last resort or rescue and used only when tissue analysis indicates a P deficiency (Westermann and Kleinkopf, 1985). Rosen et al. (2014) suggested that in-season applications should only be supplementary to soil-incorporated P applications and only if tissue analysis shows a need. Hopkins et al. (2010a) showed that preplant P fertilization resulted in significant improvements in yield. Although there were trends for yield increase, the in-season and the split (50% preplant and 50% in-season) applications did not result in significant increases over the unfertilized control. Further work showed that, although incorporation into the soil is the best option, "rescue" in-season P applications have some merit with potato when P was underapplied prior to planting (Hopkins et al., 2010a,b,c; Westermann, 2005).

Hopkins et al. (2010b) found that in-season P application gave a slight, consistent U.S. No. 1 yield increase at all preplant P levels in the study (0, 112, 224, and 336 kg P_2O_5 ha^{-1}). The response to preplant P increased steadily with rate increase, and the in-season application resulted in further increases in yield, even at the highest P rate. A similar response occurred for total yield, although the response to preplant P leveled off at the first rate of applied P. Horneck found similar results in separate trials with longer season conditions and, thus, higher yields in Oregon (Hopkins et al., 2014). Westermann (1984) found that P uptake and yields generally increased with supplemental P fertigation, although the results were mixed.

Although movement of soil P is very minimal in most cases, soluble fertilizer P can move long distances in soil, especially in soils with rapid macropore flow. Phosphorus moved to a depth of 18 cm through a loamy sand (Hergert and Reuss, 1976) and to 45 cm through a sand receiving a very high rate of irrigation (Stanberry et al., 1955). Differences in P source movement through application of P with irrigation water have been reported, with monocalcium phosphate, MAP, urea phosphate, and phosphoric acid moving downward more effectively than APP or di- and tricalcium phosphates (Lauer, 1988; O'Neill et al., 1979; Stanberry et al., 1965). Bar-Yosef et al. (1995) found no differences between broadcast and drip-injected P for sweet corn grown on sand. Carrijo and Hochmuth (2000) found that P applied in irrigation water was more effective than preplant incorporation for tomato. Instances where irrigation-applied P is more effective than broadcast likely are related to the placement of P in a concentrated form in or near to the root zone (Carrijo and Hochmuth, 2000; Mikklelsen, 1989). Application of P with irrigation can be effective if water movement through soil is adequate and the proper P source is used and placed near or in the root zone.

Direct foliar application of P has also been studied in potato. Laughlin (1962) found that 18 foliar sprays of just 11 kg P_2O_5 ha^{-1} gave a significant yield increase in potato, but not when combined with soil application of 404 kg P_2O_5 ha^{-1}. However, the soil application resulted in a significantly greater yield than foliar applications alone. Other studies showed no benefit of foliar-applied P (Allison et al., 2001; Rosen et al., 2014). Although Hiller and Koller (1987) found general responses to foliar nutrition, there was no response to foliar P nutrition at any of three field locations tested.

Foliar applications have been studied in a wide variety of other crops. In general, a limited amount of P can be delivered in this fashion but not enough to meet high demand, and in many cases, there are no or there are negative responses. Teubner et al. (1962) found that multiple foliar sprays resulted in P absorption in harvested plant parts at 12% of the total need, but that yields were unaffected and total P in the plants was not increased. Upadhyay et al. (1988) found that P fertilization applied as a foliar spray was much less effective than if all P was supplied as an incorporated soil application. Silberstein and Wittwer (1951) evaluated organic and inorganic P foliar sprays on vegetable crops, finding that orthophosphoric acid was the most effective source, but the responses were very minimal and some compounds resulted in toxicity with P concentrations as low as 0.16%. However, Barel and Black (1979a,b) found that several polyphosphate and some other phosphate fertilizers could be applied at rates up to threefold greater than orthophosphate without causing leaf toxicity and that yields of corn and soybean were higher with tri- and tetrapolyphosphate than with orthophosphate. In many cases, foliar application of P along with other nutrients (N, K, S, etc.) often resulted in maturity delays and no or negative yield responses (Batten and Wardlaw, 1987; Garcia and Hanway, 1976; Harder et al., 1982a,b; Parker and Boswell, 1980; Robertson et al., 1977). In summary, timing is not a critical factor for P fertilizer application, but incorporation into the soil is relatively more efficient than canopy ground or surface applications.

Nutrient placement can increase PUE (Kissel et al., 1985; Stevens et al., 2007). Fertilization can impact P availability through at least two avenues. First, there are more microsites with readily soluble adsorbed or precipitated P. Each site increases the likelihood of a root encounter and uptake. Broadcast fertilization greatly impacts this means of P supply to plants. The other avenue is through an increase in soil solution equilibrium P level (Kissel et al., 1985; Lindsay, 2001; Sposito, 2008). A concentrated fertilizer band or point injection greatly amplifies these effects in a small zone in the soil, providing a highly soluble pool for plant uptake. Hopkins et al. (2014) showed that there is about a 60-fold increase in the bioavailable P in the center of a fertilizer band compared with when the same amount is broadcast in the bulk soil. This increase is temporary but allows plant roots to bathe in soluble P, particularly during the critical early-season growth period. Kovar and Barber (1987) used modeling to show PUE will likely increase if P banding contacts about 5% of the soil volume, especially with high P-fixing soils low in soil test P (Kovar and Barber, 1987).

For maximum effect, the fertilizer needs to be placed in an area where roots are likely to be congregated. For corn and most other species, placement generally is recommended at 5 cm to the side and 5 cm down from the seed for interception by early roots, which tend to grow diagonally for most species. Potato is similar, although placement is generally slightly further away at 7–8 cm, with a wider range of acceptable depth ranging from 7–8 above or below the seed piece (Hopkins and Stephens, 2008, 2014). Placement too far from the main root system results in little or no P uptake, especially for species with small root systems (Lesczynski and Tanner, 1976; Opena and Porter, 1999). Moorby (1978) found no P uptake from labeled fertilizer applied in the adjacent furrow or beyond for potato. Hammes (1961, 1962) found little P uptake for banded fertilizer applied below 30 cm and that the most efficient uptake occurred 5 cm to the side of the seed piece. For sugar beet, placement should be directly below the seed in order to intercept the taproot dominant in the first few weeks of growth (Hopkins and Ellsworth, 2006; Stevens et al., 2007).

It is important to understand root morphology and architecture of individual species in order to most effectively apply a concentrated fertilizer band. Usually, these concentrated fertilizer bands are applied at planting. However, in some cases, the application is applied preplant. Preplant application is especially common for potato, with the P often applied when rows are formed. In this case, it is essential that the concentrated band be placed to the side of the seed piece and deeper than the planting depth to avoid disruption of the band at planting when the soil is disturbed. It is crucial that the concentrated band of P remains intact to realize the benefit of increased P solubility (Hopkins and Stephens, 2008, 2014).

Appropriately placed fertilizer bands increase P uptake efficiency to 25%–35% (first year recovery) compared with 1%–10% if the P is broadcast applied (Hopkins et al., 2014; Kissel et al., 1985; Mattingly and Widdowson, 1958; Randall et al., 1985; Syers et al., 2008). Using radioactively labeled P, Baerug and Steenberg (1971) showed a doubling of P recovery from a concentrated band (5 cm to the side and 2 cm down from the seed) compared with a broadcast application. Although not always a replacement for broadcast fertilizer P, adding P to soil in a concentrated band often results in additional increases in potato tuber yield and quality over a single broadcast application (Hawkins, 1954; Jackson and Carter, 1976; Kelling and Speth, 1997; Kingston and Jones, 1980; Liegel et al., 1981; Soltanpour, 1969b; Sparrow et al., 1992). Banding P increases P uptake, especially for early-season growth when P availability is most limiting due to low soil temperatures and a poorly developed root system (Marschner, 2012). Hopkins et al. (2014) stated that these concentrated bands often result in increased rates of early-season shoot and root growths and higher concentrations of potato petiole P, with the consequence being gains in yield and quality. However, early-season growth boosts due to concentrated bands do not always equate to end of season yield increases, as plants can sometimes "catch-up" if the conditions and length of growing season are optimum.

Recently reported research results show an additive response if banded fertilizer P was applied in conjunction with broadcast-incorporated P for potato grown in calcareous soil (2%–12% $CaCO_3$) with Olsen bicarbonate extractable P of 8–18 mg kg^{-1} (Hopkins et al., 2014). In moderately high testing soils, such as those that have received heavy manure applications over time, plants may respond to a band application even when the soil test recommends no additional P applications (Stark et al., 2004). The effectiveness of banded P for potato has been shown to vary with P source in calcareous soil (Stark and Ojala, 1989), with the pH of the fertilizer solution being a key factor. Banding also has been beneficial in low pH soils by concentrating P near the early developing root system (Rosen and Eliason, 2005).

Despite the benefits of applying P in a concentrated band, all plant roots require adequate P throughout the entire rooting zone. Although P is mobile in plants, it may not be translocated efficiently from one distant root to another. This inefficiency is because the P would have to be transported to the shoots and then back to the root with photosynthates; consequently, it is best to apply both broadcast and banded P to soils with low to medium soil test levels (Stark et al., 2004). When soil test values are high, it is generally not recommended to apply a broadcast P (Stark et al.,

2004). However, there are reported incidents of responses to banded P in soils with high residual P (Hopkins and Ellsworth, 2006; Hopkins and Stephens, 2008, 2014; Rosen et al., 2014). Liegel et al. (1981) found broadcasting to be as effective as banding on two sand soils, although they did find an advantage for band application on a more coarse textured soil. Rosen et al. (2014) stated that it is essential that all of the P be banded on soils that have a high potential for P fixation.

It should be noted, however, that too much of a good thing can be bad. Hegney and McPharlin (1999) found negative results when banded P was applied in direct contact with potato seed pieces. Other species show toxicity if high rates are applied in direct contact with seeds. Plants need salts in order to regulate water uptake, and all nutrients are found in salt form. However, excessive salts desiccate plant tissues if the soil osmotic potential becomes extremely negative, particularly for germinating seeds and seedlings. Fertilizer can be applied in direct seed contact as long as the rate is not too high. Orthophosphate is a salt component, but when it is applied as a fertilizer its salt effect is minimal because the majority is quickly precipitated into solid forms. As such, its direct impact on salt concentration is less than more soluble nutrients, such as N and K. Thus, P can be directly applied to seed relatively more safely than other nutrients, although accompanying cations are often soluble salts. To be safe, no fertilizer should be applied in direct seed contact without research showing that the rate applied is acceptable for the species and soil and environmental conditions. Note that because salt damage is a function of soil moisture status, dry soil conditions are relatively more likely to result in salt damage to plants. Furthermore, small seeded species tend to be more readily impacted by salts in close proximity to the seed or seedling than species with large seeds.

3.10.5 Best Management Practices Impact Fertilization

As P is poorly soluble and immobile in the soil, any factor that increases root growth should expose the plant to more P for absorption. Miller and Hopkins (2007) and Hopkins et al. (2007a) reviewed best management practices (BMPs). In general, these BMPs can be applied across crops and conditions, but specific circumstances call for specific practices. One BMP that is especially important is to avoid root damage due to tillage, insect and pathogen damage, herbicides, salt, or other toxicities to promote a healthier root system, which can greatly enhance P uptake.

In summary, P nutrition in plants has been a major contributor to societal success in the last two centuries. It is vital to understand its essential role in plants and their difficulty in obtaining it from soil, with wide differences observed across species and growing conditions. Tissue and soil analysis tools greatly help in guiding the management of P nutrition in plants. As a society, we need to follow BMPs for the growing of plants and P fertilization to sustain crop productivity simultaneously with improving environmental quality and resource conservation. Ample, but not excessive availability of water and nutrients is also important.

3.10.6 Environmental Issues

Despite the many positive roles of P, not long after its widespread use, it became known that it is an environmental contaminant (Ruark et al., 2014). In the United States, the Environmental Protection Agency (EPA) has declared that more than one-third of all water bodies are impaired with P pollution. The problem is related to water quality from point and nonpoint sources (Romkens et al., 1973; Ryden et al., 1973). Point source pollution occurs primarily from P-rich municipal and industrial wastes being dumped into surface water bodies. Even when treated to remove pathogens and other hazards, P typically has not been removed. However, even if all of the P was removed from point sources, nonpoint source contributions are massive and very difficult to identify and treat.

The problem with P entering surface water is related to nutrient enrichment (Ruark et al., 2014; Sharpley et al., 1999). Algae and other aquatic organisms are like land plants, needing P and other nutrients. In most freshwater systems, P is the primary limiting factor for algal growth. When these simple organisms are fertilized with nutrient-rich pollution, their yields increase similarly to what happens with more complex land plants when the most limiting factor for growth is overcome (Correll, 1998; Schinder, 1977). Therefore, when adequate sunlight and heat are present, algal blooms occur. By themselves, the population of algae can be unsightly and odoriferous and can interfere with recreation and aquatic-based industries. However, the most serious problem occurs upon the death of the algae. The microbes responsible for decomposing the algae also require nutrients, including O_2. In some cases, their population exceeds the carrying capacity of the water body as they deplete it of O_2 and create a hypoxic condition (Correll, 1998; Daniel et al., 1998). This condition can result in the broadscale death of fish and other aquatic organisms, which also require O_2, again causing unsightly and odoriferous problems, as well as negative impacts on fishing and other related industries and recreational activities. In addition to algal growth, eutrophication can result in cyanobacterial blooms as well and can result in poor palatability of the water for drinking and livestock and human health hazards (Kotak et al., 1993; Lawton and Codd, 1991). There are many reviews examining losses due to P pollution (Buczko and Kuchenbuch, 2007; Chien et al., 2011; McDowell et al., 2001; Ryden et al., 1973; Sharpley et al., 1993; Shi et al., 2011; Withers and Jarvie, 2008).

This increasing problem in recent years has prompted regulations and guidelines regarding P levels in surface waters and for practices to reduce P loading of water bodies. Ruark et al. (2014) stated that no official U.S. standard has been set for P loading to freshwaters; however, USEPA established the criterion of 0.001 mg total P L^{-1} for marine and estuary water (Parry, 1998). The state of Florida has adopted this same guideline for freshwater systems (Daniel et al., 1998). Other states have established a critical maximum for P, such as 0.05 mg total P L^{-1} in streams that enter lakes and 0.1 mg L^{-1} for total P in flowing waters (Ruark et al., 2014). Although the total P losses from agricultural fields are generally small compared with the total P in the soil, concentrations as low as 0.02 mg P L^{-1} can cause eutrophication (Correll, 1998; Sawyer, 1947; Sharpley et al., 1999; USEPA, 1996).

Point sources are easily identified and cleaned of P using the solubility principles known for soils to precipitate out the P for removal, although costs of doing so can be prohibitive. Nonpoint sources of P are much more difficult to identify and prevent. There are natural systems, as well as a variety of anthropogenic P sources, that result in P loading. However, agriculture is identified as the largest contributor of nonpoint source P loading into water bodies.

Phosphorus-enriched soils from application of traditional and, mostly, manure P fertilizers are the major source of the nonpoint problems. Organic P is much more mobile in soil than inorganic sources (Hannapel et al., 1964). The typical scenario is that crops are harvested from a wide-ranging geographical area and transported to concentrated animal operations. These feedstuffs are fed to the animals, with their wastes accumulating. These manure wastes generally are applied back to the land but are not transported back to where they came due to high transportation costs, which society, to this point, is unwilling to pay as a part of the cost of the food and fuel consumed. So, instead, the land immediately surrounding these concentrated animal facilities tends to become oversupplied with nutrient wastes.

Phosphorus moves from land to water via various mechanisms (Sharpley et al., 1994). Although not common, P concentration can become so high in soil that the equilibrium concentrations are atypically high. This action results in movement through the soil where P can enter subsurface water systems and drains (Brye et al., 2002; Eghball et al., 1996; Hansen et al., 1999; Kleinman et al., 2003; Mozaffari and Sims, 1994). Ruark et al. (2014) stated that while surface runoff is historically considered the dominant pathway by which P is transported from land to water, subsurface flow can be an important pathway in certain landscapes (e.g., sandy soil; well-structured, fine-textured soils with well-defined macropores; and porous organic soils with low P-sorption capacities). They also stated that subsurface movement of P is dominated by preferential flow via macropores exacerbated by the presence of artificial drainage such as tile drainage.

However, most of the P transport is a surface phenomenon. P tends to accumulate at the soil surface and, as such, precipitation events that result in overland flow of water tend to pick up soluble P, which is transported to surface water. The quantity of P transported as dissolved P is a function of desorption–sorption and dissolution–precipitation reactions that cause P to exist in soil solution

(Sharpley et al., 1993, 1994). Ruark et al. (2014) stated that transport of dissolved P movement is a function of the factors that cause surface runoff to occur (e.g., slope, surface roughness, and residue cover).

Another mechanism of P movement to surface water is due to transport of P adsorbed to soil and mineral solids via wind and water erosions. Ruark et al. (2014) cited several sources that show an average of 86% attributed to particulate P movement. Researchers found during year-round monitoring of fields in Wisconsin that, whereas the median annual dissolved and sediment-bound P losses were similar, the maximum annual particulate P loss was five times greater than maximum dissolved P loss, due to erosion with spring precipitation (Good et al., 2012).

Areas that tend to have P-enriched water bodies have an abundance of animals per unit land area along with high precipitation rates. Steep slopes and nonvegetated soils are more prone to offsite P movement through water and soil transport. Proximity to water is also a factor, with manure or fertilizer P applied to a soil close to a water body much more likely to be deposited in water than if applied further away.

Practices to control P losses target the reduction of P available for loss (source management) and reduction of movement of P to a water body (transport management) (Ruark et al., 2014; Sharpley et al., 1994). Ruark et al. (2014) stated that concerns relative to source management include (1) soil test P levels, (2) rate and manner of P applied, and (3) rate and implementation of BMPs (Daniel et al., 1998; Eghball et al., 2000; Ginting et al., 1998; Sharpley et al., 1994). The most obvious strategy is to avoid the excessive application of manures and fertilizers. Many areas have passed laws for farmers and even homeowners with regard to manure and fertilizer applications. The Natural Resource Conservation Service (NRCS) has offered various financial incentives to growers to implement strategies to prevent P pollution. One example of how this action has impacted agriculture is that there has been an influx of dairy operations into western states of the United States where soils have a high capacity for P fixation, precipitation is low, and water bodies tend to be further away from farm fields. For example, dairy and cattle productions recently have surpassed potato as the main source of agricultural income in Idaho. Moving operations to other locales is more financially agreeable than is paying to transport manure long distances, although transport is another strategy that can be employed to avoid the accumulation of P in soils.

Several studies show a positive correlation between concentrations of P in soil tests and in runoff water (Andraski and Bundy, 2003; Cox and Hendricks, 2000; Pote et al., 1996, 1999; Sims, 1998). Andraski and Bundy (2003) concluded that traditional soil P tests are effective for predicting the risk of P loss, although the relationship is not perfect (Cox and Hendricks, 2000; Daniel et al., 1993; Hart et al., 2004; Sharpley, 1995). There has been a trend for increasing levels of soil P in some locales due to P applications in excess of crop need (Bundy and Sturgul, 2001; Sharpley et al., 1994, 2001). A survey of 1928 farms in Wisconsin revealed that 80% applied excessive P to corn (Shepard, 2000). However, Bundy and Sturgul (2001) stated that, although excessive application continues, the trend is in decline. The Bray P_1 soil test concentrations increased from 34 to 51 mg kg^{-1} between about 1970 through the late 1990s, but then stabilized. The International Plant Nutrition Institute (IPNI, 2011) also reflects this trend, with some locales decreasing in soil test P. However, soil P is especially an issue in locales with an abundance of farm manure (Ginting et al., 1998; Shepard, 2000; Sims, 1998).

Ruark et al. (2014) stated that having a high test for P in soil is not enough alone to cause eutrophication of nearby water bodies. Conditions must exist where P is transported easily, especially steep slope and close proximity to water (Sharpley et al., 1992; Sims et al., 2000). As particulate P is the main source of contamination of waters, efforts to control soil erosion are the primary focus [e.g., soil type, slope, distance to surface water, crop management, conservation practices, and intensity, timing, and duration of rainfall (Gburek et al., 2000; Hart et al., 2004; Hudson, 1995; Kimmell et al., 2001; McDowell et al., 2001; Sharpley et al., 2001)].

There have been significant efforts in the United States to develop P indexes to help prevent P pollution (Sharpley et al., 2003). The P indexes tend to factor in the aforementioned soil-loss risk factors, along with bioavailable P concentration in soil. It is recommended that soils in close proximity to water, especially those that are on steep slopes, should not receive manure applications and that P fertilizer should be applied carefully and judiciously. Soils that have low soil test P can accept manure or fertilizer P applications without much risk of loss, especially if incorporation takes place. Keeping a soil vegetated greatly decreases the risk of P transport as soil erosion is much more likely to occur when soil is void of plant growth. Similarly, vegetated buffer strips between farm fields and water bodies can be used to capture soil and P in water runoff. Incorporating P into the soil can reduce the surface concentration of P, but there is a temporary increase in risk due to exposure of bare soil to the forces of erosion.

Ruark et al. (2014) stated that fields with the highest risk for P loss are those with both source and transport factors, that is, those with high P additions or soil test P and are coincident with high relative transport risk (surface runoff, erosion, and/or subsurface flow). If a site has high soil test P or high amounts of nutrient addition from manure or fertilizer but is not located near a lake or stream, the risk for P loss to water is much less. Likewise, if a field is located next to a stream but has low levels of soil test P, risk for P loss is also low.

Fortunately, there seem to be reasonable solutions to avoid problems of P pollution in most circumstances (Daniel et al., 1998; Sharpley et al., 1994, 2001; Sims et al., 2000). For most crops, the level of soil test P that is considered optimum for production and for which no added P is needed is well below that of where there is a high risk of P transport. The exception to this rule is our unique potato species. The problem has been discussed previously and is reviewed thoroughly by Ruark et al. (2014), who give an excellent review of P pollution and the unique problems with potato. Essentially, the issue is that potato, unlike most all other species, continues to respond to P fertilization at high and even exceptionally high soil test levels. In some cases, potato grown in high-rainfall areas with soil test levels that are very high may still require P applications. Furthermore, typical potato production results in the soil remaining bare for relatively longer periods of time than for most other crops. The soil receives deep tillage in the fall or early spring in order to create friable soil conducive to tuber growth, and then the planting, hilling, and harvesting operations all involve near complete turning over of the soil. These four operations result in bare soil exposure that, along with very slow early-season growth, give extremely high susceptibility to P transport by wind and water erosion of soil. However, Ruark et al. (2014) also stated that there is minimal research on P pollution from potato fields and that the risks may not be as great as suggested here. The bottom line is that potato production in high-rainfall areas in close proximity to water bodies is very problematic with regard to P pollution risk.

REFERENCES

Abbott, J.L. and T.C. Tucker. 1973. Persistence of manure phosphorus availability in calcareous soil. *Soil Sci. Soc. Am. J.* 37:60–63.

Alexander, M. 1977. *Introduction to Soil Microbiology*. New York: John Wiley & Sons.

Allison, M.F., J.H. Fowler, and E.J. Allen. 2001. Effects of soil- and foliar-applied phosphorus fertilizers on the potato (*Solanum tuberosum* L.) crop. *J. Agric. Sci. Camb.* 137:379–395.

Alt, D. 1987. Influence of P- and K-fertilization on yield of different vegetable species. *J. Plant Nutr.* 10:1429–1435.

Alva, A.K. 1992. Differential leaching of nutrients from soluble vs. controlled-release fertilizers. *Environ. Manag.* 16:769–776.

Alvarez-Tinault, M.C., A. Leal, and L. Recalde-Martinez. 1980. *Iron-Manganese Interaction and Its Relation to Boron Levels in Tomato Plants*. Granada, Spain: Zaidin Experimental Station (C.S.I.C.).

Anderson, G. 1967. Nucleic acids, derivatives, and organic phosphorus. In *Soil Biochemistry*, Volume 1, eds. A.D. McLaren and G.H. Peterson, pp. 67–90. New York: Marcel Dekker, Inc.

Anderson, G. 1975. Other organic phosphorus compounds. In *Soil Components*, Volume 1. Organic Components, ed. J.E. Gieseking, pp. 305–331. New York: Springer-Verlag.

Andrade, F.V., E.S. Mendonca, I.R. Silva, and R.F. Mateus. 2007. Dry-matter production and phosphorus accumulation by maize plants in response to the addition of organic acids in oxisols. *Commun. Soil Sci. Plant Anal.* 38:2733–2745.

Andraski, T.W. and L.G. Bundy. 2003. Relationships between phosphorus levels in soil and in runoff from corn production systems. *J. Environ. Qual.* 32:310–316.

Anghinoni, I. and S.A. Barber. 1980a. Phosphorus influx and growth characteristics of corn roots as influenced by phosphorus supply. *Agron. J.* 72:685–688.

Anghinoni, I. and S.A. Barber. 1980b. Predicting the most efficient phosphorus placement for corn. *Soil Sci. Soc. Am. J.* 44:1016–1020.

Anuradha, M. and A. Narayanan. 1991. Promotion of root elongation by phosphorus deficiency. *Plant Soil* 136:273–275.

Armstrong, D.L. (ed.). 1999. Phosphorus for agriculture. *Better Crops with Plant Food* 83(1):1–39.

Asfary, A.F., A. Wild, and P.M. Harris. 1983. Growth, mineral nutrition, and water use by potato crops. *J. Agric. Sci. Camb.* 100:87–101.

Assuero, S.G., A. Mollier, and S. Pellerin. 2004. The decrease in growth of phosphorus-deficient maize leaves is related to a lower cell production. *Plant Cell Environ.* 27:887–895.

Bacon, P.E. and B.G. Davey. 1982. Nutrient availability under trickle irrigation: Distribution of water and Bray No 1 phosphate. *Soil Sci. Soc. Am. J.* 46:981–987.

Baerug, R. and K. Steenberg. 1971. Influence of placement method and water supply on the uptake of phosphorus by early potatoes. *Potato Res.* 14:282–291.

Baker, D.E., A.E. Jarrell, L.E. Marshall, and W.I. Thomas. 1970. Phosphorus uptake from soils by corn hybrids selected for high and low phosphorus accumulation. *Agron. J.* 62:103–106.

Bar-Yosef, B., B. Sagiv, T. Markovitch, and I. Levkovitch. 1955. Phosphorus placement effects on sweet corn growth, uptake, and yield. In *Proceedings of Dahlia Greidinger International Symposium on Fertigation.* Haifa, Israel, pp. 141–154.

Barben, S.A., B.G. Hopkins, V.D. Jolley, B.L. Webb, and B.A. Nichols. 2010a. Optimizing phosphorus and zinc concentrations in hydroponic chelator-buffered nutrient solution for Russet Burbank potato. *J. Plant Nutr.* 33:557–570.

Barben, S.A., B.G. Hopkins, V.D. Jolley, B.L. Webb, and B.A. Nichols. 2010b. Phosphorus and manganese interactions and their relationships with zinc in chelator-buffered solution grown Russet Burbank potato. *J. Plant Nutr.* 33:752–769.

Barben, S.A., B.G. Hopkins, V.D. Jolley, B.L. Webb, and B.A. Nichols. 2010c. Phosphorus and zinc interactions in chelator-buffered solution grown Russet Burbank potato. *J. Plant Nutr.* 33:587–601.

Barben, S.A., B.G. Hopkins, V.D. Jolley, B.L. Webb, B.A. Nichols, and E.A. Buxton. 2011. Zinc, manganese and phosphorus relationships and their effects on iron and copper in chelator-buffered solution grown Russet Burbank potato. *J. Plant Nutr.* 34:1144–1163.

Barber, S.A. 1958. Relation of fertilizer placement to nutrient uptake and crop yield. I. Interaction of row phosphorus and the soil level of phosphorus. *Agron. J.* 50:535–539.

Barber, S.A. 1977. Application of phosphate fertilizers: Methods, rates and time of application in relation to the phosphorus status of soils. *Phosphorus Agric.* 70:109–115.

Barber, S.A. 1980. Soil-plant interactions in the phosphorus nutrition of plants. In *The Role of Phosphorus in Agriculture*, eds. F.E. Khasawneh, E.C. Sample, and E.J. Kamprath, pp. 591–615. Madison, WI: American Society of Agronomy, Crop Science Society of America and Soil Science Society of America.

Barber, S.A. 1995. *Soil Nutrient Bioavailability: A Mechanistic Approach*, 2nd edition. New York: John Wiley & Sons.

Barel, D. and C.A. Black. 1979a. Foliar application of P. I. Screening of various inorganic and organic P compounds. *Agron. J.* 71:15–21.

Barel, D. and C.A. Black. 1979b. Foliar application of P. II. Yield response of corn and soybeans sprayed with various condensed phosphates and P-N compounds in greenhouse and field experiments. *Agron. J.* 71:21–24.

Barry, D.A.J. and M.H. Miller. 1989. Phosphorus nutritional requirement of maize seedlings for maximum yield. *Agron. J.* 81:95–99.

Batten, G.D. and I.F. Wardlaw. 1987. Senescence of the flag leaf and grain yield following late foliar and root applications of phosphate on plants of differing phosphate status. *J. Plant Nutr.* 10:735–748.

Beer, K., C. Durst, C. Grundler, A. Willing, and B. Witter. 1972. Effect of lime and of physiologically different N and P fertilizers on the dynamics of manganese fractions in soil, and on manganese uptake by arable crops in various locations in the German Democratic Republic, in relation to the efficiency of manganese fertilizers applied for the production of high yields. *Archiv für Acker und Pflanzenbau und Bodenkunde* 16:471–481.

Bell, L.C. and C.A. Black. 1970. Crystalline phosphates produced by interaction of orthophosphate fertilizers with slightly acid and alkaline soils. *Proc. Soil Sci. Soc. Am.* 34:735–740.

Bennett, W.F. (ed.). 1993. *Nutrient Deficiencies and Toxicities in Crop Plants*. St. Paul, MN: APS Press.

Berger, K.C., P.E. Potterton, and E.L. Hobson. 1961. Yield quality and phosphorus uptake of potatoes as influenced by placement and composition of potassium fertilizers. *Am. Potato J.* 38:272–285.

Bieleski, R.L. 1973. Phosphate pools, phosphate transport, and phosphate availability. *Annu. Rev. Plant Physiol.* 24:225–252.

Bingham, F.T. 1963. Relation between phosphorus and micronutrients in plants. *Proc. Soil Sci. Soc. Am.* 27:389–391.

Bingham, F.T. 1966. Phosphorus. In *Diagnostic Criteria for Plants and Soils*, ed. H.D. Chapman, pp. 324–361. Riverside, CA: Division of Agricultural Sciences, University of California.

Birch, J.A., J.R. Devine, M.R.J. Holmes, and J.D. Whitear. 1967. Field experiments on the fertilizer requirement of main crop potatoes. *J. Agric. Sci. Camb.* 69:13–24.

Bishop, R.F., C.R. MacEacherne, and D.C. MacKay. 1967. The relation of soil test values to fertilizer response by the potato. IV. Available phosphorus and phosphatic fertilizer requirements. *Can. J. Soil Sci.* 47:175–185.

Boawn, L.C. and G.E. Leggett. 1963. Zinc deficiency of the Russet Burbank potato. *Soil Sci.* 95:137–141.

Boawn, L.C. and G.E. Leggett. 1964. Phosphorus and zinc concentrations in Russet Burbank potato tissues in relation to development of zinc deficiency symptoms. *Proc. Soil Sci. Soc. Am.* 28:229–232.

Bould, C. and R.I. Parfitt. 1973. Leaf analysis as a guide to the nutrition of fruit crops. X. Magnesium and phosphorus sand culture experiments with apple. *J. Sci. Food Agric.* 24:175–185.

Boyd, D.A. and W. Dermott. 1967. Fertiliser requirements of potatoes in relation to kind of soil and soil analysis. *J. Sci. Food Agric.* 18:85–89.

Bradely, D.B. and D.H. Sieling. 1953. Effect of organic anions and sugars on phosphate precipitation by iron and aluminum as influenced by pH. *Soil Sci.* 76:175–179.

Bray, R.H. and L.T. Kurtz. 1945. Determination of total, organic, and available forms of phosphorus in soils. *Soil Sci.* 59:39–45.

Broadley, M., P. Brown, I. Cakmak, Z. Rengel, and F. Zhao. 2012. Function of nutrients: Micronutrients. In *Mineral Nutrition of Higher Plants*, 3rd edition, ed. P. Marschner, pp. 191–248. London, U.K.: Academic Press.

Brown, B., J. Hart, D. Horneck, and A. Moore. 2010. Nutrient management for field corn silage and grain in the inland Pacific Northwest. *PNW 615*. Moscow, ID: University of Idaho Agricultural Communications.

Brown, J.C. and L.O. Tiffin. 1962. Zinc deficiency and iron chlorosis dependent on the plant species and nutrient-element balance in Tulare clay. *Agron. J.* 56:356–358.

Brye, K.R., T.W. Andraski, W.M. Jarrell, L.G. Bundy, and J.M. Norman. 2002. Phosphorus leaching under a restored tallgrass prairie and corn agroecosystems. *J. Environ. Qual.* 31:769–781.

Buczko, U. and R.L. Kuchenbuch. 2007. Phosphorus indices as risk-assessment tools in the USA and Europe—A review. *J. Plant Nutr. Soil Sci.* 170:445–460.

Bundy, L.G. and S.J. Sturgul. 2001. A phosphorus budget for Wisconsin cropland. *J. Soil Water Conserv.* 56:243–249.

Bundy, L.G., H. Tunney, and A.D. Halvarson. 2005. Agronomic aspects of phosphorus management. In *Phosphorus: Agriculture and the Environment*, eds. J.T. Sims and A.N. Sharpley, pp. 685–727. Madison, WI: American Society of Agronomy, Crop Science Society of America, Soil Science Society of America.

Burns, I.G. 1987. Effects of interruptions in N, P, or K supply on the growth and development of lettuce. *J. Plant Nutr.* 10:1571–1578.

Buso, G.S.C. and F.A. Bliss. 1988. Variability among lettuce cultivars grown at two levels of available phosphorus. *Plant Soil* 111:67–93.

Cakmak, I. and H. Marschner. 1987. Mechanism of phosphorus-induced zinc deficiency in cotton. III. Changes in physiological availability of zinc in plants. *Physiol. Plant.* 70:13–20.

Carpenter, P.N. 1963. Mineral accumulation in potato plants as affected by fertilizer application and potato variety. *Maine Agr. Exp. Sta. Bull. 610*, Orono, MN: University of Maine.

Carrijo, O.A. and G. Hochmuth. 2000. Tomato responses to preplant incorporated or fertigated phosphorus on soils varying in Mehlich-1 extractable phosphorus. *HortScience* 35:67–72.

Chatterjee, C. and N. Khurana. 2007. Zinc stress induced changes in biochemical parameters and oil content of mustard. *Commun. Soil Sci. Plant Anal.* 38:751–761.

Chien, S.H., D. Edmeades, R. McBride, and K.L. Sahrawat. 2014. Review of maleic–itaconic acid copolymer purported as urease inhibitor and phosphorus enhancer in soils. *Agron. J.* 106:423–430.

Chien, S.H., L.I. Pronchnow, S. Tu, and C.S. Snyder. 2011. Agronomic and environmental aspects of phosphate fertilizer varying in source and solubility: An update review. *Nutr. Cycl. Agroecosys.* 89:229–255.

Childers, N.F. (ed.). 1966. *Temperate to Tropical Fruit Nutrition*. New Brunswick, NJ: Rutgers–The State University.

Christensen, N.W. 1972. A new hypothesis to explain phosphorus-induced zinc deficiencies. PhD dissertation, Corvallis, OR: Oregon State University.

Christensen, N.W. and T.L. Jackson. 1981. Potential for phosphorus toxicity in zinc-stressed corn and potato. *Soil Sci. Soc. Am. J.* 45:904–909.

Christenson, D.R. and E.C. Doll. 1968. Effect of phosphorus source and rate on potato yields and phosphorus content of petioles and tubers. *Mich. Agric. Exp. Sta. Q. Bull.* 50(4):616–624.

Chu, C-C., H. Plate, and D.L. Matthews. 1984. Fertilizer injury to potatoes as affected by fertilizer source, rate and placement. *Am. Potato J.* 61:591–597.

Clarkson, D.T., M. Carvajal, T. Henzler et al. 2000. Root hydraulic conductance: Diurnal aquaporin expression and the effects of nutrient stress. *J. Exp. Bot.* 51:61–70.

Clarkson, D.T. and J.B. Hanson. 1980. The mineral nutrition of higher plants. *Annu. Rev. Plant Physiol.* 31:239–298.

Clarkson, D.T., A.W. Robards, J.E. Stephens, and M. Stark. 1987. Suberin lamellae in the hypodermis of maize (*Zea mays*) roots; development and factors affecting the permeability of hypodermal layers. *Plant Cell Environ.* 10:83–93.

Clemens, S., M.G. Palmgren, and U. Krämer. 2002. A long way ahead: Understanding and engineering plant metal accumulation. *Trends Plant Sci.* 7:309–315.

Cogliatti, D.H. and D.T. Clarkson. 1983. Physiological changes in, and phosphate uptake by potato plants during development of, and recovery from phosphate deficiency. *Physiol. Plant.* 58:287–294.

Cole, C.V., S.R. Olsen, and C.O. Scott. 1953. The nature of phosphate sorption by calcium carbonate. *Proc. Soil Sci. Soc. Am.* 17:352–356.

Cordell, D., J.O. Drangert, and S. White. 2009. The story of phosphorus: Global food security and food for thought. *Global Environ. Change* 19:292–305.

Correll, D.L. 1998. The role of phosphorus in the eutrophication of receiving waters: A review. *J. Environ. Qual.* 27:261–266.

Cosgrove, D.J. 1977. Microbial transformations in the phosphorus cycle. In *Advances in Microbial Ecology*, ed. M. Alexander, pp. 95–134. New York: Plenum Press.

Cox, F.R. and S.E. Hendricks. 2000. Soil test phosphorus and clay content effects on runoff water quality. *J. Environ. Qual.* 29:1582–1586.

Cumbus, I.P., D.J. Hornsley, and L.W. Robinson. 1977. The influence of phosphorus, zinc and manganese on absorption and translocation of iron in watercress. *Plant Soil* 48:651–660.

Curless, M.A., K.A. Kelling, and P.E. Speth. 2005. Nitrogen and phosphorus availability from liquid dairy manure to potatoes. *Am. J. Potato Res.* 82:287–297.

Daniel, T.C., D.R. Edwards, and A.N. Sharpley. 1993. Effect of extractable soil surface phosphorus on runoff water quality. *Trans. Am. Soc. Agr. Eng.* 36:1079–1085.

Daniel, T.C., A.N. Sharpley, and J.L. Lemunyon. 1998. Agricultural phosphorus and eutrophication: A symposium overview. *J. Environ. Qual.* 27:251–257.

Davenport, J.R., P.H. Milburn, C.J. Rosen, and R.E. Thornton. 2005. Environmental impacts of potato nutrient management. *Am. J. Potato Res.* 85:321–328.

Davis, J.R., J.C. Stark, L.H. Sorenson, and A.T. Schneider. 1994. Interactive effects of nitrogen and phosphorus on Verticillium wilt of Russet Burbank potato. *Am. Potato J.* 71:467–481.

Dechassa, N., M. Schenk, N. Claassen, and B. Steingrobe. 2003. Phosphorus efficiency of cabbage (*Brassica oleraceae* L. *var. capitata*), carrot (*Daucus carota* L.), and potato (*Solanum tuberosum* L.). *Plant Soil* 250:215–224.

Deguchi, T., E. Itoh, M. Matsumoto, K. Furukawa, and K. Iwama. 2011. Root vertical distribution and water absorption ability in Konyu potato cultivars with drought tolerance. In *Abstracts of the 18th Triennial Conference of the European Association for Potato Research*, eds. J. Santala and J.P.T Valkonen, p. 109. Oulu, Finland.

Dong, B., P.R. Ryan, Z. Rengel, and E. Delhaize. 1999. Phosphate uptake in *Arabidopsis thaliana*: Dependence of uptake on the expression of transporter genes and internal phosphate concentrations. *Plant Cell Environ.* 22:1455–1461.

Drew, M.C. and L.R. Saker. 1984. Uptake and long-distance transport of phosphate, potassium and chloride in relation to internal ion concentration in barley: Evidence for non-allosteric regulation. *Planta* 160:500–507.

Drew, M.C., L.R. Saker, S.A. Barber, and W. Jenkins. 1984. Changes in the kinetics of phosphate and potassium absorption in nutrient-deficient barley roots measured by a solution-depletion technique. *Planta* 160:490–499.

Dubetz, S. and J.B. Bole. 1975. Effects of nitrogen, phosphorus and potassium on yield components and specific gravity of potatoes. *Am. Potato J.* 52:399–405.

Ducic, T. and A. Polle. 2007. Manganese toxicity in two varieties of Douglas fir (*Pseudotsuga menziesii* var. viridis and glauca) seedlings as affected by phosphorus supply. *Funct. Plant Biol.* 34:31–40.

Dunn, D.J. and G. Stevens. 2008. Response of rice yields to phosphorus fertilizer rates and polymer coating. *Crop Manag.* doi:10.1094/CM-2008-0610-01-RS. http://www.plantmanagementnetwork.org./cm/element/sum2.aspx?id = 6946. Accessed May 13, 2013.

Dyer, B. 1894. Analytical determination of probably available mineral plant food in soils. *Trans. Chem. Soc.* 65:115–167.

Dyson, P.W. and D.J. Watson. 1971. An analysis of the effects of nutrient supply on the growth of potato crops. *Ann. Appl. Biol.* 69:47–63.

Eghball, B., G.D. Binford, and D.D. Baltensperger. 1996. Phosphorus movement and adsorption in a soil receiving long-term manure and fertilizer application. *J. Environ. Qual.* 25:1339–1343.

Eghball, B., J.E. Gilley, L.A. Kramer, and T.B. Moorman. 2000. Narrow grass hedge effects on phosphorus and nitrogen in runoff following manure and fertilizer application. *J. Soil Water Conserv.* 55:172–176.

Elias-Azar, K., A.E. Lang, and P.F. Pratt. 1980. Bicarbonate-extractable phosphorus in fresh and composted dairy manures. *Soil Sci. Soc. Am. J.* 44:435–437.

Emsley, J. 2000. *The 13th Element: The Sordid tale of Murder, Fire, and Phosphorus*. New York: John Wiley & Sons.

Engelstad, O.P. and G.L. Teramn. 1980. Agronomic effectiveness of phosphate fertilizers. In *The Role of Phosphorus in Agriculture*, eds. F.E. Khasawneh, E.C. Sample, and E.J. Kamprath, pp. 311–332. Madison, WI: American Society of Agronomy.

Erich, M.S, C.B Fitzgerald, and G.A Porter. 2002. The effect of organic amendments on phosphorus chemistry in a potato cropping system. *Agric. Ecosys. Environ.* 88:79–88.

Ernst, M., V. Römheld, and H. Marschner. 1989. Estimation of phosphorus uptake capacity by different zones of the primary root of soil-grown maize (*Zea mays* L.). *Z. Pflanzenernähr. Bodenk.* 152:21–25.

Fageria, V.D. 2001. Nutrient interactions in crop plants. *J Plant Nutr.* 24:1269–1290.

Fallah, A.S. 1979. The effect of temperature, nitrogen, foliage removal and root pruning on growth and brown center development in the potato. Dissertation. Pullman, WA: Washington State University.

Feder, J. 1973. The phosphatases. In *Environmental Phosphorus Handbook*, eds. E.J. Griffith, A. Beeton, J.M. Spencer, and D.T. Mitchell, pp. 475–508. New York: John Wiley & Sons.

Fixen, P.E. and T.W. Bruulsema. 2014. Potato management challenges created by phosphorus chemistry and plant roots. *Am. J. Potato Res.* 91:121–131.

Fixen, P.E. and J.H. Grove. 1990. Testing soil for phosphorus. In *Soil Testing and Plant Analysis*, ed. R.L. Westerman, pp. 141–180. Madison, WI: Soil Science Society of America.

Fixen, P.E., E.A. Liegel, C.R. Simson, L.M. Walsh, and R.P. Wolkowski. 1979–1981. Effect of phosphorus source and rate on potatoes. Unpublished Research Reports, Department of Soil Science, University of Wisconsin–Madison.

Föhse, D., N. Claassen, and A. Jungk. 1988. Phosphorus efficiency in plants I. External and internal P requirement and P uptake efficiency of different plant species. *Plant Soil* 110:101–109.

Föhse, D., N. Claassen, and A. Jungk. 1991. Phosphorus efficiency in plants II: Significance of root radius, root hairs and cation balance for phosphorus influx in seven plant species. *Plant Soil* 132:261–272.

Forsee, W.T. Jr. and R.V. Allison. 1944. Evidence of phosphorus interference in the assimilation of copper by citrus on the organic soils of the lower east coast of Florida. *Soil Sci. Soc. Fla. Proc.* 6:162–165.

Fox, T.C. and M.L. Guerinot. 1998. Molecular biology of cation transport in plants. *Annu. Rev. Plant Physiol. Mol. Biol.* 49:669–696.

Foy, C.D., R.L. Chaney, and M.C. White. 1978. The physiology of metal toxicity in plants. *Annu. Rev. Plant Physiol.* 29:511–566.

Fredeen, A.L., I.M. Rao, and N. Terry. 1989. Influence of phosphorus nutrition on growth and carbon partitioning in *Glycine max*. *Plant Physiol.* 89:225–230.

Freeman, K.L., P.R. Franz, and R.W. de Jong. 1998. Effect of phosphorus on the yield, quality and petiolar phosphorus concentrations of potatoes (cvv. Russet Burbank and Kennebec) grown in the krasnozem and duplex soils of Victoria. *Aust. J. Exp. Agric.* 38:83–93.

Gale, P.M., M.D. Mullen, C. Cieslik, D.D. Tyler, B.N. Deuk, M. Kirchner, and J. McClure. 2000. Phosphorus distribution and availability in response to dairy manure applications. *Commun. Soil Sci. Plant Anal.* 31:553–565.

Galvez, L., R.B. Clark, L.M. Gourley, and J.W. Maranville. 1989. Effects of silicon on mineral composition of sorghum growth with excess manganese. *J. Plant Nutr.* 12:547–561.

Garcia, R. and J.J. Hanway. 1976. Foliar fertilization of soybeans during the seed-filling period. *Agron. J.* 68:653–657.

Gardner, B.R. 1984. Effects of soil P levels on yields of head lettuce. *Agron. Abst. P* 204.

Gardner, B.R. and J. Preston Jones. 1973. Effects of temperature on phosphate sorption isotherms and phosphate desorption. *Commun. Soil Sci. Plant Anal.* 4:83–93.

Gburek, W.J., A.N. Sharpley, L. Heathwaite, and G.J. Folmar. 2000. Phosphorus management at the watershed scale: A modification of the phosphorus index. *J. Environ. Qual.* 29:130–144.

George, E., W.J. Horst, and E. Neumann. 2012. Adaptation of plants to adverse chemical soil conditions. In *Mineral Nutrition of Higher Plants*, 3rd edition, ed. P. Marschner, pp. 409–472. London, U.K.: Academic Press.

Ginting, D., J.F. Moncrief, S.E. Gupta, and S.D. Evans. 1998. Corn yield, runoff, and sediment losses from manure and tillage systems. *J. Environ. Qual.* 27:1396–1402.

Giroux, M., A. Dube, and G.M. Barnett. 1984. Effect de la fertilisation phosphate e sur la pomme de terre en relation avec l'analyse du sol et al source de phosphore utilisee. *Can. J. Soil Sci.* 64:369–381.

Good, L.W., P. Vadas, J. Panuska, C.A. Bonilla, and W.E. Jokela. 2012. Testing the Wisconsin phosphorus index with year-round, field-scale runoff monitoring. *J. Environ. Qual.* 41:1730–1740.

Gracey, H.I. 1984. Availability of phosphorus in organic manures compared with monoammonium phosphate. *Agric. Wastes* 11:133–141.

Grant, C.A., D.N. Flaten, D.J. Tomasiewicz, and S.C. Sheppard. 2001. The importance of early season phosphorus nutrition. *Can. J. Plant Sci.* 81:211–224.

Greenwood, D.J., T.J. Cleaver, M.K. Turner, J. Hunt, K.B. Niendorf, and S.M.H. Loquens. 1980. Comparison of the effects of phosphate fertilizer on the yield, phosphate content and quality of 22 different vegetable and agricultural crops. *J. Agric. Sci. Camb.* 95:457–469.

Grierson, P.F. 1992. Organic acids in the rhizosphere of *Banksia integrifolia* L.f. *Plant Soil* 44:259–265.

Griffin, R.A. and J.J. Jurinak. 1973. The interaction of phosphate with calcite. *Soil Sci. Soc. Am. Proc.* 37:847–850.

Grossl, P.R. and W.P. Inskeep. 1991. Precipitation of dicalcium phosphate dihydrate in the presence of organic acids. *Soil Sci. Soc. Am. J.* 55:670–675.

Grunes, D.L., H.R. Haise, and L.O. Fine. 1958. Proportional uptake of soil and fertilizer phosphorus by plants as affected by nitrogen fertilization: Field experiments with sugarbeets and potatoes. *Soil Sci. Soc. Am. Proc.* 22:49–52.

Grusak, M.A., J.N. Pearson, and E. Marentes. 1999. The physiology of micronutrient homeostasis in field crops. *Field Crops Res.* 60:41–56.

Gunes, A., M. Alpaslan, and A. Inal. 1998. Critical nutrient concentrations and antagonistic and synergistic relationships among the nutrients of NFT-grown young tomato plants. *J. Plant Nutr.* 21:2035–2047.

Guppy, C.N., N.W. Menzies, F.P.C. Blamey, and P.W. Moody. 2005. Do decomposing organic matter residues reduce phosphorus sorption in highly weathered soils? *Soil Sci. Soc. Am. J.* 69:1405–1411.

Haldar, M. and L.N. Mandal. 1981. Effect of phosphorus and zinc on the growth and phosphorus, zinc, copper, iron and manganese nutrition of rice. *Plant Soil* 59:415–425.

Hall, A.D. 1905. The analysis of the soil by means of the plant. *J. Agric. Sci.* 1:65–88.

Hall, A.D. 1909. *Fertilisers and Manures*. London, U.K.: John Murray. https://ia600308.us.archive.org/7/items/fertilisersmanur00hall/fertilisersmanur00hall.pdf.

Halstead, R.L. and R.B. McKercher. 1975. Biochemistry and cycling of phosphorus. In *Soil Biochemistry*, Volume 4, eds. A. Paul and A.D. McLaren, pp. 31–63. New York: Marcel Dekker, Inc.

Hambidge, G. 1941. *Hunger Sign in Crops*. Washington, DC: American Society of Agronomy and The National Fertilizer Council.

Hamblin, R.L., C. Schatz, and A.V. Barker. 2003. Zinc accumulation in Indian Mustard as influenced by nitrogen and phosphorus nutrition. *J. Plant Nutr.* 26:177–190.

Hammes, J.K. 1961–1962. Influence of fertilizer placement on the cumulative uptake of fertilizer phosphorus by Early Gem potatoes. Unpublished research reports, Department of Soil Science, Madison, WI: University of Wisconsin–Madison.

Hanafi, M.M., S.M. Eltaib, M.B. Ahmad, and S.R. Syed Omar. 2002. Evaluation of controlled-release compound fertilizers in soil. *Commun. Soil Sci. Plant Anal.* 33:1139–1156.

Hang, Z. 1993. Influence of chloride on the uptake and transport of phosphorus in potato. *J. Plant Nutr.* 16(9):1733–1737.

Hanlon, E.A. and G.V. Johnson. 1984. Bray/Kurtz, Mehlich III, AB/D, and ammonium acetate extractions of P, K, and Mg in four Oklahoma soils. *Commun. Soil Sci. Plant Anal.* 15:277–294.

Hannapel, R.J., W.H. Fuller, S. Bosma, and J.S. Bullock. 1964. Phosphorus movement in a calcareous soil. I. Predominance of organic forms of phosphorus in phosphorus movement. *Soil Sci.* 97:350–357.

Hansen, H.C.B., P.E. Hansen, and J. Magid. 1999. Empirical modeling of the kinetics of phosphate sorption to macropore materials in aggregated subsoils. *Eur. J. Soil Sci.* 50:317–327.

Hanway, J.J. and R.A. Olson. 1980. Phosphate nutrition of corn, sorghum, soybeans, and small grains. In *The Role of Phosphorus in Agriculture*, eds. F.E. Kasanweh, E.C. Sample, and E.J. Kamprath, pp. 681–692. Madison, WI: American Society of Agronomy.

Harder, H.J., R.E. Carlson, and R.H. Shaw. 1982a. Corn grain yield and nutrient response to foliar fertilization applied during grain fill. *Agron. J.* 74:106–110.

Harder, H.J., R.E. Carlson, and R.H. Shaw. 1982b. Leaf photosynthetic response to foliar fertilizer applied to corn during grain fill. *Agron. J.* 74:759–761.

Hart, M.R., B.F. Quin, and M.L. Nguyen. 2004. Phosphorus runoff from agricultural land and direct fertilizer effects: A review. *J. Environ. Qual.* 33:1954–1972.

Hash, C.T., R.E. Schaffert, and J.M. Peacock. 2002. Prospects for using conventional techniques and molecular biological tools to enhance performance of 'orphan' crop plants on soils low in available phosphorus. *Plant Soil* 245:135–146.

Hawkesford, M., W. Horst, T. Kichey et al. 2012. Functions of macronutrients. In *Marscher's Mineral Nutrition of Higher Plants*, 3rd edn, ed. P. Marschner, pp. 135–189. London, U.K.: Academic Press.

Hawkins, A. 1954. Time, method of application, and placement of fertilizer for efficient production of potatoes in New England. *Am. Potato J.* 31:106–113.

Hecht-Buchholz, C. 1967. Über die Dunkelfärbung des Blattgrüns bei Phosphormangel. *Z. Pflanzenernähr. Bodenk.* 118:12–22.

Hegney, M.A. and I.R. McPharlin. 1999. Broadcasting phosphate fertiliser produces higher yields of potatoes (*Solanum tuberosum* L.) than band placement on coastal sands. *Aust. J. Exp. Agric.* 39:495–503.

Hegney, M.A., I.R. McPharlin, and R.C. Jeffery. 2000. Using soil testing and petiole analysis to determine phosphorus fertiliser requirements of potatoes (*Solanum tuberosum* L. cv. Delaware) in the Manjimup–Pemberton region of Western Australia. *Aust. J. Exp. Agric.* 40:107–117.

Hergert, G.W. and J.O. Reuss. 1976. Sprinkler application of P and Zn fertilizer. *Agron. J.* 68:5–8.

Herlihy, M. 1970. Contrasting effects of nitrogen and phosphorus on potato tuber blight. *Plant Pathol.* 19:69–71.

Herlihy, M. and P.J. Carroll. 1969. Effects of N, P and K and their interactions on yield, tuber blight and quality of potatoes. *J. Sci. Food Agric.* 20:513–517.

Hill, M.W., B.G. Hopkins, and V.D. Jolley. 2014a. Maize in-season growth response to organic acid-bonded phosphorus fertilizer (Carbond P®). *J. Plant Nutr.* DOI: 10.1080/01904167.2014.973040.

Hill, M.W., B.G. Hopkins, and V.D. Jolley. 2014b. Phosphorus mobility through soil increased with organic acid-bonded phosphorus fertilizer (Carbond P®). *J. Plant Nutr.* DOI: 10.1080/01904167.2014.973041.

Hiller, L.K. and D.C. Koller. 1987. Foliar fertilization of potatoes: Some 1986 results. *Potato Country, U.S.A.* 32:1–2.

Hoffland, E. 1992. Quantitative evaluation of the role of organic acid exudation in the mobilization of rock phosphate by rape. *Plant Soil* 140:279–289.

Holford, I.C.R. and G.E.G. Mattingly. 1975. Phosphate sorption by Jurassic Oolitic limestones. *Geoderma* 13:257–264.

Holford, I.C.R. and W.H. Patrick Jr. 1979. Effect of reduction and pH changes on phosphate sorption and mobility in an acid soil. *Soil Sci. Soc. Am. J.* 43:292–297.

Hooker, M.L., G.A. Peterson, D.H. Sander, and L.A. Daigger. 1980. Phosphate fractions in calcareous soils as altered by time and amounts of added phosphate. *Soil Sci. Soc. Am. J.* 44:269–277.

Hopkins, B.G. 2013. Russet Burbank potato phosphorus fertilization with dicarboxylic acid copolymer additive (AVAIL®). *J. Plant Nutr.* 36:1287–1306.

Hopkins, B.G. and J.W. Ellsworth. 2006. Banded P increases sugarbeet yields. *Fluid J.* 14(1):14–16.

Hopkins, B.G., J.W. Ellsworth, T.R. Bowen, A.G. Cook, S.C. Stephens, V.D. Jolley, A.K. Shiffler, and D. Eggett. 2010a. Phosphorus fertilizer timing for Russet Burbank potato grown in calcareous soil. *J. Plant Nutr.* 33:529–540.

Hopkins, B.G., J.W. Ellsworth, A.K. Shiffler, T.R. Bowen, and A.G. Cook. 2010b. Pre-plant versus in-season application of phosphorus fertilizer for Russet Burbank potato grown in calcareous soil. *J. Plant Nutr.* 33:1026–1039.

Hopkins, B.G., J.W. Ellsworth, A.K. Shiffler, A.G. Cook, and T.R. Bowen. 2010c. Monopotassium phosphate as an in-season fertigation option for potato. *J. Plant Nutr.* 33:1422–1434.

Hopkins, B.G. and R.E. Hirnyck. 2007a. Organic potato production. In *Potato Health Management*, ed. D.A. Johnson, pp. 101–108. Minneapolis, MN: American Phytopathological Society.

Hopkins, B.G., D.A. Horneck, and A.E. MacGuidwin. 2014. Improving phosphorus use efficiency through potato rhizosphere modifications and extension. *Am. J. Potato Res.* 91:161–174.

Hopkins, B.G., D.A. Horneck, M.J. Pavek et al. 2007a. Evaluation of potato production best management practices. *Am. J. Potato Res.* 84:19–27.

Hopkins, B.G., D.A. Horneck, R.G. Stevens, J.W. Ellsworth, and D.M. Sullivan. 2007b. Managing irrigation water quality for crop production in the Pacific Northwest. *PNW* 597-E. Corvallis, OR: Oregon State University.

Hopkins, B.G., S.A. Randall, T.M. Rae, C.J. Ransom, and L.E. Sutton. 2013. Fertilizer bans coming to a city near you? *Western Turf.* Idaho Falls, ID: Harris Publishing.

Hopkins, B.G., C.J. Rosen, A.K. Shiffler, and T.W. Taysom. 2008. Enhanced efficiency fertilizers for improved nutrient management: Potato (*Solanum tuberosum*). *Crop Manag.* online. doi:10.1094/CM-2008-0317-01-RV. http://www.plantmanagementnetwork.org/cm/element/cmsum2.asp?id = 6920.

Hopkins, B.G. and J.C. Stark. 2003. Humic acid effects on potato response to phosphorus. In *Proceedings of the Winter Commodity Schools*, Volume 35, eds. L.D. Robertson et al., pp. 87–92. *Idaho Potato Conference*, Pocatello, ID: UI-Cooperative Extension System, Moscow, ID: UI-Cooperative Extension System.

Hopkins, B.G. and S.C. Stephens. 2008. Band placement critical to potato yield. *Fluid J.* 16(3):1–3.

Horneck, D.A., D. Wysocki, B.G. Hopkins, J. Hart, and R.G. Stevens. 2007. Acidifying soil for crop production: Inland Pacific Northwest. *PNW* 599-E, Corvallis, OR: Oregon State University.

Hudson, N. 1995. *Soil Conservation*, 3rd edition. London, U.K.: B.T. Batsford.

IPNI. 2011. Soil test levels summary for North America, 2010 summary. Norcross, GA: International Plant Nutrition Institute.

Itoh, S. and S.A. Barber. 1983. Phosphorus uptake by six plant species as related to root hairs. *Agron. J.* 75:457–461.

Iwama, K. 2008. Physiology of the potato: New insights into root system and repercussions for crop management. *Potato Res.* 51:333–353.

Iwama, K., K. Nakaseko, A. Isoda, K. Gotoh, and Y. Nishibe. 1981. Relations between root system and tuber yield in the hybrid population of the potato plants. *Jpn. J. Crop Sci.* 50:233–238.

Iwasa, T., A.M. Anithakumari, S. Niura et al. 2011. QTL analysis for root length and dry weight in a diploid potato population. In *Abstracts of the 18th Triennial Conference of the European Association for Potato Research*, eds. J. Santala and J.P.T Valkonen, p. 56. Oulu, Finland.

Iyamuremye, F. and R.P. Dick. 1996. Organic amendments and phosphorus sorption by soils. *Adv. Agron.* 56:139–185.

Jackson, T.L. and G.E. Carter. 1976. Nutrient uptake by Russet Burbank potatoes as influenced by fertilization. *Agron. J.* 68:9–12.

Jacob, W.C. and L.A. Dean. 1950. The utilization of phosphorus by two potato varieties on Long Island. *Am. Potato J.* 27:439–445.

Jacob, W.C., C.H. van Middelem, W.L. Nelson, C.D. Welch, and N.S. Hall. 1949. Utilization of phosphorus by potatoes. *Soil Sci.* 68:113–120.

James, D.W., C.J. Hurst, and T.A. Tindall. 1995. Alfalfa cultivar response to phosphorus and potassium deficiency: Elemental composition of the herbage. *J. Plant Nutr.* 18:2447–2464.

James, D.W., W.H. Weaver, and R.L. Rader. 1970. Chloride uptake by potatoes and the effects of potassium chloride, nitrogen and phosphorus fertilization. *Soil Sci.* 109:48–52.

Jenkins, P.D. and H. Ali. 1999. Growth of potato cultures in response to application of phosphate fertilizer. *Ann. Appl. Biol.* 135:431–438.

Johnston, A.E., P.W. Lane, G.E.G. Mattingly, P.R. Poulton, and M.V. Hewitt. 1986. Effects of soil and fertilizer P on yields of potatoes, sugar beet, barley and winter wheat on a sandy clay loam soil at Saxmundham, Suffolk. *J. Agric. Sci. Camb.* 106:155–167.

Jones, M.P., B.L. Webb, V.D. Jolley, B.G. Hopkins, and D.A. Cook. 2013. Evaluating nutrient availability in semi-arid soils with resin capsules and conventional soil tests, I. Native plant bioavailability under glasshouse conditions. *Commun. Soil Sci. Plant Anal.* 44:971–986.

Kalifa, A., N.N. Barthankur, and D.J. Donnelly. 2000. Phosphorus reduces salinity stress in micro-propagated potato. *Am. J. Potato Res.* 77:179–182.

Kalkafi, U., B. Bar-Yosef, and A. Hadas. 1978. Fertilization decision model—A synthesis of soil and plant parameters in a computerized program. *Soil Sci.* 125:261–268.

Kamprath, E.J. and M.E. Watson. 1980. Conventional soil and tissue tests for assessing the phosphorus status of soils. In *The Role of Phosphorus in Agriculture*, eds. F.E. Khasawneh, E.C. Sample, and E.J. Kamprath, pp. 433–469. Madison, WI: American Society of Agronomy.

Karamanos, R.E. and D. Puurveen. 2011. Evaluation of a polymer treatment as enhancer of phosphorus fertilizer efficiency in wheat. *Can. J. Soil Sci.* 91:123–125.

Kelling, K.A. and P.E. Speth. 1997. Influence of phosphorus rate and timing on Wisconsin potatoes. *Proc. Wisconsin Potato Meet.* 10:33–41.

Kelling, K.A., S.A. Wilner, R.F. Hensler, and L.M. Massie. 1998. Placement and irrigation effects on nitrogen fertilizer use efficiency. *Proc. Wisconsin Annu. Potato Meet.* 11:79–88.

Kelling, K.A., R.P. Wolkowski, J.G. Iyer, R.B. Corey, and W.R. Stevenson. 1992. Potato responses to phosphorus application and using petiole analysis in determining P status. *Proc. Wisconsin Annu. Potato Meet.* 5:39–50.

Khiari, L., L.E. Parent, A. Pellerin et al. 2000. An agri-environmental phosphorus saturation index for acid coarse-textured soils. *J. Environ. Qual.* 29:1561–1567.

Kimmel, R.J., G.M. Pierzynski, K.A. Janssen, and P.L. Barnes. 2001. Effects of tillage and phosphorus placement on phosphorus runoff losses in a grain sorghum-soybean rotation. *J. Environ. Qual.* 30:1324–1330.

Kingston, B.D. and R.W. Jones. 1980. Response of potatoes to phosphorus rate and placement on the Texas rolling plains. *Texas Agricultural Research Station Report PR3680*, College Station: TX.

Kissel, D.E., D.H. Sander, and R. Ellis Jr. 1985. Fertilizer-plant interactions in alkaline soils. In *Fertilizer Technology and Use*, 3rd edition, ed. O.P. Engelstad, pp. 153–196. Madison, WI: Soil Science Society of America.

Kitchen, H.B. (ed.). 1948. *Diagnostic Techniques for Soil and Crops*. Washington, DC: The American Potash Institute.

Kleinkopf, G.E., D.T. Westermann, and R.B. Dwelle. 1981. Dry matter production and nitrogen utilization by six potato cultivars. *Agron. J.* 73:799–802.

Kleinman, P.J.A., B.A. Needelman, A.N. Sharpley, and R.W. McDowell. 2003. Using soil phosphorus profile data to assess phosphorus leaching potential in manured soils. *Soil Sci. Soc. Am. J.* 67:215–224.

Ko, W.H. and F.K. Hora. 1970. Production of phospholipases by soil microorganisms. *Soil Sci.* 10:355–358.

Kotak, B.G., S.L. Kenefick, D.L. Fritz, C.G. Rousseaux, E.E. Prepas, and S.E. Hrudey. 1993. Occurrence and toxicological evaluation of cyanobacterial toxins in Alberta lakes and farm dugouts. *Water Res.* 27:495–506.

Kovar, J.L. and S.A. Barber. 1987. Placing phosphorus and potassium for greatest recovery. *J. Fertil. Issues* 4:1–6.

Kuo, S. 1996. Phosphorus. In *Methods of Soil Analysis. Part 3. Chemical Methods*, SSSA No. 5, eds. D.L. Spark, A.L. Page, P.A. Helmke, R.H. Loeppert, P.N. Soltanpour, M.A. Tabatabai, C.T. Johnston, and M.E. Sumner, pp. 869–919. Madison, WI: Soil Science Society of America.

Laboski, C.A.M. and J.A. Lamb. 2003. Changes in soil test phosphorus concentration after application of manure or fertilizer. *Soil Sci. Soc. Am. J.* 67:544–554.

Laboski, C.A.M. and J.B. Peters. 2012. *Nutrient Application Guidelines for Field, Vegetable and Fruit Crops in Wisconsin*. Madison, WI: University of Wisconsin-Extension Publication A2809.

Lambers, H., M.C. Brundrett, J.A. Raven, and S.D. Hopper. 2010. Plant mineral nutrition in ancient landscapes: High plant species diversity on infertile soils is linked to functional diversity for nutritional strategies. *Plant Soil* 334:11–31.

Lambers, H., P.M. Finnegan, E. Laliberté et al. 2011. Phosphorus nutrition of Proteaceae in severely phosphorus-impoverished soils: Are there lessons to be learned for future crops? *Plant Physiol.* 156:1058–1066.

Lambers, H., M.W. Shane, M.D. Cramer, S.J. Pearse, and E.J. Veneklaas. 2006. Root structure and functioning for efficient acquisition of phosphorus: Matching morphological and physiological traits. *Ann. Bot.* 98:693–713.

Lambert, D.H., M.L. Powelson, and W.R. Stevenson. 2005. Nutritional interactions influencing diseases of potato. *Am. J. Potato Res.* 82:309–319.

Lang, N.S., R.G. Stevens, R.E. Thornton, W.L. Pan, and S. Victory. 1999. Potato nutrient management for central Washington. *Extension Bulletin 871*, Pullman, WA: Washington State University.

Lauer, D.A. 1988. Vertical distribution in soil of sprinkler-applied phosphorus. *Soil Sci. Soc. Am. J.* 52:862–868.

Laughlin, W.M. 1962. Influence of soil and spray applications of phosphorus on potato yield, dry matter content, and chemical composition. *Am. Potato J.* 39:343–347.

Lawton, L.A. and G.A. Codd. 1991. Cyanobacterial (blue-green algae) toxins and their significance in UK and European waters. *J. Inst. Water Env. Man.* 5:460–465.

Le Mare, P.H. 1977. Experiments on effects of phosphorus on the manganese nutrition of plants I. Effects of monocalcium phosphate and its hydrolysis derivatives of manganese in ryegrass grown in two Buganda soils. *Plant Soil* 47:593–605.

Leece, D.R. 1978. Effects of boron on the physiological activity of zinc in maize. *Aust. J. Agric. Res.* 29:739–748.

Leikam, D.F., L.S. Murphy, D.E. Kissel, D.A. Whitney, and H.C. Moser. 1983. Effects of nitrogen and phosphorus application method and nitrogen source on winter wheat grain yield and leaf tissue phosphorus. *Soil Sci. Soc. Am. J.* 47:530–535.

Lesczynski, D.B. and C.B. Tanner. 1976. Seasonal variation of root distribution of irrigated, field-grown russet Burbank potato. *Am. Potato J.* 53:69–78.

Liegel, E.A., C.R. Simson, P.E. Fixen, R.E. Rand, and G.G. Weis. 1981. Potato responses to phosphorus and potassium and recommendations for P-K fertilization. *Potato Manual 81EL*, Madison, WI: University of Wisconsin.

Lindsay, W.L. 1979. *Chemical Equilibria in Soils*. New York: John Wiley & Sons, Inc.

Lindsay, W.L. 2001. *Chemical Equilibria in Soils*. Caldwell, ID: The Blackburn Press.

Lindsay, W.L., A.W. Frazier, and H.F. Stephenson. 1962. Identification of reaction products from phosphate fertilizers in soils. *Soil Sci. Soc. Am. Proc.* 26:446–452.

Lingle, J.C. and R.M. Davis. 1959. The influence of soil temperature and phosphorus fertilization on the growth and mineral absorption of tomato seedling. *Proc. Am. Soc. Hortic. Sci.* 73:312–322.

Lingle, J.C. and J.R. Wright. 1964. Fertilizer experiments with cantaloupes. *California Agricultural Exp. Bull. 807*, Berkeley, CA.

Little, S.A., P.J. Hocking, and R.S.B. Greene. 2004. A preliminary study of the role of cover crops in improving soil fertility and yield for potato production. *Commun. Soil Sci. Plant Anal.* 35:471–494.

Locasio, S.J. and R.D. Rhue. 1990. Phosphorus and micronutrient sources for potato. *Am. Potato J.* 67:217–226.

Locascio, S.J., G.F. Warren, and G.E. Wilcox. 1960. The effect of phosphorus placement on uptake of phosphorus and growth of direct seeded tomatoes. *Proc. Am. Soc. Hortic. Sci.* 76:503–514.

Lombi, E., M.J. McLaughlin, C. Johnston, R.D. Armstrong, and R.E. Holloway. 2004. Mobility and lability of phosphorus from granular and fluid monoammonium phosphate differs in a calcareous soil. *Soil Sci. Soc. Am. J.* 68:682–689.

Loneragan, J.F., Y.S. Grove, A.D. Robson, and K. Snowball. 1979. Phosphorus toxicity as a factor in zinc-phosphorus interactions in plants. *Soil Sci. Soc. Am. J.* 43:966–972.

Lorenz, O.A. 1947. Studies on potato nutrition: III. Chemical composition and uptake of nutrients by Kern County potatoes. *Am. Potato J.* 24:281–293.

Lorenz, O.A., K.B. Tyler, and O.D. McCoy. 1964. Phosphate sources and rates for winter lettuce on a calcareous soil. *Proc. Am. Soc. Hortic. Sci.* 84:348–355.

Love, S.L., R. Novy, D.L. Corsini, and P. Bain. 2003. Variety selection and management. In *Potato Production Systems*, eds. J.C. Stark and S.L. Love, 21–47. Moscow, ID: University of Idaho Agriculture Communications.

Lucas, R.E. and M.T. Vittum. 1976. Fertilizer placement for vegetables. In: *Phosphorus Fertilization-Principles and Practices of Band Application*, eds. G.E. Richards, St Louis, MO: Olin Corp.

Lynch, J. 1998. The role of nutrient efficient crops in modern agriculture. In: *Nutrient Use in Crop Production*, ed. Z. Rengel, pp. 241–264. Binghamton, NY: The Haworth Press Inc.

Lynch, J., A. Läuchli, and E. Epstein. 1991. Vegetative growth of the common bean in response to phosphorus nutrition. *Crop Sci.* 31:380–387.

Macy, P. 1936. The quantitative mineral nutrient requirements of plants. *Plant Physiol.* 11:749–764.

MacKay, D.C., J.M. Carefoot, and T. Entz. 1988. Detection and correction of midseason P deficiency in irrigated potatoes. *Can. J. Plant Sci.* 68:523–534.

MacLean, A.A. 1983. Source of fertilizer nitrogen and phosphorus for potatoes in Atlantic Canada. *Am. Potato J.* 69:913–918.

Maier, N.A., K.A. Potocky-Pacay, A.P. Dahlenburg, and C.M.J. Williams. 1989a. Effect of phosphorus on the specific gravity of potato tubers (*Solanum tuberosum* L.) of cultivars Kennebec and Coliban. *Aust. J. Exp. Agric.* 29:869–874.

Maier, N.A., K.A. Potocky-Pacay, J.M. Jacka, and C.M.J. Williams. 1989b. Effect of phosphorus fertiliser on the yield of potato tubers (*Solanum tuberosum* L.) and the prediction of tuber yield response by soil analysis. *Aust. J. Exp. Agric.* 29:419–432.

Malik, M.A., P. Marschner, and K.S. Khan. 2012. Addition of organic and inorganic P sources to soil—Effects on P pools and microorganisms. *Soil Biol. Biochem.* 49:106–113.

Mallarino, A.P. 1997. Interpretation of soil phosphorus tests for corn in soils with varying pH and calcium carbonate content. *J. Prod. Agric.* 10:163–167.

Mallarino, A.P., J.R. Webb, and A.M. Blackmer. 1991. Corn and soybean yields during 11 years of phosphorus and potassium fertilization on a high-testing soil. *J. Prod. Agric.* 4:312–317.

Marschner, P. 2012. *Mineral Nutrition of Higher Plants*, 3rd edition. San Diego, CA: Elsevier.

Marsh, K.B., L.A. Peterson, and B.H. McCown. 1989. A microculture method for assessing nutrient uptake II. The effect of temperature on manganese uptake and toxicity in potato shoots. *J. Plant Nutr.* 12:219–232.

Mattingly, G.E.G. and F.V. Widdowson. 1958. Uptake of phosphorus from P[32]-labelled superphosphate by field crops. *Plant Soil* 9:286–304.

McBeath, T., M. McLaughlin, J. Kirby, and R. Armstrong. 2011. Effect of soil water on phosphorus use in agricultural soils. *Proceedings of the 2011 Fluid Forum*. Online http://www.fluidfertilizer.com/fertilizer_fluid_forum.

McCollum, R.E. 1978a. Analysis of potato growth under differing P regimes. I. Tuber yields and allocation of dry matter and P. *Agron. J.* 70:51–57.

McCollum, R.E. 1978b. Analysis of potato growth under differing P regimes. II. Time by P-status interactions for growth and leaf efficiency. *Agron. J.* 70:58–66.

McDowell, R., A. Sharpley, and G. Folmar. 2001. Phosphorus export from an agricultural watershed: Linking source and transport mechanisms. *J. Environ. Qual.* 30:1587–1595.

McGrath, J.M. and G.D. Binford. 2012. Corn response to starter fertilizer with and without AVAIL. *Crop Management.* doi:10.1094/CM-2012-0320-02-RS. Accessed May 13, 2013.

McLaughlin, M.J., T.M. McBeath, R. Smernik, S.P. Stacey, B. Ajiboye, and C. Guppy. 2011. The chemical nature of P accumulation in agricultural soils—Implications for fertiliser management and design: An Australian perspective. *Plant Soil* 349:69–87.

McLean, E.O. and T.J. Logan. 1970. Sources of phosphorus for plants grown in soils with differing phosphorus fixation tendencies. *Soil Sci. Soc. Am. J.* 34:907–911.

McMurtrey Jr, J.E. 1948. Visual symptoms of malnutrition in plants. In: *Diagnostic Techniques for Soils and Crops*, ed. H.B. Kitchen, pp. 231–289. Washington, DC: American Potash Institute.

Meek, B.D., L.E. Graham, T.J. Donovan, and K.S. Mayberry. 1979. Phosphorus availability in a calcareous soil after high loading rates of animal manure. *Soil Sci. Soc. Am. J.* 43:741–743.

Mehlich, A. 1978. New extractant for soil test evaluation of phosphorus, potassium, magnesium, calcium, sodium, manganese, and zinc. *Commun. Soil Sci. Plant Anal.* 9:477–492.

Mehlich, A. 1984. Mehlich 3 soil test extractant: A modification of the Mehlich 2 extractant. *Commun. Soil Sci. Plant Anal.* 15:1409–1416.

Meisinger, J.J., D.R. Bouldin, and E.D. Jones. 1978. Potato yield reductions associated with certain fertilizer mixtures. *Am. Potato J.* 55:227–234.

Mikklelsen, R.L. 1989. Phosphorus fertilization through drip irrigation. *J. Prod. Agric.* 2:279–286.

Miller, J.S. and B.G. Hopkins. 2007. Checklist for a holistic potato health management plan. In *Potato Health Management*, ed. D.A. Johnson, pp. 7–10. Minneapolis, MN: American Phytopathological Society.

Mills, H.A. and J.B. Jones Jr. 1996. *Plant Analysis Handbook II*. Athens, GA: MicroMacro Publishing Inc.

Miyasaka, S.C. and M. Habte. 2001. Plant mechanisms and mycorrhizal symbiosis to increase phosphorus uptake efficiency. *Commun. Soil Sci. Plant Anal.* 32:1101–1147.

Moorby, J. 1978. The physiology of growth and tuber yield. In *The Potato Crop—The Scientific Basis for Improvement*, ed. P.M. Harris, pp. 153–194. London, U.K.: Chapman & Hill.

Moorehead, S., R. Coffin, and B. Douglas. 1998. Phosphorus needs of processing potato varieties. *Better Crops* 82(4):6–7.

Morgan, M.F. 1941. Chemical soil diagnosis by the universal soil testing system. *Connecticut Agricultural Experiment Station Bulletin 45*.

Mortvedt, J.J., L.S. Murphy, and R.H. Follet. 1999. *Fertilizer Technology and Application*. Willoughby, OH: Meister Publishing.

Motavalli, P.P., K.A. Kelling, and J.C. Converse. 1989. First-year nutrient availability from injected dairy manure. *J. Environ. Qual.* 18:180–185.

Mozaffari, M. and J.T. Sims. 1994. Phosphorus availability and sorption in an Atlantic coastal plain watershed dominated by animal-based agriculture. *Soil Sci.* 157:97–107.

Munoz, F., R.S. Mylavarapu, and C.M. Hutchinson. 2005. Environmentally responsible potato production systems: A review. *J. Plant Nutr.* 28:1287–1309.

Murphy, H.J., F.N. Carpenter, and M.J. Goven. 1967. Effect of differential rates of phosphorus, potassium and lime on yield, specific gravity and nutrient uptake of the Katahdin and Russet Burbank. *Maine Agricultural Experiment Station Bulletin* 652, Orono, MN: University of Maine.

Murphy, L.S., D.R. Leikam, R.E. Lamond, and P.J. Gallagher. 1978. Dual applications of N and P—Better agronomics and economics? *Fertilizer Solutions* 22:8–20.

Nagarajah, S., A.M. Posner, and J.P. Quirk. 1970. Competitive adsorption of phosphate with polygalacturonate and other organic acids on kaolinite and oxide surfaces. *Nature* 228:83–85.

Nagata, R.T., C.A. Sanchez, and F.J. Coale. 1992. Crisphead lettuce cultivar response to fertilizer phosphorus. *J. Am. Soc. Hort. Sci.* 117:721–724.

Neilsen, D., G.H. Neilsen, A.H. Sinclair, and D.J. Linehan. 1992. Soil phosphorus status in the manganese nutrition of wheat. *Plant Soil* 145:45–50.

Nelson, W.L. and A. Hawkins. 1947. Response of Irish potatoes to phosphorus and potassium on soils having different levels of these nutrients in Maine and North Carolina. *J. Am. Soc. Agron.* 39:1053–1067.

Nichols, B.A., B.G. Hopkins, V.D. Jolley, B.L. Webb, B.G. Greenwood, and J.R. Buck. 2012. Phosphorus and zinc interactions and their relationships with other nutrients in maize grown in chelator-buffered nutrient solution. *J. Plant Nutr.* 35:123–141.

Nishomoto, R.K., R.L. Fox, and P.E. Parvin. 1977. Response of vegetable crops to phosphorus concentrations in soil solution. *J. Am. Soc. Hort. Sci.* 102:705–709.

Nogueira, M.A., G.C. Magalhaes, and E.J.B.N. Cardoso. 2004. Manganese toxicity in mycorrhizal and phosphorus-fertilized soybean plants. *J. Plant Nutr.* 27:141–156.

Oburger, E., G.J.D. Kirk, W.W. Wenzel, M. Puschenreiter, and D.L. Jones. 2009. Interactive effects of organic acids in the rhizosphere. *Soil Biol. Biochem.* 41:449–457.

Olsen, S.R., C.V. Cole, F.S. Watanabe, and L.A. Dean. 1954. Estimation of available P in soils by extraction with NaHCO₃. *USDA Cir. 939.* Washington, DC: U.S. Government Printing Office.

Omotoso, T.I. and A. Wild. 1970. Content of inositol phosphates in some English and Nigerian soils. *J. Soil Sci.* 21:216.

O'Neill, M.K., B.R. Gardner, and R.L. Roth. 1979. Orthophosphoric acid as a phosphorus fertilizer in trickle irrigation. *Soil Sci. Soc. Am. J.* 43:283–286.

Opena, G.B. and G.A. Porter. 1999. Soil management and supplemental irrigation effects on potato: II. Root growth. *Agron. J.* 91:426–431.

Ozanne, P.G. 1980. Phosphate nutrition of plants—A general treatise. In *The Role of Phosphorus in Agriculture,* eds. F.E. Khasawneh, E.C. Sample, and E.K. Kamprath, pp. 559–589. Madison, WI: American Society of Agronomy, Crop Science Society of America, Soil Science Society of America.

Pack, J.E., C.M. Hutchinson, and E.H. Simonne. 2006. Evaluation of controlled-release fertilizers for northeast Florida chip potato production. *J. Plant Nutr.* 29:1301–1313.

Pan, W.L., R.P. Bolton, E.J. Lundquist, and L.K. Hiller. 1998. Portable rhizotron and color scanner system for monitoring root development. *Plant Soil* 200:107–112.

Panique, E., K.A. Kelling, E.E. Schulte, D.E. Hero, W.R. Stevenson, and R.V. James. 1997. Potassium rate and source effects on potato yield, quality and disease interaction. *Am. Potato J.* 74:379–398.

Parker, M.B. and F.C. Boswell. 1980. Foliage injury, nutrient uptake, and yield of soybeans as influenced by foliar fertilization. *Agron. J.* 72:110–113.

Parry, R. 1998. Agricultural phosphorus and water quality: A U.S. Environmental Protection Agency perspective. *J. Environ. Qual.* 27:258–261.

Pearson, N.J. and Z. Rengel. 1997. Mechanisms of plant resistance to nutrient deficiency stresses. In *Mechanisms of Environmental Stress Resistance in Plants,* eds. A.S. Basra and R.K. Basra, pp. 213–240. Amsterdam, the Netherlands: Harwood Academic Publishers.

Peralta, J.M. and C.O. Stockle. 2002. Dynamics of nitrate leaching under irrigated potato rotation in Washington State: A long-term simulation study. *Agric. Ecosys. Environ.* 88:23–34.

Peterson, G.A., D.H. Sanders, P.H. Grabouski, and M.L. Hooker. 1981. A new look at row and broadcast phosphate fertilizer recommendations for winter wheat. *Agron. J.* 73:13–17.

Pointing, C. 2007. *A New Green History of the World: The Environment and the Collapse of Great Civilizations.* New York: Penguin House.

Ponnamperuma, F.N. 1972. The chemistry of submerged soils. *Adv. Agron.* 24:29–96.

Pote, D.H., T.C. Daniel, D.J. Nichols et al. 1999. Relationship between phosphorus levels in three ultisols and phosphorus concentration in runoff. *J. Environ. Qual.* 28:170–175.

Pote, D.H., T.C. Daniel, A.N. Sharpley, P.A. Moore, Jr., D.R. Edwards, and D.J. Nichols. 1996. Relating extractable soil phosphorus to phosphorus losses in runoff. *Soil Sci. Soc. Am. J.* 60:855–859.

Powell, J.M., Z. Wu, and L.D. Satter. 2001. Dairy diet effects on phosphorus cycles of cropland. *J. Soil Water Conserv.* 56:22–26.

Pursglove, J.D. and F.E. Sanders. 1981. The growth and phosphorus economy of the early potato (*Solanum tuberosum*). *Commun. Soil Sci. Plant Anal.* 12:1105–1121.

Randall, G.W. and R.G. Hoeft. 1988. Placement methods for improved efficiency of P and K fertilizers: A review. *J. Prod. Agric.* 1:70–78.

Randall, G.W., T.K. Iragavarapu, and S.D. Evans. 1997. Long-term P and K applications: 1. Effect on soil test incline and decline rates and critical soil test levels. *J. Prod. Agric.* 10:565–571.

Randall, G.W., K.L. Wells, and J.J. Hanway. 1985. Modern techniques in fertilizer application. In *Fertilizer Technology and Use,* 3rd edition, ed. O.P. Engelstad, p. 526. Madison, WI: Soil Science Society of America.

Rao, I.M. and N. Terry. 1989. Leaf phosphate status, photosynthesis, and carbon partitioning in sugar beet. I. Changes in growth, gas exchange, and Calvin cycle enzymes. *Plant Physiol.* 90:814–819.

Ratjen, A.M. and J. Gerendás. 2009. A critical assessment of the suitability of phosphite as a source of phosphorus. *J. Plant Nutr. Soil Sci.* 172:821–828.

Redulla, C.A., J.R. Davenport, R.G. Evans, M.J. Hattendorf, A.K. Alva, and R.A. Boydston. 2002. Relating potato yield and quality to field scale variability in soil characteristics. *Am. J. Potato Res.* 79:317–323.

Reichman, S.M. 2002. The response of plants to metal toxicity: A review focusing on copper, manganese and zinc. Australian Minerals and Energy Environment Foundation Occasional Paper No. 14.

Rhue, R.D., D.R. Hensel, T.L. Yuan, and W.K. Robertson. 1981. Ammonium orthophosphate and ammonium polyphosphate as sources of phosphorus for potatoes. *Soil Sci. Soc. Am. J.* 45:1229–1233.

Roberts, S., H.H. Cheng, and F.O. Farrow. 1991. Potato uptake and recovery of nitrogen-15-enriched ammonium nitrate from periodic applications. *Agron. J.* 83:378–381.

Robertson, W.K., K. Hinson, and L.C. Hammond. 1977. Foliar fertilization of soybeans (*Glycine max* L) Merr in Florida. *Soil Crop Sci. Soc. Fla. Proc.* 36:77–79.

Robinson, D. 1986. Limits to nutrient inflow rates in roots and root systems. *Physiol. Plant.* 68:551–559.

Rolland, F., E. Baena-Gonzalez, and J. Sheen. 2006. Sugar sensing and signalling in plants: Conserved and novel mechanisms. *Annu. Rev. Plant Biol.* 57:675–709.

Romkens, M.J.M., D.W. Nelson, and J.V. Mannering. 1973. Nitrogen and phosphorus composition of surface runoff as affected by tillage method. *J. Environ. Qual.* 2:292–295.

Rosen, C.J. and P.M. Bierman. 2008. Potato yield and tuber set as affected by phosphorus fertilization. *Am. J. Potato Res.* 85:110–120.

Rosen, C.J. and R. Eliason. 2005. Nutrient management for commercial fruit and vegetable crops. *BU-05886*. St. Paul, MN: University of Minnesota.

Rosen, C.J., K.A. Kelling, J.C. Stark, and G.A. Porter. 2014. Optimizing phosphorus management in potato production. *Am. J. Potato Res.* 91:145–160.

Rossiter, R.C. 1978. Phosphorus deficiency and flowering in subterranean clover (*Tr. subterraneum* L.). *Ann. Bot.* 42:325–329.

Ruark, M.D., K.A. Kelling, and L.W. Good. 2014. Environmental concerns of phosphorus management in potato production. *Am. J. Potato Res.* 91:132–144.

Russell, E.J. 1961. *Soil Conditions and Plant Growth*, 9th edition. New York: John Wiley & Sons.

Ryden, J.C., J.K. Syers, and R.F. Harris. 1973. Phosphorus in runoff and streams. *Adv. Agron.* 25:1–45.

Safaya, N.M. 1976. Phosphorus-zinc interaction in relation to absorption rates of phosphorus, zinc, copper, manganese and iron in corn. *Soil Sci. Soc. Am. J.* 40:719–722.

Sanchez, C.A. 1990. Soil testing and fertilizer recommendations for crop production on organic soils in Florida. *University of Florida Agricultural Experiment Station Bulletin 876*, Gainesville, FL.

Sanchez, C.A. 2007. Phosphorus. In *Handbook of Plant Nutrition*, eds. A.V. Barker and D.J. Pilbeam, pp. 51–90. Boca Raton, FL: CRC Press.

Sanchez, C.A. and N.M. El-Hout. 1995. Response of diverse lettuce types to fertilizer phosphorus. *HortScience* 30:528–531.

Sanchez, C.A., V.L. Guzman, and R.T. Nagata. 1990. Evaluation of sidedress fertilization for correcting nutritional deficits in crisphead lettuce on Histosols. *Proc. Fla. State Hortic. Soc.* 103:110–113.

Sanchez, C.A., P.S. Porter, and M.F. Ulloa. 1991. Relative efficiency of broadcast and banded phosphorus for sweetcorn produced on Histosols. *Soil Sci. Soc. Am. J.* 55:871–875.

Sanchez, C.A., S. Swanson, and P.S. Porter. 1990. Banding P to improve fertilizer use efficiency in lettuce. *J. Am. Soc. Hortic. Sci.* 115:581–584.

Sanderson, J.B., T.W. Bruulsema, R. Coffin, B. Douglas, and J.A. MacLeod. 2002. Phosphorus sources for potato production. *Better Crops* 86(4):10–12.

Sanderson, J.B., J.A. MacLeod, B. Douglas, R. Coffin, and T. Bruulsema. 2003. Phosphorus research on potato in PEI. *Acta Hortic.* 619:409–417.

Sarkar, D., S.K. Pandey, K.C. Sud, and A. Chanemougasoundharam. 2004. In vitro characterization of manganese toxicity in relation to phosphorus nutrition in potato (*Solanum tuberosum* L.). *Plant Sci.* 167:977–986.

Sattelmacher, B., F. Klotz, and H. Marschner. 1990. Influence of the nitrogen level on root morphology of two potato varieties differing in nitrogen acquisition. *Plant Soil* 123:131–137.

Sawyer, C.N. 1947. Fertilization of lakes by agricultural and urban drainage. *J. New Engl. Water Works Assoc.* 61:109–127.

Schenk, M.K. and S.A. Barber. 1979. Root characteristics of corn genotypes as related to P uptake. *Agron. J.* 71:921–924.

Schinder, D.W. 1977. Evolution of phosphorus limitation in lakes. *Science* 195:260–262.

Shane, M.W., M.E. McCully, and H. Lambers. 2004. Tissue and cellular phosphorus storage during development of phosphorus toxicity in *Hakea prostrata* (Proteaceae). *J. Exp. Bot.* 55:1033–1044.

Sharma, U.C. and B.R. Arora. 1987. Effect of nitrogen, phosphorus and potassium application on yield of potato tubers (*Solanum tuberosum* L.). *J. Agric. Sci. Camb.* 108:321–329.

Sharpley, A.N. 1995. Dependence of runoff phosphorus on extractable soil phosphorus. *J. Environ. Qual.* 24:920–926.

Sharpley, A.N., S.C. Chapra, R. Wedepohl, J.T. Sims, T.C. Daniel, and K.R. Reddy. 1994. Managing agricultural phosphorus for protection of surface waters: Issues and options. *J. Environ. Qual.* 23:437–451.

Sharpley, A.N., T.C. Daniel, and D.R. Edwards. 1993. Phosphorus movement in the landscape. *J. Prod. Agric.* 6:492–500.

Sharpley, A.N., T.C. Daniel, J.T. Sims, J.L. Lemunyon, R.G. Stevens, and R. Parry. 1999. Agricultural phosphorus and eutrophication. Washington, DC: *U.S. Department of Agriculture, Agricultural Research Service, ARS-*149, 42 pp.

Sharpley, A.N., T.C. Daniel, J.T. Sims, J. Lemunyon, R. Stevens, and R. Parry. 2003. *Agricultural Phosphorus and Eutrophication*, 2nd edition, *ARS-*149. Washington, DC: USDA.

Sharpley, A.N., R.W. McDowell, and P.J. Kleinman. 2001. Phosphorus loss from land to water: Integrating agricultural and environmental management. *Plant Soil* 237:287–307.

Sharpley, A.N., U. Singh, G. Uehara, and J. Kimble. 1989. Modeling soil and plant phosphorus dynamics in calcareous and highly weathered soils. *Soil Sci. Soc. Am. J.* 53:153–158.

Sharpley, A.N., S.J. Smith, O.R. Jones, W.A. Berg, and G.A. Coleman. 1992. The transport of bioavailable phosphorus in agricultural runoff. *J. Environ. Qual.* 21:30–35.

Shepard, R. 2000. Nitrogen and phosphorus management on Wisconsin farms: Lessons learned for agricultural water quality programs. *J. Soil Water Conserv.* 55:63–68.

Shi, X.N., L.S. Wu, W.P. Chien, and Q.J. Wang. 2011. Solute transfer from the soil surface to overland flow: A review. *Soil Sci. Soc. Am. J.* 75:1214–1225.

Silberstein, O. and S.H. Wittwer. 1951. Foliar application of phosphatic fertilizers to vegetable crops. *Proc. Am. Soc. Hortic. Sci.* 58:179–180.

Sims, J.T. 1998. Phosphorus soil testing: Innovation for water quality protection. *Commun. Soil Sci. Plant Anal.* 29:1471–1489.

Sims, J.T., A.C. Edwards, O.F. Schoumans, and R.R. Simard. 2000. Integrating soil phosphorus testing into environmentally based agricultural management practices. *J. Environ. Qual.* 29:60–71.

Singh, C.P. and A. Amberger. 1998. Organic acids and phosphorus solubilization in straw composted with rock phosphate. *Biores. Technol.* 63:13–16.

Singh, J.P., R.E. Karamanos, and J.W.B. Stewart. 1988. The mechanism of phosphorus-induced zinc deficiency in bean (*Phaseolus vulgaris* L.). *Can. J. Soil Sci.* 68:345–358.

Sleight, D.M., D.H. Sander, and G.A. Peterson. 1984. Effect of fertilizer phosphorus placement on the availability of phosphorus. *Soil Sci. Soc. Am. J.* 48:336–340.

Smith, F.W., W.A. Jackson, and P.J. Van den Berg. 1990. Internal phosphorus flows during development of phosphorus stress in *Stylosanthes hamata. Aust. J. Plant Physiol.* 17:451–464.

Smith, J.A. and R.W. Sheard. 1957. Evaluation and calibration of phosphorus soil test methods for predicting fertilizer requirements of potatoes. *Can. J. Soil Sci.* 37:134–142.

Soltanpour, P.N. 1969. Effect of nitrogen, phosphorus and zinc placement on yield and composition of potatoes. *Agron. J.* 61:288–289.

Sparrow, L.A., K.S.R. Chapman, D. Parsley, P.R. Hardman, and B. Cullen. 1992. Response of potatoes (*Solanum tuberosum* cv. Russet Burbank) to band-placed and broadcast high cadmium phosphorus fertiliser on heavily cropped Krasnozems in north-western Tasmania. *Aust. J. Exp. Agric.* 32:113–119.

Sposito, G. 2008. *The Chemistry of Soils*, 2nd edition. New York: Oxford University Press.

Stalham, M.A. and E.J. Allen. 2001. Effect of variety, irrigation regime and planting date on depth, rate, duration, and density of root growth in the potato (*Solanum tuberosum*) crop. *J. Agric. Sci. Camb.* 137:251–270.

Stanberry, C.O., C.D. Converse, H.R. Haise, and A.J. Kelly. 1955. Effect of moisture and phosphate variables on alfalfa hay production on the Yuma mesa. *Proc. Soil Sci. Soc. Am.* 19:303–310.

Stanberry, C.O., H.A. Schreiber, L.R. Cooper, and S.D. Mitchell. 1965. Vertical movement of phosphorus in some calcareous soils of Arizona. Unpublished data. University of Arizona Agric. Exp. Stn. Tech. Paper.

Stark, J.C. and B.G. Hopkins. 2014. Fall and spring phosphorus fertilization of potato using a dicarboxylic acid polymer (AVAIL®). *J. Plant Nutr.* DOI: 10.1080/01904167.2014.983124.

Stark, J.C. and J.C. Ojala. 1989. Comparison of banded ammonium polyphosphate and acid urea phosphate as P sources for potatoes. *HortScience* 24:282–284.

Stark, J.C. and D.T. Westermann. 2003. Nutrient management. In *Potato Production Systems*, eds. J.C. Stark and S.L. Love, pp. 115–135. Moscow, ID: University of Idaho Agricultural Communications.

Stark, J.C., D.T. Westermann, and B.G. Hopkins. 2004. Nutrient management guidelines for Russet Burbank potato. *Bulletin 840*. Moscow, ID: University of Idaho Agricultural Communications.

Stevens, W.B., A.D. Blaylock, J.M. Krall, B.G. Hopkins, and J.W. Ellsworth. 2007. Sugarbeet yield and nitrogen use efficiency with preplant broadcast, banded, or point-injected nitrogen application. *Agron. J.* 99:1252–1259.

Steward, J.H. and M.E. Tate. 1971. Gel chromatography of soil organic phosphorus. *J. Chromatogr.* 60:75–78.

Summerhays, J.S., B.G. Hopkins, V.D. Jolley, and M.W. Hill. 2014. Enhanced phosphorus fertilizer (Carbond P®) supplied to maize in moderate and high organic matter soils. *J. Plant Nutr.* DOI 10.1080/01904167.2014.973039.

Sutton, C.D. 1969. Effects of low temperature on phosphate nutrition of plants—A review. *J. Sci. Food Agric.* 20:1–3.

Syers, J.K., A.E. Johnston, and D. Curtin. 2008. Efficiency of soil and fertilizer phosphorus use. Reconciling changing concepts of soil phosphorus behavior with agronomic information. *FAO Fertilizer and Plant Nutrition Bulletin No. 18.* Rome, Italy: FAO.

Tan, K.H. 2003. *Humic Matter in Soil and the Environment; Principles and Controversies.* New York: Marcel Dekker, Inc.

Tanner, C.B., G.G. Weis, and D. Curwen. 1982. Russet Burbank rooting in sandy soils with pans following deep plowing. *Am. J. Potato Res.* 59:107–112.

Terman, G.L., P.M. Giordano, and S.E. Allen. 1972. Relationship between dry matter yields and concentration of Zn and P in young corn plants. *Agron. J.* 64:686–687.

Teubner, F.G., M.J. Bukovac, S.H. Wittwer, and B.K. Guar. 1962. The utilization of foliar-applied radiophosphorus by several vegetable crops and tree fruits under field conditions. *Mich. Agric. Exp. Stn. Q. Bull.* 44:455–465.

Thien, S.J. 1976. Stabilizing soil aggregates with phosphoric acid. *Soil Sci. Soc. Am. J.* 40:105–108.

Thomas, G.W. and D.E. Peaslee. 1973. Testing soils for phosphorus. In *Soil Testing and Plant Analysis*, eds. L.M. Walsh and J.D. Beaton, pp. 115–132. Madison, WI: Soil Science Society of America Inc.

Thornton, M.K., D. Beck, J. Stark, and B. Hopkins. 2008. Potato variety response to phosphorus fertilizer. In *Proceedings of the Idaho Nutrient Management Conference*, pp. 19–23. Jerome, ID.

Thornton, M.K., R.G. Novy, and J.C. Stark. 2014. Improving phosphorus use efficiency in the future. *Am. J. Potato Res.* 91:175–179.

Truog, E. 1930. Determination of the readily available phosphorus of soils. *Agron. J.* 22:874–882.

Upadhyay, A.P., M.R. Deshmukh, R.P. Rajput, and S.C. Deshmukh. 1988. Effect of sources, levels and methods of phosphorus application on plant productivity and yield of soybean. *Indian J. Agron.* 33:14–18.

USDA. 2014. Crop production 2013 summary. *National Agricultural Statistics Service.* http://www.usda.gov/nass/PUBS/TODAYRPT/crop1014.pdf. Accessed July 21, 2014.

U.S. Environmental Protection Agency (USEPA). 1996. *Clean Water Action Plan: Restoring and Protecting America's Waters.* Washington, DC: U.S. Environmental Protection Agency.

Van Kauwenbergh, S.J., M. Stewart, and R. Mikkelsen. 2013. World reserves of phosphate rock…A dynamic and unfolding story. *Better Crops* 97(3):18–20.

Wagganman, W.H. 1969. *Phosphoric Acid, Phosphates, and Phosphatic Fertilizers*, 2nd edition. New York: Hafner Publications Company.

Wallace, T. 1961. *The Diagnosis of Mineral Deficiencies in Plants by Visual Diagnosis.* New York: Chemical Publishing Co. Inc.

Walworth, J.L. and J.E. Muniz. 1993. A compendium of tissue nutrient concentrations for field-grown potatoes. *Am. Potato J.* 70:579–597.

Watanabe, F.S., S.R. Olsen, and R.E. Danielson. 1960. Phosphorus availability as related to soil moisture. *Transactions of the 7th International Congress of Soil Science*, Volume III, pp. 450–456. Madison, MI: International Society of Soil Science.

Weaver, J.E. 1926. *Root Development of Field crops.* New York: McGraw-Hill.

Webb, J.R., A.P. Mallarino, and A.M. Blackmer. 1992. Effects of residual and annually applied phosphorus on soil test values and yields of corn and soybean. *J. Prod. Agric.* 5:148–152.

Webb, M.J. and J.F. Loneragan. 1988. Effect of zinc deficiency on growth, phosphorus concentration, and phosphorus toxicity of wheat plants. *Soil Sci. Soc. Am. J.* 52:1676–1680.

Welch, L.F., D.L. Mulvaney, L.V. Boone, G.E. McKibben, and J.W. Pendleton. 1966. Relative efficiency of broadcast versus banded phosphorus for corn. *Agron. J.* 58:283–287.

Westermann, D.T. 1984. Mid-season P fertilization effects on potatoes. In *Proceedings of the 35th Annual N.W. Fertilizer Conference*, NWISRL Publication Number 0545, 73–81. Pasco, Washington, DC: USDA. http://eprints.nwisrl.ars.usda.gov/1036/. Accessed May 16, 2013.

Westermann, D.T. 1992. Lime effects on phosphorus availability in a calcareous soil. *Soil Sci. Soc. Am. J.* 56:489–494.

Westermann, D.T. 2005. Nutritional requirements of potato. *Am. J. Potato Res.* 82:301–307.

Westermann, D.T. and G.E. Kleinkopf. 1985. Phosphorus relationships in potato plants. *Agron. J.* 77:490–494.

White, P.J. 2012a. Ion uptake mechanisms of individual cells and roots: Short-distance transport. In *Mineral Nutrition of Higher Plants*, 3rd edition, ed. P. Marschner, pp. 7–47. London, U.K.: Academic Press.

White, P.J. 2012b. Long-distance transport in the xylem and phloem. In *Mineral Nutrition of Higher Plants*, 3rd edition, ed. P. Marschner, pp. 49–70. London, U.K.: Academic Press.

White, P.J. and J.P. Hammond. 2008. Phosphorus nutrition of terrestrial plants. In *The Ecophysiology of Plant-Phosphorus Interactions*, eds. P.J. White and J P. Hammond, pp. 51–81. Dordrecht, the Netherlands: Springer-Verlag.

Withers, P.J.A. and H.P. Jarvie. 2008. Delivery and cycling of phosphorus in rivers: A review. *Sci. Total Environ.* 400:379–395.

Yamaguchi, J. 2002. Measurement of root diameter in field-grown crops under a microscope without washing. *Soil Sci. Plant Nutr.* 48:625–629.

Yamaguchi, J. and A. Tanaka. 1990. Quantitative observation on the root system of various crops growing in the field. *Soil Sci. Plant Nutr.* 36:483–493.

Yanai, J., A. Nakano, K. Kyuma, and T. Kosaki. 1997. Application effects of controlled-availability fertilizer on dynamics of soil solution composition. *Soil Sci. Soc. Am. J.* 61:1781–1786.

Young, R.D., D.G. Westfall, and G.W. Colliver. 1985. Production, marketing, and use of phosphorus fertilizers. In *Fertiliser Technology and Use*, 3rd edition, ed. O.P. Engelstad, pp. 323–376. Madison, WI: Soil Science Society of America.

Zhang, F.S., J. Ma, and Y.P. Cao. 1997. Phosphorus deficiency enhances root exudation of low-molecular weight organic acids and utilization of sparingly soluble inorganic phosphates by radish (*Raphanus sativus* L.) and rape (*Brassica napus* L.) plants. *Plant Soil* 196:261–264.

Zhong, H. 1993. Influence of chloride on the uptake and translocation of phosphorus in potato. *J. Plant Nutr.* 16:1733–1737.

Zhu, Y.-G, F.A. Smith, and S.E. Smith. 2002. Phosphorus efficiencies and their effects on Zn, Cu, and Mn nutrition of different barley (*Hordeum vulgare*) cultivars grown in sand culture. *Aust. J. Agric. Res.* 53:211–216.

4 Potassium

Nand Kumar Fageria

CONTENTS

4.1 HISTORICAL BACKGROUND

When man started cultivation of crop plants to meet food demands marks the dawn of agriculture. The exact time is not known, but certainly, it was several thousand years before the birth of Christ. Until then, man hunted almost exclusively for food and was nomadic in habits (Tisdale and Nelson, 1966). In the early days, several scientists, especially from Europe, contributed significantly to the development of agriculture science or mineral nutrition of plants. However, the most significant contribution came from Justus von Liebig (1803–1873), a German chemist, who very effectively deposed the humus myth. Liebig also emphasized the importance of inorganic plant nutrients as cycling between the living nature and the inorganic nature, mediated by plants. Hence, the contributions that Liebig made to the advancement of agriculture were monumental, and Liebig is perhaps quite rightly recognized as the father of agricultural chemistry (Tisdale and Nelson, 1966).

The *era* of field experimentation, which began in 1834 when J.B. Boussingault, a French Chemist, set up the first field experiments at Bechelbonn, Alsace (France), was placed on a modern scientific basis by Liebig's report of 1840 (Collis-George and Davey, 1960). The first field experiment in the form used today was established by Lawes and Gilbert at Rothamsted in 1843. Since then, field experiments have sought for and have confirmed the importance of the essential elements in influencing the production of field crops. However, a great deal of the evidence for discovery of the essentiality of nutrients has been in laboratory experiments in nutrient solution and not from field experiments (Collis-George and Davey, 1960; Fageria, 2007a).

The application of field trial results led to a large increase in agricultural production around the world. Research in agriculture is a complex process and demands constant efforts and experimentation due to change in weather conditions, soil heterogeneity, and release of new cultivars (Barley, 1964; Fageria, 2007a). These changes are sometimes so significant that all management practices in use to produce good yields of crops need reevaluation and adjustments to changed situations. For example, when a new cultivar of a crop is released, its nutritional requirements are different from those under cultivation due to difference in yield potential, disease and insect resistance, and change in architecture. Therefore, field experiments are the basic need in modern agriculture to evaluate nutritional requirements under different agroecological regions. It is very hard to transfer experimental results of one region to another due to differences in soil properties, climatic differences, and socioeconomic conditions of the farmers. All of these factors determine the technological development and its adaptation by the farmers. In conducting field experimentation, certain basic principles should be followed to arrive at meaningful conclusions. Some of these important principles or considerations in field experimentation are discussed by Fageria (2007a). The principles discussed in this chapter will help agricultural scientists in planning and execution of research trials. Of course, the discussion is concerned mainly with the field of soil fertility and plant nutrition, but some basic principles are applicable to other disciplines of agricultural science too. These principles are applicable everywhere with slight modification according to the circumstances of a particular situation. Most of the points discussed are the outcome of the author's practical experience of more than 40 years in the field of agriculture, in general, and soil fertility and plant nutrition, in particular (Fageria, 2007a).

In the earlier days, supply of potassium (K) was with animal manures and wood ashes. However, now, it is manufactured from ore deposits. Potassium salts were mined first commercially in Germany in 1861 after Liebig's doctrine of mineral nutrition had demonstrated their value as a fertilizer (Sheldrick, 1985). The essentiality of K for plants was established in 1890 by A. F. Z. Schimper (Fageria et al., 2011). Since then, the use of this element in crop production has increased steadily. More than 95% of world KCl consumption is by the agricultural sector (Stone, 2008). Canada, the Soviet Union, and Europe have the largest deposits of K ores. Deposits of K ores are also in some countries of Africa, Asia, and Latin America (Fageria, 2009). Most of the potassium in the world is produced as potassium chloride and is known as muriate of potash in the fertilizer industry.

World consumption of muriate of potash, fertilizer grade KCl, increased from an estimated 38 Tg in 2001 to 49 Tg in 2007 (Stone, 2008; Nelson et al., 2010). Global K removal from soils has far exceeded KCl use (Nelson et al., 2010). From 1997 to 1999, KCl applications replaced approximately 73% of K removed from soils in developed countries, whereas in developing countries, KCl applications replaced only 20% of K removed (Nelson et al., 2010). Potassium deficiency in annual crops is reported widely in many countries (Rengel and Damson, 2008; Fageria, 2009; Fageria et al., 2010a,b; Nelson et al., 2010). Presently, about one-quarter of the arable soils and three-quarters of paddy soils in China are K deficient for rice (Yang et al., 2004). Long-term field experiments across five Asian countries indicated a negative K balance in irrigated rice systems, for example, in the Mekong River delta in Vietnam, such negative balances can be as high as 86 kg K ha^{-1} year^{-1} (Nguyen et al., 2006), with removal of straw having a particularly severe impact on the K budget (Rengel and Damos, 2008). In southwestern Australia, the incidence of K deficiency in wheat has increased steadily, with two-thirds of the arable soils prone to K depletion through continued removal in hay or grain and straw (Pal et al., 2001).

Potassium plays a significant role in many physiological and biochemical processes in plants (Brady and Weil, 2002; Fageria, 2009). In addition, if applied in adequate amount, K also controls many plant diseases and provides tolerance to drought, heat, and cold (Qian et al., 1997; Fageria, 2009; Rowland et al., 2010). In Brazil, the major part of K fertilizers is imported (>85%). Looking into these scenarios, improving K nutrition of annual crops is crucial to improve yields and to reduce costs of production. The objectives of this chapter are to discuss latest advances in mineral nutrition of K by annual crops to improve its uptake and use efficiency and consequently impart higher yields.

4.2 CYCLE IN SOIL–PLANT SYSTEM

Nutrient cycling involves addition, transformation, and uptake of a nutrient by plants. Knowledge of nutrient cycles in the soil–plant system is fundamental to adopt necessary soil and crop management practices to improve its uptake and use efficiency and consequently to give higher yields. Potassium is added to soil as chemical fertilizers, crop residues, or farmyard manures. In addition, K is also available to plants by weathering of soil minerals. Its removal from the soil–plant system is by soil erosion, leaching, and uptake by plants. Figure 4.1 shows the potassium cycle in the soil–plant system.

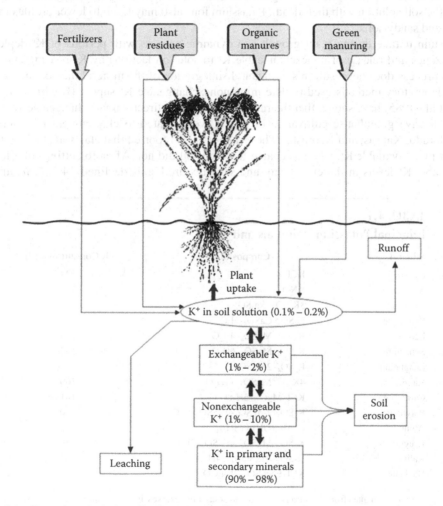

FIGURE 4.1 Potassium cycle in soil–plant system. (From Fageria, N.K., *Nutrient Uptake in Crop Plants*, CRC Press, Boca Raton, FL, 2009.)

The total soil K concentration typically ranges from 0.2% to 3.3% of the total soil mass. Up to 98% of the total K in the plow layer, however, is relatively unavailable for plant growth as it is held in primary minerals, such as micas or feldspar (Sparks and Huang, 1985). The plant-available and temporarily dynamic pools, on the other hand, include exchangeable K, nonexchangeable K, and microbial biomass K (Lorenz et al., 2010). The exchangeable K pool in soils ranges from 1% to 2% of total soil K and includes K electrostatically bound to clay mineral surfaces and organic matter and K in the soil solution. The nonexchangeable K pool represents 1%–10% of total soil K. Within this pool, K is bound primarily in the interlayers of micaceous clay minerals and is therefore slowly available (Sparks and Huang, 1985; Fageria, 2013).

Potassium is not complexed or bound up into organic matter to any degree, compared to N and P, and essentially all of it is associated with the mineral fraction (Foth and Ellis, 1988). In addition to chemical fertilizers, weathering of K minerals is the major process for supplying K in forms that plants can use. A list of principal K minerals, their composition, and their K concentration is given in Table 4.1. Soil parent materials contain K that is mainly in feldspars and micas. These minerals weather, and the K ions are released into the soil solution or exchange with other cations on soil colloids. The K that is released into solution may undergo various changes in the soil, including fixation, or uptake by plants or microorganisms. The K that is fixed by microorganism may return again to the soil solution with their death. Potassium ions also may leach to lower profiles in humid climates and sandy soils.

Potassium uptake during plant growth is a dynamic process with periods of K^+ depletion in the root zones and release of nonexchangeable K^+ to solution factions by K^+ bearing clay minerals. This process does not result in significant disintegration of the mineral matrix and can occur in some laboratory methods used to determine nonexchangeable K^+ supply (Havlin et al., 1985). Havlin et al. (1985) have shown that the rate and amount of directly nonexchangeable K^+ released during intensive greenhouse cultivation of alfalfa were related to clay content of several soils from Colorado, Kansas, and Nebraska. These authors also reported that clay soils had a long-term supply of plant-available K^+, whereas coarse-textured soil did not. After 16 cuttings of alfalfa, the exchangeable K^+ levels in the clay, loam, and sand-textured soils declined 44%, 33%, and 58%,

TABLE 4.1
Principal Potassium Minerals and Ores

Mineral	Composition	K Content (g kg^{-1})
Sylvite	KCl	524
Niter	KNO_3	386
Glaserite	$3K_2SO_4 \cdot Na_2SO_4$	353
Syngenite	$K_2SO_4 \cdot CaSO_4 \cdot H_2O$	238
Leonite	$K_2SO_4 \cdot MgSO_4 \cdot 4H_2O$	213
Schoenite	$K_2SO_4 \cdot MgSO_4 \cdot 6H_2O$	194
Langbeinite	$K_2SO_4 \cdot 2MgSO_4$	188
Kainite	$4KCl \cdot 4MgSO_4 \cdot 11H_2O$	160
Carnallite	$KCl \cdot MgCl_2 \cdot 6H_2O$	141
Polyhalite	$K_2SO_4 \cdot 2MgSO_4 \cdot 2CaSO_4 \cdot H_2O$	130
Alunite	$K_2Al_6 \cdot (OH)_{12} \, (SO_4)_4$	95
Krugite	$K_2SO_4 \cdot MgSO_4 \cdot 4CaSO_4 \cdot 2H_2O$	89
Kalinite	$K_2SO_4 \cdot Al_2(SO_4)_3 \cdot 24H_2O$	82
Hanksite	$KCl \cdot 9Na_2SO_4 \cdot 2Na_2CO_3$	25

Source: Adapted from Sheldrick, W.F., World potassium reserves, in: *Potassium in Agriculture*, D. Munson (ed.), ASA, CSSA, and SSSA, Madison, WI, 1985, pp. 3–28.

respectively. Reddy (1976) reported that medium- and fine-textured Colorado soils were capable of supplying sufficient K^+ to a 12-metric-ton (Mg) ha^{-1} annual alfalfa crop for 3 years, whereas the K^+ supply in coarse-textured soils was sufficient for only 1 year. Soil chemical properties, which influence the phytoavailability of K^+, include K^+ activity in soil solution, K^+ diffusion rate, buffering capacity, and exchange equilibrium. Activity of K^+ in the soil solution is reported to represent readily available K^+.

4.3 DEFICIENCY SYMPTOMS

Potassium deficiency limits plant growth and development. This type of deficiency can be recognized only if there are plants without K application for comparison. Figure 4.2 shows deficiency of N, P, and K in upland rice compared with rice with adequate rate N, P, and K fertilization. It is very clear from this figure that K shortage suppressed growth, but K deficiency symptoms were not as clear as N and P. Similarly, Figures 4.3 and 4.4 show the growth of two genotypes of lowland or flooded rice at low and high K levels. Growth at 0 mg K kg^{-1} was limited significantly in both genotypes. But there were no visual deficiency symptoms at low K levels. If there have been no comparative plants, it would have been impossible to recognize K deficiency in the zero mg K treatments in the genotypes. Figure 4.5 shows that K increases panicle density in rice and also that maturation was enhanced in the pot that received 300 mg K kg^{-1} compared to zero mg K kg^{-1} soil treatment. Potassium also improved root growth of crop plants if supplied in adequate amount (Figures 4.6 and 4.7).

Potassium like N and P is highly mobile in plant tissues. Hence, K^+ deficiency symptoms first appear in the old leaves. Potassium deficiency symptom shows up as scorching along leaf margins of leaves (Figure 4.8). Potassium-deficient plants grow slowly. They have poorly developed root systems. Stalks are weak, and lodging is common. Seeds and fruits are small and shriveled, and plants possess low resistance to disease. Plants under stress from short-K supplies are very susceptible to unfavorable weather. Although it cannot be detected as it is happening, stand loss in forage grass and legumes is a direct result of K^+ deficiency. In grass–legume pastures, the grass crowds out the legume when the K^+ runs short because the grass has the greater capacity to absorb K and the legume is starved out (Fageria, 2009).

FIGURE 4.2   Upland rice plants with N + P + K and without –N, –P, and –K.

FIGURE 4.3 Growth of irrigated rice genotype BRA051108 at two K levels.

FIGURE 4.4 Growth of irrigated rice genotype BRA051130 at two K levels.

FIGURE 4.5 Growth of upland rice cultivar BRS Sertaneja at different N, P, and K levels. Left ($N_1P_1K_0$ = 150 mg N kg^{-1}, 100 mg P kg^{-1}, and 0 mg K kg^{-1} soil) and right ($N_1P_1K_2$ = 150 mg N kg^{-1}, 100 mg P kg^{-1}, and 200 mg K kg^{-1} soil).

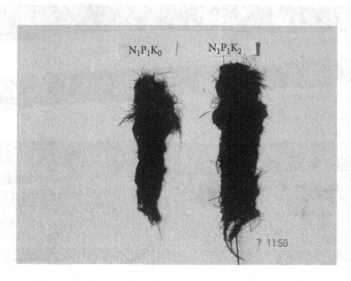

FIGURE 4.6 Root growth of upland rice cultivar BRS Sertaneja at different N, P, and K levels. Left ($N_1P_1K_0$ = 150 mg N kg^{-1}, 100 mg P kg^{-1}, and 0 mg K kg^{-1} soil) and right ($N_1P_1K_2$ = 150 mg N kg^{-1}, 100 mg P kg^{-1}, and 200 mg K kg^{-1} soil).

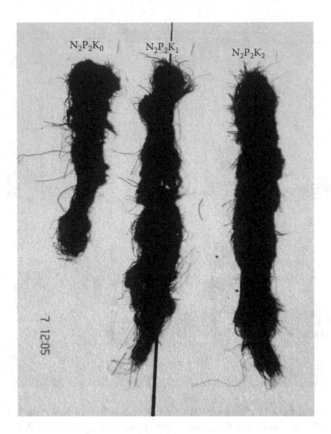

FIGURE 4.7 Influence of potassium on root growth of upland rice. Left to right ($N_2P_2K_0$ = 300 mg N kg^{-1} + 200 mg P kg^{-1} + 0 mg K kg^{-1} soil), ($N_2P_2K_1$ = 300 mg N kg^{-1} + 200 mg P kg^{-1} + 100 mg K kg^{-1} soil), and ($N_2P_2K_2$ = 300 mg N kg^{-1} + 200 mg P kg^{-1} + 300 mg K kg^{-1} soil).

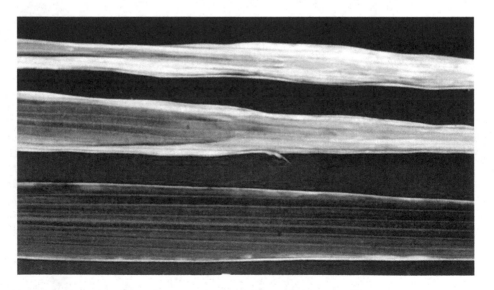

FIGURE 4.8 **(See color insert.)** Rice leaves with severe (top), moderate (middle), and mild (bottom) K deficiency symptoms. (From Fageria, N.K. and Barbosa Filho, M.P., *Nutritional Deficiencies in Rice: Identification and Correction*, EMBRAPA-CNPAF document 42, Goiania, Brazil, 1994.)

Potassium deficiency has an impact on numerous synthetic processes, such as synthesis of sugar and starch, lipid, and ascorbate and also on the formation of leaf cuticles (Mengel, 2007). The latter are developed poorly under K deficiency. Cuticles protect plants against water loss and infection by fungi. At an advanced stage of K deficiency, chloroplasts and mitochondria collapse. Potassium-deficient plants have a low energy status because K is essential for efficient energy transfer in chloroplasts and mitochondria (Mengel, 2007).

4.4 UPTAKE OF POTASSIUM BY PLANTS

Nutrient absorption by plants is referred to usually as ion uptake or ion absorption because it is the ionic form in which nutrients are absorbed by roots (Fageria, 2013). Cations and anions may be absorbed independently and may not be absorbed in equal quantities; however, electroneutrality must be maintained in the growth medium. Therefore, ionic relationship achieves major importance in plant nutrition. The majority of cations in plant tissues are in the inorganic form, predominantly K^+, Ca^{2+}, and Mg^{2+}, and the majority of anions are in the organic form. The organic ions are synthesized within the tissue, whereas inorganic ions are absorbed from the growth medium. Uptake of K^+ is very selective and closely related to metabolic activity of plants. Generally, the monovalent cations are absorbed rapidly, whereas the divalent cations, especially Ca^{2+}, are absorbed more slowly. Similarly, the monovalent anions generally are absorbed more rapidly than polyvalent anions (Fageria et al., 2006). The so-called physiological acidity or alkalinity of a salt depends upon which ion of the salt, the cation or the anion, is absorbed most rapidly from solution. Thus, a salt like K_2SO_4 would be physiologically acid, that is, it would acidify a nutrient solution or soil solution, since the K^+ would enter the roots more rapidly than the SO_4^{2-}. By the same token, $CaCl_2$ would be physiologically alkaline, raising the alkalinity of a solution, since the Ca^{2+} enters plant roots more slowly than the Cl^- (Fageria and Baligar, 2005; Fageria, 2013).

Knowledge of uptake of nutrients by crop plants is an important aspect in maintaining soil fertility for sustainable crop production. The amount of nutrients removed by a crop species or cultivar of the same species should be replenished to sustain crop productivity over a longer duration. The removal of nutrients in harvested plant biomass has been used to estimate nutrient requirements of annual crops (Fageria, 2009). The nutrient requirement of crop plants varies with soil type, crop species, and yield level. To determine nutrient uptake of a crop species or cultivar, weight of straw and grain should be measured at harvest and chemical analysis of plant materials (grain and straw separately) should be done. Results of straw and grain yield of grain crops generally are expressed in kg ha^{-1} or Mg ha^{-1}. Nutrient uptake results are expressed in kg ha^{-1} for macronutrients and g ha^{-1} for micronutrients for field experiments. When nutrient uptake is determined for a particular crop species, its yield should be above average. Lower yield does not represent an actual value of nutrient uptake or requirement (Fageria, 2013).

Accumulation of potassium by crop plants is larger than accumulation of any other mineral nutrient derived from soil with the exception of nitrogen in some crop species. Data related to uptake of macronutrients and micronutrients by lowland rice or upland rice (*Oryza sativa* L.), dry bean (*Phaseolus vulgaris* L.), corn (*Zea mays* L.), and soybean (*Glycine max* Merr.) are presented in Tables 4.2 through 4.4. In lowland rice, upland rice, and corn, accumulation of K was the highest in straw and grain among macronutrients, and in legumes nitrogen accumulation was highest in straw and grain compared to other nutrients. Among micronutrients, the highest accumulation was of Mn and Fe and the minimum accumulation was of B and Cu in cereals (lowland, upland rice, and corn). In legumes, maximum uptake of micronutrients was of Fe, and the minimum was of Cu. Data in Table 4.5 show macronutrient and micronutrient requirements to produce 1 metric ton (Mg) of grain. To produce one metric ton of grain in lowland or upland rice, K requirement was maximum followed by N. In legumes, the highest requirement to produce one metric ton of grain was for N followed by K. Overall, phosphorus had maximum use efficiency in the cereals as well as legume crops among macronutrients. In micronutrients, maximum use efficiency for grain production was for B in rice and Cu in corn, dry bean, and soybean (Fageria, 2013).

TABLE 4.2
Straw and Grain Yield and Nutrient Accumulation by Lowland and Upland Rice (*Oryza sativa* L.) Grown on Brazilian Inceptisol and Oxisol

Yield and Nutrient Accumulation	Lowland Rice		Upland Rice	
	Straw	Grain	Straw	Grain
Yield (kg ha⁻¹)	9423	6389	6343	4568
N (kg ha⁻¹)	65	86	56	70
P (kg ha⁻¹)	15	15	3	10
K (kg ha⁻¹)	156	20	150	56
Ca (kg ha⁻¹)	26	5	23	4
Mg (kg ha⁻¹)	15	7	12.5	5
Zn (g ha⁻¹)	546	224	161	138
Cu (g ha⁻¹)	77	102	35	57
Mn (g ha⁻¹)	4724	369	1319	284
Fe (g ha⁻¹)	2553	505	654	117
B (g ha⁻¹)	69	33	53	30

Source: Fageria, N.K. et al., *Growth and Mineral Nutrition of Field Crops*, 3rd edn., CRC Press, Boca Raton, FL, 2011.

TABLE 4.3
Straw and Grain Yield and Nutrients Accumulation by Dry Bean (*Phaseolus vulgaris* L.) and Corn (*Zea mays* L.) Grown on a Brazilian Oxisol

Yield and Nutrient Accumulation	Dry Bean		Corn	
	Straw	Grain	Straw	Grain
Yield (kg ha⁻¹)	1893	3356	11,873	8501
N (kg ha⁻¹)	13.47	119	72	127
P (kg ha⁻¹)	1.69	12.3	5	17
K (kg ha⁻¹)	35.2	61.5	153	34
Ca (kg ha⁻¹)	16.6	8.02	33	8
Mg (kg ha⁻¹)	7.17	6.09	21	9
Zn (g ha⁻¹)	49.1	122.5	184	192
Cu (g ha⁻¹)	8.4	37.9	53	12
Mn (g ha⁻¹)	25.7	45.6	452	82
Fe (g ha⁻¹)	897	397	2,048	206
B (g ha⁻¹)			103	43

Source: Fageria, N.K. et al., *Growth and Mineral Nutrition of Field Crops*, 3rd edn., CRC Press, Boca Raton, FL, 2011.

4.5 PHYSIOLOGICAL RESPONSES OF PLANTS TO SUPPLY OF POTASSIUM

Potassium is responsible for many physiological functions in plants. Potassium helps in pH stabilization and osmotic and ion regulation in plants. The osmotic function is not a specific one as there are numerous organic and inorganic solutes in plants. Potassium is of utmost importance for water status of plants (Mengel et al., 2001). It also regulates stomatal opening and closing in plants. An increase in the potassium concentration in the guard cells results in the uptake of water from the adjacent cells and a corresponding increase in turgor in the guard cells and thus stomatal opening. Closure of the stomata in

TABLE 4.4

Yield of Straw and Grain and Nutrients Accumulation by Soybean (*Glycine max* Merr.) Grown on a Brazilian Oxisol

Yield and Nutrient Accumulation	Straw	Grain
Yield (kg ha^{-1})	3518	4003
N (kg ha^{-1})	38	280
P (kg ha^{-1})	2	12
K (kg ha^{-1})	58	78
Ca (kg ha^{-1})	31	13
Mg (kg ha^{-1})	20	10
Zn (g ha^{-1})	29	169
Cu (g ha^{-1})	33	60
Mn (g ha^{-1})	117	120
Fe (g ha^{-1})	187	373

Source: Fageria, N.K. et al., *Growth and Mineral Nutrition of Field Crops*, 3rd edn., CRC Press, Boca Raton, FL, 2011.

TABLE 4.5

Macronutrient and Micronutrient Requirement to Produce 1 Metric Ton of Grain of Important Food Crops

Nutrient	Lowland Rice	Upland Rice	Dry Bean	Corn	Soybean
N (kg)	24	28	37	23	79
P (kg)	5	3	5	3	4
K (kg)	28	45	27	22	34
Ca (kg)	5	6	7	5	11
Mg (kg)	3	4	4	4	8
Zn (g)	121	65	48	44	50
Cu (g)	28	20	13	8	23
Mn (g)	797	351	20	63	59
Fe (g)	479	169	365	265	140
B (g)	16	18	18	17	30

Source: Fageria, N.K. et al., *Growth and Mineral Nutrition of Field Crops*, 3rd edn., CRC Press, Boca Raton, FL, 2011.

the dark is correlated with K efflux and a corresponding decrease in the osmotic pressure of the guard cells (Marschner, 1995). Potassium functions as a cofactor or activator for many enzymes of carbohydrate and protein metabolism (Fageria et al., 2011). Suelter (1970) and Marschner (1995) reported that there are more than 50 enzymes that either completely depend on or are stimulated by potassium ions.

Potassium also activates membrane-bound ATPases. The activation of ATPases by K not only facilitates its own transport from the external solution across the plasma membrane into the root cells but also makes K the most important mineral element in cell extension and osmoregulation (Marschner, 1995). Potassium also is required for protein synthesis in plants. It plays a significant role in photosynthesis. Fageria et al. (2006) reported that supply of K in adequate amount reduces lodging in crops, decreases water stress, increases photosynthetic rates, and improves grain quality.

Responses of 15 upland rice genotypes to K fertilization in relation to grain yield, plant height, and panicle density are presented in Table 4.6. Grain yield varied from 6.4 g per 4 plants to 24.6 g per 4 plants, with an average value of 16.7 g per 4 plants at low K level. Similarly, at high K level,

TABLE 4.6
Grain Yield, Plant Height, and Panicle Density of 15 Upland Rice Genotypes as Influenced by Potassium Fertilization

Genotype	Grain Yield (g pot⁻¹)		Plant Height (cm)		Panicle Number per Pot	
	K_0	K_{200}	K_0	K_{200}	K_0	K_{200}
Rio Paranaíba	19.8abcd	39.5a	106abcd	143[a]	10.3abcde	14.0bc
CNA6975-2	6.4e	9.4de	91ef	107d	4.3e	6.7de
CNA7690	13.7de	29.2abc	93def	130ab	13.0abcd	13.7bc
L141	18.9abcd	38.7a	95cdef	118bcd	10.7abcd	12.3bcd
CNA7460	13.7de	26.1abcd	87f	112cd	10.0bcde	16.3b
CNA6843-1	14.9bcde	36.7ab	100bcde	117bcd	9.3cde	14.3bc
Guarani	23.9ab	25.1abcd	95bcdef	115bcd	16.3a	15.0bc
CNA7127	16.6abcd	31.9abc	90ef	120bcd	9.7bcde	11.3bcd
CNA6187	23.1abc	32.0abc	107abc	128abc	12.7abcd	14.3bc
CNA7911	24.6a	20.3bcde	108ab	120bcd	15.7ab	9.0cde
CNA7645	14.1cde	38.9a	108ab	145[a]	9.3cde	15.7bc
CNA7875	13.3de	6.2e	117[a]	117bcd	7.7de	4.7e
CNA7680	17.5abcd	19.2bcde	102bcde	131ab	11.3abcd	15.7bc
CNA6724-1	16.5abcd	19.2bcde	106abcd	130ab	10.3abcde	9.0cde
CNA7890	13.3de	17.3cde	104abcd	120bcd	14.3abc	23.7a
Mean	16.7	26.0	101	124	10.9	13.0

Source: Fageria, N.K., *Pesq. Agropec. Bras.*, 35, 2115, 2000a.

Note: There were 4 plants per pot. Means followed by the same letter in the same column are not statistically different at 5% probability level by Duncan's multiple range test.

the grain yield varied from 6.2 g per 4 plants to 39.5 g per 4 plants with an average value of 26 g per 4 plants. Overall, grain yield increase was 56% at the higher K level (200 mg K kg⁻¹) than at the lowest K level (0 mg K kg⁻¹). Plant height varied from 87 to 117 cm, with an average value of 100.6 cm at the low K level. Similarly, at high K level, plant height varied from 107 to 145 cm, with an average value of 123.5 cm. Overall, increase in plant height was 23% at the higher K level compared to the lower K level. Overall, increase in panicle density was 19% at the higher K level compared to the lower K level. Plant height had a highly significant correlation with grain yield (Fageria, 2000a). Similarly, panicle number or density also had a highly significant correlation with grain yield. Hence, it can be concluded that K fertilization improved grain yield due to increase in plant height and panicle density in upland rice genotypes grown on a Brazilian Oxisol. Figures 4.9 and 4.10 show response of lowland rice and dry bean to K fertilization applied to a Brazilian Inceptisol. The growth of these two crops increased in a quadratic fashion with increasing K levels in the range of 0–600 mg kg⁻¹.

Data in Table 4.7 show root length and root dry weight of 12 lowland rice genotypes as influenced by K fertilization grown on a Brazilian Inceptisol. Root length and root dry weight were increased with the addition of K fertilizer. In addition, the K × genotype interaction was significant for these two growth parameters, indicating variation in root length as well as root dry weight at two K levels. Hence, it is possible to select lowland rice genotypes for root length and root dry weight at low as well as at high K levels. Root length at low K level varied from 8.7 cm produced by genotype BRA 051083 to 36 cm produced by genotype BRA 051077, with an average value of 17.5 cm. Root length at high K level varied from 14.0 cm produced by genotype BRA 051129 to 39 cm produced by genotype BRS Jaçanã, with an average value of 25.9 cm. Overall, increase in root length was 48% with the addition of 300 mg K kg⁻¹ compared to the unfertilized treatment (Fageria, 2013).

FIGURE 4.9 Response of lowland rice grown on a Brazilian Inceptisol to potassium fertilization. (From Fageria, N.K., *The Role of Plant Roots in Crop Production*, CRC Press, Boca Raton, FL, 2013.)

FIGURE 4.10 Response of dry bean grown on a Brazilian Inceptisol to potassium fertilization. (From Fageria, N.K., *The Role of Plant Roots in Crop Production*, CRC Press, Boca Raton, FL, 2013.)

Root dry weight varied from 0.34 g per plant produced by genotype BRS 051250 to 2.11 g plant^{-1} produced by genotype BRS Jaçanã, with an average value of 0.89 g plant^{-1} at low K level. At high K level, root dry weight varied from 1.04 g plant^{-1} produced by genotype BRA 051134 to 6.58 g plant^{-1} produced by genotype BRA 051108, with an average value of 3.08 g plant^{-1}. Overall, increase in root dry weight at high K level was 246% compared to low level of K. There was also difference among genotypes in root growth at low as well as at high K levels (Fageria, 2013). Figure 4.11 shows root growth of upland rice at two K levels.

Adequate K in plant tissues suppresses incidence of diseases. Potassium not only increases plant resistance to diseases but also reduces fungal populations in the soil, reduces fungal pathogenicity, and promotes more rapid healing of injuries. Although there is a large amount of literature on the relationship between K and plant diseases, there is little quantitative information available on the concentration of K in soil or plant tissues that result in the observed effects on disease expression (Huber and Arny, 1985; Prabhu et al., 2007). Generally, K fertilization restricts the intensity of several infectious diseases, and this suppression occurs with diseases caused by obligate as well as by facultative parasites (Kiraly, 1976; Prabhu et al., 2007). Effects of K on bacterial, viral, fungal, and nematode diseases are summarized in Table 4.8.

TABLE 4.7

Maximum Root Length and Root Dry Weight of 12 Lowland Rice (*Oryza sativa* L.) Genotypes as Influenced by K Fertilization

Genotype	Maximum Root Length (cm)		Root. Dry Weight, (g Plant⁻¹)	
	0 mg K kg⁻¹	300 mg K kg⁻¹	0 mg K kg⁻¹	300 mg K kg⁻¹
BRS Tropical	23.3bc	33.0abc	0.44cd	5.48a
BRS Jaçanã	26.0b	39.0a	2.11a	6.18
BRS 02654	17.3cd	25.7cde	1.14bc	2.02
BRA 051077	36.0a	35.0ab	2.01a	6.57a
BRA 051083	8.7f	26.0cde	0.51cd	1.34b
BRA 051108	25.7b	31.3abc	1.62ab	6.58a
BRA 051126	10.7ef	21.0def	0.49cd	1.81b
BRA 051129	14.0def	14.0f	0.56cd	1.13b
BRA 051130	10.3ef	29.0bcd	0.48cd	1.36b
BRA 051134	11.7def	14.7f	0.49cd	1.04b
BRA 051135	15.0de	20.3ef	0.58cd	2.17b
BRS 051250	11.0ef	21.7def	0.34d	1.25b

Source: Fageria, N.K., *The Role of Plant Roots in Crop Production*, CRC Press, Boca Raton, FL, 2013.

Note: Means followed by the same letter in the same column (in the same line for averages) are not significantly different at the 5% probability level by Tukey's test.

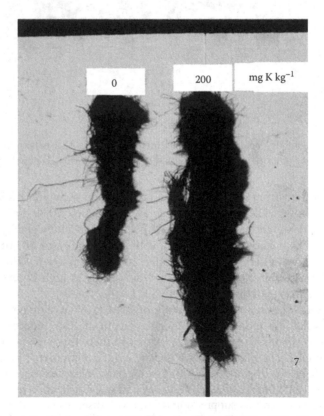

FIGURE 4.11 Root growth of upland rice at two K levels. (From Fageria, N.K., *The Role of Plant Roots in Crop Production*, CRC Press, Boca Raton, FL, 2013.)

TABLE 4.8

Influence of Potassium on Bacterial, Fungal, Viral, and Nematode Diseases on Principal Field Crops

Crop Species	Disease	Pathogen	K Effect
Bacterial Diseases			
Bean, lima (*Phaseolus lunatus* L.)	Bacterial blight	*Pseudomonas syringae*	Decrease
Cassava (*Manihot esculenta* Crantz)	Bacterial blight	*Xanthomonas manihotis*	Decrease
Cotton (*Gossypium hirsutum* L.)	Angular leaf spot	*Xanthomonas malvacearum*	Decrease
Potato (*Solanum tuberosum* L.)	Scab	*Streptomyces scabies*	Increase
Rice (*Oryza sativa* L.)	Bacterial blight	*Xanthomonas oryzae*	Decrease
Corn (*Zea mays* L.)	Stewarts wilt	*Erwinia stewartii*	Decrease
Fungal Diseases			
Corn (*Zea mays* L.)	Stalk rot	*Fusarium moniliforme*	Decrease
Alfalfa (*Medicago sativa* L.)	Leaf spot	*Pseudopeziza medicaginis*	Decrease
Barley (*Hordeum vulgare* L.)	Powdery mildew	*Erysiphe graminis*	Decrease
Dry bean or common bean (*Phaseolus vulgaris* L.)	Leaf spot	*Mycosphaerella cruenta*	Increase
	Root rot	*Rhizoctonia solani*	Increase
Wheat (*Triticum aestivum* L.)	Leaf rust	*Puccinia triticina*	Decrease
	Bunt	*Tilletia* spp.	Decrease
	Take-all	*Gaeumannomyces graminis*	Increase or decrease
	Powdery mildew	*Erysiphe graminis*	Decrease
Peanut (*Arachis hypogaea* L.)	Pod rot	*Rhizoctonia solani*	Increase
	Leaf spot	*Mycosphaerella arachidis*	Decrease
Cotton (*Gossypium hirsutum* L.)	Leaf blight	*Cercospora gossypina*	Decrease
	Root rot	*Phymatotrichum omnivorum*	Decrease
Potato (*Solanum tuberosum* L.)	Canker	*Rhizoctonia solani*	Decrease or increase
	Late blight	*Phytophthora infestans*	Decrease
Rice (*Oryza sativa* L.)	Sheath blight	*Corticium sasakii*	Decrease
	Stem rot	*Leptosphaeria salvinii*	Decrease
	Blast	*Pyricularia oryzae*	Decrease or increase
Soybean (*Glycine max* Merr.)	Pod rot	*Diaporthe sojae*	Decrease
	Root rot	*Phytophthora megasperma*	Increase
Sugarcane (*Saccharum officinarum* L.)	Eyespot	*Helminthosporium sacchari*	Decrease
Viral Diseases			
Barley (*Hordeum vulgare* L.)	Barley yellow dwarf	*Barley yellow dwarf virus*	Decrease
Dry bean (*Phaseolus vulgaris* L.)	Mosaic	*Tobacco mosaic virus*	Decrease
Cassava (*Manihot esculenta* Crantz)	Mosaic	*African cassava mosaic virus*	None
Oat (*Avena sativa* L.)	Barley yellow dwarf	*Barley yellow dwarf virus*	Increase or none
Potato (*Solanum tuberosum* L.)	Mosaic	*Potato mosaic virus*	Decrease
Soybean (*Glycine max* L.)	Mosaic	*Soybean mosaic virus*	Increase
Pea (*Pisum sativum* L.)	Leaf roll	*Pea leaf roll virus*	Increase
Nematode Diseases			
Lima bean (*Phaseolus lunatus* L.)	Root knot	*Meloidogyne incognita*	Decrease
Sugar beet (*Beta vulgaris* L.)	Sugar beet nematode	*Heterodera schachtii*	Decrease
Cotton (*Gossypium hirsutum* L.)	Root knot	*Meloidogyne incognita*	None
Rice (*Oryza sativa* L.)	White tip	*Aphelenchoides oryzae*	Increase
Soybean (*Glycine max* L.)	Root knot	*Meloidogyne incognita*	Increase
	Soybean cyst	*Heterodera glycines*	Increase

Source: Compiled from Prabhu, A.S. et al., Potassium and plant disease, in: *Mineral Nutrition and Plant Disease*, L.E. Datnoff, W.H. Elmer, and D.M. Huber, (eds.), The American Phytopathological Society, St. Paul, MI, 2007, pp. 57–78.

Potassium is reported to be beneficial for nitrogen (N_2) fixation in legume crops. The effect of K on legume N_2 fixation is reported to be indirect due to effects on the legume plant itself and not to direct effect on the nodule (Duke and Collins, 1985). Duke and Collins (1985) reported that positive effects of K fertilization in legume N_2 fixation are due to K fertilization–induced increases in photosynthesis, the transport of photosynthetic products to the legume nodule, increased root growth, or a combination of these factors.

4.6 GENETICS OF ACQUISITION OF POTASSIUM BY PLANTS

In the last three decades of the twentieth century until today (2014), much progress has been made in investigations on absorption of nutrients by higher plants. By synthesizing this progress, scientists in the 1970s developed mathematical models to describe the relationship between availability and plant nutrient absorption. These models have had a major impact on the understanding of nutrient absorption and have contributed profoundly both conceptually and practically (Okajima, 2001). By the 1990s, knowledge of ion uptake by plants had reached the stage that the term nutrient acquisition had become common usage. Okajima (2001) reported that plants have an autocontrolled feedback system in nutrient absorption, whereby the nutrient absorption power is induced and increased by nutrient stress and roots actively absorb nutrients. Plants also have the ability to adjust or adapt to adverse environmental conditions to satisfy nutrient requirements. The mechanisms of plant adaption to adverse conditions in the absorption of nutrients are as follows: (1) root growth or extension for increasing the root surface to gather nutrients, (2) selective nutrient absorption by plants, (3) root action to make rhizosphere conditions favorable for root growth, and (4) root action to solubilize more of the limiting nutrients (Okajima, 2001).

Great genetic variability occurs among plant species and genotypes of the same species in absorption and utilization of nutrients. Fageria (1989) defined plant genetic variability as the heritable characters of a particular crop species or cultivar that shows differences in growth or production with other species, or cultivars of the same species, under favorable or unfavorable growth conditions. In the last few decades, it has been shown that large differences exist among species or genotypes of the same species in the absorption and utilization on mineral nutrients. Similarly, differences also have been observed among plant species and cultivars in their tolerance to nutrient or elemental toxicities (Fageria, 2009, 2013).

Potassium-use efficiencies varied in 14 upland rice genotypes (Table 4.9). Agronomic efficiency (AE) (crop yield/nutrient applied) varied from 2.7 to 24.8 mg mg^{-1}, physiological efficiency (PE) (yield increase/nutrient absorbed) varied from 27.5 to 63.2 mg mg^{-1}, agrophysiological efficiency (APE) (yield increase/nutrient absorbed) varied from 3.1 to 37.7 mg kg^{-1}, recovery efficiency (nutrient absorbed/nutrient applied) varied from 44.3% to 87.7%, and utilization efficiency (yield increase/nutrient applied) varied from 16.48 to 46.31 mg mg^{-1}. Overall, genotype Rio Paranaíba had higher K-use efficiency than other genotypes. The variation in K accumulation among upland rice genotypes suggests that genetic variation occurs among these genotypes in the uptake and utilization of K. There are several possible mechanisms, which are responsible for this variation. These mechanisms are listed in Table 4.10.

Hogh-Jensen and Pedersen (2003) discussed the ability of crops to adapt to low-potassium conditions and to their ability to utilize sparingly soluble K sources. These authors also studied K uptake by different crop species at two K levels. In this study, they used pea (*Pisum sativum* L.), red clover (*Trifolium pratense* L.), alfalfa (*Medicago sativa* L.), barley (*Hordeum vulgare* L.), rye (*Secale cereale* L.), perennial ryegrass (*Lolium perenne* L.), and oilseed rape (*Brassica napus oleifera* L.). The K levels used were 26 and 60 mg kg^{-1}. The legumes (pea, red clover, alfalfa) accumulated larger amounts of nitrogen but lower amounts of K than rye, ryegrass, barley, or oilseed rape. The differences in K accumulation correlated with root hair length. Rye had an outstanding root surface that in total as well as per unit root dry matter was twice that of the other crops. The ranking of K-use efficiency in decreasing order was rye, ryegrass, oilseed rape, alfalfa, barley, pea, and

TABLE 4.9
Potassium-Use Efficiency of 14 Upland Rice Genotypes

Genotype	AE (mg mg⁻¹)	PE (mg mg⁻¹)	APE (mg mg⁻¹)	RE (%)	EU (mg mg⁻¹)
Rio Paranaíba	19.6abc	52.8b	19.2bcde	87.7[a]	46.31a
CNA6975-2	6.2bc	34.2b	6.3de	76.3ab	26.09abc
CNA7690	15.6abc	46.9b	20.7bcde	63.6abc	29.83abc
L141	19.8abc	63.2a	26.5abc	64.4abc	40.70ab
CNA7460	12.5abc	43.3b	12.8cde	79.1ab	34.25abc
CNA6843-1	21.8ab	48.9b	21.8abcd	76.4ab	37.36abc
Guarani	20.7abc	40.7b	37.7a	44.3c	18.03c
CNA7127	15.2abc	46.8b	19.3bcde	61.6abc	28.83abc
CNA6187	8.9abc	33.4b	11.2cde	68.2abc	22.79bc
CNA7911	10.8abc	27.5b	14.2bcde	60.5abc	16.64c
CNA7645	24.8a	53.6b	30.9ab	80.6ab	43.20ab
CNA7680	6.6abc	30.4b	11.2cde	54.2bc	16.48c
CNA6724-1	2.7c	29.5b	3.1e	59.2bc	17.46c
CNA7890	4.0c	28.8b	7.4de	57.7bc	16.62c

Source: Fageria, N.K., *Pesq. Agropec. Bras.*, 35, 2115, 2000a.

Notes: Means followed by the same letter in the same column are not statistically different at 5% probability level by Duncan's multiple range test.

AE, agronomic efficiency; PE, physiological efficiency; APE, agrophysiological efficiency; RE, recovery efficiency; UE, utilization efficiency. Potassium rates used in this experiments were 0 mg kg⁻¹ and 200 mg K kg⁻¹.

$$AE \text{ (mg mg}^{-1}) = \frac{\text{GY at 200 mg K kg}^{-1} \text{ in mg} - \text{GY at 0 mg K kg}^{-1} \text{ in mg}}{\text{K rate applied in mg}}$$

$$PE \text{ (mg mg}^{-1}) = \frac{\text{BY at 200 mg K kg}^{-1} \text{ in mg} - \text{BY at 0 mg K kg}^{-1} \text{ in mg}}{\text{K uptake in BY at 200 mg K kg}^{-1} \text{ in mg} - \text{K uptake in BY at 0 mg K kg}^{-1} \text{ in mg}}$$

$$APR \text{ (mg mg}^{-1}) = \frac{\text{GY at 200 mg K kg}^{-1} \text{ in mg} - \text{GY at 0 mg K kg}^{-1} \text{ in mg}}{\text{K uptake in BY at 200 mg K kg}^{-1} \text{ in mg} - \text{K uptake in BY at 0 mg K kg}^{-1} \text{ in mg}}$$

$$UE \text{ (mg mg}^{-1}) = PE \times RE$$

where
  GY is the grain yield
  BY is the biological yield (grain plus straw)

red clover. Variation in root morphology and K uptake and utilization have been reported by many workers with various plant species (Drew and Nye, 1969; Mengel and Kirkby, 1980; Barber, 1995; Marschner, 1995). Fageria et al. (2008) gave a detailed discussion on mechanisms involved in the variation in nutrient uptake and use efficiency in crop plants.

4.7 CONCENTRATION OF POTASSIUM IN PLANTS

Concentration of potassium in the plant tissue is normally expressed in g kg⁻¹ on a dry weight basis. The nutrient concentration generally is used to identify deficient, sufficient, or toxic level in the plant tissues. Potassium toxicity rarely occurs in plant tissues, but K may create nutrient imbalance

TABLE 4.10
Soil and Plant Mechanisms and Processes and Other Factors that Influence Genotypic Differences in Nutrient-Use Efficiency

Nutrient acquisition

1. Diffusion and mass flow (buffer capacity, ionic concentration, ionic properties, tortuosity, soil moisture, bulk density, and temperature)
2. Root morphological factors (number, length, root hair density, root extension, and root density)
3. Physiological (root–shoot, root microorganisms such as mycorrhizal fungi, nutrient status, water uptake, nutrient influx and efflux, rate of nutrient transport in roots and shoots, affinity for uptake [K_m], threshold concentration [C_{min}])
4. Biochemical (enzyme secretion as phosphate, chelating compounds, and phytosiderophore, proton exudes, organic acid production such as citric, trans-aconitic, malic acid exudates)

Nutrient movement in root

1. Transfer across endodermis and transport within root
2. Compartmentation/binding within roots
3. Rate of nutrient release to xylem

Nutrient accumulation and remobilization in shoot

1. Demand at cellular level and storage in vacuoles
2. Retransport from older to younger leaves and from vegetative to reproductive parts
3. Role of chelates in xylem transport

Nutrient utilization and growth

1. Metabolism at reduced tissue concentration of nutrient
2. Lower element concentration in supporting structure, particularly the stem
3. Elemental substitution, for example, Na for K function

Other factors

1. Soil factors
   a. Soil solution (ionic equilibrium, solubility precipitation, competing ions, organic ions, pH, and phytotoxic ions)
   b. Physicochemical properties of soil (organic matter, pH, aeration, structure, texture, compaction, soil moisture)
2. Environmental effects
   a. Intensity and quality of solar radiation
   b. Temperature
   c. Moisture supply
   d. Plant diseases, insects, weeds, and allelopathy

Sources:	Baligar, V.C. et al., *Soil Sci. Plant. Anal.*, 32, 921, 2001; Fageria, N.K. and Baligar, V.C., *J. Plant Nutr.*, 26, 1315, 2003a; Fageria, N.K., *Nutrient Uptake in Crop Plants*, CRC Press, Boca Raton, FL, 2009; Fageria, N.K. et al., *Growth and Mineral Nutrition of Field Crops*, 3rd edn., CRC Press, Boca Raton, FL, 2011.

if absorbed in excess of needs by a plant. Like soil analysis, plant analysis also involves sampling, sample preparation (of plant tissues), analysis, and interpretation of analytical results. All these steps are important for a meaningful plant analysis program. Many factors such as soil, climate, plant, and their interactions affect absorption of nutrients by growing plants. However, the concentrations of the nutrients are maintained within rather narrow limits in plant tissues. Such consistency is thought to arise from the operation of delicate feedback systems, which enable plants to respond in a homeostatic fashion to environmental fluctuations (Fageria and Baligar, 2005).

Potassium concentrations (content per unit of shoot or grain dry weight) in shoots and grain of upland rice are presented in Table 4.11. Potassium concentrations in shoots at low K level (0 mg K kg^{-1}) varied from 5.5 to 9.0 g kg^{-1} depending on genotype, with an average value of 7.3 g kg^{-1}. Similarly, K concentration in shoot at high K level (200 mg K Kg^{-1}) varied from 10.5 to 13.0 g kg^{-1}, with an average value of 11.7 g kg^{-1}. Overall, K concentration in shoot at 200 mg K kg^{-1} soil was 60% higher than at 0 mg K kg^{-1} of soil. However, in the grain overall, K concentration was slightly higher at low K

TABLE 4.11
Potassium Concentration in Shoot and Grain as Influenced by K and Genotype Treatments

Genotype	K Concentration in Shoot (g kg^{-1})		K Concentration in Grain (g kg^{-1})	
	K_0	K_{200}	K_0	K_{200}
BRS Bonança	6.0c	12.0	1.5	2.0
BRS Primavera	7.5b	12.5	1.5	2.0
BRSMG Curinga	9.0a	10.5	2.5	1.0
BRA 032033	8.0ab	13.0	1.5	1.0
BRA 01596	5.5c	11.5	1.5	1.0
BRA 02582	8.0ab	12.5	1.5	1.5
Average	7.3	11.67	1.67	1.42

Source: Fageria, N.K. et al., *Soil Sci. Plant Anal.,* 41, 2676, 2010b.

Note: Means followed by the same letter in the same column are not statistically different at 5% probability level by Duncan's multiple range test.

TABLE 4.12
Potassium Accumulation Traits (X) Association with Grain Yield (Y)

K Trait	Regression Equation	R^2
K conc. in shoot	$Y = 9.4715 + 0.4339X$	0.1576*
K conc. in grain	$Y = 21.6379 - 9.8800X + 2.6465X^2$	0.1262[NS]
K total in shoot	$Y = 12.1350 + 0.0058X$	0.1568*
K total in grain	$Y = 12.9373 + 0.0319X$	0.1714*
K-use efficiency ratio (KUER) in shoot	$Y = 17.1310 - 0.0309X$	0.1303*
KUER in grain	$Y = 11.9189 + 0.0023X$	0.0372[NS]

Source: Fageria, N.K. et al., *Soil Sci. Plant Anal.,* 41, 2676, 2010b.

*Significant at the 5% probability level; [NS] not significant.

concentration than at high K concentration. This effect may happen due to high grain yield at high K level, giving what is known as a dilution effect (Fageria et al., 2011). The K concentration in shoots had a significantly positive association with grain yield (Table 4.12). However, K concentration in grain had a nonsignificant association with grain yield, indicating that increasing K in grain of upland rice did not increase grain yield of this crop.

For the interpretation of plant analysis results, a critical nutrient concentration concept was developed. This concept is used widely now in interpretation of plant analysis results for nutritional disorders for diagnostic purposes. Critical nutrient concentration is designated usually as a single point within the bend of the curve when crop yield is plotted against nutrient concentration where the plant nutrient status shifts from deficient to adequate. The critical nutrient concentration has been defined in several ways (Fageria and Baligar, 2005), including (1) the concentration that is just deficient for maximum growth, (2) the point where growth is 10% less than the maximum, (3) the concentration where plant growth begins to decrease due to a limited supply of nutrient, and (4) the lowest amount of element in the plant accompanying the highest yield. Nutrient concentration in plant tissue decreases with the advancement of plant age. Hence, for nutrient sufficiency or deficiency diagnosis purposes, plant tissue analysis should be done at different growth stages. Data in Table 4.13 show adequate concentration of K^+ in plant tissues of principal crops.

TABLE 4.13
Adequate Level of K⁺ in Plant Tissue of Principal Field Crops

Crop	Growth Stage	Plant Part	Adequate K Level (g kg⁻¹)
Barley (*Hordeum vulgare* L.)	Tillering	Leaves	42–47
	Heading	Whole tops	15–30
Wheat (*Triticum aestivum* L.)	Tillering	Leaf blade	34–42
	Heading	Whole tops	15–30
Rice (*Oryza sativa* L.)	75 days after sowing	Whole tops	15–40
	Flowering	Whole tops	12–30
Corn (*Zea mays* L.)	30–45 days after emergence	Whole tops	30–50
	Before tasseling	Leaf blade	20–25
Sorghum (*Sorghum spp.* L.)	Seedling	Whole tops	30–45
	Early vegetative	Whole tops	25–40
	Bloom	Third blade below panicle	10–15
Soybean (*Glycine max* Merr.)	Prior to pod set	Upper fully developed trifoliate	17–25
Dry bean (*Phaseolus vulgaris* L.)	Early flowering	Uppermost blade	15–35
Cowpea (*Vigna unguiculata* Walp.)	Early flowering	PUMB	1.2–6.0
Peanut (*Arachis hypogaea* L.)	Early pegging	Upper stems and leaves	17–30

Sources: Piggott, T.J., Vegetable crops. *Plant Analysis: An Interpretation Manual*, Reuter, D.J. and Robinson, J.B. (eds.), Inkata Press, Melbourne, Victoria, Australia, 1986, pp. 148–187; Reuter, D.J., Temperate and subtropical crops, *Plant Analysis: An Interpretation Manual*, D.J. Reuter and J.B. Robinson (eds.), Inkata Press, Melbourne, Victoria, Australia, 1986, pp. 38–99; Fageria, N.K., *Nutrient Uptake in Crop Plants*, CRC Press, Boca Raton, FL, 2009; Fageria, N.K. et al., *Growth and Mineral Nutrition of Field Crops,* 3rd edn., CRC Press, Boca Raton, FL, 2011.

PUMB, Petiole of uppermost mature leaf blade.

4.8   INTERACTION WITH OTHER NUTRIENTS

Nutrient interaction may be positive, negative, or neutral. Negative nutrient interaction reflects an imbalance between two nutrients or among many nutrients. Proper nutrient balance is as important as adequate supply of nutrients for maximizing crop yields. Generally, nutrient interactions are measured in terms of growth and yield. Nutrient concentration and accumulation also are considered or determined to evaluate nutrient interaction in plant tissues. Positive interactions of K with N and P have been reported (Dibb and Thomson, 1985). Dibb and Welch (1976) reported that increased K allowed for rapid assimilation of absorbed NH_4^+ ions in the plant, maintaining a low, nontoxic level of NH_4^+. Increased yield of crops with the addition of N and P requires a higher level of K in the soil (Dibb and Thompson, 1985; Fageria et al., 2011). An example of positive association between P and K is presented in Table 4.14. There was a significant quadratic increase in K uptake with the increasing P rate in the range of 0–655 kg ha⁻¹.

Antagonistic interaction between K^+ and Mg^{2+} or Ca^{2+} uptake has been reported widely (Johnson et al., 1968; Fageria, 1983; Dibb and Thompson, 1985). Fageria (1983) reported that reduction in Ca uptake with increasing K^+ concentration in the growth medium was associated closely with the increasing uptake of K^+, indicating that there may have been a competitive effect. A competition between K^+ and Ca^{2+} and Mg^{2+} due to physiological properties of these ions has been reported (Johnson et al., 1968; Fageria, 1983).

Potassium and micronutrient interactions have been observed for many crop plants. Hill and Morrill (1975) and Gupta (1979) reported that high K^+ rates reduced B accumulation and intensified B deficiency in crop plants. Dibb and Thompson (1985) reviewed interactions between K and Cu in

TABLE 4.14

Influence of P on the Accumulation of K in the Straw and Grain of Lowland Rice

P Rate (kg ha⁻¹)	K in Straw (kg ha⁻¹)	K in Grain (kg ha⁻¹)
0	58	2.9
131	118	8.9
262	121	10.8
393	134	14.6
524	128	13.6
655	132	12.9

Regression analysis

P rate vs. K accumulation in straw $(Y) = -17.79 + 0.59X - 0.00055X^2$, $R^2 = 0.62**$

P rate vs. K accumulation in grain $(Y) = -3.73 + 0.06X - 0.000048X^2$, $R^2 = 0.80**$

Source: Adapted from Fageria, N.K. and Santos, A.B., *Commun. Soil Sci. Plant Anal.*, 39, 873, 2008.

Note: Data are averages of 2 years field trial.

**Significant at the 1% probability level.

crop plants and reported that Cu^{2+} uptake increased with the addition of K^+. Matocha and Thomas (1969) reported that Fe application with K^+ increased sorghum yield. Fageria (1984) reported that Fe^{2+} toxicity in flooded rice was reduced with the addition of adequate K^+ to the soil. These authors also reported that plants with adequate K^+ have more metabolic activity in the roots and a high level of Fe^{2+} excluding power. Dibb and Thompson (1985) in a review reported that K^+ improved Mn^{2+} uptake when Mn^{2+} was in low concentration in the growth medium and decreased Mn^{2+} uptake in high concentration that might be toxic. A beneficial effect of K on the uptake of Zn^{2+} has been reported (Dibb and Thompson, 1985).

4.9 MANAGEMENT PRACTICES TO MAXIMIZE POTASSIUM-USE EFFICIENCY

Appropriate management practices can improve potassium-use efficiency in crop plants and consequently efficient crop production. Various important practices can be adapted in this respect, including (1) liming acid soils; (2) use of effective source, method, and timing of application; (3) use of adequate rate; (4) incorporation of crop residues in the soil; (5) adequate water supply; (6) use of cover crops; (7) use of farmyard manures; (8) use of crop rotation; (9) use of conservation tillage; (10) maintaining adequate K^+ saturation ratio; and (11) use of efficient crop species and genotypes within species. These practices are discussed briefly in the succeeding section.

4.9.1 LIMING ACID SOILS

Liming is the most important practice for improving fertilizer-use efficiency of acid soils. Liming of acid soils not only decreases soluble soil Al but also increases retention of applied K^+ and thereby decreases K^+ leaching. Liming increases K^+ retention in soils by replacing Al^{3+} on the exchange sites with Ca^{2+}, allowing K^+ to compete better for exchange sites and increasing the effective cation exchange capacity (CEC) (Fageria, 2009). Data in Table 4.15 show that liming of a Brazilian Oxisol increased pH, Ca^{2+}, and Mg^+ concentrations and base saturation and decreased Al^{3+} concentration. Improvement in these chemical properties improves upland rice yield and consequently gives higher K-use efficiency (Fageria, 2009). Fageria (1989) also reported that improvement in these properties improved grain yield of many crops in acid soils. Data in Table 4.16 show that improvement in soil pH with liming improves shoot dry weight grain yield and number of panicles in upland rice. Increasing yields means improving K^+-use efficiency in crop plants.

TABLE 4.15

Influence of Liming on Chemical Properties of an Oxisol

Lime Rate, g kg⁻¹ soil	pH in H₂O	Elemental Concentration, cmol_c kg⁻¹			Base Saturation (%)
		Ca²⁺	Mg²⁺	Al³⁺	
0	4.6	0.9	1.0	0.7	22
6	5.7	5.6	2.8	0.0	68
12	6.2	7.4	2.8	0.0	83
18	6.4	7.8	2.3	0.0	90
24	6.6	8.0	2.2	0.0	90
36	6.8	9.2	2.1	0.0	94
r²	0.99**	0.99**	0.96**	0.98**	0.99**

Source: Fageria, N.K., *Pesq. Agropec. Bras.*, 35, 2115, 2000b.
**Linear regression significant at the 1% probability level.

TABLE 4.16

Relationship between Soil pH (X) and Shoot Dry Weight (Y), Grain Yield (Y), and Panicle Number (Y) of Upland Rice

Plant Parameter	Regression Equation	R²	Adequate pH
Shoot dry weight	$Y = -73.64 + 35.68X - 3.53X^2$	0.99**	5.1
Grain yield	$Y = -94.26 + 39.98X - 3.72X^2$	0.91*	5.4
Panicle number	$Y = -8.19 + 5.34X - 0.54X^2$	0.94*	5.0

Source: Fageria, N.K., *Pesq. Agropec. Bras.*, 35, 2115, 2000b.
*,**Significant at the 5% and 1% probability level, respectively.

Liming also has a positive influence in preventing K⁺ leaching from soil–plant systems. Nemeth (1982a,b) analyzed Malaysian and Brazilian soils before and after lime application. In both cases, desorption, that is, soil solution K⁺, decreased when lime was applied. The occupation of selective adsorption sites by K⁺ prevents its removal by leaching. Mielniczuk (1977) also reported that K⁺ buffering capacity of Brazilian soils from Rio Grande do Sul increased with liming. The increases in K⁺ buffering capacity by liming indicate a raise in K⁺ adsorption, by which K⁺ becomes less subject to loss by leaching (Malavolta, 1985). Liming can also improve root development to increased depths due to neutralizing Al³⁺ and improving pH and Ca²⁺ and Mg²⁺ contents, which can improve the uptake of water and nutrients. Muzilli (1982) studied the response of soybean to levels of potassium fertilization in the presence and absence of lime in a Brazilian Oxisol. Soybean yield response to K was higher in limed plots compared with in unlimed plots. Leaf analysis for K also showed a significant increase in K contents in the limed plots than in unlimed plots.

4.9.2 Use of Effective Source, Method, and Timing of Application

Potassium chloride is the major source of single K⁺ fertilization, with potassium sulfate and potassium magnesium sulfate being minor sources for supplying K⁺ to field crops. The reason for widespread use of potassium chloride is associated with low cost of production and high analysis. The use of potassium sulfate may be useful in areas where S deficiency is reported. However, it should not be used in flooded rice where sulfate reduction and hydrogen sulfide toxicity are a problem

TABLE 4.17

Principal Potassium Fertilizers, Their Potassium Content, and Solubility

Common Name	Formula	K_2O (%)	Solubility
Potassium chloride	KCl	60	Water soluble
Potassium sulfate	K_2SO_4	50	Water soluble
Potassium–magnesium sulfate	$K_2SO_4 \cdot MgSO_4$	23	Water soluble
Potassium nitrate	KNO_3	44	Water soluble
Kainit	$MgSO_4 + KCl + NaCl$	12	Water soluble
Potassium metaphosphate	KPO_3	40	Low water solubility

(De Datta and Mikkelsen, 1985). Similarly, the use of KCl may be discouraged in saline soils because it may increase salt concentrations in the rhizosphere. Potassium from various salts is used in formulated, complete fertilizers, reported as %N–%P_2O–%K_2O. In Brazil, the use of a fertilizer mixture of 4–30–16 or 4–20–20 is very common for fertilization of annual crops. Principal K^+ fertilizers are presented in Table 4.17. All of the potassium fertilizers are highly water soluble, except potassium metaphosphate (Fageria, 1989). In general, crop yields are comparable whether one or the other source is used, though this result may depend on the soil, the crop, or the manner of application (Stewart, 1985). However, the fertilizer source selected by farmers depends mainly on cost of transportation and handling convenience in application.

Generally, fertilizers are applied broadcast or in bands, or furrows, below or alongside the planted seed row. As a relatively immobile ion in soil, K^+ movement to roots depends mostly on diffusion (Barber, 1995). This phenomenon means that in K^+-deficient soils, band application of K^+ may be more efficient than broadcasting, because the mean level of K^+ supply to developing roots and diffusion rates are increased. Maintaining a high level of K^+ intensity in a portion of the root system may ensure that the K^+-supplying power of the soil does not limit the rate of K^+ accumulation. Welch et al. (1966) found that less K^+ fertilizer was required to obtain a given corn yield if the K^+ was banded rather than broadcast. Randall and Hoeft (1988), in a review on fertilizer-placement studies, concluded that, in many situations, localized placement of fertilizer can improve fertilizer efficiency and economic return. Heckman and Kamprath (1992) reported that there was little benefit from band placement of K^+ (either alone or in combination with broadcasting) if corn was grown with intensive production practices on sandy soils testing relatively high in K^+. De Datta and Mikkelsen (1985) reported that the placement of K^+ fertilizers in bands in upland soils often enhances their effectiveness, since the ion is not very mobile. Placement becomes less critical, however, as soil-available K levels increase (De Datta and Mikkelsen, 1985).

Plant accumulation and stover yield were, however, increased in a dry year if a band was included with broadcast placement. According to Fageria (1982), the application of moderate amounts of K^+ (30–45 kg ha^{-1}) in the planting furrow of upland rice produced twice as much yield as K^+ broadcast and incorporated. Lopes (1983) recommended that K^+ fertilization of the Brazilian Cerrado soils should be made in two complementary ways, (1) broadcasting and incorporation to raise K^+ saturation to 3% and (2) application of maintenance fertilizer in bands 15–20 cm wide. Localized placement has little advantage over broadcasting if the soil is well supplied with K^+ (Malavolta, 1985).

In addition to application in bands or broadcasting, uniform application of fertilizer is usually considered essential for maximum yield and high K^+ utilization efficiency. Nonuniform fertilizer application can occur because of faulty machinery, faulty machine operation, or because of fertilizer properties that adversely affect the performance of the machine. Hence, uniform application of fertilizers is very important for the uptake of nutrients by plant roots and improving or maximizing fertilizer-use efficiency.

Efficient use of K^+ fertilizer for crop production is essential to maximize economic return to the grower. Timing of fertilizer K^+ application is an important management tool in this effort. Maximum efficiency is obtained if K^+ is applied so that it is available for uptake by the plants as needed. Generally, K^+ fertilizer is applied as a basal application at the time of sowing because of its relative immobility in clay soils. In contrast to N, few studies have been conducted to define the best time for K^+ application. However, in some soils, large losses of K are subjected to leaching and runoff. Oxisols and Ultisols, two predominant tropical soils groups, have a very low CEC. They do not contain K-fixing minerals, so there is little chance of large amounts of K being retained in these soils. Table 4.18 shows extractable K^+ values in an Oxisol of central Brazil after harvest of each upland rice crop under field conditions. Significant differences in extractable K^+ existed with K application rate and depth. Extractable K^+ increased with K application rate and decreased with soil depth (Table 4.18). Applied K increased extractable K^+ in the surface soil and in the subsoil. Leaching of K^+ to lower depths was especially noticeable at the higher K rates. In this situation, split application of K^+ may be an appropriate management practice to reduce K^+ losses by leaching and improving K^+ utilization efficiency by crop plants.

Split application of K^+ is recommended also if high rates are required and fertilizer is applied in the furrow. High rate (>500 kg ha^{-1} fertilizer mixture of 4–30–16) may create high salt index if the

TABLE 4.18

Influence of Potassium Fertilizer Application Rates on Extractable K (mg kg^{-1}) at Four Soil Depths in an Oxisol of Central Brazil

K Applied (kg ha^{-1})	Soil Depth (cm)	1st Crop	2nd Crop	3rd Crop
0	0–20	30	26	31
0	20–40	19	20	22
0	40–60	16	15	15
0	60–80	14	14	13
42	0–20	31	29	45
42	20–40	21	24	27
42	40–60	17	18	17
42	60–80	18	16	16
84	0–20	34	31	50
84	20–40	24	28	33
84	40–60	20	18	21
84	60–80	19	19	21
126	0–20	37	34	55
126	20–40	29	37	40
126	40–60	24	23	25
126	60–80	23	24	25
168	0–20	43	40	59
168	20–40	29	42	46
168	40–60	23	25	30
168	60–80	24	28	30
F-Test				
K rate	**			
Soil depth (D)	**			
K X D	**			

Source: Adapted from Fageria, N.K. et al., *Fert. Res.*, 21, 141, 1990c.

Note: Soil analysis was done after harvest of each upland rice crop.

**Significant at the 1% probability level.

fertilizer is placed too close to the seeds or to the root system. Generally, split application of K^+ is done along with topdressing of N in annual crops. For example, in lowland rice in Brazil, a beneficial effect of topdressing of K^+ (half of the recommended rate applied at sowing and the remaining half applied at 45 days after emergence along with N) has been observed (Fageria, 1991a).

Malavolta (1985) reported beneficial effects of split application of K^+ in several annual crops in Brazil. Similarly, Haque et al. (1982) in Bangladesh and Ismunadji et al. (1982) in Indonesia also reported yield increase in lowland rice with split application of K^+. Kim and Park (1973) reported that split application of K^+ significantly increased rice yield on heavy sulfate soils in South Korea. From experiments conducted in Japan and Taiwan, Su (1976) reported the beneficial effect to rice of split application of K^+. Similar benefits of K topdressing in rice have been reported from India (Singh and Kumar, 1981).

4.9.3 Use of Adequate Rate

Due to the introduction of high-yielding cultivars and an increase of multiple cropping indexes, yields per unit area have increased around the world, resulting in the removal of considerable quantities of K^+ from soils. Due to this fact, since the early 1970s a very rapid increase in world K^+ fertilizer consumption has occurred, and in spite of periodic market adjustment this increase will continue in the long run. However, the supplying of K^+ to soils may not be matching the removal by cropping. Adequate fertility management requires knowledge of crop response to K fertilizer. Adequate rate of K application varies with crop species and genotypes of the same species. Other factors which determine K rates are yield level, environmental factors, and crop management practices. Hence, supply of adequate rate of K^+ for crops is fundamental to improve yields and K-use efficiency (Fageria, 2009).

The most important criteria for adequate K application are crop response to applied K and a corresponding soil analysis value known as a soil test calibration study. Dahnke and Olson (1990) defined soil test calibration as the process of ascertaining the degree of limitation to crop growth or the probability of getting a growth response to applied nutrient at any soil test level. The amount of extractable nutrient usually is expressed in terms of low, medium, and high or as a range of critical concentrations. The final step is to develop fertilizer recommendations (Dahnke and Olson, 1990). Table 4.19 shows potassium recommendations based on K soil test calibration in various states of Brazil. If soil test calibration data are not available, potassium fertilizer recommendations also can be made on the basis of crop response curves to applied potassium. In this case, maximum

TABLE 4.19

Soil Test Values for Potassium Recommendations for Dry Bean in Brazil

State	Exchangeable K Level in the Soil (mg kg^{-1})		
	Low	Medium	High
Rio Grande do Sul and Santa Catarina	4–6	61–80	81–120
Paraná	<39	40–117	118–235
São Paulo	<27	48–59	60–117
Espírito Santo	<30	30–60	>60
Minas Gerais	<30	31–60	>60
Pernambuco	<15	16–30	31–45
Goiás	<25	26–50	>50

Sources: Moraes, J.F.V., Liming and fertilization, *Bean Crop: Factors Affecting Productivity*, M.J. Zimmermann, M. Rocha, and T. Yamada. (eds.), Piracicaba, Brazilian Potash and Phosphate Institute, São Paulo, Brazil, 1988, pp. 260–301; Fageria, N.K., *Soil Sci. Plant Anal.*, 33, 2301, 2002.

economic rate of K^+ is determined or defined on the basis of 90% maximum yield. Sometimes absolute yield is transformed into relative yield, and a relationship between potassium fertilizer rates and relative yield is determined. In this way it is easy to determine economic K^+ rate at 90% relative yield. Relative yield can be calculated by using the following formula (Fageria et al., 2011):

$$\text{Relative yield}(\%) = \frac{\text{Yield of unfertilized or another fertilized plot}}{\text{Maximum yield of fertilized plot}} \times 100$$

According to Fageria et al. (2011), a wide scattering of absolute yields may occur as a result of factors other than soil fertility. This scattering of absolute yields does not necessarily mean that there is poor correlation, but a better relationship may be obtained by using a relative yield to eliminate some of the climate and site influences.

The absolute minimum level of exchangeable K^+ for tropical agriculture is considered to be close to 0.10 $cmol_c$ kg^{-1} (39 mg kg^{-1}) but may vary from 0.07 to 0.20 $cmol_c$ kg^{-1} (27–78 mg kg^{-1}) depending on the kinds of soil and crop species involved (Fageria and Gheyi, 1999). A level below 0.13 $cmol_c$ K kg^{-1} (51 mg K kg^{-1}) is usually inadequate to support normal plant growth of most crop species (Fageria and Gheyi, 1999). Muns (1982) reviewed the suggested critical concentrations of exchangeable K^+ in a number of tropical and subtropical soils. Deficiencies were reported in soil having concentrations ranging from 0.15 to 0.45 $cmol_c$ kg^{-1}, depending on soil characteristics and crop requirements. According to Mahapatra and Prasad (1970), an exchangeable K^+ content of about 0.2 $cmol_c$ kg^{-1} is considered a satisfactory level for flooded rice. This value has been suggested as appropriate to upland rice in Latin America (De Datta and Mikkelson, 1985).

4.9.4 INCORPORATION OF CROP RESIDUES

The incorporation of crop residues in the soil after harvest enables a substantial amount of plant K^+ to be recycled. Fageria et al. (1990d) and Fageria and Gheyi (1999) reported that approximately 70%–80% of the total K^+ content remains in the vegetative shoot of cereals such as wheat and rice, whereas about 40%–50% remains in the shoot of legumes, such as cowpea and beans. Fageria et al. (1982) also reported that most of the K^+ was retained in the straw of upland rice at harvest, and hence it is advisable to incorporate rice straw in the soil to contribute K^+ for the succeeding crop. Fageria (1991b) determined K^+ distribution in different plant parts of cowpea and concluded that 5% was in the roots, 49% was in the tops, 36% was in the seeds, and 10% was in pod husks. Similarly, Fageria (1991c) reported distribution of K^+ in upland rice plants as 3% in the roots 86% in the tops and the remaining 10% in the grain. Fageria et al. (2011) reported that in lowland rice at the time of harvest, for the proportion of total plant K^+, 15% was in the grain, and the remaining 85% was in the shoot.

De Datta and Mikkelsen (1985) reported that recycling of K^+ occurs to a considerable extent if crop residues are left in the field and incorporated into the soil. These authors also reported that recycling rice straw returns up to about 28 kg of K^+ per metric ton of straw. Returning crop residues was essential for prolonging the residual effects from K applications under high-input systems in acid, low-CEC Oxisols (Da Silva and Ritchey, 1982). The same authors reported that one application of 150 kg K ha^{-1} provided enough K^+ for five crops if stover was returned. In another study established after clearing a virgin forest on a clayey Oxisol in the Brazilian Amazon basin, a seven-crop rotation was supplied adequately with K^+ if crop stover was returned (Smyth et al., 1987; Cox and Uribe, 1992).

4.9.5 ADEQUATE WATER SUPPLY

Soil water is an important factor affecting diffusion of K^+ in the rhizosphere and K^+ uptake. Low soil water contents reduce K^+ uptake by plant roots because K^+ diffusion rates decrease as soil water content decreases. Crop response to fertilizers also is affected by the interaction between fertilizer placement and soil water content. Placement of fertilizers in soil zones that are susceptible to drying may limit plant uptake. Kaspar et al. (1989) reported that drying of the fertilized soil layer even though water was available at greater depth suppressed soybean growth and K^+ utilization efficiency. Change in soil water content has a greater effect on K^+ diffusion rate than changes in soil temperature. Schaff and Skogley (1982) found that K^+ diffusion rate increased an average 2.8-fold as soil water was raised from 10% to 28% (w/w). This result compared with only a 1.6- to 1.7-fold increase in K^+ diffusion as soil temperature was raised from 5° C to 30°C. Bertsch and Thomas (1985) reported that as soil water is reduced, the concentration of divalent ions like Ca^{2+} and Mg^{2+} increases faster than the concentrations of K^+ in the soil solution, thereby resulting in a decreasing K^+ concentration with increasing soil moisture tension and consequently lower uptake.

4.9.6 USE OF COVER CROPS OR GREEN MANURING

Cover crops are defined as close-growing crops that provide soil protection and soil improvement between periods of normal crop production or between trees in orchards and vines in vineyards. When plowed under and incorporated into the soil, cover crops are referred to as green manure crops (Soil Science Society of America, 2008). Positive roles of cover cropping or green manuring in crop production have been known since ancient times. The importance of this soil-ameliorating practice is increasing in recent years because of the high cost of chemical fertilizers, increased risk of environmental pollution, and the need for sustainable cropping systems (Fageria, 2007b). Cover cropping or green manuring can improve soil physical, chemical, and biological properties and consequently crop yields. Cover cropping or green manuring can increase cropping system sustainability by reducing soil erosion (Fageria et al., 2011), by increasing soil organic matter (SOM) and fertility levels (Fageria, 2012), and by reducing global warming potential (Robertson et al., 2000). Furthermore, potential benefits of cover cropping or green manuring are reduced NO_3^- leaching risk and lower fertilizer N requirements for succeeding crops (Fageria et al., 2005). However, its influence may vary from soil to soil, crop to crop, environmental variables, type of green manure crop used, and its management. Beneficial effects of cover cropping or green manuring in crop production should not be evaluated in isolation, however, but rather in integration with chemical fertilizers (Fageria et al., 2005; Fageria, 2007b).

Soil compaction is a major problem for crop production around the world. Deep-rooted cover crops are one possible solution to compaction problems, especially in no-till farming systems (Unger and Kaspar, 1994; Williams and Weil, 2004). The deep-growing tap roots of the perennial alfalfa can increase infiltration rate on compacted no-till soils (Meek et al., 1990), and recolonization of root channels left by alfalfa has been shown to benefit corn (*Z. mays* L.) root systems that follow (Rasse and Smucker, 1998). Similarly, Williams and Weil (2004) reported that soybean roots were reported to take advantage of the root channels left by decomposition of cover crop roots of the cereal rye (*S. cereale* L.) and forage radish (*Raphanus sativus* L. Diachon).

4.9.7 USE OF FARMYARD MANURES

The use of livestock manure may be a complementary source of N, P, and K when applied in amounts that account for the composition of the manure and the nutrient-supplying potential of the soil (Chang et al., 1993; Fageria and Baligar, 2003b). Integration of manure with inorganic fertilizers may result in the benefits of greater residual effects of organic rather than inorganic sources and advantages of manure in addition to a supply of N, P, and K (e.g., improved soil physical properties and supply of bases) (Fageria and Baligar, 2003b).

In addition, the use of farmyard manures improves SOM content. SOM is a storehouse of nutri-ents and water, gives improved soil structure and improved activities of beneficial microorganisms, reduces elemental toxicity, and makes soil less susceptible to erosion. These benefits mean that improving or maintaining SOM is one of the most important management practices to maintain soil fertility of tropical soils for sustainable crop production (Fageria and Baligar, 2003b).

4.9.8 USE OF CROP ROTATION

Conventional monoculture agriculture systems can reduce the quality of soils by allowing loss of SOM and structure because of a low level of organic inputs and regular disturbance from tillage practices (Acosta-Martinez et al., 2004). Crop rotation may have many positive effects on soil qual-ity and consequently on crop production. Crop rotation is defined as a planned sequence of crops growing in a regularly recurring succession on the same area of land, as contrasted to continuous culture of one crop (Soil Science Society of America, 2008). Bullock (1992) defined crop rotation as a system of growing different types of crops in a recurrent succession and in an advantageous sequence on the same land. Crop rotations are a key component of successful organic arable systems (Robson et al., 2002). Rotations can be optimized to conserve and recycle nutrients and minimize pest, disease, and weed problems (Robson et al., 2002). Appropriate crop rotation has significant influence on SOM content of soils. The results of long-term field trials in Illinois (United States) showed that crop rotation influenced the content of SOM (Odell et al., 1984). The level of soil C and N was highest in the rotation of corn–oats–clover and lowest in the permanent corn rotation (Mengel et al., 2001).

Crop rotations under conventional tillage (CT) that provide residues with low C/N ratios stimu-late decomposition of native SOM to a greater extent than rotations providing residues with high C/N ratios (Fageria, 2009). Under no-tillage (NT), crop rotations have been shown to have minimal effect on native SOM decomposition (Sisti et al., 2004). Wright and Hons (2004) also reported that greater differences in SOM between crop species occurred under CT rather than NT, especially in subsurface soil. Wani et al. (1994) reported that green manures and organic amendments in crop rotations provided a measurable increase in SOM and other soil quality attributes compared with continuous cereal systems.

Crop rotations have positive effects on soil properties related to the higher C inputs and diversity of plant residues in soils in comparison with continuous systems (Fageria, 2009). Conservation tillage increases soil organic C and microbial biomass and modifies the soil microbial community (Acosta-Martinez et al., 2004).

4.9.9 USE OF CONSERVATION TILLAGE

Conservation tillage is defined as any tillage sequence, the objective of which is to minimize or reduce loss of soil and water; operationally, a tillage or tillage and planting combination that leaves a 30% or greater cover of crop residues on the surface (Soil Science Society of America, 2008). Minimum-tillage, no-tillage, or zero-tillage terms are also used in the litera-ture. According to the Soil Science Society of America (2008), minimum tillage is defined as the minimum use of primary and/or secondary tillage necessary for meeting crop production requirements under the existing soil and climatic conditions, usually resulting in fewer tillage operations for CT. Similarly, minimum tillage or zero tillage is defined as a procedure whereby a crop is planted directly into the soil with no primary or secondary tillage since harvest of the previous crop. In this process, usually a special planter is necessary to prepare a narrow shallow seedbed immediately surrounding the seed being planted. No tillage is sometimes practiced in combination with subsoiling to facilitate seeding and early root growth, whereby the surface residue is left virtually undisturbed except for a small slot in the path of the subsoil shank. Conservation, minimum tillage, or no tillage has been adopted widely in developed as well as developing countries in recent years for crop production. It is projected that conservation tillage will be practiced on 75% of cropland in the United States by 2020 (Lal, 1997).

There is a general concept that tillage decreases aggregate stability by increasing mineralization of organic matter and exposing aggregates to additional raindrop impact energies (Park and Smucker, 2005). Tillage promotes SOM loss through crop residue incorporation into soil, physical breakdown of residues, and disruption of macroaggregates (Wright and Hons, 2004). In contrast, conservation or no tillage reduces soil mixing and soil disturbance, which allows SOM accumulation (Blevins and Frye, 1993). Many studies have shown that conservation tillage improves soil aggregation and aggregate stability (Fageria, 2012). Conservation or minimum tillage promotes soil aggregation through enhanced binding of soil particles as a result of greater SOM content (Six et al., 2002). Microaggregates often form around particles of undecomposed SOM, providing protection from decomposition (Six et al., 2002; Wright and Hons, 2004). Microaggregates are more stable than macroaggregates, and, thus, tillage is more disruptive of large aggregates than smaller aggregates, making SOM from large aggregates more susceptible to mineralization (Six et al., 2002; Wright and Hons, 2004; Fageria, 2012). Since tillage often increases the proportion of microaggregates to macroaggregates, there may be less crop-derived SOM in CT than conservation or no tillage (Wright and Hons, 2004). Fungal growth and mycorrhizal fungi, which are promoted by no tillage, contribute to the formation and stabilization of macroaggregates (Beare and Bruce, 1993).

The larger SOM accumulation in conservation tillage had been observed in intensive cropping systems, where multiple crops are grown yearly (Wright and Hons, 2004). The use of conservation tillage, including no till, is being considered as part of a strategy to reduce C loss from agricultural soils (Denef et al., 2004). Crop species also influence SOM accumulation in the soil. Residue quality often plays an important role in regulating long-term SOM storage (Lynch and Bragg, 1985). Crop residues having low N concentration, such as wheat, generally decompose at slower rates than residues with higher N, such as sorghum and soybean (Franzluebbers et al., 1995; Wright and Hons, 2004), since wheat residues often persist longer and increase SOM more than sorghum or soybean (Wright and Hons, 2004).

4.9.10 Optimum Potassium Saturation in Soil Solution

Optimum K^+ saturation in soil solution is an important index for improving crop yields and potassium-use efficiency in crop plants. Dry bean yield increased linearly when K^+ saturation increased from 1% to 3% across two soils layers (Figure 4.12). This result indicates that K^+ saturation level was not sufficient for achieving optimum dry bean yield in the soil under investigation. The optimum K^+ saturation value reported by Eckert (1987) from various studies is in the range of 2%–5%.

4.9.11 Use of Efficient Crop Species and Cultivars

The use of nutrient-efficient crop species or genotypes within species is an important management strategy for maximizing potassium-use efficiency in crop plants. In the literature nutrient-efficient plants are defined in several ways. Some of these definitions are presented in Table 4.20. From a practical standpoint, crop nutrient-use efficiency reflects the ability to produce a high yield under nutrient-limited growth conditions. But to classify plant genotypes within a crop species as nutrient-use efficient or inefficient, Gerloff (1976) proposed that inefficient genotypes must yield approximately the same as efficient strains under optimum supplies of the limiting element.

This restriction avoids comparison of genotypes with widely differing growth potential and nutrient demand. Cassman et al. (1989) evaluated cultivar differences in K^+-use efficiency (defined as higher yield with a limited K^+ supply) in relation to K^+ uptake, K^+ partitioning, and critical internal and external K^+ requirements for cotton. Without K^+ addition, the yield was 29% and 35% greater in the 2-year experimentation in the K-efficient cultivar. If K^+ supply was not limited, cultivar yields were similar. Yield of both cultivars was associated closely with leaf K^+ concentration and

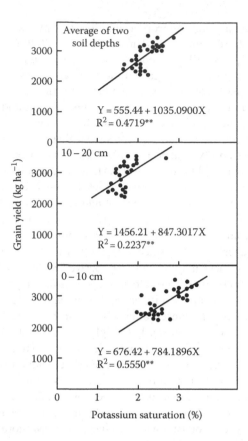

FIGURE 4.12 Relationship between potassium saturation of CEC and grain yield of dry bean. (From Fageria, N.K., *Commun. Soil Sci. Plant Anal.*, 39, 845, 2008.)

soil K^+ availability, but response curves indicated a lower leaf and soil K^+ requirement for the K^+-efficient cultivar. The K^+-efficient cultivar had a higher K^+ uptake rate during fruit development and greater total K^+ accumulation, particularly at low soil K^+ levels.

The author evaluated 20 upland rice genotypes for potassium-use efficiency in Brazilian Oxisol. Based on grain yield efficiency index (GYEI), genotypes were classified as efficient, moderately efficient, or inefficient (Figure 4.13). The GYEI was calculated by using the following formula:

$$\text{GYEI} = \frac{\text{Grain yield at low K level}}{\text{Average grain yield of 20 genotypes at low K level}}$$

$$\times \frac{\text{Grain yield at high K level}}{\text{Average grain yield of 20 genotypes at high K level}}$$

GYEI is useful in separating high-yielding, stable, nutrient-efficient genotypes from low-yielding, unstable, and nutrient-inefficient genotypes (Fageria, 2014). Genotypes having GYEI higher than 1 were considered K efficient; inefficient genotypes were in the range of 0–0.5 GYEI, and genotypes in between these two limits were considered intermediate in K-use efficiency. Primavera and BRA 1600 were the most efficient, and BRSMG Curinga and BRA 02582 were the most inefficient in K use. Cultivar differences in K^+ accumulation have been reported in a number of crop species (Epstein and Bloom, 2005; Damon and Rengel, 2007; Fageria, 2009, 2013; Fageria et al., 2011).

Potassium utilization differs among crop species. Generally, sugar beet, sugarcane, potato, tomato, and celery have a very high demand, whereas cotton and wheat take up much lower amounts

TABLE 4.20
Definitions of Nutrient-Efficient Plants

Definition	Reference
A nutrient-efficient plant is defined as a plant that absorbs, translocates, or utilizes more of a specific nutrient than another plant under conditions of relatively low nutrient availability in the soil or growth media.	Soil Science Society of America (2008)
The nutrient efficiency of a genotype (for each element separately) is defined as the ability to produce a high yield in a soil that is limiting in that element for a standard genotype.	Graham (1984)
Nutrient efficiency of a genotype or cultivar is defined as the ability to acquire nutrients from growth medium and/or to incorporate or utilize them in the production of shoot and root biomass or utilizable plant material (grain).	Blair (1993)
An efficient genotype is one that absorbs relatively high amounts of a nutrient from soil and fertilizer, produces a high grain yield per unit of absorbed nutrient, and stores relatively little nutrients in the straw.	Isfan (1993)
Efficient plants are defined as those that produce more dry matter or have a greater increase in harvested portion per unit time, area, or applied nutrient, have fewer deficiency symptoms, or have greater increment increases and higher concentrations of mineral nutrients than other plants grown under similar conditions or compared to a standard genotype.	Clark (1990)
Efficient germplasm requires less of a nutrient than an inefficient one for normal metabolic processes.	Gourley et al. (1994)
An efficient plant is defined as a plant that produces higher economic yield with a determined quantity of applied or absorbed nutrient compared to other or standard plant under similar growing conditions.	Fageria et al. (2008)

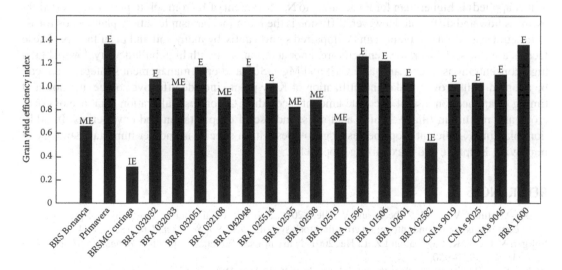

FIGURE 4.13 Classification of upland rice genotypes for K-use efficiency. Letters on bar: E, efficient; ME, moderately efficient; IE, inefficient. (From Fageria, N.K., *Nutrient Uptake in Crop Plants*, CRC Press, Boca Raton, FL, 2009.)

of K^+ (Mengel et al., 2001). Pretty and Stangel (1985) reported that root crops, sugar crops, fiber crops, and forages are among those with a high K^+ requirement. Oilseeds and grains such as soybean and corn have intermediate needs, whereas the cereal grains wheat and rice require relatively small amounts of K^+. These differences among cultivars suggested that it is possible to improve K^+ utilization and forage composition of the species through the use of efficient cultivars.

4.9.12 CONTROL OF DISEASES, INSECTS, AND WEEDS

Diseases, insects, and weeds are serious problems for crop production worldwide. Their presence in the crops not only reduces plant use of water, nutrients, and solar radiation but also reduces nutrient uptake and use efficiency. Weed seeds also reduce crop quality by mixing with crop seeds. Hence, control of weeds is an important strategy in reducing their harmful effects in crop production and improves nutrient-use efficiency, including of potassium. Plant diseases and insects can be controlled by use of appropriate fungicides and insecticides. Planting disease- and insect-resistant crop cultivars is an attractive strategy and also environmentally sound. Similarly, weeds can be controlled mechanically and by use of herbicides.

4.10 CONCLUSIONS

Potassium plays significant roles in many physiological and biochemical processes in plants. Adequate supply of K in the soil may lessen plant stresses to abiotic and biotic factors, regulate osmotic pressure, reduce plant lodging, and improve grain quality. Potassium in adequate supply also improves root growth, an action that has special significance for water and nutrient conditions for crop plants. Potassium is highly mobile in the plant; hence, its deficiency symptoms first appear in the old leaves of crop plants. Potassium toxicity symptoms rarely are observed in crop plants under field conditions. Its range of deficiency and toxicity application in the soil is high. Accumulation of potassium is higher in cereal crops than for other plant nutrients. In legumes, uptake of K is second to N. In cereals, accumulation of K is smaller in grain than in straw. For example, in rice about 85% of the K remains in the straw. Hence, incorporation of straw after harvest of a rice crop is an important strategy in recycling K in the soil. Overall, K-use efficiency (grain yield per unit K accumulation) is lower than for N and P. However, K recovery efficiency (accumulation in the plant per unit K applied) is higher than for P and equal to N. Movement of K from soil to plant roots is mainly by mass flow and diffusion. However, diffusion is the main mechanism to satisfy plant needs for K. Potassium availability to plants can be impaired significantly by many soil and plant factors. These factors are low soil water, low temperature, poor aeration, soil with high bulk density, low pH, and high concentrations of exchangeable Ca^{2+} and Mg^{2+}. Soil and crop management strategies that can be adopted to improve uptake and utilization of K by plants include effective source, method and timing of application, use of adequate amounts according to soil test calibration data, use of cover crops and minimum tillage, liming acid soils, and use of crop rotation and cover crops. In addition, planting K-efficient crop species or genotypes within species is another important strategy in improving K uptake and maximizing crop yields.

REFERENCES

Acosta-Martinez, V., T. M. Zobeck, and V. Allen. 2004. Soil microbial, chemical and physical properties in continuous cotton and integrated crop-livestock systems. *Soil Sci. Soc. Am. J.* 68:1875–1884.

Baligar, V. C., N. K. Fageria, and Z. L. He. 2001. Nutrient use efficiency in plants. *Commun. Soil Sci. Plant Anal.* 32:921–950.

Barber, S. A. 1995. *Soil Nutrient Bioavailability*. New York: John Wiley & Sons.

Barley, K. P. 1964. The utility of field experiments. *Soils Fert.* 27:267–269.

Beare, M. H. and R. R. Bruce. 1993. A comparison of methods for measuring water-stable aggregates: Implications for determining environmental effects on soil structure. *Geoderma* 56:87–104.

Bertsch, P. M. and G. W. Thomas. 1985. Potassium status of temperate region soils. In: *Potassium in Agriculture*, ed. R. D. Munson, pp. 131–162. Madison, WI: ASA, CSSA and SSSA.

Blair, G. 1993. Nutrient efficiency-what do we really mean. In: *Genetic Aspects of Mineral Nutrition*, eds. P. J. Randall, E. Delhaize, R. A. Richards, and R. Munns, pp. 205–213. Dordrecht, the Netherlands: Kluwer Academic Publishers.

Blevins, R. L. and W. W. Frye. 1993. Conservation tillage: An ecological approach to soil management. *Adv. Agron.* 51:33–78.

Brady, N. C. and R. R. Weil. 2002. *The Nature and Properties of Soils*, 13th edn. Upper Saddle River, NJ: Prentice Hall.

Bullock, D. G. 1992. Crop rotation. *Crit. Rev. Plant Sci.* 11:309–326.

Cassman, K. G., T. A. Kerby, B. A. Roberts, D. C. Bryant, and S. M. Brouder. 1989. Differential response of two cotton cultivars to fertilizer and soil potassium. *Agron. J.* 81:870–876.

Chang, C., T. G. Sommerfield, and T. Entz. 1993. Barley performance under heavy applications of cattle feedlot manure. *Agron. J.* 85:1013–1018.

Clark, R. B. 1990. Physiology of cereals for mineral nutrient uptake, use, and efficiency. In: *Crops as Enhancers of Nutrient Use*, eds. V. C. Baligar and R. R. Duncan, pp. 131–209. San Diego, CA: Academic Press.

Collis-George, N. and B. G. Davey. 1960. The doubtful utility of present day field experimentation and other determinations involving soil-plant interactions. *Soils Fert.* 23:307–310.

Cox, F. R. and E. Uribe. 1992. Management and dynamics of potassium in a humid tropical Ultisol under a rice-cowpea rotation. *Agron. J.* 84:655–660.

Da Silva, J. E. and K. D. Ritchey. 1982. Potassium fertilization of cerrado soils. In: *Potassium in Brazilian Agriculture*, ed. T. Yamada, pp. 323–338. São Paulo, Brazil: Brazilian Potassium and Phosphate Association.

Dahnke, W. C. and R. A. Olsen. 1990. Soil test correlation, calibration, and recommendation. In: *Soil Testing and Plant Analysis*, ed. R. L. Westerman, pp. 45–71. Madison, WI: Soil Science Society of America.

Damon, P. M. and Z. Rengel. 2007. Wheat genotypes differ in potassium efficiency under glasshouse and field conditions. *Aust. J. Agric. Res.* 58:816–825.

De Datta, S. K. and D. S. Mikkelsen. 1985. Potassium nutrition of rice. In: *Potassium in Agriculture*, ed. D. Munson, pp. 665–699. Madison, WI: ASA, CSSA and SSSA.

Denef, K., J. Six, R. Merckx, and K. Paustian. 2004. Carbon sequestration in microaggregates of no-tillage soils with different clay mineralogy. *Soil Sci. Soc. Am. J.* 68:1935–1944.

Dibb, D. W. and W. R. Thompson, Jr. 1985. Interaction of potassium with other nutrients. In: *Potassium in Agriculture*, ed. R. D. Munson, pp. 515–533. Madison, WI: ASA, CSSA, and SSSA.

Dibb, D. W. and L. F. Welch. 1976. Corn growth as affected by ammonium vs. nitrate absorbed from soil. *Agron. J.* 68:89–94.

Drew, M. C. and P. H. Nye. 1969. The supply of nutrient ions by diffusion to plant roots in soil. II. The effect of root hairs on the uptake of potassium by roots of ryegrass (*Lolium multiflorum*). *Plant Soil* 31:407–424.

Duke, S. H. and M. Collins. 1985. Role of potassium in legume dinitrogen fixation. In: *Potassium in Agriculture*, ed. R. D. Munson, pp. 443–465. Madison, WI: ASA, CSSA, and SSSA.

Eckert, D. J. 1987. Soil test interpretations: Basic cation saturation ratios and sufficiency levels. In: *Soil Testing: Sampling, Correlation, Calibration, and Interpretation*, ed. J. R. Brown, pp. 53–64. Madison, WI: SSSA.

Epstein, E. and A. J. Bloom. 2005. *Mineral Nutrition of Plants: Principles and Perspectives*, 2nd edn. Sunderland, MA: Sinauer Associations, Inc.

Fageria, N. K. 1982. Fertilization and potassium nutrition of rice in Brazil. In: *Potassium in Brazilian Agriculture*, ed. T. T. Yamada, pp. 421–436. São Paulo, Brazil: Brazilian Potassium and Phosphate Association.

Fageria, N. K. 1983. Ionic interactions in rice plants from dilute solutions. *Plant Soil* 70:309–316.

Fageria, N. K. 1984. *Fertilization and Mineral Nutrition of Rice*. Rio de Janeiro, Brazil: CNPAF/Campus.

Fageria, N. K. 1989. *Tropical Soils and Physiological Aspects of Crops*. Goiania, Brazil: EMBRAPA-CNPAF.

Fageria, N. K. 1991a. Response of rice to fractional applied potassium in Brazil. *Better Crops Intern.* 7:19.

Fageria, N. K. 1991b. Response of cowpea to phosphorus on an Oxisol with special reference to dry matter production and mineral ion contents. *Trop. Agric.* 68:384–388.

Fageria, N. K. 1991c. Response of rice cultivars to phosphorus fertilization on a dark red latosol of central Brazil. *Rev. Bras. Ci. Solo* 15:63–67.

Fageria, N. K. 2000a. Potassium use efficiency of upland rice genotypes. *Pesq. Agropec. Bras.* 35:2115–2120.

Fageria, N. K. 2000b. Upland rice response to soil acidity in cerrado soil. *Pesq. Agropec. Bras.* 35:2115–2120.

Fageria, N. K. 2002. Soil quality versus environmentally based agricultural management practices. *Commun. Soil Sci. Plant Anal.* 33:2301–2329.

Fageria, N. K. 2007a. Soil fertility and plant nutrition research under field conditions: Basic principles and methodology. *J. Plant Nutr.* 30:203–223.

Fageria, N. K. 2007b. Green manuring in crop production. *J. Plant Nutr.* 30:691–719.

Fageria, N. K. 2008. Optimum soil acidity indices for dry bean production on an Oxisol in no-tillage system. *Commun. Soil Sci. Plant Anal.* 39:845–857.

Fageria, N. K. 2009. *Nutrient Uptake in Crop Plants*. Boca Raton, FL: CRC Press.

Fageria, N. K. 2012. Role of soil organic matter in maintaining sustainability of cropping systems. *Commun. Soil Sci. Plant Anal.* 43:2063–2163.

Fageria, N. K. 2013. *The Role of Plant Roots in Crop Production*. Boca Raton, FL: CRC Press.

Fageria, N. K. 2014. *Mineral Nutrition of Rice*. Boca Raton, FL: CRC Press.

Fageria, N. K. and V. C. Baligar. 2003a. Methodology for evaluation of lowland rice genotypes for nitrogen use efficiency. *J. Plant Nutr.* 26:1315–1333.

Fageria, N. K. and V. C. Baligar. 2003b. Fertility management of tropical acid soil for sustainable crop production. In: *Handbook of Soil Acidity*, ed. Z. Rengel, pp. 359–385. New York: Marcel Dekker.

Fageria, N. K. and V. C. Baligar. 2005. Nutrient availability. In: *Encyclopedia of Soils in the Environment*, ed. D. Hillel, pp. 63–71. San Diego, CA: Elsevier.

Fageria, N. K., V. C. Baligar, and B. A. Bailey. 2005. Role of cover crops in improving soil and row crop productivity. *Commun. Soil Sci. Plant Anal.* 36:2733–2757.

Fageria, N. K., V. C. Baligar, and R. B. Clark. 2006. *Physiology of Crop Production*. New York: The Haworth Press.

Fageria, N. K., V. C. Baligar, and D. G. Edward. 1990a. Soil-plant nutrient relationships at low pH stress. In: *Crops as Enhancers of Nutrient Use*, eds. V. C. Baligar and R. R. Duncan, pp. 475–507. New York: Academic Press.

Fageria, N. K., V. C. Baligar, and C. A. Jones. 2011. *Growth and Mineral Nutrition of Field Crops*, 3rd edn. Boca Raton, FL: CRC Press.

Fageria, N. K., V. C. Baligar, and Y. C. Li. 2008. The role of nutrient efficient plants in improving crop yields in the twenty first century. *J. Plant Nutr.* 31:1121–1157.

Fageria, N. K., V. C. Baligar, R. J. Wright, and J. R. P. Carvalho. 1990b. Lowland rice response to potassium fertilization and its effect on N and P uptake. *Fert. Res.* 21:157–162.

Fageria, N. K. and M. P. Barbosa Filho. 1994. *Nutritional Deficiencies in Rice: Identification and Correction*. Goiania, Brazil: EMBRAPA-CNPAF document 42.

Fageria, N. K., M. P. Barbosa Filho, and J. R. P. Carvalho. 1982. Response of upland rice to phosphorus fertilization on an Oxisol of central Brazil. *Agron. J.* 74:51–56.

Fageria, N. K., J. R. P. Carvalho, V. C. Baligar, and R. J. Wright. 1990c. Upland rice response to potassium fertilization on a Brazilian oxisol. *Fert. Res.* 21:141–147.

Fageria, N. K. and H. R. Gheyi. 1999. *Efficient Crop Production*. Campina Grande, Brazil: Federal University of Paraiba.

Fageria, N. K. and A. B. Santos. 2008. Lowland rice response to thermophosphate fertilization. *Commun. Soil Sci. Plant Anal.* 39:873–889.

Fageria, N. K., A. B. Santos, and M. F. Moraes. 2010b. Yield, potassium uptake and use efficiency in upland rice genotypes. *Commun. Soil Sci. Plant Anal.* 41:2676–2684.

Fageria, N. K., A. B. Santos, A. Moreira, and M. F. Moraes. 2010a. Potassium soil test calibration for lowland rice on an Inceptisol. *Commun. Soil Sci. Plant Anal.* 41:2595–2601.

Fageria, N. K., R. J. Wright, V. C. Baligar, and J. R. P. Carvalho. 1990d. Upland rice response to potassium fertilization on an Oxisol. *Fert. Res.* 21:141–147.

Foth, H. D. and B. G. Ellis. 1988. *Soil Fertility*. New York: John Wiley & Sons.

Franzluebbers, A. J., F. M. Hons, and D. A. Zuberer. 1995. Soil organic carbon, microbial biomass, and mineralizable carbon and nitrogen in sorghum. *Soil Sci. Soc. Am. J.* 59:460–466.

Gerloff, G. C. 1976. Plant efficiencies in the use of nitrogen, phosphorus, and potassium. In: *Plant Adaptation to Mineral Stress in Problem Soils*, ed. M. J. Wright, pp. 161–173. Ithaca, NY: Cornell University Agricultural Experiment Station.

Gourley, C. J. P., D. L. Allan, and M. P. Russelle. 1994. Plant nutrient efficiency: A comparison of definitions and suggested improvement. *Plant Soil* 158:29–37.

Graham, R. D. 1984. Breeding for nutritional characteristics in cereals. In: *Advances in Plant Nutrition*, Vol. 1, eds. P. B. Tinker and A. Lauchi, pp. 57–102. New York: Praeger Publisher.

Gupta, U. C. 1979. Boron nutrition of crops. *Adv. Agron.* 31:273–307.

Haque, S. A., Z. H. Bhuya, A. K. M. Idris Ali, F. A. Choudhury, M. Jahiruddin, and M. M. Rahman. 1982. Response of HVY paddy to potash fertilization in different regions of Bangladesh. In: *Phosphorus and Potassium in the Tropics*, eds. E. Pushparajah and E. H. A. Hamid, pp. 425–430. *Proceedings of International Conference on Phosphorus and Potassium in the Tropics*, Kuala Lumpur, Malaysia, August 1981. Kuala Lumpur, Malaysia: Malaysian Society of Soil Science.

Havlin, J. L., D. G. Westfall, and S. R. Olsen. 1985. Mathematical models for potassium release kinetics in calcareous soils. *Soil Sci. Soc. Am. J.* 49:371–376.

Heckman, J. R. and E. J. Kamprath. 1992. Potassium accumulation and corn yield related to potassium fertilizer rate and placement. *Soil Sci. Soc. Am. J.* 56:141–148.

Hill, W. E. and L. G. Morrill. 1975. Boron, calcium and potassium interactions in Spanish peanuts. *Proc. Soil Sci. Soc. Am.* 39:80–83.

Hogh-Jensen, H. and M. B. Pedersen. 2003. Morphological plasticity by crop plants and their potassium use efficiency. *J. Plant Nutr.* 26:969–984.

Huber, D. M. and D. C. Arny. 1985. Interactions of potassium with plant disease. In: *Potassium in Agriculture*, ed. R. D. Munson, pp. 467–488. Madison, WI: American Society of Agronomy.

Isfan, D. 1993. Genotypic variability for physiological efficiency index of nitrogen in oats. *Plant Soil* 154:53–59.

Ismunadji, M., I. Nasution, and S. Parthohardjono. 1982. Potassium nutrition of lowland rice. In: *Phosphorus and Potassium in the Tropics*, eds. E. Pushparajah and E. H. A. Hamid, pp. 315–321. *Proceedings of International Conference on Phosphorus and Potassium in the Tropics*, Kuala Lumpur, Malaysia, August 1981. Kuala Lumpur, Malaysia: Malaysian Society of Soil Science.

Johnson, C., D. G. Edwards, and J. F. Loneragan. 1968. Interactions between potassium and calcium in their absorption by intact barley plants. I. Effects of potassium on calcium absorption. *Plant Physiol.* 43:1717–1721.

Kaspar, T. C., J. B. Zahler, and D. R. Timmons. 1989. Soybean response to phosphorus and potassium fertilizers as affected by soil drying. *Soil Sci. Soc. Am. J.* 53:1448–1454.

Kim, Y. S. and S. C. Park. 1973. Effect of split potassium application to paddy rice grown on acid sulfate soils. *Potash Rev.* 7:1–9.

Kiraly, Z. 1976. Plant disease resistance as influenced by biochemical effects on nutrients in fertilizers. In: *Fertilizer Use and Plant Health*, ed. International Potash Institute, pp. 33–46. Bern, Switzerland: International Potash Institute.

Lal, R. 1997. Residue management, conservation tillage and soil restoration for mitigating greenhouse effects by CO_2-enrichment. *Soil Tillage Res.* 43:81–107.

Lopes, A. S. 1983. *Soil under Cerrado*. São Paulo, Brazil: Potassium and Phosphate Institute.

Lorenz, N., K. Verdell, C. Ramsier, and R. P. Dick. 2010. A rapid assay to estimate soil microbial biomass potassium in agricultural soils. *Soil Sci. Soc. Am. J.* 74:512–516.

Lynch, J. M. and E. Bragg. 1985. Microorganisms and soil aggregate stability. *Adv. Soil Sci.* 2:133–171.

Mahapatra, I. C. and R. Prasad. 1970. Response of rice to potassium in relation to its transformation and availability under waterlogged condition. *Fert. News* 15:34–41.

Malavolta, E. 1985. Potassium status of tropical and subtropical region soils. In: *Potassium in Agriculture*, ed. R. D. Munson, pp. 163–200. Madison, WI: ASA, CSSA and SSSA.

Marschner, H. 1995. *Mineral Nutrition of Higher Plants*, 2nd edn. New York: Academic Press.

Matocha, J. E. and G. W. Thomas. 1969. Potassium and organic nitrogen content of grain sorghum as affected by iron. *Agron. J.* 61:425–428.

Meek, B. D., W. R. Detarr, D. Rolph, E. R. Rechel, and L. M. Carter. 1990. Infiltration rate as affected by an alfalfa and no-till cotton cropping system. *Soil Sci. Soc. Am. J.* 54:505–508.

Mengel, K. 2007. Potassium. In: *Handbook of Plant Nutrition*, eds. A. V. Barker and D. J. Pilbeam, pp. 91–120. Boca Raton, FL: CRC Press.

Mengel, K. and E. A. Kirkby. 1980. Potassium in crop production. *Adv. Agron.* 33:59–110.

Mengel, K., E. A. Kirkby, H. K. Kosegarten, and T. Appel. 2001. *Principles of Plant Nutrition*, 5th edn. Dordrecht, the Netherlands: Kluwer Academic Publishers.

Mielniczuk, J. 1977. Potassium forms in Brazilian soils. *Rev. Bras. Ci. Solo* 1:55–56.

Moraes, J. F. V. 1988. Liming and fertilization. In: *Bean Crop: Factors Affecting Productivity*, eds. M. J. Zimmermann, M. Rocha, and T. Yamada, pp. 260–301. São Paulo, Brazil: Brazilian Potash and Phosphate Institute.

Muns, R. D. 1982. *Potassium, Calcium and Magnesium in the Tropics and Subtropics*, ed. J. C. Brosheer, Tech. Bull. No. IFDC-T-23. Muscle Shoals, AL: International Fertilizer Development Center.

Muzilli, O. 1982. Fertilization and nutrition of soybean in Brazil. In: *Potassium in Brazilian Agriculture*, ed. T. Yamada, pp. 373–392. São Paulo, Brazil: Brazilian Potassium and Phosphate Institute.

Nelson, K. A., P. P. Motavalli, W. E. Stevens, D. Dunn, and C. G. Meinhardt. 2010. Soybean response to preplant and foliar-applied potassium chloride with strobilurin fungicides. *Agron. J.* 102:1657–1663.

Nemeth, K. 1982a. Potassium analysis methods in soil and their interpretation. In: *Potassium in Brazilian Agriculture*, ed. T. Yamada, pp. 77–94. São Paulo, Brazil: Brazilian Potassium and Phosphate Institute.

Nemeth, K. 1982b. Nutrient dynamics in some humid tropical soils as determined by electro-ultrafiltration. In: *Phosphorus and Potassium in the Tropics*, eds. E. Pushparajah and E. H. A. Hamid, pp. 3–14. *Proceedings of International Conference on Phosphorus and Potassium in the Tropics*, Kuala Lumpur, Malaysia, August 1981. Kuala Lumpur, Malaysia: Malaysian Society of Soil Science.

Nguyen, M. H., B. H. Janssen, O. Oenema, and A. Dobermann. 2006. Potassium budgets in rice cropping systems with annual flooding in the Mekong River Delta. *Better Crops Plant Food* 90:25–29.

Odell, R. T., S. W. Melsted, and W. M. Walker. 1984. Changes in organic carbon and nitrogen of Morrow plot soils under different treatments, 1904–1973. *Soil Sci.* 137:160–171.

Okajima, H. 2001. Historical significance of nutrient acquisition in plant nutrient research. In: *Plant Nutrient Acquisition: New Perspectives*, eds. N. Ae, J. Arihara, K. Okada, and A. Srinivasan, pp. 3–31. New York: Springer.

Pal, Y., R. J. Gilkes, and M. T. F. Wong. 2001. Soil factors affecting the availability of potassium to plants for Western Australian soils: A glasshouse study. *Aust. J. Soil Res.* 39:611–625.

Park, E. J. and A. J. M. Smucker. 2005. Saturated hydraulic conductivity and porosity within macroaggregates modified by tillage. *Soil Sci. Soc. Am. J.* 69:38–45.

Piggott, T. J. 1986. Vegetable crops. In: *Plant Analysis: An Interpretation Manual*, eds. D. J. Reuter and J. B. Robinson, pp. 148–187. Melbourne, Victoria, Australia: Inkata Press.

Prabhu, A. S., N. K. Fageria, D. M. Huber, and F. A. Rodrigues. 2007. Potassium and plant disease. In: *Mineral Nutrition and Plant Disease*, eds. L. E. Datnoff, W. H. Elmer, and D. M. Huber, pp. 57–78. St. Paul, MI: The American Phytopathological Society.

Pretty, K. M. and P. J. Stangel. 1985. Current and future use of world potassium. In: *Potassium in Agriculture*, ed. R. D. Munson, pp. 99–128. Madison, WI: ASA, CSSA and SSSA.

Qian, Y. L., J. D. Fry, and W. S. Upham. 1997. Rooting and drought avoidance of warm-season turfgrasses and tall fescue in Kansas. *Crop Sci.* 37:905–910.

Randall, G. W. and R. G. Hoeft. 1988. Placement methods for improved efficiency of P and K fertilizers: A review. *J. Prod. Agric.* 1:70–79.

Rasse, D. P. and A. J. M. Smucker. 1998. Root recolonization of previous root channels in corn and alfalfa rotation. *Plant Soil* 204:203–212.

Reddy, S. V. 1976. Availability of potassium in different Colorado soils. PhD dissertation, Colorado State University, Ft. Collins, CO (Diss. Abstr. 37-3195B).

Rengel, Z. and P. M. Damon. 2008. Crops and genotypes differ in efficiency of potassium uptake and use. *Physiol. Plant.* 133:624–636.

Reuter, D. J. 1986. Temperate and sub-tropical crops. In: *Plant Analysis: An Interpretation Manual*, eds. D. J. Reuter and J. B. Robinson, pp. 38–99. Melbourne, Victoria, Australia: Inkata Press.

Robertson, G. P., E. A. Paul, and R. R. Harwood. 2000. Greenhouse gases in intensive agriculture: Contributions of individual gases to the radiative forcing of the atmosphere. *Science* 289:1922–1925.

Robson, M. C., S. M. Fowler, N. H. Lampkin, C. Leifert, M. Leitch, D. Robinson, C. A. Watson, and A. M. Litterick. 2002. The agronomic and economic potential of break crops for ley/arable rotations in temperate organic agriculture. *Adv. Agron.* 77:369–427.

Rowland, J. H., J. L. Cisar, G. H. Snyder, J. B. Sartain, A. L. Wright, and J. E. Erickson. 2010. Optimal nitrogen and potassium fertilization rates for establishment of warm-season putting greens. *Agron. J.* 102:1601–1605.

Schaff, B. E. and E. O. Skogley. 1982. Diffusion of potassium, calcium, and magnesium in Bozeman silt loam as influenced by temperature and moisture. *Soil Sci. Soc. Am. J.* 46:521–524.

Sheldrick, W. F. 1985. World potassium reserves. In: *Potassium in Agriculture*, ed. D. Munson, pp. 3–28. Madison, WI: ASA, CSSA, and SSSA.

Singh, R. P. and A. Kumar. 1981. Effect of levels and times of potassium application on upland rice. *Indian Potash J.* 6:12–15.

Sisti, C. P. J., H. P. Santos, R. Kohhann, B. J. R. Alves, S. Urquiaga, and R. M. Boddey. 2004. Change in carbon and nitrogen stocks in soil under 13 years of conventional and zero tillage in southern Brazil. *Soil Tillage Res.* 76:39–58.

Six, J., C. Feller, K. Denef, S. M. Ogle, J. C. Moraes, and A. Albrecht. 2002. Soil organic matter, biota and aggregation in temperate and tropical soils-effects of no-tillage. *Agronomie* 22:755–775.

Smyth, J. T., M. Cravo, and J. B. Bastos. 1987. Soil nutrient dynamics and fertility management for sustained crop production on oxisols in the Brazilian Amazon. In: *Tropsoils Technical Report 1985–1986*, pp. 88–91. Raleigh, NC: Department of Soil Science, North Carolina State University.

Soil Science Society of America. 2008. *Glossary of Soil Science Terms*. Madison, WI: Soil Science Society of America.

Sparks, D. and P. M. Huang. 1985. Physical chemistry of soil potassium. In: *Potassium in Agriculture*, ed. R. E. Munson, pp. 201–276. Madison, WI: ASA, CSSA, and SSSA.

Stewart, J. A. 1985. Potassium sources, use, and potential. In: *Potassium in Agriculture*, ed. R. D. Munson, pp. 83–98. Madison, WI: ASA, CSSA and SSSA.

Stone, K. 2008. Potash. In: *Canadian Mineral Year Book*, ed. Canadian Potash Institute, 36.1–36.12. Ottawa, Ontario, Canada: Natural Resources Canada.

Su, N. R. 1976. Potassium fertilization of rice. In: *The Fertility of Paddy Soils and Fertilizer Application for Rice*, ed. Asian Pacific Food and Fertilizer Technology Center, pp. 117–148. Taipei, Taiwan, People's Republic of China: Asian Pacific Food and Fertilizer Technology Center.

Suelter, C. H. 1970. Enzymes activated by monovalent cations. *Science* 168:789–795.

Tisdale, S. L. and W. L. Nelson. 1966. *Soil Fertility and Fertilizers*, 2nd edn. New York: The Macmillan Company.

Unger, P. W. and T. C. Kaspar. 1994. Soil compaction and root growth: A review. *Agron. J.* 86:759–766.

Wani, S. P., W. B. McGill, K. L. Haugen-Kozyra, J. A. Robertson, and J. J. Thurston. 1994. Improved soil quality and barley yields with fababeans, manure, forages and crop rotation on a Gray Luvisol. *Can. J. Soil Sci.* 74:75–84.

William, S. M. and R. R. Weil. 2004. Crop cover root channels may alleviate soil compaction effects on soybean crop. *Soil Sci. Soc. Am. J.* 68:1403–1409.

Welch, L. F., P. E. Johnson, G. E. Mckibben, L. V. Boone, and J. W. Pendleton. 1966. Relative efficiency of broadcast versus banded potassium for corn. *Agron. J.* 58:618–621.

Wright, A. L. and F. M. Hons. 2004. Soil aggregation and carbon and nitrogen storage under soybean cropping sequences. *Soil Sci. Soc. Am. J.* 68:507–513.

Yang, X. E., J. X. Liu, W. M. Wang, Z. Q. Ye, and A. C. Luo. 2004. Potassium internal use efficiency relative to growth vigor, potassium distribution, and carbohydrate allocation in rice genotypes. *J. Plant Nutr.* 27:837–852.

5 Calcium

Philip J. White

CONTENTS

5.1 CHEMISTRY OF CALCIUM

Calcium (Ca) is one of the most abundant elements in the earth's crust (4.15% by weight), where it is present in minerals such as limestone (calcium carbonate, $CaCO_3$), gypsum (calcium sulfate, $CaSO_4$), apatite (various calcium phosphates), and fluorite (calcium fluoride, CaF_2). Although the agricultural benefits of adding ground limestone to acid soils to increase crop yields had been known for two millennia, it was not until 200 years ago that Carl Sprengel (1787–1859) is credited with concluding that Ca is an essential element for plants, without which they are unable to grow or complete their life cycles (Pilbeam and Morley, 2007).

Calcium behaves as a typical Group 2 alkali metal. Although the solubility of calcium salts in water is often high, notable exceptions include calcium hydroxide (lime, $Ca(OH)_2$), calcium carbonate, calcium sulfate, calcium phosphates, and calcium oxalate (CaC_2O_4). This low solubility has several important consequences for both agriculture and cellular biochemistry. In an agricultural context, the presence of calcium carbonate in soils buffers the soil solution to an alkaline pH, which reduces the phytoavailability of mineral elements essential for plant nutrition, such as iron (Fe), zinc (Zn), manganese (Mn), copper (Cu), and nickel (Section 5.2; White et al., 2012a, 2013; White and Greenwood, 2013). The low solubility of calcium phosphates can also compromise the mineral nutrition of plants by reducing the concentrations of phosphate in the soil solution (White and Hammond, 2008). In the context of cellular biochemistry, the low-solubility product of Ca^{2+} and phosphate requires cells to maintain a submicromolar cytosolic Ca^{2+} concentration ($[Ca^{2+}]_{cyt}$) to allow cellular energy metabolism to proceed, which is thought to have led to the evolution of sophisticated mechanisms to maintain low $[Ca^{2+}]_{cyt}$ homeostasis and, subsequently, to the evolution of intracellular signaling through $[Ca^{2+}]_{cyt}$ perturbations (Section 5.7.1; Sanders et al., 1999; White and Broadley, 2003).

Calcium forms a relatively large divalent cation, with a hydrated ionic radius of 0.412 nm and a hydration energy of 1577 J mol^{-1}, that has a high affinity for negatively charged surfaces. This charge attraction has important consequences for both agriculture and cellular biochemistry. In soil, Ca^{2+} binds to cation-exchange sites in both minerals and organic residues, which influences the retention and phytoavailability of other cations (White and Greenwood, 2013). It also binds to fixed negative charges in plant cell walls and lipid membranes, which confers structural stability to plant tissues and integrity to cells and cellular compartments (White and Broadley, 2003; Hepler and Winship, 2010; Aghdam et al., 2012; Hawkesford et al., 2012). In addition, since Ca^{2+} can coordinate six to eight uncharged oxygen atoms, it can bind to proteins and alter their conformation. This is the mechanism by which intracellular signaling cascades are controlled through changes in $[Ca^{2+}]_{cyt}$ (Section 5.7.2.4; White and Broadley, 2003).

5.2 CALCIUM IN SOILS AND FERTILIZERS

The pH of the soil solution is one of the most important factors influencing the phytoavailability of mineral elements (Taiz and Zeigler, 2006; White and Greenwood, 2013). Different chemical equilibria buffer the soil solutions at different pH values (Figure 5.1; George et al., 2012; White and Greenwood, 2013). Soil solutions with pH < 4 are generally buffered by hydroxyl-aluminum complexes; those with a pH between 5 and 6 are generally buffered by the replacement of Ca^{2+}, Mg^{2+}, K^+,

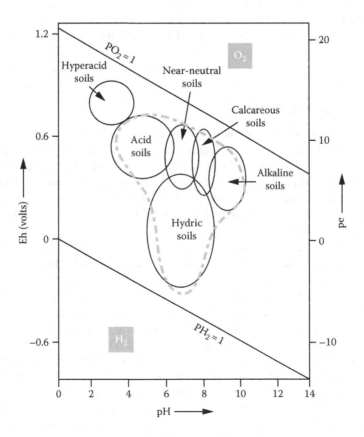

FIGURE 5.1 Characteristics of common soils. The dashed line encloses the conditions in most mineral soils. (Redrawn from Macías, F. et al., Acid soils, in *The Encyclopedia of Soil Science*, Chesworth, W. (ed.), Springer-Verlag, Dordrecht, the Netherlands, 2008, pp. 7–9. With permission.)

and Na^+ on cation-exchange sites with H^+; and soil solutions with pH > 7 are generally buffered by the reaction of H^+ on base carbonates, such as $CaCO_3$ and Na_2CO_3 (George et al., 2012; White and Greenwood, 2013). Crop production on acid soils is restricted by proton, aluminum (Al), and Mn toxicities, together with phosphorus (P), molybdenum (Mo), magnesium (Mg), Ca, and potassium (K) deficiencies, and crop production on calcareous and alkaline soils is often limited by the phytoavailability of Fe, Zn, and Cu (George et al., 2012; Ryan and Delhaize, 2012; White et al., 2012a, 2013; White and Greenwood, 2013). Growing legumes and applying ammonium-based fertilizers accelerate soil acidification (Ryan and Delhaize, 2012). It is estimated that acid soils underlie about 40% of all agricultural land, often occurring in humid environments where rainfall leads to extensive leaching and weathering of minerals, and that alkaline and calcareous soils underlie another 25%–30%, often in semiarid and arid regions (White and Broadley, 2009; George et al., 2012; White et al., 2012a; White and Greenwood, 2013).

Although plants rarely lack sufficient Ca in nature, Ca deficiency can occur in plants growing in soils with a low cation-exchange capacity (CEC), which have a low Ca content due to Ca^{2+} leaching, or in plants growing in acid soils (McLaughlin and Wimmer, 1999; White and Broadley, 2003; Pilbeam and Morley, 2007; White and Greenwood, 2013). When plants are grown hydroponically, shoot Ca concentrations generally decline as the pH of the nutrient solution decreases (Islam et al., 1980). This is thought to be a consequence of H^+ replacing Ca^{2+} on cation-binding sites in the root apoplast, which reduces Ca^{2+} concentrations at the plasma membrane of root cells, together with a decrease in the membrane potential of root cells. Both of these actions diminish the electrochemical gradient for Ca^{2+} influx to root cells (Shabala et al., 1997). In soils, the concentrations of both Al^{3+} and Mn^{2+} increase markedly as the pH of the soil solution decreases (George et al., 2012; Ryan and Delhaize, 2012; White and Greenwood, 2013). A molar ratio of Ca to total cations of about 0.15 in the soil solution is thought to be required for maximal root growth and a lower ratio than this is commonly observed in acid soils (Lynch et al., 2012). The presence of Al^{3+} in the soil solution inhibits root growth and the formation of root hairs by a variety of mechanisms (George et al., 2012; Ryan and Delhaize, 2012; White and Greenwood, 2013). This reduces the ability of plants to acquire all essential mineral elements, which compounds the direct effects of Al^{3+} on root growth. In addition, Al^{3+} and Mn^{2+} can displace cations from exchange sites in the soil, which, together with the shallow rooting depth of plants growing on acid soils, increases the leaching of essential cationic elements, including Ca^{2+} from the topsoil. Both Al^{3+} and Mn^{2+} can also compete with Ca^{2+} for binding sites in the root apoplast, which reduces Ca^{2+} uptake by root cells indirectly. In addition, Al^{3+} can block cation channels responsible for Ca^{2+} influx to root cells, and Mn^{2+} competes with Ca^{2+} for uptake by root cells (White et al., 2002; George et al., 2012). By contrast, K^+ uptake by roots is relatively unaffected by the presence of Al^{3+} in the soil solution, which results in an increase in the K/(Ca + Mg) ratio in shoots and an increase in the likelihood of Ca-deficiency symptoms in plants and the incidence of grass tetany in ruminants using grass as forage (George et al., 2012).

The relative contributions of Al inhibition of root growth and Al-induced Ca deficiency to reduced crop yields on acidic soils depend on plant species. The growth of grasses and forest trees on acid soils appears to be limited by Al inhibition of root growth and a general lack of essential mineral elements, whereas the growth of legumes on acid soils appears to be limited mainly by Al-induced Ca deficiency (Keltjens and Tan, 1993; Fageria et al., 2011; George et al., 2012). A correlation between Al resistance and the physiological efficiency of Ca utilization has been observed among genotypes of some legume species (Horst, 1987). The nodulation of legumes and, therefore, their N nutrition is particularly sensitive to both soil acidity and low Ca phytoavailability. This is thought to be a consequence of low survival of rhizobial strains, reduced abundance of root hairs, and impaired attachment of rhizobia to the root surface in acid soils, caused directly or indirectly by increased Al^{3+} concentrations (Sadowsky, 2005; Cooper and Scherer, 2012). Thus, the nodulation of legumes (and plant growth in general) on acid soils can be improved by liming (Fageria, 2009), through the application of gypsum, which shifts the ionic equilibria from Al^{3+} to the less toxic $AlSO_4^+$ cation (Lynch et al., 2012), or by cultivating genotypes that reduce Al^{3+} concentrations in the rhizosphere by secretion of organic acids or mucilage to chelate Al^{3+}, or by binding Al^{3+} to cell wall components (Ma et al., 2001; Kochian et al., 2004; Delhaize et al., 2007; Ryan and Delhaize, 2012; White et al., 2012a). It is also noteworthy that the net release of H^+ associated with N_2 fixation by nodulated legumes is considerable, and it can require the equivalent of up to 100 g of $CaCO_3$ to neutralize the acidity formed in the production of 1 kg of shoot dry matter (Jarvis and Hatch, 1985; Cooper and Scherer, 2012). The nodulation of legumes has also been found to require greater Ca phytoavailability in the rhizosphere than plant growth, and the initial stages of nodulation are particularly sensitive to low Ca phytoavailability (Sadowsky, 2005; Cooper and Scherer, 2012).

Applying Ca fertilizers in an appropriate form and at an appropriate time, rate, and place is an important factor for crop production (Kopittke and Menzies, 2007; Fageria, 2009; Kannan, 2010; White and Greenwood, 2013). Liming is the most effective agronomic practice to increase Ca concentrations in the soil solution (Fageria, 2009; White and Greenwood, 2013). Calcite limestone ($CaCO_3$) and dolomitic limestone ($CaCO_3 \cdot MgCO_3$) are the most appropriate Ca fertilizers for raising soil pH to avoid Ca deficiency and Al and Mn toxicities in acid soils, although burned lime (CaO), slaked lime ($Ca(OH)_2$), basic slag ($CaSiO_3$), and wood ash are also commonly used Ca fertilizers (Fageria, 2009; White and Greenwood, 2013). Other Ca fertilizers include gypsum, calcium nitrate ($Ca(NO_3)_2$), superphosphate (a mixture of $CaSO_4$ and $Ca(H_2PO_4)_2$), triple superphosphate ($Ca(H_2PO_4)_2$), rock phosphates, and farmyard manures (0.5%–2.3% Ca; Defra, 2010). Supplying adequate irrigation and improving the organic matter content of soil also increase Ca phytoavailability. Subsoil acidity is a significant problem throughout the world, especially in the tropics. The penetration of roots into acidic subsoils is enabled either by the application of large amounts of lime to the soil surface, and the eventual leaching of Ca^{2+} into the subsoil, or by the application of fertilizers with a greater water solubility, such as gypsum, that distribute Ca rapidly throughout the soil profile (Whitten et al., 2000; Fageria, 2009; White and Greenwood, 2013). In general, the application of Ca fertilizers to the soil, or to the rooting substrate, increases Ca concentrations in roots, leaves, and tubers but does not always increase Ca concentrations in fruits and seeds (Shear, 1975; McLaughlin and Wimmer, 1999; White and Broadley, 2003, 2009; Ho and White, 2005; White et al., 2009). Foliar applications of soluble Ca fertilizers are, therefore, often made to prevent Ca-deficiency disorders in fruits (Shear, 1975; Ho and White, 2005; Pilbeam and Morley, 2007; Fernández and Eichert, 2009; Liebisch et al., 2009; Lurie, 2009; Blanco et al., 2010; Kannan, 2010; Val and Fernández, 2011; Fernández et al., 2013).

Symptoms of Ca deficiency can also be observed when crops are grown on saline or sodic soils. Inhibition of Ca^{2+} uptake by excessive sodium (Na^+) concentrations, and reduced Ca^{2+} delivery to organs with low transpiration rates caused by low osmotic potential of the soil solution, can both contribute toward reduced yields on saline and sodic soils (Ho and White, 2005; Munns and Tester, 2008; George et al., 2012; Hodson and Donner, 2013), which underlie about 5%–15% of all agricultural land (Munns and Tester, 2008; George et al., 2012). Saline soils are defined simply as having an electrical conductivity >4 dS m^{-1} at 25°C in a saturated soil paste (George et al., 2012; Hodson and Donner, 2013). The high electrical conductivity can be caused by high concentrations of various ions, and saline soils do not always have high Na^+ concentrations and are not necessarily alkaline. By contrast, sodic soils, which are defined as having a high ratio of Na to Ca and Mg in the soil solution, are generally buffered by Na_2CO_3 and have pH > 8.5 (George et al., 2012; Hodson and Donner, 2013). The application of Ca fertilizers can improve crop yields on both saline and sodic soils by increasing the Ca^{2+}/Na^+ ratio in the soil solution, which not only reduces Na^+ uptake but also increases the uptake of both Ca^{2+} and other essential cations by plants (Pilbeam and Morley, 2007; George et al., 2012). The nodulation of legumes, which is inhibited by high Na^+ concentrations in the rhizosphere solution, can be improved by the application of Ca fertilizers (Bolanos et al., 2006). In addition, sodic soils can be remediated through the application of Ca^{2+}, generally as gypsum, followed by flushing the soil with freshwater, while saline soils can be remediated by leaching soluble salts from the soil profile by irrigation with freshwater (Hodson and Donner, 2013; White and Greenwood, 2013).

Although sensitivity to P and Fe deficiencies is likely to determine the flora of calcareous soils (Lee, 1999), excessive Ca accumulation by plants can result in Ca toxicity, even though most plants can accommodate Ca concentrations up to about 100 mg Ca g^{-1} dry matter (DM; Römheld, 2012; White et al., 2012a). When sampled from their natural environments, tissue Ca concentrations in calcicoles, which occur naturally on calcareous soils, generally exceed those of calcifuges, which occur on acid soils (White and Broadley, 2003). Calcifuges grow well when Ca^{2+} concentrations in the rhizosphere solution are low, but their growth is inhibited at the Ca^{2+} concentrations found in soil solutions from calcareous soils (Jeffries and Willis, 1964; White and Broadley, 2003). In contrast, calcicoles grow well in soils containing high-solution Ca^{2+} concentrations but do not grow well when the Ca^{2+} concentration in the soil solution is low (Jeffries and Willis, 1964; White and Broadley, 2003). The latter phenomenon might be because the protective mechanisms maintaining low [Ca^{2+}]$_{cyt}$ homeostasis in calcifuges in their natural habitat are constitutive and cause a *physiological* Ca deficiency when Ca^{2+} is restricted (Lee, 1999; White and Broadley, 2003). These mechanisms include the ability to reduce shoot apoplastic Ca^{2+} concentrations to maintain stomatal conductance (De Silva et al., 1996, 1998), to sequester Ca and P in different cell types to avoid precipitation of CaHPO$_4$ (Karley et al., 2000a,b; Conn and Gilliham, 2010), and adaptations that allow *calciotrophe* (calcium-accumulating) species in the Crassulaceae, Brassicaceae, and Fabaceae to utilize soluble Ca as an osmoticum in dry calcareous environments (Kinzel, 1982; White and Broadley, 2003). Other adaptations of calcicoles to high rhizosphere Ca^{2+} concentrations include the chelation of Ca^{2+} by organic compounds and the precipitation of Ca salts (CaCO$_3$, Ca oxalate) in the rhizosphere or root apoplast, which in addition to restricting the movement of Ca to the shoot can increase the phytoavailability of sparingly soluble essential mineral elements such as Fe, Zn, and P (White and Broadley, 2003; George et al., 2012). These plant adaptations can be complemented by associations with beneficial mycorrhizal fungi that perform similar functions (George et al., 2012).

5.3 ACQUISITION OF CALCIUM FROM THE SOIL

In most agricultural soils, the Ca concentration in the soil solution lies in the millimolar range (White and Broadley, 2003; Marschner and Rengel, 2012), and the dominant physical process delivering Ca to the root surface is transpiration-driven mass flow of the soil solution (Barber, 1995; Fageria et al., 2011; White and Greenwood, 2013; White et al., 2013). In some soils, the delivery of Ca to the root surface can exceed the capacity of plant roots to take up Ca. This leads to the accumulation of Ca at the root surface and, occasionally, the precipitation of Ca salts, such as CaCO$_3$, CaSO$_4$, or Ca oxalate, in tubular concretions (rhizocylinders) around roots (Marschner and Rengel, 2012; Neumann and Römheld, 2012). In addition, CaCO$_3$ crystals are found in cortical cells of roots of many herbaceous species growing in calcareous soils (Neumann and Römheld, 2012). These crystals, which eventually attain the shape of the cortical cells in which they form, are thought to be formed by cycles of rhizosphere acidification by plant roots, which generate cyclic fluctuations in rhizosphere Ca concentration that require the periodic precipitation of excess Ca within the vacuoles of root cortical cells (Jaillard, 1985; Jaillard et al., 1991). Consistent with this hypothesis, roots containing CaCO$_3$ crystals are generally surrounded by a decalcified rhizocylinder with a silicon–aluminum matrix (Jaillard, 1985; Jaillard et al., 1991). Since plant species differ in both their transpiration rates and their capacity for Ca uptake, Ca concentrations in the rhizospheres of plants growing in the same environment can differ markedly (Barber and Ozanne, 1970; Neumann and Römheld, 2012). This can have consequences for the spatial and temporal phytoavailability of other mineral elements, water movement through the soil, and the abundance of rhizosphere microbes.

Calcium enters the root apoplast from the rhizosphere. Most of the Ca in the apoplast is located in the middle lamellae of cell walls, where it is bound to the carboxyl groups of polygalacturonic acids (pectins) in a readily exchangeable form (Carpita and McCann, 2000). The Ca^{2+} concentration in the apoplastic solution of the root cortex generally lies in the millimolar range (White and Broadley, 2003; White, 2012a). It is determined by various factors including (1) the mass flow of the soil solution to the root, (2) the Ca^{2+} concentration in the rhizosphere solution, (3) the cation-exchange capacities of rhizosphere and cell walls, (4) the ionic composition and proportion of Ca^{2+} of the total cation complement of the rhizosphere and cell walls, (5) the pH of the rhizosphere and the apoplast, and (6) Ca^{2+} uptake by root cells. The properties of the rhizosphere and cell walls are, in turn, influenced both directly and indirectly by the physical and chemical properties, and biological activities, of the bulk soil.

Root cells must meet their Ca requirement for growth by taking up Ca^{2+} directly from the rhizosphere or apoplastic solutions. Calcium uptake by root cells occurs through cation-permeable ion channels in their plasma membranes (Table 5.1; White and Broadley, 2003; Karley and White, 2009). However, Ca^{2+} fluxes through these ion channels must be tightly controlled, and the Ca^{2+} entering the cytoplasm of root cells must be well buffered, because perturbations in $[Ca^{2+}]_{cyt}$ act as signals to initiate cellular responses to numerous developmental and environmental stimuli (Section 5.7.2). For this reason, it is thought that much of the Ca transported across the root to the xylem might not pass through the root symplast (White, 1998, 2001). Indeed, most of the Ca delivered to the shoot is loaded into the xylem at the extreme root tip and in regions where lateral roots are being initiated (Clarkson, 1984; White, 2001, 2012a). It is thought that Ca can reach the xylem solely via the root apoplast in regions where Casparian bands are absent or disrupted and via the cytoplasm of unsuberized endodermal cells where Casparian bands are present (White, 2001; Moore et al., 2002; Pilbeam and Morley, 2007).

The relative contributions of the apoplastic and symplastic pathways to the delivery of Ca to the xylem are debated (White, 2001; Cholewa and Peterson, 2004; Hayter and Peterson, 2004). Nevertheless, the rate at which Ca is delivered to the shoot increases with increasing transpiration rate (Lazaroff and Pittman, 1966; Drew and Biddulph, 1971; McLaughlin and Wimmer, 1999; Conn and Gilliham, 2010; White, 2012b), and Ca delivery to the xylem is reduced by the precipitation of Ca salts in the apoplast, the sequestration of Ca in the vacuoles of root cells, and the accelerated deposition of suberin and lignin in the endodermis (White, 2001; Baxter et al., 2009; Conn and Gilliham, 2010). Since Ca is taken up at the tips of roots, the number of root apices in moist soil is of critical importance for Ca acquisition and extensive root branching is beneficial for Ca acquisition (Pilbeam and Morley, 2007; White and Greenwood, 2013; White et al., 2013). The Ca concentrations of root tissues are generally lower than those of shoot tissues (Conn and Gilliham, 2010).

The relationship between Ca uptake by isolated root cells and the Ca^{2+} concentration in the rhizosphere solution can generally be approximated by the sum of a saturatable and a nonsaturatable component (Figure 5.2; Maas, 1969; Macklon and Sim, 1981; Huang et al., 1992). At low Ca^{2+} concentrations in the rhizosphere solution (≤ 300 μM), the relationship between Ca uptake (I) and Ca^{2+} concentration (S) can be described by a Michaelis–Menten function ($I = (I_{max} * S)/(K_m + S)$, where I_{max} is the maximal rate of Ca^{2+} uptake catalyzed by the saturatable component and $K_m = S$ when $I = 0.5I_{max}$. Calcium influx appears to be selective and closely coupled to metabolism (Epstein and Leggett, 1954; Maas, 1969). Estimates of the K_m for Ca^{2+} influx to isolated root cells range from about 50 to 200 μM (Dunlop, 1973; Macklon and Sim, 1981; Huang et al., 1992). By contrast, at higher Ca^{2+} concentrations in the rhizosphere solution, Ca influx often becomes linearly related to Ca^{2+} concentration ($I = kS$) through a proportionality parameter (k), is not very selective, and is not particularly sensitive to temperature or metabolic inhibitors (Maas, 1969; White, 2012a).

The relationship between shoot Ca concentration and the Ca^{2+} concentration in the rhizosphere solution differs between plants with contrasting physiotypes (Kinzel, 1982; Kinzel and Lechner, 1992; White, 2005). In many plants, shoot Ca concentration exhibits both saturatable and nonsaturatable components with increasing Ca^{2+} concentration in the rhizosphere solution (Figure 5.3; Lazaroff and Pitman, 1966; Baker, 1978; English and Barker, 1987; Islam et al., 1987; White, 2001). In calciotrophes, such as members of the Brassicaceae, Crassulaceae, and Fabaceae, which have high concentrations of soluble Ca, shoot Ca concentration continues to increase as the Ca concentration in the rhizosphere solution is increased in the millimolar range (Figure 5.3; Kinzel and Lechner, 1992; White, 2001). By contrast, in *potassium plants* such as the

TABLE 5.1

Protein Families Implicated in Calcium Uptake and Distribution between Plant Organs, Cell Types, and Subcellular Compartments[a]

Transport Mechanism	Protein Family[b]	Putative Function	Putative Ca Transporters	Tissue[c]
Cation channel	Annexin (7)	Ca influx to cell	AtAnn1, AtAnn2, AtAnn4	R,S,L
			AtAnn3	R,L
		Ca efflux from vacuole	AtAnn1, AtAnn3	R,S,L
Cation channel	CNGC (20)	Ca influx to cell	AtCNGC1, AtCNGC5, AtCNGC6, AtCNGC12, AtCNGC14, AtCNGC15, AtCNGC17, AtCNGC19	R,S,L
			AtCNGC2, AtCNGC3, AtCNGC8, AtCNGC9, AtCNGC10, AtCNGC11, AtCNGC13, AtCNGC20	R,L
			AtCNGC18	R
			AtCNGC4	L
Cation channel	GLR (20)	Ca influx to cell	AtGLR1.1, AtGLR1.2, AtGLR1.3, AtGLR1.4, AtGLR2.1, AtGLR2.2, AtGLR2.4, AtGLR3.2, AtGLR3.3, AtGLR3.4, AtGLR3.5, AtGLR3.6, AtGLR3.7	R,S,L
			AtGLR2.3, AtGLR2.5, AtGLR2.7, AtGLR2.8, AtGLR2.9	R,L
			AtGLR2.6	R,S
Cation channel	MCA (2)	Ca influx to cell	AtMCA1, AtMCA2	R,S,L
K+ channel	*Shaker* (9)	K efflux from cell (Ca influx to cell)	AtSKOR	R,S
Ca transporter		Ca influx to cells	TaLCT1	
Ca transporter		Ca efflux from cell	BjPCR1	R,S
Ca^{2+} channel	TPC (1)	Ca efflux from vacuole	AtTPC1	R,S,L
Ca^{2+}-ATPase	AtECA (4)	Ca efflux from cell	AtECA2	R,L
			AtECA4	S,L
		Endomembrane Ca sequestration	AtECA1, AtECA3	R,L
			AtECA4	S,L
Ca^{2+}-ATPase	AtACA (10)	Ca efflux from cell	AtACA8	R,S,L
			AtACA10, AtACA13	R,L
			AtACA7, AtACA9	L
		Endomembrane Ca sequestration	AtACA1, AtECA2	R,S,L
		Vacuolar Ca sequestration	AtACA11	R,S,L
			AtACA4, AtACA12	R,L
Cation/H+ antiporter	AtCAX (6)	Vacuolar Ca sequestration	AtCAX2	R,S,L
			AtCAX1, AtCAX3	R,L
			AtCAX4, AtCAX5	R
Cation ATPase	HMA (8)	Ca influx to nucleus and plastids	AtHMA1	S

The expression of genes encoding transport proteins in the root stele (S) was based primarily upon data obtained by Birnbaum et al., (2003), and the subcellular location of Ca transporters was based on data in the SUBA database (Heazlewood et al., 2007).

[a] Key references include Dodd et al., (2010), Edmond et al., (2009), Dietrich et al., (2010), Kudla et al., (2010), White and Karley (2010), Yamanaka et al., (2010), Michard et al., (2011), Song et al., (2011), Urquhart et al., (2011), Laohavisit et al., (2012), and Vincill et al., (2012).

[b] The number in parentheses indicates the number of genes in the gene family in *A. thaliana*.

[c] R, root; S, root stele; L, leaf.

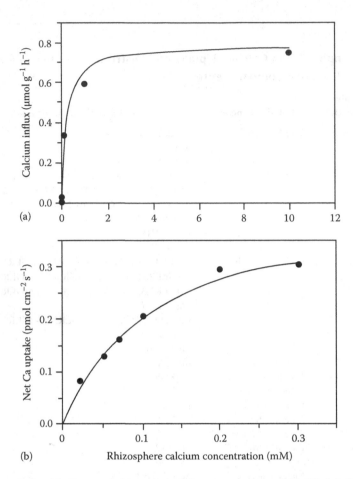

FIGURE 5.2   The relationships between rhizosphere Ca^{2+} concentration and (a) $^{45}Ca^{2+}$ influx to excised onion root segments (I_{max} = 0.79 μmol g^{-1} h^{-1}, K_m = 0.20 mM). (From Macklon, A.E.S. and Sim, A., *Planta*, 152, 381, 1981; Britto, D.T. and Kronzucker, H.J., *Planta*, 217, 490, 2003.) and (b) net Ca^{2+} uptake into cells at the root apex of intact wheat plants measured using a noninvasive, Ca^{2+}-sensitive, vibrating microelectrode (I_{max} = 4.15 pmol cm^{-1} s^{-1}, K_m = 0.10 mM). (From Huang, J.W. et al., *Planta*, 188, 414, 1992.)

Apiales and Asterales, which are characterized by high shoot K/Ca ratios and low concentrations of soluble Ca, shoot Ca concentration does not increase markedly when the Ca^{2+} concentration in the rhizosphere solution is increased above about 100 μM (Kinzel and Lechner, 1992). Among the *oxalate plants* the accumulation of Ca does not increase with increasing Ca supply in plants containing soluble oxalate, such as the Oxalidaceae, whereas in plants that precipitate Ca oxalate, such as the members of the Caryophyllaceae, Chenopodiaceae, and Polygonaceae, there is a proportional increase in both Ca and oxalate concentrations with increasing Ca supply (Libert and Franceschi, 1987; Kinzel and Lechner, 1992; Rahman and Kawamura, 2011). In all plant physiotypes, however, Ca accumulation in the shoot is increased greatly by increasing transpiration (Lazaroff and Pitman, 1966).

A large electrochemical gradient drives Ca^{2+} influx to root cells (White et al., 2002; White and Broadley, 2003). The rhizosphere and root apoplastic solutions have millimolar Ca^{2+} concentrations, whereas $[Ca^{2+}]_{cyt}$ is maintained at submicromolar concentrations. Calcium influx to root cells is mediated by a variety of Ca^{2+}-permeable ion channels (Table 5.1). These include ion channels that are regulated by the cell membrane potential, by the binding of extracellular or intracellular ligands, by mechanical stresses, or by protein modification (White et al., 2002; Demidchik and Maathuis, 2007; Roux and Steinebrunner, 2007; Karley and White, 2009; Dodd et al., 2010; Laohavisit and Davies, 2011). Many of

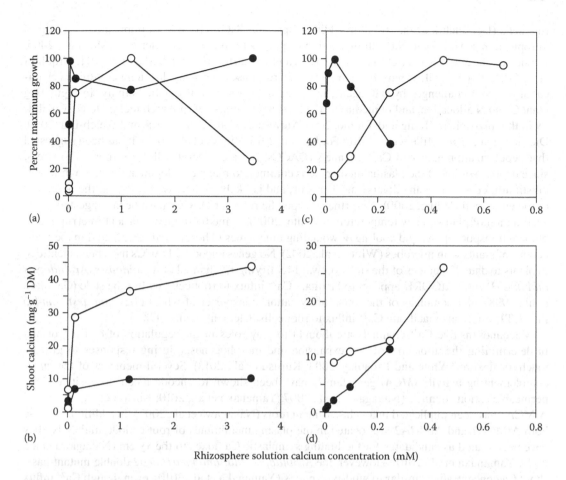

FIGURE 5.3 The relationships between (a, c) plant growth and (b, d) shoot Ca concentrations and the Ca concentration in the rhizosphere solution of (a, b) wheat (filled circles), a typical commelinid monocot species, and French bean (open circles), a typical eudicot species. (From Islam, A.K.M.S. et al., *Plant Soil*, 98, 377, 1987.) and (c, d) *Nardus stricta* (open circles), a calcifuge species, and *Origanum vulgare* (filled circles), a calcicole species. (From Jefferies, R.L. and Willis, A.J., *J. Ecol.*, 52, 691, 1964.)

these Ca^{2+}-permeable ion channels have been implicated in the coordination of cellular responses to environmental challenges, and their role in mediating *nutritional* Ca^{2+} fluxes through the symplasm to the xylem is still debated (White and Broadley, 2003; Miedema et al., 2008; Dayod et al., 2010).

Hyperpolarization-activated Ca^{2+} channels (HACCs) are thought to be formed by annexins (White et al., 2002; Mortimer et al., 2008; Laohavisit et al., 2010, 2012, 2013; Laohavisit and Davies, 2011; Clark et al., 2012). These channels mediate the Ca^{2+} influx required to elevate $[Ca^{2+}]_{cyt}$ to initiate and maintain the elongation of root cells and determine the rate and direction of elongation of root hairs (Demidchick et al., 2002; Foreman et al., 2003; Laohavisit et al., 2012). They are also implicated in initiating root tropisms, symbiotic associations, and responses to environmental factors causing oxidative stress (White and Broadley, 2003; Mortimer et al., 2008; Laohavisit et al., 2010, 2013; Laohavisit and Davies, 2011). Voltage-independent cation channels (VICCs) are thought to be formed by members of the cyclic nucleotide-gated channel (CNGC) and/or glutamate receptor (GLR) protein families (Karley and White, 2009; Dietrich et al., 2010; Dodd et al., 2010). It has been suggested that Ca^{2+} influx though VICCs might balance the constitutive Ca^{2+} efflux from root cells and, thereby, help maintain $[Ca^{2+}]_{cyt}$ homeostasis (White and Davenport, 2002) and that VICCs provide the general increase in cellular Ca^{2+} influx that initiates HACC-dependent elongation of root cells (Demidchik et al., 2002). In addition, specific CNGCs have been implicated in root gravitropism, in sensitivity to Na^+ salinity, and in responses to rhizosphere microbes (Ma et al., 2006; Borsics et al., 2007; Kugler et al., 2007; Dietrich et al., 2010; Ma and Berkowitz, 2011; Urquhart et al., 2011). Specific GLRs have been implicated in responses to gravity, low temperature, mechanical stress, and challenges by pathogens, in maintaining root meristematic activity, in regulating plant C and N allocation, and in mediating the foraging response of root architecture to N availability in the rhizosphere (Kang and Turano, 2003; Meyerhoff et al., 2005; Forde and Walch-Liu, 2009; Dietrich et al., 2010; Miller et al., 2010; Forde et al., 2013; Vincill et al., 2013). It has been suggested that depolarization-activated Ca^{2+} channels (DACCs) initiate general cellular responses to stress, since depolarization of the plasma membrane is common to many developmental and environmental stimuli, occurs by many diverse mechanisms, and is likely to increase $[Ca^{2+}]_{cyt}$ at the periphery of the entire cell (White, 2000). Nevertheless, specific roles for DACCs have been suggested in the acclimation of plants to low temperatures (White, 2009), in mediating the transient interruption of phloem transport upon rapid cooling or wounding in legumes (Thorpe et al., 2010), and in the interactions of plants with microbes (White et al., 2002). No genes encoding DACCs have been identified in plants to date. Homologs of the stelar outward-rectifying K^+ channel of *Arabidopsis* (*Arabidopsis thaliana* Heynh.) AtGORK appears to facilitate Ca^{2+} influx to root cells within the stele (Gaymard et al., 1998), and homologs of the low-affinity cation transporter of wheat (*Triticum aestivum* L.) (TaLCT1) might also facilitate Ca^{2+} influx to root cells (Clemens et al., 1998).

Mechanosensitive Ca^{2+} channels are thought to play roles in the regulation of cell turgor and in determining the allometry of cell expansion and morphogenesis during responses to gravity, touch, or flexure (White and Broadley, 2003; Kurusu et al., 2013). Several members of the *mid1*-complementing activity (*MCA*) gene family have been shown to encode mechanosensitive Ca^{2+}-permeable cation channels (Nakagawa et al., 2007; Yamanaka et al., 2010; Kurusu et al., 2013), and AtMCA1 has been implicated in touch sensing in roots (Nakagawa et al., 2007). In addition, because both AtMCA1 and AtMCA2 are located in the plasma membranes of root endodermal cells, they have been touted as candidates for facilitating symplastic Ca fluxes to the xylem (Nakagawa et al., 2007; Yamanaka et al., 2010). However, the *Arabidopsis thaliana mca1/mca2* double mutant has a shoot Ca concentration similar to wild-type plants (Yamanaka et al., 2010), even though Ca^{2+} influx to roots of mutants lacking *AtMCA1* or *AtMCA2* is less (Nakagawa et al., 2007; Yamanaka et al., 2010) and Ca^{2+} influx to roots of mutants overexpressing *AtMCA1* is more (Nakagawa et al., 2007) than that into roots of wild-type plants. Members of the mechanosensitive channel of small conductance (MscS)-like (*MSL*) gene family have also been implicated in $[Ca^{2+}]_{cyt}$ signaling in root cells (Kurusu et al., 2013). However, although AtMSL9 and AtMSL10 (and possibly AtMSL4, AtMSL5, and AtMSL6) are present on the plasma membrane of root cells, they appear to facilitate anion influx (Haswell et al., 2008; Maksaev and Haswell, 2012). It is, therefore, most likely that MSL channels affect $[Ca^{2+}]_{cyt}$ indirectly, via alterations in cell membrane potential or changes in turgor pressure (Kurusu et al., 2013).

The $[Ca^{2+}]_{cyt}$ of all plant cells is maintained at submicromolar concentrations by the activity of Ca^{2+} ATPases, encoded by members of the P_{2A}-ATPase (*ECA*) and P_{2B}-ATPase (*ACA*) gene families, and Ca^{2+}/H^+ antiporters, encoded by the Ca^{2+}/H^+-antiporter (*CAX*) gene family, which transport Ca^{2+} from the cytosol to the apoplast, endoplasmic reticulum (ER), plastids, or vacuoles (Table 5.1; White and Broadley, 2003; Shigaki et al., 2006; Boursiac and Harper, 2007; McAinsh and Pittman, 2009; Dodd et al., 2010; Bonza and De Michelis, 2011; Pittman, 2011). The ACA proteins are activated by the binding of calmodulin (CaM) to an autoinhibitory domain and regulated by Ca-dependent phosphorylation and interactions with acidic phospholipids (White and Broadley, 2003; Boursiac and Harper, 2007; Bonza and De Michelis, 2011; Pittman, 2011). The transport of Ca^{2+} into the vacuole by CAX proteins is energized by the H^+ gradient generated by vacuolar H^+ ATPases and/or H^+ pyrophosphatases, and their activity can also be regulated by phosphorylation, heteromerization, and the binding of CAX-interacting proteins (Shigaki and Hirschi, 2006; Pittman, 2011; Martinoia et al., 2012).

Within the cytosol, Ca^{2+} is chelated by various proteins, including CaM, CaM-related proteins, calcineurin-B-like (CBL) proteins, Ca^{2+}-dependent protein kinases (CDPKs), and annexins (Section 5.7.2.4; White and Broadley, 2003; McAinsh and Pittman, 2009; Dodd et al., 2010; Kudla et al., 2010; Reddy et al., 2011; Clark et al., 2012). These proteins can be present at high concentrations, and although $[Ca^{2+}]_{cyt}$ is submicromolar, the cytosol can contain between 0.1 and 15 mM Ca (White et al., 1992; Malhó et al., 1998; White and Broadley, 2003; McAinsh and Pittman, 2009). The Ca^{2+}-binding proteins calreticulin, calsequestrin, and calnexin and the lumenal binding protein (BiP) are present in the ER (White and Broadley, 2003; McAinsh and Pittman, 2009), and several Ca^{2+}-binding proteins have been purified from vacuoles (Peiter, 2011). However, most Ca in the root vacuole is present as Ca^{2+} and water-soluble Ca complexes, although $CaCO_3$ or Ca oxalate crystals can be present in some plants under particular soil conditions.

Calcium is released from the vacuole through Ca^{2+} permeable cation channels in the tonoplast (Table 5.1; White and Broadley, 2003; Pottosin and Schönknecht, 2007; Karley and White, 2009; Dodd et al., 2010; Kudla et al., 2010; Peiter, 2011). These include (1) hyperpolarization-ctivated channels, which could be formed by annexins (Mortimer et al., 2008); (2) depolarization-activated channels, such as the ubiquitous slow-vacuolar (SV) channel encoded by homologs of the *Arabidopsis thaliana AtTPC1* gene (Peiter et al., 2005; Ranf et al., 2008); and (3) channels gated by ligands such as inositol-1,4,5-triphosphate (IP_3) and cyclic ADP ribose (cADPR). No gene encoding any ligand-gated Ca^{2+} channel in the tonoplast has yet been identified (Wheeler and Brownlee, 2008; Verrett et al., 2010), although electrophysiological evidence for such channels has been available for over two decades (Alexandre et al., 1990; White, 2000).

It is thought that to prevent the large nutritional Ca flux to the shoot interfering with the minute Ca^{2+} influx required for cell signaling, symplastic Ca^{2+} transport to the xylem takes a minimal route through the cytoplasm of unsuberized endodermal cells (Figure 5.4; White, 2001). This pathway requires the spatial separation of Ca^{2+} channels, facilitating Ca^{2+} influx on the cortical side, and Ca^{2+} ATPase, catalyzing Ca^{2+} efflux on the stelar side of the Casparian band. A requirement for Ca^{2+} to enter the root symplasm *en route* to the xylem is thought to allow the root to control the rate and selectivity of Ca transport to the shoot (White, 2001). This would prevent the transport of excess Ca, or toxic solutes, to the shoot through the apoplastic pathway (White, 2001). The P_{2B} ATPases present in the plasma membranes of cells within the root stele, such as AtACA1, AtACA2, and AtACA8 in *A. thaliana*, might load Ca into the xylem (Table 5.1; Karley and White, 2009). Recently, the *Brassica juncea* PCR1 protein, which is located in the plasma membrane of root epidermal cells, has been demonstrated to be a Ca^{2+} efflux transporter necessary for the radial transfer of Ca^{2+} through the root symplast to the xylem (Song et al., 2011).

5.4 EFFECTS OF RHIZOSPHERE CALCIUM ON THE UPTAKE OF OTHER ELEMENTS

The Ca^{2+} concentration in the rhizosphere solution can exert a variety of effects on the uptake of other elements, depending upon soil conditions. At low pH, the presence of Ca^{2+} in the rhizosphere solution can increase the uptake of both anions and cations (White, 2012a). It is likely that this effect is a consequence of the displacement of H^+ from cation-binding sites in the apoplast or results from greater membrane stability. However, the presence of Ca^{2+} in the rhizosphere solution generally decreases the uptake of both monovalent and divalent cations (Hawkesford et al., 2012; White et al., 2012a). Calcium can influence the acquisition of other cations either directly, by competing for transport by membrane proteins, or indirectly by (1) competing for binding sites in the rhizosphere and apoplast, thereby altering the concentrations of cations in the apoplast, (2) inhibiting transport by membrane proteins, or (3) regulating transport processes via changes in $[Ca^{2+}]_{cyt}$.

Increasing Ca^{2+} concentration in the rhizosphere solution generally decreases the accumulation of monovalent cations by plants (White, 2012a). It also alters the relative accumulation of monovalent cations. For example, increasing Ca^{2+} concentration in the rhizosphere solution increases the

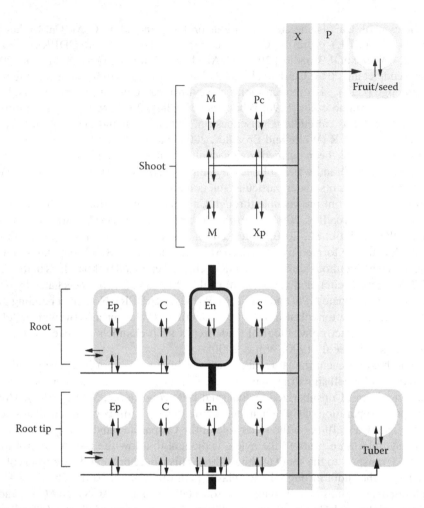

FIGURE 5.4 Calcium transport within the plant. Calcium is taken up by individual epidermal (Ep), corti-cal (C) cells, and stelar (S) cells of the root from the apoplastic solution and moves to the xylem (X) either apoplastically, in regions where the Casparian band is incomplete, or through the symplast of unsuberized endodermal cells (En) in the root tip. In the shoot, Ca follows the transpiration stream and, again, is taken up by individual xylem parenchyma (Xp), phloem (Pc), and mesophyll cells (M). Little Ca is transported in the phloem (P) and fruit/seeds and tubers must acquire most of their Ca from the xylem.

K^+/Na^+ ratio for uptake (Epstein, 1961; White, 2012a). One explanation of the latter phenomenon is that extracellular Ca^{2+} inhibits Na^+ influx through VICCs, which catalyze much of the Na^+ influx to root cells (Maathuis and Amtmann, 1999; White, 1999; White and Davenport, 2002; Munns and Tester, 2008), but has little effect on K^+ influx through the inward-rectifying K^+ channels, which con-tribute substantially to K^+ nutrition (White and Karley, 2010). Thus, high Ca^{2+} concentrations in the soil solution are beneficial for maintaining appropriate tissue K/Na ratios in saline environments.

Although Ca and Mg are likely to be taken up and transported within the plant by different membrane proteins (Karley and White, 2009), it is often observed that (1) increasing Ca concentra-tion in the rhizosphere solution decreases both Mg uptake by roots and Mg accumulation in shoots and (2) the application of Ca fertilizers can induce Mg deficiency in crops (Hawkesford et al., 2012; White, 2012a). However, and by contrast, although Ca^{2+}-permeable ion channels in the plasma membrane of root cells are permeable to Ca^{2+}, Ba^{2+}, and Sr^{2+} (White, 2001), which is consistent with competition among these cations for uptake by roots (Epstein and Leggett, 1954), and their selectiv-ity is consistent with the accumulation ratios of Ca/Ba and Ca/Sr in plant roots, it is often observed that the Ca/Ba and Ca/Sr ratios in plant shoots are identical to those in the rhizosphere solution (White, 2001). This implies that there is little competition between these cationic elements in their transport to the shoot. Similarly, there are generally strong positive correlations between shoot Ca, Ba, and Sr concentrations among plant species growing in the same environment (White, 2001, 2005; Watanabe et al., 2007; Broadley and White, 2012). This has been cited as evidence for the apoplastic transport movement of these cations across the root to the xylem (Section 5.3). Increasing Ca^{2+} concentrations in the rhizosphere solution can also reduce the uptake of other cations including cadmium (Lux et al., 2011), manganese (Löhnis, 1960), and zinc (Baker, 1978).

5.5 CALCIUM REQUIREMENTS OF PLANTS

Calcium requirements for optimal growth are related to both phylogenetic inheritance and ecological adaptation (White and Broadley, 2003; White, 2005). In general, commelinoid monocots (Arecales, Commelinales, Zingiberales) require less Ca than noncommelinoid monocots (Acorales, Alismatales, Asparagales, Dioscoreales, Liliales, Pandanales, Petrosaviales) or eudicots (Figure 5.3a,b; Table 5.2; Loneragan et al., 1968; Loneragan and Snowball, 1969; Islam et al., 1987; Pilbeam and Morley, 2007; Römheld, 2012). This difference appears to be related primarily to the demand for Ca as a structural element in cell walls. Tissue Ca concentrations of different plant species growing in the same environment are often positively correlated with the CEC of their cell walls and, specifically, the free carboxyl groups of galacturonic acid of pectins in the middle lamella (White and Broadley, 2003; White, 2005). However, there are exceptions to

TABLE 5.2
Calcium Concentrations Considered Adequate for the Growth of Selected Crops

Crop		Tissue	Calcium Concentration (g kg⁻¹ Dry Matter)
Sorghum	*Sorghum bicolor* (L.) Moench	Mature leaves, new growth	1.5–9.0
Sugarcane	*Saccharum officinarum* L.	Third leaf, 3–5 months after planting	2.0–5.0
Winter wheat	*Triticum aestivum* L.	Leaves, heading	2.0–10.0
Maize	*Zea mays* L.	Ear leaf, initial silk	2.1–10.0
Soybean	*Glycine max* (L.) Merr.	Mature leaves, new growth	3.5–20.0
Sugar beet	*Beta vulgaris* L.	Leaves, 50–80 days after planting	5.0–15.0
Cassava	*Manihot esculenta* Crantz	Mature leaves, new growth	6.0–15.0
Spinach	*Spinacia oleracea* L.	Mature leaves, new growth	6.0–15.0
Strawberry	*Fragaria* × *ananassa* Duchesne	Mature leaves, new growth	6.0–25.0
Pigeon pea	*Cajanus cajan* (L.) Millsp.	Mature leaves, new growth	7.5–15.0
Apple	*Malus domestica* Borkh.	Mature leaves, new growth	8.0–16.0
Snap beans	*Phaseolus vulgaris* L.	Mature leaves, new growth	8.0–30.0
Subterranean clover	*Trifolium subterraneum* L.	Mature leaves, new growth	10.0–15.0
Orange	*Citrus* × *sinensis* (L.) Osbeck	Mature leaves, subtending fruit	11.0–40.0
Garden pea	*Pisum sativum* L.	Mature leaflets, first bloom	12.0–20.0
Cabbage	*Brassica oleracea* L.	Wrapper leaves, mature plants	13.0–35.0
Onion	*Allium cepa* L.	Leaves, 50% maturity	15.0–22.0
Iceberg lettuce	*Lactuca sativa* L.	Wrapper leaves, mature plants	15.0–22.5
Potato	*Solanum tuberosum* L.	Mature leaves, tuber filling	15.0–25.0
Sunflower	*Helianthus annuus* L.	Mature leaves, new growth	15.0–30.0
Tomato	*Solanum lycopersicum* L.	Mature leaves, new growth	16.0–32.1

Source: Data from Mills, H.A. and Jones, J.B., *Plant Analysis Handbook II*, MicroMacro Publishing, Athens, GA, 1996.

this observation. For example, the calcifuge plant blue lupin (*Lupinus angustifolius* L.) (Fabales), which prefers acidic soils and grows poorly in calcareous soils, has an optimal tissue Ca concentration (and Ca requirement) similar to cereals and grasses, which might be related to an inability to maintain a low $[Ca^{2+}]_{cyt}$ through vacuolar Ca sequestration or cytosolic Ca buffering in environments with high Ca phytoavailability (Islam et al., 1987). Calcium requirements are often greater in plants adapted to calcareous soils, and calcifuge ecotypes achieve their growth potential at lower Ca^{2+} concentrations in the soil solution than calcicole ecotypes and are generally more tolerant of acid soils (Figure 5.3c,d; Jefferies and Willis, 1964; White and Broadley, 2003).

The Ca^{2+} concentration in the rhizosphere solution that allows for maximal plant growth is determined by (1) the Ca demand of the plant; (2) the delivery of Ca^{2+} to the rhizosphere; (3) the concentrations of other cations in the rhizosphere solution, which can compete with Ca^{2+} for cation-binding sites in the rhizosphere and apoplast and for membrane transport processes; and (4) the rates of uptake and transport of Ca within the plant. When plants are grown in flowing nutrient solutions, sufficient Ca for maximal growth can be obtained at solution Ca^{2+} concentrations as low as 2.5–100 µM, depending upon the plant species (Figure 5.3c,d; Loneragan et al., 1968; Islam et al., 1987). However, if the concentrations of other cations are increased or the solution pH is lowered, much higher-solution Ca^{2+} concentrations are required for optimal growth (Asher and Edwards, 1983). When plants are grown in soil, the Ca concentration of the rhizosphere solution required for maximal growth is greater than that in flowing nutrient solutions (Hawkesford et al., 2012).

Calcium concentrations in plant tissues range from about 1 to 50 mg g^{-1} DM, depending on the growth conditions, plant species, and plant organ sampled (Figure 5.5; Broadley et al., 2003; Watanabe et al., 2007). Much of the genetic variation in shoot Ca concentrations occurs at the taxonomic level of the order or above (Thompson et al., 1997; Broadley et al., 2003, 2004; Watanabe et al., 2007; White et al., 2012b). This implies that shoot Ca concentrations are constrained by an ancient evolutionary heritage. Commelinoid monocots have lower shoot Ca concentrations than other angiosperm species, whereas noncommelinoid monocots have shoot Ca concentrations that are less than, or similar to, most eudicot species, and among the eudicots, members of orders within the rosid (Brassicales, Cucurbitales, Malvales, and Rosales) and asterid (Apiales, Asterales,

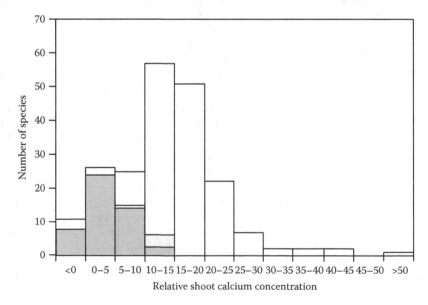

FIGURE 5.5 Mean relative shoot calcium concentrations (approximating mg Ca g^{-1} dry weight) of 49 commelinid monocot (dark gray bars), 4 noncommelinid monocot (light gray bars), and 153 eudicot (white bars) species obtained from a literature survey of 61 papers containing 244 comparative studies. (Data from Broadley, M.R. et al., *J. Exp. Bot.*, 54, 1431, 2003.)

Lamiales, and Solanales) clades typically have the highest shoot Ca concentrations (Figure 5.5; Thompson et al., 1997; Broadley et al., 2003, 2004; White and Broadley, 2003; White, 2005; Watanabe et al., 2007; White et al., 2012b; Zhang et al., 2012). However, the absolute Ca concentrations in shoots depend greatly upon the phytoavailability of Ca in the rhizosphere, the transpirational water flux, and their interactions with plant physiotype (Kinzel and Lechner, 1992; White, 2001; White and Broadley, 2003; White et al., 2012b). Considerable variation has been observed among genotypes of several plant species in their growth responses to Ca supply and their tissue Ca concentrations, which often reflect differences in their acquisition of Ca by roots, Ca transport within the plant, and ability to accumulate Ca in specific tissues (Sleper et al., 1989; Horst et al., 1992; White and Broadley, 2003, 2009; Broadley et al., 2008, 2009).

5.6 MOVEMENT OF CALCIUM WITHIN THE PLANT

Calcium is not transported readily in the phloem and its distribution within the plant is largely determined by Ca movement in the xylem, which is related to both the rate of Ca delivery to the xylem and transpirational water flows (White and Broadley, 2003, 2009; Karley and White, 2009; Gilliham et al., 2011; White, 2012b). The Ca concentration in the xylem sap can vary widely, depending upon Ca^{2+} supply and plant transpiration rate, and values from 300 µM to 16.5 mM have been reported (White et al., 1992; De Silva et al., 1998; White, 2012b). Within the xylem, Ca is transported as Ca^{2+} and as Ca complexes with organic acids such as malate and citrate (Welch, 1995). There is often a positive relationship between tissue Ca concentration and the amount of water transpired and also between the distribution of Ca within a plant and the route of the transpiration stream. Thus, leaf Ca concentration generally increases with leaf age (Pilbeam and Morley, 2007; Römheld, 2012), and Ca concentrations in leaves generally exceed those in fruits and tubers (Kärenlampi and White, 2009; White, 2012b). There is little movement of Ca from leaves to fruit or tubers during vegetative growth, and shoot Ca content often increases until leaf senescence (White, 2012a). This necessitates continued Ca uptake by roots and Ca transport to leaves and fruit via the xylem, after anthesis. Nevertheless, Ca remobilization from leaves to fruit can be substantial in some plants during the reproductive stage. For example, in white and blue lupins (*Lupinus albus* L. and *L. angustifolius* L.) at least 13%–18% of the Ca accumulated in leaves is translocated to the fruits (Hocking and Pate, 1978).

It is commonly observed that a temporary interruption in Ca delivery to the xylem leads to the development of Ca-deficiency disorders in tissues with low transpiration rates, such as enclosed leaves and fruits (Figure 5.6; Shear, 1975; McLaughlin and Wimmer, 1999; White and Broadley, 2003; Ho and White, 2005; Dayod et al., 2010). Root-pressure-driven water flow in the xylem during periods of low transpiration is of great importance for the Ca nutrition of these tissues and environmental factors that reduce root pressure, such as poor aeration, low temperatures, or salinity of the rooting medium, will reduce the Ca supply to young leaves and fruits and increase the incidence of Ca-deficiency symptoms (Ho and White, 2005; White, 2012b). Differences between genotypes in their ability to translocate Ca to tissues with low transpiration rates underlie differences in the susceptibility of the enclosed shoot meristems of leafy vegetables and fruits to Ca-deficiency disorders (Shear, 1975; Ho and White, 2005; Volz et al., 2006). The same problem with restricted mobility of Ca applies to other tissues that are fed principally by the phloem, such as tubers. For example, the lack of susceptibility of tubers of some potato genotypes to Ca-deficiency disorders has been correlated with their ability to acquire Ca directly from the soil solution (Krauss and Marschner, 1975; Busse and Palta, 2006).

Within leaves, Ca follows the apoplastic route of the transpiration stream and accumulates in specific cell types, such as mesophyll cells, trichomes, or epidermal cells adjacent to guard cells, depending upon the plant species (De Silva et al., 1996, 1998; Karley et al., 2000a,b; Franceschi and Nakata, 2005; Kerton et al., 2009; Conn and Gilliham, 2010; Conn et al., 2011; Gilliham et al., 2011; Rios et al., 2012). The removal of Ca^{2+} from the apoplastic solution, either by binding to cell wall pectins; by precipitation as Ca salts, as occurs in various gymnosperms (Kinzel, 1989; Fink, 1991); or by

FIGURE 5.6 **(See color insert.)** Calcium-related disorders in horticultural crop: (a) blackheart in celery, (b) tipburn in lettuce, (c) blossom-end rot in immature tomato fruit, and (d) cracking in mature tomato fruit. (Photographs from White, P.J. and Broadley, M.R., *Ann. Bot.*, 92, 487, 2003. Courtesy of the HRI collection.)

sequestration within specific cell types, is necessary because the $[Ca^{2+}]_{cyt}$ of guard cells is sensitive to apoplastic Ca^{2+} concentration within the range of the Ca^{2+} concentrations in the xylem sap (McAinsh et al., 1995; Webb et al., 2001; Conn et al., 2011). Changes in the guard cell $[Ca^{2+}]_{cyt}$ are central to the control of stomatal aperture, gas exchange, and water loss in plants (Amtmann and Blatt, 2008; Kudla et al., 2010). Indeed, an increased Ca^{2+} concentration in the shoot apoplast is responsible for reduced stomatal opening and impaired gas exchange in the *Arabidopsis thaliana cax1/cax3* mutant (Conn et al., 2011).

Calcium is likely to enter the cytoplasm from the apoplast of shoot cells through Ca^{2+}-permeable cation channels, such as annexins, CNGCs, GLRs, MCAs, and MSLs. However, all of these channels have also been implicated in generating $[Ca^{2+}]_{cyt}$ signals necessary for appropriate cellular responses to a variety of developmental and environmental stimuli (Section 5.7.2; White and Broadley, 2003; Conn and Gilliham, 2010; Ma and Berkowitz, 2011). In *A. thaliana*, both the ability to accumulate Ca in the shoot and the accumulation of Ca in specific cell types within the shoot correlate with the expression of *AtCAX1*, which encodes a tonoplast Ca^{2+}/H^+ antiporter (Table 5.1; Conn et al., 2011, 2012). Calcium is accumulated in the vacuoles of shoot cells as Ca^{2+}, as soluble complexes with proteins or organic acids, and in insoluble forms, including Ca carbonate, Ca oxalate, Ca malate,

Ca citrate, Ca tartrate, Ca phytate, and Ca phosphate, depending upon the plant species and the amount of Ca delivered to the shoot (Kinzel, 1982; Kinzel and Lechner, 1992; White and Broadley, 2003; Franceschi and Nakata, 2005; White, 2005; Bauer et al., 2011).

The misexpression of various genes has been shown to influence tissue Ca concentrations in plants. Mutants of *A. thaliana* lacking AtSKOR, which is thought to remove Ca^{2+} from the xylem sap in roots, have greater shoot Ca concentrations than wild-type plants (Gaymard et al., 1998). Two additional *A. thaliana* mutants (*aca4* and *aca8*) with shoot Ca concentrations that are significantly higher (Z-score >3 for shoot Ca in more than 50% of the samples analyzed) than other genotypes can be identified in the Purdue Ionomics Information Management System (PiiMS version 6/10/2011, http://www.ionomicshub.org; Baxter et al., 2007), which holds ionomic profiles of 152,601 *A. thaliana* samples (MR Broadley, LX Dupuy, A Lyseko and PJ White, unpublished observations), although this phenotype is not always observed in *aca4* (Conn et al., 2011), nor in the *aca11* or *aca4/aca11* mutants (Conn et al., 2011).

Overexpression of genes encoding a truncated version of AtCAX1 lacking its autoinhibitory domain (*sAtCAX1*), AtCAX2, or the yeast vacuolar Ca^{2+}/H^+ antiporter (VCX1), increases shoot Ca concentrations in transgenic tobacco (Hirschi, 2001), and the expression of *sAtCAX1*, a modified *AtCAX2* gene (*sAtCAX2*), or *AtCAX4*, increases Ca concentrations in edible portions of transgenic carrot (Park et al., 2004; Morris et al., 2008), lettuce (Park et al., 2009), tomato (Park et al., 2005a), and potato (Park et al., 2005b; Kim et al., 2006). However, despite having greater tissue Ca concentrations, the leaves of tobacco plants overexpressing *sAtCAX1* exhibit tipburn (Hirschi, 2001), and the fruits of tomato plants overexpressing *sAtCAX1* are prone to blossom-end rot (Park et al., 2005a; de Freitas et al., 2011), two classic horticultural *Ca-deficiency* disorders (Section 5.8; Shear, 1975; White and Broadley, 2003). It has been suggested that this phenomenon occurs because the overexpression of genes encoding vacuolar Ca^{2+}/H^+ antiporters affects $[Ca^{2+}]_{cyt}$ homeostasis or apoplastic Ca^{2+} concentration to produce a *physiological* Ca deficiency (Ho and White, 2005; de Freitas et al., 2011). Interestingly, both the tipburn in tobacco leaves and blossom-end rot in tomato fruit in transgenic plants overexpressing *sAtCAX1* can be reduced by coexpressing a gene encoding maize calreticulin (*ZmCRT*), a Ca^{2+}-binding protein located in ER that is thought to buffer $[Ca^{2+}]_{cyt}$ (Wu et al., 2012). Transgenic plants overexpressing calreticulin have greater shoot Ca concentrations than wild-type plants (Wyatt et al., 2002). Consistent with these observations, Conn et al., (2011) observed that shoot Ca concentration among 15 *A. thaliana* ecotypes was positively correlated with the expression of *AtCAX1* in shoots, but not with the expression of *AtCAX3*, *AtACA4*, or *AtACA11*, and Conn et al., (2012) observed that shoot Ca concentration among 31 *A. thaliana* ecotypes was positively correlated with the expression of *AtCAX1* and *AtTPC1*, but negatively correlated with the expression of *AtMCA1*, in shoots. Although *A. thaliana* mutants lacking AtCAX1 have occasionally been reported to have lower shoot Ca concentrations than wild-type plants (Catalá et al., 2003), this is not always observed (Cheng et al., 2003, 2005; Conn et al., 2011), and mutants lacking AtCAX3 have similar shoot Ca concentrations to wild-type plants, possibly because there is transcriptional compensation for the loss of these proteins (Cheng et al., 2005; Conn et al., 2011). Consistent with this interpretation, the shoot Ca concentration of *cax1/ cax3* double mutants is often significantly lower than that of wild-type plants (Cheng et al., 2005; Conn et al., 2011, 2012).

A. thaliana mutants lacking plasma membrane cation channels encoded by members of the *AtCNGC* or *AtGLR* gene families rarely differ in their shoot Ca concentrations from wild-type plants, although they often show altered developmental sensitivities to Ca supply (White and Broadley, 2003; Kaplan et al., 2007; Ma and Berkowitz, 2011; Urquhart et al., 2011). These observations suggest that the primary role of such channels is in $[Ca^{2+}]_{cyt}$ signaling and that the Ca-related phenotypes of mutants misexpressing *AtCNGC* or *AtGLR* genes are probably a consequence of altered $[Ca^{2+}]_{cyt}$ homeostasis. Nevertheless, *A. thaliana* cngc1 (Hampton et al., 2005; Ma et al., 2006), cngc2 (Ma et al., 2010), and cngc10 (Guo et al., 2010) have occasionally exhibited lower shoot Ca concentrations than wild-type plants, although these phenotypes are not always observed (Chan et al., 2003; Hampton et al., 2005).

Chromosomal quantitative trait loci (QTL) affecting leaf Ca concentration have been identified in *A. thaliana* (White, 2005) and genes colocalizing with QTL affecting seed Ca concentration in *A. thaliana* have been reported (Vreugdenhil et al., 2004; Waters and Grusak, 2008). These include two members of the cation/H⁺ exchanger (*CHX*) gene family (*AtCHX1*, *AtCHX16*) and several genes encoding AtCAXs (Vreugdenhil et al., 2004; Waters and Grusak, 2008). In addition, QTL affecting Ca concentrations in edible portions of rice (*Oryza sativa* L.) (Lu et al., 2008; Garcia-Oliveira et al., 2009), wheat (Peleg et al., 2009), bean (*Phaseolus vulgaris* L.) (Guzmán-Maldonado et al., 2003; Gelin et al., 2007; Casanas et al., 2013), soybean (*Glycine max* Merr.) (Zhang et al., 2009), oilseed rape (*Brassica napus* L.) (Ding et al., 2010), leafy brassicas (Broadley et al., 2008, 2009), and potato (*Solanum tuberosum* L.) (Subramanian, 2012) have been identified, but the genes responsible for these phenotypes have not yet been identified. It is noteworthy that the Ca concentrations in seeds of commelinid monocot species are less than those in seeds of eudicot species (White and Broadley, 2009). Seeds of cereals often have Ca concentrations between about 0.10 and 1.18 g kg⁻¹ DM, whereas Ca concentrations in seeds of most eudicots lie between 0.28 and 4.80 g kg⁻¹ DM (White and Broadley, 2009).

5.7 FUNCTIONS OF CALCIUM IN PLANTS

Calcium has unique roles in maintaining the expansion and structural integrity of cell walls and lipid membranes (Section 5.7.1) and as a cytosolic signal coordinating cellular responses to environmental stimuli (Section 5.7.2). It also has a nonspecific role in maintaining cation–anion balance and in osmoregulation under specific environmental conditions (Section 5.7.3).

5.7.1 CELL WALLS AND LIPID MEMBRANES

A high proportion of Ca in plant tissues is located in the cell wall, especially in eudicot species (Hawkesford et al., 2012). This distribution results from the binding of Ca^{2+} to the carboxyl groups of pectins in the cell wall. The cross-linking of pectins by Ca^{2+} not only provides strength and rigidity to plants but also determines the size of pores in the cell wall matrix and, therefore, the movements of large solutes and the ingress of microorganisms (Carpita and McCann, 2000; Pilbeam and Morley, 2007). Structural weaknesses in cell walls lacking sufficient Ca result often result in cracking during rapid extension growth (Figure 5.6; White and Broadley, 2003; Ho and White, 2005).

Calcium is also important for maintaining membrane integrity by bridging anionic groups of lipids and proteins and can promote membrane fusion by cross-linking lipids from two separate membranes (Jaiswal, 2001). When plant tissues lack sufficient Ca, there is greater leakage of low-molecular-weight solutes, such as K⁺ and sugars, from cells, which is most apparent when plants are subject to environmental stresses restricting membrane fluidity or energy metabolism, such as low temperature or waterlogging (Pilbeam and Morley, 2007; Lurie, 2009). In severely Ca-deficient tissues, a general disintegration of membrane structures and a loss of intracellular compartmentation can occur (Hawkesford et al., 2012).

5.7.2 CELL SIGNALING

5.7.2.1 Signature Perturbations in $[Ca^{2+}]_{cyt}$

Perturbations in $[Ca^{2+}]_{cyt}$ play a central role in intracellular signaling, acting as a *second messenger* linking many environmental and developmental stimuli to an appropriate response (Figure 5.7; White and Broadley, 2003; Wheeler and Brownlee, 2008; Batistič and Kudla, 2010, 2012; Dodd et al., 2010; Kudla et al., 2010; Hawkesford et al., 2012). It is thought that the $[Ca^{2+}]_{cyt}$ perturbation elicited by a specific stimulus has a unique *signature*, which is associated with the subcellular location, magnitude, and kinetics of the $[Ca^{2+}]_{cyt}$ perturbation (White and Broadley, 2003; Wheeler and

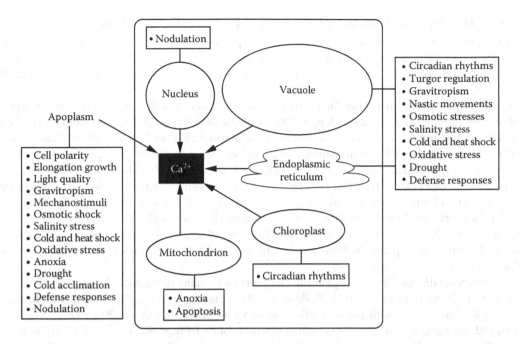

FIGURE 5.7 Principal sources for Ca^{2+} influx to the cytosol initiating cellular responses to developmental and environmental stimuli. (Adapted from Hawkesford, M. et al., 2012. Functions of macronutrients, in *Marschner's Mineral Nutrition of Higher Plants*, 3rd edn., Marschner, P. (ed.), Academic Press, London, U.K., 2012, pp. 135–189.)

Brownlee, 2008; McAinsh and Pittman, 2009; Dodd et al., 2010; Kudla et al., 2010; Batistič and Kudla, 2012). Cytosolic Ca^{2+} concentration can be increased by Ca^{2+} influx either from the apoplast, across the plasma membrane, or from intracellular compartments, such as the vacuole, ER, golgi, endosomal vesicles, mitochondria, plastids, or nucleus (Figure 5.7). Each of these compartments has a different complement of Ca^{2+} transport proteins with contrasting properties (Table 5.1). Calcium influx is mediated by Ca^{2+}-permeable ion channels, and their biochemistry, subcellular location, and abundance will influence the spatial characteristics, magnitude, and kinetics of any $[Ca^{2+}]_{cyt}$ perturbation. Since diffusion of Ca^{2+} within the cytosol is slow, because of the low $[Ca^{2+}]_{cyt}$, and $[Ca^{2+}]_{cyt}$ is strongly buffered by cytosolic proteins, any increase in $[Ca^{2+}]_{cyt}$ caused by the opening of an individual ion channel will be restricted to its vicinity and dissipate rapidly after the channel is closed (White and Broadley, 2003; Batistič and Kudla, 2010). The signature of a $[Ca^{2+}]_{cyt}$ perturbation is also related to the characteristics of the cellular $[Ca^{2+}]_{cyt}$ buffering capacity and the subcellular location and activities of Ca^{2+} ATPases and Ca^{2+}/H^+ antiporters that remove Ca^{2+} from the cytosol to the apoplast, ER, plastids, or vacuoles and restore $[Ca^{2+}]_{cyt}$ homeostasis upon a perturbation in $[Ca^{2+}]_{cyt}$ (White and Broadley, 2003; Bonza and De Michaelis, 2011). Perturbations in $[Ca^{2+}]_{cyt}$ range from a single transient rise in $[Ca^{2+}]_{cyt}$, such as often occurs in response to mechanical or temperature shocks (White and Broadley, 2003), through the oscillations in $[Ca^{2+}]_{cyt}$ that ensure stomatal closure for long durations (Allen et al., 2001; Amtmann and Blatt, 2008; Dodd et al., 2010; Kudla et al., 2010) and the interactions between bacteria and root hairs during nodulation (White and Broadley, 2003; Kudla et al., 2010) or the circadian rhythms that contribute to the diurnal regulation of photosynthesis, leaf metabolism, and solute transport within the plant (Love et al., 2004; Haydon et al., 2011), to the wave of elevated $[Ca^{2+}]_{cyt}$ that crosses the egg following fertilization (Antoine et al., 2001) and the sustained $[Ca^{2+}]_{cyt}$ elevations found in elongating root cells (Demidchik et al., 2002) or at the apex of growing root hairs and pollen tubes (Foreman et al., 2003; Konrad et al., 2011; Hepler et al., 2012).

5.7.2.2 Developmental Processes Coordinated by $[Ca^{2+}]_{cyt}$ Perturbations

Perturbations in $[Ca^{2+}]_{cyt}$ control many developmental processes including cell division, cell elongation, generation of polarity, circadian rhythms, apoptosis, senescence, and symbiotic associations with N-fixing bacteria and mycorrhizal fungi (Figure 5.7; White and Broadley, 2003; Hawkesford et al., 2012).

Cell division appears to be coordinated by Ca^{2+} fluxes to and from the ER, which is closely associated with the mitotic apparatus (Hepler, 2005; Batistič and Kudla, 2010). Cell elongation often depends upon continuous Ca^{2+} influx across the plasma membrane, and the direction of elongation of root hairs and pollen tubes is determined by a local elevation in $[Ca^{2+}]_{cyt}$, which focuses the exocytosis of cell wall material at a specific site (White and Broadley, 2003; Dodd et al., 2010; Konrad et al., 2011; Hepler et al., 2012). The Ca^{2+} influx associated with elongation growth is mediated by annexins in root hairs (Foreman et al., 2003; Laohavisit et al., 2012), and several HACCs, CNGCs, and GLRs are required for the elongation of pollen tubes (Frietsch et al., 2007; Konrad et al., 2011; Michard et al., 2011; Hepler et al., 2012; Tunc-Ozdemir et al., 2013). The secretion of mucilage from cells of the root cap and the formation of callose are also Ca^{2+}-dependent secretory processes (Kauss, 1987).

Circadian oscillations in $[Ca^{2+}]_{cyt}$ have been observed in many plant cells (Johnson et al., 1995; Wood et al., 2001; Love et al., 2004; Dodd et al., 2010; Haydon et al., 2011). They are controlled by complex interactions with the molecular circadian oscillator (Xu et al., 2007; Dalchau et al., 2010) and are thought to involve the circadian synthesis of cADPR, which triggers Ca^{2+} efflux from intracellular organelles (Dodd et al., 2007). Although the exact function of these oscillations is unknown, it has been speculated that they might synchronize photosynthesis, solute transport, and growth (Haydon et al., 2011).

Many cellular responses to phytohormones involve an initial influx of Ca^{2+} across the plasma membrane followed by complex oscillations in $[Ca^{2+}]_{cyt}$ resulting from Ca^{2+} fluxes across various cell membranes (White and Broadley, 2003; Batistič and Kudla, 2010; Hawkesford et al., 2012). These occur, for example, in remodeling root system architecture in response to auxin (Monshausen et al., 2011; Monshausen, 2012), during stomatal closure in response to abscisic acid (Amtmann and Blatt, 2008; Dodd et al., 2010; Kudla et al., 2010), in the synthesis of α-amylase during seed germination in response to gibberellin (Batistič and Kudla, 2010), and in the initiation of cellular responses to ethylene production (Zhao et al., 2007; Aghdam et al., 2012). The bolting and flowering of *A. thaliana* appears to be controlled, in part, by the interplay of Ca^{2+} signals originating from the cytoplasm and chloroplasts (Batistič and Kudla, 2010). Sustained elevations in $[Ca^{2+}]_{cyt}$ have been implicated in cell death (apoptosis), both during normal development, for example, during pollen self-incompatibility responses and tissue patterning, and in hypersensitive responses to pests and pathogens (White and Broadley, 2003; Dumas and Gaude, 2006; Lecourieux et al., 2006). Leaf senescence, which is considered to be a form of programmed cell death, also appears to be influenced by $[Ca^{2+}]_{cyt}$ perturbations and consequent signaling cascades (Ma et al., 2010).

Perturbations of $[Ca^{2+}]_{cyt}$ also coordinate cellular events during the development of symbiotic associations with N-fixing bacteria and mycorrhizal fungi (Kosuta et al., 2008; Dodd et al., 2010; Kudla et al., 2010; Sieberer et al., 2011; Oldroyd, 2013). During these interactions, there is often an initial Ca^{2+} influx across the plasma membrane of a root cell, followed by subsequent oscillations in $[Ca^{2+}]_{cyt}$ and nuclear Ca^{2+} concentrations resulting from periodic Ca^{2+} fluxes across membranes of intracellular organelles, including the vacuole, ER, and nucleus (White and Broadley, 2003; Sieberer et al., 2009, 2011; Dodd et al., 2010; Capoen et al., 2011; Granqvist et al., 2012). Several transport proteins have been implicated in oscillations in nuclear Ca^{2+} concentration including the cation channels CASTOR and POLLUX in lotus (*Lotus japonicus* L.) and the K^+ channel MtDMI1 and Ca^{2+}-ATPase MtMCA8 in barrel medic (*Medicago truncatula* Gaertn.) (Venkateshwaran et al., 2013). It is possible that annexins behave as Ca^{2+}-permeable cation channels in nuclear membranes (Clark et al., 2012). The progression of bacterial or fungal symbiont from epidermal to cortical cells is preceded by slow oscillations in $[Ca^{2+}]_{cyt}$ and nuclear Ca^{2+} concentration, which become faster as the microbes enter the cell (Sieberer et al., 2011).

5.7.2.3 Environmental Responses Coordinated by $[Ca^{2+}]_{cyt}$ Perturbations

Perturbations in $[Ca^{2+}]_{cyt}$ coordinate plant responses to a wide variety of environmental signals, including light quality and day length, extreme temperatures, drought, salinity, anoxia, oxidative stresses, mechanical stresses, and attack by pathogens (Figure 5.7; White and Broadley, 2003; Lecourieux et al., 2006; McAinsh and Pittman, 2009; Batistič and Kudla, 2010; Dodd et al., 2010; Haydon et al., 2011; Hawkesford et al., 2012).

Perturbations in $[Ca^{2+}]_{cyt}$ have an important role in coordinating osmotically driven movements in plants in response to environmental cues (Moran, 2007; Amtmann and Blatt, 2009; Kim et al., 2010). These movements include the regulation of stomatal aperture by changes in the turgor of guard cells and nyctinastic and seismonastic movements produced by dissimilar changes in turgor of cells in two oppositely located parts of the pulvinus. These turgor changes are driven mainly by fluxes of K^+, Cl^-, and malate (Moran, 2007; Amtmann and Blatt, 2009; Kim et al., 2010). Stomatal closure is produced by loss of solutes, and turgor, of guard cells. Short-term closure of stomata occurs when a threshold $[Ca^{2+}]_{cyt}$ is exceeded, whereas long-term stomatal closure is effected by Ca^{2+} oscillations within a defined range of amplitude, frequency, and duration (Dodd et al., 2010; Kudla et al., 2010). These $[Ca^{2+}]_{cyt}$ perturbations are generated by the opening of HACCs in the plasma membrane, in consort with the opening of various types of Ca^{2+}-permeable channels in the tonoplast and the activities of Ca^{2+} ATPase and Ca^{2+}/H^+ antiporters (Amtmann and Blatt, 2009; Dodd et al., 2010; Kim et al., 2010; Kudla et al., 2010). In pulvini, the opening of HACCs, together with a variety of endomembrane Ca^{2+}-permeable ion channels, generates the $[Ca^{2+}]_{cyt}$ perturbations required to coordinate the loss of solutes from shrinking cells, while the plasma membrane H^+-ATPase generates the electrochemical gradient for the accumulation of solutes in swelling cells (Moran et al., 2007). The closure of the Venus flytrap (*Dionaea muscipula* Sol ex Ellis) is thought to be initiated by the opening of mechanosensitive Ca^{2+} channels (Volkov et al., 2008). Illumination by red light, blue light, or UV-B stimulates Ca^{2+} influx across the plasma membrane, which produces a transient increase in $[Ca^{2+}]_{cyt}$ that is implicated in various tropic and developmental responses (White and Broadley, 2003; Harada and Shimazaki, 2007; Dodd et al., 2010).

Acclimation to chilling temperatures appears to be initiated by $[Ca^{2+}]_{cyt}$ perturbations generated by Ca^{2+} influx through DACCs in the plasma membrane (White, 2009; Knight and Knight, 2012), and freezing tolerance requires both Ca^{2+} influx from the apoplast and Ca^{2+}-dependent exocytosis involving synaptotagmin and SNARE proteins to reseal membranes after cold-induced rupture (Schapire et al., 2008; Yamazaki et al., 2008; Knight and Knight, 2012). Heat shock also results in Ca^{2+} influx and an elevation of $[Ca^{2+}]_{cyt}$ lasting 15–30 min, which is maintained by Ca^{2+} influx to the cytosol from both the apoplast and intracellular organelles (White and Broadley, 2003; Mittler et al., 2012). This perturbation in $[Ca^{2+}]_{cyt}$ is thought to initiate acclimatory responses that mitigate the effects of heat stress on cellular metabolism and allow plants to function at higher temperatures (Mittler et al., 2012).

Exposure to pathogens elicits biphasic perturbations in $[Ca^{2+}]_{cyt}$, which are capable of inducing both basal defense responses and the hypersensitive response, which prevents the spread of the pathogen by rapid programmed cell death around the infection site (Lecourieux et al., 2002, 2006; White and Broadley, 2003; Nemchinov et al., 2008; Ma and Berkowitz, 2011). Various Ca^{2+}-permeable cation channels have been implicated in generating these $[Ca^{2+}]_{cyt}$ signatures, including CNGCs and GLRs in the plasma membrane and IP_3- and cADPR-gated channels in the tonoplast (Batistič and Kudla, 2010; Kudla et al., 2010; Ma and Berkowitz, 2011; Reddy et al., 2011). Elevated $[Ca^{2+}]_{cyt}$ also stops cyclosis in root cells and closes plasmodesmata, both of which appear to be strategies to isolate cells to prevent the movement of pathogens (White and Broadley, 2003).

5.7.2.4 Translation of $[Ca^{2+}]_{cyt}$ Signals

Perturbations in $[Ca^{2+}]_{cyt}$ exert their effects through a network of interactions that alter solute transport, cellular biochemistry, gene expression, and physiological and morphological processes. The primary targets of $[Ca^{2+}]_{cyt}$ signals are cytosolic proteins that alter conformation upon binding Ca^{2+}. These include proteins that bind Ca^{2+} using a helix–loop–helix structure termed an "EF hand," such as CaMs, CaM-like (CML) proteins, CBL proteins, CDPKs and CDPK-related kinases (CRKs), and other proteins such as annexins and phospholipase D (Figure 5.8). CaMs and CML proteins are involved in Ca^{2+}-dependent initiation of developmental responses to environmental variables, such as light, gravity, and mechanical stimuli; acclimation to adverse environmental conditions, including salinity, anoxia, and temperature shocks; and defense responses to pathogens (White and Broadley, 2003; Batistič and Kudla, 2010, 2012; Dodd et al., 2010; Kudla et al., 2010; Reddy et al., 2011). In *A. thaliana*, CaMs and CMLs can bind over 300 different proteins to modify their activities (Popescu et al., 2007; Lee et al., 2010; Reddy et al., 2011). Important targets for CaMs include ion transport proteins, including several involved in $[Ca^{2+}]_{cyt}$ homeostasis, CaM-dependent protein kinases (CaMK), and Ca^{2+}- and CaM-dependent protein kinases (CCaMKs), which regulate the activities of numerous proteins via phosphorylation, and CaM-binding transcription factors (e.g., CAMTAs), which control gene expression (White and Broadley, 2003; Batistič and Kudla, 2010, 2012; Dodd et al., 2010; Kudla et al., 2010; Reddy et al., 2011). Similarly, the CBL proteins, together with their target proteins including the SNF-related serine/threonine protein kinases (CBL-interacting protein kinases [CIPKs]), act in a wide variety of response networks, including those initiated by low temperatures, drought, salinity, anoxia, wounding, and mineral deficiencies (White and Broadley, 2003; Amtmann and Blatt, 2009; Weinl and Kudla, 2009; Batistič and Kudla, 2010, 2012; Dodd et al., 2010; Kudla et al., 2010; Reddy et al., 2011; Kim, 2013; Sánchez-Barrena et al., 2013). The CDPKs are also involved in cellular responses to many stimuli and exert their effects through the $[Ca^{2+}]_{cyt}$-dependent phosphorylation of a multitude of target proteins, including ion transport proteins and water channels, NADPH oxidases, enzymes involved in carbon and nitrogen metabolism, cytoskeletal proteins, and DNA-binding proteins (White and Broadley, 2003; Batistič and Kudla, 2010, 2012; Dodd et al., 2010). They are implicated in seed germination, root development, stomatal movements, elongation growth, and responses to salinity, drought, and wounding (Batistič and Kudla, 2010, 2012; Dodd et al., 2010). Plant annexins are multifunctional proteins that can

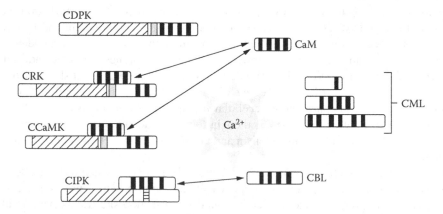

FIGURE 5.8 Signal transduction through $[Ca^{2+}]_{cyt}$-regulated kinase cascades. CaMs, CML proteins, CBL proteins, CDPKs, and CCaMKs can bind Ca^{2+} using an *EF hand* (black box). Both CaM and CBL can bind to many proteins to modify their activities. Among their target proteins, CaM binds CaMKs (CRK) and CCaMKs through a CaM-binding domain (gray box), and CBL binds to CIPKs through a CBL-binding NAF domain (striped box). These kinases regulate the activities of numerous proteins via phosphorylation. Their kinase domain is indicated by the hatched box. (Adapted from Batistic, O. and Kudla, J., *Biophys. Acta*, 1820, 1283, 2012.)

form cation channels in membranes, facilitate membrane fusion, and act as peroxidases (Mortimer et al., 2008; Laohavisit and Davies, 2011; Clark et al., 2012). Their activity has been implicated in Ca^{2+}-dependent membrane repair, secretory processes, cell elongation, and acclimatory responses to drought and salinity (Mortimer et al., 2008; Laohavisit and Davies, 2011). The activity of phospholipase D, which cleaves membrane phospholipids into a soluble head group and phosphoric acid, has been implicated in leaf senescence and drought tolerance (White and Broadley, 2003). Calcium-binding proteins in the ER include calreticulin, calsequestrin, calnexin, and BiP (White and Broadley, 2003; McAinsh and Pittman, 2009). These proteins are involved in cellular Ca^{2+} homeostasis, protein folding, and posttranslational modifications. Since cells differ in their complement of Ca^{2+}-binding proteins, this might confer appropriate tissue specificity in responses to developmental cues and environmental challenges (White and Broadley, 2003).

5.7.3 Calcium Accumulation in Vacuoles

In addition to its unique roles in maintaining tissue integrity and in intracellular signaling, Ca^{2+} also has nonspecific roles in cation–anion balance and in osmoregulation. Most of the water-soluble Ca in plants is located in the vacuole, where Ca^{2+} acts as a countercation for the accumulation of inorganic anions, such as nitrate or chloride, and organic anions, such as oxalate, malate, or citrate. Calcium salts can either be precipitated in the vacuole (or apoplast) to remove excess organic acids or Ca^{2+} can be accumulated in millimolar concentrations in the vacuole to increase osmotic pressure (White and Broadley, 2003). In plant species that synthesize oxalate in response to nitrate reduction, the formation of Ca oxalate in the vacuole, or the precipitation of Ca oxalate in the apoplast, is important for the maintenance of $[Ca^{2+}]_{cyt}$ homeostasis (Kinzel, 1989), and the precipitation of Ca oxalate in the vacuole allows the accumulation of oxalate in nitrate-fed plants without an increase in osmotic pressure (Osmond, 1967). Conversely, the ability of calciotrophe species to utilize soluble Ca as an osmoticum is thought to be an adaptation to dry calcareous environments (Kinzel, 1982).

5.8 SYMPTOMS OF PLANTS SUPPLIED INSUFFICIENT OR EXCESS CALCIUM

As with all essential mineral elements, Ca deficiency results in biochemical and physiological disorders, aberrant development, and reduced yields (White and Broadley, 2003; Pilbeam and Morley, 2007). The symptoms of Ca deficiency reflect the roles of Ca^{2+} in the cell wall structure, maintaining membrane integrity and coordinating responses to developmental signals and environmental challenges. They include chlorosis and, ultimately, necrosis of the youngest leaves and shoot apices, weakened stems, wilting, malformation, and necrosis of fruit and tubers and, in some plants, premature flower dehiscence and failure to set seed (Bould et al., 1983; Bergmann, 1992; Pilbeam and Morley, 2007). Calcium requirements differ between plant species and largely reflect the Ca requirement for cell wall structure (Section 5.5). Thus, tissue Ca concentrations considered adequate in cereals and grasses are generally lower than those considered adequate for other crops (Table 5.2; Bergmann, 1992; Mills and Jones, 1996; Pilbeam and Morley, 2007; Römheld, 2012). The tissue location and occurrence of Ca-deficiency symptoms reflect the immobility of Ca in the phloem (Section 5.6). Thus, phloem-fed tissues and enclosed tissues with a low transpiration rate are most susceptible to Ca deficiency. The susceptibility of crops to Ca-deficiency disorders is genetically determined, and plant varieties that are less susceptible to Ca-deficiency disorders can be developed through breeding programs (Clark, 1983; Hochmuth, 1984; Horst, 1987; Caines and Shennan, 1999; Ho and White, 2005; Pilbeam and Morley, 2007).

Tissue analyses can be used to assess crop Ca status, but since Ca is immobile in the phloem and leaf Ca concentrations generally increase with leaf age (Section 5.6), the diagnostic tissue for Ca analysis must be chosen wisely. Tissue analyses can indicate the general Ca nutritional status of a crop, but since most Ca-deficiency disorders occur rapidly in developing tissues in response to their immediate Ca supply (Section 5.6), it is difficult to predict these from tissue analyses.

Routine spraying of Ca salts on valuable commodities is, therefore, advocated (Defra, 2010; Fernández et al., 2013). Analysis of Ca in fruit at harvest, or even specific parts of a fruit, provides the best indication of their quality and storage properties and can be used to inform the postharvest application of Ca salts (Defra, 2010).

Although Ca deficiency is rare in field-grown crops, it can occur in crops grown on soils prone to Ca^{2+} leaching or in crops grown on acid soils (McLaughlin and Wimmer, 1999; White and Broadley, 2003; White and Greenwood, 2013). In addition, several costly Ca-deficiency disorders occur in horticulture when the immediate Ca supply is insufficient to meet the growth requirements of developing tissues (Shear, 1975; White et al., 2003). Classic Ca-deficiency disorders include tipburn in a wide range of plants, blackheart in celery (*Apium graveolens* Pers.), bract necrosis in poinsettia (*Euphorbia pulcherrima* Willd. ex Klotzsch), water core and bitter pit in apples (*Malus domestica* Borkh.), empty pod of peanut (*Arachis hypogaea* L.), blossom-end rot in tomato (*Lycopersicon esculentum* Mill. syn. *Solanum lycopersicum* L.), pepper (*Capsicum annuum* L.), and watermelon (*Citrullus lanatus* Matsum. and Nakai) and the cracking of orchard fruits after irrigation or rainfall (Figure 5.6; Shear, 1975; Bould et al., 1983; White and Broadley, 2003; Ho and White, 2005; Pilbeam and Morley, 2007).

Several of these disorders are related to the role of Ca in the cell wall structure. For example, the cracking of cherries, tomatoes, and apples, upon hypoosmotic shock in fruit lacking sufficient Ca, is directly related to a weakness in the cell wall structure (Shear, 1975; White and Broadley, 2003). The structural integrity of the cell wall also provides resistance to insects, nematodes, and many bacterial and fungal infections (Huber et al., 2012). In this context, Ca not only provides structural integrity to the cell wall, but Ca^{2+} also inhibits many pectolytic enzymes released by bacteria and fungi to degrade the middle lamella of the cell wall (Wehr et al., 2004; Huber et al., 2012). Similarly, the losses of fleshy fruits and tubers during storage, caused by accelerated softening and susceptibility to fungal and bacterial infections, are related directly to cell wall properties and can be reduced by increasing cell wall Ca concentration by postharvest treatment with Ca salts (Pilbeam and Morley, 2007; Dayod et al., 2010; Oms-Oliu et al., 2010; Aghdam et al., 2012; Hawkesford et al., 2012). The role of Ca^{2+} in maintaining the integrity of cell membranes can be illustrated by the increased susceptibility to environmental stresses that cause membrane damage and solute leakage, such as chilling injury and heat stress, of plants lacking Ca (Kleinhenza and Palta, 2002; Lurie, 2009).

Since cellular responses to many environmental challenges are coordinated by $[Ca^{2+}]_{cyt}$ perturbations (Section 5.7.2), several Ca-deficiency disorders, including tipburn of leafy vegetables and blossom-end rot of solanaceous fruit, are thought to be, at least in part, a consequence of aberrant $[Ca^{2+}]_{cyt}$ homeostasis. In these disorders, there is sometimes no absolute Ca deficiency even in the affected tissues (Ho and White, 2005; de Freitas et al., 2011; Römheld, 2012). For example, although the leaves of transgenic tobacco plants overexpressing *sAtCAX1* have greater tissue Ca concentrations than wild-type plants, they exhibit tipburn (Hirschi, 2001), and the fruits of transgenic tomato plants overexpressing *sAtCAX1* have greater tissue Ca concentrations than wild-type plants but are more prone to blossom-end rot (Park et al., 2005a; de Freitas et al., 2011). Similarly, necrosis in Ca-deficient tissues is probably also partly a consequence of aberrant $[Ca^{2+}]_{cyt}$ signaling.

Although Ca toxicity can determine the survival of plants on well-fertilized, calcareous soils, it is sensitive to P and Fe deficiencies that appear to determine the natural flora of these soils (Lee, 1999; Pilbeam and Morley, 2007). Many plants adapted to calcareous soils can restrict their Ca uptake, accumulate large amounts of Ca in particular cell types, or precipitate Ca in the apoplast to avoid compromising energy metabolism or stomatal function (Sections 5.3 and 5.6). Similarly, when horticultural plants are supplied excessive Ca, they often exhibit similar strategies to avoid Ca toxicity. A common symptom of excessive Ca supply in tomato, for example, is the occurrence of *gold specks*, which are caused by the precipitation of Ca oxalate crystals (White and Broadley, 2003).

ACKNOWLEDGMENTS

I thank Dr. Timothy George and Dr. Lionel Dupuy for their comments on a draft version of this chapter. This work was supported by the Rural and Environment Science and Analytical Services Division of the Scottish Government through Workpackage 3.3 (2011–2016).

REFERENCES

Aghdam, M.S., M.B. Hassanpouraghdam, G. Paliyath, and B. Farmani. 2012. The language of calcium in post-harvest life of fruits, vegetables and flowers. *Sci. Hortic.* 144:102–115.

Alexandre, J., J.P. Lassalles, and R.T. Kado. 1990. Opening of Ca^{2+} channels in isolated red beet root vacuole membrane by inositol 1,4,5-trisphosphate. *Nature* 343:567–570.

Allen, G.J., S.P. Chu, C.L. Harrington et al. 2001. A defined range of guard cell calcium oscillation parameters encodes stomatal movements. *Nature* 411:1053–1057.

Amtmann, A. and M.R. Blatt. 2008. Regulation of macronutrient transport. *New Phytol.* 181:35–52.

Antoine, A.F., C. Dumas, J.-E. Faure, J.A. Feijó, and M. Rougier. 2001. Egg activation in flowering plants. *Sex. Plant Reprod.* 14:21–26.

Asher, C.J. and D.G. Edwards. 1983. Modern solution culture techniques. In *Encyclopedia of Plant Physiology, New Series*, vol. 15A, eds. A. Läuchli and R.L. Bieleski, pp. 94–119. Berlin, Germany: Springer-Verlag.

Baker, A.J.M. 1978. The uptake of zinc and calcium from solution culture by zinc tolerant and non tolerant *Silene maritima* With. in relation to calcium supply. *New Phytol.* 81:321–330.

Barber, S.A. 1995. *Soil Nutrient Bioavailability: A Mechanistic Approach*, 2nd edn. New York: John Wiley & Sons.

Barber, S.A. and P.G. Ozanne. 1970. Autoradiographic evidence for the differential effect of four plant species in altering the calcium content of the rhizosphere soil. *Soil Sci. Soc. Am. Proc.* 34:635–637.

Batistič, O. and J. Kudla. 2010. Calcium: Not just another ion. In *Plant Cell Monographs 17, Cell Biology of Metals and Nutrients*, eds. R. Hell and R.R. Mendel, pp. 17–54. Berlin, Germany: Springer-Verlag.

Batistič, O. and J. Kudla. 2012. Analysis of calcium signaling pathways in plants. *Biochim. Biophys. Acta* 1820:1283–1293.

Bauer, P., R. Elbaum, and I.M. Weiss. 2011. Calcium and silicon mineralization in land plants: Transport, structure and function. *Plant Sci.* 180:746–756.

Baxter, I., P.S. Hosmani, A. Rus et al. 2009. Root suberin forms an extracellular barrier that affects water relations and mineral nutrition in *Arabidopsis*. *PLoS Genet.* 5:1–12.

Baxter, I., M. Ouzzani, S. Orcun, B. Kennedy, S.S. Jandhyala, and D.E. Salt. 2007. Purdue ionomics information management system. An integrated functional genomics platform. *Plant Physiol.* 143:600–611.

Bergmann, W. 1992. *Nutritional Disorders of Plants: Development, Visual and Analytical Diagnosis*. Jena, Germany: Gustav Fischer.

Birnbaum, K., D.E. Shasha, J.Y. Wang et al. 2003. A gene expression map of the *Arabidopsis* root. *Science* 302:1956–1960.

Blanco, A., V. Fernández, and J. Val. 2010. Improving the performance of calcium-containing spray formulations to limit the incidence of bitter pit in apple (*Malus × domestica* Borkh.). *Sci. Hortic.* 127:23–28.

Bolaños, L., M. Martín, A. El-Hamdaoui, R. Rivilla, and I. Bonilla. 2006. Nitrogenase inhibition in nodules from pea plants grown under salt stress occurs at the physiological level and can be alleviated by B and Ca. *Plant Soil* 280:135–142.

Bonza, M.C. and M.I. De Michelis. 2011. The plant Ca^{2+}-ATPase repertoire: Biochemical features and physiological functions. *Plant Biol.* 13:421–430.

Borsics, T., D. Webb, C. Andeme-Ondzighi, L.A. Staehelin, and D.A. Christopher. 2007. The cyclic nucleotide-gated calmodulin binding channel AtCNGC10 localizes to the plasma membrane and influences numerous growth responses and starch accumulation in *Arabidopsis thaliana*. *Planta* 225:563–573.

Bould, C., E.J. Hewitt, and P. Needham. 1983. *Diagnosis of Mineral Disorders in Plants*, Vol. 1: *Principles*. London, U.K.: HMSO.

Boursiac, Y. and J.F. Harper. 2007. The origin and function of calmodulin regulated Ca^{2+} pumps in plants. *J. Bioenerg. Biomembr.* 39:409–414.

Britto, D.T. and H.J. Kronzucker. 2003. Ion fluxes and cytosolic pool sizes: Examining fundamental relationships in transmembrane flux regulation. *Planta* 217:490–497.

Broadley, M.R., H.C. Bowen, H.L. Cotterill et al. 2003. Variation in the shoot calcium content of angiosperms. *J. Exp. Bot.* 54:1431–1446.

Broadley, M.R., H.C. Bowen, H.L. Cotterill et al. 2004. Phylogenetic variation in the shoot mineral concentration of angiosperms. *J. Exp. Bot.* 55:321–336.

Broadley, M.R., J.P. Hammond, G.J. King et al. 2008. Shoot calcium (Ca) and magnesium (Mg) concentrations differ between subtaxa, are highly heritable, and associate with potentially pleiotropic loci in *Brassica oleracea. Plant Physiol.* 146:1707–1720.

Broadley, M.R., J.P. Hammond, G.J. King et al. 2009. Biofortifying *Brassica* with calcium (Ca) and magnesium (Mg). In *Proceedings of the International Plant Nutrition Colloquium XVI*, August 26–30, 2009, Sacramento, USA. Paper 1256. http://escholarship.org/uc/item/9936g2vv.

Broadley, M.R. and P.J. White. 2012. Some elements are more equal than others: Soil-to-plant transfer of radiocaesium and radiostrontium, revisited. *Plant Soil* 355:23–27.

Busse, J.S. and J.P. Palta. 2006. Investigating the in vivo calcium transport path to developing potato tuber using ^{45}Ca: A new concept in potato tuber calcium nutrition. *Physiol. Plant.* 128:313–323.

Caines, A.M. and C. Shennan. 1999. Growth and nutrient composition of Ca^{2+} use efficient and Ca^{2+} use inefficient genotypes of tomato. *Plant Physiol. Biochem.* 37:559–567.

Capoen, W., J. Sun, D. Wysham et al. 2011. Nuclear membranes control symbiotic calcium signaling of legumes. *Proc. Nat. Acad. Sci. USA* 108:14348–14353.

Carpita, N. and M. McCann. 2000. The cell wall. In *Biochemistry and Molecular Biology of Plants*, eds. B.B. Buchanan, W. Gruissem, and R.L. Jones, pp. 52–108. Rockville, MD: American Society of Plant Physiologists.

Casanas, F., E. Perez-Vega, A. Almirall, M. Plans, J. Sabate, and J.J. Ferreira. 2013. Mapping of QTL associated with seed chemical content in a RIL population of common bean (*Phaseolus vulgaris* L.). *Euphytica* 192:279–288.

Catalá, R., E. Santos, J.M. Alonso, J.R. Ecker, J.M. Martínez-Zapater, and J. Salinas. 2003. Mutations in the Ca^{2+}/H^+ transporter CAX1 increase CBF/DREB1 expression and the cold-acclimation response in *Arabidopsis. Plant Cell* 15:2940–2951.

Chan, C.W., L.M. Schorrak, R.K. Smith, A.F. Bent, and M.R. Sussman. 2003. A cyclic nucleotide-gated ion channel, CNGC2, is crucial for plant development and adaptation to calcium stress. *Plant Physiol.* 132:728–731.

Cheng, N.-H., J.K. Pittman, B.J. Barkla, T. Shigaki, and K.D. Hirschi. 2003. The *Arabidopsis cax1* mutant exhibits impaired ion homeostasis, development, and hormonal responses and reveals interplay among vacuolar transporters. *Plant Cell* 15:347–364.

Cheng, N.-H., J.K. Pittman, T. Shigaki et al. 2005. Functional association of *Arabidopsis* CAX1 and CAX3 is required for normal growth and ion homeostasis. *Plant Physiol.* 138:2048–2060.

Cholewa, E. and C.A. Peterson. 2004. Evidence for symplastic involvement in the radial movement of calcium in onion roots. *Plant Physiol.* 134:1793–1802.

Clark, G.B., R.O. Morgan, M.-P. Fernandez, and S.J. Roux. 2012. Evolutionary adaptation of plant annexins has diversified their molecular structures, interactions and functional roles. *New Phytol.* 196:695–712.

Clark, R.B. 1983. Plant genotype differences in the uptake, translocation, accumulation, and use of mineral elements required for plant growth. *Plant Soil* 72:175–196.

Clarkson, D.T. 1984. Calcium transport between tissues and its distribution in the plant. *Plant Cell Environ.* 7:449–456.

Clemens, S., D.M. Antosiewicz, J.M., Ward, D.P. Schachtman, and J.I. Schroeder. 1998. The plant cDNA *LCT1* mediates the uptake of calcium and cadmium in yeast. *Proc. Nat. Acad. Sci. USA* 95:12043–12048.

Conn, S., P. Berninger, M.R. Broadley, and M. Gilliham. 2012. Exploiting natural variation to uncover candidate genes that control element accumulation in *Arabidopsis thaliana. New Phytol.* 193:859–866.

Conn, S. and M. Gilliham. 2010. Comparative physiology of elemental distribution in plants. *Ann. Bot.* 105:1081–1102.

Conn, S.J., M. Gilliham, S. Athman et al. 2011. Cell-specific vacuolar calcium storage mediated by *CAX1* regulated apoplastic calcium concentration, gas exchange, and plant productivity in *Arabidopsis. Plant Cell* 23:240–255.

Cooper, J. and H. Scherer. 2012. Nitrogen fixation. In *Marschner's Mineral Nutrition of Higher Plants*, 3rd edn, ed. P. Marschner, pp. 389–408. London, U.K.: Academic Press.

Dalchau, N., K.E. Hubbard, F.C. Robertson et al. 2010. Correct biological timing in *Arabidopsis* requires multiple light-signaling pathways. *Proc. Nat. Acad. Sci. USA* 107:13171–13176.

Dayod, M., S.D. Tyerman, R.A. Leigh, and M. Gilliham. 2010. Calcium storage in plants and the implications for calcium biofortification. *Protoplasma* 247:215–231.

de Freitas, S.T., M. Padda, Q. Wu, S. Park, and E.J. Mitcham. 2011. Dynamic alternations in cellular and molecular components during blossom-end rot development in tomatoes expressing sCAX1, a constitutively active Ca^{2+}/H^+ antiporter from *Arabidopsis. Plant Physiol.* 156:844–855.

Delhaize, E., B.D. Gruber, and P.R. Ryan. 2007. The roles of organic anion permeases in aluminium resistance and mineral nutrition. *FEBS Lett.* 581:2255–2262.

Demidchik, V., H.C. Bowen, F.J.M. Maathuis et al. 2002. *Arabidopsis thaliana* root non-selective cation channels mediate calcium uptake and are involved in growth. *Plant J.* 32:799–808.

Demidchik, V. and F.J.M. Maathuis. 2007. Physiological roles of nonselective cation channels in plants: From salt stress to signalling and development. *New Phytol.* 175:387–404.

Department for Environment, Food and Rural Affairs, UK [Defra]. 2010. *Fertiliser Manual (RB209)*, 8th edition. Norwich, U.K.: The Stationery Office.

De Silva, D.L.R., A.M. Hetherington, and T.A. Mansfield. 1996. Where does all the calcium go? Evidence of an important regulatory role for trichomes in two calcicoles. *Plant Cell Environ.* 19:880–886.

De Silva, D.L.R., A.M. Hetherington, and T.A. Mansfield. 1998. The regulation of apoplastic calcium in relation to intracellular signalling in stomatal guard cells. *Z. Pflanzenernähr. Bodenk.* 161:533–539.

Dietrich, P., U. Anschütz, A. Kugler, and D. Becker. 2010. Physiology and biophysics of plant ligand-gated ion channels. *Plant Biol.* 12:80–93.

Ding, G., M. Yang, Y. Hu et al. 2010. Quantitative trait loci affecting seed mineral concentrations in *Brassica napus* grown with contrasting phosphorus supplies. *Ann. Bot.* 105:1221–1234.

Dodd, A.N., M.J. Gardner, C.T. Hotta et al. 2007. A cADPR-based feedback loop modulates the *Arabidopsis* circadian clock. *Science* 318:1789–1792.

Dodd, A.N., J. Kudla, and D. Sanders. 2010. The language of calcium signaling. *Annu. Rev. Plant Biol.* 61:593–620.

Drew, M.C. and O. Biddulph. 1971. Effect of metabolic inhibitors and temperature on uptake and translocation of ^{45}Ca and ^{42}K by intact bean plants. *Plant Physiol.* 48:426–432.

Dumas, C. and T. Gaude. 2006. Fertilization in plants: Is calcium a key player? *Semin. Cell Dev. Biol.* 17:244–253.

Dunlop, J. 1973. The kinetics of calcium uptake by roots. *Planta* 112:159–167.

Edmond, C., T. Shigaki, S. Ewert et al. 2009. Comparative analysis of CAX2-like cation transporters indicates functional and regulatory diversity. *Biochem. J.* 418:145–154.

English, J.E. and A.V. Barker. 1987. Ion interactions in calcium-efficient and calcium-inefficient tomato lines. *J. Plant Nutr.* 10:857–869.

Epstein, E. 1961. The essential role of calcium in selective cation transport by plant cells. *Plant Physiol.* 36:437–444.

Epstein, E. and J.E. Leggett. 1954. The absorption of alkaline earth cations by barley roots—Kinetics and mechanism. *Am. J. Bot.* 41:785–791.

Fageria, N.K. 2009. *The Use of Nutrients in Crop Plants*. Boca Raton, FL: CRC Press.

Fageria, N.K., V.C. Baligar, and C.A. Jones. 2011. *Growth and Mineral Nutrition of Field Crops*. Boca Raton, FL: CRC Press.

Fernández, V. and T. Eichert. 2009. Uptake of hydrophilic solutes through plant leaves: Current state of knowledge and perspectives of foliar fertilization. *Crit. Rev. Plant Sci.* 28:36–68.

Fernández, V., T. Sotiropoulos, and P. Brown. 2013. *Foliar Fertilization: Scientific Principles and Field Practices*. Paris, France: International Fertilizer Industry Association.

Fink, S. 1991. The micromorphological distribution of bound calcium in needles of Norway spruce [*Picea abies* (L.) Karst.]. *New Phytol.* 119:33–40.

Forde, B.G., S.R. Cutler, N. Zaman, and P.J. Krysan. 2013. Glutamate signalling via a MEKK1 kinase-dependent pathway induces changes in *Arabidopsis* root architecture. *Plant J.* 75:1–10.

Forde, B.G. and P. Walch-Liu. 2009. Nitrate and glutamate as environmental cues for behavioural responses in plant roots. *Plant Cell Environ.* 32:682–693.

Foreman, J., V. Demidchik, J.H. Bothwell et al. 2003. Reactive oxygen species produced by NADPH oxidase regulate plant cell growth. *Nature* 422:442–446.

Franceschi, V.R. and P.A. Nakata. 2005. Calcium oxalate in plants: Formation and function. *Annu. Rev. Plant Biol.* 56:41–71.

Frietsch, S., Y.-F. Wang, C. Sladek et al. 2007. A cyclic nucleotide-gated channel is essential for polarized tip growth of pollen. *Proc. Nat. Acad. Sci. USA* 104:14531–14536.

Garcia-Oliveira, A.L., L. Tan, Y. Fu, and C. Sun. 2009. Genetic identification of quantitative trait loci for contents of mineral nutrients in rice grain. *J. Integr. Plant Biol.* 51:84–92.

Gaymard, F., G. Pilot, B. Lacombe et al. 1998. Identification and disruption of a plant shaker-like outward channel involved in K^+ release into the xylem sap. *Cell* 94:647–655.

Gelin, J.P., S. Forster, K.F. Grafton, P.E. McClean, and G.A. Kooas-Cifuentes. 2007. Analysis of seed zinc and other minerals in a recombinant inbred population of navy bean (*Phaseolus vulgaris* L.). *Crop Sci.* 47:1361–1366.

George, E., W.J. Horst, and E. Neumann. 2012. Adaptation of plants to adverse chemical soil conditions. In *Marschner's Mineral Nutrition of Higher Plants*, 3rd edn, ed. P. Marschner, pp. 409–472. London, U.K.: Academic Press.

Gilliham, M., M. Dayod, B.J. Hocking et al. 2011. Calcium delivery and storage in plant leaves: Exploring the link with water flow. *J. Exp. Bot.* 62:2233–2250.

Granqvist, E., D. Wysham, S. Hazledine et al. 2012. Buffering capacity explains signal variation in symbiotic calcium oscillations. *Plant Physiol.* 160:2300–2310.

Guo, K.M., O. Babourina, D.A. Christopher, T. Borsic, and Z. Rengel. 2010. The cyclic nucleotide-gated channel AtCNGC10 transports Ca^{2+} and Mg^{2+} in *Arabidopsis*. *Physiol. Plant.* 139:303–312.

Guzmán-Maldonado, S.H., O. Martínez, J.A. Acosta-Gallegos, F. Guevara-Lara, and O. Paredes-López. 2003. Putative quantitative trait loci for physical and chemical components of common bean. *Crop Sci.* 43:1029–1035.

Hampton, C.R., M.R. Broadley, and P.J. White. 2005. Short review: The mechanisms of radiocaesium uptake by *Arabidopsis* roots. *Nukleonika* 50:S3–S8.

Harada, A. and K. Shimazaki. 2007. Phototropins and blue light-dependent calcium signaling in higher plants. *Photochem. Photobiol.* 83:102–111.

Haswell, E.S., R. Peyronnet, H. Barbier-Brygoo, E.M. Meyerowitz, and J.M. Frachisse. 2008. Two MscS homologs provide mechanosensitive channel activities in the *Arabidopsis* root. *Curr. Biol.* 18:730–734.

Hawkesford, M., W. Horst, T. Kichey et al. 2012. Functions of macronutrients. In *Marschner's Mineral Nutrition of Higher Plants*, 3rd edn, ed. P. Marschner, pp. 135–189. London, U.K.: Academic Press.

Haydon, M.J., L.J. Bell, and A.R.R. Webb. 2011. Interactions between plant circadian clocks and solute transport. *J. Exp. Bot.* 62:2333–2348.

Hayter, M.L. and C.A. Peterson. 2004. Can Ca^{2+} fluxes to the root xylem be sustained by Ca^{2+}-ATPases in exodermal and endodermal plasma membranes? *Plant Physiol.* 136:4318–4325.

Heazlewood, J.L., R.E. Verboom, J. Tonti-Filippini, I. Small, and A.H. Millar. 2007. SUBA: The *Arabidopsis* subcellular database. *Nucleic Acids Res.* 35:D213–D218.

Hepler, P.K. 2005. Calcium: A central regulator of plant growth and development. *Plant Cell* 17:2142–2155.

Hepler, P.K., J.G. Kunkel, C.M. Rounds, and L.J. Winship. 2012. Calcium entry into pollen tubes. *Trends Plant Sci.* 17:32–38.

Hepler, P.K. and L.J. Winship. 2010. Calcium at the cell wall-cytoplast interface. *J. Integr. Plant Biol.* 52:147–160.

Hirschi, K. 2001. Vacuolar H^+/Ca^{2+} transport: Who's directing the traffic? *Trends Plant Sci.* 6:100–104.

Ho, L. and P.J. White. 2005. A cellular hypothesis for the induction of blossom-end rot in tomato fruit. *Ann. Bot.* 95:571–581.

Hochmuth, G.J. 1984. Variation in calcium efficiency among strains of cauliflower. *J. Am. Soc. Hortic. Sci.* 109:667–672.

Hocking, P.J. and J.S. Pate. 1978. Accumulation and distribution of mineral elements in annual lupins *Lupinus albus* L. and *Lupinus angustifolius* L. *Aust. J. Agric. Res.* 29:267–280.

Hodson, M.E. and E. Donner. 2013. Managing adverse soil chemical environments. In *Soil Conditions and Plant Growth*, eds. P.J. Gregory and S. Nortcliff, pp. 195–237. Oxford, U.K.: Blackwell Publishing.

Horst, W.J. 1987. Aluminium tolerance and calcium efficiency of cowpea genotypes. *J. Plant Nutr.* 10:1121–1129.

Horst, W.J., C. Currle, and A.H. Wissemeier. 1992. Differences in calcium efficiency between cowpea (*Vigna unguiculata* (L.) Walp.) cultivars. *Plant Soil* 146:45–54.

Huang, J.W., D.L. Grunes, and L.V. Kochian. 1992. Aluminium effects on the kinetics of calcium uptake into cells of the wheat root apex. *Planta* 188:414–421.

Huber, D., V. Römheld, and M. Weinmann. 2012. Relationship between nutrition, plant diseases and pests. In *Marschner's Mineral Nutrition of Higher Plants*, 3rd edn, ed. P. Marschner, pp. 283–298. London, U.K.: Academic Press.

Islam, A.K.M.S., C.J. Asher, and D.G. Edwards. 1987. Response of plants to calcium concentration in flowing solution culture with chloride or sulphate as the counter-ion. *Plant Soil* 98:377–395.

Islam, A.K.M.S., D.G. Edwards, and C.J. Asher. 1980. pH optima for crop growth. Results of a flowing solution culture experiment with six species. *Plant Soil* 54:339–357.

Jaillard, B. 1985. Activite racinaire et rhizostructures en milieu carbonate. *Pedologie* 35:297–313.

Jaillard, B., A. Guyon, and A.F. Maurin. 1991. Structure and composition of calcified roots, and their identification in calcareous soils. *Geoderma* 50:197–210.

Jaiswal, J.K. 2001. Calcium—How and why? *J. Biosci.* 26:357–363.

Jarvis, S.C. and D.J. Hatch. 1985. Rates of hydrogen ion efflux by nodulated legumes grown in flowing solution culture with continuous pH monitoring and adjustment. *Ann. Bot.* 55:41–51.

Jeffries, R.L. and A.J. Willis. 1964. Studies on the calcicole-calcifuge habit. II. The influence of calcium on the growth and establishment of four species in soil and sand cultures. *J. Ecol.* 52:691–707.

Johnson, C.H., M.R. Knight, T. Kondo et al. 1995. Circadian oscillations of cytosolic and chloroplastic free calcium in plants. *Science* 269:1863–1865.

Kang, J.M. and F.J. Turano. 2003. The putative glutamate receptor 1.1 (AtGLR1.1) functions as a regulator of carbon and nitrogen metabolism in *Arabidopsis thaliana*. *Proc. Nat. Acad. Sci. USA* 100:6872–6877.

Kannan, S. 2010. Foliar fertilization for sustainable crop production. In *Sustainable Agriculture Reviews*, Vol. 4: *Genetic Engineering, Biofertilisation, Soil Quality and Organic Farming*, ed. E. Lichtfouse, pp. 371–402. Dordrecht, the Netherlands: Springer-Verlag.

Kaplan, B., T. Sherman, and H. Fromm. 2007. Cyclic nucleotide-gated channels in plants. *FEBS Lett.* 581:2237–2246.

Kärenlampi, S. and P.J. White. 2009. Potato proteins, lipids and minerals. In *Advances in Potato Chemistry and Technology*, ed. J. Singh, pp. 99–126. Oxford, U.K.: Elsevier.

Karley, A.J., R.A. Leigh, and D. Sanders. 2000a. Where do all the ions go? The cellular basis of differential ion accumulation in leaf cells. *Trends Plant Sci.* 5:465–470.

Karley, A.J., R.A. Leigh, and D. Sanders. 2000b. Differential ion accumulation and ion fluxes in the mesophyll and epidermis of barley. *Plant Physiol.* 122:835–844.

Karley, A.J. and P.J. White. 2009. Moving cationic minerals to edible tissues: Potassium, magnesium, calcium. *Curr. Opin. Plant Biol.* 12:291–298.

Kauss, H. 1987. Some aspects of calcium-dependent regulation in plant metabolism. *Annu. Rev. Plant Physiol.* 38:47–72.

Keltjens, W.G. and K. Tan. 1993. Interactions between aluminium, magnesium and calcium with different monocotyledonous and dicotyledonous plant species. *Plant Soil* 155/156:458–488.

Kerton, M., H.J. Newbury, D. Hand, and J. Pritchard. 2009. Accumulation of calcium in the centre of leaves of coriander (*Coriandrum sativum* L.) is due to an uncoupling of water and ion transport. *J. Exp. Bot.* 60:227–235.

Kim, C.K., J.-S. Han, H.-S. Lee et al. 2006. Expression of an *Arabidopsis CAX2* variant in potato tubers increases calcium levels with no accumulation of manganese. *Plant Cell Rep.* 25:1226–1232.

Kim, K.-N. 2013. Stress responses mediated by the CBL calcium sensors in plants. *Plant Biotechnol. Rep.* 7:1–8.

Kim, T., M. Bohmer, H. Hu, N. Nishimura, and J.I. Schroeder. 2010. Guard cell signal transduction network: Advances in understanding abscisic acid, CO_2, and Ca^{2+} signaling. *Annu. Rev. Plant. Biol.* 61:561–591.

Kinzel, H. 1982. *Pflanzenökologie und Mineralstoffwechsel*. Stuttgart, Germany: Ulmer.

Kinzel, H. 1989. Calcium in the vacuoles and cell walls of plant tissue. Forms of deposition and their physiological and ecological significance. *Flora* 182:99–125.

Kinzel, H. and I. Lechner. 1992. The specific mineral metabolism of selected plant species and its ecological implications. *Bot. Acta* 105:355–361.

Kleinhenza, M.D. and J. Palta. 2002. Root zone calcium modulates the response of potato plants to heat stress. *Physiol. Plant.* 15:111–118.

Knight, M.R. and H. Knight. 2012. Low-temperature perception leading to gene expression and cold tolerance in higher plants. *New Phytol.* 195:737–751.

Kochian, L., O.A. Hoekenga, and M.A. Piñeros. 2004. How do crop plants tolerate acid soils? Mechanisms of aluminum tolerance and phosphorous efficiency. *Annu. Rev. Plant Biol.* 55:459–493.

Konrad, K.R., M.M. Wudick, and J.A. Feijo. 2011. Calcium regulation of tip growth: New genes for old mechanisms. *Curr. Opin. Plant Biol.* 14:721–730.

Kopittke, P.M. and N.W. Menzies. 2007. A review of the use of the basic cation saturation ratio and the "ideal" soil. *Soil Sci. Soc. Am. J.* 71:259–265.

Kosuta, S., S. Hazledine, J. Sun et al. 2008. Differential and chaotic calcium signatures in the symbiosis signaling pathway of legumes. *Proc. Nat. Acad. Sci. USA* 105:9823–9828.

Krauss, A. and H. Marschner. 1975. Einflus des Calcium-Angebotes auf Wachstumsrte und Calcium-Gehalt von Kartoffelknollen. *Z. Pflanzenernähr. Bodenk.* 138:317–326.

Kudla, J., O. Batistič, and K. Hashimoto. 2010. Calcium signals: The lead currency of plant information processing. *Plant Cell* 22:541–563.

Kugler, A., B. Köhler, K. Palme, P. Wolff, and P. Dietrich. 2009. Salt-dependent regulation of a CNG channel subfamily in *Arabidopsis*. *BMC Plant Biol.* 9:140.

Kurusu, T., K. Kuchitsu, M. Nakano, Y. Nakayama, and H. Iida. 2013. Plant mechanosensing and Ca^{2+} transport. *Trends Plant Sci.* 18:227–233.

Laohavisit, A., A.T. Brown, P. Cicuta, and J.M. Davies. 2010. Annexins: Components of the calcium and reactive oxygen signalling network. *Plant Physiol.* 152:1824–1829.

Laohavisit, A. and J.M. Davies. 2011. Annexins. *New Phytol.* 189:40–53.

Laohavisit, A., S.L. Richards, L. Shabala et al. 2013. Salinity-induced calcium signalling and root adaptation in *Arabidopsis* require the calcium regulatory protein annexin1. *Plant Physiol.* 163:253–262.

Laohavisit, A., Z. Shang, L. Rubio et al. 2012. *Arabidopsis* Annexin1 mediates the radical-activated plasma membrane Ca^{2+}- and K^+-permeable conductance in root cells. *Plant Cell* 24:1522–1533.

Lazaroff, N. and M.G. Pitman. 1966. Calcium and magnesium uptake by barley seedlings. *Aust. J. Biol. Sci.* 19:991–1005.

Lecourieux, D., C. Mazars, N. Pauly, R. Ranjeva, and A. Pugin. 2002. Analysis and effects of cytosolic free calcium increases in response to elicitors in *Nicotiana plumbaginifolia* cells. *Plant Cell* 14:2627–2641.

Lecourieux, D., R. Raneva, and A. Pugin. 2006. Calcium in plant defence-signalling pathways. *New Phytol.* 171:249–269.

Lee, J.A. 1999. The calcicole-calcifuge problem revisited. *Adv. Bot. Res.* 29:1–30.

Lee, K., D. Thorneycroft, P. Achuthan, H. Hermjakob, and T. Ideker. 2010. Mapping plant interactomes using literature curated and predicted protein-protein interaction data sets. *Plant Cell* 22:997–1005.

Libert, B. and V.R. Franceschi. 1987. Oxalate in crop plants. *J. Agric. Food Chem.* 35:926–938.

Liebisch, F., J.F.J. Max, G. Heine, and W.J. Horst. 2009. Blossom end rot and fruit cracking of tomato grown in net-covered greenhouses in central Thailand can partly be corrected by calcium and boron sprays. *J. Plant Nutr. Soil Sci.* 172:140–150.

Löhnis, M.P. 1960. Effect of magnesium and calcium supply on the uptake of manganese by various crop plants. *Plant Soil* 12:339–376.

Loneragan, J.F. and K. Snowbal 1969. Calcium requirements of plants. *Aust. J. Agric. Res.* 20:465–478.

Loneragan, J.F., K. Snowball, and W.J. Simmons. 1968. Response of plants to calcium concentration in solution culture. *Aust. J. Agric. Res.* 19:845–857.

Love, J., A.N. Dodd, and A.R.R. Webb. 2004. Circadian and diurnal calcium oscillations encode photoperiodic information in *Arabidopsis*. *Plant Cell* 16:956–966.

Lu, K., L. Li, X. Zheng, Z. Zhang, T. Mou, and Z. Hu. 2008. Quantitative trait loci controlling Cu, Ca, Zn, Mn and Fe content in rice grains. *J. Genet.* 87:305–310.

Lurie, S. 2009. Stress physiology and latent damage. In *Postharvest Handling: A Systems Approach*, eds. W.J. Florkowski, R.L. Shewfelt, B. Brueckner, and S.E. Prussia, pp. 443–459. San Diego, CA: Academic Press.

Lux, A., M. Martinka, M. Vaculík, and P.J. White. 2011. Root responses to cadmium in the rhizosphere: A review. *J. Exp. Bot.* 62:21–37.

Lynch, J., P. Marschner, and Z. Rengel. 2012. Effect of internal and external factors on root growth and development. In *Marschner's Mineral Nutrition of Higher Plants*, 3rd edn, ed. P. Marschner, pp. 331–346. London, U.K.: Academic Press.

Ma, J.F., P.R. Ryan, and E. Delhaize. 2001. Aluminium tolerance in plants and the complexing role of organic acids. *Trends Plant Sci.* 6:273–278.

Ma, W., R. Ali, and G.A. Berkowitz. 2006. Characterization of plant phenotypes associated with loss-of-function of AtCNGC1, a plant cyclic nucleotide gated cation channel. *Plant Physiol. Biochem.* 44:494–505.

Ma, W. and G.A. Berkowitz. 2011. Ca^{2+} conduction by plant cyclic nucleotide gated channels and associated signaling components in pathogen defense signal transduction cascades. *New Phytol.* 190:566–572.

Ma, W., A. Smigel, R.K. Walker, W. Moeder, K. Yoshioka, and G.A. Berkowitz. 2010. Leaf senescence signaling: The Ca^{2+}-conducting *Arabidopsis* cyclic nucleotide gated channel2 acts through nitric oxide to repress senescence programming. *Plant Physiol.* 154:733–743.

Maas, E.V. 1969. Calcium uptake by excised maize roots and interactions with alkali cations. *Plant Physiol.* 44:985–989.

Maathuis, F.J.M. and A. Amtmann. 1999. K^+ nutrition and Na^+ toxicity: The basis of cellular K^+/Na^+ ratios. *Ann. Bot.* 84:123–133.

Macías, F., M.C. Arbestain, and W. Chesworth. 2008. Acid soils. In *The Encyclopedia of Soil Science*, ed. W. Chesworth, pp. 7–9. Dordrecht, the Netherland: Springer-Verlag.

Macklon, A.E.S. and A. Sim. 1981. Cortical cell fluxes and transport to the stele in excised root segments of *Allium cepa* L. 4. Calcium as affected by its external concentration. *Planta* 152:381–387.

Maksaev, G. and E.S. Haswell. 2012. MscS-Like10 is a stretch-activated ion channel from *Arabidopsis thaliana* with a preference for anions. *Proc. Nat. Acad. Sci. USA* 109:19015–19020.

Malhó, R., A. Moutinho, A. van der Luit, and A.J. Trewavas. 1998. Spatial characteristics of calcium signalling: The calcium wave as a basic unit in plant cell calcium signalling. *Philos. Trans. R. Soc. Lond. B Biol. Sci.* 353:1463–1473.

Marschner, P. and Z. Rengel. 2012. Nutrient availability in soils In *Marschner's Mineral Nutrition of Higher Plants*, 3rd edn, ed. P. Marschner, pp. 315–330. London, U.K.: Academic Press.

Martinoia, E., S. Meyer, A. De Angeli, and R. Nagy. 2012. Vacuolar transporters in their physiological context. *Annu. Rev. Plant Biol.* 63:183–213.

McAinsh, M.R. and J.K. Pittman. 2009. Shaping the calcium signature. *New Phytol.* 161:275–294.

McAinsh, M.R., A.R.R. Webb, J.E. Taylor, and A.M. Hetherington. 1995. Stimulus-induced oscillations in guard cell cytosolic free calcium. *Plant Cell* 7:1207–1219.

McLaughlin, S.B. and R. Wimmer. 1999. Calcium physiology and terrestrial ecosystem processes. *New Phytol.* 142:373–417.

Meyerhoff, O., K. Müller, M.R.G. Roelfsema et al. 2005. *AtGLR3.4*, a glutamate receptor channel-like gene is sensitive to touch and cold. *Planta* 222:418–427.

Michard, E., P.T. Lima, F. Borges et al. 2011. Glutamate receptor-like genes form Ca^{2+} channels in pollen tubes and are regulated by pistil D-serine. *Science* 332:434–437.

Miedema, H., V. Demidchik, A.-A. Véry, J.H.F. Bothwell, C. Brownlee, and J.M. Davies. 2008. Two voltage-dependent calcium channels co-exist in the apical plasma membrane of *Arabidopsis thaliana* root hairs. *New Phytol.* 179:378–385.

Miller, N.D., T.L. Durham Brooks, A.H. Assadi, and E.P. Spalding. 2010. Detection of a gravitropism phenotype *in glutamate receptor-like 3.3* mutants of *Arabidopsis thaliana* using machine vision and computation. *Genetics* 186:585–593.

Mills, H.A. and J.B. Jones. 1996. *Plant Analysis Handbook II*. Athens, GA; MicroMacro Publishing.

Mittler, R., A. Finka, and P. Goloubinoff. 2012. How do plants feel the heat? *Trends Biochem. Sci.* 37:118–125.

Monshausen, G.B. 2012. Visualizing Ca^{2+} signatures in plants. *Curr. Opin. Plant Biol.* 15:677–682.

Monshausen, G.B., N.D. Miller, A.S. Murphy, and S. Gilroy. 2011. Dynamics of auxin-dependent Ca^{2+} and pH signaling in root growth revealed by integrating high-resolution imaging with automated computer vision-based analysis. *Plant J.* 65:309–318.

Moore, C.A., H.C. Bowen, S. Scrase-Field, M.R. Knight, and P.J. White. 2002. The deposition of suberin lamellae determines the magnitude of cytosolic Ca^{2+} elevations in root endodermal cells subjected to cooling. *Plant J.* 30:457–466.

Moran, N. 2007. Osmoregulation of leaf motor cells. *FEBS Lett.* 581:2337–2347.

Morris, J., K.M. Hawthorne, T. Hotze, S.A. Abrams, and K.D. Hirschi. 2008. Nutritional impact of elevated calcium transport activity in carrots. *Proc. Nat. Acad. Sci. USA* 105:1431–1435.

Mortimer, J.C., A. Laohavisit, N. Macpherson et al. 2008. Annexins: Multifunctional components of growth and adaptation. *J. Exp. Bot.* 59:533–544.

Munns, R. and M. Tester. 2008. Mechanisms of salinity tolerance. *Ann. Rev. Plant Biol.* 59:651–681.

Nakagawa, Y., T. Katagiri, K. Shinozaki et al. 2007. *Arabidopsis* plasma membrane protein crucial for Ca^{2+} influx and touch sensing in roots. *Proc. Nat. Acad. Sci. USA* 104:3639–3644.

Nemchinov, L.G., L. Shabala, and S. Shabala. 2008. Calcium efflux as a component of the hypersensitive response of *Nicotiana benthamiana* to *Pseudomonas syringae*. *Plant Cell Physiol.* 49:40–46.

Neumann, G. and V. Römheld. 2012. Rhizosphere chemistry in relation to plant nutrition. In *Marschner's Mineral Nutrition of Higher Plants*, 3rd edn, ed. P. Marschner, pp. 347–368. London, U.K.: Academic Press.

Oldroyd, G.E.D. 2013. Speak, friend, and enter: Signalling systems that promote beneficial symbiotic associations in plants. *Nat. Rev. Microbiol.* 11:252–263.

Oms-Oliu, G., M.A. Rojas-Grau, L.A. Gonzalez et al. 2010. Recent approaches using chemical treatments to preserve quality of fresh-cut fruit: A review. *Postharvest Biol. Technol.* 57:139–148.

Osmond, C.B. 1967. Acid metabolism in *Atriplex*. I. Regulation in oxalate synthesis by the apparent excess cation absorption. *Aust. J. Biol. Sci.* 20:575–587.

Park, S., N.H. Cheng, J.K. Pittman et al. 2005a. Increased calcium levels and prolonged shelf life in tomatoes expressing *Arabidopsis* H^+/Ca^{2+} transporters. *Plant Physiol.* 139:1194–1206.

Park, S., M.P. Elless, J. Park et al. 2009. Sensory analysis of calcium-biofortified lettuce. *Plant Biotechnol. J.* 7:106–117.

Park, S., T.S. Kang, C.K. Kim et al. 2005b. Genetic manipulation for enhancing calcium content in potato tuber. *J. Agric. Food Chem.* 53:5598–5603.

Park, S., C. Kim, L. Pike, R. Smith, and K. Hirschi. 2004. Increased calcium in carrots by expression of an *Arabidopsis* H^+/Ca^{2+} transporter. *Mol. Breed.* 14:275–282.

Peiter, E. 2011. The plant vacuole: Emitter and receiver of calcium signals. *Cell Calcium* 50:120–128.

Peiter, E., F.J.M. Maathuis, L.M. Mills et al. 2005. The vacuolar Ca^{2+}-activated channel TPC1 regulates germination and stomatal movement. *Nature* 434:404–408.

Peleg, Z., I. Cakmak, L. Ozturk et al. 2009. Quantitative trait loci conferring grain mineral nutrient concentrations in durum wheat × wild emmer wheat RIL population. *Theor. Appl. Genet.* 119:353–369.

Pilbeam, D.J. and P.S. Morley. 2007. Calcium. In *Handbook of Plant Nutrition*, eds. A.V. Barker and D.J. Pilbeam, pp. 121–144. Boca Raton, FL: CRC Press.

Pittman, J.K. 2011. Vacuolar Ca^{2+} uptake. *Cell Calcium* 50:139–146.

Popescu, S.C., G.V. Popescu, S. Bachan et al. 2007. Differential binding of calmodulin-related proteins to their targets revealed through high-density *Arabidopsis* protein microarrays. *Proc. Nat. Acad. Sci. USA* 104:4730–4735.

Pottosin, I.I. and G. Schönknecht. 2007. Vacuolar calcium channels. *J. Exp. Bot.* 58:1559–1569.

Rahman, M.M. and O. Kawamura. 2011. Oxalate accumulation in forage plants: Some agronomic, climatic and genetic aspects. *Asian Aust. J. Anim. Sci.* 24:439–448.

Ranf, S., P. Wünnenberg, J. Lee et al. 2008. Loss of the vacuolar cation channel, AtTPC1, does not impair Ca^{2+} signals induced by abiotic and biotic stresses. *Plant J.* 53:287–299.

Reddy, A.S.N., G.S. Ali, H. Celesnik, and I.S. Day. 2011. Coping with stresses: Roles of calcium- and calcium/calmodulin-regulated gene expression. *Plant Cell* 23:2010–2032.

Rios, J.J., S. Ó Lochlainn, J. Devonshire et al. 2012. Distribution of calcium (Ca) and magnesium (Mg) in the leaves of *Brassica rapa* under varying exogenous Ca and Mg supply. *Ann. Bot.* 109:1081–1089.

Römheld, V. 2012. Diagnosis of deficiency and toxicity of nutrients. In *Marschner's Mineral Nutrition of Higher Plants*, 3rd edn, ed. P. Marschner, pp. 299–312. London, U.K.: Academic Press.

Roux, S.J. and I. Steinebrunne. 2007. Extracellular ATP: An unexpected role as a signaler in plants. *Trends Plant Sci.* 12:522–527.

Ryan, P.R. and E. Delhaize. 2012. Adaptations to aluminium toxicity. In *Plant Stress Physiology*, ed. S. Shabala, pp. 171–193. Wallingford, U.K.: CABI.

Sadowsky, M.J. 2005. Soil stress factors influencing symbiotic nitrogen fixation. In *Nitrogen Fixation in Agriculture, Forestry, Ecology, and the Environment*, eds. D. Werner and W.E. Newton, pp. 89–112. Dordrecht, the Netherlands: Springer-Verlag.

Sánchez-Barrena, M.J., M. Martínez-Ripoll, and A. Albert. 2013. Structural biology of a major signaling network that regulates plant abiotic stress: The CBL-CIPK mediated pathway. *Int. J. Mol. Sci.* 14:5734–5749.

Sanders, D., C. Brownlee, and J.F. Harper. 1999. Communicating with calcium. *Plant Cell* 11:691–706.

Schapire A.L., B. Voigt, J. Jasik et al. 2008. *Arabidopsis* synaptotagmin 1 is required for the maintenance of plasma membrane integrity and cell viability. *Plant Cell* 20:3374–3388.

Shabala, S.N., I.A. Newman, and J. Morris. 1997. Oscillations in H^+ and Ca^{2+} ion fluxes around the elongation region of corn roots and effects of external pH. *Plant Physiol.* 113:111–118.

Shear, C.B. 1975. Calcium-related disorders of fruits and vegetables. *HortScience* 10:361–365.

Shigaki, T. and K.D. Hirschi. 2006. Diverse functions and molecular properties emerging for CAX cation/H^+ exchangers in plants. *Plant Biol.* 8:419–429.

Shigaki, T., I. Rees, L. Nakhleh, and K.D. Hirschi. 2006. Identification of three distinct phylogenetic groups of CAX cation/proton antiporters. *J. Mol. Evol.* 63:815–825.

Sieberer, B.J., M. Chabaud, J. Fournier, A.C.J. Timmers, and D.G. Barker. 2011. A switch in Ca^{2+} spiking signature is concomitant with endosymbiotic microbe entry into cortical root cells of *Medicago truncatula*. *Plant J.* 69:822–830.

Sieberer, B.J., M. Chabaud, A.C. Timmers, A. Monin, J. Fournier, and D.G. Barker. 2009. A nuclear-targeted cameleon demonstrates intranuclear Ca^{2+} spiking in *Medicago truncatula* root hairs in response to rhizobial nodulation factors. *Plant Physiol.* 151:1197–1206.

Sleper, D.A., K.P. Vogel, K.H. Asay, and H.F. Mayland. 1989. Using plant breeding and genetics to overcome the incidence of grass tetany. *J. Anim. Sci.* 67:3456–3462.

Song, W.-Y., K.-S. Choi, D.A. Alexisa, E. Martinoia, and Y. Lee. 2011. *Brassica juncea* plant cadmium resistance 1 protein (BjPCR1) facilitates the radial transport of calcium in the root. *Proc. Nat. Acad. Sci. USA* 108:19808–19813.

Subramanian, N.K. 2012. Genetics of mineral accumulation in potato tubers. PhD thesis. University of Nottingham, Nottingham, U.K.

Taiz, L. and E. Zeigler. 2006. *Plant Physiology*, 4th edn. Sunderland, MA: Sinauer Associates.

Thompson, K., J.A. Parkinson, S.R. Band, and R.E. Spencer. 1997. A comparative study of leaf nutrient concentrations in a regional herbaceous flora. *New Phytol.* 136:679–689.

Thorpe, M.R., A.C.U. Furch, P.E.H. Minchin, J. Föller, A.J.E. Van Bel, and J.B. Hafke. 2010. Rapid cooling triggers forisome dispersion just before phloem transport stops. *Plant Cell Environ.* 33:259–271.

Tunc-Ozdemir, M., C. Rato, E. Brown et al. 2013. Cyclic nucleotide gated channels 7 and 8 are essential for male reproductive fertility. *PLoS ONE* 8:e55277.

Urquhart, W., K. Chin, H. Ung, W. Moeder, and K. Yoshioka. 2011. The cyclic nucleotide-gated channels AtCNGC11 and 12 are involved in multiple Ca^{2+}-dependent physiological responses and act in a synergistic manner. *J. Exp. Bot.* 62:3671–3682.

Val, J. and V. Fernández. 2011. In-season calcium-spray formulations improve calcium balance and fruit quality traits of peach. *J. Plant Nutr. Soil Sci.* 174:465–472.

Venkateshwaran, M., J.D. Volkening, M.R. Sussman, and J.M. Ané. 2013. Symbiosis and the social network of higher plants. *Curr. Opin. Plant Biol.* 16:118–127.

Verret, F., G. Wheeler, A.R. Taylor, G. Farnham, and C. Brownlee. 2010. Calcium channels in photosynthetic eukaryotes: Implications for evolution of calcium-based signalling. *New Phytol.* 187:23–43.

Vincill, E.D., A.M. Bieck, and E.P. Spalding. 2012. Ca^{2+} conduction by an amino acid-gated ion channel related to glutamate receptors. *Plant Physiol.* 159:40–46.

Vincill, E.D., A.E. Clarin, J.N. Molenda, and E.P. Spalding. 2013. Interacting glutamate receptor-like proteins in phloem regulate lateral root initiation in *Arabidopsis*. *Plant Cell* 25:1304–1313.

Volkov, A.G., T. Adesina, V.S. Markin, and E. Jovanov. 2008. Kinetics and mechanism of *Dionaea muscipula* trap closing. *Plant Physiol.* 146:694–702.

Volz, R.K., P.A. Alspach, D.J. Fletcher, and I.B. Ferguson. 2006. Genetic variation in bitter pit and fruit calcium concentrations within a diverse apple germplasm collection. *Euphytica* 149:1–10.

Vreugdenhil, D., M.G.M. Aarts, M. Koornneef, H. Nelissen, and W.H.O. Ernst. 2004. Natural variation and QTL analysis for cationic mineral content in seeds of *Arabidopsis thaliana*. *Plant Cell Environ.* 27:828–839.

Watanabe, T., M.R. Broadley, S. Jansen et al. 2007. Evolutionary control of leaf element composition in plants. *New Phytol.* 174:516–523.

Waters, B.M. and M.A. Grusak. 2008. Quantitative trait locus mapping for seed mineral concentrations in two *Arabidopsis thaliana* recombinant inbred populations. *New Phytol.* 179:1033–1047.

Webb, A.A.R., M.G. Larman, L.T. Montgomery, J.E. Taylor, and A.M. Hetherington. 2001. The role of calcium in ABA-induced gene expression and stomatal movements. *Plant J.* 26:351–362.

Wehr, J.B., N.W. Menzies, and F.P.C. Blamey. 2004. Inhibition of cell-wall autolysis and pectin degradation by cations. *Plant Physiol. Biochem.* 42:485–492.

Weinl, S. and J. Kudla. 2009. The CBL–CIPK Ca^{2+}-decoding signalling network: Function and perspectives. *New Phytol.* 184:517–528.

Welch, R.M. 1995. Micronutrient nutrition of plants. *Crit. Rev. Plant Sci.* 14:49–82.

Wheeler, G.L. and C. Brownlee. 2008. Ca^{2+} signalling in plants and green algae—Changing channels. *Trends Plant Sci.* 13:506–514.

White, P.J. 1998. Calcium channels in the plasma membrane of root cells. *Ann. Bot.* 81:173–183.

White, P.J. 1999. The molecular mechanism of sodium influx to root cells. *Trends Plant Sci.* 4:245–246.

White, P.J. 2000. Calcium channels in higher plants. *Biochim. Biophys. Acta* 1465:171–189.

White, P.J. 2001. The pathways of calcium movement to the xylem. *J. Exp. Bot.* 52: 891–899.

White, P.J. 2005. Calcium. In *Plant Nutritional Genomics*, ed. M.R. Broadley and P.J. White, pp. 66–86. Oxford, U.K.: Blackwell.

White, P.J. 2009. Depolarisation-activated calcium channels shape the calcium signatures induced by low-temperature stress. *New Phytol.* 183:6–8.

White, P.J. 2012a. Ion uptake mechanisms of individual cells and roots: Short distance transport. In *Marschner's Mineral Nutrition of Higher Plants*, 3rd edn, ed. P. Marschner, pp. 7–47. London, U.K.: Academic Press.

White, P.J. 2012b. Long-distance transport in the xylem and phloem. In *Marschner's Mineral Nutrition of Higher Plants*, 3rd edn, ed. P. Marschner, pp. 49–70. London, U.K.: Academic Press.

White, P.J. and M.R. Broadley. 2003. Calcium in plants. *Ann. Bot.* 92:487–511.

White, P.J. and M.R. Broadley. 2009. Biofortification of crops with seven mineral elements often lacking in human diets—Iron, zinc, copper, calcium, magnesium, selenium and iodine. *New Phytol.* 182:49–84.

White, P.J. and R.J. Davenport. 2002. The voltage-independent cation channel in the plasma membrane of wheat roots is permeable to divalent cations and may be involved in Ca^{2+} homeostasis. *Plant Physiol.* 130:1386–1395.

White, P.J. and D.J. Greenwood. 2013. Properties and management of cationic elements for crop growth. In *Soil Conditions and Plant Growth*, eds. P.J. Gregory and S. Nortcliff, pp. 160–194. Oxford, U.K.: Blackwell Publishing.

White, P.J. and J.P. Hammond. 2008. Phosphorus nutrition of terrestrial plants. In *The Ecophysiology of Plant-Phosphorus Interactions*, eds. P.J. White and J.P. Hammond, pp. 51–81. Dordrecht, the Netherlands: Springer-Verlag.

White, P.J., J. Banfield, and M. Diaz. 1992. Unidirectional Ca²⁺ fluxes in roots of rye (*Secale cereale* L.). A comparison of excised roots with roots of intact plants. *J. Exp. Bot.* 43:1061–1074.

White, P.J., H.C. Bowen, V. Demidchik, C. Nichols, and J.M. Davies. 2002. Genes for calcium-permeable channels in the plasma membrane of plant root cells. *Biochim. Biophys. Acta* 1564:299–309.

White, P.J., J.E. Bradshaw, M.F.B. Dale, G. Ramsay, J.P. Hammond, and M.R. Broadley. 2009. Relationships between yield and mineral concentrations in potato tubers. *HortScience* 44:6–11.

White, P.J., M.R. Broadley, and P.J. Gregory. 2012a. Managing the nutrition of plants and people. *Appl. Environ. Soil Sci.* 2012:104826.

White, P.J., M.R. Broadley, J.A. Thompson et al. 2012b. Testing the distinctness of shoot ionomes of angiosperm families using the Rothamsted Park Grass Continuous Hay Experiment. *New Phytol.* 196:101–109.

White, P.J., T.S. George, L.X. Dupuy et al. 2013. Root traits for infertile soils. *Front. Plant Sci.* 4:193.

White, P.J. and A.J. Karley. 2010. Potassium. In *Plant Cell Monographs 17. Cell Biology of Metals and Nutrients*, eds. R. Hell and R.R. Mendel, pp. 199–224. Berlin, Germany: Springer-Verlag.

Whitten, M.G., M.T.F. Wong, and A.W. Rate. 2000. Amelioration of subsurface acidity in the south-west of Western Australia: Downward movement and mass balance of surface-incorporated lime after 2–15 years. *Aust. J. Soil Res.* 38:711–728.

Wood, N.T., A. Haley, M. Viry-Moussaïd, C.H. Johnson, A.H. van der Luit, and A.J. Trewavas. 2001. The calcium rhythms of different cell types oscillate with different circadian phases. *Plant Physiol.* 125:787–796.

Wu, Q., T. Shigaki, J.-S. Han, C.K. Kim, K.D. Hirschi, and S. Park. 2012. Ectopic expression of a maize calreticulin mitigates calcium deficiency-like disorders in *sCAX1*-expressing tobacco and tomato. *Plant Mol. Biol.* 80:609–619.

Wyatt, S.E., P.-L. Tsou, and D. Robertson. 2002. Expression of the high capacity calcium-binding domain of calreticulin increases bioavailable calcium stores in plants. *Transgenic Res.* 11:1–10.

Xu, X., C.T. Hotta, A. Dodd et al. 2007. Distinct light and clock modulation of cytosolic free Ca²⁺ oscillations and rhythmic CHLOROPHYLL A/B BINDING PROTEIN 2 promoter activity in *Arabidopsis*. *Plant Cell* 19:3474–3490.

Yamazaki, T., Y. Kawamura, A. Minami, and M. Uemura. 2008. Calcium-dependent freezing tolerance in *Arabidopsis* involves membrane resealing via synaptotagmin SYT1. *Plant Cell* 20:3389–3404.

Yamanaka, T., Y. Nakagawa, K. Mori et al. 2010. MCA1 and MCA2 that mediate Ca²⁺ uptake have distinct and overlapping roles in *Arabidopsis*. *Plant Physiol.* 152:1284–1296.

Zhang, B., P. Chen, A. Shi, A. Hou, T. Ishibashi, and D. Wang. 2009. Putative quantitative trait loci associated with calcium content in soybean seed. *J. Hered.* 100:263–269.

Zhang, S.-B., J.-L. Zhang, J.W.F. Slik, and K.-F. Cao. 2012. Leaf element concentrations of terrestrial plants across China are influenced by taxonomy and the environment. *Global Ecol. Biogeogr.* 21:809–818.

Zhao, M.-G., Q.-Y. Tian, and W.-H. Zhang. 2007. Ethylene activates a plasma membrane Ca²⁺-permeable channel in tobacco suspension cells. *New Phytol.* 174:507–515.

6 Magnesium

Witold Grzebisz

CONTENTS

6.1   ORIGIN AND DISCOVERY OF MAGNESIUM

The origin of magnesium is rooted in astrophysics processes taking place in the universe since the Big Bang (Schlesinger, 1997). It is assumed that in the second phase of star development, the combustion of helium yielded $^{12}_{6}C$, which subsequently underwent fusion, leading to the formulation as follows:

$$^{12}_{6}C + ^{12}_{6}C \rightarrow ^{24}_{12}Mg + \gamma$$

where γ is a photon.

As a result of different fusion processes, magnesium occurs in three stable isotopes, namely, ^{24}Mg (78.99% of relative contribution), ^{25}Mg (10%), and ^{26}Mg (11.01%) (Rosman and Taylor, 1998).

The history of the discovery of magnesium lasted about one century (102 years). The first step was to distinguish magnesia from lime. In 1729, Hoffman (1667–1742), a German chemist, found that *magnesia alba* (magnesium carbonate) treated with sulfuric acid resulted in a bitter water-soluble salt, but lime yielded a non-water-soluble compound. This difference was corroborated by Black in 1754 (Joseph Black, 1728–1799). In the conducted experiment, the heated magnesia alba yielded CO_2 (termed the fixed air) and a solid residue reacting with sulfuric acid to form a water-soluble salt. He called the discovered salt *magnesia*, being, in fact, a magnesium oxide. However, this compound was not recognized as a metal until the beginning of the nineteenth century.

Pure magnesium was first isolated from magnesium oxide about 50 years later, in 1808 by Davy (Humphry Davy, 1778–1829). The magnesium metal was obtained from a magnesium oxide–mercury amalgam by distilling mercury off. A few years later, in 1833, Faraday (Michael Faraday, 1791–1867) obtained free magnesium metal by electrolysis of anhydrous magnesium chloride (Page, 2002; Shand, 2006). The next step in the history of investigating magnesium of major importance for human existence and knowledge was the study of chlorophyll. The structure of this molecule, including the presence of magnesium as a central metal, was elaborated by Richard Willstätter (1872–1942), the Nobel Prize winner in 1915 (Sourkes, 2009).

The present name of the element refers to the Magnesia district located in Thessaly (Greece). This region is naturally rich in iron and manganese oxides and magnesium carbonates. Therefore, the latter minerals are enriched with magnetite, which exhibits magnetic properties. The physical and morphological similarities resulted in the same name for different minerals, that is, magnesia alba and magnesia nigra (manganite) and two elements: magnesium and manganese. Both minerals, due to their lustrous surface, were highly important for alchemists and were considered for centuries as being philosopher stones. This consideration was the key reason for the high interest in their chemical recognition and transformation (Shand, 2006).

Magnesium belongs to the alkaline earth metal family, with an atomic number of 12 and stable oxidation valence of 2^{+}. Its atomic weight is 24.305 g mol^{-1}, with a density of 1.74 g cm^{-3}, which is two-thirds that of aluminum (2.7 g cm^{-3}). Melting and boiling points of 649°C and 1091°C, respectively, make this metal easy for physical transformation. Pure magnesium does not exist under natural conditions in the elemental form, due to a high reactivity with oxygen. In air, the solid form of magnesium metal undergoes oxidation, thus creating a patina at the surface. These physical properties of pure, free magnesium are broadly exploited in industry for constructing light materials and alloys with aluminum. Finely powdered magnesium metal burns immediately, when exposed to oxygen. Therefore, magnesium powder is applied as a source of illumination in flashlight photography. It has been exploited for making incendiary bombs, broadly used during World War II, but nowadays, it is used for making flares (Halka and Nordstrom, 2010).

Based on the solubility of magnesium salts in water, two key groups of compounds can be distinguished, water soluble and water insoluble. The first group comprises compounds like halides, nitrates, and sulfates. Halides, which dominate in seawater, brines, and wells, are exploited broadly by electrolysis as a source of pure metal:

$$MgCl_2 \rightarrow Mg^{2+}_{(cathode - molten\ metal)} + Cl_{2(anode - gas)}$$

Under conditions of high-water evaporation, water-soluble magnesium compounds undergo transformation into crystals, as currently observed in the Dead Sea or in the Great Salt Lake, for example. The presence of soluble sulfate compounds, even in small amounts, leads to water having a tart taste. Therefore, its regular presence in natural wells led to magnesium sulfate ($MgSO_4 \cdot 7H_2O$) being called *the bitter salt*. Its English name, Epsom salt, refers to the English village of Epsom, where it was first recognized in the seventeenth century. This salt is used broadly as a health remedy. Magnesium sulfates are the water-soluble compounds most exploited from geological deposits for making magnesium fertilizers. The classical example of a solid form of water-soluble sulfate minerals is kieserite ($KCl \cdot NaCl \cdot MgSO_4 \cdot H_2O$), being at present the key source for producing magnesium fertilizers. Other Mg minerals like carnallite ($KMgCl_3 \cdot 6H_2O$) are used for obtaining pure magnesium metal. Among water-insoluble compounds such as carbonates and phosphates, hydroxides prevail in compounds representing carbonates, like magnesite ($MgCO_3$) and dolomite [$MgCa(CO_3)_4$]. Magnesium oxide is produced using the thermal reduction method (Eliezer et al., 1998; Halka and Nordstrom, 2010):

$$MgCO_3 \rightarrow MgO + CO_{2(gas)}$$

The global resources of geological deposits of water-soluble magnesium salts are still sufficiently high to cover the demand for industry and agriculture (Eliezer et al., 1998; Shand, 2006). Unfortunately, the human population suffers from inadequate supply of this element. The main reason is the steady-state soil magnesium *mining* by crop plants. The current situation was summarized 80 years ago by Charles Northen, who stated: *It is simpler to cure sick soils than sick people—which shall we choose* (Beach, 1936).

6.2 MAGNESIUM UPTAKE

6.2.1 MECHANISMS OF UPTAKE

Nutrient uptake by crop plants is described using two approaches, nutrient supply and nutrient demand (Mankin and Fynn, 1996). Magnesium, as a mobile nutrient in the soil solution, is considered the typical example of the *nutrient demand* uptake approach. This concept relies on the assumption that the growth rate of a crop, primarily induced by the amount of absorbed sunlight energy, determines its current demand for a particular amount of a given nutrient (Rengel, 1993). The first concept, assuming the uptake of magnesium under conditions of limited supply, also requires attention.

There are two groups of factors that prevent an adequate uptake of magnesium by plants. The first group, called *the absolute magnesium deficiency*, results from too low concentrations of soil magnesium in forms that are available to currently cultivated crops. The second group, called *the induced deficiency*, comprises all factors that disturb the flow of Mg^{2+} ions to the root (Metson, 1974). It is also necessary to consider the fact that crop requirements for magnesium are not constant during the season. This variability in requirement is driven by the growth rate and by developmental changes. Therefore, uptake of magnesium by crop plants should be considered to be a series of consecutive events that are responsible for the following:

1. Variability of the crop plant requirement throughout the growing season
2. Transportation of ions from the soil solution at the root surface
3. Ion transfer through the plasma membrane
4. Transport and redistribution of Mg ions within a cell and plant tissues

TABLE 6.1
Magnesium Accumulation for Selected Crop Plants, kg t⁻¹ Harvested Yield

Crops	Germany[a]	Poland[b]	USA[c]	WFM[d]
Wheat (*Triticum aestivum* L.)	1.5–3.0	3.0	5.0	n.s.
Maize (*Zea mays* L.)	3.0–5.0	4.0	5.5	4.5
Rice (*Oryza sativa* L.)	n.s.	n.s.	4.1	4.0
Oilseed rape (*Brassica napus* L.)	4.0–7.0	8.0	n.s.	n.s.
Soybean (*Glycine max* Merr.)	n.s.	8.0	13.8	n.s.
Pea (*Pisum sativum* L.)	3.0–5.0	5.0	n.s.	3.5
Potato (*Solanum tuberosum* L.)	0.5–0.9	0.8	n.s.	0.2–0.3
Sugar beet (*Beta vulgaris* L.)	0.6–1.2	0.8	0.7–0.9	1.0–2.0

n.s., not specified.
[a] Finck (1992).
[b] Grzebisz (2009).
[c] Jones (2003).
[d] Wichmann (1992).

Total crop requirement for any nutrient, including magnesium, is calculated based on the critical nutrient content in the main yield unit plus its adequate amount in the respective portion of any harvested byproducts. This index is called the unit magnesium uptake (UMgUp). The key difficulty in the calculation procedure is, however, a relatively high variability in the published ranges of UMgUp indices (Table 6.1). This variability results from the method used for the index calculation. In many cases, certain plant parts, such as chaff in cereals, or even straw, are omitted. In the case of magnesium, all vegetative parts should be included in the calculation procedure.

The dynamics of accumulation of a particular element in the plant canopy during growth is the key nutrient uptake characteristic. In general, magnesium accumulation progresses in plant parts over the entire growth period, independent of the species, up to the end of development. This specific trend, following the logistic regression model, is typical for a wide variety of crops, such as wheat (*Triticum aestivum* L.) (Barraclough, 1986; Preez and Bennie, 1991), lupine (*Lupinus angustifolius* L.) (Barłóg, 2000), oilseed rape (*Brassica napus* L.) (Barłóg et al., 2005b), and okra (*Abelmoschus esculentus* Moench) (Moustakas et al., 2011). Magnesium accumulation in potato (*Solanum tuberosum* L.), as reported by Nunes et al. (2006), followed the linear regression model. As illustrated in Figure 6.1, the pattern of magnesium accumulation in sugar beet is slightly different, following a quadratic regression model, with a significant peak. This specific trend is due to successive decreases, or to concomitant decreases, in magnesium content in older leaves at the onset of maturity (Grzebisz et al., 1998).

It has been well recognized that increasing nitrogen supply leads to increased magnesium content in crop plants. The extended study by Osaki et al. (1996) revealed the linear relationship between nitrogen content and magnesium content in various crop species (cereals, corn [*Zea mays* L.] and rice [*Oryza sativa* L.]; pulses, soybean [*Glycine max* Merr.] and field bean [*Vicia faba* L.]; tuber and root crops, potato and sugar beet [*Beta vulgaris* L.]), independent of growth stage:

$$Mg_c = Mg_0 + MgNi \cdot N_c$$

where
Mg_c, N_c are magnesium, nitrogen content (mg m⁻²)
$MgNi$, Mg_0 are coefficients of the linear regression function

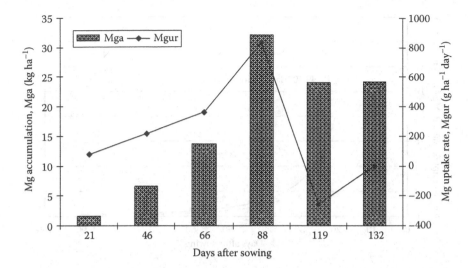

FIGURE 6.1 Characteristics of magnesium accumulation (Mga) and rate of uptake (Mgur) by high-yielding sugar beet (*B. vulgaris* L.). (Based on Grzebisz, W. et al., *Bibliotheka Frag. Agron.*, 3(98), 242, 1998.)

The primary goal of the well-defined course of magnesium accumulation in the crop is to determine the critical stage of its uptake during the season. For example, in the high-yielding crop species sugar beet, the maximum uptake, amounting to 0.83 kg Mg ha^{-1} day^{-1}, is achieved by the canopy approximately 80 days after germination (Figure 6.1). For comparison, the calculated maximum Mg uptake by winter wheat, amounting to 0.21 kg Mg ha^{-1} day^{-1}, is one-quarter of that for sugar beet (Barraclough, 1986). It is revealed at heading and is related to a significant increase in developing ear biomass (Miller et al., 1994). These two examples implicitly indicate the higher sensitivity of all leafy crops to a shortage of magnesium.

The rate of magnesium uptake by the root varies with crop age. It has been well documented that the highest flow of water occurs in the youngest immature zone of the growing root. This phenomenon can be explained by the fact that the zone, extending from a few millimeters to some centimeters behind the root apex, is composed of loosely and weakly differentiated cells. Therefore, this part of the root is easily permeable to water and nutrients. The first border limiting the rate of water flow is the endodermis. In this root zone, the flow of water and divalent ions is strongly reduced by the Casparian strip (Clarkson, 1985; Steudle, 2000; Taiz and Zeiger, 2006). This phenomenon was described extensively for corn roots by Barber (1984). As illustrated in Figure 6.2, the highest rate of any nutrient uptake, as a rule, occurs during the early stages of plant growth and declines shortly thereafter. The degree of the decrease is nutrient specific. In the 10-day period extending from the 20th to the 30th day of corn vegetative growth, the phosphorus uptake rate declined almost 92%, magnesium 89%, nitrogen 86%, and potassium 75%. Among these four major nutrients, phosphorus and magnesium showed the best fit of the experimental data to the developed power function. The close resemblance of uptake patterns of both nutrients has also been observed for winter wheat (Barraclough, 1986) and sugar beet (Grzebisz et al., 1998).

A key point of scientific discussion since the 1950s has been focused on transportation pathways of the Mg^{2+} ion toward the root surface (Barber, 1984). In general, mass flow is considered the main mechanism that is responsible for magnesium movement from any place in the soil solution to the root surface. The quantity of magnesium arriving at the root surface is driven by two principal factors: (1) the rate of water transpiration and (2) Mg^{2+} concentration in the soil solution. The main cause of water movement through the plant is the vapor-pressure deficit in the atmosphere. This factor is highly variable during the growing season and over the course of the day. A temporary water deficit is most likely the most important determinant of magnesium shortage to the cultivated crop.

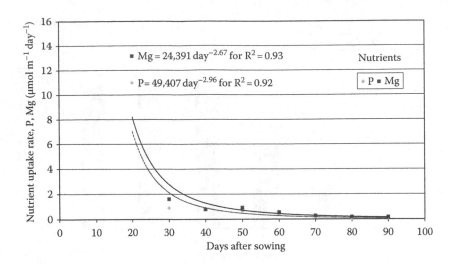

FIGURE 6.2 The course of phosphorus and magnesium uptake rate by unit root length of corn (*Z. mays* L.) during the growing season. (Based on Barber, S., *Soil Nutrient Bioavailability: A Mechanistic Approach*, Wiley & Sons Interscience, New York, 1984.)

Maize and sugar beet respond to fertilizer magnesium, mostly in years with a mild water deficit (Grzebisz, 2013). As reported by Barłóg and Grzebisz (2004), the unit amount of accumulated magnesium, at the same level of available magnesium in the soil, amounted to 0.6 kg Mg t^{-1} of storage roots (plus an adequate amount of Mg in leaves) in years with optimal weather (1997 and 1998) but increased threefold to 1.5 Mg t^{-1} under water shortage (1999).

All academic textbooks follow the extended review by Clarkson (1985), who documented that the amount of magnesium transported by mass flow toward the root is several times higher than the corn magnesium requirement. In general, magnesium concentration at the level of 1 g kg^{-1} DM in plant tissue is generally considered to be the lowest critical concentration for most crops. The upper range is highly variable, depending on the crop, ranging from 3 g Mg kg^{-1} DM as sufficient for cereals and up to 6 g Mg kg^{-1} DM for other crops. Based on these values, the required Mg concentration in the growth medium can be calculated using the procedure proposed by Barber (1984), assuming 300 g of transpired water per gram plant dry matter. As the Mg concentration in plants increases, the required concentration in the solution of the growth medium increases, as follows:

1. 1 g Mg kg^{-1} DM → 1/300 · 1000 = 3.33 mg dm^{-3} = 0.138 mM = 138 µmol dm^{-3}.
2. 3 g Mg kg^{-1} DM → 3/300 · 1000 = 10 mg dm^{-3} = 0.417 mM = 417 µmol dm^{-3}.
3. 6 g Mg kg^{-1} DM → 6/300 · 1000 = 20 mg dm^{-3} = 0.833 mM = 833 µmol dm^{-3}.

However, if the transpiration water exceeds 300 g g^{-1} dm, the concentration in the medium need not be as high as these values. For example, if the transpiration is 600 g g^{-1} dm, the concentration in the solution of the medium need be only half of these values.

Magnesium concentration in the soil solution, according to Karley and White (2009), ranges from 0.125 to 8.5 mM. The optimum concentration of Mg in the tissue of cereals during critical stages of growth varies from 1 to 3 g Mg kg^{-1} DM (Bergmann, 1992). The top requirement can be fulfilled by maintaining a magnesium concentration in the soil solution at the level of 0.3 to 0.400 mM. Therefore, magnesium supply from soil resources can be considered the critical factor for a crop during intensive stages of growth.

The extent to which magnesium uptake by a crop depends on other transportation mechanisms, such as diffusion, is still not fully understood. It is well known that diffusion of a particular nutrient in the soil–plant continuum is driven by its effective diffusion coefficient, which is related to temperature and water content within the soil body. This general assumption has been fully corroborated by Schaff and Skogley (1982), who stated that temperature and moisture interaction significantly increased the diffusion of Mg in soil developed from a silt loam. The study of Mullins and Edwards (1987) showed that the addition of salts, for example, $Ca(NO_3)_2$, to moist soil considerably increases the diffusion of magnesium. This phenomenon can be explained by salt-induced exchange processes, leading to higher concentrations of Mg^{2+} in the soil solution (Baligar, 1984).

During the period of elevated demand of the aboveground canopy for magnesium, its rate of supply is frequently too low to maintain adequate Mg in plants. Therefore, roles of other transport mechanisms, such as diffusion, especially under the highest requirement or unfavorable growth conditions, cannot be excluded. The present approach is supported by Mota Oliveira et al. (2010), who studied magnesium transportation mechanisms in soil cropped with maize. The pot experiment studies with soil originated from two agronomic systems that are typical for Brazil, that is, Cerrado and long-term maize monoculture; these studies showed that diffusion was responsible for approximately 30%–43% of magnesium supply to growing plants. Magnesium concentration was 5.5 times higher in the soil solution for the Cerrado compared with the monoculture. Consequently, its concentration in the xylem exudates of maize plants was also differentiated by soil origin, being 2.5 times higher and resulting in a 60% higher biomass.

6.2.2 MOLECULAR BACKGROUND OF MAGNESIUM UPTAKE AND REDISTRIBUTION

The amount of magnesium ions accumulated at the root surface exceeds crop requirements (Clarkson, 1985). In fact, magnesium ions arriving at the root surface undergo two key processes, fixation in the extracellular space within the root (apoplast) and transportation across the plasmalemma into the intracellular space (cytosol, symplasm). The apoplast reserves can be used during the period of the highest requirement of the aboveground plant parts or in response to magnesium stress due to the sudden decrease of its soil supply. Any investigation of magnesium transport mechanisms into and within the plant should always consider the pattern of root system development during the growing season. The root system is dynamic due to continuous growth and simultaneous mortality of aged parts. In seed crops, the growth of new roots prevails until flowering and then undergoes a reduction due to a high rate of old root mortality (Fageria and Moreira, 2011).

Magnesium inflow into the root xylem, considering this tissue as a transportation passage, is critical for its subsequent redistribution within plant parts. The first route, which is the major one, exists in the root extension zone, which is an undifferentiated part within the root. The inner tissue of young roots is immature, allowing fast transportation of nutrients in the water transpiration stream. The main limiting physical barrier is the oxydermis. Magnesium inflow into the adult part of the root, called the minor route, is impaired by the endodermis (Marschner, 2012). This magnesium transportation route exists in the mature root, where the endodermis with extended suberin lamellae builds up the inner biophysical barrier. For the purposes of this chapter, these two mechanisms will be, respectively, called *free-limited flow* and *endodermal-limited flow* of magnesium into the plant. Consequently, a free flow of water and nutrients dominates during the vegetative period of plant development. This basic difference, resulting from the root anatomy, is frequently underestimated in studies concerning magnesium uptake by the plant.

The pathway of Mg^{2+} transportation from the root into the leaves as the final physiological sink comprises several steps. Among them, three seem to be crucial: (1) Mg^{2+} arrival into the cytosol and redistribution among the cellular constituents, (2) xylem loading, and (3) xylem unloading. There are key questions about mechanisms of magnesium ion transportation from the cell apoplast into the cytoplasm through the plasma membrane. To date, the Mg^{2+}-specific transporters have not been recognized precisely. The moving ions can exploit two mechanisms: channels (permeable to mono- and divalent cations) and transporters. The putative transport channels, for example, the *rca* channel determined in the wheat root, are permeable to a broad range of ions, such as K^+ and Ca^{2+} and Mg^{2+} (Bose et al., 2011; Karley and White, 2009; Shaul, 2002). The second mechanism of magnesium transfer through the plasma membrane assumes the existence of specific carrier proteins. The extended work on arabidopsis (*Arabidopsis thaliana* Heynh.) as the model plant revealed the presence of the *MRS2* (mitochondrial RNA splicing2)-type gene. This subfamily of genes is within the *CorA* (cobalt resistance) gene family that was identified in bacteria and yeast as coding for magnesium transporters (Gardner, 2003). The genes of the *MRS2* group can activate the synthesis of plasma membrane protein carriers with the ability to move Mg^{2+} ions into and out of the cell. Among the ten plant *MRS2* genes identified in arabidopsis, three (*AtMGT1*, *AtMGt10*, and *AtMGt7*) express their activity in the plasma membrane of the root cell. The first two have been identified as the activator of a high-affinity Mg transporter, and the third was identified as the activator of a low-affinity Mg transporter (Bose et al., 2011; Chen and Ma, 2013). In the rice genome, which is also widely investigated, among the nine *CorA* homologs, the *OsMGT1* gene has been identified as being involved in the activation of Mg transporters. The study by Cai et al. (2012) on *OsMGT1* gene expression showed its sensitivity to a shortage of magnesium in shoots.

Magnesium ions entering the cytoplasm of the xylem parenchyma cell undergo bonding by negatively charged compounds and/or partitioning among intracellular organelles. The major portion of Mg throughout the plant, which ranges from a few mM to 100 mM, is fixed in cellular structures. The majority of cell Mg is bound to adenosine triphosphate (ATP). The free Mg is present only in the cytoplasm, achieving a level of 0.4–0.5 mM in well-nourished plants, which constitutes approximately 10% of its total content in the root. This range is recognized as optimal for the nondisturbed course of plant processes (Karley and White, 2009; Shaul, 2002). The key transport processes, within the parenchyma cells of the endodermis, include magnesium exchange between the cytoplasm and the vacuole (storage–release exchange) and between the cytoplasm and xylem (xylem loading). Magnesium concentration in the root vacuole ranges from 20 to 120 mM. It is also higher in the xylem than in the cytosol, ranging from 0.5 to 1 mM. Therefore, any transport of Mg^{2+} ions requires a significant input of metabolic energy, which is necessary to move and redistribute magnesium absorbed from the soil within the plant physiological sinks (Karley and White, 2009; Shaul, 2002).

The vacuole, as a storage organelle, plays a crucial role in the management of internal magnesium resources. Magnesium concentration in this pool is much higher than in other organelles and has a high daily and day-to-day variability. Magnesium transportation from the cytoplasm through the tonoplast into the vacuole is mediated by Mg^{2+}/H^+ exchangers with the stoichiometry of 1:3. To date, a single *AtMHX* (*A. thaliana* magnesium proton exchanger) gene, expressed in the xylem parenchyma, has been discovered to encode the synthesis of proteins involved in magnesium transport through the tonoplast. The AtMHX protein, and its homologs in other plants have been recently considered as affecting Mg^{2+} intracellular redistribution and xylem loading (Gardner, 2003; Shaul, 2002).

Magnesium, transported from roots via the xylem into leaves, undergoes redistribution primarily among leaf tissues, and thereafter among cell organelles. Shabala and Hariadi (2005), working with broad bean (*V. faba* L.), discovered two main routes of magnesium flux into the cytoplasm within the mesophyll cell. The first route, operating at high magnesium concentration in the cell apoplast (>30 μM), uses a nonselective cation channel (also permeable to Ca^{2+} and K^+). The second transportation route, operating at low concentrations of Mg in the apoplast, (<30 μM), exploits the Mg^{2+}/H^+ exchanger, with a stoichiometry of 1:1 or 1:2. It has been recently documented that *MRS2-1* and *MRS2-5* genes encode the synthesis of specific tonoplast transport proteins. These genes are responsible for increased Mg accumulation in the mesophyll compared with the epidermal and bundle sheath cells. The excessive amount of supplied Mg moves into the vacuole. Consequently, vacuoles of these cells become the primary storage pool of leaf magnesium (Chen and Ma, 2013).

6.2.3 External Factors Affecting Magnesium Uptake

There are numerous factors affecting nutrient uptake by crop plants. For sessile plants, soil is a standard milieu of growth, impacting the spatial distribution of nutrients considered as a feeding source and the roots as a physiological sink. The environmental factors affecting magnesium uptake by crop plants should be divided into three groups: (1) ordinary, largely independent of farmer activity, such as water and temperature; (2) natural and agronomic, responsible for root system size, partly dependent on farmer activity; and (3) nutritional, amount of plant-available magnesium and its interactions with other nutrients.

Water content and its availability in the growth medium is the decisive factor in the rate of nutrient supply to a plant during its life cycle. Water shortage is a key factor driving the root size proliferation, in turn impacting nutrient uptake. A continuously reduced water availability decreases the rate of delivery of nitrate to the root and to the whole plant, resulting in the development of local and systemic hormone signaling (Forde and Lorenzo, 2001). Consequently, carbohydrate partitioning between plant organs and tissues changes, thereby favoring root system growth. The induced plant adaptation strategy depends on potassium supply, which is required by new root cells for accelerating their longitudinal growth (Grzebisz et al., 2013). Magnesium, due to its impact on phloem loading, is a decisive factor in adequate root feeding (Verbruggen and Hermans, 2013). Plant response to water stress is partly under genetic control. As illustrated in Figure 6.3, four of five cultivars of cherry tomato (*Lycopersicon esculentum* Mill.) showed a drastic but cultivar-dependent decrease in the rate of magnesium flux in response to induced water stress. The cultivar Katalina exhibited the highest drop in magnesium uptake, down to 70% compared with the well-watered plant. In addition, magnesium concentration in plant tops decreased much less or increased under similar conditions, as in the case of the resistant cultivar Zarina. The year-to-year variability of magnesium uptake by plants can be explained by water-induced variability of Mg concentration in the soil solution, subsequently increasing its concentration in the xylem exudate under conditions of water shortage. This process is highly dependent on soil available magnesium content, as reported by Mota Oliveira et al. (2010).

Metabolic energy is a basis for nutrient uptake by a plant, but its level depends on temperature, water, and nutrient supply. Magnesium uptake by plants is highly sensitive to temperature fluctuations during the growing period. Temperature is a key factor affecting the activity of the enzyme

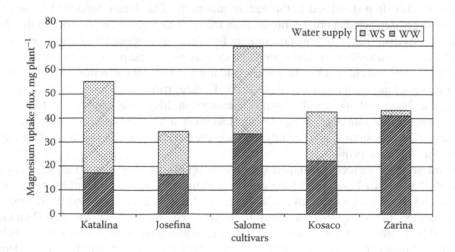

FIGURE 6.3 Uptake flux of magnesium in plants of cherry tomato (*L. esculentum* Mill.) cultivars under imposed water stress. *Legend*: water supply; well-watered plants (WW), water-stressed plants (WS). (Based on Sanchez-Rodriguez, E. et al., *Plant Soil*, 335, 339, 2010.)

ribulose-1,5-bisphosphate (RuBisco), which is consequently crucial for net photosynthesis (Taiz and Zeiger, 2006). In addition, temperature indirectly accelerates the rate of plant transpiration, in turn affecting soil water management. These series of temperature-induced events include changes in the rates of nutrient exchange between the soil cation exchange complex (CEC) and the soil solution, mineralization of organic matter, root respiration, and ion movement across the plasma membrane. Any increase in soil temperature leads to the relative increase in nitrate over ammonium ion uptake (BassiriRad, 2000). These events can explain, at least partly, the tight relationship between magnesium and nitrogen uptake during the main growing season (Grzebisz et al., 2010; Osaki et al., 1996). In many regions around the world, even in temperate areas, heat stress impairs plant growth due to an accelerated formation of reactive oxygen species (ROS), which in turn leads to growth disturbance. As reported by Mengutay et al. (2013), these negative outcomes can be partially overcome via an adequate magnesium supply.

Magnesium transportation in soil is sensitive to soil temperature, with magnesium showing much higher availability with increase in temperature than potassium (Pregitzer and King, 2005). One of the most important soil characteristics, affected by temperature, is water viscosity, which behaves inversely in response to temperature change, decreasing linearly with temperature increases. Under low soil temperatures, the viscosity of water is high, in turn reducing the rate of nutrient flow to the root surface. The classical examples that illustrate this concept are temperate grasses. It has been well documented that in the spring, the rate of magnesium uptake is much lower but increases in the summer (Barber, 1984). The low uptake rate of magnesium is compensated for by an increased rate of potassium uptake, leading to the risk of grass tetany in cows (Whitehead, 2000). Any increase in temperature results in increase in magnesium uptake, especially when connected with the simultaneous increase in nitrate nitrogen concentration in the growth medium. This effect is most important for monocots during early stages of growth, as reported by Huang and Grunes (1992) for wheat seedlings.

A study conducted on various crop plants exposed to artificial or naturally toxic levels of Al^{3+} implicitly showed a high disturbance in the size and morphology of the root system. It has been estimated that plant growth is reduced seriously if the pH of the nutrient solution drops below 4.0 (Gloser and Gloser, 2000). Kobayashi et al. (2013a), studying the uptake of magnesium by rice plants under conditions of a decreased pH of the nutrient solution, found the largest reduction in the zone that was proximate to the apex. In a period lasting 15 min, the amount of magnesium ion flux into the rice plant was 4.5, 3.0, and 2.0 μmol plant^{-1} for pH values of 6.5, 5.6, and 4.5, respectively. This sudden influx drop stabilized in the mature root parts. The drastic reduction in root length, followed by decreased nutrient uptake, results from the inhibition of root meristem cell division and the simultaneous disturbance in transport protein functions, as proposed by Rengel and Robinson (1989). The aluminum toxicity, as related to Al^{3+} ions, may be, at least partly, overcome by the application of Mg^{2+}. Both ions have a similar radius, 0.48 nm for Al^{3+} and 0.476 nm for Mg^{2+}, competing therefore for root-binding sites (Bose et al., 2011). To date, there are a few scientific reports corroborating the theoretical applicability of this phenomenon. Magnesium applied in higher rates at least partly restored the rate of seedling growth. Hossain et al. (2004), working with wheat, showed that the applied magnesium can efficiently alleviate the harmful concentration of aluminum, if a low supply of calcium is provided.

The third group of factors that impair magnesium uptake by crop plants refers to imbalances between Mg^{2+} and other basic ions in the growth medium. The ionic radius of hydrated ions that predominate in the soil solution, such as Ca^{2+}, Mg^{2+}, and K^+ is 0.295, 0.476, and 0.232 nm, respectively. In addition, the *transport numbers* of these ions were 8–12, 12–14, and 4–6 (Maguire and Cowan, 2002). Therefore, among these ions, the magnesium ions are the most mobile within the soil solution. Consequently, they arrive at the root surface much faster than the others. However, the actual competition for uptake between the basic ions occurs at the root surface. It is necessary to assume a strong competition between Mg^{2+} against K^+ and Ca^{2+} ions for the transportation sites

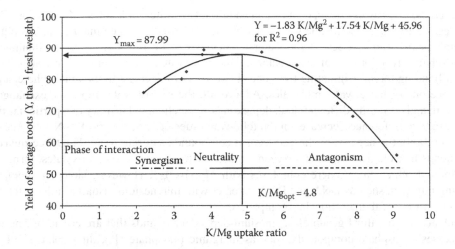

FIGURE 6.4 Response of sugar beet yield of storage roots to uptake ratios of potassium to magnesium. (Based on Barłóg, P. et al., *Biul. IHAR*, 222, 119, 2002a; Barłóg, P. et al., *Biul. IHAR*, 222, 127, 2002b.)

in the nonselective channels and/or transporters (Horie et al., 2011). The main reason for the lower uptake rate of magnesium is the quantity of energy required for dehydration of its water shell. It is well documented that the hydrated radius and strength of hydration water shell are inversely related to the rate of ion transportation through the plasma membrane (Marschner, 2012).

Based on field experiments, the dominating type of relationship between the basic cations and magnesium ions is not easily defined. The analysis of some crop characteristics, such as the rate of uptake, concentration, and content of magnesium in response to increasing concentration of calcium or potassium in the growth solution, shows a very specific course. The K/Mg curve, as illustrated in Figure 6.4, can be divided into three distinct parts, indicating the existence of different types of relationship between potassium and magnesium uptake: (1) synergism, (2) neutrality, and (3) antagonism. There are numerous factors that correct the potential antagonism, such as the initial concentration of competitive ions, the actual concentration as affected by the applied fertilizer magnesium, the stage of plant growth, and the plant part. The study of the impact of a particular relationship on the yield of storage roots of sugar beet has underlined the universality of this model. The course of the curve implicitly indicates a fixed range of each nutrient uptake, which is required for a functional K/Mg balance and is a prerequisite for the maximum growth rate and/or the highest yield of the cultivated crop. These assumptions have been corroborated by Ding and Xu (2011), who showed that the impact of the imbalanced concentration ratios of Mg and K in the nutrient solution culture on magnesium concentration in rice plants was much stronger for roots than for shoots in an effect that progressed with plant age. However, in the treatment with a K to Mg solution ratio of 1:1, shoot and root Mg concentrations during the entire growth period peaked, resulting in the greatest final biomass production.

6.3 BIOPHYSICAL FUNCTION OF MAGNESIUM

6.3.1 BASIC BIOCHEMISTRY

The impact of magnesium on biological processes in plants is a consequence of its basic physical properties. The Mg^{2+} cation shows a high affinity to numerous ligands and is surrounded in aqueous solution by two layers of water molecules. Therefore, under full hydration, its volume is approximately 400 times larger than its nonhydrated size. The first water layer, called the *inner water layer*, consists of six molecules and creates an octahedral magnesium complex $[Mg(H_2O)_6]^{2+}$. The second layer, called the *outer water layer*, contains an additional 12 water molecules (Maguire and Cowan, 2002; Weston, 2009). These double water coordination spheres implicate magnesium's uniqueness in all biological systems. Each of the water molecules can be replaced by select biomolecules. Ligand-exchange reactions occur in the outer and the inner water layers. First, a substrate is fixed by hydrogen-bonding forces originating from one or two water molecules. In the inner sphere, a ligand may replace one or more water molecules. As a result, the rate of water molecule exchange with biological ligands is highly differentiated, depending on the chemical affinity of a particular ligand. The ion-binding preference decreases in the following order of ligands: $O > P > N > S$. According to Weston (2009), the negatively charged oxygen compounds saturate more than three-quarters of the Mg bonds in biological systems. Oxygen ligands, such as water and carboxylates, form strong, mostly monodentate, inner-sphere complexes with $Mg(H_2O)_6^{2+}$. However, biological compounds containing nitrogen, such as DNA and RNA, interact with magnesium, usually in the outer sphere and are much weaker than carboxylate complexes.

Phosphates are the third group of magnesium-dependent ligands that are crucial for numerous life processes. Phosphate compounds, such as inorganic phosphate (P_i), diphosphates (PP_i, adenosine diphosphate, ADP), and triphosphate (ATP), usually form monodentate complexes in the inner sphere. Phosphorus compounds, together with magnesium, are required for the activation of enzymes that are decisive for production, storage, and transfer of metabolic energy into other cell compartments. The key metabolic process that requires both elements is the synthesis of ATP:

$$ADP + P_i \leftrightarrow ATP$$

There are two different pathways of magnesium action in this process. In the first pathway, which is the predominant one, the Mg^{2+} ion interacts as a cofactor with ATP yielding a Mg-ATP molecule, which subsequently binds to the active site of the enzyme. The hydrolysis of the terminal phosphate group of ATP yields ADP and releases energy ($\Delta G \approx -30.5$ kJ mol^{-1}), which is later used for conducting enzymatic process. In the second pathway of magnesium action, the ion is strictly attached to the enzyme, changing its structure and leading to its activation (Maguire and Cowan, 2002).

Most (¾) of the intracellular magnesium is bound by ATP. The remaining portion acts as the counteranion that neutralizes and stabilizes numerous, negatively charged organic ligands (mostly phosphates and carboxylates) in structural components within the cell, including other membranes, nucleic acids, ribozymes, enzyme complexes, and the mitochondria. Consequently, the concentration of the Mg^{2+} ion in the cell cytosol is generally low at 0.4 mM, representing approximately 10% of the total magnesium content of the plant. Some of the most important structures of the plant cell that are stabilized by magnesium are the ribosomes. The key role of the Mg^{2+} ion is to fix ribozymes into the ribosome and subsequently aggregating them into polysomes. This cellular structure becomes unstable at Mg^{2+} concentrations below 10 mM. The stability of ribosomes is crucial for protein synthesis (Hawkesford et al., 2012).

The biophysical properties of magnesium affect numerous biochemical and physiological processes in living organisms. The group of magnesium-dependent enzymes comprises 250 EC numbers, contributing to approximately 40% of all metal-dependent enzymes. Among the Mg-activated enzymes, transferases (45% of all Mg-dependent enzymes, EC2 group number) are the most common, followed by hydrolases (23%, EC3), ligases (14%, EC6), lyases (9%, EC4), oxidoreductases (5%, EC1), and isomerases (4%, EC5) (Andreini et al., 2008; Weston, 2009).

6.3.2 Photosynthesis

Human life on earth depends on magnesium. The element is critical for the biosynthesis of chlorophyll, which in turn creates a biological basis for absorption of sunlight energy, resulting in production of oxygen and carbohydrate. The stroma of chloroplasts comprises sets of 20–100 stacked discs, which are grouped into specific structures called grana. The individual unit disc, named the thylakoid, is a biological membrane that holds chlorophylls (Sarafis, 1998). Chloroplasts are the core of photosynthesis, a process with the following key functions:

Absorption of the basic unit quantum of energy—photons
Conversion of light energy into its metabolic forms such as ATP and NADPH
Use of these newly generated compounds to fix CO_2 and synthesize sugars as primary organic molecules

The first two processes occur in the thylakoid membrane, and the third occurs in the stroma, which contains a set of enzymes that fix CO_2 and produce sugars (Sarafis, 1998; Taiz and Zeiger, 2006).

Chlorophyll is composed of two parts, a hydrophilic head and a lipophilic tail. The first part is a porphyrin, which is a compound that contains a four-pyrrole ring surrounding a central metal ion. Chlorophyll contains Mg^{2+}, as the inner metal ion, which cannot be replaced by other ions, even those with similar physical properties (Mn^{2+}, Zn^{2+}, Cd^{2+}, Cu^{2+}.). A chlorophyll phytol chain, which is a C_{20} polyisoprene alcohol, is attached to the seventh carbon of the porphyrin ring, resulting in the lipophilic character of the other part of the chlorophyll molecule. In green plants, there are two forms of chlorophyll, namely, *a* and *b*. The key difference between the forms is the composition of the side group that is attached to the third carbon. A methyl ($-CH_3$) group is contained in chlorophyll *a*, whereas a formyl ($-CHO$) group is contained in chlorophyll *b*. Green plants possess both forms of chlorophyll, which enables the utilization of a broader range of wavelengths of sunlight energy. The key function of chlorophyll *a* is to bind a photon. The optimal ratio of chlorophyll *a* to *b* ranges from 3 to 1. The importance of chlorophyll *b* for photosynthesis is discussed broadly. It is assumed that under ample growth conditions, chlorophyll *b* accounts for 15%–25% of total plant chlorophyll (Beale, 1999). In the opinion of Tanaka and Tanaka (2011), chlorophyll *b* stabilizes proteins in the light-harvesting complex. The chlorophyll *a/b* ratio is under strict plant control because it determines the structure of the photosynthetic proteins, which are synthesized in a fixed stoichiometric ratio (Tanaka and Tanaka, 2006, 2011).

Biosynthesis of chlorophyll is a process that is driven by specialized enzymes in the tetrapyrrole pathway. Based on the main product, there are three principal phases in the chlorophyll life cycle: (1) synthesis of chlorophyll *a*; (2) chlorophyll cycle, including conversion of chlorophyll *a* into chlorophyll *b*; and (3) chlorophyll degradation. The primary source of tetrapyrrole compounds is glutamate, which is a substrate for δ-amino-levulinic acid (ALA) synthesis and an initial substrate for monopyrrole synthesis. During the tetrapyrrole route, the ALA undergoes numerous transformations, driven by hierarchically ordered enzymes, in a process that finally results in the synthesis of protoporphyrin IX. This compound is the critical one that is used as a substrate for two biologically important molecules, that is, chlorophyll and heme. Chlorophyll is magnesium dependent, but heme, which contains iron, is iron dependent (Beale, 1999; Tanaka and Tanaka, 2006).

Magnesium chelatase (MgChl) is the enzyme responsible for the insertion of the Mg^{2+} ion into protoporphyrin IX. The enzyme is composed of three subunits, which differ in molecular weight: CHLI (38–42 kDa), CHLD (60–74 kDa), and CHLH (150 kDa). The process governing Mg^{2+} ion insertion into protoporphyrin IX occurs in the stroma and requires certain physicobiochemical conditions for regulation. The enzyme undergoes diurnal cycling, depending on both Mg^{2+} concentration and the ATP/ADP ratios in the stroma. The level of free Mg^{2+} concentration in the stroma affects the activity of MgChl. The calculated increase of Mg^{2+} during the day ranges from 1 to 5 mM. The upper level of Mg^{2+} concentration in the stroma is required for insertion and stabilization of the MgChl molecules within the thylakoid membrane, which is a required step for its subsequent chelation. At low concentrations of Mg^{2+} in the stroma, MgChl molecules move freely in this milieu. A unique function is attributed to ATP concentration. First, it inhibits the activity of FeCh, an enzyme that is necessary for heme synthesis. Second, the MgChl enzyme requires a certain ATP/ADP ratio and pH of 8 in the stroma for activation (Beale, 1999; Masuda, 2008).

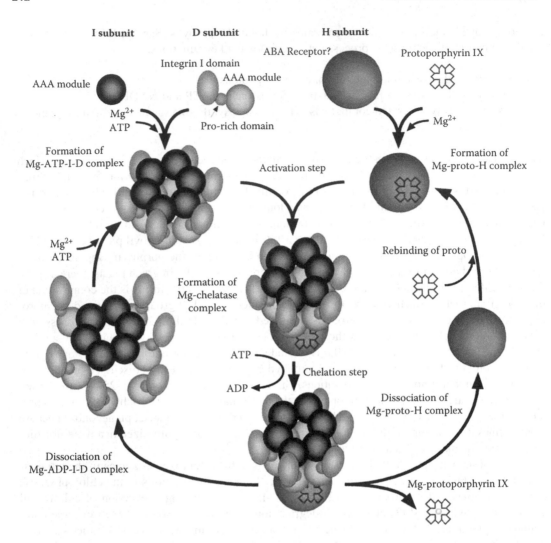

FIGURE 6.5 **(See color insert.)** Model of the catalytic cycle of Mg-Chelatase. (With kind permission from Springer Science+Business Media: *Photosynth. Res.*, Recent overview of the branch of the tetrapyrrole biosynthesis leading to chlorophylls, 96, 2008, 121–143, Masuda, T.)

Activation step: In a Mg^{2+} and ATP-dependent process, six I subunits are assembled into a hexameric ring structure, while six D subunits form a hexameric ring in an ATP-independent manner. The two-tiered hexameric ring forms the Mg-ATP-I-D complex. It is suggested that the ATPase activity of subunit I is inhibited in this complex due to binding of the integrin I domain of subunit D to the integrin I binding domain of subunit I. Meanwhile, subunit H binds to protoporphyrin IX and most likely also the Mg^{2+} substrate—as this subunit is considered the catalytic subunit—to form the Mg-H-protoporphyrin IX complex.

Chelatation step: ATP hydrolysis is triggered upon switching the binding of the integrin I domain of subunit D to the integrin I binding domain of subunit H. After the formation of Mg-protoporphyrin IX, the complex disassembles. Here, subunit H is shown docking to the D subunits, although it remains unknown if subunit H docks to the I or D side of the complex.

The insertion of the Mg^{2+} ion into the protoporphyrin IX is a two-step process, composed of an initial step, *activation*, and the main step, *chelation* (Figure 6.5). According to a recently proposed concept, two complexes are formed during the activation step. The first complex is composed of four subunits: I and D of the MgChl enzyme, the free Mg^{2+} ion, and ATP (complex: I-D-Mg-ATP). The activity of the MgChl-I also depends on sulfur supply in the form of cysteine, which stimulates ATP activity. In this step, ATP is included within the complex but does not undergo hydrolysis.

The ATP is responsible for the formation of a complex between the Chl-I and Chl-D units of the magnesium chelatase. The critical concentration of ATP is high at 1 mM. The second complex formed in this step also consists of four units: Mg^{2+}; the MgChl-H, which is a sulfur-dependent enzyme; protoporphyrin IX; and the GUN4 protein. In the chelation step, the Mg^{2+} is incorporated into protoporphyrin IX by the Chl-H. This process requires an elevated input of energy, which is necessary for the removal of a water sphere from the Mg^{2+} ion (Masuda, 2008). The GUN4 protein, through binding to both primary substrates, accelerates the rate of Mg chelation. The GUN4 protein, which has been recently discovered, is an important factor that significantly affects the activity of MgChl (Stenbaek and Jensen, 2010). According to Davison et al. (2005), the MgChl becomes inactive in the absence of the GUN4 protein.

The goal of photosynthesis is to convert sunlight energy into metabolic energy. The primary reaction, taking place in photosystem II, yields oxygen, which is released into the surrounding atmosphere. The second product, protons released during the splitting of a water molecule, causes acidification of the stroma. The optimum pH of this chloroplast space is 8.0. To maintain this pH, protons are continuously transferred from the stroma through the thylakoid membrane, into the lumen, lowering its pH by 2–3 units. This process occurring across the thylakoid membrane yields two gradients, (1) pH and (2) electrical, which triggers the compensatory flux of Cl^-, K^+, and Mg^{2+} ions. Increase in concentration of magnesium in the stroma is the key factor affecting MgChl activity. However, too high an increase in Mg^{2+} concentration can disturb the K^+ flux from the cytosol into the stroma, which would consequently decrease the MgChl activity (Shaul, 2002).

Chlorophyll concentration in the leaves of crop plants is affected by numerous factors. It seems obvious, based on the biochemical functions of chlorophyll pigments, that magnesium, phosphorus, and nitrogen are crucial for the synthesis of these pigments (Masuda, 2008). However, some questions arise: (1) To what extent does the chlorophyll concentration in leaf respond to the supply of key elements? (2) Does the chlorophyll concentration follow the same pattern in response to nitrogen supply? (3) Does the chlorophyll concentration show a seasonal trend? Despite the importance of these questions, there are only a few recent field studies concerning the first. Sabo et al. (2002), working with wheat, showed that increasing magnesium concentration significantly affected the concentration of chlorophyll a but not chlorophyll b. Studies on magnesium supply in tea plants showed a slight increase in magnesium concentration in harvested leaves following magnesium fertilizer application, independent of the chemical form of the magnesium (Jayaganesh et al., 2011a). This experiment also confirmed a significant role of other nutrients, such as K and S, in chlorophyll concentration in the flush leaves. The greatest chlorophyll concentration was obtained in fertilizer treatments containing magnesium sulfate.

The effect of the nitrogen supply (fertilizer rates) on chlorophyll concentration has been more extensively studied. It has been documented that the photosynthetic potential of leaves depends on the content of proteins, which are constitutive components of both the thylakoid and the ribulose bisphosphate carboxylase (RuBisco). This enzyme shows a conservative response to increasing nitrogen supply. This response is due to the fixed proportion of the thylakoid N to chlorophyll N, which is 50 mol of the thylakoid N per 1 mol of chlorophyll N. The concentration of RuBisco in plant responds in a different manner to an increased supply of nitrogen. Under conditions of low irradiance or insufficient supply of nitrogen, chlorophyll and nitrogen concentrations in the leaves follow a linear relationship. This trend is because the effect of baseline or moderate irradiation results in higher investment of the plant in the leaves, which in turn decreases RuBisco concentration per unit leaf area (Evans, 1989; Makino et al., 1997). An optimum supply of nitrogen leads to progressive accumulation of RuBisco up to the saturation level. Under an elevated supply of nitrogen leading to an accumulation of soluble proteins, the rate of carbon fixation is limited by the thylakoid nitrogen (Evans, 1989).

The optimum level of RuBisco in the leaves of crop plants is the key factor for this discussion in terms of food production and nitrogen-use efficiency (Murchie et al., 2009). Under field conditions,

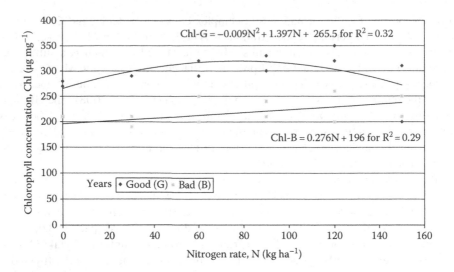

FIGURE 6.6 Impact of fertilizer nitrogen on chlorophyll *a* concentration in the corn (Z. *mays* L.) ear leaf at anthesis. *Legend*: G, good years with normal distribution of precipitation; B, bad years with prolonged drought. (Based on Szulc, P. et al., *Zemdirbyste-Agriculture*, 99(3), 247, 2012.)

the nitrogen supply to the photosynthetic apparatus depends on numerous environmental factors that have an impact on its uptake. In general, there is a linear relationship between nitrogen rates and chlorophyll concentrations in the indicative parts of crop plants (Baranisrinivasan, 2011; Szulc, 2009). Pot and field experiments, as a rule, confirm the additive increase in chlorophyll concentration in leaves in response to increasing rates of applied nitrogen, as documented for black gram (*Vigna mungo* Hepper) (Baranisrinivasan, 2011) and soybean (Ahmed, 2011). A field study on maize showed a much different pattern of chlorophyll *a* response to increasing rates of nitrogen (Szulc et al., 2012). Chlorophyll concentration in maize leaves at anthesis, as illustrated in Figure 6.6, was variable year to year. In years with prevailing drought, chlorophyll concentration increased in accordance with the progressively increased amount of fertilizer nitrogen. In years with ample water supply, this relationship followed a quadratic regression model.

Chlorophyll synthesis is significantly dependent on potassium and magnesium relationships during the light phase of photosynthesis (Shaul, 2002). This relationship is very difficult to demonstrate in crop production practice. As reported by Abu-Zinada (2009), a positive impact of potassium has been observed for potatoes. In contrast, a study by Jayaganesh et al. (2011a) in tea (*Camellia sinensis* Kuntze) plants showed that excessive application of potassium caused chlorophyll to decrease by as much as 50%. Among other nutrients, the supply of molybdenum seems to be of great interest due to its impact on the synthesis of chlorophyll precursors. As documented by Yu et al. (2006) for wheat, a lack of molybdenum decreased the concentration of protoporphyrin IX in addition to other molecules.

The seasonal trend of chlorophyll concentration in cereals and in other seed crops shows, as a rule, the highest values at the beginning of anthesis, which is the critical stage of yield formation. As documented by Sabo et al. (2002), the concentration of chlorophyll in the wheat canopy peaked at the beginning of flowering. In addition, the concentrations of both chlorophyll molecules were significantly correlated with magnesium and nitrogen, which also reached maximum concentrations just before the onset of anthesis. As a result, at this stage of seed crop development, magnesium reaches its greatest uptake rate (Barraclough, 1986; Miller et al., 1994). Its high uptake should be coupled with the maximum growth rate of the whole plant assimilation area. The process of chlorophyll degradation begins as the first seed appears and the plant accelerates toward maturity (Barry, 2009).

6.3.3 Magnesium Deficiency: Physiological Background

6.3.3.1 Dry Matter Partitioning

Magnesium deficiency in crop plants is, in general, easily recognized by the pale color of the leaves, which is called classical chlorosis (Bergmann, 1992). A recent physiological study showed that non-visible magnesium deficiency symptoms can be observed much earlier than the leaf color changes. These hidden symptoms are important for developing appropriate and practical diagnostic tools considering the significance of Mg for maintaining crop production. The order of metabolic events, postulated as the early magnesium deficiency symptom, is as follows:

An elevated accumulation of sugars in leaves
A disturbance of dry matter partitioning among plant organs
Chlorophyll degradation processes—chlorosis
An accelerated increase in concentration of ROS
Destruction of leaf green area—necrosis

In the last 20 years, two main concepts related to indices of *hidden* markers of magnesium deficiency have been proposed. Both approaches are based on studies conducted mainly in nutrient solution cultures. The first concept assumes growth disturbance of a crop plant in medium that is lacking or has extremely low magnesium after germination. In accordance with this concept, the primary symptom of magnesium deficiency is the increased accumulation of sugars in the plant assimilatory organs, the leaves, which are considered the physiological source of sugars. The elevated accumulation of sugars in magnesium-poor leaves due to reduced phloem transport can be significantly higher compared with the leaves from plants fed adequate Mg. As documented for numerous crops, the growth of the root system, for example, in bean (*Phaseolus vulgaris* L.) stops with low Mg (Cakmak et al., 1994). The key reason for the limited export of sugars produced by Mg-deficient leaves is the disturbed action of phloem loading, the efficiency of which depends on the concentration of magnesium (Verbruggen and Hermans, 2013). The second concept of magnesium deficiency assumes a disturbance in dry matter partitioning due to a sudden decrease in magnesium supply during the full vegetative growth period. This concept more closely resembles natural conditions of crop plant growth and is based on data from sugar beet, arabidopsis (Hermans et al., 2004, 2005; Verbruggen and Hermans, 2013), and rice (Ding et al., 2006; Ding and Xu, 2011).

Based on the concepts presented earlier, scientific reports, and observations, it is possible to formulate a strategy for crop plants to cope with magnesium deficiency. Four key stages in the adaptation of plants to the shortage of magnesium during vegetative growth have been proposed:

1. *Early adaptation*: plant growth depends on the root Mg uptake activity from the available resources in the soil.
2. *Advanced adaptation*: plant growth depends on the internal magnesium resources in the root.
3. *Critical adaptation*: plant growth depends on the internal magnesium resources both in the roots and in older leaves.
4. *Growth rate degradation*: plant growth depends on the recycling of the whole, but limited, Mg-plant resources.

In the earliest stages of magnesium deficiency, a plant increases the export of sugars from the leaves to the roots. As a consequence of this adaptation strategy, an increase in the root absorption area extends the capability of the plant to take up extra water and dissolved nutrients from soil resources, including mobile nutrients such as nitrogen and magnesium but also immobile nutrients such as phosphorus and potassium. As reported by Ding and Xu (2011), the translocation ratio of magnesium from the roots to shoots in rice was very high, reaching 95% at 15 days and 97% at 30 days after supply of magnesium; these ratios were slightly higher in Mg-deficient plants. As a result, the rate of shoot growth decreased, which in turn increased the root to shoot ratios. Therefore, during the early stages of growth of crop plants under field conditions, visible symptoms of magnesium deficiency appear only under extremely unfavorable growth conditions such as low pH, poor soil Mg availability, and drought (Grzebisz, 2013).

The advanced stage of magnesium deficiency appears to be due to an insufficient rate of magnesium flow toward the roots in the growth milieu. Under these conditions, the root magnesium resources are explored. It has been documented that the endodermis and the xylem parenchyma cells are the temporary storage areas of magnesium. Therefore, these resources are the first to be explored during a shortage of magnesium. The observed phenomenon is the result of two simultaneously occurring processes. The first, which is the dominant one, is due to a fast dry matter yield increase, which in turn results in a Mg concentration decrease, called the *dilution effect*. The second, which is the minor process, is due to easy remobilization of Mg^{2+} ions from root resources (Ding and Xu, 2011; Hermans et al., 2004; Kobayashi et al., 2013b).

In the third stage of plant crop adaptation to Mg shortage, plant growth depends on root and shoot reserves. The root resources of magnesium are of minor importance due to intensive exploitation in the early stages of Mg deficiency. Therefore, under prolonged magnesium stress, crop plant growth depends much more on its aboveground resources, accumulated previously in older leaves. As reported by Ding and Xu (2011), the Mg-deficient rice plants showed the first perceptible symptoms of magnesium chlorosis on the 35th day after the imposed Mg treatment, that is, after 65 days of rice growth. The visible symptoms of chlorosis, occurring on old leaves, indicate an accelerated rate of chlorophyll destruction, imposed by a strong demand of new growing tissues for magnesium. In the earlier study in rice, Ding et al. (2006) showed that magnesium concentrations below 1.1 g kg^{-1} DW resulted in biomass decrease, which is consistent with a reduction in net photosynthesis and the simultaneous appearance of perceptible deficiency symptoms. This study on rice is consistent with a previous investigation on leafy crops, such as sugar beet (Hermans et al., 2004). In that study, the significant reduction of total biomass of beet plants was first noted after 20 days of growth in the Mg-free solution. This stage of growth corresponds to the full rosette stage, which is crucial for the final yield of sugar beet (Grzebisz, 2013). Magnesium decrease was noted earlier in the true roots and storage roots (fourth day of treatment) compared with the leaves (eighth day). In the first two plant parts, the magnesium concentration drop was sudden, but in the leaves, the decrease was gradual, following an exponential regression model.

The fourth stage of magnesium shortage leads to the deficiency of magnesium in the young, fully expanded leaves. The key biochemical–physiological symptoms are reflected by an elevated accumulation of sugars in older leaves, leading to insufficient food supply to the other plant parts (Figure 6.7). The main reason for the decreased transport of sugars from the source leaf to the sink, which are the new leaves, is dysfunction of the phloem due to magnesium shortage. Transport of sucrose is an active process that requires an input of metabolic energy. Magnesium is essential for H$^+$-ATPase activity within the plasma membrane in cells within the sieve tube (Verbruggen and Hermans, 2013). Prolonged Mg deficiency during the vegetative plant growth leads to a reduction of both root and aboveground biomass. This process results from the sharp suppression in the growth of aerial plant parts as indicated by the stunting of magnesium-deficient plants. As reported by Kumar Tewari et al. (2004), 30-day-old corn plants, grown for 15 days in the Mg-free solution, had reductions in biomass of 40% and reductions of concentration of total chlorophyll of 31%, compared with the plants well fed with magnesium. The reduced but balanced biomass of the basic plant parts indicates plant growth dependence on internal magnesium cycling. A prolonged magnesium deficiency due to a limited supply of carbohydrates results in the accelerated decrease in root system size, in turn leading to decline of productivity in the whole crop plant. An excellent example that supports this observation is the mulberry (*Morus alba* L.) plant. As reported by Kumar Tewari et al. (2006), the shoot biomass of 62-day-old mulberry plants that were deficient in magnesium was reduced to 18% of the well Mg-fed plants; an even larger decrease was recorded for roots (to 13%). In addition,

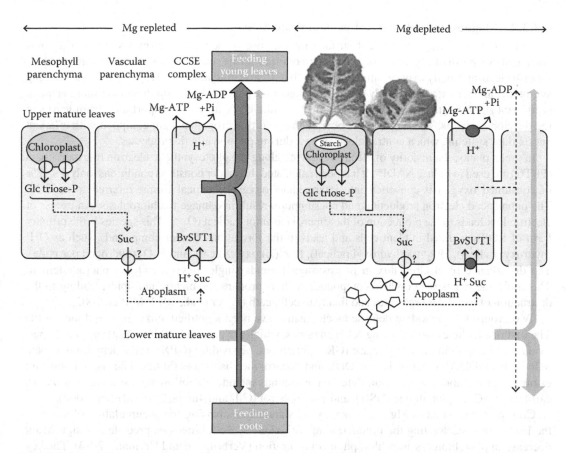

FIGURE 6.7 Allocation of sugar sink organs during Mg deficiency in sugar beet. Model of sucrose (Suc) phloem loading strategy II (apoplastic loading) species. (With kind permission from Springer Science+Business Media: *Plant Soil*, Physiological and molecular response to magnesium nutritional imbalance in plants, 368, 2013, 87–90, Verbruggen, N. and Hermans, C.)

After synthesis in the cytoplasm of mesophyll cells, sucrose is exported (through uncharacterized transport systems) to the apoplasm for uptake into companion cell-sieve element (CCSE) complex. Sucrose transport across the plasma membrane of companion cell is operated through symport with protons (SUT1). That step is energized by H$^+$-motive force generated by proton-pumping APTase in the plasma membrane of the CCSE complex, which requires Mg-ATP for their activity. Impact is here illustrated in an experimental setting inducing Mg deficiency with mature plants (Hermans et al., 2005). Lower Mg metabolic pool, particularly in most upper Mg-deficient leaves, is thought to impact on ATPase activity in the CCSE complex. Upon Mg deficiency, sucrose export to the roots is proportionally less affected than that to the youngest leaves. The upper mature leaves, where the deficiency symptoms first manifest, have the highest starch (revealed here by iodine staining) and sucrose contents and the lowest Mg concentration among all plant organs. Those leaves provide less carbon resources to young leaves. Consequently, the overall aerial biomass is decreased compared with control plants. At later stages of Mg deficiency induction, intermediate and lower leaves also accumulate higher starch amount; and finally, the root and taproot growth is affected when treatment is prolonged. The width of the arrows is proportional to the flow of assimilates delivered to the sink (immature leaves or roots) organs.

the magnesium concentration decreased sixfold in young leaves and 14.5-fold in old leaves. As a result, the Mg-deficient mulberry plants were not able to produce enough sugars to maintain the required growth rate. The same response has been observed for a lemon (*Citrus volkameriana* Tan. and Pasq.) tree that was exposed to magnesium deficiency for seven months (Lavon et al., 1999). The yield of the Mg-deficient plants was 37% lower than the well-fed plants. The almost unchanged leaf/root and shoot/root ratios compared with well-fed plants indicate the stunted growth pattern.

6.3.3.2 Magnesium-Induced Antioxidant Protection

The necrosis appearing on the leaf surface in response to acute magnesium stress is the most advanced symptom of Mg deficiency. The main reason for this reaction is oxidative damage to chloroplasts, leading finally to programmed cell death (Gill and Tuteja, 2010). These visible and striking symptoms appear extremely rarely during the vegetative part of plant growth and are more acute at seed plant ripening, that is, at stages of a very high rate of assimilate transportation from leaves to the developing fruits. Therefore, it can be hypothesized that crop plants are equipped with efficient antioxidant systems, which control electron flow during photosynthetic processes.

In green plants, the majority of electrons flowing along the photosynthetic electron transport chain (PETC) are used to reduce $NADP^+$. It has been calculated that under nonstress conditions, only 1%–2% of consumed oxygen is converted into intermediate oxygen chemical forms, referred to as ROS. The unbalanced electron production and consumption result in damage to chloroplasts. An excess of electron flux leads to the production of the superoxide anion radical ($O_2^{\cdot-}$). This species is the primary form of ROS generated in plant cells and leads to the formation of other compounds, such as $OH^{\cdot}$ (hydroxyl radical), $HO_2^{\cdot}$ (perhydroxyl radical), $RO^{\cdot}$ (alkoxy radical), and H_2O_2 (hydrogen peroxide). The disturbance of electron flux in photosystem I yields singlet oxygen ($^1O_2$) in photosystem II. This molecule reacts with organic compounds such as proteins, pigments, and lipids, leading to the destruction of a cellular organelles and finally to cell death (Foyer et al., 2012; Mittler, 2002).

Two groups of antioxidant defense mechanisms have been identified: enzymatic and metabolic. The first one includes the following set of enzymes, which are present in chloroplasts: ascorbate peroxidase (APX), glutathione reductase (GR), glutathione peroxidase (GPX), monodehydroascorbate reductase (MDHAR), peroxidase (POD), and superoxide dismutase (SOD). The most important examples of metabolic antioxidant defense compounds include the following: ascorbic acid (AsA), carotenoids (Car), glutathione (GSH), and γ-tocopherol (Gill and Tuteja, 2010; Mittler, 2002).

Crop plants respond to Mg deficiency by substantially increasing the accumulation of sugar in the leaves and accelerating the production of antioxidants. Both processes precede any significant decrease in plant biomass and chlorophyll concentration (Verbruggen and Hermans, 2013). The key question is not whether but to what extent are these two processes interrelated. It is well recognized that sugars are transiently stored as starch in chloroplasts during the day, undergoing hydrolysis during the night. The key question that therefore remains is to what extent does an elevated sugar accumulation in the leaf due to a shortage of magnesium act as the stimulus for ROS production. Concepts addressing the role of sugars as signaling molecules have been recently proposed (Bolouri-Moghaddam et al., 2010). These authors assumed an impact of sugar metabolism on hormone status of plants, in turn inducing the synthesis of ROS. Hermans et al. (2010) supported this concept by showing that magnesium deficiency results in enhanced production of ethylene by arabidopsis.

As shown in Table 6.2, magnesium shortage significantly induces the production of various enzymes and metabolites, considerably protecting plant tissues against photooxidative damage. The elevated concentration of certain enzymes in response to low magnesium supply has been documented only for a few crops grown under controlled conditions (nutrient solution). The effective control of superoxide radicals and hydrogen peroxide levels in plant cell depends on a balance between SOD and APX or SOD and CAT activities (Mittler, 2002). As reported by Hermans et al. (2010), magnesium shortage leads to an extra uptake of calcium, iron, and copper by plants. In the chloroplast stroma, the elevated concentration of transition metals, such as copper and iron, activates a series of ROS-generating processes. For example, the Fenton reaction yields $OH^{\cdot}$, which is the most reactive chemical species in the biological world (Gill and Tuteja, 2010). Two-week magnesium-deficient maize, as reported by Kumar Tewari et al. (2004), exhibited SOD activity that was increased by 28% compared with plants grown with an ample supply of magnesium. A long-term experiment (95 days) with pepper (*Capsicum annuum* L.) showed a progressive increase in SOD concentration in all Mg-deficient leaves, independent of their age. In this case, the lowest difference in the SOD elevation was attributed to the youngest leaf (Anza et al., 2005).

TABLE 6.2
Magnesium-Induced Activity of Antioxidant Enzymes and Accumulation of Antioxidants in Selected Crops

Antioxidant Enzymes and Metabolites	Response to Magnesium			
	Corn[a]	Pepper[b]	Mulberry[c]	Citrus[d]
Superoxide dismutase (SOD)	++	++	+	++
Catalase (CAT)	0	n.d.	–	–
Ascorbate oxidase (APX)	++	++	++	++
Glutathione reductase (GR)	n.d.	++	n.d.	+
Ascorbic acid (AsA)	n.d.	++	++	++
Carotenoids (Car)	–	n.d.	–	–
H_2O_2 concentration	0	n.d.	++	n.d.

Responses: –, decrease; 0, no significant response; +, more than 10%; ++, more than 25% increase compared to insufficient magnesium growth medium; n.d., not determined.
[a] Kumar Tewari et al. (2004).
[b] Anza et al. (2005).
[c] Kumar Tewari et al. (2006).
[d] Tang et al. (2012).

The highest increase of this enzyme (of 250%) in Mg deficiency has been reported by Tang et al. (2012) for Mandarin orange (*Citrus reticulata* Blanco) trees.

APX plays an important role in scavenging ROS compounds in chloroplasts. This enzyme has a high affinity for H_2O_2, which is effectively decomposed (Mittler, 2002). As presented in Table 6.2, an elevated increase in APX was noted for all tested plants. In corn, for example, the activity of APX in Mg-deficient plants increased by one-third, whereas under P shortage the APX activity doubled (Kumar Tewari et al., 2004). In Chile piquin, the time course of APX was leaf and stage dependent and was significantly increased in older leaves (Anza et al., 2005). GR occurs in great amounts in chloroplasts. The high level of GSH in plant cells is required for the regulation of the ASH–GSH cycle. The second component of this cycle, AsA, accumulates mostly in chloroplasts (30%–40% of total plant content). The main role of AsA is to donate electrons to numerous enzymatic and nonenzymatic reactions. This metabolite can directly neutralize $O_2^{\cdot-}$ and $OH^{\cdot}$ radicals and acts as a donor of electrons to APX. As presented in Table 6.2, GR activities and AsA concentrations were much higher in Mg-deficient plants. The concentration of AsA and the activities of APX, MDAR, SOD, and GPX were much higher in Mandarin orange (*C. reticulata* Blanco) plants that were Mg deficient than in plants well fed with magnesium (Tang et al., 2012). The study of GR by Anza et al. (2005) showed that the mechanisms responsible for protection of Chile piquin (*C. annuum*) relied much more on AsA than on GR. Hermans et al. (2010) studied the expression of antioxidants in the model crop *A. thaliana* under magnesium shortage and did not find any differences in the concentration of AsA and GS in mature leaves, but recorded a significant increase in the roots. In addition, the ratio of GSSG/GSH was much higher for magnesium-stressed plants.

6.4 MAGNESIUM PROFILE OF CROP PLANTS

6.4.1 Concentration and Distribution

Magnesium is an important nutrient among another 25 elements in the whole food chain. Its content in the edible part of crop plants is a significant characteristic of food quality and consequently of human health. The requirement of humans for magnesium varies based on age, weight, and gender from 80 to 420 mg day^{-1} (Murphy et al., 2008). The daily intake of magnesium is covered by various types of products, which are grouped based on magnesium concentration (expressed in mg kg^{-1} raw or final product) into four classes (Gebhardt and Thomas, 2002):

1. Very high (>1000 mg kg^{-1} DM): buckwheat grain, cocoa, almonds, pumpkin seeds, and bread made from whole grains of wheat, rye, and oats
2. High (500–1000 mg kg^{-1} DM): spinach and boiled soybeans
3. Medium (250–500 mg kg^{-1} DM): artichokes, potatoes with skin, and green beans
4. Low (<250 mg kg^{-1} DM): apples, lettuce, and potatoes (boiled, without skin)

In the modern lifestyle, the key sources of regular magnesium needs are products made from cereals. The daily magnesium requirements vary in cereal products: 45% for wheat (corn, rice, rye) to 63% for oats. This high contribution exists because cereals are the source of approximately 50% and 60% of commonly consumed carbohydrates and proteins, respectively (Cordain, 1999). Therefore, a rational choice of edible products is a decisive factor to cover daily magnesium requirements.

6.4.1.1 Magnesium Trends during the Life Cycle of a Cereal Crop

The general course of magnesium concentration in vegetative parts of cereal crops shows a progressive declining trend from tillering to reach its lowest concentration at full ripening (Hundt and Kerschberger, 1991; Preez and Bennie, 1991). In the high-yielding seed crop plantation, the steepness of the Mg concentration curve is relatively low up to heading, declining precisely from anthesis onward (Grzyś et al., 2012). The important period of photosynthetic activation of the canopy of a cereal crop occurs just during booting and heading (Shearman et al., 2005). In high-yielding winter wheat, a critical concentration of magnesium at booting is 1.5 g kg^{-1} DM, whereas the optimum ranges from 1.5 to 3.0 g kg^{-1} DM (Preez and Bennie, 1991). Hundt and Kerschberger (1991) studied the magnesium distribution among parts of winter wheat during ripening and stressed the importance of grain as the dominating physiological sink. Under conditions of poor magnesium supply to plants during the growing season, approximately 50% of the total content of this element accumulates in grain. Under ample magnesium supply, grain contributes only 25% of the entire plant magnesium. Therefore, the magnesium harvest index (MgHI) varies very broadly depending mostly on environmental factors and agricultural practice (Figure 6.8).

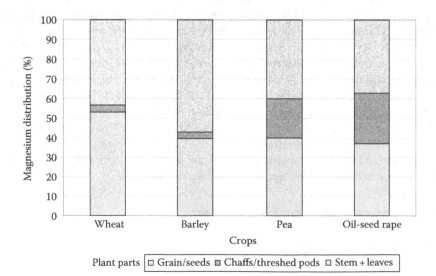

FIGURE 6.8 Magnesium partition between plant parts of selected crops at harvest. (From Grzebisz, W., *J. Elem.*, 16(2), 299, 2011. With kind permission.)

6.4.1.2 Concentrations of Magnesium in C3 Cereals and Pseudocereals

The whole-grain magnesium concentration, in general, is driven by genotype (G); it is significantly modified by environmental factors (E). Therefore, its concentration varies between species and regions, that is, origin of the variety and production site (Table 6.3). A study by Oury et al. (2006) on 175 bread wheat genotypes from many regions of the world showed extremely high differences in magnesium concentration in grain, ranging from 600 to 1400 mg kg^{-1} DM for modern varieties and up to 1890 mg kg^{-1} DM for exotic genotypes. The ranking of the regions in descending order of magnesium concentration is as follows: North and South America (1550 mg kg^{-1} DM) > Asia (1543 mg kg^{-1} DM) > Mediterranean (1504 mg kg^{-1} DM) > France (1443 mg kg^{-1} DM) > Europe (1337 mg kg^{-1} DM). These investigations showed that the G and E interaction can be considered a source of alleles that are potentially available for breeding Mg-rich varieties.

One of the most important events in human civilization was the Green Revolution, which reduced the number of hungry people. The key success was rooted in breeding new generations of intensive, short-stem varieties of wheat and rice. Therefore, the yield potential of current varieties is much higher than that of the older ones, provided that there is an adequate supply of fertilizers. Consequently, the key attribute of modern agriculture is high consumption of nitrogen fertilizer, leading to significant increases in the yields of many crop species, including forages. However, there are some recently discovered negative aspects of this progress. In the United States, between 1940 and 2002, as reported by Thomas (2007), the concentration of magnesium in many food products significantly decreased. Data presented for select products showed the largest decrease in cheeses (−26%), followed by vegetables (−24%), fruits (−16%), and meat (−15%). These unwanted changes generally have resulted from poorer quality plant edible parts and fodder (Davis, 2009; Thomas, 2007).

Grain is the staple source of magnesium for humans, and grains therefore require significant attention from farmers, breeders, and consumers. A study by Fan et al. (2008) showed a much higher concentration of magnesium in the grain of old wheat cultivars compared with modern wheat cultivars. The average Mg concentration in the grain of current wheat varieties cultivated from 1968

TABLE 6.3
Magnesium Concentration in Grains of Cereals and Pseudocereals

Crop	mg kg^{-1} DM	Region, Remarks	Reference
Soft wheat (*Triticum aestivum* L.)	1008–1318	Spain, landraces	Rodríguez et al. (2011)
	1049–1284	Poland, grain test	Kowieska et al. (2011)
	610–850	Australia, field trial	Craighead and Martin (2001)
Hard wheat (*Triticum aestivum* L.)	1490–1640	Canada, variety trial	Boila et al. (1993)
	1285–1289	Canada, field trial	Nelson et al. (2011)
Spelt wheat (*Triticum spelta* L.)	1260–1420	United States, genotype trial	Ranhotra et al. (1996)
	1630–1910	Poland, genotype trial	Suchowilska et al. (2012)
Rye (*Secale cereale* L.)	970–1006	Poland, grain test	Kowieska et al. (2011)
	1110–1250	Canada, variety trial	Boila et al. (1993)
Triticale (×*Triticosecale* Wittm. ex A. Camus.)	1050–1146	Poland, grain test	Kowieska et al. (2011)
Barley (*Hordeum vulgare* L.)	1202–1378	Poland, grain test	Kowieska et al. (2011)
	1280–1410	Canada, variety trial	Boila et al. (1993)
Oat (*Avena sativa* L.)	1430–1450	Canada, variety trial	Boila et al. (1993)
Rice (*Oryza sativa* L.)	896–1480	China, introgression lines	Garcia-Oliveira et al. (2008)
	1640–1760	Nigeria, variety trial	Oko et al. (2012)
Buckwheat (*Fagopyrum esculentum* Moench)	2500–2780	Japan, variety test	Ikeda et al. (2004)

to the present decreased to 924 mg kg^{-1} DM compared with 1138 mg kg^{-1} DM for old varieties (1845–1967). The wheat varieties cultivated in the United States between 1842 and 1965 on average were richer in Mg (1403 mg Mg kg^{-1} DM) compared with new ones (1308 mg Mg kg^{-1} DM) (Murphy et al., 2008). Consequently, the importance of wild wheat species is increasingly emphasized in breeding programs. The most promising species are einkorn (*Triticum monococcum* L.) and emmer (*Triticum dicoccum* L.). Both species are much richer in magnesium compared with commercial cultivars (Suchowilska et al., 2012). In this study, the magnesium concentration in einkorn varied from 1170 to 1800 mg Mg kg^{-1} DM and for emmer from 1480 to 2010 mg Mg kg^{-1} DM. Chatzav et al. (2010), working with wild emmer in the Near Eastern Fertile Crescent, reported a Mg concentration of 1800–1900 mg kg^{-1} DM in 25% of the studied genotype. This high frequency of Mg-rich genotypes can be considered a broad source of alleles that are potentially applicable to improving Mg concentration in cultivars.

The search for cereals that are rich in grain magnesium should also consider other species. The set of data presented in Table 6.3 implicitly indicates that the domesticated spelt wheat (*Triticum spelta* L. syn. *T. aestivum* L. subsp. *spelta* Thell.) is an important source of magnesium, which can be successfully used in bread production. Magnesium concentration in rye grain, a second bread species, is much lower compared with other cereals. This crop is cultivated on sandy soils, which are typically poor in available magnesium (Grzebisz, 2011). Triticale, barley (*Hordeum vulgare* L.), and oats (*Avena sativa* L.) are typical fodder plants but are also used for food production. One of the most promising grain crops is common buckwheat (*Fagopyrum esculentum* Moench), which belongs to the pseudocereals. As shown in Table 6.3, magnesium concentration in this grain is twice as high as in modern wheat varieties. A study by Ikeda et al. (2004) on the mineral composition of common and tartary buckwheat (*Fagopyrum tartaricum* Gaertn.) grain showed concentrations ranging from 2500 to 2800 and 1900 to 2700 mg kg^{-1} DM, respectively.

The magnesium concentration in grain is highly variable and responds strongly to weather conditions during the growing season. This characteristic is as important as the genetic mineral profile. As reported by Boila et al. (1993), environment and genotype interaction was responsible for 34% of the magnesium concentration variability in wheat, 20% in barley, 17% in rye, and 12% in oat grain. The latter crop is highly sensitive to weather conditions during the growing period. As reported by Pisulewska et al. (2009), in the dry season of 2000, Mg concentration declined, independent of the crop variety, by 33% compared with years with ample water supply. The question remains, to what extent do agronomy practices affect magnesium concentration? Nelson et al. (2011), using the *side-by-side comparison* approach, tested the response of five modern and old red spring wheat varieties to organic and conventional management systems. In general, grain produced in the former system was richer in magnesium.

Milling is the most important external process that affects Mg concentration in grain flour. For example, rice grain is composed of bran layers (6%–7% by weight), which are surrounded by the endosperm (90%) and embryo (2%–3%). The content of magnesium among the basic anatomical structures of brown rice is as follows in decreasing order: bran > outer endosperm > middle endosperm > core endosperm. As documented by Wang et al. (2011), the genotypic-induced distribution shows a cultivar dependence, mostly for bran and outer endosperm. The content of Mg in the bran of the ZN7 cultivar contributed approximately 34% of its total content, but in the ZN60 cultivar, this increased to 57%. Magnesium distribution in the outer endosperm was 52% and 26%, respectively. Therefore, the first cultivar is much less suitable for strong milling.

6.4.1.3   Corn, Pulses, and Oil Plants

The course of magnesium concentration in corn during vegetative growth is very similar to that in C3 cereals. The crop reaches the maximum magnesium concentration at the 6–8 leaf stage, declining thereafter toward maturity (Hundt and Kerschberger, 1991). At this stage, corn is sensitive to magnesium supply because it is critical for inflorescence initiation. Szulc et al. (2012) found that early application of magnesium, leading to higher nitrogen and magnesium uptake, resulted

TABLE 6.4
Magnesium Concentration in Maize Kernels and Seeds of Pulses and Oil Crops

Crop	mg kg⁻¹ DM	Region, Remarks	Reference
Corn (*Zea mays* L.)	1480–1620	United States, field trial	Heckman et al. (2003)
	1320–1710	Brazil, variety trial	Ferreira et al. (2012)
	1373–1380	India, genotype trial	Shobha et al. (2010)
	1148–1350	Croatia, genotype trial	Brkić et al. (2003)
	1115–1565	Nigeria, genotype trial	Menkir (2008)
Common bean (*Phaseolus vulgaris* L.)	1020–2050	United States, variety trial	Moraghan et al. (2006)
Mung bean (*Vigna radiata* R. Wilczek)	1290–1660	India, variety trial	Dahiya et al. (2013)
Cowpea (*Vigna unguiculata* Walp)	1300–2400	S. Africa, genotype trial	Belane and Dakora (2012)
Soybean (*Glycine max* Merr.)	2234–2750	Brazil, genotype trial	Destro et al. (2002)
Oilseed rape (*Brassica napus* L.)	3080–3270	Poland, field trial	Barłóg et al. (2005a)

in a significant increase in seedling biomass. The differences were sufficiently high to impose the pattern of plant growth and grain yield. In maize, the critical stage of magnesium uptake is in two parts. The first occurs just before tasselling, and the second occurs during the early dough stage of cob growth. At the latter stage, the magnesium concentration varies greatly among plant parts in decreasing order as follows: ear leaf (1800 mg kg⁻¹ DM) > stover (1200 mg kg⁻¹ DM) > immature ear (1000 mg kg⁻¹ DM) > husk = grain (900 mg kg⁻¹ DM) (Bruns and Ebelhar, 2006).

Corn products, as a staple food, cover 15% and 20% of the world's consumed proteins and carbohydrates, respectively (Nuss and Tanumihardjo, 2010). Therefore, magnesium concentration in kernels is an important indicator of food product quality (Table 6.4). Breeding for Mg-rich varieties is the first step in increasing food quality. To date, scientific reports about this matter are not consistent. As illustrated in Figure 6.9, magnesium concentration in the grain of 28 maize genotypes was negatively correlated with grain yield. Based on Davis's approach (2009), this phenomenon can be called *the genetically induced magnesium dilution effect*. Heckman et al. (2003), studying 23 cornfields, documented a site-induced variability within the range of 0.88–2.18 g Mg kg⁻¹ DM. In addition, the yield of grain varied from 4.9 to 16.7 t ha⁻¹. In spite of this huge variability, magnesium concentration did not show any significant impact on grain yield, in turn stressing that its

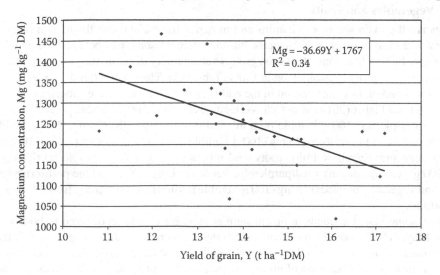

FIGURE 6.9 Impact of corn grain (85% dry matter) yield on magnesium concentration in grain—*the syndrome of Mg dilution*. (Based on Brig, I. et al., *Maydica*, 48, 293, 2003.)

supply to plants was sufficient. Ferreira et al. (2012), testing a wide range of varieties (commercial, landrace, and double-, triple-, and single-cross hybrids), concluded that in spite of high genetic differences, water supply to plants was the key factor affecting the variability of magnesium concentration in corn grain.

The consumption of pulses has occurred throughout the history of human civilization, including peas and lentils in the Fertile Crescent and kidney beans in Mesoamerica (Phillips, 1993). Throughout the world, the production of a few species dominates; these include common bean, followed by peas, chickpea (*Cicer arietinum* L.), cowpea (*Vigna unguiculata* Walp.), broad bean (*V. faba* L.), and lentils (*Lens culinaris* Medikus) (Campos-Vega et al., 2010). All of these species are rich in carbohydrate, protein, and magnesium. The seeds of pulses typically exhibit a much higher Mg concentration compared with grain of cereals. To date, studies on mineral variability in the genotypes of legumes deliver a broad field for successful breeding, as presented in Table 6.4. Destro et al. (2002), searching 72 food-type genotypes of soybean in Brazil, showed a great diversity in magnesium grain concentration. The authors found six genotypes that were high in concentration of both protein and several nutrients, including magnesium. As reported by Moraghan et al. (2006) for the common bean, the key source of variability in this species is the seed coat. A high correlation between the dry matter content in the seed coat of bean and magnesium concentration (r = 0.91) has been found. Cowpea is an important legume crop in Africa because both the leaves and the seeds are consumed. A 2-year study by Belane and Dakora (2012) documented a higher magnesium concentration in the leaves than in the grain of cowpea.

Seeds of oil-bearing crops, such as oilseed rape, can be used for oil production, and their residues can be used as fodder. Certain other crops, such as soybean, have much broader functionality, including use as food products. The comparative analyses of magnesium concentration in oil-bearing seeds or kernels cultivated in Turkey reported significant differences between species (Özcan, 2006). The five Mg-richest species in descending order are as follows: cotton (2620 mg kg^{-1} DM) > stone pine (2420 mg kg^{-1} DM) > poppy = linen ($\approx$2330 mg kg^{-1} DM) > sunflower (2250 mg kg^{-1} DM).

Corn with 1154 mg kg^{-1} DM was among four other species, turpentine, pistachio, hazelnut, and walnut, with the poorest source of magnesium. Seeds of peanuts with 1377 mg kg^{-1} DM can be classified into the low magnesium subgroup. Oilseed rape with 1898 mg kg^{-1} DM represents the medium class of Mg concentration. However, as shown in Table 6.4, Mg concentration in seeds is much higher based on other published studies.

6.4.1.4 Vegetables and Fruits

Vegetables are the main source of vitamins and minerals for maintaining the human body in good health. Fruits are minor sources of minerals, but they contain some phytochemicals, which significantly improve human functions. Both food groups substantially differ in magnesium content, which is several times higher in vegetables than in fruits (Table 6.5). The general trend for magnesium content, which is based on its concentration in the edible portion for each of the subgroups, as proposed by Pennington and Fisher (2010), is as follows: legumes (49 mg · 100 g^{-1} edible portion) > dark green leafy vegetables (44 mg · 100 g^{-1} edible portion) > allium family bulbs (21 mg · 100 g^{-1} edible portion) $\geq$ other fruits and vegetables (20 mg · 100 g^{-1} edible portion) > cabbages (18 mg · 100 g^{-1} edible portion) > deep orange/yellow fruits, roots, and tubers (15 mg · 100 g^{-1} edible portion) $\geq$ lettuce (14 mg · 100 g^{-1} edible portion) = red/purple/blue berries (14 mg · 100 g^{-1} edible portion) $\geq$ tomatoes/red fruits and vegetable tomatoes (13 mg · 100 g^{-1} edible portion) > citrus family fruits (11 mg · 100 g^{-1} edible portion).

Recently, some negative trends in magnesium content have been recognized in both vegetables and fruits. Mayer (1997) examined the concentrations of seven nutrients in vegetables and fruits that were consumed in Great Britain in the 1930s and 1980s and observed a significant decline in magnesium concentrations in edible parts of numerous species. The most striking decrease was noted for carrots (fourfold decrease from 12 to 4 mg 100 g^{-1}), followed by cabbage, radishes, Brussels sprouts (more than twofold decrease), onions, and celery (twofold decrease). The magnesium contents,

TABLE 6.5
Mean Magnesium Concentration in Vegetables and Fruits

Crop	Concentration	Region, Samples Source	Reference
Vegetables, mg kg⁻¹ DM			
Potato (*Solanum tuberosum* L.)	1160	Finland, food samples	Ekholm et al. (2007)
	1340	United States, field trial	Clough (1994)
Cabbage (*Brassica oleracea* var. *capitata* L.)	1300	Poland, field trial	Majkowska-Gadomska and Wierzbicka (2008)
Carrot (*Daucus carota* L.)	1500	Finland, food samples	Ekholm et al. (2007)
	1600	Ghana, food samples	Adotey et al. (2009)
Lettuce (*Lactuca sativa* L.)	2310	Czech, variety trial	Koudela and Petříková (2008)
Onion (*Allium cepa* L.)	590	Ghana, food samples	Adotey et al. (2009)
Tomato (*Lycopersicon esculentum* Mill.)	1450	Finland, food samples	Ekholm et al. (2007)
	1600	Ghana, food samples	Adotey et al. (2009)
Spinach (*Spinacia oleracea* L.)	1370	Pakistan, food samples	Bangash et al. (2011)
Sweet pepper (*Capsicum annuum* L.)		Poland, variety trial	Jadczak et al. (2010)
Fruits (mg 100 g⁻¹ edible part)			
Apple (pulp) (*Malus domestica* Borkh.)	320	Pakistan, variety trial	Manzoor et al. (2012)
Avocado (*Persea gratissima* L.)	27–40	Spain, variety trial	Hardisson et al. (2001)
Banana (pulp) (*Musa* spp.)	36	Hawaii, variety trial	Wall (2006)
Olives, green (*Olea* spp.)	96–149	Spain, variety trial	López et al. (2008)
Papaya (*Carica papaya* L.)	25	Hawaii, variety trial	Wall (2006)
Peach (pulp) (*Prunus persica* Batsch)	62	Turkey, variety trial	Başar (2006)

which were averaged over 26 studied species, decreased by 20% during this 50-year period. Two approaches have been postulated to explain the observed long-term dilution. The first, called the *environmental dilution effect*, indicates that the fertility of the growth medium increased primarily through intensive application of fertilizers. The faster growth of edible plant parts relative to nutrient uptake leads to a decrease in magnesium concentrations. An extended study on broccoli indicated a significant difference in calcium and magnesium contents between species, that is, representing *genetically induced* decline (Davis, 2009).

The first two subgroups of the aforementioned vegetables are of key importance as sources of magnesium in the daily human diet. However, some other species can provide a substantial amount of magnesium. An extended study by Ekholm et al. (2007) on nutrient concentrations in raw vegetables and fruits provided the following ratings of magnesium sources:

1. *Leafy and fruit-like vegetables*: squash (3180 mg Mg kg⁻¹ DM) > cucumber (2650 mg Mg kg⁻¹ DM) > green beans (2080 mg Mg kg⁻¹ DM) > cauliflower (1950 mg Mg kg⁻¹ DM)
2. *Root vegetables*: radish (2490 mg Mg kg⁻¹ DM) > parsnip (1770 mg Mg kg⁻¹ DM) > turnip (1660 mg Mg kg⁻¹ DM) > beetroot (1500 mg Mg kg⁻¹ DM)
3. *Fruits and berries*: raspberry = peach (1600 mg Mg kg⁻¹ DM) > strawberry (1400 mg Mg kg⁻¹ DM) > blackcurrant (1300 mg Mg kg⁻¹ DM)

The magnesium concentrations in spinach and cabbages are sufficiently high to cover a major part of the daily human magnesium requirement (Table 6.5). Most of the vegetables are sensitive to the supply of nutrients, which, in turn, leads to contrasting results, that is, an elevated yield increase and concomitant magnesium dilution. The genetic potential for the improvement of the magnesium content in vegetables is substantial. In a study by Jadczak et al. (2010), the magnesium content of sweet peppers ranged between varieties from 750 to 1250 mg kg⁻¹ DM. The results were significantly modified by the course of the weather. The genotype–environment interaction impact on magnesium content in edible plant parts was also noted for other crops, such as broccoli, potatoes, and sweet peppers (Farnham et al., 2000; White et al., 2009). Special attention should be given to the potato, which is a key staple food throughout the world. A negative trend of magnesium content in tubers in response to increased yield has also been documented. The magnesium content, which is evaluated by the dry matter content, is only slightly lower compared with that of Mg-rich vegetables. There remains a genetic space for the improvement of the magnesium content in potato tubers. As reported by Gugała et al. (2012), the magnesium content of tubers displayed a significant year-to-year variability (11%); however, the impact of cultivated varieties was much higher (25%).

The magnesium content in the edible portion of fruits is several times lower compared with black-green vegetables (Pennington and Fisher, 2010). A seasonal trend of magnesium concentrations in developing fruits is weakly recognized. Storey and Treeby (2000), who studied oranges (*Poncirus trifoliata* Raf.), demonstrated a two-stage trend in magnesium concentrations during fruit development. In the first stage, which lasts approximately 40 days, the magnesium concentration increases and reaches its peak concentration. Once the magnesium concentration peaks, there is a progressive decrease. In the case of the edible parts (ep) of fruits, it is important to recognize the nutritional value of the plant's anatomical parts. As reported by Juranović Cindrić et al. (2012), the mean Mg concentration in apple peels was 6060 mg kg^{-1} DM, and, in apple flesh, the mean Mg concentration was only 2710 mg kg^{-1} DM. The tissue-induced difference is typical for all apple (*Malus domestica* Borkh.) cultivars; however, some cultivars are richer in magnesium, such as Golden Delicious and Red Delicious (Manzoor et al., 2012). The magnesium concentration in citrus fruit ranges from 8 mg 100 g^{-1} for lemon to 12 mg 100 g^{-1} (ep) for tangerine. There is much room for genetic improvement, as determined by studying magnesium concentration variability in fruits of different orange varieties (Topuz et al., 2005). The magnesium concentration in banana fruit is also highly variable; this variability is induced both genetically and environmentally. The highest content of 64 mg Mg 100 g^{-1} fresh weight has been recorded for banana (*Musa* x *paradisiaca* L.), which originated from Southeast Asia, and the lowest concentration was 14 mg Mg 100 g^{-1} fresh weight for the Indian variety Harichal (Hardisson et al., 2001). As reported by Wall (2006), the magnesium concentration in banana flesh is highly sensitive to the environment and changes have been demonstrated within an identical variety that was grown in Hawaii, with 32.4–45.8 mg 100 g^{-1} fresh weight depending on the field location. The significant impact of the genotype (variety, hybrid, inbreed) is a concern for other fruit species, such as avocado, apples, and plums (Hardisson et al., 2001; Manzoor et al., 2012; Milošević and Milošević, 2012).

6.4.1.5 Fodder Crops

The group of plants that deliver fodder for domesticated animals is extremely broad and includes numerous species that are grown in natural meadows, in pastures, and in fields, for example, alfalfa and clovers. A dairy cow, which yields 25–45 kg milk per day, requires a fodder containing 1.8–2.0 g Mg kg^{-1} DM. Beef cattle require an approximately half this magnesium concentration (McDowell, 1996). In fodder plants, magnesium concentrations in the range of 1.0–1.3 g kg^{-1} DM are considered low and result in a significant yield decrease (Whitehead, 2000). As shown in Table 6.6, the average concentration of magnesium in different fodder plants is highly variable, depending on species, but significantly modified by the environment. In general, the magnesium concentration in grasses is much poorer than that in any other leafy plant. Timothy (*Phleum pratense* L.) represents a grass species that is typically poor in magnesium. Other species, such as ryegrass (*Lolium perenne* L.), are much richer; however, the magnesium concentration undergoes high variability, depending on the soil, the course of the weather during vegetation, and the consecutive cutting. As a rule, magnesium concentrations increase throughout vegetative growth and reach the highest values in summer (Aydin and Uzun, 2008; Pirhofer-Walzl et al., 2011). Leafy fodder plants, such as alfalfa (*Medicago sativa* L.) and clovers (*Trifolium* spp. L.), are

TABLE 6.6

Magnesium Concentration in Selected Fodder and Pasture Plants

Crop	mg kg⁻¹ DM	Region, Remarks	Reference
Ryegrass, perennial or annual	1500	Lithuania, plant test	Juknevičius and Nomeda (2007)
(*Lolium* spp. L.)	1730	New Zealand, plant test	Harrington et al. (2006)
	2900	Botswana, field trial	Aganga et al. (2004)
Orchardgrass (*Dactylis* spp. L.)	2400	UDA, pasture trial	Soder and Stout (2003)
Timothy (*Phleum pratense* L.)	1000	Lithuania, plant test	Juknevičius and Nomeda (2007)
Alfalfa (*Medicago sativa* L.)	2600	Spain, field trial	Lloveras et al. (2004)
	2800	Denmark, field trial	Pirhofer-Walzl et al. (2011)
Red clover (*Trifolium pratense* L.)	2400	Poland, pasture trial	Grzegorczyk et al. (2013)
	3800	Denmark, field trial	Pirhofer-Walzl et al. (2011)
White clover (*Trifolium repens* L.)	2370	New Zealand, plant test	Harrington et al. (2006)
	2800	Poland, pasture trial	Grzegorczyk et al. (2013)
Dandelion (*Taraxacum officinale*	3530	New Zealand, plant test	Harrington et al. (2006)
F.H. Wigg)	3200	Poland, pasture trial	Grzegorczyk et al. (2013)

naturally richer in magnesium and simultaneously present much lower variability during the growing season. In natural meadows, one of the most important components that deliver magnesium is herbs. In this group, valuable sources of magnesium include caraway (*Carum carvi* L.), chicory (*Cichorium intybus* L.), dandelion (Taraxacum officinale F.H. Wigg), cow vetch (*Vicia cracca* L.), common yarrow (*Achillea millefolium* L.), and plantain (*Plantago major* L.) (Grzegorczyk et al., 2013; Harrington et al., 2006; Pirhofer-Walzl et al., 2011).

6.4.2 INTERACTIONS OF MAGNESIUM WITH OTHER NUTRIENTS

The importance of the degree of any imbalance between pairs of plant nutrients depends on its impact on crop plant growth during critical stages of yield formation and on the final yield. Interactions of magnesium with other nutrients are driven by genetic traits; however, these interactions are modified, to some extent, by environmental factors, particularly deficiencies of availability from the soil, or by factors that disturb nutrient uptake processes.

6.4.2.1 Nitrogen

The nitrogen and magnesium interaction results from functions of both elements in plants that relate to chlorophyll biosynthesis, carbon fixation, and to the rate of growth during critical stages of yield formation (Gerendás and Führs, 2013; Grzebisz, 2013; Shaul, 2002). In general, irrespective of growth conditions, there is an assumed positive correlation between nitrogen and magnesium (Osaki et al., 1996). The data that have been presented in published materials are controversial, however, and depend on plant species, genotypes, growth conditions, and fertilizing systems. A study by Shobha et al. (2010) of the chemical composition of 11 corn genotypes, which were grown under the same conditions (water, nutrient supply), demonstrated that a small variability in nitrogen and magnesium concentrations was present in kernels. Simultaneously, no significant correlation between N and Mg was found. Chatzav et al. (2010) demonstrated, however, a positive ($r = 0.59$, $P \leq 0.01$) correlation between protein and magnesium concentrations in grains of wild emmer genotypes. In addition, concentrations of both nutrients were 100% and 60% higher for N and Mg, respectively, compared with domesticated varieties. This same trend has been observed for genotypes of spelt wheat (Figure 6.10). The course of the developed curve simply indicates that the optimum concentration of nitrogen, which is 14.9 g N kg⁻¹ DM, is achievable, when provided

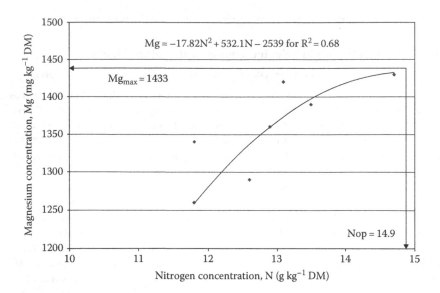

FIGURE 6.10 Magnesium concentration as a function of nitrogen concentration in grain of spelt wheat (*T. spelta* L.). *Note*: Nop, N optimum. (Based on Ranhotra, G.S. et al., *J. Food Compost. Anal.*, 9, 81, 1996.)

with the maximum Mg concentration in grain, which is 1433 mg Mg kg^{-1} DM. As documented by Hussaini et al. (2008) and by Szulc (2010) for corn, a low rate of N supply can lead to slight increases in both nitrogen and magnesium concentrations in maize grain. However, a further N rate increase resulted in N concentration stabilization and in the concomitant decrease in Mg concentration. The weak correlation coefficients for nitrogen and magnesium in modern high-yielding varieties of cereals can be explained by considering differences in the distributions of both nutrients among plant parts. Nitrogen primarily accumulates, that is, with 70%–80% of total crude protein content, in grain. For magnesium, its grain contribution is much lower and rarely exceeds 50% (Figure 6.8).

Vegetables and fruits require special attention due to the parts that are consumed and to the importance of the correlation between N and Mg in the diet index for humans. Plants within these groups are generally harvested at stages of high photosynthetic activity. For example, spinach is an important source of easily available plant magnesium. Therefore, any shortage of water, nitrogen, and magnesium results in reduced yields, and its quality decreases. As reported by Sheikhi and Ronaghi (2012), an increasing nitrogen rate led to a fast rise in magnesium concentration in the spinach rosette. The nitrogen supply to fruit plants can significantly change the magnesium concentration; however, the number of reports is limited. For example, melon (*Cucumis melo* L.) fruit shows a considerable cultivar-dependent variability in nitrogen and magnesium concentrations (Majkowska-Gadomska, 2009). In general, an increase in nitrogen concentration resulted in a concomitant increase in magnesium concentration (r = 0.55, P ≤ 0.01).

Magnesium concentrations in forage plants respond in various manners to increasing N rates. As observed by Kering et al. (2011) for Bermuda grass, this correlation follows a quadratic regression model. This trend was typical during their 5-year study, irrespective of the course of the weather. The level of magnesium concentration in response to the optimum N fertilizer rate was, however, year-to-year specific and varied from 2.5 g kg^{-1} DM under ample water conditions to 4.5 g kg^{-1} DM in years with water shortages. Significantly higher concentrations of Mg were attributes of yield of forage harvested in years with relatively low precipitation during summer months. This type of compensation has also been noted for other crops, such as maize and sugar beet (Szulc et al., 2011; Grzebisz, 2013).

6.4.2.2 Phosphorus and Sulfur

Magnesium and phosphorus are involved in basic metabolic processes, whose genesis is deeply rooted in ancient forms of life (Holm, 2012). It would be, therefore, logical to expect a close and, simultaneously, positive correlation between P and Mg concentrations in seeds or grain and/or vegetative plant parts. Menkir (2008), working with maize inbreed lines that were adapted to the lowlands of Nigeria, found a reasonable linear correlation for both elements (r = 0.63, P ≤ 0.05). Ferreira et al. (2012) showed a year-to-year variability of the P and Mg correlations in maize cultivars cultivated in Brazil. However, the correlation coefficient (r) varied from year to year and reached 0.85 (P ≤ 0.01) in 2005/2006, but only 0.58 in 2007 (P ≤ 0.01). In both years, the amount of each nutrient that accumulated in the grain significantly affected the yield. The positive interaction between P and Mg has also been corroborated in studies that were conducted in temperate regions around the world, such as the Mid-Atlantic region in the United States (Heckman et al., 2003). As indicated by the authors, yields of maize (cultivar Pioneer Hybrid Brand 3394), which was cultivated in six different site years, were significantly related to phosphorus and magnesium concentrations in kernels.

A broad study by Garcia-Oliveira et al. (2008) on variability of mineral grain composition of 85 crosses between the cultivar Tequing (an Indica type of domesticated rice) and the wild rice *Oryza rufipogon* Griff. showed a high correlation between P and Mg, as indicated by a correlation coefficient of 0.77 (P ≤ 0.01). This value was the highest "r" value between magnesium and all nutrients studied, which indicated the importance of wild rice as a source of alleles for magnesium concentration increases in grain. In contrast, a study by Oko et al. (2012) of 15 indigenous and 5 hybrid rice varieties in Nigeria did not show any significant interaction between phosphorus and magnesium. The lack of important differences can be explained by the degree of plant saturation with one or both nutrients. Nigerian varieties of grain were rich in phosphorus but, simultaneously, were variable for magnesium and ranged from 5.0 to 5.5 g kg^{-1} DM for P and from 0.7 to 2.3 g kg^{-1} DM for Mg. For comparison, the average concentrations of both nutrients in rice grains in a study by Garcia-Oliveira et al. (2008) were twice as low. Therefore, a significant correlation between both nutrients, under good supply conditions, is weakly predictable. The postulated hypothesis can be explained by a study by Rodriguez et al. (2011) of the chemical composition of wheat landraces from the Canary Islands. As illustrated in Figure 6.11, the critical concentration of P (P_c) was 2.58 g kg^{-1} DM, which resulted in the Mg maximum concentration of 1.18 g kg^{-1} DM. It can be concluded, therefore, that the linear form of the P and Mg concentration correlation represents the unsaturated status of both nutrient concentrations, and the quadratic regression model represents the saturated status of both nutrient concentrations. The positive trend between phosphorus and

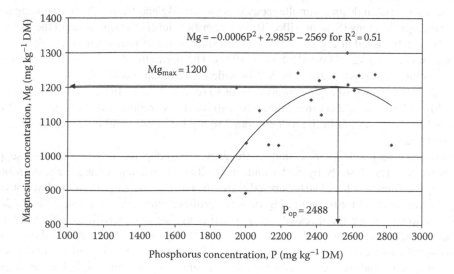

FIGURE 6.11 Magnesium concentration as a function of phosphorus concentration in wheat grain (*T. aestivum* L.). (Based on Rodríguez, L.H. et al., *J. Food Compost. Anal.*, 24, 1081, 2011.)

magnesium concentrations has been found for legumes (Fabaceae). A study by Destro et al. (2002) demonstrated that magnesium concentration was affected by phosphorus concentration. Bananas, as documented by Wall (2006), also followed a linear regression model, similar to the dominating type of phosphorus and magnesium correlations.

Correlations between sulfur and magnesium are due to the impact of both nutrients on basic plant processes that are important for the growth and yields of crops. To date, the amount of available data is too low to indicate the dominating trend. As reported by Menkir (2008) for maize, the sulfur and magnesium correlation generally follows a linear model. As reported by Shobha et al. (2010), maize genotypes in India are insufficiently supplied with Mg and S, as indicated by a highly significant additive correlation (r = 0.81, P ≤ 0.001). In contrast, a study by Boila et al. (1993) of the chemical composition of cereals that were cultivated in Manitoba did not show any significant correlation between both nutrients in wheat and barley grain. A study by Chatzav et al. (2010) of emmer wheat indicated a high interdependence of sulfur and magnesium (r = 0.71, P ≤ 0.001). Pulses are important sources of both minerals; however, sulfur-containing amino acids limit their nutritional quality (Iqbal et al., 2006). The analysis of S and Mg correlations for 72 soybean genotypes by Destro et al. (2002) in Brazil implicitly indicated the unsaturated status of both nutrients (r = 0.14, P ≤ 0.001). The authors concluded that some of the tested genotypes could be considered rich sources of alleles that are required for improving both sulfur and magnesium concentrations in soybean.

6.4.2.3 Potassium and Calcium

Nutrients such as potassium and calcium act as magnesium antagonists, which disturbs its functions in plants. It is well documented that potassium, which is a nutrient that is taken by most crop plants in the highest amounts, exerts a significant impact on the management of nitrogen, which in turn affects the rate of growth and components of the yield (Grzebisz et al., 2013). Calcium, which is an element that influences basic processes of root growth, is required for the stability of plant tissues and organs (White and Broadley, 2003). Therefore, correlations between these three nutrients are complex and have resulted in the development of a series of diagnostic indices.

The most frequently used index is the equivalent ratio of potassium to the sum of calcium and magnesium (K/Ca + Mg). This index is useful for forage crops. These crops are an important source of energy, proteins, and minerals for ruminants and horses. One of the most important threats to their use is an imbalance of potassium and divalent cations (McDowell, 1996). An increase in potassium concentrations in forage grasses leads to decreases in the concentration of both divalent ions. A diet that is rich in K and, simultaneously, is poor in Mg and Ca results in hypomagnesemia, which is called grass tetany (Schonewille, 2013). The critical level of magnesium in the diet of dairy cows has been established at 2 g kg^{-1} DM. The threshold value of potassium in the dairy cow diet should not exceed 10 g kg^{-1} DM (McDowell, 1996). The optimum K concentration for the maximum growth rate of a grass sward, which is provided under ample water supply, is 25 g kg^{-1} DM (Whitehead, 2000). Any lower concentration leads to reduced yields. In general, it is assumed that a K/Ca + Mg index above 2.2 creates a threat for ruminants. As reported by Schonewille (2013), an increase in Mg concentration in the fodder from 2 to 4 g kg^{-1} DM could decrease the risk of milk fever by up to 62%.

In practice, the accumulation of cations in the grass sward depends not only on their supply but also on nitrogen rates. A study by Soder and Stout (2003) of nutrient concentrations in orchard-grass (*Dactylis glomerata* L.) clearly showed a significant increase in potassium, but not calcium and magnesium, in response to increasing rates of fertilizer nitrogen. As a result, the K/Ca + Mg equivalent ratio was above the threshold value in all plots that were fertilized with N rates above 336 kg ha^{-1}, which, in turn, created a grass tetany risk (Figure 6.12). The value of the K/Ca + Mg index was significantly driven by potassium (r = 0.95). The counterbalancing impact of divalent ions was nutrient specific and was reasonably high for calcium (r = 0.85) but negligible for magnesium (r = 0.24). In this particular case, adjustment to give an adequate nitrogen rate was the best practical option for the effective control of the K/Ca + Mg ratio. This hypothesis has been fully corroborated

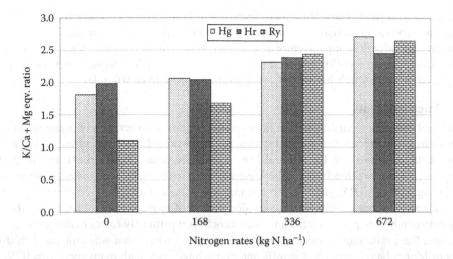

FIGURE 6.12 Effect of increasing nitrogen rates on the K/Ca + Mg equivalent ratio in grass sward. *Legend*: Experimental sites; Hg, Hagerstown; Ha, Hartleton; Ry, Rayne. (Based on Soder, K.J. and Stout, W.L., *J. Anim. Sci.*, 81, 1603, 2003.)

by Aydin and Uzun (2008), who documented that the application of 120 kg N and 100 kg K per ha to rangelands in Turkey resulted in a K/Ca + Mg ratio above 2.2. In both cases, the effect of magnesium fertilizer was of secondary importance compared with the controlling effect of the N rate.

Legumes accumulate much lower amounts of potassium compared with grass and are frequently included in the forage sward. A study by Kuusela (2006) of various grass–clover mixtures showed a high stability of the K/Ca + Mg index below 2.0. However, this ratio fluctuated during the grazing season, from 2.4 in the first cut to 1.58 in the fifth cut. A study by Peters et al. (2005) of yields and the chemical composition of alfalfa that was cultivated under conditions of rising pH levels and simultaneously increased K rates implicitly showed a significant impact of K/Mg and/K/Ca + Mg indices on sward yield; however, these results were dependent on the initial soil pH level. Any significant correlations between these three ions were found in soil of pH below 5.8. As illustrated in Figure 6.13, yields of the third cut of alfalfa were substantially affected by the K/Mg ratio.

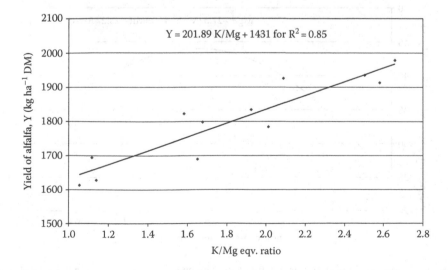

FIGURE 6.13 Effect of potassium to magnesium equivalent ratio on yield of the third cut of alfalfa (*M. sativa* L.). (Based on Peters, J.B. et al., *Commun. Soil Sci. Plant Anal.*, 36, 583, 2005.)

It was observed that the K/Ca + Mg indices increased following an increasing K rate (r = 0.79) but decreased in response to increasing concentrations of both divalent ions in harvested forage. The counterbalancing effect of magnesium was much stronger, which resulted in a higher value of its correlation coefficient (r = 0.97) compared with calcium (r = 0.78). The highest yield of the third cut was harvested in treatments with the K/Ca + Mg ratio ranging from 0.6 to 0.8.

6.4.2.4 Micronutrients

The published evidence regarding magnesium and micronutrient synergism or antagonism is insufficient to draw a simple conclusion. It can, however, be assumed that both synergism and antagonism can occur, as illustrated for Mg and Mn ratios in bananas (Figure 6.14). In this particular case, a synergistic correlation between both elements occurred, with manganese concentration of up to 0.87 mg Mn kg^{-1} DM, yielding the Mg maximum concentration of 2070 mg kg^{-1} DM. An extended review of literature sources shows a dominance of synergism or a lack of significant correlations between binary pairs of nutrients. This conclusion primarily refers to domesticated crops. For example, the grain magnesium concentration in spring wheat that was cultivated in different locations in Manitoba (Canada) was significantly correlated only with manganese (r = 0.79) (Boila et al., 1993). A study by Nelson et al. (2011) of wheat revealed that the current state of magnesium and micronutrient correlations in grain also depends on the cultivation system. The authors documented that spring wheat varieties, which were grown in an organic system, showed a positive and significant correlation between magnesium and zinc in grain. Notably, yields of wheat in the conventional system were almost double compared with yields in the biological system (5.0 vs. 2.7 t ha^{-1}). In both systems, the grain yield negatively affected magnesium and zinc concentrations, resulting in a dilution effect.

The key challenge for increasing magnesium concentration in grain is to cultivate species with naturally high concentrations of minerals. As reported by Ranhotra et al. (1996), spelt wheat grain is rich in magnesium. In addition, the magnesium concentration of spelt wheat grain highly correlated with manganese (r = 0.75) and iron (r = 0.71). Wild genotypes are an alternative to modern, cultivated genotypes for improvement of mineral profile. For example, emmer wheat ecotypes are broadly exploited in breeding programs, in spite of weak correlations between magnesium and micronutrients (Chatzav et al., 2010). In this case, the most stable binary pair, which is irrespective of the year-to-year variability, was documented for Mg and Fe and for Mg and Zn. Correlations between these

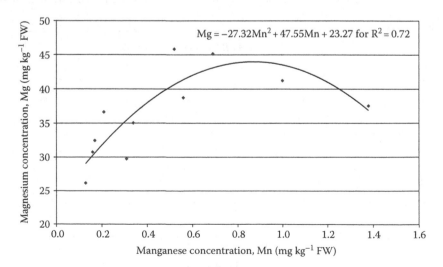

FIGURE 6.14 Magnesium concentration as a function of manganese concentration in banana (*Musa* spp. L.) edible part. (Based on Wall, M.M., *J. Food Compost. Anal.*, 19, 434, 2006.)

pairs of nutrients have also been reported for wheat landraces in other regions, such as the Canary Islands (Rodriguez et al., 2011). The authors of this report found a high correlation between Mg and Cu (r = 0.82) and between Mg and Fe (r = 0.67).

Maize, due to its increasing importance as food and fodder, requires more attention by farmers concerning its nutritional profile. The attribute of high-yielding maize cultivars and genotypes is magnesium dilution (Figure 6.9). As in the case of C3 cereals, all three states of correlations between magnesium and micronutrients have been found. An extended study on the genotype variability of magnesium concentrations in maize grain, against the background of micronutrient concentration, which was conducted by Menkir (2008), followed a quadratic regression model. In this particular case, a synergistic relationship between zinc and magnesium occurred at zinc concentrations of up to 41.9 mg kg^{-1} DM, yielding an Mg maximum concentration of 1674 mg kg^{-1} DM. Among other studied micronutrients, only iron significantly and positively affected the magnesium concentration. A study by Ferreira et al. (2012) of the mineral profile of various maize genotypes, which was carried out in the subtropics, corroborated a positive, but not significant, impact of micronutrients (Fe, Mn, Zn, Cu) on magnesium concentration, except for in the dry 2006 season.

6.4.2.5 Toxic Elements

Toxic elements, which negatively affect magnesium concentrations in edible plant parts, refer to sodium, aluminum, and heavy metals. The depression of crop yield due to salinity is well recognized and has been broadly described in scientific journals and experienced by farmers in agricultural practice. As reported by Saleethong et al. (2013), rice plants that are exposed to NaCl stress at the early booting stage have reduced magnesium concentrations in the grain. In this particular case, a salt-tolerant species under nonstressed conditions was richer in Mg compared with the salt-sensitive species. The salt-tolerant species responded to salinity with a slightly higher decrease in Mg of 15%, whereas the salt-sensitive species only decreased by 9%. Simultaneously, grain yield losses were 32% and 56% lower, respectively. Spinach (*Spinacia oleracea* L.), which is a rich source of magnesium, is considered a crop with moderate sensitivity to salinity. However, the key player, which is decisive for magnesium concentrations in leaf blades of spinach, is nitrogen. As documented by Sheikhi and Ronaghi (2012), the interactional effect of nitrogen on both magnesium and sodium concentrations was dependent on the N fertilizer rate. Plants that were grown in the nitrogen control responded positively to increasing rates of sodium from 0 to 3000 mg NaCl kg^{-1} soil. Increasing N rates of up to 150 mg kg^{-1} soil resulted in magnesium concentration increases in plants with 1000 mg NaCl kg^{-1}. A further increase in the N rate led to a progressive decrease in the magnesium concentration, which corresponded with an increasing rate of applied NaCl. Generally, fodder legumes are sensitive to increasing salinity; however, the degree of magnesium concentration decrease depends on the species. For example, alfalfa and white melilot (*Melilotus albus* Medik.) significantly differ in response to salinity (Guerrero-Rodríguez et al., 2011). The authors documented that alfalfa plants that were irrigated with salt-free water maintained the Mg concentration at 5.8 g kg^{-1} DM. This level dropped to 2.3 g kg^{-1} DM when the NaCl concentration in applied water increased to 110 mM. The second crop, white melilot, did not show any response to increasing salinity and maintained the magnesium concentration at a much lower level, 2.5 g kg^{-1} DM.

The effect of aluminum on the mineral profile of crop plants, in spite of its severity, is weakly recognized. The main reason is a tremendous drop in the rate of plant growth, which leads to a significant reduction in biomass production (Rout et al., 2001). A great decrease in the dominating ion concentration, such as magnesium and calcium, would also occur. A study by Keltjens and Tan (1993) of monocot and dicot responses to acid growth medium (pH of 4.2) showed a high decrease in magnesium concentration, below the critical deficient level. The authors of this study reported that calcium application could break down the toxicity of aluminum in dicots, whereas in monocots magnesium was also required.

6.4.3 Crop Responses to Fertilizer Magnesium

Crop responses to application of magnesium fertilizer depend on many factors, including the content of soil available magnesium, other nutrients, toxic elements, and the course of the weather. There are three areas of magnesium fertilizer action. The first refers to the quantity of the harvested yield. The second includes processes that are responsible for yield development. Consequently, a reasonable explanation of the yield increase due to improved magnesium supply requires an insight into basic processes of yield formation. The third is connected with nitrogen, whose use efficiency is induced by the magnesium supply to crops (Grzebisz et al., 2010; Grzebisz, 2013).

6.4.3.1 C3 Cereals

The period of cereal sensitivity to the application of magnesium is long and extends from tillering to grain maturity. The key yield-forming processes during this period are as follows: (1) growth of reproductive tillers, (2) ear development, and (3) grain growth. Yield-forming processes that are responsible for the number of ears per unit area extend from tillering to heading. The second component, which is the number of grains in the ear, develops during three consecutive stages, including booting, heading, and anthesis. The third component, the weight of the individual grain, which is usually considered as the 1000 grain weight (TGW), develops during the period that extends from the watery stage of the growing grain to the fully ripe stage (Shearman et al., 2005). As illustrated in Figure 6.15, the grain yield, which was averaged over seven species, showed a progressive increase in accordance with the development stage and the applied rate of magnesium. The highest yield was achieved in the treatment with 5 kg Mg ha^{-1}, which was applied during heading. The relative sensitivity of cereal species to Mg application followed the following order: winter barley = winter triticale = spring triticale (8.5%) < spring barley = winter rye (11%) < spring wheat = winter wheat (12%). This question remains: to what extent does magnesium application affect nitrogen uptake, which is the basic factor of plant biomass increase? As reported by Grzebisz (2013), the net increase in nitrogen accumulation was in accordance with the net increase in magnesium uptake, which reached, on average, 19.8 kg N ha^{-1} at heading. The net accumulated nitrogen considerably improved all components of the yield structure, which decreased in the following order: number of ears (r = 0.98) > TGW (r = 0.85) > number of grains per ear (r = 0.76).

As documented by Szczepaniak et al. (2013), split application of Mg fertilizer, that is, directly before sowing and additionally during early stages of vegetation, resulted in an increase of the number of grains per ear, which led to an increase in the yield of spring barley and winter wheat.

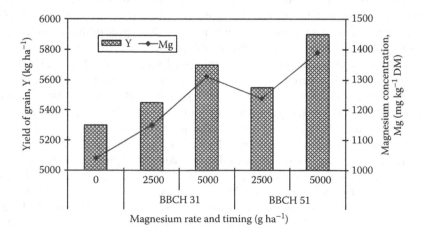

FIGURE 6.15 Effect of magnesium application technology on grain yield averaged over seven species and magnesium concentration. (From Grzebisz, W., *J. Elem.*, 16(2), 299, 2011. With kind permission.)

The aforementioned order of yield-forming components implicitly corroborates the principal effect of magnesium on the yield of grain. There remains, however, a question regarding the effect of magnesium supply on quality traits of grain. The starch content increase, which is simply related to TGW, has been recorded in many studies of cereals (Gerendás and Führs, 2013). A study by Grzebisz (2011, 2013) showed that foliar magnesium applications on cereals, in spite of increasing TGW, resulted in both an increased concentration of crude proteins and magnesium grain density. In this particular case, an application of 5 kg Mg ha^{-1} to winter wheat during heading resulted in a protein concentration increase from 9.9% to 11.5% and in a magnesium density increase from 1040 to 1390 mg kg DM (Grzebisz, 2011). Therefore, the foliar application of magnesium to cereals is a simple and practical way to overcome both the *dilution effect* of protein and the magnesium grain concentration in modern cultivars of cereals.

6.4.3.2 Corn

In corn, in contrast to C3 cereal crops, the number of cobs is fixed at sowing. Therefore, the crucial grain yield component is the number of kernels per cob (NKC) and 1000 kernel weight (TKW). The critical period for the first component of development extends from tasselling to the watery stage of kernel development (Grzebisz et al., 2008; Otegui and Bonhomme, 1998; Subedi and Ma, 2005). In fact, this period begins much earlier, at the fifth leaf stage, when the initial inflorescence appears. Reports concerning maize responses to magnesium fertilizer application indicate an advantage of early treatments. A study by Szulc et al. (2011) showed the highest effect of 10 kg Mg ha^{-1} applied directly prior to sowing. Each kg of magnesium resulted in a net yield increase of 127 kg. The same rate of productivity by foliar-applied magnesium was lower and amounted to 97 kg grain per 1 kg Mg. According to Szczepaniak et al. (2013), maize responds well to the split application of magnesium, that is, before sowing and during early stages of plant growth, to foliage. In this case, the net yield increase correlated with NKC.

The key challenge for breeders is to develop a cultivar that is highly efficient in nitrogen use and resistant to abiotic stresses, such as water shortage. Corn cultivars that slightly differ in FAO number, such as Veritis, 240, and Splendis, 260, showed a year-to-year variability due to water stress, as illustrated in Figure 6.16. In the dry 2006 season, the Splendis yield was significantly lower compared with Veritis. Potassium was the key factor and decreased the size of the yield gap that was caused by water stress. The effect of applied magnesium was low. In 2007, with ample water supply to maize during vegetative growth, the impact of magnesium on the final yield was cultivar specific. For Veritis and Splendis, the effect of magnesium was higher than that of potassium, contributing to 21% and 9% yield increases, respectively. Contrasting results have been reported by Szulc et al. (2011). In this case, the highest impact of magnesium on maize yield was achieved in the dry 2006 season. Plants that were fertilized with soil-applied magnesium (NPK Mg) produced 5.6 t ha^{-1}, which was one-third higher compared with 4.2 t ha^{-1}, which was harvested from the control treatment (NPK). In 2008, the effect of magnesium was nonsignificant, and the respective yields were 12.4 and 11.8 t ha^{-1}.

The corn response to magnesium fertilizers depends on both water and nitrogen supplies (Potarzycki, 2010, 2011). In years with ample water supply, magnesium fertilizer can increase the grain yield when a low rate of N application is provided. This finding was corroborated by Szulc (2010), who documented a much higher N recovery under suboptimum N fertilizer rate conditions when magnesium fertilizer application was provided. Under unfavorable growth conditions, conditions affected by water shortage, magnesium fertilizing results in a positive yield increase, irrespective of the N fertilizer rate (Potarzycki, 2010). The yield-forming effect of magnesium can be explained only by its effect on nitrogen-use efficiency, which is related to a higher uptake of nitrogen or an increase in the physiological productivity of accumulated nitrogen. This conclusion has been corroborated by Potarzycki (2011), who showed that the nitrogen recovery in response to Mg soil application in a plot with 80 kg N ha^{-1} increased from 78% to 92%. In a treatment with 140 kg N ha^{-1}, the nitrogen recovery ranged from 65% to 67%.

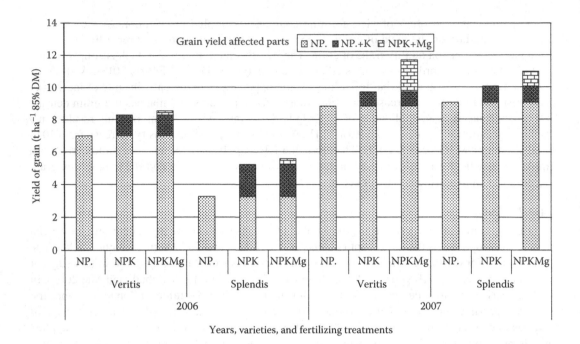

FIGURE 6.16 Effect of corn balanced fertilization with potassium and magnesium on yield of kernels (author's study nonpublished).

6.4.3.3 Root and Tuber Crops

The response of root and tuber crops, such as sugar beet and potatoes, to magnesium fertilizer is highly variable and depends on the weather course during the growing season. In years with ample water supply, as documented for sugar beet, the effect of magnesium application is generally low. This conclusion is illustrated in Figure 6.17 for sugar beet, which yielded at the potential level of the cultivar that was used in the study. In the first two years, with normal precipitation in summer

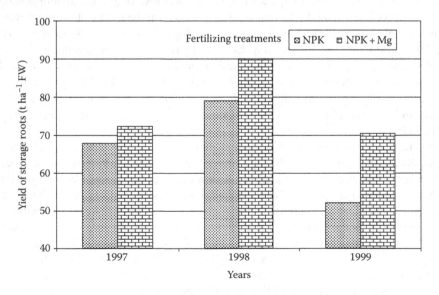

FIGURE 6.17 Effect of NPK fertilization and NPK + magnesium on yield of sugar beets on the background of different production years. (Based on Barłóg, P. and Grzebisz, W., *Biul. IHAR*, 234, 93, 2004.)

months, the magnesium-induced yield of storage root was below 15%. In 1999, with drought in the summer months, the harvested yield of plants that were fertilized only with NPK was much lower compared with 1997. The yield of storage roots increased by one-third in response to Mg fertilizer. This crop prefers soil application to foliar application of magnesium fertilizer (Barłóg and Grzebisz, 2001; Grzebisz et al., 2001). This phenomenon can be explained by the rate of leaf growth. The key factor accelerating this process is the rate of nitrogen uptake by plants during early stages of growth. The positive response results from an additional intake of nitrogen due to an ample supply of magnesium (Grzebisz et al., 2010).

Data on the impact of magnesium fertilizers on potato yield are scarce. The most cited study by Allison et al. (2001), which was conducted in Great Britain, showed a broad range of potato responses to soil or foliarly applied magnesium. The highest was 12.5% compared with the NPK treatment. In new areas of potato production, such as Bangladesh, studies on this crop are important for meeting the food requirements of a growing population. Bari et al. (2001) studied different traits of applied nutrients, which showed a low increase in tuber yield in response to foliarly spread magnesium (6.3%), but it increased to 22% when magnesium was applied in combination with sulfur, boron, and zinc.

6.4.3.4 Other Crops

Oilseed rape is an important oil-bearing crop in temperate regions. A double foliar application, at the rosette stage and repeated at budding, can increase seed yield up to 15% compared with NPK (Barłóg and Potarzycki, 2000). Therefore, in practice, magnesium fertilizer is applied both before sowing, broadcasted directly before spring regrowth, and sprayed on foliage during vegetative growth. A three-year study by Barłóg and Grzebisz (1994) of the effect of the rate and timing of magnesium foliar application on narrow lupine showed an advantage of spraying conducted at budding. The rate of 0.5 kg Mg ha^{-1} was sufficient to increase the seed yield by 0.13 t ha^{-1} on average. As reported by Reinbott and Blevins (1995), the response of soybean to the foliar application of magnesium and boron was both site and fertilizer specific. A single application of a particular nutrient was not effective. The mixture of both nutrients, which was applied four times during vegetation, resulted in a positive trend; however, the degree of the increase was site dependent. The increased number of pods correlated with increased yields.

6.4.4 Diagnosis of Magnesium Status in Crop Plants

Information regarding nutrient deficiencies, including of magnesium, in crop plants is essential to optimize productivity. This section focuses on the methods that are used in the diagnosis of magnesium status of crop plants, such as visual observation, chemical foliar analysis, and leaf green color assessment.

It has been well documented that magnesium concentration, which is measured in the whole plant biomass during vegetative growth, generally declines. This phenomenon has been documented for numerous crops, such as maize by Hundt and Kerschbergr (1991), wheat by Barraclough (1986) and by du Preez and Bennie (1991), winter barley by Barczak (2008), and orange by Storey and Treeby (2000). The trend of magnesium concentration in indicative plant parts, however, has not been clearly defined. Young, but fully developed leaves, maintain a constant level of Mg concentration during the period of greatest growth, as has been reported by Jimenez et al. (1996) for soybean. This finding underlines some important physiological events during the vegetative period of crop plant growth. First, most magnesium that has accumulated in old leaves undergoes remobilization and is consecutively transported from older leaves to developing leaves. Therefore, aged leaves are used as a first indicator of visible symptoms of magnesium deficiency (Bergmann, 1992). Second, the onset of seed development induces a sudden change in magnesium concentration, particularly in plant parts, as imposed by the requirement of developing seeds for carbohydrates, proteins, and minerals. Consequently, its content suddenly drops in vegetative tissues (Barczak, 2008; Barłóg, 2000; Bruns and Ebelhar, 2006).

FIGURE 6.18 **(See color insert.)** Magnesium-deficient grape (*V. vinifera* L.).

Visual symptoms of magnesium deficiency are not specific, but are easily recognized. In general, visual symptoms appear on blades of older, fully developed leaves, which are revealed as classical interveinal chlorosis. The physiologically natural stage of magnesium visual deficiency appearance is the onset of plant ripening. The most striking example is the grape (*Vitis vinifera* L.) (Figure 6.18) or blackberry (*Rubus* spp.) bush, for example, where classical symptoms are apparent on leaves next to ripening fruit clusters (Figure 6.19). Potatoes, when cultivated on sandy soils, experience both very early, and at the same time deep, symptoms of this deficiency (Figure 6.20). Symptoms appearing in early stages of crop plant development can disturb its growth rate and components of yield development. For example, oilseed rape plants are highly sensitive to early deficiency in magnesium, which leads to reduced rosette size (Figure 6.21). The first symptoms in cereal plants appear at the two-leaf stage and disappear shortly thereafter (Bergmann, 1992). These temporary hidden deficiencies result in the thinning of the number of tillers per plant (Figure 6.22). In corn,

FIGURE 6.19   Magnesium-deficient blackberry (*Rubus* sp. L.).

FIGURE 6.20 Magnesium-deficient potato (*S. tuberosum* L.).

FIGURE 6.21 Magnesium-deficient oilseed rape (*B. napus* L.).

acute symptoms on older leaves lead to stunted growth (Figure 6.23). Sugar beet is a crop that is sensitive to magnesium deficiency, which is imposed by a high rate of leaf growth. In spite of the elevated requirement for this nutrient, visual symptoms are rarely apparent. This crop experiences magnesium deficiency late in the season (Figure 6.24). This deficiency appears in response to the accelerated rate of sugar accumulation in the storage root.

The yield of crop plants may be limited unless adequate amounts of all required nutrients accumulate in the plant tissues during important stages of yield formation. Therefore, any operative evaluation of crop plant Mg nutritive status requires a strict definition of the critical stage of yield formation. There is an ongoing debate regarding the most appropriate stage of plant growth for magnesium nutrient status evaluation. In cereals, for example, magnesium that is applied during vegetative growth differentially affects components of grain yield. The fertilizer that is applied at tillering results in an increased number of ears, which consequently affects grain yield the most. Magnesium that is applied at later stages, for instance, at heading, affects both numbers of grains per ear and their weight (Grzebisz, 2011). The same rules have been documented for oilseed rape and pulses (Fabaceae) (Barłóg et al., 2005a). It can therefore be concluded that magnesium shortage in seed crops during the decisive period for the basic yield component of development leads

FIGURE 6.22   Magnesium-deficient wheat (*T. aestivum* L.).

FIGURE 6.23   **(See color insert.)** Magnesium-deficient corn (*Z. mays* L.).

FIGURE 6.24 (**See color insert.**) Magnesium-deficient sugar beet (*B. vulgaris* L.).

to reduced yields (Grzebisz, 2011). Some monocots, such as maize, or dicots, such as sugar beet, respond much more to early, mostly soil-, than to late, usually foliar-applied magnesium fertilizer (Barłóg and Grzebisz, 2001; Szulc et al., 2011). According to Grzebisz (2013), nitrogen accumulation in the canopy of cereal crops increases in response to the external supply of magnesium fertilizer. Therefore, the appropriate stage of plant sampling, which is focused on correction of plant nutritional status, should be coupled with evaluation of nitrogen status because nitrogen is the key component of yield formation.

Visible deficiency symptoms are not, in general, an accurate guide for detecting magnesium deficiency because these symptoms do not appear until the deficiency is acute. Therefore, the mineral analysis of plant indicative parts, mostly leaves, is an important procedure for the detection of nonvisible symptoms of magnesium deficiency. Two terms are significant for any evaluation of the magnesium nutritive status in crop plants. The first, the critical value, is defined as the Mg concentration below which deficiency occurs, which results in a growth rate reduction by more than 10% from the optimum. For example, for wheat, the required Mg leaf concentration at tillering is 1.3 g kg^{-1} DM, whereas at booting, this concentration is approximately 1.5 mg kg^{-1} (Bergmann, 1992). The second term, the sufficiency range, is given for plant parts of the crop at stages that have been determined appropriate for the evaluation of yield component formation. The term optimal concentration range refers to Mg concentrations in leaves that are required for 95%–100% yield in response to an adequate supply of magnesium during critical stages of yield formation (Jones, 2003). The set of sufficiency ranges that are shown in Table 6.7 are inconsistent for the same crop, considering the stage of growth and the presented range. The key reason for these inconsistencies, which lead to inaccuracy in evaluation of plant nutritional status, is that these values are primarily valid under conditions where the study was performed and for the tested cultivar. For modern cultivars, a practical applicability of critical or sufficiency nutrient ranges that were established two or three decades ago is questionable. As reported by Walworth and Ceccotti (1990), maize produced 18 t ha^{-1}, which required an optimum Mg concentration in corn ear leaf that extended from 1300 to 1800 mg kg^{-1} DM. The referenced sufficiency ranges are much wider, extending from 2000 to 4000 mg Mg kg^{-1} DM (Jones, 2003). Older studies may have been relevant for the cropping system at that particular time and environment; however, these studies are not necessarily helpful for understanding nutritional principles for cropping modern varieties and for achieving high yield.

TABLE 6.7

Optimal Ranges of Magnesium Concentration in Parts of Selected Crop Plants in Critical Phase/Stages of Development

Crop	Phase/Stage of Development	Optimal Range mg kg^{-1} DM	Reference
Cereals or small grains	Beginning of shooting	1,200–3,000	Bergmann (1992)
	Shooting	1,000–3,000	
	Booting/heading	1,500–10,000	Jones (2003)
		1,500–5,000	Seefeldt (2013)
Rice (*Oryza sativa* L.)	Tillering—Y stage	1,500–3,000	Jones (2003)
	Midtillering	1,200–2,100	Campbell (2000)
Corn (*Zea mays* L.)	6th leaf	2,000–6,000	Campbell (2000)
	Tasselling/silking—ear leaf	1,500–6,000	
		2,000–4,000	Jones (2003)
Soybean (*Glycine max* Merr.)	Flowering—fully developed trifoliate	2,500–7,000	Campbell (2000)
		3,000–10,000	Prado and Caione (2012)
Potato (*Solanum tuberosum* L.)	Recently matured leaves	3,600–4,900	Seefeldt (2013)
Sugar beet (*Beta vulgaris* L.)	Sixth leaf	1,200–4,500	Barłóg (2009)
	Rosette stage	2,500–8,000	Bergmann (1992)
Cabbage (*Brassica oleracea* var. *capitata* L.)	2–6 weeks old, whole top	5,000–20,000	Seefeldt (2013)
	Head formation	2,400–7,500	
Apple (*Malus domestica* Borkh.)	Recently matured leaves	2,500–4,000	Prado and Caione (2012)
Orange (*Citrus* spp.)	Recently matured leaves	2,500–4,000	Uchida (2000)
	4–6-month-old spring flash leaves	3,000–4,000	Mattos Junior et al. (2012)
Banana (*Musa* spp.)	Third leaf from the apex	2,700–6,000	Prado and Caione (2012)
	Midsection leaf strips	2,500–8,000	Uchida (2000)
Coffee (*Coffea arabica* L.)	Third–fourth leaf pair from the apex	3,100–4,500	Prado and Caione (2012)
	Leaves, 90 days after bloom	2,800–4,000	Martinez et al. (2003)
Red clover (*Trifolium pratense* L.)	Early bloom, top (5 cm)	2,500–6,000	Bergmann (1992)
	Prior to bloom	2,100–6,000	Seefeldt (2013)
Alfalfa (*Medicago sativa* L.)	Early bloom, top (10–15 cm)	1,600–10,000	Seefeldt (2013)
		2,500–10,000	Campbell (2000)
Timothy (*Phleum pratense* L.)	Early bloom, top (5 cm)	600–2,500	Seefeldt (2013)
		1,500–5,000	Bergmann (1992)

The interactional effects of other nutrients on the absorption and functions of magnesium in crop plants have been a matter of extensive debate by both scientists and farmers. The diagnosis and recommendation integrated system (DRIS) approach was the first concept that allowed the extension of an interpretation of crop plant data. The nutrient binary ratios in the indicative plant part, which was usually leaves, are the base of this balance approach. The corresponding ratio in the high-yielding subpopulation is considered the norm (standard) and can be used to determine optimum ranges, as reported for maize (Elwali et al., 1985), for lettuce (*Lactuca sativa* L.) by Hartz and Johnstone (2007), and for pineapple (*Ananas comosus* Merr.) by Sema et al. (2010). Hence, the distance of a particular element concentration ratio in the low-yielding subpopulation compared with the high-yielding subpopulation is considered a nutritious index (Serra et al., 2013). Bailey et al. (1997), who studied nutrient balances in the grassland sward, expressed the opinion of a slight physiological basis for Mg compared with N, P, K, and S ratios, which are involved in the balance ratio evaluation. According to these authors, the DRIS indices, which were developed for Mg, reflect only the degree of its uptake limitation through the impact of environmental factors.

The third group of methods, which were used to determine the magnesium status of crop plants, considers its key importance in chlorophyll synthesis. A close link between leaf chlorophyll and leaf nitrogen concentrations has been documented (Shaahan et al., 1999). As reported by Ayala-Silva and Beyl (2005), a deficiency of magnesium or nitrogen leads to the significant reduction of chlorophyll concentration and spectral characteristics of wheat. Traditionally, plant tissue analysis has been used for assessing nitrogen availability for crops. At present, the chlorophyll meter (SPAD-502) is frequently applied to monitor plant N status by measuring the transmittance of radiation through a leaf in two wavelength bands, which are centered near 650 and 940 nm. However, at high N levels, the linearity between chlorophyll and nitrogen concentrations undergoes a change from a linear to a nonlinear function, as presented for coffee leaves by Netto et al. (2005). The main reason for this inconsistency is the accumulation of nonchlorophyll nitrogen compounds. Despite this inconsistency, the SPAD-502 is a useful tool for the nutritional assessment of crop plants and for grain yield prediction. As presented by Ramesh et al. (2002), SPAD values that were measured in leaves of rice at five consecutive dates after sowing significantly correlated with the yield of grain. A portable chlorophyll meter has also been successfully applied to determine the nutritional status of magnesium, as reported by Shaahan et al. (1999) for some fruit trees, such as mango, guava, and grapevine, but not for mandarin. Shaaban et al. (2002) documented the applicability of this meter for dicots, such as potatoes, snap bean, and sugar beet, but not for monocots, such as wheat and maize.

6.5 SOIL MAGNESIUM

6.5.1 Soil Magnesium Natural Sources

Magnesium, which constitutes 22 g kg^{-1}, is one of the most abundant elements (sixth) of the continental Earth crust. In the bedrock, that is, a rock layer that is directly exposed to the impact of the atmosphere, its average concentration is lower, amounting to 16.4 g kg^{-1}. In the soil, magnesium content is several times lower, amounting to approximately 5 g kg^{-1}. The drastically declined content of magnesium in arable soils compared with the bedrock composition emphasizes the elevated rate of weathering processes that are occurring in the soil. On a global scale, magnesium losses are estimated at 180 kg ha^{-1} $year^{-1}$. In comparison with calcium, this loss is 4.45-fold lower but, simultaneously, is 9-fold higher compared with potassium (Schlesinger, 1997). The mineral composition of the soil parent material provides insight into the potential sources and supply of magnesium to plants. Igneous basic rocks (<66% of SiO_2), which are represented by basalt, gabbro, and norite, undergo weathering processes much faster compared with acid rocks (>66% of SiO_2), which are represented by granite and rhyolite. Therefore, based on the mineral composition of the soil parent material, one can conclude that soils that contain minerals such as olivine, serpentine, or dolomite are naturally rich in total magnesium (Metson, 1975; Schlesinger, 1997). One of the main reasons for an elevated rate of magnesium release from minerals is the activity of plants. The key plant processes that are involved in the dissolution of magnesium from the bedrock are organic acids and chelates that are extruded by roots into the rhizosphere (Hinsinger et al., 2001).

Magnesium-bearing minerals represent two basic groups of minerals, primary and secondary. The first group is composed of several silicate minerals, of which some are an essential source of soil magnesium:

1. Primary minerals of the ferromagnesian series: amphiboles, olivine, pyroxenes, hornblendes; these minerals are important components of igneous rocks and of some metamorphic rocks (serpentine, talc).
2. The micas (muscovite, biotite); these are identified by a specific 2:1 layer-type silicate structure.

The solubility of magnesium-bearing silicates is inherently related to the silicon/oxygen ratio. Under natural conditions, this process significantly depends on many other factors; however, the most important factor is pH, as presented in the following (Lindsay, 1979; Schlesinger, 1997):

1. Olivine: $Mg_{1.6}Fe(II)_{0.4}SiO_4 + 4H^+ \leftrightarrow 1.6Mg^{2+} + 0.4Fe^{2+} + H_4SiO_4{}^\circ$ log $K^\circ = 26.18$
2. Serpentine: $Mg_6Si_4O_{10}(OH)_8 + 12H^+ \leftrightarrow 6Mg^{2+} + 4H_4SiO_4{}^\circ + 2H_2O$ log $K^\circ = 61.75$

Serpentine, which is a product of the metamorphosis of primary minerals, primarily olivine, is less sensitive to weathering compared with other silicates. A high content of ferromagnesian minerals in the soil bedrock is, therefore, the prerequisite of a high cation exchange content (CEC) and an increased content of exchangeable magnesium (EX_{Mg}) (Chu and Johnson, 1985; Metson and Brooks, 1975).

The second group of natural sources of magnesium is composed of secondary minerals of different geological origins, which represents three main groups as follows:

1. The carbonates
2. The silicates: secondary silicate minerals: vermiculites, chlorites, clay minerals (hydrous mica, illite, smectite, montmorillonite)
3. Soluble salts

The carbonate family is composed of many minerals; however, dolomite and magnesite are the key sources of magnesium in soils. The main factor affecting solubility is the soil pH (Lindsay, 1979):

1. Dolomite:

$$CaMg(CO_3)_2 + 4H^+ \leftrightarrow Mg^{2+} + Ca^{2+} + 2CO_{2(g)} + 2H_2O \text{ log } K^\circ = 18.46$$

2. Magnesite:

$$MgCO_3 + 2H^+ \leftrightarrow Mg^{2+} + CO_{2(g)} + H_2O \text{ log } K^\circ = 10.69$$

Dolomite, in natural conditions, is an important source of available magnesium, as documented for chernozem and semigley soils (Jakovljević et al., 2003). All water-soluble magnesium salts, such as sulfates, chlorides, or nitrates, are not detectable in the soil solution due to their immediate dissolution. This fact is important for agronomic practice because any addition of these salts to soil results in the direct increase in magnesium concentration.

Magnesium content in soils is highly differentiated due to pedogenic processes, which lead to the formation of the dominant soil type. Most soils that cover the world land surface are the result of soil forming processes that took place during the Quaternary. The mineralogical composition of bedrocks, which has been affected by geological processes during the last 10,000 years, has been crucial for current soil total magnesium content. This hypothesis is supported by three examples of soil formation in different climatological zones. The Pleistocene, similar to the present Mediterranean climate zone, was humid. Soil formation, which was dominated by the leaching of carbonates, illuviation, and rubification, resulted in the well-developed B_t horizon (Ortiz et al., 2002). Consequently, the total magnesium content is low, ranging from 156 to 1120 mg kg^{-1}. In accordance with the rating, which is presented in Table 6.8, these soils belong to the low or medium class of potential reserves. The vertical profile of magnesium distribution is not consistent in the studied soils, but, for most profiles, declines. Soils that developed in the Pleistocene permafrost zone, for example, in Serbia, present both a high content of total magnesium and, simultaneously, a huge variability among soil types (Jakovljević et al., 2003). In Serbia, the declining order of the entire magnesium content in soils is as follows: fluvisol (7500 mg kg^{-1}), semigley (7400 mg kg^{-1}) > chernozem (6700 mg kg^{-1}) $\geq$ ranker (6400 mg kg^{-1}) $\geq$ humogley (6100 mg kg^{-1}) > eutric cambisol

TABLE 6.8

Rating of Exchangeable and Reserve Magnesium in Soils

Rating[a]	Exchangeable Mg (mg kg⁻¹)	Reserve Mg (mg kg⁻¹)
Very high	>850	>3600
High	360–850	1800–3600
Medium	120–360	840–1800
Low	60–150	360–840
Very low	<60	<360

[a] According to Metson and Brooks (1975).

(4800 mg kg⁻¹) > pseudogley (4200 mg kg⁻¹) > luvisol (3600 mg kg⁻¹) ≥ vertisol (3500 mg kg⁻¹). The main magnesium-bearing minerals are micas and illites. The total concentration of these dominating minerals ranges from 13.5% for luvisols to 22.4% for fluvisols. The concentration of soil available magnesium shows a slightly different order: humogley (580 mg kg⁻¹) > fluvisol (510 mg kg⁻¹) > semigley and eutric cambisol (430 mg kg⁻¹) ≥ (420 mg kg⁻¹) ≥ pseudogley (320 mg kg⁻¹) ≥ luvisol (300 mg kg⁻¹) > vertisol (270 mg kg⁻¹) > ranker (230 mg kg⁻¹). With respect to the total magnesium rating, the first five types represent the high class. However, the content of available magnesium in most soils is classified as high and in all other cases as medium, which indicates a high-potential supply for growing plants (Table 6.8). The youngest soils were formed in the Holocene. As reported by Orzechowski and Smólczynski (2010), alluvial soils, which contain 8900 mg Mg kg soil⁻¹ on average, are rich sources of magnesium. For comparison, deluvial soils that originated from moraine deposits on the riverine plain are twice as low, containing only 4700 mg Mg kg⁻¹ soil. Based on Table 6.8, both soils are rich in both total and available magnesium.

6.5.2 Soil Magnesium Fractions

Magnesium occurs in soils in numerous chemical forms with different potential availabilities to plants. The direct source of ions to growing plants is the soil solution containing magnesium-soluble salts and/or its complexes with inorganic ions and/or organic ligands (Lindsay, 1979). The ionic forms are easily determined using water; however, the amount of extracted magnesium is low, mostly below 10 mg kg⁻¹ soil. This amount is not sufficient to cover crop plant requirements. In soil that is rich in magnesium, its removal is compensated via exchange processes. The most proximate magnesium reserves are various complexes and in the CEC as the EX_{Mg}. These forms are extracted using different extractors, for example, ammonium acetate (extracted by 1 M NH_4OAc at pH = 7.0). Both forms, water soluble (WS) and exchangeable (EX_{Mg}), create the available pool of magnesium. The third pool, called non-EX_{Mg} (NEX_{Mg}), represents forms of different chemical origin that are not convertible (extracted by 1 M HNO_3). The fourth is a residual pool that is composed of minerals and/or lattice-bound magnesium (extracted by the sodium carbonate fusion method or by boiling in 1 M HNO_3 [Metson and Brooks, 1975]).

The content of water-soluble magnesium is weakly related to soil properties. There are, however, some controversies regarding the EX_{Mg}. The first question refers to its correlations with other forms, primarily the NEX_{Mg}. It is generally assumed that the flow of magnesium to the soil solution pool is controlled by the size and strength of Mg binding in the NEX_{Mg}. Okpamen et al. (2013), who studied soils in Nigeria, documented a high correlation between EX_{Mg} and NEX_{Mg} pools. The second question refers to the impact of basic soil properties on the particular pool size and on the rate of exchange processes. As reported by Chu and Johnson (1985), the content of EX_{Mg} in Pennsylvania soils was significantly dependent on its content in the whole soil or in sand and silt, but not in clay. Orzechowski and Smólczynski (2010) found a significant impact of soil type on the EX_{Mg} pool.

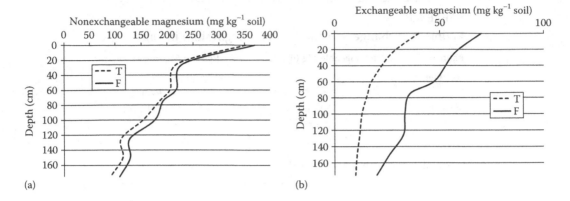

FIGURE 6.25 (a) Effect of the laterite soil land cover on the vertical distribution of NEX_{Mg}. (b) Effect of the laterite soil land cover on the vertical distribution of exchangeable EX_{Mg}. *Legend*: T, tea crop; F, natural forest. (Based on Jayaganesh, S.V. et al., *Int. J. Soil Sci.*, 6(1), 67, 2011b.)

The EX_{Mg} pool was dependent on the clay content in alluvial and deluvial soils, whereas the EX_{Mg} pool was dependent on silt in deluvial ones. The effect of organic matter was positive in mineral soils, but negative in peat-muck soils.

Plants take up magnesium ions directly from the soil solution; however, the degree of cover of the current requirement is affected by the size of its available pool, which, in turn, depends on the quantity of water-soluble and exchangeable forms and on the depth of plant rooting. The vertical profile of total magnesium is affected by many factors, including the type of soil parent material, land usage, and growing crops. The key factor that affects the magnesium soil profile is bedrock composition. As reported by Okpanem et al. (2013), the content of EX_{Mg} showed a declining trend with depth, which indicated strong leaching. The NEX_{Mg} content and distribution throughout the soil profile were significantly influenced by soil parent material and were the lowest in alluvium soil. In surface layers, extending from 15 to 75 cm, magnesium content was lower in comparison to deeper layers of the soil profile. This phenomenon indicates advanced exhaustion processes, which are modified by soil type and land usage.

The extent of magnesium depletion in arable soils is difficult to evaluate. As illustrated in Figure 6.25a and b for laterite soil, land usage (in this case forest or tea plantation) significantly affects Mg content. Both soils can be classified as poor in available Mg content, but as medium in total reserve contents. The impact of cultivated tea plants was highly specific. A significant decrease in WS, EX_{Mg}, and lattice forms, but not the NEX_{Mg} content, along the whole soil profile has been documented. The vertical pattern of the distribution of available magnesium is significantly affected by cultivated crops, as has been observed for winter rye and potato in the long-static fertilizing experiment on the typically temperate albic luvisol (Figure 6.26a and b). The amount of attainable magnesium in the soil profile to 120 cm varies in accordance with the crop and applied fertilizers. The magnesium content in the surface-plow layer was much lower compared with subsoil layers. The amount of exchangeable Mg that was recorded in the NPK treatment indicates Mg exhaustion over the whole profile, independent of the crop. Farmyard manure that was applied at a rate of 30 t ha^{-1} year^{-1} was the main source of magnesium and was allowed to saturate the subsoil layers, as indicated by its much higher content, preceding a substantial decline with soil depth. In accordance with the rating, which is presented in Table 6.9, all these plots represent the low class, even when systematically fertilized with manure.

6.5.3 SOIL TESTING

The procedure for calculating the magnesium fertilizer rate requires data regarding (1) the predicted crop yield, (2) the unit Mg uptake, and (3) the current content of soil available magnesium.

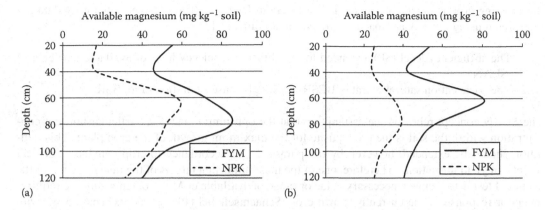

FIGURE 6.26   (a) Effect of long-term winter rye monoculture on the vertical distribution of available magnesium. (b) Effect of long-term potato monoculture on the vertical distribution of available magnesium. *Legend*: FYM, farm manure; NPK, mineral fertilizer. (Based on Piechota, T. et al., *Folia Univ. Agric. Stetin.*, 84, 393, 2000.)

TABLE 6.9
Magnesium Sufficiency Level for Selected Soil Tests

| Extracting Solution | Mg Sufficiency Level (mg kg Soil^{-1}) | | Country |
	Sandy Soils	Other Soils	
Calcium chloride	50–70	70–140	Poland[a]
Ammonium nitrate		51–100	England[b]
Doppel lactate	110–150	170–300	Germany[c]
Mehlich 1	30–60	50–100	United States[d]
Mehlich 3	60–120	70–140	United States[d]
NH$_4$OAc (pH 7.0)	100–200	140–250	United States[d]
		100–125	Denmark[b]
NH$_4$OAc (pH 4.65)	120–200	120–400	Finland[b]

[a] Jadczyszyn (2009).
[b] Ristimäki (2007).
[c] Finck (1992).
[d] Carrow et al. (2004).

It should be emphasized, however, that a particular soil test, even if it is reliable or accurate, can be considered only as the key factor supporting a farmer's decisions regarding the need for magnesium fertilizer application. There are many other supporting factors, such as the soil type, environmental conditions, applied agronomic technology, and economic conditions of production. Therefore, one of the most important challenges in agriculture is to improve soil productivity through increasing the volume that is occupied by the root system of the currently cultivated crop. The procedure of determination of the available magnesium content should be coupled with the concomitant assessment of soil acidity and lime requirement.

A cyclic determination of plant attainable nutrient, as in the case of available magnesium, is the core of magnesium fertilizer management. There are several laboratory methods for determining soil magnesium status by soil testing. The laboratory tests are interpreted based on the classical agrochemical rating of plant-available magnesium. Magnesium deficiency is shown to occur when the growing crop responds to the applied magnesium fertilizer. The obtained laboratory data are interpreted using two key approaches (Gransee and Führs, 2013):

1. The sufficiency level (SL), extended to the phrase: sufficiency level of available nutrients (SLAN)
2. The basic cation saturation ratio, BCSR, or, simply, base saturation ratio, BSR

The SLAN concept relies on the assumption that the optimum range of a particular nutrient concentration within the soil is the prerequisite for the maximum growth of the crop plant. This definition should be extended, however, by the phrase: ..."in accordance to crop demand in critical stage(s) of yield formation." Therefore, one of the most important targets of the interpretation of the chemical test is to couple a necessary value or range of available or Mg_{ex} contents with the requirements or response of the currently grown crop. Schachtschabel (1954), who extracted Mg_{ex} with 0.0125 M $CaCl_2$, defined its critical content as the level of 50, 70, and 120 mg kg^{-1} soil for light, medium, and heavy soils, respectively. These ranges, to some extent, have been corroborated by other researchers, as reported by Metson (1974). The same author published the magnesium rating a year later, evaluating both EX_{Mg} and reserve magnesium (Table 6.8). In light of this rating, the expected Mg deficiency is related to convertible magnesium concentration lower than 60 mg kg^{-1} soil and to reserves below 360 mg kg^{-1} soil. It is assumed that a deficiency of magnesium is highly probable in soils with values of CEC that are below 15 $cmol_c$ kg^{-1}, which is an attribute of most soils that have been formed from sandy parent materials (Carrow et al., 2004). Therefore, the SLAN for a particular test should be coupled with the soil agronomic type, as shown in Table 6.9.

The second concept of available magnesium interpretation, the BSR, relies on the assumption that optimum plant growth does not depend on the amount of attainable magnesium, but on the ratios between main basic ions in the growth medium. The BSR concept is rooted in the study that was first conducted during the last decade of the nineteenth century and continued to the present day. The core of this approach is to define the ideal binary ratio between the exchangeable content of basic ions in the soil for achieving the maximum growth of crop plants. Based on many studies, the perfect cation composition has been proposed: Ca^{2+}, 65%; Mg^{2+}, 10%; K^+, 5%; and H^+, 20%. This structure was established in 1940s by Bear et al. (Kopittke and Menzies, 2007). In other studies, a 6% contribution of Mg_{ex} in the CEC is considered sufficient to cover the requirement for magnesium by crop plants throughout vegetative growth (Metson and Brooks, 1975). The *ideal* ratios of Ca/Mg, Ca/K, and Ca/H have been defined as 6.5:1, 13:1, and 3.25:1, respectively. This concept has been applied for soil test interpretation in some states of the United States and Australia (Kopittke and Menzies, 2007). In spite of extended studies, which have lasted approximately one century, the possible recommendations, considering a positive crop yield response to the ideal Ca/Mg ratio, are controversial. The recently published study by Osemwota et al. (2007) on maize response to differentiated soil Ca/Mg ratios within the range from 1:1 to 8:1 has indicated the optimum value, but only in pot experiments. In this particular case, a ratio of exchangeable Ca/Mg of 3:1 was required to reach the maximum yield. The parallel field study, which was based on the same experimental assumptions, did not corroborate the hypothesis of the existence of the ideal Ca/Mg ratio. This conclusion also has been corroborated by Stevens et al. (2005), who worked with cotton. These authors did not find any response of lint cotton yield to Ca/Mg ratios, which extended from 2.5:1 to 7.6:1 (% base saturation). Kopittke and Menzies (2007) stated that the *ideal ratio* did not exist. In the opinion of these authors, the response of crop plants is much more affected by total nutrient availability and by conditions that determine the rate of magnesium uptake. The failure of the BSR philosophy, considering field experiments, is not easy to explain. First, the hypothesis assumes the yield increase but, simultaneously, neglects the course of nutrient uptake during crop plant growth. Second, in spite of the well-defined functions of magnesium, calcium, and potassium in plants, their yield-forming effects should be coupled with the action of nitrogen (Grzebisz, 2013; Rubio et al., 2003). All the discussed ratios should be extended, therefore, from binary to tertiary forms, for example, N/Ca/Mg or N/Mg + Ca.

In addition, each nutritive ratio presents a state of dynamic equilibrium throughout the period of crop growth. The most important seems to be nutrient ratios, which affect crop plant growth at critical stages of yield formation. Third, the impact of fertilizers, which contain magnesium or calcium, on aluminum activity is frequently overlooked. It has been documented that both nutrients improve the root system growth, which, in turn, changes element ratios through increasing the accessibility of both mobile and immobile nutrients to growing plants (Keltjens and Tan, 1993).

6.6 SOIL MAGNESIUM MANAGEMENT

6.6.1 SOIL MAGNESIUM BALANCE

The primary reason for magnesium deficiency in crop plants is the low content of magnesium in the soil solution, which, in turn, decreases the rate of Mg^{2+} ion flow to the root (Hundt and Kerschberger, 1991). Two main reasons for this deficiency can be distinguished. The first results from low soil reserves, which are primarily related to the extractable magnesium content. The second is composed of factors that disturb the growth of roots into soil patches rich in available magnesium and its uptake due to the elevated concentration of antagonistic ions. The emphasis of farmers should be focused, therefore, not only on increasing the content of soil attainable magnesium but also on plant crop accessibility to soil magnesium. Therefore, the management of soil magnesium requires an adequate crop supply in accordance with the requirements of crops and soil conditions. A three-step procedure of magnesium supply control has been proposed:

1. The balance sheet of magnesium in a particular crop rotation
2. The analysis of the extractable magnesium trend in a particular crop rotation
3. The cyclic testing of soil available magnesium content

The potential threat of magnesium-deficient supply from soil resources to crops that have been cultivated in a fixed rotation can be easily controlled by implementing a balance sheet method, which allows the indication of current trends of this element change. The main components of the classical balance sheet, in accordance with the *field surface balance*, are as follows:

1. Input: (a) plant residues, (b) organic fertilizers, (c) mineral fertilizers, and (d) other sources, for example, the amounts of nutrients that are delivered in rainfall
2. Output: (a) crop removal, (b) leaching, and (c) nonexchangeable Mg^{2+} fixation

The first step of quantification of soil magnesium output is to make a reliable determination of the removal of the magnesium quota from the field. These values are primarily evaluated based on the utilizable part of the growing crop. In the light of the data that are presented in Figure 6.8, 40%–55% of magnesium accumulated in the final biomass is included in the main product. The rest remains in crop residues and undergoes different forms of recycling. Another indicator, which is known as unit nutrient uptake (in the described case, UMgUp), allows a simple estimation of magnesium removal to be performed. This estimation considers total magnesium in harvested crop biomass, which is recalculated on the unit of harvestable plant parts (Table 6.1). The calculated UMgUp index for a particular crop is a useful tool in determining its sensitivity to the net magnesium supply as a function of the assumed yield.

The second method of evaluating the soil magnesium trend relies on the measurement of extractable magnesium at the beginning and at the end of a three- or four-course rotation or during an extended period of arable land cultivation. As illustrated in Figure 6.26, a long-term cultivation of rye and potatoes on the background of a constant rate of application of farmyard manure or NPK fertilizers resulted in the high differentiation of extractable magnesium contents in the whole soil profile. Plots that are fertilized annually with NPK are in a state of extended mining, irrespective of

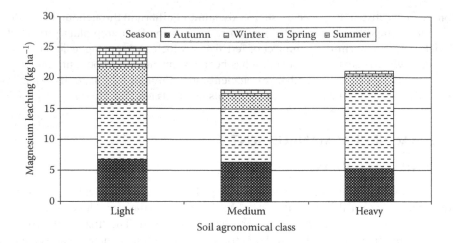

FIGURE 6.27 Magnesium leaching from soils of different textures during the key parts (seasons) of the year. (From Grzebisz, W., *J. Elem.*, 16(2), 299, 2011. With kind permission.)

the crop cultivated. This process can be mitigated by manure application. Even in a mixed farm, the amount of manure produced is too low to be managed in this manner. The soil balance of extractable magnesium can also be calculated by the difference method, considering its content at the beginning and at the end of the observation period. As documented by Kaniuczak (1999), the balance of extractable magnesium after 8 years of study was negative, in spite of magnesium fertilizers that were applied at a total rate of 288 kg Mg ha^{-1}. The net magnesium losses increased significantly in accordance with the raising NPK rates. Lime positively affected magnesium management, decreasing its losses. The maximum calculated magnesium losses amounted to 50 kg ha^{-1} year^{-1}.

The quantity of magnesium that leaches out of a field is the main part of its total losses. Losses of extractable magnesium in a lysimeter experiment were affected by soil type (Schweiger and Amberger, 1979). The quantity of lost magnesium varied from 72 to 92 kg ha^{-1} for a sandy or a medium-textured soil, respectively. Under field conditions, the amount of leached magnesium depends upon many factors, including site, soil magnesium content, the rate of weathering, and the amount and distribution of precipitation. As illustrated in Figure 6.27, Mg leaching from arable fields, measured by its concentration in the water outflow, ranged from 18 to 25 kg Mg ha^{-1}, respectively, for medium and light soil. Over the course of the year, these values were highly variable, extending from a few to more than 40 kg Mg ha^{-1} (Szymczyk et al., 2005). The highest quantitative leaching of Mg in temperate regions around the world is an attribute of the winter part of the season.

6.6.2 Magnesium Fertilizers

There are many magnesium-bearing sources for the manufacture of magnesium fertilizers. All of these sources can be divided into three groups, which are based on the following criteria:

1. Chemical origin: (a) organic and (b) mineral
2. Mineral group: (a) soluble salts, (b) oxides and carbonates, and (c) silicates
3. Solubility: (a) water soluble, (b) weakly water soluble, and (c) water nonsoluble

In spite of the rate of release of Mg^{2+} ions that is realized, each of the aforementioned groups of magnesium sources plays an important role in soil magnesium soil balance (Table 6.10). The primary sources, such as plant residues, which are indicated in Figure 6.8, contain from 45% to 60%

TABLE 6.10
Current and Potential Sources of Magnesium Fertilizers

Fertilizer	Chemical Composition	Magnesium Concentration (%)	Water Solubility (g dm^{-3})
Organic, fresh weight			
Cattle manure[a]	Mineral and organic matter	0.05–0.15	Variable
Cattle slurry[a]		0.03–0.15	
Pig manure[a]		0.25	
Pig slurry[a]		0.05–0.18	
Poultry manure[a]		0.12–0.42	
Water-soluble salts, dry weight			
Magnesium nitrate	$Mg(NO_3)_2 \cdot 6H_2O$	10	1250
Magnesium chloride	$MgCl_2 \cdot 6H_2O$	25	1570
Epsom salt	$MgSO_4 \cdot 7H_2O$	10	335
Kieserite	$MgSO_4 \cdot H_2O$	18	360
Sulfate of potash magnesia	$2MgSO_4 \cdot K_2SO_4$	12	240
Schoenite	$MgSO_4 \cdot K_2SO_4 \cdot 6H_2O$	6	330
Magnesium chelates	Variable	3–5	High
Oxides and carbonates			
Magnesium oxide	MgO	50–55	0.009
Dolomite	$MgCO_3CaCO_3$	8–20	0.006
Magnesite	$MgCO_3$	27	0.034
Other mineral			
FCMP[b]	Unknown	9	90% CAS[c]
Struvite	$MgNH_4PO_4 \cdot 6H_2O$	10	Low
Magnesium–ammonium–phosphate	$MgNH_4PO_4$	16	0.14
Serpentine	$Mg_6SiO(OH)_4$	20–40	Low

Source: Grzebisz, W., *J. Elem.*, 16(2), 299, 2011.
[a] European Communities (2001).
[b] Fused calcium–magnesium–phosphate.
[c] Citric acid soluble.

of the total magnesium in crop plants at harvest. Plant residues of dicotyledonous crops, such as sugar beet (*B. vulgaris* L.) or oilseed rape (*B. napus* L.), are naturally richer in magnesium than plant residues of cereals (Hundt and Kerschberger, 1991). These residues are considered natural, recycled sources of basic ions, including magnesium. The second group of organic sources, farmyard manures, contains recycled fodder magnesium, such as silage, straw, and hay, but also any other organic fodder or mineral additives that were used for animal nutrition. Therefore, the content of magnesium in manures is highly variable, depending on the type of farm, method of nutrition, and type of manure. Irrespective of the type, all organic sources of manure can be classified as slow-release magnesium fertilizers and the main objective of their application is to increase the content of total soil magnesium.

The second main group of magnesium-bearing fertilizers is minerals, such as soluble salts, which are easily dissolved in water. The first group of minerals, such as sulfates, is used widely as magnesium fertilizers in pure or processed forms. These fertilizers enrich the concentration of Mg^{2+} ions within the soil solution. The second group of minerals, such as carbonates and oxides, is dissolved weakly in water. However, these minerals release Mg^{2+} ions when incorporated into the soil. In light of soil geochemistry rules, soil can be considered a weak acid, which, in turn, significantly affects magnesium mineral dissolution from sparingly soluble fertilizers, as presented in Section 6.5 for dolomite and magnesite. Some other groups of magnesium fertilizers are highly specific. The first is fused calcium–magnesium–phosphate (FCMP), whose production is based on phosphate rock and serpentine. Both substrates are fused in an electric furnace. The obtained product contains 18%–20% of P_2O_5, 12% of MgO, and also silicon and lime (Ranawat et al., 2009). Other potential sources of magnesium are ammonium–magnesium–phosphates, including a mineral called struvite. This fertilizer is a product of municipal and animal manure wastewater purifying. Another, but weakly recognized, magnesium fertilizer source is serpentine ($Mg_6Si_4O_{10}(OH)_8$). This mineral was first used in the 1940s in New Zealand to produce magnesium fertilizer (Metson, 1974).

The agricultural evaluation of usage of the main magnesium fertilizers depends on the production target. This evaluation is defined by crop magnesium accumulation, crop plant yield and/or quality increase, and leaching. In humid regions of the world, the leaching of magnesium from soluble salts is high, approaching approximately 50% of the applied magnesium (Hanly et al., 2005). The expected soil leaching of magnesium, which originates from soluble salts, can be overcome by application of this type of fertilizer at lower rates or by splitting the main rate into sub-rates, which are applied in consecutive stages of the currently cultivated crop (Härdter et al., 2004). The second option assumes a foliar application of water-soluble fertilizers. This strategy of magnesium supply to the crop requires an exact determination of three variables. These variables are as follows: (1) time of application, adequately related to the critical stage of plant growth and to components of yield formation; (2) amount of applied magnesium, effectively used by plants; and (3) salt concentration in the spraying solution, below the toxic amount. Considering the main objective of this conceptual review, application of foliar magnesium spray seems to be the simplest agronomic way to increase nitrogen-use efficiency, which, in turn, results in increased yield, and magnesium density increases in edible plant parts.

REFERENCES

Abu-Zinada, I.A.I. 2009. Potato response to potassium and nitrogen fertilization under Gaza strip conditions. *J. Al Azhar Univ. Gaza* 11:15–30.

Adotey, D.K., Y. Serfor-Armah, J.R. Fianko, and P.O. Yeboah. 2009. Essential elements content in core vegetables grown and consumed in Ghana by instrumental neutron activation analysis. *Afr. J. Food Sci.* 3(9):243–249.

Aganga, A.A., U.J. Omphile, T. Thema, and L.Z. Wilson. 2004. Chemical composition of ryegrass (*Lolium multiflorum*) at different stages of growth and ryegrass silages with additives. *J. Biol. Sci.* 4(5):645–649.

Ahmed, S.U. 2011. Effects of soil deficit on leaf nitrogen, chlorophylls and SPAD chlorophyll meter reading on growth stages of soybean. *Bangladesh J. Bot.* 40(2):171–175.

Allison, M.F., J.H. Fowler, and E.J. Allen. 2001. Factors affecting the magnesium nutrition of potatoes (*Solanum tuberosum*). *J. Agric. Sci.* 137:397–409.

Andreini, C., I. Bertini, G. Cavallaro, G.L. Holliday, and J.M. Thorthon. 2008. Metal ions in biological catalysis: From enzyme databases to general principles. *J. Biol. Inorg. Chem.* 12(8):1205–1218.

Anza, M., P. Riga, and C. Garbisu. 2005. Time course of antioxidant responses of *Capsicum annuum* subjected to a progressive magnesium deficiency. *Ann. Appl. Biol.* 146:123–134.

Ayala-Silva, T. and C.A. Beyl. 2005. Changes in spectral reflectance of wheat leaves in response to specific macronutrient deficiency. *Adv. Space Res.* 35:305–317.

Aydin, I. and F. Uzun. 2008. Potential decrease of grass tetany risk in rangelands combining N and K fertilization with MgO treatments. *Eur. J. Agron.* 29:33–37.

Bailey, J.S., J.A.M. Beattie, and D.J. Kilpatrick. 1997. The diagnosis and recommendation integrated system (DRIS) for diagnosing the nutrient status of grassland swards: I. Model establishment. *Plant Soil* 197:127–135.

Baligar, V.C. 1984. Effective diffusion coefficient of cations as influenced by physical and chemical properties of selected soils. *Commun. Soil Sci. Plant Anal.* 15(11):1367–1376.

Bangash, J.A., M. Arif, F. Khan, F. Khan, Amin-Ur-Rahman, and I. Hussain. 2011. Proximite composition, minerals and vitamins content of selected vegetables grown in Peshawar. *J. Chem. Soc. Pak.* 33(1):118–122.

Baranisrinivasan, P. 2011. Chlorophyll and nutrient content in black gram using organic manures under drought stress. *BELPS* 11:24–29.

Barber, S. 1984. *Soil Nutrient Bioavailability: A Mechanistic Approach.* New York: Wiley & Sons Interscience.

Barczak, B. 2008. Contents and ratios of mineral components of winter barley biomass cultivated under conditions of different nitrogen fertilization. *J. Elem.* 13(3):291–300.

Bari, M.S., M.G. Rabbani, M.Sq. Rahman, M.J. Islam, and A.T. Hoque. 2001. Effect of zinc, boron, sulphur and magnesium on the growth and yield of potato. *Pak. J. Biol. Sci.* 4(9):1090–1093.

Barłóg, P. 2000. Effect of Kieserite and nitrogenous fertilizers interaction on the magnesium and nitrogen content and uptake dynamics by narrow-leafed lupine. *Rocz. AR Pozn. CCCXX, Agric.* 57:129–138 (in Polish with English summary).

Barłóg, P. 2009. Study on sugar beet macronutrients with special emphasis to sodium. *Fertil. Fertil.* 35:1–147 (in Polish with English summary).

Barłóg, P. and W. Grzebisz. 1994. Effect of oil-seed rape foliar feeding on yield and seeds quality. *Biul. Magnezol.* 4:9–12 (in Polish with English summary).

Barłóg, P. and W. Grzebisz. 2001. Effect of magnesium foliar application on the yield and quality of sugar beet roots. *Rostl. Výroba* 47(9):418–422.

Barłóg, P. and W. Grzebisz. 2004. Sugar beets fertilization with potassium, sodium and magnesium—Yielding and diagnostic evaluation. Part III. Prognosis of root yield and its quality. *Biul. IHAR* 234:93–105 (in Polish with English summary).

Barłóg, P., W. Grzebisz, and J. Diatta. 2005a. Effect of timing and nitrogen fertilizers on nutrients content and uptake of winter oilseed rape. Part I. Dry matter production and nutrients content. In *Development in Production and Use of New Agrochemicals*, eds. H. Górecki, Z. Dobrzański, and P. Kafarski, pp. 102–112. Prague, Czech Republic: Czech-Pol Trade.

Barłóg, P., W. Grzebisz, and J. Diatta. 2005b. Effect of timing and nitrogen fertilizers on nutrients content and uptake of winter oilseed rape. Part II. Dynamics of nutrients uptake. In *Development in Production and Use of New Agrochemicals*, eds. H. Górecki, Z. Dobrzański, and P. Kafarski, pp. 113–123. Prague, Czech Republic: Czech-Pol Trade.

Barłóg, P., W. Grzebisz, and A. Paradowski. 2002a. Effect of potassium, sodium and magnesium fertilization on yielding of three sugar beet varieties. Part I. Yield of root and sugar. *Biul. IHAR* 222:119–126 (in Polish with English summary).

Barłóg, P., W. Grzebisz, and A. Paradowski. 2002b. Effect of potassium, sodium and magnesium fertilization on yielding of three sugar beet varieties. Part I. Content and uptake of macronutrients. *Biul. IHAR* 222:127–133 (in Polish with English summary).

Barłóg, P. and J. Potarzycki. 2000. Yield and economic efficiency of magnesium applied to oil-seed rape foliage. In *The Balanced Fertilization of Oil-Seed Rape—Current Problems*, ed. W. Grzebisz, pp. 151–156. Poznań, Poland: Agricultural University (in Polish with English summary).

Barraclough, P.B. 1986. The growth and activity of winter wheat roots in the field: Nutrient uptakes of high yielding crops. *J. Agric. Sci.* 106:45–52.

Barry, C.S. 2009. The stay-green revolution: Recent progress in deciphering the mechanisms of chlorophyll degradation in higher plants. *Plant Sci.* 176:325–333.

Başar, H. 2006. Elemental composition of various peach cultivars. *Sci. Hortic.* 107:259–263.

BassiriRad, H. 2000. Kinetics of nutrient uptake by roots: Responses to global change. *New Phytol.* 147:155–169.

Beach, R. 1936. Modern miracle men. Senate, *74th Congress 2nd Session*, Document no. 264, Washington, DC. www.bioelectrichealth.org/MMM%201936.htm.

Beale, S.I. 1999. Enzymes of chlorophyll biosynthesis. *Photosynth. Res.* 60:43–73.

Belane, A.K. and F.D. Dakora. 2012. Elevated concentrations of dietarily-important trace elements and macronutrients in edible leaves and grain of 267 cowpea (*Vigna unguiculata* L. Walp.) genotypes: Implications for human nutrition and health. *Food Nutr. Sci.* 3:377–386.

Bergmann, W. 1992. *Nutritional Disorders of Plants*. Jena, Germany: Gustav Fisher Verlag.

Boila, R.J., L.D. Campbell, S.C. Stothers, G.H. Crow, and E.A. Ibrahim. 1993. Variation in the mineral content of cereal grains grown at selected location throughout Manitoba. *Can. J. Anim. Sci.* 73:421–429.

Bolouri-Moghaddam, M.R., K. Le Roy, L. Xiang, F. Rolland, and W. Ende. 2010. Sugar signaling and antioxidant network connections in plant cells. *FEBS J.* 277:2022–2037.

Bose, J., O. Babourina, and Z. Rengel. 2011. Role of magnesium in alleviation of aluminium toxicity in plants. *J. Exp. Bot.* 62(7):2251–2264.

Brkić, I., Z. Zdunić, A. Jambrović, T. Ledencan, V. Kovacević, and I. Kadar. 2003. Combining abilities of corn-belt inbreed lines of maize for mineral content in grain. *Maydica* 48:293–297.

Bruns, H.A. and M.W. Ebelhar. 2006. Nutrient uptake of maize affected by nitrogen and potassium fertility in a humid subtropical environment. *Commun. Soil Sci. Plant Anal.* 37(1–2):275–293.

Cai, J., L. Chen, H. Qu, J. Lian, W. Liu, Y. Hu, and G. Xu. 2012. Alternation of nutrient allocation and transporter genes expression in rice under N, P, K and Mg deficiencies. *Acta Physiol. Plant.* 34:939–946.

Cakmak, I., C. Hengeler, and H. Marschner. 1994. Changes in phloem export of sucrose in leaves in response to phosphorus, potassium and magnesium deficiency in bean plants. *J. Exp. Bot.* 45(9):1251–1257.

Campbell, C.R. 2000. Reference sufficiency ranges for plant analysis in the southern region of the United States. Updated and reformatted July 2009. www.ncagr.gov/agronomi/saaesd/scsb394.pdf (accessed October 16, 2013).

Campos-Vega, R., G. Loarca-Pina, and B.D. Oomah. 2010. Minor components of pulses and their potential impact on human health. *Food Res. Int.* 43:461–482.

Carrow, R.N., L. Stowell, W. Gelernter, S. Davis, R.R. Duncan, and J. Skorulski. 2004. Clarifying soil testing: III. SLAN sufficiency ranges and recommendations. *Golf Course Manag.* 72(1):194–198.

Chatzav, M., Z. Peleg, L. Ozturk, A. Yazici et al. 2010. Genetic diversity for grain nutrients in wild emmer wheat: Potential for wheat improvement. *Ann. Bot.* 105:1211–1220.

Chen, Z.C. and J.F. Ma. 2013. Magnesium transporters and their role in Al tolerance in plants. *Plant Soil* 368:51–56.

Chu, C.H. and L.J. Johnson. 1985. Relationship between exchangeable and total magnesium in Pennsylvania soils. *Clays Clay Miner.* 33(4):340–344.

Clarkson, D. 1985. Factors affecting mineral nutrient acquisition by plants. *Annu. Rev. Plant Physiol.* 16:77–115.

Clough, G.H. 1994. Potato tuber yield, mineral concentration, and quality after calcium fertilization. *J. Am. Soc. Hortic. Sci.* 119(2):175–179.

Cordain, L. 1999. Cereal grains: Humanity's double-edged sword. *World Rev. Nutr. Diet.* 84:19–73.

Craighead, M.D. and R.J. Martin. 2001. Responses to magnesium fertilizers in wheat in mid Canterbury. *Agron. N. Z.* 31:63–70.

Dahiya, P.K., A.R. Linnemann, M.J.R. Nout, M.A.J. van Boekel, and R.B. Grewal. 2013. Nutrient composition of selected newly bred and established mung bean varieties. *LWT Food Sci. Technol.* 54:249–256.

Davis, D. 2009. Declining fruit and vegetable nutrient composition: What is the evidence? *HortScience* 44(1):15–19.

Davison, P.A., H.L. Schubert, J.D. Reid et al. 2005. Structural and biochemical characterization of Gun4 suggests a mechanism for its role in chlorophyll biosynthesis. *Biochemistry* 44(21):7603–7612.

Ding, Y., W. Luo, and G. Xu. 2006. Characterization of magnesium nutrition and interaction of magnesium and potassium in rice. *Ann. Appl. Biol.* 149:111–123.

Ding, Y. and G. Xu. 2011. Low magnesium with high potassium supply changes sugar partitioning and root growth pattern prior to visible magnesium deficiency in leaves of rice (*Oryza sativa* L.). *Am. J. Plant Sci.* 2:601–608.

de Mello Prado, R. and G. Caione. 2012. Plant analysis. In *Soil Fertility*, ed. R.N. Issaka, pp. 115–134. Rijeka, Republic of Croatia: InTech.

Destro, D., Z. Miranda, A.L. Martins et al. 2002. Agronomic and chemical characterization of soybean genotypes for human consumption. *Crop Breed. Appl. Biotechnol.* 2(4):599–608.

du Preez, C.C. and A.T.P. Bennie. 1991. Concentration, accumulation and uptake rate of macro-nutrients by winter wheat under irrigation. *South Afr. J. Plant Soil* 8(1):31–37.

Ekholm, P., H. Reinivou, P. Mattila et al. 2007. Changes in the mineral and trace element contents in cereals, fruits and vegetables in Finland. *J. Food Compost. Anal.* 20:487–495.

Eliezer, D., E. Aghion, and F.H. Froes. 1998. Magnesium science, technology and application. *Adv. Perform. Mater.* 5:201–212.

Elwali, A.M.O., G.J. Gasho, and M.E.E. Summer. 1985. Dris norms for 11 nutrients in corn leaves. *Agron. J.* 77:506–508.

European Commission. 2001. *Survey of Wastes Spread on Land*. Luxembourg, Europe: European Communities.

Evans, J.R. 1989. Photosynthesis and nitrogen relationships in leaves of C_3 plants. *Oecologia* 78:9–19.

Fageria, N.K. and A. Moreira. 2011. The role of mineral nutrition on root growth of crop plants. *Adv. Agron.* 110:251–331.

Fan, M.-S., F.-J. Zhao, S. Fairweather-Tait, P. Poulton, S. Dunham, and S. McGrath. 2008. Evidence of decreasing mineral density in wheat grain over the last 160 years. *J. Trace Elem. Med. Biol.* 22:315–324.

Farnham, M.W., M.A. Grusak, and M. Wang. 2000. Calcium and magnesium concentration of inbred and hybrid broccoli heads. *J. Am. Soc. Hortic. Sci.* 125(3):344–349.

Ferreira C.F., A.C. Varga Motta, S.A. Prior, C.B. Reissman, N.Z. dos Santos, and J. Gabardo. 2012. Influence of corn (*Zea mays* L.) cultivar development in grain nutrient concentration. *Int. J. Agron.* 12(5):1–7. http://dx.doi.org/10.1155/2012/842582.pdf.

Finck, A. 1992. *Dünger und düngung*. Weinheim, Germany: VCH Verlagsgesellschaft.

Forde, B. and H. Lorenzo. 2001. The nutritional control of root development. *Plant Soil* 232:51–68.

Foyer, C.H., J. Neukermans, G. Queval, G. Noctor, and J. Harbinson. 2012. Photosynthetic control of electron transport and the regulation of gene expression. *J. Exp. Bot.* 63(4):1637–1661.

Garcia-Oliveira, A., L. Tan, Y. Fu, and C. Sun. 2008. Genetic identification of quantitative trait loci for contents of mineral nutrients in rice grain. *J. Integr. Plant Biol.* 51(1):84–92.

Gardner, R.C. 2003. Genes for magnesium transport. *Curr. Opin. Plant Biol.* 6:263–267.

Gebhardt, S. and T. Thomas. 2002. Nutritive value of food. *Home Garden Bull.* 72:1–103.

Gerendás, J. and H. Führs. 2013. The significance of magnesium fro crop quality. *Plant Soil* 368:101–128.

Gill, S.S. and N. Tuteja. 2010. Reactive oxygen species and antioxidant machinery in abiotic stress tolerance in crop plants. *Plant Physiol. Biochem.* 48:909–930.

Gloser, V. and J. Gloser. 2000. Nitrogen and base cation uptake in seedlings of *Acer pseudoplatanus* and *Calamagrostis villosa* exposed to an acidified environment. *Plant Soil* 226:71–77.

Gransee, A. and H. Führs. 2013. Magnesium mobility in soil as a challenge for soil and plant analysis, magnesium fertilization and root uptake under adverse growth conditions. *Plant Soil* 368:5–21.

Grzebisz, W. 2009. *Crop Plant Fertilization: Fertilizing Systems*. Poznań, Poland: Państwowe Wydawnictwo Rolniczo i Leśne (in Polish).

Grzebisz, W. 2011. Magnesium—Food and human health. *J. Elem.* 16(2):299–323.

Grzebisz, W. 2013. Crop response to magnesium fertilization as affected by nitrogen supply. *Plant Soil* 368:23–39.

Grzebisz, W., P. Barłóg, and M. Feć. 1998. The dynamics of nutrient uptake by sugar beet and its effect on dry matter and sugar yield. *Bibliotheka Frag. Agron.* 3(98):242–249.

Grzebisz, W., P. Barłóg, and R. Lehrke. 2001. Effect of the interaction between the method of magnesium application and amount nitrogen fertilizer on sugar recovery and technical quality of sugar beet. *Zuckerindustrie* 126:956–960.

Grzebisz, W., A. Gransee, W. Szczepaniak, and J. Diatta. 2013. The effects of potassium fertilization on water-use efficiency in crop plants. *J. Plant Nutr. Soil Sci.* 176:355–374.

Grzebisz, W., K. Przygocka-Cyna, W. Szczepaniak, J. Diatta, and J. Potarzycki. 2010. Magnesium as a nutritional tool of nitrogen efficient management—Plant production and environment. *J. Elem.* 15(4):771–788.

Grzebisz, W., M. Wrońska, J. Diatta, W. Szczepaniak, and J. Diatta. 2008. Effect of zinc application at early stages of maize growth on the patterns of nutrients and dry matter accumulation by canopy. Part II. Nitrogen uptake and dry matter accumulation patterns. *J. Elem.* 13(1):29–39.

Grzegorczyk, S., J. Alberski, and M. Olszewska. 2013. Accumulation of potassium, calcium and magnesium by selected species of grassland legumes and herbs. *J. Elem.* 18(1):69–78.

Grzyś, E., A. Demczuk, E. Sacała, and G. Kulczycki. 2012. Effect of retardants on the content of selected macroelements in winter wheat cultivars. *Prog. Plant Prot.* 52(4):893–897.

Guerrero-Rodríguez, J.D., D.K. Revell, and W.D. Bellotti. 2011. Mineral composition of lucerne (*Medicago sativa*) and white melilot (*Melilotus albus*) is affected by NaCl salinity of the irrigation water. *Anim. Feed Sci. Technol.* 170:97–104.

Gugała, M., K. Zarzecka, and I. Mystkowska. 2012. Potato tuber content of magnesium and calcium depending on weed control method. *J. Elem.* 17(2):247–254.

Halka, M. and B. Nordstrom. 2010. *Periodic Table of the Elements: Alkali & Alkaline Earth Metals*. New York: Facts on File, Inc.

Hanly, J., P. Loganathan, and L. Currie. 2005. Effect of serpentine rock and its acidulated products as magnesium fertilizers for pasture, compared with magnesium oxide and Epsom salts, on a Pumice Soil. Part 2. Dissolution and estimated leaching losses of fertilizer magnesium. *N. Z. J. Agric. Res.* 48:461–471.

Hardisson, A., C. Rubio, A. Báez, M.M. Martin, and R. Alvarez. 2001. Mineral composition of four varieties of avocado (*Persea gratissima*, L.). *Eur. Food Res. Technol.* 213:225–230.

Härdter, R., M. Rex, and K. Orlovius. 2004. Effect of different Mg fertilizer sources on the magnesium availability in soils. *Nutr. Cycl. Agroecosyst.* 70:249–259.

Harrington, K.C., A. Thatcher, and P.D. Kemp. 2006. Mineral composition and nutritive value of some common pasture weeds. *N. Z. Plant Prot.* 59:261–265.

Hartz, T. and P.R. Johnstone. 2007. Establishing lettuce leaf nutrient optimum ranges through DRIS analysis. *HortScience* 42(1):143–146.

Hawkesford, M., W. Horst, T. Kichey et al. 2012. Functions of macronutrients. In *Mineral Nutrition of Higher Plants*, 3rd edn., ed. P. Marschner, pp. 135–189. Amsterdam, the Netherlands: Elsevier Ltd.

Heckman, J.R., J.T. Sims, D.B. Beegle et al. 2003. Nutrient removal by corn grain harvest. *Agron. J.* 95:587–591.

Hermans, C., F. Bourgis, M. Faucher, R.J. Strasser, S. Delrot, and N. Verbruggen. 2005. Magnesium deficiency in sugar beets alters sugar partitioning and phloem loading in young mature leaves. *Planta* 220:541–549.

Hermans, C., G.N. Johnson, R.J. Strasser, and N. Verbruggen. 2004. Physiological characterisation of magnesium deficiency in sugar beet: Acclimation to low magnesium differentially affects photosystems I and II. *Planta* 220:344–355.

Hermans, C., M. Vuylsteke, F. Coppens et al. 2010. Systems analysis of the responses to long-term magnesium deficiency and restoration in *Arabidopsis thaliana*. *New Phytol.* 187:132–144.

Hinsinger, P., O.N. Fernandes Barros, M.F. Benedetti, Y. Noack, and G. Callot. 2001. Plant-induced weathering of a basaltic rock: Experimental evidence. *Geochim. Cosmochim. Acta* 65(1):137–152.

Holm, N.G. 2012. The significance of Mg in prebiotic geochemistry. *Geobiology* 10:269–279.

Horie, T., D.E. Brodsky, A. Costa et al. 2011. K$^+$ transport by the OsHK2;4 transporter from rice with atypical Na$^+$ transport properties and competition in permeation of K$^+$ over Mg^{2+} and Ca^{2+} ions. *Plant Physiol.* 156:1493–1507.

Hossain, M.A., K. Ban, A.K.M. Zakir Hossain, H. Koyama, and T. Hara. 2004. Combined effects of Mg and Ca supply on alleviation of Al toxicity in wheat plants. *Soil Sci. Plant Nutr.* 50(2):283–286.

Huang, J.W. and D.L. Grunes. 1992. Effects of root temperature and nitrogen form on magnesium uptake and translocation by wheat seedlings. *J. Plant Nutr.* 15(6–7):991–1005.

Hundt, I. and M. Kerschberger. 1991. Magnesium und Pflanzenwachstum. *Kali-Briefe* 20(7/8):539–552.

Hussaini, M.A., V.B. Ogunlela, A.A. Ramalan, and A.M. Falaki. 2008. Mineral composition of dry season maize (*Zea mays* L.) in response to varying levels of nitrogen, phosphorus and irrigation at Kadawa, Nigeria. *World J. Agric. Sci.* 4(6):775–780.

Ikeda, S., K. Tomura, L. Lin, and I. Kreft. 2004. Nutritional characteristics of minerals in Tartary buckwheat. *Fagopyrum* 21:79–84.

Iqbal, A., I.A. Khalil, N. Ateeq, and M.S. Khan. 2006. Nutritional quality of important food legumes. *Food Chem.* 97:331–335.

Jadczak, D., M. Grzeszczuk, and D. Kosecka. 2010. Quality characteristics and content of mineral compounds in fruit of some cultivars of sweet pepper (*Capsicum annuum* L.). *J. Elem.* 15(3):509–515.

Jadczyszyn, T. 2009. Polish fertilizer recommendations system Naw-Sald. *Fertil. Fertil.* 37:195–203.

Jakovljević, M.D., N.M. Kostić, and S.B. Antić-Mladenović. 2003. The availability of base elements (Ca, Mg, Na, K) in some important soil types in Serbia. *Proc. Nat. Sci., Matica Srpska Novi Sad*, 104:11–21.

Jayaganesh, S.V. and V.K. Senthurpandian. 2011a. Impact of different sources and doses of magnesium fertilizer on biochemical constituents and quality parameters of black tea. *Asian J. Biochem.* 6(3):273–281.

Jayaganesh, S.V., V.K. Senthurpandian, and K. Poobathiraj. 2011b. Vertical distribution of magnesium in the laterite soils of South India. *Int. J. Soil Sci.* 6(1):67–76.

Jiménez, M.P., D. Effrón, A.M. de la Horra, and R. Defrieri. 1996. Foliar potassium, calcium, magnesium, zinc and manganese content in soybean cultivars at different stages of development. *J. Plant Nutr.* 19(6):807–816.

Jones, J.B., Jr. 2003. *Agronomic Handbook: Management of Crops, Soils and Their Fertility*. Boca Raton, FL: CRC Press.

Juknevičius, S. and S. Nomeda. 2007. The content of mineral elements in some grasses and legumes. *Ekologija* 53(1):44–52.

Juranović Cindrić, I., I. Krizman, M. Zeiner, Š. Kampić, G. Medunić, and G. Stingeder. 2012. ICP-AES determination of minor- and major elements in apples after microwave assisted digestion. *Food Chem.* 135:2675–2680.

Kaniuczak, J. 1999. Content of some magnesium forms in grey-brown-podzolic soil depending on liming and mineral fertilization. *Zesz. Probl. Postęp. Nauk Rol.* 467:307–316 (in Polish with English summary).

Karley, A. and P.H. White. 2009. Moving cationic minerals to edible tissues: Potassium, magnesium, calcium. *Curr. Opin. Plant Biol.* 12:291–298.

Keltjens, W.G. and K. Tan. 1993. Interactions between aluminium, magnesium and calcium with different monocotyledonous and dicotyledonous plant species. *Plant Soil* 155:485–488.

Kering, M.K., J. Guretzky, E. Funderburg, and J. Mosali. 2011. Effect of nitrogen fertilizer rate and harvest season on forage yield, quality, and macronutrient concentrations in Midland Bermuda grass. *Commun. Soil Sci. Plant Anal.* 42:1958–1971.

Kobayashi, N.I., N. Iwata, T. Saito et al. 2013a. Application of ^{28}Mg for characterization of Mg uptake in rice seedling under different pH conditions. *J. Radioanal. Nucl. Chem.* 296:531–534.

Kobayashi, N.I., T. Saito, N. Iwata et al. 2013b. Leaf senescence in rice due to magnesium deficiency mediated defect in transpiration rate before sugar accumulation and chlorosis. *Physiol. Plant.* 148:490–501.

Kopittke, P.M. and N.W. Menzies. 2007. A review of the use of the basic cation saturation ratios and the "ideal". *Soil Sci. Soc. Am. J.* 71:259–265.

Koudela, M. and K. Petřiková. 2008. Nutrients content and yield of selected cultivars of leaf lettuce (*Lactuca sativa* L. var. *crispa*). *HortScience (Prague)* 35(3):99–106.

Kowieska, A., R. Lubowicki, and I. Jaskowska. 2011. Chemical composition and nutritional characteristics of several cereal grain. *Acta Sci. Pol. Zootechnica* 10(2):37–50.

Kumar Tewari, R., P. Kumar, and P. Nand Sharma. 2006. Magnesium deficiency induced oxidative stress and antioxidant responses in mulberry plants. *Sci. Hortic.* 108:7–14.

Kumar Tewari, R., P. Kumar, N. Tewari, S. Srivastava, and P.N. Sharma. 2004. Macronutrient deficiencies and differential antioxidant responses—Influence on the activity and expression of superoxide dismutase in maize. *Plant Sci.* 166:687–694.

Kuusela, E. 2006. Annual and seasonal changes in mineral contents (Ca, Mg, P, K and Na) of grazed clover-grass mixtures in organic farming. *Agric. Food Sci.* 15:23–34.

Lavon, R., R. Salomon, and E.E. Goldschmidt. 1999. Effect of potassium, magnesium, and calcium deficiencies on nitrogen constituents and chloroplast components in *Citrus* leaves. *J. Am. Soc. Hortic. Sci.* 124(2):158–162.

Lindsay, W.L. 1979. *Chemical Equilibria in Soils.* New York: Wiley & Sons Interscience.

Lloveras, J., M. Aran, P. Villar et al. 2004. Effect of swine slurry on Alfalfa production and on tissue and soil nutrient concentration. *Agron. J.* 96:986–991.

López, A., P. Garcia, and A. Garrido. 2008. Multivariate characterization of table olives according to their mineral nutrient composition. *Food Chem.* 106:369–378.

Maguire, M.E. and J.A. Cowan. 2002. Magnesium chemistry and biochemistry. *Biometals* 15:203–210.

Majkowska-Gadomska, J. 2009. Mineral composition of melon fruit (*Cucumis melo* L.). *J. Elem.* 14(4):717–727.

Majkowska-Gadomska, J. and B. Wierzbicka. 2008. Content of basic nutrients and minerals in heads of selected varieties of red cabbage (*Brassica oleracea* var. *capitata f. rubra*). *Pol. J. Environ. Stud.* 17(2):295–298.

Makino, A., T. Sato, H. Nakono, and T. Mae. 1997. Leaf photosynthesis, plant growth, and nitrogen allocation in rice under different irradiances. *Planta* 203:390–398.

Mankin, K.R. and R.P. Fynn. 1996. Modeling individual nutrient uptake by plants: Relating demand to microclimate. *Agric. Syst.* 50:101–114.

Manzoor, M., F. Anwar, N. Saari, and M. Ashraf. 2012. Variations of antioxidant characteristics and mineral contents in pulp and peel of different apple (*Malus domestica* Borkh.) cultivars from Pakistan. *Molecules* 17:390–407.

Marschner, P. 2012. *Marchner's Mineral Nutrition of Higher Plants.* Amsterdam, the Netherlands: Elsevier.

Martinez, H.E.P., R.B. Souza, J.A. Bayona, V.H. Alvarez Venegas, and M. Sanz. 2003. Coffee-tree floral analysis as a mean of nutritional diagnosis. *J. Plant Nutr.* 26(7):1467–1482.

Masuda, T. 2008. Recent overview of the branch of the tetrapyrrole biosynthesis leading to chlorophylls. *Photosynth. Res.* 96:121–143.

Mattos Junior, D., J.A. Quaggio, H. Cantarella, R.M. Boaretto, and F.C. Zambrosi. 2012. Nutrient management for high citrus fruit yield in tropical soils. *Better Crops* 96(1):4–7.

Mayer, A.-M. 1997. Historical changes in the mineral content of fruits and vegetables. *Br. Food J.* 99(6):207–211.

McDowell, L.R. 1996. Feeding minerals to cattle on pasture. *Anim. Feed Sci. Technol.* 60:247–271.

Mengutay, M., Y. Ceylan, U.B. Kutman, and I. Cakmak. 2013. Adequate magnesium nutrition mitigates adverse effects of heat stress on maize and wheat. *Plant Soil* 368:57–72.

Menkir, A. 2008. Genetic variation for grain mineral content in tropical-adapted maize inbred lines. *Food Chem.* 110:454–464.

Metson, A. 1974. Magnesium in New Zealand soils. Part I. Some factors governing the availability of soil magnesium. *N. Z. J. Exp. Agric.* 2:277–319.

Metson, A. and J. Brooks. 1975. Magnesium in New Zealand soils. Part I. Distribution of exchangeable and "reserve" magnesium in the main soil groups. *N. Z. J. Agric. Res.* 18:317–335.

Miller, R.O., J.S. Jacobsen, and E.O. Skogley. 1994. Aerial accumulation and partitioning of nutrients by hard red spring wheat. *Commun. Soil Sci. Plant Anal.* 25(11):1891–1911.

Milošević, T. and N. Milošević. 2012. Factors influencing mineral composition of plum fruits. *J. Elem.* 17(3):453–464.

Mittler, R. 2002. Oxidative stress, antioxidants and stress tolerance. *Trends Plant Sci.* 7(9):405–411.

Moraghan, J.T., J.D. Etchevers, and J. Padilla. 2006. Contrasting accumulations of calcium and magnesium in seed coats and embryos of common bean and soybean. *Food Chem.* 95:554–561.

Mota Oliveira, E.M., H.A. Ruiz, V.H. Alvarez et al. 2010. Nutrient supply by mass flow and diffusion to maize plants in response to soil aggregate size and water potential. *Rev. Bras. Ciênc. Solo* 34:317–327.

Moustakas, N.K., K.A. Akoumianakis, and H.C. Passam. 2011. Patterns of dry matter accumulation and nutrient uptake by okra (*Abelmoschus esculentus* L.) under different rates of nitrogen application. *Aust. J. Crop Sci.* 5(8):993–1000.

Mullins, G.L. and J.H. Edwards. 1987. Effect of fertilizer amendments, bulk density, and moisture on calcium and magnesium diffusion. *Soil Sci. Soc Am. J.* 51:1219–1224.

Murchie, E.H., M. Pinto, and Horton P. 2009. Agriculture and the challenges for photosynthesis research. *New Phytol.* 181:532–552.

Murphy, K.M., Ph.G. Reeves, and S.S. Jones. 2008. Relationship between yield and mineral nutrient concentrations in historical and modern spring wheat cultivars. *Euphytica* 163:381–390.

Nelson, A.G., S.A. Quideau, B. Frick et al. 2011. The soil microbial community and grain micronutrient concentration in historical and modern hard red spring wheat cultivars grown organically and conventionally in the black soil zone of the Canadian Prairies. *Sustainability* 3:500–517.

Netto, A.T., E. Campostrini, J. Gonçalves de Oliveira, and R.E. Bressan-Smith. 2005. Photosynthetic pigments, nitrogen, chlorophyll a fluorescence and SPAD-502 readings in coffee leaves. *Sci. Hortic.* 104:199–209.

Nunes, S.J.C., P.C.R. Fontes, E.F. Araújo, and C. Sediyama. 2006. Potato plant growth and macronutrient uptake as affected by soil tillage and irrigation systems. *Pesq. Agropec. Bras., Brasilia* 41(12):1787–1792.

Nuss, E.T. and S.A. Tanumihardjo. 2010. Maize: A paramount staple crop in the context of global nutrition. *Compr. Rev. Food Sci. Food Saf.* 9(4):417–436.

Oko, A.O., B.E. Ubi, A.A. Efisue, and N. Dambaba. 2012. Comparative analysis of the chemical nutrient composition of selected local and newly introduced rice varieties grown in Ebonyi State of Nigeria. *Inter. J. Agric. Forest.* 2(2):16–23.

Okpamen, S.U., E.G. Ilori, I. Agho, A. Nkechika A., F.U., Maidoh, and P.N. Okonjo. 2013. Influence of depth and soil pH on forms of magnesium in soils of four parent materials (*Rhodic paleudults, Rhodic tropudalfs, Oxic tropudalfs* and *Aquic tropossamment*). *J. Soil Sci. Environ. Manag.* 4(4):71–76.

Ortiz I., M. Simón, C. Dorronsoro, F. Martín, and I. García. 2002. Soil evolution over the quaternary period in a Mediterranean climate (SE Spain). *Catena* 48:131–148.

Orzechowski, M. and S. Smólczyński. 2010. Content of Ca, Mg, Na, K, P, Fe, Mn, Zn, Cu in the soil developed from the Holocene deposits in N-E Poland. *J. Elem.* 5(1):149–159.

Osaki, M., T. Zheng, and K. Konno. 1996. Carbon-nitrogen interaction related to P, K, Ca, and Mg nutrients in field crops. *Soil Sci. Plant Nutr.* 42(3):539–552.

Osemwota, I.O., J.A.I. Omueti, and A.I. Ogboghodo. 2007. Effect of calcium/magnesium ratios in soil on magnesium availability, yield, and yield components of maize. *Commun. Soil Sci. Plant Anal.* 38:2849–2860.

Otegui, M. and R. Bonhomme. 1998. Grain yield components in maize. I. Ear growth and kernel set. *Field Crop. Res.* 56:247–256.

Özcan, M.M. 2006. Determination of the mineral compositions of some selected oil-bearing seeds and kernels using inductively coupled plasma atomic emission spectrometry (ICP-AES). *Grasas Aceites* 57(2):211–218.

Page, G. 2002. Francis Home and Joseph Black: The chemistry and testing of alkaline salts in the early bleaching and alkali trade. *Bull. Hist. Chem.* 27(2):107–113.

Pennington, J.A.T. and R.A. Fisher. 2010. Food component profiles for fruit and vegetables subgroups. *J. Food Compost. Anal.* 23:411–418.

Peters, J.B., K.A. Kelling, P.E. Speth, and S.M. Offer. 2005. Alfalfa yield and nutrient uptake as affected by pH and applied K. *Commun. Soil Sci. Plant Anal.* 36:583–596.

Phillips, R.D. 1993. Starchy legumes in human nutrition, health and culture. *Plant Food Hum. Nutr.* 44:195–211.

Piechota, T., A. Blecharczyk, and I. Małecka. 2000. Effect of long-term organic and mineral fertilization on nutrients content in soil profile. *Folia Univ. Agric. Stetin.* 84:393–398 (in Polish with English summary).

Pirhofer-Walzl, K., K. Søegaard, H. Høgh-Jensen et al. 2011. Forage herbs improve mineral composition of grassland herbage. *Grass Forage Sci.* 66:415–423.

Pisulewska, E., R. Poradowski, J. Antonkiewicz, and R. Witkowicz. 2009. The effect of variable mineral fertilization on yield and grain mineral composition of covered and naked oat cultivars. *J. Elem.* 14(4):763–772.

Potarzycki, J. 2010. Yield forming effect of combined application of magnesium, sulfur, and zinc in maize fertilization. *Fertil. Fertil.* 39:44–59.

Potarzycki, J. 2011. Effect of magnesium or zinc supplementation at the background of nitrogen rate on nitrogen management by maize canopy cultivated in monoculture. *Plant Soil Environ.* 57(1):19–25.

Pregitzer, K.S. and J.S. King. 2005. Effects of soil temperature on nutrient uptake. *Ecol. Stud.* 181:277–310.

Oury, F.-X., F. Leenhardt, C. Remesy et al. 2006. Genetic variability and stability of grain magnesium, zinc and iron concentrations in bread wheat. *Eur. J. Agron.* 25:177–185.

Ramesh, K., B. Chandrasekaran, T.N. Balusubramanian, U. Bangarusamy, R. Sivasamy, and N. Sankaran. 2002. Chlorophyll dynamics in rice (*Oryza sativa*) before and after flowering based on SPAD (chlorophyll) meter monitoring and its relation with grain yield. *J. Agron. Crop Sci.* 188:102–105.

Ranawat, P., K. Mohan Kumar, and N. Sharma. 2009. A process for making slow-release phosphate fertilizer from low-grade rock phosphate and siliceous tailings by fusion with serpentine. *Curr. Sci.* 96(6):843–848.

Ranhotra, G.S., J.A. Gelroth, B.K. Glaser, and K.J. Lorenz. 1996. Nutrient composition of spelt wheat. *J. Food Compost. Anal.* 9:81–84.

Reinbott, T.M. and D.G. Blevins. 1995. Response of soybean to foliar-applied boron and magnesium and soil-applied boron. *J. Plant Nutr.* 18(1):179–200.

Rengel, Z. 1993. Mechanistic simulation models of nutrient uptake: A review. *Plant Soil* 152:161–173.

Rengel, Z. and D.L. Robinson. 1989. Competitive Al^{3+} inhibition of net Mg^{2+} uptake by intact *Lolium multiflorum* roots. *Plant Physiol.* 91:1407–1413.

Ristimäki, L.M. 2007. Potassium and magnesium fertiliser recommendations in some European countries. *Proceedings of the International Fertiliser Society*, No. 620, York, U.K.

Rodríguez, L.H., D.A. Morales, E.R. Rodríguez, and C.D. Romero. 2011. Minerals and trace elements in a collection of wheat landraces from the Canary Islands. *J. Food Compost. Anal.* 24:1081–1090.

Rosman, K.J.R. and P.D.T. Taylor. 1998. Isotopic compositions of the elements 1997. *Pure Appl. Chem.* 70:217–235.

Rout, G.R., S. Samantaray, and P. Das. 2001. Aluminium toxicity in plants: A review. *Agronomie* 21:3–21.

Rubio, G., J. Zhu, and J. Lynch. 2003. A critical test of the prevailing theories of plant response to nutrient availability. *Am. J. Bot.* 90(1):143–152.

Sabo, M., T. Teklić, and I. Vidović. 2002. Photosynthetic productivity of two winter wheat varieties (*Triticum aestivum* L.). *Rostl. Výroba* 48(2):80–86.

Saleethong, P., J. Sanitchon, K. Kong-ngern, and P. Theerakulpisut. 2013. Effects of exogenous spermidine (Spd) on yield, yield-related parameters and mineral composition of rice (*Oryza sativa* L. ssp. *indica*) grains under salt stress. *AJCS* 7(9):1293–1301.

Sanchez-Rodriguez, E., M. del mar Rubio-Wilhelmi, L.M. Cervilla et al. 2010. Study of the ionome and water fluxes in cherry tomato plants under moderate water stress conditions. *Plant Soil* 335:339–347.

Sarafis, V. 1998. Chloroplasts: A structural approach. *J. Plant Physiol.* 152:248–264.

Schachtschabel, P. 1954. Das pflanzenverfügbare Magnesium des Bodens und seine Bestimmung. *J. Plant Nutr. Soil Sci.* 67:9–23.

Schaff, B.E. and E.O. Skogley. 1982. Diffusion of potassium, calcium and magnesium in Bozeman silt loam as influenced by temperature and moisture. *J. Soil Sci. Soc. Am.* 46:521–524.

Schlesinger, W.H. 1997. *Biogeochemistry: An Analyses of Global Change.* San Diego, CA: Academic Press.

Schonewille, J.T. 2013. Magnesium in dairy cow nutrition: An overview. *Plant Soil* 358:167–178.

Schweiger, P. and A. Amberger. 1979. Mg-Auswaschung und Mg-Bilanz in einem langjärigen Lysimeterversuch. *Z. Acker. Pflanzenbau* 148:403–410.

Seefeldt, S. 2013. Plant tissue testing. FGV-00244. University of Alaska. www.uaf.edu/ces.

Sema, A., C.S. Maiti, A.K. Singh, and A. Bendangsengla. 2010. DRIS nutrient norms for pineapple on Alfisols of India. *J. Plant Nutr.* 33:1384–1399.

Serra, A.P., M.E. Marchetti, D.J. Bungenstab et al. 2013. Diagnosis and Recommendation Integrated System (DRIS) to assess the nutritional state of plants. In *Biomass Now—Sustainable Growth and Use*, ed. M.D. Matović, pp. 129–145. Rijeka, Republic of Croatia: InTech.

Shaaban, M.M., M.A. El-Nabarawy, and E.A.A. Abou El-Nour. 2002. Evaluation of magnesium and iron nutritional status in some monocot and dicot crop using a portable chlorophyll meter. *Pak. J. Biol. Sci.* 5(10):1014–1016.

Shaahan, M.M., A.A. El-Sayed, and E.A.A. Abou El-Nour. 1999. Predicting nitrogen, magnesium and iron nutritional status in some perennial crops using a portable chlorophyll meter. *Sci. Hortic.* 82:339–348.

Shabala, S. and Y. Hariadi. 2005. Effects of magnesium availability on the activity of plasma membrane ion transporters and light-induced responses from broad bean leaf mesophyll. *Planta* 221:56–65.

Shand, M.A. 2006. *The Chemistry and Technology of Magnesia.* Hoboken, NJ: John Wiley.

Shaul, O. 2002. Magnesium transport and function in plants: The tip of the iceberg. *Biometals* 15:309–323.

Shearman, V., R. Sylvester-Bradley, R. Scott, and M. Foulkes. 2005. Physiological processes associated with wheat progress in the UK. *Crop Sci.* 45(1):175–185.

Sheikhi, J. and A. Ronaghi. 2012. Growth and macro and micronutrients concentration in spinach (*Spinacia oleracea* L.) as influenced by salinity and nitrogen rates. *Int. Res. J. Appl. Basic Sci.* 3(4):770–777.

Shobha, D., T.A. Sreeramasetty, Puttaramanaik, and K.T. Pandurange Gowda. 2010. Evaluation of maize genotypes for physical and chemical composition at silky and hard stage. *Karnataka J. Agric. Sci.* 23(2):311–314.

Soder, K.J. and W.L. Stout. 2003. Effect of soil type and fertilization level on mineral nutrient concentration of pasture: Potential relationships to ruminant performance and health. *J. Anim. Sci.* 81:1603–1610.

Sourkes, T.L. 2009. The discovery and early history of carotene. *Bull. Hist. Chem.* 34(1):32–38.

Stenbaek, A. and P.E. Jensen. 2010. Redox regulation of chlorophyll biosynthesis. *Phytochemistry* 71:853–859.

Steudle, E. 2000. Water uptake by roots: Effects of water deficit. *J. Exp. Bot.* 51(350):1531–1542.

Stevens, G., T. Gladbach, P. Motavalli, and D. Dunn. 2005. Soil calcium: magnesium ratios and lime recommendations for cotton. *The J Cotton Sci.* 9:65–71.

Storey, R. and M.T. Treeby. 2000. Seasonal changes in nutrient concentrations of naval orange fruit. *Sci. Hortic.* 84:67–82.

Subedi, K. and B. Ma. 2005. Nitrogen uptake and partitioning in stay-green and leafy maize hybrids. *Crop Sci.* 45:740–747.

Suchowilska, E., M. Wiwart, W. Kandler, and R. Krska. 2012. A comparison of macro- and microelement concentrations in the whole grain of four *Triticum* species. *Plant Soil Environ.* 58(3):141–147.

Szczepaniak, W., P. Barłóg, R. Łukowiak, and K. Przygocka-Cyna. 2013. Effect of balanced nitrogen fertilization in four-year crop rotation on plant productivity. *J. Cent. Eur. Agric.* 14(1):64–77.

Szulc, P. 2009. Effect of nitrogen fertilization and methods of magnesium application on chlorophyll content, accumulation of mineral compounds and morphology of two maize hybrid types in the initial growth period. Part I. Content of chlorophyll and mineral components. *Acta Sci. Pol. Agric.* 8(2):43–50.

Szulc, P. 2010. Effects of differentiated levels of nitrogen fertilization and the method of magnesium application on the utilization of nitrogen by two different maize cultivars for grain. *Pol. J. Environ. Stud.* 19(2):407–412.

Szulc, P., J. Bocianowski, and M. Rybus-Zając. 2011. The reaction of stay green maize hybrid (*Zea mays* L.) to various methods of magnesium application. *Fresenius Environ. Bull.* 20(8):216–234.

Szulc, P., J. Bocianowski, and M. Rybus-Zając. 2012. The effect of soil supplementation with nitrogen and elemental sulphur on chlorophyll content and grain yield of maize (*Zea mays* L.). *Zemdirbyste-Agriculture* 99(3):247–254.

Szymczyk, S., U. Szyperek, A. Rochweger, and M. Rafałowska. 2005. Influence of precipitation on calcium and magnesium outflows from soils in young glacial areas. *J. Elem.* 10(1):155–166.

Taiz, L. and E. Zeiger 2006. *Plant Physiology*. Sunderland, MA: Sinauer Associates.

Tanaka, A. and R. Tanaka. 2006. Chlorophyll metabolism. *Curr. Opin. Plant Biol.* 9:248–255.

Tanaka, R. and A. Tanaka. 2011. Chlorophyll cycle regulates the construction and destruction of the light-harvesting complexes. *Biochim. Biophys. Acta* 1807(8):968–976.

Tang, N., Y. Li, and L.-S. Chen. 2012. Magnesium deficiency-induced impairment of photosynthesis in leaves of fruiting *Citrus reticulata* trees accompanied by up-regulation of antioxidant metabolism to avoid photo-oxidative damage. *J. Plant Nutr. Soil Sci.* 175:784–793.

Thomas, D. 2007. The mineral depletion of foods available to US as nations (1940–2002)—A review of the sixth edition of the McCance and Widdowson. *Nutr. Health* 19:21–55.

Topuz, A., M. Topakci, M. Canakci, I. Akinci, and F. Ozdemir. 2005. Physical and nutritional properties of four orange varieties. *J. Food Eng.* 66:519–523.

Uchida, R. 2000. Recommended plant tissue nutrient levels for some vegetable, fruit, and ornamental foliage and flowering plants in Hawaii. In *Plant Nutrient Management in Hawaii's Soils, Approaches for Tropical and Subtropical Agriculture*, eds. J.A. Silva and R. Uchida, pp. 57–65. Honolulu, HI: University of Hawaii.

Verbruggen, N. and C. Hermans. 2013. Physiological and molecular response to magnesium nutritional imbalance in plants. *Plant Soil* 368:87–90.

Wall, M.M. 2006. Ascorbic acid, vitamin A, and mineral composition of banana (*Musa* sp.) and papaya (*Carica papaya*) cultivars grown in Hawaii. *J. Food Compost. Anal.* 19:434–445.

Walworth, J.L. and S. Ceccotti. 1990. A re-examination of optimum foliar magnesium levels in corn. *Commun. Soil Sci. Plant Anal.* 21(13–16):1457–1473.

Wang, K.M., J.G. Wu, G. Li, D.P. Zhang, Z.W. Yang, and C.H. Shi. 2011. Distribution of phytic acid and mineral elements in three indica rice (*Oryza sativa* L.) cultivars. *J. Cereal Sci.* 54:116–121.

Weston, J. 2009. Biochemistry of magnesium. In *PATAI's Chemistry of Functional Groups in 2009*. Chichester, U.K.: John Wiley. doi:10.1002/9780470682531.pat0407.

White, P.H. and M. Broadley. 2003. Calcium in plants. *Ann. Bot.* 92:1–25.

White, P.J., J.E. Bradshaw, M. Finlay, B. Dale, and G. Ramsay. 2009. Relationships between yield and mineral concentrations in potato tubers. *HortScience* 44(1):6–11.

Whitehead, W.C. 2000. *Nutrient Elements in Grasslands*. Wallingford, U.K.: CABI Publishing.

Wichmann, W. 1992. *World Fertilizer Use Manual*. Paris, France: BASF, IFA.

Yu, M., Ch-X. Hu, and Y.-H. Wang. 2006. Effects of molybdenum on the intermediates of chlorophyll biosynthesis in winter wheat cultivars under low temperature. *Agric. Sci. China* 5(9):670–677.

7 Sulfur

Cynthia Grant and Malcolm J. Hawkesford

CONTENTS

7.1 HISTORICAL BACKGROUND

Sulfur (S) has been recognized as a plant nutrient since the time of Liebig, who indicated that S was accessed by plants from the soil solution (Meidner, 1985). Sachs also determined that S was an essential element from his own and previous research (Sachs, 1865; cited by Epstein, 2000). In the past, deficiencies were relatively rare in industrialized areas due to inadvertent inputs of S from industrial pollution (Lehmann et al., 2008) and in the superphosphate and ammonium sulfate fertilizers that were used frequently to supply P and N. In regions with limited industrialization and low atmospheric input of S, such as the Northern Great Plains of North America, S deficiencies were recognized early. For example, S deficiencies were identified in the Canadian prairies as a risk for legume production in the 1930s and for canola (rapeseed, *Brassica napus* L.) in the late 1960s and early 1970s (Hamm, 1967; Hamm et al., 1973; Beaton and Soper, 1986). In the Prairie Provinces, there are more than 4 million ha of agricultural soils deficient in plant-available S and substantially greater areas are potentially deficient (Bettany and Janzen, 1984; Doyle and Cowell, 1993). With movement to intensive crop production practices, use of high-yielding cultivars, movement away from S-containing fertilizers such as ammonium sulfate and superphosphate, and decreases in aerial deposition of S due to increased air quality standards, S deficiencies have been identified on a broad range of soils in North America, Europe, Asia, and Australia (Karamanos, 1988; Ali et al., 1996; McGrath and Zhao, 1996b; Jackson, 2000; Grant et al., 2004; Malhi et al., 2004; Brennan and Bolland, 2006; Solberg et al., 2007; Brennan and Bolland, 2008; Haneklaus et al., 2008; Egesel et al., 2009; Eriksen, 2009). In Europe, S deficiencies have been noted in rapeseed crops since the mid-1980s, with the occurrence of deficiencies related to the reduction in atmospheric S deposition that previously provided a significant source of plant-available S (Haneklaus et al., 2008). Similarly, effects of S deficiency on the quality of wheat (*Triticum aestivum* L.) grain have been observed increasingly (Zhao et al., 1995, 1997, 1999a). Application of S fertilizer to rapeseed crops is a recommended and widely followed commercial production practice (Good and Glendinning, 1998; Thomas, 2003; Haneklaus et al., 2008).

7.2 UPTAKE OF SULFUR BY PLANTS

Plants absorb S primarily through their roots from the soil as sulfate. The uptake and transport of sulfate in plants is an energy-dependent process, mediated through a number of different sulfate transporters (Section 7.4). Sulfate is taken up across cell membranes against a concentration gradient, probably coupled to a proton gradient through a cotransporter system (Hawkesford et al., 1993; Smith et al., 1995; Hawkesford, 2010). Excessive sulfate can be stored as free sulfate ions, mainly in the vacuoles, but sulfate taken up by the roots is transported primarily to the leaves, where it is reduced in the chloroplasts or stored in the vacuoles. The biochemistry and molecular biology of the assimilation of sulfate into sulfide have been extensively reviewed elsewhere and are summarized only briefly in the following text (Kopriva, 2006; Takahashi et al., 2011). Reduction from sulfate to sulfide can occur in most tissues and in multiple subcellular compartments but mainly occurs in the plastids of green tissues, where ATP and redox equivalent are abundant. The reduction from sulfate to sulfide is a three-stage process.

Sulfate is activated first to adenosine 5′-phosphosulfate (APS), catalyzed by ATP sulfurylase (Figure 7.1). The K_m for this enzyme is about 1 mM, so the sulfate concentration in the chloroplast may be rate limiting for S reduction. The APS is reduced by APS reductase to form sulfite. Sulfite is then reduced to sulfide by sulfite reductase. Sulfide is subsequently incorporated into cysteine in a reaction catalyzed by the cysteine synthase complex, composed of *O*-acetylserine(thiol)lyase (OASTL) and serine acetyl transferase (SAT). SAT catalyzes the acetylation of serine by acetyl-CoA to form *O*-acetylserine (OAS). Subsequently, OASTL catalyzes the conjugation of sulfide with OAS to form cysteine. The SAT is active only when in a complex with OASTL, and in the presence of an excess of OAS, for example, under a limiting supply of sulfide, the complex dissociates, preventing further OAS production. Conversely, OASTL catalyzes sulfide incorporation as a free dimer of OASTL, and excess sulfide on the other hand may promote OASTL–SAT complex stabilization. The OASTL–SAT complex is therefore said to have a sensing and regulatory function for the synthesis of cysteine, controlling the balance of substrates and cysteine formation. As described later (Section 7.4), the levels of these substrates and products influence the expression of the uptake and primary assimilation components of the pathway. The synthesis of cysteine and mechanisms of regulation have been comprehensively reviewed elsewhere (Hell and Wirtz, 2008). In addition, serine supply depends on having sufficient C and N present, so S assimilation is interconnected at this point with C and N metabolism. There is also a cysteine allosteric feedback loop that restricts the production of excess cysteine if serine and sulfide are both present in large amounts (Hawkesford, 2000; Saito, 2000). Cysteine is the reduced sulfur donor for the synthesis of the other major S-containing amino acid, methionine, and is the precursor for other various S-containing compounds including glutathione (GSH) and phytochelatins.

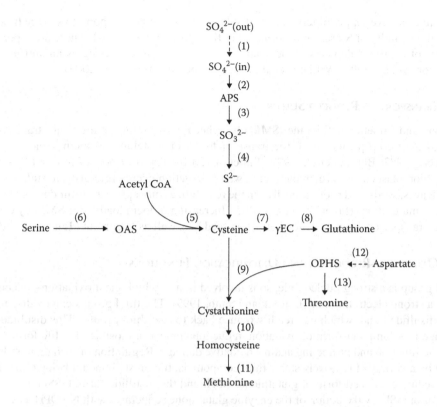

FIGURE 7.1 The pathways of cysteine, glutathione, and methionine biosynthesis. Key to enzymes involved: (1) sulfate transporters; (2) ATP sulfurylase (enzyme class [EC] 2.7.7.4); (3) adenosine phosphosulfate (APS) reductase (EC 1.8.4.9); (4) sulfite reductase (EC 1.8.7.1); (5) OASTL (EC 2.5.1.47); (6) serine acetyltransferase (EC 2.3.1.30); (7) γ-glutamylcysteine synthetase (EC 6.3.2.2); (8) glutathione synthetase (EC 6.3.2.3); (9) cystathionine γ-synthase (EC 4.2.99.2); (10) cystathionine β-lyase (EC 4.4.1.8); (11) methionine synthase (EC 2.1.1.14); (12) aspartate kinase (EC 2.7.2.4), aspartate semialdehyde dehydrogenase (EC 1.2.1.11), homoserine dehydrogenase (EC 1.1.1.3), and homoserine kinase (EC 2.7.1.39); (13) threonine synthase (EC 4.2.99.2). CoA, coenzyme A; γEC, γ-glutamylcysteine; OPHS, O-phosphohomoserine. (From Hawkesford, M.J. and De Kok, L.J., *Plant Cell Environ.*, 29(3), 382, 2006.)

7.3 PHYSIOLOGICAL RESPONSES OF PLANTS TO SUPPLY OF SULFUR

Sulfur plays an important role in many plant processes. It is a component of the essential amino acids cysteine and methionine and so is present in all proteins. The thiol groups on the cysteine residues in proteins can oxidize to form disulfide bridges that contribute to the structure of proteins. Sulfur is a component of ferredoxin and therefore is involved in photosynthesis, glutamate synthesis, N_2 fixation, and nitrate reduction (Mills and Jones, 1996). Sulfur is required for chlorophyll production and for the synthesis of various enzyme cofactors such as biotin, thiamine, glutathione, and coenzyme A. Coenzyme A is involved in the Krebs cycle and in lipid and fatty acid metabolism (Mills and Jones, 1996). Sulfur is also a component of critical secondary S compounds including glucosinolates in Cruciferae and alliins in Liliaceae as well as phytochelatins.

7.3.1 FORMS OF SULFUR IN TISSUE

Sulfur occurs in plant tissue primarily as sulfate or in proteins, with smaller amounts being present as cysteine and methionine, glutathione, sulfolipids, and other secondary compounds such as glucosinolates (Hawkesford, 2000). The amount of sulfate is very dependent upon nutritional conditions, and with an excessive supply of fertilizer, leaf sulfate can be the largest pool. In storage tissues such as seeds, the quantity of S present in forms other than protein is minimal. There are reports of the occurrence of localized deposits of elemental S in vascular tissues, acting as an antifungal compound (Cooper et al., 1996; Williams et al., 2002; Cooper and Williams, 2004).

7.3.2 TRANSPORT OF REDUCED SULFUR

Glutathione and S-methylmethionine (SMM) have been proposed to be the major transport forms of reduced S in crop plants, being the transport forms of cysteine and methionine, respectively (Rennenberg, 1982; Bourgis et al., 1999; Tan et al., 2010). The transport of reduced S is required from the chloroplasts in mature green tissues, where reduction occurs primarily, to sink tissues such as young leaves, seeds, and roots, as well as in the remobilization of protein sulfur during plant development and maturation (Haneklaus et al., 2007). Increasing phloem loading of SMM by expression of specific transporters has a beneficial effect on seed S as well as N content (Tan et al., 2010).

7.3.3 OXIDATION–REDUCTION AND DETOXIFICATION FUNCTIONS

The thiol group is a strong nucleophile, so is involved in many biological oxidation–reduction processes as a strong electron donor (Leustek and Saito, 1999). The thiol groups can oxidize to form a covalent disulfide bond, which can readily reduce back to two thiol groups. This disulfide–dithiol interchange is a dominant form of oxidation–reduction control in most aerobic life forms, regulating enzyme function and protecting against oxidative damage. Regulation of oxidation–reduction is mediated by a variety of relatively stable thiol compounds, the most abundant being GSH. The balance between the reduced form of glutathione, GSH, and the disulfide form, GSSG, is maintained in the favor of GSH by the action of the enzyme glutathione reductase, with NADPH as an electron source. This action buffers the plant cytoplasm, chloroplast stroma, and mitochondrial matrix in the reducing state, which is necessary for the proper functions of a range of intracellular enzymes. The reducing state also suppresses the formation of disulfide bonds between cysteine residues in proteins, preventing a disruption of structure and loss of activity (Leustek and Saito, 1999). Although some proteins require disulfide bonds for structural formation, this need is not common in soluble intracellular proteins.

Glutathione also plays an important role in stress mitigation, acting as a source of electrons for the enzyme glutathione peroxidase to deal with reactive oxygen species. Glutathione reacts to inactivate toxins through a reaction mediated by glutathione S-transferase, allowing them to be bound in the extracellular matrix or stored as metabolites in the vacuole (Kreuz et al., 1996; Leustek and Saito, 1999). Activity of glutathione S-transferase can be increased by the presence of natural or synthetic xenobiotics to enhance detoxification. Similarly, the presence of pathogens or other stressors that can lead to the formation of hydroperoxides can also induce glutathione S-transferase, reflecting its role in the detoxification of lipid hydroperoxidases (Haneklaus et al., 2007). Glutathione appears to play an important role in drought-stress tolerance through its roles in reducing free radicals (Chan et al., 2013). Under drought conditions, glutathione reductase activity increases, allowing the recycling of GSSG to GSH at the expense of NADPH in order to maintain GSSG/GSH ratios at an adequate level.

Heavy metal detoxification also may be mediated by phytochelatins or metallothioneins, where thiol groups are the metal ion ligand (Leustek and Saito, 1999). It appears that phytochelatin can create a complex with the heavy metals that is then transported to the vacuole, reducing the cytoplasmic concentration and potential toxicity of the metal. The role of S-containing compounds appears to be important in response to high concentrations of cadmium, as cadmium stress induces higher sulfate uptake. Cadmium-induced sulfate uptake was related to higher mRNA encoding for a high-affinity sulfate transport in roots, as well as with a decrease in sulfate and glutathione content and production of high levels of phytochelatins (Nocito et al., 2002; Gill and Tuteja, 2011).

7.3.4 FUNCTIONS OF SECONDARY SULFUR COMPOUNDS

Plants produce a wide range of bioactive secondary compounds. Many of the S-containing phytochemicals are important in protection against pests, diseases, and other environmental stresses. Glucosinolates are important S-containing compounds present in a variety of dicotyledonous families, with Brassicaceae being the most important agriculturally. Glucosinolates and their volatile hydrolysis products are responsible for the characteristic smell and flavor of cruciferous vegetables such as cabbage and broccoli and crops such as mustard. Although some degradation products of glucosinolates, such as isothiocyanates, may have anticarcinogenic properties, others can cause health problems, including the goitrogenic effect of a degradation product of 2-hydroxy-3-butenyl glucosinolate, the major glucosinolate in rapeseed. To improve the quality of rapeseed meal for inclusion in animal diets, standard breeding techniques were used to develop forms of rapeseed, referred to as canola, where the seed contains less than 30 μmole of glucosinolates per gram of air-dried, oil-free solid (http://www.canolacouncil.org/oil-and-meal/what-is-canola, accessed July 12, 2013).

A major function of glucosinolates in the plant is in defense against insects, fungi, and microorganisms (Redovniković et al., 2008). When the plant tissues are disrupted by wounding, insect, or pathogen attack, the enzyme myrosinase comes into contact with the glucosinolates in the plant, catalyzing their hydrolysis to compounds with varying biological activities (Redovniković et al., 2008). Glucosinolates function to inhibit growth or deter feeding by a range of generalist pests, including insects, slugs, and birds, but may serve as attractants for insects or other pests that selectively target crucifers (Giamoustaris and Mithen, 1995, 1997; Redovniković et al., 2008). Glucosinolates and their hydrolysis products can inhibit the activity of soil-borne pathogens, an effect that may persist to protect subsequent crops in the rotation (Angus et al., 1991, 1994; Kirkegaard et al., 1994, 1997, 2004). Defense signal molecules that are derivatives of jasmonic acid, salicylic acid, and ethylene may influence glucosinolate metabolism to bolster the defense mechanisms of the plant (Smetanska et al., 2007; Redovniković et al., 2008; Schreiner et al., 2011). Elemental S deposits within plants have also been implicated as a defense mechanism against pathogens, although the pathways that result in these accumulations are unclear (Cooper et al., 1996; Williams et al., 2002; Cooper and Williams, 2004).

Increasing S supply will generally increase the glucosinolate content of crops, whereas decreasing S supply reduces free sulfate and glucosinolates and increases the activity of the enzyme myrosinase (Falk et al., 2007). Glucosinolates also may serve as an S storage source to be remobilized under conditions of S deficiency (Blake-Kalff et al., 1998). In rapeseed that is adequately supplied with S, the youngest leaves contained about 2% of their S as glutathione and about 6% as glucosinolates, with the proportions being even lower in older leaves (Blake-Kalff et al., 1998). Insoluble protein S was about 50% in young leaves, whereas sulfate was about 42%. In the middle and oldest leaves, glutathione and glucosinolates made up less than 1% of the total S, and 70%–90% was present as sulfate. As glucosinolates make up only a small portion of the S reserves in the plant, it is unlikely that they are a major source of remobilizable S for the plant (Blake-Kalff et al., 1998; Redovniković et al., 2008). When S supply is removed, the youngest leaves convert sulfate, glutathione, and glucosinolates into insoluble S, with sulfate being the most important contributor, indicating that glucosinolates are not a major contributor of S under S-deficient situations (Blake-Kalff et al., 1998).

Allium species store a large proportion of their total plant S as bioactive secondary compounds, known as alliins. The thiosulfinates or alkane(ene) thial-S-oxides formed from S-alk(en)yl cysteine sulfoxides and thiosulfinates can decompose to form additional S constituents (Benkeblia and Lanzotti, 2007). In garlic (*Allium sativum* L.), the major compound is alliin, whereas in onion (*Allium cepa* L.), it is isoalliin. As with glucosinolates, alliins appear to contribute to the defense of plants against various pests and pathogens (Curtis et al., 2004). In intact cells, the alliinase enzyme is located in the vacuole, while the alliin and alkyl cysteine sulfoxides are located in the cytoplasm

(Lancaster and Collin, 1981). When the cells are disrupted, the enzyme catalyzes the hydrolysis of the cysteine sulfoxides to form pyruvate, ammonia, and low-molecular-weight volatile organosulfur compounds that produce the characteristic flavor of onions and garlic (Lancaster and Collin, 1981). Alliin also acts as an antioxidant and possibly contributes to the pool of antioxidant compounds, including glutathione and other thiols, as well as inhibiting enzymes that promote prooxidant status through thiol exchange (Rabinkov et al., 1998; Chung, 2006; Haneklaus et al., 2007).

7.3.5 RESPONSE OF PLANTS TO SULFUR DEFICIENCY

Due to the essential nature of S for plant metabolism, plants have developed a complex system for regulation of S supply (Hawkesford and DeKok, 2006). The sulfate assimilation pathway is regulated strongly by demand, being activated under conditions of high demand and depressed when supplies of reduced S are high. Details of the regulation of sulfate transporters are provided in Section 7.4.

Uptake of sulfate by the plant is reduced when the supply of sulfate is high but increases when plants are S deficient. When plants are supplied adequately with sulfate, the excess is stored in the vacuoles of cells. Storage in the vacuoles allows the cytoplasmic concentrations of sulfate to be kept fairly constant in spite of fluctuating sulfate supply. However, in some cases, it appears that sulfate stored in mesophyll vacuoles always is not accessed rapidly during periods of S deficiency (Bell et al., 1994, 1995).

For many nutrients, including N, remobilization from old or senescing leaves to young leaves (or seeds) is a major source of nutrients to support new growth. Senescence can lead to mobilization of more than 70% of leaf protein, especially Rubisco, increasing N supply to younger leaves. Rubisco also contains significant cysteine and methionine and so also is likely to be an important S reserve (Gilbert et al., 1997). If there are adequate reserves of S in the plant, S present in the flag leaf can be mobilized and moved to the developing grain of wheat (*Triticum aestivum* L.) if sulfate supply at anthesis is restricted, thus maintaining the S content of the grain. However, if the plants are S deficient from an earlier stage, S accumulation in the flag leaf is low, thereby limiting the amount of S available for mobilization and restricting the concentration of S in the grain (Hawkesford, 2000).

In hydroponic studies with rapeseed, transient sulfate deficiencies (initiated after 51 days of growth and kept up for 35 days) after the rosette stage had limited effects on biomass production (Abdallah et al., 2010), indicating that growth could be maintained for a period of time with relatively low amounts of available S if the plant content was initially high. When S supply was adequate, S taken up after the rosette stage was allocated mainly to the leaves (55%) and the roots (27%) (Abdallah et al., 2010). When S was applied at low concentration in hydroponic studies, the newly absorbed S that was available was kept mainly in the roots (65%) with less than 23% moving to the leaves. Of the S that moved to the leaves, most went to younger leaves.

Mobilization of sulfate from the old leaves of rapeseed is significant (Abdallah et al., 2010). Nearly 60% of the mobilized S seemed to come from sulfate with the rest coming from organic S. The sulfate mobilization from the vacuole was related to upregulation of *BnSultr4;1* and *BnSultr4;2* genes coding for vacuolar transporters, with the regulation depending on adequate N availability (Dubousset et al., 2009). During S deficiency, therefore, sulfate appears to be the most important source for rapeseed, and glucosinolates and glutathione were relatively unimportant (Blake-Kalff et al., 1998).

Most (60%) of the remobilized S was moved to the root under deficient conditions. In total, about five times as much S was allocated to the roots in S-deficient than in S-sufficient plants, so the roots were the major sink for newly absorbed and mobilized S in S-deficient plants (Abdallah et al., 2010). A similar pattern was seen in previous studies, where after 6 days of S starvation, S mobilization occurred from old leaves, primarily as sulfate, but there was no increase in S in the middle and youngest leaves (Blake-Kalff et al., 1998). The sink strength of the roots for sulfate was enhanced, an action that may decrease the transport of newly absorbed sulfate to young leaves via xylem, thereby increasing the impact on young leaves (Lee et al., 2013). Therefore, under short-term S deficiency, the plant maintains growth by recycling foliar S compounds, especially sulfate, from mature leaves and maintains its root growth by moving more of the S that is taken up or mobilized to the roots.

The limited mobilization from old to young leaves leads to a distinct differential in the form of S present in old as compared to young leaves in rapeseed as well as to a differential between old and young leaves in their response to S deficiency. If sulfate supply was sufficient, about half of the total S in the youngest leaves was in insoluble S, such as proteins, about 42% was in the form of sulfate, and the remainder present as glutathione and glucosinolates (Blake-Kalff et al., 1998). When sulfate supply was removed, the youngest leaves converted sulfate, glutathione, and glucosinolates into insoluble protein S, with sulfate being the biggest net contributor. Therefore, under S deficiency, most of the S in the young leaves was in the form of protein, with the S from the proteins not being remobilized unless N was also deficient (Blake-Kalff et al., 1998). In contrast, in the middle and oldest leaves, about 70%–90% of the S was present as sulfate, whereas glutathione and glucosinolates together comprised only about 1%. The accumulation of sulfate mainly in mature leaves in rapeseed, even after S supply was withdrawn, may indicate inefficient xylem to phloem transfer of sulfate (Blake-Kalff et al., 1998).

Sulfur deficiency also leads to a large decrease in newly assimilated amino acids compared with situations where S is in adequate supply, with the greatest decrease occurring in young leaves (Lee et al., 2013). However, S deficiency leads to an increase in total free amino acids but a decrease in cysteine and glutathione, especially in roots and young leaves, when compared with control plants. The accumulation of amino acids is due mainly to hydrolysis of previously synthesized proteins, with the mature leaves acting as the source organ and the roots being the main sinks. The pools of amino acids likely increase, because of excessively reduced N availability associated with the decreased *de novo* synthesis of proteins. Excessive amino acids may be long-distance signals and act as a negative feedback for nitrate uptake. Asparagine is well documented to accumulate upon S starvation (Shewry et al., 1983), and an unfortunate consequence is the tendency to form toxic acrylamide in a reaction with sugars upon cooking (Muttucumaru et al., 2008; Curtis et al., 2009).

Although plants have the ability to mobilize sulfate reserves under periods of S deficiency, the degree of mobilization may be insufficient to support plant growth during extended S starvation. With the onset of sulfate deficiency, the sulfate reserves in the vacuole are not an adequate source of sulfate to support the growing plant. The mobilization of sulfate from the vacuole appears to be relatively slow. The sulfate that is stored in the vacuoles appears to be released only after prolonged S stress, with the release being too slow to support plant growth adequately.

Ultimately, sustained S deficiency leads to a decline in Rubisco enzyme activity and reduced chlorophyll content, particularly in the young leaves, indicated by yellow coloration and a reduction in photosynthetic activity. Protein synthesis declines (Lee et al., 2013), whereas soluble nitrate pools increase due to the N/S imbalance and reduced nitrate assimilation (Zhao et al., 1996). Plant productivity can be restricted severely.

7.3.6 Symptoms of Sulfur Deficiency

Symptoms of S deficiency tend to be less specific than symptoms of some of the other nutrients and should be confirmed by tissue analysis. *Brassica* crops such as rapeseed and mustard tend to show the most distinct symptoms of common crop species, with deficiency symptoms occurring earlier and at a higher tissue S concentration (below ~3.5 mg g^{-1}) than in most other crops. Crops such as potatoes (*Solanum tuberosum* L.) and sugar beets (*Beta vulgaris* L.) show deficiencies at lower tissue levels (below ~1.7 to 2.1 mg S g^{-1}) than the *Brassica* crops but at higher levels than field beans (*Phaseolus vulgaris* L.), soybean (*Glycine max* Merr.), field pea (*Pisum sativum* L.), or cereal crops such as wheat, barley, or corn, where the critical tissue level ranges from 1.0 to 1.2 mg S g^{-1} (Mills and Jones, 1996; Haneklaus et al., 2007).

In general, young leaves are the first to show S deficiency symptoms, because of the preferential allocation of sulfate and remobilized S to roots. Therefore, fewer new leaves are produced and the growth of the leaves is slower. Young leaves are chlorotic and photosynthetic capacity is reduced.

Oilseed rape and similar *Brassica* species develop very distinctive S deficiency symptoms, with the expression being similar whether the crop is the low glucosinolate canola type or high glucosinolate forms. Sulfur deficiency symptoms can begin at early growth stages in canola, with interveinal chlorosis (Figure 7.2) appearing initially in the younger leaves (Haneklaus et al., 2005b). The areas near the veins remain green, leading to a mottled appearance of the leaves. This development is in contrast to symptoms in crops such as peas, beans, and sugar beets, in which the chlorosis is more evenly spread over the leaves. Unlike the case for N deficiency, chlorosis related to S deficiency does not normally proceed to necrosis. Sulfur deficiency symptoms can be accentuated by N applications. In *Brassica* crops, severe S deficiency can lead to anthocyanin accumulation in the chlorotic areas, producing a purple discoloration (Figure 7.3). The accumulation of anthocyanates as a detoxification product for accumulated carbohydrates related to the disruption of protein metabolism can occur with a range of environmental stress factors and so is not a definitive symptom of S deficiency. In *Brassica* species, a distinctive downward cupping of the leaves can occur due to the reduced rate of cellular growth in the edges of the leaves relative to that near the veins; the purple, cupped leaves are quite characteristic of S deficiency in canola.

Sulfur deficiency that persists until flowering leads to distinct lightening of blossoms so that S-deficient canola crops exhibit pale, whitish flowers in contrast to the normal strong yellow petal coloration (Figure 7.4) (Haneklaus et al., 2005b). The mechanism behind the lightening of the flower color appears to be similar to that behind the purpling of the leaves, in that excess carbohydrate caused by disruption of the protein metabolism leads to the formation of colorless leucoanthocyanins. The petals are smaller and more oval in shape than the normally roundish-shaped petals of plants well supplied with S. Higher photosynthetic activity leads to greater expression of the symptoms. Although the fertility of the flowers is not affected by S deficiency, they may be less attractive to honeybees. Although canola primarily is self-pollinated, insect pollination may have some impact on crop yield, particularly in unrestored hybrids or male-sterile lines that rely

FIGURE 7.2 Interveinal chlorosis due to S deficiency occurs initially in the younger leaves. (Photograph courtesy of Malcolm Hawkesford.)

FIGURE 7.3   Severe S deficiency can lead to purpling and cupping in rapeseed (*Brassica napus* L.) leaves. (Photograph courtesy of Malcolm Hawkesford.)

FIGURE 7.4   **(See color insert.)** Sulfur deficiency can cause distinct whitening of canola (*Brassica napus* L.) flowers. (Photographs courtesy of Malcolm Hawkesford.)

on cross-pollination (Steffan-dewenter, 2003; Shakeel and Inayatullah, 2013). Branching and pod number are restricted, but the greatest influence is on the number of seeds per pod, with a large increase in the number of small, seed-free pods occurring under severe deficiency. Branches and pods also may display a purple discoloration due to anthocyanin accumulation. Limited branching, pod production, and seed production lead to a severe reduction in seed yield.

Sulfur deficiency symptoms in cereal crops such as wheat, barley, and corn are less distinctive than in canola and may often be confused with N, Mg, Mn, or Zn deficiencies. Symptoms begin

FIGURE 7.5 **(See color insert.)** Sulfur deficiency in (a) wheat (*Triticum aestivum* L.), (b) barley (*Hordeum vulgare* L.), (c) corn (*Zea mays* L.), and (d) sugarcane (*Saccharum officinarum* L.). (Photographs courtesy of International Plant Nutrition Institute.)

with a general slight chlorosis, possibly accompanied by a green striping along the veins (Figure 7.5) (Camberato et al., 2012). Plants are slightly stunted, with shorter and narrower leaves than unaffected plants (Anonymous, 2011). Again, because S is relatively immobile in the plant, symptoms tend to occur first in young rather than old leaves, which may help to differentiate between S and N deficiency. Sulfur deficiency can result in low seed set and limited yield, related to reduction in the number of fertilized florets leading to fewer kernels per head.

Sulfur deficiency, especially in cereals, may be difficult to distinguish from N deficiency. The pattern of deficiency symptoms in a field may provide an indication of whether problems are due to S deficiency or some other stress factor. As sulfate is mobile in the soil solution, S deficiencies can occur in well-drained, coarse-textured fields or areas of the field that are subject to leaching (Franzen and Grant, 2008). Mineralization from soil organic matter (SOM) is a major source of available sulfate, so S deficiencies tend to occur in portions of the field that have low organic matter content. In arid regions, exposed gypsum on eroded knolls or sulfate-based salinity in lower slope positions may reduce the likelihood of S deficiencies. Therefore, S deficiencies generally occur in irregular shapes in the field, related to special variability in soil texture, organic matter content, and water movement. Stratification of sulfate supply also may occur by depth, with accumulations of sulfate occurring deeper in the soil from leaching of sulfate or from subsurface gypsum or salt deposits. Crops displaying deficiency symptoms due to low S levels in the upper soil horizon may recover when the roots contact available sulfate in the lower soil layers or when mineralization of SOM and crop residues releases sulfate for crop uptake. As well as spatial differences in S deficiency, temporal differences may also occur, with deficiencies being more frequent after wet periods that encourage leaching.

7.4 MOLECULAR GENETICS OF SULFUR ACQUISITION BY PLANTS

Sulfate uptake and transport in plants occurs through the action of sulfate transporters that are encoded for by multiple genes of the SulP family (Hawkesford, 2010). Uptake of sulfate from the soil and its distribution through the plant tissue require the transport of charged sulfate against a concentration gradient across lipid membranes. This action involves high-affinity and low-affinity transport systems, mediated by the operation of proton-coupled cotransporter proteins (Hawkesford et al., 1993; Smith et al., 2000; Smith, 2001; Hawkesford, 2010). Multiple transporters are required for the transport across the cell membranes of roots and other organs as well as across intracellular membranes of organelles during plant growth and development.

There are two broad classes of transporters that have been identified, high-affinity and low-affinity saturable systems. However, the plant sulfate transporter family can be further broken down into five main groups. Group 1 refers to high-affinity sulfate transporters that are primarily responsible for the uptake of sulfate from the soil solution into the cells of the root but are also expressed in other tissues (Buchner et al., 2010). Group 1 is responsible for uptake into cells throughout the plant including probably having a role in phloem loading (Hawkesford, 2010). Group 2 is low-affinity transporters ($K_m > 100$ μM) expressed solely in the plasma membranes of cells of vascular tissue. The high K_m value supports the idea of a role in cell-to-cell transfer in vascular tissues where the sulfate concentration would be high. Group 3 is poorly characterized and appears to have multiple functions, but most plant species contain several forms. One report indicates a role for an arabidopsis (*Arabidopsis thaliana* Heynh.) Group 3 protein in transporter heterodimer function for optimal activity (Kataoka et al., 2004a). Group 4 transporters are localized in the tonoplast. Group 4 transporters mediate the efflux of sulfate from the vacuole (Kataoka et al., 2004b), likely in a proton cotransport mechanism similar to that for plasma membrane influx. Group 5 appears to be localized in vesicular or mitochondrial membranes and recently has been implicated in Mo transport (Tomatsu et al., 2007; Baxter et al., 2008; Shinmachi et al., 2010).

All plants studied so far possess multiple related genes that can be divided into these groups, with some variation occurring among species (Table 7.1). For example, arabidopsis has 14 related gene members of the SulP family that are divided into the five groups, with Groups 1–3 representing the plasma membrane sulfate transporters. Multiple members occur because of historical gene duplication. While this originally led to duplication of function, often there was a subsequent development of specialized properties or location and conditions of expression (Hawkesford, 2010). The initial cDNA library screening resulted in the isolation of the *ShST*1, *ShST*2, and *HvST*1 genes that encode high-affinity sulfate transporters with K_m values for sulfate of 10, 11.2, and 6.8 μM, and *ShST*3 that encodes a lower-affinity sulfate transporter with a K_m of 100 μM

TABLE 7.1

Occurrence and Characteristics of Clades of Transporters within the SulP Family for Five Plant Species

Group	Numbers of Genes					Substrate	Typical K_m (µM)	Location	Inducibility
	Arabidopsis	Brassica	Rice	Wheat	Brachypodium				
1	3	3	3	2[a]	3	Sulfate	<10	PM	+++
2	2	2	2	1	1	Sulfate	>100	PM	+
3	5	6	6	5	5	Sulfate		PM	No
4	2	2	1	1	1	Sulfate		V	++
5	2	2	2	2	2	Molybdate		M/V	No

Source: Hawkesford, M.J., *Plant Cell Monogr.*, 19, 291, 2010.

Notes: PM, plasma membrane; V, vacuole; M, mitochondrion.

[a] Gene numbers were obtained by examination of sequence databases. For wheat, see Buchner et al. (2010).

(Smith et al., 1995, 1997; Smith, 2001). Subsequently, several members of Groups 1 and 2 from arabidopsis were characterized functionally and generally may be ascribed to the high- and low-affinity types, respectively (see Hawkesford, 2010).

Uptake, distribution, and assimilation of sulfate are strongly controlled by the plant to optimize vegetative growth, reproductive potential, and seed vigor (Hawkesford and DeKok, 2006). Control of S influx is apparently by allosteric regulation of enzyme activity and by regulation of gene expression at several points in the pathway (Hawkesford, 2000). Regulation of the sulfate transporter system plays a key role in this process (Hawkesford, 2010).

The generally accepted model is that the S nutritional status of the plant regulates transcription of the sulfate transporter, allowing the plant to respond to varying sulfate levels in the environment. Expression of the genes that encode the sulfate transporters is transcriptionally regulated by signals related to the S nutritional status of the plant (Figure 7.6). By controlling the gene transcription, the plant regulates the number of transporters that are acting in the membranes.

Positive regulation of transcription of the genes encoding high-affinity sulfate transporters has been described (Smith, 2001). If the sulfate supply is restricted, the mRNA transcripts in the root that correspond to the high-affinity sulfate transporters increase rapidly, increasing the capacity of the plant to take up sulfate. When sulfate is resupplied to the plant, mRNA levels for the high-affinity sulfate transporters decline rapidly, reducing the ability of the plant to accumulate sulfate. The very rapid rate that mRNA transcripts for high-affinity sulfate transporters decay and that sulfate uptake declines if sulfate supply is reintroduced to S-starved plants indicates that the transcripts and the transporter proteins must turn over very rapidly, with the proteins in the membrane continuously forming and degrading. The low-affinity sulfate transporters ShST3 and Sultr2;1 also are regulated transcriptionally by the sulfate status of the plant, with feedback loops resulting in increased levels of expression when roots are deprived of sulfate. The feedback mechanism maintains sulfate levels in the plant within physiologically suitable levels.

If plants are deprived of sulfate, tissue concentrations of sulfate, glutathione, and cysteine decline (Hawkesford and DeKok, 2006). The S supply of the plant leads to regulation of S reduction at several points in the pathway, with the two main points of regulation apparently being sulfate uptake and APS reduction (Davidian and Kopriva, 2010). The regulation is from systemic rather than localized signals, responding to the overall sulfate status of the plant. Metabolic intermediates are part of a signal transduction pathway that regulates the expression of genes that encode for the proteins for key steps in the pathway. Thus, the intermediates help to regulate the movement of S through the pathway, preventing excess uptake and reduction of S and

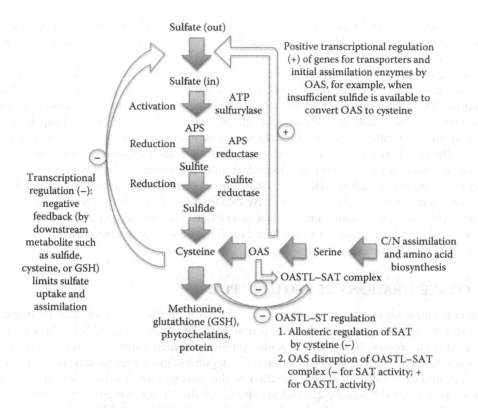

FIGURE 7.6 Metabolite control of sulfate uptake and assimilation. A series of feedback loops are proposed in which cellular concentrations of pathway intermediates may act as part of a signal transduction pathway to repress or activate expression of the genes encoding the proteins controlling some of the individual steps in the pathway. In addition, there is also allosteric feedback regulation of SAT by cysteine and of the OASTL–SAT complex by OAS. Solid arrows represent metabolite fluxes; open arrows are feedback control loops. (Modified and updated from Hawkesford, M.J., *J. Exp. Bot.*, 51(342), 131, 2000.)

so avoiding accumulation of sulfide in situations where OAS is limiting. The main regulatory point is APS reductase. Regulation occurs by allosteric inhibition and by activation or repression of gene expression for APS reductase by metabolites. Expression and activity of APS reductase respond rapidly to S deficiency or exposure to reduced S compounds, including sulfide, OAS, cysteine, or glutathione. Where S is in short supply, serine accumulates, and there is no sulfide present for biosynthesis of cysteine or for allosteric inhibition of OAS synthesis, leading to the accumulation of OAS. The OAS may regulate the transcription of transporter expression positively and also may override the negative feedback partly provided by the reduced S compound. These actions would maximize the production of cysteine when S was limited but also provide controls to prevent overactivity of the system.

If S supply is reintroduced, transcription is repressed, mRNA abundance declines, and sulfate influx decreases. Excessive S compounds suppress the expression of the transporter and APS reductase. Regulation of expression occurs for many of the steps in the pathway, especially with APS reductase, but the greatest control seems to be at the transporter level, where the mRNA pool sizes and activity are regulated by the availability of S. If sulfate is provided to S-deficient plants, the mRNA coding for the transporters decreases within an hour, and the activity and protein abundance decrease within 2–4 h, indicating a rapid turnover of the transcript and the protein. Because the response is so rapid, it indicates that the plant is responding to signals that are tied closely to the pools of S in the root (Hawkesford, 2010). The feedback may be due to repression by cysteine, glutathione, or possibly sulfide and serves to prevent excessive accumulation of sulfate, with a secondary effect on reduction of sulfate (Hawkesford, 2000). Sulfate uptake therefore is mediated by sulfate availability, the demand for reduced S, and the supply of C/N skeletons (Hawkesford, 2000).

Exposure of barley plants with roots adequately supplied with sulfite to OAS, a cysteine precursor, resulted in the derepression of transcription of the gene encoding the HvST1 high-affinity transporter. This effect led to rapid increases in the levels of HvST1 mRNA transcripts, sulfate uptake rate, and concentration of cysteine and glutathione in the plant. Enhanced supply of OAS could counteract partially the negative feedback system, indicating that the genes that encode for the high-affinity sulfate transporters are regulated transcriptionally by signals corresponding to the demand for sulfate in the synthesis of reduced S compounds and to the supply of specific precursors in the assimilatory pathway. However, studies with potato did not support the role of OAS in regulation (Hopkins et al., 2005). An alternative explanation may be that OAS supplies additional substrate, which otherwise would limit cysteine production. The enhanced cysteine synthesis would increase the demand for reduced S in the form of sulfide, depleting the repressive metabolite pool and inducing sulfate transporter expression.

7.5 CONCENTRATIONS OF SULFUR IN PLANTS

The S concentration of plants varies substantially with species, cultivar, growth stage, and environmental conditions, particularly S availability. The major pool of S in plant tissues is associated with protein, in the amino acids cysteine and methionine. A ratio of around 30:1 N to S is common, but this ratio can be larger in legumes. In storage tissues, the ratio is very dependent upon the nature and concentration of the storage proteins. In wheat, mature grain S can vary between 0.1% and 0.2% and nationwide in the United Kingdom between 1981 and 1993 was shown to vary depending on the nutrient status of the crop (Zhao et al., 1999a). The N/S ratio varied from 12:1 in 1981/1982 to 16:1 in 1992/1993, a change that was considered to reflect an increasing incidence of S deficiency in the U.K. wheat over this period (Zhao et al., 1999a). Secondary compounds, although being a significant S pool in all tissues, generally represent a much lower total amount, suggested to be around 0.1%–6.4%, depending on leaf maturity, and also are strongly influenced by N and S nutrition (Blake-Kalff et al., 1998). Generally, sulfate and other nonprotein pools are low in seed tissues, with the exception of some oilseed crops, particularly those with high glucosinolate content (Zhao et al., 1993); such varieties are not optimal for animal feed quality.

The dicotyledonous crops normally have higher S concentrations than graminaceous plants, but considerable variation occurs within the genus. Plants that produce high concentrations of S-containing secondary metabolites such as many plants in the Cruciferae and Liliaceae families accumulate high levels of S. For example, mustard (*Brassica juncea* L.) had a fivefold higher S concentration in roots, leaves, and seeds than did groundnut (*Arachis hypogaea* L.) (Lakkineneni and Abrol, 1992).

Concentrations of S in plants tend to be highest in the stems and leaves and slightly lower in the seed (Lakkineneni and Abrol, 1992; Tabe and Droux, 2001). Generally, the photosynthetically active leaves have the highest concentrations. Total S concentration in leaves will generally increase with increasing sulfate supply (Kowalska, 2005).

In S-deficient crops, S in leaf tissue can be remobilized to sink tissues, as described in Section 7.3 (Blake-Kalff et al., 1998; Howarth et al., 2008; Dubousset et al., 2009). During grain filling, a substantial fraction of the vegetative tissue protein S can be remobilized to the seed, although not to the extent to which N is remobilized, at least in an S-fertilized crop (Howarth et al., 2008). Some of this S is from the breakdown of photosynthetic proteins such as Rubisco (Gilbert et al., 1997). Sulfate-S varies considerably depending on the condition of supply and growth, as described elsewhere in this chapter.

7.6 RATIOS AND INTERACTIONS OF SULFUR WITH OTHER ELEMENTS

Interactions between S and other nutrients can influence crop growth and quality. Generalized interactions may occur because of concentration–dilution effects, growth restrictions due to the first limiting nutrient, or effects of soil pH or ionic strength of the soil solution. However, S also has specific interactions with N, P, Se, and several of the essential micronutrients.

7.6.1 NITROGEN

Nitrogen and S are both important constituents of protein, and adequate supplies of both nutrients are important for optimum crop yield. Lack of adequate S will restrict crop response to N fertilizer applications (Figure 7.7). Excess N in relation to the S supply can lead to a nutrient imbalance that can restrict protein synthesis and reduce crop growth. Therefore, the demand for S will increase with increasing crop yield and N application (Janzen and Bettany, 1984). On soils that are deficient in S, N application can accentuate S deficiency and decrease crop yield (Nyborg et al., 1974; Janzen and Bettany, 1984; McGrath and Zhao, 1996b; Brennan and Bolland, 2008). In studies conducted in northern Saskatchewan on a highly S-deficient soil, increasing N application from 0 to 150 kg ha^{-1} without S application increased the S deficiency of rapeseed, leading to a large decrease in seed yield (Malhi and Gill, 2002, 2007). When N and S were applied at optimum rates, seed yield increased to 1880 kg ha^{-1}, indicating the importance of balanced nutrition. Similar results were observed with winter oilseed rapeseed in England, where N applied alone decreased seed yield, whereas N applied with S led to large increases in seed yield (McGrath and Zhao, 1996b). Plants grown without S but with low rather than high N showed less loss of chlorophyll than when N supply was high (Blake-Kalff et al., 1998). The negative effect of N on plant growth when S was restricted may have been because high-N plants grew faster and had a higher S demand and hence a more rapid decrease in the concentration of available S than plants grown with low N.

FIGURE 7.7 Sulfur deficiency can restrict nitrogen responses. Nitrogen rate increases from right to left, with S at the bottom and without S at the top. (Photograph courtesy of Cynthia Grant.)

The interaction between N and S also relates to their joint function in protein production. Regulatory interactions between S assimilation and N metabolism function to coordinate the flow of the two elements because plants tend to utilize a relatively fixed ratio of N/S for protein assimilation (Hesse et al., 2004; Carfagna et al., 2011). Restrictions in S supply can affect N metabolism and protein synthesis (Prosser et al., 2001; Carfagna et al., 2011). Sulfur deficiency can lead to a decrease in protein synthesis and an increase in glutamine, asparagine, serine, and soluble N compounds, such as nitrate and amides (Prosser et al., 2001; Matula, 2004; Carfagna et al., 2011). Nitrate enrichment increases with N supply.

Similarly, if rapeseed plants have high S and low N content, there may be an accumulation of glucosinolates (Blake-Kalff et al., 1998; Kim et al., 2002). If N and S are deficient, adding N fertilizer to rapeseed may decrease glucosinolate concentration by increasing seed number and sink demand when S supply is limited (Haneklaus et al., 2007). Alternately, applications of N to N-deficient plants that have an adequate S supply can increase glucosinolate concentration, because there will be increased production of S-containing amino acids that are glucosinolate precursors.

In *Allium* species, N/S interactions can affect the yield and quality. In pot experiments using soilless growth medium, yield and pungency of several types of onion were reduced by inadequate N or S supply or excess N supply (Liu et al., 2009). Pot studies with onion and garlic also showed an interaction between N and S, with S content of leaves decreasing with increasing N fertilization and total N content of bulbs decreasing with increasing S fertilization (Bloem et al., 2005). Alliin concentration of bulbs increased with S fertilization, with the alliin content of bulbs being correlated with leaf S concentration and showing a tendency to decrease with N supply. In hydroponic studies, N and S deficiencies lowered amounts of different flavor components of onions, indicating that varying N and S supply would interact to affect biosynthetic pathways that determine the flavor quality in the bulb (Coolong and Randle, 2003).

The N/S balance is also important in ensuring protein quality in wheat. Sulfur is an important component of protein in wheat, being a constituent of the SH-containing amino acids that contribute the enhanced quality of protein for baking (Zhao et al., 1999a–c). Interchain disulfide bonds formed by the S-rich subunits of the gliadins and glutenins, which comprise gluten, influence the elasticity of dough by stabilizing the polymer network formed by the gluten molecules (Shewry and Tatham, 1997; Shewry et al., 2002). If the S supply is low relative to the N supply, amounts of S-poor proteins such as ω-gliadins increase and S-rich proteins such as γ-gliadins and low-molecular-weight subunits of glutenin decrease (Tea et al., 2004, 2007; Wieser et al., 2004). The reduced proportion of S-rich compounds leads to tough, less extensible dough and lower bread volume even with a similar protein content (Thomason et al., 2007; Reinbold et al., 2008; Zörb et al., 2009). Therefore, an adequate N/S balance must be present for protein quantity and quality (Zhao et al., 1999a–c; Flæte et al., 2005; Thomason et al., 2007). Adding high amounts of N fertilizer to S-deficient fields may aggravate S deficiencies by widening the N/S ratio. A lack of improved loaf quality after urea application could sometimes be related to insufficient grain S concentrations, high N/S ratios, and associated changes in the proportions of protein fractions (Gooding and Davies, 1992). In studies with wheat in England, application of N in the absence of S produced lower N concentration than applications of N with S, consistent with a need for a balance between N and S to support protein synthesis (Godfrey et al., 2010). Dough from grain grown with high N, but without applied S, showed lower dough strength than samples grown with S (Wooding et al., 2000; Godfrey et al., 2010). An adequate supply of S must be available through grain fill to optimize crop quality because remobilization of S from vegetative tissue to the grain is limited. Under highly S-restricted greenhouse conditions, late S fertilization resulted in a higher loaf volume than early S fertilization (Figure 7.8) (Flæte et al., 2005; Zörb et al., 2009). Sulfur may interact also with foliar N applications, with a greater increase in protein occurring with foliar N fertilizer when applied to wheat that had previously received S applications (Thomason et al., 2007).

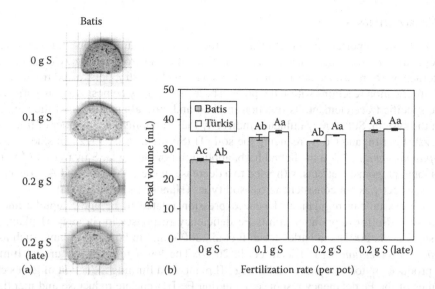

FIGURE 7.8 Effect of different S fertilization rates on the baking quality of two wheat cultivars as measured by a microscale baking test using 10 g of whole meal flour. (a) Images of micro bread slices of the cultivar Batis at a comparable scale. (b) Histograms of bread volumes; different letters represent significant differences of the mean values. Error bars represent standard errors of five independent pot replicates. Statistical significance ($p < 0.05$) is indicated by small letters for the S rates and capitals for the cultivars. (From Zörb, C. et al., *J. Agric. Food Chem.*, 57(9), 3877, 2009.)

The observed interaction between N and S has given rise to the idea that a specific plant-available N/S ratio in the soil is needed to avoid yield restrictions, with values from 5:1 to 8:1 being suggested for rapeseed (Janzen and Bettany, 1984; Bailey, 1986; Saskatchewan Agriculture, 2005; Manitoba Agriculture, 2009). However, if S is deficient, increasing the N/S ratio with N application can limit crop yield, but if S supply is increased above deficiency levels, adjusting the N/S ratio will not have an impact (Blake-Kalff et al., 2000, 2001; Malhi and Gill, 2002; Karamanos et al., 2007). Therefore, optimum crop yields can occur across a range of N/S ratios if the individual requirements for the nutrients are satisfied (Karamanos et al., 2004, 2005, 2007).

7.6.2 Phosphorus

Sulfur fertilization may influence P solubility because of the acidifying effect of S on soil, and may increase P availability on calcareous soils (Soliman et al., 1992). However, in studies with mung bean (*Phaseolus aureus* L.) under growth-chamber conditions, S and P showed antagonistic effects with S reducing P concentration and P reducing S concentrations in tissue and grain. There was also a negative interaction between S and P on grain yield (Aulakh and Pasricha, 1977). The authors suggested that sulfate and phosphate might compete for uptake and translocation pathways, affecting influx and movement of the nutrients by the plant; however, this is an unlikely explanation given the subsequent knowledge on the independent molecular nature of the transporters. They also suggested that P may reduce the S availability in the soil and that high applications of P fertilizer could aggravate S deficiencies if S level in the soil is low. In contrast, in experiments on subterranean clover (*Trifolium subterraneum* L.) in pot studies using soil that was highly S and P deficient, P or S applied alone did not increase forage yields, but yields increased substantially if the two nutrients were applied together (Jones et al., 1970). Similar synergistic interactions occurred in a range of crops including wheat (Marok and Dev, 1980), chickpea (*Cicer arietinum* L.) (Islam et al., 2001), cluster bean (*Cyamopsis tetragonoloba* Taub.) (Yadev, 2011), toria (*Brassica napus* L.) (Bharose et al., 2011), and raya (*Brassica juncea* L.) (Jaggi and Sharma, 1999), indicating the importance of balanced fertilization.

7.6.3 MICRONUTRIENTS

Micronutrients are important essential elements for plants and animals but may also have toxic effects, depending on the type, concentration, and time of exposure (Mendoza-Cózatl et al., 2005). Sulfur interacts with trace elements through effects on soil chemistry, uptake and transport dynamics, and detoxification reactions within the plant. These effects may be positive or negative, depending on the specific trace element, its function in plant and animal nutrition, and the concentration present in the system. Sulfur may influence indirectly the availability of a number of trace elements including Zn, Fe, Mn, and Cd by reducing the soil pH (Soliman et al., 1992). In studies on nasturtium (*Tropaeolum majus* L.), a medicinal herb, application of elemental S increased Mn and Cu in the leaves, but antagonistic effects with S led to a decrease in the B, Mo, Se, and As concentrations, although the effect was not consistent across soil types (Haneklaus et al., 2005a).

There is cross talk in regulation of gene expression related to Fe and S uptake and assimilation pathways. Sulfur deprivation limits Fe deficiency responses in Strategy II plants such as maize (Astolfi et al., 2003) and barley (Astolfi et al., 2006) and in Strategy I plants such as tomato (*Lycopersicon esculentum* Mill.) (Zuchi et al., 2009). The lower response occurs by limiting the ability to produce phytosiderophores in Strategy II plants and limiting induction of genes encoding key enzymes of the Fe deficiency response, including Fe(III)-chelate reductase and nicotianamine synthase, in Strategy I plants. On the other hand, Fe deficiency induces genes involved in sulfate uptake and assimilation as a response to an increased need for S metabolites such as methionine and its derivatives (Ciaffi et al., 2013; Paolacci et al., 2014).

Molybdenum is essential for all plants, being a component of several key enzymes including nitrate reductase and sulfite oxidase and for N fixation in legumes. Excessive Mo can interfere with Cu uptake by animals, producing a Cu deficiency referred to as molybdenosis (Gupta and Lipsett, 1981; Shinmachi et al., 2010). Sulfate and molybdate anions have similar physicochemical characteristics and so may compete for the same transport-binding sites. Sulfate is normally present at much higher concentrations in the soil solution than Mo, but when sulfate levels are low under S-deficient conditions, the competition is reduced allowing greater Mo uptake and translocation. Wheat grown on plots not fertilized with S had a yield penalty of between 1% and 10%, and grain S concentration 4%–20% lower than S-fertilized plants, but Mo content increased up to 7.5-fold (Stroud et al., 2010). In tomato plants grown in water culture, Mo concentration in plant sap was over 10 times greater in the absence of sulfate than when sulfate was available (Alhendawi et al., 2005). Restoration of sulfate supply led to a rapid decrease in Mo in the sap. The repressive effect of S on Mo uptake has been reported in a range of crops, including tomato (Stout et al., 1951), pea (Stout et al., 1951; Reisenauer, 1963), soybean (Singh and Kumar, 1979; Kumar and Singh, 1980), timothy (*Phleum pratense* L.), alfalfa (*Medicago sativa* L.), and red clover (*Trifolium pratense* L.) (Gupta and Macleod, 1975).

Trace element exposure can increase sulfate uptake and upregulate a number of enzymes that are involved in S assimilation (Na and Salt, 2011). Glutathione appears to be important in trace element tolerance and increases in concentration after exposure to Ni, Zn, or Cd (Na and Salt, 2011). Similarly, phytochelatin synthase is upregulated by a range of trace elements including Cd, Zn, Fe, Cu, Hg, Al, and Pb (Wagner and Krotz, 1989; Lee and Korban, 2002; Schat et al., 2002; Stolt et al., 2003; Na and Salt, 2011). Cadmium in particular serves to upregulate a number of enzymes involved in glutathione and phytochelatin metabolism, leading to enhanced Cd complexation and sequestration (Mendoza-Cózatl et al., 2005; Tamás et al., 2008). In nonaccumulators, trace element detoxification in the cytosol is mainly via complexation to S-derived compounds such as phytochelatin, while in hyperaccumulators, tolerance is partly due to the action of glutathione to reduce metal-induced oxidative stress (Na and Salt, 2011). Sequestration of phytochelatin compounds into the vacuole may play a role in detoxification (Cobbett, 2000).

Selenium is important to human nutrition, with deficiencies and excesses having negative impacts on human health (Gissel-Nielsen et al., 1984; Combs, 2001; Gupta and Gupta, 2002; Hawkesford and Zhao, 2007; Shinmachi et al., 2010). Selenium is similar to S in its ionic size and chemistry and so can substitute for S in many metabolic processes (Barillas et al., 2011). Also based on their similar chemical properties, Se and S are taken up and assimilated into the plant through the same initial transporters. In the same study where Mo accumulation was observed in S-unfertilized plots, substantial Se accumulation in the grain was observed (Stroud et al., 2010). An analysis of transporter expression indicated that root-expressed transporters had increased expression and, combined with a presumed favorable ratio of Se and Mo/S in the soil, contributed to the enhanced uptake and final partitioning of both Se and Mo (Shinmachi et al., 2010). Plants can be divided into nonaccumulator, selenium indicator, and selenium accumulator crops based on their ability to accumulate Se (White et al., 2004, 2007). Selenium is not viewed as a nutrient for plant growth and may lead to toxicity in many nonaccumulator plants (White et al., 2004). For example, selenocysteine and selenomethionine can be incorporated into proteins in place of cysteine and methionine, leading to loss of protein structure and function. Selenium accumulators appear to tolerate Se by methylating selenocysteine to prevent its incorporation into protein (Galeas et al., 2007).

In arabidopsis grown on agar, increasing S concentration increased shoot fresh weight and S concentration in the shoot but decreased Se concentration in the shoot (White et al., 2004). Increasing Se in the agar increased Se and S in the shoot but limited shoot weight. Selenium and S may enter the plant by multiple transport pathways that respond differentially to nutritional status. Sulfate appears to inhibit Se uptake, but Se may enhance S uptake by S transporters. Barley or rice grown in solution culture demonstrated decreased shoot and root growth if Se was added at low sulfate concentration but not when sulfate was present at high concentrations (Mikkelsen and Wan, 1990). Sulfate concentration in the shoot also increased with increasing Se concentrations at low S but not at high S. Similarly, under hydroponic conditions, Se enhanced S uptake in rapid cycling *Brassica oleracea* but reduced the production of some glucosinolates; the lower production of glucosinolates in spite of an adequate S level indicated that Se may interfere with glucosinolate metabolism (Toler et al., 2007).

There are differences between Se accumulator and nonaccumulator plants in their Se and S dynamics (Barillas et al., 2011). It may be that in selenium accumulator plants, the high-affinity sulfate transporters are selective for selenate, whereas in other species, they are more selective for sulfate (White et al., 2004, 2007). Accumulators preferentially take up selenate over sulfate and translocate more Se to the shoot, leading to a higher Se/S ratio than nonaccumulators. Selenium and S tissue concentrations tend to be negatively correlated in hyperaccumulators and positively correlated in non-hyperaccumulators (Galeas et al., 2007).

7.7 DIAGNOSIS OF PLANT SULFUR STATUS

Plant tissue testing can be used to assess the S sufficiency of crops (Maynard et al., 1983b; Jones, 1986; Blake-Kalff et al., 2000, 2001). The sufficiency range for tissue S depends on the type of crop, the portion of the plant assessed, and the age of the plant. For example, the sufficiency range for rapeseed is considered to be between 0.17% and 1.04% in mature leaves without petioles between the rosette and pod development stages (Mills and Jones, 1996). The Canola Council of Canada suggests that S should be above 0.24% in the whole aboveground plant at flowering (http://www.canolacouncil.org/crop-production/canola-grower's-manual-contents/chapter-9-soil-fertility/chapter-9#plantandtissuetesting, accessed October 2, 2013), whereas European work indicates that deficiency symptoms may occur if the S value in the younger leaves at the start of stem elongation falls below 0.35% (Haneklaus et al., 2007). In spring wheat, the suggested sufficiency range is 0.15%–0.40% in 25 whole tops as the head emerges from the boot, while that for maize is 0.15%–0.50% in 12 leaves below the whorl prior to tasselling (Mills and Jones, 1996). Critical concentrations of S in various selected crops are given in Table 7.2.

As with any tissue testing, care must be taken to select samples from representative areas of the field. Plants that have been under climatic or nutritional stress, those from saline portions of the

TABLE 7.2
Critical Sulfur Concentrations in Plant Tissue of Various Crop Species

Crop	Part Sampled[a]	Time of Sampling	Deficient	Low	Sufficient	High
			Deficient	Low	Sufficient	High
Alfalfa	Top 15 cm	Early bud	<0.20	0.20–0.25	0.26–0.50	>0.50
Barley	Whole top	Heading			0.15–0.40	
Canola/rape	YMB	Before flowering			0.35–0.47	
Cassava	YML	Vegetative			0.30–0.40	
Cotton	YMB	Early flowering			0.20–0.25	
Cowpea	YML	Early bloom			0.17–0.22	
Maize	Ear leaf	Initial silk	<0.10	0.10–0.20	0.21–0.50	>0.50
Oats	Top leaves	Boot stage	<0.15	0.15–0.20	0.21–0.40	>0.40
Onion	Whole top	Half maturity			0.50–1.0	
Peanut	YML	Preflowering			0.20–0.35	
Potato	YML	Plants 30 cm tall			0.19–0.36	
Rice	Whole top	Max. tillering		0.10–0.20	0.20–0.30	>0.30
Ryegrass	Young herbage	Active growth			0.10–0.25	
Soybean	First trifoliate	Early flower	<0.15	0.15–0.20	0.21–0.40	>0.40
Sugarcane	Third leaf from tip	12–15 weeks after planting			0.14–0.20	
Sunflower	YML	Midseason			0.30–0.55	
Tea	Third leaf from tip	Active growth			0.10–0.30	
White clover	Young herbage	Active growth			0.18–0.30	
Wheat	YEB/YMB	Mid–late tillering			0.15–0.40	

The header "Critical Concentration at Various Accumulation Levels (% Dry Wt.)" spans the Deficient, Low, Sufficient, and High columns.

Sources: Adapted from Mills, H.A. and Jones, J.B., *Plant Analysis Handbook II*, MicroMacro Publishing, Inc., Jefferson City, MO, 1996; Dick, W.A. et al., Availability of sulfur to crops from soil and other sources, in *Sulfur: A Missing Link between Soils, Crops, and Nutrition*, ed. J. Jez, American Society of Agronomy, Crop Science Society of America, Soil Science Society of America, Madison, WI, 2008, pp. 59–82; Kovar, J.L. and Grant, C.A., Nutrient cycling in soils: Sulfur, in *Soil Management: Building a Stable Base for Agriculture*, eds. J.L. Hatfield and T.J. Sauer, American Society of Agronomy, Crop Science Society of America, Soil Science Society of America, Madison, WI, 2011, pp. 103–116.

[a] YEB, youngest emerged leaf blade; YMB, youngest mature leaf blade; and YML, youngest mature leaf.

field, those that are damaged by insects or disease, those that have been mechanically damaged, or those covered with soil, dust, or sprays should be avoided. Avoid dead plant tissue and plants from borders or headlands. Samples should be collected from multiple plants, with it being preferable to select samples from more plants than more tissue from fewer plants (Mills and Jones, 1996).

Plant tissue analysis may be conducted to diagnose if symptoms in affected plants are due to a S deficiency. In this case, a comparison would be made between the nutrient concentration in affected and unaffected plants. The comparison may be challenging because stress that reduces crop growth can also influence plant nutrient composition through concentration/dilution effects. Therefore, when interpreting analytical results, it is important to consider the growth stage and relative dry matter yield of the plants being sampled, other stress factors that may be affecting crop growth, and concentration of other nutrients in the tissue, in addition to the plant tissue concentration of S to get a broader understanding of the nutrient status of the affected plants. Combining plant tissue tests with soil samples from the areas of concern can aid in interpretation.

It has been suggested that N/S ratio of the tissue may provide a better measure of S status than total S, with ratios of greater than 10–12 possibly being associated with S deficiencies in rapeseed (Spencer et al., 1984; Bailey, 1986; McGrath and Zhao, 1996b). However, McGrath and Zhao (1996) reported that the concentration of total S at early flowering was a superior indicator of plant S status than the N/S ratio. Alternatively, the ratio of malate to sulfate has been proposed as a diagnostic indicator of S deficiency (Blake-Kalff et al., 2000). This assessment has practical advantages in being a single analysis and, as a ratio, requires little calibration. For rapeseed, a malate/sulfate ratio of one or below is indicative of S sufficiency, whereas greater than one indicates deficiency. Assessment of plant S status at flowering either using tissue S concentration or the N/S ratio would be too late for application of S fertilizers to completely restore crop yield (McGrath and Zhao, 1996a; Malhi and Gill, 2002).

7.8 FORMS AND CONCENTRATIONS OF SULFUR IN SOILS AND AVAILABILITY TO PLANTS

Sulfate is the form of S most important from a plant nutrition viewpoint as it is the ion that is directly taken up by the plant root. However, S is present in the soil in a wide range of organic and inorganic forms, assuming several valencies and a large range in oxidation state, from 2^- to 6^+ (Solomon et al., 2011). The transformations among these various forms have an important effect on the global S cycle and on the availability of S for plant nutrition.

7.8.1 ORGANIC SULFUR POOLS

In nonsaline soils, most S in the soil is in the organic pool, with frequently more than 90% of the S being present in the organic form in the upper soil horizons (Schoenau and Germida, 1992; Wang et al., 2006; Eriksen, 2009). Although plants absorb S primarily in the sulfate form, mineralization of the organic pool can release sulfate that can then be utilized for crop growth (Freney et al., 1975; McGill and Cole, 1981; Janzen and Kucey, 1988; Zhao et al., 2006). In a study of soils from across North America, the concentration of soil organic S ranged from 194 to 853 mg kg^{-1} soil and from 135 to 441 mg kg^{-1} soil in the native grassland and cultivated soils assessed, representing on average 96% of the total soil S (Wang et al., 2006).

SOM is a mix of organic molecules including the soil microbial biomass, compounds such as chelating compounds, enzymes or proteins released from living plants and animals, and plant and animal residues in various states of decomposition. The microbial biomass comprises approximately 1%–5% of the organic S in the soil (Banerjee et al., 1993) and is an active pool for S turnover in the soil. The SOM is a source or a sink for S compounds involved in the S cycle and the ability of SOM to contribute to plant-available S depends on the forms of organic S present.

Traditionally, organic S in the soil has been divided into hydriodic acid (HI)-reducible and nonreducible S, according to its susceptibility to reduction by reducing agents (Freney et al., 1975; Zhao et al., 2006). These two pools are interpreted as being the sulfate ester-S (C–O–S) or sulfamates (C–N–S) and the C-bonded S, with the C-bonded S being further divided into Raney Ni-reducible and nonreducible fractions (Freney et al., 1975; McGill and Cole, 1981; Zhao et al., 2006; Churka Blum et al., 2013). Recent work using x-ray absorption near edge structure (XANES) spectroscopy has differentiated these S species in a number of soils (Zhao et al., 2006; Solomon et al., 2009, 2011) identifying distinct organosulfur compounds in soils, varying in oxidation and availability. These can be grouped into three major groups:

1. Strongly reduced (S^0 to S^{1+}) organic S including monosulfides, thiols, disulfides, polysulfides, and thiophenes
2. Organic S in intermediate oxidation (S^{2+} to S^{5+}) states including sulfoxides and sulfonates
3. Highly oxidized (S^{6+}) organic S comprising ester SO_4–S

The strongly reduced and intermediate oxidation states represent the C-bonded S, directly linked to C in C–S or C–S–O linkage such as in S-containing amino acids and sulfonates. The S in the highly oxidized form represents the ester sulfates with S linked mainly to O atoms through C–O–S, as well as compounds that have N–S linkages such as sulfamates and N–O–S linkages such as glucosinolates (Solomon et al., 2011). The C-bonded S is more likely to be incorporated into humic material than are the sulfate esters. Humic acids contain from 0.1% to 1.5% S, primarily derived from C-bonded forms, which are likely to play a role in determining the 3D structure (McGill and Cole, 1981). In contrast, the HI-reducible S is likely not an integral part of the humic component but rather is related to the periphery of the humic acid, possibly in an adsorbed form (McGill and Cole, 1981).

Sulfur mineralization in the soil can proceed by either biochemical or biological means. Biological mineralization refers to the utilization of C-bonded S compounds by soil microorganisms as an energy source, with S released as a by-product, while the biochemical process refers to the release of S through enzymatic hydrolysis of ester sulfates (McGill and Cole, 1981). Sulfur bonded directly to C (the reduced and intermediate oxidation states) will be mineralized as a result of biological oxidation of C to provide energy, while the more strongly oxidized S may be mobilized by extracellular enzymes, in a biochemical mineralization process that is controlled by the end product (McGill and Cole, 1981). The presence of two pathways of S mineralization helps to explain why N and S mineralization are not more tightly correlated (McGill and Cole, 1981).

Mineralization of organic S in the soil is affected by the form of S present, which is in turn affected by factors such as the SOM origin, chemical composition, age, and degree of decomposition (Churka Blum et al., 2013). Early research indicated that mineralization and release of both N and S were strongly affected by the N concentration of residues, with more rapid mineralization occurring from residues containing higher concentrations of N (Janzen and Kucey, 1988). The S concentration of the residue is also important in determining S release through residue decomposition. For example, canola residues are high in readily mineralizable S, whereas crop residues with a higher C/S ratio may lead to temporary S immobilization by soil microorganisms (Niknahad-Gharmakher et al., 2012). Plant residues contain S in a range of oxidation states, and the most highly oxidized forms are the most immediate source for mineralization of inorganic S (Churka Blum et al., 2013). Freshly incorporated crop residue appears to contain organic S primarily as reduced S (thiols, sulfides, thiophenic S), although it also contains some highly labile inorganic sulfate (Zhao et al., 2006; Solomon et al., 2011). In residues that contain highly oxidized forms of S, mineralization is controlled mainly by the enzymatic hydrolysis of ester-S, which is stimulated by the depletion of sulfate (McGill and Cole, 1981; Maynard et al., 1983a; Churka Blum et al., 2013). In contrast, with residues that are higher in C-bonded S, the process is controlled by biological mineralization (Churka Blum et al., 2013), with mineralization driven by the need for C as an energy source for the soil microorganisms (McGill and Cole, 1981). The reduced S forms oxidize relatively rapidly with aerobic incubation, so the C-bonded S moieties represented by the reduced and intermediate oxidation states appear to be the major short-term source of mineralizable S in aerobic soils (Zhao et al., 2006).

Initial stages of residue decomposition in the soil lead to a decrease in the reduced and intermediate C-bonded S and an increase in ester-S (Freney et al., 1975; McGill and Cole, 1981; Zhao et al., 2006; Churka Blum et al., 2013). Therefore, the composition of SOM, as a mixture of differing organic matter components, differs from those of the freshly applied plant residues. An evaluation of North American soils using HI-reduction methodology showed that carbon-bonded S (reduced and intermediate S) comprised about 2/3 or more of the organic S in the soil (49%–68% and 62%–71% in native grassland and cultivated areas, respectively) with only a small proportion (32%–51% and 29%–38% in native grassland and cultivated areas, respectively) of the soil organic sulfur being present in ester linkages, equivalent to the oxidized S (Wang et al., 2006). Similarly, Solomon et al. (2011) reported that the C-bonded S accounted for 80% of the total organic S detected by XANES in the undisturbed soils that they studied. Zhao et al. (2006), also using XANES, found ranges of 14.4%–32.4% for the reduced, 32.8%–49.7% for the intermediate, and 21.5%–52.5% for the oxidized S, with the S form being affected by soil type, land use, and manure application (Table 7.3)

TABLE 7.3

Percentage Distribution of Different Oxidation States of Organic Sulfur Determined Using XANES Spectroscopy

Soil	Reduced S (0.17 – 0.26) + (0.96 – 1.24)[a]	Intermediate (2.42 – 2.47) + (5.0)	Oxidized S −6
Broadbalk woodland	24.8	46	29.1
Broadbalk grassland	18.4	37.4	44.3
Broadbalk arable control	14.7	32.8	52.5
Broadbalk arable CFYM	28.7	49.7	21.5
Scottish pasture	24.3	35.7	40
Scottish arable	14.4	34.6	51
New Zealand pasture	19	34.4	46.6
New Zealand arable	19.7	34.3	45.9
Chinese paddy A	27.7	38.2	34.1
Chinese paddy B	32.4	40.2	27.4

Source: Adapted from Zhao, F.J. et al., *Soil Biol. Biochem.*, 38(5), 1000, 2006.

[a] Values in parentheses are electronic oxidation state in eV.

(Zhao et al., 2006). Paddy soils were more dominated by reduced and intermediate species as compared to the aerobic soils, reflecting the redox potential of the systems.

Conversion of soils from uncultivated grassland to cultivated crop production will have a major effect on the forms of organic S present. In established grasslands, Solomon et al. (2011) found mainly strongly reduced and intermediate S (directly bound to C), largely as sulfide (C–Sn–C) and thiol moieties (C–SH). The dominance of C–S relates to the input of highly reduced S in plant materials in the form of S-containing amino acids, sulfolipids, and sulfocarbohydrates in plant material. The small amount of highly oxidized S (C–OSO₃) in undisturbed grasslands may have come from direct input in the form of aryl, alkyl, phenol, and polysaccharide sulfates that occur in small amounts in plant tissue. However, it commonly is believed to be a transitory product of the biochemical processes of the microflora in the presence of inorganic sulfate.

If native grassland is converted to long-term cultivation, mineralization of SOM leads to a loss of organic S, but to a lesser extent than loss of soil organic C and N (Wang et al., 2006). This differential decreased the C/S and N/S ratios, indicating that soil organic S may be more resistant to mineralization than soil organic C and N. When grassland soils are cultivated, the dynamics of the organic S are shifted to more strongly oxidized forms (Solomon, 2011). Therefore, cultivated soils tend to have a higher relative proportion of oxidized S and a lower proportion of the reduced or intermediate S than native soils. The largest loss of organic S upon cultivation comes from the strongly reduced oxidation states (thiols, monosulfides, disulfides, polysulfides, and thiopens) and to a lesser extent from the intermediate oxidation states of sulfoxides and sulfonates. Meanwhile, cultivation greatly increases the proportion of organic S present in the strongly oxidized state. The oxidation of the reduced forms of S is likely due to increased soil aeration, exposure of physically protected organic S, increased microbial activity, and stimulation of the aerobic decomposition process as commonly occurs when grasslands are converted into cultivation. This result agrees with the concepts of Zhao et al. (2006), McGill and Cole (1981), and Solomon et al. (2005, 2009, 2011) that the S–C in highly reduced oxidation states is most reactive to long-term disturbance, and so the highly reduced and intermediate states are the major source of biologically mineralizable S in grassland systems. To restate, the ester sulfate forms are more labile than the C-bonded S. The sulfate added to the soil quickly incorporates into the ester fraction. The ester fraction can mineralize rapidly, and so a large portion may come from this pool when it is present. However, over time, the ester pool is incorporated into the more stable C-bonded S pool (McGill and Cole, 1981; McLaren et al., 1985; Scherer, 2009).

Climatic conditions influence the amount of organic matter and organic S present in a soil. Cool summer soil temperatures and freezing of soils over winter reduce SOM mineralization, leading to higher total SOM and concentration of organic S. Wang et al. (2006) reported that the average concentration of total organic S in either native grassland or cultivated sites declined in the following order: cryic > frigid > mesic > thermic > hyperthermic. In contrast, precipitation seemed to have little effect on native grasslands, but increasing precipitation increased organic S loss from cultivated soils.

7.8.2 Inorganic Sulfur

Although less than 10% of the S in the soil normally occurs in an inorganic form, it is the inorganic sulfate-S form that is directly available for plant uptake. Inorganic S can occur in reduced (elemental S [S^0], thiosulfate [$S_2O_3^{2-}$], and sulfite [SO_3^{2-}]) and oxidized (sulfate [SO_4^{2-}]) states (Dick et al., 2008). Reduced S compounds such as sulfides, elemental S, and thiosulfates may occur in soils where previous flooding produced sulfides, in soils that formed from parent materials that contain sulfide, or where reduced S forms such as elemental S have been added as a fertilizer. In anaerobic soils, reducing conditions can lead to the microbial conversion of sulfates into sulfides by bacteria when other electron acceptors such as oxygen and nitrate are depleted (Schoenau and Malhi, 2008). The reduced forms such as sulfide and elemental S may dominate in anaerobic soils but will rapidly be oxidized under aerobic conditions (Scherer, 2009). Although the reduced inorganic S forms such as pyrite may persist in soils over time even after soils are oxidized (Prietzel et al., 2009), they are normally transitory in soils and their concentration is usually negligible in aerobic soils (Kowalenko, 1993; Solomon et al., 2005). In paddy systems, release of oxygen from rice roots may lead to oxidation of reduced S compounds, even under flooded conditions (Wind and Conrad, 1995).

Reduced S compounds must be oxidized to sulfate before they are available for crop uptake. The S oxidation pathway is as follows:

$$S^{2-} \rightarrow S^0 \rightarrow [S_2O_3^{2-} \rightarrow S_4O_6^{2-}] \rightarrow SO_3^{2-} \rightarrow SO_4^{2-}$$

During the oxidation of reduced S compounds, acidity is produced along with the sulfate.

Oxidation can be chemical or microbiological, although normally microbiological oxidation occurring by heterotrophic microorganisms predominates, with autotrophic organisms playing a lesser role (Lawrence and Germida, 1991; Germida and Janzen, 1993). Therefore, in freely drained aerobic soils, inorganic S occurs primarily as sulfate, the S form that normally is taken up by plants. Sulfate can be free in the soil solution or adsorbed to the soil particles. It can also be present as Ca, Mg, or Na sulfates. Sulfate can be cocrystallized or coprecipitated with calcium carbonate and may be present in large amounts in calcareous soils.

Sulfate may bind to anion-exchange sites on the soil colloids just as phosphate does, and the adsorbed sulfate will be in equilibrium with the sulfate in the soil solution. Soil pH is an important factor affecting sulfate adsorption, with adsorption decreasing as pH increases from 3 to 7 and being extremely low above pH 7.0 (Bohn et al., 1986; Scherer, 2009). Therefore, above a pH of 6.5–7.0, the majority of soil sulfate is present in the soil solution, whereas in acid soils, sulfate may be adsorbed on the edges of aluminosilicate clay minerals on the surfaces of Al or Fe hydrous oxides. The adsorbed S can be an important source of highly available sulfate for crop uptake on highly weathered acidic soils and can also help to protect sulfate-S from leaching. Sulfate adsorption is affected by factors influencing the anion adsorption potential of the soil such as the proportion of Al and Fe oxides, clay content, and mineralogy (Dick et al., 2008). Adsorption sites on Fe and Al hydroxides can be blocked by the anionic groups of organic matter. The presence of competing ions such as nitrate, phosphate, and chloride can also influence adsorption. Phosphate is held more tightly to the exchange site than is sulfate, so application of phosphate fertilizer may increase sulfate availability by reducing adsorption (Bohn et al., 1986). Liming may also mobilize adsorbed sulfate by increasing phosphate solubility as pH increases up to a level of approximately 6.5 or by the competition supplied by the OH$^-$ ions (Korentajer et al., 1983; Scherer, 2009).

The sulfate in the soil solution is the form immediately available for crop uptake but is normally present in very small concentrations, often only a few micrograms per milliliter. The concentration is affected by the interactions among mineralization, immobilization, inputs from fertilizers, aerial deposition, irrigation, and crop uptake (Scherer, 2009).

Sulfate is mobile in the soil and so moves with water, influencing the distribution of S within a field. Soil sulfate concentration tends to be higher in lower than upper landscape positions. The differential is due partly to higher organic matter content in lower slope positions but also is affected by the presence of sulfate associated with water movement to the lower slope positions (Kovar and Grant, 2011). Leaching also may remove sulfate from the surface soil profile, particularly on coarse-textured soils, with sulfate salts and gypsum accumulating lower in the soil. Leaching losses of sulfate depend on the intensity of water movement through the soil and on the capacity of the soil to retain the sulfate. Risk of leaching is greater in coarse than in fine-textured soils, because there is greater movement of water through the coarse soil and because the coarse soils have a lower ability to adsorb sulfate (Scherer, 2009).

7.9 SOIL TESTING

Soil tests have been used for many years to determine the need for fertilizer S applications and the likelihood of attaining an economic crop response (Fox et al., 1964; Hamm et al., 1973). The most common base measurements of soil S concentrations are total, organic, inorganic, and extractable S (Kowalenko and Grimmett, 2008). Total S determination generally requires conversion of all S in the soil to a single form and then quantification of the resulting concentration. Conversion can be done by dry ashing or wet digestions, using either alkaline or acidic reactants. The resulting S can be measured either as a gas or in solution, with gases measured by infrared spectroscopy, chemiluminescence, coulometry, flame photometry, or other methods, and solutions usually are analyzed as sulfide or sulfate by spectrometry, ion-selective electrodes, titration, gravimetry, or chromatography (Kowalenko and Grimmett, 2008).

Most of the S in the surface horizons of nonsaline soils is generally in the form of organic S. Methods of directly determining organic S in the soil have not been universally accepted, and the difference between total S and hydriodic acid–reducible S may provide a better estimate of carbon-bonded S than direct measurement (Kowalenko and Grimmett, 2008). However, difference methods require accurate measurement of all fractions being used for the calculation.

Inorganic S generally is present as sulfate in aerobic soils and as elemental S or sulfide in anaerobic soils. Methods of analysis of reduced forms have not been evaluated to the same extent as of oxidized forms because reduced forms are less common in agricultural soils (Kowalenko and Grimmett, 2008).

Sulfate can be present in the soil solution, adsorbed or bound to soil surfaces, as gypsum, or associated with calcium carbonate. The solution and adsorbed sulfate are considered to be immediately available for plant uptake and therefore are the forms that are considered most relevant for agricultural soil analysis. Most soil tests measure the sulfate that can be extracted from the soil solution and adsorbed phase using water or a dilute solution of salts such as calcium chloride or lithium chloride (Fox et al., 1964; Hamm et al., 1973; Curtin and Syers, 1990; Blair et al., 1991; Kowalenko and Grimmett, 2008; Ketterings et al., 2011). Methods of quantifying the sulfate must be compatible with the extracting solution. The most frequently used methods use precipitation with barium or sulfide analysis after hydriodic acid reduction. Quantification in the barium precipitation technique can include turbidimetric, gravimetric, titrimetric, and colorimetric methods, all of which are subject to interference (Kowalenko and Grimmett, 2008). Hydriodic acid reduction is costly, time-consuming, and difficult to automate but is sensitive and relatively free of interference. The hydriodic acid reagent is quite specific for sulfate, but the analysis includes inorganic and organic sulfate and also may extract some elemental S. Ion chromatography can be used for inorganic sulfate analysis, whereas inductively coupled plasma spectrophotometry and x-ray fluorescence measure organic and inorganic forms of S. Detailed information on methodology for chemical analysis of soil S is provided in Kowalenko and Grimmett (2008).

Anion-exchange resins also can be used to assess the S-supplying power of soils (Schoenau et al., 1993; Grant et al., 2003b; Qian and Schoenau, 2007). Positively charged resins inserted directly in the field or in a moist soil sample adsorb sulfate ions present in the solution. Resins can then be desorbed and the sulfate present in the solution measured.

Regardless of the measurement method used, recommendations for application of S fertilizer require calibration of the measured S content in the soil with the likelihood of an economic crop response. Fox et al. (1964) suggested, based on a relatively limited data set, that yield increases in response to S applications would be possible if extractable sulfate-S was less than 8 mg kg^{-1} for corn and 10 mg kg^{-1} for alfalfa. Early studies in Western Canada showed that a seed yield response of rapeseed could be expected if soils contained less than 22 kg sulfate-S ha^{-1} in the top 60 cm (Ridley, 1972). However, there are many factors that interfere with the ability of the sulfate-S test to predict response to S application, whether using chemical extractants or ion-exchange resins. Subsoil S availability or inputs of available S in precipitation or irrigation water may reduce S response (Fox et al., 1964). Sulfate in the soil is highly variable, spatially and temporally, because of microbial processes and physical actions such as leaching that can affect the availability and distribution of sulfate in the field (Schoenau and Germida, 1992; Karamanos et al., 2007). Presence of salinity in portions of the field may be associated with thousands of kilogram of S ha^{-1} in the form of sulfate salts. The large amount of sulfate present in samples taken from a saline area can increase the concentration of sulfate in a soil sample composed of numerous cores taken from across the field, possibly masking a deficiency occurring in the bulk of the field. Similarly, exposed gypsum on eroded knolls also may elevate the average of a composite test. Therefore, when soil sampling, care must be taken to avoid saline or eroded portions of the field, with samples being taken on areas that are representative of the overall S status of the field. However, gypsum crystals randomly distributed in the soil profile may still generate false-positive results when soil samples are analyzed (Karamanos, 2001). In addition, sulfate salts or gypsum located at depth in the field may serve as a supply of S when the plant roots contact the layer late in the growing season. Plants may experience early-season S stress but recover when the roots access a sulfate supply deep in the soil.

Mineralization of organic matter also will serve as a source of S for the crop, and agricultural advisory services in some regions include organic matter when determining the need for S application or suggest that crops are more likely to respond to S applications on low organic matter soils (Franzen and Grant, 2008). Ion-exchange resins may be useful in measuring supply of sulfate from mineralization of organic matter or from dissolution of sparingly soluble inorganic forms of S (Qian et al., 1992; Schoenau et al., 1993; Qian and Schoenau, 2007). In growth-chamber studies, rapeseed S uptake was better correlated with the supply rate of sulfate measured using ion-exchange resins than with water-soluble sulfate (Schoenau et al., 1993), whereas under field conditions ion-exchange resins provided a good indication of the plant availability of elemental and sulfate forms of the S fertilizer (Grant et al., 2003b). Use of the N/S ratio in soils has also been suggested as an indication of the S-supplying ability of a soil, as high N levels may increase the likelihood of S deficiency at low S concentrations (Janzen and Bettany, 1984; Bailey, 1985).

As described in Section 7.7, plant tissue testing also can be used to assess the S sufficiency of crops (Maynard et al., 1983b; Jones, 1986; Mills and Jones, 1996; Blake-Kalff et al., 2001). Combination of soil and plant testing may provide a more complete understanding of the S status of a crop than either practice in isolation.

7.10 SULFUR FERTILIZERS

Effective fertilizer management requires selection of a combination of source, rate, timing, and placement to optimize fertilizer use efficiency for the crop and environmental conditions in the field. A range of S fertilizer sources are available, containing elemental S, sulfate, thiosulfate, or a blend of these sources. Plants absorb S primarily through the roots from the soil solution as sulfate, so elemental or thiosulfate sources must be converted to sulfate before they can be utilized by the plant (Janzen and Bettany, 1987a; Janzen, 1990). Therefore, optimum management practices in terms of rate, timing, and placement will vary substantially, depending on the source of S used.

Sulfate fertilizer sources include ammonium sulfate [$(NH_4)_2SO_4$], gypsum ($CaSO_4 \cdot 2H_2O$), kieserite ($MgSO_4 \cdot H_2O$), Epsom salts ($MgSO_4 \cdot 7H_2O$), langbeinite ($K_2SO_4 \cdot 2MgSO_4$), and potassium sulfate (K_2SO_4). Sulfur also is present in single superphosphate fertilizer [$CaSO_4 \ Ca(H_2PO_4)_2$]. In numerous trials assessing crop response to S applications, sulfate sources have been shown to be effective at correcting S deficiencies in the year of application (Table 7.4) (Nuttall et al., 1990, 1993; Dumanski et al., 1998; Malhi, 1998, 2005; Grant et al., 2003a,b, 2004; Karamanos and Poisson, 2004; Karamanos et al., 2007; Malhi and Gill, 2007). As sulfate does not require time for oxidation, sulfate fertilizer sources will be available for crop uptake upon dissolution and movement into the soil solution in the rooting zone (Malhi, 1998, 2005; Grant et al., 2004). Therefore, sulfate fertilizers can be applied near the time of crop demand and still provide available S to meet crop requirements. Fertilizer use efficiency may decrease with increasing time between application and crop uptake, depending on the environmental conditions. Sulfate present in the soil solution may be lost by leaching under conditions of high water input, especially in sandy soils (Malhi, 1998, 2005; Grant et al., 2004; Malhi et al., 2009). Losses also may occur through immobilization, although evidence from long-term Swedish studies indicates that immobilization of S may be relatively low (Kirchmann et al., 1996). Efficiency of autumn applications in studies conducted in northern Saskatchewan (Malhi, 2005) and Manitoba (Grant et al., 2004), where soils are frozen for several months between fertilizer application and the resumption of crop growth, was generally similar to or only slightly lower than with spring applications, indicating that autumn to spring losses of S were relatively low. In contrast, studies under higher moisture and warmer conditions in England suggest that there can be significant loss of sulfate where leaching conditions exist (Riley et al., 2002).

In-crop applications of sulfate-S sources also may be effective, depending on the timing of supply relative to crop demand. For example, the demand for S by rapeseed is greatest during flowering and seed set (Malhi et al., 2007). Therefore, although yield normally is optimized when adequate S is present throughout crop growth, rapeseed can recover to a great extent from early-season S deficiencies and applications of an available S source as late as rosette to early bolting can increase crop yield (Hocking et al., 1996; Malhi, 1998). Similar results occurred with pot studies with wheat, where S applications at ear emergence, when deficiency symptoms are observed commonly, increased tiller fertility and kernels per plant but did not lead to full crop recovery (Haneklaus et al., 1995). Therefore, top dressing of a sulfate form of S in-crop is a viable option if deficiency symptoms are observed during early crop development, although yield may be lower than if adequate S was available throughout crop growth. Adequate water will be required, however, to move the fertilizer into the rooting zone where it will be available for crop uptake.

Thiosulfate fertilizers combine thiosulfate with various cations to form products including ammonium thiosulfate [$(NH_4)_2S_2O_3$], potassium thiosulfate ($K_2S_2O_3$), calcium thiosulfate (CaS_2O_3), and magnesium thiosulfate (MgS_2O_3). The $S_2O_3^{2-}$ ion rapidly oxidizes to sulfate and is almost immediately available for plant uptake (Lettl et al., 1981; Janzen and Bettany, 1986). Effectiveness of ammonium thiosulfate in increasing rapeseed seed yield and S uptake was comparable to that of ammonium sulfate (Grant et al., 2003b, 2004).

Either sulfate or thiosulfate fertilizer forms may have residual benefits for several years after application, depending on the environmental conditions. Applications of 20–30 kg S ha^{-1} as ammonium sulfate limited S deficiency in crops for 2–4 years after application in the Canadian prairies

TABLE 7.4

Seed Yield (Mg ha⁻¹) of Canola at Melfort, Saskatchewan, as Affected by Tillage and Sulfur Fertilizer Source, Timing, and Placement in Year of Application

	1996		1997		1998	
Fertilizer	CT[a]	RT	CT	RT	CT	RT
Control	1.01	1.39	2.27	1.86	0.07	0.07
Fall surface						
Tiger 90[b]	—[c]	—	2.06	2.01	0.09	0.08
AS	—	—	2.25	2.10	1.15	0.76
ATS	—	—	2.17	2.19	1.02	1.17
Spring surface						
Tiger 90	0.85	1.17	1.86	1.74	0.2	0.08
Elemental S	1.14	1.42	1.96	2.12	0.07	0.07
AS	1.92	2.18	2.14	2.05	1.32	1.36
ATS	1.80	2.05	2.13	2.22	1.35	1.07
Blend	1.72	1.98	2.3	2.04	1.26	1.24
Spring banded						
Tiger 90	0.89	0.96	2.05	1.55	0.53	0.09
AS	1.98	2.06	2.13	2.26	—	—
Seed placed						
Tiger 90	0.92	1.48	1.95	1.92	0.14	0.12
AS	1.98	2.19	1.93	1.9	1.53	1.26
Contrast	F Value	Significance	F Value	Significance	F Value	Significance
CT vs. RT	0.01	ns	1.36	ns	4.19	ns
Control vs. AS	32.92	0.0001	0.09	ns	135.34	0.0001
Tiger 90 vs. AS surface	32.62	0.0001	6.59	0.0124	211.67	0.0001
AS fall vs. spring	—	—	0.45	ns	15.49	0.0002
Tiger 90 fall vs. spring	—	—	3.77	ns	0.2	ns
Elemental S vs. Tiger 90	2.27	ns	4.02	0.0488	0.42	ns
Tiger 90 surface vs. seed placed	1.09	ns	1.06	ns	0	ns
AS surface vs. seed placed	0.03	ns	2.1	ns	0.37	ns
Tiger 90 banded vs. surface	0.22	ns	0	ns	2.98	ns
AS band vs. surface	0.02	ns	0.8	ns	152.87	0.0001
Tiger 90 vs. AS	91.71	0.0001	11.23	0.0013	258.59	0.0001
AS vs. ATS	0.49	ns	0.29	ns	0.01	ns
Tiger 90 vs. AS seed placed	23.48	0.0001	0.02	ns	172.16	0.0001

Source: Grant, C.A. et al., *Can. J. Plant Sci.*, 83(4), 745, 2003a.

[a] CT, conventional tillage; RT, reduced tillage; AS, ammonium sulfate; ATS, ammonium thiosulfate.

[b] Tiger 90 is an elemental S-bentonite clay formulation.

[c] Treatment not available.

where leaching losses are low (Table 7.5) (Grant et al., 2003b, 2004; Malhi and Leach, 2003; Karamanos and Poisson, 2004). Ammonium thiosulfate also had residual benefits on S availability and crop yield similar to that of ammonium sulfate (Grant et al., 2003a, 2004; Malhi and Leach, 2003; Karamanos and Poisson, 2004). Residual benefits from sulfate fertilizers may be due to carry-over of sulfate, as ammonium sulfate has been shown to increase soil sulfate concentration after harvest particularly under drier conditions where leaching is minimal (Malhi and Leach, 2003; Karamanos and Poisson, 2004; Malhi et al., 2009). In wet environments, leaching of sulfate can be substantial, and the residual value of sulfate sources will be low (Riley et al., 2002). Mineralization

TABLE 7.5

Seed Yield (kg ha⁻¹) of Canola as Influenced by Tillage and Residual Effects of Sulfur Fertilizer Source, Timing, and Placement at Melfort, Saskatchewan

	1997		1998		1998	
	S Added to Wheat in 1996		S Added to Wheat in 1997		S Added to Canola in 1996	
Fertilizer Treatment	CT	RT	CT	RT	CT	RT
Control	0.52	0.39	1.67	1.61	0.85	0.37
Fall surface						
Tiger 90[b]	—[c]	—	1.64	1.86	—	—
AS	—	—	1.76	1.9	—	—
ATS	—	—	1.76	1.73	—	—
Spring surface						
Tiger 90	0.82	0.49	1.83	1.6	1.34	1.44
Elemental S	0.33	0.3	1.84	1.76	0.51	0.67
AS	1.74	1.7	1.51	1.89	1.32	1.36
ATS	1.75	1.76	1.72	1.86	1.41	1.55
Blend	0.92	0.51	1.45	1.76	0.63	0.99
Spring banded						
Tiger 90	0.58	0.2	1.72	1.79	0.79	0.58
AS	1.75	1.71	1.60	1.71	1.57	1.55
Seed placed						
Tiger 90	0.76	0.31	1.36	1.76	0.94	1.13
AS	0.86	0.24	1.42	1.64	1.28	1.62

	1997 Crop Year		1998 Crop Year		1998 Crop Year	
Contrast	F Value	Significance	F Value	Significance	F Value	Significance
CT vs. RT[a]	11.43	0.0013	2.71	ns	0.18	ns
Control vs. AS	46.13	<0.0001	0.22	ns	38.58	<0.0001
Tiger 90 vs. AS surface	44.51	<0.0001	0.18	ns	0.11	ns
AS fall vs. spring	—[b]	—	1.02	ns	—	—
Tiger 90 fall vs. spring	—	—	0.05	ns	—	—
Elemental S vs. Tiger 90	4.59	0.0365	0.32	ns	23.38	<0.0001
Tiger 90 surface vs. seed placed	0.56	ns	1.15	ns	4.63	0.0358
AS surface vs. seed placed	53.63	<0.0001	0.87	ns	0.46	ns
Tiger 90 banded vs. surface	2.76	ns	0.07	ns	18.28	<0.0001
AS band vs. surface	0	ns	0.11	ns	1.84	ns
Tiger 90 vs. AS	76.69	<0.0001	0.00	ns	18.7	<0.0001
AS vs. ATS	0.04	ns	0.01	ns	0.78	ns
Tiger 90 vs. AS seed placed	0.01	ns	0.00	ns	6.28	0.0152

Source: Grant, C.A. et al., *Can. J. Plant Sci.*, 84(2), 453, 2004.

[a] CT, conventional tillage; RT, reduced tillage; AS, ammonium sulfate; ATS, ammonium thiosulfate.

[b] Tiger 90 is an elemental S-bentonite clay formulation.

[c] Treatment not available.

of high S crop residues resulting from ammonium sulfate fertilization also may contribute available S to subsequent crops (Jackson, 2000; Grant et al., 2003b).

Selection of method of placement for sulfate fertilizers depends on soil characteristics and environmental conditions. In studies conducted in the black and gray soils of the northern Canadian prairies, rapeseed yield was equal when ammonium sulfate was surface broadcast, in-soil banded, or seed placed, under reduced or conventional tillage (Grant et al., 2004). In studies conducted over a 3-year period in northern Saskatchewan, seed row–placed or side-banded fertilizer produced higher seed yield than broadcast and incorporated S in 1 year, whereas yields were similar among the treatments in the remaining 2 years of the studies (Malhi and Gill, 2002). When moisture was limited, S in bands provided better root access for crop uptake than did shallow placement in surface soil. Therefore, in relatively drier areas, it may be preferable to side-band or preseed band ammonium sulfate to prevent any loss of soil water due to incorporation, avoid stranding of the S fertilizer in dry surface soil, and increase the accessibility of applied S to plants early in the growing season. Placement of sulfate-S fertilizer in the seed row is another option, but excessive application rates may lead to seedling damage in sensitive crops such as rapeseed, delay maturity, and reduce potential yield (Malhi and Gill, 2002). Safe rates for seed-row placement are influenced by crop type, soil and environmental conditions, and seed-bed utilization, which is a function of opener width and row spacing (http://www.agriculture.gov.sk.ca/Default.aspx?DN = e42316e3-15ea-4249-ac0e-369212b23131, accessed August 28, 2013).

Elemental S must be oxidized to sulfate before it is available for crop uptake (Janzen and Bettany, 1986). Oxidation of elemental S to sulfate is mediated by soil microorganisms, including chemolithotrophs, such as *Thiobacillus* sp.; photoautotrophs, including species of purple and green S bacteria; and heterotrophs, encompassing a range of bacteria and fungi (Germida and Janzen, 1993). Therefore, oxidation rate is influenced by factors affecting microbial activity, such as soil temperature, water, pH, and organic matter content, as well as dispersion of elemental S particles from fertilizer granules as affected by product formulation and placement (Janzen and Bettany, 1986, 1987a,b; Solberg, 1986; Solberg et al., 1987, 2003, 2005, 2007; Janzen, 1990; Boswell and Friesen, 1993; Riley et al., 2000).

In many environments, the oxidation rate of elemental S is not rapid enough to release sufficient available sulfate in the year of application to support crop production on S-deficient soils (Bettany and Janzen, 1984; Janzen and Bettany, 1986, 1987b; Janzen and Karamanos, 1991; Riley et al., 2000; Grant et al., 2003b, 2004; Solberg et al., 2007). In studies in the Northern Great Plains comparing ammonium sulfate with elemental S forms, elemental S consistently produced lower yields in the year of application than ammonium sulfate on S-responsive soils (Ukrainetz, 1982; Solberg and Nyborg, 1983; Swan et al., 1986; Karamanos and Janzen, 1991; Malhi and Leach, 2003; Solberg et al., 2003, 2007; Grant et al., 2004; Karamanos and Poisson, 2004; Malhi, 2005). Similarly, in Australia, elemental S applied at sowing was oxidized too slowly over the winter growing season to meet crop demand (Hocking et al., 2011). In contrast, elemental sources were usually as effective as sulfate sources in studies in West Africa and had a greater residual benefit (Friesen, 1991). Effectiveness of elemental S depends on local climatic conditions as well as the time frame when the product must be plant available (Boswell and Friesen, 1993).

Availability of sulfate from applications of elemental S increases with time as oxidation proceeds (Janzen and Karamanos, 1991; Wen et al., 2003; Solberg et al., 2007; Malhi et al., 2009). Application of elemental S as a surface broadcast application well in advance of crop demand allows more time for oxidation to occur (Grant et al., 2003b, 2004; Malhi, 2005; Malhi et al., 2009). Increased plant-available S and yield may occur in crops grown 2–4 years after initial application (Malhi and Leach, 2003; Solberg et al., 2003, 2007; Wen et al., 2003; Grant et al., 2004; Karamanos and Poisson, 2004). In studies conducted on the Canadian prairies, seed yield and S uptake of rapeseed were still lower from elemental S than from ammonium sulfate when both sources were applied in the previous spring, 12 months prior to rapeseed seeding (Grant et al., 2003b, 2004; Malhi, 2005), but by the third year after application oxidation from the elemental S could be sufficient to optimize crop growth, if surface applied (Grant et al., 2003b, 2004; Wen et al., 2003; Karamanos and Poisson, 2004). However, even after four annual applications, seed yield and S uptake of rapeseed were still less in some cases with elemental S than with ammonium sulfate (Mahli, 2005).

Elemental S supply may give available sulfate over extended periods of time since its gradual conversion to sulfate might reduce leaching losses (Solberg and Nyborg, 1983; Friesen, 1991; Boswell and Friesen, 1993; Riley et al., 2002; Solberg et al., 2007). However, once the elemental S converts to sulfate, it will be as susceptible to leaching as sulfate that originated from ammonium sulfate. Under Canadian prairie conditions, residual benefits from elemental S and ammonium sulfate were similar 3 years after application although substantially more of the S from ammonium sulfate was recovered by the preceding crops in the first 1–2 years after application (Grant et al., 2004; Karamanos and Poisson, 2004; Malhi et al., 2009). In contrast, in England, on a soil with very high leaching potential, use of elemental S decreased the leaching loss over a 3-year period as compared to ammonium sulfate (Riley et al., 2002), and in West Africa elemental sources had a greater residual availability than did sulfate sources (Friesen, 1991).

Availability of elemental S may be enhanced by fertilizer formulations designed to reduce the particle size and increase the surface area exposed to microbial activity. Oxidation rate increases as particle size of elemental S decreases (Janzen and Bettany, 1987a; Boswell and Friesen, 1993). In incubation studies, oxidation to sulfate was much more rapid with finely dispersed particles as compared to larger granules, with an average conversion to sulfate of 1%, 4%, and 8% of the S in the granules and 11%, 26%, and 56% in the finely dispersed elemental S after 10 weeks, at 5°C, 10°C, and 20°C, respectively (Wen et al., 2003). Elemental S oxidation under field conditions was also much greater with finely ground S particles mixed into the soil as compared to granules (Solberg et al., 2003). To increase the dispersion and reduce the particle size of elemental S distributed in the soil, finely divided elemental S is often combined with bentonite clay and formulated into prills, which make the product easier to handle (Germida and Janzen, 1993). As the prills absorb water, the bentonite expands and the prills disintegrate, distributing the fine elemental S particles in the soil. Dispersion depends on wetting and drying, weathering, and physical transport with seeding and tillage implements (Solberg et al., 1987; Nuttall et al., 1993).

Even with formulations designed to promote dispersion, elemental S oxidation may not be adequate to optimize plant-available S and crop yield of high S-demand crops such as rapeseed in the year of application under some environmental conditions (Malhi and Leach, 2003; Wen et al., 2003; Grant et al., 2004; Karamanos and Poisson, 2004). In field studies on the Canadian prairies, neither elemental S nor elemental S formulated with bentonite increased available sulfate measured in the soil solution using an ion-exchange resin in place for 2 weeks after seeding, nor did they increase tissue concentration of the rapeseed crop at flowering. Sulfate-S was generally numerically, but not significantly, higher with the bentonite treatment than elemental S fertilizer (Grant et al., 2003b). Therefore, oxidation of the bentonite-treated elemental S or elemental S fertilizer to sulfate-S was generally not sufficient to significantly increase plant-available sulfate in the year of application in the cool, short-season prairie environment. In contrast, finely divided elemental S applied as a suspension in water sprayed on the soil surface was much more effective than bentonite formulations, with 40 kg S ha^{-1} eliciting a comparable yield response to an application of 20 kg sulfate-S ha^{-1}, whereas application of up to 120 kg S ha^{-1} of S-bentonite forms produced a yield response equivalent to only 5 kg sulfate-S ha^{-1} (Karamanos and Janzen, 1991). Similarly, micronized elemental S was oxidized more rapidly than a bentonite clay elemental product in lysimeter studies in England (Riley et al., 2002).

Effectiveness of elemental S is increased by practices that increase the rate of oxidation by enhancing the exposure of the product to weathering and microbial activity. Broadcasting elemental S on the soil surface and leaving the granules on the surface for an extended time prior to incorporation can increase the physical breakdown of the particles through freezing and thawing and wetting and drying cycles, and increasing time between time of application and crop demand will allow more time for dispersion of S particles and oxidation to occur before plant uptake. Shallow cultivation can be used to incorporate the S particles in the soil, thus increasing their dispersion into the soil and their exposure to microbiological activity (Nyborg et al., 1980; Solberg et al., 1987, 2003). Broadcast application of elemental S should be used to maximize dispersion (Grant et al., 2003a,b). However, there is a conflict between this application and the ever-increasing practice of direct seeding or no-till where fertilizer is placed in bands. Banding or seed placement should be avoided if the goal of application is to provide available S for the current crop, as banding restricts dispersion and limits the exposure of the fertilizer to microbial oxidation (Swan et al., 1986; Germida and Janzen, 1993). In studies comparing various placement options for an elemental S-bentonite clay product, availability of the banded or seed placed was low, even 3 years after application (Grant et al., 2003a,b).

Elemental S has been used for centuries as a fungicide, as elemental S is reduced in fungi to produce H_2S, causing death of the fungus (Legris-Delaporte et al., 1987). A small proportion, in the range of 1%–2%, of micronized elemental S applied as a foliar application may be taken up and metabolized by the plant directly, most likely via a pathway including oxidation to sulfate, particularly if the plants were initially S deficient (Legris-Delaporte et al., 1987; Landry et al., 1991). In studies applying foliar elemental S at anthesis in wheat, S was taken up by the leaves, assimilated and a proportion translocated to the grain, with a synergistic effect occurring when both foliar N and S were applied together (Tea et al., 2007). However, a major route for the incorporation of S from applied elemental sulfur is likely to be rain-driven wash down to the soil, followed by microbial oxidation to sulfate and uptake via the root system.

7.11   PROSPECTS

Sulfur is a macronutrient and is a major structural component of proteins and also participates in the catalytic activities of many enzymes. Other S-containing compounds have a diverse set of biological functions and impart many functional properties on plant tissues, including important structural, nutritional, and organoleptic characteristics of crops. There are many interactions between S and other nutrients, although much is yet to be learned with respect to the mechanisms involved and implications for crop production. Sulfur deficiencies in crops have significant effects on yield and quality but are easily rectified once identified. In recent years, S deficiencies have become more widely recognized, and the calculated inclusion of S in fertilizer formulations is now commonplace. In the future, key developments will be the optimization of fertilizer practice, a deeper understanding of soil S cycles, and the development of crop germplasm optimized for S use.

ACKNOWLEDGMENTS

Rothamsted Research receives support from the Biotechnology and Biological Sciences Research Council (BBSRC) of the United Kingdom, and MJH is supported as part of the 20:20 Wheat® project.

REFERENCES

Abdallah, M., L. Dubousset, F. Meuriot, P. Etienne, J.C. Avice, and A. Ourry. 2010. Effect of mineral sulphur availability on nitrogen and sulphur uptake and remobilization during the vegetative growth of *Brassica napus* L. *J. Exp. Bot.* 61(10):2635–2646.

Alhendawi, R.A., E.A. Kirkby, and D.J. Pilbeam. 2005. Evidence that sulfur deficiency enhances molybdenum transport in xylem sap of tomato plants. *J. Plant Nutr.* 28(8):1347–1353.

Ali, M.H., S.M.H. Zaman, and S.M. Altaf Hossain. 1996. Variation in yield, oil and protein content of rapeseed (*Brassica campestris*) in relation to levels of nitrogen, sulphur and plant density. *Indian J. Agron.* 41(2):290–295.

Angus, J.F., P.A. Gardner, J.A. Kirkegaard, and J.M. Desmarchelier. 1994. Biofumigation: Isothiocyanates released from Brassica roots inhibit growth of the take-all fungus. *Plant Soil* 162(1):107–112.

Angus, J.F., A.F. van Herwaarden, and G.N. Howe. 1991. Productivity and break crop effects of winter-growing oilseeds. *Aust. J. Exp. Agric.* 31(5):669–677.

Anonymous. 2011. Sulfur deficiency symptoms in wheat. [Online] Available: www.agronomy.ksu.edu/doc3579.ashx, accessed August 12, 2013.

Astolfi, S., S. Cesco, S. Zuchi, G. Neumann, and V. Römheld. 2006. Sulfur starvation reduces phytosiderophores release by iron-deficient barley plants. *Soil Sci. Plant Nutr.* 52(1):43–48.

Astolfi, S., S. Zuchi, C. Passera, and S. Cesco. 2003. Does the sulfur assimilation pathway play a role in the response to Fe deficiency in maize (*Zea mays* L.) plants? *J. Plant Nutr.* 26:2111–2121.

Aulakh, M.S. and N.S. Pasricha. 1977. Interaction effect of sulphur and phosphorus on growth and nutrient content of moong (*Phaseolus aureus* L.). *Plant Soil* 47(2):341–350.

Bailey, L.D. 1985. The sulphur status of eastern Canadian prairie soils: The relationship of sulphur, nitrogen and organic carbon. *Can. J. Soil Sci.* 65(1):179–186.

Bailey, L.D. 1986. The sulphur status of eastern Canadian prairie soils: Sulphur response and requirements of alfalfa (*Medicago sativa* L.), rape (*Brassica napus* L.) and barley (*Hordeum vulgare* L.). *Can. J. Soil Sci.* 66(2):209–216.

Banerjee, M.R., S.J. Chapman, and K. Killham. 1993. Factors influencing the determination of microbial biomass sulphur in soil. *Commun. Soil Sci. Plant Anal.* 24:939–950.

Barillas, J.R.V., C.F. Quinn, and E.A.H. Pilon-Smits. 2011. Selenium accumulation in plants—Phytotechnological applications and ecological implications. *Int. J. Phytorem.* 13(suppl. 1):166–178.

Baxter, I., B. Muthukumar, C.P. Hyeong et al. 2008. Variation in molybdenum content across broadly distributed populations of *Arabidopsis thaliana* is controlled by a mitochondrial molybdenum transporter (MOT1). *PLoS Genet.* 4(2):e1000004.

Beaton, J.D. and R.J. Soper. 1986. Plant response to sulfur in Western Canada. In *Sulfur in Agriculture*, ed. M.A. Tabatabai, pp. 375–403. Madison, WI: American Society of Agronomy, Crop Science Society of America, Soil Science Society of America.

Bell, C.I., D.T. Clarkson, and W.J. Cram. 1995. Partitioning and redistribution of sulphur during S-stress in *Macroptilium atropurpureum* cv. Siratro. *J. Exp. Bot.* 46(1):73–81.

Bell, C.I., W.J. Cram, and D.T. Clarkson. 1994. Compartmental analysis of $^{35}SO_4^{2-}$ exchange kinetics in roots and leaves of a tropical legume *Macroptilium atropurpureum* cv. Siratro. *J. Exp. Bot.* 45:879–886.

Benkeblia, N. and V. Lanzotti. 2007. Allium thiosulfinates: Chemistry, biological properties and their potential utilization in food preservation. *Food* 1(2):193–201.

Bettany, J.R. and H.H. Janzen. 1984. Transformations of sulfur fertilizers in prairie soils. *Proceedings of International Sulphur '84 Conference*, Sulphur Development Institute of Canada, Calgary, Alberta, Canada, pp. 817–822.

Bharose, R., S. Chandra, T. Thomas, and D. Dhan. 2011. Effect of different levels of phosphorus and sulphur on yield and availability of N P K, protein and oil content in toria (*Brassica* sp.) VAR. P.T.–303. *J. Agric. Biol. Sci.* 6(2):31–33.

Blair, G., N. Chinoim, R. Lefroy, G. Anderson, and G. Crocker. 1991. A soil sulfur test for pastures and crops. *Soil Res.* 29(5):619–626.

Blake-Kalff, M.M.A., K.R. Harrison, M.J. Hawkesford, F.J. Zhao, and S.P. McGrath. 1998. Distribution of sulfur within oilseed rape leaves in response to sulfur deficiency during vegetative growth. *Plant Physiol.* 118(4):1337–1344.

Blake-Kalff, M.M.A., M.J. Hawkesford, F.J. Zhao, and S.P. McGrath. 2000. Diagnosing sulfur deficiency in field-grown oilseed rape (*Brassica napus* L.) and wheat (*Triticum aestivum* L.). *Plant Soil* 225(1–2):95–107.

Blake-Kalff, M.M.A., F.J. Zhao, M.J. Hawkesford, and S.P. McGrath. 2001. Using plant analysis to predict yield losses caused by sulphur deficiency. *Ann. Appl. Biol.* 138(1):123–127.

Bloem, E., S. Haneklaus, and E. Schnug. 2005. Influence of nitrogen and sulfur fertilization on the alliin content of onions and garlic. *J. Plant Nutr.* 27(10):1827–1839.

Bohn, H.L., N.J. Barrow, S.S.S. Rajan, and R.L. Parfitt. 1986. Reactions of inorganic sulfur in soils. *Sulfur Agric.* 27:233–249.

Boswell, C.C. and D.K. Friesen. 1993. Elemental sulfur fertilizers and their use on crops and pastures. *Fert. Res.* 35(1–2):127–149.

Bourgis, F., S. Roje, M.L. Nuccio et al. 1999. S-methylmethionine plays a major role in phloem sulfur transport and is synthesized by a novel type of methyltransferase. *Plant Cell* 11:1485–1497.

Brennan, R.F. and M.D.A. Bolland. 2006. Soil and tissue tests to predict the sulfur requirements of canola in south-western Australia. *Aust. J. Exp. Agric.* 46(8):1061–1068.

Brennan, R.F. and M.D.A. Bolland. 2008. Significant nitrogen by sulfur interactions occurred for canola grain production and oil concentration in grain on sandy soils in the Mediterranean-type climate of southwestern Australia. *J. Plant Nutr.* 31(7):1174–1187.

Buchner, P., S. Parmar, A. Kriegel, M. Carpentier, and M.J. Hawkesford. 2010. The sulfate transporter family in wheat: Tissue-specific gene expression in relation to nutrition. *Mol. Plant* 3(2):374–389.

Camberato, J., S. Maloney, and S. Casteel. 2012. Sulfur deficiency in corn. [Online] Available: www.agry. purdue.edu/ext/corn/news/timeless/sulfurdeficiency.pdf, accessed August 12, 2013.

Carfagna, S., V. Vona, V. Di Martino, S. Esposito, and C. Rigano. 2011. Nitrogen assimilation and cysteine bio-synthesis in barley: Evidence for root sulphur assimilation upon recovery from N deprivation. *Environ. Exp. Bot.* 71(1):18–24.

Chan, K.X., M. Wirtz, S.Y. Phua, G.M. Estavillo, and B.J. Pogson. 2013. Balancing metabolites in drought: The sulfur assimilation conundrum. *Trends Plant Sci.* 18(1):18–29.

Chung, L.Y. 2006. The antioxidant properties of garlic compounds: Alyl cysteine, alliin, allicin, and allyl disul-fide. *J. Med. Food* 9(2):205–213.

Churka Blum, S., J. Lehmann, D. Solomon, E.F. Caires, and L.R.F. Alleoni. 2013. Sulfur forms in organic substrates affecting S mineralization in soil. *Geoderma* 200–201:156–164.

Ciaffi, M., A.R. Paolacci, S. Celletti, G. Catarcione, S. Kopriva, and S. Astolfi. 2013. Transcriptional and physi-ological changes in the S assimilation pathway due to single or combined S and Fe deprivation in durum wheat (*Triticum durum* L.) seedlings. *J. Exp. Bot.* 64(6):1663–1675.

Cobbett, C.S. 2000. Phytochelatins and their roles in heavy metal detoxification. *Plant Physiol.* 123(3):825–832.

Combs Jr, G.F. 2001. Selenium in global food systems. *Br. J. Nutr.* 85(5):517–547.

Coolong, T.W. and W.M. Randle. 2003. Sulfur and nitrogen availability interact to affect the flavor biosynthetic pathway in onion. *J. Am. Soc. Hortic. Sci.* 128(5):776–783.

Cooper, R.M., M.L.V. Resende, J. Flood, M.G. Rowan, M.H. Beale, and U. Potter. 1996. Detection and cel-lular localization of elemental sulphur in disease-resistant genotypes of *Theobroma cacao*. *Nature* 379:159–162.

Cooper, R.M. and J.S. Williams. 2004. Elemental sulphur as an induced antifungal substance in plant defence. *J. Exp. Bot.* 55:1947–1953.

Curtin, D. and J.K. Syers. 1990. Extractability and adsorption of sulphate in soils. *J. Soil Sci.* 41(2):305–312.

Curtis, H., U. Noll, J. Stormann, and A.J. Slusarenko. 2004. Broad-spectrum activity of the volatile phyto-anticipin allicin in extracts of garlic (*Allium sativum* L.) against plant pathogenic bacteria, fungi and Oomycetes. *Physiol. Mol. Plant Pathol.* 65(2):11–11.

Curtis, T.Y., N. Muttucumaru, P.R. Shewry et al. 2009. Effects of genotype and environment on free amino acid levels in wheat grain: Implications for acrylamide formation during processing. *J. Agric. Food Chem.* 57:1013–1021.

Davidian, J.C. and S. Kopriva. 2010. Regulation of sulfate uptake and assimilation—The same or not the same? *Mol. Plant* 3(2):314–325.

Dick, W.A., D. Kost, and I. Chen. 2008. Availability of sulfur to crops from soil and other sources. In *Sulfur: A Missing Link between Soils, Crops, and Nutrition*, ed. J. Jez, pp. 59–82. Madison, WI: American Society of Agronomy, Crop Science Society of America, Soil Science Society of America.

Doyle, P.J. and L.E. Cowell. 1993. Sulphur. Chapter 6. In *Impact of Macronutrients on Crop Responses and Environmental Sustainability on the Canadian Prairies*. Eds. D.A. Rennie, C.A. Campbell and T.L. Roberts. Ottawa, Ontario, Canada: Canadian Society Soil Science.

Dubousset, L., M. Abdallah, A.S. Desfeux et al. 2009. Remobilization of leaf S compounds and senescence in response to restricted sulphate supply during the vegetative stage of oilseed rape are affected by mineral N availability. *J. Exp. Bot.* 60(11):3239–3253.

Dumanski, J., R.L. Desjardins, C. Tarnocai et al. 1998. Possibilities for future carbon sequestration in Canadian agriculture in relation to land use changes. *Clim. Change* 40(1):81–103.

Egesel, C.Ö., M.K. Gül, and F. Kahrlman. 2009. Changes in yield and seed quality traits in rapeseed genotypes by sulphur fertilization. *Eur. Food Res. Technol.* 229(3):505–513.

Epstein, E. 2000. The discovery of the essential elements. In *Discoveries in Plant Biology*, vol. 3, eds. S.-D. Kung and S.-F. Yang, pp. 1–16. Singapore: World Scientific.

Eriksen, J. 2009. Chapter 2. Soil sulfur cycling in temperate agricultural systems. *Adv. Agron.* 102:55–89.

Falk, K.L., J.G. Tokuhisa, and J. Gershenzon. 2007. The effect of sulfur nutrition on plant glucosinolate con-tent: Physiology and molecular mechanisms. *Plant Biol.* 9(5):573–581.

Flæte, N.E.S., K. Hollung, L. Ruud et al. 2005. Combined nitrogen and sulfur fertilisation and its effect on wheat quality and protein composition measured by SE-FPLC and proteomics. *J. Cereal Sci.* 41(3):357–369.

Fox, R.L., R.A. Olson, and H.F. Rhoades. 1964. Evaluating the sulfur status of soils by plant and soil tests 1. *Soil Sci. Soc. Am. J.* 28(2):243–246.

Franzen, D.W. and C.A. Grant. 2008. Sulfur response based on crop, source, and landscape position. In *Sulfur: A Missing Link between Soils, Crops, and Nutrition*, ed. J. Jez, pp. 105–116. Madison, WI: American Society of Agronomy.

Freney, J.R., G.E. Melville, and C.H. Williams. 1975. Soil organic matter fractions as sources of plant-available sulphur. *Soil Biol. Biochem.* 7(3):217–221.

Friesen, D.K. 1991. Fate and efficiency of sulfur fertilizer applied to food crops in West Africa. In *Alleviating Soil Fertility Constraints to Increased Crop Production in West Africa*, ed. A.U. Mokwunye, pp. 59–68. Dordrecht, the Netherlands: Springer.

Galeas, M.L., L.H. Zhang, J.L. Freeman, M. Wegner, and E.A.H. Pilon-Smits. 2007. Seasonal fluctuations of selenium and sulfur accumulation in selenium hyperaccumulators and related nonaccumulators. *New Phytol.* 173(3):517–525.

Germida, J.J. and H.H. Janzen. 1993. Factors affecting the oxidation of elemental sulfur in soils. *Fert. Res.* 35(1–2):101–114.

Giamoustaris, A. and R. Mithen. 1995. The effect of modifying the glucosinolate content of leaves of oilseed rape (*Brassica napus* ssp. *oleifera*) on its interaction with specialist and generalist pests. *Ann. Appl. Biol.* 126(2):347–363.

Giamoustaris, A. and R. Mithen. 1997. Glucosinolates and disease resistance in oilseed rape (*Brassica napus* ssp. *oleifera*). *Plant Pathol.* 46(2):271–275.

Gilbert, S.M., D.T. Clarkson, M. Cambridge, H. Lambers, and M.J. Hawkesford. 1997. SO_4^{2-} deprivation has an early effect on the content of ribulose-1,5-bisphosphate carboxylase/oxygenase and photosynthesis in young leaves of wheat. *Plant Physiol.* 115(3):1231–1239.

Gill, S.S. and N. Tuteja. 2011. Cadmium stress tolerance in crop plants: Probing the role of sulfur. *Plant Signal. Behav.* 6(2):215–222.

Gissel-Nielsen, G., U.C. Gupta, M. Lamand, and T. Westermarck. 1984. Selenium in soils and plants and its importance in livestock and human nutrition. *Adv. Agron.* 37:397–460.

Godfrey, D., M.J. Hawkesford, S.J. Powers, S. Millar, and P.R. Shewry. 2010. Effects of crop nutrition on wheat grain composition and end use quality. *J. Agric. Food Chem.* 58(5):3012–3021.

Good, A.J. and J.S. Glendinning. 1998. Canola needs sulphur. Better crops international 12. [Online] Available: http://anz.ipni.net/ipniweb/regions/anz/regionalPortalANZ.nsf/4d69fbe2e4abc746852574ce0 05c0150/d6dd95ec92bd94de852576ff0082d61c/$FILE/BC%20sulfur%20and%20canola.pdf, accessed September 14, 2011.

Gooding, M.J. and W.P. Davies. 1992. Foliar urea fertilization of cereals: A review. *Fert. Res.* 32(2):209–222.

Grant, C.A., G.W. Clayton, and A.M. Johnston. 2003a. Sulphur fertilizer and tillage effects on canola seed quality in the Black soil zone of western Canada. *Can. J. Plant Sci.* 83(4):745–758.

Grant, C.A., A.M. Johnston, and G.W. Clayton. 2003b. Sulphur fertilizer and tillage effects on early season sulphur availability and N:S ratio in canola in western Canada. *Can. J. Soil Sci.* 83(4):451–463.

Grant, C.A., A.M. Johnston, and G.W. Clayton. 2004. Sulphur fertilizer and tillage management of canola and wheat in western Canada. *Can. J. Plant Sci.* 84(2):453–462.

Gupta, U.C. and S.C. Gupta. 2002. Quality of animal and human life as affected by selenium management of soils and crops. *Commun. Soil Sci. Plant Anal.* 33(15–18):2537–2555.

Gupta, U.C. and J. Lipsett. 1981. Molybdenum in soils, plants and animals. *Adv. Agron.* 34:73–115.

Gupta, U.C. and L.B. Macleod. 1975. Effects of sulfur and molybdenum on the molybdenum, copper, and sulfur concentrations of forage crops. *Soil Sci.* 119(6):441–447.

Hamm, J.W. 1967. Sulfur on rapeseed and cereals. *Tenth Annual Manitoba Soil Science Meeting*, Winnipeg, Manitoba, Canada, pp. 91–108.

Hamm, J.W., J.R. Bettany, and E.H. Halstead. 1973. A soil test for sulphur and interpretative criteria for Saskatchewan. *Commun. Soil Sci. Plant Anal.* 4:219–231.

Haneklaus, S., E. Bloem, S. Hayfa, and E. Schnug. 2005a. Influence of elemental sulphur and nitrogen fertilisation on the concentration of essential micro-nutrients and heavy metals in *Tropaeolum majus* L. *Landbauforschung Volkenrode*, Special issue 286:25–36.

Haneklaus, S., E. Bloem, and E. Schnug. 2008. History of sulfur deficiency in crops. In *Sulfur: A Missing Link between Soils, Crops, and Nutrition*, ed. J. Jez, pp. 45–58. Madison, WI: American Society of Agronomy, Crop Science Society of America, Soil Science Society of America.

Haneklaus, S., E. Bloem, E. Schnug, L.J. de Kok, and I. Stulen. 2007. Sulfur. In *Handbook of Plant Nutrition*, eds. A.V. Barker and D.J. Pilbeam, pp. 183–238. Boca Raton, FL: CRC Press.

Haneklaus, S., A. Brauer, E. Bloem, and E. Schnug. 2005b. Relationship between sulfur deficiency in oilseed rape (*Brassica napus* L.) and its attractiveness to honeybees. *Landbauforschung Volkenrode Sonderheft*, Special issue 283:37–43.

Haneklaus, S., D.P.L. Murphy, G. Nowak, and E. Schnug. 1995. Effects of the timing of sulphur application on grain yield and yield components of wheat. *Zeitschrift für Pflanzenernährung und Bodenkunde* 158(1):83–85.

Hawkesford, M.J. 2000. Plant responses to sulphur deficiency and the genetic manipulation of sulphate transporters to improve S-utilization efficiency. *J. Exp. Bot.* 51(342):131–138.

Hawkesford, M.J. 2010. Sulfate transport. *Plant Cell Monogr.* 19:291–301.

Hawkesford, M.J., J.C. Davidian, and C. Grignon. 1993. Sulphate/proton cotransport in plasma-membrane vesicles isolated from roots of *Brassica napus* L.: Increased transport in membranes isolated from sulphur-starved plants. *Planta* 190(3):297–304.

Hawkesford, M.J. and L.J. De Kok. 2006. Managing sulphur metabolism in plants. *Plant Cell Environ.* 29(3):382–395.

Hawkesford, M.J. and F.J. Zhao. 2007. Strategies for increasing the selenium content of wheat. *J. Cereal Sci.* 46(3):282–292.

Hell, R. and M. Wirtz. 2008. Metabolism of cysteine in plants and phototrophic bacteria. In *Advances in Photosynthesis and Respiration*, vol. 27, eds. R. Hell, C. Dahl, D.B. Knaff, and T. Leustek, pp. 59–91. Dordrecht, the Netherlands: Springer.

Hesse, H., V. Nikiforova, B. Gakière, and R. Hoefgen. 2004. Molecular analysis and control of cysteine biosynthesis: Integration of nitrogen and sulphur metabolism. *J. Exp. Bot.* 55(401):1283–1292.

Hocking, P., R. Norton, and A. Good. 2011. Canola nutrition. [Online] Available: http://aof.clients.squiz.net/__data/assets/pdf_file/0013/2704/Chapter_4_-_Canola_Nutrition.pdf, accessed July 19, 2011.

Hocking, P.J., A. Pinkerton, and A. Good. 1996. Recovery of field-grown canola from sulfur deficiency. *Aust. J. Exp. Agric.* 36(1):79–85.

Hopkins, L., S. Parmar, A. Błaszczyk, H. Hesse, R. Hoefgen, and M.J. Hawkesford. 2005. O-acetylserine and the regulation of expression of genes encoding components for sulfate uptake and assimilation in potato. *Plant Physiol.* 138(1):433–440.

Howarth, J.R., S. Parmar, J. Jones et al. 2008. Co-ordinated expression of amino acid metabolism in response to N and S deficiency during wheat grain filling. *J. Exp. Bot.* 59(13):3675–3689.

Islam, M., S. Mohsan, S. Afzal, S. Ali, M. Akmal, and R. Khalid. 2001. Phosphorus and sulfur application improves the chickpea productivity under rainfed conditions. *Int. J. Agric. Biol.* 13:713–718.

Jackson, G.D. 2000. Effects of nitrogen and sulfur on canola yield and nutrient uptake. *Agron. J.* 92(4):644–649.

Jaggi, R.C. and R.K. Sharma. 1999. Sulphur-phosphorus interaction in raya (*Brassica juncea* var. Varuna) in acid Alfisols of western Himalaya. *Trop. Agric.* 76(3):157–163.

Janzen, H.H. 1990. Elemental sulfur oxidation as influenced by plant growth and degree of dispersion within soil. *Can. J. Soil Sci.* 70(3):499–502.

Janzen, H.H. and J.R. Bettany. 1984. Sulfur nutrition of rapeseed. 1. Influence of fertilizer nitrogen and sulfur rates. *Soil Sci. Soc. Am. J.* 48(1):100–107.

Janzen, H.H. and J.R. Bettany. 1986. Release of available sulfur from fertilizers. *Can. J. Soil Sci.* 66(1):91–103.

Janzen, H.H. and J.R. Bettany. 1987a. The effect of temperature and water potential on sulfur oxidation in soils. *Soil Sci.* 144(2):81–89.

Janzen, H.H. and J.R. Bettany. 1987b. Oxidation of elemental sulfur under field conditions in central Saskatchewan. *Can. J. Soil Sci.* 67(3):609–618.

Janzen, H.H. and R.E. Karamanos. 1991. Short-term and residual contribution of selected elemental S fertilizers to the S fertility of two Luvisolic soils. *Can. J. Soil Sci.* 71(2):203–211.

Janzen, H.H. and R.M.N. Kucey. 1988. C, N, and S mineralization of crop residues as influenced by crop species and nutrient regime. *Plant Soil* 106(1):35–41.

Jones, M.B. 1986. Sulphur availability indexes. In *Sulfur in Agriculture*, ed. M.A. Tabatabai, pp. 549–566. Madison, WI: American Society of Agronomy, Crop Science Society of America, Soil Science Society of America.

Jones, M.B., J.H. Oh, and J.E. Ruckman. 1970. Effect of phosphorus and sulphur fertilization on the nutritive value of subterranean clover. *Proceedings of the New Zealand Grassland Association*, Dunedin, New Zealand, vol. 32, pp. 69–75.

Karamanos, R.E. 1988. The effect of ammonium sulphate placement on canola yields. *Proceedings of the 25th Alberta Soil Science Workshop*, Edmonton, Alberta, Canada, pp. 117–124.

Karamanos, R.E. 2001. Soil sampling to optimize fertilizer responses. *Proceedings of Western Canada Agronomy Workshop*, Lethbridge, Alberta, Canada.

Karamanos, R.E., T.B. Goh, and D.N. Flaten. 2007. Nitrogen and sulfur fertilizer management for growing canola on sulfur sufficient soils. *Can. J. Plant Sci.* 87(2):201–210.

Karamanos, R.E., T.B. Goh, and D.P. Poisson. 2004. Is there a need to provide N and S to canola in any given ratio? *Proceedings of the Great Plains Soil Fertility Conference*, Denver, CO.

Karamanos, R.E., T.B. Goh, and D.P. Poisson. 2005. Nitrogen, phosphorus, and sulfur fertility of hybrid canola. *J. Plant Nutr.* 28(7):1145–1161.

Karamanos, R.E. and H.H. Janzen. 1991. Crop response to elemental sulfur fertilizers in central Alberta. *Can. J. Soil Sci.* 71(2):213–225.

Karamanos, R.E. and D.P. Poisson. 2004. Short- and long-term effectiveness of various sulfur products in prairie soils. *Commun. Soil Sci. Plant Anal.* 35(13–14):2049–2066.

Kataoka, T., N. Hayashi, T. Yamaya, and H. Takahashi. 2004a. Root-to-shoot transport of sulfate in *Arabidopsis*. Evidence for the role of SULTR3;5 as a component of low-affinity sulfate transport system in the root vasculature. *Plant Physiol.* 136(4):4198–4204.

Kataoka, T., A. Watanabe-Takahashi, N. Hayashi et al. 2004b. Vacuolar sulfate transporters are essential determinants controlling internal distribution of sulfate in *Arabidopsis*. *Plant Cell* 16(10):2693–2704.

Ketterings, Q., C. Miyamoto, R.R. Mathur, K. Dietzel, and S. Gami. 2011. A Comparison of soil sulfur extraction methods. *Soil Sci. Soc. Am. J.* 75(4):1578–1583.

Kim, S.-J., T. Matsuo, M. Watanabe, and Y. Watanabe. 2002. Effect of nitrogen and sulphur application on the glucosinolate content in vegetable turnip rape (*Brassica rapa* L.). *Soil Sci. Plant Nutr.* 48(1):43–49.

Kirchmann, H., F. Pichlmayer, and M.H. Gerzabek. 1996). Sulfur balances and sulfur-34 abundance in a long-term fertilizer experiment. *Soil Sci. Soc. Am. J.* 60(1):174–178.

Kirkegaard, J.A., P.A. Gardner, J.F. Angus, and E. Koetz. 1994. Effect of Brassica break crops on the growth and yield of wheat. *Aust. J. Agric. Res.* 45(3):529–545.

Kirkegaard, J.A., P.J. Hocking, J.F. Angus, G.N. Howe, and P.A. Gardner. 1997. Comparison of canola, Indian mustard and Linola in two contrasting environments. II. Break-crop and nitrogen effects on subsequent wheat crops. *Field Crops Res.* 52(1–2):179–191.

Kirkegaard, J.A., S. Simpfendorfer, J. Holland, R. Bambach, K.J. Moore, and G.J. Rebetzke. 2004. Effect of previous crops on crown rot and yield of durum and bread wheat in northern NSW. *Aust. J. Agric. Res.* 55(3):321–334.

Kopriva, S. 2006. Regulation of sulfate assimilation in *Arabidopsis* and beyond. *Ann. Bot.* 97:479–495.

Korentajer, L., B.H. Byrnes, and D.T. Hellums. 1983. The effect of liming and leaching on the sulfur-supplying capacity of soils 1. *Soil Sci. Soc. Am. J.* 47(3):525–530.

Kovar, J.L. and C.A. Grant. 2011. Nutrient cycling in soils: Sulfur. In *Soil Management: Building a Stable Base for Agriculture*, eds. J.L. Hatfield and T.J. Sauer, pp. 103–116. Madison, WI: American Society of Agronomy, Crop Science Society of America, Soil Science Society of America.

Kowalenko, C.G. 1993. Extraction of available sulfur. *Soil Sampling and Methods of Analysis*, ed. M.R. Carter, pp. 65–74. Boca Raton, FL: Lewis Publishers.

Kowalenko, C.G. and M. Grimmett. 2008. Chemical characterization of soil sulfur. In *Soil Sampling and Methods of Analysis*, 2nd edn., eds. M.C. Carter and E.G. Gregorich, pp. 252–263. Boca Raton, FL: CRC Press.

Kowalska, I. 2005. Effects of sulphate level in the nutrient solution on plant growth and sulphur content in tomato plants. *Folia Horti.* 17(1):91–100.

Kreuz, K., R. Tommasini, and E. Martinoia. 1996. Old enzymes for a new job (herbicide detoxification in plants). *Plant Physiol.* 111(2):349–353.

Kumar, V. and M. Singh. 1980. Interactions of sulfur, phosphorus, and molybdenum in relation to uptake and utilization of phosphorus by soybean. *Soil Sci.* 130(1):26–31.

Lakkineneni, K.C. and Y.P. Abrol. 1992. Effect of sulfur fertilization on rapeseed-mustard and groundnut. *Phyton (Hom, Austria)* 32(3):75–78.

Lancaster, J.E. and H.A. Collin. 1981. Presence of alliinase in isolated vacuoles and of alkyl cysteine sulphoxides in the cytoplasm of bulbs of onion (*Allium cepa*). *Plant Sci. Lett.* 22(2):169–176.

Landry, J., S. Legris-Delaporte, and F. Ferron. 1991. Foliar application of elemental sulphur on metabolism of sulphur and nitrogen compounds in leaves of sulphur-deficient wheat. *Phytochemistry* 30(3):729–732.

Lawrence, J.R. and J.J. Germida. 1991. Enumeration of sulfur-oxidizing populations in Saskatchewan agricultural soils. *Can. J. Soil Sci.* 71(1):127–136.

Lee, B.R., S. Muneer, K.Y. Kim, J.C. Avice, A. Ourry, and T.H. Kim. 2013. S-deficiency responsive accumulation of amino acids is mainly due to hydrolysis of the previously synthesized proteins—Not to de novo synthesis in *Brassica napus*. *Physiol. Plant.* 147(3):369–380.

Lee, S. and S.S. Korban. 2002. Transcriptional regulation of *Arabidopsis thaliana* phytochelatin synthase (AtPCS1) by cadmium during early stages of plant development. *Planta* 215(4):689–693.

Legris-Delaporte, S., F. Ferron, J. Landry, and C. Costes. 1987. Metabolization of elemental sulfur in wheat leaves consecutive to its foliar application. *Plant Physiol.* 85(4):1026–1030.

Lehmann, J., D. Solomon, F.-J. Zhao, and S.P. McGrath. 2008. Atmospheric SO_2 emissions since the late 1800s change organic sulfur forms in humic substance extracts of soils. *Environ. Sci. Technol.* 42(10):3550–3555.

Lettl, A., O. Langkramer, and V. Lochman. 1981. Some factors influencing production of sulphate by oxidation of elemental sulphur and thiosulphate in upper horizons of spruce forest soils. *Folia Microbiol.* 26(2):158–163.

Leustek, T. and K. Saito. 1999. Sulfate transport and assimilation in plants. *Plant Physiol.* 120(3):637–644.

Liu, S., H. He, G. Feng, and Q. Chen. 2009. Effect of nitrogen and sulfur interaction on growth and pungency of different pseudostem types of Chinese spring onion (*Allium fistulosum* L.). *Sci. Hortic. (Amsterdam)* 121(1):12–18.

Malhi, S.S. 1998. Restoring canola yield by applying sulphur fertilizer during the growing season. *Proceedings Agrium Symposium: Sulfur Fertility and Fertilizers*, Calgary, Alberta, Canada.

Malhi, S.S. 2005. Influence of four successive annual applications of elemental S and sulphate-S fertilizers on yield, S uptake and seed quality of canola. *Can. J. Plant Sci.* 85(4):777–792.

Malhi, S.S. and K.S. Gill. 2002. Effectiveness of sulphate-S fertilization at different growth stages for yield, seed quality and S uptake of canola. *Can. J. Plant Sci.* 82(4):665–674.

Malhi, S.S. and K.S. Gill. 2007. Interactive effects of N and S fertilizers on canola yield and seed quality on S-deficient Gray Luvisol soils in northeastern Saskatchewan. *Can. J. Plant Sci.* 87(2):211–222.

Malhi, S.S., A.M. Johnston, J.J. Schoenau, Z.H. Wang, and C.L. Vera. 2007. Seasonal biomass accumulation and nutrient uptake of canola, mustard, and flax on a black chernozem soil in Saskatchewan. *J. Plant Nutr.* 30(4):641–658.

Malhi, S.S. and D. Leach. 2003. Effectiveness of elemental S fertilizers on canola after four annual applications. *Proceedings of Soils and Crop Workshop (Disc Copy)*, Saskatoon, Saskatchewan, Canada.

Malhi, S.S., D. Leach, and Z.H. Wang. 2004. Factors affecting yield variations of various crops in north-eastern Saskatchewan. *Proceedings of Soils and Crops Workshop (Disc Copy)*, Saskatoon, Saskatchewan, Canada.

Malhi, S.S., J.J. Schoenau, and C.L. Vera. 2009. Influence of six successive annual applications of sulphur fertilizers on wheat in a wheat-canola rotation on a sulphur-deficient soil. *Can. J. Plant Sci.* 89(4):629–644.

Manitoba Agriculture, Food and Rural Initiatives. 2009. Canola production and management. [Online], http://www.gov.mb.ca/agriculture/crops/production/canola.html), accessed April 21, 2009.

Marok, A.S. and G. Dev. 1980. Phosphorus and sulphur inter-relationships in wheat (*Triticum aestivum* L.). *J. Indian Soc. Soil Sci.* 28(2):184–188.

Matula, J. 2004. The effect of chloride and sulphate application to soil on changes in nutrient content in barley shoot biomass at an early phase of growth. *Plant Soil Environ.* 50(7):295–302.

Maynard, D.G., J.W.B. Stewart, and J.R. Bettany. 1983a. Sulfur and nitrogen mineralization in soils compared using two incubation techniques. *Soil Biol. Biochem.* 15(3):251–256.

Maynard, D.G., J.W.B. Stewart, and J.R. Bettany. 1983b. Use of plant analysis to predict sulfur deficiency in rapeseed (*Brassica napus* and *B. campestris*). *Can. J. Soil Sci.* 63:387–396.

McGill, W.B. and C.V. Cole. 1981. Comparative aspects of cycling of organic C, N, S and P through soil organic matter. *Geoderma* 26(4):267–286.

McGrath, S.P. and F.J. Zhao. 1996a. Sulphur uptake, yield responses and the interactions between nitrogen and sulphur in winter oilseed rape (*Brassica napus*). *J. Agric. Sci.* 126(01):53–62.

McGrath, S.P. and F.J. Zhao. 1996b. Sulphur uptake, yield responses and the interactions between nitrogen and sulphur in winter oilseed rape (*Brassica napus*). *J. Agric. Sci.* 126(1):53–62.

McLaren, R.G., J.I. Keer, and R.S. Swift. 1985. Sulphur transformations in soils using sulphur-35 labelling. *Soil Biol. Biochem.* 17(1):73–79.

Meidner, H. 1985. Historical sketches 5: Progress in mineral nutrition. *J. Exp. Bot.* 36(5):848–849.

Mendoza-Cózatl, D., H. Loza-Tavera, A. Hernández-Navarro, and R. Moreno-Sánchez. 2005. Sulfur assimilation and glutathione metabolism under cadmium stress in yeast, protists and plants. *FEMS Microbiol. Rev.* 29(4):653–671.

Mikkelsen, R.L. and H.F. Wan. 1990. The effect of selenium on sulfur uptake by barley and rice. *Plant Soil* 121(1):151–153.

Mills, H.A. and J.B. Jones Jr. 1996. *Plant Analysis Handbook II*. Jefferson City, MO: MicroMacro Publishing, Inc.

Muttucumaru, N., J.S. Elmore, T. Curtis, D.S. Mottram, M.A.J. Parry, and N.G. Halford. 2008. Reducing acrylamide precursors in raw materials derived from wheat and potato. *J. Agric. Food Chem.* 56:6167–6172.

Na, G. and D.E. Salt. 2011. The role of sulfur assimilation and sulfur-containing compounds in trace element homeostasis in plants. *Environ. Exp. Bot.* 72(1):18–25.

Niknahad-Gharmakher, H., S. Piutti, J.M. Machet, E. Benizri, and S. Recous. 2012. Mineralization-immobilization of sulphur in a soil during decomposition of plant residues of varied chemical composition and S content. *Plant Soil* 360(1–2):391–404.

Nocito, F.F., L. Pirovano, M. Cocucci, and G.A. Sacchi (2002). Cadmium-induced sulfate uptake in maize roots. *Plant Physiol.* 129(4):1872–1879.

Nuttall, W.F., C.C. Boswell, A.G. Sinclair, A.P. Moulin, L.J. Townley-Smith, and G.L. Galloway. 1993. The effect of time of application and placement of sulphur fertilizer sources on yield of wheat, canola, and barley. *Commun. Soil Sci. Plant Anal.* 24(17–18):2193–2202.

Nuttall, W.F., C.C. Boswell, and B. Swanney. 1990. Influence of sulphur fertilizer placement, soil moisture and temperature on yield response of rape to sulphur-bentonite. *Fert. Res.* 25(2):107–114.

Nyborg, M., A. Ayala, P. Yeung, S.S. Malhi, and M. Shier. 1980. The rate of oxidation of elemental sulphur fertilizers. *Proceedings of Soils and Crops Workshop*, Saskatoon, Saskatchewan, Canada, pp. 216–223.

Nyborg, M., C.F. Bentley, and P.B. Hoyt. 1974. Effect of sulphur deficiency. *Sulphur Inst. J.* 10:14–15.

Paolacci, A.R., S. Celletti, G. Catarcione, M.J. Hawkesford, S. Astolfi, and M. Ciaffi. 2014. Iron deprivation results in a rapid but not sustained increase of the expression of genes involved in iron metabolism and sulfate uptake in tomato (*Solanum lycopersicum* L.) seedlings. *J. Integr. Plant Biol.* 56:88–100.

Prietzel, J., J. Thieme, N. Tyufekchieva, D. Paterson, I. McNulty, and I. Kögel-Knabner. 2009. Sulfur speciation in well-aerated and wetland soils in a forested catchment assessed by sulfur K-edge X-ray absorption near-edge spectroscopy (XANES). *J. Plant Nutr. Soil Sci.* 172(3):393–403.

Prosser, I.M., J.V. Purves, L.R. Saker, and D.T. Clarkson. 2001. Rapid disruption of nitrogen metabolism and nitrate transport in spinach plants deprived of sulphate. *J. Exp. Bot.* 52(354):113–121.

Qian, P. and J.J. Schoenau. 2007. Using an anion exchange membrane to predict soil available N and S supplies and the impact of N and S fertilization on canola and wheat growth. *Pedosphere* 17(1):77–83.

Qian, P., J.J. Schoenau, and W.Z. Huang. 1992. Use of ion exchange membranes in routine soil testing. *Commun. Soil Sci. Plant Anal.* 23(15–16):1791–1804.

Rabinkov, A., T. Miron, L. Konstantinovski, M. Wilchek, D. Mirelman, and L. Weiner. 1998. The mode of action of allicin: Trapping of radicals and interaction with thiol containing proteins. *Biochim. Biophys. Acta (BBA)—General Subjects* 1379(2):233–244.

Redovniković, I.R., T. Glivetić, K. Delonga, and J. Vorkapić-Furač. 2008. Glucosinolates and their potential role in plant. *Period. Biol.* 110(4):297–309.

Reinbold, J., M. Rychlik, S. Asam, H. Wieser, and P. Koehler. 2008. Concentrations of total glutathione and cysteine in wheat flour as affected by sulfur deficiency and correlation to quality parameters. *J. Agric. Food Chem.* 56(16):6844–6850.

Reisenauer, H.M. 1963. The effect of sulfur on the absorption and utilization of molybdenum by peas. *Soil Sci. Soc. Am. J.* 27(5):553–555.

Rennenberg, H. 1982. Glutathione metabolism and possible biological roles in higher plants. *Phytochemistry* 21:2771–2781.

Ridley, A.O. 1972. Effect of nitrogen and sulfur fertilizers on yield and quality of rapeseed. *Seventeenth Annual Manitoba Soil Science Meeting*, Winnipeg, Manitoba, Canada, pp. 182–187.

Riley, N.G., F.J. Zhao, and S.P. McGrath. 2000. Availability of different forms of sulphur fertilisers to wheat and oilseed rape. *Plant Soil* 222(1–2):139–147.

Riley, N.G., F.J. Zhao, and S.P. McGrath. 2002. Leaching losses of sulphur from different forms of sulphur fertilizers: A field lysimeter study. *Soil Use Manage.* 18(2):120–126.

Saito, K. 2000. Regulation of sulfate transport and synthesis of sulfur-containing amino acids. *Curr. Opin. Plant Biol.* 3(3):188–195.

Saskatchewan Agriculture, Food and Rural Revitalization. 2005. Nutrient requirement guidelines for field crops in Saskatchewan. *Saskatchewan Agriculture, Food and Rural Revitalization*. Saskatchewan Institute of Pedology, Saskatoon, Saskatchewan, Canada.

Schat, H., M. Llugany, R. Vooijs, J. Hartley-Whitaker, and P.M. Bleeker. 2002. The role of phytochelatins in constitutive and adaptive heavy metal tolerances in hyperaccumulator and non-hyperaccumulator metallophytes. *J. Exp. Bot.* 53(379):2381–2392.

Scherer, H.W. 2009. Sulfur in soils. *J. Plant Nutr. Soil Sci.* 172(3):326–335.

Schoenau, J. and S. Malhi. 2008. Sulfur forms and cycling processes in soil and their relationship to sulfur fertility. In *Sulfur: A Missing Link between Soils, Crops, and Nutrition*, ed. J. Jez, pp. 1–10. Madison, WI: American Society of Agronomy, Crop Science Society of America, Soil Science Society of America.

Schoenau, J., P. Qian, and W.Z. Huang. 1993. Assessing sulphur availability in soil using ion exchange membranes. *Sulphur Agric.* 17:13–17.

Schoenau, J.J. and J.J. Germida. 1992. Sulphur cycling in upland agricultural systems. In *Sulphur Cycling on the Continents: Wetlands, Terrestrial Ecosystems and Associated Water Bodies, SCOPE 48*, eds. R.W. Howarth, J.W.B. Stewart, and M.V. Ivanov, pp. 261–277. Chichester, U.K.: John Wiley & Sons.

Schreiner, M., A. Krumbein, D. Knorr, and I. Smetanska. 2011. Enhanced glucosinolates in root exudates of *Brassica rapa* ssp. *rapa* mediated by salicylic acid and methyl jasmonate. *J. Agric. Food Chem.* 59(4):1400–1405.

Shakeel, M. and M. Inayatullah. 2013. Impact of insect pollinators on the yield of canola (*Brassica napus*) in Peshawar, Pakistan. *J. Agric. Urban Entomol.* 29:1–5.

Shewry, P.R., J. Franklin, S. Parmar, S.J. Smith, and B.J. Miflin. 1983. The effects of sulfur starvation on the amino-acid and protein compositions of barley-grain. *J. Cereal Sci.* 1:21–31.

Shewry, P.R., N.G. Halford, P.S. Belton, and A.S. Tatham. 2002. The structure and properties of gluten: An elastic protein from wheat grain. *Phil. Trans. R. Soc. B* 357(1418):133–142.

Shewry, P.R. and A.S. Tatham. 1997. Disulphide bonds in wheat gluten proteins. *J. Cereal Sci.* 25(3):207–227.

Shinmachi, F., P. Buchner, J.L. Stroud et al. 2010. Influence of sulfur deficiency on the expression of specific sulfate transporters and the distribution of sulfur, selenium, and molybdenum in wheat. *Plant Physiol.* 153(1):327–336.

Singh, M. and V. Kumar. 1979. Sulfur, phosphorus, and molybdenum interactions on the concentration and uptake of molybdenum in soybean plants (*Glycine max*). *Soil Sci.* 127(5):307–312.

Smetanska, I., A. Krumbein, M. Schreiner, and D. Knorr. 2007. Influence of salicylic acid and methyl jasmonate on glucosinolate levels in turnip. *J. Hort. Sci. Biotechnol.* 82(5):690–694.

Smith, F.W. 2001. Sulphur and phosphorus transport systems in plants. *Plant Soil* 232(1–2):109–118.

Smith, F.W., P.M. Ealing, M.J. Hawkesford, and D.T. Clarkson. 1995. Plant members of a family of sulfate transporters reveal functional subtypes. *Proc. Natl. Acad. Sci. USA* 92(20):9373–9377.

Smith, F.W., M.J. Hawkesford, P.M. Ealing et al. 1997. Regulation of expression of a cDNA from barley roots encoding a high affinity sulphate transporter. *Plant J.* 12(4):875–884.

Smith, F.W., A.L. Rae, and M.J. Hawkesford. 2000. Molecular mechanisms of phosphate and sulphate transport in plants. *Biochim. Biophys. Acta—Biomembr.* 1465(1–2):236–245.

Solberg, E.D. 1986. Oxidation of elemental S fertilizers in agricultural soils of northern Alberta and Saskatchewan. MSc thesis, University of Alberta, Edmonton, Alberta, Canada.

Solberg, E.D., D.H. Laverty, and M. Nyborg. 1987. Effects of rainfall, wet-dry, and freeze-thaw cycles on the oxidation of elemental sulphur fertilizers. *Proceedings of 24th Annual Alberta Soil Science Workshop*, Edmonton, Alberta, Canada, pp. 120–126.

Solberg, E.D., S.S. Malhi, M. Nyborg, and K.S. Gill. 2003. Fertilizer type, tillage, and application time effects on recovery of sulfate-S from elemental sulfur fertilizers in fallow field soils. *Commun. Soil Sci. Plant Anal.* 34(5–6):815–830.

Solberg, E.D., S.S. Malhi, M. Nyborg, and K.S. Gill. 2005. Temperature, soil moisture, and antecedent S application effects on recovery of elemental sulfur as SO_4-S in incubated soils. *Commun. Soil Sci. Plant Anal.* 36:1–12.

Solberg, E.D., S.S. Malhi, M. Nyborg, B. Henriquez, and K.S. Gill. 2007. Crop response to elemental S and sulfate-S sources on S-deficient soils in the Parkland Region of Alberta and Saskatchewan. *J. Plant Nutr.* 30(2):321–333.

Solberg, E.D. and M. Nyborg. 1983. Comparison of sulphate and elemental sulphur fertilizers. *Proceedings of the Sulphur-82 Conference*, The British Sulphur Corporation Limited, London, U.K., vol. 2, pp. 843–852.

Soliman, M.F., S.F. Kostandi, and M.L. van Beusichem. 1992. Influence of sulfur and nitrogen fertilizer on the uptake of iron, manganese, and zinc by corn plants grown in calcareous soil. *Commun. Soil Sci. Plant Anal.* 23(11–12):1289–1300.

Solomon, D., J. Lehmann, K.K. de Zarruk et al. 2011. Speciation and long- and short-term molecular-level dynamics of soil organic sulfur studied by x-ray absorption near-edge structure spectroscopy. *J. Environ. Qual.* 40(3):704–718.

Solomon, D., J. Lehmann, J. Kinyangi et al. 2009. Anthropogenic and climate influences on biogeochemical dynamics and molecular-level speciation of soil sulfur. *Ecol. Appl.* 19(4):989–1002.

Solomon, D., J. Lehmann, I. Lobe et al. 2005. Sulphur speciation and biogeochemical cycling in long-term arable cropping of subtropical soils: Evidence from wet-chemical reduction and S K-edge XANES spectroscopy. *Eur. J. Soil Sci.* 56(5):621–634.

Spencer, K., J. Freney, and M. Jones. 1984. A preliminary testing of plant analysis procedures for the assessment of the sulfur status of oilseed rape. *Aust. J. Agric. Res.* 35(2):163–175.

Steffan-dewenter, I. 2003. Seed set of male-sterile and male-fertile oilseed rape (*Brassica napus*) in relation to pollinator density. *Apidologie* 34(3):227–235.

Stolt, J.P., F.E.C. Sneller, T. Bryngelsson, T. Lundborg, and H. Schat. 2003. Phytochelatin and cadmium accumulation in wheat. *Environ. Exp. Bot.* 49(1):21–28.

Stout, P.R., W.R. Meagher, G.A. Pearson, and C.M. Johnson. 1951. Molybdenum nutrition of crop plants. *Plant Soil* 3(1):51–87.

Stroud, J.L., F.J. Zhao, P. Buchner et al. 2010. Impacts of sulphur nutrition on selenium and molybdenum concentrations in wheat grain. *J. Cereal Sci.* 52:111–113.

Swan, M., R.J. Soper, and G. Morden. 1986. The effect of elemental sulfur, gypsum and ammonium thiosulfate as sulfur sources on yield of rapeseed. *Commun. Soil Sci. Plant Anal.* 17:1983–11390.

Tabe, L.M. and M. Droux. 2001. Sulfur assimilation in developing lupin cotyledons could contribute significantly to the accumulation of organic sulfur reserves in the seed. *Plant Physiol.* 126(1):176–187.

Takahashi, H., S. Kopriva, M. Giordano, K. Saito, and R. Hell. 2011. Sulfur assimilation in photosynthetic organisms: Molecular functions and regulations of transporters and assimilatory enzymes. *Annu. Rev. Plant Biol.* 62:157–184.

Tamás, L., J. Dudíková, K. Ďurčeková, L. Halušková, J. Huttová, I. Mistrík, and M. Ollé. 2008. Alterations of the gene expression, lipid peroxidation, proline and thiol content along the barley root exposed to cadmium. *J. Plant Physiol.* 165(11):1193–1203.

Tan, Q., L. Zhang, J. Grant, P. Cooper, and M. Tegeder. 2010. Increased phloem transport of S-methylmethionine positively affects sulfur and nitrogen metabolism and seed development in pea plants. *Plant Physiol.* 154(4):1886–1896.

Tea, I., T. Genter, N. Naulet, V. Boyer, M. Lummerzheim, and D. Kleiber. 2004. Effect of foliar sulfur and nitrogen fertilization on wheat storage protein composition and dough mixing properties. *Cereal. Chem.* 81(6):759–766.

Tea, I., T. Genter, N. Naulet, M. Lummerzheim, and D. Kleiber. 2007. Interaction between nitrogen and sulfur by foliar application and its effects on flour bread-making quality. *J. Sci. Food Agric.* 87(15):2853–2859.

Thomas, P. 2003. *Canola Growers Manual.* Winnipeg, Manitoba, Canada: Canola Council of Canada.

Thomason, W.E., S.B. Phillips, T.H. Pridgen et al. 2007. Managing nitrogen and sulfur fertilization for improved bread wheat quality in humid environments. *Cereal Chem.* 84(5):450–462.

Toler, H.D., C.S. Charron, C.E. Sams, and W.R. Randle. 2007. Selenium increases sulfur uptake and regulates glucosinolate metabolism in rapid-cycling *Brassica oleracea*. *J. Am. Soc. Hortic. Sci.* 132(1):14–19.

Tomatsu, H., J. Takano, H. Takahashi et al. 2007. An *Arabidopsis thaliana* high-affinity molybdate transporter required for efficient uptake of molybdate from soil. *Proc. Natl. Acad. Sci. USA* 104(47):18807–18812.

Ukrainetz, H. 1982. Oxidation of elemental sulphur fertilizers and response of rapeseed to sulphur on Gray Wooded soils. *Proceedings of the 19th Annual Alberta Soil Science Workshop*, pp. 278–307.

von Sachs, J. 1865. *Handbuch der Experimental-Physiologie der Pflanzen.* Leipzig, Germany: Verlag von Wilhelm Engelmann.

Wagner, G.J. and R.M. Krotz. 1989. Perspectives on Cd and Zn accumulation, accommodation and tolerance in plant cells: The role of Cd-binding peptide versus other mechanisms. In *UCLA Symposia on Molecular and Cellular Biology*, Alan R. Liss, New York, vol. 98, pp. 325–336.

Wang, J., D. Solomon, J. Lehmann, X. Zhang, and W. Amelung. 2006. Soil organic sulfur forms and dynamics in the Great Plains of North America as influenced by long-term cultivation and climate. *Geoderma* 133(3–4):160–172.

Wen, G., J.J. Schoenau, S.P. Mooleki et al. 2003. Effectiveness of an elemental sulfur fertilizer in an oilseed-cereal-legume rotation on the Canadian prairies. *J. Plant Nutr. Soil Sci.* 166(1):54–60.

White, P.J., H.C. Bowen, B. Marshall, and M.R. Broadley. 2007. Extraordinarily high leaf selenium to sulfur ratios define "Se-accumulator" plants. *Ann. Bot.* 100(1):111–118.

White, P.J., H.C. Bowen, P. Parmaguru et al. 2004. Interactions between selenium and sulphur nutrition in *Arabidopsis thaliana*. *J. Exp. Bot.* 55(404):1927–1937.

Wieser, H., R. Gutser, and S. von Tucher. 2004. Influence of sulphur fertilisation on quantities and proportions of gluten protein types in wheat flour. *J. Cereal Sci.* 40(3):239–244.

Williams, J.S., S.A. Hall, M.J. Hawkesford, M.H. Beale, and R.M. Cooper. 2002. Elemental sulfur and thiol accumulation in tomato and defense against a fungal vascular pathogen. *Plant Physiol.* 128(1):150–159.

Wind, T. and R. Conrad. 1995. Sulfur compounds, potential turnover of sulfate and thiosulfate, and numbers of sulfate-reducing bacteria in planted and unplanted paddy soil. *FEMS Microbiol. Ecol.* 18(4):257–266.

Wooding, A.R., S. Kavale, F. MacRitchie, F.L. Stoddard, and A. Wallace. 2000. Effects of nitrogen and sulfur fertilizer on protein composition, mixing requirements, and dough strength of four wheat cultivars. *Cereal Chem.* 77(6):798–807.

Yadev, B.K. 2011. Interaction effect of phosphorus and sulfur on yield and quality of clusterbean in Typic Haplustept. *World J Agric. Sci.* 7(5):556–560.

Zhao, F.J., E.J. Evans, P.E. Bilsborrow, and J.K. Syers. 1993. Sulphur uptake and distribution in double and single low varieties of oilseed rape (*Brassica napus*, L.). *Plant Soil* 150(1):69–76.

Zhao, F.J., M.J. Hawkesford, and S.P. McGrath. 1999a. Sulphur assimilation and effects on yield and quality of wheat. *J. Cereal Sci.* 30(1):1–17.

Zhao, F.J., M.J. Hawkesford, A.G.S. Warrilow, S.P. McGrath, and D.T. Clarkson. 1996. Responses of two wheat varieties to sulphur addition and diagnosis of sulphur deficiency. *Plant Soil* 181(2):317–327.

Zhao, F.J., J. Lehmann, D. Solomon, M.A. Fox, and S.P. McGrath. 2006. Sulphur speciation and turnover in soils: Evidence from sulphur K-edge XANES spectroscopy and isotope dilution studies. *Soil Biol. Biochem.* 38(5):1000–1007.

Zhao, F.J., S.P. McGrath, A.R. Crosland, and S.E. Salmon. 1995. Changes in the sulphur status of British wheat grain in the last decade, and its geographical distribution. *J. Sci. Food Agric.* 68(4):507–514.

Zhao, F.J., S.E. Salmon, P.J.A. Withers et al. 1999b. Responses of breadmaking quality to sulphur in three wheat varieties. *J. Sci. Food Agric.* 79(13):1865–1874.

Zhao, F.J., S.E. Salmon, P.J.A. Withers et al. 1999c. Variation in the breadmaking quality and rheological properties of wheat in relation to sulphur nutrition under field conditions. *J. Cereal Sci.* 30(1):19–31.

Zhao, F.J., P.J.A. Withers, E.J. Evans et al. 1997. Sulphur nutrition: An important factor for the quality of wheat and rapeseed. *Soil Sci. Plant Nutr.*, Special issue 43:1137–1142.

Zörb, C., D. Steinfurth, S. Seling et al. 2009. Quantitative protein composition and baking quality of winter wheat as affected by late sulfur fertilization. *J. Agric. Food Chem.* 57(9):3877–3885.

Zuchi, S., S. Cesco, Z. Varanini, R. Pinton, and S. Astolfi. 2009. Sulphur deprivation limits Fe-deficiency responses in tomato plants. *Planta* 230(1):85–94.

Section III

Essential Elements: Micronutrients

8 Boron

Monika A. Wimmer, Sabine Goldberg, and Umesh C. Gupta

CONTENTS

8.1   HISTORICAL BACKGROUND

In the early twentieth century, boron (B) was discovered to be ubiquitously present in plant tissue (Agulhon, 1910) and to be beneficial for the growth of corn (*Zea mays* L.) (Mazé, 1919). It was, however, the work of Warington (1923) that secured strong experimental evidence of the essentiality of B for broad bean (*Vicia faba* L.), and later, research showed the essentiality for other species (Sommer and Lipman, 1926; Sommer, 1927). It rapidly became clear that the B requirement is highly variable among plant species, and plants were categorized into three main groups: (1) graminaceous species (especially of the order Poales) with the lowest B demand, (2) nongrass monocots and most dicots with an intermediate requirement, and (3) some latex-producing species with a very high B demand (Goldbach, 1997).

Fungi do not have an established B requirement, but diatoms (Lewin, 1996), cyanobacteria depending on heterocyst formation (Mateo et al., 1986; Bonilla et al., 1990), actinomycetes (Bolanos et al., 2002), yeast (Bennett et al., 1999), and some algae (Carrano et al., 2009) seem to need B. In several marine bacteria, B plays an unusual role as a structural component of the quorum sensing molecule autoinducer-2 (AI-2), which is involved in monitoring cell density and serves as a universal signal for communication between bacterial species (Chen et al., 2002). Interestingly, B is bound to AI-2 in marine, but not in terrestrial, bacteria (Chen et al., 2002). Whether or not B is also essential, or only beneficial, for animals and humans is still a matter of debate. Essentiality is established for the embryonic development of zebra fish, *Danio rerio* F. Hamilton (Rowe and Eckhert, 1999), mouse, *Mus* spp. L. (Lanoue et al., 2000), and African clawed frog, *Xenopus laevis* Daudin (Fort et al., 1998), and beneficial effects were observed in several nutritional studies with different animal species (Hunt, 2012).

8.2   UPTAKE OF BORON BY PLANTS

In the soil solution, boron is usually present in a pH-dependent equilibrium between boric acid and borate, with a pK_a of 9.24. The uncharged boric acid ($B(OH)_3$) molecule is the dominant B form in neutral or acidic soils, whereas the borate anion ($B(OH)_4^-$) can only be found in relevant amounts in alkaline soils. B is the only essential element taken up by plants predominantly as an uncharged molecule.

8.2.1 Boron Transport Mechanisms

For many years, passive diffusion was thought to be the only uptake mechanism for B, based on the high permeability of lipid membranes to boric acid (Raven, 1980). However, the last decade has revealed that B can be taken up by plants through three different pathways: (1) passive diffusion, (2) facilitated uptake through channels, and (3) active uptake through carriers, where the latter mechanism is especially relevant under limited B supply (for review, see Miwa and Fujiwara, 2010).

8.2.1.1 Passive Uptake

As an uncharged molecule, boric acid can diffuse easily through lipid bilayers following a concentration gradient (Dordas et al., 2000). The permeability of boric acid through cell membranes is, however, significantly lower than that observed in artificial lipid bilayers (Raven, 1980; Dordas and Brown, 2000). Passive uptake is likely the dominant uptake pathway under conditions of adequate or high B supply, but it is not sufficient to satisfy the plant B requirement when supply is limited (Brown et al., 2002).

8.2.1.2 Facilitated Absorption

Major intrinsic proteins (MIPs) have been suggested as candidates for facilitating B uptake (Dordas et al., 2000; Dordas and Brown, 2001). Takano et al. (2006) identified NIP5;1 as a boric acid channel that facilitated B influx into root cells of arabidopsis (*Arabidopsis thaliana* Heynh.). *NIP5;1* transcripts were increased 15-fold after 24 h of low B treatment. Under limited B supply, *A. thaliana* T-DNA insertion mutants exhibited reduced B uptake into roots, while expression of NIP5;1 enhanced B uptake into *Xenopus laevis* oocytes (Takano et al., 2006). Nine NIP genes are present in *A. thaliana*, and it was suggested that other members of the NIP family also are involved in facilitated B uptake, but so far only OsNIP3;1, a close homologue of AtNIP5;1, was shown to transport boric acid under B limitation in rice (*Oryza sativa* L.) (Hanaoka and Fujiwara, 2007). Another channel, NIP6;1, was proposed to function in xylem–phloem transfer of B and thus in B distribution within the shoot (Tanaka et al., 2008).

Other members of the MIP superfamily also have been suggested to be involved in B uptake (Dordas and Brown, 2000, 2001), even though their functions are not yet fully understood. Most prominent are plasma-membrane intrinsic proteins (PIPs), also known as aquaporins, which are responsible for the transport of water and also for some small uncharged molecules such as urea, arsenite, or silicic acid (Fitzpatrick and Reid, 2009). Indications for B transport through PIPs are an increased sensitivity to high B supply after expression of barley PIP1;3 and PIP1;4 in yeast cells (Fitzpatrick and Reid, 2009) and enhanced B permeability of *Xenopus laevis* oocytes after expression of maize PIP1 (but not Zm-PIP3) (Dordas and Brown, 2000; Dordas et al., 2000) and of *A. thaliana* PIP1b, PIP2a, and PIP2b (Nuttall, 2000).

8.2.1.3 Active Uptake

Under limited B supply, active mechanisms additionally improve the B supply for plants (Dannel et al., 2000; Stangoulis et al., 2001). The first active B transporter identified in plants was BOR1, a homologue of the yeast anion transporter YNL275w (Sc-BOR1) (Takano et al., 2002). Expression of *A. thaliana* BOR1 in yeast lacking the yeast *BOR1* gene reduced cellular B concentrations, indicating its function as a B efflux transporter (Takano et al., 2002). BOR1 is localized predominantly in the root pericycle cells, is involved in loading of B (in the form of borate) into the xylem, and allows transport of B against a concentration gradient (Takano et al., 2002). Consistently, the *A. thaliana bor1-1* mutant was more susceptible to B deficiency than the wild-type plants (Noguchi et al., 1997). Similar efflux-type B transporters also were identified in rice (OsBOR1: Nakagawa et al., 2007) and in mammals (NaBC1: Park et al., 2004).

8.2.1.4 Active Efflux

Maintenance of low cellular B concentrations is an important trait for B tolerance, and efflux-type B transporters were suggested to be candidates for removing excessive B from plant cells (Hayes and Reid, 2004). However, while BOR1 is necessary for B transport into the xylem in roots under limited B supply, it is degraded rapidly under high B supply (Takano et al., 2005). Unlike BOR1, the paralog BOR4 was not broken down under high B supply, and its overexpression greatly increased B tolerance in *A. thaliana* (Miwa et al., 2007).

In barley (*Hordeum vulgare* L.) and wheat (*Triticum aestivum* L.), positive correlations were observed between mRNA levels of HvBOR2 (also named Bot1) or TaBOR2 and the tolerance of different cultivars to high B levels (Reid, 2007). A B-tolerant barley cultivar contained multiple copies of Bot1 compared to a B-susceptible cultivar (Sutton et al., 2007). It was suggested that efflux-type B transporters may export B from root cells, and move B from the cells into the apoplast of leaf cells (Miwa and Fujiwara, 2010).

8.2.2 SOIL FACTORS AFFECTING BORON UPTAKE BY PLANTS

8.2.2.1 Soil pH

Soil pH is one of the most important factors affecting B availability for plants, since the equilibrium between boric acid and borate in the soil solution is pH dependent, and only boric acid can be taken up by passive diffusion. At soil pH values below 7, 99% of the B is present as boric acid, which is readily available, but can be leached easily. In alkaline soils, however, more B is present in the form of the borate anion, which can be adsorbed on organic matter, oxides, and clay minerals. Liming of acidic soils reduced plant B accumulation, for example, in corn (Jansen van Rensburg et al., 2010), tobacco (*Nicotiana tabacum* L.) (Tsadilas et al., 2005), spruce (Lehto and Malkonen, 1994), or pea (Dwivedi et al., 1992), and accelerated the occurrence of B deficiency symptoms (Gupta and MacLeod, 1977). At high B levels in the soil, liming delayed the occurrence of B toxicity symptoms in carnation (*Dianthus caryophyllus* L.) and reduced soluble soil B reserves (Eck and Campbell, 1962). The liming effect appeared to be caused by the increased pH rather than increased Ca concentrations in Norway spruce (*Picea abies* [L.] Karst.) (Lehto and Malkonen, 1994) and in rutabaga grown under low B conditions (Gupta and MacLeod, 1977).

8.2.2.2 Soil Organic Matter

Soil organic matter affects B uptake because it can bind B either by ligand exchange (Yermiyahu et al., 1988) or by formation of B-diol complexes with polyhydroxy compounds (Coddington and Taylor, 1989). Hot-water-soluble B is correlated significantly and positively with the organic matter content of the soil (Gupta, 1968), indicating that organically bound B is available for plants. Therefore, addition of organic matter, for example, compost, can increase B uptake especially in acidic or sandy soils under humid climate conditions (Berger and Truog, 1946). Application of compost can induce phytotoxic effects if too much B is supplied (Purves and MacKenzie, 1973). The influence of organic material on B uptake is amplified by increasing pH and clay content of the soil (Goldberg, 1997).

8.2.2.3 Soil Texture

Boron deficiency is more common in coarse-textured soils than in fine-textured soils (Gupta, 1968). This occurrence is due to the higher adsorption of B on some clay minerals (Goldberg and Glaubig, 1986) and a higher occurrence of leaching of B from sandy soils. In agricultural lands, a higher water-soluble soil B concentration is necessary in fine textured than in sandy soils to produce similar plant B concentrations because of the adsorption to the clays (Singh et al., 1976).

8.2.2.4 Soil Salinity

Boron toxicity often occurs in saline soils, either because plants are irrigated with water containing high levels of salts and B or because soils are derived from marine sediments naturally high in both components (Nable et al., 1997). In alkaline soils, high pH further prevents leaching of B due to the presence of sparingly soluble borates. Ample evidence indicates an interaction between salt and high B stress, but results often are inconsistent between different experiments, sometimes indicating an antagonistic (Holloway and Alston, 1992; Alpaslan and Gunes, 2001; Ismail, 2003; Diaz and Grattan, 2009; Masood et al., 2012) and sometimes a synergistic interaction (Ismail, 2003, maize; Alpaslan and Gunes, 2001, cucumber, *Cucumis sativus* L.). The nature of this interaction is still not fully clear, but the emerging picture is that it is different under marginal/low and adequate/high B supply. In the presence of adequate or high B levels, when B uptake is predominantly passive by diffusion or channel-mediated via aquaporins, soil salinity generally reduces B uptake rates and transpiration-driven B accumulation in aerial plant parts. When B supply is limited, however, and a significant portion of B can be taken up via active pathways, other factors seem to dominate the B–salinity interaction (Wimmer and Goldbach, 2012). Suggested mechanisms include the control of B uptake by salt-exclusion mechanisms (Alpaslan and Gunes, 2001), especially in the presence of chloride (Yermiyahu et al., 2008), or by downregulation of aquaporins (Bastias et al., 2004a; Martinez-Ballesta et al., 2008). Alternatively, salt-induced damage to membrane components (Masood et al., 2012) and promotion of B uptake and transport to aerial parts (Bastias et al., 2004b) also were suggested. Genotypic differences in salt sensitivity are additional modifying factors for B–salt interactions (Wimmer and Goldbach, 2012).

8.3 PHYSIOLOGICAL RESPONSES OF PLANTS TO BORON SUPPLY

Inadequate B supply represents one of the economically most important micronutrient disorders worldwide (Gupta, 1979; Shorrocks, 1997; Rerkasem and Jamjod, 2004). Under high rainfall conditions, in sandy, heavily weathered, or acidic soils, B can be leached easily and B deficiency severely limits yield (Shorrocks, 1997). On the other hand, under low rainfall conditions or in fields dominated by alkaline or organic soils, B often accumulates to levels that are toxic for plants. Irrigation with high B-containing water or without proper drainage also can increase topsoil B levels and reduce crop yields (Camacho-Cristobal et al., 2008b). The window between insufficient and toxic B levels can be quite small (Reid et al., 2004).

Plants respond to an inadequate B supply with a large number of anatomical, physiological, and biochemical alterations (reviews by Blevins and Lukaszewski, 1998; Brown et al., 2002; Bolanos et al., 2004a; Camacho-Cristobal et al., 2008b; see Sections 7.1 and 7.2), some of which may be secondary effects of growth inhibition (Goldbach, 1997). The so far only known B function in plants, namely, the cross-linking of the cell wall component rhamnogalacturonan II, is based on the unique chemistry of boric acid and borate, which can form reversible diester bonds with a range of *cis*-diol-containing molecules (Bassil et al., 2004). Whether all other physiological responses to B deficiency or toxicity are also based on this unique cross-linking ability is still under debate (Brown et al., 2002; Bolanos et al., 2004a), since the molecular basis for most of these effects is still not fully understood.

8.3.1 LOW BORON SUPPLY

8.3.1.1 Low Boron and Plant Growth

A constant supply of B is necessary for plant growth (O'Neill et al., 2001). Cessation of growth of apical meristems of both roots and shoots and the development of deformed and brittle leaves are the most obvious symptoms of B deficiency in plants (Goldbach, 1997). This malady is related clearly to the role of B in cross-linking two apiose moieties of the cell wall pectin rhamnogalacturonan II

(Ishii and Matsunaga, 1996; Kobayashi et al., 1996; O'Neill et al., 1996). Under low B supply, cell walls become unstable, abnormally thick, deformed, and less extensible (Loomis and Durst, 1992; Findeklee and Goldbach, 1996). The lack of B-RGII cross-links not only reduces cell wall stability but also results in altered cell wall pore size (Fleischer et al., 1999), which may disturb access of wall-modifying enzymes, cell wall deposition, and thus growth (Brown et al., 2002). On a molecular level, the recycling (internalization) of dimeric B-RGII pectins is inhibited under B-deficient conditions (Yu et al., 2002). Altered patterns of polymerization of the cytoskeleton in corn (Yu et al., 2003) may reflect an adaptive response to strengthen the weakened cell wall under B deficiency.

Growth processes also depend on the formation, transport, and excretion of cell material through vesicles. Functional membranes are important for these processes, and there is increasing evidence that B deficiency also hampers the functionality of cell membranes (see Brown et al., 2002; Camacho-Cristobal et al., 2008b), but molecular mechanisms are not yet understood. It is possible that B stabilizes membranes or affects membrane composition and function by cross-linking membrane-bound diol-containing molecules, such as glycoproteins or glycolipids (Bolanos et al., 2004a; Goldbach and Wimmer, 2007). Several putative B-binding membrane proteins, including arabinogalactan proteins (AGPs), have been identified (Wimmer et al., 2009), even though complexes of these proteins with B have not yet been isolated. In pea (*Pisum sativum* L.) plants, three putative B-binding membrane glycoproteins were absent under B deficiency (Redondo-Nieto et al., 2007), and several genes for AGPs were downregulated under B deficiency in *A. thaliana* roots (Camacho-Cristobal et al., 2008a).

8.3.1.2 Low Boron and Plant Development

B deficiency strongly inhibits reproductive yield and/or seed and fruit quality, even if no deficiency symptoms are expressed during the previous vegetative growth (Dell and Huang, 1997). Yield loss due to male sterility was observed in rice (Garg et al., 1979; Lordkaew et al., 2013), corn (Lordkaew et al., 2011), barley (*Hordeum vulgare* L.) (Ambak and Tadano, 1991), and wheat (*Triticum aestivum* L.), if B supply was inadequate during the sensitive phase of male gametogenesis (Rerkasem and Jamjod, 2004). The critical period for normal pollen development in wheat extends from after emergence of the flag leaf tip to shortly after the flag leaf is expanded fully (Rawson, 1996). Whether or not this development is related to an impairment of meiosis, or rather to an abnormal development of the tapetum as observed in oilseed rape (*Brassica napus* L.) (Zhang et al., 1994), is still not clear (Dell and Huang, 1997; Huang et al., 2000). Pollen sterility can be increased in plants grown under conditions of reduced transpiration, indicating that B transport to the reproductive tissues mainly occurs via xylem elements (Dell and Huang, 1997). However, some of the B supplied to the anther can also be transported via the phloem, as was demonstrated for broccoli (*Brassica oleracea* var. *italica* Plenck.) (Liu et al., 1993). The extent of phloem or xylem B transport to floral organs may vary among species and is not well known. Considerable uncertainty also exists with regard to B requirements of female fertility, but pistil development appears to be less sensitive to low B supply than anther development (Dell and Huang, 1997).

In addition to pollen development, pollen tube growth is known to have a high B requirement (Blevins and Lukaszewski, 1998). Under B deficiency, pollen tube growth was arrested in oilseed rape (Shen et al., 1994), and pollen tubes showed abnormal morphology or even burst explosively within minutes of B deficiency (reviewed in Loomis and Durst, 1992).

Other observations of B deficiency symptoms related to yield include a decrease in pod number in black gram (*Vigna mungo* Hepper) (Rerkasem et al., 1988), a reduced number of seeds per pod in soybean (*Glycine max* Merr.) (Rerkasem et al., 1993) and green gram (*Vigna radiata* R. Wilczek) (Bell et al., 1990), an enhanced seed size in wheat (Rerkasem and Jamjod, 2004), or reduced seed size in peanut (*Arachis hypogaea* L.) (Keerati-Kasikorn et al., 1991). Seeds can also abort or abscise (Dell and Huang, 1997). At a later stage, insufficient B supply induces abnormal growth and malformations of seeds or fruits (e.g., peanut, Harris and Brolmann, 1966; avocado, *Persea americana* Mill., Harkness, 1959). Reduced fruit quality occurs especially in rapidly growing or subterranean fruits, where B supply via xylem vessels cannot meet the B demand either because of very rapid tissue expansion or because of insufficient transpiration. Only species producing sugar alcohols, such as some in the Rosaceae, are able to provide larger amounts of B via phloem transport to fruits (Brown and Hu, 1996). What is often ignored is the fact that B deficiency structurally damages xylem and phloem vessels (reviewed in Dell and Huang, 1997; Wimmer and Eichert, 2013). This may limit translocation of photosynthates to flowers and/or fruits and may explain why flower initials are aborted and fruits shed in B-deficient plants (Brown et al., 2002).

8.3.1.3 Low Boron and Water Relations

Boron deficiency can affect plant water relations in several ways. Inhibited root growth limits the soil volume that can be exploited for available water and reduces the available surface for water and nutrient uptake. Water movement through the plant and water transport to aerial plant parts can be inhibited substantially because of structural damage to xylem and phloem vessels, a problem that is observed often under low B supply (Dell and Huang, 1997; Wimmer and Eichert, 2013). Additionally, severe B deficiency reduces transpiration rates in well-watered plants (e.g., Baker et al., 1956; Pinho et al., 2010) and likely is related to a reduced density of stomata, structural damage of guard cells, and thus stomatal limitation of photosynthesis (Rosolem and Leite, 2007; Will et al., 2011). Stomatal responsiveness to environmental conditions (e.g., drought) also can be impaired (Zhao and Oosterhuis, 2003), possibly explaining why B-deficient plants were more susceptible to drought stress than B-sufficient plants (Mottonen et al., 2001). Overall, B deficiency seems to aggravate abiotic stresses, which by themselves impair water relations of plants, due to a combination of damage to the structural integrity of the vascular system, damage to stomata, and inhibition of root growth (Wimmer and Eichert, 2013).

8.3.1.4 Low Boron and Oxidative Stress

It was proposed that oxidative damage is involved in B deficiency stress (Kobayashi et al., 2004). Undoubtedly, membrane leakiness and accumulation of products of lipid peroxidation are observed commonly under B deficiency (Dordas and Brown, 2005; Koshiba et al., 2009; Tewari et al., 2010). Once membranes are damaged, ions, sugars, and phenolic compounds can leak from the cells (Cara et al., 2002; Dordas and Brown, 2005), and phenolics can be oxidized enzymatically or nonenzymatically, resulting in the production of quinones and ultimately reactive oxygen species (ROS) (Shkolnik, 1984; Cakmak and Römheld, 1997). Under low B supply, the activity of polyphenoloxidase is increased (Goldbach, 1997; Camacho-Cristobal et al., 2002), and the composition of the phenolic pool can be changed (Camacho-Cristobal et al., 2002, 2004; Liakopoulos and Karabourniotis, 2005; Karioti et al., 2006). Accumulation of phenols is likely a secondary event occurring after cells have been damaged and is not related to a direct interaction between B and phenolic compounds (Ruiz et al., 1998; Brown et al., 2002; Cara et al., 2002). However, there is also evidence that oxidative damage can occur rapidly and may be the cause of cell death, as was shown for B-deficient cultured tobacco BY2 cells (Koshiba et al., 2009). The authors suggest that ROS production may be triggered by rapid cell wall alterations and function as a signal for downstream responses (Koshiba et al., 2009). Boron deficiency induced root growth inhibition of squash correlated with reductions in the ascorbate pool and supply of ascorbate restored root growth (Lukaszewski and Blevins, 1996). This result could indicate a direct interaction between B and the ascorbate/glutathione cycle, by a yet unknown mechanism (Brown et al., 2002). Early occurrence of ROS under B deficiency is also in line with an upregulation of 13 genes responsive to oxidative stress in low B-acclimated tobacco cells, of which two genes responded to B deficiency within 30 min (Kobayashi et al., 2004).

Even though it is not yet clear whether ROS are accumulated under B deficiency as a consequence of membrane damage and phenol leakage (suggested by Cakmak and Römheld, 1997) or by some yet unknown process involving the reduction of ascorbate (suggested by Lukaszewski and Blevins, 1996), it is certain that elevated levels of ROS can cause an impairment of the capacity of photosystem II (El-Shintinawy, 1999; Pinho et al., 2010) and an overall inhibition of photosynthesis

(Han et al., 2008; Tewari et al., 2010). A reduced efficiency of PSII under B deficiency was observed in sunflower (*Helianthus annuus* L.) leaves (Kastori et al., 1995). Cessation of growth also can cause an overreduction of the electron transport chain leading to a secondary oxidative stress under insufficient B supply (Dordas and Brown, 2005).

8.3.1.5 Low Boron and Nitrogen Fixation

The essentiality of B for N_2 fixation was first demonstrated in heterocysts of the cyanobacterium *Anabaena* (Mateo et al., 1986; Bonilla et al., 1990), where B seems to stabilize the inner glycolipid envelope and to maintain low O_2 diffusion (Garcia-González et al., 1988, 1991). Later, it was shown that nodule development and N_2 fixation in legumes were also highly dependent on B supply (Bolanos et al., 1994, 1996; Yamagishi and Yamamoto, 1994). All stages of the development of the rhizobia–legume interaction are affected under B deficiency, including recognition and rhizobial infection (Bolanos et al., 1996; Redondo-Nieto et al., 2001; Reguera et al., 2010a,b), development of the symbiosome (Bolanos et al., 2001; Redondo-Nieto et al., 2007), and nodule organogenesis (Reguera et al., 2009). It was suggested that the interaction between the plant matrix glycoproteins and the bacterial infection thread was impaired under B deficiency (Bolanos et al., 1996; Redondo-Nieto et al., 2008), that failure to deliver a hydroxyproline-rich protein (ENOD2) into the cell wall results in disturbance of the nodule O_2 barrier (Bonilla et al., 1997), and that some plant glycoproteins, which are crucial signals for bacteroid differentiation, are not targeted correctly into nodules (Bolanos et al., 2001; Bolanos, 2004b). Recently, transcriptomic analysis of barrel medic (*Medicago truncatula* Gaertn.) nodules showed that >70% of the genes analyzed were altered by B nutrition during nodule organogenesis (Redondo-Nieto et al., 2012), suggesting that B may have a function in control of gene expression or in signaling mechanisms, especially during cell differentiation (Reguera et al., 2009).

8.3.2 EXCESSIVE BORON SUPPLY

Boron can form complexes with a number of molecules containing several hydroxyl groups in *cis*-configuration, such as ribose or sugar alcohols (Woods, 1996; Hu et al., 1997). Most of these complexes are, however, rather weak under physiological pH conditions (Loomis and Durst, 1992). Since cellular concentrations of soluble boric acid or borate are rather low under adequate or low B supply (Brown et al., 2002), it is hard to perceive how B complexes with these molecules could interfere with metabolism under "adequate" B supply. Under excessive B supply, though, cellular B levels can increase strongly, because passive B uptake is concentration dependent and hardly can be controlled. Under these conditions, it is conceivable that a significant amount of the metabolic pool of complexing molecules could be bound and that cellular metabolic processes could be inhibited (Reid et al., 2004; Reid, 2010). Ribose, which has a very strong B complexing ability under physiological conditions, is a component of ATP, NADH, and NADPH, and of RNA, and is thus involved in all major metabolic events.

8.3.2.1 High Boron and Plant Growth

In concordance with the importance of B as a structural component of the cell wall, high B supply also inhibits plant growth. In solution-cultured barley, root and shoot growth was inhibited by B concentrations above 1 mM, and leaf necrosis occurred (with concentrations of B in the necrotic areas being above approximately 23 mM in the water content of the tissue) (Reid et al., 2004). Meristematic cells of barley root tips were more sensitive to B toxicity than mature tissues, implying that specific processes of growing cells, such as cell division or cell expansion, were affected first under high B levels (Reid et al., 2004). The molecular basis for growth inhibition is not yet clear, but cell division could be inhibited by the formation of complexes between B and ribose, either in DNA or in mRNA or tRNA involved in the synthesis of proteins that are essential for mitosis (Liu et al., 2000). It has also been argued that B could affect enzyme activities, either by direct binding to the enzyme or by complexing of the substrates such as NAD(P)$^+$ or NAD(P) H (Ralston and Hunt, 2001). Determination of K_m and V_{max} values for several enzymes indicates that this may be relevant for enzymes using NAD$^+$ and only to a lesser extent NADH or NADP$^+$ as substrate, especially under conditions of high energy demand for biosynthesis, such as in meristems (Reid et al., 2004). It is also possible that B interferes with transcription and translation, as suggested by Nozawa et al. (2006). Several ribosomal proteins and transcription factors have been identified in *A. thaliana*, which increase B tolerance when expressed in yeast (Nozawa et al., 2006). Additionally, inhibition of cell wall synthesis (Reid et al., 2004) or accumulation of suberin and lignin and a concomitant stiffening of the cell wall (Ghanati et al., 2005) could disrupt cell expansion. So far, there is no evidence indicating that high B concentrations cause osmotic stress or inhibit leaf expansion by reducing turgor at the leaf base (Karabal et al., 2003; Reid et al., 2004).

8.3.2.2 High Boron and Photosynthesis

Even though several papers clearly show that B toxicity has an effect on photosynthesis, this effect seems to occur at B concentrations much higher than those sufficient to reduce meristematic growth (Reid et al., 2004). It is, therefore, unlikely that inhibition of photosynthesis is the primary cause for growth disruption. More likely, inhibition of growth would result in a reduced demand for photosynthates, an oversupply of sugars in the source cells, and a concomitant increase in (photo-)oxidative stress, which has indeed been observed under B toxicity. This result is in line with the observation of negative effects on photosynthetic rate only after severe or long-term B toxicity (Sotiropoulos et al., 2002; Han et al., 2009; Sheng et al., 2010; Guidi et al., 2011; Landi et al., 2013). Inhibition of CO_2 assimilation was reduced due to both stomatal and nonstomatal limitations (orange, *Citrus* spp. L., Papadakis et al., 2004; sweet basil, *Ocimum basilicum* L., Landi et al., 2013), but to nonstomatal limitations only in kiwi (*Actinidia* spp. Lindl., Sotiropoulos et al., 2002). At a later stage of B toxicity, reduced photosynthesis can have a strong negative impact on yield, because it will further inhibit the growth of leaf tissue and limit the supply of sugars to developing fruits and grains.

8.3.2.3 High Boron and Oxidative Stress

Membrane permeability and peroxidation are observed commonly under B toxicity (Gunes et al., 2006; Molassiotis et al., 2006; Landi et al., 2013). Antioxidative enzymes respond in different ways to high B stress (Karabal et al., 2003; Gunes et al., 2006; Sotiropoulos et al., 2006). However, a general picture is not recognizable. At least in barley, the observed membrane damage was not caused by accumulation of H_2O_2, and B toxicity tolerance was not correlated with the activity of antioxidant enzymes (Karabal et al., 2003). On the other hand, H_2O_2 was increased substantially under severe B toxicity in tomato (Cervilla et al., 2007) and apple (*Malus domestica* Borkh.) (Molassiotis et al., 2006), and occurrence of oxidative stress was indicated by induction of antioxidant enzymes (Molassiotis et al., 2006; Cervilla et al., 2007; Han et al., 2009; Landi et al., 2013) and a reduction of glutathione (Ruiz et al., 2003). All reports indicate that oxidative stress occurs after prolonged high B stress, but it remains unclear whether this stress is a direct effect of toxic cellular B levels or a secondary event caused by growth inhibition. Reported discrepancies might be caused by differences in the level and duration of the high B treatment and thus depend on the severity of the stress perceived by different plant species.

8.4 GENETICS OF ACQUISITION OF BORON BY PLANTS

There is considerable variation with respect to B efficiency and B tolerance among species and genotypes within species (Xue et al., 1998; Jamjod and Rerkasem, 1999; Jamjod et al., 2004; Ahmed et al., 2007; Zeng et al., 2008; Mei et al., 2011; Punchana et al., 2012). Many studies indicate that both B efficiency under limited supply and B tolerance under excessive supply depend on the ability of the plants to increase or limit B absorption and translocation within the plant (see Zeng et al., 2008; Punchana et al., 2012). Incorporating traits for B efficiency into breeding programs is therefore one of the promising approaches to reduce yield losses caused by B imbalances (Reid et al., 2004; Zeng et al., 2008; Punchana et al., 2012).

Some genes relevant for B uptake and transport (*BOR1* and *NIP5;1*) have been identified. In *A. thaliana*, overexpression of *BOR1* increased seed yield (Miwa et al., 2006), whereas enhanced expression of *NIP5;1* improved root growth under deficient B supply (Kato et al., 2009). Expression of a NIP5-like gene (*CiNIP5*) was increased under low B supply and correlated with the tolerance of two citrus species to B deficiency (An et al., 2012). Quantitative trait loci (QTL) mapping has been used to determine genomic regions responsible for differences in B efficiency in rape, wheat, and *A. thaliana* (Xu et al., 2001; Jamjod et al., 2004; Zeng et al., 2008; Zhao et al., 2008, 2012). One major and three minor QTL controlling B efficiency were identified in oilseed rape (Xu et al., 2001), and five and three QTL were found for B efficiency coefficient and seed yield under low B conditions, respectively, in *A. thaliana* (Zeng et al., 2008). In wheat, B efficiency is thought to be controlled by two major genes, Bo_d1 and Bo_d2 (Jamjod et al., 2004), and a QTL for improved grain set under low B supply was identified on chromosome 4D, which was associated with the Bo_d2 gene.

More recently, the discovery of B efflux transporters in plants has stimulated research regarding B tolerance. Although it has long been known that tolerant varieties maintain lower B concentrations in their tissues (Nable, 1988; Paull et al., 1988), this tolerance can now be assigned at least in part to a higher ability to efflux B, as was shown, for example, for barley (Hayes and Reid, 2004). The B-tolerant barley landrace cv. Sahara contains 3.8 times more gene copies of *Bot1*, an ortholog of *BOR1*, than the intolerant cultivar Clipper, which is correlated with the maintenance of lower shoot B concentrations (Sutton et al., 2007). *Bot1* underlies the 4H QTL identified as one of four genetic loci associated with B tolerance of barley (Jefferies et al., 1999). In wheat and barley, B tolerance might be related to the expression of BOR2, another efflux-type boron transporter (Reid, 2007). In *A. thaliana*, accumulation of the efflux transporter BOR4 has been shown to correlate with B tolerance, and plants overexpressing BOR4 exhibited improved growth under high B supply (Miwa et al., 2007). In addition, limitation of B uptake by a reduced expression of the aquaporin HvNIP2;1, which underlies a QTL for B tolerance on chromosome 6H, also determines B toxicity tolerance in barley (Jefferies et al., 1999; Schnurbusch et al., 2010). The genes underlying two other QTL for B tolerance have not yet been identified (Schnurbusch et al., 2010).

Boron tolerance can now be transferred either by conventional breeding or by molecular techniques, for example, by overexpressing genes encoding efflux-type boron transporters (Reid, 2010). However, even though B tolerance has been increased substantially in several species under controlled conditions, field trials with these cultivars have so far given highly variable results and indicate only very moderate, if any, improvements in yield (Emebiri et al., 2009; McDonald et al., 2010). This result might be due to the cotransfer of other, less desirable genes or to the occurrence of additional constraints in the field, such as salinity or nutrient deficiencies, significantly limiting yield in addition to the B toxicity (Reid, 2010).

8.5   CONCENTRATIONS OF BORON IN PLANTS

Boron is not distributed evenly within plants, and B levels in plant parts depend on its phloem mobility or immobility. Although B was long considered a phloem-immobile element (Oertli, 1993), it is now known that B mobility and thus B levels in different plant parts vary greatly among species (Brown and Shelp, 1997).

8.5.1   BORON CONCENTRATIONS IN SPECIES WHERE BORON IS IMMOBILE

In species that use sucrose as a primary photosynthetic metabolite, B is transported to the leaves with the transpiration stream, but cannot be remobilized via the phloem, and is therefore considered phloem immobile (Oertli and Richardson, 1970; Brown and Shelp, 1997). However, some reports indicate a limited B mobility in species that use sucrose as primary photoassimilate such as canola (*Brassica napus* L.) or wheat, and recently, a bis-sucrose borate complex was identified in canola, and a *bis-N*-acetyl-serine borate complex in wheat phloem sap (Stangoulis et al., 2010). In most of these species, however, mobility is low, and B tends to accumulate at the sites of termination of leaf veins, that is, in leaf tips and margins (Oertli, 1994; Brown and Shelp, 1997). A steep gradient in B concentration often is encountered from leaf tip to leaf base (Kohl and Oertli, 1961; Brown and Shelp, 1997; Wimmer et al., 2003) and from leaf margins to the midrib section (Touchton and Boswell, 1975). Typically, B concentrations are lowest in fruits, seed, and stems and highest in leaves (Gupta, 1991; Brown and Shelp, 1997), where B levels are higher in old or mature than in young leaves (Clark, 1975).

8.5.2 BORON CONCENTRATIONS IN SPECIES WHERE BORON IS MOBILE

Boron is highly phloem mobile in species that use certain polyols as primary photosynthetic metabolites (Brown and Hu, 1996; Brown and Shelp, 1997). In many of these species, B is distributed more evenly throughout the plant, and no steep concentration gradients are observed within leaves especially when B supply is limited (Brown and Shelp, 1997; Marentes et al., 1997). Fruit tissues of these species can contain higher B levels than leaves (Brown and Shelp, 1997). Indicative of B mobility are (1) decreasing B concentrations of older leaves, as observed in broccoli (Benson et al., 1961; Shelp, 1988), and unchanging concentrations in fruits, as seen in peanut and subterranean clover (Campbell et al., 1975), upon the onset of B deficiency, and (2) export of foliar-applied B (see Table 8.1).

TABLE 8.1
Crop Plant Species Producing Sugar Alcohols and Species with an Observed Apparent Boron Mobility

Sugar Alcohol	Species	References	Observed Apparent B Mobility[b]
Mannitol	Onion, asparagus, cabbage, cauliflower, fennel, carrot, coffee, bean, pea, pomegranate	1	
	Celery[a]	1, 2, 3	
	Olive	1, 4	4, 9, 10, 11
Sorbitol	Almond	1	12, 13
	Apple, cherry, pear, plum	1, 5	12, 14, 15, 16, 17
	Apricot	5	18
	Loquat, quince, nectarine	5	
	Peach[a]	3, 5	13, 19
	Plum and hybrid species	1	13
	Rice	6	
Perseitol	Avocado	7	
Myo-inositol	Kiwifruit	8	

References: (1) Brown and Shelp (1997): based on Bourne (1958), Plouvier (1963), Bieleski (1982); (2) Davis et al. (1988); (3) Hu et al. (1997); (4) Perica et al. (2001b); (5) Wallaart (1980); (6) Bellaloui et al. (2003); (7) Minchin et al. (2012); (8) Sotomayor et al. (2012); (9) Delgado et al. (1994); (10) Liakopoulos et al. (2005); (11) Liakopoulos et al. (2009); (12) Brown and Hu (1996); (13) El-Motaium et al. (1994); (14) Woodbridge et al. (1971); (15) Crandall et al. (1981); (16) Hanson (1991); (17) Picchioni et al. (1995); (18) Dye et al. (1983); (19) Kamali and Childers (1970).
[a] B-polyol determined *in vivo*.
[b] Based on B concentration pattern and/or observed export of foliar applied B.

Polyols are produced by a range of important crop species (Table 8.1), typically in the families Rosaceae, Apiaceae, Rubiaceae, and Brassicaceae (Bourne, 1958; Plouvier, 1963; Wallaart, 1980; Bieleski, 1982; Brown and Shelp, 1997). However, B mobility does not occur in all members of these families (Brown et al., 1999). The most widespread polyols in higher plants are mannitol and sorbitol, but less common polyols such as dulcitol, ribitol, erythritol, and perseitol are present in single genera or species (Lewis and Smith, 1967).

8.6 INTERACTION OF BORON WITH UPTAKE OF OTHER ELEMENTS

Plant nutrient concentrations can change in response to low or high B supply (Camacho-Cristobal and González-Fontes, 1999; Lopez-Lefebre et al., 2002). The fact that nutrient decreases were observed in tobacco under long-term B deficiency (Camacho-Cristobal and González-Fontes, 1999) but not under short-term B deficiency (Camacho-Cristobal et al., 2005) may indicate that they were likely the results of B-induced growth inhibition, or other external factors, and not of a primary effect of B on the uptake of other nutrients. Relatively little is known about direct effects of B supply on the uptake of other elements, and even less is known about underlying mechanisms.

Severe B deficiency causes a decline in root and leaf nitrate concentrations and nitrate reductase activity (Kastori and Petrovic, 1989; Ramon et al., 1989; Shen et al., 1993). After only a few days of B deficiency, net nitrate uptake levels were reduced substantially, most likely by a reduction of nitrate transporter activity (Camacho-Cristobal and González-Fontes, 2007).

Interactions between B and Ca have been observed in plants, animals, and humans, but evidence indicates that these interactions are related to the formation of complexes in the cell wall rather than a direct effect of B on Ca uptake (for review, see Bolanos et al., 2004a). Although increasing B supply significantly reduced Ca concentrations in roots and suspension-cultured cells of rape (Wang et al., 2003), it increased Ca translocation into the shoots of grafted rose (*Rosa* spp. L.) plantlets (Ganmore-Neumann and Davidov, 1993) and of tobacco (Lopez-Lefebre et al., 2002). In micropropagated potato (*Solanum tuberosum* L.) plantlets, higher B levels reduced, whereas lower B levels increased, Ca concentrations in plant leaves (Abdulnour et al., 2000). On the other hand, deficient B supply inhibited Ca transport into tomato leaves (Yamauchi et al., 1986). In both rape and potato, the responses were opposite for two cultivars, indicating that genotypic differences may play an important role for B–Ca interactions (Abdulnour et al., 2000; Wang et al., 2003). Overall, the interaction between B and Ca is not well understood, and unequivocal evidence for a B effect on Ca uptake is missing.

8.7 DIAGNOSIS OF BORON STATUS IN PLANTS

Crop plants have been classified into three groups: Graminaceae with a very low B demand (optimum leaf concentrations 2–5 µg B g^{-1} DW), other monocot and all dicot species with an intermediate demand (optimum leaf concentrations 20–80 µg B g^{-1} DW), and latex-producing species with a very high B demand (optimum leaf concentrations >80 µg B g^{-1} DW) (Goldbach, 1997). Different requirements of graminaceous and dicot species can be largely assigned to differences in their cell wall composition, especially the content of pectins and galacturonans containing *cis*-diols with B-binding capacities (Carpita and Gibeaut, 1993; Hu and Brown, 1994; Hu et al., 1996). The molecular basis for the extraordinarily high demand of latex-producing plant species has not yet been unraveled.

Since B is an indispensable component of the plant cell wall, it needs to be available at low concentrations, but, permanently, for all growth processes in plants (O'Neill et al., 2001). Whereas short interruptions of B supply can be compensated by species with high phloem B mobility, species that have no means to remobilize B from older plant parts will rapidly suffer from B deficiency.

The occurrence of B deficiency and B toxicity symptoms is accordingly quite different between B-mobile and B-immobile species.

8.7.1 Deficiency Symptoms

The most common symptoms of B deficiency in shoots include the inhibition of apical growth (deformation, discoloration, and death of meristematic tissues such as terminal buds and young leaves), reduced leaf expansion and deformation of leaf blades, and breaking of tissues due to brittleness (Goldbach, 1997; Brown et al., 2002). Young leaves remain small and often dark green, before chlorosis develops (Dell and Huang, 1997) (Figure 8.1). There seems to be a difference between species that exhibit chlorotic and finally necrotic leaves, probably due to the accumulation of toxic metabolites such as phenolics, and those species where chlorosis and necrosis do not develop (Broadley et al., 2012). Production of lateral shoots and shorter internodes can lead to a bushy shoot appearance (Broadley et al., 2012). More rapidly, but usually unnoticed in the field, root growth is inhibited severely by low B supply (Dell and Huang, 1997). Since root elongation, but not initiation of lateral roots, is inhibited, deficient plants often exhibit a bushy root phenotype (Dell and Huang, 1997). During reproductive growth, typical deficiency symptoms include poor flower production, abortion of flower initials, infertility, reduced fruit and seed set, premature shedding of fruits, malformations of seeds and fruits, and breakdown of storage tissues (Figures 8.2 and 8.3) (Dell and Huang, 1997; Goldbach, 1997; Brown et al., 2002).

8.7.2 Toxicity Symptoms

The occurrence of toxicity symptoms depends on the mobility of B. In species where B is immobile, toxicity symptoms always occur as leaf tip and edge burn, mostly of older leaves (Oertli, 1993; Shelp et al., 1995). Before leaf necrosis is observed, shoot growth often is inhibited, especially of

FIGURE 8.1 Symptoms of boron deficiency in alfalfa (*Medicago sativa*) are expressed as red and yellow discoloration of young leaves, as noted by the light-colored leaves in the photograph. (From Gupta, U.C., Boron, in *Handbook of Plant Nutrition*, 1st edn., Barker, A.V. and Pilbeam, D.J., (eds.), CRC Press: Boca Raton, FL, 2007, pp. 242–268.)

FIGURE 8.2 Symptoms of boron deficiency in cauliflower (*Brassica oleracea* var. *botrytis*) showing as waterlogged patches, and rotting in the center of the head. (From Gupta, U.C., Boron, in *Handbook of Plant Nutrition*, 1st edn., Barker, A.V. and Pilbeam, D.J., (eds.), CRC Press: Boca Raton, FL, 2007, pp. 242–268.)

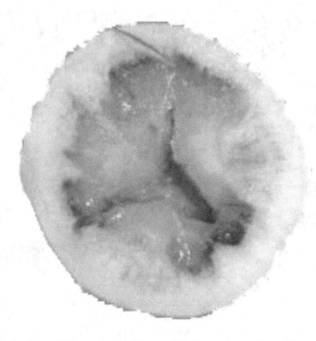

FIGURE 8.3 **(See color insert.)** Symptoms of boron deficiency (brown heart) in rutabaga (*Brassica napobrassica* Mill.) showing a soft, watery center of a cut root. (From Gupta, U.C., Boron, in *Handbook of Plant Nutrition*, 1st edn., Barker, A.V. and Pilbeam, D.J., (eds.), CRC Press: Boca Raton, FL, 2007, pp. 242–268.)

FIGURE 8.4 **(See color insert.)** Boron toxicity in peach (*Prunus persica* Stokes). Note meristematic dieback on side shoot (top left). (Photograph courtesy of Patrick H. Brown, UC Davis, CA.)

expanding tissues (Nable et al., 1997; Reid et al., 2004; Reid and Fitzpatrick, 2009). These "classical" B toxicity symptoms are absent in species in which B is mobile. Those species exhibit toxicity as dieback of young shoots (meristematic dieback), profuse gumming in the leaf axil, and the appearance of brown, corky lesions along the stems and petioles (Figure 8.4) (Brown and Shelp, 1997) (Table 8.2).

8.7.3 Deficiency, Sufficiency, and Toxicity Levels of Boron in Crop Plants

To determine the B status of a plant, total B levels in different plant parts usually are measured after tissue digestion, and a range of B levels associated with either expression of deficiency or toxicity symptoms or with yield suppressions is summarized for different plant species in Table 8.3. For some crops, deficiency and optimum levels seem to differ markedly, whereas for others, deficiency and toxicity ranges may overlap. This occurrence might be due to the well-known fact that cultivars and genotypes can differ markedly in their B efficiency as well as B tolerance. In addition, differences in sampling and determination methods, and external influencing factors such as climate or soil conditions, cannot be ruled out.

Sampling techniques and symptom descriptions have long been based on the assumption that B is not mobile in the plant (Oertli and Richardson, 1970). However, B mobility needs to be taken into account when selecting tissues for the determination of the B status of a plant. Generally, the B level of growing tissues should be tested when assessing the B status of crop plants (Brown and Hu, 1996; Brown and Shelp, 1997). In B-immobile species, old leaves will not represent the B status of growing tissues and are therefore inadequate for the diagnosis of B deficiency. Rather, growing tissues should be sampled. On the other hand, mature or old leaves are suitable plant parts for sampling in species where B is mobile. For the diagnosis of B toxicity, old leaves are a suitable indicator in species where B is immobile. In species with B mobility, young apical leaves or fruits may be more suitable for the assessment of B toxicity.

TABLE 8.2
Species Specific Symptoms of Boron Deficiency and Toxicity

Species	Main Symptoms	References
Deficiency symptoms		
Alfalfa (*Medicago sativa* L.)	Lower leaves remain green; tips and young leaves with red and yellow color development; buds appear as white or light-brown tissues; no flowers and internodes; and stems short.	1, 2
Almond (*Prunus amygdalus* Batsch)	Decreased seed set, premature flower and fruitlet drop.	3
Barley (*Hordeum vulgare* L.)	No ear formation; partial sterility causes swelling of ovaries and opening of flowers; increased appearance of ergot.	4, 5
Broccoli (*Brassica oleracea* var. *italica* L.)	Midrib cracking, stem corkiness, necrotic lesions, hollowing in the stem pith (*hollow stem disorder*), water-soaked areas inside heads, callus formation slower on the cut end of the stems after head harvest.	6, 7
Carrot (*Daucus carota* L.)	Longitudinal splitting of roots, roots small, rough, with yellow tops and white core in the center, browning of plant tops.	8, 9, 10
Cauliflower (*Brassica oleracea* var. *botrytis* L.)	Small heads with brown, waterlogged patches and rotting of the core; outer and inner portions of head with bitter flavor; *hollow stem*: cracking of stems, stiff stems with hollow cores, roots rough, and dwarfed.	11
Clover (*Trifolium* spp.)	Plants weak, with thick stems that are swollen close to the growing point; leaf margins look burnt or show red coloration that gradually spreads over the leaves; leaf tip of younger leaves red; may die.	2
Corn (*Zea mays* L.)	Youngest leaves with white spots scattered between veins; spots coalesce and form white, waxy, or raised stripes 2.5–5.0 cm long; disturbed development of ear when B supply was interrupted from 1 week prior to tasseling until maturity.	12, 13
Cotton (*Gossypium hirsutum* L.)	Death of the terminal bud, retarded internodal growth, enlarged nodes, bushy appearance, rosette condition, root growth inhibited, secondary roots stunted, bolls deformed and small.	14
Mango (*Mangifera indica* L.)	Misshapen, cracked, or bleeding fruit.	15
Olive (*Olea europaea* L.)	Abnormal fruit development (*monkey face*) and fruit drop, leaf-tip yellowing, shoot dieback, bark abnormalities, death of the terminal bud, resulting in a "bushy" appearance.	16
Pea (*Pisum sativum* L.)	New leaves small and shriveled, leaves with yellow or white veins, followed by changes in interveinal areas; growing points die; short internodes; blossoms shed.	17
Peanut (*Arachis hypogaea* L.)	*Hollow heart*, decreased seed size.	18, 19
Potato (*Solanum tuberosum* L.)	New leaves malformed and chlorotic, leaves thicken and margins roll upward, death of growing points, short internodes, rosetting of terminal buds and shoot.	20
Radish (*Raphanus sativus* L.)	*Brown heart*: dark spots on the thickest parts of the roots, roots with thick periderm, browning when cut.	8, 21
Rutabaga (*Brassica napobrassica* Mill.)	*Brown heart*: soft, watery area of cut roots, root surface rough, netted, fibrous, bitter, often elongated and with corky and somewhat leathery skin.	22, 23
Snap bean (*Phaseolus vulgaris* L.)	Yellowing of tops, slow flowering, and pod formation.	8
Soybean (*Glycine max* Merr.)	Necrosis of apical growing point and young growth, localized depression on the internal surface of one or both cotyledons of some seeds, lamina thick and brittle, reduced number of seeds per pod, floral buds wither before opening.	24

(Continued)

TABLE 8.2 (*Continued*)
Species Specific Symptoms of Boron Deficiency and Toxicity

Species	Main Symptoms	References
Sugar beet (*Beta vulgaris* L.)	Retarded growth; young leaves curl and turn black; old leaves show cupping; curling and surface cracking; beets are rough; scabby and off-color; *heart rot*.	8, 25
Sunflower (*Helianthus annuus* L.)	Basal fading and distortion of young leaves with soaked areas and tissue necrosis.	26
Tobacco (*Nicotiana tabacum* L.)	Interveinal chlorosis, dark and brittle new leaves, water-soaked areas in leaves, tissue breakdown at base of leaves, stalk toward the top distorted or twisted, delayed flowering, formation of seedless pods, death of terminal bud.	17
Tomato (*Lycopersicon esculentum* Mill.)	Growing point injured, flower injury, failure to set fruit, fruits ridged, with corky patches, imperfectly filled, unevenly ripening.	27
Wheat (*Triticum aestivum* L.)	Ear forms, but fails to flower; male sterility; development of inflorescences and grain setting restricted; shift in spikelets with seed fill to higher orders; increased seed size.	28, 29
Toxicity symptoms		
Alfalfa (*Medicago sativa* L.)	Burnt edges on older leaves.	2
Almond (*Prunus amygdalus* Batsch)	Stem dieback, gum formation along main stem, no leaf-tip burn.	30
Red clover (*Trifolium pratense* L.)	Burnt edges on older leaves.	2
Barley (*Hordeum vulgare* L.)	Older leaf tips with dark-brown blotches, browning, spotting, and burning; marginal chlorotic spots developing into brown necrotic lesions; eventually the whole leaf blade is affected; increased leaf senescence.	31, 32, 33
Celery (*Apium graveolens* (Mill.) Pers.)	Deformed young leaves, "bitter" and misshapen stems.	34
Corn (*Zea mays* L.)	Leaves with tip and marginal burn, especially older leaves, yellowing between the veins.	8
Cowpea (*Vigna sinensis* Savi.)	Marginal chlorosis and spotted necrosis.	35
Durum wheat (*Triticum durum* Desf.)	Retarded growth, aborted tillers, delayed heading, lower grain yield per tiller.	36
Faba bean (*Vicia faba* L.)	Reduced stem growth, yellowing of mature foliage, later marginal necrosis, young leaves wrinkled, thick, dark blue.	37
Grapevine (*Vitis vinifera* L.)	Smaller leaves, yellowing of leaf edges, interveinal chlorosis followed by necrosis, epical internodes.	38
Oat (*Avena sativa* L.)	Light-yellow-bleached leaf tips.	32
Onion (*Allium cepa* L.)	Burning of leaf tips, no development of bulb.	39
Orange (*Citrus sinensis* (L.) Osbeck)	Tip chlorosis of old leaves only, extending from tip to base.	40, 41
Pea (*Pisum sativum* L.)	Reduced plant height and number of nodes, edge burn of older leaves.	42
Potato (*Solanum tuberosum* L.)	Necrosis of leaf margins, arching midrib and downward cupping of leaves.	20
Prunus spp.	Shoot dieback, gummy exudates, and necrotic spots along the lower or middle part of the stem, whole shoot death, no leaf symptoms.	43

(*Continued*)

TABLE 8.2 (*Continued*)
Species Specific Symptoms of Boron Deficiency and Toxicity

Species	Main Symptoms	References
Rutabaga (*Brassica napobrassica* Mill.)	Marginal bleaching of cotyledons and first leaves; leaf margins are yellow and tend to curl and wrinkle; water-soaked appearance of tissues in the center of roots.	44
Snap bean (*Phaseolus vulgaris* L.)	Reduced growth, marginal chlorosis, and burning of older leaves.	8, 35
Strawberry (*Fragaria x ananassa* Duchesne)	Reduced plant growth, marginal curling and interveinal bronzing and necrosis, leaf margins distorted and cracked; death of apex.	30, 45
Wheat (*Triticum aestivum* L.)	Light browning of older leaf tips converging into light greenish-blue spots.	32

Note: Color plates showing species specific symptoms can be found in Sprague (1964), Bergmann (1983), Zorn et al. (2006).

References: (1) Berger (1962); (2) Gupta (1971a); (3) Nyomora et al. (1997); (4) Simojoki (1991); (5) Wongmo et al. (2004); (6) Shattuck and Shelp (1987); (7) Shelp et al. (1992); (8) Gupta (1983); (9) Gupta and Cutcliffe (1985); (10) Hole and Scaife (1993); (11) Shelp and Shattuck (1987a); (12) Lordkaew et al. (2011); (13) Mozafar (1987); (14) van de Venter and Currier (1977); (15) Ram et al. (1989); (16) Perica et al. (2001b); (17) Bergmann (1983); (18) Rerkasem et al. (1988); (19) Keerati-Kasikorn et al. (1991); (20) Roberts and Rhee (1990); (21) Shelp et al. (1987); (22) Shelp and Shattuck (1987b); (23) Gupta and Cutcliffe (1972); (24) Rerkasem et al. (1993); (25) Vlamis and Ulrich (1971); (26) Dube et al. (2000); (27) Francois (1984); (28) Rerkasem and Jamjod (2004); (29) Dell and Huang (1997); (30) Brown et al. (1999); (31) Riley (1987); (32) Gupta (1971b); (33) McDonald et al. (2010); (34) Francois (1988); (35) Francois (1989); (36) Yau and Saxena (1997); (37) Poulain and Almohammad (1995); (38) Yermiyahu et al. (2006); (39) Francois (1991); (40) Sheng et al. (2010); (41) Papadakis et al. (2004); (42) Bagheri et al. (1993); (43) El-Motaium et al. (1994); (44) Muller and McSweeney (1976); (45) Haydon (1981).

8.7.4 DETERMINATION OF BORON IN PLANT TISSUES

Boron in plant tissues usually is determined after digestion of the dried material. Dry ashing of tissue in a muffle furnace with subsequent heating to solubilize the ash in dilute HNO_3, followed by filtration (Wimmer et al., 2005), is common, but includes a risk of B loss during the heating step. Alternatively, wet digestion in HNO_3, using either pressure digestion or a microwave system, can be carried out (Evans and Krahenbuhl, 1994).

A range of photometric methods have been developed for the determination of B, the most common being the curcumin (De la Chevallerie-Haaf et al., 1986) and the azomethine-H (Lohse, 1982) methods. Both methods are prone to interferences from other nutrients (Zarcinas, 1995). To overcome this problem, B can be extracted from sample solutions using extraction with a mixture of 2-ethyl-1,3-hexanediol/chloroform (10% v/v) (Wikner and Uppström, 1980; Wimmer and Goldbach, 1999). Analytical methods with a higher sensitivity, or at least less vulnerability to interferences, include inductively coupled plasma atomic emission spectroscopy (ICP-AES) and inductively coupled plasma mass spectrometry (ICP-MS). Both methods are relatively fast, require small sample volumes, and can be used without interferences for a large number of samples types. Detection limits are in the order of 0.5 mg L^{-1} for ICP-AES, 0.001 mg L^{-1} for ICP-MS (Carrano et al., 2009), and 0.01 mg L^{-1} for the spectrometric curcumin method (Wimmer and Goldbach, 1999). ICP-MS has the additional advantage that the stable isotopes ^{10}B and ^{11}B can be distinguished and has been used successfully to determine B uptake rates and B distribution within plants (Brown and Hu, 1994; Marentes et al., 1997; Dannel et al., 2000; Nachiangmai et al., 2004; Wimmer et al., 2005). If B concentrations in the plant material are sufficiently high, ^{11}B nuclear magnetic resonance (NMR) can be used for B studies, with the benefit of providing additional structural information (Chuda et al., 1997).

TABLE 8.3
Deficiency, Sufficiency, and Toxicity Levels of Boron in Field and Horticultural Crops

Crop	(mg B kg⁻¹ in Dry Matter)			Plant Part Sampled	References
	Deficiency	Sufficiency	Toxicity		
Alfalfa (*Medicago sativa* L.)	<15	20–40	200	Whole tops at early bloom	1
		15–20[a]			1
	<20	31–80	>100	Top one-third of plant shortly before flowering	2
		30[a]		Upper stem cuttings in early flower stage	3
		17–18[a]		Whole tops in early bud	4
	8–12	39–52	>99	Whole tops at 10% bloom	5
	61			46 days old, leaves	6
	39			46 days old, whole tops	6
Barley (*Hordeum vulgare* L.)	1.9–3.5	10	>20	Boot-stage tissue	7
			50–70[a]	Boot-stage tissue	8
	7.1–8.6	21	>46	Straw	7
			>2–15	Grain	9
			50–420	Whole shoots at maturity	9
Beans (*Phaseolus* spp.):		12	>160	43-day-old plants	10
		36–94	144	Aerial portion of plants 1 month after planting	11
		28	43	Pods	12
			109	Recently matured leaves at prebloom	13
	<12	42	>125	Plant tops at prebloom	14
Blueberry (*Vaccinium corymbosum* L.)	<15			Leaves	15
Brassica oleracea:					
Broccoli (var. *italica* Plenck)		70		Leaves	16
	2–9	10–71		Leaf tissue when 5% heads formed	17, 18
Brussels sprouts (var. *gemmifera* Zenker)	6–10	13–101		Leaf tissue when sprouts begin to form	17, 18
			161[b]	Leaf tissue when sprouts begin to form	19
Cabbage (var. *capitata* L.)			132[b]	Mature leaf blade prior to head formation	13
Cauliflower (var. *botrytis* L.)	3	12–23		Whole tops before the appearance of curd	20
	23	36		Leaves	16
	4–9	11–97		Leaf tissue when 5% heads formed	17, 18
Carrots (*Daucus carota* L.)	<16	32–103	175–307	Mature leaf lamina	21
	<28	54		Whole plants at swelling of roots	22
Corn (*Zea mays* L.)		8–38	>98	Whole plants when 25 cm tall	14
		10[a]		Leaf at or opposite and below ear level at tassel stage	3
	<9	15–90	>100	Total aboveground plant material at vegetative stage until ear formation	2

(Continued)

TABLE 8.3 (*Continued*)

Deficiency, Sufficiency, and Toxicity Levels of Boron in Field and Horticultural Crops

| Crop | (mg B kg⁻¹ in Dry Matter) | | | Plant Part Sampled | References |
	Deficiency	Sufficiency	Toxicity		
Cotton (*Gossypium hirsutum* L.)	45			Leaves, 30 days old (squaring), mature leaf	23
	38			Leaves, 30 days old (squaring), youngest open leaf (YOL)	23
	55			Leaves, 60 days old (flowering), mature leaf	23
	43			Leaves, 60 days old (flowering), YOL	23
			>198	Leaves	24
	17–20			Uppermost fully expanded main-stem leaf blades, during squaring and fruiting	25
Cucumber (*Cucumis sativus* L.)	<20	40–120	>300	Mature leaves from center of stem 2 weeks after first picking	2
Eucalyptus globulus	12–16			Youngest fully expanded leaf, up to 3 years old	26
Faba bean (*Vicia faba* L.)		25–100		Whole plants	27
Mung bean (*Vigna radiata* L.)	<25	25		Whole shoot	28
Oats (*Avena sativa* L.)			>105	47-day-old plants	10
		15–50	44–400	Boot-stage tissue	29
	<1	8–30	>30[b]	Boot-stage tissue	2
	1.1–3.5		>35	Boot-stage tissue	7
	3.5–5.6	14–24	>50	Straw	7
Olive (*Olea europaea* L.)		19–150			30
Orange (*Citrus sinensis* cv. Navelina)		55	>444	Old leaves	31
Pasture grass (Gramineae family)		10–50	>800	Aboveground part at first bloom at first cut	2
Peanuts (*Arachis hypogaea* L.)		29		Shoot terminals	32
		54–65	>250	Young leaf tissue from 30-day-old plants	33
		18–20[a]			33
Peas (*Pisum sativum* L.)	10.5	23	110	Young leaves	34
	7.6	10.5	51	Seeds	34
Potatoes (*Solanum tuberosum* L.)		12	>180	32-day-old plants	10
	<15	21–50	>50[b]	Fully developed first leaf at 75 days after planting	2
	<15	37–48	82–220	Shoots	35
Radish (*Raphanussativus* L.)	<9	96–217		Whole plant when roots began to swell	14
Rape (*Brassica napus* L.)	20–25			Youngest open leaf before flowering	36
	18			75 days after transplanting, youngest open leaf	37

(*Continued*)

TABLE 8.3 (*Continued*)

Deficiency, Sufficiency, and Toxicity Levels of Boron in Field and Horticultural Crops

Crop	(mg B kg⁻¹ in Dry Matter)			Plant Part Sampled	References
	Deficiency	Sufficiency	Toxicity		
Red clover	12–20	21–45	>59	Whole tops at bud stage	5, 20
(*Trifolium*		20–60	>60[b]	Top one-third of plant at bloom	2
pratense L.)		15–18[a]		Whole tops at rapid growth	4
Rice (*Oryza*	<7.3			Flag leaves	38
sativa L.)	<3.6			Shoots	38
Rutabaga (*Brassica*	20–38[c]	38–140	>250	Leaf tissue at harvest	39
napobrassica	<12[d]				39
Mill.)	32–40[c]	40		Leaf tissue when roots begin to swell	40, 41
	<12[d]				40, 41
Ryegrass (*Lolium*		9–38	>39–42	Whole plants at rapid growth	42
perenne L.)					
Sorghum		17–18		Whole shoots	43
(*Sorghum bicolor*		25–31		Recently matured leaves	43
Moench)					
Soybean (*Glycine*	14–40		63	Mature trifoliate leaves at early bloom	44
max Merr.)					
		34–137		Shoot 20 days after sowing	45
Sugar beets (*Beta*	12–40	35–200		Blades of recently matured leaves	46
vulgaris L.)	<20	31–200	>800	Middle fully developed leaf without stem taken at the end of June or early July	2
Sunflower	12.5	27	89	Leaves	47
(*Helianthus*	19.7			Early vegetative, youngest open leaf blades	48
annuus L.)					
	25			75 days after transplanting	37
	46–63			Whole shoots, 4 weeks old	49
	<36			Whole shoots, 8 weeks old	49
Sweet pepper	69			Whole shoots, 3 weeks old	50
(*Capsicum*	49			Whole shoots, 6 weeks old	50
annuum L.)	20			Whole shoot	51
	24			Matured leaf	51
	15			Seed	51
Timothy (*Phleum*		3–93	>102	Whole plants at heading stage	52
pratense L.)		11–46	47	Whole plants at rapid growth	42
Tomatoes	<10	30–75	>200	Mature young leaves from top of the plant	2
(*Lycopersicon*			>125	63-day-old plants	10
esculentum Mill.)	<12	51–88	>172	Whole plants when 15 cm tall	14
Wheat (*Triticum*	2.1–5.0	8	>16	Boot-stage tissue	7
aestivum L.)	4.6–6.0	17	>34	Straw	7
			>400	Leaves	53
	1.2			Early vegetative growth, youngest open leaf	48
	4–6			Young whole shoots	54

(Continued)

TABLE 8.3 (*Continued*)
Deficiency, Sufficiency, and Toxicity Levels of Boron in Field and Horticultural Crops

Crop	(mg B kg⁻¹ in Dry Matter)			Plant Part Sampled	References
	Deficiency	Sufficiency	Toxicity		
	5–7			Flag leaves	54
	<0.3	2.1–10.1	>10[b]	Aboveground vegetative plant tissue when plants 40 cm high	2
White clover		13–16[a]		Whole tops at rapid growth	4
(*Trifolium*			53	Whole plants at 6 weeks	55
repens L.)					

References: (1) Meyer and Martin (1976); (2) Neubert et al. (1970); (3) Melsted et al. (1969); (4) Sherrell (1983a); (5) Gupta (1972); (6) dos Santos et al. (2004); (7) Gupta (1971b); (8) Riley (1987); (9) Riley and Robson (1994); (10) MacKay et al. (1962); (11) Robertson et al. (1975); (12) Purves and MacKenzie (1973); (13) Gupta and Cutcliffe (1984); (14) Gupta (1983); (15) Wojcik (2005); (16) Wallace (1951); (17) Gupta and Cutcliffe (1973); (18) Gupta and Cutcliffe (1975); (19) Gupta et al. (1987); (20) Gupta (1971a); (21) Kelly et al. (1952); (22) Gupta and Cutcliffe (1985); (23) Ahmed et al. (2013); (24) Ahmed et al. (2008); (25) Zhao and Oosterhuis (2002); (26) Sakya et al. (2002); (27) Poulain and Almohammad (1995); (28) Naeem et al. (2013); (29) Jones and Scarseth (1944); (30) Perica et al. (2001a); (31) Papadakis et al. (2004); (32) Rashid et al. (1997a); (33) Morrill et al. (1977); (34) Sinha et al. (2000); (35) Roberts and Rhee (1990); (36) Wei et al. (1998); (37) Asad et al. (2002); (38) Yu and Bell (1998); (39) Gupta and Munro (1969); (40) Gupta and Cutcliffe (1971); (41) Gupta and Cutcliffe (1972); (42) Sherrell (1983b); (43) Rashid et al. (1997b); (44) Woodruff (1979); (45) Rahman et al. (1999); (46) Hills and Ulrich (1976); (47) Dube et al. (2000); (48) Asad (2002); (49) Rashid and Rafique (2005); (50) Nabi et al. (2006); (51) Rafique et al. (2012); (52) Gupta and MacLeod (1973); (53) Grieve and Poss (2000); (54) Rashid et al. (2011); (55) Prasad and Byrne (1975).

[a] Considered critical.
[b] Considered high.
[c] Moderately deficient.
[d] Severely deficient.

8.8   FORMS AND CONCENTRATIONS OF BORON IN SOILS AND AVAILABILITY TO PLANTS

8.8.1   SOLID FORMS

8.8.1.1   Total Boron

The total amount of B in soils varies from 1 mg kg⁻¹ for leached spodosols to 250 mg kg⁻¹ for peat soils (Aubert and Pinta, 1997). The total amount of soil B ranges widely depending on the soil parent material, mineralogy, and weathering. Whetstone et al. (1942) found total B content to range from 4 to 98 mg kg⁻¹ for 300 soils from the United States. Soils derived from igneous rocks and unconsolidated sediments are low in total B, whereas soils from alluvium, limestone, shale, and glacial drift are high in total B (Whetstone et al., 1942). Soils derived from marine sediments are particularly rich in total B (Bingham, 1973). Total B content in most minerals is not well correlated with B availability, and therefore, total soil B content is not a good indicator of availability of soil B (Bradford, 1966).

8.8.1.2   Boron Minerals

Common B-containing minerals in soils are tourmaline and the various hydrated calcium, magnesium, and sodium boron minerals. These mineral groups seldom control the concentration of B in soil solution (Goldberg, 1997).

8.8.1.2.1 Hydrated Boron Minerals

Hydrated B minerals are formed as evaporites by chemical precipitation following concentration of brine waters in arid regions or evaporation of seawater (Watanabe, 1967). Examples of these minerals are borax, colemanite, ulexite, inyoite, and inderite (Christ et al., 1967). Upon evaporation, saline lakes that have a high pH and are low in Ca and high in Na content produce hydrated Na borates, whereas lakes having a higher Ca content produce hydrated Ca borates. Such hydrated minerals are all highly soluble and are therefore generally not present in agricultural soils.

8.8.1.2.2 Tourmaline

Tourmaline is the most common B-containing mineral in soils (Whetstone et al., 1942). The tourmaline group consists of various minerals that are complex borosilicates having rhombohedral symmetry. The tourmaline structure contains linked sheets of island units where the B atoms form strong covalent B–O bonds within BO_3 triangles (Tsang and Ghose, 1973). Tourmalines are highly resistant to weathering processes and virtually insoluble. For this reason, they generally do not control the soluble B concentrations in soil solutions.

8.8.1.3 Boron in Coprecipitates

Boron can be found coprecipitated into various minerals. It can be incorporated into the crystal structure of Mg hydroxide (Rhoades et al., 1970) as well as be coprecipitated with Ca carbonate (Kitano et al., 1978). Boron can incorporate into the structure of phyllosilicate clay minerals (Stubican and Roy, 1962) and is considered to be incorporated into the clay lattice substituting for tetrahedral silicon and aluminum (Harder, 1961). This fourfold coordination of B was observed spectroscopically for phlogopite, saponite, illite, and montmorillonite (Stubican and Roy, 1962; Jasmund and Lindner, 1973).

8.8.1.4 Soil Boron Fractions

Soil B has been partitioned into various forms using chemical fractionation techniques (Jin et al., 1987; Hou et al., 1994, 1996; Datta et al., 2002; Raza et al., 2002). Also called selective extractions, these sequential procedures assume that specific reagents selectively can attack and dissolve discrete soil components thereby releasing associated B. These methods determine the distribution of total soil B into various B pools consisting of readily soluble, specifically adsorbed, oxide-bound, organically bound, and residual form. Table 8.4 shows chemical extractants used to sequentially determine the various B fractions in the four fractionation methods.

Readily soluble B includes nonspecifically adsorbed B and has been determined as water soluble, Ca chloride extractable, ammonium acetate extractable, or anion membrane exchangeable. Specifically adsorbed B was measured as phosphate extractable or mannitol exchangeable. Residual B was determined by Na carbonate fusion, extracted with aqua regia/hydrofluoric acid or a mixture of sulfuric/perchloric/hydrofluoric acids (Jin et al., 1987; Hou et al., 1994, 1996; Datta et al., 2002; Raza et al., 2002). For most of the soils studied, the vast majority of the soil B was in the residual fraction. Boron concentration in corn tissue was correlated positively with readily available and exchangeable B, indicating that B in these fractions was available for plant uptake (Jin et al., 1987).

8.8.2 SOIL SOLUTION BORON

8.8.2.1 Soluble Boron Species

The element B exhibits properties that place it on the borderline between metals and nonmetals. However, because it is a semiconductor, it must be classified chemically as a nonmetal (Cotton and Wilkinson, 1980). Boron has a high ionization potential and therefore does not lose electrons to form B^{3+} cation species. Instead, covalent bond formation is of major importance, causing B chemistry to resemble that of the group IV element silicon more closely than that of the group III elements

TABLE 8.4

Sequential Extraction Procedures for Boron in Soils

Fraction	Jin et al. (1987)	Hou et al. (1994, 1996)	Datta et al. (2002)	Raza et al. (2002)
Readily soluble		0.01 M $CaCl_2$	0.01 M $CaCl_2$	
Water soluble	Deionized water			Hot water or 0.01 M $CaCl_2$
Nonspecifically adsorbed	0.02 M $CaCl_2$			1 M NH_4 acetate or anion exchange membrane
Specifically adsorbed	0.01 M mannitol/ 0.02 M $CaCl_2$	0.05 M KH_2PO_4	0.05 M KH_2PO_4	0.05 M KH_2PO_4
Oxide bound		0.2 M NH_4 oxalate pH 3	0.2 M NH_4 oxalate pH 3.25	0.2 M NH_4 oxalate pH 3
Mn oxyhydroxides	0.1 M $NH_2OH \cdot HCl$/ 0.01 M HNO_3			
Noncrystalline Al and Fe oxyhydroxides	0.175 M NH_4 oxalate pH 3.25 (dark)			
Crystalline Al and Fe oxyhydroxides	0.175 M NH_4 oxalate pH 3.25 (UV light)			
Organically bound		0.02 M HNO_3/ 30% H_2O_2	0.5 M NaOH	0.02 M HNO_3/ 30% H_2O_2
Residual	$NaCO_3$ fusion	Aqua regia/HF	H_2SO_4/$HClO_4$/HF	Aqua regia/HF

(Cotton and Wilkinson, 1980). Since their octet is incomplete, trigonal B compounds act as Lewis acids (electron acceptors) toward Lewis bases to form tetrahedral structures.

Boric acid is the predominant form of B in soil solution. It is a very weak, monobasic, trigonal Lewis acid that accepts a hydroxyl ion to form the tetrahedral borate anion. The hydrolysis constant for this reaction is 5.75×10^{-10} or $pK_a = 9.24$ (Bassett, 1980). This property means that boric acid exists predominantly in undissociated form over most agricultural solution pH values. At B concentrations ≥ 0.025 M, polymeric B species begin to form (Cotton and Wilkinson, 1980). However, such concentrations seldom are reached under agricultural conditions. Although alkali and alkaline earth B complexes may be present in significant concentrations in seawaters and brines, concentrations of these complexes in saline soil solutions even at pH values >9 constitute less than 5% of total solution B (Keren and Bingham, 1985).

8.8.2.2 Available Boron

The term "available boron" is generally used to describe the amount of B that can be taken up by plants. This amount consists of B in soil solution plus B on various soil surfaces that is in sufficiently labile form that it can be released to renew solution phase B concentrations during the course of the growing season. Only water-soluble B is immediately available to plants (Saarela, 1985). Plants respond exclusively to the B activity in the soil solution and not to B present on or in the solid phase (Ryan et al., 1977; Keren et al., 1985a,b). Measurement of available B is usually done using various extraction techniques and forms the basis for B soil tests. Available B measurements for agricultural soils range from 0.4 to 5 mg kg^{-1} using various extraction methods (Gupta, 1968).

8.9 MEASUREMENT OF BORON IN SOILS

Many diverse analytical methods are available for B determination. Boron determination methods for plant and soil analyses have been primarily colorimetric and spectrometric (Keren, 1996; Sah and Brown, 1997). The most common colorimetric methods used for soil solutions are carmine

(Hatcher and Wilcox, 1950), curcumin (Dible et al., 1954), and azomethine-H (Wolf, 1971). In the carmine method, B is complexed with carmine, an anthraquinone dye, in concentrated sulfuric acid, and its absorbance is measured at a wavelength of 585 nm. The primary drawback of this method is the hazard of using concentrated sulfuric acid. In the curcumin method, B forms a reddish-brown complex with curcumin, a phenol dye. After evaporation to dryness, the B–curcumin complex is determined in ethanol at a wavelength of 540 nm. This method is time-consuming, and some B may be lost during the evaporation step. Boron concentrations for standard curves are suggested to range from 2 to 10 mg L^{-1} for both the carmine and the curcumin methods (Keren, 1996).

The azomethine-H method has become the most commonly used colorimetric method for B determination because it is fast, simple, and sensitive and does not require concentrated acids. A yellow B/azomethine-H complex is formed and determined at a wavelength of 420 nm. The method is automated readily for use with flow injection analyzers. A major drawback for soil extracts is the interference resulting from the yellow color of dissolved organic matter. Keren (1996) recommends a standard calibration curve of 0.4–2 mg B L^{-1} for the azomethine-H method.

The development of plasma source spectrometric techniques has revolutionized B analysis. One reason is because colored extracts do not cause interference in ICP analyses as they sometimes do for colorimetric techniques. The detection limits and precision for B analyses using inductively coupled plasma optical emission spectrometry (ICP-OES) and ICP-MS are better than for all colorimetric methods (Sah and Brown, 1997). The detection limit for ICP-OES is 10 μg L^{-1}, and that for ICP-MS is one order of magnitude lower at 1 μg L^{-1}. In contrast to the colorimetric methods, colored solutions do not result in interferences because the samples are ionized. High iron concentrations, as may be present in some acid soil extracts, may cause interferences in B determination with ICP-OES (Sah and Brown, 1997). Precision can be improved by use of a calibration procedure with an internal standard or standard addition and isotope dilution methods. Beryllium is considered the ideal internal standard because its mass is close to that of B (Sah and Brown, 1997).

8.9.1 Soil Sampling Protocols

When sampling agricultural fields, it is important to obtain a representative soil sample. A composite sample of about 500 g usually is taken to represent a uniform field. Such a homogenized sample consists of subsamples from 15 to 20 locations across the field. At least 15 subsamples per hectare are necessary to obtain a representative composite sample. Soil corings are usually taken to a plow depth of 15–20 cm (Tisdale and Nelson, 1975). Individual soil testing laboratories have their own specific protocols.

8.9.2 Extractable Boron

Extraction methods are used to measure the amount of B available for plant uptake. Many methods have been developed and are used to diagnose conditions of B deficiency. However, the tests can also be applied to evaluate the potential for inducing B toxicity symptoms, especially in sensitive crops.

8.9.2.1 Hot Water Soluble

Historically, the most common method for estimating plant-available soil B has been the hot-water-soluble procedure developed by Berger and Truog (1940). Over the years, this method has been modified and simplified by various researchers. In the modification of Gupta (1967), B was extracted by boiling soil slurries on a hot plate. Reproducibility and sample handling efficiency of the method were improved by use of a block digester (John, 1973).

8.9.2.2 Calcium Chloride–Mannitol

A Ca chloride–mannitol extract was recommended as the best extractant for Australian soils ranging in B concentration from potentially deficient to potentially toxic (Cartwright et al., 1983).

These authors considered 0.01 M $CaCl_2$ and 0.05 M mannitol to be the optimal concentrations. The amount of B determined using Ca chloride–mannitol was significantly correlated to hot-water-soluble B. A Ca chloride–mannitol extract was used to represent specifically adsorbed B in the sequential extraction procedure of Jin et al. (1987). Boron content of barley leaves was highly correlated with Ca chloride–mannitol extractable soil B (Tsadilas et al., 1994).

8.9.2.3 DTPA–Sorbitol

A diethylene triamine pentaacetic acid (DTPA)–sorbitol extract is recommended currently by the North American Proficiency Testing Program to estimate soil availability of B simultaneously with the micronutrients Zn, Cu, Mn, and Fe (Miller et al., 2000). Despite being correlated significantly with hot-water-soluble B, DTPA–sorbitol extractable B was correlated with alfalfa yield only on alkaline and not on acid soils (Schiffler et al., 2005a,b). Highly significant correlations occurred between DTPA–sorbitol extractable B and B content of melon fruits (Goldberg et al., 2003).

8.9.2.4 Ammonium Bicarbonate–DTPA

Ammonium bicarbonate–DTPA allows for the extraction of the major nutrient elements phosphorus and potassium as well as the micronutrients Zn, Fe, Cu, and Mn (Soltanpour and Schwab, 1977). This soil test was extended to include B (Gestring and Soltanpour, 1984). The ammonium bicarbonate–DTPA extract was correlated highly with B content of wheat but was inferior to the hot-water-soluble B soil test. Its advantages are simplicity and multielement extraction. Gestring and Soltanpour (1984) recommended that the extractant be used only if the soil variables clay content, organic matter content, and pH are measured and included in the regression equation for plant B content. Ammonium bicarbonate–DTPA extractable B was correlated highly with yield, B content, and B uptake of alfalfa (360). However, Gestring and Soltanpour (1987) recommended that ammonium acetate extractable Ca be included in the regression equation. Matsi et al. (2000) found that ammonium bicarbonate–DTPA extractable B was correlated highly with hot-water-soluble B and that both soil tests extracted similar amounts of B. Cation-exchange capacity was included in the regression equation for B concentration.

8.9.2.5 Saturation Extracts

Saturation extracts are used to determine water-soluble B concentrations that might be conducive to B toxicity effects in plants. In this procedure, the soil is saturated with water, and, after 24 h of equilibration, the solution is vacuum extracted. Saturation extract B concentrations have been considered to be comparable to soil solution B concentration (Bingham, 1973). However, this expression is a simplification because amounts of adsorbed B were neglected. Since saturation extract B concentrations are much lower than for other soil extractants (Matsi et al., 2000), this soil test is used primarily to diagnose toxicity conditions. Unlike hot-water-soluble B and Ca chloride–mannitol extractable B, saturation extract B was not correlated significantly with leaf B content of kiwifruits (Tsadilas et al., 1997).

8.9.3 EVALUATION OF BORON SOIL TESTS

Three requirements for a successful soil test outlined by Bray (1948) are applicable to B: (1) the extracting solution should extract either the total amount or a proportional part of the available form(s) of B from soils having variable chemical and physical properties; (2) the amount of B in the extracting solution should be measured with reasonable speed and accuracy; and (3) the amount of B extracted should be correlated with crop growth and response under various conditions. Berger and Truog (1940) evaluated the hot-water-soluble B test in this manner with regard to B deficiency in beets. As new soil tests for B have been developed, they have generally been compared to the hot-water-soluble B test. In the evaluation of a new B soil test, it is important to establish that the new test meets the following three criteria: (1) it has practical advantages of analysis; (2) it is at least as good at predicting plant response as the old test; and (3) there is a mathematical relationship between the soil B values of the old test and those of the new test. In the evaluation of new soil tests, criteria (1) and (3) are usually satisfied, but criterion (2) may be neglected because it is much more time-consuming to carry out. This case occurs for the DTPA–sorbitol method, which was recommended by the North American Proficiency Testing Program based on its high correlation with hot-water-soluble B before it had been correlated with plant B uptake (Miller et al., 2000).

8.10   BORON FERTILIZERS

Boron fertilizer is one of the most common micronutrient fertilizers applied in agricultural production. Table 8.5 provides a list of common B fertilizers in the order of decreasing solubility. All of these B fertilizers are inorganic materials. Solubor and boric acid are the most soluble, the solid Na borates are soluble, ulexite and colemanite are slightly soluble, and the B frits are only very slightly soluble (Mortvedt and Woodruff, 1993). In general, it is recommended that B fertilizers be applied to the soil prior to sowing and seedling emergence (Follett et al., 1981) and that the foliar applications of B be made during early plant growth because of the poor phloem mobility of B in most species (Shorrocks, 1997).

8.10.1   APPLICATION AS SOLIDS

The solid Na borates differ in particle size and number of waters of hydration (Follett et al., 1981). Solubility is a function of hydration state with borax, the decahydrate, being more soluble than the pentahydrate, which, in turn, is more soluble than the anhydrous material. These solubility differences are not considered to be agronomically significant (Shorrocks, 1997). Borax historically has been the most popular B-containing fertilizer. Soil application is the most common method of supplying B. Boron can be mixed with bulk N–P–K–S fertilizer blends or added to granular formulations. Broadcast applications are preferred over banding to ensure uniform application and to minimize the possibility of toxic effects at high application rate (Follett et al., 1981).

The Na–Ca borate mineral, ulexite, is less soluble than the Na borates, and the Ca borate mineral, colemanite, is even less soluble. For this reason, these products are used primarily in sandy soils in regions of high rainfall where B leaching from the soil is a problem (Mortvedt and Woodruff, 1993).

Boron frits are manufactured by mixing powdered Na borates with silicates and melting and fusing in a furnace. Subsequently, they are quenched, dried, and milled. Because of their very low

TABLE 8.5
Common Boron Fertilizers

Boron Compound	Chemical Formula	% B Content
Solubor[a]	$Na_2B_8O_{13} \cdot 4H_2O$	17.5
Boric acid	H_3BO_3	20.5
Borax	$Na_2B_4O_7 \cdot 10H_2O$	11.3
Fertilizer borate	$Na_2B_4O_7 \cdot 5H_2O$	14.9
Anhydrous borax	$Na_2B_4O_7$	21.5
Ulexite	$NaCaB_5O_9 \cdot 8H_2O$	13.3
Colemanite	$Ca_2B_6O_{11} \cdot 5H_2O$	15.8
Boron frits	Borosilicate · glass	2–11

[a]   Disodium octaborate tetrahydrate, registered trademark of 20 Mule Team Borax, Boron, California.

solubility, B frits are used in sandy soil in high rainfall areas where leaching is a problem. They are used to supply a relatively insoluble, slowly available source of B and are therefore appropriate for maintenance programs more so than for correcting B deficiencies (Mortvedt and Woodruff, 1993).

8.10.2 APPLICATION IN SOLUBLE FORMS

8.10.2.1 Surface Watering

The completely soluble sources of B, Na borates and boric acid, can be applied to soils in liquid form. This can occur as part of a program of inclusion in the irrigation water (fertigation). Borates can also be applied in solution in combination with herbicides and insecticides (Shorrocks, 1997).

8.10.2.2 Foliar Application

Solubor, a highly soluble form of Na borate, and boric acid are dissolved in water and sprayed directly onto the plant leaves. Foliar application of B is common in perennial crops, such as fruit trees, where it is often combined with pesticide application (Mortvedt and Woodruff, 1993). Foliar applications can be timed for a particular stage in the crop growth cycle and are most appropriate for remediating B deficiency that develops during the growing season. The effectiveness of foliar fertilization depends on the phloem B mobility. Repeated applications are often necessary and are usually more effective than a single treatment containing a larger amount of B (Follett et al., 1981) in species where B is phloem immobile. Foliar applications are also appropriate in soils where micronutrient elements such as Zn, Mn, and Cu are not readily available to crops due to high pH.

REFERENCES

Abdulnour, J.E., D.J. Donnelly, and N.N. Barthakur. 2000. The effect of boron on calcium uptake and growth in micropropagated potato plantlets. *Potato Res.* 43:287–295.
Agulhon, H. 1910. Présence et utilité du bore chez les végéteaux. *Ann. Inst. Pasteur* 24:321–329.
Ahmed, N., M. Abid, and F. Ahmad. 2008. Boron toxicity in irrigated cotton (*Gossypium hirsutum* L.). *Pakistan J. Bot.* 40:2443–2452.
Ahmed, N., M. Abid, A. Rashid, M.A. Ali, and M. Ammanullah. 2013. Boron requirement of irrigated cotton in a typic haplocambid for optimum productivity and seed composition. *Commun. Soil Sci. Plant Anal.* 44:1293–1309.
Ahmed, M., M. Jahiruddin, and M.H. Mian. 2007. Screening of wheat genotypes for boron efficiency. *J. Plant Nutr.* 30:1127–1138.
Alpaslan, M. and A. Gunes. 2001. Interactive effects of boron and salinity stress on the growth, membrane permeability and mineral composition of tomato and cucumber plants. *Plant Soil* 236:123–128.
Ambak, K. and T. Tadano. 1991. Effect of micronutrient application on the growth and occurrence of sterility in barley and rice in a Malaysian deep peat. *Soil Sci. Plant Nutr.* 37:715–724.
An, J., Y. Liu, C. Yang, G. Zhou, Q. Wei, and S. Peng. 2012. Isolation and expression analysis of *CiNIP5*, a citrus boron transport gene involved in tolerance to boron deficiency. *Scientia Hortic.* 142:149–154.
Asad, A. 2002. Boron requirements for sunflower and wheat. *J. Plant Nutr.* 25:885–899.
Asad, A., F.P.C. Blamey, and D.G. Edwards. 2002. Dry matter production and boron concentrations of vegetative and reproductive tissues of canola and sunflower plants grown in nutrient solution. *Plant Soil* 243:243–252.
Aubert, H. and M. Pinta (eds.). 1997. *Trace Elements in Soils.* Developments in Soil Science 7, Amsterdam, the Netherlands: Elsevier Scientific Publishing Company.
Bagheri, A., J.G. Paull, A.J. Rathjen, S.M. Ali, and D.B. Moody. 1993. Genetic variation in the response of pea (*Pisum sativum* L.) to high soil concentrations of boron. In: P.J. Randall, ed. *Genetic Aspects of Plant Mineral Nutrition.* Dordrecht, the Netherlands: Kluwer Academic Publishers, pp. 377–385.
Baker, J.E., H.G. Gauch, and W.M. Dugger. 1956. Effects of boron on the water relations of higher plants. *Plant Physiol.* 31:89–94.
Bassett, R.L. 1980. A critical evaluation of the thermodynamic data for boron ions, ion pairs, complexes, and polyanions in aqueous solution at 298.15 K and 1 bar. *Geochim. Cosmochim. Acta* 44:1151–1160.

Bassil, E., H.N. Hu, and P.H. Brown. 2004. Use of phenylboronic acids to investigate boron function in plants. Possible role of boron in transvacuolar cytoplasmic strands and cell-to-wall adhesion. *Plant Physiol.* 136:3383–3395.

Bastias, E., N. Fernandez-Garcia, and M. Carvajal. 2004a. Aquaporin functionality in roots of *Zea mays* in relation to the interactive effects of boron and salinity. *Plant Biol.* 6:415–421.

Bastias, E.I., M.B. Gonzalez-Moro, and C. Gonzalez-Murua. 2004b. *Zea mays* L. amylacea from the Lluta Valley (Arica-Chile) tolerates salinity stress when high levels of boron are available. *Plant Soil* 267:73–84.

Bell, R.W., L. McLay, D. Plaskett, B. Dell, and J.F. Loneragan. 1990. Internal boron requirements of green gram (*Vigna radiata*). In: M.L. van Beusichem, ed. *Plant Nutrition—Physiology and Application*. Dordrecht, the Netherlands: Kluwer Academic Publishers, pp. 275–280.

Bellaloui, N., R.C. Yadavc, M.S. Chern et al. 2003. Transgenically enhanced sorbitol synthesis facilitates phloem-boron mobility in rice. *Physiol. Plant.* 117:79–84.

Bennett, A., R.I. Rowe, N. Soch, and C.D. Eckhert. 1999. Boron stimulates yeast (*Saccharomyces cerevisiae*) growth. *J. Nutr.* 129:2236–2238.

Benson, N.R., I.C. Chmelir, and E.S. Degman. 1961. Translocation and re-use of boron in broccoli. *Plant Physiol.* 36:296–307.

Berger, K.C. 1962. Micronutrient shortages—Micronutrient deficiencies in United States. *J. Agric. Food Chem.* 10:178–181.

Berger, K.C. and E. Truog. 1940. Boron deficiencies as revealed by plant and soil tests. *J. Am. Soc. Agron.* 32:297–301.

Berger, K.C. and E. Truog. 1946. Boron availability in relation to soil reaction and organic matter content. *Soil Sci. Soc. Am. Proc.* 10:113–116.

Bergmann, W. 1983. *Farbatlas: Ernährungsstörungen bei Kulturpflanzen*. Jena, Germany: VEB Gustav Fischer Verlag, pp. 110–145.

Bieleski, R.L. 1982. Sugar alcohols. In: F. Loewus, and W. Tanner, eds. *Encyclopedia of Plant Physiology*. Berlin, Germany: Springer Verlag, pp. 158–192.

Bingham, F.T. 1973. Boron and cultivated soils and irrigation waters. *Adv. Chem. Ser.* 123:130–138.

Blevins, D.G. and K.M. Lukaszewski. 1998. Boron in plant structure and function. *Annu. Rev. Plant Physiol. Plant Mol. Biol.* 49:481–500.

Bolanos, L., N.J. Brewin, and I. Bonilla. 1996. Effects of boron on Rhizobium-legume cell-surface interactions and nodule development. *Plant Physiol.* 110:1249–1256.

Bolanos, L., A. Cebrian, M. Redondo-Nieto, R. Rivilla, and I. Bonilla. 2001. Lectin-like glycoprotein PsNLEC-1 is not correctly glycosylated and targeted in boron deficient pea nodules. *Mol. Plant Microbe Interact.* 14:663–670.

Bolanos, L., E. Esteban, C. Delorenzo et al. 1994. Essentiality of boron for symbiotic dinitrogen fixation in pea (*Pisum sativum*) rhizobium nodules. *Plant Physiol.* 104:85–90.

Bolanos, L., K. Lukaszewski, I. Bonilla, and D. Blevins. 2004a. Why boron? *Plant Physiol. Biochem.* 42:907–912.

Bolanos, L., M. Redondo-Nieto, I. Bonilla, and L.G. Wall. 2002. Boron requirement in the *Discaria trinervis* (Rhamnaceae) and *Frankia* symbiotic relationship. Its essentiality for *Frankia* BCU110501 growth and nitrogen fixation. *Physiol. Plant.* 115:563–570.

Bolanos, L., M. Redondo-Nieto, R. Rivilla, N.J. Brewin, and I. Bonilla. 2004b. Cell surface interactions of Rhizobium bacteroids and other bacterial strains with symbiosomal and peribacteroid membrane components from pea nodules. *Mol. Plant Microbe Interact.* 17:216–223.

Bonilla, I., M. Garcia-González, and P. Mateo. 1990. Boron requirement in cyanobacteria. Its possible role in the early evolution of photosynthetic organisms. *Plant Physiol.* 94:1554–1560.

Bonilla, I., C. Mergold-Villasenor, M.E. Campos et al. 1997. The aberrant cell walls of boron-deficient bean root nodules have no covalently bound hydroxyproline-/proline-rich proteins. *Plant Physiol.* 115:1329–1340.

Bourne, E.J. 1958. The polyhydric alcohols. Acyclic polyhydric alcohols. In: W. Ruhland, ed. *Encyclopedia of Plant Physiology*. Berlin, Germany: Springer Verlag, pp. 345–361.

Bradford, G. 1966. Boron. In: H.D. Chapman (ed.) *Diagnostic Criteria for Plants and Soils*. Berkeley, CA: University of California, pp. 33–61.

Bray, R.H. 1948. Requirements for successful soil tests. *Soil Sci.* 66:83–89.

Broadley, M., P.H. Brown, I. Cakmak, Z. Rengel, and F. Zhao. 2012. Function of nutrients: Micronutrients. In: P. Marschner, ed. *Mineral Nutrition of Higher Plants*. Amsterdam, the Netherlands: Elsevier Academic Press, pp. 233–243.

Brown, P.H., N. Bellaloui, M.A. Wimmer et al. 2002. Boron in plant biology. *Plant Biol.* 4:205–223.

Brown, P.H. and H.N. Hu. 1994. Boron uptake by sunflower, squash and cultured tobacco cells. *Physiol. Plant.* 91:435–441.

Brown, P.H. and H. Hu. 1996. Phloem mobility of boron is species dependent: Evidence for phloem mobility in sorbitol-rich species. *Ann. Bot.* 77:497–505.

Brown, P.H., H.N. Hu, and W.G. Roberts. 1999. Occurrence of sugar alcohols determines boron toxicity symptoms of ornamental species. *J. Am. Soc. Hortic. Sci.* 124:347–352.

Brown, P.H. and B.J. Shelp. 1997. Boron mobility in plants. *Plant Soil* 193:85–101.

Cakmak, I. and V. Römheld. 1997. Boron deficiency-induced impairments of cellular functions in plants. *Plant Soil* 193:71–83.

Camacho-Cristobal, J.J., D. Anzellotti, and A. Gonzalez-Fontes. 2002. Changes in phenolic metabolism of tobacco plants during short-term boron deficiency. *Plant Physiol. Biochem.* 40:997–1002.

Camacho-Cristobal, J.J. and A. Gonzalez-Fontes. 1999. Boron deficiency causes a drastic decrease in nitrate content and nitrate reductase activity, and increases the content of carbohydrates in leaves from tobacco plants. *Planta* 209:528–536.

Camacho-Cristobal, J.J. and A. Gonzalez-Fontes. 2007. Boron deficiency decreases plasmalemma H+-ATPase expression and nitrate uptake, and promotes ammonium assimilation into asparagine in tobacco roots. *Planta* 226:443–451.

Camacho-Cristobal, J.J., M.B. Herrera-Rodriguez, V.M. Beato et al. 2008a. The expression of several cell wall-related genes in *Arabidopsis* roots is down-regulated under boron deficiency. *Environ. Exp. Bot.* 63:351–358.

Camacho-Cristobal, J.J., L. Lunar, F. Lafont, A. Baumert, and A. Gonzalez-Fontes. 2004. Boron deficiency causes accumulation of chlorogenic acid and caffeoyl polyamine conjugates in tobacco leaves. *J. Plant Physiol.* 161:879–881.

Camacho-Cristobal, J.J., J.M. Maldonado, and A. Gonzalez-Fontes. 2005. Boron deficiency increases putrescine levels in tobacco plants. *J. Plant Physiol.* 162:921–928.

Camacho-Cristobal, J.J., J. Rexach, and A. Gonzalez-Fontes. 2008b. Boron in plants: Deficiency and toxicity. *J. Integr. Plant Biol.* 50:1247–1255.

Campbell, L.C., M.H. Miller, and J.F. Loneragan. 1975. Translocation of boron to plant fruits. *Aust. J. Plant Physiol.* 2:481–487.

Cara, F.A., E. Sanchez, J.M. Ruiz, and L. Romero. 2002. Is phenol oxidation responsible for the short-term effects of boron deficiency on plasma-membrane permeability and function in squash roots? *Plant Physiol. Biochem.* 40:853–858.

Carpita, N. and D.M. Gibeaut. 1993. Structural models of primary cell walls in flowering plants: Consistency of molecular structure with the physical properties of the walls during growth. *Plant J.* 3:1–30.

Carrano, C.J., S. Schellenberg, S.A. Amin, D.H. Green, and F.C. Kuepper. 2009. Boron and marine life: A new look at an enigmatic bioelement. *Mar. Biotechnol.* 11:431–440.

Cartwright, B., K.G. Tiller, B.A. Zarcinas, and L.R. Spouncer. 1983. The chemical assessment of the boron status of soils. *Aust. J. Soil Res.* 21:321–332.

Cervilla, L.M., B. Blasco, J.J. Rios, L. Romero, and J.M. Ruiz. 2007. Oxidative stress and antioxidants in tomato (*Solanum lycopersicum*) plants subjected to boron toxicity. *Ann. Bot.* 100:747–756.

Chen, X., S. Schauder, N. Potier et al. 2002. Structural identification of a bacterial quorum-sensing signal containing boron. *Nature* 415:545–549.

Christ, C.L., A.H. Truesdell, and R.C. Erd. 1967. Borate mineral assemblages in the system Na_2O-CaO-MgO-B_2O_3-H_2O. *Geochim. Cosmochim. Acta* 31:313–337.

Chuda, Y., M. Ohnishikameyama, and T. Nagata. 1997. Identification of the forms of boron in seaweed by B-11 NMR. *Phytochem.* 46:209–213.

Clark, R.B. 1975. Mineral element concentrations in corn leaves by position on plant and age. *Commun. Soil Sci. Plant Anal.* 6:439–450.

Coddington, J.M. and M.J. Taylor. 1989. High field ^{11}B and ^{13}C NMR investigations of aqueous borate solutions and borate-diol complexes. *J. Coord. Chem.* 20:27–38.

Cotton, F.A. and G. Wilkinson. 1980. *Advanced Inorganic Chemistry*, 4th ed. New York: John Wiley & Sons.

Crandall, P.C., J.D. Chamberlain, and J.K.L. Garth. 1981. Toxicity symptoms and tissue-levels associated with excess boron in pear trees. *Commun. Soil Sci. Plant Anal.* 12:1047–1057.

Dannel, F., H. Pfeffer, and V. Römheld. 2000. Characterization of root boron pools, boron uptake and boron translocation in sunflower using the stable isotopes [10]B and [11]B. *Aust. J. Plant Physiol.* 27:397–405.

Datta, S.P., R.K. Rattan, K. Suribabu, and S.S. Datta. 2002. Fractionation and colorimetric determination of boron in soils. *J. Plant Nutr. Soil Sci.* 165:179–184.

Davis, J.M., J.K. Fellman, and W.H. Loescher. 1988. Biosynthesis of sucrose and mannitol as a function of leaf age in celery (*Apium graveolens* L). *Plant Physiol.* 86:129–133.

De la Chevallerie-Haaf, U., A. Meyer, and G. Henze. 1986. Photometrische Bestimmung von Bor im Grund- und Oberflächenwasser. *Fresenius Z. Anal. Chem.* 323:266–270.

Delgado, A., M. Benlloch, and R. Fernandezescobar. 1994. Mobilization of boron in olive trees during flowering and fruit development. *HortScience* 29:616–618.

Dell, B. and L. Huang. 1997. Physiological response of plants to low boron. *Plant Soil* 193:103–120.

Diaz, F.J. and S.R. Grattan. 2009. Performance of tall wheatgrass (*Thinopyrum ponticum* cv. 'Jose') irrigated with saline-high boron drainage water: Implications on ruminant mineral nutrition. *Agr. Ecosyst. Environ.* 131:128–136.

Dible, W.T., E. Truog, and K.C. Berger. 1954. Boron determination in soils and plants. A simplified curcumin method. *Anal. Chem.* 26:418–421.

Dordas, C. and P.H. Brown. 2000. Permeability of boric acid across lipid bilayers and factors affecting it. *J. Membr. Biol.* 175:95–105.

Dordas, C. and P.H. Brown. 2001. Evidence for channel mediated transport of boric acid in squash (*Cucurbita pepo*). *Plant Soil* 235:95–103.

Dordas, C. and P.H. Brown. 2005. Boron deficiency affects cell viability, phenolic leakage and oxidative burst in rose cell cultures. *Plant Soil* 268:293–301.

Dordas, C., M.J. Chrispeels, and P.H. Brown. 2000. Permeability and channel-mediated transport of boric acid across membrane vesicles isolated from squash roots. *Plant Physiol.* 124:1349–1361.

dos Santos, A.R., W.T. de Mattos, A.A.D. Almeida, F.A. Monteiro, B.D. Correa, and U.C. Gupta. 2004. Boron nutrition and yield of alfalfa cultivar Crioula in relation to boron supply. *Sci. Agric.* 61:496–500.

Dube, B.K., P. Sinha, and C. Chatterjee. 2000. Boron stress affects metabolism and seed quality of sunflower. *Trop. Agric.* 77:89–92.

Dwivedi, B.S., M. Ram, B.P. Singh, M. Das, and R.N. Prasad. 1992. Effect of liming on boron nutrition of pea (*Pisum sativum* L.) and corn (*Zea mays* L.) grown in sequence in an acid alfisol. *Fert. Res.* 31:257–262.

Dye, M.H., L. Buchanan, F.D. Dorofaeff, and F.G. Beecroft. 1983. Die-back of apricot trees following soil application of boron. *N. Z. J. Exp. Agric.* 11:331–342.

Eck, P. and F.J. Campbell. 1962. Effect of high calcium application on boron tolerance of carnation *Dianthus caryophyllus*. *Proc. Amer. Soc. Hortic. Sci.* 81:510–517.

El-Motaium, R., H. Hu, and P.H. Brown. 1994. The relative tolerance of six *Prunus* rootstocks to boron and salinity. *J. Am. Soc. Hortic. Sci.* 119:1169–1175.

El-Shintinawy, F. 1999. Structural and functional damage caused by boron deficiency in sunflower leaves. *Photosynthetica* 36:565–573.

Emebiri, L.C., P. Michael, and D.B. Moody. 2009. Enhanced tolerance to boron toxicity in two-rowed barley by marker-assisted introgression of favourable alleles derived from Sahara 3771. *Plant Soil* 314:77–85.

Evans, S. and U. Krahenbuhl. 1994. Boron analysis in biological material—Microwave digestion procedure and determination by different methods. *Fresen. J. Anal. Chem.* 349:454–459.

Findeklee, P. and H.E. Goldbach. 1996. Rapid effects of boron deficiency on cell wall elasticity modulus in *Cucurbita pepo* roots. *Bot. Acta* 109:463–465.

Fitzpatrick, K.L. and R.J. Reid. 2009. The involvement of aquaglyceroporins in transport of boron in barley roots. *Plant Cell Environ.* 32:1357–1365.

Fleischer, A., M.A. O'Neill, and R. Ehwald. 1999. The pore size of non-graminaceous plant cell walls is rapidly decreased by borate ester cross-linking of the pectic polysaccharide rhamnogalacturonan II. *Plant Physiol.* 121:829–838.

Follett, R.H., L.S. Murphy, and R.L. Donahue (eds.). 1981. *Fertilizers and Soil Amendments.* Englewood Cliffs, NJ: Prentice-Hall, Inc.

Fort, D.J., T.L. Propst, E.L. Stover, and P.L. Strong. 1998. Adverse reproductive and developmental effects in *Xenopus* from insufficient boron. *Biol. Trace Elem. Res.* 66:237–259.

Francois, L.E. 1984. Effect of excess boron on tomato yield, fruit size, and vegetative growth. *J. Am. Soc. Hortic. Sci.* 109:322–324.

Francois, L.E. 1988. Yield and quality responses of celery and crisphead lettuce to excess boron. *J. Am. Soc. Hortic. Sci.* 113:538–542.

Francois, L.E. 1989. Boron tolerance of snap bean and cowpea. *J. Am. Soc. Hortic. Sci.* 114:615–619.

Francois, L.E. 1991. Yield and quality responses of garlic and onion to excess boron. *HortScience* 26:547–549.

Ganmore-Neumann, R. and S. Davidov. 1993. Uptake and distribution of calcium in rose plantlets as affected by calcium and boron concentration in culture solution. *Plant Soil* 155/156:151–154.

Garcia-González, M., P. Mateo, and I. Bonilla. 1988. Boron protection for O_2 diffusion in heterocysts of *Anabaena* sp. PCC 7119. *Plant Physiol.* 87:785–789.

Garcia-Gonzalez, M., P. Mateo, and I. Bonilla. 1991. Boron requirement for envelope structure and function in *Anabaena* PCC 7119 heterocysts. *J. Exp. Bot.* 42:925–929.

Garg, O.K., A.N. Sharma, and G. Kona. 1979. Effect of boron on the pollen vitality and yield of rice plants (*Oryza sativa* L. var. Jaya). *Plant Soil* 52:591–594.

Gestring, W.D. and P.N. Soltanpour. 1984. Evaluation of the ammonium bicarbonate–DTPA soil test for assessing boron availability to alfalfa. *Soil Sci. Soc. Am. J.* 48:96–100.

Gestring, W.D. and P.N. Soltanpour. 1987. Comparison of soil tests for assessing boron toxicity to alfalfa. *Soil Sci. Soc. Am. J.* 51:1214–1219.

Ghanati, F., A. Morita, and H. Yokota. 2005. Deposition of suberin in roots of soybean induced by excess boron. *Plant Sci.* 168:397–405.

Goldbach, H.E. 1997. A critical review on current hypotheses concerning the role of boron in higher plants: Suggestions for further research and methodological requirements. *J. Trace and Microprobe Techniques* 15:51–91.

Goldbach, H.E. and M.A. Wimmer. 2007. Boron in plants and animals: Is there a role beyond cell-wall structure? *J. Plant Nutr. Soil Sci.* 170:39–48.

Goldberg, S. 1997. Reactions of boron with soils. *Plant Soil* 193: 35–48.

Goldberg, S. and R.A. Glaubig. 1986. Boron adsorption and silicon release by the clay minerals kaolinite, montmorillonite, and illite. *Soil Sci. Soc. Am. J.* 50:1442–1448.

Goldberg, S., P.J. Shouse, C.M. Grieve, J.A. Poss, H.S. Forster, and D.L. Suarez. 2003. Effect of high boron application on boron content and growth of melons. *Plant Soil* 256:403–411.

Grieve, C.M. and J.A. Poss. 2000. Wheat response to interactive effects of boron and salinity. *J. Plant Nutr.* 23:1217–1226.

Guidi, L., E. Degl'Innocenti, G. Carmassi, D. Massa, and A. Pardossi. 2011. Effects of boron on leaf chlorophyll fluorescence of greenhouse tomato grown with saline water. *Environ. Exp. Bot.*73:57–63.

Gunes, A., G. Soylemezoglu, A. Inal, E.G. Bagci, S. Coban, and O. Sahin. 2006. Antioxidant and stomatal responses of grapevine (*Vitis vinifera* L.) to boron toxicity. *Scientia Hortic.* 110:279–284.

Gupta, U.C. 1967. A simplified method for determining a hot water-soluble boron and podzol soils. *Soil Sci.* 103:424–428.

Gupta, U.C. 1968. Relationship of total and hot-water soluble boron, and fixation of added boron, to properties of podzol soils. *Soil Sci. Soc. Am. Proc.* 32:45–48.

Gupta, U.C. 1971a. Boron requirement of alfalfa, red clover, brussels sprouts and cauliflower grown under greenhouse conditions. *Soil Sci.* 112:280–281.

Gupta, U.C. 1971b. Boron and molybdenum nutrition of wheat, barley and oats grown in Prince Edward Island soils. *Can. J. Soil Sci.* 51:415–422.

Gupta, U.C. 1972. Effects of boron and lime on boron concentration and growth of forage legumes under greenhouse conditions. *Commun. Soil Sci. Plant Anal.* 3:355–365.

Gupta, U.C. 1979. Boron nutrition of crops. *Adv. Agron.* 31:273–307.

Gupta, U.C. 1983. Boron deficiency and toxicity symptoms for several crops as related to tissue boron levels. *J. Plant Nutr.* 6:387–395.

Gupta, U.C. 1991. Boron, molybdenum and selenium status in different plant-parts in forage legumes and vegetable crops. *J. Plant Nutr.* 14:613–621.

Gupta, U.C. 2007. Boron. In: A.V. Baker and D.J. Pilbeam, eds. *Handbook of Plant Nutrition* 1st edn. Boca Raton, FL: CRC Press, pp. 242–268.

Gupta, U.C., R. Cormier, and J.A. Cutcliffe. 1987. Tolerance of brussels-sprouts to high boron levels. *Can. J. Soil Sci.* 67:205–207.

Gupta, U.C. and J.A. Cutcliffe. 1971. Determination of optimum levels of boron in rutabaga leaf tissue and soil. *Soil Sci.* 111:382–385.

Gupta, U.C. and J.A. Cutcliffe. 1972. Effects of lime and boron on brown heart, leaf tissue, calcium/boron ratios, and boron concentrations of rutabaga. *Soil Sci. Soc. Am. Proc.* 36:936–939.

Gupta, U.C. and J.A. Cutcliffe. 1973. Boron nutrition of broccoli, brussels-sprouts, and cauliflower grown on Prince Edward Island soils. *Can. J. Soil Sci.* 53:275–279.

Gupta, U.C. and J.A. Cutcliffe. 1975. Boron deficiency in cole crops under field and greenhouse conditions. *Commun. Soil Sci. Plant Anal.* 6:181–188.

Gupta, U.C. and J.A. Cutcliffe. 1984. Effects of applied and residual boron on the nutrition of cabbage and field beans. *Can. J. Soil Sci.* 64:571–576.

Gupta, U.C. and J.A. Cutcliffe. 1985. Boron nutrition of carrots and table beets grown in a boron deficient soil. *Commun. Soil Sci. Plant Anal.* 16:509–516.

Gupta, U.C. and J.A. MacLeod. 1973. Boron nutrition and growth of timothy as affected by soil pH. *Commun. Soil Sci. Plant Anal.* 4:389–395.

Gupta, U.C. and J.A. MacLeod. 1977. Influence of calcium and magnesium sources on boron uptake and yield of alfalfa and rutabaga as related to soil pH. *Soil Sci.* 124:279–284.

Gupta, U.C. and D.C. Munro. 1969. Boron content of tissues and roots of rutabagas and of soil as associated with brown-heart condition. *Soil Sci. Soc. Am. Proc.* 33:424–426.

Han, S., L.-S. Chen, H.-X. Jiang, B.R. Smith, L.-T. Yang, and C.-Y. Xie. 2008. Boron deficiency decreases growth and photosynthesis, and increases starch and hexoses in leaves of citrus seedlings. *J. Plant Physiol.* 165:1331–1341.

Han, S., N. Tang, H.-X. Jiang, L.-T. Yang, Y. Li, and L.-S. Chen. 2009. CO_2 assimilation, photosystem II photochemistry, carbohydrate metabolism and antioxidant system of citrus leaves in response to boron stress. *Plant Sci.* 176:143–153.

Hanaoka, H. and T. Fujiwara. 2007. Channel-mediated boron transport in rice. *Plant Cell Physiol.* 48:S227.

Hanson, E.J. 1991. Movement of boron out of tree fruit leaves. *HortScience* 26: 271–273.

Harder, H. 1961. Einbau von Bor in detritische Tonminerale. Experimente zur Erklärung des Borgehaltes toniger Sedimente. *Geochim. Cosmochim. Acta* 21:284–294.

Harkness, R.W. 1959. Boron deficiency and alternate bearing in avocados. *Proc. Fla. State Hortic. Soc.* 72:311–317.

Harris, H.C. and J.B. Brolmann. 1966. Comparison of calcium and boron deficiencies of peanut. 1. Physiological and yield differences. *Agron. J.* 58:S 575.

Hatcher, J.T. and L.V. Wilcox. 1950. Colorimetric determination of boron using carmine. *Anal. Chem.* 22:567–569.

Haydon, G.F. 1981. Boron toxicity of strawberry. *Commun. Soil Sci. Plant Anal.* 12:1085–1091.

Hayes, J.E. and R.J. Reid. 2004. Boron tolerance in barley is mediated by efflux of boron from the roots. *Plant Physiol.* 136:3376–3382.

Hills, F.J. and A. Ulrich. 1976. Plant analysis as a guide for mineral-nutrition of sugar-beets. *Calif. Agric. Exp. Stat. Bull.* 1879:18–21.

Hole, C.C. and A. Scaife. 1993. An analysis of the growth-response of carrot seedlings to deficiency in some mineral nutrients. *Plant Soil* 150:147–156.

Holloway, R.E. and A.M. Alston. 1992. The effects of salt and boron on growth of wheat. *Aust. J. Agric. Res.* 43:987–1001.

Hou, J., L.J. Evans, and G.A. Spiers. 1994. Boron fractionation in soils. *Commun. Soil Sci. Plant Anal.* 25:1841–1853.

Hou, J., L.J. Evans, and G.A. Spiers. 1996. Chemical fractionation of soil boron: I. Method development. *Can J. Soil Sci.* 76:485–491.

Hu, H. and P.H. Brown. 1994. Localization of boron in cell walls of squash and tobacco and its association with pectin. *Plant Physiol.* 105:681–689.

Hu, H., P.H. Brown, and J.M. Labavitch. 1996. Species variability in boron requirement is correlated with cell wall pectin. *J. Exp. Bot.* 47:227–232.

Hu, H., S.G. Penn, C.B. Lebrilla, and P.H. Brown. 1997. Isolation and characterization of soluble boron complexes in higher plants. *Plant Physiol.* 113:649–655.

Huang, L., J. Pant, B. Dell, and R. Bell. 2000. Effects of boron deficiency on anther development and floret fertility in wheat (*Triticum aestivum* L. 'Wilgoyne'). *Ann. Bot.* 85:493–500.

Hunt, C.D. 2012. Dietary boron: Progress in establishing essential roles in human physiology. *J. Trace Elem. Med. Biol.* 26:157–160.

Ishii, T. and T. Matsunaga. 1996. Isolation and characterization of a boron-rhamnogalacturonan-II complex from cell walls of sugar beet pulp. *Carbohydr. Res.* 284:1–9.

Ismail, A.M. 2003. Response of maize and sorghum to excess boron and salinity. *Biol. Plant.* 47:313–316.

Jamjod, S., S. Niruntrayagul, and B. Rerkasem. 2004. Genetic control of boron efficiency in wheat (*Triticum aestivum* L.). *Euphytica* 135:21–27.

Jamjod, S. and B. Rerkasem. 1999. Genotypic variation in response of barley to boron deficiency. *Plant Soil* 215:65–72.

Jansen van Rensburg, H.G., A.S. Claassens, and D.J. Beukes. 2010. Evaluation of the effect of soil acidity amelioration on maize yield and nutrient interrelationships using stepwise regression and nutrient vector analysis. *S. Afr. J. Plant Soil* 27:117–125.

Jasmund, K. and B. Lindner. 1973. Experiments on the fixation of boron by clay minerals. *Proc. Int. Clay Conf.* 1972:399–412.

Jefferies, S.P., A.R. Barr, A. Karakousis et al. 1999. Mapping of chromosome regions conferring boron toxicity tolerance in barley (*Hordeum vulgare* L.). *Theor. Appl. Genet.* 98:1293–1303.

Jin, J.Y., D.C. Martens, and L.W. Zelazny. 1987. Distribution and plant availability of soil boron extractions. *Soil Sci. Soc. Am. J.* 51:1228–1231.

John, M.K. 1973. A batch-handling technique for hot-water extraction of boron from soils. *Soil Sci. Soc. Am. Proc.* 37:332–333.

Jones, H.E. and G.D. Scarseth. 1944. The calcium-boron balance in plants as related to boron needs. *Soil Sci.* 57:15–24.

Kamali, A.R. and N.F. Childers. 1970. Growth and fruiting of peach in sand culture as affected by boron and a fritted form of trace elements. *J. Am. Soc. Hortic. Sci.* 95:652–656.

Karabal, E., M. Yucel, and H.A. Oktem. 2003. Antioxidant responses of tolerant and sensitive barley cultivars to boron toxicity. *Plant Sci.* 164:925–933.

Karioti, A., A. Chatzopoulou, A.R. Bilia, G. Liakopoulos, S. Stavrianakou, and H. Skaltsa. 2006. Novel secoiridoid glucosides in *Olea europaea* leaves suffering from boron deficiency. *Biosci. Biotech. Bioch.* 70:1898–1903.

Kastori, R. and N. Petrovic. 1989. Effect of boron on nitrate reductase activity in young sunflower plants. *J. Plant Nutr.* 12:621–632.

Kastori, R., M. Plesnicar, D. Pankovic, and Z. Sakac. 1995. Photosynthesis, chlorophyll fluorescence and soluble carbohydrates in sunflower leaves as affected by boron deficiency. *J. Plant Nutr.* 18:1751–1763.

Kato, Y., K. Miwa, J. Takano, M. Wada, and T. Fujiwara. 2009. Highly boron deficiency-tolerant plants generated by enhanced expression of NIP5;1, a boric acid channel. *Plant Cell Physiol.* 50:58–66.

Keerati-Kasikorn, P., R.W. Bell, and J.F. Loneragan. 1991. Response of 2 peanut (*Arachis hypogaea* L.) cultivars to boron and calcium. *Plant Soil* 138:61–66.

Kelly, W.C., G.F. Somers, and G.H. Ellis. 1952. The effect of boron on the growth and carotene content of carrots. *Proc. Am. Soc. Hortic. Sci.* 59:352–360.

Keren, R. 1996. Boron. In: D.L. Sparks, ed. *Methods of Soil Analysis. Part 3—Chemical Methods*, SSSA Book Series: 5. Madison, WI: ASA and SSSA, pp. 603–626.

Keren, R. and F.T. Bingham. 1985. Boron in water, soils, and plants. *Adv. Soil Sci.* 1:229–276.

Keren, R., F.T. Bingham, and J.D. Rhoades. 1985a. Plant uptake of boron as affected by boron distribution between liquid and solid phases in soil. *Soil Sci. Soc. Am. J.* 49:297–302.

Keren, R., F.T. Bingham, and J.D. Rhoades. 1985b. Effect of clay content in soil on boron uptake and yield of wheat. *Soil Sci. Soc. Am. J.* 49:1466–1470.

Kitano, Y., M. Okamura, and M. Idogaki. 1978. Coprecipitation of borate-boron with calcium carbonate. *Geochem. J.* 12:183–189.

Kobayashi, M., T. Matoh, and J.-I. Azuma. 1996. Two chains of rhamnogalacturonan II are cross-linked by borate-diol ester bonds in higher plant cell walls. *Plant Physiol.* 110:1017–1020.

Kobayashi, M., T. Mutoh, and T. Matoh. 2004. Boron nutrition of cultured tobacco BY-2 cells. IV. Genes induced under low boron supply. *J. Exp. Bot.* 55:1441–1443.

Kohl, H.C., and J.J. Oertli. 1961. Distribution of boron in leaves. *Plant Physiol.* 36:420–424.

Koshiba, T., M. Kobayashi, and T. Matoh. 2009. Boron nutrition of tobacco BY-2 cells. V. Oxidative damage is the major cause of cell death induced by boron deprivation. *Plant Cell Physiol.* 50:26–36.

Landi, M., A. Pardossi, D. Remorini, and L. Guidi. 2013. Antioxidant and photosynthetic response of a purple-leaved and a green-leaved cultivar of sweet basil (*Ocimum basilicum*) to boron excess. *Environ. Exp. Bot.* 85:64–75.

Lanoue, L., D.R. Trollinger, P.L. Strong, and C.L. Keen. 2000. Functional impairments in preimplantation mouse embryos following boron deficiency. *FASEB J.* 14 A:539.

Lehto, T. and E. Malkonen. 1994. Effects of liming and boron fertilization on boron uptake of *Picea abies*. *Plant Soil* 163:55–64.

Lewin, J. 1966. Physiological studies of the boron requirement of the diatom *Cylindrotheca fusiformis* Reimann and Lewin. *J. Exp. Bot.* 17:473–479.

Lewis, D.H. and D.C. Smith. 1967. Sugar alcohols (polyols) in fungi and green plants. I. Distribution, physiology and metabolism. *New Phytol.* 66:143–184.

Liakopoulos, G. and G. Karabourniotis. 2005. Boron deficiency and concentrations and composition of phenolic compounds in *Olea europaea* leaves: A combined growth chamber and field study. *Tree Physiol.* 25:307–315.

Liakopoulos, G., S. Stavrianakou, M. Filippou et al. 2005. Boron remobilization at low boron supply in olive (*Olea europaea*) in relation to leaf and phloem mannitol concentrations. *Tree Physiol.* 25:157–165.

Liakopoulos, G., S. Stavrianakou, D. Nikolopoulos et al. 2009. Quantitative relationships between boron and mannitol concentrations in phloem exudates of *Olea europaea* leaves under contrasting boron supply conditions. *Plant Soil* 323:177–186.

Liu, D.G., W.S. Jiang, L.X. Zhang, and L.F. Li. 2000. Effects of boron ions on root growth and cell division of broadbean (*Vicia faba* L.). *Isr. J. Plant Sci.* 48:47–51.

Liu, L., B.J. Shelp, and G.A. Spiers. 1993. Boron distribution and retranslocation in field-grown broccoli (*Brassica oleracea* var. Italica). *Can. J. Plant Sci.* 73:587–600.

Lohse, G. 1982. Microanalytical azomethine-H method for boron determination in plant tissue. *Commun. Soil Sci. Plant Anal.* 13:127–134.

Loomis, W.D. and R.W. Durst. 1992. Chemistry and biology of boron. *BioFactors* 3:229–239.

Lopez-Lefebre, L.R., R.M. Rivero, P.C. Garcia, E. Sanchez, J.M. Ruiz, and L. Romero. 2002. Boron effect on mineral nutrients of tobacco. *J. Plant Nutr.* 25:509–522.

Lordkaew, S., B. Dell, S. Jamjod, and B. Rerkasem. 2011. Boron deficiency in maize. *Plant Soil* 342:207–220.

Lordkaew, S., S. Konsaeng, J. Jongjaidee, B. Dell, B. Rerkasem, and S. Jamjod. 2013. Variation in responses to boron in rice. *Plant Soil* 363:287–295.

Lukaszewski, K.M. and D.G. Blevins. 1996. Root growth inhibition in boron-deficient or aluminum-stressed squash may be a result of impaired ascorbate metabolism. *Plant Physiol.* 112:1135–1140.

MacKay, D.C., W.M. Langille, and E.W. Chipman. 1962. Boron deficiency and toxicity in crops grown on sphagnum peat soil. *Can. J. Soil Sci.* 42:302–310.

Marentes, E., B.J. Shelp, R.A. Vanderpool, and G.A. Spiers. 1997. Retranslocation of boron in broccoli and lupin during early reproductive growth. *Physiol. Plant.* 100:389–399.

Martinez-Ballesta, M.D., E. Bastias, C. Zhu et al. 2008. Boric acid and salinity effects on maize roots. Response of aquaporins ZmPIP1 and ZmPIP2, and plasma membrane H+-ATPase, in relation to water and nutrient uptake. *Physiol. Plant.* 132:479–490.

Masood, S., M.A. Wimmer, K. Witzel, C. Zoerb, and K.H. Muehling. 2012. Interactive effects of high boron and NaCl stresses on subcellular localization of chloride and boron in wheat leaves. *J. Agron. Crop Sci.* 198:227–235.

Mateo, P.I., Bonilla, E. Fernandezvaliente, and E. Sanchezmaeso. 1986. Essentiality of boron for dinitrogen fixation in *Anabaena* sp. PCC 7119. *Plant Physiol.* 81:430–433.

Matsi, T., V. Antoniadis, and N. Barbayiannis. 2000. Evaluation of the NH$_4$HCO$_3$-DTPA soil test for assessing boron availability to wheat. *Commun. Soil Sci. Plant Anal.* 31:669–278.

Mazé, P. 1919. Recherche d'une solution purement minérale capable d'assurer l'évolution complete du mais cultivé à l'abri des microbes. *Ann. Inst. Pasteur* 33:139–173.

McDonald, G.K., J.K. Eglinton, and A.R. Barr. 2010. Assessment of the agronomic value of QTL on chromosomes 2H and 4H linked to tolerance to boron toxicity in barley (*Hordeum vulgare* L.). *Plant Soil* 326:275–290.

Mei, L., O. Sheng, S.-A. Peng, G.-F. Zhou, Q.-J. Wei, and Q.-H. Li. 2011. Growth, root morphology and boron uptake by citrus rootstock seedlings differing in boron-deficiency responses. *Scientia Hortic.* 129:426–432.

Melsted, S.W., H.L. Motto, and T.R. Peck. 1969. Critical plant nutrient composition values useful in interpreting plant analysis data. *Agron. J.* 61:17–20.

Meyer, R.D. and W.E. Martin. 1976. Plant analysis as a guide for fertilization of alfalfa. *Calif. Agric. Exp. Stat. Bull.* 1879:26–29.

Miller, R.O., B. Vaughan, and J. Kutoby-Amacher. 2000. Extraction of soil boron with DTPA-sorbitol. *Soil-Plant Anal.* 4–5:10.

Minchin, P.E.H., T.G. Thorp, H.L. Boldingh et al. 2012. A possible mechanism for phloem transport of boron in 'Hass' avocado (*Persea americana* Mill.) trees. *J. Hortic. Sci. Biotechnol.* 87:23–28.

Miwa, K. and T. Fujiwara. 2010. Boron transport in plants: Coordinated regulation of transporters. *Ann. Bot.* 105:1103–1108.

Miwa, K., J. Takano, and T. Fujiwara. 2006. Improvement of seed yields under boron-limiting conditions through overexpression of BOR1, a boron transporter for xylem loading, in *Arabidopsis thaliana*. *Plant J.* 46:1084–1091.

Miwa, K., J. Takano, H. Omori, M. Seki, K. Shinozaki, and T. Fujiwara. 2007. Plants tolerant of high boron levels. *Science* 318:1417.

Molassiotis, A., T. Sotiropoulos, G. Tanou, G. Diamantidis, and I. Therios. 2006. Boron-induced oxidative damage and antioxidant and nucleolytic responses in shoot tips culture of the apple rootstock EM 9 (*Malus domestica* Borkh). *Environ. Exp. Bot.* 56:54–62.

Morrill, L.G., W.E. Hill, W.W. Chrudimsky et al. 1977. Boron requirements of Spanish peanuts in Oklahoma—Effects on yield and quality and interaction with other nutrients. *Okla. Agric. Exp. Stat. Misc. Publ.* 99:1–20.

Mortvedt, J.J. and J.R. Woodruff. 1993. Technology and application of boron fertilizers for crops. In: U.C. Gupta, ed. *Boron and Its Role in Crop Production*. Boca Raton, FL: CRC Press, pp. 157–176.

Mottonen, M., P.J. Aphalo, and T. Lehto. 2001. Role of boron in drought resistance in Norway spruce (*Picea abies*) seedlings. *Tree Physiol.* 21:673–681.

Mozafar, A. 1987. Effect of boron on ear formation and yield components of 2 maize (*Zea mays* L.) hybrids. *J. Plant Nutr.* 10:319–332.

Muller, F.B. and G. McSweeney. 1976. Toxicity of borates to turnips. *N. Z. J. Exp. Agric.* 4:451–455.

Nabi, G., E. Rafique, and M. Salim. 2006. Boron nutrition of four sweet pepper cultivars grown in boron-deficient soil. *J. Plant Nutr.* 29:717–725.

Nable, R.O. 1988. Resistance to boron toxicity amongst several barley and wheat cultivars: A preliminary examination of the resistance mechanism. *Plant Soil* 112:45–52.

Nable, R.O., G.S. Banuelos, and J.G. Paull. 1997. Boron toxicity. *Plant Soil* 193:181–198.

Nachiangmai, D., B. Dell, R. Bell, L.B. Huang, and B. Rerkasem. 2004. Enhanced boron transport into the ear of wheat as a mechanism for boron efficiency. *Plant Soil* 264:141–147.

Naeem, M.A., M.A. Maqsood, S. Hussain, M.K. Khan, and S. Kanwal. 2013. Irrigation with brackish water modifies the boron requirement of mungbean (*Vigna radiata* L.) on typic calciargid. *Arch. Agron. Soil Sci.* 59:133–145.

Nakagawa, Y., H. Hanaoka, M. Kobayashi, K. Miyoshi, K. Miwa, and T. Fujiwara. 2007. Cell-type specificity of the expression of Os *BOR1*, a rice efflux boron transporter gene, is regulated in response to boron availability for efficient boron uptake and xylem loading. *Plant Cell* 19:2624–2635.

Neubert, P., W. Wrazidlo, H.P. Vielemeyer, I. Hundt, F. Gollmick, and W. Bergmann. 1970. *Tabellen zur Pflanzenanalyse—Erste orientierende Übersicht*. Jena, Germany: Inst. für Pflanzenernährung, pp. 1–40.

Noguchi, K., M. Yasumori, T. Imai et al. 1997. *bor 1–1*, an *Arabidopsis thaliana* mutant that requires a high level of boron. *Plant Physiol.* 115:901–906.

Nozawa, A., K. Miwa, M. Kobayashi, and T. Fujiwara. 2006. Isolation of *Arabidopsis thaliana* cDNAs that confer yeast boric acid tolerance. *Biosci. Biotech. Biochem.* 70:1724–1730.

Nuttall, C.Y. 2000. Boron tolerance and uptake in higher plants, Thesis, University of Cambridge, UK: University of Cambridge.

Nyomora, A.M.S., P.H. Brown, and M. Freeman. 1997. Fall foliar-applied boron increases tissue boron concentration and nut set of almond. *J. Am. Soc. Hort. Sci.* 122:405–410.

Oertli, J.J. 1993. The mobility of boron in plants. *Plant Soil* 155/156:310–304.

Oertli, J.J. 1994. Nonhomogeneity of boron distribution in plants and consequences for foliar diagnosis. *Commun. Soil Sci. Plant Anal.* 25:1133–1147.

Oertli, J.J. and W.F. Richardson. 1970. The mechanism of boron immobility in plants. *Physiol. Plant.* 23:108–116.

O'Neill, M., S. Eberhard, P. Albersheim, and A. Darvill. 2001. Requirement of borate cross-linking of cell wall rhamnogalacturonan II for Arabidopsis growth. *Science* 294:846–849.

O'Neill, M., D. Warrenfeltz, K. Kates et al. 1996. Rhamnogalacturonan-II, a pectic polysaccharide in the walls of growing plant cell, forms a dimer that is covalently cross-linked by a borate ester. *J. Biol. Chem.* 271:22923–22930.

Papadakis, I.E., K.N. Dimassi, A.M. Bosabalidis, I.N. Therios, A. Patakas, and A. Giannakoula. 2004. Effects of B excess on some physiological and anatomical parameters of 'Navelina' orange plants grafted on two rootstocks. *Environ. Exp. Bot.* 51:247–257.

Park, M., Q. Li, N. Shcheynikov, W.Z. Zeng, and S. Muallem. 2004. NaBC1 is a ubiquitous electrogenic Na$^+$-coupled borate transporter essential for cellular boron homeostasis and cell growth and proliferation. *Mol. Cell* 16:331–341.

Paull, J.G., B. Cartwright, and A.J. Rathjen. 1988. Responses of wheat and barley genotypes to toxic concentrations of soil boron. *Euphytica* 39:137–144.

Perica, S., N. Bellaloui, C. Greve, H.N. Hu, and P.H. Brown. 2001a. Boron transport and soluble carbohydrate concentrations in olive. *J. Am. Soc. Hort. Sci.* 126:291–296.

Perica, S., P.H. Brown, J.H. Connell et al. 2001b. Foliar boron application improves flower fertility and fruit set of olive. *HortScience* 36:714–716.

Picchioni, G.A., S.A. Weinbaum, and P.H. Brown. 1995. Retention and the kinetics of uptake and export of foliage-applied, labeled boron by apple, pear, prune, and sweet cherry leaves. *J. Am. Soc. Hort. Sci.* 120:28–35.

Pinho, L.G.R., E. Campostrini, P.H. Monnerat et al. 2010. Boron deficiency affects gas exchange and photochemical efficiency (Jpi test parameters) in green dwarf coconut. *J. Plant Nutr.* 33:439–451.

Plouvier, B. 1963. Distribution of aliphatic polyols and cyclitols. In: T. Swain, ed. *Chemical Plant Taxonomy*. New York: Academic Press, pp. 313–336.

Poulain, D. and H. Almohammad. 1995. Effects of boron deficiency and toxicity on faba bean (*Vicia faba* L.). *Eur. J. Agron.* 4:127–134.

Prasad, M. and E. Byrne. 1975. Boron source and lime effects on yield of 3 crops grown in peat. *Agron. J.* 67:553–556.

Punchana, S., M. Cakir, B. Rerkasem, and S. Jamjod. 2012. Mapping the Bo(d)2 gene associated with boron efficiency in wheat. *Scienceasia* 38:235–243.

Purves, D. and E.J. MacKenzie. 1973. Effects of applications of municipal compost on uptake of copper, zinc and boron by garden vegetables. *Plant Soil* 39:361–371.

Rafique, E., M. Mahmood-ul-Hassan, K.M. Khokhar, M. Ishaq, M. Yousra, and T. Tabassam. 2012. Boron requirement of chili (*Capsicum annuum* L.): Proposed diagnostic criteria. *J. Plant Nutr.* 35:739–749.

Rahman, M.H.H., Y. Arima, K. Watanabe, and H. Sekimoto. 1999. Adequate range of boron nutrition is more restricted for root nodule development than for plant growth in young soybean plant. *Soil Sci. Plant Nutr.* 45:287–296.

Ralston, N.V.C. and C.D. Hunt. 2001. Diadenosine phosphates and S-adenosylmethionine: novel boron binding biomolecules detected by capillary electrophoresis. *Biochim. Biophys. Acta* 1527:20–30.

Ram, S., L.D. Bist, and S.C. Sirohi. 1989. Internal fruit necrosis of mango and its control. *Acta Hort.* 231:805–813.

Ramon, A.M., R.O. Carpena Ruiz, and A. Garate. 1989. In vitro stabilization and distribution of nitrate reductase in tomato plants: Incidence of boron deficiency. *J. Plant Physiol.* 135:126–128.

Rashid, A. and E. Rafique. 2005. Internal boron requirement of young sunflower plants: Proposed diagnostic criteria. *Commun. Soil Sci. Plant Anal.* 36:2113–2119.

Rashid, A., E. Rafique, and N. Ali. 1997a. Micronutrient deficiencies in rainfed calcareous soils of Pakistan.2. Boron nutrition of the peanut plant. *Commun. Soil Sci. Plant Anal.* 28:149–159.

Rashid, A., E. Rafique, A.U. Bhatti, J. Ryan, N. Bughio, and S.K. Yau. 2011. Boron deficiency in rainfed wheat in Pakistan: Incidence, spatial variability and management strategies. *J. Plant Nutr.* 34:600–613.

Rashid, A., E. Rafique, and N. Bughio. 1997b. Micronutrient deficiencies in rainfed calcareous soils of Pakistan.3. Boron nutrition of sorghum. *Commun. Soil Sci. Plant Anal.* 28:441–454.

Raven, J.A. 1980. Short- and long-distance transport of boric acid in plants. *New Phytol.* 84:231–249.

Rawson, H.M. 1996. The developmental stage during which boron limitation causes sterility in wheat genotypes and the recovery of fertility. *Aust. J. Plant Physiol.* 23:709–717.

Raza, M., A.R. Mermut, J.J. Schoenau, and S.S. Malhi. 2002. Boron fractionation in some Saskatchewan soils. *Can J. Soil Sci.* 82:173–179.

Redondo-Nieto, M., N. Maunoury, P. Mergaert, E. Kondorosi, I. Bonilla, and L. Bolanos. 2012. Boron and calcium induce major changes in gene expression during legume nodule organogenesis. Does boron have a role in signalling? *New Phytol.* 195:14–19.

Redondo-Nieto, M., L. Pulido, M. Reguera, I. Bonilla, and L. Bolanos. 2007. Developmentally regulated membrane glycoproteins sharing antigenicity with rhamnogalacturonan II are not detected in nodulated boron deficient Pisum sativum. *Plant Cell Environ.* 30:1436–1443.

Redondo-Nieto, M., M. Reguera, I. Bonilla, and L. Bolanos. 2008. Boron dependent membrane glycoproteins in symbiosome development and nodule organogenesis: A model for a common role of boron in organogenesis. *Plant Signal. Behav.* 3:298–300.

Redondo-Nieto, M., R. Rivilla, A. El-Hamdaoui, I. Bonilla, and L. Bolanos. 2001. Boron deficiency affects early infection events in the pea-Rhizobium symbiotic interaction. *Aust. J. Plant Physiol.* 28:819–823.

Reguera, M., I. Abreu, N.J. Brewin, I. Bonilla, and L. Bolanos. 2010a. Borate promotes the formation of a complex between legume AGP-extensin and Rhamnogalacturonan II and enhances production of Rhizobium capsular polysaccharide during infection thread development in *Pisum sativum* symbiotic root nodules. *Plant Cell Environ.* 33:2112–2120.

Reguera, M., A. Espi, L. Bolanos, I. Bonilla, and M. Redondo-Nieto. 2009. Endoreduplication before cell differentiation fails in boron-deficient legume nodules. Is boron involved in signaling during cell cycle regulation? *New Phytol.* 183:9–12.

Reguera, M., M. Wimmer, P. Bustos, H.E. Goldbach, L. Bolanos, and I. Bonilla. 2010b. Ligands of boron in *Pisum sativum* nodules are involved in regulation of oxygen concentration and rhizobial infection. *Plant Cell Environ.* 33:1039–1048.

Reid, R. 2007. Identification of boron transporter genes likely to be responsible for tolerance to boron toxicity in wheat and barley. *Plant Cell Physiol.* 48:1673–1678.

Reid, R. 2010. Can we really increase yields by making crop plants tolerant to boron toxicity? *Plant Sci.* 178:9–11.

Reid, R.J. and K.L. Fitzpatrick. 2009. Redistribution of boron in leaves reduces boron toxicity. *Plant Signal. Behav.* 4:1091–1093.

Reid, R.J., J.E. Hayes, A. Post, J.C.R. Stangoulis, and R.D. Graham. 2004. A critical analysis of the causes of boron toxicity in plants. *Plant Cell Environ.* 27:1405–1414.

Rerkasem, B., R.W. Bell, S. Lodkaew, and J.F. Loneragan. 1993. Boron deficiency in soybean (*Glycine max* L. Merr.), peanut (*Arachis hypogaea* L.) and black gram (*Vigna mungo* L. Hepper): Symptoms in seeds and differences among soybean cultivars in susceptibility to boron deficiency. *Plant Soil* 150:289–294.

Rerkasem, B. and S. Jamjod. 2004. Boron deficiency in wheat: A review. *Field Crops Res.* 89:173–186.

Rerkasem, B., R. Netsangtip, R.W. Bell, J.F. Loneragan, and N. Hiranburana. 1988. Comparative species responses to boron on a typic-tropaqualf in Northern Thailand. *Plant Soil* 106:15–21.

Rhoades, J.D., R.D. Ingvalson, and J.T. Hatcher. 1970. Adsorption of boron by ferromagnesian minerals and magnesium hydroxide. *Soil Sci. Soc. Am. Proc.* 34:938–941.

Riley, M.M. 1987. Boron toxicity in barley. *J. Plant Nutr.* 10:2109–2115.

Riley, M.M. and A.D. Robson. 1994. Pattern of supply affects boron toxicity in barley. *J. Plant Nutr.* 17:1721–1738.

Roberts, S. and J.K. Rhee. 1990. Boron utilization by potato in nutrient cultures and in field plantings. *Commun. Soil Sci. Plant Anal.* 21:921–932.

Robertson, L.S., B.D. Knezek, and J.O. Belo. 1975. Survey of Michigan soils as related to possible boron toxicities. *Commun. Soil Sci. Plant Anal.* 6:359–373.

Rosolem, C.A., and V.M. Leite. 2007. Coffee leaf and stem anatomy under boron deficiency. *Revista Brasileira De Ciencia Do Solo* 31:477–483.

Rowe, R.I. and C.D. Eckhert. 1999. Boron is required for zebrafish embryogenesis. *J. Exp. Biol.* 202:1649–1654.

Ruiz, J.M., G. Bretones, M. Baghour, A. Belakbir, and L. Romero. 1998. Relationship between boron and phenolic metabolism in tobacco leaves. *Phytochemistry* 48:269–272.

Ruiz, J.M., R.M. Rivero, and L. Romero. 2003. Preliminary studies on the involvement of biosynthesis of cysteine and glutathione concentration in the resistance to B toxicity in sunflower plants. *Plant Sci.* 165:811–817.

Ryan, J., S. Miyamoto, and J.L. Stroehlein. 1977. Relation of solute adsorbed boron to the boron hazard in irrigation water. *Plant Soil* 47:253–256.

Saarela, I. 1985. Plant-available boron in soils and the boron requirement of spring oilseed rapes. *Ann. Agric. Fenn.* 24:183–265.

Sah, R.N. and P.H. Brown. 1997. Techniques for boron determination and their application to the analysis of plant and soil samples. *Plant Soil* 193:15–33.

Sakya, A.T., B. Dell, and L. Huang. 2002. Boron requirements for *Eucalyptus globulus* seedlings. *Plant Soil* 246:87–95.

Schiffler, A.K., V.D. Jolley, J.E. Christopherson, and B.L. Webb. 2005a. Pressurized hot water and DTPA-sorbitol as viable alternatives for soil boron extractions. I. Boron-treated soil incubation and efficiency of extraction. *Commun. Soil Sci. Plant Anal.* 36:2179–2187.

Schiffler, A.K., V.D. Jolley, J.E. Christopherson, and B.L. Webb. 2005b. Pressurized hot water and DTPA-sorbitol as viable alternatives for soil boron extractions. II. Correlation of soil extraction to responses of boron-fertilized alfalfa. *Commun. Soil Sci. Plant Anal.* 36:2189–2207.

Schnurbusch, T., J. Hayes, M. Hrmova et al. 2010. Boron toxicity tolerance in barley through reduced expression of the multifunctional aquaporin HvNIP2;1. *Plant Physiol.* 153:1706–1715.

Shattuck, V.I. and B.J. Shelp. 1987. Effect of boron nutrition on hollow stem in broccoli (*Brassica oleracea* var. italica). *Can. J. Plant Sci.* 67:1221–1225.

Shelp, B.J. 1988. Boron mobility and nutrition in broccoli (*Brassica oleracea* var. italica). *Ann. Bot.* 61:83–91.

Shelp, B.J., E. Marentes, A.M. Kitheka, and P. Vivekanandan. 1995. Boron mobility in plants. *Physiol. Plant.* 94:356–361.

Shelp, B.J., R. Penner, and Z. Zhu. 1992. Broccoli (*Brassica oleracea* var italica) cultivar response to boron deficiency. *Can. J. Plant Sci.* 72:883–888.

Shelp, B.J. and V.I. Shattuck. 1987a. Boron nutrition and mobility, and its relation to hollow stem and the elemental composition of greenhouse grown cauliflower. *J. Plant Nutr.* 10:143–162.

Shelp, B.J. and V.I. Shattuck. 1987b. Boron nutrition and mobility, and its relation to the elemental composition of greenhouse grown root crops.1. Rutabaga. *Commun. Soil Sci. Plant Anal.* 18:187–201.

Shelp, B.J., V.I. Shattuck, and J.T.A. Proctor. 1987. Boron nutrition and mobility, and its relation to the elemental composition of greenhouse grown root crops.2. Radish. *Commun. Soil Sci. Plant Anal.* 18:203–219.

Shen, Z., X. Zhang, Z. Wang, and K. Shen. 1994. On the relationship between boron nutrition and development of anther (pollen) in rapeseed plant. *Scientia Agric. Sin.* 27:51–56.

Shen, Z.G., Y.C. Liang, and K. Shen. 1993. Effect of boron on the nitrate reductase-activity in oilseed rape plants. *J. Plant Nutr.* 16:1229–1239.

Sheng, O., G. Zhou, Q. Wei, S. Peng, and X. Deng. 2010. Effects of excess boron on growth, gas exchange, and boron status of four orange scion-rootstock combinations. *J. Plant Nutr. Soil Sci.* 173:469–476.

Sherrell, C.G. 1983a. Boron deficiency and response in white and red clovers and lucerne. *New Zealand J. Agric. Res.* 26:197–203.

Sherrell, C.G. 1983b. Boron nutrition of perennial ryegrass, cocksfoot, and timothy. *N. Z. J. Agric. Res.* 26:205–208.

Shkolnik, M.Y. (ed.). 1984. *Trace Elements in Plants—Developments in Crop Science*, Volume 6. Amsterdam, the Netherlands: Elsevier, pp. 1–463.

Shorrocks, V.M. 1997. The occurrence and correction of boron deficiency. *Plant Soil* 193:121–148.

Simojoki, P. 1991. Boron deficiency in barley. *Ann. Agric. Fenn.* 30:389–405.

Singh, D.V., R.P.S. Chauhan, and R. Charan. 1976. Safe and toxic limits of boron for grain in sandy loam and clay loam soils. *Indian J. Agron.* 21:309–310.

Sinha, P, R. Jain, and C. Chatterjee. 2000. Interactive effect of boron and zinc on growth and metabolism of mustard. *Commun. Soil Sci. Plant Anal.* 31:41–49.

Soltanpour, P.N. and A.P. Schwab. 1977. A new soil-test for simultaneous extraction of macro- and micronutrients in alkaline soils. *Commun. Soil Sci. Plant Anal.* 8:195–207.

Sommer, A.L. 1927. The search for elements essential in only small amounts for plant growth. *Science* 66:482–484.

Sommer, A.L. and C.B. Lipman. 1926. Evidence on the indispensable nature of zinc and boron for higher green plants. *Plant Physiol.* 1:231–249.

Sotiropoulos, T.E., A. Molassiotis, D. Almaliotis et al. 2006. Growth, nutritional status, chlorophyll content, and antioxidant responses of the apple rootstock MM 111 shoots cultured under high boron concentrations in vitro. *J. Plant Nutr.* 29:575–583.

Sotiropoulos, T.E., I.N. Therios, K.N. Dimassi, A. Bosabalidis, and G. Kofidis. 2002. Nutritional status, growth, CO_2 assimilation, and leaf anatomical responses in two kiwifruit species under boron toxicity. *J. Plant Nutr.* 25:1249–1261.

Sotomayor, C., R. Ruiz, and L. Munoz. 2012. Phloematic mobility of (10)Boron in kiwifruit (*Actinidia deliciosa*) mixed shoots. *Cienc. E Inv. Agr.* 39:563–567.

Sprague, H.B. (ed.). 1964. *Hunger Signs in Crops*, 3rd edn. New York: David McKay Co.

Stangoulis, J., M. Tate, R. Graham et al. 2010. The mechanism of boron mobility in wheat and canola phloem. *Plant Physiol.* 153: 76–881.

Stangoulis, J.C.R., R.J. Reid, P.H. Brown, and R.D. Graham. 2001. Kinetic analysis of boron transport in Chara. *Planta* 213:142–146.

Stubican, V. and R. Roy. 1962. Boron substitution in synthetic micas and clays. *Am. Miner.* 47:1166–1173.

Sutton, T., U. Baumann, J. Hayes et al. 2007. Boron-toxicity tolerance in barley arising from efflux transporter amplification. *Science* 318:1446–1449.

Takano, J., M. Kobayashi, K. Noguchi et al. 2002. Role of BOR1 in B transport in *Arabidopsis thaliana*. *Plant Cell Physiol.* 43:S191.

Takano, J., K. Miwa, L.X. Yuan, N. von Wiren, and T. Fujiwara. 2005. Endocytosis and degradation of BOR1, a boron transporter of *Arabidopsis thaliana*, regulated by boron availability. *Proc. Natl. Acad. Sci. USA* 102:12276–12281.

Takano, J., M. Wada, U. Ludewig, G. Schaaf, N. von Wiren, and T. Fujiwara. 2006. The Arabidopsis major intrinsic protein NIP5;1 is essential for efficient boron uptake and plant development under boron limitation. *Plant Cell* 18:1498–1509.

Tanaka, M., I.S. Wallace, J. Takano, D.M. Roberts, and T. Fujiwara. 2008. NIP6;1 is a boric acid channel for preferential transport of boron to growing shoot tissues in Arabidopsis. *Plant Cell* 20:2860–2875.

Tewari, R.K., P. Kumar, and P.N. Sharma. 2010. Morphology and oxidative physiology of boron-deficient mulberry plants. *Tree Physiol.* 30:68–77.

Tisdale, S.L. and W.L. Nelson (eds.). 1975. *Soil Fertility and Fertilizers*, 3rd ed. New York: Macmillan Publishing Co., Inc.

Touchton, J.T. and F.C. Boswell. 1975. Boron application for corn grown on selected Southeastern soils. *Agron. J.* 67:197–200.

Tsadilas, C.D., D. Dimoyiannis, and V. Samaras. 1997. Methods of assessing boron availability to kiwifruit plants growing on high boron soils. *Commun. Soil Sci. Plant Anal.* 28:973–987.

Tsadilas, C.D., T. Kassioti, and I.K. Mitsios. 2005. Influence of liming and nitrogen forms on boron uptake by tobacco. *Commun. Soil Sci. Plant Anal.* 36:701–708.

Tsadilas, C.D., N. Yassoglou, C.S. Kosmas, and Ch. Kallianou. 1994. The availability of soil boron fractions to olive trees and barley and their relationships to soil properties. *Plant Soil* 167:211–217.

Tsang, T. and S. Ghose. 1973. Nuclear magnetic resonance of 1H, 7Li, 11B, 23Na, 27Al in tourmaline (elbaite). *Am. Mineral.* 58:224–229.

van de Venter, H.A. and H.B. Currier. 1977. Effect of boron deficiency on callose formation and C-14 translocation in bean (*Phaseolus vulgaris* L) and cotton (*Gossypium hirsutum* L). *Am. J. Bot.* 64:861–865.

Vlamis, J. and A. Ulrich. 1971. Boron nutrition in growth and sugar content of sugar beets. *J. Am. Soc. Sugar Beet Technol.* 16:428–439.

Wallaart, R.A.M. 1980. Chemotaxonomy of Rosaceae.1. Distribution of sorbitol in Rosaceae. *Phytochem.* 19:2603–2610.

Wallace, T. 1951. *The Diagnosis of Mineral Deficiencies in Plants.* London, U.K.: HM Stationery Office, pp. 26–35, 61–75.

Wang, H.Y., Y.H. Wang, C.W. Du, F.S. Xu, and Y.H. Yang. 2003. Effects of boron and calcium supply on calcium fractionation in plants and suspension cells of rape cultivars with different boron efficiency. *J. Plant Nutr.* 26:789–806.

Warington, K. 1923. The effect of boric acid and borax on the broad bean and certain other plants. *Ann. Bot.* 37:629–672.

Watanabe, T. 1967. Geochemical cycling concentration of boron in the earth's crust, reprinted from Chemistry of the Earth's Crust, Vol. 2, Israel Program for Scientific Translations, Ltd. In: C.T. Walker (ed.) *Geochemistry of Boron*, Benchmark Paper in Geology 23, Stroudsburg, PA: Dowden, Hutchinson and Ross, Inc.

Wei, Y.Z., R.W. Bell, Y. Yang, Z.Q. Ye, K. Wang, and L.B. Huang. 1998. Prognosis of boron deficiency in oilseed rape (*Brassica napus*) by plant analysis. *Aust. J. Agric. Res.* 49:867–874.

Whetstone, R.R., W.O. Robinson, and H.G. Byers. 1942. Boron distribution in soils and related data. *U.S. Dep. Agric. Tech. Bull.* No. 797, Washington, DC.

Wikner, B. and L. Uppström. 1980. Determination of boron in plants and soils with a rapid modification of the curcumin method utilizing different 1,3-diols to eliminate interferences. *Commun. Soil Sci. Plant Anal.* 11:105–126.

Will, S., T. Eichert, V. Fernández, J. Möhring, T. Müller, and V. Römheld. 2011. Absorption and mobility of foliar-applied boron in soybean as affected by plant boron status and application as a polyol complex. *Plant Soil* 344:283–293.

Wimmer, M.A., E.S. Bassil, P.H. Brown, and A. Läuchli. 2005. Boron response in wheat is genotype dependent and related to boron uptake, translocation, allocation, plant phenological development and growth rate. *Funct. Plant Biol.* 32:507–515.

Wimmer, M.A. and T. Eichert. 2013. Review: Mechanisms for boron deficiency-mediated changes in plant water relations. *Plant Sci.* 203:25–32.

Wimmer, M.A. and H.E. Goldbach. 1999. A miniaturized curcumin method for the determination of boron in solutions and biological samples. *Z. Pflanzenernähr. Bodenkd.* 162:15–18.

Wimmer, M.A. and H.E. Goldbach. 2012. Boron-and-salt interactions in wheat are affected by boron supply. *J. Plant Nutr. Soil Sci.* 175:171–179.

Wimmer, M.A., G. Lochnit, E. Bassil, K.H. Muehling, and H.E. Goldbach. 2009. Membrane-associated, boron-interacting proteins isolated by boronate affinity chromatography. *Plant Cell Physiol.* 50:1292–1304.

Wimmer, M.A., K.H. Mühling, A. Läuchli, P.H. Brown, and H.E. Goldbach. 2003. The interaction between salinity and boron toxicity affects the subcellular distribution of ions and proteins in wheat leaves. *Plant Cell Environ.* 26:1267–1274.

Wojcik, P. 2005. Response of 'bluecrop' highbush blueberry to boron fertilization. *J. Plant Nutr.* 28:1897–1906.

Wolf, B. 1971. The determination of boron in soil extracts, plant materials, compost, manures, water and nutrient solutions. *Commun. Soil Sci. Plant Anal.* 2:363–374.

Wongmo, J., S. Jamjod, and B. Rerkasem. 2004. Contrasting responses to boron deficiency in barley and wheat. *Plant Soil* 259:103–110.

Woodbridge, C.G., A. Venegas, and P.C. Crandall. 1971. Boron content of developing pear, apple and cherry flower buds. *J. Am. Soc. Hortic. Sci.* 96:613–615.

Woodruff, J.R. 1979. Soil boron and soybean leaf boron in relation to soybean yield. *Commun. Soil Sci. Plant Anal.* 10:941–952.

Woods, W.G. 1996. Review of possible boron speciation relating to its essentiality. *J. Trace Elem. Exp. Med.* 9:153–163.

Xu, F.S., Y.H. Wang, and J. Meng. 2001. Mapping boron efficiency gene(s) in *Brassica napus* using RFLP and AFLP markers. *Plant Breed.* 120:319–324.

Xue, J.M., M.S. Lin, R.W. Bell, R.D. Graham, X. Yang, and Y. Yang. 1998. Differential response of oilseed rape (*Brassica napus* L.) cultivars to low boron supply. *Plant Soil* 204:155–163.

Yamagishi, M. and Y. Yamamoto. 1994. Effects of boron on nodule development and symbiotic nitrogen fixation in soybean plants. *Soil Sci. Plant Nutr.* 40:265–274.

Yamauchi, T., T. Hara, and Y. Sonoda. 1986. Effects of boron deficiency and calcium supply on the calcium metabolism in tomato plant. *Plant Soil* 93:223–230.

Yau, S.K. and M.C. Saxena. 1997. Variation in growth, development, and yield of durum wheat in response to high soil boron.1. Average effects. *Aust. J. Agric. Res.* 48:945–949.

Yermiyahu, U., A. Ben-Gal, R. Keren, and R.J. Reid. 2008. Combined effect of salinity and excess boron on plant growth and yield. *Plant Soil* 304:73–87.

Yermiyahu, U., A. Ben-Gal, and P. Sarig. 2006. Boron toxicity in grapevine. *HortScience* 41:1698–1703.

Yermiyahu, U., R. Keren, and Y. Chen. 1988. Boron sorption on composted organic-matter. *Soil Sci. Soc. Am. J.* 52:1309–1313.

Yu, Q., F. Baluska, F. Jasper, D. Menzel, and H.E. Goldbach. 2003. Short-term boron deprivation enhances levels of cytoskeletal proteins in maize, but not zucchini, root apices. *Physiol. Plant.* 117:270–278.

Yu, Q., A. Hlavacka, T. Matoh et al. 2002. Short-term boron deprivation inhibits endocytosis of cell wall pectins in meristematic cells of maize and wheat root apices. *Plant Physiol.* 130:415–421.

Yu, X.H. and P.F. Bell. 1998. Nutrient deficiency symptoms and boron uptake mechanisms of rice. *J. Plant Nutr.* 21:2077–2088.

Zarcinas, B.A. 1995. Suppression of iron interference in the determination of boron using the azomethine-H procedure. *Commun. Soil Sci. Plant Anal.* 26:713–729.

Zeng, C., Y. Han, L. Shi et al. 2008. Genetic analysis of the physiological responses to low boron stress in *Arabidopsis thaliana. Plant Cell Environ.* 31:112–122.

Zhang, X., Z. Shen, and K. Shen. 1994. The effect of boron on the development of floral organs and seed yield of rape. *Acta Ped. Sin.* 31:146–151.

Zhao, D.L. and D.M. Oosterhuis. 2002. Cotton carbon exchange, nonstructural carbohydrates, and boron distribution in tissues during development of boron deficiency. *Field Crops Res.* 78:75–87.

Zhao, D.L. and D.M. Oosterhuis. 2003. Cotton growth and physiological responses to boron deficiency. *J. Plant Nutr.* 26:855–867.

Zhao, H., L. Shi, X. Duan, F. Xu, Y. Wang, and J. Meng. 2008. Mapping and validation of chromosome regions conferring a new boron-efficient locus in *Brassica napus. Mol. Breed.* 22:495–506.

Zhao, Z., L. Wu, F. Nian et al. 2012. Dissecting quantitative trait loci for boron efficiency across multiple environments in *Brassica napus. Plos One* 7, 2012.

Zorn, W., G. Marks, H. Heß, and W. Bergmann. 2006. *Handbuch zur visuellen Diagnose von Ernährungsstörungen bei Kulturpflanzen.* Heidelberg, Germany: Elsevier Spektrum Akademischer Verlag, pp. 197–239.

9 Chlorine

David E. Kopsell and Dean A. Kopsell

CONTENTS

9.1 HISTORICAL BACKGROUND

Chlorine (from the Greek "chloros" meaning yellowish green, which describes the color of the gaseous diatomic molecule it exists as under standard conditions) is a member of the halogen family and is a strong oxidizing agent. Chlorine combines with elements to create chlorides (Winterton, 2000). Chlorine exists predominantly in soils and plants as chloride, one of the most common anions in nature. This anionic form contributes greatly to the behavior of chlorine in the environment and its function in plants (Winterton, 2000). Although plants can take up chloride in amounts of 50–500 $\mu mol \cdot kg^{-1}$, which is similar to macronutrient accumulation, the requirement of chloride for plant growth is considerably lower, clearly establishing it as a micronutrient (Mengel and Kirkby, 1987). Chlorine deficiency in plants seldom is observed in agriculture or nature due to the plentiful supplies of chlorine in the environment and its redistribution from natural occurrences such as rainfall, marine aerosols, and volcanic emissions. Although reports of chloride being beneficial to plant growth date back to 1862, it was not until 1954 that chlorine was proven to be a micronutrient for the growth of plants (Broyer et al., 1954; Hewitt, 1966). Ironically, the essentiality of chlorine resulted from a study of cobalt nutrition in tomato (*Lycopersicon esculentum* Mill.) where there was increased plant growth when cobalt was supplied as its chloride salt (Broyer et al., 1954). Little concern for chlorine nutrition existed, however, until research in several states of the United States in the 1980s demonstrated that crops grown on soils low in chlorine content responded to chlorine fertilizer applications (Lamond and Leikam, 2002). Even though chlorine has gained the attention of agronomists, much of the focus on chlorine in terms of crop production continues to be over the presence of excessive levels of chloride salts in soils, water, and fertilizers (Xu et al., 2000).

9.2   UPTAKE OF CHLORINE BY PLANTS

Most plant species take up chloride rapidly, and the rate of uptake depends primarily on the concentration in the soil solution. In six-day-old barley (*Hordeum vulgare* L.) seedlings in solution culture, Blevins et al. (1974) reported that chloride uptake is rapid initially but declines within a few hours after exposure to chloride salts; moreover, chloride uptake exceeded nitrate uptake during the first 4 h after nutrient treatment. Chloride uptake is a metabolically controlled, active process which is sensitive to variations in temperature and metabolic inhibitors (Xu et al., 2000). Some plant species seem to require higher concentrations of chloride than others, for example, onion (*Allium cepa* L.) in which chloride functions in stomatal movement (Schnabl and Raschke, 1980). Recent field observations and scientific research with sweet onions demonstrate that the chloride requirement for onions may be higher than previously thought, and research in which chloride was supplied at concentrations of up to 500 mg·L^{-1} in hydroponic culture demonstrates that the uptake of chloride was exceeded only by nitrogen, potassium, and phosphorus (Randle, 2004).

Chloride ions entering a root can pass through apoplastic or symplastic routes (White and Broadley, 2001). In the apoplastic route, the ions move in intercellular spaces and cell walls, but cannot move all the way to the vascular tissue, except in very young root tissues, as it is impeded by the Casparian band of the endodermis. In the symplastic route, ions cross the plasmalemma to enter cells initially and move from cell to cell through plasmodesmata. To enter the vacuole, chloride crosses the tonoplast. Movement of chloride from the soil solution into the xylem occurs against a concentration gradient. There is discrimination between chloride and similar ions such as bromide, and its translocation is impeded by inhibitors of anion transport. Therefore, chloride movement into and within plants appears to be carried out by selective transport systems (White and Broadley, 2001). These include proteinaceous carriers and channels, with different transporters being present in different membranes and in different tissues. A variety of different channels that can allow for the uptake of chloride have been identified in plant roots, and in many cases, these channels are also able to take up nitrate and in some cases sulfate (White and Broadley, 2001; Roberts, 2006).

The plasmalemma is fairly permeable to chloride, and under soil conditions where external chloride concentrations are lower than internal concentrations, uptake occurs against an electrochemical gradient. This action seems to occur via a symport of chloride and H$^+$, with the protons being generated by hydrolysis of ATP (White and Broadley, 2001; Saleh and Plieth, 2013). The observation that low soil pH values (i.e., abundant H$^+$) promote anion and chloride uptake (Mengel and Kirkby, 1987) can be seen to be due to an effect on this uptake system. Rubinstein (1974) reported a linear uptake of ^{36}Cl$^-$ in excised oat (*Avena sativa* L.) coleoptile sections if external pH was lowered from 5 to 3.

Under saline conditions, with high external chloride concentrations, the ion moves down an electrochemical gradient (White and Broadley, 2001; Saleh and Plieth, 2013). This transport is initially a fast process linked to membrane depolarization on exposure to salinity and probably through transport channels. Elgharably (2011) reported the shoot chloride concentration in wheat (*Triticum aestivum* L.) increased 5–6 mg·L^{-1} each time when soil EC increased from 3 to 6 dS·m^{-1} to 9 to 12 dS·m^{-1}, regardless of the level of phosphorus or nitrogen fertilization. Later, a slower influx may occur through a variety of mechanisms, possibly including cation-chloride cotransport, and maybe even cation channels (Saleh and Plieth, 2013). A cation-chloride cotransporter has been identified in the root plasmalemma of rice (*Oryza sativa* L.) (Kong et al., 2011). Two cation-chloride cotransporters have also been identified in arabidopsis (*Arabidopsis thaliana* Heynh.), and it seems that they are involved in regulating long-distance ion transport through controlling xylem loading or unloading. Genetic expression of these cotransporters was observed with xylem parenchyma cells (Colmenero-Flores et al., 2007).

Two different fluxes initially influence the rate of chloride uptake from the soil solution. Flux across the plasmalemma into root cells approximately equals the flux across the tonoplast into the vacuole at an external chloride concentration below approximately 5 mM, and external concentration determines the rate of uptake (White and Broadley, 2001). Above this concentration, the flux from cytosol to vacuole continues to increase, but the flux across the plasmalemma does not, so the direct relationship between external concentration and rate of overall uptake is broken (White and Broadley, 2001). There is also efflux of chloride from the cells if tissue chloride concentrations are high.

Chloride taken up by the roots is moved to the shoots in the xylem, with only a small proportion returning to the roots in the phloem and a very small proportion remaining in the roots (Peuke, 2010). This movement is particularly efficient in young root tissues, where the Casparian band has not developed, but some movement to the xylem occurs even in mature parts of the root system (White and Broadley, 2001). In general, a correlation occurs between the concentration of chloride in the soil solution and the amount that reaches the xylem, but this action is influenced by other factors, such as the availability of competitive ions (White and Broadley, 2001; Peuke, 2010).

Leaves absorb chlorine as either chloride in solution or as gaseous Cl_2 (Mills and Jones, 1996). Simini and Leone (1982) demonstrated that the foliage of a mixture of plants (bean, *Phaseolus vulgaris* L.), tomato, white pine (*Pinus strobus* L.), Douglas fir (*Pseudotsuga menziesii* Franco), mountain laurel (*Kalmia latifolia* L.), Norway spruce (*Picea abies* H. Karst.), red maple (*Acer rubrum* L.), arborvitae (*Thuja occidentalis* L.), holly (*Ilex opaca* Aiton), hybrid poplar (*Populus maximowiczii* Henry × *P. trichocarpa* Torr. and Gray), and dogwood (*Cornus florida* L.) absorbed more chloride from a simulated salt spray when exposed to short photoperiods, low temperatures, and high relative humidity than with long photoperiods, high temperatures, and low relative humidity. Although plants, especially woody ornamentals, can absorb chlorine gas, it can be toxic at concentrations above $0.1 \text{ mL} \cdot \text{L}^{-1}$, causing such abnormalities as foliar chlorosis, flecking and stippling, glazing, scorching, and necrosis (Coder, 2011).

9.3 PHYSIOLOGICAL RESPONSES OF PLANTS TO SUPPLY OF CHLORINE

With wheat, chlorine fertilization has been associated with accelerated plant development, improved kernel weight, reduced late-season foliar lodging, yield increases, and reduced disease damage (Murphy et al., 2007). However, Thomason et al. (2001) reported that chlorine fertilization increased yield with wheat only in 3 years of an 8-year trial in Oklahoma, USA. Younts and Musgrove (1958) reported that incidence of stalk rot in corn (*Zea mays* L.) decreased if potassium was supplied as KCl instead of K_2SO_4. The effect of applied chlorine fertilizers to suppress take-all disease in wheat over an 8-year study in Oklahoma was inconsistent (Thomason et al., 2001).

9.3.1 ROLE IN PHOTOSYNTHESIS

Chlorine is known to be a cofactor in the catalysis of water oxidation in the Hill reaction of Photosystem II since the relationship was first reported by Warburg and Lüttgens in 1944 (Homann, 2002). Despite being an essential cofactor, the amount of chlorine required by plants for photosynthesis is very small (Kafkafi et al., 2001). However, chlorine depletion in isolated chloroplastic membranes of spinach (*Spinacia oleracea* L.) decreased oxygen evolution by up to 50% (Olesen and

Andréasson, 2003). Chlorine deficiency in plants does not inhibit photosynthesis because leaves exhibiting chloride deficiency concentrate chloride in chloroplasts (Kafkafi et al., 2001). However, chloride deficiency lowers leaf cell expansion, producing a smaller leaf area in which photosynthesis can occur (Terry, 1977; Marschner, 1995). It is thought that chloride anions interact with the oxidation–reduction chemistry of the four-Mn^{2+} ion cluster in the oxygen-evolving complex (water-splitting complex), but this interaction is understood poorly (Popelková and Yocum, 2007). Fine and Frasch (1992) reported that chloride prevents hydrogen peroxide formation. Olesen and Andréasson (2003) proposed that chloride works together with lysine side chains and other charged amino acids of proteins to maintain a proton-relay network that allows the transport and release of protons in the water-splitting reaction.

9.3.2 ROLE IN STOMATAL REGULATION

Stomatal guard-cell movement is the result of the movement of K^+ and corresponding anions, such as malate or chloride, increasing cell turgor in the guard cells and thus opening the stomatal pore (Kafkafi et al., 2001). Stomata open at dawn as the guard cells become turgid and close at dusk when the guard cells lose this turgidity. This diurnal cycle can be cancelled by imposition of water-deficit stress, which causes open stomata to close. The increase in turgidity with the onset of daylight is brought about by a rapid inward flux of K^+ ions, which lowers water potential of the guard cells, and the decreased turgidity as darkness falls is brought about by a rapid flux of K^+ ions from the guard cells. The flux of K^+ ions is brought about by the action of ATPase enzymes in the plasma membrane generating an H^+ gradient, so that movement of the positively charged cations in the opposite direction to the protons maintains electroneutrality. However, in many plant species, the inward flux of K^+ ions is balanced in part by accumulation of malate, and the K^+ flux continues. Malate that accumulates during stomatal opening is synthesized in the guard cells, and this synthesis is affected by the concentration of chloride and its effect on cellular pH and phosphoenolpyruvate (PEP) carboxylase activity (Raschke and Schnabl, 1978; Du et al., 1997). An influx of chloride causes inhibition of PEP carboxylase, where the absence of chloride increases malate synthesis through a stimulation of PEP carboxylase (Kafkafi et al., 2001). When the K^+ ions leave the guard cells, malate is metabolized. However, the uptake of K^+ also can be balanced by the uptake of anions, and chloride is the anion most commonly used as a counterion in this process. In some plants, for example, onion, malate is not formed in the guard cells, so movement of chloride is very important in stomatal opening and closing (Marschner, 1995). In fact, onion guard cells contain equivalent amounts of K^+ and Cl^- ions (Kafkafi et al., 2001). Kiwifruit (*Actinidia chinensis* Planch.) is another example of a plant that uses chloride rather than organic anions for charge balance (Buwalda and Smith, 1991). In arabidopsis, aluminum-activated malate transporter 9 functions as a chloride channel in the guard-cell tonoplast and has a major role in stomatal opening (De Angeli et al., 2013). It is likely that the rate of synthesis of malate in the guard cells, or the uptake of malate from the apoplast, is not as fast as the inward movement of K^+ ions and of Cl^- ions into the cytoplasm and then into the vacuole (De Angeli et al., 2013). Chloride concentrations in the vacuoles of guard cells increase when stomata open and decrease when they close (Kafkafi et al., 2001). It is not clear what transporters are involved in uptake of chloride through the plasmalemma into the guard-cell cytoplasm, although various channels involved in efflux of anions during stomatal closure have been identified (Tavares et al., 2011).

9.3.3 ROLE IN VACUOLAR TRANSPORT

Salt-tolerant plants take up salts and store them in the vacuole if exposed to high salt concentrations, and this action enables the plant to maintain turgor without the synthesis of organic osmotica (Martinoia et al., 1986). However, the mobility of various anions and cations across the tonoplast appears to differ (Martinoia et al., 1986). In plasmalemma, ATPases hydrolyze water and generate a movement of H^+ into the apoplast and vacuole. This process results in an increase of cytoplasmic pH and a decrease of apoplastic and vacuolar pH, which provides energy for the movement of cations into the cell and anions into the vacuole (Marschner, 1995). This proton gradient gives rise to the movement of chloride into the vacuoles of guard cells during stomatal opening. It has been known for many years that chloride stimulates tonoplast ATPases. Stimulation of H^+ movement into vacuoles is caused partly by chloride dissipating the tonoplast membrane potential, but the action is mainly a direct stimulation arising from interaction between chloride and protein (Churchill and Sze, 1984). Potassium chloride can have an additional effect since K^+ stimulates the plasmalemma-bound ATPases. In the process, K^+ is accumulated in the cytoplasm, and chloride accumulates in the vacuole. Other anions, including sulfate and nitrate, only slightly stimulate the activity of tonoplast ATPases. 4,4'-Diisothiocynao-2,2'-stilbenedisulfonic acid and MgATP partially inhibited chloride vacuole accumulation in young barley leaves, whereas p-chloromercuribenzene sulfonate and $HgCl_2$ stimulated accumulation (Martinoia et al., 1986; Wissing and Smith, 2000). Although Martinoia et al. (1986) showed that the tonoplast is very permeable to anions such as nitrate and chloride, they proposed the theory of a specific chloride channel that regulated chloride movement to and from the vacuole. The K_m for chloride uptake at low external chloride concentrations was 2.3 mM and was by active absorption, but at elevated chloride concentrations, uptake was characteristic of a diffusional process (Martinoia et al., 1986). Wissing and Smith (2000) reported that vacuolar chloride transport was not affected by sulfate, malate, or nitrate in the halophytic common ice plant (*Mesembryanthemum crystallinum* L.).

9.3.4 OSMOTIC ADJUSTMENT IN HALOPHYTES

In the classic scheme of Greenway and Munns (1980) describing responses of plants to salinity, different species are positioned on a spectrum from very salt sensitive to very salt tolerant. Species at any point on the spectrum can be subdivided into includers, those species that take up the ions causing salinity, or excluders, those species that select against the ions and do not take them up, or at least have efflux of these ions that equals or exceeds influx. In a salt-tolerant (halophyte) plant species, the imposition of salinity by NaCl gives rise to considerable uptake of Na^+ and Cl^- ions in includers, but selection against their uptake in excluders (Greenway and Munns, 1980).

In includers, potential for phytotoxicity occurs due to the presence of excessive chloride, but the action of tonoplast ATPases results in the chloride accumulating in the vacuole, where it does not interfere with cellular metabolism. The accumulation of chloride lowers the water potential of the cells below that in the saline soil solution and allows water to enter the plant. Thus, in halophytes with the includer mechanism, the accumulation of chloride in the vacuoles enables the plants to acquire water (Greenway and Munns, 1980).

In excluder halophytes, the potential toxicity of chloride (and Na^+) is avoided by excluding these ions from shoots (Garthwaite et al., 2005). These plants take up water due to a lowered water potential in the cells from synthesis and accumulation in the vacuoles of malate and other soluble organic anions. The salt tolerance of wild barley (*Hordeum marinum* Huds.) is due to exclusion of sodium and chloride from shoots. When wild barley species was grown in 300 mM NaCl, sodium and chloride ions accounted for 6% and 10% of leaf osmotic potential, respectively, whereas in domestic barley grown in 300 mM NaCl, sodium and chloride ions accounted 25% and 21% of leaf osmotic potential (Garthwaite et al., 2005).

9.3.5 ACTIVATION OF ENZYMES

Several plant enzymes appear to require chloride for optimal activity (Kafkafi et al., 2001). They include asparagine synthetase (Rognes, 1980) and ATPase (Churchill and Sze, 1984). Chloride increased the affinity of asparagine synthetase 30-fold for its substrate glutamate in cotyledons of European yellow lupine (*Lupinus luteus* L.) seedlings (Rognes, 1980). Hence, in plants in which asparagine is the major compound in the long-distance transport of organic nitrogen, chloride might play a role in nitrogen metabolism (Marschner, 1995). ATPase in the tonoplast of oat roots (*Avena sativa* L.) increased asymptotically with increasing chloride concentrations up to 50 mM (Churchill and Sze, 1984). Chloride also stimulated the activity of vacuolar-type H^+-ATPase from guard-cell protoplasts of Asiatic dayflower (*Commelina communis* L.) (Willmer et al., 1995). Although mammalian and bacterial amylase enzymes require chloride as a cofactor (Metzler, 1979), Swain and Dekker (1966) reported that α-amylase enzyme in cotyledons of germinating peas (*Pisum sativum* L.) did not require chloride. Wedding (1989) reported that chloride exhibited inhibitory effects on malic enzymes in plant tissues.

9.3.6 OTHER FUNCTIONS OF CHLORIDE IN PLANTS

Addition of chlorine fertilizers during crop growth reduced the severity of 15 different foliar and root diseases on 11 crops (Heckman, 2007). Chlorine applied as KCl reduced foliar and root diseases in small grains if soil chloride was greater than 33.6 kg·ha^{-1} (Fixen et al., 1986). Heckman (1998) reported the specific effect of chloride in controlling the incidence of corn stalk rot (*Gibberella zeae* Petch, and *Diplodia maydis* Sacc.) and anthracnose (*Colletotrichum graminicola* G.W. Wilson). Although the direct mechanism of chloride on disease suppression is not fully understood (Xu et al., 2000), it appears to originate from a suppression of nitrification by chloride in the soil solution. If chloride is added to acidic (pH 5.0–5.5) soils, nitrification is suppressed and results in an increase in uptake of NH_4^+, a decrease in the uptake of NO_3^-, and a decrease in rhizosphere pH. This acidification creates a favorable environment for Mn^{2+} uptake, which may be beneficial in disease control (Huber and Wilhelm, 1988). It should be noted that balanced and adequate fertility for any crop reduces plant stress and improves resistance to diseases (Krupinsky et al., 2002).

More than 130 natural chlorine-containing compounds have been isolated from plants including polyacetylenes, thiophenes, iridoids, sesquiterpene lactones, pterosinoids, diterperenoids, steroids and gibberellins, maytansinoids, alkaloids, chlorinated chlorophyll, chloroindoles and amino acids, phenolics, and fatty acids (Engvild, 1986). Some chlorine-containing compounds may behave as hormones in the plant, or they may have a function in plant protection (Heckman, 2007). A chlorinated indole-3-acetic acid (IAA) in the seeds of some legume species enhanced hypocotyl elongation at ten times the rate of IAA alone (Hofinger and Bottger, 1979). Chloride seems to be essential for pollen germination and pollen-tube growth (Tavares et al., 2011).

9.4 GENETICS OF ACQUISITION OF CHLORINE BY PLANTS

The reported differences in tolerance to and uptake of chloride salts among plant genera, species, and cultivars demonstrate genetic regulation (Xu et al., 2000; Loudet et al., 2003; Munns and Tester, 2008; Darvishzadeh and Alavi, 2011). Garthwaite et al. (2005) reported that wild barley was more tolerant of high chloride and sodium ion concentrations than cultivated barley. Darvishzadeh and Alavi (2011) performed a genetic analysis of eight oriental tobacco (*Nicotiana tabacum* L.) genotypes exhibiting variability for chloride leaf concentration in Iran. The tobacco genotypes were crossed in a diallel-mating system without reciprocal crosses to produce 28 F_1 hybrid combinations. Although general combining ability and specific combining ability effects for chloride leaf concentration were highly significant, indicating additive and dominant genetic effects, the greater mean square of specific combining ability in the analysis of variance indicated a nonadditive effect in controlling chloride leaf concentration. The analysis of variance indicated that the effect of location or genotype was significant on the chloride concentration in tobacco leaves, but the location by genotype interaction was not (Darvishzadeh and Alavi, 2011).

Loudet et al. (2003) reported highly significant differences between 415 genotypes of an F_7 generation of Bay-0 × Shahdara arabidopsis population in accumulation of chloride, when supplied with either a low or high supply of nitrate. The chloride accumulation was heritable in arabidopsis grown under low nitrate, and a result that Loudet et al. (2003) reasoned showed that chloride is a very important osmoticum if nitrate is deficient. Furthermore, all of the quantitative trait loci (QTLs) for water content in the plants grown in low nitrate colocated with QTLs for chloride concentration in these plants. QTLs for water content are linked to time of flowering, and this linkage is further proof that the chloride is acting as an osmoticum (Loudet et al., 2003). Eight QTLs for chloride concentration were identified in the high-nitrate arabidopsis, and six were in the low-nitrate plants. One gene locus where four QTLs overlap (QTLs for nitrate concentration in high-nitrate plants, water content in high-nitrate plants, water content in low-nitrate plants, and chloride concentration in low-nitrate plants) seemed to overlap with an aquaporin gene and two QTLs on chromosome 5 for chloride concentration in high-nitrate arabidopsis colocated to two chloride channel genes (Loudet et al., 2003).

9.5 CONCENTRATIONS OF CHLORINE IN PLANTS AND PLANT PARTS

The amount of chloride in plants varies with habitat and production practices Table 9.1 (Xu et al., 2000). There are also differences among plant genera, species, and cultivars in tolerance to chloride salts (Xu et al., 2000). Normal plant leaf content of chlorine is 50–200 mg·kg^{-1}, but some plants can contain up to 20,000 mg·kg^{-1} without adverse effects (Shulte, 1999). Deficiencies generally occur when levels are in the range of 2–20 mg·kg^{-1}, whereas toxicity occurs in the range of 0.5%–2.0% for sensitive plants and greater than 4% for tolerant plants. Darvishzadeh and Alavi (2011) reported the leaf tissue chloride content of tobacco leaves, a plant sensitive to excessive chloride, ranged from 1.61% to 2.81%, and a leaf chloride content greater than 2.0% results in poor burning of tobacco leaves. However, sufficient chloride ranges for onion tissue are reported from 0.25% to 5%.

Crops can remove substantial amounts of chloride from soils, especially when available chloride levels are high, whereas the removal of chloride in grains is very limited (Xu et al., 2000). Chloride concentration is usually highest in the leaf blades, followed by the petioles, stems, fruit stems, and fruit. In cereal crops, the highest chloride content is usually in stems and leaves, followed by grain and chaff (Mills and Jones, 1996). In rice plants, the highest concentration of chloride was in the roots (4800 mg·kg^{-1} dry weight) followed by the straw (3800 mg·kg^{-1} dry weight) and the rice hull and bran (1270 mg·kg^{-1} dry weight), and the lowest was in the polished rice (140 mg·kg^{-1} dry weight) (Tsukada and Takeda, 2008).

Chloride is stored partially in the vacuoles of leaf cells, and its negative charge is balanced by Ca^{2+} and Mg^{2+} (Beringer et al., 1990). Lloyd et al. (1989) reported differences in salinity tolerance between species of citrus (Citrus sp. L.) depended on the compartmentalization of chloride anions under saline conditions. In leaf cells of NaCl-treated halophytes, Na^+ tended to concentrate in the cytoplasm, whereas chloride was distributed evenly in the cytosol and the vacuole (Eshel and Waisel, 1979). Chloride has high mobility in short-distance or long-distance transport. The control of chloride transport to shoots may be due to reduced loading of chloride via anion channels but also may be due to increased active retrieval of chloride from the xylem stream (Munns and Tester, 2008). Leaf chloride content is controlled not only by root uptake but also by restricted translocation of chloride from the roots to the leaves (Kafkafi et al., 2001). Sites of tissue chloride accumulation indicate that chloride can be retrieved from the xylem in petioles, stems, and roots (Munns and Tester, 2008).

For determination of chloride in plant tissue, techniques such as atomic absorption spectrometry (AAS), inductively coupled plasma mass spectrometry (ICP-MS), inductively coupled plasma optical emission spectrometry (ICP-OES), instrumental neutron activation analysis (INAA), and x-ray fluorescence (XRF) have been used (Mello et al., 2013).

TABLE 9.1
Plant Species Variation for Chloride Concentrations

Plant Species by Crop Type	Plant Part and Treatment	Chlorine Concentration (mg kg^{-1} Dry Mass)	Reference
Agronomic crops			
Alfalfa (*Medicago sativa* L.)	Shoot	900–2,700	Eaton (1966)
	Shoot	6,910	Wheal and Palmer (2010)
Barley (*Hordeum vulgare* L.)	Heads	1,900	Sheppard et al. (1999)
	Stalks	3,600	
Canola (*Brassica napus* L.)	Leaves	6,970	Wheal and Palmer (2010)
Corn (*Zea mays* L.)	Cob	420	Sheppard et al. (1999)
	Leaves and stalk	370	
	Silk and ear leaves	250	
Pasture grasses	Leaves	4,100–38,000	Tsukada et al. (2007)
Peanut (*Arachis hypogaea* L.)	Shoot	3,900	Wang et al. (1989)
Rice (*Oryza sativa* L.)	Polished rice	127–220	Tsukada and Takeda (2008)
	Rice bran	670	
	Hull	600	
	Straw	3,800	
	Root	4,800	
Wheat (*Triticum aestivum* L.)	Shoot	200–22,000	Elgharably (2011)
Vegetable crops			
Green bean (*Phaseolus vulgaris* L.)	Stem and leaf	270	Sheppard et al. (1999)
	Pod	370	
Beet (*Beta vulgaris* L.)	Petiole and Leaves	2,800	Sheppard et al. (1999)
	Root	1,100	
Cabbage (*Brassica oleracea* var. *capitata* L.)	Inner leaf	1,000	Sheppard et al. (1999)
	Outer leaf	820	
	Stalk	520	
	Head	2,700–10,400	Tsukada et al. (2007)
Chinese cabbage (*Brassica chinensis* L.)	Shoot	4,520–24,800	Tsukada et al. (2007)
Carrot (*Daucus carota* L.)	Shoot	1,630–5,430	Sahin et al. (2012)
	Root	4,000–14,800	
	Petiole and Leaf	640	Sheppard et al. (1999)
	Root	1,200	
	Root	3,910–8,560	Tsukada et al. (2007)
Chives (*Allium schoenoprasum* L.)	Leaves	1,200	Sheppard et al. (1999)
Cucumber (*Cucumis sativus* L.)	Flesh	590	Sheppard et al. (1999)
	Vine	300	
	Fruit	1,430–9,690	Tsukada et al. (2007)
Dill (*Anthethum graveolens* L.)	Plant	640	Sheppard et al. (1999)
Garlic (*Allium sativum* L.)	Clove	125–588	Tsukada et al. (2007)
Lettuce (*Lactuca sativa* L.)	Leaves	790	Sheppard et al. (1999)
Mung bean (*Vigna radiata* R. Wilczek)	Leaves	1,480 ± 13	Wheal and Palmer (2010)
Onion (*Allium cepa* L.)	Leaves	450	Sheppard et al. (1999)
	Bulb	300	
Pea (*Pisum sativum* L.)	Stem and leaves	450	Sheppard et al. (1999)
	Pod	320	

(Continued)

TABLE 9.1 (*Continued*)
Plant Species Variation for Chloride Concentrations

Plant Species by Crop Type	Plant Part and Treatment	Chlorine Concentration (mg kg⁻¹ Dry Mass)	Reference
Potato (*Solanum tuberosum* L.)	Stem and leaves	1,300	Sheppard et al. (1999)
	Tuber	620	
	Leaves	16,600 ± 220	Wheal and Palmer (2010)
Pumpkin (*Cucurbita pepo* L.)	Fruit and shell	400	Sheppard et al. (1999)
	Fruit	809–3020	Tsukada et al. (2007)
Tomato (*Lycopersicon esculentum* Mill.)	Stem and Leaf	690	Sheppard et al. (1999)
	Fruit	550	
	Fruit	3,650–9,440	Tsukada et al. (2007)
	Leaves	6,340 ± 40	Wheal and Palmer (2010)
Fruit crops			
Avocado (*Persea americana* Mill.)	Leaves	1,500–4,000	Bar et al. (1997)
Citrus (*Citrus* spp. L.)	Leaves	716	Wheal and Palmer (2010)
Cantaloupe (*Cucumis melo* L.)	Fruit	420	Sheppard et al. (1999)
Grape (*Vitis vinifera* L.)	Petiole	4,100	Wheal and Palmer (2010)
Kiwifruit (*Actinidia deliciosa* A. Chev.)	Leaves	6,000–13,000	Smith et al. (1987)
Melon (*Cucumis melo* L.)	Leaves	1,310–2,700	Tsukada et al. (2007)
Peach (*Prunus persica* Batsch.)	Leaves	900–3,900	Robinson (1986)
Strawberry (*Fragaria* × *ananassa* Duchesne)	Shoot	1,000–5,000	Robinson (1986)
Ornamental Plants			
Zinnia elegans Jacq.	Shoots	2,400–5,700	Carter and Grieve (2010)
Lupine (*Lupinus* spp. L.)	Leaves	9,570 ± 44	Wheal and Palmer (2010)

9.6 RATIOS OF CHLORINE AND INTERACTIONS WITH OTHER ELEMENTS

Chloride has reported uptake interactions with other negatively charged anions such as nitrate, sulfate, and phosphate, and interactions with cations such as ammonium, potassium, calcium, and micronutrients have been reported (Kalkafi et al., 2001).

Increasing concentrations of nitrate in solution decreased chloride concentrations in avocado (*Persea americana* Mill.; Bar et al. 1997), broccoli (*Brassica oleracea* L. var. *italica* Plenck.; Liu and Shelp, 1996), citrus seedlings (Chapman and Liebig, 1940), kiwifruit (*Actinidia* sp., Smith et al., 1987), melon (*Cucumis melo* L.), tomato (Feigin et al., 1987), strawberry (*Fragaria* x *ananassa* Duchesne), and spring wheat (*Triticum aestivum* L.; Wang et al., 1989). Loudet et al. (2003) reported the mean values of chloride concentration in recombinant inbred lines of arabidopsis were 1.34 µg·kg⁻¹ dry weight in plants given 10 mM nitrate and 14.07 µg·kg⁻¹ dry weight in plants given 3 mM nitrate. Influx of chloride in barley is inhibited by nitrate in the rootzone (Glass and Siddiqi, 1985). Bar et al. (1997) suggested that after nitrate is assimilated following its uptake, the chloride electrochemical potential gradient builds up because chloride maintains its negative charge, thus reducing chloride uptake. Conversely, the inhibition of nitrate uptake by chloride depends on the plant species and the concentrations of nitrate and chloride in the medium (Cerezo et al., 1982). The addition of moderate amounts of chloride to the medium of broccoli plants decreased the nitrate content by increasing nitrate reduction (Liu and Shelp, 1996). Barber et al. (1989) reported that nitrate reductase activity is inhibited by chloride in spinach (*Spinacia oleracea* L.). The competition of nitrate and chloride was stronger in salt-sensitive plants, such as peanut (*Arachis hypogaea* L.), than in salt-tolerant plants, such as cotton (*Gossypium hirsutum* L.) (Leidi et al., 1992).

Plants fertilized with ammonium (NH_4^+) usually contain much more chloride than plants fertilized with nitrate or a combination of ammonium and nitrate, irrespective of the chloride level in the nutrient solution (White and Broadley, 2001). Following uptake of ammonium, anions have to be taken up to maintain electrical neutrality in the nutrient solution (Kafkafi et al., 2001). Chloride suppresses nitrification in soils and leads to higher NH_4^+/NO_3^- ratios. Meador and Fisher (2013) reported that free chloride from sodium hypochlorite was significantly decreased in nutrient solutions containing ammonium fertilizers, whereas it increased with nitrate or urea fertilizers.

The interaction of chloride with phosphorus is complex (Kafkafi et al., 2001). In spring wheat, phosphorus uptake was stimulated by small amounts of chloride but was inhibited by high amounts of chloride (Wang et al., 1989). Although James et al. (1970) showed that the chloride content of potato (*Solanum tuberosum* L.) petioles was not affected by phosphate fertilizers, increasing the supply of phosphate decreased the uptake of chloride by wheat from saline soil (Elgharably, 2011). Kafkafi et al. (2001) reasoned that these conflicting reports were due to the effect of chloride concentrations on nitrate uptake, soil rhizosphere pH, and phosphorus availability.

Chloride applications can increase potassium uptake and content in citrus seedlings (Chapman and Liebig, 1940), avocado (*Persea americana* Mill.), strawberry (Wang et al., 1989), potato (Jackson and McBride, 1986), and kiwifruit (*Actinidia deliciosa* C.F. Liang et A.R. Ferguson) (Buwalda and Smith, 1991). If grown in nutrient solutions containing equivalent amounts of K^+, plants generally take up more potassium when KCl rather than K_2SO_4 is the potassium source (Kafkafi et al., 2001). On the other hand, more K^+ was taken up by barley plants from KNO_3 treatments than from KCl treatments (Blevins et al., 1974). In chloride-deficient soils in the Magdalena River Valley, Columbia, chloride supplied as KCl or NaCl increased the uptake of potassium, calcium, and magnesium in rachises, petioles, and stems of oil palm (*Elaeis guineensis* Jacq.) (Dubos et al., 2011).

Banuls et al. (1997) demonstrated that accumulation of chloride in orange (*Citrus sinensis* Osbeck) leaves was reduced by increasing calcium in nutrient solution culture, whereas the chloride content of roots increased slightly. Kafkafi et al. (2001) explained that this effect might be due to withdrawal of chloride from the xylem stream by calcium.

Boron deficiency reduces chloride plasmalemma and tonoplast transport. It appears that boron is required to maintain inward transport across the plasmalemma and the tonoplast but has no effect on the processes of efflux. Bromide in nutrient solution inhibits the uptake of chloride competitively, but fluoride and iodide have little effect (White and Broadley, 2001). The concentration of bromide in carrot (*Daucus carota* L.) shoots and storage roots (applied as bromate, but probably converted to Br^- before uptake) was lowered by supply of chloride, and the concentration of chloride in shoots and roots was lowered by supply of bromate (Sahin et al., 2012). Under saline conditions, added silicon increased net photosynthesis, stomata conductance, and transpiration, yet inhibited the accumulation of chloride in shoots but not roots of rice (Shi et al., 2013). Saleh and Plieth (2013) showed inhibition of uptake of chloride from high concentration by La^{3+} (lanthanum), Gd^{3+} (gadolinium), and Al^{3+}. Uptake of chloride can be stimulated by the hormones indole-3-acetic acid (IAA) and 1-napthaleneacetic acid (NAA) in oat coleoptiles (Rubinstein, 1974) due to the action of auxin in plant cells on the plasma membrane chloride transport system mediating increased chloride uptake (Babourina et al., 1998).

9.7 DIAGNOSIS OF CHLORINE STATUS IN PLANTS

Chloride is mobile within plants, and this mobility is an important criterion for diagnosing deficiency symptoms. Under deficient conditions, available chloride will be moved from old leaves to new leaves; therefore, old leaves will show deficiency symptoms (Peuke, 2010). Plant tissue analysis has proven to be a valuable means of assessing chloride status in crops. Research on wheat, corn, and grain sorghum (*Sorghum bicolor* Moench) demonstrated that leaf chloride concentrations of 0.10%–0.12% are good indicators of low soil chloride levels (Lamond and Leikam, 2002).

9.7.1 DEFICIENT CONDITIONS

The physiological symptoms of chloride deficiency in plants grown in chloride-free nutrient solutions have been well characterized (Kafkafi et al., 2001). Xu et al. (2000) reported that chloride deficiency symptoms occur in plants when the tissue concentration is between 0.1 and 5.7 $mg \cdot kg^{-1}$ dry weight. Typical chloride deficiency symptoms include wilting of leaves, curling of leaflets, leaf bronzing and chlorosis, root growth inhibition, and a restriction in fruit number and size (Ozanne et al., 1957; Smith et al., 1987; Kafkafi et al., 2001; White and Broadley, 2001). Deficiency in kiwifruit produced chlorosis at the leaf margin, interveinal chlorosis, and a downward cupping of some of the old leaves, but not tissue necrosis (Smith et al., 1987). Chloride deficiency causes wilting of the leaflet blade tips, followed progressively by chlorosis, bronzing, and necrosis in tomato (Broyer et al., 1954). The least sensitive plants to chloride deficiency are beans (*Phaseolus vulgaris* L.), squash (*Cucurbita maxima* Duchesne), barley, corn (*Zea mays* L.), and buckwheat (*Fagopyrum esculentum* Moench) (White and Broadley, 2001). Onion leaves appear chlorotic and wilted under chloride deficiency. Deficient amounts of chloride in onions also will disrupt water movement, create heat stress under hot conditions, and decrease photosynthesis. In chlorine-deficient wheat, the symptoms are expressed as chlorotic or necrotic lesions on leaf tissue. The symptoms that result from chlorine deficiency have been named "chloride-deficient leaf spot syndrome" (Engel et al., 1997). Chloride deficiency is usually difficult to observe in field environments unless direct comparison can be made to similar plants with sufficient chloride supply (Kafkafi et al., 2001).

9.7.2 EXCESSIVE CONDITIONS

Chloride toxicity in agronomic, fruit, and vegetable crops is much more common in field situations than chloride deficiency (Eaton, 1966; Kafkafi et al., 2001). The most commonly induced cause of chloride toxicity in crop plants is from the application of irrigation water high in chloride. Determining chloride toxicity in plants is difficult due to separating the effects of chloride from accompanying cations, such as sodium, and distinguishing between what is a toxic effect within tissue and what is cellular dehydration caused by excessive external salt concentration (Xu et al., 2000). Analysis of plant tissues is used commonly to confirm chloride toxicity. Greenway and Munns (1980) grouped salinity tolerance in plants as (1) halophytes which either continue to grow rapidly at 200–500 mM NaCl, which are mostly native flora of saline soils or grow very slowly above 200 mM NaCl; (2) halophytes and nonhalophytes whose growth is substantially reduced by >100 mM NaCl, which can be subdivided into tolerant (e.g., red fescue *Festuca rubra* L., cotton, and barley), intermediate (e.g., tomatoes), and sensitive (e.g., beans and soybeans (*Glycine max* Merr.); and (3) very salt-sensitive nonhalophytes, such as citrus. For sensitive crops, toxicity will occur at foliar levels between 0.5% and 2% dry weight, whereas halophytes may contain up to 4% foliar chloride dry weight (Mills and Jones, 1996; White and Broadley, 2001). In fact, chloride is a more sensitive indicator of salt damage than Na^+ since it is stored in plants, whereas Na^+ is absorbed in smaller quantities despite high Na^+ concentrations in the soil (Alam, 1999). Reuter and Robinson (1997) reported that chloride toxicity in wheat resulted when shoot tissue concentrations were $\geq 1.5\%$. Curling of the leaf margins, marginal leaf scorch, leaf necrosis, and leaf drop are typical chloride toxicity symptoms (Mills and Jones, 1996). Older leaves are usually the first to exhibit symptoms, which may progress upward affecting the entire foliage. Dieback of the terminal axis and small branches may occur in cases of severe chloride toxicity (Heckman, 2007). Symptoms of chloride toxicity in onions include leaf chlorosis, leaf-tip burn, and stunting of the plant. Plant species differ in their sensitivity to excessive chloride; sugar beet (*Beta vulgaris* L.), barley, corn, spinach, and tomato are highly tolerant, whereas tobacco, bean, citrus, potato, lettuce (*Lactuca sativa* L.), and some legumes are very prone to toxicity and are called chlorophobic crops (Mengel and Kirkby, 1987). Chlorophobic crops should be fertilized with sulfate-based carriers instead of chloride-based carriers (Mengel and Kirkby, 1987). Under saline soil conditions, excessive chloride can produce toxicity by direct competition with other essential anions such as nitrate and sulfate and also through the induction of water stress due to the increased water potential in the growing environment (Mills and Jones, 1996).

9.8 FORMS AND CONCENTRATIONS OF CHLORINE IN SOILS AND AVAILABILITY TO PLANTS

9.8.1 FORMS AND CONCENTRATIONS OF CHLORINE IN SOIL

Chloride is one of the first elements removed from minerals by weathering during soil formation; therefore, most of the chloride of the earth is in oceans or in salt deposits left from the evaporation of ancient inland seas (Kafkafi et al., 2001; Matucha et al., 2010). Chlorine is in the primary minerals chlorapatite ($Ca_5(PO_4)Cl$), carnallite ($K, MgCl_3 \cdot 6H_2O$), sodalite ($Na_4Al_3Si_3O_{12}Cl$), halite (NaCl), and sylvite (KCl) (Mills and Jones, 1996). A large fraction of the total chlorine in common igneous rocks occurs in hydroxyl minerals such as biotite and hornblende, where chloride replaces the hydroxyl ion (Kuroda and Sandell, 1953). However, because igneous rocks and parent materials contain minor amounts of chlorine, there is little contribution to chlorine content from soil weathering.

Chlorine occurs predominantly as the chloride anion in soils, although over 2000 different organochlorine compounds have been identified in soil and marine environments (Fleming, 1995). The concentration of organic chlorine was two to four times higher than inorganic chlorine in Swedish forest soils (Johansson et al., 2001). Öberg (1998) concluded that the chlorine content of organic matter in the soil is similar to that of phosphorus. In the upper 60 cm of a temperate evergreen forest soil, the reservoir of organically bound chlorine could be as high as 0.6 t·ha^{-1} (Öberg, 1998). Chloride does not form complexes readily. Above pH 7.0, chloride is not absorbed by minerals and absorbed only weakly in kaolinitic and oxidic soils that have positive charges under acidic conditions (Mortvedt, 2000). Kasparov et al. (2005) showed that there was no sorption of ^{36}Cl in four soil types and the entire amount of applied isotope remained in the soil solution.

Although chlorine is the 21st most abundant element in the crust of the earth, the content of chloride in soils varies greatly from region to region. The chloride content of the crust of the earth is similar to sulfur (~500 mg·L^{-1}) and slightly less than half that of phosphorus (Flowers, 1988). Concentrations of reported chloride content in soils range from 20 to 900 mg·kg^{-1}, with a mean concentration of 100 mg·kg^{-1} (Mortvedt, 2000). Soils are considered low in chloride when concentrations are below 2 mg·kg^{-1} (James et al., 1970). Sheppard et al. (1999) surveyed multiple soil types in and around the lower English River and Manitoba Lowlands, Canada, and reported forest soils ranged from 22 to 23 mg Cl·kg^{-1} and wet land soils ranged from 94 to 180 mg Cl·kg^{-1}. An outcropping soil had a 56 mg Cl·kg^{-1} content; a field crop soil had a 29 mg Cl·kg^{-1} content; and a garden soil had a 64 mg Cl·kg^{-1} content. Total chlorine concentrations in France ranged from 19 to 100 mg·kg^{-1} for agricultural soils, from 13 to 1248 mg·kg^{-1} for grassland soils, and from 34 to 340 mg·kg^{-1} for forest soils (Redon et al., 2013). Raji and Jimba (1999) reported that the chloride content of 32 soils representing four savanna zones in Nigeria varied from 47.2 to 296.5 mg·kg^{-1}. However, chloride deficiencies are widespread in the coconut (*Cocos nucifera* L.)-growing areas in the Philippines and South Sumatra, Indonesia (Potash & Phosphate Institute, 1995). Although chloride deficiencies in agronomic grain crops in Kansas and the Great Plains, USA, have been reported, chlorine deficiencies are rare in the United States. (Lamond and Leikam, 2002). Soil testing in the sandy soils of southeastern Georgia, USA, revealed that extractable chloride was less than 12 mg·kg^{-1}, near the lower limit of detection (Randle, 2004). It is estimated that 80% of the soils in Nigeria may be chloride deficient (Raji and Jimba, 1999). Fine-textured soils with a history of fertilization with potassium chloride and soils irrigated with high-chloride-content irrigation water likely will be high in chloride (Murphy et al., 2007).

9.8.2 Factors Affecting Chlorine Availability to Plants

The availability of chloride in soils for plant uptake is determined by the chloride concentration in the soil solution. Chlorine inputs to soils occur mainly as the deposition of chloride from rainwater, fertilizer applications, irrigation waters, sea spray, dust, and air pollution (White and Broadley, 2001). Marschner (1995) reported that the minimal leaf concentration for crop growth is 1 g Cl·kg^{-1} dry weight. Approximately 4–8 kg Cl·ha^{-1} are sufficient to achieve this concentration and maintain average crop yields, and often this amount can be supplied by rainfall (Marschner, 1995). Chloride concentrations in rainwater have been reported from 20 to 50 mg·L^{-1} in coastal areas, but this content decreases rapidly with distance from the ocean (Hewitt and Smith, 1975; Xu et al., 2000). Anthropogenic sources of organochlorine also contribute to chloride soil levels, namely, through the intentional chlorination of soils, the disposal of household waste, and the dechlorination of herbicides, pesticides, medicinal pharmaceuticals, and artificial sweeteners (White and Broadley, 2001).

Under field conditions, atmospheric chlorine is usually adequate to prevent deficiency symptoms from occurring, especially near ocean coastal regions as the evaporation of seawater and its subsequent precipitation as onshore rainfall introduce chloride into agricultural systems. Most chloride salts are water soluble, and because chloride is supplied regularly to soil in rainwater, it is rarely deficient in plants (Marschner, 1995). The amount of chloride supplied from precipitation ranges from 1 to 175 kg·ha^{-1} per year (Junge, 1963; Yaalon, 1963; Reynolds et al., 1997), with higher amounts in coastal regions and lower amounts in midcontinental areas (Xu et al., 2000). Irrigation water of low to medium salinity contains 100–300 g Cl$^-$·m^{-3}, and high to very high saline irrigation water contains 300–1200 g Cl$^-$·m^{-3} (Xu et al., 2000).

The solubility and mobility of chloride in soil is similar to nitrate. Chloride is one of the most mobile ions and is easily lost from soils by leaching caused by rainfall or irrigation. Hence, the movement of chloride within soil is determined by water fluxes (White and Broadley, 2001). Burns (1974) reported that the movement of chloride in an uncropped, unirrigated sandy loam with CaCl$_2$ as the chloride source incorporated to a depth of 15 cm. If soil evaporation exceeded rainfall, an upward movement of chloride occurred; after a heavy rain, a downward movement of chloride occurred; and when rainfall equalled evapotranspiration, chloride movement was small (Burns, 1974).

9.9 ASSESSMENT OF CHLORINE STATUS IN SOILS

Historically, soil chloride analysis has been conducted primarily for the purpose of salinity characterization and irrigation management (Gelderman et al., 2012), but currently, there is increasing interest in assessing chloride soil status for fertilization purposes. Chloride concentration in soils can be determined in an extracted soil solution (Kafkafi et al., 2001). However, accurate sampling is the most important step in soil testing to ensure a representative sample of the production environment. In 1995, Huang et al. reported that a large amount of chloride was leached to a depth of 40–60 cm by precipitation in a rice field after one season (Xu et al., 2000). Fixen et al. (1987) reported that the depth of soil sampling for chloride was recommended to be 60 cm for spring wheat and barley on the North American Great Plains. When testing for mobile nutrients in soils (such as N, S, or Cl), a 60 cm sampling depth is recommended (Lamond and Leikam, 2002).

9.9.1 Extraction of Chloride from Soils

Most soil testing laboratories offer a chloride soil test (Lamond and Leikam, 2002). Chloride salts are highly water soluble and display limited adsorption to soil particles; therefore, chloride can be extracted with water or a weak electrolyte, such as 0.1 M Ca(NO$_3$)$_2$, 0.5 M K$_2$SO$_4$, 100 g·L^{-1} concentrated acetic acid, or CaO-saturated solutions (Kafkafi et al., 2001). Theoretically, each extractant should give similar results, but the method of determination may make some extractants more useful than others (Gelderman et al., 2012). Time of extraction can range upward from 5 (Gaines et al., 1984) to 60 min (Bolton, 1971), but minimum extraction time periods should be determined through recovery studies on the soil to be analyzed (Gelderman et al., 2012). Care should be taken to avoid chloride contamination from dust, water, filter paper, paper bags, perspiration, and many common cleaning agents during extraction and analysis (Johnson and Fixen, 1990). Plastic gloves should be worn when handling filter paper for soil chloride determination (Gelderman et al., 2012).

9.9.2 DETERMINATION OF CHLORIDE IN SOIL EXTRACTS

Although many methods have been developed for chloride determination in soils, several are not suitable for routine soil testing (Gelderman et al., 2012). Depending on the extractant, chloride can be measured in soil extracts by chemical analyses, ion-specific electrodes, ion chromatography, or spectrophotometry.

A classic volumetric procedure for chloride determination is titration with silver nitrate ($AgNO_3$) using a chromate (K_2CrO_4) indicator (Chapman and Pratt, 1961). The chloride in the extract first forms a white precipitate of AgCl, and then the colorless K_2CrO_4 indicator forms a brown-red precipitate of Ag_2CrO_4 that indicates the end of titration. The amount of chloride in a sample is obtained from the volume and normality of the standard $AgNO_3$ solution used (Chapman and Pratt, 1961).

The chloride-specific ion electrode can be used to measure the chloride extracted from soils and plant tissues, but its use has not been reliable across diverse soils and may be more suited as an endpoint indicator in titrations (Hipp and Langdale, 1971; Adriano and Doner, 1982; Gelderman et al., 2012). The chloride electrode has limitations due to ions or molecules that damage the electrode or that are measured as chloride (Van Loon, 1968). Serious interferences with the use of the chloride-specific ion electrode are caused by hydroxide, sulfide, cyanide, thiosulfate, bromide, iodide, and ammonia (Van Loon, 1968; Adriano and Doner, 1982). Of these interferences, only bromide and iodide are encountered often in soil solutions or plant ash. The extent of interferences depends on their kind and concentration and the concentration of chloride (Kafkafi et al., 2001). Instruction manuals accompanying the electrodes provide procedures for measuring chloride in the presence of the interfering substances.

Chlorine, along with bromine, iodine, and other halogens, can be analyzed in plant tissues using ICP-MS after the tissue is leached with tetramethylammonium hydroxide $(CH_3)_4N \cdot OH$ (Tagami et al., 2006). ICP-MS provides excellent limits of detection, suitable sample throughput, suitable linear dynamic range, and multielemental capacity (Wheal and Palmer, 2010; Mello et al., 2013). As a general rule, halogens such as chlorine cannot be directly measured by AAS since their analytical lines lie at wavelengths shorter than 190 nm (Mello et al., 2013). However, indirect determination can be carried out by the use of precipitation reactions or the formation of metal complexes. Chloride in soil extracts can be determined spectrophotometrically by complexation with mercury(II) thiocyanate $Hg(SCN)_2$ (Gelderman et al., 2012). This spectrophotometric method is very sensitive and has a limit of detection of approximately 1 µg $Cl \cdot g^{-1}$ (on a dry soil basis) and is generally reproducible at $\pm 7\%$ (Gelderman et al., 2012). Nitrate, sulfide, cyanide (CN^-), thiocyanate (SCN^-), bromide, and iodide cause interferences but are not usually present in sufficient amounts to cause problems (Gelderman et al., 2012). For halogens, the respective absorption spectra can also be used for quantification of elements by molecular absorption spectrometry (MAS) (Welz et al., 2010). Other reported analytical determinations of chloride include total reflection XRF and INAA (Mello et al., 2013; Redon et al., 2013).

9.10   CHLORINE FERTILIZERS

The most common chlorine fertilizer sources are potassium chloride or muriate of potash (KCl), ammonium chloride (NH_4Cl), calcium chloride ($CaCl_2$), and magnesium chloride ($MgCl_2$) (Lamond and Leikam, 2002) (Table 9.2). Over 90% of global muriate of potash production is used for crop fertilization, mainly as a potassium source. Potassium chloride can vary in color from deep red to white, the color coming from contamination with minor amounts of iron oxides that are occluded in the crystals of sylvinite ores (Stewart, 1985). Grades of KCl are defined by their K_2O content

TABLE 9.2

Source, Chemical Formula, and Concentration for Various Chlorine Fertilizers

Source	Chemical Formula	Chlorine Content (%)	Other Nutrient Content
Ammonium chloride	NH_4Cl	67	25%–26% N
Calcium chloride	$CaCl_2$	64	36% Ca
Magnesium chloride	$MgCl_2$	74	26% Mg
Potassium chloride	KCl	47	50%–53% K

and particle size. Agricultural grades are those with 40%–60% K_2O, and industrial grades are >62% K_2O. Agricultural grades of KCl are usually treated with additives such as amines and oils to prevent caking and dusting during shipping, handling, and processing (Eatock, 1985). Potassium chloride is produced in five agricultural decreasing grades: granular (0.841–3.36 mm particle size), coarse (0.595–2.38 mm), standard (0.210–1.19 mm), fine (0.105–0.420 mm), and soluble/suspension (0.105–0.420 mm) (Kafkafi et al., 2001). Granular, coarse, or standard grade is suitable for direct application as a single nutrient (K) fertilizer or a component in a fertilizer blend. Fine grade KCl is used as a raw material in the granulation of NPK fertilizers and for the production of K_2SO_4. Soluble grade KCl is used in the production of clear liquid and suspension solution fertilizers (Kafkafi et al., 2001). Grades of KCl are either white or pink/red in color. The solubility characteristics of white muriate of potash make it a popular potassium fertilizer for application with irrigation, but the red forms contain Fe oxides, which may clog irrigation equipment. Calcium chloride, potassium chloride, and magnesium chloride also are available as liquid formulations, but the chloride content is lower than their solid forms (Lamond and Leikam, 2002; Murphy et al., 2007).

9.10.1 METHODS OF CHLORINE APPLICATION

9.10.1.1 Soil Applications

Chlorine can be soil incorporated as a broadcasted application before planting, at seedling or transplanting, or top- or side-dressed during crop growth (Potash & Phosphate Institute, 1995). The relatively high salt index of chloride fertilizers precludes application in direct seed contact for small grains or row crops (Murphy et al., 2007). Small grain research in Kansas and Oregon, USA, has not shown significant yield differences in the timing of chlorine applications, but results in Texas, USA, show that heavy winter rainfall can reduce chlorine content in sandy soils due to its soil mobility (Potash & Phosphate Institute, 1995). Approximately 60 kg $Cl^- \cdot ha^{-1}$ per 60 cm of soil depth are reported to be adequate for target yields of small grain crops. For soils in North America that are high in potassium and not regularly fertilized with KCl, crops such as wheat, corn, grain sorghum, sunflower (*Helianthus annuus* L.), and forage grasses could benefit from chlorine fertilization (Murphy et al., 2007). Chlorine applications in the range of 20–30 kg·ha have been beneficial in crop production, and supplementation can increase wheat yields from 270 to 375 kg·ha in the central United States (Murphy et al., 2007).

9.10.1.2 Foliar Applications

Much of the foliar application of chlorine to plants occurs with fruits and vegetables as calcium chloride to delay aging or ripening, reduce postharvest decay, and control the development of physiological disorders. Foliar calcium chloride has been reported to delay ripening and retard mold development in strawberry (Chéour et al., 1991) and raspberry (*Rubus idaeus* L.) and increase fruit quality in peach (*Prunus persica* Batsch.) (Robson et al., 1989) and oranges. Calcium chloride at concentrations of 2% and 4% significantly increased average fruit weight and ascorbic acid (AA) content of pomegranate (*Punica granatum* L.) (Ramezanian et al., 2009). Even though most of the beneficial effects of calcium chloride foliar applications can be attributed to calcium (Chéour et al., 1991), Subbiah and Perumal (1990) demonstrated that calcium chloride was better than calcium oxide, calcium sulfate, and calcium nitrate for improving lycopene and AA content and firmness in tomato. Foliar applications of potassium chloride (0.27–1.34 M) can reduce powdery mildew on wheat (Kettlewell et al., 2000).

Mepiquat chloride (N,N-dimethylpiperidinium chloride) is a plant growth regulator used exclusively on cotton. It is intended to increase yield by inhibiting gibberellic acid synthesis (Rosolem et al., 2013). Registered product formulations include soluble concentrate/liquid (SC/L) and dry flowable (DF) forms. It is applied aerially or using ground boom equipment. Application rates are limited to 0.049 kg·ha and seasonally to 0.148 kg·ha (U.S. EPA, 1997).

9.11 CONCLUSIONS

Chlorine is an essential element that often is taken up by plants in large quantities. Soluble chlorine exists in nature as the chloride anion, and deficiency in crop plants seldom is observed due to the plentiful supplies of chloride in the environment and its redistribution from natural occurrences such as rainfall, marine aerosols, and volcanic emissions. Concentrations of reported chloride content in soils range from 20 to 900 mg·kg^{-1}, with a mean concentration of 100 mg kg^{-1}. Plants take up chlorine as the chloride anion through an active uptake process, and the rate of uptake depends on the chloride concentration in the soil solution, which is affected by the atmospheric deposition of chloride, the concentration of chloride in irrigation waters, and the content of chloride in fertilizers and manures. Physiological responses of plants to chloride include essential roles in photosynthesis, stomatal regulation, vacuolar transport, osmotic adjustment, and activation of certain enzymes. There is an interaction of the uptake of chloride with other anions, such as nitrate, sulfate, and phosphate, but also with cations, such as ammonium, potassium, calcium, and other micronutrients. Chloride is usually in the highest concentration in plants in leaf blades, followed by petioles, stems, and fruit or seeds. The chloride ion is mobile in plants, and tissue analysis has proven to be a valuable means of assessing chloride status in crops. Although chloride toxicity is more common than deficiency in crop plants, typical deficiency symptoms include wilting and chlorosis of leaves, curling of leaflets, root growth inhibition, and a restriction in fruit number and size. Chloride toxicity is hard to diagnose, and plants are grouped according to their tolerance to salinity and chloride salts. Chloride concentration in soils can be determined directly in an extract of the soil solution, and most soil testing laboratories offer a chloride soil test. Depending on the extractant, chloride can be measured by chemical analyses, ion-specific electrodes, ion chromatography, or spectrophotometry. The most common chloride fertilizer sources are potassium chloride, ammonium chloride, calcium chloride, and magnesium chloride and can be soil incorporated as a broadcast preplant, applied at seedling or transplanting, top or side dressed during crop growth, or applied foliarly.

REFERENCES

Adriano, D.C. and H.E. Doner. 1982. Bromine, chlorine, and fluorine. In *Agronomy Monograph, Methods of Soil Analysis. Part 2. Chemical and Microbiological Properties 9.2*, ed. A. Klute, pp. 449–483. Madison, WI: American Society of Agronomy and Soil Science Society of America.

Alam, S.M. 1999. Nutrient uptake by plants under stress conditions. In *Handbook of Plant and Crop Stress*, 2nd edn., ed. M. Pessarakli, pp. 285–314. New York: Marcel Dekker.

Babourina, O., S. Shabala, and I. Newman. 1998. Auxin stimulates Cl$^-$ uptake by oat coleoptiles. *Ann. Bot.* 82(3):331–336.

Banuls, J.E., M.D. Serna, F. Legaz, M. Talon, and E. Primo-Millo. 1997. Growth and gas exchange parameters of citrus plants stressed with different salts. *J. Plant Physiol.* 150:194–199.

Bar, Y., A. Apelbaum, U. Kafkafi, and R. Goren. 1997. Relationship between chloride and nitrate and its effect on growth and mineral composition of avocado and citrus plants. *J. Plant Nutr.* 20:715–731.

Barber, M.J., B.A. Notton, C.J. Kay, and L.P. Solomonson. 1989. Chloride inhibition of spinach nitrate reductase. *Plant Physiol.* 90:70–94.

Beringer, H., K. Hoch, and M.G. Lindauer. 1990. Source:sink relationship in potato as influenced by potassium chloride or potassium sulphate nutrition. In *Plant Nutrition—Physiology and Application*, ed. M.L. van Beusichem, pp. 639–642. Dordrecht, the Netherlands: Kluwer.

Blevins, D.G., A.J. Hiatt, and R.H. Lowe. 1974. The influence of nitrate and chloride uptake on expressed sap pH, organic acid synthesis, and potassium accumulation in higher plants. *Plant Physiol.* 54:82–87.

Bolton, J. 1971. The chloride balance in a fertilizer experiment on sandy soil. *J. Sci. Food Agric.* 22:292–294.

Broyer, T.C., A.B. Carlton, C.M. Johnson, and P.R. Stout. 1954. Chlorine—A micronutrient element for higher plants. *Plant Physiol.* 29:526–532.

Burns, I.G. 1974. A model for predicting the redistribution of salts applied to fallow soils after excessive rainfall or evaporation. *J. Soil Sci.* 25:165–178.

Buwalda, J.G. and G.S. Smith. 1991. Influence of anions on the potassium status and productivity of kiwifruit (*Actinidia deliciosa*) vines. *Plant Soil* 133:209–218.

Carter, C.T. and C.M. Grieve. 2010. Growth and nutrition of two cultivars of *Zinnia elegans* under saline conditions. *HortScience* 45:1058–1063.

Cerezo, M., P. Garcia-Agustin, M.D. Serna, and E. Primo-Millo. 1997. Kinetics of nitrate uptake by citrus seedlings and inhibitory effects of salinity. *Plant Sci.* 126:105–112.

Chapman, H.D. and G.F. Liebig. 1940. Nitrogen concentration and ion balance in relation to citrus nutrition. *Hilgardia* 13:141–173.

Chapman, H.D. and P.F. Pratt. 1961. *Methods of Analysis for Soils, Plants and Waters*. Riverside, CA: University California Division of Agricultural Sciences.

Chéour, F., C. Willemot, J. Arul, J. Makhlouf, and Y. Desjardines. 1991. Postharvest response of two strawberry cultivars to foliar application of $CaCl_2$. *HortScience* 26(9):1186–1188.

Churchill, K.A. and H. Sze. 1984. Anion-sensitive, H^+-pumping ATPase of oat roots. Direct effects of Cl^-, NO_3^-, and a disulfonic stilbene. *Plant Physiol.* 76:490–497.

Coder, K. 2011. Chlorine gas exposure and trees. *Warnell School Outreach Monograph WSFNR11-11*. Athens, GA: The University of Georgia.

Colmenero-Flores, J.M., G. Martínez, G. Gamba et al. 2007. Identification and functional characteristics of cation-chloride cotransporters in plants. *Plant J.* 50(2):278–292.

Darvishzadeh, R. and R. Alavi. 2011. Genetic analysis of chloride concentration in oriental tobacco genotypes. *J. Plant Nutr.* 34(5–8):1070–1078.

De Angeli, A., J. Zhang, S. Meyer, and E. Martinoia. 2013. AtALMT9 is a malate-activated vacuolar chloride channel required for stomatal opening in *Arabidopsis*. *Nature Commun.* 4:1804.

Du, Z., K. Aghoram, and W.H. Outlaw, Jr. 1997. In vivo phosphorylation of phosphoenolpyruvate carboxylase in guard cells of *Vicia faba* L. is enhanced by fusicoccin and suppressed by abscisic acid. *Arch. Biochem. Biophys.* 337:345–350.

Dubos, B., W.H. Alarcón, J.E. López, and J. Ollivier. 2011. Potassium uptake and storage in oil palm organs: The role of chlorine and the influence of soil characteristics in the Magdalena valley, Colombia. *Nutr. Cycl. Agroecosyst.* 89:219–227.

Eatock, W.H. 1985. Advances in potassium mining and refining. In *Potassium in Agriculture*, ed. R.D. Munson, pp. 29–48. Madison, WI: ASA/CSSA/SSSA.

Eaton, F.M. 1966. Chlorine. In *Diagnostic Criteria for Plants and Soils*, ed. H.D. Chapman, pp. 98–135. Riverside, CA: University of California.

Elgharably, A. 2011. Wheat response to combined application of nitrogen and phosphorus in a saline sandy loam soil. *Soil Sci. Plant Nutr.* 57:396–402.

Engel, R.E., P.L. Bruckner, D.E. Mathre, and S.K.Z. Brumfield. 1997. A chloride-deficient leaf spot syndrome of wheat. *Soil Sci. Soc. Am. J.* 61:176–184.

Engvild, K.C. 1986. Chlorine-containing natural compounds in higher plants. *Phytochemistry* 25:781–791.

Eshel, A. and Y. Waisel. 1979. Distribution of sodium and chloride in leaves of *Suaeda monoica* halophyte. *Physiol. Plant.* 46:151–154.

Feigin, A., I. Rylski, A. Meiri, and J. Shalhevet. 1987. Response of melon and tomato plants to chloride-nitrate ratio in saline nutrient solutions. *J. Plant Nutr.* 10:1787–1794.

Fine, P.L. and W.D. Frasch. 1992. The oxygen-evolving complex requires chloride to prevent hydrogen peroxide formation. *Biochemistry* 31(48):12204–12210.

Fixen, P.E., R.H. Gelderman, J.R. Gerwing, and F.A. Cholick. 1986. Response of spring wheat, barley, and oats to chloride in potassium chloride fertilizers. *Agron. J.* 78:664–668.

Fixen, P.E., R.H. Gelderman, J.R. Gerwing, and B.G. Faber. 1987. Calibration and implementation of a soil Cl test. *J. Fert. Issues* 4:91–97.

Fleming, B.I. 1995. Organochlorines in perspective. *Tappi J.* 78:93–98.

Flowers, T.J. 1988. Chloride as a nutrient and an osmoticum. *Adv. Plant Nutr.* 3:55–78.

Gaines, T.P., M.B. Parker, and G.J. Gascho. 1984. Automated determination of chloride ion in soil and plant tissue by sodium nitrate. *Agro. J.* 76:371–374.

Garthwaite, A.J., R. von Bothmer, and T.D. Colmer. 2005. Salt tolerance in wild *Hordeum* species is associated with restricted entry of Na^+ and Cl^- into the shoots. *J. Exp. Bot.* 56(419):2365–2378.

Gelderman, R.H., J.L. Denning, and R.J. Goos. 2012. Chlorides. In *Recommended Chemical Soil Test Procedures for the North Central Region*, pp. 49–52. Columbia, MO: Missouri Agricultural Experiment Station.

Glass, A.D.M. and M.Y. Siddiqi. 1985. Nitrate inhibition of chloride influx in barley: Implications for a proposed chloride homeostat. *J. Exp. Bot.* 36:556–566.

Greenway, H. and R. Munns. 1980. Mechanisms of salt tolerance in nonhalophytes. *Annu. Rev. Plant Physiol.* 31:149–190.

Heckman, J.R. 1998. Corn stalk rot suppression and grain yield response to chloride. *J. Plant Nutr.* 21(1):149–155.

Heckman, J.R. 2007. Chlorine. In *Handbook of Plant Nutrition*, ed. A.V. Barker and D.J. Pilbeam, pp. 279–291. Boca Raton, FL: CRC Press.

Hewitt, E.J. 1966. *Sand and Water Culture Methods Used in the Study of Plant Nutrition*, 2nd edn. Farnham Royal, U.K.: Commonwealth Agricultural Bureau.

Hewitt, E.J. and T.A. Smith. 1975. *Plant Mineral Nutrition*. London, U.K.: Hodder & Stoughton Ltd.

Hipp, B.W. and G.W Langdale. 1971. Use of solid-state chloride electrode for chloride determination in soil extractions. *Commun. Soil Sci. Plant Anal.* 2:237–240.

Hofinger, M. and M. Bottger. 1979. Identification by GC-MS of 4-chloroindolylacetic acid and its methyl ester in immature *Vicia faba* broad bean seeds. *Phytochemistry* 18:653–654.

Homann, P.H. 2002. Chloride and calcium in photosystem II: From effects to enigma. *Photosyn. Res.* 73:169–175.

Huber, D.M. and N.S. Wilhelm. 1988. The role of manganese in resistance to plant disease. In *Manganese in Soils and Plants*, eds. R.D. Graham, J. Hannam, and N.C. Uren, pp. 155–173. Boston, MA: Kluwer.

Jackson, T.L. and R.E. McBride. 1986. Yield and quality of potatoes improved with potassium and chloride fertilization. In *Special Bulletin on Chloride and Crop Production*, ed. T.L. Jackson, pp. 73–83. Norcross, GA: Potash & Phosphate Institute.

James, D.W., W.H. Weaver, and R.L. Reeder. 1970. Chloride uptake by potatoes and the effects of potassium chloride, nitrogen and phosphorus fertilization. *Soil Sci.* 109:48–53.

Johannson, E., G. Ebenå, P. Sandén, T. Svensson, and G. Öberg. 2001. Organic and inorganic chlorine in Swedish spruce forest soils: Influence of nitrogen. *Geoderma* 101:1–13.

Johnson, G.V. and P.E. Fixen. 1990. Testing soils for sulfur, boron, molybdenum and chlorine. In *Soil Testing and Plant Analysis*, 3rd edn., ed. R.L.Westerman, pp. 265–273. Madison, WI: Soil Science Society of America.

Junge, C.E. 1963. *Air Chemistry and Radioactivity*. New York: Academic Press.

Kafkafi, U., G. Xu, P. Imas, H Magen, and J. Tarchitzky. 2001. *Potassium and Chloride in Crops and Soils: The Role of Potassium Chloride Fertilizer in Crop Nutrition*. Basel, Switzerland: International Potash Institute.

Kashparov, V., C. Colle, S. Zvarich, V. Yoschenko, S. Levchuk, and S. Lundin. 2005. Soil-to-plant halogens transfer studies 2. Root uptake of radiochlorine by plants. *J. Environ. Radioact.* 79:233–253.

Kettlewell, P.S., J.W. Cook, and D.W. Perry. 2000. Evidence for an osmotic mechanism in the control of powdery mildew disease of wheat by foliar-applied potassium chloride. *Euro. J. Plant Path.* 106:297–2000.

Kong, X.-Q., X.-H. Gao, W. Sun, J. An, Y.-X. Zhao, and H. Zhang. 2011. Cloning and functional characterization of a cation-chloride cotransporter gene *OsCCC1*. *Plant Mol. Biol.* 75:567–578.

Krupinsky, J.M., K.L. Bailey, M.P. McMullen, B.D. Gossen, and T.K. Turkington. 2002. Managing plant disease risk in diversified cropping systems. *Agron. J.* 94(2):198–209.

Kuroda, P.K. and E.B. Sandell. 1953. Chlorine in igneous rocks: Some aspects of the geochemistry of chlorine. *Geol. Soc. Am. Bull.* 64(8):879–896.

Lamond, R.E. and D.F. Leikam. 2002. Chloride in Kansas: Plant, soil, and fertilizer considerations. Bulletin MF-2570. Manhattan, KS: Kansas State University Agriculture Experiment Station and Extension Service.

Leidi, E.O., M. Silberbush, M.I.M. Soares, and S.H. Lips. 1992. Salinity and nitrogen nutrition studies on peanut and cotton plants. *J. Plant Nutr.* 15:591–604.

Liu, L. and B.J. Shelp. 1996. Impact of chloride on nitrate absorption and accumulation by broccoli (*Brassica oleracea* var. *italica*). *Can. J. Plant Sci.* 76:367–377.

Lloyd, J., P.E. Kriedemann, and D. Aspinall. 1989. Comparative sensitivity of Prior Lisbon lemon and Valencia orange trees to foliar sodium and chloride concentrations. *Plant Cell Environ.* 12(5):529–540.

Loudet, O., S. Chaillou, A. Krapp, and F. Daniel-Vedele. 2003. Quantitative trait loci analysis of water and anion contents in interaction with nitrogen availability in *Arabidopsis thaliana*. *Genetics* 163(2):711–722.

Marschner, H. 1995. *Mineral Nutrition of Higher Plants*, 2nd ed. San Diego, CA: Academic Press.

Martinoia, E., M.J. Schramm, G. Kaiser, W.M. Kaiser, and U. Heber. 1986. Transport of anions in isolated barley vacuoles. *Plant Physiol.* 80(4):895–901.

Matucha, M., N. Clarke, Z. Lachmanová, S.T. Forczek, K. Fuksová, and M. Gryndler. 2010. Biogeochemical cycles of chlorine in the coniferous forest ecosystem: Practical implications. *Plant Soil Environ.* 56(8):357–367.

Meador, D.P. and P.R. Fisher. 2013. Ammonium in nutrient solutions decreases free chlorine concentration from sodium hypochlorite. *HortScience* 48(10):1304–1308.

Mello, P.A., J.S. Barin, F.A. Duarte et al. 2013. Analytical methods for the determination of halogens in bioanalytical sciences: A review. *Anal. Bioanal. Chem.* 405:7615–7642.

Mengel, K. and E.A. Kirkby. 1987. Further elements of importance. In *Principles of Plant Nutrition*, pp. 573–588. Bern, Switzerland: International Potash Institute.

Metzler, D.E. 1979. *Biochemistry-the Chemical Reactions of Living Cells*. New York: Academic Press.

Mills, H.A. and J.B. Jones, Jr. 1996. Part II-Utilization of plant tissue analysis. In *Plant Analysis Handbook II*, pp. 155–414. Athens, GA: Micro-Macro Publishing.

Mortvedt, J.J. 2000. Bioavailability of micronutrients. In *Handbook of Soil Science*, ed. M.E. Sumner, pp. D-71–D-88. New York: CRC Press.

Munns, R. and M. Tester. 2008. Mechanisms of salinity tolerance. *Annu. Rev. Plant Biol.* 59:651–681

Murphy, L., B. Gordon, and B. Evans. 2007. More profits with chloride. *Fluid J.* Winter 20–23.

Öberg, G. 1998. Chloride and organic chlorine in soil. *Acta Hydro. Hydrobiol.* 26(3):137–144.

Olesen, K. and L. Andréasson. 2003. The function of the chloride ion in photosynthetic oxygen evolution. *Biochemistry* 42:2025–2035.

Ozanne, P.G., J.T. Woolley, and T.C. Broyer. 1957. Chloride and bromine in the nutrition of higher plants. *Aust. J. Biol. Sci.* 10:66–79.

Peuke, A.D. 2010. Correlations in concentrations, xylem and phloem flows, and partitioning of elements and ions in intact plants. A summary and statistical re-evaluation of modelling experiments in *Ricinus communis*. *J. Exp. Bot.* 61:635–655.

Popelková, H. and C.F. Yocum. 2007. Current status of the role of Cl$^-$ ion in the oxygen-evolving complex. *Photosynth. Res.* 93:111–121.

Potash & Phosphate Institute. 1995. *International Soil Fertility Manual*. Norcross, GA.

Raji, B.A. and B.W. Jimba. 1999. A preliminary chlorine survey of the savannah soils of Nigeria. *Nutr. Cycl. Agroecosys.* 55:29–34.

Ramezanian, A., M. Rahemi, and M.R. Vazifehshenas. 2009. Effects of foliar application of calcium chloride and urea on quantitative and qualitative characteristics of pomegranate fruit. *Sci. Hort.* 121:171–175.

Randle, W.M. 2004. Chloride requirements in onion: Clarifying a widespread misunderstanding. *Better Crops* 88:10–11.

Raschke, K. and H. Schnabl. 1978. Availability of chloride affects the balance between potassium chloride and potassium malate in guard cells of *Vicia faba* L. *Plant Physiol.* 62:84–87.

Redon, P.-O., C. Jolivet, N.P.A. Saby, A. Abdelouas, and Y. Thiry. 2013. Occurrence of natural organic chlorine in soils of different land uses. *Biogeochemistry* 114:413–419.

Reuter, D.J. and J.B. Robinson. 1997. *Plant Analysis, an Interpretation Manual*, 2nd edn. Melbourne, Victoria, Australia: CSIRO Publications.

Reynolds, B.D., R. Fowler, I. Smith, and J.R. Hall. 1997. Atmospheric inputs and catchment solute fluxes for major ions in five Welsh upland catchments. *J. Hydrol.* 195:305–329.

Roberts, S.K. 2006. Plasma membrane anion channels in higher plants and their putative functions in roots. *New Phytol.* 169:647–666.

Robinson, J.B. 1986. Fruits, vines, and nuts. In *Plant Analysis: An Interpretation Manual*, eds. D.J. Reuter and J.B. Robinson, pp. 120–147. Sydney, New South Wales, Australia: Inkata Press.

Robson, M.G., J.A. Hopfinger, and P. Eck. 1989. Postharvest sensory evaluation of calcium treated peach fruit. *Acta Hort.* 254:173–177.

Rognes, S.E. 1980. Anion regulation of lupin asparagine synthetase: Chloride activation of the glutamine-utilizing reaction. *Phytochemistry* 19:2287–2293.

Rosolem, C.A., D.M. Oosterhuis, and F.S. de Souza. 2013. Cotton response to mepiquat chloride and temperature. *Sci. Agric.* 70(2):82–87.

Rubinstein, B. 1974. Effect of pH and auxin on chloride uptake into *Avena* coleoptile cells. *Plant Physiol.* 54:835–839.

Sahin, O., M.B. Taskin, Y.K. Kadioglu, A. Inal, A. Gunes, and D.J. Pilbeam. 2012. Influence of chloride and bromate interaction on oxidative stress in carrot plants. *Sci. Hort.* 137:81–86.

Saleh, L. and C. Plieth. 2013. A9C sensitive Cl⁻—Accumulation in *A. thaliana* root cells during salt stress is controlled by internal and external calcium. *Plant Signal. Behav.* 8(6):e24259.

Schnabl, H. and K. Raschke. 1980. Potassium chloride as stomatal osmoticum in *Allium cepa* L., a species devoid of starch in guard cells. *Plant Physiol.* 65(1):88–93.

Schulte, E.E. 1999. Soil and applied chlorine. Fact Sheet A3556. Madison, WI: University of Wisconsin Cooperative Extension Service.

Sheppard, S.C., W.G. Evenden, and C.R. Macdonald. 1999. Variation among chlorine concentration ratios for native and agronomic plants. *J. Environ. Radioact.* 43:65–76.

Shi, Y., Y. Wang, T.J. Flowers, and H. Gong. 2013. Silicon decreases chloride transport in rice (*Oryza sativa* L.) in saline conditions. *J. Plant Physiol.* 170(9):847–853.

Simini, M. and I.A. Leone. 1982. Effect of photoperiod, temperature, and relative humidity on chloride uptake of plant exposed to salt spray. *Phytopathology* 72(9):1163–1166.

Smith, G.S., C.J. Clark, and P.T. Holland. 1987. Chlorine requirement of kiwifruit (*Actinidia deliciosa* L.). *New Phytol.* 106:71–80.

Stewart, J.A. 1985. Potassium source, use and potential. In *Potassium in Agriculture*, ed. R.D. Munson, pp. 83–98. Madison, WI: ASA/CSSA/SSSA.

Subbiah, K. and R. Perumal. 1990. Effect of calcium sources, concentrations, stages and number of sprays on physic-chemical properties of tomato fruits. *South Indian Hort.* 38(1):20–27.

Swain, R.R. and E.E. Dekker. 1966. Seed germination studies. I. Purification and properties of an α-amylase from the cotyledons of germinating peas. *Biochem. Biophy. Acta* 122:75–86.

Tagami, K., S. Uchida, I. Hirai, H. Tsukada, and H. Takeda. 2006. Determination of chlorine, bromine and iodine in plant samples by inductively coupled plasma-mass spectrometry after leaching with tetramethyl ammonium hydroxide under a mild temperature condition. *Anal. Chim. Acta* 570(1):88–92.

Tavares, B., P. Domingos, P. Nuno Dias, J.A. Feijó, and A. Bicho. 2011. The essential role of anionic transport in plant cells: The pollen tube as a case study. *J. Exp. Bot.* 62:2273–2298.

Terry, N. 1977. Photosynthesis, growth, and the role of chloride. *Plant Physiol.* 60(1):69–75.

Thomason, W.E., K.J. Wynn, K.W. Freeman et al. 2001. Effect of chloride fertilizers and lime on wheat grain yield and take-all disease. *J. Plant Nutr.* 24(4–5):683–692.

Tsukada, H., H. Hasegawa, A. Takeda, and S. Hisamatsu. 2007. Concentrations of major and trace elements in polished rice and paddy soils collected in Aomori, Japan. *J. Radioanal. Nucl. Chem.* 273(1):199–203.

Tsukada, H. and A. Takeda. 2008. Concentration of chlorine in rice plant components. *J. Radioanal. Nucl. Chem.* 278(2):387–390.

United States Environmental Protection Agency. 1997. Mepiquat chloride. R.E.D. Facts 7508W. Washington, DC: United States Environmental Protection Agency.

Van Loon, J.C. 1968. Determination of chloride in chloride-containing materials with a chloride membrane electrode. *Analyst* 93:788–791.

Wang, D.Q., B.C. Guo, and X.Y. Dong. 1989. Toxicity effects of chloride on crops. *Chin. J. Soil Sci.* 30:258–261.

Wedding, R.T. 1989. Malic enzymes in higher plants. *Plant Physiol.* 90:367–371.

Welz, B., S. Mores, E. Carasek, M.G.R. Vale, M. Okruss, and H. Becker-Ross. 2010. High-resolution continuum source atomic and molecular absorption spectrometry-A review. *Appl. Spectrosc. Rev.* 45(5):327–354.

Wheal, M.S. and L.T. Palmer. 2010. Chloride analysis of botanical samples by ICP-OES. *J. Anal. Atom. Spectrom.* 25:1946–1952.

White, P.J. and M.R. Broadley. 2001. Chloride in soils and its uptake and movement within the plant: A review. *Ann. Bot.* 88:967–988.

Willmer, C.M., G. Grammatikopoulos, G. Lasceve, and A. Vavasseur. 1995. Characterization of the vacuolar-type H⁺-ATPase from guard cell protoplasts of *Commelina*. *J. Exp. Bot.* 46:383–389.

Winterton, N. 2000. Chlorine: The only green element—Towards a wider acceptance of its role in natural cycles. *Green Chem.* 2:173–225.

Wissing, F. and J.A.C. Smith. 2000. Vacuolar chloride transport in *Mesembryanthemum crystallinum* L. measured using the fluorescent dye lucigenin. *J. Membr. Biol.* 177(3):199–208.

Xu, G., H. Magen, J. Tarchitzky, and U. Kafkafi. 2000. Advances in chloride nutrition of plants. *Adv. Agron.* 68:97–150.

Yaalon, D.H. 1963. The origin and accumulation of salts in groundwater and in soils in Israel. *Bull. Res. Counc. Isr. Sect. G* 11:105–131.

Younts, S.E. and R.B. Musgrave. 1958. Chemical composition, nutrient absorption, and stalk rot incidence in corn as affected by chloride in potassium fertilizer. *Agron. J.* 50:426–429.

10 Copper

Inmaculada Yruela

CONTENTS

10.1 HISTORICAL BACKGROUND

Copper (Cu) is a $3d^{10}$ transition metal that belongs to Group IB. It has an atomic number of 29 and an atomic mass of 63.546 amu. The melting point is 1083.0°C (1356.15 K, 1981.4°F), the boiling point is 2567.0°C (2840.15 K, 4652.6°F), and the density at 293 K is 8.96 g cm^{-3}. Copper is an active oxidation–reduction transition metal that can exist as Cu^{2+} (Cu(II), cupric) and Cu^+ (Cu(I), cuprous) ions. A single electron in the outer 4s orbital ($4s^1$) confers this electronic characteristic, and this $4s^1$ electron is difficult to remove. The first and second ionization potentials are 7.72 and 20.29 eV, respectively. The fact that the second ionization potential is much higher that the first one means that several Cu^+ species can exist. Copper is one of the few metals that exist in nature in elemental form, although the quantities in which it occurs in its free state are very small.

Copper is an essential element for all living organisms since it is a component of a variety of Cu-containing proteins involved in biological reactions such as oxidation of iron (Fe), insertion of oxygen in organic substrates, and disproportionation of oxygen free radicals. These reactions are important in redox electron transport chains and oxidative stress responses. In particular, in plants, Cu functions in photosynthesis and respiratory electron transport, and also it is important for cell wall formation, detoxification of oxygen species, ethylene sensing, and synthesis of polyphenols. The biological role of Cu started to be relevant in biology with the accumulation of oxygen in the atmosphere and the oceans, which changed the earth from aerobic to anaerobic conditions (Crowe et al., 2013). The development of primitive oxygenic photosynthetic organisms (i.e., cyanobacteria) and ancestors of bryophytes and land plants was responsible for this environmental change. This fact led to a decrease in Fe solubility and a shift from the use of Fe toward Cu in similar biological systems (Crichton and Pierre, 2001; Yruela, 2013).

Copper has diverse roles in biology due to its redox properties, which led it to participate in numerous interactions with proteins to drive diverse structures and reactions. Monovalent Cu^+ has affinity for thiol and thioether groups (found in cysteine or methionine amino acids), and Cu^{2+} normally exhibits coordination to oxygen or imidazole nitrogen groups (found in aspartic and glutamic acid or histidine, respectively). According to spectroscopic and magnetic features, the Cu centers in proteins have been classified as type I, type II, type III, and binuclear Cu_A and binuclear Cu_B centers. In type I centers (also named blue copper centers), the Cu ion is coordinated to nitrogens (N) of two histidine (His) residues and sulfurs (S) of cysteine (Cys) and methionine (Met) (or oxygen [O] of a glutamine, Gln), exhibiting a square planar coordination. It shows an intense absorption maximum at around 600 nm and narrow hyperfine splittings in electron paramagnetic resonance (EPR) spectroscopy. In plants, plastocyanin and multicopper oxidases belong to this type of Cu-containing protein. Type II centers (also named nonblue copper centers) are formed by only one Cu atom. They do not give strong absorption at 600–700 nm since they lack S atoms as ligands and show larger hyperfine coupling constants in the EPR spectrum. In this case, the Cu ion is coordinated to four or five ligands, which can be N atoms of His residues, or O atoms of tyrosine (Tyr) residues and water molecules. They exhibit a square planar coordination. Normally, these centers are present in enzymes involved in oxidations or oxygenations. In plants, copper–zinc superoxide dismutase (Cu/ZnSOD) and amine oxidase belong to this group. Type III Cu centers are constituted by two Cu atoms, each coordinated to N atoms of three His residues. It is not detected in the EPR spectrum because of the exchange interaction between the two Cu^{2+} (S = ½) metal ions caused by bridging ligation, leading to strong antiferromagnetic coupling. They play important roles in O_2 binding, activation, and reduction to H_2O. In plants, enzymes such as catechol oxidase, ascorbate oxidase, and laccase contain this type of Cu center.

The Cu_A and the Cu_B centers are characterized by a binuclear and mononuclear geometry, respectively. In the binuclear Cu_A center, two Cu atoms are bound by two S atoms of Cys residues, and each is bound to one N atom of a His residue. The coordination is completed with an O atom from a glutamic acid (Glu) residue and an S from a Met residue. In the Cu_B center, the Cu atom is coordinated to three His ligands in trigonal pyramidal geometry. In plants, cytochrome c oxidase is an enzyme that contains both Cu_A and Cu_B centers. Finally, the tetranuclear copper Z center (CuZ), found in nitrous-oxide reductases, is constituted by four Cu atoms coordinated by seven His residues and bridged by an S atom.

In the oxidation–reduction process between Cu^+ and Cu^{2+} states, Cu ions can generate harmful hydroxyl radicals and other reactive oxygen species (ROS) via the Haber–Weiss and Fenton reactions (Halliwell and Gutteridge, 1984; Yruela et al., 1996). These can damage proteins, nucleic acids, and lipids and interfere with the synthesis of Fe–S clusters or the activity of enzymes. On the other hand, ROS have a physiological role in signaling and regulatory processes, but the details of the relationship between Cu-sensing mechanisms and ROS-sensing mechanisms in plants so far are understood poorly (Ravet and Pilon, 2013).

10.2   UPTAKE OF COPPER BY PLANTS

Cu availability is a prerequisite for plant growth and development. The Cu content of soils ranges from 2 to 100 mg kg^{-1}, with an average value of ca. 30 mg kg^{-1}, but most of this Cu is not available for plants. Copper concentration in plant tissues varies depending on plant species or ecotypes, developmental stage, and environmental factors such as nitrogen supply and soil chemical properties (Ginocchio et al., 2002). For instance, Cu availability in soils decreases above pH 7.0 due to slightly basic pH conditions favoring the binding of Cu to soil chemical components. On the contrary, Cu availability increases under acidic soil conditions due to the increase of Cu ions in the soil solution (Kopsell and Kopsell, 2007). Nitrogen supply also can affect the availability of

Cu (Jarvis and Whitehead, 1981). Conversely, the presence of Cu can affect the development of some legume symbioses and the effectiveness of nitrogen fixation and the productivity of legumes (O'Hara, 2001).

Normally, the Cu concentration in various plant tissues is between 2 and 50 μg g^{-1} dry weight (Epstein and Bloom, 2005; Marschner, 2012), and the average composition of Cu in leaves is 10 μg g^{-1} dry weight (ranging from 5 to 20 μg g^{-1} dry weight) (Baker and Senef, 1995), but these concentrations can vary among plant species and varieties. The amount of free Cu ions in the cytosol probably is limited to less than one ion per cell (Rae et al., 1999). Copper-deficient plants (<5–20 μg g^{-1} in vegetative tissue) exhibit reduced growth and development, with reproductive organs and youngest leaves displaying the most severe symptoms (Burkhead et al., 2009). The rate of Cu uptake in plants is low in comparison with other micronutrients (Kabata-Pendias and Pendias, 1992). This occurrence is explained partly by the fact that (1) plants grown under high nitrogen supply require significantly more Cu and (2) the bioavailability of Cu tends to be greater in acidic soils, meaning that in most agricultural soils supply is comparatively low.

Copper concentrations in cells need to be maintained at low levels since it is extremely toxic because of its high oxidation–reduction properties. The critical free Cu concentration in nutrient media (below which Cu deficiency occurs) ranges from 10^{-14} to 10^{-16} M. Plants usually have a variable but adequate supply of Cu in the soil since typically soil solution concentrations range from 10^{-6} to 10^{-9} M (Marschner, 2012), but plants may still need to solubilize and reduce the metal.

Concentrations of free metal ions or metal chelates in the soil solution are generally rather low, although they depend on soil properties (Kochian, 1991; Marschner, 2012). Either the soil solution or the solid phase Cu is associated mainly with inorganic and organic matter by complexation or adsorption. Copper ions have a high affinity for binding sites of soil components and can also be adsorbed onto surfaces of clays and Fe or Mn oxides, coprecipitated with carbonates and phosphates, or be present in the lattice of primary silicate minerals. Copper ions also can be bound to cell walls and to the outer membrane surface of plant root cells. The distribution of Cu among these various solid and plant components will influence the chemical mobility greatly and hence the amount of Cu potentially taken up by plants. At acidic pH, dissolved Cu increases because of its weaker adsorption, and the free Cu ion activity is higher. Additionally, with increasing pH, competitive adsorption arises between organic matter in the solid phase and dissolved organic carbon, generally leading to an increase in Cu concentration in the soil solution due to a higher dissolved organic carbon content (Carrillo-González and González-Chávez, 2006). Thus, upon increasing pH, the Cu ion activity considerably decreases at the expense of organically bound complex species in the soil solution (Sauvé et al., 1997).

On the other hand, in the rhizosphere, root and microbial activities can influence the chemical mobility of metal ions and ultimately their uptake by plants as a consequence of alterations of soil pH or dissolved organic carbon content (Hinsinger and Courchesne, 2007). For instance, in the case of graminaceous species, the increased root secretion of Fe-chelating compounds (phytosiderophores [PS]) under Fe deficiency has been reported to increase Cu uptake in a calcareous soil (Chaignon et al., 2002b). It is noticeable that soil chemical properties can differ between the bulk soil and the rhizosphere, so considering only properties in the bulk soil might be a poor predictor of Cu bioavailability and ultimately Cu uptake, which is influenced by the particular properties induced in the rhizosphere by roots. Accordingly, contradictory results concerning the effect of pH on Cu uptake by plants are found in the literature. In very acidic soils, plant Cu concentration increased compared with calcareous soils in rape (*Brassica napus* L.) and tomato (*Solanum lycopersicum* L., syn. *Lycopersicon esculentum* Mill.) (Chaignon et al., 2002a, 2003; Cornu et al., 2007). In contrast, Cu accumulation in corn (*Zea mays* L.) was as high in calcareous soils as in acidic soils (Brun et al., 2001). Michaud et al. (2007) did not find a clear relationship between Cu uptake and soil pH in durum wheat (*Triticum turgidum* subsp. *durum* Husn.) in Cu-contaminated soils, probably due to the implication of root-induced changes of pH and dissolved organic carbon in the rhizosphere. At low pH, alkalization was observed in the rhizosphere compared with the bulk soil, which may result in a reduced Cu bioavailability. In calcareous soils, a larger chemical mobility may be related to phytosiderophore secretion leading to greater Cu uptake by plants.

The molecular mechanisms of Cu mobilization and uptake by roots from soil solutions remain unclear. Iron reduction and complexation mechanisms have been shown to affect Cu speciation in controlled environments. For instance, ferric reductase oxidase 2 (FRO2) is capable of reducing Cu^{2+}, and FRO3 is upregulated during Cu deficiency (Robinson et al., 1999; Burkhead et al., 2009; Palmer and Guerinot, 2009; Bernal et al., 2012). The fractionation of stable Cu isotopes ($^{65}Cu/^{63}Cu$) during uptake into plant roots and translocation to shoots can provide information on Cu acquisition mechanisms (Ryan et al., 2013). The heavier isotope was translocated preferentially to shoots in tomato (strategy I plant), whereas oat (*Avena sativa* L) (strategy II plant) showed no significant fractionation during translocation and with no effect of Fe supply on Cu accumulation in either species. The majority of Cu in the roots and leaves of both species existed as sulfur-coordinated Cu^+ species resembling glutathione–cysteine-rich proteins. The presence of isotopically light Cu in tomato is attributed to a reductive uptake mechanism, and the isotopic shifts within various tissues are attributed to redox cycling during translocation. The lack of isotopic discrimination in oat plants suggested that Cu uptake and translocation are not oxidation–reduction selective.

The Cu concentration in shoots of plants varies considerably among species (Beeson et al., 1947; Thomas et al., 1952; Alloway, 2008). This variation suggests different abilities to either absorb Cu from the soil or to translocate Cu from root to shoot. The factors that allow some plants to take up more Cu than others are unclear. Low uptake of Cu into shoots of plants is due partly to Cu being strongly bound to soil organic matter or partly because Cu remains in the roots in high amounts. In both cases, Cu is translocated to the shoot poorly (Jarvis and Whitehead, 1981; Whitehead, 1987). On the other hand, divalent cations can alter the permeability of the plasma membrane depending on their concentration and affect the trans-root potential and H^+ efflux of excised roots (Kennedy and Gonsalves, 1987). One of the physiological responses to excessive Cu is K^+ efflux from the roots, which is interpreted as a symptom of toxicity resulting from Cu-induced oxidative damage to the plasma membrane. Cu retention by roots impairs Cu translocation to xylem and shoots and increases the risks of membrane damage in the roots themselves.

10.3 PHYSIOLOGICAL RESPONSES OF PLANTS TO SUPPLY OF COPPER

Copper is a micronutrient but is highly phytotoxic above micromolar concentrations (Marschner, 2012; Yruela, 2005, 2008) and consequently induces physiological responses when supplied in excess. Copper-derived chemicals have been used as broad-specificity fungicides since 1882, when the French botanist and mycologist Pierre-Marie-Alexis Millardet developed a formulation, consisting of hydrated lime (calcium hydroxide), copper sulfate, and water (Bordeaux mixture), which controlled and protected the vineyards of France infested by the destructive phylloxera (*Daktulosphaira vitifoliae* Fitch). It was used subsequently on late potato blight caused by *Phytophthora infestans* de Bary (the agent of the Irish potato famine) (Large, 1940; Dixon, 2004). It was the first fungicide to be used worldwide, and this agronomic practice can increase the Cu concentration in soils (Lepp et al., 1984; de Loland and Singh, 2004; Pietrzak and McPhail, 2004; Fan et al., 2011; Ruyters et al., 2013; Wightwick et al., 2013).

Copper phytotoxicity symptoms normally are characterized by interveinal chlorosis, reduction of root and shoot volume, and reduction of stem size and leaf size (Ouzounidou et al., 1995; Prasad and Strzalka, 1999; Yruela, 2005, 2008; Kopsell and Kopsell, 2007; Marschner, 2012). It is noted that this toxicity is dependent on plant species, concentration of metal supplied, exposure time, culture conditions, and soil properties (Rooney et al., 2006; Li et al., 2010). Differences in Cu phytotoxicity on vineyard and cereal (barley, *Hordeum vulgare* L., corn, rice, *Oryza sativa* L.) crops grown in acidic or calcareous soils have been reported (Delas, 1963; Reichman, 2002; Michaud et al., 2007; Guo et al., 2010; Li et al., 2010). Normally, Cu toxicity decreases as soil pH increases. Guo et al. (2010) reported that the critical concentration of Cu added to soils that decrease corn grain yield by 10% is higher (711 mg kg^{-1}) for calcareous soil with a pH of 8.9 than for acidic soil (23 mg kg^{-1}) with a pH of 5.3. Rhizosphere alkalization can restrict Cu bioavailability in acidic soils. Bravin et al. (2009) found that Cu bioavailability was 2.4- to 4.2-fold higher when durum wheat plants were fed with mixed NH_4^+–NO_3^- than plants fed NO_3^- alone.

The early effect of Cu toxicity in plants is rhizotoxicity (Sheldon and Menzies, 2005; Kopittke and Menzies, 2006; Marshner, 2012). Copper translocation toward shoots is restricted efficiently by the large accumulation of Cu in roots. Thus, important Cu rhizotoxicity and deleterious physiological effects (i.e., altered root growth and nutrient uptake) are expected to occur before shoot Cu concentration reaches abnormal values. To evaluate ecotoxicological risks caused by excessive Cu, simple and sensitive indicators of Cu phytotoxicity based on early and primarily Cu rhizotoxic effects are necessary. Few references of such Cu phytotoxicity are available in the literature, and most of these were established for shoots (i.e., critical Cu concentration), not for roots, probably due to difficulties encountered in measuring Cu concentration in roots of soil-grown plants because of potential contamination of roots with soil particles (Beckett and Davis, 1978; MacNicol and Beckett, 1985; Reuter and Robinson, 1997; Kopittke and Menzies, 2006).

Some of these responses are common to Cu deficiency, and this similarity makes it difficult to distinguish the origin of such symptoms. Either Cu deficiency or Cu toxicity cause changes in the expression of genes, which activate morphological changes in the plant, mainly concerning root and leaf architecture. Altogether, this action limits plant biomass and crop productivity.

10.4 GENETICS OF ACQUISITION AND DISTRIBUTION OF COPPER BY PLANTS

Plants are characterized by a vascular transport system that requires a complex Cu homeostasis machinery and that provides the metal to Cu-dependent proteins in chloroplasts, endoplasmic reticulum, and mitochondria. Different types of transporter proteins involved in Cu uptake and distribution have been identified, based on their homology with bacteria, yeast, and mammals. These transporters include integral membrane proteins and soluble proteins named metallochaperones. Examples are (1) the high-affinity Cu transport proteins named copper transporter proteins (COPTs) in plants (Ctr in yeast, *Saccharomyces cerevisiae*, and humans) (Kampfenkel et al., 1995; Puig and Thiele, 2002; Pole and Schützdendübel, 2004), (2) the P-type ATPase pump transporters (William et al., 2000; Williams and Mills, 2005; Yruela, 2005, 2008), (3) the ZRT/ITR-like protein (ZIP) zinc (Zn) transporter family (Puig et al. 2007), (4) yellow stripe-like (YSL) protein family, and (5) copper chaperones (CCHs) (Table 10.1).

10.4.1 COPT TRANSPORTERS

Copper enters the cytosol of plant cells mediated by members of the COPT family of transporters. The COPT proteins share 35%–64% sequence identity and 47%–73% sequence similarity with each other and have a similar structure to COPT/Ctr proteins in other species. Members of this family have been investigated at the molecular and functional level in arabidopsis (*Arabidopsis thaliana* Henyh.) (Puig and Thiele, 2002; Puig et al., 2007) and rice (Yuan et al., 2011). In the arabidopsis genome, six members of the COPT/Ctr family (COPT1-6) have been identified (Sancenón et al., 2003; Peñarrubia et al., 2010; Perea-García et al., 2013). In rice, the COPT/Ctr family consists of seven members (COPT1-7) (Yuan et al., 2011). They could act alone or cooperatively to mediate Cu transport in different plant tissues.

All transporters belonging to the COPT/Ctr family contain three predicted transmembrane (TM) segments, and most possess an N-terminus methionine- and histidine-rich putative metal-binding domain localized in the extracellular space. The C-terminus and the loop between TM1 and TM2 segments are exposed to the cytosol (Puig and Thiele, 2002; Klomp et al., 2003). Genetic data and in vivo uptake experiments have demonstrated that an extracellular methionine residue, located approximately 20 amino acids before TM1, and a MxxxM motif within TM2 are essential for

TABLE 10.1
COPTs in Plants

Family	Name	Description	Subcellular Localization	Tissue Expression	Code	References
COPT	AtCOPT1	High-affinity Cu^+ transporter	Plasma membrane	Root apex, lateral root, embryo, trichomes, guard cells, and pollen grains	At5g59030	1, 2, 3, 4, 5
	OsCOPT1				Os01g0770700	6
	AtCOPT2	High-affinity Cu^+ transporter	Plasma membrane	Leaves (high), roots, stems, and flowers	At3g46900	2, 4, 7
	OsCOPT2				Os05g0424700	6
	AtCOPT3	High-affinity Cu^+ transporter	Plasma membrane	Stems (high), leaves, and flowers (low)	At5g59040	2
	OsCOPT3				Os01g0770800	6
	AtCOPT4	High-affinity Cu^+ transporter	Plasma membrane	Roots (high), leaves (low), stems, and flowers	At2g37925	2
	OsCOPT4				Os03g0370800	6
	AtCOPT5	High-affinity Cu^+ transporter	Secretory pathway	Leaves (high), stems, roots, and flowers (low)	At5g20650	2, 8, 9
	OsCOPT5				Os09g0440700	6
	AtCOPT6	High-affinity Cu^+ transporter	Plasma membrane	Leaves (high)	At2g26975	10
	OsCOPT6				Os04g0415600	6
	OsCOPT7	High-affinity Cu^+ transporter			Os09g0440700	6
ZIP	AtZIP1	Divalent cation transporter	Plasma membrane	Roots	At3g12750	11, 12, 13, 14
	OsZIP1			Roots and leaves	Os01g0972200	
	AtZIP2	Divalent cation transporter	Plasma membrane	Roots	At5g59520	11, 12, 15
	OsZIP2				Os03g0411800	
	AtZIP3	Divalent cation transporter	Plasma membrane	Roots	At2g32270	11
	OsZIP3				Os04g0613000	
	AtZIP4	Divalent cation transporter	Thylakoid membrane	Leaf and chloroplasts	At1g10970	11, 12
	OsZIP4			Roots and shoots phloem)	Os08g0207500	16
	MtZIP4			Root and leaf	Q6VM18	17
P_{1B} ATPase	AtHMA1	Cu^{2+}–P_{1B}-ATPase transporter	Chloroplast envelope	Root and shoot	At4g37270	18, 19
	AtHMA5	Cu^+–P_{1B}-ATPase transporter	Plasma membrane	Root, flower, and pollen	At1g63440	20
	AtHMA6(PAA1)	Cu^+–P_{1B}-ATPase transporter	Chloroplast envelope	Root and shoot	At4g33520	21, 22
					At5g44790	23, 24

(Continued)

TABLE 10.1 (Continued)
COPTs in Plants

Family	Name	Description	Subcellular Localization	Tissue Expression	Code	References
	AtHMA7(RAN1)	Cu⁺-P₁B-ATPase transporter	Trans-Golgi network	Root, shoot, and leaf		25, 26
	OsHMA9			Vascular tissue (xylem, phloem) and anthers		27
	BnHMA7(BnRAN1)					
	AtHMA8(PAA2)	Cu⁺-P₁B-ATPase transporter	Thylakoid membrane	Shoot and leaf	At5g21930	22, 28
	GmHMA8			Leaf		29
Chaperones	AtCCH	ATX1-like Cu chaperone	Cytosol	Stem and vascular tissue	At3G56240	30, 31
	LeCCH					32
	AtATX1	ATX1-like Cu chaperone	Cytosol	Root	At1G66240	20
	AtCCS	Chaperone for Cu/ZnSOD	Cytosol and chloroplast	Leaf	At1g12520	21, 33, 34
	LeCCS				Q9ZSC1	35
	StCCS				Q6XZF8	36
	ZmCCS					37
	GmCCS				Q9BBU5	38
	AtCOX17	COX17-like Cu chaperone	Mitochondria		At3g15352	33, 39
YSL	AtYSL1	Fe³⁺-phytosiderophore/NA transporter	Plasma membrane	Leaf, shoots, and pollen	At4G24120	40
	ZmYS1				Q9AY27	41, 42, 43
	HvYSL1				J7QZU1	42, 44
	BdYSL1A, BdYSL1B					45
	OsYSL1				Os01g0238700	46, 47, 48
	AtYSL2	Fe³⁺/Cu²⁺-NA transporter	Plasma membrane	Leaf, roots, shoots (low), pollen, petals, and sepals	At5g24380	12, 49, 50
	OsYSL2				Os02g0649900	51, 52
	AtYSL3	Fe³⁺/Cu²⁺-NA transporter	Plasma membrane	Leaf, anther, and pollen	At5g53550	12, 53
	TcYSL3				Q2XPY3	54
	AtYSL4	Metal-NA complex transporter	Vacuolar membrane and endoplasmic reticulum	Leaf	At5g41000	55, 56
	OsYSL4				Os05g0252000	
	AtYSL6	Metal-NA complex transporter	Vacuolar membrane and endoplasmic reticulum	Leaf	At3g27020	56
	OsYSL6				Os04g0390500	

(Continued)

TABLE 10.1 (Continued)
COPTs in Plants

Family	Name	Description	Subcellular Localization	Tissue Expression	Code	References
	AtYSL7	Metal–NA complex transporter	Plasma membrane		At1g65730	57
	OsSYSL7				Os02g0116300	
	AtYSL8	Metal–NA complex transporter	Plasma membrane		At1g48370	57
	OsYSL8				Os02g0116400	
	OsYSL14	Metal–NA complex transporter	Plasma membrane	Leaf and roots (low)	Os02g0633300	46, 47, 48
	OsYSL15	Metal–NA complex transporter	Plasma membrane	Roots and shoots (low)	Os02g0650300	46
	OsYSL16	Metal–NA complex transporter	Plasma membrane	Roots	Os04g0542800	47
	OsYSL18	Metal–NA complex transporter	Plasma membrane	Leaf and reproductive organs	Os01g0829900	58

References: (1) Kampfenkel et al. (1995); (2) Sancenón et al. (2003); (3) Sancenón et al. (2004); (4) Andrés-Colás et al. (2010); (5) Andrés-Colás et al. (2013); (6) Yuan et al. (2011); (7) Perea-García et al. (2013); (8) García-Molina et al. (2011); (9) Klaumann et al. (2011); (10) Jung et al. (2012); (11) Grotz et al. (1998); (12) Wintz et al. (2003); (13) Connolly et al. (2002); (14) Vert et al. (2002); (15) Vert et al. (2001); (16) Ishimaru et al. (2007); (17) López-Millán et al. (2004); (18) Seigneurin-Berny et al. (2006); (19) Kim et al. (2009); (20) Andrés-Colás et al. (2006); (21) Shikanai et al. (2003); (22) Abdel-Ghany et al. (2005); (23) Woeste and Kieber (2000); (24) Chen et al. (2002); (25) Lee et al. (2007); (26) Sichul et al. (2007); (27) Southron et al. (2004); (28) Weigel et al. (2003); (29) Bernal et al. (2007b); (30) Mira et al. (2001a); (31) Mira et al. (2001b); (32) Company and González-Bosch (2003); (33) Wintz and Vulpe (2002); (34) Chu et al. (2005); (35) Zhu et al. (2000); (36) Trindade et al. (2003); (37) Ruzsa and Scandalios (2003); (38) Sagasti et al. (2011); (39) Balandin and Castresana (2002); (40) Le Jean et al. (2005); (41) Curie et al. (2001); (42) Ueno et al. (2009); (43) Contre and Walker (2012); (44) Murata et al. (2006); (45) Yordem et al. (2011); (46) Inoue et al. (2009); (47) Lee et al. (2009); (48) Zheng et al. (2012); (49) DiDonato et al. (2004); (50) Schaaf et al. (2005); (51) Koike et al. (2004); (52) Colangelo and Guerinot (2006); (53) Waters et al. (2006); (54) Gendre et al. (2007); (55) Conte et al. (2013); (56) Divol et al. (2013); (57) Hofstetter et al. (2013); (58) Aoyama et al. (2009).

Cu acquisition and probably mediate metal coordination during transport. The structure of human Ctrl transporter, a homolog of COPT proteins, in a phospholipid bilayer has shown a compact and symmetrical trimer organization with a novel channel-like architecture where a conserved GxxxG motif within TM3 is essential for trimerization (Aller et al., 2004; Aller and Unger, 2006). The expression of members of the COPT/Ctr gene family is controlled by environmental Cu level in different species. In general, they are transcriptionally upregulated in response to Cu deficiency and downregulated in response to Cu excess. COPT1 and COPT2 localized in the plasma membrane are involved in Cu acquisition and transport toward the cytosol (Sancenón et al., 2003; Andrés-Colás et al., 2010) and are highly specific for Cu+ ions. COPT1 is activated under Cu-limiting conditions and regulates root elongation and pollen development. COPT1 plays a predominant role in Cu uptake from soil through the root tips (Sancenón et al., 2004). *COPT1* expression in roots has been corroborated in transgenic plants where the COPT1 promoter drives GUS expression, and it has been assigned to specific peripheral cells in a limited narrow root apical zone. Reduction of Cu uptake was observed in *COPT1* antisense plants. COPT2 participates in the attenuation of Cu-deficiency responses driven by Fe limitation, possibly to minimize further Fe consumption. In arabidopsis, *COPT2* expression is upregulated in roots by both Cu and Fe deficiencies (Sancenón et al., 2003; Colangelo and Guerinot, 2006; Waters et al., 2012; Perea-García et al., 2013), indicating links between Cu and Fe transport in arabidopsis. Furthermore, *AtCOPT1* and *AtCOPT2* genes display a similar expression pattern under slight Cu deficiency in several plant aerial tissues and organs (including cotyledons from young seedlings, trichomes, anthers, and mature pollen), suggesting a partial functional redundancy (Perea-García et al., 2013). On the contrary, *AtCOPT1* and *AtCOPT2* present notable differences in root expression patterns. *AtCOPT1* is exclusively expressed in primary and secondary root tips, whereas *AtCOPT2* is expressed in subapical root regions, indicating local and specific functions and signaling in roots. Moreover, *AtCOPT1* and *AtCOPT2* differ in their response under Fe deficiency. In these conditions, *AtCOPT1* is downregulated in restricted Cu medium and *AtCOPT2* is upregulated. Additionally, a role of AtCOPT2 transport activation has been reported in Pi starvation conditions. Connections between Pi starvation responses and the regulation of other metal ion transporters have been suggested (Abel, 2011; Chiou and Lin, 2011).

The AtCOPT3 and AtCOPT5 proteins are involved in intracellular Cu distribution. In particular, AtCOPT5 participates in the mobilization of Cu from the vacuole or prevacuolar compartments toward the cytosol under extreme Cu deficiency (García-Molina et al., 2011; Klaumann et al., 2011). AtCOPT6 is localized in the plasma membrane, interacts with AtCOPT1, and regulates the response to either Cu limitation or excess (Jung et al., 2012). Its transcript is highly expressed in leaves and is upregulated under Cu limitation, indicating that probably its role consists in maintaining optimum Cu level for the photosynthetic apparatus. In *O. sativa*, it has been described that COPT2, COPT3, and COPT4 physically interact with COPT6, indicating that they may cooperate with COPT6 for Cu transport. COPT7 would function alone in different rice tissues, except in root (Yuan et al., 2011). The *OsCOPT2*, *OsCOPT3*, *OsCOPT4*, and *OsCOPT6* genes are all expressed in stem, sheath, leaf, and panicle tissues.

10.4.2   P₁ᵦ-Type ATPase Pump Transporters

The P-type heavy metal ATPases (HMAs) are a subgroup of the large superfamily of P-type ATPases, which use ATP to pump a variety of charged substrates across biological membranes and are distinguished by the formation of a phosphorylated intermediate during the reaction cycle. This family of transporters (HMAs) is diverse in terms of tissue distribution, subcellular localization, and metal specificity. In plants, at least eight members have been identified in arabidopsis. The rice genome contains nine P_{1B}-type ATPase genes and ten members of this subfamily have been identified in barley. The P-type HMAs are implicated in the transport of a range essential and potentially toxic metals (i.e., Cu^+, Cu^{2+}, Zn^{2+}, Cd^{2+}, Pb^{2+}) across cell membranes. Solioz and Vulpe (Solioz and

Vulpe, 1996) defined the P-type HMAs as CPx ATPases because they share the common feature of a conserved intramembranous cysteine–proline–cysteine, cysteine–proline–histidine, or cysteine–proline–serine motif (CPx motif), which is thought to function in heavy metal transduction. Structurally, P_{1B}-type ATPases contain eight TM segments with various cytoplasmic domains involved in enzyme phosphorylation (P-domain), nucleotide binding (N-domain), and energy transduction (A-domain), domains that are common for all P-type ATPases. Additionally, P_{1B} ATPases show different features associated with their singular function in heavy metal transport (Argüello, 2003; Argüello et al., 2007).

Functional studies of HMAs have shown that these transporters can be divided into two subgroups based on their metal-substrate specificity: a copper (Cu)/silver (Ag) group, which transports monovalent cations, and a zinc (Zn)/cobalt (Co)/cadmium (Cd)/lead (Pb) group, which is involved with divalent cations. In arabidopsis, AtHMA1 to AtHMA4 are divalent cation transporters, involved in the export of Cu^{2+}, Zn^{2+}, and Cd^{2+}. In contrast, AtHMA5 to AtHMA8 act in the transport of monovalent Cu^+ ions. In rice, OsHMA1 to OsHMA3 belong to the divalent Zn/Co/Cd/Pb subgroup (Takahashi et al., 2012a). The first member cloned in plants was PAA1(AtHMA6) (P1B-type ATPase of arabidopsis 1) (Tabata et al., 1997), which is responsible for the delivery of Cu to chloroplasts and provides the cofactor for the stromal Cu/ZnSOD enzyme and for the thylakoid lumen protein plastocyanin, two proteins involved in antioxidant enzymatic activity and photosynthetic electron transport function, respectively (Shikanai et al., 2003). PAA2(AtHMA8), closely related to PAA1(AtHMA6), transports Cu into the thylakoid lumen to supply plastocyanin (Abdel-Ghany et al., 2005). PAA1(AtHMA6) is expressed in both roots and shoots, while PAA2(AtHMA8) is only detected in shoots. A double *paa1/paa2* mutant resulted in seedling lethality, a more severe phenotype than that observed for plants defective for either gene separately, underlying the importance of Cu to photosynthesis (Weigel et al., 2003; Abdel-Ghany et al., 2005). The homolog of PAA2(AtHMA8) in soybean (*Glycine max* Merr.), named GmHMA8, was identified and localized in the thylakoid membrane (Bernal et al., 2007b). AtHMA1 is localized in the chloroplast envelope and contributes to the detoxification of excess Zn and Cu (Seigneurin-Berny et al., 2006; Kim et al., 2009). AtHMA3 is localized in the vacuolar membrane and plays a role in detoxifying Zn and Cd through vacuolar sequestration (Gravot et al., 2004; Morel et al., 2009). HMA3 is recognized as the major locus responsible for the variation in leaf Cd accumulation in arabidopsis (Chao et al., 2012). AtHMA2 and AtHMA4 are localized in the plasma membrane and function in Zn and Cd efflux from cells (Mills et al., 2003; Eren et al., 2004; Hussain et al., 2004; Mills et al., 2005; Verret et al., 2005). AtHMA2 and AtHMA4 are expressed in tissues surrounding the vascular vessels of roots (Hussain et al., 2004; Verret et al., 2004). Additionally, it has been suggested that HMA4 acts in Zn loading to the xylem and that HMA2 and HMA4 could be involved in Cd translocation in arabidopsis (Wong and Cobbett, 2009; Mills et al., 2010). Furthermore, HMAs are involved in metal hyperaccumulation and hypertolerance. In *Arabidopsis halleri* L., greater HMA4 expression in the roots contributes to the high efficiency of Zn translocation from roots to shoots (Talke et al., 2006; Courbot et al., 2007). Furthermore, expression of *HMA4* contributes to Zn and Cd hyperaccumulation in *A. halleri* and alpine pennycress (*Noccaea caerulescens* F. K. Mey; formerly *Thlaspi caerulescens* J. & C. Presl) (Ó Lochlainn et al., 2012). NcHMA4/TcHMA4 is present in vascular tissue and is thought to function in metal distribution (Craciun et al., 2012). TcHMA3 in the leaves also contributes to Cd hyperaccumulation and hypertolerance via the high sequestration of Cd in leaf vacuoles (Ueno et al., 2011).

The responsive to antagonist (RAN1)/AtHMA7 transporter is responsible for the biogenesis of ethylene receptors by delivering Cu to ETR1 through the endoplasmic reticulum, where it is required for the formation of functional ethylene receptors (Woeste and Kieber, 2000; Chen et al., 2002). The plant hormone ethylene is an important signal in many abiotic stress situations and also in plant pathogen interaction. RAN1 (AtHMA7) has also been found in rape (BnRAN1) (Southron et al., 2004). Among the rice P_{1B} ATPases, OsHMA9 was found to form a subclass with RAN1 (AtHMA7), which might transport Zn, Cd, and Pb, although OsHMA9 belongs to the Cu/Ag subgroup phylogenetically (Lee et al., 2007). It plays a role in Cu detoxification, acting as an efflux pump in the plasma membrane (Sichul et al., 2007). Plants mutant in the *HMA9* gene, such as *oshma9-1* and *oshma9-2* exhibited the phenotype of increased sensitivity to high levels of Cu and also Zn and Pb. The *OsHMA9* gene was expressed mainly in vascular tissues, including xylem and phloem, and weakly expressed in mesophyll tissues. In developing tissues, expression was strong in anthers, suggesting a putative role in metal delivery to rice anthers. The importance of metal transport in anthers has been reported previously. The AtHMA5 transporter, the closest homolog of RNA1(AtHMA7) in the P_{1B}-type ATPase subfamily, is strongly and specifically induced by Cu in whole plants. The *hma5* T-DNA insertion mutants are hypersensitive to Cu, and HMA5-defective plants accumulate Cu in roots to a greater extent than wild-type plants, suggesting its key role in TM transport, and particularly in root Cu detoxification (Andrés-Colás et al., 2006). This phenotype is the opposite of that observed for the COPT antisense lines, supporting the notion that COPT1 and AtHMA5 transport Cu in opposite directions. AtHMA5 is expressed mostly in roots, flowers, and pollen. The specific interaction of AtHMA5 with two different antioxidant protein1 (ATX1)–type chaperones, ATX1 and CCH, in arabidopsis has been demonstrated. Although further experiments are necessary to confirm the fact, it has been proposed that AtHMA5 could be involved in Cu efflux from specific root cells and its overexpression in plants could be a strategy for improving Cu detoxification under Cu excess (Figure 10.1).

OsHMA3 transports only Cd and plays a role in the sequestration of Cd into vacuoles in root cells (Ueno et al., 2010; Miyadate et al., 2011). On the contrary, there is little information on the role of OsHMA1, which is thought to be involved in Zn transport. *OsHMA1* expression is highly upregulated by Zn deficiency in shoot tissue (Suzuki et al., 2012). *OsHMA1* is highly expressed in the leaf blade but is also expressed in the root, inflorescence, anther, pistil, lemma, palea, ovary, embryo, and endosperm. This occurrence suggests that OsHMA1 plays a role in Zn transport in the entire plant through all growth and developmental stages. OsHMA1, just like AtHMA1, may play a role in Zn efflux from plastids and may contribute to the detoxification of excess Zn.

OsHMA2 is localized in the plasma membrane and transports Zn and Cd (Nocito et al., 2011; Satoh-Nagasawa et al., 2012; Takahashi et al., 2012b). The expression of *OsHMA2* was observed mainly in the roots, where *OsHMA2* transcripts were abundant in vascular bundles (Takahashi et al., 2012b). OsHMA2 could play a role in loading Zn and Cd into the xylem and participates in root-to-shoot translocation of these metals in rice. It also could participate in Zn transport during flowering and seed maturing.

10.4.3 ZIP Transporters

ZIP transporters refers to ZRT-IRT-like proteins, which were named for the two first members found: the high-affinity yeast plasma membrane Zn-uptake transporter, ZRT1, and the high-affinity arabidopsis plasma membrane Fe-uptake transporter, IRT1 (Eide et al., 1996; Zhao and Eide, 1996). Subsequent studies of other plant ZIP family members demonstrated that not all the members of this family are involved in plasma membrane micronutrient uptake. The ZIP membrane proteins are involved in the transport of four essential micronutrients: Zn, Fe, Mn, and Cu (Eide et al., 1996; Grotz et al., 1998; Wintz et al., 2003; Cohen et al., 2004; Pedas et al., 2008; Lin et al., 2009). They act as influx carriers, participating in the uptake from the soil (similar to ZRTs in yeast). In plant roots, as in yeast, Zn enters the cell via ZIP proteins, which are divalent metal transporters (Grotz et al., 1998; Colangelo and Guerinot, 2006; Puig et al., 2007). ZIP family members have also been shown to transport heavy metals such as Cd and may also play a significant role in how various heavy metals, either essential or toxic, are taken up and translocated throughout the plant (Guerinot, 2000; Pence et al., 2000; Rogers et al., 2000). ZIP proteins contain eight TM domains and a histidine-rich variable loop between TM3 and TM4.

In arabidopsis, 15 ZIP family members have been identified. In general, these transporters are highly expressed under conditions of Zn deficiency, whereas their expression decreases quickly

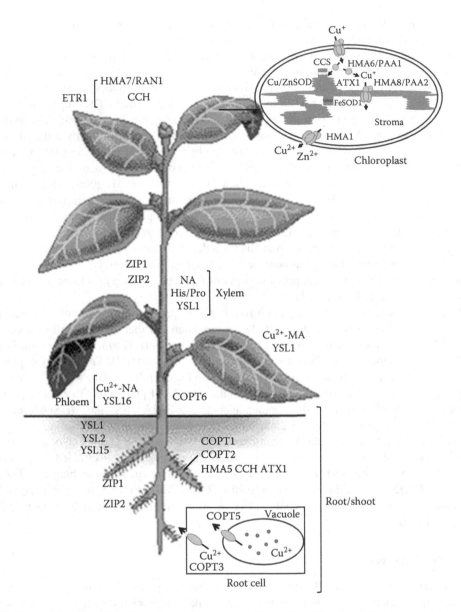

FIGURE 10.1 **(See color insert.)** Scheme of transport pathways identified for Cu in a generic plant. Arrows indicate the proposed direction for metal transport. CCH, copper chaperone; ATX1, antioxidant 1; CCS, copper chaperone for Cu/Zn SOD; COPT, copper transporter; ERT1, endoplasmic reticulum; HMA, heavy metal P-type ATPase; MA, mugineic acid; NA, nicotianamine; PAA, P-type ATPase of *Arabidopsis*; RAN1, responsive to antagonist 1; SOD, superoxide dismutase; YSL, yellow stripe-like protein; ZIP, ZRT/IRT-like protein.

when Zn is added to the media (Talke et al., 2006) although they could transport other metals. Six of the arabidopsis ZIP genes complemented a yeast Zn-uptake–deficient mutant, one was able partially to complement a yeast Fe-uptake–deficient mutant, six ZIP family members complemented a Mn-uptake–deficient mutant, and none complemented the Cu-uptake–deficient mutant (Milner et al., 2013). The most characterized members of this family are the three arabidopsis ZIP transporters: AtZIP1(IRT1), AtZIP2(IRT2), and AtZIP3(IRT3), with AtZIP1(IRT1) being by far the most well studied based on its seminal role in root Fe uptake and transport (Eide et al., 1996; Rogers et al., 2000; Vert et al., 2001; Connolly et al., 2002; Vert et al., 2002; Lin et al., 2009;

Vert et al., 2009; Milner et al., 2013). AtZIP1 is localized in the vacuole. Very little or nothing is known about the function of the other 12 arabidopsis ZIPs.

AtZIP2 and *AtZIP4* complement growth defects of yeast Cu and Zn transport mutants (Grotz et al., 1998; Wintz et al., 2003), and their transcript expression is upregulated in arabidopsis by deficiency of Cu and Zn, but not of Fe. AtZIP1 may play a role in remobilizing Mn from the vacuole to the cytoplasm in root stelar cells and may contribute to radial movement to the xylem parenchyma. AtZIP2, on the other hand, may mediate Mn (and possibly Zn) uptake into root stelar cells and thus also may contribute to Mn/Zn movement in the stele to the xylem parenchyma, for subsequent xylem loading and transport to the shoot. Although the role of these proteins in plant Cu transport still requires further characterization, the preference that ZIP family members show for divalent metals suggests that ZIP2 and ZIP4 proteins may transport Cu^{2+} ions.

Expression of *AtZIP1* and *AtZIP2* genes are localized in the root stele, although *AtZIP1* expression was also in the leaf vasculature. It also was found that AtZIP1 is a vacuolar transporter, whereas AtZIP2 is localized in the plasma membrane. The ZIP4 family member is localized in the chloroplast (Guerinot, 2000). *AtZIP1* transcript levels increase in the roots as the arabidopsis plant ages, and its expression decreases in the shoot during the same developmental time sequence. *AtZIP1* transcript levels were higher in the roots under both Zn and Fe deficiency. Fe acquisition in arabidopsis roots under Fe deficiency mostly depends on AtIRT1, which is considered the major Fe transporter at the root surface in arabidopsis. On the other hand, root and shoot *AtZIP2* transcript abundance decreased in response to Zn, Fe, and Mn deficiency. However, the observed lower transcript levels under Zn and Mn deficiency suggest that AtZIP2 is not a primary transporter involved in Zn and Mn uptake from the soil under Zn- and Mn-limiting conditions. Six cDNA encoding ZIP family members have been identified in the model legume, barrel clover (*Medicago truncatula* L.), and have been tested for the ability to complement yeast metal-uptake mutants (López-Millán et al., 2004).

The exact mechanism of this regulation is still unknown. It has been shown that at least ZIP4 in arabidopsis is regulated by transcription factors of the basic-region leucine zipper (bZIP) family: bZIP19 and bZIP23. These factors bind to a zinc-deficiency response element, which has been found in the upstream region not only of *ZIP4* but also of *ZIP1*, *ZIP3*, *ZIP9*, and *IRT3*. Therefore, it is reasonable to assume similar regulation for these ZIP transporters.

10.4.4 YSL TRANSPORTERS

The YSL transporters belong to the oligopeptide transporter (OPT) superfamily (Curie et al., 2001, 2009) and transport tri-, tetra-, penta-, and hexapeptides (Yen et al., 2001). Although the 3D structure and topology of these transporters is unknown, structural modelling predicts between 11 and 16 TM domains and loops between TM domains containing a higher number of charged residues, which may participate in substrate recognition of metal–chelator complexes (Conte and Walker, 2012). The YSL proteins play an important role in the long-distance transport of metals complexed with PS and nicotianamine (NA) (Colangelo and Guerinot, 2006; Conte and Walker, 2012). They are responsible for primary uptake of Fe–PS complexes into plant roots. It has been reported that *ys1* mutants were defective in the uptake of Fe–PS and Zn–deoxymugineic acid (DMA) (von Wirén et al., 1994; 1996). The plants that use strategy II for Fe uptake, which involves the synthesis and secretion of PS to increase Fe solubility in the rhizosphere, typically take up Fe–PS complexes at the root epidermis by YSL transporters. Members of this family have been investigated in *Z. mays*, maize (ZmYS1) (Curie et al., 2001), barley (HvYS1) (Murata et al., 2006), arabidopsis (AtYS1, AtYS2, AtYS3), rice (OsYS1, OsYS14, OsYS15, OsYS16) (Inoue et al., 2009; Lee et al., 2009; Zheng et al., 2012), and brachypodium (*Brachypodium distachyon* P. Beauv) (BdYS1A) (Yordem et al., 2011). ZmYS1 protein, which was the first member of this family characterized, accumulates in roots and leaves of Fe-deficient plants and transports Fe–PS. It is localized on the distal side of epidermal cells of the crown and lateral roots (Ueno et al., 2009). In the leaves, ZmYS1 is localized in mesophyll but not epidermal cells, implicating it in intracellular transport of Fe in maize (Ueno et al., 2009). ZmYS1 has been characterized extensively on a biochemical level indicating that it plays a role in the homeostasis of Cu, Zn, Ni, or Mn (Conte and Walker, 2012). ZmYS1 has a broad substrate specificity and can transport Cu(II)–MA, Zn(II)–MA, and Fe(II)–NA but has a lower affinity for Ni(II)–MA, Mn(II)–MA, and Co(II)–MA. The ortholog HvYS1 transporter is localized in the plasma membrane, and it is expressed more strongly in roots, where it is upregulated by Fe deficiency but not by deficiency of Mn, Zn, or Cu (Murata et al., 2006; Ueno et al., 2009).

Arabidopsis has eight predicted YSL proteins. AtYSL1, AtYSL2, and AtYSL3 have been studied is some detail. The *AtYSL1* gene was expressed in the xylem parenchyma of leaves, pollen, and young siliques, and it was induced by Fe excess in shoots. AtYSL2 is the most similar transporter to ZmYS1, and its transcript accumulates under conditions of Fe sufficiency or Fe resupply, and the transcript levels also respond to Cu and Zn (DiDonato et al., 2004; Le Jean et al., 2005; Schaaf et al., 2005). Localization of AtYSL2 in root endodermis and pericycle cells facing the xylem tubes has suggested its participation in lateral movement of Fe and Cu within the veins (Schaaf et al., 2005). These proteins seem to be involved in the unloading of metal–NA from vasculature into developing tissues, in immobilization of metal–NA from senescent leaves, and in an efficient loading of metal–NA into seeds. AtYSL1 and AtYSL3 are upregulated during leaf senescence and could function in delivery of Cu, among other metals, from vascular tissues, as well as in Fe–NA delivery to seeds (Waters et al., 2006). *AtYSL2* and *AtYSL3* are expressed differentially under metal deficiencies, and their products can transport Cu^{2+}, Mn^{2+}, and Fe^{2+} (Wintz et al., 2003). AtYSL4 and AtYSL6 are involved in managing chloroplastic Fe. They are localized in the vacuolar membrane and endoplasmic reticulum (Conte et al., 2013; Divol et al., 2013). *YSL4* and *YSL6* expression patterns support the physiological role of YSL4 and YSL6 in detoxifying Fe during plastid dedifferentiation occurring in embryogenesis and senescence.

The rice genome contains 18 putative *YSL* genes. *Os*YSL2 has been shown to transport Fe^{2+}–NA and Mn^{2+}–NA complexes but not Fe^{3+}–NA. A role in the transport of divalent cations in the phloem has been suggested (Koike et al., 2004; Colangelo and Guerinot, 2006). The member most closely related to ZmYS1 is OsYSL15, which similarly is localized in the plasma membrane. It is upregulated in roots under Fe deficiency and transports Fe(III)–DMA and probably Fe(II)–NA based on the *fet3fet4* complementation (Murata et al., 2006; Inoue et al., 2009). Overexpression of *OsYSL15* increases the Fe concentration in leaves and seeds. On the other hand, the tissue expression profile of *OsYSL18* suggests that its protein products are involved in translocation of Fe in reproductive organs and phloem in leaf joints (Aoyama et al., 2009). By contrast, OsYSL16 is a phloem-localized transporter involved in Cu–NA distribution and redistribution. It is present in the roots, leaves, and unelongated nodes at the vegetative growth stage. It has been suggested that OsYSL16 is required for delivering Cu to developing young tissues and seeds (Lee et al., 2009).

The genome of brachypodium contains 19 protein homologs to ZmYS1, and two of them are most likely YS1 orthologs: BdYS1A and BdYS1B (Yordem et al. 2011). Studies on YSL genes have also been carried out in metal hyperaccumulator plants such as alpine pennycress. *TcYSL3* is expressed throughout the plant body, and *TcYSL5* and *TcYSL7* are expressed in shoots and the central cylinder in the roots. *TcYSL5* is highly expressed in shoots and *TcYSL7* is highly expressed in flowers (Gendre et al., 2007).

10.4.5 Copper Chaperones

The Cu chaperones are low-molecular-weight metal-receptor proteins involved in the intracellular trafficking of metal ions. These proteins contain Cu-binding domain(s), to assist Cu intracellular homeostasis by their Cu-chelating ability. The limited solubility and high reactivity of Cu^+ inside the cell requires the participation of these specialized proteins. Consequently, Cu chaperones bind and deliver Cu ions to intracellular compartments and insert the Cu into the active sites of specific partners, Cu-dependent enzymes (O'Halloran and Culotta, 2000; Huffman and O'Halloran, 2001). These proteins prevent inappropriate Cu interaction with other cellular components. Arabidopsis has at least three Cu chaperones, including the Cu chaperone for SOD (CCS) and two homologs of yeast ATX1, the CCH and ATX1 (Casareno et al., 1998; Chu et al., 2005; Puig et al., 2007; Shin et al., 2012). CCH has been the most extensively studied of the Cu chaperones in plants (Mira et al., 2001a,b). The CCH chaperone exhibits the conserved features of the ATX1-type metallochaperone family such as typical lysine residues, overall βαββαβ-fold structure, and a MxCxxC Cu⁺-binding motif in the N-terminus (Pufahl et al., 1997). However, CCH also has a plant-specific C-terminal domain with special structural characteristics (Mira et al., 2001a,b, 2004). In arabidopsis, the ATX1- and ATX1-like CCHs share high sequence homology. Both CCH and ATX1 chaperones complement the yeast *atx1* mutant and interact with the N-terminus of AtHMA5 (Andrés-Colás et al., 2006). However, the C-terminus of CCH has a negative effect on its interaction with AtHMA5. The plant *CCH* gene expression has been related to oxidative stress and senescence, when the plant reallocates nutrient resources. High levels of *CCH* expression were found in arabidopsis stems and vascular cells that lack nuclei. A plant-specific role in Cu symplasmic transport through the plasmodesmata during senescence, associated with nutrient mobilization, has been proposed for this extra C-terminus domain of CCH. Expression of *CCH* increases with oxidative stress, senescence, and Cu deficiency. The activities of antioxidant enzymes in *atx1* and *cchatx1* mutants were markedly regulated in response to excess Cu.

A CCH chaperone has been also identified by a differential display in tomato (*L. esculentum*; LeCCH) infected with the fungal pathogen *Botrytis cinerea* (Company and González-Bosch, 2003), suggesting an interesting relationship between Cu homeostasis and plant defense responses.

The COX17 chaperone shares sequence similarity to COX17 from yeast that may mediate the delivery of Cu to the mitochondria for the assembly of a functional cytochrome c oxidase complex (Balandin and Castresana, 2002). In this manner, COX17 would contribute to the increase in activity of specific enzymes that are required to preserve organelle functionality in a number of biotic and abiotic stress situations.

Despite their role in Cu homeostasis, neither CCH nor RAN1 (AtHMA7) is induced by Cu treatment, indicating that they may be more important in helping cells cope with Cu deficit than Cu excess. In contrast, activation of *AtCOX17* gene expression in response to Cu treatment might be an indication of a function-like metallothioneins, which also are induced by high concentrations of metals (Zhou and Goldsbrough, 1994). Nevertheless, further experimental support is necessary to establish the function of these proteins.

The *CCS* gene, homolog of the yeast *Ccs1p/Lys7p* gene, encodes a protein that delivers Cu to the Cu/ZnSOD by a protein–protein interaction. It has been identified in tomato (*LeCCS*) (Zhu et al., 2000), arabidopsis (Wintz and Vulpe, 2002; Chu et al., 2005), potato (*Solanum tuberosum* L.; *StCCS*) (Trindade et al., 2003), corn (*ZmCCS*) (Ruzsa and Scandalios, 2003), and soybean (*GmCCS*) (Sagasti et al., 2011). AtCCS has a predicted chloroplast-targeting sequence, but dual localization in both cytosol and plastids (Chu et al., 2005). Therefore, it is possible that AtCCS delivers Cu to cytosolic and chloroplastic Cu/ZnSOD enzymes, perhaps using an alternative translation start site. It has been shown that AtCCS is Cu upregulated and coregulated with cytosolic and chloroplastic Cu/ZnSOD targets, indicating an important role in the regulation of oxidative stress protection. An upregulation of *AtCCS* mRNA also has been found in response to senescence. Additionally, AtCCS and cytosolic and chloroplastic Cu/ZnSODs were downregulated in response to Cu deficiency. It has been also proposed that *AtCCS* expression is regulated to allow the most optimal use of Cu for photosynthesis (Shikanai et al., 2003).

StCCS gene expression is induced by auxin, which is known to play a role in different stages of plant development. Auxins have a promoting effect on cell elongation–expansion. Surprisingly, potato plants sprayed with $CuSO_4$ did not respond with a significant change in *StCCS* expression (Trindade et al., 2003). This action is consistent with the inhibition of *StCCS* gene expression observed when potato plants were grown in vitro in media supplemented with 10 mM $CuSO_4$. This surprised finding may be explained if the presence of a chaperone would not be required for the incorporation of Cu in the Cu/ZnSOD when Cu is present at high concentrations in leaves.

10.4.6 REGULATION OF COPPER PROTEINS

The arabidopsis genome encodes a 17-member zinc finger plant-specific transcription factor family named SPL (for SQUAMOSA-promoter binding-like proteins) (Birkenbihl et al., 2005). The most studied member of this family is SPL7, which has been shown to be essential for the transcriptional activation in response to Cu deficiency (Bernal et al., 2012). It mediates the substitution of Cu/ZnSOD by FeSOD1 in chloroplasts. SPL7 activates the expression of *FeSOD1* and promotes the degradation of Cu/ZnSOD mRNA through its binding to GTAC motifs within the promoter region of target genes (Yamasaki et al., 2009). Several Cu-regulated microRNAs such as *miR398*, *miR397*, *miR408, and miR857* have been described to be involved in the degradation of the mRNAs encoding cuproteins (Yamasaki et al., 2007; Abdel-Ghany and Pilon, 2008). The regulation by *miRNA* is a widespread response to Cu deficiency. SPL7 also activates the Cu-responsive genes *COPT1*, *COPT2*, and *COPT6* (Gayomba et al., 2013). The comparison of global responses to mild deficiency and excess Cu has been investigated in arabidopsis (Andrés-Colás et al., 2013), and regulatory elements in the promoter regions of the Cu-deficiency overrepresented gene were proposed. The CuAtDB database lists the cuproteins and Cu homeostasis factors identified (http://www.uv.es/cuatlab; Andrés-Colás et al., 2013). Regulators that mediate the response to excess Cu are still known.

10.5 DISTRIBUTION AND SPECIATION OF COPPER IN PLANTS

The distribution and speciation of Cu in plants has been investigated using different techniques such as inductively coupled plasma mass spectrometry (ICP-MS), electron dispersive x-ray spectroscopy (EDXS) (Monni et al., 2002; Ni et al., 2005), electron energy-loss spectroscopy (Turnau et al., 1993; Neumann et al., 1995; Lichtenberger and Neumann, 1997), x-ray absorption spectroscopy (XAS) (Tao et al., 2004), x-ray absorption fine structure, synchrotron-based micro-x-ray fluorescence (Punshon et al., 2013), micro-x-ray absorption near-edge structure (Shi et al., 2004; Song et al., 2013), and mass spectrometric imaging techniques (LA-ICP-MS imaging) (Wu and Becker, 2012; Wu et al., 2013). In the last few years, significant improvement has been seen in the usefulness of synchrotron-based techniques and XAS for biological samples, including in the study of localization of metals and metalloids (West et al., 2012). Comparatively, it has been reported that synchrotron radiation x-ray fluorescence (SRXRF) microprobe is a more sensitive technique and is less injurious to cells (Song et al., 2013).

Copper normally is accumulated in root tissues and in particular in cell walls (Nishizono et al., 1987; Shi et al., 2008; Kopittke et al., 2011; Shi et al., 2011). For instance, in wild rye (*Leymus chinensis* Hohst.), Cu concentration varies between plant parts in the order of roots > rhizomes > stems > leaves > litter (Zhou et al., 2013). The dynamic distribution of nutrients during germination of rapeseeds revealed a relatively rapid allocation of Cu to roots (Eggert and von Wirén, 2013). However, the results derived from these studies are mostly difficult to compare since Cu distribution can vary depending on the nature of chemical reagent used, the concentration applied, the time of exposure, and the handling of samples (Kopittke et al., 2011). Long-term exposure may not give a real status of the initial metal uptake or toxicity. Differences in trace elements' distribution and speciation have been found in the use of fresh, frozen-hydrated, freeze-dried, or oven-dried plant materials (Kopittke et al., 2011). Furthermore, the type of plant also can influence these kinds of analyses; differences can be found between a sensitive, tolerant, or hyperaccumulator plant since in nonhyperaccumulator plants, trace elements are in much lower concentrations. Studies on agronomic crops and vegetables are comparatively more limited than those referring to tolerant plants and hyperaccumulators (Callahan et al., 2006).

A comparison of Cu localization by EDXS in shiny elsholtzia (*Elsholtzia splendens* Nakai ex F. Maek.), a Cu-tolerant plant growing in mine areas in the south of China, and the nontolerant Chinese milkvetch (*Astragalus sinicus* L.) showed that the majority of Cu in the tolerant plant was localized primarily in the cell wall and vacuole, but in the nontolerant species, Cu precipitates on the plasma membrane and in the chloroplasts and cytoplasm under levels of Cu supply that were toxic to both species (Ni et al., 2005). The spatial distribution and speciation of Cu in different zones of the root tip and meristematic zone of cucumber (*Cucumis sativus* L.) have been investigated using μ-SRXRF and μ-XANES and freeze-dried samples (Song et al., 2013). Control roots exhibited a higher content of Cu in the root cap and the front of the meristematic zone. These are the most active regions for the absorption of trace elements (Walker et al., 2003). After exposure of roots to 100 μM Cu over 72 h, the distribution of elements, especially Cu and Fe, changed in these regions. The content of Cu in the root cap and meristematic zone increased sharply, but no accumulation was observed in the maturation and elongation zones. By contrast, Fe content decreased in these regions (Song et al., 2013). XANES analysis indicated that after 72 h of treatment, Cu in the root tip bound mainly to alginate, citrate, and cysteine-like ligands, and little was deposited as CuO. Distribution dynamics indicated that Cu-alginate-like ligands were accumulated more as a proportion of the Cu bound with ligands with distance from the root cap to the maturation zone (from 25.7% to 71.2%), whereas proportions of Cu bound with citrate-like ligands decreased along the same direction (from 53.1% to 17.7%). The cysteine-Cu-like species gradually increased in proportion from the root cap to the elongation zone (from 17.7% to 28.7%), but this proportion sharply declined in the maturation zone (5.1%). CuO-like species accounted for only small proportion but also tended to increase from the root cap to the maturation zone (from 3.4% to 6.0%). In Asiatic dayflower (*Commelina communis* L.), a Cu-tolerant plant grows in Cu mine areas; the metal distribution after 100 μM Cu treatment was analyzed by SRXRF (Shi et al., 2011). A high level of Cu was found in the root meristem and epidermis, it being lower in the cortex than in the vascular cylinder. In the cross section of elongation tissue, Cu concentration decreased from the epidermis to the endodermis and reached the highest level in the vascular cylinder. In situ analysis with μ-SRXRF of hydrated roots of cowpea (*Vigna unguiculata* Walp.) after 24 h of exposure to copper revealed that Cu was located mostly in the rhizodermis (cell wall) and the outer cortex bound to polygalacturonic acid (60%), which is the skeleton component of pectin, and its concentration was substantially lower within the inner cortex and stele. However, after only a short time (3 h) of exposure, Cu was mainly associated with cysteine (57%) or citric acid (43%) (Kopittke et al., 2011). In comparison, the distribution of Zn was different, showing the highest content in the meristematic region and the lowest content in the cortex. In roots, stems, and leaves of *E. splendens*, a Cu-tolerant plant, most Cu was bound to O-containing ligands in the cell wall when plants were grown in 300 μM Cu for 10–60 days (Shi et al., 2008). Similarly, in swamp stonecrop (*Crassula helmsii* Cockayne), a Cu-tolerant amphibious water plant, Cu was accumulated in its shoots bound almost exclusively by oxygen ligands, like organic acids, with no contribution of sulfur ligands or Cu–Cu interactions (Küpper et al., 2009). This finding contrasts with observations in nonaccumulator plants (Mijovilovich et al., 2009). It is thought that Cu is bound by weak ligands (i.e., sulfur ligands) in Cu-hyperaccumulator plants like swamp stonecrop but is bound by strong ligands (i.e., O-ligands) in Cu nonaccumulators such as alpine pennycress.

The distribution of Cu in leaf epidermis, and cross section of the stem of *E. splendens*, was analyzed by μ-SRXRF (Shi et al., 2004). The Cu concentration in leaves of this species reached 1000 mg kg^{-1} in solution culture (Yang et al., 2002). The highest Cu levels were measured in the vascular tissues of the stems and petioles, whereas Cu levels in the mesophyll were higher than in the leaf epidermis. A significant correlation between distribution of Cu and distribution of P, S, and Ca was observed, suggesting that P, S, and Ca can play an important role in Cu accumulation in *E. splendens*. Based on the significant correlation between Cu and distribution of the elements Mn, Fe, and Zn, it seemed that Cu, Mn, Fe, and Zn could be transported by the same transporters, with these having a broad substrate range.

A comparison between Cu ions and Cu chelate was assayed in Ethiopian mustard (*Brassica carinata* A. Braun) plants. Chelates make Cu more available for plant uptake and translocation to the shoots. The plants were treated with 30 μM or 150 μM $CuSO_4$ or Cu–EDDS ((S,S)-*N,N'*-ethylenediamine disuccinic acid) in hydroponic solution, and Cu distribution was determined by microproton-induced x-ray emission (Cestone et al., 2012). Differences depending on Cu concentrations were found between both treatments. In roots, the 30 µM treatments with either CuSO$_4$ or Cu–EDDS (ethylenediamine-N,N′-disuccinic acid) resulted in higher Cu concentrations in epidermal/cortical regions. With the 150 µM CuSO$_4$ treatment, the Cu was concentrated mainly in the vascular tissues, indicating increased uptake into the symplast and further into the xylem. Similar effects were not observed in the 30 µM Cu treatment, supporting the concept that this treatment was not as harmful to the root membranes (Cestone et al., 2010), and thus a lower level of passive symplast Cu influx was maintained. With 150 µM Cu–EDDS, the highest Cu concentration was detected in the endodermis and the adjacent inner cortical cell layer, which indicates that the endodermis prevented the translocation of Cu into the vascular tissues and that an efficient apoplastic barrier was still preserved, in spite of the high root Cu concentrations. The incubation of plants with 150 µM Cu–EDDS enhanced metal translocation to shoots, in comparison with the corresponding CuSO$_4$ treatment. The transport of Cu–EDDS was active and dependent on ATPase since inhibition of H$^+$-ATPase activity resulted in a reduction of Cu accumulation in 30 µM Cu–EDDS-treated roots and 150 µM Cu–EDDS-treated leaves and induced changes in Cu distribution in the leaves. These results indicate that active mechanisms are involved in retaining Cu in the leaf vascular tissues.

10.6 INTERACTION OF COPPER WITH UPTAKE OF OTHER ELEMENTS

Metal ions can bind to organic ligands in a metal-binding site of a metalloprotein, metal chaperone, or metal transporter with different affinities (Fraústo da Silva and Williams, 2001). These differential metal-binding affinities are determined by diverse factors such as their different chemical properties (i.e., redox potential, coordination geometry, charge and thermodynamic and kinetic properties of ligand exchange), the size of metal-binding-site cavity in a protein, and the geometry of ligand atoms, among others. Accordingly, in a given metalloenzyme, a specific metal ion is used for a specific function. However, according to the Irving–Williams series ($Zn^{2+} < Cu^+ > Cu^{2+} > Ni^{2+} > Co^{2+} > Fe^{2+} > Mn^{2+} > Mg^{2+} > Ca^{2+}$), cations chemically similar to each other can compete in uptake pathways. Thus, one major mechanism of toxic action of all transition metal in plants is the efficient competition of metal ions for specific binding sites (Fraústo da Silva and Williams, 2001; Yruela, 2008). For instance, the central ion Mg^{2+} in chlorophyll can be substituted by Cu and other toxic metals under metal excess conditions, resulting in an impairment of photosynthesis (Küpper and Kroneck, 2005).

It has been discussed that metal toxicity may be caused by a disturbance of nutritional balance, resulting in a deficiency of essential elements, an action that increases toxicity. Antagonistic interactions between essential micro- and macronutrients and nonessential elements can take place. Copper and Fe antagonism often occurs in plants grown under Cu toxicity (Foy et al., 1978; Wallace and Cha, 1989; Lombardi and Sebastiani, 2005). Excess Cu in hydroponic medium induces an Fe deficiency in bean plants (Pätsikkä et al., 2002). In *C. communis* Fe deficiency induces Cu accumulation (Chen et al., 2004). In leaf blades of sugar beet grown hydroponically, Fe deficiency increases the Cu content and decreases the Zn content (Rombolà et al., 2005). In bean (*Phaseolus vulgaris* L.) plants, concentrations of Fe, Zn, and K were reduced significantly simultaneously when CuSO$_4$ was administrated (Bouazizi et al., 2010). Copper competed with Fe, Mn, and Zn uptake in the Mn-hyperaccumulator American pokeweed (*Phytolacca americana* L.) plants supplemented with 25 µM Cu (Zhao et al., 2012). Zn decreased concentrations of micro- and macroelements such as Fe, Mg, Mn, Ca, and K. Another interesting aspect is that micronutrient interactions can vary depending on how they are supplied. Copper can interact differently with Fe and Zn depending on whether excessive Cu is supplied through a foliar treatment or in a hydroponic medium (Bernal et al., 2007a). Soybean plants showed no antagonist interaction between Cu and Fe when excessive Cu was supplied through leaves, but Cu competed with Fe uptake in plants grown with excessive Cu in the hydroponic medium. Concerning Zn uptake, soybean plants exhibited a decrease in Zn content upon Cu treatment of leaves, whereas the opposite was observed upon Cu supply through the roots. The different plant response observed following these two Cu treatments might be explained assuming different Cu-uptake strategies in leaf and root cells or different compartmentalization mechanisms that prevent the metals from being transported.

In metal-polluted areas, toxic metal ions can enter into most plants, since the metal homeostasis network is not equipped to avoid the entry of nonessential metal transitions at high concentration. The Mn and Zn content of roots decreased in stone pine (*Pinus pinea* L.), maritime pine (*Pinus pinaster* Ait.), and narrow-leafed ash (*Fraxinus angustifolia* Vahl) at increasing Cu and Cd concentrations (Arduini et al., 1998). The absorption and translocation of different elements can be conditioned by the total metal composition in the soil. For instance, the cooccurrence of metals such as Cu, Zn, and Cr resulted in a greater reduction of the biomass of the hyperaccumulator wand riverhemp (*Sesbania virgata* Pers.) than the presence of a single metal, suggesting a synergistic or additive response. In the binary mixture of Cu and Zn, *Sesbania* plants absorbed the highest concentrations of these metals. In contrast, Cr was more absorbed in the individual treatments (Branzini et al., 2012). In particular, the highest concentration of Cu in shoot leaves and roots was observed when Zn was added simultaneously at high doses. The simultaneous presence of Cu and Zn increases the extraction capacity of *S. virgata*, indicating synergistic effects between them. This finding is in agreement with observations by Luo and Rimmer (1995), who demonstrated that the increase in Zn uptake due to the addition of Cu is approximately 20% and that Cu uptake also increases with the addition of Zn. Total Zn concentration had a pattern of variation different from that of Cu. At high doses, the Zn concentration in shoots/leaves was higher than at low doses only with Cu. In contrast, when Cr was in binary mixtures, the concentration in roots was lower (Branzini et al., 2012). Cr was more absorbed in the individual treatments, suggesting a possible antagonistic relationship between the mixture constituents. A possible explanation for this trend could be that the sorption capacity of each metallic cation of the mixture might decrease in competitive processes (Flogeac et al., 2007). Accumulation of metals in rattlebush (*Sesbania drummondii* Cory) seedlings was dependent on the combination of metals in the medium (Israr et al., 2011). Rattlebush can accumulate higher concentrations of metals such as Pb, Cu, Ni, and Zn, but the uptake of these elements is affected not only by the elements in single applications but also by the combinations of the elements. For all different combinations of metal accumulation studied with rattlebush seedlings, bioaccumulation of a single metal in the roots as well in the shoots was affected by the presence of a second metal, resulting in the inhibition or increase in the bioaccumulation of one metal over another. Uptake of a single metal by rattlebush was affected by the presence of a second metal, suggesting an antagonistic effect or competition between metals at the plant uptake site (Israr et al., 2011). The uptake of metals followed the orders Pb > Cu > Zn > Ni in roots and Pb > Zn > Cu > Ni in shoots.

10.7 DIAGNOSIS OF COPPER STATUS IN PLANTS

A preliminary diagnosis of nutrient status can be carried out by the observation of changes in the appearance of the leaves. The lack of green color in leaves is a symptom associated with alterations in Cu status by either deficiency or toxicity but can be caused also by changes in other nutrients. Morphological changes in leaves and roots are also indicative of nutrient alterations. The visual method is not diagnostic for any specific nutrient stress symptoms but is an extremely valuable tool for the rapid evaluation of the nutrient status of a plant. The main advantage of the visual diagnosis of symptoms is that symptoms are observed readily and provide an immediate evaluation of nutrient status. However, the fact that these visual symptoms do not develop until after there is a major effect on growth and development constitutes a disadvantage. More precise tools include microscopic studies, spectral analysis, tissue and soil analysis, and enzymatic assays. These methods all vary in their precision, rapidity, and their ability to predict future Cu status. Enzymatic analysis normally gives a more specific diagnosis (i.e., the determination of the CuZnSOD activity).

It is known that Cu and Fe uptakes are balanced and in equilibrium within the chloroplast in order to preserve the photosynthetic process. The chloroplast enzymes involved in the defense against ROS, FeSOD, and Cu/ZnSOD are synthesized depending on the Cu status. Cu/ZnSOD and FeSOD are downregulated and upregulated in response to Cu deficiency, respectively (Abdel-Ghany and Pilon, 2008; Pilon et al., 2011; Andrés-Colás et al., 2013). The downregulation of the chloroplast Cu/ZnSOD upon occurrence of low Cu content contributes to maintaining a Cu pool for plastocyanin, allowing plants to save Cu for essential functions such as photosynthetic electron transport (Yamasaki et al., 2007). The availability of Cu affects Cu/ZnSOD enzymes, diminishing their expression and activity. Therefore, the analysis of these proteins and enzymatic activity levels or their transcript expression provides information of the Cu status of the plant. These chloroplastic enzymes can be considered as markers of Cu demand.

TABLE 10.2
Copper Concentrations in Crop Plants

Crop Species	Plant Part	Cu Concentration (mg kg^{-1} Dry Matter)	Reference
Alfalfa (*Medicago sativa* L.)	Aerial parts	8.8	Kubota (1983)
Bluegrass (*Poa* sp.)	Aerial parts	5.5	Kubota (1983)
Clover, ladino (*Trifolium repens* L.)	Aerial parts	7.9	Kubota (1983)
Clover, red (*Trifolium pratense* L.)	Aerial parts	10.0	Kubota (1983)
Durum wheat (*Triticum turgidum durum* L.)	Roots	11–705	Michaud et al. (2007)
	Roots of plants showing interveinal chlorosis due to excess Cu supply	128–705	
	Shoots	6–39	
	Shoots of plants showing interveinal chlorosis due to excess Cu supply	11–39	
Fescue (*Festuca* sp.)	Aerial parts	4.4	Kubota (1983)
Lettuce (*Lactuca sativa* L.)	Whole plant	23–235	Ginocchio et al. (2002)
Maize (*Zea mays* L.)	Roots	32	Ouzounidou et al. (1995)
	Roots of plants with Cu supply that gives lower biomass	299–7790	
	Roots	23–584	Brun et al. (2001)
	Aerial parts	7–17	
	Leaves	10–21	Guo et al. (2010)
	Stems	8–12	
	Grains	2.5–3	
Oilseed rape (*Brassica napus* L.)	Roots	52–107	Chaignon et al. (2002a)
	Shoots	9–14	
Onion (*Allium cepa* L.)	Whole plant	8–45	Ginocchio et al. (2002)
Orchard grass (*Dactylis glomerata* L.)	Aerial parts	5.2	Kubota (1983)
Perennial ryegrass (*Lolium perenne* L.)	Roots	14–42	Jarvis and Whitehead (1981)
	Shoots	3.2–9.5	
Timothy grass (*Phleum pratense* L.)	Aerial parts	4.6	Kubota (1983)
Tomato (*Solanum lycopersicum* L.)	Roots	48–157	Chaignon et al. (2002a)
	Shoots	6–10	
	Whole plants	15–92	Ginocchio et al. (2002)
	Roots	14–42	Cornu et al. (2007)
	Shoots	3–7	

The chloroplast Ca^{2+} transducer *CAS* gene and the gene for the flavoprotein subunit of succinate dehydrogenase in mitochondria *SDH1-2* also have been proposed as markers for Cu deficiency and excess Cu, respectively (Andrés-Colás et al., 2013).

The accumulation of Cu in parts of various plants is used often in diagnosis of sufficiency of Cu for crop growth. Researchers and growers can develop concepts of amounts of Cu in leaves, in particular, and in other organs to assess the nutritional status of crops. The concentrations of Cu in organs of several crop plants are reported in Table 10.2.

10.8   FORMS AND CONCENTRATIONS OF COPPER IN SOILS AND ITS AVAILABILITY TO PLANTS

The average concentration of total Cu in soils ranges from 2 to 200 mg kg^{-1} (Mortvedt, 2000). The distribution of Cu concentrations in soils around the world has been reviewed (Kubota, 1983; Adriano, 1986; Kabata-Pendias and Pendias, 1992; Kopsell and Kopsell, 2007). Kastanozems, chernozems, ferrasols, and fuvisols contain the highest levels of copper, whereas podzols and histosols contain the lowest levels. Plants usually cannot take up the total amount of a metal micronutrient present in the growth medium or soil. Only the soluble fraction of soil solution, where Cu can exist as ionic species or complexed to soluble compounds, is directly available for plant uptake, whereas metal precipitates, complexes with organic matter, metals adsorbed on clays, oxides, and the matrix of soil minerals are less available (Barber, 1995). The proportion of total metal that is in the soil solution is determined by factors such as pH, organic matter, clay, and oxidation–reduction conditions. The fraction of the metal that plants can absorb is known as the available or bioavailable fraction.

Copper is present as sulfide minerals, stable oxides, silicates, sulfates, and carbonates (Krauskopf, 1972; Mortvedt, 2000). The most abundant copper-containing mineral is chalcopyrite ($CuFeS_2$). The oxidation–reduction conditions of a soil can play a role in the availability of Cu. The oxidation–reduction status of the soil can be affected by many factors, including waterlogging and compaction (Patrick and Jugsujinda, 1992; Evangelou, 1998). High P levels can alter the surface properties of soil colloids, possibly resulting in a redistribution of trace metals among various forms in soils.

10.8.1   Distribution of Copper in Soils

The pH in culture solution and soils affects Cu speciation, solubility, complexation, and adsorption (Payne and Pickering, 1975; Msaky and Calvet, 1990; Reddy et al., 1995). However, some soil studies have found little relationship between soil pH and Cu concentration in the soil solution (Jeffery and Uren, 1983; McGrath et al., 1988; Sauvé et al., 1997). The reason for this lack of relationship is the strong affinity of Cu for organic matter (Norvell, 1991). Therefore, the amount of organic matter dissolved in the soil solution, especially in soils with high organic matter content, can be a more important determining factor on Cu solubility than pH. Cu ions form strong coordination complexes with organic matter, altering their availability to plants (Stevenson, 1976, 1991). Hence, Cu is often bound predominantly to the organic matter fraction in the soil, and soil organic matter can be the most important soil factor in determining Cu bioavailability (del Castilho et al., 1993). In a chernozem, between 37% and 91% of the total soil Cu was present in the organic fraction, depending on level of Cu contamination (Pampura et al., 1993). In a range of Cu-contaminated soils, greater than 98% of the Cu in the soil solution was bound to organic complexes, irrespective of pH (Sauvé et al., 1997). Also, in a different range of soils, approximately 95% of soil solution Cu was complexed, irrespective of pH (Fotovat et al., 1997). Reddy et al. (1995) found the proportion of Cu bound to organic matter in the soil solution increased from 37% to 95% as the pH decreased. In addition, Cu applied as sewage sludge was retained in the soil solution in greater quantities than Cu applied as a sulfate because it was bound to dissolved organics from the sludge (Miller et al., 1987), and the activity of the highly available Cu^{2+} ion has been inversely correlated with soil organic matter (McBride et al., 2003).

10.9 SOIL TESTING

Soil testing methods can be used to predict the soil capacity to provide essential micronutrients to a crop. These methods were developed mostly more than 40 years ago and have changed little in recent decades, despite changing demands. They are based on a chemical extraction of the soil, which reflects the levels of nutrients that will be available to the plant during its growth, combined with a determination by using atomic absorption spectrometry or ICP-MS. Historically, soil testing was developed to determine if soil nutrients were deficient or adequate. However, more recently, they are being extended to evaluate excess or toxic levels of nutrients and trace elements. Unfortunately, in these cases, the validation of soil tests presents limitations.

Various extraction methods have been used to assess the availability of metals in soils. In particular, for Cu, the most common chemical reagents used are acids such as 0.1 M HCl, complexing solutions such as Mehlich-1, Mehlich-2, and Mehlich-3 (Mehlich, 1984), diethylenetriaminepentaacetic acid (DTPA), (Lindsay and Norvell, 1978), and salts like sodium acetate trihydrate (Morgan, 1941), ammonium acetate (McIntosh, 1969; Ure et al., 1993), or 0.01 mol L^{-1} $CaCl_2$. Mehlich-3 is composed of 0.2 M CH_3COOH, 0.25 M NH_4NO_3, 0.015 M NH_4F, 0.013 M HNO_3, and 0.001 M ethylenediaminetetraacetic acid (EDTA). It has been demonstrated that in comparison with Mehlich-2, the combination of EDTA and acids increases Cu extraction by 170% (Mehlich, 1984). The DTPA extractant consists of 0.005 M DTPA, 0.1 M triethanolamine, and 0.01 M $CaCl_2$, with a pH of 7.3.

De Abreu et al. (1996) compared the efficiency of several reagents in the determination of available Cu in 31 soils from the state of Sao Paulo, Brazil, using wheat and beans as test plants. The results showed that the extraction of the available Cu decreased in the order Mehlich-3 > TEA-DTPA > Mehlich-1. In pasture soils from New Zealand, the efficiency of various chemical reagents in extracting the Cu from the soil followed the order TEA-DTPA > Mehlich-3 > Mehlich-1 > 0.02 M $SrCl_2$ > 0.1 M HCl > 1.0 M NH_4NO_3 > 0.01 M $CaCl_2$ > 0.1 M $NaNO_3$ > 0.01 M $Ca(NO_3)_2$ (Khan et al., 2005). Field experiments carried out in a Typic Eutrorthox also have indicated that Mehlich-3 yields the highest concentrations of Cu (Nogueirol et al., 2013). Testing experiments in acidic soils of Venezuela using five Cu reagents, DTPA, DTPA-HCl, EDTA, HCl, and Mehlich-1, in a greenhouse experiment with corn as a test crop indicated that DTPA-HCl and DTPA extractants could be useful for producing an index for predicting Cu deficiency in acidic soils (Rodríguez and Ramírez, 2005).

The main difficulties in choosing a soil test derive from the extraction capacity of the reagent, since a substantial fraction of the extracted element may not be available for plants, especially when aggressive reagents are used. On the other hand, the proportion of Cu that can be extracted by soil test extractants can vary with the soil matrix (Khan et al., 2005). McBride et al. (2009) reported that the fraction of total Cu and Zn extracted by aggressive tests (i.e., Mehlich-3, DTPA) was much higher than the fraction extracted by $CaCl_2$, with the Morgan tests being intermediate. Less aggressive reagents, such as 0.01 M $CaCl_2$, are considered frequently better for predicting plant availability of excess trace elements in soils than the traditional aggressive tests (Lebourg et al., 1996; Houba et al., 2000; McBride et al., 2003; Meers et al., 2007; Menzies et al., 2007). However, success in prediction of plant uptake by dilute $CaCl_2$ tends to diminish as the diversity of soils included in the analysis increases (McLaughlin et al., 2000). This problem is encountered with all soil tests and presumably reflects the various biological, chemical, and physical factors that influence plant uptake that are not measured in the laboratory soil test itself. The main disadvantage of a nonaggressive reagent such as 0.01 M $CaCl_2$ is that it yields low extractable concentrations for some trace metals, particularly when soil metal concentrations are at or near background levels.

The accumulation of organic matter in soils can affect the efficiency of chemical extractants by causing structural and chemical alterations including ion-exchange reactions, reactions with soil minerals, increased liberation of organic anions, complexing of metallic cations, and oxidation–reduction reactions. Certain investigations have demonstrated that extractants developed for predicting positive crop responses to fertilization do a poor job of predicting toxicity or excessive uptake (Nogueirol et al., 2013). Accordingly, the soil tests have difficulties when sewage sludge has been applied for several years to crops, since extractants were originally developed for conventional systems and not to predict environmental risks. It has been observed that the efficiency of Cu and Zn extraction from field-contaminated soils was much lower than that from laboratory-spiked aged soils. For Mehlich-3 and DTPA tests, Cu and Zn in field-contaminated soils were less extractable by a factor of about 2 compared with the spiked soils. For less aggressive tests, the difference in extractability was even greater. It has been suggested that chemically nonaggressive neutral salts may be the most appropriate extractants where phytotoxicity is the concern in metal-contaminated soils (McBride et al., 2009).

REFERENCES

Abdel-Ghany, S.E., P. Müller-Moulé, K.K. Niyogi, M. Pilon, and T. Shikanai. 2005. Two P-type ATPases are required for copper delivery in *Arabidopsis thaliana* chloroplasts. *Plant Cell* 17:1–19.

Abdel-Ghany, S.E. and M. Pilon. 2008. MicroRNA-mediated systemic down-regulation of copper protein expression in response to low copper availability in *Arabidopsis*. *J Biol. Chem.* 283:15932–15945.

Abel, S. 2011. Phosphate sensing in root development. *Curr. Opin. Plant Biol.* 14:303–309.

Adriano, D.C. 1986. *Trace Elements in the Terrestrial Environment.* New York: Springer.

Aller, S.G., E.T. Eng, C.J. de Feo, and V.M Unger. 2004. Eukaryotic CTR copper uptake transporters require two faces of the third transmembrane domain for helix packing, oligomerization, and function. *J. Biol. Chem.* 279:53435–53441.

Aller, S.G. and V.M. Unger. 2006. Projection structure of the human copper transporter CTR1 at 6-Å resolution reveals a compact trimer with a novel channel-like architecture. *Proc. Natl. Acad. Sci. U.S.A.* 103:3627–3632.

Alloway, B.J. 2008. *Micronutrient Deficiencies in Global Crop Production.* Heidelberg, Germany: Springer.

Andrés-Colás, N., A. Perea-García, S. Mayo de Andrés et al. 2013. Comparison of global responses to mild deficiency and excess copper levels in *Arabidopsis* seedlings. *Metallomics* 5:1234–1246.

Andrés-Colás, N., A. Perea-García, S. Puig, and L. Peñarrubia. 2010. Deregulated copper transport affects *Arabidopsis* development especially in the absence of environmental cycles. *Plant Physiol.* 153:170–184.

Andrés-Colás, N., V. Sancenón, S. Rodríguez-Navarro et al. 2006. The *Arabidopsis* heavy metal P-type ATPasa HMA5 interacts with metallochaperones and functions in copper detoxification of roots. *Plant J.* 45:225–236.

Aoyama, T., T. Kobayashi, M. Takahashi et al. 2009. OsYSL18 is a rice iron(III)-deoxymugineic acid transporter specifically expressed in reproductive organs and phloem of lamina joints. *Plant Mol. Biol.* 70:681–692.

Arduini, L., D.L. Godbold, A. Onnis, and A. Stefani. 1998. Heavy metals influence mineral nutrition of tree seedlings. *Chemosphere* 36:739–744.

Argüello, J.M. 2003. Identification of ion-selectivity determinants in heavy-metal transport P1B-type ATPases. *J. Membr. Biol.* 195:93–108.

Argüello, J.M., E. Eren, and M. González-Guerrero. 2007. The structure and function of heavy metal transport P1B-ATPases. *Biometals* 20:233–248.

Baker, D.E. and J.P. Senef. 1995. Copper. In *Heavy Metals in Soils*, ed. B.J. Alloway, pp. 179–205. London, U.K.: Blackie Academic & Professionals.

Balandin, T. and C. Castresana. 2002. AtCOX17, an *Arabidopsis* homolog of the yeast copper chaperone COX17. *Plant Physiol.* 129:1852–1857.

Barber, S.A. 1995. *Soil Nutrient Bioavailability*, 2nd edn. New York: Wiley.

Beckett, P.H.T. and R.D. Davis. 1978. The additivity of the toxic effects of Cu, Ni, and Zn in young barley. *New Phytol.* 81:155–173.

Beeson, K.C., L. Gray, and M.B. Adams. 1947. The absorption of mineral elements by forage plants: I. The phosphorus, cobalt, manganese and copper content of some common grasses. *J. Am. Soc. Agron.* 39:356–362.

Bernal, M., D. Casero, V. Singh et al. 2012. Transcriptome sequencing identifies SPL7-regulated copper acquisition genes FRO4/FRO5 and the copper dependence of iron homeostasis in *Arabidopsis*. *Plant Cell* 24:738–761.

Bernal, M., R. Cases, R. Picorel, and I. Yruela. 2007a. Foliar and root Cu supply affect differently Fe and Zn uptake and photosynthetic activity in soybean plants. *Environ. Exp. Bot.* 60:145–150.

Bernal, M., P. Sánchez-Testillano, M. Alfonso, M.C. Risueño, R. Picorel, and I. Yruela. 2007b. Identification and subcellular localization of the soybean copper P_{1B}-ATPase GmHMA8 transporter. J. Struct. Biol. 158:46–58.

Birkenbihl, R.P., G. Jach, H. Saedler, and P. Huijser. 2005. Functional dissection of the plant-specific SBP-domain: Overlap of the DNA-binding and nuclear localization domains. J. Mol. Biol. 352:585–596.

Bouazizi, H., H. Jouili, A. Geitmann, and E. El Ferjani. 2010. Copper toxicity in expanding leaves of Phaseolus vulgaris L.: Antioxidant enzyme response and nutrient element uptake. Ecotoxicol. Environ. Saf. 73:1304–1308.

Branzini, A., R. Santos González, and M. Zubillaga. 2012. Absorption and translocation of copper, zinc and chromium by Sesbania virgata. J. Environ. Manage. 102:50–54.

Bravin, M.N., A.L. Martí, M. Clairotte, and P. Hinsinger. 2009. Rhizosphere alkalisation a major driver of copper bioavailability over a broad pH range in an acidic, copper-contaminated soil. Plant Soil 318:257–268.

Brun, L.A., J. Maillet, P. Hinsinger, and M. Pépin. 2001. Evaluation of copper availability to plants in copper-contaminated vineyard soils. Environ. Pollut. 111:293–302.

Burkhead, J.L., K.A. Reynolds, S.E. Abdel-Ghany, C.M. Cohu, and M. Pilon. 2009. Copper homeostasis. New Phytol. 182:799–816.

Callahan, D.L., A.J.M. Baker, S.D. Kolev, and A.G. Wedd. 2006. Metal ion ligands in hyperaccumulating plants. J. Biol. Inorg. Chem. 11:2–12.

Carrillo-González, R. and M.C.A. González-Chávez. 2006. Metal accumulation in wild plants surrounding mining wastes. Environ. Pollut. 144:84–92.

Casareno, R.L., D. Waggoner, and J.D. Gitlin. 1998. The copper chaperone CCS directly interacts with copper/zinc superoxide dismutase. J. Biol. Chem. 273:23625–23628.

Cestone, B., M.F. Quartacci, and F. Navari-Izzo. 2010. Uptake and translocation of CuEDDS complexes by Brassica carinata. Environ. Sci. Technol. 44:6403–6408.

Cestone, B., K. Vogel-Mikuš, M.F. Quartacci et al. 2012. Use of micro-PIXE to determine spatial distributions of copper in Brassica carinata plants exposed to $CuSO_4$ or CuEDDS. Sci. Total Environ. 427–428:339–346.

Chaignon, V., F. Bedin, and P. Hinsinger. 2002a. Copper bioavailability and rhizosphere pH changes as affected by nitrogen supply for tomato and oilseed rape cropped on an acidic and a calcareous soil. Plant Soil 243:219–228.

Chaignon, V., D. Di Malta, and P. Hinsinger. 2002b. Fe-deficiency increases Cu acquisition by wheat cropped in a Cu-contaminated vineyard soil. New Phytol. 154:121–130.

Chaignon, V., I. Sanchez-Neira, P. Herrmann, B. Jaillard, and P. Hinsinger. 2003. Copper bioavailability and extractability as related to chemical properties of contaminated soils from a vine-growing area. Environ. Pollut. 123:229–238.

Chao, D.-Y., A. Silva, I. Baxter et al. 2012. Salt. Genome-wide association studies identify heavy metal ATPase3 as the primary determinant of natural variation in leaf cadmium in Arabidopsis thaliana. PLoS Genet. 8:e1002923.

Chen, Y., J. Shi, G. Tian, S. Zheng, and Q. Lin. 2004. Fe deficiency induces Cu uptake and accumulation in Commelina communis. Plant Sci. 166:1371–1377.

Chen, Y.F., M.D. Randlett, J.L. Findell, and G.E. Schaller. 2002. Localization of the ethylene receptor ETR1 to the endoplasmic reticulum of Arabidopsis. J. Biol. Chem. 277:19861–19866.

Chiou, T.J. and S.I. Lin. 2011. Signaling network in sensing phosphate availability in plants. Annu. Rev. Plant Biol. 62:185–206.

Chu, C.C., W.C. Lee, W.Y. Guo et al. 2005. Copper chaperone for superoxide dismutase that confers three types of copper/zinc superoxide dismutase activity in Arabidopsis. Plant Physiol. 139:425–436.

Cohen, C.K., D.F. Garvin, and L.V. Kochian. 2004. Kinetic properties of a micronutrient transporter from Pisum sativum indicate a primary function in Fe uptake from the soil. Planta 218:784–792.

Colangelo, E.P. and M.L. Guerinot. 2006. Put metal to petal: Metal uptake and transport throughout plants. Curr. Opin. Plant Biol. 9:322–330.

Company, P. and C. González-Bosch. 2003. Identification of a copper chaperone from tomato fruits infected with Botrytis cinerea by differential display. Biochem. Biophys. Res. Commun. 304:825–830.

Connolly, E.L., J.P. Fett, and M.L. Guerinot. 2002. Expression of the IRT1 metal transporter is controlled by metals at the levels of transcript and protein accumulation. Plant Cell 14:1347–1357.

Conte, S.S., H.H. Chu, D.C. Rodríguez et al. 2013. Arabidopsis thaliana Yellow Stripe1-Like4 and Yellow Stripe1-Like6 localize to internal cellular membranes and are involved in metal ion homeostasis. Front. Plant Sci. 4:283.

Conte, S.S. and E.L. Walker. 2012. Genetic and biochemical approaches for studying the yellow stripe-like transporter family in plants. Curr. Top. Membr. 69:295–322.

Cornu, J.Y., S. Staunton, and P. Hinsinger. 2007. Copper concentration in plants and in the rhizosphere as influenced by the iron status of tomato (*Lycopersicon esculentum* L.). *Plant Soil* 292:63–77.

Courbot, M., G. Willems, P. Motte et al. 2007. A major quantitative trait locus for cadmium tolerance in *Arabidopsis halleri* colocalizes with HMA4, a gene encoding a heavy metal ATPase. *Plant Physiol.* 144:1052–1065.

Craciun, A.R., C.-L. Meyer, J. Chen et al. 2012. Variation in HMA4 gene copy number and expression among *Noccaea caerulescens* populations presenting different levels of Cd tolerance and accumulation. *J. Exp. Bot.* 63:4179–4189.

Crichton, R.R. and J.L. Pierre. 2001. Old iron, young copper: From Mars to Venus. *Biometals* 14:99–112.

Crowe, S.A., L.N. Døssing, and N.J. Beukes. 2013. Atmospheric oxygenation three billion years ago. *Nature* 501:535–538.

Curie, C., G. Cassin, D. Couch, F. Divol, K. Higuchi, and M. Le Jean. 2009. Metal movement within the plant: Contribution of nicotianamine and yellow stripe 1-like transporters. *Ann. Bot.* 103:1–11.

Curie, C., Z. Panaviene, C. Loulergue, S.L. Dellaporta, J.F. Briat, and E.L. Walker. 2001. Maize yellow stripe1 encodes a membrane protein directly involved in Fe(III) uptake. *Nature* 409:346–349.

de Abreu, C.A., B. van Raij, M.F. de Abreu, W.R. de Santos, and J.C. de Andrade. 1996. Efficiency of multinutrient extractants for the determination of available copper in soils. *Commun. Soil Sci. Plant Anal.* 27:763–771.

de Loland, J.O. and B.R. Singh. 2004. Copper contamination of soil and vegetation in coffee orchards after long-term use of Cu fungicides. *Nutr. Cycl. Agroecosyst.* 69:203–211.

del Castilho, P., W.J. Chardon, and W. Salomons. 1993. Influence of cattle manure slurry application on the solubility of cadmium, copper and zinc in a manure acidic, loamy-sandy soil. *J. Environ. Qual.* 22:689–697.

Delas, J. 1963. La toxicité du cuivre accumulé dans les sols. *Agrochimica* 7:258–288.

DiDonato, R.J. Jr., L.A. Roberts, T. Sanderson, R.B. Eisley, and E.L. Walker. 2004. *Arabidopsis* yellow stripe-like2 (YSL2): A metal-regulated gene encoding a plasma membrane transporter of nicotianamine-metal complexes. *Plant J.* 39:403–414.

Divol, F., D. Couch, G. Conéjéro, H. Roschzttardtz, S. Mari, and C. Curie. 2013. The *Arabidopsis* Yellow Stripe LIKE4 and 6 transporters control iron release from the chloroplast. *Plant Cell* 25:1040–1055.

Dixon, B. 2004. Pushing bordeaux mixture. *Lancet Infec. Dis.* 4:594.

Eggert, K. and N. von Wirén. 2013. Dynamics and partitioning of the ionome in seeds and germinating seedlings of winter oilseed rape. *Metallomics* 5:1316–1325.

Eide, D., M. Broderius, J. Fett, and M.L. Guerinot. 1996. A novel iron-regulated metal transporter from plants identified by functional expression in yeast. *Proc. Natl. Acad. Sci. U.S.A.* 93:5624–5628.

Epstein, E. and A.J. Bloom. 2005. *Mineral Nutrition of Plants: Principles and Perspectives*, 2nd edn. Sunderland, MA: Sinauer Associates, Inc.

Eren, E. and J.M. Argüello. 2004. *Arabidopsis* HMA2, a divalent heavy metal-transporting P(IB)-type ATPase, is involved in cytoplasmic Zn^{2+} homeostasis. *Plant Physiol.* 136:3712–3723.

Evangelou V.P. 1998. *Environmental Soil and Water Chemistry Principles and Applications*. New York: John Wiley & Sons.

Fan, J., Z. He, L.Q. Ma, and P.J. Stoffella. 2011. Accumulation and availability of copper in *citrus* grove soils as affected by fungicide application. *J. Soils Sediments* 11:639–648.

Flogeac, K., E. Guillon, and M. Aplincourt. 2007. Competitive sorption of metal ions onto a north-eastern France soil. Isotherms and XAFS studies. *Geoderma* 139:180–189.

Fotovat, A., R. Naidu, and M.E. Sumner. 1997. Water: Soil ratio influences aqueous phase chemistry of indigenous copper and zinc in soils. *Austr. J. Soil Res.* 35:687–709.

Foy, C.D., R.L. Chaney, and M.C. White. 1978. The physiology of metal toxicity in plants. *Annu. Rev. Plant Physiol.* 29:511–566.

Fraústo da Silva, J.J.R., and R.J.P. Williams. 2001. *The Biological Chemistry of the Elements*, 2nd edn. Oxford, U.K.: Clarendon Press.

García-Molina, A., N. Andrés-Colás, A. Perea-García, S. Del Valle-Tascón, L. Peñarrubia, and S. Puig. 2011. The intracellular *Arabidopsis* COPT5 transport protein is required for photosynthetic electron transport under severe copper deficiency. *Plant J.* 65:848–860.

Gayomba, S.R., H.I. Jung, J. Yan et al. 2013. The CTR/COPT-dependent copper uptake and SPL7-dependent copper deficiency responses are required for basal cadmium tolerance in *A. thaliana*. *Metallomics* 5:1262–1275.

Gendre, D., P. Czernic, G. Conéjéro et al. 2007. TcYSL3, a member of the YSL gene family from the hyperaccumulator *Thlaspi caerulescens*, encodes a nicotianamine-Ni/Fe transporter. *Plant J.* 49:1–15.

Ginocchio, R., P.H. Rodríguez, R. Badilla-Ohlbaum, H.E. Allen, and G.E. Lagos. 2002. Effect of soil copper content and pH on copper uptake of selected vegetables grown under controlled conditions. *Environ. Toxicol. Chem.* 21:1736–1744.

Gravot, A., A. Lieutaud, F. Verret, P. Auroy, A. Vavasseur, and P. Richaud. 2004. *At*HMA3, a plant P1B-ATPase, functions as a Cd/Pb transporter in yeast. *FEBS Lett.* 561:22–28.

Grotz, N., T. Fox, E. Connolly, W. Park, M.L. Guerinot, and D. Eide. 1998. Identification of a family of zinc transporter genes from *Arabidopsis* that respond to zinc deficiency. *Proc. Natl. Acad. Sci. U.S.A.* 95:7220–7224.

Guerinot, M.L. 2000. The ZIP family of metal transporters. *Biochim. Biophys. Acta* 1465:190–198.

Guo, X.Y., Y.B. Zuo, B.R. Wang, J.M. Li, and Y.B. Ma. 2010. Toxicity and accumulation of copper and nickel in maize plants cropped on calcareous and acidic field soils. *Plant Soil* 333:365–373.

Halliwell, B. and J.M.C. Gutteridge. 1984. Oxygen toxicity, oxygen radicals, transition metals and disease. *Biochem. J.* 219:1–14.

Hinsinger, P. and F. Courchesne. 2007. Biogeochemistry of metals and metalloids at the soil–root interface. In *Biophysico-Chemical Processes of Heavy Metals and Metalloids in Soil Environments*, eds. A. Violante, P.M. Huang, and G.M. Gadd, pp. 265–311. Hoboken, NJ: John Wiley & Sons, Inc.

Hofstetter, S.S., A. Dudnik, H. Widmer, and R. Dudler. 2013. *Arabidopsis* YELLOW STRIPE-LIKE7 (YSL7) and YSL8 transporters mediate uptake of Pseudomonas virulence factor syringolin A into plant cells. *Mol. Plant Microbe Interact.* 26:1302–1311.

Houba, V.J.G., E.J.M. Temminghoff, G.A. Gaikhorst, and W. van Vark. 2000. Soil analysis procedures using 0.01 M calcium chloride as extraction reagent. *Comm. Soil Sci. Plant Anal.* 31:1299–1396.

Huffman, D.L. and T.V. O'Halloran. 2001. Function, structure, and mechanism of intracellular copper trafficking proteins. *Annu. Rev. Biochem.* 70:677–701.

Hussain, D., M.J. Haydon, Y. Wang et al. 2004. P-type ATPase heavy metal transporters with roles in essential zinc homeostasis in *Arabidopsis*. *Plant Cell.* 16:1327–1339.

Inoue, H., T. Kobayashi, T. Nozoye et al. 2009. Rice OsYSL15 is an iron-regulated iron(III)-deoxymugineic acid transporter expressed in the roots and is essential for iron uptake in early growth of the seedlings. *J. Biol. Chem.* 284:3470–3479.

Ishimaru, Y., H. Masuda, M. Suzuki et al. 2007. Overexpression of the OsZIP4 zinc transporter confers disarrangement of zinc distribution in rice plants. *J. Exp. Bot.* 58:2909–2915.

Israr, M., A. Jewell, D. Kumar, and S.V. Sahi. 2011. Interactive effects of lead, copper, nickel and zinc on growth, metal uptake and antioxidative metabolism of *Sesbania drummondii*. *J. Hazard. Mater.* 186:1520–1526.

Jarvis, S.C. and D.C. Whitehead. 1981. The influence of some soil and plant factors on the concentration of copper in perennial ryegrass. *Plant Soil* 60:275–286.

Jeffery, J.J. and N.C. Uren. 1983. Copper and zinc species in the soil solution and the effects of soil pH. *Aust. J. Soil Res.* 21:479–488.

Jung, H.I., S.R. Gayomba, M.A. Rutzke, E. Craft, L.V. Kochian, and O.K. Vatamaniuk. 2012. COPT6 is a plasma membrane transporter that functions in copper homeostasis in *Arabidopsis* and is a novel target of SQUAMOSA promoter-binding protein-like 7. *J. Biol. Chem.* 287:33252–33267.

Kabata-Pendias, A. and H. Pendias. 1992. *Trace Elements in Soils and Plants*, 2nd edn. Boca Raton, FL: CRC Press.

Kampfenkel, K., S. Kushnir, E. Babiychuk, D. Inzé, and M. van Montagu. 1995. Molecular characterization of a putative *Arabidopsis thaliana* copper transporter and its yeast homologue. *J. Biol. Chem.* 270:28479–28486.

Kennedy, C.D. and F.A.N. Gonsalves. 1987. The action of divalent zinc, cadmium, mercury, copper and lead on the trans-root potential and H+ efflux of excised roots. *J. Exp. Bot.* 38:800–817.

Khan, M.A.R., N.S. Bolan, and A.D. Mackay. 2005. Soil test to predict the copper availability in pasture soils. *Commun. Soil Sci. Plant Anal.* 36:2601–2624.

Kim, Y.-Y., H. Choi, S. Segami et al. 2009. *At*HMA1 contributes to the detoxification of excess Zn(II) in *Arabidopsis*. *Plant J.* 58:737–753.

Klaumann, S., S.D. Nickolaus, S.H. Fürst et al. 2011. The tonoplast copper transporter COPT5 acts as an exporter and is required for interorgan allocation of copper in *Arabidopsis thaliana*. *New Phytol.* 192:393–404.

Klomp, A.E., J.A. Juijn, L.T. van der Gun, I.E. van den Berg, R. Berger, and L.W. Klomp. 2003. The N-terminus of the human copper transporter 1 (hCTR1) is localized extracellularly, and interacts with itself. *Biochem. J.* 370:881–889.

Kochian, L.V. 1991. Mechanisms of micronutrient uptake and translocation in plants. In *Micronutrients in Agriculture*, eds. J.J. Mortvedt, F.R. Cox, L.M. Shuman, and R.M. Welch, pp. 229–296, Madison, WI: Soil Science Society of America.

Koike, S., H. Inoue, D. Mizuno et al. 2004. OsYSL2 is a rice metal-nicotianamine transporter that is regulated by iron and expressed in the phloem. *Plant J.* 39:415–424.

Kopittke, P.M. and N.W. Menzies. 2006. Effect of Cu toxicity on growth of cowpea (*Vigna unguiculata*). *Plant Soil* 279:287–296.

Kopittke, P.M., N.W. Menzies, M.D. de Jonge et al. 2011. In situ distribution and speciation of toxic copper, nickel, and zinc in hydrated roots of cowpea. *Plant Physiol.* 156:663–673.

Kopsell, D.E. and D.A. Kopsell. 2007. Copper. In *Handbook of Plant Nutrition*, eds. A.V. Barker and D.J. Pilbeam, pp. 293–328. Boca Raton, FL: CRC Press.

Krauskopf, K.B. 1972. Geochemistry of micronutrients. In *Micronutrients in Agriculture*, eds. J.J. Mortvedt, P.M. Giordano, and W.L. Lindsay, pp. 7–40. Madison, WI: Soil Science Society of America.

Kubota, J. 1983. Copper status of United States soils and forage plants. *Agron. J.* 75:913–918.

Küpper, H., B. Götz, A. Mijovilovich, F.C. Küpper, and W. Meyer-Klaucke. 2009. Complexation and toxicity of copper in higher plants. I. Characterization of copper accumulation, speciation, and toxicity in *Crassula helmsii* as a new copper accumulator. *Plant Physiol.* 151:702–714.

Küpper, H. and P.M.H. Kroneck. 2005. Heavy metal uptake by plants and cyanobacteria. *Met. Ions Biol. Syst.* 44:97–144.

Large, E.C. 1940. *The Advance of the Fungi*. New York: Holt.

Lebourg, A., T. Sterckeman, H. Ciesielski, and N. Proix. 1996. Intérêt de différents reactifs d'extraction chimique pour l'évaluation de la biodisponibilité des métaux en traces du sol. *Agronomie* 16:201–215.

Le Jean, M., A. Schikora, S. Mari, J.F. Briat, and C. Curie. 2005. A loss-of-function mutation in *AtYSL1* reveals its role in iron and nicotianamine seed loading. *Plant J.* 44:769–782.

Lee, S., J.C. Chiecko, S.A. Kim et al. 2009. Disruption of *OsYSL15* leads to iron inefficiency in rice plants. *Plant Physiol.* 150:786–800.

Lee, S., Y.-Y. Kim, Y. Lee, and G. An. 2007. Rice P1B-type heavy-metal ATPase, OsHMA9, is a metal efflux protein. *Plant Physiol.* 145:831–842.

Lepp, N.W., N.M. Dickinson, and K.I. Ormand. 1984. Distribution of fungicide-derived copper in soils, litter and vegetation of different aged stands of coffee (*Coffea arabica* L.) in Kenya. *Plant Soil* 77:263–270.

Li, B., Y. Ma, M.J. McLaughlin, J.K. Kirby, G. Cozens, and J. Liu. 2010. Influences of soil properties and leaching on copper toxicity to barley root elongation. *Environ. Toxicol. Chem.* 29:835–842.

Lichtenberger, O. and D. Neumann. 1997. Analytical electron microscopy as a powerful tool in plant cell biology: Examples using electron energy loss spectroscopy and x-ray microanalysis. *Eur. J. Cell Biol.* 73:378–386.

Lin, Y.-F., H.-M. Liang, S.-Y. Yang et al. 2009. *Arabidopsis* IRT3 is a zinc-regulated and plasma membrane localized zinc/iron transporter. *New Phytol.* 182:392–404.

Lindsay, W.L. and W.A. Norvell. 1978. Development of a DTPA soil test for zinc, iron, manganese and copper. *Soil Sci. Soc. Am. J.* 42:421–428.

Lombardi, L. and L. Sebastiani. 2005. Copper toxicity in *Prunus cerasifera*: Growth and antioxidant enzymes responses of in vitro grown plants. *Plant Sci.* 168:797–802.

López-Millán, A.F., D.R. Ellis, and M.A. Grusak. 2004. Identification and characterization of several new members of the ZIP family of metal ion transporters in *Medicago truncatula*. *Plant Mol. Biol.* 54:583–596.

Luo, Y. and D.L. Rimmer. 1995. Zinc-copper interaction affecting plant growth on metal-contaminated soil. *Environ. Pollut.* 88:79–83.

MacNicol, R.D. and P.H.T. Beckett. 1985. Critical tissue concentrations of potentially toxic elements. *Plant Soil* 85:107–129.

Marschner, P. (ed.) 2012. *Marschner's Mineral Nutrition of Higher Plants*. 3rd edn. London, U.K.: Academic Press.

McBride, M.B., E.A. Nibarger, B.K. Richards, and T. Steenhuis. 2003. Trace metal accumulation by red clover grown on sewage sludge-amended soils and correlation to Mehlich 3 and calcium chloride-extractable metals. *Soil Sci.* 168:29–38.

McBride, M.B., M. Pitiranggon, and B. Kim. 2009. A comparison of tests for extractable copper and zinc in metal-spiked and field-contaminated soil. *Soil Sci. J.* 174:439–444.

McGrath, S.P., J.R. Sanders, and M.H. Shalaby. 1988. The effects of soil organic matter levels on soil solution concentrations and extractabilities of manganese, zinc and copper. *Geoderma* 42:177–188.

McIntosh, J.L. 1969. Bray and Morgan soil test extractants modified for testing acid soils from different parent materials. *Agron. J.* 61:259–265.

McLaughlin, M.J., B.A. Zarcinas, D.P. Stevens, and N. Cook. 2000. Soil testing for heavy metals. *Commun. Soil Sci. Plant Anal.* 31:1661–1700.

Meers, E., R. Samson, F.M.G. Tack et al. 2007. Phytoavailability assessment of heavy metals in soils by single extractions and accumulation by *Phaseolus vulgaris*. *Environ. Exp. Bot.* 60:385–396.

Mehlich, A. 1984. Mehlich-3 soil test extractant: A modification of Mehlich-2 extractant. *Commun. Soil Sci. Plant Anal.* 15:1409–1416.

Menzies, N.W., M.J. Donn, and P.M. Kopittke. 2007. Evaluation of extractants for estimation of the phytoavailable trace metals in soils. *Environ. Pollut.* 145:121–130.

Michaud, A.M., M.N. Bravin, M. Galleguillos, and P. Hinsinger. 2007. Copper uptake and phytotoxicity as assessed in situ for durum wheat (*Triticum turgidum durum* L.) cultivated in Cu-contaminated, former vineyard soils. *Plant Soil* 298:99–111.

Mijovilovich, A., B. Leitenmaier, W. Meyer-Klaucke, P.M. Kroneck, B. Götz, and H. Küpper. 2009. Complexation and toxicity of copper in higher plants. II. Different mechanisms for copper versus cadmium detoxification in the copper-sensitive cadmium/zinc hyperaccumulator *Thlaspi caerulescens* (Ganges ecotype). *Plant Physiol.* 151:715–731.

Miller, W.P., D.C. Martens, and L.W. Zelazny. 1987. Short-term transformations of copper in copper-amended soils. *J. Environ. Qual.* 16:176–180.

Mills, R.F., A. Francini, P.S.C. Ferreira da Rocha et al. 2005. The plant P1B-type ATPase AtHMA4 transports Zn and Cd and plays a role in detoxification of transition metals supplied at elevated levels. *FEBS Lett.* 579:783–791.

Mills, R.F., G.C. Krijger, P.J. Baccarini, J.L. Hall, and L.E. Williams. 2003. Functional expression of AtHMA4, a P1B-type ATPase of the Zn/Co/Cd/Pb subclass. *Plant J.* 35:164–176.

Mills, R.F., B. Valdes, M. Duke et al. 2010. Functional significance of AtHMA4 C-terminal domain in planta. *PLoS One* 5:e13388.

Milner, M.J., J. Seamon, E. Craft, and L.V. Kochian. 2013. Transport properties of members of the ZIP family in plants and their role in Zn and Mn homeostasis. *J. Exp. Bot.* 64:369–381.

Mira, H., F. Martínez-García, and L. Peñarrubia. 2001a. Evidence for the plant-specific intercellular transport of the *Arabidopsis* copper chaperone CCH. *Plant J.* 25:521–528.

Mira, H., M. Vilar, V. Esteve et al. 2004. Ionic self-complementarity induces amyloid-like fibril formation in an isolated domain of a plant copper metallochaperone protein. *BMC Struct. Biol.* 4:7.

Mira, H., M. Vilar, E. Pérez-Raya, and L. Peñarrubia. 2001b. Functional and conformational properties of the exclusive C-domain from the *Arabidopsis* copper chaperone (CCH). *Biochem. J.* 357:545–549.

Miyadate, H., S. Adachi, A. Hiraizumi et al. 2011. OsHMA3, a P1B-type of ATPase affects root-to-shoot cadmium translocation in rice by mediating efflux into vacuoles. *New Phytol.* 189:190–199.

Monni, S., H. Bucking, and I. Kottke. 2002. Ultrastructural element localization by EDXS in *Empetrum nigrum*. *Micron* 33:339–351.

Morel, M., J. Crouzet, A. Gravot et al. 2009. AtHMA3, a P1B-ATPase allowing Cd/Zn/Co/Pb vacuolar storage in *Arabidopsis*. *Plant Physiol.* 149:894–904.

Morgan, M.F. 1941. Chemical soil diagnosis by the universal soil testing system. In *Connecticut Agriculture Experiment Station Bulletin*, University of Connecticut Agriculture Experiment Station, New Haven, CT, p. 450.

Mortvedt, J.J. 2000. Bioavailability of micronutrients. In *Handbook of Soil Science, D71-D88*, ed. M.E. Sumner. Boca Raton, FL: CRC Press.

Msaky, J.J. and R. Calvet. 1990. Adsorption behavior of copper and zinc in soils: Influence of pH on adsorption characteristics. *Soil Sci.* 150:513–522.

Murata, Y., J.F. Ma, N. Yamaji, D. Ueno, K. Nomoto, and T.A. Iwashita. 2006. A specific transporter for iron(III)–phytosiderophore in barley roots. *Plant J.* 46:563–572.

Neumann, D., U.Z. Nieden, O. Lichtenberger, and I. Leopold. 1995. How does *Armeria maritima* tolerate high heavy metal concentrations? *J. Plant Physiol.* 146:704–717.

Ni, C.Y., Y.X. Chen, Q. Lin, and G.M. Tian. 2005. Subcellular localization of copper in tolerant and nontolerant plant. *J. Environ. Sci. (China)* 17:452–456.

Nishizono, H., H. Ichikawa, S. Suziki, and F. Ishii. 1987. The role of the root cell-wall in the heavy-metal tolerance of *Athyrium yokoscense*. *Plant Soil* 101:15–20.

Nocito, F.F., C. Lancilli, B. Dendena, G. Lucchini, and G.A. Sacchi. 2011. Cadmium retention in rice roots is influenced by cadmium availability, chelation and translocation. *Plant Cell Environ.* 34:994–1008.

Nogueirol, R.C., J.de M. Wanderley, and L.R. Ferracciú. 2013. Testing extractants for Cu, Fe, Mn and Zn in tropical soils treated with sewage sludge for 13 consecutive years. *Water Air Soil Pollut.* 224:1557.

Norvell, W.A. 1991. Reactions of metal chelates in soils and nutrient solutions. In *Micronutrients in Agriculture*, eds. J.J. Mortvedt, F.R. Cox, L.M. Shuman, and R.M. Welch, 2nd edn., pp. 187–227. Madison, WI: Soil Science Society of America.

O'Halloran, T.V. and V.C. Culotta. 2000. Metallochaperones, an intracellular shuttle service for metal ions. *J. Biol. Chem.* 275:25057–25060.

O'Hara, G.W. 2001. Nutritional constraints on root nodule bacteria affecting symbiotic nitrogen fixation: A review. *Aust. J. Exp. Agr.* 41:417–433.

Ó Lochlainn, S., H.C. Bowen, R.G. Fray et al. 2012. Tandem quadruplication of HMA4 in the zinc (Zn) and cadmium (Cd) hyperaccumulator *Noccaea caerulescens*. *PLoS One* 6:e17814.

Ouzounidou, G., M. Mousbakas, and S. Karataglis. 1995. Responses of maize (*Zea mays* L.) plants to copper stress: Growth, mineral content and ultrastructure of roots. *Environ. Exp. Bot.* 35:167–176.

Palmer, C.M. and M.L. Guerinot. 2009. Facing the challenges of Cu, Fe and Zn homeostasis in plants. *Nat. Chem. Biol.* 5:333–340.

Pampura, T.B., D.L. Pinskiy, V.G. Ostroumov, V.D. Gershevich, and V.N. Bashkin. 1993. Experimental study of the buffer capacity of a Chernozem contaminated with copper and zinc. *Eur. Soil Sci.* 25:27–38.

Patrick, W.H. Jr. and A. Jugsujinda. 1992. Sequential reduction and oxidation of inorganic nitrogen, manganese, and iron in flooded soil. *Soil Sci. Soc. Am. J.* 56:1071–1073.

Pätsikkä, E., M. Kairavuo, F. Sersen, E.-M. Aro, and E. Tyystjärvi. 2002. Excess copper predisposes photosystem II to photoinhibition in vivo by outcompeting iron and causing decrease in leaf chlorophyll. *Plant Physiol.* 129:1359–1367.

Payne, K. and W.F. Pickering. 1975. Influence of clay-solute interactions on aqueous copper ion levels. *Water Air Soil Pollut.* 5:63–69.

Pedas, P., C.K. Ytting, A.T. Fuglsang, T.P. Jahn, J.K. Schjoerring, and S. Husted. 2008. Manganese efficiency in barley: Identification and characterization of the metal ion transporter HvIRT1. *Plant Physiol.* 148:455–466.

Peñarrubia, L., N. Andrés-Colás, J. Moreno, and S. Puig. 2010. Regulation of copper transport in *Arabidopsis thaliana*: A biochemical oscillator? *J. Biol. Inorg. Chem.* 15:29–36.

Pence, N.S., P.B. Larsen, S.D. Ebbs et al. 2000. The molecular basis for heavy metal hyperaccumulation in *Thlaspi caerulescens*. *Proc. Natl. Acad. Sci. U.S.A.* 97:4956–4960.

Perea-García, A., A. García-Molina, N. Andrés-Colás et al. 2013. *Arabidopsis* copper transport protein COPT2 participates in the cross talk between iron deficiency responses and low-phosphate signaling. *Plant Physiol.* 162:180–194.

Pietrzak, U. and D.C. McPhail. 2004. Copper accumulation, distribution and fractionation in vineyard soils of Victoria, Australia. *Geoderma* 122:151–166.

Pilon M., K. Ravet, and W. Tapken. 2011. The biogenesis and physiological function of chloroplast superoxide dismutases. *Biochim. Biophys. Acta* 1807:989–998.

Pole, A. and A. Schützdendübel. 2004. Plant responses to abiotic stress. In *Topics in Current Genetics*, eds. H. Hirt, and K. Shinozaki, Vol. 4, pp. 187–215. Berlin/Heidelberg, Germany: Springer-Verlag.

Prasad, M.N.V. and K. Strzalka. 1999. Impact of heavy metals on photosynthesis. In *Heavy Metal Stress in Plants*, eds. M.N.V. Prasad and J. Hagemeyer, pp. 117–138. Berlin, Germany: Springer Publishers.

Pufahl, R.A., C.P. Singer, K.L. Peariso et al. 1997. Metal ion chaperone function of the soluble Cu(I) receptor Atx1. *Science* 278:853–856.

Puig, S., N. Andrés-Colas, A. García-Molina, and L. Peñarrubia. 2007. Copper and iron homeostasis in *Arabidopsis*: Response to metal deficiencies, interactions and biotechnological applications. *Plant Cell Environ.* 30:271–290.

Puig, S. and D.J. Thiele. 2002. Molecular mechanisms of copper uptake and distribution. *Curr. Opin. Chem. Biol.* 6:171–180.

Punshon, T., F.K. Ricachenevsky, M.N. Hindt, A.L. Socha, and H. Zuber. 2013. Methodological approaches for using synchrotron x-ray fluorescence (SXRF) imaging as a tool in ionomics: Examples from *Arabidopsis thaliana*. *Metallomics* 5:1133–1145.

Rae, T.D., P.J. Schmidt, R.A. Pufahl, V.C. Culotta, and T.V. O'Halloran. 1999. Undetectable intracellular free copper: The requirement of a copper chaperone for superoxide dismutase. *Science* 284:805–808.

Ravet, K. and M. Pilon. 2013. Copper and iron homeostasis in plants. The challenges of oxidative stress. *Antioxidants and Redox Signaling* 19:919–932.

Reddy, K.J., L. Wang, and S.P. Gloss. 1995. Solubility and mobility of copper, zinc and lead in acidic environments. *Plant Soil* 171:53–58.

Reichman, S. 2002. *The Responses of Plants to Metal Toxicity: A Review Focusing on Copper, Manganese and Zinc*, Melbourne, Victoria, Australia: Australian Minerals & Energy Environment Foundation.

Reuter, D.J. and J.B. Robinson. 1997. *Plant Analysis and Interpretation Manual*, 2nd edn. Collingwood, Victoria, Australia: Australian Soil Analysis and Plant Analysis Council Inc.

Robinson, N.J., C.M. Procter, E.L. Connolly, and M.L. Guerinot. 1999. A ferric- chelate reductase for iron uptake from soils. *Nature* 397:694–697.

Rodríguez, B. and R. Ramírez. 2005. A soil test for determining available copper in acidic soils of Venezuela. *Interciencia* 30:361–364.

Rogers, E.E., D.J. Eide, and M.L. Guerinot. 2000. Altered selectivity in an *Arabidopsis* metal transporter. *Proc. Natl. Acad. Sci. U.S.A.* 97:12356–12360.

Rombolà, A.D., Y. Gogorcena, A. Larbi et al. 2005. Iron deficiency-induced changes in carbon fixation and leaf elemental composition of sugar beet (*Beta vulgaris*) plants. *Plant Soil* 271:39–45.

Rooney, C.P., F.-J. Zhao, and S.P. McGrath. 2006. Soil factors controlling the expression of copper toxicity to plants in a wide range of European soils. *Environ. Toxicol. Chem.* 25:726–732.

Ruyters, S., P. Salaets, K. Oorts, and E. Smolders. 2013. Copper toxicity in soils under established vineyards in Europe: A survey. *Sci. Total Environ.* 43:470–477.

Ruzsa, S.M. and J.G. Scandalios. 2003. Altered Cu metabolism and differential transcription of Cu/ZnSod genes in a Cu/ZnSOD-deficient mutant of maize: Evidence for a Cu-responsive transcription factor. *Biochemistry* 42:1508–1516.

Ryan, B.M., J.K. Kirby, F.H. Degryse, H. Harris, M.J. McLaughlin, and K. Scheiderich. 2013. Copper speciation and isotopic fractionation in plants: Uptake and translocation mechanisms. *New Phytol.* 199:367–378.

Sagasti, S., I. Yruela, M. Bernal et al. 2011. Characterization of the recombinant copper chaperone (CCS) from the plant *Glycine (G.) max*. *Metallomics* 3:169–175.

Sancenón, V., S. Puig, I. Mateu-Andres, E. Dorcey, D.J. Thiele, and L. Peñarrubia. 2004. The *Arabidopsis* copper transporter COPT1 functions in root elongation and pollen development. *J. Biol. Chem.* 279:15348–15355.

Sancenón, V., S. Puig, H. Mira, D.J. Thiele, and L. Peñarrubia. 2003. Identification of a copper transporter family in *Arabidopsis thaliana*. *Plant Mol. Biol.* 51:577–587.

Satoh-Nagasawa, N., M. Mori, N. Nakazawa et al. 2012. Mutations in rice (*Oryza sativa*) heavy metal ATPase 2 (OsHMA2) restrict the translocation of zinc and cadmium. *Plant Cell Physiol.* 53:213–224.

Sauvé, S., M.B. McBride, W.A. Norvell, and W.H. Hendershot. 1997. Copper solubility and speciation of in situ contaminated soils: Effects of copper level, pH and organic matter. *Water Air Soil Pollut.* 100:133–149.

Schaaf, G., A. Schikora, J. Harberle et al. 2005. A putative function for the *Arabidopsis* Fe phytosiderophore transporter homolog AtYSL2 in Fe and Zn homeostasis. *Plant Cell Physiol.* 46:762–774.

Seigneurin-Berny, D., A. Gravot, P. Auroy et al. 2006. HMA1, a new Cu-ATPase of the chloroplast envelope, is essential for growth under adverse light conditions. *J. Biol. Chem.* 28:2882–2892.

Sheldon, A.R. and N.W. Menzies. 2005. The effect of copper toxicity on the growth and root morphology of Rhodes grass (*Chloris gayana* Knuth.) in resin buffered solution culture. *Plant Soil* 278:341–349.

Shi, J.Y., Y.X. Chen, Y.Y. Huang, and W. He. 2004. SRXRF microprobe as a technique for studying elements distribution in *Elsholtzia splendens*. *Micron* 35:557–564.

Shi, J.Y., B. Wu, X.F. Yuan et al. 2008. An x-ray absorption spectroscopy investigation of speciation and biotransformation of copper in *Elsholtzia splendens*. *Plant Soil* 302:163–174.

Shi, J.Y., X. Yuan, X. Chen, B. Wu, Y. Huang, and Y.X. Chen. 2011. Copper uptake and its effect on metal distribution in root growth zones of *Commelina communis* revealed by SRXRF. *Biol. Trace Elem. Res.* 141:294–304.

Shikanai, T., P. Müller-Moulé, Y. Munekage, K.K. Niyogi, and M. Pilon. 2003. PPA1, a P-type ATPase of *Arabidopsis*, functions in copper transport in chloroplasts. *Plant Cell* 15:1333–1346.

Shin, L.J., J.C. Lo, and K.C. Yeh. 2012. Copper chaperone antioxidant protein1 is essential for copper homeostasis. *Plant Physiol.* 159:1099–1110.

Sichul, L., K. Yu-Young, L. Youngsook, and A. Gynheung. 2007. Rice P1B-ATPase, OsHMA9, is a metal efflux protein. *Plant Physiol.* 145:831–842.

Solioz, M. and C. Vulpe. 1996. CPx-type ATPases: A class of P-type ATPases that pump heavy metals. *Trends Biochem. Sci.* 21:237–241.

Song, J., Y.Q. Yang, S.H. Zhu et al. 2013. Spatial distribution and speciation of copper in root tips of cucumber revealed by μ-XRF and μ-XANES. *Biol. Plant.* 57:581–586.

Southron, J.L., U. Basu, and G.J. Taylor. 2004. Complementation of *Saccharomyces cerevisiae* ccc2 mutant by a putative P1B-ATPase from *Brassica napus* supports a copper-transporting function. *FEBS Lett.* 566:218–222.

Stevenson, F.J. 1976. Stability constants of Cu^{2+}, Pb^{2+}, and Cd^{2+} complexes with humic acids. *Soil Sci. Soc. Am. J.* 40:665–672.

Stevenson, F.J. 1991. Organic matter-micronutrient reactions in soil. In *Micronutrients in Agriculture*, eds. J.J. Mortvedt, F.R. Cox, L.M. Shuman, and R.M. Welch, 2nd edn., pp. 145–186. Madison, WI: Soil Science Society of America.

Suzuki, M., K. Bashir, H. Inoue, M. Takahashi, H. Nakanishi, and N.K. Nishizawa. 2012. Accumulation of starch in Zn-deficient rice. *Rice* 5:9.

Tabata, K., S. Kashiwagi, H. Mori, C. Ueguchi, and T. Mizuno. 1997. Cloning of a cDNA encoding a putative metal-transporting P-type ATPase from *Arabidopsis thaliana*. *Biochim. Biophys. Acta* 1326:1–6.

Takahashi, R., K. Bashir, Y. Ishimaru, N.K. Nishizawa, and H. Nakanishi. 2012a. The role of heavy-metal ATPases, HMAs, in zinc and cadmium transport in rice. *Plant Signal Behav.* 7:1605–1607.

Takahashi, R., Y. Ishimaru, H. Shimo et al. 2012b. The OsHMA2 transporter is involved in root-to-shoot translocation of Zn and Cd in rice. *Plant Cell Environ.* 35:1948–1957.

Talke, I.N., M. Hanikenne, and U. Krämer. 2006. Zinc-dependent global transcriptional control, transcriptional deregulation, and higher gene copy number for genes in metal homeostasis of the hyperaccumulator *Arabidopsis halleri*. *Plant Physiol.* 142:148–167.

Tao, S., W.X. Liu, Y.J. Chen et al. 2004. Evaluation of factors influencing root-induced changes of copper fractionation in rhizosphere of a calcareous soil. *Environ. Pollut.* 129:5–12.

Thomas, B., A. Thompson, V.A. Oyenuga, and R.H Armstrong. 1952. The ash constituents of some herbage plants at different stages of maturity. *Emp. J. Exp. Agric.* 20:10–22.

Trindade, L.M., B.M. Horváth, M.J.E. Bergervoet, and R.G.F. Visser. 2003. Isolation of a gene encoding a copper chaperone for copper/zinc superoxide dismutase and characterization of its promoter in potato. *Plant Physiol.* 133:618–629.

Turnau, K., I. Kottke, and F. Oberwinkler. 1993. Element localization in mycorrhizal roots of *Pteridium aquilinum* (L.) Kuhn collected from experimental plots treated with cadmium dust. *New Phytol.* 123:313–324.

Ueno, D., M.J. Milner, N. Yamaji et al. 2011. Elevated expression of *TcHMA3* plays a key role in the extreme Cd tolerance in a Cd-hyperaccumulating ecotype of *Thlaspi caerulescens*. *Plant J.* 66:852–862.

Ueno, D., N. Yamaji, I. Kono et al. 2010. Gene limiting cadmium accumulation in rice. *Proc. Natl. Acad. Sci. U.S.A.* 107:16500–16505.

Ueno, D., N. Yamaji, and J.F. Ma. 2009. Further characterization of ferric-phytosiderophore transporters ZmYS1 and HvYS1 in maize and barley. *J. Exp. Bot.* 60:3513–3520.

Ure, A.M., P. Quevauviller, H. Muntau, and B. Griepink. 1993. Speciation of heavy metals in soils and sediments—An account of the improvement and harmonization of extraction techniques undertaken under the auspices of the BCR of the Commission-of-the-European-Communities. *Int. J. Environ. Anal. Chem.* 51:135–151.

Verret, F., A. Gravot, P. Auroy et al. 2004. Overexpression of AtHMA4 enhances root-to-shoot translocation of zinc and cadmium and plant metal tolerance. *FEBS Lett.* 576:306–312.

Verret, F., A. Gravot, P. Auroy et al. 2005. Heavy metal transport by AtHMA4 involves the N-terminal degenerated metal binding domain and the C-terminal His11 stretch. *FEBS Lett.* 579:1515–1522.

Vert, G., M. Barberon, E. Zelazny, M. Séguéla, J.F. Briat, and C. Curie. 2009. *Arabidopsis* IRT2 cooperates with the high-affinity iron uptake system to maintain iron homeostasis in root epidermal cells. *Planta* 229:1171–1179.

Vert, G., J.F. Briat, and C. Curie. 2001. *Arabidopsis* IRT2 gene encodes a root–periphery iron transporter. *Plant J.* 26:181–189.

Vert, G., N. Grotz, F. Dédaldéchamp et al. 2002. IRT1, an *Arabidopsis* transporter essential for iron uptake from the soil and for plant growth. *Plant Cell* 14:1223–1233.

von Wirén, N., H. Marschner, and V. Römheld. 1996. Roots of iron-efficient maize also absorb phytosiderophore-chelated zinc. *Plant Physiol.* 111:1119–1125.

von Wirén, N., S. Mori, H. Marschner, and V. Römheld. 1994. Iron inefficiency in maize mutant ys1 (*Zea mays* L. cv Yellow-Stripe) is caused by a defect in uptake of iron phytosiderophores. *Plant Physiol.* 106:71–77.

Walker, T.S., H.P. Bais, E. Grotewold, and J.M. Vivanco. 2003. Root exudation and rhizosphere biology. *Plant Physiol.* 132:44–51.

Wallace, A. and J.W. Cha. 1989. Interactions involving copper toxicity and phosphorus deficiency in bush bean plants grown in solutions of low and high pH. *Soil Sci.* 147:430–431.

Waters, B.M., H.-H. Chu, R.J. DiDonato et al. 2006. Mutations in *Arabidopsis* Yellow Stripe-Like1 and Yellow Stripe-Like3 reveal their roles in metal ion homeostasis and loading of metal ions in seeds. *Plant Physiol.* 141:1446–1458.

Waters, B.M., S.A. McInturf, and R.J. Stein. 2012. Rosette iron deficiency transcript and microRNA profiling reveals links between copper and iron homeostasis in *Arabidopsis thaliana*. *J. Exp. Bot.* 63:5903–5918.

Weigel, M., C. Varotto, P. Pesaresi et al. 2003. Plastocyanin is indispensable for photosynthetic electron flow in *Arabidopsis thaliana*. *J. Biol. Chem.* 278:31286–31289.

West, M., A.T. Ellis, P.J. Potts et al. 2012. Atomic spectrometry update—X-ray fluorescence spectrometry. *J. Anal. Atom. Spectrom.* 27:1603–1644.

Whitehead, D.C. 1987. Some soil–plant and root–shoot relationships of copper, zinc and manganese in white clover and perennial ryegrass. *Plant Soil* 97:47–56.

Wightwick, A.M., S.A. Salzman, S.M. Reichman, G. Allinson, and N.W. Menzies. 2013. Effects of copper fungicide residues on the microbial function of vineyard soils. *Environ. Sci. Pollut. Res.* 20:1574–1585.

Williams, L.E. and R.F. Mills. 2005. P_{1B}-ATPases-an ancient family of transition metal pumps with diverse function in plants. *Trends Plant Sci.* 10:491–502.

Williams, L.E., J.K. Pittman, and J.L. Hall. 2000. Emerging mechanisms for heavy metal transport in plants. *Biochim. Biophys. Acta: Biomembr.* 1465:104–126.

Wintz, H., T. Fox, Y.Y. Wu et al. 2003. Expression profiles of *Arabidopsis thaliana* in mineral deficiencies reveal novel transporters involved in metal homeostasis. *J. Biol. Chem.* 278:47644–47653.

Wintz, H. and C. Vulpe. 2002. Plant copper chaperones. *Biochem. Soc. Trans.* 30:732–735.

Woeste, K.E. and J.J. Kieber. 2000. A strong loss of function mutation in RAN1 results in constitutive activation of the ethylene response pathway as well as rosette-lethal phenotype. *Plant Cell* 12:443–455.

Wong, C.K.E. and C.S. Cobbett. 2009. HMA P-type ATPases are the major mechanism for root-to-shoot Cd translocation in *Arabidopsis thaliana*. *New Phytol.* 181:71–78.

Wu, B., F. Andersch, W. Weschke, H. Weber, and J.S. Becker. 2013. Diverse accumulation and distribution of nutrient elements in developing wheat grain studied by laser ablation inductively coupled plasma mass spectrometry imaging. *Metallomics* 5:1276–1284.

Wu, B. and J.S. Becker. 2012. Imaging techniques for elements and element species in plant science. *Metallomics* 4:403–416.

Yamasaki, H., S.E. Abdel-Ghany, C.M. Cohu, Y. Kobayashi, T. Shikanai, and M. Pilon. 2007. Regulation of copper homeostasis by micro-RNA in *Arabidopsis*. *J. Biol. Chem.* 282:16369–16378.

Yamasaki, H., M. Hayashi, M. Fukazawa, Y. Kobayashi, and T. Shikanai. 2009. SQUAMOSA promoter binding protein-like7 is a central regulator for copper homeostasis in *Arabidopsis*. *Plant Cell* 21:347–361.

Yang, M.J., X.E. Yang, and V. Römheld. 2002. Growth and nutrient composition of *Elsholtzia splendens* Nakai under copper toxicity. *J. Plant Nutr.* 25:1359–1375.

Yen, M.-R., Y.-H. Tseng, and M.H. Saier Jr. 2001. Maize Yellow Stripe1, an iron-phytosiderophore uptake transporter, is a member of the oligopeptide transporter (OPT) family. *Microbiology* 147:2881–2883.

Yordem, B.K., S.S. Conte, J.F. Ma et al. 2011. *Brachypodium distachyon* as a new model system for understanding iron homeostasis in grasses: Phylogenetic and expression analysis of Yellow Stripe-Like (YSL) transporters. *Ann. Bot.* 108:821–833.

Yruela, I. 2005. Copper in plants. *Braz. J. Plant Physiol.* 17:145–156.

Yruela, I. 2008. Copper in plants: Acquisition, transport and interactions. *Funct. Plant Biol.* 36:409–430.

Yruela, I. 2013. Transition metals in plant photosynthesis. *Metallomics* 5:1090–1109.

Yruela, I., J.J. Pueyo, P.J. Alonso, and R. Picorel. 1996. Photoinhibition of photosystem II from higher plants. Effect of copper inhibition. *J. Biol. Chem.* 271:27408–27415.

Yuan, M., X. Li, J. Xiao, and S. Wang. 2011. Molecular and functional analyses of COPT/Ctr type copper transporter-like gene family in rice. *BMC Plant Biol.* 11:69.

Zhao, H. and D. Eide. 1996. The yeast ZRT1 gene encodes the zinc transporter of a high affinity uptake system induced by zinc limitation. *Proc. Natl. Acad. Sci. U.S.A.* 93:2454–2458.

Zhao, H., L. Wu, T. Chaia, Y. Zhang, J. Tana, and S. Ma. 2012. The effects of copper, manganese and zinc on plant growth and elemental accumulation in the manganese-hyperaccumulator *Phytolacca americana*. *J. Plant Physiol.* 169:1243–1252.

Zheng, L., N. Yamaji, K. Yokosho, and J.F. Ma. 2012. YSL16 is a phloem-localized transporter of the copper-nicotianamine complex that is responsible for copper distribution in rice. *Plant Cell* 24:3767–3782.

Zhou, J. and P.B. Goldsbrough. 1994. Functional homologs of fungal metallothionein genes from *Arabidopsis*. *Plant Cell* 6:875–884.

Zhou, X., Z. Zuo, Y. Li, and B. Liu. 2013. The distribution dynamic of Cu element between soil and plant in northeast *Leymus chinensis* grassland. *J. Food Agric. Environ.* 11:1310–1314.

Zhu, H., E. Shipp, R.J. Sánchez et al. 2000. Cobalt^{2+} binding to human and tomato copper chaperone for superoxide dismutase: Implications for the metal ion transfer mechanism. *Biochemistry* 39:5413–5421.

FIGURE 2.3 Nitrogen-deficient young pepper (*Capsicum* sp. L.) plants showing chlorosis (paler coloration) of oldest leaves. (Photograph by A.V. Barker.)

FIGURE 3.2 Phosphorus-deficient corn (*Zea mays* L.) showing discoloration, chlorosis, and necrosis of leaves. (Photograph by A.V. Barker.)

FIGURE 3.3 Phosphorus-deficient cucumber (*Cucumis sativus* L.) showing early necrosis of leaves. (Photograph by A.V. Barker.)

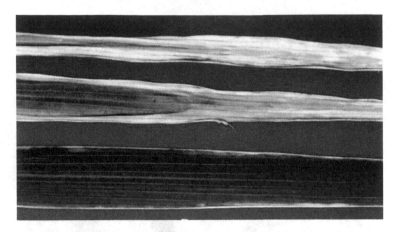

FIGURE 4.8 Rice leaves with severe (top), moderate (middle), and mild (bottom) K deficiency symptoms. (From Fageria, N.K. and Barbosa Filho, M.P., *Nutritional Deficiencies in Rice: Identification and Correction*, EMBRAPA-CNPAF document 42, Goiania, Brazil, 1994.)

FIGURE 5.6 Calcium-related disorders in horticultural crop: (a) blackheart in celery, (b) tipburn in lettuce, (c) blossom-end rot in immature tomato fruit, and (d) cracking in mature tomato fruit. (Photographs from White, P.J. and Broadley, M.R., *Ann. Bot.*, 92, 487, 2003. Courtesy of the HRI collection.)

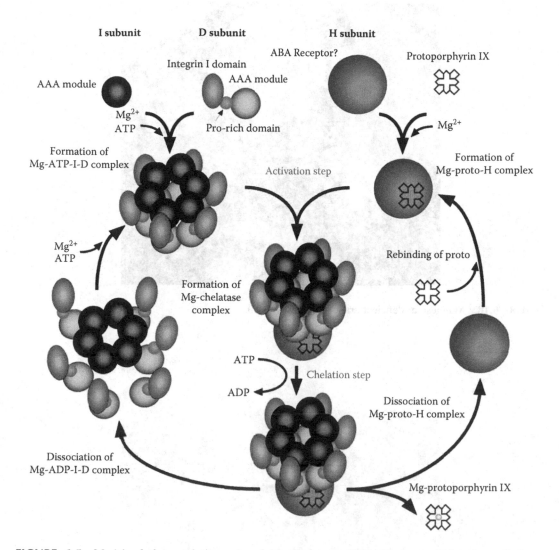

FIGURE 6.5 Model of the catalytic cycle of Mg-Chelatase. (With kind permission from Springer Science+Business Media: *Photosynth. Res.*, Recent overview of the branch of the tetrapyrrole biosynthesis leading to chlorophylls, 96, 2008, 121–143, Masuda, T.)

Activation step: In a Mg^{2+} and ATP-dependent process, six I subunits are assembled into a hexameric ring structure, while six D subunits form a hexameric ring in an ATP-independent manner. The two-tiered hexameric ring forms the Mg-ATP-I-D complex. It is suggested that the ATPase activity of subunit I is inhibited in this complex due to binding of the integrin I domain of subunit D to the integrin I binding domain of subunit I. Meanwhile, subunit H binds to protoporphyrin IX and most likely also the Mg^{2+} substrate—as this subunit is considered the catalytic subunit—to form the Mg-H-protoporphyrin IX complex.

Chelatation step: ATP hydrolysis is triggered upon switching the binding of the integrin I domain of subunit D to the integrin I binding domain of subunit H. After the formation of Mg-protoporphyrin IX, the complex disassembles. Here, subunit H is shown docking to the D subunits, although it remains unknown if subunit H docks to the I or D side of the complex.

FIGURE 6.18 Magnesium-deficient grape (*Vitis vinifera* L.).

FIGURE 6.23 Magnesium-deficient corn (*Zea mays* L.).

FIGURE 6.24 Magnesium-deficient sugar beet (*Beta vulgaris* L.).

FIGURE 7.4 Sulfur deficiency can cause distinct whitening of canola (*Brassica napus* L.) flowers. (Photographs courtesy of Malcolm Hawkesford.)

(a)

(b)

(c)

(d)

FIGURE 7.5 Sulfur deficiency in (a) wheat (*Triticum aestivum* L.), (b) barley (*Hordeum vulgare* L.), (c) corn (*Zea mays* L.), and (d) sugarcane (*Saccharum officinarum* L.). (Photographs courtesy of International Plant Nutrition Institute.)

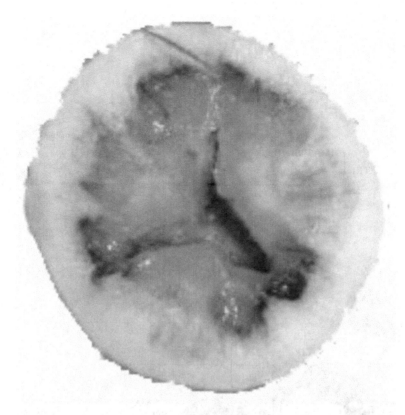

FIGURE 8.3 Symptoms of boron deficiency (brown heart) in rutabaga (*Brassica napobrassica* Mill.) showing a soft, watery center of a cut root. (From Gupta, U.C., Boron, in *Handbook of Plant Nutrition*, 1st edn., Barker, A.V. and Pilbeam, D.J., (eds.), CRC Press: Boca Raton, FL, 2007, pp. 242–268.)

FIGURE 8.4 Boron toxicity in peach (*Prunus persica* Stokes). Note meristematic dieback on side shoot (top left). (Photograph courtesy of Patrick H. Brown, UC Davis, CA.)

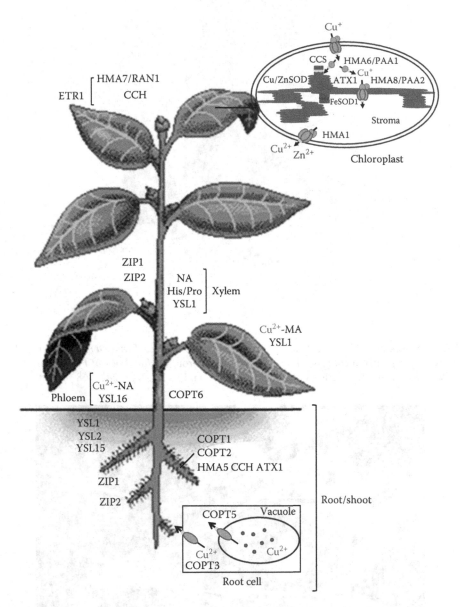

FIGURE 10.1 Scheme of transport pathways identified for Cu in a generic plant. Arrows indicate the proposed direction for metal transport. CCH, copper chaperone; ATX1, antioxidant 1; CCS, copper chaperone for Cu/Zn SOD; COPT, copper transporter; ERT1, endoplasmic reticulum; HMA, heavy metal P-type ATPase; MA, mugineic acid; NA, nicotianamine; PAA, P-type ATPase of *Arabidopsis*; RAN1, responsive to antagonist 1; SOD, superoxide dismutase; YSL, yellow stripe-like protein; ZIP, ZRT/IRT-like protein.

FIGURE 11.2   Iron-deficient marigold (*Tagetes patula* L.) showing chlorosis of young leaves.

FIGURE 11.3   Iron-deficient borage (*Borago officinalis* L.) showing chlorosis of young leaves.

FIGURE 11.4   Iron-deficient ixora (*Ixora chinensis* Lam.).

FIGURE 11.5   Iron-deficient corn (*Zea mays* L.) showing interveinal chlorosis of leaves.

FIGURE 12.2   Manganese deficiency in tomato (*Lycopersicon esculentum* Mill.).

FIGURE 12.4   Manganese toxicity in soybean (*Glycine max* Merr.) showing necrotic spots.

FIGURE 13.2 Molybdenum deficiency of poinsettia (*Euphorbia pulcherrima* Willd.). (Photograph courtesy of Dr. Douglas A. Cox, University of Massachusetts, Amherst, MA.)

(a)　　　　　(b)

(c)　　　　　(d)

FIGURE 14.1 Nickel deficiency symptoms for pecan (*Carya illinoinensis* K. Koch). (a) Reduced leaf and leaflet size as severity of deficiency increases; (b) altered shape of leaflet blade to produce mouse-ear-like or little-leaf foliage; (c) blunting, crinkling, dark green zone, and necrosis of the apical portion of the leaf blade; and (d) absence of leaf lamina development in severe cases, plus deformed and pointed buds, and rosetting (loss of apical dominance).

FIGURE 14.2 Influence of Ni deficiency on pecan (*Carya illinoinensis* K. Koch). (a) Loss of apical dominance and rosetting; (b) delayed bud break (degree of Ni deficiency decreasing from left to right); (c) short internodes, dwarfed and weakened shoots possessing weak pointed buds (Ni deficient on left and Ni sufficient on right); and (d) dwarfed tree (tree is 1 m tall at age 10 years) typical of the replant form of Ni deficiency.

FIGURE 15.2 Zinc-deficient cucumber (*Cucumis sativus* L.).

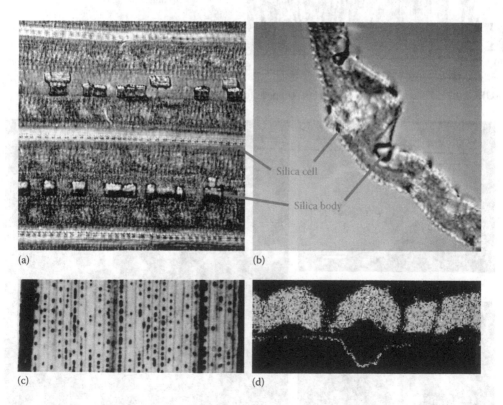

FIGURE 20.5 Silicon accumulation in rice leaf blade. (a, b) Silica cell and silica body. (c) Silica body detected by soft x-ray. (d) Silica deposition on epidermal cells detected by SEM–EDX.

FIGURE 20.7 Effect of silicon on (a) blast, (b) pest, (c) overtranspiration, and (d) fertility in rice.

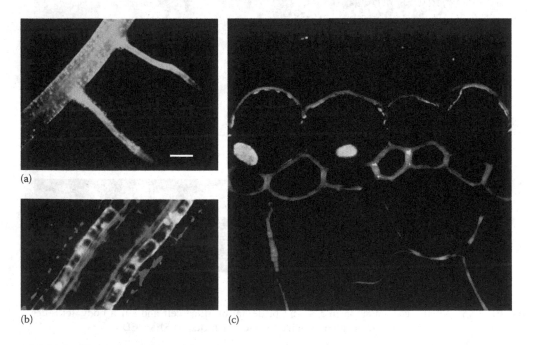

FIGURE 20.9 Tissue localization of (a) Lsi1 (green), (b) Lsi2 (red), and (c) polar localization of Lsi1 (green) and Lsi2 (red) in rice.

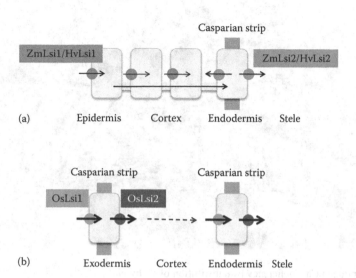

(a) Epidermis   Cortex   Endodermis   Stele

Casparian strip

ZmLsi1/HvLsi1   ZmLsi2/HvLsi2

Casparian strip   Casparian strip

OsLsi1   OsLsi2

(b) Exodermis   Cortex   Endodermis   Stele

FIGURE 20.10   Lsi1- and Lsi2-mediated silicon uptake system in (a) upland crop (barley/maize) and (b) rice.

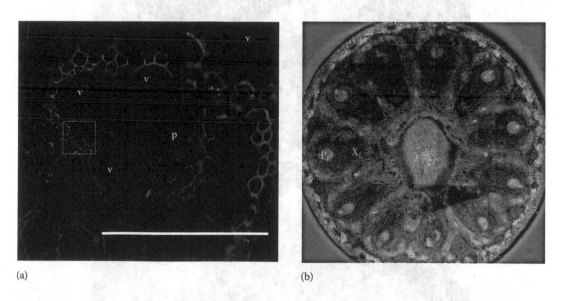

FIGURE 20.11   Localization of Lsi6 protein in (a) leaf blade and (b) node I in rice. Red color shows signal of Lsi6 protein. v, vessel; p, phloem; xylem (X_L) and phloem regions (P_L) of large vascular bundles.

FIGURE 21.3 Induced Ca^{2+} deficiency by substitution of K^+ by Na^+.

FIGURE 21.4 Corn (*Zea mays* L.) plant showing typical Na^+ toxicity symptoms.

11 Iron

Allen V. Barker and Margie L. Stratton

CONTENTS

11.1 HISTORICAL BACKGROUND

Iron (Fe) has been known from prehistoric times. It is the fourth most common element in the crust of the earth and is the most common element in the entire mass of the earth. According to dictionaries, the name iron derives from Middle English and Old English words of unknown meaning, and the chemical symbol Fe comes from the Latin *ferrum* for firmness. In crop production, Gris (1843) and Wallihan (1965) noted that growth and appearance of chlorotic grape vines were improved by foliar application of Fe salts. The essentiality of Fe as a nutrient for plants was demonstrated in about 1860 by Julius von Sachs, who included Fe in nutrient solutions for hydroponics (Römheld and Nikolic, 2007). In plants, Fe deficiency produces easily recognized symptoms that appeared readily in research on essentiality of nutrients so that Fe was recognized as an essential element at the same time as the macronutrients.

11.2 FUNCTION OF IRON IN PLANTS

The ability of Fe to undergo oxidation–reduction reactions ($Fe^{+2} \leftrightarrow Fe^{+3}$ + electron) is important in its functions in metabolism in heme, nonheme, or coordinated protein structures (Table 11.1). Heme proteins have Fe in a porphyrin ring complex. Cytochromes are well-known heme proteins and participate in electron transport in photosynthesis between the two photosystems in chloroplasts and in respiration in oxidative phosphorylation in mitochondria (Broadley et al., 2012). Other heme-containing proteins are catalase, peroxidase, and leghemoglobin. Catalase catalyzes the formation of water and oxygen through the decomposition of hydrogen peroxide and protects cells from oxidative damage (Mhamdi et al., 2010; Su et al., 2014). Iron deficiency suppresses the activity of catalase (Broadley et al., 2012; Dasgan et al., 2003). For many peroxidases, the substrate is hydrogen peroxide, but other peroxidases are more active with organic hydroperoxides. Peroxides are created as by-products of various biochemical reactions within organisms and can cause damage as they are oxidizing agents. Peroxidases break these compounds down into harmless substances in oxidation–reduction reactions in which one molecule of peroxide is reduced to water and another molecule

TABLE 11.1

Some Enzymes and Other Proteins in Which Iron Functions in Plant Metabolism

Protein	Structure	Function
Catalase	Heme	Catalyzes the decomposition of hydrogen peroxide to water and oxygen
Peroxidases	Heme	A family of enzymes that metabolize reactive oxygen species including organic peroxides or hydrogen peroxide to water and an oxidized compound
Leghemoglobin	Heme	Oxygen carrier in nitrogen fixation
Ferredoxin	Nonheme	Electron transport in photosynthesis, nitrate reduction, sulfite reduction, and nitrogen fixation
Aconitase	Nonheme	Catalyzes isomerization of citrate to isocitrate via aconitate
Superoxide dismutase	Coordinated	Catalyzes decomposition of superoxide to hydrogen peroxide and oxygen
Succinic dehydrogenase	Heme	Catalyzes oxidation of succinate to fumarate
Cytochromes	Heme	Electron transport in photosynthesis and in oxidative phosphorylation in respiration
Aminolevulinic acid synthetase	Catalytic	Condenses glycine and succinyl-CoA into δ-aminolevulinic acid, the first common precursor of tetrapyrroles in heme biosynthesis

is oxidized. There are a large number of these enzymes in plants and animals. Peroxidases are abundant in cell walls of the epidermis or of aerial plant organs or of roots (Broadley et al., 2012; Codignola et al., 1989; Hendriks and van Loon, 1990). In addition to their detoxifying roles in ridding cells of reactive oxygen species, peroxidases are required for synthesis of lignin and suberin (Broadley et al., 2012).

Dasgan et al. (2003) reported that measurement of activities of Fe-containing enzymes in leaves is more reliable than measuring total concentration of Fe for characterization of the Fe nutritional status of plants and for assessing genotypical differences in responses to Fe deficiency. Tewari et al. (2005) reported that total Fe concentration in young leaves of mulberry (*Morus* spp. L.), corn (*Zea mays* L.), and cauliflower (*Brassica oleracea* var. *botrytis* L.) plants supplied with a deficient amount of Fe did not show any significant change from leaves with an adequate supply of Fe; however, a deficient supply of Fe decreased activities of catalase, peroxidase, and ascorbate peroxidase and increased activity and isoforms of superoxide dismutase, accumulation of superoxide (O_2^-), and concentration of H_2O_2. With young corn plants, grown hydroponically, in vitro activities of catalase, peroxidase, and nitrate reductase in leaves decreased with intense Fe deficiency (Nenova and Stoyanov, 1995). With peanuts (*Arachis hypogaea* L.), the Fe concentration in leaves increased with foliar or soil applications of chelated Fe ethylenediamine-di-O-hydroxyphenylacetic acid (FeEDDHA). Either soil or leaf applications of Fe had more positive effects on the activities of peroxidase and catalase than on superoxide dismutase (Panjtandoust et al., 2013).

Glutathione and ascorbic acid are multifunctional metabolites playing important roles in oxidation–reduction balancing. In Fe deficiency, both metabolites increased the activity of ascorbate peroxidase to a level similar to that in Fe-sufficient seedlings (Ramirez et al., 2013). Glutathione content decreased in Fe-deficient arabidopsis (*Arabidopsis thaliana* Heynh.) seedlings, whereas ascorbic acid was not affected. The results showed that glutathione and ascorbic acid supplementations protect arabidopsis seedlings from Fe deficiency, preserving cell oxidation–reduction status and improving internal Fe availability.

Iron deficiency in pea (*Pisum sativum* L.) leaves caused a large decrease in chlorophyll and carotenoids and smaller decreases in soluble protein and net photosynthesis (Iturbe-Ormaetxe et al., 1995). Also, catalase and peroxidase declined sharply in young Fe-deficient leaves, whereas monodehydroascorbate reductase, dehydroascorbate reductase, and glutathione reductase activities were unaffected. Ascorbate peroxidase activity was correlated highly with the Fe content of leaves, thereby allowing its use as an indicator of the Fe nutritional status of the plant. Iron deficiency resulted in an increase of Cu/Zn-containing superoxide dismutase but not of Mn superoxide dismutase.

Excess of free Fe is considered to harm plant cells by enhancing the intracellular production of reactive oxygen intermediates (ROIs). With bean plants (*Phaseolus vulgaris* L.), Pekker et al. (2002) reported that cytosolic ascorbate peroxidase, which is an Fe-containing, ROI-detoxifying enzyme induced in response to Fe overload or oxidative stress, increased rapidly in response to Fe overload as noted by an increase in Fe content in leaves. The results suggested that high intracellular levels of free Fe in plants lead to the enhanced production of ROI, which in turn induces the expression of cytosolic ascorbate peroxidase. Fang et al. (2001) reported that excess Fe from toxic levels of $FeSO_4$ enhanced lipid peroxidation but not the content of H_2O_2. Superoxide dismutase activity was reduced by the excessive $FeSO_4$, but ascorbate peroxidase and glutathione reductase activities were increased by excessive $FeSO_4$.

Herbik et al. (1996) reported that the nicotianamine-deficient tomato (*Lycopersicon esculentum* Mill.) mutant, *chloronerva*, resembled an Fe-deficient plant despite the high accumulation of Fe in the leaves. In root tips of the Fe-deficient wild-type 'Bonner Beste' and the Fe-sufficient as well as the Fe-deficient mutant, the activity of ascorbate peroxidase was increased compared with the Fe-sufficient wild type. In leaves of the Fe-sufficient or Fe-deficient mutant, chloroplastic and cytosolic Cu/Zn-containing superoxide dismutase and Cu-containing plastocyanin were nearly absent.

Babalakova et al. (2011) reported that Fe-deficiency stress in hydroponically grown cucumber (*Cucumis sativus* L.) induced an increase of Fe(III) citrate reductase activity and Fe reduction in roots. Hydrogen peroxide content in leaves increased about 20% after Fe exclusion.

Abdelmajid et al. (2008) showed that nitrogen fixation and leghemoglobin accumulation decreased with limited Fe availability in common bean (*Phaseolus vulgaris* L.). A cultivar tolerant to Fe deficiency was less affected than a sensitive one. A significant stimulation of peroxidase, superoxide dismutase, and catalase activities was observed in the tolerant cultivar under Fe deficiency. The Fe for leghemoglobin accumulation, superoxide dismutase, catalase, and peroxidase was critical for the protection of the symbiotic system against oxidative burst and for the maintenance of an optimal functioning of the N_2 fixing system.

Iron, although not a constituent of chlorophyll, is required for chlorophyll synthesis and functioning of the photosynthetic apparatus of plants (Marsh et al., 1963; Yadavalli et al., 2012). Iron activates ∂-aminolevulinic acid synthetase and coproporphyrinogen oxidase in the synthesis of precursors of chlorophyll (Römheld and Nikolic, 2007). Römheld and Nikolic (2007) suggested also that during Fe deficiency, the decreased chloroplast volume and protein content indicate that the protein–chlorophyll complex of chloroplasts is not synthesized adequately and results in chlorosis. In Fe-deficient corn leaves, the total protein decreases by 25% relative to nondeficient plants, but the protein content of chloroplasts declines by 82% (Broadley et al., 2012; Perur et al., 1961).

Ferredoxin is a nonheme protein that has involvement in photosynthesis. Ferredoxin is reduced in Photosystem I and can transfer electrons to a number of acceptors. In these transfers, reduced nicotinamide adenine dinucleotide phosphate (NADP) is produced (Green et al., 1991). Ferredoxin is required for reduction of nitrite to ammonia in nitrate reduction reactions (Knaff and Hirasawa, 1991; Zafrilla et al., 2011).

11.3 UPTAKE OF IRON BY PLANTS AND PHYSIOLOGICAL RESPONSES OF PLANTS TO SUPPLY OF IRON

In well-aerated soils, Fe is predominately in the oxidized ferric form (Fe^{3+}), with only a small amount being in solution since most Fe in soil is in combination with oxides. An estimated concentration of Fe in soil solution is less than 10^{-15} M Fe^{3+} (Marschner, 1995). Iron absorption into roots of dicotyledonous plants is mainly in the ferrous (Fe^{2+}) form (Lindsay and Schwab, 1982; Römheld and Nikolic, 2007), whereas in grasses, absorption is as a ferric (Fe^{3+}) complex

(Stephan, 2002). Consequently, plants have developed two distinct strategies of uptake to deal with low or deficient concentrations of Fe in soil solutions and with reduction of Fe to the ferrous form (Brumbarova and Bauer, 2008; Charlson and Shoemaker, 2006; Graham and Stangoulis, 2003; Schmidt, 1999; Thomine and Lanquar, 2011). The path of iron from the root surface up to the point of the xylem vessels within the stele may be symplasmic; however, a part of this route also may be apoplastic, through the free space (apoplast) to cells of the rhizodermis and cortex that absorb ions (Stephan, 2002).

Dicots and nongraminaceous monocots have developed a reduction strategy, and graminaceous plants, which include the important grain crops of the world, such as rice (*Oryza sativa* L.), corn, and wheat, have developed a chelation strategy for absorbing Fe from the soil solution (Kobayashi and Nishizawa, 2012). Mechanisms for these strategies reside in the plasma membranes of root cells exposed to soil or nutrient solutions. Iron reductases and high-affinity transporters are located in roots and increase in response to Fe deficiency (Colangelo and Guerinot, 2004). A family of ferric chelate reductases (FCRs) exist to form soluble ferrous ion for transport across the plasma membrane but also may function in membranes of subcellular compartments such as mitochondria, chloroplasts, and vacuoles (Jeong and Connolly, 2009). With dicotyledonous and nongraminaceous monocotyledonous species, plants reduce chelated Fe by a FCR that is bound to the root epidermis. The reduced Fe is released and is absorbed by a Fe^{2+} transporter in the plasmalemma of root cells (Römheld and Nikolic, 2007). Plants with this mode of action are Strategy I plants. Grasses (Poaceae) are classified as Strategy II. The main components of the Strategy II mechanism for Fe uptake are secretion of chelating compounds, phytosiderophores, and the uptake of a Fe^{3+}–phytosiderophore complex (Roberts et al., 2004; Walker and Connolly, 2008; Zuchi et al., 2011). Strategy II plants secrete phytosiderophores, compounds of the mugineic acid family that form stable Fe^{3+} chelates in soil (Mori, 2001). Uptake of Fe–phytosiderophores occurs through specific transporters at the root surface and represents the primary route of Fe entry into Strategy II roots (Roberts et al., 2004). However, rice contains a Fe^{2+} transporter, indicating that, in addition to absorbing a Fe^{3+}–phytosiderophore, rice possesses a novel Fe uptake system that directly absorbs the Fe^{2+}, a strategy that is advantageous for growth in submerged conditions (Ishimaru et al., 2006; Walker and Connolly, 2008).

With several species of Strategy I and Strategy II plants, Fe^{2+} uptake rates varied within a narrow range for Strategy I and Strategy II plants adequately supplied with Fe (Zaharieva and Römheld, 2000). For plants grown in preculture under Fe deficiency, the Fe^{2+} uptake rates markedly increased (4–10-fold) with a short-term resupply of Fe in the Strategy I plants (tomato and cucumber). But, with graminaceous species (barley, *Hordeum vulgare* L., and corn) as well as in an Fe-inefficient tomato (*Tfer*), the increase in uptake was less (1.3–1.6-fold). A lack of Fe and a low supply of 3 μM chelated Fe (FeEDDHA) induced an increase in the Fe-reducing capacity and rhizosphere acidification by roots of citrus seedlings (*C. taiwanica* Tanaka and Shimada or *C. volkameriana* V. Ten. and Pasq.) roots (Chouliaras et al., 2004). Citrus seedlings treated with adequate Fe (20 μM FeEDDHA) showed a decline in the Fe-reducing capacity and in rhizosphere acidification.

These strategies function during conditions of Fe deficiency in soils. For example, with Fe-deficient cucumber (*Cucumis sativus* L.), a Strategy I plant, roots readily reduced Fe chelate (ferric hydroxyethylenediaminetriacetic acid [FeHEDTA]) via FCR, whereas Fe-sufficient roots had low activity of the enzyme (Johnson et al., 2002). Other Fe chelates and Fe siderophores did not function as well as FeHEDTA to supply Fe to the cucumbers. The transcription intensity of encoding of a Fe^{3+} chelate reductase protein in roots was dependent on the Fe status of leaves (Li et al., 2004). In another investigation (Johnson et al., 2002), Fe-deficient cucumber roots readily reduced chelated Fe, whereas Fe-sufficient roots had low ferric chelate reduction activity. Genes encode a root FCR and a high-affinity Fe transporter involved in the Fe-deficiency-induced uptake system (Vert et al., 2003). Recovery from Fe-deficient conditions and modulation of apoplastic Fe pools indicate that Fe itself plays a major role in the regulation of root Fe-deficiency responses at the mRNA and protein levels and may be under diurnal regulation (Vert et al., 2003).

Agnolon et al. (2002) showed that the development of the Fe-deficiency response in cucumber roots can be regulated by the amount of Fe supply. Acidification of the nutrient solutions occurred with plants grown at 1 µM or lesser concentrations of ferric ethylenediaminetetraacetic acid (EDTA) and was related inversely to the external Fe concentration. Ferric EDTA reduction by intact roots also was depressed gradually by increasing Fe supply.

Another property of Strategy I plants is the acidification of the medium to increase the solubility of Fe. Protein-extruding H^+-ATPases function in the solubilization of Fe through rhizosphere acidification by Strategy I dicotyledonous plants (Aranto et al., 2003; Mlodzinska, 2012; Rabotti et al., 1995). These ATPases mediate H^+ extrusion to the extracellular space and create H^+ gradients and potential differences across the plasma membrane, potentials that activate ion and metabolite transport (Aranto et al., 2003; Morsomme and Boutry, 2000; Palmgren, 2001). Chickpea (*Cicer arietinum* L.) acidified the nutrient solution in response to Fe deficiency, with subsequent regreening of chlorotic leaves (Ohwaki and Sugahara, 1997). No recovery occurred if the nutrient solution was buffered at pH 6.3. During the period of acidification induced by Fe deficiency, the roots exuded more carboxylic acids than roots supplied with sufficient Fe. The Fe-deficiency stress response of two cultivars of subterranean clover species, Koala (*Trifolium brachycalycinum* Katzn. and Morley) (Fe-deficiency resistant) and Karridale (*T. subterraneum* L.) (Fe-deficiency susceptible), were evaluated for their effects on acidity of nutrient solutions (Wei et al., 1997). After 6 days of Fe treatment, the Fe-deficient Koala and Karridale decreased the pH of the nutrient solution by 1.83 and 0.79 units, respectively, whereas the Fe-sufficient plants increased the pH of the nutrient solution. The H^+ release rate of the Fe-deficient Koala determined 7 days after Fe treatment initiation was more than three times that of the Fe-deficient Karridale. The Fe-deficient plants had a significantly enhanced Fe^{3+} reduction rate compared to the Fe-sufficient plants for each species, but the resistant species did not exhibit a higher root Fe^{3+} reduction rate than the susceptible species cultivar at each Fe treatment. The increased acidification and reductase capacity are present in the subapical region of the roots. Resupply of Fe after 2 weeks partially reversed the tendency of the roots to acidify the nutrient solution and to reduce ferric EDTA (Vizzotto et al., 1997).

Iron-deficient plants also undergo anatomical changes to improve their capacity to absorb Fe (Briat et al., 1995; Landsberg, 1994; Marschner and Römheld, 1994). Roots develop several morphological changes that disappear when the plants acquire enough Fe (Agnolon et al., 2002; Lopez-Millan et al., 2001). With Strategy I plants, these changes include subapical swelling with abundant root hairs and formation of specialized cells called transfer cells, which have a distinct wall labyrinth (Broadley et al., 2012; Kramer et al., 1980). However, the enhanced root tips and root hairs sometimes die with prolonged Fe deficiency (Egilla et al., 1994). Transfer cells develop from peripheral cells within 24–48 h of Fe deficiency (Kramer et al., 1980). The wall labyrinth is situated on the walls that face the external medium. The cytoplasm has numerous mitochondria, extensive rough endoplasmic reticulum, and large leucoplasts that contain protein bodies. Transfer cells are likely sites of Fe-deficiency-induced responses—excretion of protons, increased ferric-reducing capacity, and release of phenolic compounds.

The regulation of these responses is not understood fully, but some evidence suggests the involvement of ethylene or nitric oxide in this process (Chen et al., 2010; Garcia et al., 2010, 2011; Kabir et al., 2012; Lingam et al., 2011; Lynch and Brown, 1997; Romera and Alcantara, 2004; Walker and Connolly, 2008; Waters and Blevins, 2000). With tomato, extranumerary root hairs and transfer cell-like structures in the epidermis developed in response to phosphorus or Fe deficiency; however, this research did not show the involvement of ethylene in the processes (Schikora and Schmidt, 2002). Roots from Fe-deficient cucumber, tomato, and pea produced more ethylene than roots of Fe-sufficient plants (Romera et al., 1999). The higher production of ethylene in Fe-deficient cucumber and pea plants occurred before Fe-deficient plants showed chlorosis symptoms and occurred with Fe-deficiency stress responses. The addition of the ethylene precursor, 1-aminocyclopropane-1-carboxylic acid, or the ethylene-releasing substance, ethephon, to several Fe-sufficient Strategy I plants promoted some Fe-deficiency stress responses, such as enhanced root ferric-reducing capacity and swollen root tips. However, Fe-deficient roots from several Strategy II plants (corn; wheat, *Triticum aestivum* L.; and barley) did not produce more ethylene than the Fe-sufficient ones. Research by Schikora and Schmidt (2001) with pea and tomato suggested that FCR is regulated by a shoot-originated signal molecule, communicating the Fe status of the shoot to the roots, but that formation of transfer cells depended on local concentrations of Fe. Similar to transfer cells, formation of extra root hairs in an arabidopsis mutant was regulated by the Fe concentration of the growth medium and was unaffected by interorgan signaling. Schikora and Schmidt (2001) suggested that auxin has no primary function in inducing transfer cell or root hair development with tomato or arabidopsis mutants. Schmidt et al. (2000a,b) studying tomato and arabidopsis mutants suggested that the morphological and physiological components of the Fe stress syndrome are regulated separately.

Peanut–corn intercropping evidently enhances the Fe nutrition of peanuts in calcareous soils. Iron-deficiency chlorosis occurred in the young leaves of peanut in monoculture and particularly at the flowering stage, whereas leaves of peanut grown in a mixture with corn remained green (Zuo et al., 2003). The chlorophyll and HCl-extractable Fe concentrations in young leaves of peanut grown in the mixture were much higher than those in monoculture. Peanut roots in the mixture with corn produced more lateral roots and had more root length than plants in monoculture. Peanut grown together with corn had rhizodermal transfer cells in the subapical root zone, but transfer cells were poorly developed in peanut in monoculture. Xiong et al. (2013) noted that the corn *YS3* mutant, which cannot release phytosiderophores, did not improve Fe nutrition of peanut, whereas the corn *YS1* mutant, which can release phytosiderophores, prevented Fe deficiency. Hydroponic experiments revealed that phytosiderophores released by Fe-deficient wheat promoted Fe acquisition in nearby peanuts. The phytosiderophore deoxymugineic acid was detected in the roots of intercropped peanuts.

11.4   GENETICS OF ACQUISITION OF IRON BY PLANTS

Plants initiate a set of responses that allow them to obtain Fe from the dilute concentrations of soil solutions and to restrict Fe absorption in excess of cellular needs (Schmidt, 2003; Stephan, 2002) (Table 11.2). These responses are under genetic control in plants and microorganisms (Curie and Briat, 2003; Curie et al., 2001; O'Rourke et al., 2007; Peiffer et al., 2012; Toker et al., 2010; Urzica et al., 2012). Therefore, a feasible method for alleviating Fe deficiency in crops may be in the selection of suitable cultivars resistant to Fe-deficiency chlorosis (Lin et al., 1997; Rengel, 2002; Toker et al., 2010). Strategy I plants respond with actions localized in the plasma membrane of root cells to dissolve and transport of Fe into roots. These actions include acidification of the root zone, generation of a FCR that converts Fe^{3+} to Fe^{2+}, and a transporter that carries Fe^{2+} across the plasma membrane (Walker and Connolly, 2008). In arabidopsis, a ferric reductase oxidase gene (*FRO2*) encodes FCR that reduces Fe^{3+} at the root surface, and another gene, *IRT1*, encodes a high-affinity transporter

TABLE 11.2
Some Genes That Regulate Iron Absorption and Transport by Plants

Gene	Process Regulated
Ferric reductase oxidase (*FRO2*)	Encodes FCR that reduces ferric Fe at the root surface
Fe regulated transporter (*IRT1*)	Encodes a high-affinity transporter of Fe^{2+} across the plasma membrane
ATPase genes (*HA1*)	Encode a H⁺-ATPase that functions in acidification of the root zone
Fe-deficiency response transcription factor (*FIT*)	Encodes a transcription factor necessary for *IRT1* and *FRO2* induction under Fe deficiency
Yellow-stripe gene, yellow-stripe like (*YS1*, *YSL*)	Encode production of plasma membrane protein that transports ferric–phytosiderophore complexes
Nicotianamine synthase (*NAS*)	Long-distance transport of Fe

of Fe^{2+} across the plasma membrane (Varotto et al., 2002; Vert et al., 2002; Walker and Connolly, 2008). The *CsHA1* gene of cucumber encodes a H^+-ATPase that functions in acidification of the root zone (Santi et al., 2005). Santi et al. (1995) also noted acidification of the root zone of cucumber with membrane-associated H^+-ATPase involved in nitrate absorption. Santi and Schmidt (2008) further investigated the expression of genes in structural and physiological responses involved in the mobilization of Fe (*CsHA1*), the reduction of ferric chelates (*CsFRO1*), and the uptake of Fe^{2+} (*CsIRT1*) in epidermal cells of Fe-sufficient and Fe-deficient cucumber roots. Growing plants hydroponically in media deprived of Fe induced the differentiation of almost all epidermal cells into root hairs, whereas no root hairs were formed under Fe-sufficient conditions. The formation of root hairs in response to Fe starvation was associated with a dramatic increase in message levels of *CsFRO1*, *CsIRT1*, and the *CsHA1* gene that codes for Fe-inducible H^+-ATPase relative to activities in epidermal cells of Fe-sufficient plants.

A protein FIT (Fe-deficiency-induced transcription factor) is a central regulator of Fe acquisition in roots and is regulated by environmental cues and internal requirements (Bauer et al., 2007). Lingam et al. (2011) demonstrated a direct molecular link between ethylene signaling and FIT. Lingam et al. (2011) noted that increased FIT abundance led to the high level of expression of genes required for Fe acquisition. Chen et al. (2010) reported that an Fe-deficiency-induced increase of auxin and nitric oxide (NO) levels in wild-type arabidopsis was accompanied by upregulation of root FCR activity and the expression of the basic transcription factor (FIT) and the ferric reduction oxidase (*FRO2*) genes. Garcia et al. (2010) showed that ethylene and NO are involved in the upregulation of many important Fe-regulated genes of arabidopsis and showed that Fe deficiency upregulates genes involved in ethylene synthesis and signaling in roots. Yuan et al. (2008) reported that transcriptions of FCR *FRO2* and ferrous-transporter *IRT1* are directly regulated by a complex of FIT/basic helix-loop-helix (bHLH) proteins.

Graminaceous plants have evolved a unique mechanism to acquire Fe through the secretion of molecules, the mugineic acid family of phytosiderophores (Bashir et al., 2006; Kobayashi et al., 2005). L-methionine is a precursor for these phytosiderophores (Bashir et al., 2006; Mori and Nishizawa, 1987). Bashir et al. (2006) isolated genes for deoxymugineic acid synthase (*DMAS*) from rice (*OsDMAS1*), barley (*HvDMAS1*), wheat (*TaDMAS1*), and corn (*ZmDMAS1*). The expression of each of these DMAS genes was upregulated under Fe-deficient conditions in root tissue. All grasses do not have the same capacities to secrete phytosiderophores. For example, young rice plants excrete a very low amount of deoxymugineic acid even under Fe deficiency, and the secretion ceased within 10 days under Fe deficiency, whereas barley secreted mugineic acid during a period of more than 1 month (Mori et al., 1991). Murata et al. (2006) identified a gene (*HvYS1*) encoding an Fe–phytosiderophore transporter in barley, reportedly the most tolerant species to Fe deficiency among graminaceous plants. This *HvYS1* was suggested by Murata et al. (2006) to encode a polypeptide having a high identity with the yellow-stripe transporter encoded by the yellow-stripe gene *ZmYS1* of corn (Roberts et al., 2004; Schaaf et al., 2004; von Wirén et al., 1994).

Nicotianamine, a ubiquitous molecule in plants, is an important metal ion chelator and also a key polyamine for phytosiderophore biosynthesis in Poaceae (Curie et al., 2001; Higuchi et al., 1996). Recent research has indicated a role for nicotianamine in metal homeostasis, through the chelation and transport of heavy metal complexes, including Fe (Gendre et al., 2007). The function of transport of nicotianamine–metal chelates was assigned to the Yellow Stripe 1–like family of proteins. A nicotianamine-deficient mutant of tomato (*chloronerva*) phenotypically resembled an Fe-deficient plant despite the high accumulation of Fe in the leaves, and the plant also suffered copper deficiency (Herbik et al., 1996). Research by Ling et al. (1999) proposed that the *chloronerva* gene encodes the enzyme nicotianamine synthase. During symplasmic transport, the intracellular environment is protected against the reactive species of Fe that may accumulate with Fe in excess of plant metabolic needs by holding of Fe in chelated forms. A chelating compound for this purpose may be plant-endogenous nicotianamine (Stephan, 2002). Nicotianamine is a chelator for Fe^{2+} and is essential for the proper function of Fe^{2+}-dependent processes, such as the regulation of Fe-deficiency response mechanisms and phloem loading/unloading of Fe (Scholz et al., 1992).

Dicotyledonous plants do not synthesize phytosiderophores but do synthesize nicotianamine, which is essential for maintenance of Fe homeostasis and copper translocation (Le Jean et al., 2005). Plants that do form phytosiderophores also synthesize nicotianamine, and Zhou et al. (2013) identified nine genes (*ZnNAS*) for nicotianamine synthase in corn and determined their expression patterns in different organs. They suggested that the two classes of *ZmNAS* genes may be regulated in response to demands for Fe uptake, translocation, and homeostasis.

11.5 INTERACTIONS OF IRON WITH OTHER ELEMENTS

An adequate supply of potassium is critical to proper function of the Fe-stress response mechanisms of monocotyledonous and dicotyledonous plants. The diminished capacity to develop specific Fe-deficiency responses in the absence of potassium resulted in intense chlorosis and low leaf Fe concentrations for Fe-deficiency stressed muskmelon (*Cucumis melo* L.), soybean (*Glycine max* Merr.), tomato, and oat (*Avena byzantina* C. Koch.) (Hughes et al., 1992). Judicious application of nitrogen, phosphorus, and potassium fertilization was recommended for the prevention of Fe-induced malnutrition in rice (Panda et al., 2012). With three rice cultivars, nitrogen, phosphorus, and potassium applications improved plant biomass and grain. Among the nutrients, nitrogen was most effective in increasing leaf Fe concentration. However, high doses of these nutrients adversely affected grain yield and Fe content of leaf and grain. Raising nitrogen supply substantially enhanced shoot and grain concentrations of Fe and zinc of durum wheat (*Triticum durum* Desf.) (Aciksoz, 2011). Improving nitrogen status of plants from low to sufficient resulted in a threefold increase in shoot Fe accumulation, whereas the increase was only 42% for total shoot dry weight. Kutman et al. (2011a) reported that Fe and zinc uptakes in durum wheat were enhanced by a high supply of nitrogen compared to a low supply of nitrogen. Distribution of Fe and zinc into grain was enhanced also by the high supply of nitrogen. With corn, nitrogen fertilization increased Fe in the ear leaf (Losak et al., 2011). With pot experiments with rice, optimum application of nitrogen alone on rice crops increased the concentration of Fe in the polished grain (Zhang et al., 2008). In greenhouse experiments, depending on foliar zinc supply, high nitrogen supply elevated the Fe concentration of durum wheat endosperm up to 100% and also increased the Fe in the total grain (Kutman et al., 2011b). Hence, nitrogen fertilization may be a factor to enrich rice or wheat grain for human consumption. On the other hand, dry matter mass of corn plants increased with increasing ammonium nitrate and zinc fertilization in a greenhouse experiment, and Fe, copper, and manganese concentrations decreased likely from a dilution effect of the enhanced growth (Adiloglu, 2007). Celik and Katkat (2010) reported that increasing potassium and Fe concentrations in nutrient solutions increased dry weight of corn shoots and roots and increased total Fe concentrations. Increasing concentrations of potassium and Fe decreased manganese, zinc, and copper concentrations in the shoots and roots.

Form of nitrogen has an effect on Fe accumulation in plants. Ammonium absorption increases the acidity of media and increases the solubility of Fe. Leaves of petunia (*Petunia* spp. L.) grown with ammonium nutrition showed an increase in green color, reportedly due to an increase in Fe concentration in leaves (Ramos et al., 2013). In comparison to ammonium-fed plants, young leaves of nitrate-fed corn showed severe chlorosis (Zou et al., 2001). Nitrate supply caused Fe accumulation in roots, whereas ammonium supply gave a high Fe concentration in young leaves and a low Fe concentration in roots. Seeds of sorghum (*Sorghum bicolor* Moench) dressed with variable additions of FeEDDHA showed that at all Fe additions, ammonium nutrition approximately doubled the Fe, zinc, copper, or manganese concentrations in shoots compared with nitrate nutrition (Alhendawi et al., 2008). With nitrate nutrition, increasing rates of Fe dressing had little effect on Fe concentrations of shoots, which were about 40 mg/kg and below the critical range for sufficiency. Differences in solution acidity apparently affected the availability of Fe in this study. At the end of the growth period, the pH of the sand around the roots had fallen by about 0.5 unit with ammonium-fed plants and increased by about 0.2 unit with nitrate-fed plants. In an experiment with arabidopsis, leaf growth and chlorophyll content were highest in nitrate nutrition with sufficient Fe but were diminished strongly by Fe deficiency under nitrate nutrition or ammonium nutrition (Karray-Bouraoui et al., 2010). However, with sufficient Fe, the leaves of ammonium-fed plants had a higher Fe concentration than leaves of Fe-sufficient, nitrate-fed plants. The medium was acidified under the ammonium regime and alkalinized under the nitrate regime regardless of Fe supply. Stimulation of FCR activity in response to Fe deficiency occurred only with nitrate-fed plants. The ammonium regime increased peroxidase expression and anthocyanin accumulation, whereas Fe deficiency enhanced superoxide dismutase expression. Waters et al. (2012) reported that Fe deficiency led to upregulation of Cu/Zn superoxide dismutase genes and to a downregulation of Fe superoxide genes. With grapes susceptible (*Vitis riparia* Michx.) or resistant (*Vitis vinifera* L.) to lime-induced chlorosis and cultured in Fe-free solutions with nitrate as the sole nitrogen source, typical Fe-deficiency response reactions as acidification of the growth medium and enhanced FCR activity in the roots were observed only in the tolerant genotype. With ammonium nutrition, the sensitive genotype displayed some decrease in pH of the growth medium and an increase in FCR activity (Jimenez et al., 2007).

Ladouceur et al. (2006) reported that growth, chlorophyll content, and Fe accumulation of barley cultured under low phosphorus conditions were higher than in plants receiving high phosphorus supplies. These results indicated that low phosphorus depressed phytosiderophore release from and accumulation in barley roots, an action possibly resulting from the higher Fe content in shoots and the alleviation of chlorosis with low phosphorus treatment of plants.

An initial effect of short-term (2 days) sulfur deprivation of young Fe-inefficient corn plants caused low Fe concentration in bottom leaves (Bouranis et al., 2003). With prolonged starvation (6 days), plants showed a complex constraint consisting of sulfur depletion, Fe deficiency, and induced nitrogen deficiency. In hydroponic culture, high sulfur (0.0025 M) supply increased the concentration of Fe in the shoots of durum wheat (Zuchi et al., 2012) over two lower concentrations of sulfur. The effect of sulfur nutrition on Fe accumulation was explained by an increased production of phytosiderophores and nicotianamine possibly due to increased methionine synthesis induced by the abundance of sulfur. Siderophore release rate increased with increasing sulfur supply in wheat plants under Fe-limiting conditions. In sulfur-sufficient tomato, Fe deficiency caused an increase in FCR activity, Fe uptake rate, and ethylene production in roots, but when sulfur-deficient plants were transferred to an Fe-free solution, no induction of FCR activity or ethylene production occurred (Zuchi et al., 2009).

Metallic micronutrients may interact with Fe to suppress Fe accumulation in crops. With durum wheat, copper and Fe concentrations in shoots varied inversely, suggesting that an antagonism between copper and Fe could lead to Fe deficiency (Michaud et al., 2008). If either Zn^{2+} or Mn^{2+} were added to culture media containing FeEDDHA, Fe concentrations in roots and shoots of bitter orange (*Citrus aurantium* L.) were reduced with respect to the seedlings cultured in the same medium lacking Zn^{2+} or Mn^{2+} (Martinez-Cuenca et al., 2013). Also, the presence of Fe chelate in the medium lowered zinc and manganese levels in the plant. Either Zn^{2+} or Mn^{2+} stimulated the activities of genes (*FRO2, IRT1*), which also were induced by Fe deficiency, but these effects appeared to be independent of the inhibition of Fe uptake by these cations. With tomato, zinc excess also imparted Fe deficiency in leaves and transcriptional activation of Fe uptake systems in roots (Barabasz et al., 2012). With pot experiments with peanut, Fe fertilization induced symptoms suggestive of manganese deficiency, a diagnosis that was supported by decreasing concentration of manganese in leaf blades with increasing Fe levels (Ali, 1998). The concentration of copper in tomato roots was enhanced under Fe deficiency in acidic soils, but copper in shoots did not vary with Fe status of the plant (Cornu et al., 2007). Manganese concentrations in shoots and roots of excessive-Mn-stressed barley were reduced significantly with 100 μM Fe but remained above critical toxicity levels, suggesting that elevation of Fe could help to ameliorate manganese toxicity (Alam et al., 2001).

With soybean, the addition of 0.5 mM soluble silicon to the nutrient solution without Fe prevented chlorophyll degradation, abated the growth suppression due to the Fe deficiency, and maintained the Fe content in leaves (Gonzalo et al., 2013). However, with cucumber, silicon addition delayed suppression of growth and suppression of Fe content in stems and roots, but no effect was observed in alleviation of leaf chlorosis. Work with cucumber suggested a beneficial role of silicon in plant nutrition, indicating that silicon-mediated alleviation of Fe deficiency includes an increase of the apoplastic Fe in roots and an enhancement of Fe acquisition (Pavlovic et al., 2013). For two genotypes of soybean, an antagonism occurred between the Fe and manganese concentrations in leaves, but no relationship occurred between Fe and zinc, phosphorus, and copper concentrations (Izaguirre-Mayoral and Sinclair, 2005).

11.6   CONCENTRATIONS OF IRON IN PLANTS

Plants and animals cannot acquire Fe easily from soils although it is abundant in nature. Thus, Fe deficiency is a major limiting factor affecting crop yields, food quality, and human nutrition. Therefore, approaches need to be developed to increase Fe uptake by roots, accumulation in edible plant portions, and availability to humans from plant food sources (Blair et al., 2010; Geurinot, 2010; Imtiaz et al., 2010; von Wirén, 2004; Zuo and Zhang, 2011). Recent research progress in soil and crop management has provided means to address plant nutritional problems with Fe through fertilization and water regulation and managing of cropping systems and screening for Fe-efficient species and varieties. Changes in production of rice in flooded conditions to aerobic conditions have required modifications in strategies in crop production (Fan et al., 2012; Zuo and Zhang, 2011). Intercropping strategies of Fe-efficient and Fe-deficient crops help to alleviate Fe deficiency in Fe-inefficient plants (Zuo and Zhang, 2011).

Major factors affecting Fe accumulation in plants are soil acidity and oxidation–reduction status of Fe in soils. Accumulation of Fe in several plant species increases with an increase in soil acidity (Fageria and Zimmerman, 1998). An interaction of high concentrations of Fe in acid soils during waterlogging led to elevated Fe concentrations that approached toxic levels in plants (Becker and Asch, 2005). Iron concentrations in leaves of waterlogged rice greatly exceeded the critical concentration of 300 mg/kg dry weight (Sahrawat, 2000). Iron toxicity is a widely distributed nutritional disorder in flooded and irrigated rice and is derived from ferrous ions generated by the reduction of Fe oxides (Schmidt et al., 2013). Iron toxicity in rice was ameliorated by drainage of paddies to permit reoxidation of the Fe during vegetative growth. Yaduvanshi et al. (2012) noted, however, that reaeration, which resulted in an increase in oxidation–reduction potential and a decrease in diethylenetriaminepentaacetic acid (DTPA)-extractable (see Section 11.10) Fe in soil solutions, occurred slowly, taking 15–25 days. In waterlogged acidic soil, shoot concentrations of Fe increased, and in some wheat (*Triticum aestivum* L.) varieties, they were above critical concentrations; however, Fe decreased or remained the same in shoots of plants grown in waterlogged neutral soil concentrations compared with plants in drained soil (Khabaz-Saberi et al., 2006). Research by Fu et al. (2012) suggested that silicon could detoxify Fe^{2+} and ameliorate Fe toxicity in rice roots, likely by restricting Fe uptake by roots and translocation to shoots. Khabaz-Saberi et al. (2012) noted that wheat genotypes differed in tolerance to Fe toxicities imparted by waterlogging.

With increasing water shortages, rice production cultivation in some regions is shifting from continuously flooded conditions to partly or even completely aerobic conditions in upland production. Results of Fan et al. (2012) suggest that a shift from flooded to aerobic cultivation will increase Fe deficiency in rice and will increase the problem of Fe deficiency in humans who depend on rice for nutrition.

Due to the alkalinity that is imparted by bicarbonate, lime-induced chlorosis occurs in soils with high bicarbonate concentrations (Covarrubias and Rombola, 2013; Nikolic and Kastori, 2000; Römheld and Nikolic, 2007). Poor growth of white lupin (*Lupinus albus* L.) in alkaline soils may result from its sensitivity to Fe deficiency, resulting in poor nodulation (Tang et al., 2006). Likewise, nitrate nutrition can lead to alkaline conditions as nitrate absorption exceeds cation absorption from a soil or nutrient solution or in leaf apoplast (Bar and Kafkafi, 1992; Kosegarten and Englisch, 1994; Lucena, 2000; Mengel, 1994; Smolders, 1997). The resulting chlorosis may be related to insufficient Fe accumulation in leaves or to immobilization of Fe into physiologically inactive forms (Zohlen and Tyler, 1997). Temporary flooding may be used to increase Fe availability in calcareous soils (Sanchez-Alcala, 2011).

Grasses have been suggested to solubilize sparingly soluble Fe sources by release of phyto-siderophores (Cesco et al., 2006). Chlorosis-susceptible citrus trees growing on calcareous soils recovered in the presence of grass cover species, which may improve the Fe nutrition of the trees by enhancing Fe availability (Cesco et al., 2006). Likewise, visual observations indicated that guava (*Psidium guajava* L.) seedlings had fewer symptoms of Fe-deficiency chlorosis and better growth when grown with corn or sorghum (Kamal et al., 2000). Also, extractability of Fe and zinc in calcareous soils and the concentrations of these two micronutrients in guava leaves were increased. With a peanut–corn intercropping system, corn improved the Fe nutrition of peanut, an effect that was suggested as being due to phytosiderophores being released from Fe-deficient corn (Zhang et al., 2004). Soil-applied aqueous extracts of pigweed (*Amaranthus retroflexus* L.) shoots improved Fe nutritional status of pear trees (*Pyrus communis* L.), likely due to the Fe-chelating capacity of compounds released from the extract (Sorrenti et al., 2011). Plant breeding and selection for genotypes that are Fe efficient in calcareous soils may help to solve the problem of lime-induced chlorosis (Brancadoro et al., 2001; Covarrubias and Rombola, 2013; Decianzio, 1991; Peiffer et al., 2012). A bicarbonate-resistant line of tobacco (*Nicotiana tabacum* L.) was able to adapt to conditions that restricted Fe uptake, such as high bicarbonate concentration, high pH, and low Fe conditions (Kiriiwa et al., 2004).

Sufficiency of Fe concentrations in plant leaves varies rather narrowly from about 30 to over 300 mg/kg (Figure 11.1). The critical Fe level in taro (*Colocasia esculenta* L. Schott) leaf blades was estimated to be between 55 and 70 mg/kg dry weight (Ares et al., 1996). Alhendawi et al. (2008) reported that for sorghum, the critical concentration was about 40 mg/kg. Sahrawat (2000) reported that Fe concentrations in rice grown in waterlogged soils exceeded 300 mg/kg and approached toxic concentrations. Burns et al. (2012) reported Fe concentrations ranging from about 140 to 680 mg/kg in cassava leaves and from 8 to 24 mg/kg in roots. These concentrations seem to be in the magnitude of ranges of concentrations reported by Bryson et al. (2014) for a diversity of agronomic, horticultural, and forest species (Table 11.3).

The concentrations of Fe in seeds and fruits are lower than in leaves. In seeds, concentrations range from about 50 to just over 100 mg Fe/kg (Table 11.4). Concentrations in fruit flesh vary from

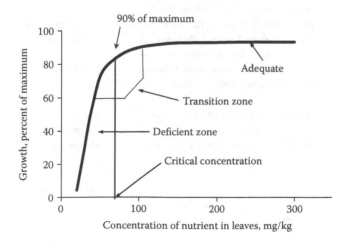

FIGURE 11.1   Model of growth of plants in response to concentrations of iron in leaves.

TABLE 11.3

Concentrations of Iron in Leaves or Shoots of Various Agricultural Crops

Crop Class	Selected Genera[a]	Plant Part[b]	Common Range,[c] mg/kg Dry Weight
Agronomic			
Grains			
Legumes	*Arachis, Glycine, Phaseolus*	Mature leaves	25–300
Grasses	*Avena, Oryza, Hordeum, Secale, Sorghum*	Mature leaves	20–200
Broadleaf, nonlegume	*Beta, Brassica, Gossypium, Helianthus, Nicotiana*	Mature leaves	30–300
Hay and forage crops	*Bromus, Cynodon, Dactylon, Lotus, Medicago, Phleum, Setaria, Sorghum, Trifolium*	Mature leaves or whole tops	30–350
Bedding plants	*Calibrachoa, Celosia, Cleome, Dianthus, Impatiens, Pelargonium, Primula, Salvia, Tagetes, Verbena, Viola, Zinnia*	Mature leaves	50–400
Forest and landscape trees			
Softwoods	*Abies, Cedrus, Ginkgo, Juniperus, Larix, Picea, Pinus, Taxodium, Taxus, Thuja, Tsuga*	Terminal cuttings or mature leaves	20–300
Hardwoods	*Acer, Aesculus, Alnus, Betula, Carpinus, Carya, Cercis, Cornus, Eucalyptus, Fagus, Fraxinus, Lagerstroemia, Liquidambar, Liriodendron, Magnolia, Populus, Prunus, Quercus, Salix, Ulmus*	Mature leaves	30–300
Plantation crops	*Camellia, Coffea, Manihot, Theobroma*	Mature leaves	
Ferns and related plants	*Adiantum, Athyrium, Dennstaedtia, Equisetum, Nephrolepis Polystichum, Osmunda, Rumohra*	Mature fronds	30–300
Fruit and nuts	*Anacardium, Ananas, Carya, Carica, Citrus, Corylus, Diospyros, Fragaria, Juglans, Malus, Musa, Olea, Persea, Pistacia, Prunus, Rubus, Vaccinium, Vitis*	Mature leaves	35–400
Herbaceous perennials	More than 100 species	Mature leaves	35–200
Herbs	*Allium, Anethum, Mentha, Origanum, Petroselinum, Rosmarinus, Thymus*	Mature leaves, terminal cuttings, whole tops	60–250
Ornamental monocots	*Acorus, Carex, Cyperus, Miscanthus, Pennisetum, Phalaris, Phyllostachys*	Mature leaves	35–150
Ornamental vines and groundcovers	*Ajuga, Campsis, Clematis, Euonymus, Ficus, Hedera, Liriope, Lonicera, Pachysandra, Parthenocissus, Trachelospermum, Vinca*	Mature leaves and terminal cuttings	40–400
Turfgrass	*Agrostis, Cynodon, Eremochloa, Festuca, Lolium, Poa, Stenotaphrum, Zoysia*	Clippings	50–500
Vegetables			
Leafy	*Amaranthus, Apium, Brassica, Lactuca, Rheum, Spinacia*	Mature leaves or wrapper leaves	50–500
Fruit	*Abelmoschus, Capsicum, Citrullus, Cucurbita, Lycopersicon, Solanum, Zea*	Mature leaves	50–350
Bulb	*Allium*	Whole tops or mature leaves	60–300
Root	*Brassica, Daucus, Ipomoea, Raphanus*	Mature leaves	50–350
Tuber	*Solanum*	Mature leaves	50–150

(Continued)

TABLE 11.3 (*Continued*)

Concentrations of Iron in Leaves or Shoots of Various Agricultural Crops

Crop Class	Selected Genera[a]	Plant Part[b]	Common Range,[c] mg/kg Dry Weight
Stems	*Asparagus, Brassica*	Mature leaves	50–300
Woody ornamentals	*Abelia, Aesculus, Buddleja, Buxus, Camellia, Cornus, Cotoneaster, Forsythia, Gardenia, Ilex, Kalmia, Leucothoe, Mahonia, Pieris, Rhododendron, Rhus, Spirea, Viburnum*	Mature leaves from new growth	35–200

[a] Genera of crops that are commonly known or commonly tested for Fe accumulation from Bryson et al. (2014).
[b] Parts tested for Fe accumulation.
[c] Sufficiency range or survey range from Bryson et al. (2014).

TABLE 11.4

Iron Concentration in Some Seeds That Are Used as Foods

Seed	Iron Concentration, mg/kg Dry wt	Reference
Amaranth (*Amaranthus caudatus* L.)	96	Nascimento et al. (2014)
Bambara groundnut (*Vigna subterranea* Verdc.)	15–48	Ijarotimi and Esho (2009)
Bean (*Phaseolus vulgaris* L.)	60–80	Meyer et al. (2013)
Chickpea (*Cicer arietinum* L.)	46–67	Thavarajah and Thavarajah (2012)
	77–112	Bueckert et al. (2011)
Corn (*Zea mays* L.)	20–45	Sokrab et al. (2012)
	55	Nascimento et al. (2014)
Field peas (*Pisum sativum* L.)	46–54	Amarakoon et al. (2012)
Quinoa (*Chenopodium quinoa* Willd.)	55	Nascimento et al. (2014)
Pearl millet (*Pennisetum glaucum* R. Br.)	>80	Rai et al. (2012)
Rice (*Oryza sativa* L.)	5.4–6.5	Wei et al. (2012)
	9–45	Liang et al. (2007)
	2.2	Nascimento et al. (2014)
Wheat (*Triticum aestivum* L.)	40 (flour)	Sharma et al. (2012)

about 10 to 50 mg/kg. For example, in common bean, concentrations in seeds ranged from about 60 to 80 mg/kg with only about 3% or 4% of that value being bioavailable (Meyer et al., 2013). Iron concentration in polished rice grains was about 5.4 mg/kg in an unfertilized crop but was increased to a maximum of 6.5 mg/kg with foliar fertilization with iron sulfate or iron chelate (Wei et al., 2012). Fertilization with Fe increased the bioavailability of Fe as expressed by ferritin formation by 34%–67% compared with unfertilized rice. Other research showed that Fe in rice varied from 9 to 45 mg/kg (Liang et al., 2007). Soaking and cooking of grains or seeds often increase the bioavailability of Fe by lowering the phytate and phenolic contents (ElMaki et al., 2007; Sokrab et al., 2012).

Fe in seeds may have poor bioavailability for human nutrition. The phytate content of seeds forms compounds with Fe and makes the Fe poorly digestible and is a major barrier to success-ful iron biofortification of foods (Ariza-Nieto et al., 2007; Bodnar et al., 2013; Doria et al., 2012; Ma et al., 2005; Meyer et al., 2013). The bioavailability to some extent depends on food processing

TABLE 11.5
Iron Concentration in Fruits That Are Used as Foods

Fruit	Iron Concentration, mg/kg, Dry wt.
Orange (*Citrus sinensis* Osbeck)	6
Papaya (*Carica papaya* L.)	6
Mango (*Mangifera indica* L.)	6
Kiwifruit (*Actinidia deliciosa* C.F. Liang et A.R. Ferguson)	8
Peach (*Prunus persica* L.)	8
Apple (*Malus domestica* Borkh.)	9
Plum (*Prunus domestica* L.)	10
Banana (*Musa paradisiaca* var. *sapientum* Kuntze)	14
Cantaloupe melon (*Cucumis melo* var. *cantalupensis* Naud.)	14
Grape (*Vitis vinifera* L.)	22
Pineapple (*Ananas comosus* L.)	30
Pear (*Pyrus communis* L.)	15
Grapefruit (*Citrus × paradisi* Macf.)	23
Avocado (*Persea americana* Mill.)	40
Raisins (*Vitis vinifera* L.)	44
Raspberry (*Rubus occidentalis* L.)	44
Watermelon (*Citrullus lanatus* Matsum. and Nakai)	60
Strawberry (*Fragaria × ananassa* Duchesne)	69

and cooking methods and crop genotype (Aluru et al., 2011; Borg et al., 2012; Eagling et al., 2014; Luo et al., 2012). Milling generally has no effect on phytate concentrations (Kayode et al., 2007).

Concentration of iron in the flesh of fruits varies with species (Table 11.5), ranging among some selected fruits from under 10 mg/kg in apple (*Malus domestica* Borkh.), orange (*Citrus sinensis* Osbeck), kiwifruit, mango (*Mangifera indica* L.), and peach (*Prunus persica* L.) to 60 or higher in strawberry (*Fragaria × ananassa* Duchesne) and watermelon (*Citrullus lanatus* Matsum. and Nakai) (Gebhardt and Thomas, 2002).

11.7 DIAGNOSIS OF IRON STATUS IN PLANTS

Iron is only slightly mobile in plants, and deficiency symptoms appear first on the youngest leaves. Tiffin (1966, 1970) concluded that citrate carries Fe in the xylem of intact plants. The stability of the Fe–citrate complex might limit the transport of Fe in the xylem, hence leading to poor movement of Fe into young leaves. Photodestruction of citrate, catalyzed by Fe, results in increase of pH in the solution and in the formation of a nondialyzable form of Fe and thus can lead to deposition of so-called inactive Fe in leaves (Bienfait and Scheffers, 1992). The nonproteinogenic amino acid nicotianamine could control Fe transport via phloem including mobilization or recirculation (Stephan et al., 1996; Nishiyama et al., 2012).

Plants need Fe to produce chlorophyll (Broadley et al., 2012; Marsh et al., 1963). Lack of Fe results in chlorosis of young leaves of herbaceous or woody dicotyledonous plants (Figures 11.2 through 11.4). With grasses, interveinal chlorosis shown by striping of young leaves is common (Figure 11.5). Because of the appearance of symptoms on young leaves, mild Fe-deficiency symptoms especially on broad-leafed species may be confused with symptoms of manganese or zinc deficiency (Wallihan, 1965). Tissue testing is used sometimes to confirm deficiencies (Bryson et al., 2014; Garcia-Escudero et al., 2013; Sumner, 1979).

With prolonged deficiency, Fe can be mobilized to young leaves, and some recovery of health often is noted. But, eventually, the Fe supply of the plant is exhausted, and the plant expresses symptoms on all of the foliage, and the plant dies.

FIGURE 11.2 **(See color insert.)** Iron-deficient marigold (*Tagetes patula* L.) showing chlorosis of young leaves.

FIGURE 11.3 **(See color insert.)** Iron-deficient borage (*Borago officinalis* L.) showing chlorosis of young leaves.

11.8 FORMS AND CONCENTRATIONS OF IRON IN SOILS

Total Fe in soil ranges from 7,000 to 500,000 mg/kg (Römheld and Nikolic, 2007) or commonly about 3%–5% of the soil weight (Bryson et al., 2014; Loeppert and Inskeep, 1996). Several primary minerals such as ferromagnesian silicates may be present (Bryson et al., 2014; Loeppert and Inskeep, 1996). Iron oxides are common in soils with hematite (Fe_2O_3) and goethite (FeOOH) being principal oxide minerals (Bohn et al., 1979; Loeppert and Inskeep, 1996). The oxides occur as coatings on particles of soil thereby imparting characteristic red colors to soils (Rossel et al., 2010). With aging,

FIGURE 11.4   **(See color insert.)** Iron-deficient ixora (*Ixora chinensis* Lam.).

FIGURE 11.5   **(See color insert.)** Iron-deficient corn (*Zea mays* L.) showing interveinal chlorosis of leaves.

the amorphous coatings become crystalline. The composition and crystalline nature of Fe oxides determines the reactivity through oxidation and reduction and solubility reactions. In laterite (soil rich in iron and aluminum oxides and with a name derived from Latin for brick), the crystallization upon drying forms stonelike structures that are hardened irreversibly. These soils have a number of physical and chemical problems, including low permeability, sodicity, and structural decline and hardsetting in the A-horizon, and require special management strategies to address these problems (Belford et al., 1992; Cochrane and Aylmore, 1991; Tennant et al., 1992).

Plant-available Fe depends on factors such as form of Fe in the soil and the ability of plant roots to solubilize Fe compounds. Dabkowska-Naskret (2000) noted that most of the Fe in soil was in Fe oxide (over 12% of the total) and fixed in silicate structures (more than 80% of the total) with substantial amounts of Fe being associated with soil organic matter. Iron associated with manganese oxides and acid-soluble Fe compounds were minor components. In six soil series in Florida representing spodosols, alfisols, and entisols, in soils with pH ≤ 6.5, more than 60% of the total Fe was in amorphous and crystalline Fe oxides, and with pH > 6.5, 80% of the total Fe was in this fraction (Zhang et al., 1997). Mathematical modeling showed that siderophores (mugineic acid) mobilize Fe from goethite at a rate that was too low to meet the demands of plants (Gerke, 2000). With ferrihydrite ($Fe_2O_3 \cdot 0.5H_2O$) as the principal form of Fe in soil, the mobilized Fe met the demand of the plants. With humic–Fe complexes, the influx of chemically mobilized Fe was much higher than with inorganic forms. Total sesquioxides (principally Fe_2O_3 and Al_2O_3) of the soil have a high correlation with phosphorus-fixing capacity of soils (Metzger, 1941). Ruiz et al. (1997) suggested that fertilizer phosphorus may be occluded in Fe oxides, particularly in the poor-crystalline forms.

Some Fe in soil is exchangeable into the soil solution from inorganic sites of soil colloids, oxide surfaces, and organic matter (Sharma et al., 2008). Sharma et al. (2008) noted that these forms were correlated with DTPA-extractable (see Section 11.10) Fe but not with total Fe. Though some forms are interrelated, none of the forms had any relationship with the total Fe.

11.9   SOIL TESTING

In measuring Fe in soil, determinations of total Fe, Fe oxides, or extractable Fe are common. Interests in total Fe relate to need for knowledge of the amount of Fe in soils or other silicate materials. For total Fe, decomposition of a soil sample is by sodium carbonate fusion or dissolution with hydrofluoric acid (Loeppert and Inskeep, 1996; Olson, 1965). The resultant residue is dissolved in an acid such as $3M\ H_2SO_4$ or $6M\ HCl$ for analysis. Interest in determining of Fe oxides occurs because of their importance in soil-forming processes and in fixation of phosphates. Oxides may exist as coatings on soil minerals, cement between soil particles, or discrete particles. Iron in oxides is reduced chemically by zinc powder and ammonium tartrate or by sodium dithionite (NaS_2O_4) for measurement (Jackson et al., 1986; Olson, 1965). This procedure is used also to strip Fe oxides from particles for mineralogical studies. The amount of Fe extracted from soils by solvents varies widely. Solvents remove water-soluble and exchangeable Fe. Neutral 1 M ammonium acetate extracts very little or no Fe from aerated, neutral, or alkaline soils but may remove considerable Fe from waterlogged or acidic soils (Olson, 1965). Universal extracting solutions such as Mehlich 3 contain acetic acid, nitric acid, ammonium nitrate, ammonium fluoride, and EDTA to provide a solvent in which the EDTA chelates micronutrients including Fe (Mehlich, 1984). Morgan Solution or Modified Morgan also is a universal extraction solution formulated from sodium acetate or ammonium acetate and glacial acetic acid (Morgan, 1941; Wolf and Beegle, 2009). Chelating agents such as DTPA are used specifically to extract metallic micronutrients (Lindsay and Norvell, 1978; Samourgiannidis and Masti, 2013).

Iron from decomposition or extraction of soil can be determined gravimetrically, volumetrically, colorimetrically, or spectrophotometrically (Olson, 1965; Soil and Plant Analysis Council, 1992).

Comparisons of universal extracting solutions with specific chelate-based solutions for Fe and other micronutrient metals have been studied extensively with different soils and cropping systems. Research with Mehlich 3 solution has been more extensive than with Morgan Solution. Wendt (1996) highly recommended the Mehlich 3 universal extracting method because this procedure was well correlated with the ammonium acetate, DPTA, and Bray 1 methods and because it was a rapid procedure for testing Malawi upland soils. Mehlich 3 extraction correlated well to exchangeable and soluble Fe but not to noncrystalline Fe (Marcos et al., 1998). A citrate–ascorbate extraction method was suggested for routine use in estimating poor-crystalline Fe oxides in soils (Reyes and Torrent, 1997). Garcia et al. (1997) evaluated Mehlich 3, DTPA-triethanolamine (TEA), and other extracting solutions for determinations of copper, zinc, manganese, and Fe. Mehlich 3 yielded the highest extractions for the four micronutrients, and correlations of the different tests for the metals were significant. The investigators suggested that for the determination of the bioavailable status of micronutrients, any of the studied tests could be applied with considerations of soil edaphic properties as factors to improve the correlations between them and to standardize the methods. Mehlich 3 or ammonium bicarbonate-diethylenetriaminepentaacetic acid (ABDTPA) was an appropriate extractant for fifteen elements including Fe in assessments of these solutions against widely accepted and conventional soil tests of thirty acidic and twenty alkaline U.S. soils from twenty-one states (Elrashidi et al., 2003). The Mehlich 3 extractant was correlated poorly with DTPA-extractable Fe and Mn but was adequate for extraction of phosphorus, potassium, and zinc across concentration ranges normally found in one hundred agricultural soils tested in North Dakota (Schmisek et al., 1998). Zbiral and Nemec (2000) suggested that although significant correlations occurred between the use of Mehlich 3 and DPTA extraction methods, determination of Fe and manganese by Mehlich 3 extraction would need more precise calibration of the method based on field and pot experiments. But, for all other elements, conversion equations can be used instead of expensive and time-consuming field calibrating studies. Gartley et al. (2002) tested several extracting solutions including Mehlich 3 and developed regression equations that allowed for conversions of results with one solution with another. Ostatek-Boczynski and Lee-Steere (2012) tested the Mehlich 3 procedure against a sulfuric acid–based extraction method and concluded that Mehlich 3 is a suitable extractant to assess the basic nutrient status of Australian sugarcane soils.

The consensus from the evaluations of universal solutions and specific chelating solutions for extracting Fe from soils was that the universal methods are adequate but that for good correlations of Fe extraction and Fe availability, solutions such as DTPA-based solvents are best.

Simultaneous extraction of nutrients, including Fe, with ABDTPA extractant was successful for lowland soils for rice production in that highly significant correlations occurred between plant uptake and extractable nutrients (Madurapperuma and Kumaragamage, 2008). Extractions with DPTA solutions formulated with other extracting agents such as ammonium bicarbonate, calcium carbonate, or triethanolamine were suggested for the determination of the available Fe of brown, forest soils in Turkey (Adiloglu, 2006). In the United States, ABDPTA is used widely for the determinations of Fe in soil tests (Rodriguez et al., 1999; Self, 1999).

The use of the chelating agent EDTA was among the most suitable extraction methods to indicate the available Fe status of calcareous soils (Celik and Katkat, 2010). Basar (2009) compared fourteen extraction methods, including Morgan's extraction solution, and recommended 0.01 M Na_2EDDHA for determination of plant-available Fe for peaches grown on alkaline alluvial soils. In other evaluations, the concentration of Fe in hydroxylamine extracts was related to poor-crystalline Fe oxides in the soil and was suggested as being useful to predict the incidence of Fe-deficiency chlorosis in several crops in Spain (de Santiago et al., 2008).

11.10   FERTILIZERS

Fertilization of crops with Fe in fields is difficult because of the rapid precipitation of Fe in soil. Therefore, supplying of Fe to deficient crops is largely by foliar applications. Iron shows a low mobility from leaves, but because of the relatively small amounts needed, foliar applications of Fe are effective in plant nutrition (Mengel, 2001). Ferrous sulfate is used commonly in a 2% solution (w/v) applied in several applications of 0.1–0.15 kg/ha in 80 L of water during a growing season (Bryson et al., 2014; Schulte, 2004). Other salts for foliar sprays are ferrous ammonium sulfate and various Fe chelates such as FeEDTA, FeDPTA, and FeEDDHA. These materials are applied similarly to use of Fe sulfate. The chelates can be applied directly to soils at 0.5–2 kg Fe/ha. FeEDTA is suitable for use in acid soils, whereas FeDTPA or FeEDDHA are better suited than FeEDTA for alkaline soils (Bryson et al., 2014). Several different formulations of FeEDDHA are available for fertilization of soil (Schenkeveld et al., 2008). In studies with soybean, Nadal et al. (2012a) suggested the use of an Fe chelate, Fe-N,N'-bis(2-hydroxybenzyl)ethylenediamine-N,N'-diacetate (HBED), as a longer-lasting alternative for FeEDDHA for dicotyledonous crops in calcareous soils. Nadal et al. (2012b) reported that another reagent, Fe 2-{2-[(2-hydroxybenzyl)amino]ethylamino}-2-{2-hydroxyphenyl} acetic (DCHA), corrected Fe chlorosis of soybean and cucumber in hydroponics or soil conditions. In nutrient solutions and in soil cultures, humic substances such as commercial preparations of humic acid or fulvic acid or extracts of peat have been used successfully to supply chelated Fe to grasses or dicotyledonous plants (Bocanegra et al., 2004; Chen et al., 2004; Garcia-Mina et al., 2004; Pinton et al., 1999). The efficiency of humic substances complexed with Fe in preventing Fe chlorosis was similar to and may substitute partially and economically for FeEDDHA (Cerdan et al., 2007; de Santiago and Delgado, 2007).

Two chelates (FeEDTA and Fe ethylenediamine disuccinic acid [FeEDDS]), Fe lignosulfonate, or Fe sulfate was applied foliarly to mature grapevines (*Vitis vinifera* L.) grown on a soil with a high active lime content (Yunta et al., 2013). All treatments resulted in greater leaf Fe concentration than in the untreated leaves, but leaves sprayed with FeEDTA or Fe sulfate showed the greatest concentrations. Fernandez et al. (2006) evaluated eighty spray formulations of various Fe compounds ($FeSO_4$, Fe citrate, FeEDTA, FeDTPA, Fe iminodisuccinic acid [FeIDHA]) with various surfactants and adjuvants. Formulations, in decreasing order, of $FeSO_4$, Fe citrate, FeEDTA, FeIDHA, and FeDTPA gave best results. The effects of amendments of surfactants or adjuvants generally were positive but varied with each Fe-containing compound.

REFERENCES

Abdelmajid, K., B.H. Karim, and A. Chedly. 2008. Symbiotic response of common bean (*Phaseolus vulgaris* L.) to Fe deficiency. *Acta Physiol. Plant.* 30(1):27–34.

Aciksoz, S.B., A. Yazici, L. Ozturk, and I. Cakmak. 2011. Biofortification of wheat with Fe through soil and foliar application of nitrogen and Fe fertilizers. *Plant Soil* 349(102):215–225.

Adiloglu, S. 2007. The effect of increasing nitrogen and zinc doses on the Fe, copper and manganese contents of maize plant in calcareous and zinc deficient soils. *Agrochimica* 51(2–3):114–120.

Agnolon, F., S. Santi, Z. Varanini, and R. Pinton. 2002. Enzymatic responses of cucumber roots to different levels of Fe supply. *Plant Soil* 241(1):35–41.

Alam, S., S. Kamei, and S. Kawai. 2001. Amelioration of manganese toxicity in barley with Fe. *J. Plant Nutr.* 24(9):1421–1433.

Alhendawi, R.A., E.A. Kirkby, D.J. Pilbeam, and V. Römheld. 2008. Effect of Fe seed dressing and form of nitrogen-supply on growth and micronutrient concentration in shoots of sorghum grown in a calcareous sand culture. *J. Plant Nutr.* 31(10):1855–1865.

Ali, Z.I., E.M.A. Malik, H.M. Babiker et al. 1998. Fe and nitrogen interactions in groundnut nutrition. *Comm. Soil Sci. Plant Anal.* 29(17–18):2619–2630.

Aluru, M.R., S.R. Rodermel, and M.B. Reddy. 2011. Genetic modification of low phytic acid 1–1 maize to enhance iron content and bioavailability. *J. Agr. Food Chem.* 59(24):12954–12962.

Amarakoon, D., D. Thavarajah, K. McPhee, and P. Thavarajah. 2012. Iron-, zinc-, and magnesium-rich field peas (*Pisum sativum* L.) with naturally low phytic acid: A potential food-based solution to global micronutrient malnutrition. *J. Food Comp. Anal.* 27(1):8–13.

Aranto, M., F. Gevaudant, M. Oufattole, and M. Boutry. 2003. The plasma membrane proton pump ATPase: The significance of gene subfamilies. *Planta* 216(3):355–365.

Ares, A., S.G. Hwang, and S.C. Miyasaka. 1996. Taro response to different Fe levels in hydroponic solution. *J. Plant Nutr.* 19(2):281–292.

Ariza-Nieto, M., M.W. Blair, R.M. Welch, and R.P. Glahn. 2007. Screening of iron bioavailability patterns in eight bean (*Phaseolus vulgaris* L.) genotypes using the Caco-2 cell in vitro model. *J. Agr. Food Chem.* 55(19):7950–7956.

Babalakova, N., Z. Salama, S. Rocheva, and M. El-Fouly. 2011. Root redox-system activities and H_2O_2-metabolizing enzymes in cucumber cultivars grown in conditions of Fe deficit. *J. Food Agric. Environ.* 9(3–4):533–537.

Bar, Y. and U. Kafkafi. 1992. Nitrate-induced Fe-deficiency chlorosis in avocado (*Persea americana* Mill.) rootstocks and its prevention by chloride. *J. Plant Nutr.* 15(10):1739–1746.

Barabasz, A., A. Wilkowska, A. Ruszczyńska et al. 2012. Metal response of transgenic tomato plants expressing P1B-ATPase. *Physiol. Plant.* 145(2):315–331.

Bashir, K., H. Inoue, S. Nagasaka et al. 2006. Cloning and characterization of deoxymugineic acid synthase genes from graminaceous plants. *J. Biol. Chem.* 281(43):32395–32402.

Bauer, P., H.W. Ling, and M.L. Guerinot. 2007. FIT, the Fer-like Fe deficiency induced transcription factor in *Arabidopsis. Plant Physiol. Biochem.* 45(5):260–261.

Becker, M. and F. Asch. 2005. Fe toxicity in rice-conditions and management concepts. *J. Plant Nutr. Soil Sci.* 168(4):558–573.

Belford, R.K., M. Dracup, and D. Tennant. 1992. Limitations to growth and yield of cereal and lupin crops on duplex soils. *Austral. J. Exp. Agric.* 32(7):929–945.

Bienfait, H.F. and M.R. Scheffers. 1992. Some properties of ferric citrate relevant to the Fe nutrition of plants. *Plant Soil* 143(1):141–144.

Blair, M.W., S.J.B. Knewtson, C. Astudillo et al. 2010. Variation and inheritance of Fe reductase activity in the roots of common bean (*Phaseolus vulgaris* L.) and association with seed Fe accumulation QTL. *BMC Plant Biol.* 10:Article 215.

Bocanegra, M.P., J.C. Lobartini, and G.A. Orioli. 2004. Fe-humate as a source of Fe for plants. *Commun. Soil Sci. Plant Anal.* 35(17–18):2567–2576.

Bodnar, A., A.K. Proulx, M.P. Scott, A. Beavers, and M.B. Reddy. 2013. Iron bioavailability of maize hemoglobin in a Caco-2 cell culture model. *J. Agr. Food Chem.* 61(30):7349–7356.

Bohn, H.L., B.L. McNeal, and G.A. O'Connor. 1979. *Soil Chemistry.* New York: Wiley-Interscience.

Borg, S., H. Brinch-Pedersen, B. Tauris et al. 2012. Wheat ferritins: Improving the iron content of the wheat grain. *J. Cereal Sci.* 56(2):204–213.

Bouranis, D.L., S.N. Chorianopoulou, V.E. Protonotarios et al. 2003. Leaf responses of young Fe-inefficient maize plants to sulfur deprivation. *J. Plant Nutr.* 26(6):1189–1202.

Brancadoro, L., G. Tamai, G. Gocchi, and O. Failla. 2001. Adaptive responses of *Vitis* spp. and *Prunus* spp. to Fe-deficiency induced by HCO_3^-. In *Proceedings of the Fourth International Symposium on Mineral Nutrition of Deciduous Fruit Crops,* Penticton, British Columbia, Canada, eds. D. Neilsen, B. Fallahi, G. Neilsen, and F. Peryea. Acta Horticulturae Book Series 564, pp. 359–364

Briat, J.-F., I. Fobis-Loisy, N. Grignon et al. 1995. Cellular and molecular aspects of Fe metabolism in plants. *Biol. Cell* 84(1–2):69–81.

Broadley, M., P. Brown, I. Cakmak, Z. Rengel, and F. Zhao. 2012. Function of nutrients: Micronutrients. In *Mineral Nutrition of Higher Plants,* 3rd edn., ed. P. Marschner, pp. 191–248. London, U.K.: Academic Press.

Brumbarova, T. and P. Bauer. 2008. Iron uptake and transport in plants. In *Plant Membrane and Vacuolar Transporters,* eds. P.K. Jaiwal, R.P. Singh, and O.P. Dhankher, pp. 149–172. Devon, U.K.: CABI Publishing.

Bryson, G.M., H.A. Mills, D.N. Sasseville, J.B. Jones, Jr., and A.V. Barker. 2014. *Plant Analysis Handbook III.* Athens, GA: Micro-Macro Publishing.

Bueckert, R.A., D. Thavarajah, P. Thavarajah, and J. Pritchard. 2011. Phytic acid and mineral micronutrients in field-grown chickpea (*Cicer arietinum* L.) cultivars from western Canada. *Eur. Food Res. Technol.* 233(2):203–212.

Burns, A.E., R.M. Gleadow, A.M. Zacarias, C.E. Cuambe, R.E. Miller, and T.R. Cavagnaro (2012). Variations in the chemical composition of cassava (*Manihot esculenta* Cranz) leaves and roots as affected by genotypic and environmental variation. *J. Agr. Food Chem.* 60(19):4946–4956.

Celik, H. and A.V. Katkat. 2010. Comparison of various chemical extraction methods used for determination of the available Fe amounts of calcareous soils. *Commun. Soil Sci. Plant Anal.* 41(3):290–300.

Cerdan, M., A. Sanchez-Sanchez, M. Juarez et al. 2007. Partial replacement of Fe(o.o-EDDHA) by humic substances for Fe nutrition and fruit quality of citrus. *J. Plant Nutr. Soil Sci.* 170(4):474–478.

Cesco, S., A.D. Rombola, M. Tagliavini, Z. Varanini, and R. Pinton. 2006. Phytosiderophores released by graminaceous species promote Fe-59-uptake in citrus. *Plant Soil* 287(1–2):223–233.

Charlson, D.V. and R.C. Shoemaker. 2006. Evolution of iron acquisition in higher plants. *J. Plant Nutr.* 29(6):1109–1125.

Chen, Y., C.E. Clapp, and H. Magen. 2004. Mechanisms of plant growth stimulation by humic substances: The role of organo-iron complexes. *Soil Sci. Plant Nutr.* 50(7):1089–1095.

Chen, W.W., J.L. Yang, C. Qin et al. 2010. Nitric oxide acts downstream of auxin to trigger root ferric-chelate reductase activity in response to iron deficiency in *Arabidopsis. Plant Physiol.* 154(2):810–819.

Chouliaras, V., K. Dimassi, I. Therios, A. Molassiotis, and G. Diamantidis. 2004. Root-reducing capacity, rhizosphere acidification, peroxidase and catalase activities and nutrient levels of *Citrus taiwanica* and *C-volkameriana* seedlings, under Fe deprivation conditions. *Agronomie* 24(1):1–6.

Cochrane, H.R. and L.A.G. Aylmore. 1991. Assessing management-induced changes in the structural stability of hardsetting soils. *Soil Tillage Res.* 20(1):123–132.

Codignola, A., L. Verotta, P. Panu et al. 1989. Cell wall bound-phenols in roots of vesicular-arbuscular mycorrhizal plants. *New Phytol.* 112(2):221–228.

Colangelo, E.P. and M.L. Guerinot. 2004. The essential basic helix-loop-helix protein FIT1 is required for the iron deficiency response. *Plant Cell* 16(12):3400–3412.

Cornu, J.Y., S. Staunton, and P. Hinsinger. 2007. Copper concentration in plants and in the rhizosphere as influenced by the iron status of tomato (*Lycopersicon esculentum* L.). *Plant Soil* 292(1–2):63–77.

Curie, C. and J.F. Briat. 2003. Iron transport and signaling in plants. *Annu. Rev. Plant Biol.* 54:183–206.

Curie, C., Z. Panaviene, C. Loulergue et al. 2001. Maize yellow stripe1 encodes a membrane protein directly involved in Fe(III) uptake. *Nature* 409(6818):346–349.

Dabkowska-Naskret, H. 2000. Status of iron in alluvial soils from the Wisla River Valley, Poland. *J. Plant Nutr.* 23(11–12):1549–1557.

Dasgan, H.Y., L. Ozturk, K. Abak, and I. Cakmak. 2003. Activities of iron-containing enzymes in leaves of two tomato genotypes differing in their resistance to Fe chlorosis. *J. Plant Nutr.* 26(10–11):1997–2007.

Decianzio, S.R.R. 1991. Recent advances in breeding for improving iron utilization by plants. *Plant Soil* 130(1–2):63–68.

de Santiago, A., I. Diaz, M.D. del Campillo, J. Torrent, and A. Delgado. 2008. Predicting the incidence of Fe deficiency chlorosis from hydroxylamine-extractable iron in soil. *Soil Sci. Soc. Am. J.* 72(5): 1493–1499.

Doria, E., B. Campion, F. Sparvoli, A. Tava, and E. Nielsen. 2012. Anti-nutrient components and metabolites with health implications in seeds of 10 common bean (*Phaseolus vulgaris* L. and *Phaseolus lunatus* L.) landraces cultivated in southern Italy. *J. Food Comp. Anal.* 26(1–2):72–80.

Eagling, T., A.L. Neal, S.P. McGrath et al. 2014. Distribution and speciation of iron and zinc in grain of two wheat genotypes. *J. Agr. Food Chem.* 62(3):708–716.

Egilla, J.N., D.H. Byrne, and D.W. Reed. 1994. Iron stress response of 3 peach cultivars—Ferric iron reduction capacity. *J. Plant Nutr.* 17(12):2079–2103.

ElMaki, H.B., S.M. AbelRahaman, W.H. Dris et al. 2007. Content of antinutritional factors and HCl-extractability of minerals from white bean (*Phaseolus vulgaris*) cultivars: Influence of soaking and/or cooking. *Food Chem.* 100(1):362–368.

Elrashidi, M.A., M.D. Mays, and C.W. Lee. 2003. Assessment of Mehlich 3 and ammonium bicarbonate-DTPA extraction for simultaneous measurement of fifteen elements in soils. *Commun. Soil Sci. Plant Anal.* 34(19–20):2817–2838.

Fageria, N.K. and F.J.P. Zimmerman. 1998. Influence of pH on growth and nutrient uptake by crop species in an oxisol. *Commun. Soil Sci. Plant Anal.* 29(17–18):2675–2682.

Fan, X.Y., M.R. Karim, X.P. Chen et al. 2012. Growth and iron uptake of lowland and aerobic rice genotypes under flooded and aerobic cultivation. *Commun. Soil Sci. Plant Anal.* 43(13):1811–1822.

Fang, W.C., J.W. Wang, C.C. Lin, and C.H. Kao. 2001. Iron induction of lipid peroxidation and effects on antioxidative enzyme activities in rice leaves. *Plant Growth Reg.* 35(1):75–80.

Fernandez, V., V. Del Rio, J Abadia, and A. Abadia. 2006. Foliar iron fertilization of peach (*Prunus persica* (L.) Batsch): Effects of iron compounds, surfactants and other adjuvants. *Plant Soil* 289(1–2):239–252.

Fu, Y.Q., H. Shen, D.M. Wu, and K.Z. Cai. 2012. Silicon-mediated amelioration of Fe^{2+} toxicity in rice (*Oryza sativa* L.) roots. *Pedosphere* 22(6):795–802.

Garcia, A., A.F. deIorio, M. Barros, M. Bargiela, and A. Rendina. 1997. Comparison of soil tests to determine micronutrients status in Argentina soils. *Commun. Soil Sci. Plant Anal.* 28(10–20):1777–1792.

Garcia, M., C. Lucena, F.J. Romera, E. Alcantara, and R. Perez-Vicente. 2010. Ethylene and nitric oxide involvement in the up-regulation of key genes related to iron acquisition and homeostasis in *Arabidopsis*. *J. Exp. Bot.* 61(14):3885–3899.

Garcia, M., V. Suarez, F.J. Romera, E. Alcantara, and R. Perez-Vicente. 2011. A new model involving ethylene, nitric oxide and Fe to explain the regulation of Fe-acquisition genes in Strategy I plants. *Plant Physiol. Biochem.* 49(5):537–544.

Garcia-Escudero, E., I. Romero, A. Benito, N. Dominguez, and I. Martin. 2013. Reference levels for leaf nutrient diagnosis of cv. Tempranillo grapevine in the Rioja Appellation. *Commun. Soil Sci. Plant Anal.* 44(1–4):645–654.

Garcia-Mina, J.M., M.C. Antolin, and M. Sanchez-Diaz. 2004. Metal-humic complexes and plant micronutrient uptake: A study based on different plant species cultivated in diverse soil types. *Plant Soil* 258(1–2):57–68.

Gartley, K.L., J.T. Sims, C.T. Olsen, and P. Chu. 2002. Comparison of soil test extractants used in Mid-Atlantic United States. *Commun. Soil Sci. Plant Anal.* 33(5–6)873–895.

Gebhardt, S.E. and R.G. Thomas. 2002. Nutritive value of foods. U.S. Department of Agriculture, Agricultural Research Service, *Home and Garden Bulletin 72*. Available at http://www.ars.usda.gov/SP2UserFiles/Place/12354500/Data/hg72/hg72_2002.pdf, accessed January 12, 2015.

Gendre, D., P. Czernic, G. Conejero et al. 2007. TcYSL3, a member of the YSL gene family from the hyper-accumulator *Thlaspi caerulescens*, encodes a nicotianamine-Ni/Fe transporter. *Plant J.* 49(1):1–15.

Gerke, J. 2000. Mathematical modelling of iron uptake by graminaceous species as affected by Fe forms in soil and phytosiderophore efflux. *J. Plant Nutr.* 23(11–12):1579–1587.

Geurinot, M.L. 2010. Iron. In *Cell Biology of Metals and Nutrients*, eds. R. Hell and R.R. Mendel. Plant Cell Monographs 17, pp. 75–94. Berlin, Germany: Springer.

Gonzalo, M.J., J.J. Lucena, and L. Hernandez-Apaolaza. 2013. Effect of silicon addition on soybean (*Glycine max*) and cucumber (*Cucumis sativus*) plants grown under Fe deficiency. *Plant Physiol. Biochem.* 70:455–461.

Graham, R.D. and J.C.R. Stangoulis. 2003. Trace element uptake and distribution in plants. *J. Nutr.* 133(5):1502S–1505S.

Green, L.S., B.C. Lee, B.B. Buchanan et al. 1991. Ferredoxin and ferredoxin-NADP reductase from photosynthetic and nonphotosynthetic tissues of tomato. *Plant Physiol.* 96(4):1207–1213.

Gris, E. 1843. New results on the action of soluble iron compounds applied to foliage and especially on the treatment of chlorosis and diseases of plants [French]. *Compt. Rend. Acad. Sci. Paris* 19:1118–1119.

Hendriks, T. and L.C. van Loon. 1990. Petunia peroxidase a is localized in the epidermis of aerial plant organs. *J. Plant Physiol.* 136(5):519–515.

Herbik, A., A. Giritch, C. Horstmann et al. 1996. Iron and copper nutrition-dependent changes in protein expression in a tomato wild type and the nicotianamine-free mutant chloronerva. *Plant Physiol.* 111(2):533–540.

Higuchi, K., K. Kanazawa, and N.K. Nishizawa. 1996. The role of nicotianamine synthase in response to Fe nutrition status in Gramineae. *Plant Soil* 178(2):171–177.

Hughes, D.F., V.D. Jolley, and J.C. Brown. 1992. Roles for potassium in the iron-stress response mechanisms of Strategy I and Strategy II plants. *J. Plant Nutr.* 15(10):1821–1839.

Ijarotimi, O.S. and T.R. Esho. 2009. Comparison of nutritional composition and anti-nutrient status of fermented, germinated and roasted bambara groundnut seeds (*Vigna subterranea*). *Br. Food J.* 111(4–5):376–386.

Imtiaz, M., A. Rashid, P. Khan, M.Y. Memon, and M. Aslam. 2010. The role of micronutrients in crop production and human health. *Pakistan J. Botany* 41(4):2565–2578.

Ishimaru, Y, M. Suzuki, T. Tsukamoti et al. 2006. Rice plants take up iron as an Fe^{3+}-phytosiderophore and as Fe^{2+}. *Plant J.* 45(3):335–346.

Iturbe-Ormaetxe, I., J.F. Moran, C. Arrese-Igor et al. 1995. Activated oxygen and antioxidant defences in iron-deficient pea plants. *Plant Cell Environ.* 18(4):421–429.

Izaguirre-Mayoral, M.L. and T.R. Sinclair. 2005. Soybean genotypic difference in growth, nutrient accumulation and ultrastructure in response to manganese and iron supply in solution culture. *Ann. Bot.* 96(1):149–158.

Jackson, M.L., C.H. Lim, and L.W. Zelazny. 1986. Oxides, hydroxides, and aluminosilicate. In *Methods of Soil Analysis. Part 1—Physical and Mineralogical Methods*, 2nd edn., ed. A. Klute, pp. 101–150. Madison, WI: American Society of Agronomy, Soil Science Society of America.

Jeong, J. and E.L. Connolly. 2009. Iron uptake mechanisms in plants: Functions of the FRO family of ferric reductases. *Plant Sci.* 176(6):709–714.

Jimenez, S., Y. Gogorcena, C. Hevin, A.D. Rombola, and N. Ollat. 2007. Nitrogen nutrition influences some biochemical responses to iron deficiency in tolerant and sensitive genotypes of *Vitis*. *Plant Soil* 290(1–2):343–355.

Johnson, G.V., A. Lopez, and N.L. Foster. 2002. Reduction and transport of Fe from siderophores—Reduction of siderophores and chelates and uptake and transport of iron by cucumber seedlings. *Plant Soil* 241(1):27–33.

Kabir, A.H., N.G. Paltridge, A. Able, J. Paull, and J.C.R Stangoulis. 2012. Natural variation for Fe-efficiency is associated with upregulation of Strategy I mechanisms and enhanced citrate and ethylene synthesis in *Pisum sativum* L. *Planta* 235(6):1409–1419.

Kamal, K., L. Hagagg, and F. Awad. 2000. Improved Fe and Zn acquisition by guava seedlings grown in calcareous soils intercropped with graminaceous species. *J. Plant Nutr.* 23(11–12):2071–2080.

Karray-Bouraoui, N., H. Attia, M. Maghzaoui et al. 2010. Physiological responses of *Arabidopsis thaliana* to the interaction of Fe deficiency and nitrogen form. *Acta Biol. Hung.* 61(2):204–213.

Kayode, A.P.P., A.R. Linnemann, M.J.R. Nout, and M.A.J.S. Van Boekel. 2007. Impact of sorghum processing on phytate, phenolic compounds and in vitro solubility of iron and zinc in thick porridges. *J. Sci. Food Agr.* 87(5):832–838.

Khabaz-Saberi, H., S.J. Barker, and Z. Rengel. 2012. Tolerance to ion toxicities enhances wheat (*Triticum aestivum* L.) grain yield in waterlogged acidic soils. *Plant Soil* 354(1–2):371–381.

Khabaz-Saberi, H., T.L. Setter, and I. Waters. 2006. Waterlogging induces high to toxic concentrations of iron, aluminum, and manganese in wheat varieties on acidic soil. *J. Plant Nutr.* 29(5):899–911.

Kiriiwa, Y., S. Osumi, M. Endo et al. 2004. Iron uptake in a bicarbonate-resistant cell line of tobacco. *Soil Sci. Plant Nutr.* 50(7):1119–1124.

Knaff, D.B. and M. Hirasawa. 1991. Ferredoxin-dependent chloroplastic enzymes. *Biochim. Biophys. Acta* 1056(2):93–925.

Kobayashi, T. and N.K. Nishizawa. 2012. Iron uptake, translocation, and regulation in higher plants. *Annu. Rev. Plant Biol.* 63:131–152.

Kobayashi, T., M. Suzuki, H. Inoue et al. 2005. Expression of iron-acquisition-related genes in iron-deficient rice is co-ordinately induced by partially conserved iron-deficiency-responsive elements. *J. Exp. Bot.* 56(415):1305–1316.

Kosegarten, H. and G. Englisch. 1994. Effect of various nitrogen forms on the pH of leaf apoplast and on iron chlorosis of *Glycine max* L. *Zeitsch. Pflanzen. Boden.* 157(6):401–405.

Kramer, D., V. Römheld, E. Landsberg, and H. Marschner. 1980. Induction of transfer cells by iron deficiency in the root epidermis of *Helianthus annuus*. *Planta* 147(4):335–339.

Kutman, U.B., B. Yildiz, and I. Cakmak. 2011a. Effect of nitrogen on uptake, remobilization and partitioning of zinc and iron throughout the development of durum wheat. *Plant Soil* 342(10):149–164.

Kutman, U.B., B. Yildiz, and I. Cakmak. 2011b. Improved nitrogen status enhances zinc and iron concentrations both in the whole grain and the endosperm fraction of wheat. *J. Cereal Sci.* 53(1):118–125.

Ladouceur, A., S. Tozawa, S. Alam, S. Kamei, and S. Kawai. 2006. Effect of low phosphorus and iron-deficient conditions on phytosiderophore release and mineral nutrition in barley. *Soil Sci. Plant Nutr.* 52(2):203–210.

Landsberg, E.-C. 1994. Transfer cell formation in sugar beet roots induced by latent Fe deficiency. *Plant Soil* 165(2):197–205.

Le Jean, M., A. Schikora, S. Mari, J.-F. Briat, and C. Curie. 2005. A loss-of-function mutation in *AtYSL1* reveals its role in Fe and nicotianamine seed loading. *Plant J.* 44(5):769–782.

Li, L., X. Cheng, and H.-Q. Ling. 2004. Isolation and characterization of Fe(III)-chelate reductase gene *LeFRO1* in tomato. *Plant Mol. Biol.* 54(1):125–136.

Liang, J., B.Z. Han, L.Z. Han, M.J.R. Nout, and R.J. Hammer. 2007. Iron, zinc and phytic acid content of selected rice varieties from China. *J. Sci. Food Agr.* 87(3):504–510.

Lin, S., S. Cianzio, and R. Shoemaker. 1997. Mapping genetic loci for iron deficiency chlorosis in soybean. *Mol. Breeding* 3(3):219–220.

Lindsay, W.L. and W.A. Norvell. 1978. Development of a DPTA test for zinc, iron, manganese, and copper. *Soil Sci. Soc. Am. J.* 42(3):421–428.

Lindsay, W.L. and A.P. Schwab. 1982. The chemistry of iron in soils and its availability to plants. *J. Plant Nutr.* 10:821–840.

Ling, H.-Q., G. Koch, H. Bäumlein, and M.W. Ganal. 1999. Map-based cloning of *chloronerva*, a gene involved in iron uptake of higher plants encoding nicotianamine synthase. *Proc. Natl. Acad. Sci. U.S.A.* 96(12):7098–7103.

Lingam, S., J. Mohrbacher, T. Brumbarova et al. 2011. Interaction between the bHLH Transcription Factor FIT and ETHYLENE INSENSITIVE3/ETHYLENE INSENSITIVE3-LIKE1 reveals molecular linkage between the regulation of iron acquisition and ethylene signaling in *Arabidopsis*. *Plant Cell* 23(5):1815–1829.

Loeppert, R.H. and W.P. Inskeep. 1996. Iron. In *Methods of Soil Analysis. Part 3. Chemical Methods*, ed. D.L. Sparks, pp. 639–664. Book Series No. 5. Madison, WI: Soil Science Society of America.

Lopez-Millan, A.F., F. Morales, Y. Gogorcena, A. Abadia, and J. Abadia. 2001. Fe resupply-mediated deactivation of Fe-deficiency stress responses in roots of sugar beet. *Aust. J. Plant Physiol.* 28(3):171–180.

Losak, T., J. Hlusek, J. Martinec et al. 2011. Nitrogen fertilization does not affect micronutrient uptake in grain maize (*Zea mays* L.). *Acta Agr. Scand. Sect. B. Soil Plant Sci.* 61(6):543–550.

Lucena, J.J. 2000. Effects of bicarbonate, nitrate and other environmental factors on iron deficiency chlorosis. A review. *J. Plant Nutr.* 23(11–12):1591–1606.

Luo, Y.W., W.H. Xie, and F.X. Luo. 2012. Effect of several germination treatments on phosphatases activities and degradation of phytate in faba bean (*Vicia faba* L.) and azuki bean (*Vigna angularis* L.). *J. Food Sci.* 77(10):C1023-C1029.

Lynch, J. and K.M. Brown. 1997. Ethylene and plant responses to nutritional stress. *Physiol. Plant.* 100(3):613–619.

Ma, G.S., Y. Jin, J. Piao et al. 2005. Phytate, calcium, iron, and zinc contents and their molar ratios in foods commonly consumed in China. *J. Agr. Food Chem.* 53(26):10285–10290.

Madurapperuma, W.S. and D. Kumaragamage. 2008. Evaluation of ammonium bicarbonate-diethylene tri-amine penta acetic acid as a multinutrient extractant for acidic lowland rice soils. *Commun. Soil Sci. Plant Anal.* 39(11–12):1773–1790.

Marcos, M.L.F., E. Alvarez, and C. Monterroso. 1998. Aluminum and iron estimated by Mehlich-3 extractant in mine soils in Galicia, northwest Spain. *Commun. Soil Sci. Plant Anal.* 29(5–6):599–612.

Marschner, H. 1995. *Mineral Nutrition of Higher Plants*, 2nd edn. London, U.K.: Academic Press.

Marschner, H. and V. Römheld. 1994. Strategies of plants for acquisition of iron. *Plant Soil* 165(2):261–274.

Marsh, H.V., G. Matrone, and H.J. Evans. 1963. Investigations of role of iron in chlorophyll metabolism. 2. Effect of iron deficiency on chlorophyll synthesis. *Plant Physiol.* 38(6):638–642.

Martinez-Cuenca, M.R., A. Quinones, D.J. Inglesias et al. 2013. Effects of high levels of zinc and manganese ions on Strategy I responses to Fe deficiency in citrus. *Plant Soil* 373(1–2):943–953.

Mehlich, A. 1984. Mehlich 3 soil test extractant: A modification of Mehlich 2 extractant. *Commun. Soil Sci. Plant Anal.* 15(12):1409–1416.

Mengel, K. 1994. Iron availability in plant tissues—Iron chlorosis on calcareous soils. *Plant Soil* 165(2):275–283.

Mengel, K. 2001. Alternative or complementary role of foliar supply in mineral nutrition. In *International Symposium on Foliar Nutrition of Perennial Fruit Plants*, Meran, Italy, eds. M. Tagliavini, M. Toselli, L. Bertschinger, D. Neilsen, and M. Thalheimer. Acta Horticulturae Book Series 594, pp. 33–47.

Metzger, W.H. 1941. Phosphorus fixation in relation to the iron and aluminum of the soil. *J. Am. Soc. Agron.* 33(12):1093–1099.

Meyer, M.R.M., A. Rojas, A. Santanen, and F.L. Stoddard. 2013. Content of zinc, iron and their absorption inhibitors in Nicaraguan common beans (*Phaseolus vulgaris* L.). *Food Chem.* 136(1):87–93.

Mhamdi, A., G. Queval, S. Chanouch et al. 2010. Catalase function in plants: A focus on *Arabidopsis* mutants as stress-mimic models. *J. Exp. Bot.* 61(15):4197–4220.

Michaud, A.M., C. Chappellaz, and P. Hinsinger. 2008. Copper phytotoxicity affects root elongation and iron nutrition in durum wheat (*Triticum turgidum durum* L.). *Plant Soil* 310(1–2):151–165.

Mlodzinska, E. 2012. Alteration of plasma membrane H^+-ATPase in cucumber roots under different Fe nutrition. *Acta Physiol. Plant.* 34(6):2125–2133.

Morgan, M.F. 1941. Chemical soil diagnosis by the universal soil testing system. *Conn. Agric. Exp. Stn. Bull.* No. 450.

Mori, S. 2001. The role of mugineic acid in iron acquisition: Progress in cloning the genes for transgenic rice. In *Plant Nutrient Acquisition: New Perspectives*, ed. N. Ae, J. Arihara, K. Okada, and A. Srinivasan, pp. 120–139. Tsukuba, Japan: International Workshop on Plant Nutrient Acquisition, March 24–27, 1998.

Mori, S. and N. Nishizawa. 1987. Methionine as a dominant precursor of phytosiderophores in Graminaceae plants. *Plant Cell Physiol.* 28(6):1081–1092.

Mori, S., N. Nishizawa, H. Hayashi, M. Chino, E. Yoshimura, and J. Ishihara. 1991. Why are young rice plants highly susceptible to iron deficiency? *Plant Soil* 130(1–2):143–156.

Morsomme, P. and M. Boutry. 2000. The plant plasma membrane H^+-ATPase: Structure, function and regulation. *Biochim. Biophys. Acta—Biomembranes* 1465(1–2):1–16.

Murata, Y., J.F. Ma, N. Yamaji, D. Ueno, K. Nomoto, and T. Iwashita. 2006. A specific transporter for Fe(III)-phytosiderophore in barley roots. *Plant J.* 46(4):563–572.

Nascimento, A.C., C. Mota, I. Coelho et al. 2014. Characterisation of nutrient profile of quinoa (*Chenopodium quinoa*); amaranth (*Amaranthus caudatus*), and purple corn (*Zea mays* L.) consumed in the North of Argentina: Proximates, minerals and trace elements. *Food Chem.* 148:420–426.

Nadal, P., C. Garcia-Delgado, D. Hernandez, S. Lopez-Rayo, and J.J. Lucena. 2012a. Evaluation of Fe-N,N'-Bis(2-hydroxybenzyl)ethylenediamine-N,N'-diacetate (HBED/Fe^{3+}) as Fe carrier for soybean (*Glycine max*) plants grown in calcareous soil. *Plant Soil* 360(1–2):349–362.

Nadal, P., S. Garcia-Marco, R. Escudero, and J.J. Lucena. 2012b. Fertilizer properties of DCHA/Fe^{3+}. *Plant Soil* 356(1–2):367–379.

Nenova, V. and I. Stoyanov. 1995. Physiological and biochemical changes in young maize plants under Fe deficiency 2. Catalase, peroxidase, and nitrate reductase activities in leaves. *J. Plant Nutr.* 18(10):2081–2091.

Nikolic, M. and R. Kastori. 2000. Effect of bicarbonate and Fe supply on Fe nutrition of grapevine. *J. Plant Nutr.* 23(11–12):1619–1627.

Nishiyama, R., M. Kato, S. Nagata, S. Yanagisawa, and T. Yoneyama. 2012. Identification of Zn-nicotianamine and Fe-2'-deoxymugineic acid in the phloem sap from rice plants (*Oryza sativa* L.). *Plant Cell Physiol.* 53(2):381–390.

Ohwaki, Y. and K. Sugahara. 1997. Active extrusion of protons and exudation of carboxylic acids in response to iron deficiency by roots of chickpea (*Cicer arietinum* L). *Plant Soil* 189(1):49–55.

Olson, R.V. 1965. Iron. In *Methods of Soil Analysis, Part 2. Chemical and Microbiological Properties*, ed. C.A. Black, pp. 963–971. Madison, WI: American Society of Agronomy.

O'Rourke, J.A., D.V. Charlson, D.O. Gonzalez et al. 2007. Microarray analysis of iron deficiency chlorosis in near-isogenic soybean lines. *BMC Genomics* 8:Article 476.

Ostatek-Boczynski, Z.A. and P. Lee-Steere. 2012. Evaluation of Mehlich 3 as a universal nutrient extractant for Australian sugarcane soils. *Commun. Soil Sci. Plant Anal.* 43(4):623–630.

Palmgren, M.G. 2001. Plant plasma membrane H^+-ATPases: Powerhouses for nutrient uptake. *Ann. Rev. Plant Physiol. Plant Mol. Biol.* 52:817–845.

Panda, B.B., S. Sharma, P.K. Mohapatra, and A. Das. 2012. Application of excess nitrogen, phosphorus, and potassium fertilizers leads to lowering of grain iron content in high-yielding tropical rice. *Commun. Soil Sci. Plant Nutr.* 43(20):2590–2602.

Panjtandoust, M., A. Sorooshzadeh, and F. Ghanati. 2013. Influence of iron application methods on seasonal variations in antioxidant activity of peanut. *Commun. Soil Sci. Plant Nutr.* 44(14):2118–2126.

Pavlovic, J., J. Samardzic, V. Maksimovic et al. 2013. Silicon alleviates iron deficiency in cucumber by promoting mobilization of iron in the root apoplast. *New Phytol.* 198(4):1096–1107.

Peiffer, G.A., K.E. King, A.J. Severin et al. 2012. Identification of candidate genes underlying an iron efficiency quantitative trait locus in soybean. *Plant Physiol.* 158(4):1745–1754.

Pekker, I., E. Tel-Or, and R. Mittler. 2002. Reactive oxygen intermediates and glutathione regulate the expression of cytosolic ascorbate peroxidase during iron-mediated oxidative stress in bean. *Plant Mol. Biol.* 49(5):429–438.

Perur, N.G., R.L Smith, and H.H. Wiebe. 1961. Effect of iron chlorosis on protein fraction of corn leaf tissue. *Plant Physiol.* 36:736–739.

Pinton, R., S. Cesco, S. Santi, F. Agnolon, and Z. Varanini. 1999. Water-extractable humic substances enhance iron deficiency responses by Fe-deficient cucumber plants. *Plant Soil* 210(2):145–147.

Rabotti, G., P. Denisi, and G. Zocchi. 1995. Metabolic implications in the biochemical responses to iron-deficiency in cucumber (*Cucumis sativus* L.) roots. *Plant Physiol.* 107(4):1195–1199.

Rai, K.N., M. Govindaraj, and A.S. Rao. 2012. Genetic enhancement of grain iron and zinc content in pearl millet. *Qual. Assur. Safe. Crops Foods* 4(3):119–125.

Ramirez, L., G.C. Bartoli, and L. Lamattina. 2013. Glutathione and ascorbic acid protect *Arabidopsis* plants against detrimental effects of iron deficiency. *J. Exp. Bot.* 64(11):3169–3178.

Ramos, L., A. Bettin, B.M.P. Herrada et al. 2013. Effects of nitrogen form and application rates on the growth of petunia and nitrogen content in the substrate. *Commun. Soil Sci. Plant Anal.* 44(1–4):473–479.

Rengel, Z. 2002. Genetic control of root exudation. *Plant Soil* 245(1):59–70.

Reyes, I. and J. Torrent. 1997. Citrate-ascorbate as a highly selective extractant for poorly crystalline iron oxides. *Soil Sci. Soc. Am. J.* 61(6):1647–1654.

Roberts, L.A., A.J. Pierson, Z. Panaviene, and E.L. Walker. 2004. Yellow stripe1. Expanded roles for the maize iron-phytosiderophore transporter. *Plant Physiol.* 135(1):112–120.

Rodriguez, J.B., J.R. Self, G.A. Peterson, and D.G. Westfall. 1999. Sodium bicarbonate-DTPA test for macro- and micronutrient elements in soils. *Commun. Soil Sci. Plant Anal.* 30(7–8):957–970.

Romera, F.J. and E. Alcantara. 2004. Ethylene involvement in the regulation of Fe-deficiency stress responses by Strategy I plants. *Funct. Plant Biol.* 31(4):315–328.

Romera, F.J., E. Alcantara, and M.D. De la Guardia. 1999. Ethylene production by Fe-deficient roots and its involvement in the regulation of Fe-deficiency stress responses by strategy I plants. *Ann. Bot.* 83(1): 51–55.

Römheld, V. and M. Nikolic. 2007. Iron. In *Handbook of Plant Nutrition*, eds. A.V. Barker and D.J. Pilbeam, pp. 329–350. Boca Raton, FL: CRC Press.

Rossel, R.A.V., E.N. Bui, P. de Caritat, and N.J. McKenzie. 2010. Mapping iron oxides and the color of Australian soil using visible-near-infrared reflectance spectra. *J. Geophys. Res.* 115:Article F04031.

Ruiz, J.M., A. Delgado, and J. Torrent. 1997. Iron-related phosphorus in overfertilized European soils. *J. Environ. Qual.* 26(6):1548–1554.

Sahrawat, K.L. 2000. Elemental composition of the rice plant as affected by iron toxicity under field conditions. *Commun. Soil Sci. Plant Anal.* 31(17–18):2819–2827.

Samourgiannidis, G. and T. Masti. 2013. Comparison of two sequential extraction methods and the DTPA method for the extraction of micronutrients from acidic soils. *Commun. Soil Sci. Plant Anal.* 44(1–4):38–49.

Sanchez-Alcala, I., M.C. del Campillo, J. Torrent, K.L. Straub, and S.M. Kraemer. 2011. Fe(III) reduction in anaerobically incubated suspensions of highly calcareous agricultural soils. *Soil Sci. Soc. Am. J.* 75(6):2136–2146.

Santi, S., S. Cesco, Z. Varanini, and R. Pinton. 2005. Two plasma membrane H(+)-ATPase genes are differentially expressed in Fe-deficient cucumber plants. *Plant. Physiol. Biochem.* 43(3):287–292.

Santi, S., G. Locci, R. Pinton, S. Cesco, and Z. Varanini. 1995. Plasma-membrane H⁺-ATPase in maize roots induced for NO_3^- uptake. *Plant Physiol.* 109(4):1277–1283.

Santi, S. and W. Schmidt. 2008. Laser microdissection-assisted analysis of the functional fate of iron deficiency-induced root hairs in cucumber. *J. Exp. Bot.* 59(3):697–704.

Schaaf, G., U. Ludewig, B.E. Erenoglu et al. 2004. ZmYS1 functions as a proton-coupled symporter for phyto-siderophore- and nicotianamine-chelated metals. *J.Biol. Chem.* 279(10):9091–9096.

Schenkeveld, W.D.C., R. Dijcker, A.M. Reichwein, E.J.M. Temminghoff, and W.H. van Riemsdijk. 2008. The effectiveness of soil-applied FeEDDHA treatments in preventing iron chlorosis in soybean as a function of the o,o-FeEDDHA content. *Plant Soil* 303(1–2):161–176.

Schikora, A. and W. Schmidt. 2001. Iron stress-induced changes in root epidermal cell fate are regulated independently from physiological responses to low iron availability. *Plant Physiol.* 125(4):1679–1687.

Schikora, A. and W. Schmidt. 2002. Formation of transfer cells and H⁺-ATPase expression in tomato roots under P and Fe deficiency. *Planta* 215 (2):304–311.

Schmidt, F., M.D. Fortes, J. Wesz, G.L. Buss, and R.O. de Sousa. 2013. The impact of water management on iron toxicity in flooded rice. *Rev. Brasil. Cienc. Solo.* 37(5):1226–1235.

Schmidt, W. 1999. Mechanisms and regulation of reduction-based iron uptake in plants. *New Phytol.* 141(1):1–26.

Schmidt, W. 2003. Iron solutions: Acquisition strategies and signaling pathways in plants. *Trends Plant Sci.* 8(4):188–193.

Schmidt, W., A. Schikora, A. Pich, and M Bartels. 2000a. Hormones induce an Fe-deficiency-like root epidermal cell pattern in the Fe-inefficient tomato mutant fer. *Protoplasma* 213(1–2):67–73.

Schmidt, W., J. Tittel, and A. Schikora. 2000b. Role of hormones in the induction of iron deficiency responses in *Arabidopsis* roots. *Plant Physiol.* 122(4):1109–1118.

Schmisek, M.E., L.J. Cihacek, and L.J. Swenson. 1998. Relationships between the Mehlich-III soil test extraction procedure and standard soil test methods in North Dakota. *Commun. Soil Sci. Plant Anal.* 29(11–14):1719–1729.

Scholz, G., R. Becker, A. Pich, and U.W. Stephan. 1992. Nicotianamine—A common constituent of Strategy I and Strategy II iron acquisition by plants—A review. *J. Plant Nutr.* 15(10):1647–1665.

Schulte, E.E. 2004. Soil and applied Fe. Understanding plant nutrients cooperative extension publications, University of Wisconsin-Extension Fact Sheet A3554. http://www.soils.wisc.edu/extension/pubs/A3554.pdf, accessed January 12, 2015.

Self, J.R. 1999. Soil test explanation. Fact Sheet 0.502. Colorado State University Extension, Fort Collins, CO.

Sharma, B.D., D.S. Chahal, P.K. Singh, and Raj-Kumar. 2008. Forms of iron and their association with soil properties in four soil taxonomic orders of arid and semi-arid soils of Punjab, India. *Commun. Soil Sci. Plant Anal.* 39(17–18):2550–2567.

Sharma, S., J.P. Gupta, H.P.S. Nagi, and R. Kumar. 2012. Effect of incorporation of corn byproducts on quality of baked and extruded products from wheat flour and semolina. *J. Food Sci.Tech.* 49(5):580–586.

Soil and Plant Analysis Council. 1992. *Handbook on Reference Methods for Soil Analysis.* Athens, GA: Georgia University Station.

Sokrab, A.W., I.A. Mohamed Ahmed, and E.E. Babiker. 2012. Effect of malting and fermentation on antinutrients, and total and extractable minerals of high and low phytate corn genotypes. *Intern. J. Food Sci. Tech.* 47(5):1037–1043.

Stephan, U.W. 2002. Intra- and intercellular iron trafficking and subcellular compartmentation within roots. *Plant Soil* 241(1):19–25.

Stephan, U.W., I. Schmidke, V.W. Stephan, and G. Scholz. 1996. The nicotianamine molecule is made-to-measure for complexation of metal micronutrients in plants. *Biometals* 9(1):84–90.

Su, Y.C., J. Guo, H. Ling et al. 2014. Isolation of a novel peroxisomal catalase gene from sugarcane, which is responsive to biotic and abiotic stresses. *PLoS ONE* 9(1):Article e84426.

Sumner, M.E. 1979. Interpretation of foliar analyses for diagnostic purposes. *Agron. J.* 71(2):343–348.

Tang, C., S.J. Zheng, Y.F. Qiao, G.H. Wang, and X.Z. Han. 2006. Interactions between high pH and iron supply on nodulation and iron nutrition of *Lupinus albus* L. genotypes differing in sensitivity to iron deficiency. *Plant Soil* 279(1–2):153–162.

Tennant, D., G. Scholz, J. Dixon, and B. Purdie. 1992. Physical and chemical characteristics of duplex soils and their distribution in the south-west of western Australia. *Austral. J. Exp. Agric.* 32(7):827–843.

Tewari, R.K., P. Kumar, and P.N. Sharma. 2005. Signs of oxidative stress in the chlorotic leaves of iron starved plants. *Plant Sci.* 169(6):1037–1045.

Thavarajah, D. and P. Thavarajah. 2012. Evaluation of chickpea (*Cicer arietinum* L.) micronutrient composition: Biofortification opportunities to combat global micronutrient malnutrition. *Food Res. Intern.* 49(1):99–104.

Thomine, S. and V. Lanquar. 2011. Iron transport and signaling in plants. In *Transporters and Pumps in Plant Signaling*, eds. M. Geisler and K. Venema, pp. 99–131, Book Series Signaling and Communication in Plants, Vol. 7. Berlin, Germany: Springer.

Tiffin, L.O. 1966. Iron transport I. Plant culture, exudate sampling, iron citrate analysis. *Plant Physiol.* 41(3):510–514.

Tiffin, L.O. 1970. Translocation of iron citrate and phosphorus in xylem exudates of soybean. *Plant Physiol.* 45(3):280–283.

Toker, C., T. Yildirim, H. Huseyin, N.E. Inci, and F.O. Ceylan. 2010. Inheritance of resistance to iron deficiency chlorosis in chickpea (*Cicer arietinum* L.). *J. Plant Nutr.* 33(9):1366–1373.

Urzica, E.I., D. Casero, H. Yamasaki et al. 2012. Systems and trans-system level analysis identifies conserved iron deficiency responses in the plant lineage. *Plant Cell* 24(10):3921–3948.

Varotto, C., D. Maiwald, P. Pesaresi et al. 2002. The metal ion transporter IRT1 is necessary for iron homeostasis and efficient photosynthesis in *Arabidopsis thaliana*. *Plant J.* 31(5):589–599.

Vert, G., N. Grotz, F. Dédaldéchamp et al. 2002. IRT1, an *Arabidopsis* transporter essential for iron uptake from the soil and for plant growth. *Plant Cell* 14(6):1223–1233.

Vert, G.A., J.F. Briat, and C. Curie. 2003. Dual regulation of the *Arabidopsis* high-affinity root iron uptake system by local and long-distance signals. *Plant Physiol.* 132(2):796–804.

Vizzotto, G., I. Matosevic, R. Pinton, Z. Varanini, and G. Costa. 1997. Iron deficiency responses in roots of kiwi. *J. Plant Nutr.* 20(2–3):327–334.

von Wirén, N. 2004. Progress in research on iron nutrition and interactions in plants. *Soil Sci. Plant Nutr.* 50(7):955–964.

von Wirén, N., S. Mori, H. Marshner, and V. Römheld. 1994. Iron inefficiency in maize mutant ys1 (*Zea mays* L. cv Yellow-Stripe) is caused by a defect in uptake of Fe phytosiderophores. *Plant Physiol.* 106(1):71–77.

Walker, E.L. and E.L. Connolly. 2008. Time to pump iron: Iron-deficiency-signaling mechanisms of higher plants. *Curr. Opinion Plant Biol.* 11:530–535.

Wallihan, E.F. 1965. Iron. In *Diagnostic Criteria for Plants and Soils*, ed. H.D. Chapman, pp. 203–212. Riverside, CA: Homer D. Chapman.

Waters, B.M. and D.G. Blevins. 2000. Ethylene production, cluster root formation, and localization of iron(III) reducing capacity in Fe deficient squash roots. *Plant Soil* 225(1–2):21–31.

Waters, B.M., S.A. McInturf, and R.J. Stein. 2012. Rosette iron deficiency transcript and microRNA profiling reveals links between copper and iron homeostasis in *Arabidopsis thaliana*. *J. Exp. Bot.* 63(16):5903–5918.

Wei, L.C., R.H. Loeppert, and W.R. Ocumpaugh. 1997. Fe-deficiency stress response in Fe-deficiency resistant and susceptible subterranean clover: Importance of induced H^+ release. *J. Exp. Bot.* 48(307):239–246.

Wei, Y.Y., M.J.I. Shohag, X.E. Yang, and Y.B. Zhang. 2012. Effects of foliar iron application on iron concentration in polished rice grain and its bioavailability. *J. Agr. Food Chem.* 60(45):11433–11439.

Wolf, A. and D. Beegle. 2009. Recommended Soil Testing Procedures for the Northeastern United States. *Cooperative Bulletin No. 493*, University of Delaware, Newark, DE. http://extension.udel.edu/lawngarden/files/2012/10/CHAP5.pdf, April 29, 2014.

Xiong, H.C., Y. Kakei, T. Kobayashi et al. 2013. Molecular evidence for phytosiderophore-induced improvement of iron nutrition of peanut intercropped with maize in calcareous soil. *Plant Cell Environ.* 36(10):1888–1902.

Yadavalli, V., S. Neelam, A.S.V.C. Rao, A.R. Reddy, and R. Subramanyam. 2012. Differential degradation of photosystem I subunits under iron deficiency in rice. *J. Plant Physiol.* 169(8):753–759.

Yaduvanshi, N.P.S., T.L. Setter, S.K. Sharma, K.N. Singh, and N. Kulshreshtha. 2012. Influence of waterlogging on yield of wheat (*Triticum aestivum*), redox potentials, and concentrations of microelements in different soils in India and Australia. *Soil Res.* 40(6):489–499.

Yuan, Y.X., H.L. Wu, N. Wang et al. 2008. FIT interacts with AtbHLH38 and AtbHLH39 in regulating iron uptake gene expression for iron homeostasis in *Arabidopsis*. *Cell Res.* 18(3):385–397.

Yunta, F., I Martin, J.J. Lucena, and A. Garate. 2013. Iron chelates supplied foliarly improve the iron translocation rate in Tempranillo grapevine. *Commun. Soil Sci. Plant Anal.* 44(1–4):794–804.

Zafrilla, B., R.M. Martinez-Espinose, M.J. Bonete et al. 2011. A haloarchaeal ferredoxin electron donor that plays an essential role in nitrate assimilation. *Biochem. Soc. Trans.* 39(6):1844–1848.

Zaharieva, T. and V. Römheld. 2000. Specific Fe^{2+} uptake system in strategy I plants inducible under Fe deficiency. *J. Plant Nutr.* 23(11–12):1733–1744.

Zbiral, J. and P. Nemec. 2000. Integrating of Mehlich 3 extractant into the Czech soil testing scheme. *Commun. Soil Sci. Plant Anal.* 31(11–14):2171–2182.

Zhang, F., J. Shen, L. Li, and X. Liu. 2004. An overview of rhizosphere processes related with plant nutrition in major cropping systems in China. *Plant Soil* 260(1–2):89–99.

Zhang, J., L.H. Wu, and M.Y. Wang. 2008. Iron and zinc biofortification in polished rice and accumulation in rice plant (*Oryza sativa* L.) as affected by nitrogen fertilization. *Acta Agr. Scand. Sect. B. Soil Plant Sci.* 58(3):267–272.

Zhang, M., A.K. Alva, Y.C. Li, and D.V. Calvert. 1997. Fractionation of iron, manganese, aluminum, and phosphorus in selected sandy soils under citrus production. *Soil Sci. Soc. Am. J.* 61(3):794–801.

Zhou, X, S.Z. Li, Q.Q. Zhao et al. 2013. Genome-wide identification, classification and expression profiling of nicotianamine synthase (NAS) gene family in maize. *BMC Genomics* 14:Article 238.

Zohlen, A. and G. Tyler. 1997. Differences in iron nutrition strategies of two calcifuges, *Carex pilulifera* L. and *Veronica officinalis* L. *Ann. Bot.* 80(4):553–559.

Zou, C., J. Shen, F. Zhang et al. 2001. Impact of nitrogen form on iron uptake and distribution in maize seedlings in solution culture. *Plant Soil* 235(2):143–149.

Zuchi, S., S. Cesco, and S. Astolfi. 2012. High S supply improves Fe accumulation in durum wheat plants grown under Fe limitation. *Environ. Exp. Bot.* 77:25–32.

Zuchi, S., S. Cesco, S. Gottardi, R. Pinton, V. Römheld, and S. Astolfi. 2011. The root-hairless barley mutant *brb* used as model for assessment of role of root hairs in Fe accumulation. *Plant Physiol. Biochem.* 49(5):506–512.

Zuchi, S., S. Cesco, Z. Varanini, R. Pinton, and S. Astolfi. 2009. Sulphur deprivation limits Fe-deficiency responses in tomato plants. *Planta* 230(1):84–94.

Zuo, Y.M., X.L. Li, Y.P. Cao, F.S. Zhang, and P. Christie. 2003. Fe nutrition of peanut enhanced by mixed cropping with maize: Possible role of root morphology and rhizosphere microflora. *J. Plant Nutr.* 26(10–11):2093–2110.

Zuo, Y.M. and F.S. Zhang. 2011. Soil and crop management strategies to prevent iron deficiency in crops. *Plant Soil* 339(1–2):83–95.

12 Manganese

Touria E. Eaton

CONTENTS

12.1 HISTORICAL BACKGROUND

In 1774, Scheele discovered manganese (Bryson et al., 2014). Since then, investigators interested in plant nutrition and in soil fertility have given considerable attention to the role of Mn in plant nutrition. McHargue (1922) is credited with showing that Mn is an essential element for plants. Arnon and Stout (1939) confirmed the essentiality of Mn in plant growth, according to the following criteria: the element is not considered essential unless (1) a deficiency of it makes it impossible for the plant to complete the vegetative or the reproductive stage of its life cycle; (2) such deficiency is specific to the element in question and can be prevented or corrected only by supplying this element; and (3) the element is directly involved in the nutrition of the plant, apart from its possible effects in correcting some unfavorable microbiological or chemical condition of the soil or other growth medium.

The determination of essentiality of Mn in plants led the way for further groundbreaking studies. Research into concentrations of Mn that confer deficiency or toxicity and the variation among and within species of plants in their tolerance or susceptibility to these afflictions has proliferated. The symptoms of Mn toxicity and deficiency also have received much attention owing to their variation among species and their similarity to other nutrient-related anomalies. In addition to research on Mn diagnostics, investigators have focused on the role of Mn in plant metabolism and in resistance to pests and diseases, revealing economically important interactions that further highlight the importance of this nutrient in optimal plant production.

This chapter reviews literature dealing with the understanding of the role of Mn as an essential element and as a toxic element to plants, the long-distance and cellular transport in plants, as well as the mechanisms or strategies involved to resist an overload of this metal. The forms and dynamics of this element in soils and the importance of acidity for availability to plants also are given. This chapter also includes literature dealing with identification of Mn deficiency and toxicity in various crops of economic importance, the physiology of Mn uptake and transport, and the interaction between manganese and diseases.

12.2 MANGANESE AS AN ESSENTIAL ELEMENT IN PLANT METABOLISM

Manganese plays an important role in many metabolic processes, including photosynthesis, respiration, protein synthesis, and hormone activation (Burnell, 1988).

12.2.1 Photosynthesis and Oxygen Evolution

The essentiality of Mn in photosynthesis was first discovered in green algae (*Chlorella vulgaris* Beijerinck) (Pirson, 1937). Eyster et al. (1958) reported that the Mn requirement for optimal growth of chlorella under heterotrophic conditions (darkness and external supply of carbohydrates) was one-thousandth of the requirement under autotrophic conditions (availability of light for carbohydrate synthesis through photosynthesis). A study on subterranean clover (*Trifolium subterraneum* L.) indicated that photosynthesis in general and photosynthetic oxygen (O_2) evolution in photosystem II (PSII) in particular are depressed when the plants are deficient in Mn (Broadley et al., 2012; Husted et al., 2010; Nable et al., 1984; Shenker et al., 2004). In this study, photosynthetic O_2 evolution was restricted by more than 50% with low Mn concentration in young leaves (Nable et al., 1984). Resupplying Mn to the deficient leaves restored photosynthetic O_2 evolution. Similar results were reported in wheat (*Triticum aestivum* L.) (Kriedemann et al., 1985) and corn (*Zea mays* L.) (Gong et al., 2010).

Due to its oxidation–reduction properties ($Mn^{2+} \leftrightarrow Mn^{3+} + e^-$), Mn plays an important role in the oxygen-evolving photosynthetic machinery, catalyzing the water-splitting reaction in PSII (Amesz, 1993; Millaleo et al., 2010; Prince, 1986). In photolysis of water, a group of four atoms of Mn (Mn cluster) is associated with the oxygen-evolving complex bound to the reaction center protein of PSII

(Goussias et al., 2002). This Mn cluster accumulates four positive charges, which oxidize two water molecules, releasing one O_2 molecule and four protons. Therefore, this metal cluster is considered a catalytic compound for water oxidation (Zouni et al., 2001), where Mn ions are close to a redox-active tyrosine residue (Goussias et al., 2002).

Husted et al. (2010) reported that Mn deficiency–induced alterations in O_2 evolution were correlated with changes in the ultrastructure of thylakoid membranes and loss of PSII functional units in the stacked areas of thylakoid membranes. Resupplying Mn restored the number of the PSII protein pigment units in the thylakoid membranes (Gong et al., 2010; Simpson and Robinson, 1984).

12.2.2 INVOLVEMENT OF MANGANESE IN PLANT ENZYMATIC REACTIONS

12.2.2.1 Manganese-Activated Enzymes

Most studies on Mn activation of enzymes have been carried out *in vitro*, and in many cases, Mn^{2+} can be replaced by magnesium (Mg^{2+}) or vice versa (Broadley et al., 2012). Given that the concentration of Mg^{2+} in the cell is on average about 50–100 times higher than that of Mn^{2+}, activation of enzymes by Mn^{2+} *in vivo* presumably is important only for those enzymes where Mn^{2+} is a more effective cofactor than Mg^{2+} (Broadley et al., 2012). An example of the higher effectiveness of Mn^{2+} over Mg^{2+} in catalyzing enzymes is the case of activation of chloroplast RNA polymerase. Activation of chloroplast RNA polymerase requires about one-tenth the concentration of Mn^{2+} than Mg^{2+} (Ness and Woolhouse, 1980).

Phosphoenolpyruvate (PEP) carboxykinase is an enzyme that catalyzes the decarboxylation of oxaloacetate in the bundle sheath chloroplasts of C_4 plants (Broadley et al., 2012). The enzyme PEP carboxykinase has an absolute requirement for Mn^{2+} that cannot be replaced by Mg^{2+} (Burnell, 1986). Maximum activity of the enzyme occurs at a Mn/ATP ratio of one, suggesting that the substrate for the enzyme is the Mn/ATP complex, rather than Mg/ATP complex, as in most other enzymatic reactions (Burnell, 1986). Another Mn-dependent enzyme is allantoate amidohydrolase, an enzyme responsible for the degradation of allantoin and allantoate in the leaves and seed coats of leguminous plants (Winkler et al., 1985, 1987). In legumes, such as soybean (*Glycine max* Merr.), nitrogen (N) is transported as allantoin or allantoate to the shoots (Broadley et al., 2012).

In addition to chloroplast RNA polymerase, PEP carboxykinase, and allantoate amidohydrolase, which depend on Mn as a sole catalyst, Mn^{2+} activates about 32 other enzymes, most of which catalyze oxidation–reduction, decarboxylation, and hydrolytic reactions (Broadley et al., 2012; Brunell, 1988). Some of these enzymes may be catalyzed by Mn^{2+} or Mg^{2+} and do not depend solely on Mn^{2+}. Manganese has a role in the tricarboxylic acid cycle in oxidative and nonoxidative decarboxylation reactions (Broadley et al., 2012). For example, Mn^{2+} catalyzes the NADPH-specific decarboxylating malate dehydrogenase, malic enzyme, and isocitrate dehydrogenase (Broadley et al., 2012).

$$\text{Malate} + \text{NADP}^+ \xrightarrow[\text{Mn}^{2+},\ \text{Mg}^{2+}]{\text{Malic enzyme}} \text{Pyruvate} + \text{NADPH} + \text{H}^+ + \text{CO}_2$$

$$\text{Isocitrate} + \text{NADP}^+ \xrightarrow[\text{Mn}^{2+},\ \text{Mg}^{2+}]{\text{Isocitarte dehydrogenase}} \text{Oxalosuccinate} + \text{NADPH}$$

Manganese activates several enzymes of the shikimic acid pathway and subsequent pathways, leading to the biosynthesis of aromatic amino acids (such as tyrosine), various secondary products (such as lignin and flavonoids), and indoleacetic acid (Broadley et al., 2012; Burnell, 1988; Hughes and Williams, 1988). For example, Mn stabilizes the active conformation of phenylalanine ammonia lyase (Wall et al., 2008), stimulates peroxidases, and works as a distributing redox vehicle in combination with peroxidases in lignin biosynthesis (Broadley et al., 2012; Önnerud et al., 2002). Morgan et al. (1976) reported that IAA oxidase activity was high in cotton (*Gossypium hirsutum* L.) leaves suffering from Mn deficiency or Mn toxicity. Manganese also catalyzes phytoene synthetase, an enzyme involved in the biosynthesis of isoprenoids (precursors of carotenoids, sterols, and gibberellic acid (GA) (Broadley et al., 2012; Wilkinson and Ohki, 1988).

A role of Mn in the activity of nitrate reductase has been suggested because of an increase of nitrate concentration in Mn-deficient leaves (Broadley et al., 2012). Leidi and Gómez (1985) attribute the increase of nitrate concentration in the Mn-deficient leaves to the adverse effects of Mn deficiency on plant growth rather than to a direct role of Mn in nitrate reductase activity.

12.2.2.2 Manganese-Containing Enzymes

Although a relatively large number of enzymes are activated by Mn^{2+}, only a few enzymes actually contain Mn. The Mn-containing enzymes include the Mn superoxide dismutase (SOD), oxalate oxidase, and the Mn protein in PSII (Broadley et al., 2012). The most well-known and best documented Mn-containing enzyme is the 33 kDa polypeptide of the water-splitting system in PSII. In this system, four Mn atoms and one atom of Ca are arranged as a cluster (Mn_4Ca), which stores positive charges prior to the four-electron oxidation of two molecules of water as follows:

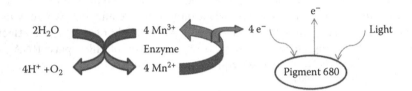

The work of Mn atoms, in storing transient electron charge and transmitting electrons, is coupled with fluctuations in the oxidation state of Mn between Mn^{2+} and Mn^{3+} (Rutherford, 1989). The Mn_4Ca catalytic cluster cycles through five oxidation states coupling the one-electron photochemistry of the reaction center with the four-electron redox chemistry of water oxidation (Yano, 2010). In photosynthesizing cells, PSII is the most sensitive metabolic utility impaired by Mn deficiency (Broadley et al., 2012).

Another Mn-containing enzyme is oxalate oxidase, which is multimeric, glycosylated, and homohexameric (Broadley et al., 2012; Dunwell et al., 2001; Kanautchi et al., 2009). This enzyme belongs to the large family of germin-like proteins termed cupins because of their conserved beta-barrel fold (Broadley et al., 2012). The active site of oxalate oxidase is in the center of the barrel and contains a Mn ion (Dunwell et al., 2001).

Manganese superoxide dismutase (MnSOD) also is a Mn-containing enzyme. SODs are present in all aerobic organisms and play an essential role in plant defense against oxidative stress, produced by elevated levels of activated forms of oxygen and free radicals (reactive oxygen species, ROS) (Elsner, 1982; Fridovich, 1975, 1983). They protect tissues from the deleterious effect of the oxygen radical (O_2^-) formed in various enzyme reactions, such as in photosynthesis, in which a single electron is transmitted to O_2 (Elsner, 1982; Fridovich, 1975, 1983):

$$O_2 + e^- \longrightarrow O_2^- \text{ (superoxide)}$$

$$O_2^- + O_2^- \xrightarrow{\text{Superoxide dismutase (SOD)}} 2H_2O_2 \text{ (hydrogen peroxide)} + O_2$$

$$H_2O_2 \xrightarrow{\text{Catalase}} 2H_2O + O_2$$

The conversion of O_2^- is catalyzed by SOD, and the subsequent dismutation of H_2O_2 into O_2 is facilitated by peroxidases, catalase, ascorbate-specific peroxidase, or ascorbate free radical reductase (Broadley et al., 2012; Elstner, 1982; Kröniger et al., 1992). Because of the role of Mn in MnSOD activity, it has been proposed that Mn can act as a scavenger of superoxide (O_2^-) and hydrogen peroxide (H_2O_2). MnSOD is not distributed widely in plants and is located mainly in mitochondria and in peroxisomes (Alscher et al., 2002; Broadley et al., 2012; Clemens et al., 2002; Sandmann and Böger, 1980). Bowler et al. (1991), however, reported the presence of MnSOD in the chloroplasts of cotton plants. In addition to MnSOD, there are other SOD isoenzymes that include iron (FeSOD) or copper and zinc (CuZnSOD) instead of Mn. The FeSOD is confined mainly to chloroplasts, and CuZnSOD is in chloroplasts and in peroxisomes and mitochondria (Palma et al., 1986).

Numerous transgenic plants have been produced over the last two decades with MnSOD targeted in the chloroplasts (Broadley et al., 2012). Such transgenic plants have showed increased tolerance to a range of abiotic stresses such as high temperature, salinity, and Mn deficiency (Silber et al., 2009; Tanaka et al., 1999; Wang et al., 2005, 2007; Waraich et al., 2012; Yu et al., 1999). Wang et al. (2004, 2007) studied the tolerance of transgenic tomatoes (*Lycopersicon esculentum* Mill.) and arabidopsis (*Arabidopsis thaliana* Heynh.) to salt (NaCl) stress. They reported that the plants responded to increased salt stress by an overexpression of MnSOD and that MnSOD increased the plants' tolerance to salt stress. The tolerance of plants to salt stress was associated with improvement in seed germination and root development (Wang et al., 2007).

Other enzymes reported to involve Mn include Mn catalase, malic enzyme, nitrate reductase, PEP carboxykinase, PEP carboxylase, pyruvate carboxylase, pyruvate kinase, RNA polymerases, and RuBP carboxylase (Burnell, 1988; Ducic and Polle, 2005; Graham, 1983; Houtz et al., 1988; Jackson et al., 1978; Marschner, 1995; Millaleo et al., 2010; Mousavi et al., 2011; Mukhopadhyay and Sharma, 1991; Uehara et al., 1974). Also through enzymatic activities, Mn is involved in the biosynthesis of chlorophyll, aromatic amino acids (tyrosine), lignin, flavonoids, isoprenoids, fatty acids, and acyl lipids (Burnell, 1988; Graham, 1983; Lidon et al., 2000; Ness and Woolhouse, 1980) and in nitrate assimilation (Burnell, 1983, 1988; Ducic and Polle, 2005; Jackson et al., 1978; Marschner, 1995; Millaleo et al., 2010; Mousavi et al., 2011; Mukhopadhyay and Sharma, 1991; Uehara et al., 1974).

12.2.3 INVOLVEMENT OF MANGANESE IN SYNTHESIS OF PROTEINS, LIPIDS, AND LIGNIN

Although Mn activates RNA polymerase (Ness and Woolhouse, 1980), protein synthesis is not especially reduced in Mn-deficient plants (Broadley et al., 2012). The protein concentration of deficient plants is either similar to or higher than that of plants adequately supplied with Mn (Lerer and Bar-Akiva, 1976). The accumulation of soluble N in Mn-deficient plants is due to a lower demand for reduced N and to limited nitrate reduction caused by a shortage of reducing equivalents and carbohydrates (Broadley et al., 2012). Indeed, Mn deficiency has a severe effect on the concentration of nonstructural carbohydrates (Broadley et al., 2012; Vielemeyer et al., 1969). A decrease in carbohydrate concentration is especially marked in the roots and is most likely a key factor responsible for the suppression of root growth in Mn-deficient plants (Mascar and Graham, 1987).

The role of Mn in lipid metabolism is complex. In Mn-deficient leaves, the concentration of thylakoid-membrane constituents such as glycolipids and polyunsaturated fatty acids may be decreased by up to 50% (Constantopoulus, 1970). This depression in lipid concentration in chloroplasts can be attributed to the role of Mn in biosynthesis of fatty acids and carotenoids (Broadley et al., 2012).

Manganese nutrition affects lipid concentration and composition in seeds (Wilson et al., 1982). In the deficiency range of leaf Mn (<20 mg Mn/kg dry weight), the Mn concentration in the leaves, the seed yield, and the seed oil concentration are all positively correlated (Wilson et al., 1982). The fatty acid composition of the oil also was affected markedly, with the concentration of linoleic acid increasing and the concentration of oleic acid decreasing with decreased leaf Mn (Wilson et al., 1982). The low oil concentrations in the seeds of Mn-deficient plants may have resulted from the direct involvement of Mn in the synthesis of fatty acid or from lower rates of photosynthesis and a consequent decreased supply of carbon skeletons for fatty acid synthesis or a combination of both (Broadley et al., 2012).

Manganese is involved in the biosynthesis of lignin through the activation of the enzyme peroxidase (Broadley et al., 2012). Peroxidase is involved in the polymerization of cinnamyl alcohols into lignin. Manganese deficiency restricts lignin concentration in plants and particularly in roots (Brown et al., 1984a,b; Rengel et al., 1993). Lignin is considered an important defense component against fungal infections (Rengel, 1993). Reduced lignin concentration in the roots of Mn-deficient plants is an important factor responsible for the low resistance of Mn-deficient plants to root-infecting pathogens (Broadley et al., 2012).

12.2.4 INVOLVEMENT OF MANGANESE IN CELL DIVISION AND ELONGATION

Manganese has a direct effect on cell division and elongation (Broadley et al., 2012). Abbott (1967) studied the effect of Mn on growth of excised roots of tomato and reported that the growth of the main axis was suppressed and that the formation of lateral roots ceased completely under Mn-deprived conditions. Compared to Mn-sufficient roots, there was a greater abundance of small, nonvacuolated cells in Mn-deficient roots, indicating that Mn deprivation impairs cell elongation more strongly than cell division, an observation also supported by research (Neumann and Steward, 1968).

12.3 AVAILABILITY OF MANGANESE TO PLANTS

Plant growth depends on the availability of water and nutrients in the rhizosphere, of which thickness varies between 0.1 mm and a few millimeters depending on the length of root hairs. Some plants such as blue lupin (*Lupinus angustifolius* L.) have an exceptionally large rhizosphere of up to 18 mm in diameter (Papavizas and Davey, 1961). Availability of Mn in the rhizosphere is controlled by the combined effects of Mn content in the soil, soil properties, plant characteristics, and the interactions of plant roots with microorganisms and the surrounding soil (Bowen and Rovira, 1992).

12.3.1 MANGANESE CONTENT, FORMS, AND DYNAMICS IN SOILS

Manganese is the eleventh most common element in the crust of the earth, with an average concentration of total Mn of 900 mg/kg (Barber, 1995). Manganese biogeochemistry in soils is complex because it is present in several oxidation states (0, 2, 3, 4, 6, and 7) in the soil (Guest et al., 2002). Divalent manganese (Mn^{2+}) is the most available form of Mn in the soil (Guest et al., 2002). Manganese bioavailability depends on the soil pH, redox reactions, microbial activity, temperature, water, and the crops (Marschner, 1995; Porter et al., 2004).

The relationship between soil pH and Mn availability is complex. Theoretically, in aerated soils, Mn^{2+} concentration in soil solution should decrease by a divisor of 100 for every unit of pH increase (Barber, 1995). However, with various organic compounds capable of complexing Mn and changing solubility equilibria, a decrease in Mn^{2+} concentration with an increase in pH is not that severe (Neilsen et al., 1992). In fact, a 0.5-unit increase in soil pH results in no more than 10%–20% decrease in Mn^{2+} concentration in the soil solution (Neilsen et al., 1992). The complexity of the relationship between Mn concentration in soil solution and soil pH was illustrated by Rule and Graham (1976) who used ^{45}Mn to determine the soil Mn pool. With an increase in pH, soil Mn pools under white clover (*Trifolium repens* L.) actually increased, whereas soil Mn pools under tall fescue (*Festuca elatior* L.) decreased. So soil supply of Mn is a complex variable that depends not only on soil chemistry but on plant responses, as well as the activity of microorganisms.

In most acid soils (pH < 5.5), Mn oxides are reduced in the soil exchange sites (Kogelmann and Sharpe, 2006), increasing the concentration of soluble Mn^{2+} (Adriano, 2001; Watmough et al., 2007) and the most available Mn form for plants (Marschner, 1995). At higher soil pH (pH up to 8), chemical Mn^{2+} auto-oxidation is favored over MnO_2, Mn_2O_3, Mn_3O_4, and even Mn_2O_7, which are sparingly soluble and not normally available to plants (Ducic and Polle, 2005; Gherardi and Rengel, 2004; Humpries et al., 2007). Furthermore, high pH allows Mn adsorption onto soil particles, decreasing their availability (Fageria et al., 2002). Nevertheless, some reports have suggested that an excess of available Mn is produced under reduced soil conditions, even at high soil pH values (Hue, 1988). Under anaerobic conditions, MnO_2 and also NO_3^- and Fe^{3+} serve as alternative electron acceptors for microbial respiration and are transformed into reduced states. This process increases the solubility and availability of Mn and Fe to plants (Ghiorse, 1988). A reducing environment can be produced with poor drainage or applications of organic material (El-Jaoual and Cox, 1998; Hue, 1988). Organic molecules can dissolve solid Mn oxides through transfer of electrons, transforming them into an available Mn form for plants (Laha and Luthy, 1990). Liming decreases the availability of Mn^{2+} in acid soils (Hue and Mai, 2002).

Environmental conditions also affect soil Mn contents. High soil temperature was reported to increase the availability of Mn^{2+} in the soil, due to inhibition of Mn-oxidizing organisms, thereby allowing the chemical reduction of Mn oxides in the soil (Conyers et al., 1997; Sparrow and Uren, 1987).

Mobilization and immobilization of nutrients, including Mn, occur in the rhizosphere (Marschner, 1995). The mobilization of Mn^{2+} is produced by the rhizosphere acidification due to the release of H^+ or low-molecular-weight organic acids from plants (Rengel and Marschner, 2005). Organic acids released in anionic forms from roots chelate Mn^{2+} released from Mn oxides (Ryan et al., 2001). Neumann and Römheld (2001) reported that mobilization of Mn into the rhizosphere is due mainly to its acidification and complexation with the organic acids (e.g., citrate) in various plant species. It has been reported that organic amendments, wood chip compost and pine bark, released organic compounds such as arabinose and malic acid (Tsuji et al., 2006). Arabinose and malic acid dissolve manganese oxides (MnO_x) and release Mn^{2+} into the rhizosphere (Tsuji et al., 2006).

Soil microorganisms also can cause mobilization or immobilization of Mn, depending on soil conditions (Marschner, 1995). In aerated soils, microorganisms may mobilize Mn through reduction of manganese oxides, and this process is favored by exudation of H^+ from roots (Marschner, 1995). In contrast, Mn-oxidizing bacteria can decrease Mn availability in aerated, poorly aerated, or calcareous soils (Marschner, 1995). Another key factor in the Mn dynamics in soil is organic matter (OM). Given that OM is negatively charged, it has a great Mn adsorption capacity, forming nonexchangeable Mn complexes (Marschner, 1995). However, exchangeable Mn can be released by the H^+ donated from roots (Bradl, 2004).

12.4 UPTAKE OF MANGANESE BY PLANTS

Manganese is absorbed preferentially by plants as the free Mn^{2+} ion from the soil solution (Geering et al., 1969; Kochian, 1991; Page et al., 1962). It also may be absorbed in complexes, with plant and microbial organic ligands or with synthetic chelates (Rengel, 2000). Kinetics of uptake differ with the two forms. Complexed forms of Mn are absorbed generally more slowly by roots than the free cation (Barber and Lee, 1974; Webb et al., 1993). In a study on the kinetics of Mn uptake by roots as free ion or complexed with ethylenediaminetetraacetic acid (EDTA), Rengel (2000) reported that free Mn^{2+} was taken up across the root-cell plasma membrane up to 50 times faster than the Mn–EDTA complex, even when all Mn was supplied as Mn–EDTA. In this study, it was not certain whether EDTA was transported across the plasma membrane as a complex with Mn or as Mn-independent entity (Rengel, 2000).

Manganese uptake depends on the nutritional status of the plant as well as the proportion of Mn that is complexed (Pearson and Renget, 1994). Under nutrient deficiency conditions, chelation of Mn with EDTA (up to 98% of Mn present in the chelated form) actually increased Mn uptake (Laurie et al., 1991). Increased complexation of Mn with EDTA in the nutrient solution (around 45% of total Mn present as a complex) increased accumulation of Mn in barley (*Hordeum vulgare* L.), indicating that not only Mn^{2+} but also the Mn–EDTA complex can be taken up across the plasma membrane (Pearson and Rengel, 1994).

12.4.1 KINETICS OF MANGANESE UPTAKE

Although there is a relative paucity of information regarding kinetics and regulation of Mn uptake by intact plants grown at realistically low Mn concentrations, Mn uptake has been intensively studied in algae and yeast. In microorganisms, transport of Mn from the external medium across the plasma membrane frequently has been studied together with the transport to cell organelles (Duerr et al., 1998).

Manganese uptake by roots occurs in phases. Initial absorption into plants not previously exposed to Mn represents transport into the apoplasm of roots (Graham et al., 1976; Page, 1964). This phase is rapid, reversible, and nonmetabolic with a half time of 8 min. In this phase, Mn^{2+} is adsorbed by the negatively charged cell wall constituents of the root-cell apoplastic spaces, and absorbed Mn is exchanged freely with Mn^{2+} or Ca^{2+} ions in the rhizosphere (Clarkson, 1988; Graham et al., 1976; Humphries et al., 2007; Page, 1964; Page and Dainty, 1964). Another phase of Mn^{2+} uptake is into the cytosol and has a half time of up to 2 h (Munns et al., 1963). A fraction with a longer half time (up to 28 h) represents transport into the vacuole (Munns et al., 1963; Page and Dainty, 1964). This phase is slow and metabolic, with Mn^{2+} being less readily exchanged (Maas and Moore, 1968). The energy dependence of Mn^{2+} uptake has been shown in a number of lower and higher plant species (Garnham et al., 1992; Maas and Moore, 1968; Parkin and Ross, 1986, Ratkovic and Vucinic, 1990, Roby et al., 1988), but it is likely to be due to an indirect energy requirement that powers H^{+}-ATPase and allows maintenance of electrochemical potential difference across the plasma membrane, which then drives the uptake of cations into the negatively charged cytoplasm (Rengel, 2000).

In an experiment on Mn^{2+} uptake by transgenic tobacco (*Nicotiana tabacum* L.), transformed with a tomato root protein with a metal-binding site at its N-terminus (LeGlpl), Takahashi and Sugiura (2001) reported that Mn^{2+} binds to the protein. This finding strongly suggests the involvement of LeGlpl in Mn^{2+} uptake from the soil (Takahashi and Sugiura, 2001).

Kinetic measurements have demonstrated 100–1000 times higher rates of Mn^{2+} transport than the estimated plant requirement for this element (Clarkson, 1988).

Uptake kinetics of Mn^{2+} range from linear to saturable and may include low- and high-affinity plasma membrane transport, depending on the experimental system (Rengel, 2000). Webb et al. (1993) reported a linear relationship between ionic activity of Mn^{2+} in solution and the Mn accumulation rate in barley. From experiments on Mn^{2+} uptake, using pH-buffered nutrient solutions, they reported the presence of two separate Mn uptake systems in barley mediating high- and low-affinity Mn^{2+} influx.

Estimates of K_m for low-affinity Mn uptake by roots range between 4 and 400 μM, which are relatively high concentrations in comparison with those in the soil solution, which indicates that uptake of Mn from most soils occurs in the concentration range where the relationship between soil solution Mn^{2+} concentration and uptake is approximately linear (Barber, 1995). The calculated supply of Mn^{2+} from soil by mass flow and root interception is reported to be linearly related to the measured total Mn accumulation by species with vastly different types of root system, such as soybean, wheat (*T. aestivum* L.), tomato, and lettuce (*Lactuca sativa* L.) (Halstead et al., 1968).

12.4.2 UPTAKE MECHANISM

Photosynthetic autotrophs have evolved specific routes for the entry of Mn^{2+} into the cell. A so-called ABC-type permease has been reported to be responsible for the uptake of Mn^{2+} in cyanobacteria (Bartsevich and Pakrasi, 1995). In yeast (*Saccharomyces cerevisiae* Meyen ex E.C.Hansen), the accumulation of Mn^{2+} is mediated by the natural resistance–associated macrophage protein (NRAMP) family transporter SMF1 and by the high-affinity phosphate (Pi) transporter PHO84 (Jensen et al., 2003). Similarly, in an alga (*Chlamydomonas reinhardtii* P.A. Dang), a NRAMP protein was identified as the main component of a Mn^{2+}-selective uptake pathway (Allen et al., 2007). No Mn^{2+}-specific transporter has been identified in plants to date, probably because Mn shares the same entry route as Fe (Yang et al., 2008). In arabidopsis, Fe is taken up by IRT1, a member of the ZIP transporter family (Eide et al., 1996; Henriques et al., 2002; Varotto et al., 2002; Vert et al., 2002). IRT1 has a relatively broad substrate spectrum and is reportedly capable of transporting other metal ions including Mn^{2+} (Korshunova et al., 1999). Recently, a transporter of the cation diffusion facilitator (CDF) family, MTP11, was shown to be crucial for maintaining Mn homeostasis in plants (Delhaize et al., 2007; Peiter et al., 2007). However, MTP11 is not expressed in the root epidermis, nor is its transcript level increased upon Mn deficiency, indicating that this protein functions in Mn tolerance rather than in Mn acquisition from the soil (Peiter et al., 2007). This assumption is supported by its localization in prevacuolar compartments (Delhaize et al., 2007). A Golgi-localized P2A-type ATPase, ECA3, was important for Mn homeostasis under Mn-depleted conditions, probably by mediating the loading of Mn^{2+} into the Golgi (Mills et al., 2008). Pedas et al. (2008) reported that a gene of the ZIP family with high similarity to *OsIRT1*, designated as *HvIRT1*, controls manganese uptake in barley roots.

In yeast, the *SMF1* gene codes for the plasma membrane proteins involved in the transport of Mn^{2+} (Pinner et al., 1997; Supek et al., 1996). The yeast *SMF1* gene is homologous to the *NRAMP2* gene that encodes the mammalian macrophage membrane protein, whereas the *OsNRAMP* genes in rice (*Oryza sativa* L.) encode membrane proteins that share high homology in the hydrophobic membrane-spanning region with the NRAMP transporter (Belouchi et al., 1997); therefore, the three genes from the rice *OsNRAMP* family may code for the first Mn^{2+} membrane transporters cloned from the higher plants (Rengel, 2000).

Ion channels embedded in the root-cell plasma membrane, even when selective for ions other than Mn^{2+}, can still facilitate uptake of Mn^{2+} (e.g., the Ca^{2+}-selective ion channel from the wheat plasma membrane) (Piñeros and Tester, 1995). Such an uptake may decrease the flux of the ion for which the channel is selective (e.g., Mn^{2+} decreased Ca^{2+} uptake through the Ca^{2+}-selective channel in corn plasma membrane by 70%) (Marchall et al., 1994), probably because of binding of Mn^{2+} to the Ca^{2+} binding site on the apoplastic side of the channel (Schumaker and Sze, 1985).

12.4.3 REGULATION OF MANGANESE UPTAKE

The physiological requirement of plants for Mn is low, and Mn^{2+} uptake capacity greatly exceeds plant requirements, indicating a poor regulation of Mn uptake. Kinetic experiments have estimated absorption to be 100–1000 times greater than the need of plants (Clarkson, 1988). This high absorption may be due to the high capacity of ion carriers and channels in the transportation of Mn^{2+} through the plasma membrane, at a speed of several hundred to several million ions per second per protein molecule (Tester, 1990; Tyerman, 1992).

12.4.3.1 Genetics of Acquisition of Manganese by Plants

Plants differ in their ability to take up Mn^{2+}. Manganese concentrations in plant tissue range from 4 mg/kg dry weight in bacon and eggs (*Oxylobium capitatum* L.) plants to 2180 mg/kg dry weight in frogfruit (*Phyla nodiflora* L.) (Foulds, 1993). Some plants in the Proteaceae accumulate Mn to over 300 mg/kg dry weight while growing on soil that is considered deficient in plant-available Mn (<2 µg Mn/g soil in the DTPA extract) (Foulds, 1993). The ability of some plants to accumulate high concentrations of Mn has been associated with proteoid (cluster) roots that Proteaceae species produce and to the chemistry and biology of the rhizosphere (Renzel, 2000).

Plant species (Lemanceau et al., 1995, Wiehe and Hoflich, 1995) and genotypes within a species (Khanna et al., 1993, Timonin, 1946) differentially influence the quantitative and qualitative composition of microbial populations in the rhizosphere. Therefore, rhizosphere microorganisms may have a role in the expression of genotypic differences in tolerance to Mn deficiency. Timonin (1946) reported greater numbers of Mn-oxidizing microbes in the seed coats of oat (*Avena sativa* L.) cultivars sensitive to Mn deficiency than in tolerant ones. They concluded that this difference in the number of Mn-oxidizing microbes in the coats was the basis of genotypic differences in Mn uptake.

Similarly, the rhizosphere of "Aroona" wheat, a plant tolerant to Mn deficiency, contained more Mn-reducing microorganisms under Mn deficiency than under Mn sufficiency (Rengel et al., 1996).

Under Mn deficiency, sensitive genotypes did not have an increase in the ratio of Mn reducers to Mn oxidizers in the rhizosphere (Rengel, 1997; Rengel et al., 1996). The genetic control of tolerance to Mn deficiency may be expressed through the qualitative or quantitative composition of root exudates encouraging a more favorable balance of Mn reducers to Mn oxidizers in the rhizosphere (Rengel, 1997).

12.4.3.2 Manganese Uptake as Affected by Other Nutrients

The Mn^{2+} ion has similar properties to alkaline cations such as Ca^{2+} and Mg^{2+} and heavy metals such as Zn^{2+} and Fe^{2+}; thus, these ions affect the uptake and transport of manganese (Aref, 2011; Hewitt, 1948; Marschner, 1995; Spiers, 1993).

Deficiency of one nutrient could substantially increase uptake of one or more other nutrients (Kochian, 1991). The compensatory absorption of micronutrients under deficiency stress has been reported between Mn^{2+} and Cu^{2+} or Zn^{2+} (del Rio et al., 1978), Zn^{2+} and Fe^{2+} or Cu^{2+} (Rengel and Graham, 1995, 1996; Rengel et al., 1998; Welch and Norvell, 1993), and Mn^{2+} and Fe^{2+} (Brown and Olsen, 1980; Iturbe-Ormaetxe et al., 1995; Welch et al., 1993).

12.4.3.2.1 Uptake Interactions with Iron

The interaction between Mn and Fe in their absorption by plants has been known for quite some time (Johnson, 1917). The nature of the interaction varies among plant species and is not always consistent among studies. The interaction between Mn and Fe was absent in tomato, sorghum (*Sorghum* spp. L.), rice, or wheat (Foy et al., 1978; Jaurigui and Reisenauer, 1982; Kuo and Mikkelson, 1981; Sanchez-Raya et al., 1974; Van Der Vorm and Van Diest, 1979); positive in oats, wheat, or barley (Mashhady and El-Damary, 1981; Shuman and Anderson, 1976; Singh and Pathak, 1968); and negative in pineapple (*Ananas comosus* Merr.), soybean, mustard (*Brassica* spp. L.), tobacco, rice, barley, or flax (*Linum usitatissimum* L.) (Beauchamp and Rossi, 1972; Dekock and Inkson, 1962; Heenan and Campbell, 1983; Hiatt and Ragland, 1963; Johnson, 1917; Moraghan, 1979; Somers and Shive, 1942; Vlamis and Williams, 1964). Beauchamp and Rossi (1972) reported that high levels of Mn in the solution culture depressed Fe absorption by barley.

Morris and Pierre (1948) reported that the beneficial action of Fe in reducing Mn toxicity appeared to be due to some mechanism that prevents the excessive absorption and subsequent accumulation of Mn in the plant rather than to an increase in Fe^{2+} absorption and corrective action of Fe^{2+} within the plant. They reported that alleviation of Mn toxicity by higher Fe^{2+} concentration (1 mg/L compared with 0.2 mg/L) was not due to an increase of total iron in the plant but to an approximate 50% reduction in the Mn content of the plant.

Michael and Beckg (2001) and Malakouti and Tehrani (1999) reported a negative correlation between Mn^{2+} and Fe^{2+} absorption. Clarkson (1988) attributed this negative correlation to the plasma membrane–embedded ferric reductase of root cells. Later research (Norvell et al., 1993) showed that the plasma membrane ferric reductase of peas and soybeans effectively reduced Mn^{3+} to Mn^{2+} and that this reduction was increased more than 10-fold under Fe deficiency that stimulated reductase activity. This conclusion was supported by a study on beans (*Phaseolus vulgaris* L.), where plants were subjected to Fe deficiency to induce ferric reductase activity (Norvell et al., 1993). In this study, the increase in ferric reductase activity resulted in increased Mn uptake (Norvell et al., 1993).

12.4.3.2.2 Uptake Interactions with Molybdenum and Phosphorus

Millikan (1948) reported that an essential function of molybdenum is associated intimately with the regulation of the deleterious effects of Mn on the physiological availability of Fe to the plants. Kirch et al. (1960) reported that Mn absorption in tomatoes was influenced by interactions among Fe, Mn, and Mo. Iron supply counteracted the yield-depressive effects of excess Mn, but the amount of Fe needed for maximum yield was increased as the Mo supply increased. Molybdenum suppressed yields at low Fe levels and increased yields at higher Fe levels (Kirch et al., 1960).

Foy et al. (1978) reported that low P levels in the medium reduced the effect of Mn toxicity in chickpeas (*Cicer arietinum* L.) because low P increased Fe^{2+} uptake.

12.4.3.2.3 Uptake Interactions with Calcium

The effectiveness of Ca in influencing Mn uptake varies depending on the growing conditions, plant species, and plant age (Löhnis, 1960; White, 1970). Bekker et al. (1994) reported that liming (31.1% Ca and 1.7% Mg) of acid soil alleviated Mn toxicity in pea (*Pisum sativum* L.) and explained that this alleviation of Mn toxicity likely was due not to decreased Mn solubility, because the lime application increased soil pH by less than 0.1 unit, but rather to an increase in Ca availability in the soil. It has been reported that application of Ca to the growth media decreased shoot Mn concentration of plants, probably due to an antagonistic relationship between Mn^{2+} and Ca^{2+} absorption (Fageria and Baligar, 1999; Williams and Vlamis, 1957c). These findings suggested that the growth inhibition of plants grown in Mn toxic conditions may be overcome by higher Ca supply (Fageria and Baligar, 1999; Williams and Vlamis, 1957c). Alam et al. (2006) investigated the mechanism by which high Ca supply to plants alleviated Mn toxicity in barley and reported that Ca^{2+} inhibited Mn^{2+} translocation to shoots but did not affect Mn^{2+} uptake.

The choice of Ca salt to suppress plant Mn is influenced strongly by the accompanying anion. According to Vlamis and Williams (1962), the order of accompanying anions in decreasing Mn in the leaves of barley is phosphate > nitrate > sulfate. Calcium chloride may be used as a Ca source considering its high solubility and low price (Alam et al., 2000).

12.5   MANGANESE TRANSPORT AND ACCUMULATION IN PLANTS

Symptoms of Mn deficiency normally do not appear in old leaves, suggesting that Mn is an immobile element (El-Baz et al., 1990; Hill et al., 1979; Nable and Loneragan, 1984a,b; Pearson and Rengel, 1994). However, symptoms of Mn deficiency regularly appear on fully expanded young leaves rather than on the newest leaves. This symptom may indicate an internal requirement in the fully expanded new leaves beyond that of the newest leaves (Grundon et al., 1997), or it may simply be a matter of supply and demand in what is the fastest growing tissue (Humphries et al., 2007).

12.5.1   Long-Distance Transport of Manganese

The location of Mn in plants is a significant factor in the expression of deficiency symptoms and is affected by its mobility in the xylem and phloem (Humphries et al., 2007). Romani and Kannan (1987) reported that Mn moves easily from the root to the shoot in the xylem transpiration stream. In contrast, translocation of Mn^{2+} within the phloem is complex, with leaf Mn being immobile, but root and stem Mn being able to be remobilized (Loneragan, 1988).

Moussavi-Nik et al. (1997) measured remobilization of Mn from the grain coat and the endosperm of germinating wheat grains. They reported that remobilized Mn was directed to roots during the first 3 days after commencement of imbibition and to roots and shoots in the later stages. Grain coat represented the largest accumulation site of Mn in the grain, which was depleted by 30%–55% during germination depending on genotype (Moussavi-Nick et al., 1997).

According to Humphries et al. (2007), Mn transport from the roots to the whole plant involves transport in the xylem, transfer from the xylem to the phloem, and subsequent retranslocation from the phloem. Xylem transport from roots to the aboveground parts of plants is performed by the transpiration stream, whereas phloem transport is more selective, taking place from sources to sinks (Marschner, 1995). Nonetheless, low mobility in phloem has been reported, and Mn redistribution depends on the plant species and stages of development (Herrén and Feller, 1994).

Page and Feller (2005) studied transport and accumulation of radioactively labeled [54]Mn in wheat at an early stage of development. They reported that Mn accumulated in shoots more than in roots and reported that Mn^{2+} was transported rapidly from roots to shoots in the xylem and was essentially immobile in the phloem. Similar transport patterns of Mn^{2+} were shown, by the same technique, in young (28 days) white lupine (*Lupinus albus* L.) (Page et al., 2006). However, in older lupine plants, Mn was present in a large amount in the root system, hypocotyls and stem, immediately after the labeling phase (day 0). Seven days later (day 7), almost all of the [54]Mn had moved to the youngest fully expanded leaves and only a small fraction to the other leaves. However, Mn accumulation was detected in the periphery of the oldest leaves. Page et al. (2006) concluded that Mn was released rapidly into the xylem, reaching photosynthetically active leaves via the transpiration stream. They also discussed that the low mobility of Mn via phloem may be due to a restricted loading of soluble Mn into the phloem or to precipitation of Mn. Page and Feller (2005) emphasized that little is known about the mechanisms involved in the loading of Mn into the phloem and the chemical transport forms.

Studies on the mobility of Mn in wheat, lupines (*Lupinus* spp. L.), and subterranean clover indicate no mobilization of Mn from the old leaves to the younger ones (Hannam et al., 1985a; Nable and Loneragan, 1984a; Single and Bird, 1958a,b). Further support for this lack of mobility of Mn in the phloem from the old leaves to the young leaves was given in a study by Nable and Loneragan (1984a). In this study, plants provided with an early supply of $^{54}Mn^{2+}$ failed to remobilize Mn after roots were placed in a solution with a low concentration of nonradioactive Mn^{2+}.

12.5.1.1 Mobilization of Manganese

There are differences among species and genotypes within species in the capacity to remobilize Mn from different plant parts. For example, white lupine can remobilize Mn from leaves, whereas subterranean clover, blue lupine (*L. angustifolius* L.), or soybean cannot (Drossopoulos et al., 1994; Nable and Loneragan, 1984a; Reay, 1987). In contrast to leaves, mobilization of Mn from stems of blue lupine and shoots and roots of wheat enhanced grain loading of Mn substantially (Bowen and Rovira, 1961; Pearson and Rengel, 1994, 1995). Manganese in petioles of blue lupine also is mobilized, but the size of that Mn pool is too small to contribute to grain nutrition in a major way (Bowen and Rovira, 1961; Rengel, 2000).

Rengel (2000) reported that the majority of Mn from vegetative tissues of wheat was mobilized within the first 2 weeks following anthesis. However, Mn was mobilized only from the roots and shoots, which have a direct xylem connection to the ear and not from the leaves. Mobilization was greater under Mn deficiency than under Mn sufficiency (Pearson and Rengel, 1994). Some species, such as rice, subterranean clover, and blue lupine also can mobilize Mn from roots (Bowen and Rovira, 1961; Munns et al., 1963; Nable and Loneragan, 1984a).

Although the stems in blue lupine and the shoots in wheat, during early stages of seed development, can represent stocks of Mn for later transport into developing seeds, it is not known where in the stems or shoots Mn is stored and which mechanisms allow for this reversible loading of Mn into the cells of stems and shoots (Bowen and Rovira, 1961; Pearson and Rengel, 1995; Rengel, 2000). It has been suggested that the pod walls in lupines can represent stores of Mn for developing seeds (Hocking et al., 1977).

A small proportion of Mn (from 1% of the applied amount in corn to 10% in rice) externally applied to leaves can be translocated from treated leaves (El Baz et al., 1990; Rengel, 2000; Vose, 1963). However, Mn that accumulated in leaves in a physiological way via the xylem sap did not appear to mobilize in most plant species (Loneragan, 1988). In addition, Mn did not move out of senescing leaves of subterranean clover or blue lupine (Hannam et al., 1985b; Nable and Loneragan, 1984b). An absence of Mn mobilization from the leaves indicates that Mn is present in the leaves in insoluble complexes (Rengel, 2000). However, some Mn can be leached out of the leaves by water (Bowen and Rovira, 1961; Nable and Loneragan, 1984a).

12.5.1.2 Loading of Manganese into Developing Grain

Mineral nutrient reserves in the seed must be adequate to sustain growth until the root system can take over the nutrient supply function (Asher, 1987; Moussavi-Nick et al., 1997). Large seed reserves of Mn are important for crops grown in soils that are deficient in Mn (Longnecker et al., 1991a, 1998; Moussavi-Nick et al., 1997). During germination and early growth in Mn-deficient soil, high Mn content in seeds will improve seedling vigor, tolerance to soil-borne root and crown diseases, canopy development, and yield potential (Rengel et al., 1993, 1994, 1999).

Mineral nutrient content determines the quality of seed (Welch, 1986). Indeed, in blue lupine, seeds with low Mn concentration split, and their cotyledons and embryos protrude through the seed coats, a response that limits the commercial value of the seeds and their capacity to germinate (Crosbie et al., 1994; Longnecker, et al., 1996; Walton, 1978). A different species of the same genus, white lupine, does not suffer from the split-seed disorder. White lupine forms cluster roots that solubilize plant-unavailable MnO_2 into plant-available Mn^{2+} (Gardner et al., 1982). Therefore, white lupine accumulates more Mn than blue lupine, which does not form cluster roots (Gladstones, 1962; White et al., 1981).

In some plant species, Mn is transported toward developing seeds via the phloem, whereas in other species, the xylem is the most important pathway of Mn translocation to the seeds (Rengel, 2000). Phloem transport of Mn appears to be important in loading Mn into developing seeds of white or blue lupine, two species with a close relationship between Mn concentration in seeds and Mn concentration in the phloem sap (Bowen and Rovira, 1961; Hannam et al., 1985b; Hocking et al., 1977; Longnecker et al., 1998). In contrast, Mn transported in the xylem toward developing grains of wheat (*T. aestivum* L.) was not loaded into the phloem within the peduncle or the rachis but reached the floret via the xylem (Herren and Feller, 1994; Pearson and Rengel, 1995). As much as 75% of Mn transported toward the developing wheat grain ended up in transpiring floral structures (palea, lemma, and glumes) and was not translocated into the grain (Pearson et al., 1995; Rengel, 2000). This lack of translocation of Mn deprives the mature grain of Mn (Pearson and Rengel, 1994). It therefore appears that utilization of Mn in the grain loading process in wheat is fairly poor. O'Brien et al. (1985) reported that a relatively small portion of Mn remaining, after most of it was partitioned into palea, lemma, and glumes, was taken up by transfer cells in the grain stalk. This portion of Mn was transferred to the grain vascular system, with transport across the plasma membrane into the phloem being a limiting step (Pearson et al., 1995, 1996; Rengel, 2000). Thereafter, Mn was transferred from the phloem into the vascular system of the pericarp before being transported relatively slowly to the embryo of the maturing wheat grain (Rengel, 2000). At maturity, the grain embryo has up to 50-fold higher concentration of Mn than the endosperm, a condition that is important in grain germination and seedling establishment (Hannam et al., 1985b; Longnecker et al., 1991a; Lott and Spitzer, 1980; Moussavi-Nick et al., 1997; Peasrson et al., 1996, 1998; Wada and Lott, 1997).

12.5.2 SHORT-DISTANCE TRANSPORT OF MANGANESE

In addition to the long-distance transport of Mn, short-distance transport mechanisms are important for the translocation of this metal into cells and cell organelles. These mechanisms involve Mn translocation throughout the plasma membrane and the biomembranes of organelles (Ducic and Polle, 2005; Pittman, 2005). Delhaize et al. (2007) and Reddi et al. (2009) discussed possible mechanisms of Mn homeostasis and transport in yeast and in arabidopsis cells. They stated that transport proteins play an important role in the maintenance of adequate Mn concentrations in the cytoplasm. Moreover, a family of metal-transporter proteins with broad specificity, such as Fe^{2+} or Ca^{2+} transporters, also transports Mn into the cells (Pittman, 2005). Migocka and Klobus (2007) demonstrated activation of an antiport system in cucumber (*Cucumis sativus* L.) with different affinities for lead, Mn, nickel, or cadmium in the root plasma membrane. This antiport system acts as part of the general defense mechanism of the plant against the harmful effects of heavy metals (Migocka and Klobus, 2007). Using tonoplast-enriched vesicles, Shigaki et al. (2003) reported that a Cd/H antiporter might also be involved in Mn^{2+} accumulation in vacuoles.

Transporter proteins are described according to their localization. In the plasma membrane, four Mn^{2+} uptake transporters are identified: AtlRTl, NRAMP, AtYSL, and PHO84 (Ducic and Polle, 2005; Pittman, 2005). The transporter protein IRT1 can transport Mn when expressed in yeast (Korshunova et al., 1999). The transporter protein NRAMP also is considered a metal transporter protein that can transport Mn and may be localized in the tonoplast rather than in the plasma membranes (Thomine et al., 2003). The transporter AtYSL of arabidopsis is a yellow stripe-like (YSL) protein, also involved in metal-complex transport in the plasma membrane (Roberts et al., 2004). This complex can be formed by nicotianamine, which is a strong chelator of metals including Mn^{2+} (Pittman, 2005). The transporter PHO84 is a transporter protein identified in yeast and has a high affinity for phosphate (Mitsukawa et al., 1997). Luk et al. (2003) reported a new form in Mn transport as $MnHPO_4$. However, there is no evidence for Mn^{2+} or $MnHPO_4$ accumulation by a plant phosphate transporter (Pittman, 2005).

Some transporter proteins have been related to the transport and accumulation of Mn^{2+} into the intracellular compartments, such as the vacuole. It has been reported that a metal transporter (specifically antiporter CAX2, calcium exchanger 2), originally identified as a Ca^{2+} transporter (which can also transport Cd^{2+}) and located in the cytosol, has the ability to transport Mn to the vacuole in tobacco plants and yeast (Hirschi et al., 2000; Pittman, 2005). The protein transporters, ATP-binding cassette (ABC), are considered to be involved in detoxification processes (Martinoia et al., 2002). Studies on cyanobacteria also suggested the putative role of these proteins in Mn^{2+} transport (Bartsevich and Pakrasi, 1996).

Another protein, considered indirectly as a metal transporter protein, is ShMTPl, which can sequester metal ions within cells or efflux them out of the cells (Delhaize et al., 2003). Therefore, this transporter protein is considered a metal-tolerant protein (Delhaize et al., 2003). Ducic and Polle (2005) and Pittman (2005) highlighted that, despite the available information about Mn^{2+} transport across membranes in plant cells, the Mn^{2+} transport and efflux strategies into the mitochondria, chloroplasts, and Golgi are not understood completely. Nonetheless, Mills et al. (2008) have recently identified a Ca-ATPase that also transports Mn^{2+} into the Golgi apparatus.

12.6 DIAGNOSIS OF MANGANESE STATUS IN PLANTS

12.6.1 MANGANESE DEFICIENCY

12.6.1.1 Prevalence

Soils that impart Mn deficiency to susceptible crops are usually siliceous and calcareous sandy soils of neutral or alkaline pH that favor chemical and microbial oxidation and immobilization of plant-available Mn^{2+} (Rengel, 2000). However, even these soils contain large reserves of total Mn relative to the amounts removed in crop harvests (Rengel, 2000). Therefore, resulting Mn deficiency of susceptible crops is due to insufficient availability of soil Mn to plants rather than to an absolute shortage of soil Mn (Rengel, 2000).

Bioavailability of Mn depends on its oxidation state, which in turn depends on soil physical and chemical conditions. Conditions diverting Mn^{2+} to an oxide can cause Mn deficiency in susceptible crops. In sandy soils, soils high in OM, or in dry well-aerated soils with high pH, the bioavailability of Mn decreases far below the level that is required for normal plant growth (Rengel, 2000; Humphries et al., 2007).

Lindsay (1979) reported that Mn deficiency is prevalent in calcareous soils (pH 7.3–8.3) due to precipitation. The pH of calcareous soils is well buffered by calcium carbonate (Jauregui and Reisenauer, 1982).

Humphries et al. (2007) reported that susceptible crops in soils with low bulk densities are prone to Mn deficiency, possibly due to restricted root growth. They also reported that conditions of cool and temperate climates are associated with Mn deficiency, although Mn deficiency has been reported in tropical and arid areas. Dry seasons have been reported to relieve or to intensify Mn deficiency (Graham et al., 1983; Hannam et al., 1984). Manganese deficiency is more severe in the cold and wet seasons, due to reduced root metabolic activity that lowers Mn^{2+} uptake (Batey, 1971; Marschner, 1995; Michael and Beckg, 2001). Humphries et al. (2007) and Broadley et al. (2012) reported occurrence of deficiency resulting from Mn depletion in highly leached tropical or podzolic soils.

12.6.1.2 Critical Deficiency Concentrations of Manganese in Plants

Despite differences in susceptibility to Mn deficiency among plant species and varieties within species, the critical deficiency concentrations of Mn in plants are similar. Except for blue lupine that has a critical Mn deficiency concentration, twice as high as that of other plants, most plants have a critical Mn deficiency concentration between 10 and 20 mg Mn/kg dry weight in fully expanded leaves, regardless of plant species or cultivar or environmental conditions (Brennan et al., 2001; Broadley et al., 2012; Hannam and Ohki, 1988).

Ward (1977) reported that symptoms of Mn deficiency in tomato were related to Mn content in the plant tissue and that the highest concentration of Mn in tissue showing deficiency symptoms was 25 mg/kg dry weight.

12.6.1.3 Tolerance to Manganese Deficiency and Indicator Plants

Plants with notable sensitivity to Mn deficiency include apple (*Malus domestica* Borkh.), cherry (*Prunus avium* L.), citrus (*Citrus* spp. L.), oat, pea, beans, soybeans, raspberry (*Rubus* spp. L.), and sugar beet (*Beta vulgaris* L.) (Chapman, et al., 1939, 1940; Labanauskas, 1965; Lucas and Knezek, 1972). Among cereals, oats are reported to be the most sensitive to Mn deficiency, and rye (*Secale cereale* L.) is the least sensitive (Humphries et al., 2007).

Tolerance to Mn deficiency usually is conferred by an ability to extract efficiently available Mn from soils (Humphries et al., 2007). Plants have developed sophisticated mechanisms to improve the acquisition of immobile nutrients, comprising biochemical, physiological, and morphological responses (Yang et al., 2008). Tolerance to Mn deficiency may be attributed to one or more of the following adaptive mechanisms: (1) superior internal utilization or lower functional requirement for Mn, (2) optimum internal distribution of Mn, (3) fast rate of absorption from low Mn concentrations at the root–soil interface, (4) superior root geometry or strong extrusion of substances from roots into the rhizosphere to mobilize insoluble Mn utilizing (a) H^+, (b) reductants, (c) manganese-binding ligands, or (d) microbial stimulants (Graham, 1988a,b).

Pedas et al. (2005) reported that two barley genotypes with differential tolerance to low availability of Mn in the growth medium differed only with respect to high-affinity Mn influx. The Mn-efficient genotype had almost four times higher V_{max} than the inefficient, whereas the K_m values were similar between the two genotypes. Yang et al. (2008) reported a unique root hair phenotype that is induced specifically by Mn deficiency and that differs from the root hair phenotypes observed in response to a lack of P or Fe. The Mn-deficient phenotype is evident at the seedling stage. The Mn deficiency–induced changes include the differentiation of trichoblasts into root hair–forming cells, as shown by the formation of root hairs in ectopic positions (Yang et al., 2008). This formation of new hairs indicates that Mn deficiency does not simply promote the elongation of hairs from trichoblasts but alters the developmental program of rhizodermal cells.

12.6.1.4 Effect of Manganese Deficiency on Crop Yield and Quality

Mn deficiency is a widespread plant nutritional disorder in agriculture and is difficult to overcome because Mn^{2+} is oxidized rapidly when supplemented as fertilizers (Rengel, 2000; Yang et al., 2007; Zhang et al., 2009).

12.6.1.4.1   Physiological and Biochemical Effects of Manganese Deficiency
Manganese plays major roles in plant growth and development and therefore in crop yield and quality, as noted by effects on photosynthesis, actions of plant hormones, synthesis of carbohydrates, and resistance to diseases.

12.6.1.4.1.1   Effect of Manganese Deficiency on Photosynthesis   Because of the key role of Mn in photosynthesis, net photosynthesis and amounts of chlorophyll decrease with Mn deficiency (Ahangar et al., 1995; Honann, 1967; Ndakidemi et al., 2011; Ohki et al., 1981; Polle et al., 1992). Kriedemann et al. (1985) reported that inhibition of photosynthesis occurs even at moderate Mn deficiency and that Mn deficiency does not affect chloroplast ultrastructure or cause chloroplast breakdown until severe deficiency is reached. Ohki et al. (1979) and Shenker et al. (2004) reported that, in Mn-deficient plants, dry matter production, net photosynthesis, and chlorophyll content decline rapidly, whereas rates of respiration and transpiration remain unaffected.

12.6.1.4.1.2   Effect of Manganese Deficiency on Plant Hormones   Critical to production of nursery crops is the role of Mn as a precursor of the plant hormone auxin (Altland, 2006). Manganese activates the auxin oxidase system (Russell, 1988). Manganese deficiency reduces auxin levels and causes hormone imbalance. A decrease in the ratio of auxin to other plant hormones causes suppressed lateral root development and root extension (Landis, 1998).

12.6.1.4.1.3   Effect of Manganese Deficiency on the Synthesis of Carbohydrates   Manganese deficiency has very serious effects on nonstructural carbohydrates and root carbohydrates especially (Altland, 2006; Marschner, 2012). During winter when plants are dormant, carbohydrates are stored in stem and root tissue (Altland, 2006). Stored carbohydrates in the root system are important for root regeneration the following year. Manganese deficiency limits the ability of plants to produce carbohydrates, thus reducing the ability of harvested perennial crops to regenerate roots and grow vigorously in the following year (Altland, 2006).

12.6.1.4.1.4   Effect of Manganese Deficiency on Plant Resistance to Diseases   Huber (1978) reported that the Mn status of a plant can affect, and be affected by, disease infections. Manganese-deficient plants are susceptible to a range of soil-borne root-rotting fungal diseases (Huber and McCay-Buis, 1993; Longnecker et al., 1991b). Huber and Wilhelm et al. (1988) reported that Mn concentration in diseased tissues decreased as the disease progressed. This occurrence may be due to the reduction of the root system in the case of root pathogens, leading to a reduction in the absorptive surface with a resultant decrease in nutrient concentrations in the plants (Wilhelm et al., 1985, 1988). Additionally, microbially induced changes in Mn status, such as grey-speck disease of Mn-deficient oats, have been reported to be due to oxidizing bacteria in the rhizosphere causing Mn to become unavailable (Bromfield, 1978; Wilhelm et al., 1990). Silber et al. (2005) reported a negative correlation between the nonpathogenic disease of blossom-end rot in pepper (*Capsicum annuum* L.) and fruit Mn. Bar-Tal and Aloni (2008) confirmed that blossom-end rot in vegetable fruits can be caused by Mn deficiency.

The most notable interaction between disease resistance and Mn is that of the wheat disease, take-all root rot caused by the pathogen *Gaeumannomyces graminis* var. *tritici*, commonly referred to as *Ggt*. The importance of Mn in the defense against infection by *Ggt* was demonstrated by Graham (1983). Manganese is the unifying factor in the susceptibility of varieties to *Ggt* under several soil conditions, including changing pH and nitrogen forms (Graham and Webb, 1991). The role of Mn fertilizers in the resistance against *Ggt* has been reported in numerous papers (Graham and Rovira, 1984; Rovira et al., 1985; Wilhelm et al., 1988). Manganese fertilizers have been reported to limit take-all disease if used before the onset of foliar symptoms (Graham and Rovira, 1984; Wilhelm et al., 1988).

Several mechanisms have been proposed for the interaction between Mn and disease resistance. These include lignification, as the concentration of lignin in the plant declines due to Mn deficiency, and this reduction is more severe in the roots, a response that reduces resistance to root fungi infections (Anderson and Pyliotis, 1996; Marschner, 1995; Mukhopadhyay and Sharma, 1991; Ness and Woolhouse, 1980).

Mechanisms proposed for the interaction between Mn and disease resistance also include the concentration of soluble phenols, where Mn deficiency leads to a decrease in their concentration (Brown et al., 1984a,b); inhibition of aminopeptidase, which supplies essential amino acids for fungal growth, under Mn-deficient conditions (Huber and Keeler, 1977); and inhibition of pectin methylesterase, a fungal enzyme causing degradation of host cell walls under Mn-deficient conditions (Sadasivan, 1965).

Another mechanism proposed for the interaction between Mn and disease resistance is the inhibition of photosynthesis in Mn-deficient plants, leading to a decrease in root exudates, an action that causes the roots to become susceptible to invasion by root pathogens (Graham and Rovira, 1984). However, this mechanism has been shown not to be important in controlling take-all because of the lack of effect of foliar-applied Mn on *Ggt* (Reis et al., 1982; Wilhelm et al., 1988).

Graham and Webb (1991) reported that a plant capable of mobilizing high concentrations of Mn^{2+} that are toxic to pathogens but not to plants, in the rhizosphere, may inhibit pathogenic attack directly.

12.6.1.4.1.5   Other Effects of Manganese Deficiency   In cereals, yield and crop quality decrease with Mn deficiency. A decrease in grain number and grain yield in Mn-deficient plants is presumably a combination of low pollen fertility and shortage of carbohydrate supply for grain filling (Longnecker et al., 1991b; Sharma et al., 1991).

12.6.1.4.2   Morphological Effects of Manganese Deficiency

Appearance of symptoms of Mn deficiency on the affected plants is a result of the metabolic and physiological effects of Mn deficiency. Characteristic foliar symptoms of Mn deficiency become unmistakable only when the growth rate is restricted significantly and include diffuse interveinal chlorosis on young expanded leaf blades (Grundon et al., 1997; Hannan and Ohki, 1988). Severe necrotic spots or streaks may also form on the leaves of deficient plants (Humphries et al., 2007) (Figures 12.1 and 12.2). Symptoms often occur first on the middle leaves (Humphries et al., 2007).

FIGURE 12.1   Manganese deficiency in bean (*Phaseolus vulgaris* L.).

FIGURE 12.2 **(See color insert.)** Manganese deficiency in tomato (*Lycopersicon esculentum* Mill.).

Symptoms of Mn deficiency differ with the plant species and cultivars within species. In mono-cotyledonous plants, symptoms of Mn deficiency appear as gray-green spots on bases of leaves (Longnecker et al., 1991a; Marschner, 1995; Sharma et al., 1991; Wilson et al., 1982). Broadley et al. (2012) reported that the symptoms of Mn deficiency in cereals appear on the old leaves of the plant and consist of greenish gray spots called "gray speck." Hannam and Ohki (1988) reported that chlorosis develops first on the leaf base in Mn-deficient cereals. Symptoms of Mn deficiency also appear as striping on grasses with symptoms that closely resemble iron or zinc deficiency (Figure 12.3).

In dicotyledonous plants, interveinal chlorosis of the younger leaves is the most distinct symptom of Mn deficiency (Broadley et al., 2012). The distal portions of the leaf blade are affected first, and symptoms often consist of small yellow or necrotic spots on leaves (Figures 12.2 and 12.4). (Hannam and Ohki, 1988; Longnecker et al., 1991a; Marschner, 1995; Sharma et al., 1991; Wilson et al., 1982). In legumes, necrotic areas, known as "marsh spots," developed on the cotyledons are the major symptom (Robson and Snowball, 1986; Uchida, 2000). In eucalyptus (*Eucalyptus* spp. L. Her.), the tip margins of juvenile and adult expanding leaves become pale green, and chlorosis extends between the lateral veins and the midrib (Grundon et al., 1997). In citrus, dark-green bands form along the midrib and main veins, with lighter green areas between the bands (Labanauskas, 1965). In mild cases of Mn deficiency, the symptoms appear on young leaves of citrus and disappear as the leaf matures (Labanauskas, 1965). Young leaves of Mn-deficient citrus often show a network of green veins in a lighter green background, closely resembling iron chlorosis (Labanauskas, 1965). Manganese deficiency is manifested as malformed seeds in lupines (Walton, 1978).

Ward (1977) reported that typical foliar symptoms of mild Mn deficiency began early in the growth of tomato and consisted of interveinal chlorotic mottling with darker green bands along the veins and lighter green spots or patches between them. As growth continued, these symptoms progressed from the bottom leaves up the plant and became more severe (Ward, 1977). Eventually, the lighter areas coalesced and the whole plant assumed a pale chlorotic appearance.

FIGURE 12.3   Manganese deficiency in corn (*Zea mays* L.).

FIGURE 12.4   **(See color insert.)** Manganese toxicity in soybean (*Glycine max* Merr.) showing necrotic spots.

In extreme deficiency, the center of the chlorotic areas died, giving the leaf an appearance of fine necrotic spotting. These spots or lesions also enlarged and coalesced. The plants deficient in Mn had greatly reduced vigor, smaller stems and leaves, but no stunting. Fruit set appeared to be normal, but fruit sizing, that is, enlargement with maturation, particularly on the upper part of the plant, was very slow.

Altland (2006) reported that Mn-deficient red maple (*Acer rubrum* L.) trees often developed foliar chlorosis. Initially, the condition appeared as mild interveinal chlorosis in newly developing foliage.

As symptoms intensified, entire leaves turned chlorotic. Under severe conditions, interveinal necrosis occurred, and chlorotic plants often had reduced height and caliper.

Visual symptoms of Mn deficiency are very similar to those associated with some stages of iron, zinc, magnesium, or sulfur deficiency, or even some expressions of tobacco mosaic virus (Römheld and Marschner, 1991; Ward, 1977). However, they are a valuable basis for the determination of nutrient imbalance and, combined with chemical analysis, can lead to a correct diagnosis. Robson and Snowball (1986) designated the youngest leaves as the most useful for chemical analysis to confirm Mn deficiency. Römheld and Marschner (1991) reported that symptoms of Mn deficiency differ from those of Fe deficiency in that chlorosis induced by Mn deficiency is not distributed uniformly over the entire leaf blade and tissue may become necrotic rapidly.

Marschner (1995) reported that symptoms of Mn deficiency are similar to those of Mg deficiency in that they both appear in the intercostal veins. However, symptoms of Mn deficiency appear first on younger leaves, and symptoms of Mg deficiency appear in older leaves primarily, because of the different dynamics of Mn and Mg in plants (Mn is essentially immobile and Mg is mobile in the plant tissue) (Humphries et al., 2007; Marschner, 1995).

A major expression of Mn deficiency in plants is a reduction in the efficiency of photosynthesis leading to a general decline in dry matter productivity and yield (Broadley et al., 2012).

12.6.1.5 Management of Manganese Deficiency

Because Mn^{2+} is not stable in the soil and because Mn^{2+} is slowly mobile in soil and in plants, prevention of Mn deficiency is more effective than its correction. Tolerant plant varieties and cultivars and appropriate management practices can be used to prevent the occurrence of Mn deficiency. Site selection and soil preparation prior to planting susceptible crops are key to preventing Mn deficiency.

12.6.1.5.1 Site Selection for Planting

Using historical knowledge, planting sites should be narrowed down to those fields where the pH is lowest but still appropriate for plant growth. For example, in the case of red maple, Altland (2006) reported that Mn availability in soil and absorption by plants decreased dramatically as the soil pH increased above 5.6, and they recommended a site with a pH of 5.6 to prevent Mn deficiency in red maple. The planting site should also be well drained (Altland, 2006). For large areas that drain poorly, tile drainage should be used to improve drainage, or crops that are more tolerant of wet soils should be used.

12.6.1.5.2 Crop Selection

Plants tolerant of Mn deficiency may be selected for production in sites with low levels or availability of Mn. Plant species and genotypes within species differ considerably in susceptibility to Mn deficiency (Rengel, 2001). For example, oak (*Quercus* spp. L.), soybean, and peaches (*Prunus persica* L.) are susceptible to Mn deficiency, whereas corn and rye are not (Reuter et al., 1988). Differential susceptibility to Mn deficiency was reported among genotypes of bread wheat, durum wheat (*Triticum durum* Desf.), and barley (Hebbern et al., 2005; Khabaz-Saberi et al., 2000; Sadana et al., 2002). Because of their sensitivity to shortage of the element, several species considered susceptible to Mn deficiency have been the focus of breeding for tolerant varieties (Humphries et al., 2007).

12.6.1.5.3 Adjusting Soil Nutrient Levels

Soil nutrient levels should be within the range of sufficiency for the crop (Altland, 2006). Fertilizers should be incorporated into the soil before planting (Altland, 2006). Phosphorus, secondary nutrients, and micronutrients move slowly in the soil and are more readily available to plants if incorporated throughout the root zone of the crop (Altland, 2006). Sulfate often is limiting to growth of some plants and influences Mn absorption (Altland, 2006).

12.6.1.5.4 Adjusting Soil Acidity

Soil acidity (pH) should be managed to improve Mn availability to plants. For example, for red maple, the pH should be adjusted to a target level of about 5.0–5.6 (Altland, 2006). If pH is suitable but Ca or Mg is deficient, gypsum ($CaSO_4$), Epsom salts ($MgSO_4$), or both can be added to raise Ca and Mg levels in the soil without affecting the pH (Altland, 2006). If pH is below 5.0, lime should be applied during field preparation (before planting) and incorporated into the soil at rates necessary to adjust the pH to the desired value (Altland, 2006; Hart, 1998).

If soil pH is greater than 5.6, granular elemental sulfur can be applied to lower soil pH to the appropriate value (Altland, 2006; Horneck et al., 2004). Elemental S is oxidized to sulfate by soil bacteria (*Thiobacillus* spp.), a process that results in lowered soil pH (Altland, 2006). Soil temperatures need to be above 10°C for this reaction to occur, and optimum reaction rates occur at temperatures between 21°C and 27°C (Altland, 2006). Applications can be made during any time of the year, but changes in soil pH will not occur until soil temperatures have warmed to above 10°C (Altland, 2006). Elemental S is not soluble in water, so it does not leach away with rain, but it typically takes at least a year for S to be oxidized (Altland, 2006). Because of the long time needed for the resulting pH change to occur, it is best to incorporate S 6–12 months before planting susceptible crops such as red maple (Altland, 2006). Elemental S topdressed on the soil surface will not be accessible to soil bacteria and thus will not lower soil pH (Altland, 2006).

12.6.1.5.5 Promoting Healthy Root Growth

Similar to deficiencies in other immobile mineral nutrients such as Fe and phosphate, the low bioavailability of Mn rather than the concentration in the soil causes Mn deficiency. An efficient means of improving the acquisition of immobile nutrients is to enlarge the soil–root interface. Root epidermal cells can differentiate into root hairs, thereby substantially increasing the absorptive surface of the root (Yang et al., 2008). These authors investigated the acclimation of seedlings of arabidopsis to low Mn availability in the medium. They reported that Mn deficiency caused a total change in the arrangement and characteristics of the root epidermal cells. A proportion of the extra hairs formed upon Mn deficiency were located in trichoblast positions, indicative of a postembryonic reprogramming of the cell fate acquired during embryogenesis (Yang et al., 2008).

To manage Mn deficiency in red maple, Altland (2006) recommends planting healthy liners with vigorous root systems and providing the plants with sufficient water and fertility so that other factors, aside from low Mn levels, do not impede root growth. Healthy, fibrous root systems release more root exudates, resulting in higher levels of Mn available for plant growth, and factors that negatively affect root growth exacerbate Mn deficiency in red maple (Altland, 2006).

12.6.1.5.6 Manganese Applications

Manganese deficiency can be corrected by treatment of seeds with Mn fertilizers or by applying the fertilizers to soil or to the foliage of plants.

12.6.1.5.6.1 Seed Treatment with Manganese
Longnecker et al. (1991a) reported that high Mn concentration in seeds of barley, either supplied naturally from the parent plants or artificially by soaking seeds in manganese sulfate ($MnSO_4$), can improve plant growth and grain yield on soils with low availability of Mn. With wheat, high Mn in seeds was reported to be more effective than Mn fertilization of land in achieving good yield in soils where Mn is poorly available (Khabaz-Saberi et al., 2000; Moussavi-Nik et al., 1997). Also, wheat produced from seeds with high Mn concentration had increased tolerance to take-all root rot disease (McCay-Buis et al., 1995).

12.6.1.5.6.2 Manganese Applications to Soil
Manganese absorption increases with increasing soil Mn levels; however, Mn availability usually can be regulated more effectively by adjusting soil pH than by Mn applications (Altland, 2006). Manganese is sparingly mobile in plants or in soil, conditions that mean, in general, that plants cannot translocate Mn from a well-supplied part of the

shoot to a deficient part (Altland, 2006). If Mn fertilizers are topdressed after planting, only those roots near the soil surface will have sufficient Mn, and the remainder of the root system will remain deficient (Altland, 2006). For best results, Altland (2006) recommends combining pH adjustment with soil fertilization with Mn.

Research on the effectiveness of Mn application to soil indicates there is a rapid (on a scale of seconds or minutes) decrease of Mn to deficiency levels after addition of Mn^{2+} fertilizers to the growth medium (Savvas et al., 2003; Silber et al., 2005; Sonneveld and Voogt, 1997). This decrease of Mn to a deficiency level might be attributed to fast adsorption of Mn^{2+} on negatively charged surfaces of soil constituents (Benhammou et al., 2005; Davies and Morgan, 1989; Norvell and Lindsay, 1969). Silber et al. (2008) reported that Mn^{2+} solubility was controlled, on a time scale of seconds to a few hours after application to perlite media, by adsorption reactions, mainly onto the organic constituents accumulated from root excretions or decomposition and rhizosphere biota. With advancing time, biotic oxidation became the predominant mechanism of Mn^{2+} removal from solution (Silber et al., 2008).

Manganese fertilizers exist in different forms (Table 12.1) (Malakouti and Tehrani, 1999; Schulte and Kelling, 1999). The most commonly used Mn fertilizer is manganese sulfate ($MnSO_4$), with about 29% Mn, and it is useful in acidic or alkaline soils (Malakouti and Tehrani, 1999). Manganese sulfate is banded or broadcasted, although banding the fertilizer is more efficient as Mn^{2+} reacts quickly with soil components and changes into unavailable forms (Malakouti and Tehrani, 1999). Manganese sulfate also is effective for foliar applications (Malakouti and Tehrani, 1999). Manganese oxide, with about 70% Mn, is usable only in acidic soils, due to its limited solubility (Malakouti and Tehrani, 1999).

Chelated Mn fertilizers contain 12% Mn and consist of Mn in organic molecules. Chelated Mn fertilizers are soluble over a wider range of soil pH conditions, thus allowing fertilizer applications in high pH soils (Altland, 2006); also, they are used often as foliar sprays. In agricultural fertilizers, the organic molecule most often used is EDTA. Chelated Mn applied as a soil fertilizer is not effective at raising available Mn levels in soil. Norvell and Lindsay (1969) reported that when Mn–EDTA was applied to a soil with pH of 5.7, less than 3% of the Mn^{2+} remained in the soil solution at 20 h after application. The chelated Mn^{2+} ion is replaced by Fe^{2+} in low pH soils and by Ca^{2+} in high pH soils (Russell, 1988). High Fe concentrations in soil solution exacerbate Mn deficiency by outcompeting Mn for uptake. When chelated Mn is applied and Fe displaces it in the chelated molecule, Mn uptake is not improved, and the additional absorbed Fe further depresses Mn uptake (Altland, 2006). This effect was demonstrated in research on a loam, where chelated Mn application worsened Mn deficiency (Schulte and Kelling, 1999). Manganese chelates are, however, recommended for foliar application of crops (Bryson et al., 2014; Malakouti and Tehrani, 1999).

TABLE 12.1
Main Sources of Manganese Fertilizer

Manganese Fertilizer	Chemical Formula	Mn (%)
Manganese sulfate	$MnSO_4 \cdot 3H_2O$	27[a]
Manganese oxide	MnO	77
Manganese carbonate	$MnCO_3$	48
Manganese chelate	Mn-EDTA	12
Manganese chloride	$MnCl_2$	44
Manganese dioxide	MnO_2	63

Sources: Malakouti, M.J. and Tehrani, M., *Effect of Micronutrients in Yield Increase and Improvement of Crops Quality*, p. 299. Tehran, Iran: Tarbiat Modarres University Press, 1999; Schulte, E.E. and Kelling, K.A. Soil and applied manganese. Understanding plant nutrients. University of Wisconsin Extension Publication, A2526, 1999.

[a] The percentage of Mn is for the chemical compound. Purity of fertilizer grade may differ.

12.6.1.5.6.3 Manganese Applications to Foliage Mousavi et al. (2007) reported that potato (*Solanum tuberosum* L.) yield increased and storage dry matter was improved by the application of Mn and Zn. In separate investigations, Hiller (1995) and Walworth (1998) reported that yield and quality of potato increased with foliar applications of micronutrients including Mn. Bansal and Nayyar (1994) investigated the effect of Mn foliar applications on 10 cultivars of soybean and observed a significant increase in the economic and biological yield. Mahler et al. (1992) examined the effects of $MnSO_4$ on wheat yield and quality and concluded that yield increased significantly with the use of $MnSO_4$. Increased yield with foliar applications of Mn is due to increased photosynthetic efficiency and synthesis of carbohydrates (Malakouti and Tehrani, 1999).

12.6.2 MANGANESE TOXICITY

Although Mn is essential for plant growth, it can be detrimental when available in excess in the growth medium. Concentrations as low as 1 mg Mn/L in solution culture can be toxic to tobacco, potato, or bush clover (*Lespedeza* spp. Michx) (Berger and Gerloff, 1947; Jacobson and Swanback, 1932; Morris and Pierre, 1949). Excessive Mn in the growth medium can interfere with the absorption, translocation, and utilization of other mineral elements such as Ca, Mg, Fe, and P (Clark, 1982), and high Mn concentration in plant tissues can alter activities of enzymes and hormones, so that other Mn-requiring processes become less active or nonfunctional (Epstein, 1961; Horst, 1988).

12.6.2.1 Prevalence

In many parts of the world, Mn toxicity is more important than Mn deficiency in crop plants (Welch et al., 1991). Manganese toxicity is encountered usually in poorly drained, acid soils (below pH 5.5) (Chesworth, 1991; Messing 1965; Morris and Pierre, 1949). Such soils usually contain relatively low concentrations of available Ca and Mg and high concentrations of soluble Fe and Mn. Many crops are sensitive to concentrations as low as 1 mg Mn/kg in the growth medium, a concentration occurring in many acid soils (Morris and Pierre, 1949). Morris and Pierre (1949) reported that some acid soils contain as high as 50 mg Mn/kg in the soil solution. Manganese toxicity also can occur in near-neutral or alkaline soils under reducing conditions created by flooding, compaction, or OM accumulation (Elliott and Blaylock 1975; Grasmanis and Leeper 1966; Kamprath and Foy, 1971). Toxicity may occur after steam sterilization of greenhouse soils, particularly those of high clay content (Roorda van Eysinga 1969).

Tropical, subtropical, and temperate soils have been reported to be sources of Mn at concentrations high enough to produce visible symptoms of toxicity. In the tropics, Mn toxicity has been reported in grasses grown in the Catalina (basalt) and the Fajardo (moderately permeable) clayey soils of Puerto Rico (Abruna et al., 1964), and in ryegrass (*Lolium* spp. L.) grown on red-brown clayey loam and granite-mica schists in Uganda (Lemare, 1977). Among the subtropical regions, toxicity has been reported in subtropical United States in poorly drained soils and soils on limestone and on ultisols (Bortner, 1935a,b). However, the impermeability of soils does not seem essential for Mn toxicity (Foy and Campbell, 1984). In southeastern Australia, toxicity has been reported in fruit trees grown in neutral-pH duplex soils (Grasmanis and Leeper, 1966), in beans grown in Mn-rich basaltic soil (Siman et al., 1971), and in pasture legumes (Siman et al., 1974). In temperate regions, toxicity was reported in Illinois, United States, on soils characterized by low pH and high concentrations of readily exchangeable manganese (Snider, 1943).

12.6.2.2 Critical Toxicity Concentrations of Manganese in Plants

Ohki (1981) defined critical toxicity concentration of manganese as Mn concentration in the leaves associated with 10% restriction in dry weight due to toxicity. Symptoms of Mn toxicity, however, may occur at a different concentration than the one that causes restricted weight (Morris and Pierre, 1949). In some cases, the accumulation of Mn may be very high. Gupta et al. (1970) reported that 2600 mg Mn/kg in carrot (*Daucus carota* L.) shoots caused symptoms (bronzing) to occur but that higher concentrations (7100–9600 mg/kg) were required to limit yields.

TABLE 12.2
Critical Toxicity Levels of Manganese in the Leaves of Various Plant Species[a]

Species	Concentration for Toxicity (mg Mn/kg Dry Weight)	Reference
Soybean	160	Foy et al. (1978)
Corn	200	Horst and Marschner (1990)
Pea	300	Horst and Marschner (1990)
Soybean	600	Horst and Marschner (1990)
Cotton	750	Horst and Marschner (1990)
Sweet potato	1380	Horst and Marschner (1990)
Sunflower	5300	Horst and Marschner (1990)
Carrot	7100–9600	Gupta et al. (1970)

[a] Critical toxicity levels are associated with a 10% suppression in dry matter production (Ohki, 1981).

In contrast to the narrow range of critical deficiency concentration of Mn in plants, the critical toxicity concentration of Mn is quite variable among plant species and varieties within species, due to wide differences in Mn tolerance (Edwards and Asher, 1982; Horst, 1988; Ohki, 1981; Khabaz-Saberi et al., 2010; Wang et al., 2002). Differences among plant species in regard to critical toxicity concentrations are shown in Table 12.2.

Based on the Mn concentration in the growth medium associated with injury, the critical Mn was 1 mg Mn/kg for tobacco (Jacobson and Swanback, 1932); 3.5 mg/kg for peas; and 10 mg/kg for soybean, peanut (*Arachis hypogaea* L.) (Morris and Pierre, 1949), and barley (Aso, 1902). El-Jaoual Eaton et al. (2012) reported that with marigold (*Tagetes erecta* L.) grown in nutrient solution, symptoms from Mn toxicity occurred at 6.5 mg Mn/L solution, but a restriction in dry weight with no symptoms appearing occurred at 4.5 mg Mn/L.

Critical toxicity concentration of Mn also depends on environmental factors, particularly temperature and the presence of silicon (Si) in the growth medium. Leaf critical toxicity concentration of Mn is higher at high temperatures than at low temperatures or when the plants are supplied with Si (Doncheva et al., 2009; Führs et al., 2009; Heenan and Carter, 1977; Iwasaki et al., 2002; Rufty et al., 1979).

12.6.2.3 Tolerance to Manganese Toxicity and Indicator Plants

The biochemical responses of higher plants to toxic doses of heavy metals, including Mn, are very complex, and several defense strategies have been suggested. These defense strategies include reduced uptake and transport of metal, sequestration and compartmentalization of the metal, or enhanced production of antioxidants that detoxify reactive oxidative responses of plants to toxic metals (Schiltzendilbel and Polle, 2002; Van Assche and Clisters, 1990).

Plant species and cultivars differ widely in tolerance to excessive Mn. In general, water plants are able to accumulate higher amounts of Mn in their tissues without injury than land plants (Gossl, 1905). Brenchley (1914) reported that barley is more tolerant to high concentrations of Mn than wheat. Benac (1976) reported that, in nutrient solution, about 20 times as much Mn was required to produce toxicity in corn than in peanut.

Morris and Pierre (1949) reported that Japanese clover (*Kummerowia striata* Thunb.) and sweet clover (*Melilotus* spp. Mill.) were the most sensitive legumes, cowpeas (*Vigna unguiculata* Walp) and soybeans were intermediate, and peanuts were the most tolerant.

Crops particularly sensitive to Mn toxicity include alfalfa (*Medicago sativa* L.), cabbage (*Brassica oleracea* var. *capitata* L.), cauliflower (*Brassica oleracea* var. *botrytis* L.), clover (*Trifolium* spp. L.), pineapple, potato, and sugar beet (Humphries et al., 2007).

Resistance to Mn toxicity can be achieved through different mechanisms, largely grouped into avoidance or tolerance. Avoidance involves a protective role that prevents the metal ions from entering the cytoplasm of plant cells (Blamey et al., 1986; Marschner, 1991). Tolerance strategy implies a detoxification of metal ions after they have crossed the plasma cell membrane or internal organelle biomembranes (Macfie et al., 1994). Mechanisms of plant resistance to Mn toxicity include restricted absorption of Mn, restricted translocation of excessive Mn to the shoots, and greater tolerance to high Mn levels within plant tissues (Andrew and Hegarty, 1969; Foy et al., 1995; Morris and Pierre, 1949).

12.6.2.3.1 Tolerance to Manganese Toxicity through Restricted Absorption of Manganese

The superior tolerance of peanut (compared with cowpea and soybean) and "Siam 29" rice (compared with "Pebifun") was attributed to limited Mn absorption (Morris and Pierre, 1949; Foy et al., 1978). Robson and Loneragan (1970) reported that the sensitivity of *Medicago* species of tropical legumes (compared with subterranean clover) was associated with a high rate of Mn absorption and less retention of Mn in the roots or to high transport into the shoots.

Plants can modify the soil chemistry within their root zone (Foy et al., 1978). Certain Mn-tolerant cultivars of wheat, barley, rice, peas, and corn increase the pH of their nutrient solutions (Barber, 1974). Such increase in the growth medium pH limits the availability of Mn to the plants. In contrast, Mn-sensitive cultivars of the same species decrease or have no effect on the pH of their growth medium (Barber, 1974) and thus are exposed to higher concentrations of available Mn. In some wheat genotypes, these differential pH changes have been demonstrated in thin layers of soil removed directly from the root zone and even in the bulk soil in pot culture (Barber, 1974). Possible causes of pH lowering in the plant root zone are release of H^+ resulting from excessive cation over anion absorption, release of H^+ ions as a result of NH_4^+ nutrition, release of H^+ ions from carboxyl groups of polygalacturonic acid residues of pectic acid, or excretion of H^+ protons from microorganisms associated with the roots (Barber, 1974).

The superior Mn tolerance of rice over soybean coincides with greater oxidizing power of rice roots (Doi, 1952a,b; Engler and Patrick, 1975). The rice roots are able to oxidize Mn from the divalent (available) to the tetravalent (unavailable) form of Mn (MnO_2), an action that decreases Mn absorption by the plants (Engler and Patrick, 1975). Doi (1952a,b) reported that soybean was Mn injured if planted alone in paddy soils, but this injury was reduced if soybean was planted with rice. The beneficial effect was probably due to the oxidation and detoxification of Mn by rice roots.

The exudation of organic acid anions into the rhizosphere also may restrict absorption of Mn by roots. Mora et al. (2009) reported that root exudates of oxalate and citrate may decrease Mn availability in the rhizosphere of ryegrass (*Lolium* spp. L.), thereby enhancing plant tolerance to excessive Mn in the growth medium.

12.6.2.3.2 Tolerance to Toxicity through Reduced Translocation of Manganese

The distribution of Mn between roots and shoots depends on plant species and genotype. Early research attributed Mn tolerance in tropical and temperate legumes to a greater retention of Mn in the roots (Andrew and Hegarty, 1969). The root retention of heavy metals, including Mn, has been attributed to the formation of metal complexes in the roots (Foy et al., 1978). The oxidation of Mn with concomitant deposition of oxidized Mn in the roots reduces its translocation to the shoots and enhances the tolerance of the plant to excessive amounts of Mn (Foy et al., 1978). Horiguchi (1987) observed deposition of oxidized Mn on the roots of rice, which displayed a dark brown color and accumulated a large amount of the element. They also reported that a peroxidase–malate dehydrogenase system in the root cell walls oxidized Mn.

The superior tolerance of corn to Mn toxicity, compared with peanuts, was associated with limited translocation of Mn from the roots and stems to the leaves (Benac, 1976). High sensitivity of peanuts to excessive Mn was associated with the accumulation of Mn in its leaf blades with little or no accumulation in the roots, stem, or petioles (Morris and Pierre, 1949). Ouellette and Dessureaux (1958) reported that tolerant clones of alfalfa contained lower concentrations of Mn in shoots and higher concentrations of Mn in roots than did the sensitive clones. Kamprath and Foy (1971) reported that excessive Mn may be detoxified by accumulating in the woody portions of azalea (*Rhododendron obtusum* Planch.).

12.6.2.3.3 Tolerance to Manganese Toxicity through Internal Tolerance to Excessive Manganese

In some species, the tolerance to Mn toxicity is associated closely with the internal tolerance of the shoots to Mn. The higher Mn tolerance of "Lee" soybean, compared with "Bragg" and "Forrest" (Brown and Jones, 1977a), and "Rex Smooth Leaf" cotton, compared with "Coher 100-A" (Foy et al., 1969), was attributed to a greater ability of the plants to tolerate high levels of Mn in the shoots. Horiguchi (1987) reported that rice retained less Mn in its roots than other species and that the shoots were able to tolerate a high Mn concentration in the tissues. On the other hand, the shoots of alfalfa were sensitive to a high Mn concentration in the tissues although the roots oxidized Mn and retained it in high concentrations.

The tolerance of plants to high concentrations of Mn in their shoots varies among species and genotypes within species (Foy et al., 1978; Morris and Pierre, 1949). Winterhalder (1963) reported that seedlings of Sydney blue gum (*Eucalyptus saligna* Sm.) were more tolerant to excessive soil Mn than those of red blood wood (*Eucalyptus gummifera* Hochr.), although the leaves of the former contained higher concentrations of Mn. Foy et al. (1995) reported that the differential Mn tolerance in cotton genotypes was associated with differential internal tolerance to Mn.

Tea (*Camellia thea* L.) and tomato are more tolerant of high internal levels of Mn than many other species (Dennis, 1968; Foy et al., 1978). Rice may tolerate 5–10 times as much Mn as many other grasses such as oat, barley, wheat, and ryegrass (*Lolium* spp. L.) (Vlamis and Williams, 1967). Sorghum (*Sorghum propinguum* Hitoho.) generally tolerates high levels of Mn. However, some genotypes such as NB 9040 tolerate higher concentrations of Mn than others (IS 7173c) (Mgema and Clark, 1995). Different mechanisms of tolerance of plants to excessive Mn in the plants have been identified.

12.6.2.3.3.1 Oxidation of Excessive Manganese
A possible explanation for the internal tolerance of some plants to high levels of Mn in the shoots is the oxidation of Mn (Horigushi, 1987). Blamey et al. (1986) reported that the shoots of sunflower (*Helianthus* spp. L.) tolerate high Mn concentrations in the tissues through the presence of Mn in a metabolically inactive form. Horiguchi (1987) reported that when cucumber absorbed an excessive amount of Mn, a part of the Mn absorbed was changed to oxidized Mn. They suggested that oxidized Mn is probably metabolically inactive since the dry weight of the plants showed a relatively high Mn concentration in the tissues. The mechanism of the Mn tolerance of rice does not appear to be associated with Mn oxidation because oxidized Mn could not be detected in the leaves (Horigushi, 1987).

12.6.2.3.3.2 Sequestration of Excessive Manganese in the Apoplastic Areas of Tissues
Horst et al. (1999) attributed the tolerance of cowpeas to excessive Mn to the sequestration of excessive Mn in the apoplast, where Mn is complexed with organic acids. In this species, symptoms such as brown leaf cell spots (identified as oxidized Mn and phenolic compounds in the cell walls) are considered an expression of a Mn tolerance mechanism (Wissemeier and Horst, 1992). Horiguchi (1987) suggested that oxidized Mn depositions in the plant tissues correspond to a tolerance mechanism to Mn toxicity, with cucumber being more tolerant to high Mn deposition in tissues than melon (*Cucumis melo* L.).

In proteomics studies, a comparison between Mn-sensitive and Mn-tolerant cultivars of cowpea indicated that Mn toxicity in sensitive plants was associated with the development of calluses and brown spots and the formation and enhanced release of phenols and peroxidases into the apoplastic areas (Fecht-Christoffers et al., 2003a,b, 2006). Specific proteins involved in the regulation processes, such as CO_2 fixation, stabilization of the Mn cluster of photosystem II, pathogenesis-response reactions, and protein degradation, were affected at low or high Mn levels, mainly in the Mn-sensitive cowpea cultivar (Führs et al., 2008). Chloroplastic proteins, which are important for CO_2 fixation and photosynthesis, were of lower abundance upon Mn-induced stress, suggesting the scavenging of metabolic energy for a specific stress response (Führs et al., 2008). Führs et al. (2008) concluded that a coordinated interplay of apoplastic and symplastic reactions seems to be important during the Mn stress response in plants.

Depending on the plant species, different organelles can serve as storage areas for this accumulation (Lidon et al., 2004). Under high Mn levels, apart from vacuoles, chloroplasts are important sinks of this metal in citrus (*Citrus volkameriana* V. Ten. & Pasq.). This feature, together with the large size of this organelle, is considered to be an adaptive response of citrus to excessive Mn (Papadakis et al., 2007). In general, rice is considered tolerant of Mn toxicity (Lidon, 2001) because the plant leaf tissues can accumulate from 5 to 10 times more Mn than other grasses (Foy et al., 1978). The Mn tolerance mechanisms in this species include the inhibition of apoplastic influx from the cortex to the stele and symplastic Mn assimilation in the shoot protoplast, where the chloroplast is the main target (Lidon, 2001).

Excessive Mn results in apoplastic deposition of oxidized Mn and phenolics (Fecht-Christoffers et al., 2003a,b). There is evidence that peroxidases are involved in this reaction (Fecht-Christoffers et al., 2003a,b). Manganese also induces pathogenesis-related and thaumatin-like proteins in the apoplast (Horst et al., 1999). However, it is unclear whether these responses belong to the activation of protection against Mn or whether these typical defense reactions occur nonspecifically due to Mn-induced H_2O_2 production and injury (Horst et al., 1999).

Maintaining sufficient concentrations of ascorbic acid in leaf apoplast can contribute to the tolerance to Mn toxicity in cowpea and common bean cultivars (Fecht-Christoffers et al., 2006; Führs et al., 2009; Gangwar et al., 2010; Gonzalea et al., 1998; Mora et al., 2009; Rosas et al., 2007).

12.6.2.3.3.3 Sequestration of Manganese in Cellular Organelles One of the main mechanisms of Mn tolerance is the sequestration of excessive Mn by organic compounds in metabolically less active cells or organelles. The vacuole is considered the biggest and most important compartment, because it can store many toxic compounds (Pittman, 2005). Baldisserotto et al. (2004) reported that the level of phenolic compounds increased in the leaves of water chestnut (*Trapa natans* L.) exposed to high concentrations of Mn (7 mg/L) in the growth medium. Phenolic compounds chelate Mn inside the vacuole, segregating the metal ion in the protoplasm and thus limiting the damage (Davis et al., 2001). A similar role in tolerance to Mn toxicity has been assigned to oxalic acid in the vacuoles of tolerant plants. Dou et al. (2008) attributed the tolerance of a Mn hyperaccumulator plant, American pokeweed (*Phytolacca americana* L.), to sequestration of internal Mn in the plant vacuoles, where oxalic acid chelates excessive Mn.

Studies on the effect of excessive Mn in the growth medium on different varieties of Douglas fir (*Pseudotsuga menziesii* Mirb.) showed that excessive Mn (>137 mg/L) in the nutrient solution suppressed root weight and elongation in the variety *glauca* and not in the variety *viridis* (Ducic and Polle, 2007). These studies also showed the presence of dark deposits of insoluble complexes of Mn in the vacuoles of the roots of tolerant plants.

Another mechanism of plant tolerance to excessive Mn inside plant cells is associated with several metal transporter proteins identified in the Mn transport mechanisms. Metal transporter proteins located in the tonoplast (CAX2 in tobacco and ShMTP1 in arabidopsis) conferred greater tolerance to Mn due to the internal sequestering of the element (Delhaize et al., 2003; Hirschi et al., 2000). Other transporter proteins (ECA1) can maintain low cytosolic Mn, since it moves into the endoplasmic reticulum (Wu et al., 2002). In this research, ECA1 limited cytosolic Mn^{2+}, thereby preventing an interference with the internal distribution of Mg^{2+}, Fe^{2+}, or Ca^{2+} (Wu et al., 2002).

12.6.2.3.3.4 Antioxidant Systems   In alleviating metal toxicity in plants, the antioxidant systems are considered an important tolerance mechanism. The antioxidant systems include enzyme scavengers such as SOD, catalase (CAT), peroxidases (phenol peroxidase [POX], ascorbate peroxidase [APX], glutathione peroxidase [GPX]), and nonenzymatic molecules, ascorbate, α-tocopherol, carotenoids, flavonoids, and glutathione (Apel and Hirt, 2004; Foyer and Noctor, 2003). High activities of antioxidant enzymes occur in response to Mn excess in woody plants, white clover, and in ryegrass (*Lolium perenne* L.) (Lei et al., 2007; Mora et al., 2009; Rosas et al., 2007).

Another strategy that plants use to prevent the toxic effects of Mn is the efflux of Mn from the cell. In this process, the cell Mn is delivered into the Golgi apparatus and finally exported from the cell via secretory pathway vesicles that carry the metal to the cell surface (Ducic and Polle, 2005). Blamey et al. (1986) reported the accumulation and secretion of Mn^{2+} in and around the trichomes of sunflower (*Helianthus annuus* L.) as a Mn tolerance mechanism.

12.6.2.3.4 Tolerance to Manganese Toxicity through Genetic Control

Genetic variations in the ability of plants to accumulate Mn have been reviewed by many investigators (Brown et al., 1972; El-Jaoual and Cox, 1998; Gerloff and Gabelman, 1983; Graham, 1984). Foy et al. (1978) reported that Mn tolerance in alfalfa was due to additive genes. Lettuce genotypes, "Neptune" (Mn-sensitive) and "Plenos" and "Troppo" (Mn-tolerant), differ greatly in tolerance to available Mn in steam-sterilized soils (Eenink and Garresten, 1977). Studies of their progenies indicated that Mn tolerance was determined by different genes in the parents (Eenink and Garresten, 1977).

12.6.2.4 Effect of Manganese Toxicity on Crop Yield

Manganese toxicity is one of the major factors limiting crop production on acid soil, which comprises about 30% of the world's total land area (Kochian, 2000). Toxicity of Mn in crop plants also occurs in calcareous soils low in available Fe (Moraghan, 1979) and in compacted and waterlogged soils with pH higher than 6.0, if the soil parent materials contain sufficient total Mn (Foy et al., 1988; Weil et al., 1997).

The geographical origin of the species and climatic conditions affect the degree of toxicity in plants. Lei et al. (2007) investigated Mn toxicity in two populations of poplar (*Populus cathayana* Rehder), coming from wet or dry climates and cultivated in acid solutions with increasing Mn concentrations. They reported that the wet-climate population accumulated more Mn in the plant tissues, especially in the leaves, than the dry-climate population. Increased Mn in the plant tissue of the wet-climate population limited their growth, chlorophyll contents, and activities of antioxidant enzymes (Lei et al., 2007).

12.6.2.4.1 Physiological and Biochemical Effects

Epstein (1961) pointed out that an element present in excess can interfere with metabolism through competition for absorption, inactivation of enzymes, or displacement of essential elements from functional sites.

12.6.2.4.1.1 Effect of Manganese Toxicity on Enzyme Activities   With cotton, Mn toxicity has been associated with an increased activity of indoleacetic acid oxidase and polyphenol oxidase; low activities of catalase, ascorbic acid oxidase, and glutathione oxidase; low ATP contents; and low respiration rates (Morgan et al., 1976). With sugar beets, excessive Mn restricted leaf cell number and volume (Terry et al., 1975). Manganese toxicity has been associated with an increase in indoleacetic acid oxidase activity, which results in a destruction of auxin (IAA) (Morgan et al., 1966, 1976). Bhatt et al. (1976) reported that 36 mg Mn/kg in the growth medium reversed the growth inhibition of sorghum roots caused by 100 mg/L GA. GA enhances auxin production. The inhibition of root growth by GA was due to elevated auxin synthesis to a level that was inhibitory for root growth and stimulatory for shoot growth.

Manganese toxicity also has been associated with an increase in peroxidase activity (Horigushi and Fukumoto, 1987; Morgan et al., 1966). Horigushi and Fukumoto (1987) demonstrated that the increase in peroxidase activity was followed by the appearance of the necrotic browning (visual symptoms of Mn toxicity). They added that the increase in the peroxidase activity may, however, not be the only factor for the browning because the contact of peroxidase with the substrates, which are oxidized to form brown substances, is necessary. It has been reported that peroxidase is localized in the cell walls and plays a significant role in lignification (Swain, 1977). When peroxidase oxidizes the precursors of lignin, the necrotic browning does not occur (Horigushi, 1987). Engelsma (1972) and Brown et al. (1984a,b) reported that Mn stimulates phenolic metabolism. If peroxidase comes into contact with substrates such as caffeic acid or chlorogenic acid, which are easily oxidized to form brown substances, necrotic browning occurs (Horigushi, 1987).

12.6.2.4.1.2 Effect of Manganese Toxicity on Photosynthesis A reduction in photosynthesis, in chlorophyll *a* and *b* contents, and their biosynthesis and a reduction in carotenoids occur frequently in plants and algae grown under Mn excess (Hauck et al., 2003; Macfie and Taylor, 1992). With rice, cultivated at different Mn concentrations (from 0.1 to 32 mg/kg growth medium), a significant decrease in the content of chlorophyll *a* was reported at the highest Mn concentration (Lidon and Teixeira, 2000a). Nable et al. (1988) reported an early inhibition of photosynthesis concomitant with a high Mn accumulation in the leaves of tobacco cultivated in nutrient solutions with excessive Mn (55 mg/L). The authors concluded that inhibition of photosynthesis is an early indicator of Mn toxicity in tobacco leaves. Lidon et al. (2004) also observed a decline in net CO_2 fixation and O_2 evolution by rice plants subjected to 0.5 or 2 mg Mn/L. However, an increase in the photochemical quenching and the quantum yield of noncyclic electron transport occurred at levels up to 2 mg Mn/L. Since Mn is accumulated in thylakoids, the element may interfere with thylakoid stacking, decreasing net photosynthesis (Lidon and Teixeira, 2000a).

Macfie and Taylor (1992) reported that high Mn concentrations (549 mg Mn/kg) in wheat leaves inhibited the biosynthesis of chlorophyll and carotenoids, inducing a decrease in photosynthetic electron transport and therefore a decrease in the rate of photosynthesis. In a study on the effect of high concentrations of Mn in the growth medium on the Mn accumulator plant, yellow tuft (*Alyssum murale* subsp. *murale* Waldst. & Kit.), Abou et al. (2002) reported that high Mn concentrations shorten plant roots and shoots and decrease the concentration of chlorophyll. Recently, Amao and Ohashi (2008) suggested that high Mn amounts in spinach leaves inhibited the activity of the oxygen-evolving complex of PSII.

12.6.2.4.1.3 Effect of Manganese Toxicity on Oxidative Stress Manganese toxicity can trigger oxidative stress in plant cells (Demirevska-Kepova et al., 2004). As a toxic metal, Mn can cause metabolic alterations and macromolecular damage that disrupts the cell homeostasis (Hegedüs et al., 2001; Polle, 2001). According to Clair and St Lynch (2004), Mn toxicity in plants generates ROS, mainly the ˙OH radical, the most reactive oxidant and harmful species in cells (Lidon and Henriques, 1993).

In a study on the oxidative stress responses of rice to Mn excess, Lidon and Teixeira (2000b) reported that, at the first growth stages of rice, two kinetic phases can be distinguished. In the first, there is an increase in Mn accumulation in the thylakoid lamellae, an action that inhibits electron leakage from the Hill and Mehler reactions, limiting ROS formation (McCain and Markley, 1989). In the second kinetic phase, higher Mn amounts inhibit noncyclic photophosphorylation, promoting an increase in ROS production that parallels an injury increase in the thylakoid peroxidase system (Lidon and Teixeira, 2000b). Lidon and Teixeira (2000b) concluded that Mn excess increases the disorganization in chloroplast lamellae but that elevated activity of SOD still limits cell damage. Shi et al. (2006) reported that, in cucumber, excessive Mn and light intensity determine an enhancement in oxidative stress by increased Mn content in the tissues concomitant with an inhibition of plant growth.

12.6.2.4.1.4   Other Effects of Manganese Toxicity   Panda et al. (1986) reported that lipid peroxidation occurred in isolated chloroplasts of wheat grown under excessive Mn conditions. González
et al. (1998) reported that lipid peroxidation was not induced by Mn toxicity stress in the mature
leaves of common beans. They also noted that the damage caused by excessive Mn could be related
to the developmental stage of leaves, with the damage being more intense in immature than in
mature leaves.

Excessive Mn in the plant tissue precipitates nucleic acids and inhibits the replication of the
nuclear deoxyribonucleic acid (DNA) and the synthesis of protein (Baranowska et al., 1977; Foy
et al., 1978; Trim, 1959). Guerin (1897) isolated a nucleic precipitate of Mn from the woody tissues
of the plants, and Foy et al. (1978) reported that excessive Mn may act as an error-producing factor
during mitochondrial replication, thereby causing mitochondrial mutations that are up to 4500 base-
pair long and that inhibit total protein synthesis by up to 90%.

Excessive Mn also may predispose plants to certain diseases. Singh et al. (1974) found that high
concentrations of Mn (1900 mg/kg) in scopolia (*Scopolia sinensis* Hemsl.), which is host to the
potato spindle tuber virus, increased the replication of the virus ribonucleic acid (RNA).

12.6.2.4.2   Morphological Effects
The morphological symptoms of Mn toxicity are the result of the physiological and biochemical
effects and are diverse among plant species. Symptoms include marginal chlorosis and necrosis of
leaves of alfalfa, rape (*Brassica napus* L.), kale (*Brassica oleracea* var. *acephala* DC), or lettuce
(Foy et al., 1978); interveinal and marginal chlorosis along with brown necrotic spotting in sweet
potato (*Ipomoea batatas* L.), snap bean, lettuce, barley, or cotton (Foy et al., 1995, 1969); small
dark spots surrounded by irregular areas of chlorotic tissues in marigold (Albano and Miller, 1996;
El-Jaoual Eaton et al., 2012) and geranium (*Pelagonium* × *hortorum* Bailey) (Bachman and Miller,
1995); dark, reddish-brown leaf spots with chlorotic margins in bush clover; crumbled, distorted
small-sized leaves with irregular interveinal chlorosis in soybean; small reddish purple spots on the
underside of the leaves in cowpeas; and marginal chlorosis and crimping of the leaves with no spotting in sweet clover (Morris and Pierre, 1949). Loss of apical dominance and enhanced formation of
auxiliary shoots (witches' broom) constitute another symptom of Mn toxicity (Kang and Fox, 1980).
Specific physiological disorders such as "bronze speckle" in marigold or geranium, "crinkle leaf"
of cotton, *stem streak necrosis* in potato, *internal bark necrosis* of apple (*Malus sylvestris* P. Mill.),
tip burn in carnation (*Dianthus caryophyllus* L.), and *fruit cracking* in muskmelon (*C. melo* L.) have
been associated with excessive Mn in the tissues (El-Jaoual and Cox, 1998; El-Jaoual Eaton et al.,
2012; Foy, 1983).

Although expression of Mn toxicity varies considerably among plant species, brown spots on
older leaves surrounded by chlorotic zones are typical symptoms of Mn toxicity. The necrotic brown
spots were reported to be localized accumulations of oxidized Mn (Horiguchi, 1987). Horst (1988)
reported that the spots contain precipitated Mn compounds. Quite often, however, Mn-induced
symptoms of deficiencies of other mineral nutrients, such as Fe, Mg, and Ca, are dominant (Fleming,
1989; Foy et al., 1981; Lee, 1972).

In many plant species, symptoms of Mn toxicity are speckles (brown or bronze) on mature
leaves (El-Jaoual and Cox, 1998; El-Jaoual Eaton et al., 2012; Wissemeier and Horst, 1987). Some
researchers maintain that, although the brown speckles contain oxidized Mn, the brown color
derives not from Mn, but from oxidized polyphenols (Führs et al., 2009; Wissemeier and Horst,
1987). The formation of brown speckles is preceded by enhanced callus formation in the same area,
indicating toxic effects of Mn on the plasma membrane and enhanced Ca^{2+} influx as a signal for callus formation (Horst et al., 1999; Wissemeier and Horst, 1987). The intensity of formation of brown
speckles can be used as a simple and rapid method for screening different cultivars for tolerance to
Mn toxicity (Doncheva et al., 2009; Wissemeier and Horst, 1991). The brown necrotic spots have
been associated with the accumulation and deposition of oxidized Mn. Horiguchi (1987) reported
that a peroxidase–malate dehydrogenase system in the root cell walls oxidized Mn. The presence of

oxidized Mn in plant tissues has been reported in the stems and leaves of pea and sunflower; in the leaves of tobacco and bean; and in the stems, petioles, and leaf blades of cucumber, localized around the trichomes (Horiguchi, 1987). Aso (1902) examined the brown spots of the Mn-affected leaves and reported that the membranes of epidermal cells, and in some cases the nuclei, were stained deeply brown. In Mn-tolerant plants, such as sunflower or stinging nettle (*Urtica dioica* L.) growing in high concentrations of Mn, the brown spots are developed around the base of the trichomes of the leaves (Blamey et al., 1986; Hughes and Williams, 1988). These brown spots contain Mn oxides and may therefore be considered as a tolerance mechanism to reduce the concentration of soluble Mn (Blamey et al., 1986; Broadley et al., 2012; Hughes and Williams, 1988).

Interveinal chlorosis and necrosis are further symptoms of Mn toxicity (González and Lynch, 1999; Horigushi, 1988; Nable et al., 1988). In dicots such as bean, soybean, cotton, and blueberry (*Vaccinium* spp. L.), these symptoms are combined with deformation of young leaves (crinkle leaf), which is a typical symptom of Ca deficiency (Bañados et al., 2009; Foy et al., 1981; Heenan and Campbell, 1980; Horst and Marschner, 1978a; Neal, 1937). Hence, Mn toxicity is accompanied by induced deficiencies of other nutrients such as Ca, Mg, Fe, and Zn (Horst, 1988; de Varennes et al., 2001).

Alam et al. (2006) reported that, in barley, visible symptoms of Mn toxicity consisted of desiccation of old leaves, brown spots on old leaves and stems, mild interveinal chlorosis on younger leaves, and brown-colored roots. The brown discoloration is indication of accumulation of MnO_2 on the root surface (Horiguchi, 1987; Horst, 1988), and brown spots appeared to be associated with oxidized phenolic substances (Horiguchi, 1987). In contrast to the observations in this study, Vlamis and Williams (1964) observed no visible symptoms of Mn toxicity on the roots of barley.

In some Mn-induced disorders, such as leaf marginal chlorosis in mustard, leaf speckling in barley, and internal bark necrosis in apple, the toxicity is related to localized accumulations of Mn (Fisher et al., 1977; Williams and Vlamis, 1957b, 1971). For example, Williams et al. (1971) reported that the yellow leaf margins of mustard affected by Mn toxicity contained 2300 mg Mn/kg compared to 570 mg/kg for green portions of the same leaves.

Symptoms of Mn toxicity also include Mn-induced symptoms of deficiencies of mineral nutrients, such as Fe, Mg, and Ca (Fleming, 1989; Foy et al., 1981; Lee, 1972). For example, Mn-induced Fe deficiency symptoms were reported in barley grown in excessive concentrations of Mn in a nutrient solution (Agarwala et al., 1977; Alam et al., 2000; Rees and Sidrak, 1961). However, other researchers reported no symptoms of Fe deficiency in barley grown in nutrient solutions with high Mn (Beauchamp and Rossi, 1972; Löhnis, 1960; Vlamis and Williams, 1964). This discrepancy may be related to the use of different cultivars in the experiments.

Other reported effects of Mn toxicity include restricted number and size of nodules (Foy et al., 1978; Vose and Jones, 1963), retarded germination (Brenchley, 1914; Millikan, 1948), retarded growth (Aso, 1902), and delayed ripening (Brenchley, 1914).

In many cases, Mn toxicity causes symptoms much before the suppression in dry weight occurs. Foy et al. (1978) reported that plant symptoms of Mn toxicity often are detectable at stress levels that produce little or no suppression in vegetative growth. Morris and Pierre (1949) reported that the dry weight of soybean and cowpeas was limited significantly at 10 mg Mn/kg in the plant tissue but the symptoms occurred at 2.5 mg/kg. The reduced dry weight may be caused by the shedding of the affected leaves since chlorosis may cover as much as 50% of the leaf area and up to 100% of the leaves may be affected in severe cases of toxicity (Morris and Pierre, 1949).

12.6.2.4.3 Losses in Yield and Crop Quality

Losses in crop yield and quality are the consequences of the morphological, physiological, and biochemical effects of Mn toxicity on plants. Galvez et al. (1989) reported growth restriction of sorghum grown with high levels of Mn in the nutrient solution. Morris and Pierre (1949) reported that Mn toxicity caused 63% suppression in the dry weight of lespedeza.

12.6.2.5 Management of Manganese Toxicity

Factors leading to Mn toxicity should be managed properly before losses in crop yield and quality occur. Because plant roots absorb Mn as the divalent cation, any factors that change its concentration in the soil solution potentially can affect the accumulation of Mn in the plant. These factors include soil physical and chemical properties (Graham, 1988a,b) including soil water, OM, and pH (Christenson et al., 1951) and the concentrations of other elements that interact with Mn absorption, translocation, and use (Welch et al., 1991). Lowering the pH below about 5.5 increases the level of the Mn^{2+} in the soil solution and increases the likelihood of toxicity (Kamprath and Foy, 1971). Air drying may greatly increase the level of exchangeable Mn in acid soils rich in easily reducible Mn, and flooding and compaction of soils limit soil aeration and favor the reduction of Mn (Kamprath and Foy, 1971).

12.6.2.5.1 Soil Management Factors

12.6.2.5.1.1 Application of Other Elements Application of other elements can help minimize the effects of Mn toxicity. For example, application of Si significantly decreases lipid peroxidation caused by excessive Mn, thereby decreasing the symptoms of Mn phytotoxicity and improving plant growth (Iwasaki et al., 2002; Shi et al., 2005). There is also an association between Mn toxicity and the decrease in Ca concentration in barley, indicating a competition and a specific interaction during the absorption of these elements, their translocation, or both (Alam et al., 2001). Another study, in which Ca was applied to reduce Mn toxicity, showed that Ca additions inhibited Mn translocation from roots to shoots in barley plants but did not affect Mn absorption in the roots (Alam et al., 2006). These findings suggested that supply of Ca could prevent Mn accumulation in the shoots, protecting the photosynthetic apparatus from the dangerous effect of excessive Mn. With barley, Mn toxicity could be suppressed by high K contents, which inhibit the absorption and the translocation of Mn in plants (Alam et al., 2005).

12.6.2.5.1.2 Application of Fertilizers and Pesticides Manganese toxicity can appear on previously productive cropland that has been treated repeatedly with Mn-containing pesticides and fertilizers (Chaney and Giodano, 1977). Excessive Mn was reported in certain Florida citrus-producing soils, in which the accumulation resulted from repeated spray and fertilizer applications (Foy et al., 1978). Manganese toxicity was reported in plants in soil treated with sewage sludge (Chaney and Giodano, 1977). Peterson et al. (1971) reported that Mn concentration in corn leaves was correlated directly with sludge fertilization.

Nitrogen form affected Mn toxicity in muskmelons (Elamin and Wilcox, 1986c). Plants grown with nitrate (NO_3^-) developed Mn toxicity symptoms and had high concentrations of Mn in the shoot tissue after exposure to high concentrations of Mn. However, plants grown with ammonium (NH_4^+) had less Mn in their shoot tissue and developed no symptoms of Mn toxicity when exposed to the same Mn concentration treatments. Arnon (1937) reported that NH_4^+ sources of N decreased Mn absorption and toxicity in barley. Reddy and Mills (1991) reported that NO_3^- source of N enhanced Mn uptake while NH_4NO_3 reduced it. They also reported that the growth of marigold was suppressed with high concentration of Mn in the nutrient solution (10 mg/L) and that the suppression was greater with NO_3^- nutrition than with NH_4NO_3 nutrition. Since NO_3^- fertilization increases soil pH, use of this N source should result in a decreased Mn concentration in the soil solution and consequently uptake by plants. However, NO_3^- or NH_4^+ influences not only the rhizosphere pH but also the uptake of other ions by plants (Barker and Mills, 1980), and this latter factor must also be considered independent of Mn availability to plants as a consequence of rhizosphere pH changes.

Application of divalent Mn sources with acid-forming fertilizers results in increased Mn absorption (Hossner and Richards, 1968). Williams et al. (1971) reported that fertilizers increased Mn concentrations in plants in the following order: $(NH_4)_2SO_4 > NH_4NO_3 > Ca(NO_3)_2$. This series coincides with the declining order of soil acidification by these fertilizers. There were no symptoms above soil pH 5.5, and yields were correlated inversely with Mn in plants (Williams et al., 1971).

Larson (1964) reported increased Mn absorption by plants if high rates of superphosphate were applied to neutral and alkaline soils. They suggested chemical mobilization of soil Mn as a result of soil reactions of the superphosphate.

12.6.2.5.1.3 Liming   The absorption of excessive amounts of Mn and the resultant toxicity in sensitive crops can be reduced by liming acid media to pH 5.5 or above (Fleming, 1989; Jackson et al., 1996). However, Mn toxicity can also occur in calcareous soils and flooded neutral soils (Moraghan and Freeman, 1978). Moraghan and Freeman (1978) reported that flax developed Mn toxicity symptoms at pH 8.1, on a calcareous soil with low available Fe.

Pre-plant addition of dolomite to a soilless peat-based medium maintained the pH within the range of 5.8–6.2 and lowered Fe and Mn concentrations in the tissue of geranium (*Pelargonium* × *hortorum* L.H. Bailey) plants suffering from Fe and Mn toxicity. Liquid dolomite drenches during plant growth had little effect on tissue Fe or Mn (Bachman and Miller, 1995).

It is generally considered that the role of liming in amelioration of Mn toxicity is due primarily to reduced Mn availability at high soil pH. However, such beneficial effects could be, in part, due to increased Ca availability (Bekker et al., 1994). Indeed, Mn toxicity is associated with a decrease in Ca concentration in plants (Alam et al., 2001; Horst and Marschner, 1978a), thus indicating specific interaction and competition during absorption of the minerals, their translocation, or both. Bekker et al. (1994) reported that liming the soil with coarse coralline lime material containing 31.1% Ca and 1.7% Mg alleviated Mn toxicity in peanut. They explained that this alleviation was not likely due to decreased Mn solubility because lime application increased the soil pH by less than 0.1 unit, but that it was due rather to an increase of Ca availability. Calcium availability reduced Mn toxicity through a Ca–Mn antagonism.

Morris and Pierre (1948) attributed the effect of liming in promoting growth of lespedeza on acid soils to removal of toxic concentrations of soluble Mn and not to the addition of Ca to the soil, because lespedeza requires very low Ca. Liming acid soils to pH values of 5.5–6.0 reduces water-soluble Mn in the soil (Morris and Pierre, 1948). The availability of Mn to plants generally decreases as soil pH increases. A sharp yield depression may occur when strongly acid soils are limed to pH 6.0 or above, because of micronutrient deficiencies that may occur under limed conditions (Kamprath and Foy, 1971).

12.6.2.5.1.4 Water Level and Growth Medium   Waterlogging of soil promotes the reduction of Mn to the divalent (available) form. Tolerance of some plant species to certain wet soils coincides with their tolerance to excessive Mn^{2+} in nutrient cultures. For example, the Mn-tolerant subterranean clover is more tolerant of waterlogging than the Mn-sensitive barrel clover (*Medicago truncatula* Gaertner) (Robson and Loneragan, 1970). Graven et al. (1965) suggested that the sensitivity of alfalfa to wet soils might be due in part to Mn sensitivity.

Gupta and Chipman (1976) and Gupta et al. (1970) reported Mn toxicity of carrots in a sphagnum peat medium. The toxicity was attributed to an increased solubility of a humic fraction of peat that rendered Mn more available to plants. Goh and Haynes (1978) reported that a different proportion of peat resulted in a different growth rate of marigold (*T. erecta* L.), petunia (*Petunia hybrida nana compacta* Juss), and forget-me-not (*Myosotis alpestris* L.). Generally, poorest growth occurred when peat represented more than 25% of the growth medium. Increased peat in the medium increased the amount of micropore water and decreased its air capacity, conditions that favor Mn reduction and make Mn^{2+} available to plants (Goh and Haynes, 1978). Poor growth also occurred if peat was mixed with soil, sawdust, or bark. Welch (1991) reported that, in England, Mn toxicity occurred in crops grown in peaty soils or mineral soils with high levels of OM. The biological reduction of Mn in the soil is enhanced by the addition of fresh OM (Cotter and Mishra, 1968) and by flooding (Ponnamperuma et al., 1969). The availability of the divalent form of Mn upon flooding is usually greater when soils contain high levels of OM (Mortvedt and Cunningham, 1971).

The availability of Mn is closely related to the transformations of organic materials in soils (Kamprath and Foy, 1971). Microorganisms influence the availability of Mn by changing the pH of the medium (Kamprath and Foy, 1971) and by changing the oxidation state of Mn (Alexander, 1961). Soil microorganisms appear to play a major role in determining soil levels of reduced Mn especially at the higher soil pH levels (Douka, 1977; Nambier, 1975). Christenson et al. (1951) reported that soil pH was of greater importance than OM content and moisture level in affecting the status of available soil Mn.

12.6.2.5.2 Environmental Interactions

The incidence of Mn toxicity strongly depends on the environmental factors and on the status of nutrients, other than Mn, in the growth medium. For example, in some cases, Mn concentrations around 700 mg/kg in potato foliage have induced Mn toxicity, but in other cases, levels up to 4000 mg/kg Mn in field-grown potato foliage did not cause toxicity (Marsh et al., 1989; Ouellette and Genereaux, 1965). This differential tolerance to Mn appeared to depend on nutritional and environmental interactions. It is apparent that several of these nutritional and environmental interactions could depend on changes in growth rate. Plant tolerance to Mn toxicity is, to some extent, dependent on growth rate, cell size, and particularly the size of the vacuole (Rufty et al., 1979), which is generally the site of Mn accumulation (Horst and Marschner, 1978b; Roby et al., 1987). The relative tolerance of plants to excessive Mn is affected particularly by climatic factors such as temperature and light intensity (Rufty et al., 1979; McCool, 1935).

12.6.2.5.2.1 Temperature
Plant tolerance to Mn toxicity increases with increasing temperature in tobacco or soybeans despite high Mn absorption (Heenan and Carter, 1975; Rufty et al., 1979). This increased tolerance to Mn toxicity with increasing temperature is attributed mainly to faster growth rate due to high temperature. Fast-growing leaves form large vacuoles, thereby sequestering the potentially toxic Mn. In the case of potato, high temperature reduces the growth rate and increases Mn absorption (Marsh et al., 1989). The combination of these two facts results in accentuated symptoms of Mn toxicity when the plants are grown under high Mn concentration in the medium. Marsh et al. (1989) attributed the severity of Mn toxicity at high temperature to an interaction between growth rate and Mn tolerance.

12.6.2.5.2.2 Light
There are conflicting views on the effect of high light intensity on Mn toxicity, with reports of it increasing the severity of toxicity symptoms (Horiguchi, 1988; González et al., 1998; Nable et al., 1988) or reducing them (Wissemeier and Horst, 1987). High light intensity was reported to stimulate Mn absorption by plants and to accentuate the severity of Mn toxicity (Horiguchi, 1988; McCool, 1935). Horiguchi (1988) reported that increased light intensity at high levels of Mn in the nutrient solution decreased the chlorophyll content of the leaves, causing chlorosis, and precipitated the oxidized Mn in the medium. High light intensity increased the severity of Mn-induced chlorosis even at similar levels of Mn concentration within the leaves. According to Horiguchi (1988), the Mn-induced chlorosis in the leaves is due to the oxidation of chlorophyll by the oxidized Mn in the medium. Gerretsen (1950) attributed Mn-induced chlorosis to photooxidation of chlorophyll.

12.6.2.5.3 Nutritional Interactions

Several investigators reported that plant tolerance to soluble Mn was affected by the concentration of other nutrient elements in the growth medium (McCool, 1913; Morris and Pierre, 1948; Somers and Shive, 1942; Williams and Vlamis, 1957a). Excessive Mn can interfere with the absorption, translocation, and utilization of other mineral elements such as Ca, Mg, Fe, and P (Clark, 1982). Galvez et al. (1989) reported that high levels of Mn in the nutrient solution decreased shoot concentrations of Si, K, Ca, Mg, Zn, and Cu; increased shoot concentrations of Mn and P; decreased root concentrations of K and Mg; and increased root concentrations of Mn, Si, P, Fe, Zn, and Cu.

Under some conditions, the addition of Si, Fe, Ca, or Mg alleviates Mn toxicity (Heenan and Carter, 1975; Horst and Marschner, 1978b; Löhnis, 1960; Osawa and Ikeda, 1976).

12.6.2.5.3.1 Nutritional Interaction with Iron

The effect of Fe in lowering the uptake and accumulation of Mn by plants (see Section 12.4.3.2.1) indicates that there is potential interaction between the two elements in crop nutrition. However, the studies in which such interactions have been demonstrated often give contradictory results. The growth medium used in the studies seems to have a major influence on the nature of the relationship between Fe and Mn. In studies conducted by Mashhady and El-Damary (1981), solution cultures tended to give a negative interaction between Mn and Fe; sand cultures gave no interaction; and soil-grown plants gave positive interactions. The authors suspected that the dramatic effect of the medium on the nature of the Mn–Fe interaction involved the presence or absence of solid phases and their interaction with plant roots. To evaluate Mn–Fe interactions for soil-grown plants, Warden and Reisenauer (1991) suggested determining not only the composition of Mn and Fe in the plant but also the levels of Mn and Fe in the rhizosphere and the length of the roots.

Most of the investigations on the interaction between Fe and Mn in plants showed a negative correlation between Fe and Mn accumulation in the shoots of susceptible cultivars. Ohki (1975) reported reciprocal Fe–Mn relationships in cotton. Manganese concentrations 4 and 247 mg/kg in plant shoots were associated with Fe concentrations of 270 and 51 mg/kg, respectively (Ohki, 1975). Kohno and Foy (1983) reported that high concentrations of Fe decreased the absorption of Mn by some plants.

12.6.2.5.3.1.1 Manganese-Induced Iron Deficiency

Osawa and Ikeda (1976) described two types of manganese toxicity in plants. The first type was identical to Fe deficiency symptoms of interveinal chlorosis on young leaves and was observed in spinach (*Spinacia oleracea* L.), tomato, pepper, and beans. The second type of was characterized by marginal chlorosis of old leaves, often with brown necrotic spots, and was observed in lettuce, celery (*Apium graveolens* L.), and cabbage (*Brassica oleracea* var. *capitata* L.). In general, increasing the Fe supply prevented or reduced the severity of the interveinal chlorosis but had little or no effect on reducing the marginal chlorosis and brown spots. Leaf analyses indicated that the Mn-induced interveinal chlorosis was due to Fe deficiency, whereas the marginal chlorosis and brown necrotic spots were due to accumulation of excessive Mn.

Foy et al. (1978) also described two types of Mn toxicity in chickpeas. The first type was pronounced in genotype T-1 and was characterized by marginal chlorosis and brown necrotic spots on leaflets of middle and young leaves. The second type of Mn toxicity was pronounced in genotype T-3 and was Mn-induced Fe deficiency chlorosis. Manganese treatment of T-3 decreased Fe concentrations of young leaves and caused chlorosis. The reduction of chlorophyll in chlorotic plants was counteracted by high Fe applications. Millikan (1948) reported that flax grown at excessive concentrations of Mn (25 mg/kg) developed lower leaf necrosis in addition to chlorosis and necrosis of the top of the plant, which are symptoms identical to those of Fe deficiency. Supplying the plants with Fe eliminated the symptoms of Fe deficiency but did not have any effect on the lower leaf necrosis.

Lee (1972) reported that excessive Mn in the medium induced Fe deficiency in potato. Manganese-induced Fe deficiency was related to the ratio Fe/Mn (Leach and Taper, 1954) as well as to the concentration of these two nutrients in the medium (Beauchamp and Rossi, 1972; Foy et al., 1995). Foy et al. (1995) reported that Fe concentrations and Fe/Mn ratio were higher in the Mn-tolerant cultivar of cotton (307) than in the Mn-sensitive cultivar (517). Beauchamp and Rossi (1972) reported that high levels of Mn in the solution culture depressed Fe absorption by barley.

In various plant species, Mn toxicity was decreased by increasing the concentration of Fe in the nutrient solution (Somers and Shive, 1942). Somers and Shive (1942) reported that the Fe/Mn ratio in culture solution, rather than the total amounts of these elements, controlled the growth of soybeans. They also reported that the optimum ratio was around 2.0, and any appreciable variation in this ratio would result in appearance of Fe or Mn toxicity symptoms. Hophins et al. (1944), however, obtained normal growth of beans in culture solutions with Fe/Mn ratios much higher than 2 and reported no toxicity from Mn regardless of the ratio. Morris and Pierre (1948) reported that 1 mg Mn/Land 0.5 mg Fe/L allowed excellent growth and that the plants grown with the same ratio (2.0) but with concentrations five times higher than the former treatment resulted in poor growth and severe Mn toxicity symptoms. Leach and Taper (1954) reported that the optimum Fe/Mn ratios in plants ranged from 1.5 to 3 for beans and from 0.5 to 5 for tomato. Iron deficiency developed at lower ratios. Osawa and Ikeda (1976) reported that, in spinach, the Fe/Mn ratio in nutrient solution was related closely to plant growth and that decreasing this ratio resulted in Mn-induced chlorosis (Mn-induced Fe deficiency) and suppressed growth. They also reported that in tomato, pepper, bean, and eggplant (*Solanum melongena* L.), both Fe/Mn ratio and Mn concentration in the nutrient solution affected plant growth and the occurrence of Mn-induced chlorosis. In celery, cabbage, and Welsh onion (*Allium fistulosum* L.), the absolute Mn concentration appeared to be the dominant factor (Osawa and Ikeda, 1976).

The actual means by which Mn induces Fe deficiency in some plant species was investigated by many researchers. Rippel (1923), Chapman (1931), and Millikan (1949) reported that plants manifesting symptoms of Fe deficiency associated with excessive Mn may actually contain the same or higher content of Fe than unaffected plants, indicating that it is not Fe absorption but its action in the tissues that is affected unfavorably by excessive Mn. Somers and Shive (1942) suggested that this unfavorable effect is probably due to the Mn catalysis of the oxidation of the physiologically active form of Fe (Fe^{2+}) to the inactive form (Fe^{3+}). They also suggested the probable formation of an insoluble ferric phosphate organic complex. Benette (1945) reported that Mn does not interfere with Fe utilization in the leaf but produces chlorosis by depressing the absorption of Fe. Johnson (1917) also explained the Mn-induced Fe deficiency by inhibiting Fe absorption. In Fe-deficient soils, Mn-induced Fe deficiency may be related to excessive mobilization of soil Mn as a result of reductant and proton release by plant roots responding to Fe deficiency (Horst, 1988). Broadley et al. (2012) reported that Mn-induced Fe deficiency was caused by inhibited Fe uptake across the plasma membrane and competition (or imbalance) at the cellular level.

Other explanations of the effect of Mn in inducing Fe deficiency could be that the plant species susceptible to Mn-induced Fe deficiency assimilate Mn preferentially over Fe and are more susceptible to interference of Mn with Fe function inside the plant or that Fe is immobilized more readily within tissues of these plants (Somers and Shive, 1942).

There may also be genotypic variations in the absorption and utilization of Mn and Fe by plants (Brown et al., 1972; Graham, 1988a,b). The abilities of Fe to alleviate Mn toxicity depend on whether or not the toxicity is due to Mn-induced Fe deficiency or to accumulation of excessive Mn in the plant tissues (Foy et al., 1978; Osawa and Ikeda, 1976). Iron supply has a marked effect on alleviating Mn toxicity in many crops (Morris and Pierre, 1948; Sideris and Young, 1949). Iron treatments corrected Mn toxicity in eucalyptus (*Eucalyptus* spp. L'her) (Winterhalder, 1963), soybean (Heenan and Campbell, 1983), flax (Moraghan and Freeman, 1978), and pineapples (Sideris and Young, 1949). Morris and Pierre (1948) reported that the beneficial action of Fe in reducing Mn toxicity appeared to be due to some mechanism that prevents the excessive absorption and subsequent accumulation of Mn in the plant rather than to an increase in Fe absorption and corrective action of Fe within the plant itself. They reported that alleviation of Mn toxicity by higher Fe concentration (1 mg/L compared with 0.2 mg/L) was not due to an increase of total Fe in the plant but to an approximate 50% reduction in the Mn content of the plant. Limitations as to the effect of Fe in overcoming Mn toxicity are evident from the research of Morris and Pierre (1948), who showed that an increase in the concentration of Fe in culture solution up to 1 mg/L resulted in marked reduction of the Mn toxicity from a given concentration of Mn. A further increase in the Fe concentration from 1 to 2.5 mg/L produced no additional decrease in Mn absorption, and higher Fe concentration resulted in suppressed growth regardless of the concentration of Mn used.

Supply of Mo or P may be effective in alleviating Mn-induced Fe deficiency. Millikan (1948) reported that an essential function of Mo is intimately associated with the regulation of the deleterious effects of Mn on the physiological availability of Fe to plants. Kirch et al. (1960) reported that Mn absorption in tomatoes was influenced by interactions among Fe, Mn, and Mo. Iron supply counteracted the yield-depressive effects of excessive Mn, but the amount of Fe needed for maximum yield was increased as the Mo supply increased. Molybdenum suppressed yields at low Fe levels and increased yields at higher Fe levels. Foy et al. (1978) reported that low P levels in the medium reduced the effect of Mn toxicity in chickpeas (variety T-3) because low P increased Fe uptake by the plants.

Manganese toxicity may develop under Fe-deficient conditions (Brown and Jones, 1977a). Iron deficiency stress enhances Mn absorption in some species (Heenan and Campbell, 1983; Sideris and Young, 1949; Somers and Shive, 1942). Brown and Jones (1977a–c) reported increased Mn absorption by soybean, cotton, or sorghum, grown under Fe deficiency stress. The absorption of Fe and Mn was greater in "Hawkeye" (Fe-efficient) than in "PI" (Fe-inefficient) soybean. When these cultivars were grown under Fe-sufficient conditions, they both absorbed low levels of Fe and Mn (Brown and Jones, 1962).

Dicots respond to Fe deficiency by an increased potential for Fe absorption and translocation (Brown and Ambler, 1973; Fleming, 1989). When Fe response is initiated (roots release reductants), the absorption of Mn and Fe is enhanced in Fe-efficient species such as sunflower and soybean (Brown and Ambler, 1973; Fleming, 1989).

The magnitude of the enhanced Mn absorption is related to the intensity of Fe stress as well as the cultivar sensitivity to Mn. In a study on bean, Fleming (1989) reported that, under alkaline conditions, Mn absorption by "Bush Blue Lake (BBL) 290" snap bean was increased dramatically with Fe stress, whereas a moderate increase was reported for "BBL 274." Fleming (1989) also reported that foliar symptoms of Mn toxicity, observed on Fe stressed "BBL 290," increased in severity at higher concentrations of Mn.

12.6.2.5.3.1.2 Manganese–Iron Toxicity Contrary to the preceding findings, susceptibility to Mn toxicity and Fe toxicity seemed to coincide in "Bragg" and "Forrest" soybeans (Brown and Jones, 1977a), marigold (Albano and Miller, 1996; El-Jaoual Eaton et al., 2012), and geranium (Bachman and Miller, 1995). Brown and Jones (1977a) suggested that high levels of Fe in the shoots of "Bragg" soybean may accentuate Mn toxicity or create Fe toxicity *per se*. Bachman and Miller (1995) reported that 1–60 mg/kg Fe in the growth medium caused Fe/Mn toxicity in "Aurora" geranium plants. Symptomatic tissues of affected geraniums had higher concentrations of Fe and Mn than asymptomatic tissues. Iron concentration increased with increasing Fe level in the growth medium, in both symptomatic and asymptomatic tissues. Manganese, however, increased in symptom-showing tissues as Fe level increased and remained the same in asymptomatic tissue, among all Fe treatments. Vlamis and Williams (1964) also reported that the chlorotic spots of lettuce contained high concentrations of Fe and Mn. Morris and Pierre (1948) reported that Fe content of lespedeza plants increased with increased Mn concentration in the nutrient solution and that high Mn concentration in the medium may thus cause Fe deficiency in the plants if the medium is not sufficient in Fe.

12.6.2.5.3.2 Nutritional Interaction with Silicon Abundant evidence shows that a soluble source of Si in the growth medium can protect plants against Mn toxicity (Bowen, 1972; Lewin and Reimann, 1969; Peaslee and Frink, 1969; Williams and Vlamis, 1957b). The higher absorption of Si by monocots than dicots may explain the higher tolerance of monocots to Mn toxicity compared with dicots (Foy et al., 1978). Lewin and Reimann (1969) reported that plants grown in low Si concentrations in the growth medium accumulated higher concentrations of Mn than those grown in high Si concentrations. Galvez et al. (1989) reported that acid-soil-tolerant genotypes had higher leaf Si concentrations than the acid-soil-susceptible genotypes grown at higher-acid soil stresses. They added that Si might be the cause of the plant tolerance to high acidity and hence to Mn toxicity.

Silicon reduced or prevented Mn toxicity in barley, rice, rye, ryegrass (Vlamis and Williams, 1967), sudangrass (*Sorghum* × *drummondi* Millsp. & Chase) (Bowen, 1972), and sorghum (*Sorghum bicolor* Moench) (Galvez et al., 1989).

Van Der Vorm and Van Diest (1979) attributed the effect of Si in alleviating Mn toxicity to a decreased rate of Mn absorption. Peaslee and Frink (1969) reported that treating acid soils with H_2SiO_3 significantly decreased Mn concentrations in the plant tissues. In contrast to these reports, Williams and Vlamis (1957b) reported that the effect of Si in suppressing Mn toxicity does not necessary result from decreased Mn absorption. In an experiment with barley, they reported that Si prevented the formation of necrotic spots of high Mn concentrations in the leaves although it did not significantly alter the absorption of Mn by plants.

The effect of Si in alleviating Mn toxicity also was attributed to decreased Mn transport from the roots to the shoots (Horigushi, 1988a) and to homogeneous distribution of Mn throughout the leaves instead of concentrating in some areas more than in others (Vlamis and Williams, 1967). Vlamis and Williams (1967) reported that the distribution of Mn was homogeneous in the leaves of plants grown in high concentrations of Si, whereas it was not homogeneous in those grown in low Si concentrations. They also reported that Mn was more concentrated in the affected areas than in the green areas when the plants were grown in low concentrations of Si.

Silicon increases the Mn critical toxicity concentration (Horst and Marschner, 1978b; Galvez et al., 1989). Horst and Marschner (1978b) reported that the critical toxicity concentration of Mn in beans increased from about 100 mg/kg dry weight in the leaves to about 1000 mg/kg in the presence of Si. Galvez et al. (1989) reported that Si enhanced the tolerance of rice to Mn toxicity by increasing root-oxidizing conditions to decrease Mn (and Fe) reducing capacities at root surfaces, so that toxic levels of Mn (and Fe) in the reduced form would not be available to plants. Foy et al. (1978) reported that Mn oxidation was greater in rice plants supplied with Si than in Si-deprived plants and that Si decreased excessive absorption of Mn and Fe.

Other reported effects of Si applications in alleviating Mn toxicity are moderated respiration rate and decreased peroxidase activity (Horigushi, 1988a). Silicon in the cell is deposited in close association with other constituents of the cell wall (Lewin and Reimann, 1969). Silicon may play a role in maintaining the compartmentalization of the cell wall and preventing peroxidase from coming into contact with the precursors of the brown substrates (Horigushi, 1988a). Silicon application also limited the increase of P in plants caused by excessive Mn in the nutrient solution (Galvez et al., 1989). In a corn genotype tolerant of Mn toxicity, silicon substantially increased the thickness of the epidermal leaf layers where excessive Mn was stored (Doncheva et al., 2009). Limitations as to the effect of Si in overcoming Mn toxicity were reported (Galvez et al., 1989). If Mn was sufficiently high to inhibit dry matter yields of sorghum by about 80% or more, added Si had no beneficial effect on overcoming Mn toxicity.

12.6.2.5.3.3 Nutritional Interaction with Calcium   Manganese-induced suppressions in Ca concentrations in plants have been reported (Clark et al., 1981; Galvez et al., 1989). The Ca content of peanut decreased markedly as the Mn concentration of culture solution increased (Bekker et al., 1994). The higher tolerance of strain L_6 of lespedeza (compared with strain L_{39}) to excessive Mn appears to be attributed to its greater tolerance of soluble Mn and to its lower requirement for Ca (Morris and Pierre, 1949). Increased Ca levels in the growth medium often decreased Mn absorption and toxicity (Heenan and Carter, 1975; Robson and Loneragan, 1970; Shuman and Anderson, 1976; Hewitt, 1945). Vose and Jones (1963) reported that increasing the Ca level in solution culture reduced the adverse effects of high levels of Mn on suppressing the number of nodules on white clover roots. Foy et al. (1978) reported that the ratio of Ca/Mn is important in nodulation and in tolerance of soybean to Mn toxicity. Bekker et al. (1994) reported the importance of a Ca/Mn ratio above 80 for a desirable Ca/Mn balance in peanut tissue. In a study on alfalfa, Ouellette and Dessureaux (1958) reported that Mn-tolerant clones contained lower concentrations of Mn in their shoots and higher concentrations of Mn and Ca in their roots than sensitive clones. They concluded that Ca absorption regulated Mn toxicity by reducing Mn transport to the shoots. Contrary to these findings, Chapman (1931) and Morris and Pierre (1948) reported that Mn toxicity was not alleviated by additional Ca and was even greater at high Ca concentrations.

The conflicting results between Hewitt (1945) and Chapman (1931) and Morris and Pierre (1948) may be due to the different crops used. Morris and Pierre (1948) pointed out that high Ca levels (300 mg/L) in the nutrient solution can limit plant growth regardless of the concentration of Mn in the solution, probably because of unbalanced nutrient solution. They concluded that optimal Ca concentration (60 mg/L) was crucial for optimal plant growth and that Ca addition to the nutrient solution may cause a nutritional imbalance to the plants.

A symptom of Ca deficiency called crinkle leaf is another well-known malady induced by Mn toxicity in dicots, such as cotton (Foy et al., 1981; Horst and Marschner, 1978a). When the supply of Mn is excessive, the translocation of Ca into the shoot apex, especially, is inhibited (Marschner, 1995). This inhibition might be related to the fact that high Mn levels decrease the IAA levels in the areas of new growth (Morgan et al., 1966; Morgan et al., 1976). Auxin was reported to be responsible for the formation of new binding sites for the transport of Ca to the apical meristems (Horst and Marschner, 1978a).

In contrast to Mn-induced Fe and Mg deficiencies, where Mn competes with Fe or Mg for uptake, Mn-induced Ca deficiency is most likely an indirect effect of Mn on Ca transport to expanding leaves (Broadley et al., 2012). Acropetal Ca transport is mediated by a basipetal countertransport of IAA, and high IAA oxidase activity or polyphenoloxidase activity, in general, is measured frequently in tissues with high Mn concentration (Horst, 1988; Fecht-Christoffers et al., 2007). Calcium deficiency symptoms induced by Mn toxicity are therefore most likely caused by enhanced degradation of IAA, a process that is aggravated by high light intensity (Horst, 1988). Loss of apical dominance and enhanced formation of auxiliary shoots (witches' broom) is another symptom of Mn toxicity (Kang and Fox, 1980, Bañados et al., 2009), further supporting the hypothesis of a relationship between impaired basipetal IAA transport and Mn toxicity (Gangwar et al., 2010).

As discussed in Section 12.4.3.2.3, calcium can be supplied to plants in order to lower the uptake of Mn, minimizing its toxicity.

12.6.2.5.3.4 Nutritional Interaction with Magnesium High concentrations of Mn in the growth medium can induce Mg deficiency in the plant (Mn-induced Mg deficiency) (Heenan and Campbell, 1981). Excessive Mn may reduce Mg uptake by up to 50% (Kazda and Znacek, 1989) due to competition (Heenan and Campbell, 1981). The competition between Mg and Mn for binding sites in the roots, during absorption, inhibits Mg absorption, since Mn competes more effectively than Mg and even blocks the binding sites for Mg (Horst and Marschner, 1990).

Broadley et al. (2012) reported that Mn-induced Mg deficiency is caused by inhibited Mg uptake across the plasma membrane and competition (or imbalance) at the cellular level. Accordingly, Mn toxicity can often be counteracted by a large supply of Mg (Bachman and Miller, 1995; Davis, 1996; El-Jaoual Eaton et al., 2012; Elamin and Wilcox, 1986a,b; Löhnis, 1960; Rezai and Farbodnia, 2008; Terry et al., 1975; Wu, 1994). Magnesium reduced Mn uptake by excised or intact roots of several plant species (Harrison and Bergman, 1981; Löhnis, 1960; Maas et al., 1969).

In some cases, Mg application is not a practical method for the avoidance of Mn toxicity (Davis, 1996). The ability of Mg to reduce Mn uptake depends on the concentration of Mn in the medium. Elamin and Wilcox (1986a) reported that, at high Mn concentration, Mg had little effect on Mn uptake and the plants were able to accumulate toxic levels of Mn at all levels of Mg supply. In addition, using Mg to prevent Mn toxicity would require large Mg applications, which could lead to serious nutritional imbalance because Mg would interfere with Ca uptake.

12.6.2.5.3.5 Nutritional Interaction with Potassium In comparison to Mg, the absorption of K is affected only slightly by increasing Mn concentrations in the growth medium (Heenan and Campbell, 1981). Brown and Jones (1977b) speculated that high K levels in the shoots of Mn-tolerant "Lee" soybean alleviated the harmful effects of high internal Mn concentrations. Alam et al. (2005) reported that high levels of K in the nutrient solution alleviated Mn toxicity and Mn-induced Fe deficiency in barley. They also reported that high K in the feeding solution reduced Mn absorption and translocation but did not affect Fe absorption and translocation.

12.6.2.5.3.6   Nutritional Interaction with Phosphorus   Bortner (1935a,b) suggested that P may reduce the toxicity of excessive Mn by rendering it inactive within the plant. Phosphorus may detoxify excessive Mn in the plant roots through precipitation (Heintze, 1968). However, P does not seem to detoxify excessive Mn in the stems and the leaves since no precipitation was found in those parts of the plant (Morris and Pierre, 1948). Applications of soluble phosphates to soils or nutrient solutions containing high amounts of Mn increased crop yields and suppressed development of Mn toxicity symptoms (Bortner, 1935a,b). Bortner (1935a,b) attributed these effects to the additional available P and to a decrease in soluble Mn in the medium by precipitation. Contrary to these findings, Morris and Pierre (1948) reported that the solubility of Mn in the nutrient solution was not affected by the additional phosphate and that P did not have any effect in preventing Mn toxicity and may even accentuate it if added at high concentrations to the growth medium. Le Mare (1977) reported that P application to the soil increased P content in the plants and increased Mn absorption. Phosphorus also increased Mn absorption by potato and accentuated the severity of visual symptoms of Mn toxicity at high temperature (Marsh et al., 1989). Millikan (1948) reported an indirect negative effect of P on Mn toxicity and reported that P may precipitate Mo in the medium making it unavailable to plants, an action that may accentuate Mn toxicity.

12.6.2.5.3.7   Nutritional Interaction with Molybdenum   Anderson and Thomas (1946) reported general similarities between Mo deficiency and Mn toxicity. Molybdenum-deficient plants showed chlorosis with a necrosis of the margins of the lower leaves, which are also symptoms of Mn toxicity.

Molybdenum is poorly available in strongly acidic soils (pH <5.5), whereas Mn availability is high at this acidity. Liming of Mo-responsive soils improves growth. This beneficial effect of lime has been attributed to an increase in the availability of Mo, but concomitantly with such an improvement in the Mo status in the soil, there would be a decrease in the availability of Mn. Millikan (1948) reported that small applications of Mo to the growth medium prevented Mn toxicity symptoms (lower leaf necrosis) of flax grown in strongly acid soil rich in $MnSO_4$. The report of Davies (1945) that whiptail disease of cauliflower may be prevented by Mo is of interest here as Mn toxicity symptoms in cauliflower appear to be similar to those of whiptail disease (Wallace, 1946; Wallace et al., 1945). Anderson and Thomas (1946) and Anderson and Oertel (1946) found that Mo alleviated the symptoms of Mn toxicity in the case of clover but not in the case of grass. This difference in responsiveness to Mo is probably due to a difference in the normal Mo requirements of the two species of plants as clover has been shown to have much higher content of Mo than grass (Lewis, 1943; Millikan, 1948).

One of the functions of Mo is to regulate the deleterious effects of Mn on the physiological availability of Fe to plants (Millikan, 1948). The addition of 5–25 mg Mo/L to the nutrient solution reduced the severity of Mn-induced Fe deficiency in plants (Millikan, 1948). With excessive Mn (25–150 mg/L), the addition of 5–25 mg Mo/L retarded the appearance of symptoms and reduced the severity of lower leaf necrosis, which is a characteristic symptom of Mn toxicity (not Fe deficiency) (Millikan, 1948). Manganese toxicity of plants may still occur with the presence of fair quantity of Mo in the growth medium, if the pH is low and the concentration of phosphate is high in the growth medium, probably because of the precipitation of Mo under these conditions (Millikan, 1948).

12.6.2.5.3.8   Other Nutritional Interactions
Plant tolerance to soluble Mn may also be affected by the concentration of sulfur (S), aluminum (Al), zinc (Zn), or copper (Cu) in the growth medium. Wallace et al. (1974) reported that S fertilization significantly decreased the yields of soybeans suffering from Mn toxicity. Aluminum reduced and prevented Fe chlorosis caused by excessive Mn in flax (Millikan, 1949) but also increased Mn absorption and toxicity symptoms in soybeans (Foy et al., 1978). Zinc (Hewitt, 1948) and Cu (Landi and Fagioli, 1983) were reported to reduce Mn absorption and toxicity in plants.

REFERENCES

Abou, M., L. Symeonidis, E. Hatzistavrou, and T. Yupsanis. 2002. Nucleolytic activities and appearance of a new DNase in relation to nickel and manganese accumulation in *Alyssum múrale*. *J. Plant Physiol.* 159:1087–1095.

Abruna, F., J. Vicente-Chandler, and R.W. Pearson. 1964. Effects of liming on yields and composition of heavily fertilized grasses and on soil properties under humid tropical conditions. *Proc. Soil Sci. Soc. Am.* 28:657–661.

Adriano, D.C. 2001. *Trace Elements in Terrestrial Environments. Biogeochemistry, Bioavailability and Risks of Metals.* New York: Springer-Verlag.

Agarwala, S.C., S.S. Bisht, and C.P. Sharma. 1977. Relative effectiveness of certain heavy metals in producing toxicity and symptoms of iron deficiency in barley. *Can. J. Bot.* 55:1299–1307.

Ahangar, A.G., A. Karimian, M. Assad, and Y. Emam. 1995. Growth and manganese uptake by soybean in highly calcareous soil as affected by native and applied manganese and predicted by nine different extractants. *J. Plant Nutr.* 17:117–125.

Alam, S., F. Akiha, S. Kamei, S. Huq, and S. Kawai. 2005. Mechanism of potassium alleviation of manganese phytotoxicity in barley. *J. Plant Nutr.* 28:889–901.

Alam, S., S. Kamei, and S. Kawai. 2000. Phytosiderophore release from manganese-induced iron deficiency in barley. *J. Plant Nutr.* 23:1193–1207.

Alam, S., S. Kamei, and S. Kawai. 2001. Amelioration of manganese toxicity in barley with iron. *J. Plant Nutr.* 28:1421–1433.

Alam, S., R. Kodama, F. Akiha, S. Kamei, and S. Kawai. 2006. Alleviation of manganese phytotoxicity in barley with calcium. *J. Plant Nutr.* 29:59–74.

Albano, J.P. and W.B. Miller. 1996. Iron deficiency stress influences the physiology of iron acquisition in marigold (*Tagetes erecta* L.). *J. Am. Soc. Hortic. Sci.* 121:438–441.

Albott, A.J. 1967. Physiological effects of micronutrient deficiencies in isolated roots of *Lycopersicon esculentum*. *New Phytol.* 66:419–437.

Alexander, M. 1961. *Introduction to Soil Microbiology.* New York: John Wiley.

Allen, M.D., J. Kropat, S. Tottey, J.A. Del Campo, and S.S. Merchant. 2007. Manganese deficiency in *Chlamydomonas* results in loss of Photosystem II and MnSOD function, sensitivity to peroxides, and secondary phosphorus and iron deficiency. *Plant Physiol.* 143:263–277.

Alscher, R., N. Erturk, and L. Heath. 2002. Role of superoxide dismutase (SODs) in controlling oxidative stress in plants. *J. Exp. Bot.* 53:1331–1341.

Altland, J. 2006. Managing manganese deficiency in nursery production of red maple. Extension faculty (nursery crops), pp: 8. Corvallis, OR: North Willamette Research and Extension Center (NWREC), Oregon State University.

Amao, Y., and A. Ohashi. 2008. Effect of Mn ion on the visible light induced water oxidation activity of photosynthetic organ grana from spinach. *Catal. Commun.* 10:217–220.

Amesz, J. 1993. The role of manganese in photosynthetic oxygen evolution. *Biochim. Biophys. Acta* 726:1–12.

Anderson, A.J. and A.C. Oertel. 1946. Plant responses to molybdenum as a fertilizer. II. Factors affecting the response of plants to molybdenum. *Coun. Sci. Ind. Res. (Aust.) Bull.* 198:25–44.

Anderson, A.J., and M.P. Thomas. 1946. Plant response to molybdenum as a fertilizer: 1. Molybdenum and symbiotic nitrogen fixation. *Australia Council Sci. Ind. Research Bull.* 198(1)7–24.

Anderson, J.M. and N.A. Pyliotis. 1996. Studies with manganese deficient chloroplasts. *Biochim. Biophys. Acta* 189:280–293.

Andrew, C.S. and M.P Hegarty. 1969. Comparative responses to manganese excess of eight tropical and four temperate pasture legume species. *Aust. J. Agric. Res.* 20:687–696.

Apel, K. and H. Hirt. 2004. Reactive oxygen species: Metabolism, oxidative stress, and signal transduction. *Annu. Rev. Plant Biol.* 55:373–399.

Aref, F. 2011. Influence of zinc and boron nutrition on copper, manganese and iron concentration in the maize leaf. *Aust. J. Basic and Appl. Sci.* 5(7):52–62.

Arnon, D.I. 1937. Ammonium and nitrate nitrogen nutrition of barley at different seasons in relation to hydrogen-ion-concentration, manganese, copper and oxygen supply. *Soil Sci.* 44:91–113.

Arnon, D.I. and P.R. Stout. 1939. The essentiality of certain elements in minute quantity for plants with special reference to copper. *Plant Physiol.* 14(2):371–375.

Asher, C.J. 1987. Crop nutrition during the establishment phase: Role of seed reserves. In *Crop Establishment Problems in Queensland: Recognition, Research and Resolution*, eds. I.M. Wood, W.H. Hazard, and F.R. From, pp. 88–106. Brisbane, Australia: Australian Institute of Agricultural Science.

Aso, K. 1902. On the physiological influence of manganese compounds on plants. *Tokyo Imp. Univ. Coll. Agric. Bull.* 5:177–185.

Bachman, G.R. and W.B, Miller. 1995. Iron chelate inducible iron/manganese toxicity in zonal geranium. *J. Plant Nutr.* 18:1917–1929.

Baldisserotto, C., L. Ferroni, and V. Medici, et al. 2004. Specific intra-tissue responses to manganese in the floating lamina of *Trapa natans* L. *Plant Biol.* 6:578–589.

Bañados, M.P., F. Ibáñez, and A.M. Toso. 2009. Manganese toxicity induces abnormal shoot growth in 'O' Neal'bluberry. *Acta Hortic.* 810:509–512.

Bansal, R.L. and V.K. Nayyar. 1994. Differential tolerance of soybean (*Glycine max*) to manganese in Mn deficient soil. *Indian J. Agric. Sci.* 64:604–607.

Baranowska, H., A. Ejchart, and A. Putrment. 1977. Manganese mutagenesis in yeast. V. On mutation and conversion induction in nuclear DNA. *Mutat. Res.* 42:343–345.

Barber, S.A. 1974. Influence of the plant root on ion movement in soil. In *The Plant Root and Its Environment*, ed. E.W. Carson, pp. 525–565. Charlottesville, VA: University of Virginia Press.

Barber, S.A. 1995. *Soil Nutrient Bioavailability: A Mechanistic Approach*, 2nd edn. New York: John Wiley.

Barber, D.A. and R.B. Lee. 1974. The effect of microorganisms on the absorption of manganese by plants. *New Phytol.* 73:97–106.

Barker, A.V. and H.A. Mills. 1980. Ammonium and nitrate nutrition of horticultural crops. *Hortic. Rev.* 2:395–423.

Bar-Tal, A. and B. Aloni. 2008. Effect of fertigation regime on blossom end rot of vegetable fruits. In *Fertigation: Optimizing the Utilization of Water and Nutrients. Fertigation Proceedings*, eds. P. Imas and R. Price, pp. 130–145. Horgen, Switzerland: International Potash Institute.

Bartsevich, V.V. and H.B. Pakrasi. 1995. Molecular identification of an ABC transporter complex for manganese: Analysis of a cyanobacterial mutant strain impaired in the photosynthetic oxygen evolution process. *EMBO J.* 14:1845–1853.

Bartsevich, V.V. and H.B. Pakrasi. 1996. Manganese transport in the cyanobacterium *Synechocystis* sp. PCC 6803. *J. Biol. Chem.* 271:26057–26061.

Batey, T. 1971. Manganese and boron deficiency. In *Trace Elements in Soils and Crops*. Ministry of Agriculture Fisheries and Food, Techn. Bull., Vol. 21, pp. 137–149. London: Her Majesty's Stationery Office.

Beauchamp, E.G. and N. Rossi. 1972. Effects of Mn and Fe supply on the growth of barley in nutrient solution. *Can. J. Plant Sci.* 52:575–581.

Bekker, A.W., N.V. Hue, L.G.G. Yapa, and R.G. Chase. 1994. Peanut growth as affected by liming, Ca-Mn interactions, and Cu plus Zn applications to oxidic Samoan soils. *Plant Soil* 164:203–211.

Belouchi, A., T. Kwan, and P. Gros. 1997. Cloning and characterization of the OsNRAMP family from *Oryza sativa*, a new family of membrane proteins possibly implicated in the transport of metal ions. *Plant Mol. Biol.* 33:1085–1092.

Benac, R. 1976. Response of a sensitive (*Arachis hypogaea*) and a tolerant (*Zea mays*) species to different concentrations of manganese in the environment (Fr.). *Can. ORSTOM. Ser. Biol.* 11(1):43–51.

Benette, J.P. (1945). Iron in leaves. *Soil Sci.* 60:91–106.

Benhammou, A., A. Yaacoubi, L. Nibou, and B. Tanouti. 2005. Adsorption of metal ions onto Moroccan Stevensite: Kinetics and isotherm studies. *J. Colloid Interface Sci.* 282:320–326.

Berger, K.C. and G.C. Gerloff. 1947. Manganese toxicity of potatoes in relation to strong soil acidity. *Proc. Soil Sci. Soc. Am.* 12:310–314.

Bhatt, K.C., P.P. Vaishnav, Y.D. Sigh, and J.J. Chinoy. 1976. Reversal of gibberellic acid-induced inhibition of root growth by manganese. *Biochem. Physiol. Pflanz.* 170:453–455.

Blamey, F.P.C., D.C. Joyce, D.G. Edwards, and C.J. Asher. 1986. Role of trichomes in sunflower tolerance to manganese toxicity. *Plant Soil* 91:171–180.

Bortner, C.E. 1935. Manganese toxicity in tobacco. *Soil Sci.* 39:15–33.

Bowen, J.E. 1972. Manganese-silicon interaction and its effect on growth of Sudan grass. *Plant Soil* 37:577–588.

Bowen, G.D. and A.D. Rovira. 1961. The effects of micro-organisms on plant growth. *Plant Soil* 15:166–188.

Bowen, G.D. and A.D. Rovira. 1992. The rhizosphere: The hidden half of the hidden half. In *Plant Roots: The Hidden Half*, eds. Y. Waisel, A. Eshel, and U. Kafkafi, pp. 641–669. New York: Marcel Decker.

Bowler, C., L. Slooten, S. Vandenbraden et al. 1991. Manganese superoxide dismutase can reduce cellular damage mediated by oxygen radicals in transgenic plants. *EMBO J.* 10:1723–1732.

Bradl, H.B. 2004. Adsorption of heavy metal ions on soils and soils constituents. *J. Colloid Interface Sci.* 277:1–18.

Brennan, R.F., M.D.A. Bolland, and G. Shea. 2001. Comparing how lupinus angustifolius and lupinus luteus use zinc fertilizer for seed production. *Nutr. Cycl. Agroecosyst.* 59:209–217.

Brenshley, W.E. 1914. *Inorganic Plant Poisons and Stimulants*, pp. 78–107. Cambridge, UK: Cambridge University Press.

Broadley M., P. Brown, I. Cakmak, Z. Rengel, and F. Zhao. 2012. Function of nutrients: Micronutrients. In *Marschner's Mineral Nutrition of Higher Plants*, 3rd edn., ed. P. Marschner, pp. 191–243. Boston, MA: Academic Press.

Bromfield, S.M. 1978. The effect of manganese-oxidising bacteria and pH on the availability of manganous ions and manganese oxides to oats in nutrient solutions. *Plant Soil* 49:23–39.

Brown, J.C. and J.E. Ambler. 1973. "Reductants" released by roots of Fe-deficient soybeans. *Agron. J.* 65:311–314.

Brown, J.C., J.E. Ambler, R.L. Chaney, and C.D. Foy. 1972. Differential responses of plant genotypes to micronutrients. In *Micronutrients in Agriculture,* eds. J.J. Morvedt, P.M. Giordano, and W.L. Lindsay, pp. 389–418. Madison, WI: American Society of Agronomy.

Brown, J.C. and W.E. Jones. 1962. Absorption of Fe, Mn, Ca, Rb and phosphate ions by soybean roots that differ in their reductive capacity. *Soil Sci.* 94:173–179.

Brown, J.C. and W.E. Jones. 1977a. Fitting plants nutritionally to soils. I. Soybeans. *Agron. J.* 69:399–404.

Brown, J.C. and W.E. Jones. 1977b. Fitting plants nutritionally to soils. II. Cotton. *Agron. J.* 69:405–409.

Brown, J.C. and W.E. Jones. 1977c. Fitting plants nutritionally to soils. III. Sorghum. *Agron. J.* 69:410–414.

Brown, J.C. and Olsen R.A. 1980. Factors related to iron uptake by dicotyledonous and monocotyledonous plants III. Competition between root and external factors for Fe. *J. Plant Nutr.* 2:661–682.

Brown, P.H., R.D. Graham, and D.J.D. Nicholas. 1984. The effects of manganese and nitrate supply on the levels of phenolics and lignin in young wheat plants. *Plant Soil* 81:437–440.

Bryson, G.M., H.A. Mills, D.N. Sasseville, J.B. Jones, Jr., and A.V. Barker. 2014. *Plant Analysis Handbook III.* Athens, GA.: Micro-Macro Publishing.

Burnell, J.N. 1986. Purification and properties of phosphoenolpyruvate carboxykinase from C_4 plants. *Aust. J. Plant Physiol.* 13:577–585.

Burnell, J.N. 1988. The biochemistry of manganese in plants. In *Manganese in Soils and Plants*, eds. R.D. Graham, R.J. Hannam, and N.C. Uren, pp. 125–137. Dordrecht, the Netherlands: Kluwer Academic Publishers.

Chaney, R.L. and P.M. Giodano. 1977. Microelements as related to plant deficiencies and toxicities. In *Soil for Management and Utilization of Organic Wastes and Waste Waters*, eds. L.F. Elliot and F.J. Stevenson, pp. 233–280. Madison, WI: American Society of Agronomy.

Chapman, G.W. 1931. The relation of iron and manganese to chlorosis in plants. *New Phytol.* 30:266–284.

Chapman, H.D., G.F.J. Liebig, and E.R. Parker. 1939. Manganese studies on Californian soils and citrus leaf symptoms of deficiency. *Calif. Citrograph.* 24(12):427–454.

Chapman, H.D., G.F.J. Liebig, and A.P. Vanselow. 1940. Some nutritional relationships as revealed by a study of mineral deficiency and excess symptoms on citrus. *Proc. Soil Sci. Soc. Am.* 4:196–200.

Chesworth, W. 1991. Geochemistry of micronutrients. In *Micronutrients in Agriculture*, 2nd edn., eds. J.J Mortvedt, F.R Cox, L.M. Shuman, and R.M Welch, pp. 96–99. Madison, WI: Soil Science Society of America.

Christenson, P.O., S.J. Toth, and F.E. Bear. 1951. The status of soil manganese as influenced by moisture, organic matter, and pH. *Proc. Soil Sci. Soc. Am.* 15:279–283.

Clair, S.B. and J.P. St Lynch. 2004. Photosynthetic and antioxidant enzyme responses of sugar maple and red maple seedlings to excess manganese in contrasting light environments. *Funct. Plant Biol.* 31:1005–1014.

Clark, R.B. 1982. Plant response to mineral element toxicity and deficiency. In *Breeding Plants for Less Favorable Environments*, eds. M.N. Christiansen and C.F. Lewis, pp. 71–142. New York: John Wiley.

Clark, R.B., P.A. Pier, D. Knudsen, and J.W. Maranville. 1981. Effect of trace element deficiencies and excesses on mineral nutrients in sorghum. *J. Plant Nutr.* 3:357–373.

Clarkson, D.T. 1988. The uptake and translocation of manganese by plant roots. In *Manganese in Soils and Plants*, eds. R.D. Graham, R.J. Hannam, and N.J. Uren, pp. 101–111. Dordrecht, the Netherlands: Kluwer Academic Publishers.

Clemens, K.L., A.A. Force, and R.D. Britt. 2002. Acetate binding at the photosystem II oxygen evolving complex: An S2-State multiline signal ESEEM study. *J. Am. Chem. Soc.* 124:10921–10933.

Constantopoulus, G. 1970. Lipid metabolism of manganese-deficient algae. I. Effect of manganese deficiency on the greening and the lipid composition of *Euglena gracilis. Z. Plant Physiol.* 45:76–80.

Conyers, M., N. Uren, K. Helyar, G. Poile, and B. Cullis. 1997. Temporal variation in soil acidity. *Aust. J. Soil Res.* 35:1115–1129.

Cotter, D.J. and U.N. Mishra. 1968. The role of organic matter in soil manganese equilibrium. *Plant Soil* 29:439–448.

Crosbie, J., N.E. Longnecker, and A.D. Robson. 1994. Seed manganese affects the early growth of lupins in manganese-deficient conditions. *Aust. J. Agric. Res.* 45:1469–1482.

Davies, E.B. 1945. A case of molybdenum deficiency in New Zealand. *Nature* 156:392–393.

Davies, S.H.R. and J.J. Morgan. 1989. Manganese (II) oxidation kinetics on metal oxide surfaces. *J. Colloid Interface Sci.* 129:63–77.

Davis, C.A., S.S. Nick, and A. Agarwal A. 2001. Manganese superoxide dismutase attenuates cisplatin-induced renal injury: Importance of superoxide. *J. Am. Soc. Nephrol.* 12:2683–2690.

Davis, J.G. 1996. Soil pH and magnesium effects on manganese toxicity in peanuts. *J. Plant Nutr.* 19:535–550.

de Varennes, A., J.P. Carneiro, and M.J. Goss, 2001. Characterization of manganese toxicity in two species of annual medics. *J. Plant Nutr.* 24:1947–1955.

Dekock, P.C. and R.H.E. Inkson. 1962. Manganese contents of mustard leaves in relation to iron and major nutrient supply. *Plant Soil* 18:183–191.

del Río, L.A., F. Sevilla, M. Gómez, J. Yáñez, and J. López-Gorgé. 1978. Superoxide dismutase: An enzyme system for the study of micronutrient interactions in plants. *Planta* 140:221–225.

Delhaize E., B.D. Gruber, J.K. Pittman et al. 2007. A role for the AtMTP11 gene of arabidopsis in manganese transport and tolerance. *Plant J.* 51:198–210

Delhaize E., T. Kataoka, D.M. Hebb, R.G. White, and P.R. Ryan. 2003. Genes encoding proteins of the cation diffusion facilitator family that confer manganese tolerance. *Plant Cell* 15:1131–1142.

Demirevska-Kepova, K., L. Simova-Stoilova, Z. Stoyanova, R. Holzer, and U. Feller. 2004. Biochemical changes in barley plants after excessive supply of copper and manganese. *Environ. Exp. Bot.* 52:253–266.

Dennis, D. 1968. Manganese toxicity in tomatoes. *N. Z. J. Agric.* 117:118–119.

Doi, Y. 1952a. Studies on the oxidizing power of roots of crop plants. I. The differences of crop plants and wild grasses. II. Interrelation between paddy rice and soybean. *Proc. Crop Sci. Soc. Jpn.* 21:12–13.

Doi, Y. 1952b. Studies on the oxidizing power of roots of crop plants. II. Interrelation between paddy rice and soybean. *Proc. Crop Sci. Soc. Jpn.* 21:14–15.

Doncheva, S., C. Poschenrieder, A. Stoyanova et al. 2009. Silicon amelioration of manganese toxicity in Mn-sensitive and Mn-tolerant maize varieties. *Environ. Exp. Bot.* 65:189–197.

Dou, C., X. Fu, X. Chen, J. Shi, and Y. Chen. 2008. Accumulation and detoxification of manganese in hyperaccumulator *Phytolacca Americana. Plant Biol.* 11:664–670.

Douka, C.E. 1977. Study of bacteria from manganese concentrations, precipitation of manganese by whole cells and cell free extracts of isolated bacteria. *Soil Biol. Biochem.* 9:89–97.

Drossopoulos, J.B., D.L. Bouranis, and B.D. Bairaktari. 1994. Patterns of mineral nutrient fluctuations in soybean leaves in relation to their position. *J. Plant Nutr.* 17:1017–1035.

Ducic, T. and A. Polle. 2005. Transport and detoxification of manganese and copper in plants. *Braz. J. Plant Physiol.* 17:103–112.

Ducic, T. and A. Polle. 2007. Manganese toxicity in two varieties of Douglas fir (*Pseudotsuga menziesii* var. viridis and glauca) seedlings as affected by phosphorus supply. *Funct. Plant Biol.* 34:31–40.

Duerr, G., J. Strayle, R. Plemper, S. Elbs et al. 1998. The medial-Golgi ion pump Pmr1 supplies the yeast secretory pathway with Ca^{2+} and Mn^{2+} required for glycosylation, sorting, and endoplasmic reticulum-associated protein degradation. *Mol. Biol. Cell* 9:1149–1162.

Dunwell, J.M., A. Culham, C.E. Carter, C.R. Sosa-Aguirre, and P.W. Goodenough. 2001. Evolution of functional diversity in the cupin superfamily. *Trends Biochem. Sci.* 26:740–746.

Edwards, D.G. and C.J. Asher. 1982. Tolerance of crop and pasture species to manganese toxicity. In *Proceedings of the Ninth International Plant Nutrition Colloquium*, ed. A. Scaife, pp. 145–150. Warwick University, England; Commonwealth Agricultural Bureaux.

Eenink, A.H. and G. Garresten. 1977. Inheritance of insensitivity of lettuce to a surplus of exchangeable manganese in steam sterilized soils. *Euphytica* 26:47–53.

Eide, D., M. Broderius, J. Fett, and M.L. Guerinot.1996. A novel iron-regulated metal transporter from plants identified by functional expression in yeast. *Proc. Natl. Acad. Sci. U. S. A.* 93:5624–5628.

Elamin, O.M. and G.E. Wilcox. 1986a. Effects of magnesium and manganese on muskmelon growth and manganese toxicity. *J. Am. Soc. Hortic. Sci.* 111:582–587.

Elamin, O.M. and G.E. Wilcox. 1986b. Effects of magnesium and manganese nutrition on watermelon growth and manganese toxicity development. *J. Am. Soc. Hortic. Sci.* 111:588–593.

Elamin, O.M. and G.E. Wilcox. 1986c. Manganese toxicity development in muskmelons as influenced by nitrogen form. *J. Am. Soc. Hortic. Sci.* 111:323–327.

El-Baz, F.K., P. Maier, A.H. Wissemeier, and W.J. Horst. 1990. Uptake and distribution of manganese applied to leaves of *Vicia faba* (cv. Herzfreya) and *Zea mays* (cv. Regent) plants. *Z. Pflanzenernähr. Bodenk.* 153:279–282.

El-Jaoual Eaton, T., D.A. Cox, and A.V. Barker. 2012. Relationship of iron-manganese toxicity disorder in marigold to manganese and magnesium nutrition. *J. Plant Nutr.* 35:142–164.

El-Jaoual, T. and D. Cox. 1998. Manganese toxicity in plants. *J. Plant Nutr.* 21:353–386.

Elliott, L.F. and J.W. Blaylock. 1975. Effects of wheat straw and alfalfa amendments on solubilization of manganese and iron in soil. *Soil Sci.* 120(3):205–211.

Elsner, E.F. 1982. Oxygen activation and oxygen toxicity. *Annu. Rev. Plant Physiol.* 33:73–96.

Engelsma, G. 1972. A possible role of divalent manganese ion in the photoinduction of phenylalanine ammonia-lyase. *Plant Physiol.* 50:599–602.

Engler, R.M. and W.H. Patrick. 1975. Stability of sulfides of manganese, iron, zinc, copper, and mercury in flooded and non-flooded soil. *Soil Sci.* 119:217–221.

Epstein, E. 1961. Mineral metabolism of halophytes. In *Ecological Aspects of the Mineral Nutrition of Plants*, ed. I.H. Rorison, pp. 345–353. Oxford, UK: Blackwell.

Eyster, C., T.E. Brown, H.A. Tanner, and S.L. Hood. 1958. Manganese requirement with respect to growth, Hill reaction and photosynthesis. *Plant Physiol.* 33(4):235–241.

Fageria, N.K. and V.C. Baligar. 1999. Phosphorus use efficiency in wheat genotypes. *J. Plant Nutr.* 22:331–340.

Fageria, N.K., V.C. Baligar, and R.B. Clark. 2002. Micronutrients in crop production. *Adv. Agron.* 77:185–268.

Fecht-Christoffers, M.M., H.P. Braun, C. Lemaitre-Guillier, A. VanDorsselear, and W.J. Horst. 2003a. Effect of Mn toxicity on the proteome of the leaf apoplast in cowpea. *Physiol. Plant.* 133:1935–1946.

Fecht-Christoffers M.M., H. Führs, H.P. Braun, and W.J. Horst. 2006. The role of hydrogen peroxide-producing and hydrogen peroxide-consuming peroxidases in the leaf apoplast of cowpea in manganese tolerance. *Plant Physiol.* 140:1451–1463.

Fecht-Christoffers, M.M., P. Maier, and W.J. Horst. 2003b. Apoplastic peroxidase and ascorbate are involved in manganese toxicity and tolerance of Vigna unguiculata. *Physiol. Plant.* 117:237–244.

Fecht-Christoffers, M.M., P. Maier, K. Iwasaki, H.P. Braun, and W.J. Horst. 2007. The role of the leaf apoplast in manganese toxicity and tolerance in cowpea (*Vigna unguiculata* L. Walp). In *The Apoplast of Higher Plants: Compartment of Storage, Transport, and Reactions*, eds. B. Sattelmacher and W.J. Horst, pp. 307–322. Dordrecht, The Netherlands: Springer.

Fisher, A.G., G.W. Eaton, and S.W. Porrit. 1977. Internal bark necrosis of Delicious apple in relation to soil pH and leaf manganese. *Can. J. Plant Sci.* 57:297–299.

Fleming, A.L. 1989. Enhanced Mn accumulation by snapbean cultivars under Fe stress. *J. Plant Nutr.* 12:715–731.

Foulds, W. 1993. Nutrient concentrations of foliage and soil in South-western Australia. *New Phytol.* 125:529–546.

Foy, C., B. Scott, and J. Fisher. 1988. Genetic differences in plant tolerance to manganese toxicity. In *Manganese in Soil and Plants*, eds. R.D. Graham, R.J. Hannam, and N.J. Uren, pp. 293–307. Dordrecht, the Netherlands: Kluwer Academic Publishers.

Foy, C.D. 1983. The physiology of plant adaptation to mineral stress. *Iowa State J. Res.* 57:355–391.

Foy, C.D. and T.A. Campbell. 1984. Differential tolerances of Amaranth's strains to high levels of aluminum and manganese in acid soils. *J. Plant Nutr.* 7:1365–1388.

Foy, C.D., R.L. Chaney, and M.C. White. 1978. The physiology of metal toxicity in plants. *Annu. Rev. Plant Physiol.* 29:511–566.

Foy, C.D., A.L. Fleming, and W.H. Armiger. 1969. Differential tolerance of cotton varieties to excess manganese. *Agron. J.* 61:690–692.

Foy, C.D., H.W. Webb, and J.E. Jones. 1981. Adaptation of cotton genotypes to an acid, manganese toxic soil. *Agron. J.* 73:107–111.

Foy, C.D., R.R. Weil, and C.A. Coradetti. 1995. Differential manganese tolerances of cotton genotypes in nutrient solution. *J. Plant Nutr.* 18:685–706.

Foyer, C.H. and G. Noctor. 2003. Redox sensing and signaling associated with reactive oxygen in chloroplasts, peroxisomes and mitochondria. *Physiol. Plant.* 119:355–364.

Fridovich, I. 1975. Superoxide dismutase. *Annu. Rev. Biochem.* 44:147–159.

Fridovich, I. 1983. Superoxide radical: An endogenous toxicant. *Annu. Rev. Pharmacol. Toxicol.* 23:239–257.

Führs, H., M. Hartwig, L.E.B. Molina et al. 2008. Early manganese-toxicity response in *Vigna unguiculata* L.: A proteomic and transcriptomic study. *Proteomics* 8:149–159.

Führs, H., S. Götze, A. Specht et al. 2009. Characterization of leaf apoplastic peroxidases and metabolites in *Vigna unguiculata* in response to toxic manganese supply and silicon. *J. Exp. Bot.* 60:1663–1678.

Galvez, L., R.B. Clark, L.M. Gourley, and J.W. Maranville. 1989. Effect of silicon on mineral composition of sorghum growth with excess manganese. *J. Plant Nutr.* 12:547–561.

Gangwar, S., V.P. Singh, S.M. Prasad, and J.N. Maurya. 2010. Differential responses of pea seedlings to indole acetic acid under manganese toxicity. *Acta Physiol. Plant* 33:451–462.

Gardner, W.K., D.G. Parbery, and D.A. Barber. 1982. The acquisition of phosphorus by *Lupinus albus* L. *Plant Soil* 68:19–32.

Garnham, G.W., G.A. Codd, and G.M. Gadd. 1992. Kinetics of uptake and intracellular location of cobalt, manganese and zinc in the estuarine green alga, *Chlorella salina. Appl. Microbiol. Biotechnol.* 37:270–276.

Geering, H.R., J.R. Hodgson, and C. Sdano. 1969. Micronutrient cation complexes in soil solution. IV. The chemical state of manganese in soil solution. *Proc. Soil Sci. Soc. Am.* 33:81–85.

Gerloff, G.C. and W.H. Gabelman. 1983. Genetic basis of inorganic plant nutrition. In *Inorganic Plant Nutrition*, eds. A. Lauchli and R.L. Bieleski, pp. 453–480. New York: Springer-Verlag.

Gerretsen, F.C. 1950. Manganese in relation to photosynthesis. III. Uptake of oxygen by illuminated crude chloroplast suspensions. *Plant Soil* 2:323–342.

Gherardi, M. and Z. Rengel. 2004. The effect of manganese supply on exudation of carboxylates by roots of Lucerne (*Medicago sativa*). *Plant Soil* 260:271–282.

Ghiorse, W.C. 1988. The biology of manganese transforming micro-organisms in soil. In *Manganese in Soils and Plants*, eds. R.D. Graham, R.J. Hannam, and N.C. Uren, pp. 75–84. Dordrecht, the Netherlands: Kluwer Academic Publishers.

Gladstones, J.S. 1962. The mineral composition of lupins 2. A comparison of the copper, manganese, molybdenum, and cobalt contents of lupins and other species at one site. *Aust. J. Exp. Agric. Anim. Husb.* 2:213–220.

Goh, K.M. and R.J. Haynes. 1978. The growth of *Tagetes erecta, Petunia hybrida nana compacta* and *Mysotis* in a range of peat-based container media. *Commun. Soil Sci. Plant Anal.* 9:373–375.

Gong, X., Y. Wang, C. Liu et al. 2010. Effect of manganese deficiency on spectral characteristics and oxygen evolution in maize chloroplasts. *Biol. Trace Elem. Res.* 136:372–382.

González, A. and J. Lynch. 1999. Subcellular and tissue Mn compartmentation in bean leaf under Mn toxicity stress. *Aust. J. Plant Physiol.* 26, 811–822.

González, A., K.L. Steffen, and J.P. Lynch. 1998. Light and excess manganese: Implications for oxidative stress in common bean. *Plant Physiol.* 118:493–504.

Gossl, J. 1905. About the content of manganese in plants and about its influence on mildew (Ger.). *Beih. Bot. Centr.* 18:119–132.

Goussias, C., A. Boussac, and W. Rutherford. 2002. Photosystem II and photosynthetic oxidation of water: An overview. *Phil. Trans. R. Soc. Lond. B* 357:1369–1381.

Graham, R.D. 1983. Effect of nutrient stress on susceptibility of plants to disease with particular reference to the trace elements. *Adv. Bot. Res.* 10:221–276, 1983.

Graham, R.D. 1984. Breeding for nutritional characteristics in cereals. *Adv. Plant Nutr.* 1:57–102.

Graham, R.D. 1988a. Development of wheats with enhanced nutrient efficiency: Progress and potential. *Wheat Production Constraints in Tropical Environments, Proceedings of the CIMMYT/INDP International Symposium*, pp. 305–320. Chiang Mai, Thailand.

Graham, R.D. 1988b. Genotypic differences in tolerance to manganese deficiency. In *Manganese in Soils and Plants*, eds. R.D. Graham, R.J. Hannam, and N.C. Uren, pp. 261–276. Dordrecht, the Netherlands: Kluwer Academic Publishers.

Graham, R.D. and A.D. Rovira. 1984. A role for manganese in the resistance of wheat plants to take-all. *Plant Soil* 78:441–444.

Graham, R.D. and M.J. Webb. 1991. Micronutrients and disease resistance and tolerance in plants. In *Micronutrients in Agriculture*, eds. J.J. Mortvedt, F.R. Cox, L.M. Shuman, and R.M. Welch, pp. 333–339. Madison, WI: Soil Science Society of America.

Graham, R.D., W.J. Davies, D.H.B. Sparrow, and J.S. Ascher. 1983. Tolerance of barley and other cereals to manganese-deficient soils of South Australia. In *Genetic Aspects of Plant Nutrition*, eds. M.R. Saric and B.C. Loughman, pp. 339–345. The Hague, the Netherlands: Martinus Nijhoff/Dr W Junk Publishers.

Graham, W.F., M.L. Bender, and G.P. Klinkhammer. 1976. Manganese in Narragansett Bay. *Limnology Oceanography* 21:665–673.

Grasmanis, V.O. and G.W. Leeper. 1966. Toxic manganese in near-neutral soils. *Plant Soil* 25:41–48.

Graven, E.H., O.J. Attoe, and D. Smith. 1965. Effect of liming and flooding on manganese toxicity in alfalfa. *Proc. Soil Sci. Soc. Am.* 29:702–706.

Grundon, N.J., A.D. Robson, M.J. Lambert, and K. Snowball. 1997. Nutrient deficiency and toxicity symptoms. In *Plant Analysis: An Interpretation Manual*, eds. D.J. Reuter and J.B. Robinson, pp. 37–51. Collingwood, Australia: CSIRO Publishing.

Guerin, G. 1897. About organic compounds rich in manganese in lignin (Fr.). *Comp. Rend. Chim.* 25:311–312.

Guest, C., D. Schulze, I. Thompson, and D. Huber. 2002. Correlating manganese X-ray absorption near-edge structure spectra with extractable soil manganese. *Soil Sci. Soc. Am. J.* 66:1172–1181.

Gupta, U. and E.W. Chipman. 1976. Influence of iron and pH on the yield and iron, manganese, zinc and sulfur concentrations of carrots grown on sphagnum peat soil. *Plant Soil* 44:559–566.

Gupta, U., E.W. Chipman, and D.C. Mackay. 1970. Influence of manganese and pH on chemical composition, bronzing of leaves, and yield of carrots grown on acid sphagnum peat soil. *Proc. Soil Sci. Soc. Am.* 34:762–764.

Halstead, E.H., S.A. Barber, D.D. Warnecke, and J.B. Bole. 1968. Supply of Ca, Sr, Mn, and Zn to plant roots growing in soil. *Proc. Soil Sci. Soc. Am.* 32:69–72.

Hannam, R.J., R.D. Graham, and J.L. Riggs. 1985a. Diagnosis and prognosis of manganese deficiency in *Lupinus angustifolius* L. *Aust. J. Agric. Res.* 36:765–777.

Hannam R.J., R.D. Graham, and J.L. Riggs. 1985b. Redistribution of manganese in maturing Lupinus angustifolius cv Illyarrie in relation to levels of previous accumulation. *Ann. Bot.* 56:821–834.

Hannam, R.J. and K. Ohki. 1988. Detection of manganese deficiency and toxicity in plants. In *Manganese in Soils and Plants*, eds. R.D. Graham, R.J. Hannam, and N.J. Uren, pp. 243–259. Dordrecht, the Netherlands: Kluwer Academic Publishers.

Hannam, R.J., W.J. Davies, R.D. Graham, and J.L. Riggs. 1984. The effect of soil- and foliar-applied manganese in preventing the onset of manganese deficiency in *Lupinus angustifolius*. *Aust. J. Agric. Res.* 35:529–538.

Harrison, H.C. and E.L. Bergman. 1981. Calcium, magnesium, and potassium interrelationships affecting cabbage production. *J. Am. Soc. Hortic. Sci.* 106:500–503.

Hart, J. 1998. Fertilizer and lime materials. Oregon State University Extension Fertilizer Guide 52. http://ir.library.oregonstate.edu/xmlui/bitstream/handle/1957/20620/fg52-e.pdf, accessed June 1, 2014.

Hauck, M., A. Paul, S. Gross, and M. Raubuch. 2003. Manganese toxicity in epiphytic lichens: Chlorophyll degradation and interaction with iron and phosphorus. *Environ. Exp. Bot.* 49:181–191.

Hebbern, C.A., P. Pedas, J.K. Schjoerring, L. Knudsen, and S. Husted. 2005. Genotypic differences in manganese efficiency: A field trial with winter barley (*Hordeum vulgare* L.). *Plant Soil* 272:233–244.

Heenan, D.P. and L.C. Campbell. 1980 Transport and distribution of manganese in two cultivars of soybean (*Glycine max* (L.) Merr.). *Aust. J. Agric. Res.* 31:943–949.

Heenan, D.P. and L.C. Campbell. 1981. Influence of potassium and manganese on growth and uptake of magnesium by soybeans (*Glycine max* L. Merr. cv Bragg). *Plant Soil* 61:447–456.

Heenan, D.P. and L.C. Campbell. 1983. Manganese and iron interactions on their uptake and distribution in soybeans (*Glycine max* L. Merr.). *Plant Soil* 70:317–326.

Heenan, D.P. and D.G. Carter. 1975. Response of two soybean cultivars to manganese toxicity as affected by pH and calcium levels. *Aust. J. Agric. Res.* 26:967–974.

Heenan, D.P. and O.G. Carter. 1977. Influence of temperature in the expression of manganese toxicities by two soybean varieties. *Plant Soil* 47:219–227.

Hegedüs, A., Erdei, S., and Horváth, G. 2001. Comparative studies of H_2O_2 detoxifying enzymes in green and greening barley seedlings under cadmium stress. *Plant Sci.* 160:1085–1093.

Heintze, J.G. 1968. Manganese phosphate reactions in aqueous systems and the effects of application of monocalcium phosphate on the availability of manganese to oats in alkaline soils. *Plant Soil* 24:407–423.

Henriques, R., J. Jasik, M. Klein et al. 2002. Knock out mutant of Arabidopsis metal transporter gene IRTI results in iron deficiency accompanied by cell differentiation defects. *Plant Mol. Biol.* 50:587–597.

Herren, T. and U. Feller. 1994. Transfer of zinc from xylem to phloem in the peduncle of wheat. *J. Plant Nutr.* 17:587–1598.

Hewitt, E.J. 1945. The resolution of the factors in soil acidity. Long Ashton Res. Sta. Annu. Rpt. pp. 51–60.

Hewitt, E.J. 1948. Relation of manganese and other metal toxicities to the iron status of plants. *Nature* 161:489–490.

Hiatt, A.J. and J.L. Ragland. 1963. Manganese toxicity of burley tobacco. *Agron. J.* 55:47–49.

Hill, J., A.D. Robson, and J.F. Loneragan. 1979. The effect of copper supply on the senescence and retranslocation of nutrients of the oldest leaf of wheat. *Ann. Bot.* 44:279–287.

Hiller, L.K. 1995. Foliar fertilization bumps potato yields in Northwest. *Fluid Journal*. Washington State University. http://www.fluidfertilizer.com/pastart/pdf/10P28-30.pdf, accessed January 22, 2015.

Hirschi, K.D., V.D. Korenkov, N.L. Wilganowski, and G.J. Wagner. 2000. Expression of Arabidopsis CAX2 in tobacco: Altered metal accumulation and increased manganese tolerance. *Plant Physiol.* 124:125–133.

Hocking, P.J., J.S. Pate, S.C. Wee, and A.J. McComb. 1977. Manganese nutrition of *Lupinus* spp., especially in relation to developing seeds. *Ann. Bot.* 41:677–688.

Honann, P.E., 1967. Studies on the manganese of the chloroplast. *Plant Physiol.* 42:997–1007.

Hophins, E.F., V. Pagan, and F.I. Ramirez Silva. 1944. Iron and Mn in relation to plant growth and its importance in Puerto Rico. *J. Agric. Univ. Puerto Rico* 28:43–101.

Horiguchi, T. 1987. Mechanism of manganese toxicity and tolerance of plants. II. Deposition of oxidized manganese in plant tissues. *Soil Sci. Plant Nutr.* 33:595–606.

Horiguchi, T. 1988a. Mechanism of manganese toxicity and tolerance of plants. IV. Effect of silicon on alleviation of manganese toxicity of rice plants. *Soil Sci. Plant Nutr.* 34:65–73.

Horiguchi, T. 1988b. Mechanism of manganese toxicity and tolerance of plants. VII. Effect of light intensity on manganese-induced chlorosis. *J. Plant Nutr.* 11:235–246.

Horiguchi, T. and T. Fukumoto. 1987. Mechanism of manganese toxicity and tolerance of plants. III. Effect of excess manganese on respiration rate and peroxidase activity of various plant species. *J. Soil Sci. Plant Nutr.* 58:713–716.

Horneck, D., J. Hart, R. Stevens, S. Petrie, and J. Altland. 2004. Acidifying soil for crop production west of the cascade mountains. Oregon State University Extension Publication EM 8857. http://www.agrisk.umn.edu/cache/arl03227.pdf, accessed June 1, 2014.

Horst, W.J. 1988. The physiology of Mn toxicity. In *Manganese in Soils and Plants*, eds. R.D. Graham, R.J. Hannam, and N.C. Uren, pp. 175–188. Dordrecht, the Netherlands: Kluwer Academic.

Horst, W.J. and H. Marschner. 1978a. Effect of excessive manganese supply on uptake and translocation of calcium in bean plants (*Phaseolus vulgaris* L.). *Z. Pflanzenphysiol.* 87:137–148.

Horst, W.J. and H. Marschner. 1978b. Effect of silicon on manganese tolerance of bean plants (*Phaseolus vulgaris* L.). *Plant Soil* 50:287–303.

Horst, W.J., M. Fecht, A. Naumann, A.H. Wissemeir, and P. Maier. 1999. Physiology of manganese toxicity and tolerance in *Vigna unguiculata* (L.) Walp. *J. Plant Nutr. Soil Sci.* 162:263–274.

Hossner, L.R. and G.E. Richards. 1968. The effect of phosphorus sources on the movement and uptake of band applied manganese. *Proc. Soil Sci. Soc. Am.* 32:83–85.

Houtz, R.L., R.O. Nable, and G.M. Cheniae. 1988. Evidence for effects on the in vivo activity of ribulose-bisphosphate carboxylase/oxygenase during development of Mn toxicity in tobacco. *Plant Physiol.* 86:1143–1149.

Huber, D.M. 1978. Disturbed mineral nutrition. In *Plant Pathology—An Advanced Treatise*, eds. J. Horsfall and E.B. Cowling, pp. 163–181. New York: Academic Press.

Huber, D.M. and R.R. Keeler. 1977. Alteration of wheat peptidase activity after infection with powdery mildew. *Proc. Am. Phytopathol. Soc., 69th Annual Meeting of the American Phytopathological Society*, Vol. 4, pp. 163. East Lansing, MI, August 14–18, 1977.

Huber D.M. and T.S. McCay-Buis. 1993. A multiple component analysis of the take-all disease of cereals. *Plant Dis.* 77:437–447.

Huber, D.M. and N.S. Wilhelm. 1988. The role of manganese in resistance to plant diseases. In *Manganese in Soils and Plants*, eds. R.D. Graham, R.J. Hannam, and N.C. Uren, pp. 155–173. Dordrecht, the Netherlands: Kluwer Academic.

Hue, N. 1988. A possible mechanism for manganese toxicity in Hawaii soils amended with a low-Mn sewage sludge. *J Environ. Qual.* 17:473–479.

Hue, N.V. and Y. Mai. 2002. Manganese toxicity in watermelon as affected by lime and compost amended to a Hawaiian soil. *HortScience* 37:656–661.

Hughes, N.P. and R.J.P. Williams. 1988. An introduction to manganese biological chemistry. In *Manganese in Soils and Plants*, eds. R.D. Graham, R.J. Hannam, and N.C. Uren, pp. 7–19. Dordrecht, the Netherlands: Kluwer Academic.

Humphries, J.M., J.C.R. Stangoulis, and R.D. Graham. 2007. Manganese. In *Handbook of Plant Nutrition*, eds. A.V. Barker and D.J. Pilbeam, pp. 352–374. Boca Raton, FL: CRC Press.

Husted, S., K.H. Laursen, C.A. Hebbern et al. 2010. Manganese deficiency leads to genotype specific changes in fluorescence induction kinetics and state transitions. *Plant Physiol.* 150:825–833.

Iturbe-Ormaetxe, I., J.F. Moran, C. Arrese-Igor, Y. Gogorcena, R.V. Klucas, and M. Becana. 1995. Activated oxygen and antioxidant defenses in iron-deficient pea plants. *Plant Cell Environ.* 18(4):421–429.

Iwasaki, K., P. Maier, M. Fecht, and W.J. Horst. 2002. Leaf apoplastic silicon enhances manganese tolerance of cowpea (*Vigna unguiculata*). *J. Plant Physiol.* 159:167–173.

Jackson, C., J. Dench, A.L. Moore, B. Halliwell, C.H. Foyer, and D.O. Hall. 1978. Subcellular localization and identification of superoxide dismutase in the leaves of higher plants. *Eur. J. Biochem.* 91:339–344.

Jackson, T.L., D.T. Westermann, and D.P. Moore. 1996. The effect of chloride and lime on the manganese uptake by bush beans and sweet corn. *Proc. Soil Sci. Soc. Am.* 30:70–73.

Jacobson, H.G.M. and T.R. Swanback. 1932. Manganese content of certain Connecticut soils and its relation to the growth of tobacco. *J. Am. Soc. Agron.* 24:237–245.

Jauregui, M.A. and H.M. Reisenauer. 1982. Calcium carbonate and manganese dioxide as regulators of available manganese and iron. *Soil Sci.* 134:105–110.

Jensen, L.T., M. Ajua-Alemanji, and V.C. Culotta. 2003. The Saccharomyces cerevisiae high affinity phosphate transporter encoded by PHO84 also functions in manganese homeostasis. *J. Biol. Chem.* 278:42036–42040.

Johnson, M.O. 1917. Manganese as a cause of the depression of the assimilation of iron by pineapple plants. *Ind. Eng. Chem.* 9:47–49.

Khanna, R., S. Chandra, and K.K. Khanna. 1993. Rhizosphere microflora of triticale. *Biol. Mem.* 19:111–121.

Kamprath, E.J. and C.D. Foy. 1971. Lime-fertilizer-plant interactions in acid soils. In *Fertilizer Technology and Use*, eds. R.A. Olson, T.J. Army, J.J. Hanway, and V.J. Kilmer, pp. 105–141. Madison, WI: Soil Science Society of America.

Kanauchi, M., J. Milet, and C.W. Bamforth. 2009. Oxalate and oxalate oxidase in malt. *J. Inst. Brew.* 115:232–237.

Kang, B.T. and R.L. Fox. 1980. A methodology for evaluating the manganese tolerance of cowpea (*Vigna unguiculata*) and some preliminary results of field trials. *Field Crops Res.* 3:199–210.

Kazda, M. and L. Znacek. 1989. Aluminum and manganese and their relation to calcium in soil solution and needle in three Norway spruce (*Picea abies* L. Karst.) stands of upper Australia. *Plant Soil* 114:257–267.

Khabaz-Saberi, H., R.D. Graham, J.S. Ascher, and A.J. Rathjen. 2000. Quantification of the confounding effect of seed manganese content in screening for manganese efficiency in durum wheat (*Triticum turgidum* L. var. *durum*). *J. Plant Nutr.* 23:855–866.

Khabaz-Saberi, H., Z. Rengel, R. Wilson, and T.L. Setter. 2010. Variation for tolerance to high concentration of ferrous iron (Fe^{2+}) in Australian hexaploid wheat. *Euphytica* 172:275–283.

Kirsh, R.K., M.E. Howard, and R.G. Peterson. 1960. Interrelationships among iron, manganese, and molybdenum in the growth and nutrition of tomato grown in nutrient solution. *Plant Soil* 12:257–275.

Kochian, L.V. 1991. Mechanism of micronutrient uptake and translocation in plants. In *Micronutrients in Agriculture*, eds. J.J. Mortvedt, F.R. Cox, L.M. Shuman, and R.M. Welch, pp. 29–296. Madison, WI: Soil Science Society of America.

Kochian, L.V. 2000. Molecular physiology of mineral nutrient acquisition, transport and utilization. In *Biochemistry and Molecular Biology of Plants*, eds. B.B. Buchanan, W. Gruissem, and R.L. Jones, pp. 1204–1249. Rockville, MD: American Society of Plant Physiologists.

Kogelmann, W.J. and W.E. Sharpe. 2006. Soil acidity and manganese in declining and non-declining sugar maple stands in Pennsylvania. *J. Environ. Qual.* 35:433–441.

Kohno, Y. and C.D. Foy. 1983. Manganese toxicity in bush bean as affected by concentration of manganese and iron in the nutrient solution. *J. Plant Nutr.* 6:353–386.

Korshunova, Y.O., D. Eide, W.G. Clark, M.L. Guerinot, and H.B. Pakrasi. 1999. The IRT1 protein from *Arabidopsis* thaliana is a metal transporter with broad specificity. *Plant Mol. Biol.* 40:37–44.

Kriedemann, P.E., R.D. Graham, and J.T. Wiskich. 1985. Photosynthetic dysfunction and in vivo changes in chlorophyll a fluorescence from manganese-deficient wheat leaves. *Aust. J. Agric. Res.* 36:157–169.

Kröniger, W., H. Rennenberg, and A. Polle. 1992. Purification of two superoxide dismutase isoenzymes and their subcellular localization in needles and roots of Norway spruce (*Picea abies* L.) trees. *Plant Physiol.* 100:334–340.

Kuo, S. and D.S. Mikkelson. 1981. Effect of P and Mn on growth response and uptake of Fe, Mn, and P by sorghum. *Plant Soil* 62:15–22.

Labanauskas, C.K. 1965. Manganese. In *Diagnostic Criteria for Plants and Soils*, ed. H.D. Chapman, pp. 264–285. Riverside, CA: University of California Division of Agricultural Science.

Laha, S. and R.G. Luthy. 1990. Oxidation of aniline and other primary aromatic amines by manganese dioxide. *Environ. Sci. Technol.* 24(3):363–373.

Landi, S. and F. Fagioli. 1983. Efficiency of manganese and copper uptake by excised roots of maize genotypes. *J. Plant Nutr.* 6:957–970.

Landis T.D. Summer 1998. *Mosaic-Pattern Stunting*, pp. 14–17. Portland, OR: USDA Forest Service, State and Private Forestry Cooperative Program. Forest Nursery Notes.

Larson, S. 1964. The effect of phosphate application on Mn content of plants grown on neutral and alkaline soils. *Plant Soil* 21:37–42.

Laurie, S.H., N.P. Tancock, S.P. McGrath, and J.R. Sanders. 1991. Influence of complexation on the uptake by plants of iron, manganese, copper, and zinc. I. Effect of EDTA in a multi-metal and computer simulation study. *J. Exp. Bot.* 42:509–513.

Le Mare, P.H. 1977. Experiments on effects of phosphorus on the manganese nutrition of plants. I. Effects of mono-calcium phosphate and its hydrolysis derivatives on manganese in ryegrass grown in two Buganda soils. *Plant Soil* 47:593–605.

Leach, W. and C.D. Taper. 1954. Studies in plant mineral nutrition. II. The absorption of Fe and Mn by dwarf kidney beans, tomato, and onion from culture solutions. *Can. J. Bot.* 32:561–570.

Lee, C.R. 1972. Interrelationships of aluminum and manganese on the potato plant. *Agron. J.* 64:546–547.

Lei, X.G., J.M. Porres, E.J. Mullaney, and H. Brinch-Pedersen. 2007. Phytase source, structure and applications. In *Industrial Enzymes. Structure, Function and Applications*, eds. J. Polaina and A.P MacCab, pp. 505–529. Dordrecht, the Netherlands: Springer.

Leidi, E.O. and M. Gómez. 1985. A role for manganese in the regulation of soybean nitrate reductase activity? *J. Plant Physiol.* 118:335–342.

Lemanceau, P., T. Corberand, L. Gardan et al. 1995. Effect of two plant species, flax (*Linum usitatissimum* L.) and tomato (*Lycopersicon esculentum* Mill.), on the diversity of soil-borne populations of fluorescent pseudomonads. *Appl. Environ. Microbiol.* 61:1004–1012.

Lerer, M. and A. Bar-Akiva. 1976. Nitrogen constituents in manganese deficient lemon leaves. *Physiol. Plant.* 38:13–18.

Lewin, J. and B.E.F. Reimann. 1969. Silicon and plant growth. *Annu. Rev. Plant Physiol.* 20:289–304.

Lewis, A.H. 1943. The teart pastures of Somerset. III. Reducing the teartness of pasture herbage. *J. Agric. Sci.* 33:58–63.

Lidon, F.C. 2001. Tolerance of rice to excess manganese in the early stages of vegetative growth. Characterisation of manganese accumulation. *J. Plant Physiol.* 158:1341–1348.

Lidon, F.C., H.G. Azinheira, M.G. Barreiro. 2000. Al toxicity in maize: Modulation of biomass production and nutrients uptake and translocation. *J. Plant Nutr.* 23:151–160.

Lidon, F.C. and F.S. Henriques. 1993. Effects of copper toxicity on growth and the uptake and translocation of metals in rice plants. *J. Plant Nutr.* 16:1449–1464.

Lidon, F.C., M.G. Barreiro, and J.C. Ramalho. 2004. Manganese accumulation in rice: implications for photosynthetic functioning. *J. Plant Physiol.* 161:1235–1244.

Lidon, F.C. and M.G. Teixeira. 2000a. Rice tolerance to excess Mn: Implications in the chloroplast lamellae and synthesis of a novel Mn protein. *Plant Physiol. Biochem.* 38:969–978.

Lidon, F.C. and M.G. Teixeira. 2000b. Oxy radicals' production and control in the chloroplast of Mn-treated rice. *Plant Sci.* 152:7–15.

Lindsay, W.L. 1979. *Chemical Equilibria in Soils*, pp. 150–161. New York: Wiley.

Loneragan, J.F. 1988. Distribution and movement of manganese in plants. In *Manganese in Soils and Plants*, eds. R.D.Graham, R.J. Hannam, and N.C. Uren, pp. 113–124. Dordrecht, the Netherlands: Kluwer Academic Publishers.

Longnecker, N., R. Brennan, and A. Robson. 1998. Lupin nutrition. In *Lupins as crop plants: Biology, production and utilization*, eds. J.S. Gladestones, C.A. Atkins, and J. Hamblin, pp. 121–148. Wallingford, UK: CAB International.

Longnecker, N., J. Crosbie, F. Davies, and A. Robson. 1996. Low seed manganese concentration and decreased emergence of *Lupinus angustifolius*. *Crop Sci.* 36:355–361.

Longnecker, N.E., R.D. Graham, and G. Card. 1991b. Effects of manganese deficiency on the pattern of tillering and development of barley (Hordeum vulgare c.v. Galleon). *Field Crops Res.* 28:85–102.

Longnecker, N.E., N.E. Marcar, and R.D. Graham. 1991a. Increased manganese contents of barley seeds can increase grain yield in manganese-deficient conditions. *Aust. J. Agric. Res.* 42:1065–1074.

Lott, J.N.A. and E. Spitzer. 1980. X-ray analysis studies of elements stored in protein body globoid crystals of Triticum grains. *Plant Physiol.* 66:494–499.

Lucas, R.E. and B.D. Knezek. 1972. Climatic and soil conditions promoting micronutrients deficiencies in plants. In *Micronutrients in Agriculture*, eds. J.J. Mortvedt, P.M. Giordano, and W.L. Lindsay, pp. 265–288. Madison, WI: Soil Science Society of America.

Luk, E., M. Carroll, M. Baker, and V.C. Culotta. 2003. Manganese activation of superoxide dismutase 2 in Saccharomyces cerevisiae requires MTM1, a member of the mitochondrial carrier family. *Proc. Natl. Acad. Sci. U. S. A.* 100(18):10357–10357.

Löhnis, M.P. 1960. Effect of magnesium and calcium supply on the uptake of manganese by various crop plants. *Plant Soil* 12:339–376.

Maas, E.V. and D.P. Moore.1968. Manganese absorption by excised barley roots. *Plant Physiol.* 43:527–530.

Maas, E.V., D.P. Moore, and B.J. Mason. 1969. Influence of calcium and magnesium on manganese absorption. *Plant Physiol.* 44:796–800.

Macfie, S., E. Cossins, and G. Taylor. 1994. Effects of excess manganese on production of organic acids in Mn-tolerant and in Mn-sensitive cultivars of *Triticum aestivum* L. (wheat). *J. Plant Physiol.* 143:135–144.

Macfie, S.M. and G.J. Taylor. 1992. Effects of excess manganese on production of organic acids in Mn-tolerant and Mn-sensitive cultivars of *Triticum aestivum* L. (wheat). *J. Plant Physiol.* 143:135–144.

Mahler, R.L., G.D. Li, and D.W. Wattenbarger. 1992. Manganese relationship in spring wheat and spring barley production in northern Idaho. *Commun. Soil Sci. Plant Anal.* 23:1671–1692.

Malakouti, M.J. and M. Tehrani. 1999. *Effect of Micronutrients in Yield Increase and Improvement of Crops Quality*, p. 299. Tehran, Iran: Tarbiat Modarres University Press.

Marcar, N.E. and R.D. Graham. 1987. Tolerance of wheat, barley, triticale and rye to manganese deficiency during seedling growth. *Aust. J. Agric. Res.* 38:501–511.

Marshall, J., A. Corzo, R.A. Leigh, and D. Sanders. 1994. Membrane potential-dependent calcium transport in right-side-out plasma membrane vesicles from *Zea mays* L. roots. *Plant J.* 5:683–694.

Marschner, H. 1991. Root-induced changes in the availability of micronutrients in the rhizosphere. In *The Plant Roots, the Hidden Half*, eds. Y. Waisel, A. Eshel, and U. Kafkafi, pp. 503–528. New York: Marcel Dekker.

Marschner, H. 1995. *Mineral Nutrition of Higher Plants*, 2nd edn. Boston, MA: Academic Press.

Marsh, K.B., L.A. Peterson, and B.H. McCown. 1989. A microculture method for assessing nutrient uptake. II. The effect of temperature on manganese uptake and toxicity in potato shoots. *J. Plant Nutr.* 12:219–232.

Martinoia, E., M. Klein, M. Geisler et al. 2002. Multi-functionality of plant ABC transporters: More than just detoxifiers. *Planta* 214:345–355.

Mashhady, A. and A.H. El-Damaty. 1981. Biological availability of iron and manganese in some representative soils of Saudi Arabia. *Agrochimica* 25:9–19.

McCain, D.C. and J.L. Markley. 1989. More manganese accumulates in maple sun leaves than in shade leaves. *Plant Physiol.* 90:1417–1421.

McCay-Buis, T.S., D.M. Huber, R.D. Graham, J.D. Phillips, and K.E. Miskin. 1995. Manganese seed content and take-all of cereals. *J. Plant Nutr.* 18:1711–1721.

McCool, M.M. 1913. The action of certain nutrient and non nutrient bases on plant growth. II. The toxicity of manganese and the antidotal relations between this and various other cations with respect to green plants. *Cornell Univ. Agric. Exp. Sta. Mem.* 2:171–200.

McCool, M.M. 1935. Effect of light intensity on the manganese content of plants. *Boyce Thompson Inst. Contrib.* 7:427–437.

McHargue, J.S. 1922. The role of manganese in plants. *J. Am. Chem. Soc.* 44:1592–1598.

Messing, J.H.L. 1965. The effects of lime and superphosphate on manganese toxicity in steam-sterilized soil. *Plant Soil* 23:1–16.

Mgema W.G. and R.B. Clark. 1995. Sorghum genotypic differences in tolerance to excess manganese. *J. Plant Nutr.* 18:983–993.

Michael, W.S. and S.C. Beckg. 2001. Manganese deficiency in pecan. *HortScience* 36:1075–1076.

Migocka, M. and G. Klobus. 2007. The properties of the Mn, Ni and Pb transport operating at plasma membranes of cucumber roots. *Physiol. Plant.* 129:578–587.

Millaleo, R., M. Reyes-Díaz, A.G. Ivanov, M.L. Mora, and M. Alberdi. 2010. Manganese as essential and toxic element for plants: Transport, accumulation and resistance mechanisms. *J. Soil Sci. Plant Nutr.* 10:476–494.

Millikan, C.R. 1948. Effect of molybdenum on the severity of toxicity symptoms in flax induced by excess of either manganese, zinc, copper, nickel or cobalt in the nutrient solution. *J. Aust. Inst. Agric. Sci.* 5:180–186.

Millikan, C.R. 1949. Effect on flax of a toxic concentration of boron, iron, molybdenum, aluminum, copper, zinc, manganese, cobalt, or nickel in the nutrient solution. *R. Soc. Vict. Proc.* 61:25–42.

Mills, R.F., M.L. Doherty, R.L. López-Marqués et al. 2008. ECA3, a Golgi-localized P2A-Type ATPase, plays a crucial role in manganese nutrition in arabidopsis. *Plant Physiol.* 146(1):116–128.

Mitsukawa, N., S. Okumura, Y. Shirano et al. 1997. Overexpression of an *Arabidopsis thaliana* high-affinity phosphate transporter gene in tobacco cultured cells enhances cell growth under phosphate-limited conditions. *Proc. Natl. Acad. Sci. U. S. A.* 94:7098–7102.

Mora, M.M., A. Rosas, A. Ribera, and Z. Rengel. 2009. Differential tolerance to mn toxicity in perennial ryegrass genotypes: Involvement of antioxidative enzymes and root exudation of carboxylates. *Plant Soil* 320:79–89.

Moraghan, J.T. 1979. Manganese toxicity on flax growing on certain calcareous soils low in available iron. *Soil Sci. Soc. Am. J.* 43:1177–1180.

Moraghan, J.T. and T.J. Freeman. 1978. Influence of FeEDDHA on growth and manganese accumulation in flax. *Soil Sci. Soc. Am. J.* 42:455–459.

Morgan, P.W., H.E. Joham, and J. Amin. 1966. Effect of manganese toxicity on the Indoleacetic acid oxidase system of cotton. *Plant Physiol.* 41:718–724.

Morgan, P.W., D.M. Taylor, and H.E. Joham. 1976. Manipulations of IAA-oxidase activity and auxin deficiency symptoms in intact cotton plants with manganese nutrition. *Physiol. Plant.* 37:149–156.

Morris, H.D. and W.H. Pierre. 1949. Minimum concentrations of manganese necessary for injury to various legumes in culture solutions. *Agron. J.* 41:107–112.

Morris, H.D. and W.H. Pierre. 1948. The effect of calcium, phosphorus, and iron on the tolerance of Lespedeza to manganese toxicity in culture solutions. *Proc. Soil Sci. Soc. Am.* 12:382–386.

Mortvedt, J.J. and H.G. Cunningham. 1971. Production, marketing, and use of other secondary and micro-nutrient fertilizers. In *Fertilizer Technology and Use*, eds. R.A. Olson, T.J. Army, J.J. Hanway, and V.J. Kilmer, pp. 413–455. Madison, WI: Soil Science Society of America.

Mousavi, S.R., M. Galavi, and G. Ahmadvand. 2007. Effect of zinc and manganese foliar application on yield, quality and enrichment on potato (*Solanum tuberosum* L.). *Asian J. Plant Sci.* 6:1256–1260.

Mousavi, S.R., M. Shahsavari, and M. Rezaei. 2011. A general overview on manganese (Mn) importance for crops production. *Aust. J. Basic Appl. Sci.* 5:1799–1803.

Moussavi-Nik, M., Z. Rengel, N.J. Pearson, and G. Hollamby. 1997. Dynamics of nutrient remobilization from seed of wheat genotypes during imbibition, germination and early seedling growth. *Plant Soil* 197:271–280.

Mukhopadhyay, M. and A. Sharma. 1991. Manganese in cell metabolism of higher plants *Bot. Rev.* 57:117–149.

Munns, D.N., L. Jacobsen, and C.M. Johnson. 1963. Uptake and distribution of manganese in oat plants. II. A kinetic model. *Plant Soil* 19:193–204.

Nable, R.O., A. Bar-Akiva, and J.F. Loneragan. 1984. Functional manganese requirement and its use as a criti-cal value for diagnosis of manganese deficiency in subterranean clover (*Trifolium subterraneum* L. cv. Seaton Park). *Ann. Bot.* 54:39–49.

Nable, R.O., R.L. Houtz, and G.M. Cheniae. 1988. Early inhibition of photosynthesis during development of Mn toxicity in tobacco. *Plant Physiol.* 86:1136–1142.

Nable, R.O. and J.F. Loneragan. 1984a. Translocation of manganese in subterranean clover. I. Redistribution during vegetative growth. *Aust. J. Plant Physiol.* 11:101–111.

Nable, R.O. and J.F. Loneragan. 1984b. Translocation of manganese in subterranean clover. II. The effects of leaf senescence and of restricting supply of manganese to part of a split root system. *Aust. J. Plant Physiol.* 11:113–118.

Nambier, E.K.S. 1975. Mobility and plant uptake of micronutrients in relation to soil water content. In *Trace Elements in Soil-Plant-Animal Systems*, eds. D.J.D. Nicholas and A.R. Egan, pp. 151–163. New York: Academic Press.

Ndakidemi, P.A., S.J. Bambara, and H.J.R. Makoi. 2011. Micronutrient uptake in common bean (*Phaseolus vulgaris* L.) as affected by rhizobium inoculation, and the supply of molybdenum and lime. *Plant OMICS: J. Plant Bio. Omics* 4(1):40–52.

Neal, D.C. 1937. Crinkle leaf, a new disease of cotton in Louisiana. *Phytopathology* 27:1171–1175.

Neilsen, D., G.H. Neilsen, A.H. Sinclair, and D.J. Linehan. 1992. Soil phosphorus status, pH and the manga-nese nutrition of wheat. *Plant Soil* 145:45–50.

Ness, P.J. and H.W. Woolhouse. 1980. RNA synthesis in Phaseolus chloroplasts. I. Ribonucleic acid synthesis and senescing leaves. *J. Exp. Bot.* 31:223–233.

Neumann, G. and V. Römheld. 2001. The release of root exudates as affected by the plants physiological status. In *The Rhizosphere: Biochemistry and Organic Substances at the Soil-Plant Interface*, eds. R. Pinto, Z. Varanini and P. Nannipibri, pp. 41–93. New York: Marcel Dekker.

Neumann, K.H. and F.C. Steward. 1968. Investigations on the growth and metabolism of cultured explants of Daucus carota. I. Effects of iron, molybdenum and manganese on growth. *Planta* 81:333–350.

Norvell, W.A. and W.L. Lindsay. 1969. Reactions of EDTA complexes of Fe, Zn, Mn, and Cu in soils. *Proc. Soil Sci. Soc. Am.* 33:86–91.

Norvell, W.A., R.M. Welch, M.L. Adams, and L.V. Kochian. 1993. Reduction of Fe(III), Mn(III), and Cu(II) chelates by roots of pea (*Pisum sativum* L.) or soybean (*Glycine max*). *Plant Soil* 155/156:123–126.

O'Brien, T.P., M.E. Sammut, J.W. Lee, and M.G. Smart. 1985. The vascular system of the wheat spikelet. *Aust. J. Plant Physiol.* 12:487–511.

Ohki, K. 1975. Mn and B effects on micronutrients and P in cotton. *Agron. J.* 67:204–207.

Ohki, K. 1981. Manganese critical levels for soybean growth and physiological processes. *J. Plant Nutr.* 3:271–284.

Ohki, K., F.C. Boswell, M.B. Parker, L.M. Shuman and D.O. Wilson. 1979. Critical manganese deficiency level of soybean related to leaf position. *Agron. J.* 71:233–234.

Ohki, K., D.O. Wilson, and O.E. Anderson. 1981. Manganese deficiency and toxicity sensitivities of soybean cultivars. *Agron. J.* 72:713–716.

Önnerud, H., L. Zhang, G. Gellerstedt, and G. Henriksson. 2002. Polymerization of monolignols by redox shuttle-mediated enzymatic oxidation: A new model in lignin biosynthesis I. *Plant Cell* 14:1953–1962.

Osawa, T. and H. Ikeda. 1976. Heavy metal toxicities in vegetable crops. I. The effect of iron concentrations in the nutrient solution on manganese toxicities in vegetable crops. *J. Jpn. Soc. Hortic. Sci.* 45:50–58.

Ouellette, G.J. and L. Dessureaux. 1958. Chemical composition of alfalfa as related to degree of tolerance to manganese and aluminum. *Can. J. Plant Sci.* 38:206–214.

Ouellette, G.J. and H. Genereaux. 1965. Effect of manganese toxicity on six varieties of potato. *Can. J. Soil Sci.* 45:24–32.

Page, E.R. 1964. The extractable manganese of soil. *J. Soil Sci.* 15:93–102.

Page, E.R. and J. Dainty. 1964. Manganese uptake by excised oat roots. *J. Exp. Bot.* 15:428–443.

Page, V. and U. Feller. 2005. Selective transport of zinc, manganese, nickel, cobalt and cadmium in the root system and transfer to the leaves in young wheat plants. *Ann. Bot.* 96:425–434.

Page, E.R., E.K. Schofield-Palmer, and A.J. McGregor. 1962. Studies in soil and plant manganese. I. Manganese in soil and its uptake by oats. *Plant Soil* 16:238–246.

Page, V., L. Weisskopf, and U. Feller. 2006. Heavy metals in white lupin: Uptake, root to shoot. Transfer and redistribution within the plant. *New Phytol.* 171:329–341.

Palma, J.M., L.M. Sandalio, and L.A. del Rio. 1986. Manganese superoxide dismutase and higher plant chloroplast: A reappraisal of a controverted cellular localization. *J. Plant Physiol.* 125:427–439.

Panda, S., A.K. Mishra, and U.C. Biswal. 1986. Manganese-induced modification of membrane lipid peroxidation during aging of isolated wheat chloroplasts. *Photobiochem. Photobiophys.* 13:53–61.

Papadakis, I., A. Giannakoula, I. Therios, A. Bosabalidis, M. Moustakas, and A. Nastou. 2007. Mn-induced changes in leaf structure and chloroplast ultrastructure of *Citrus volkameriana* (L.) plants. *J. Plant Physiol.* 164:100–103.

Papavizas, G.C. and C.B. Davey. 1961. Extent and nature of the rhizosphere of *Lupinus*. *Plant Soil* 14:215–236.

Parkin, M.J. and I.S. Ross. 1986. The specific uptake of manganese in the yeast *Candida utilis*. *J. Gen. Microbiol.* 132:2155–2160.

Pearson, J.N., C.F. Jenner, Z. Rengel, and R.D. Graham. 1996. Differential transport of Zn, Mn, and sucrose along the longitudinal axis of developing wheat grains. *Physiol. Plant.* 97:332–338.

Pearson, J.N. and Z. Rengel. 1994. Distribution and remobilisation of Zn and Mn during grain development in wheat. *J. Exp. Bot.* 45:1829–1835.

Pearson, J.N. and Z. Rengel. 1995. Uptake and distribution of ^{65}Zn and ^{54}Mn in wheat grown at sufficient and deficient levels of Zn and Mn. II. During grain development. *J. Exp. Bot.* 46:841–845.

Pearson, J.N., Z. Rengel, C.F. Jenner, and R.D. Graham. 1995. Transport of zinc and manganese to developing wheat grains. *Physiol. Plant.* 95:449–455.

Pearson, J.N., Z. Rengel, C.F. Jenner, and R.D. Graham. 1998. Dynamics of zinc and manganese movement in developing wheat grains. *Aust. J. Plant Physiol.* 25:139–144.

Peaslee, D.E. and G.R. Frink. 1969. Influence of silicic acid on uptake of Mn, Al, Zn, and Cu by tomato (*Lycoperiscon esculentum*) grown on an acid soil. *Proc. Soil Sci. Soc. Am.* 33:569–571.

Pedas, P., C.A. Hebbern, J.K. Schjoerring, P.E. Holm, and S. Husted. 2005. Differential capacity for high-affinity manganese uptake contributes to differences between barley genotypes in tolerance to low manganese availability. *Plant Physiol.* 139:1411–1420.

Pedas, P., C.K. Ytting, A.T. Fuglsang, T.P. Jahn, J.K. Schjoerring, and S. Husted. 2008. Manganese efficiency in barley: Identification and characterization of the metal ion transporter HvIRT1. *Plant Physiol.* 148:455–466.

Peiter, E., B. Montanini, A. Gobert et al. 2007. A secretory pathway-localized cation diffusion facilitator confers plant manganese tolerance. *Proc. Natl. Acad. Sci. U. S. A.* 104:8532–8537.

Peterson, J.R., T.M. McCalla, and G.E. Smith. 1971. Human and animal wastes as fertilizers. In *Fertilizer Technology and Use*, eds. R.A. Olson, T.J. Army, J.J. Hanway, and V.J. Kilmer, pp. 557–597. Madison, WI: Soil Science Society of America.

Piñeros, M. and M. Tester. 1995. Characterization of a voltage-dependent Ca^{2+}-selective channel from wheat roots. *Planta* 195:478–488.

Pinner, E., S. Gruenheid, M. Raymond, and P. Gros. 1997. Effect of host cell lipid metabolism on alphavirus replication, virion morphogenesis, and infectivity. *J. Biol. Chem.* 272: 28933–28938.

Pirson, A. 1937. Physiological investigation on the nutrition and metabolism of Fontinalis and Chlorella. *Z. Bot.* 31:193–267.

Pittman, J.K. 2005. Managing the manganese: Molecular mechanisms of manganese transport and homeostasis. *New Phytol.* 157:733–742.

Polle, A. 2001. Dissecting the superoxide dismutase-ascorbate-glutathione-pathway in chloroplasts by metabolic modeling. Computer simulations as a step towards flux analysis. *Plant Physiol.* 126:445–462.

Polle, A., K. Chakrabarti, S. Chakrabarti et al. 1992. Antioxidants and manganese deficiency in needles of Norway spruce (*Picea abies* L.) trees. *Plant Physiol.* 99:1084–1089.

Ponnamperuma, F.N., T.A. Loy, and E.M. Tianco. 1969. Redox equilibria in flooded soils: II. The manganese oxide systems. *Soil Sci.* 108:48–57.

Porter, G., J. Bajita-Locke, N. Hue, and S. Strand. 2004. Manganese solubility and phytotoxicity affected by soil moisture, oxygen levels, and green manure additions. *Commun. Soil Sci. Plant. Anal.* 35:99–116.

Prince, R.C. 1986. Manganese at the active site of the chloroplast oxygen-evolving complex. *Trend Biochem. Sci.* 132:491–492.

Ramani, S. and S. Kannan. 1987. Manganese absorption and transport in rice. *Physiol. Plant.* 33:133–137.

Ratkovic, S. and Z. Vucinic. 1990. The H NMR relaxation method applied in studies of continual absorption of paramagnetic Mn^{2+} ions by roots of intact plants. *Plant Physiol. Biochem.* 28:617–622.

Reay, P.F. 1987. The distribution of nine elements in shoots of *Lupinus albus* L. and *Lupinus angustifolius* L. in relation to manganese accumulation. *Ann. Bot.* 59:219–226.

Reddi, A.R., L.T. Jensen, and V.C. Culotta. 2009. Manganese homeostasis in *Saccharomyces cerevisiae*. *Chem. Rev.* 109:4722–1732.

Reddy, K.S. and H.A. Mills. 1991. Interspecific responses of marigold to manganese as influenced by nitrogen source. *HortScience* 26:1281–1282.

Rees, W.J. and G.H. Sidrak. 1961. Inter-relationship of aluminum and manganese toxicities towards plants. *Plant Soil* 14(2):101–117.

Reis, E.M., R.J. Cook, and B.L. McNeal. 1982. Effect of mineral nutrition on take-all of wheat. *Phytopathology* 72:224–229.

Rengel, Z. 1997. Root exudation and microflora populations in rhizosphere of crop genotypes differing in tolerance to micronutrient deficiency. *Plant Soil* 196:255–260.

Rengel, Z. 2000. Uptake and transport of manganese in plants. In *Metal Ions in Biological Systems, Vol. 37, Manganese and Its Role in Biological Processes,* eds. A. Sigel and H. Sigel, pp. 57–87. Boca Raton, FL: CRC Press.

Rengel, Z. 2001. Xylem and phloem transport of micronutrients. In *Plant Nutrition—Food Security and Sustainability of Agro-Ecosystems*, eds. W.J. Horst et al., pp. 628–629. Dordrecht, The Netherlands: Kluwer Academic Publishers.

Rengel, Z., G. Batten, and D. Crowley. 1999. Agronomic approaches for improving the micronutrient density in edible portions of field crops. *Field Crops Res.* 60:28–40.

Rengel, Z. and R.D. Graham. 1995. Wheat genotypes differ in Zn efficiency when grown in chelate-buffered nutrient solution. II. Nutrient uptake. *Plant Soil* 176:317–324.

Rengel, Z. and R.D. Graham. 1996. Uptake of zinc from chelate-buffered nutrient solutions by wheat genotypes differing in zinc efficiency. *J. Exp. Bot.* 47:217–226.

Rengel, Z., R.D. Graham, and J.F. Pedler. 1993. Manganese nutrition and accumulation of phenolics and lignin as related to differential resistance of wheat genotypes to the take-all fungus. *Plant Soil* 151:255–263.

Rengel, Z., R. Guterridge, P. Hirsch, and D. Hornby. 1996. Plant genotype, micronutrient fertilization and take-all infection influence bacterial populations in the rhizosphere of wheat. *Plant Soil* 183:269–277.

Rengel, Z. and P. Marschner. 2005. Nutrient availability and management in the rhizosphere: Exploiting genotypic differences. *New Phytol.* 168:305–331.

Rengel, Z., J.F. Pedler, and R.D. Graham. 1994. Control of Mn status in plants and rhizosphere: Genetic aspects of host and pathogen effects in the wheat take-all interactions. In *Biochemistry of Metal Micronutrients in the Rhizosphere*, eds. J.A. Manthey, D.E. Crowley, and D.G. Luster, pp. 125–145. Boca Raton, FL: Lewis Publishers/CRC Press.

Rengel, Z., V. Römheld, and H. Marschner. 1998. Uptake of zinc and iron by wheat genotypes differing in tolerance to zinc deficiency. *J. Plant Physiol.* 152:433–438.

Reuter, D.J., A.M. Alston, and J.D. McFarlane. 1988. Occurrence and correction of manganese deficiency in plants. In *Manganese in Soils and Plants*, eds. R.D. Graham, R.J. Hannan, and N.C. Uren, pp. 205–224. Dordrecht, the Netherlands: Kluwer Academic Publishers.

Rezai, K. and T. Fabodnia. 2008. The response of pea plants to manganese toxicity in solution culture. *J. Agric. Sci.* 3:248–251.

Rippel, A. 1923. About iron-induced chlorosis caused by manganese in green plants (Ger.). *Biochem. Z.* 140:315–323.

Roberts, L., A. Pierson, Z. Panaviene, and E. Walker. 2004. Yellow Stripe 1. Expanded roles for the maize iron-phytosiderophore transporter. *Plant Physiol.* 135:112–120.

Robson, A.D. and J.F. Loneragan. 1970. Sensitivity of annual Medicago species to manganese toxicity as affected by calcium and pH. *Aust. J. Agric. Res.* 21:223–232.

Robson, A.D. and K. Snowball. 1986. Nutrient deficiency and toxicity symptoms. In *Plant Analysis: An Interpretation Manual*, eds. D.J. Reuter and J.B. Robinson, pp. 13–19. Melbourne, Australia: Inkata Press.

Roby, C., R. Bligny, R. Douce, and P.E. Pfeffer. 1987. Studies of uptake and suppression of Mn^{2+} migration in highly vacuolated sycamore (*Acer pseudoplatanus* L.) cells by 31P-NMR. *HortScience* 21:176.

Roby, C., R. Bligny, R. Douce, S.I. Tu, and P.E. Pfeffer. 1988. Facilitated transport of Mn^{2+} in sycamore (*Acer pseudoplatanus*) cells and excised maize root tips. A comparative ^{31}P n.m.r. study *in vivo*. *Biochem. J.* 252:401–408.

Römheld, V. and H. Marschner. 1991. Function of micronutrients in plants. In *Micronutrients in Agriculture*, eds. J.J. Mortvedt, F.R. Cox, L.M. Shuman, and R.M. Welch, pp. 297–328. Madison, WI: Soil Science Society of America.

Roorda van Eysinga, J.P.N.L. 1969. *Nutritional Disorders in Cucumbers and Gherkins Under Glass*. Wageningen, the Netherlands: Centre for Agricultural Publications and Documents.

Rosas, A., Z. Rengel, and M. Mora. 2007. Manganese supply and pH influence growth, carboxylate exudation and peroxidase activity of ryegrass and white clover. *J. Plant Nutr.* 30:253–270.

Rovira, A.D., R.D. Graham, and J.S. Ascher. 1985. Reduction in infection of wheat roots by *Gaeumannomyces graminis* var. *tritici* with application of manganese to soil. In *Ecology and Management of Soil-Borne Plant Pathogens*, ed. C.A. Parker, pp. 212–214. St. Paul, MN: American Phytopathological Society.

Rufty, T.W., G.S. Miner, and C.D. Raper, Jr. 1979. Temperature effects on growth and manganese tolerance in tobacco. *Agron. J.* 71:638–644.

Rule, J.H. and E.R. Graham. 1976. Soil labile pools of manganese, iron, and zinc as measured by plant uptake and DTPA equilibrium. *Soil Sci. Soc. Am. J.* 40:853–857.

Russell, E.W. 1988. *Soil Conditions and Plant Growth*, 11th edn. Essex, UK: Longman Scientific and Technical.

Rutherford, A.W. 1989. Photosystem II, the water-splitting enzyme. *Trends Biochem. Sci.* 14:227–232.

Ryan, P., E. Delhaize, and D. Jones. 2001. Function and mechanism of organic anion exudation from plant roots. *Annu. Rev. Plant Physiol. Plant Mol. Biol.* 52:527–560.

Sadana, U.S., L. Kusum, and N. Claassen. 2002. Manganese deficiency of wheat cultivars as related to root growth and internal manganese requirement. *J. Plant Nutr.* 25:2677–2688.

Sadasivan, T.S. 1965. Effect of mineral nutrients on soil microorganisms and plant disease. In *Ecology of Soil-Borne Pathogens. Proceedings of the International Symposium*, eds. C.A. Baker and W.C. Snyder, pp. 460–470. Berkeley, CA: University of California Press.

Sanchez-Raya, A.L., M. Gomez-Ortega, and L. Recalde. 1974. Effect of iron on the absorption and translocation of manganese. *Plant Soil* 41:429–434.

Sandmann, G. and P. Böger. 1980. Copper-mediated lipid peroxidation processes in photosynthetic membranes. *Plant Physiol.* 66:797–800.

Savvas, D., V. Karagianni, A. Kotsiras, V. Demopoulus, I. Karkamisi, and I. Pakou. 2003. Interactions between ammonium and pH of the nutrient solution supplied to gerbera (*Gerbera jamesonii*) grown in pumice. *Plant Soil* 254:393–402.

Schulte, E.E. and K.A. Kelling. 1999. Soil and applied manganese. Understanding plant nutrients. University of Wisconsin Extension Publication, A2526. http://www.soils.wisc.edu/extension/pubs/A2526.pdf, accessed June 1, 2014.

Schumaker, K.S. and H. Sze. 1985. A Ca/H antiport system driven by the proton electrochemical gradient of a tonoplast H-ATPase from oat roots. *Plant Physiol.* 79:1111–1117.

Schützendübel, A. and A. Polle. 2002. Plant responses to abiotic stresses: Heavy metal-induced oxidative stress and protection by mycorrhization. *J. Exp. Bot.* 53:1351–1365.

Sharma, C.P. and P.N. Sharma, C. Chatterjee, and S.C. Agarwala. 1991. Manganese deficiency in maize effects pollen viability. *Plant Soil* 138:139–142.

Shenker M., O.E. Plessner, and E. Tel-Or. 2004. Manganese nutrition effects on tomato growth, chlorophyll concentration, and superoxide dismutase activity. *J. Plant Physiol.* 161:197–202.

Shi, Q., Z. Bao, Z. Zhu, Y. He, Q. Qian, and J. Yu. 2005. Silicon-mediated alleviation of Mn toxicity in *Cucumis sativus* in relation to activities of superoxide dismutase and ascorbate peroxidase. *Phytochemistry* 66:1551–1559.

Shi, Q., Z. Zhu, M. Xu, Q. Qian, and J. Yu. 2006. Effect of excess manganese on the antioxidant system in *Cucumis sativus* L. under two light intensities. *Environ. Exp. Bot.* 58:197–205.

Shigaki, T., J. Pittman, and K. Hirschi. 2003. Manganese specificity determinants in *Arabidopsis* metal/H+ antiporter CAX2. *J. Biol. Chem.* 21:6610–6617.

Shuman, L.M. and O.E. Anderson. 1976. Interactions of Mn with other ions in wheat and soybeans. *Commun. Soil Sci. Plant Anal.* 7:547–557.

Sideris, C.P. and H.Y. Young. 1949. Growth and chemical composition of *Ananas comosus* (L.) Merr., in solution cultures with different iron-manganese ratios. *Plant Physiol.* 24:416–440.

Silber, A., A. Bar-Tal, I. Levkovitch et al. 2009. Manganese nutrition of pepper (*Capsicum annuum* L.): Growth, Mn uptake and fruit disorder incidence. *Sci. Hortic.* 123:197–203.

Silber, A., B. Bar-Yosef, I. Levkovitch, L. Kautzky, and D. Minz. 2008. Kinetics and mechanisms of pH-dependent Mn(II) reactions in plant-growth medium. *Soil Biol. Biochem.* 40:2787–2795.

Silber, A., M. Bruner, E. Kenig et al. 2005. High fertigation frequency and phosphorus level: Effects on summer grown bell pepper growth and blossom-end rot incidence. *Plant Soil* 270:135–146.

Siman, A., F.W. Cradock, and A.W. Hudson. 1974. The development of manganese toxicity in pasture legumes under extreme climatic conditions. *Plant Soil* 41:129–140, 1974.

Siman, A., F.W. Cradock, P.J. Nicholls, and H.C. Kirton. 1971. Effects of calcium carbonate and ammonium sulphate on manganese toxicity in an acid soil. *Aust. J. Agric. Res.* 22:201–214.

Simpson, D.J. and S.P. Robinson. 1984. Freeze-fracture ultrastructure of thylakoid membranes in chloroplasts from manganese-deficient plants. *Plant Physiol.* 74:735–741.

Singh, R.P., C.R. Lee, and M.C. Clark. 1974. Manganese effect on the local lesion symptoms of potato spindle tuber virus in *Scopolia sinensis. Phytopathology* 64:1015–1018.

Singh, M. and A.N. Pathak. 1968. Effect of manganese and iron application on their solubility, absorption and growth of oat plants. *Agrochimica* 12:382–388.

Single, W.V. and I.F. Bird. 1958a. The mobility of manganese in the wheat plant II. Redistribution in relation to foliar application. *Ann. Bot.* 22:479–488.

Single, W.V. and I.F. Bird. 1958b. The mobility of Mn in the wheat plant II. Redistribution in relation to manganese concentration and chemical state. *Ann. Bot.* 22:489–502.

Snider, H.J. 1943. Manganese in some Illinois soils and crops. *Soil Sci.* 56:187–195.

Somers, I.I. and J.W. Shive. 1942. The iron-manganese relation in plant metabolism. *Plant Physiol.* 17:582–602.

Sonneveld, C. and W. Voogt. 1997. Effect of pH value and Mn application on yield and nutrient absorption with rockwool grown gerbera. *Acta Hortic.* 450:139–147.

Sparrow, L. and N. Uren. 1987. Oxidation and reduction of Mn in acidic soils: Effect of temperature and soil pH. *Soil Biol. Biochem.* 19:143–148.

Spiers, J.M. 1993. Nitrogen, calcium and magnesium fertilizer affects growth and leaf element content of dormered raspberry. *J. Plant Nutr.* 16:2333–2339.

Supek, F., L. Supekova, H. Nelson, and N. Nelson. 1996. A yeast manganese transporter related to the macrophage protein involved in conferring resistance to mycobacteria. *Proc. Natl. Acad. Sci. U. S. A.* 93:5105–5110.

Swain, T. 1977. Secondary compounds as protective agents. *Annu. Rev. Plant Physiol.* 28:479–501.

Takahashi, M. and M. Sugiura. 2001. Strategies for uptake of a soil micronutrient, manganese, by plant roots. *RIKEN Rev.* 35:76–77.

Tanaka, Y., T. Hibino, and Y. Hayashi et al. 1999. Salt tolerance of transgenic rice overexpressing yeast mitochondrial Mn-SOD in chloroplasts. *Plant Sci.* 148:131–138.

Terry, N., P.S. Evans, and D.E. Thomas. 1975. Manganese toxicity effects on leaf cell multiplication and expansion and on dry matter yield of sugar beets. *Crop Sci.* 15:205–208.

Tester, M. 1990. Plant ion channels: Whole-cell and single-channel studies. *New Phytol.* 114:305–340.

Thomine, S., F. Lelièvre, E. Debarbieux, J.I. Schroeder, and H. Barbier-Brygo. 2003. AtNRAMP3, a multi-specific vacuolar metal transporter involved in plant responses to iron deficiency. *Plant J.* 34:685–695.

Timonin, M.I. 1946. Microflora of the rhizosphere in relation to the manganese-deficiency disease of oats. *Proc. Soil Sci. Soc. Am.* 11:284–292.

Trim, A.R. 1959. Metal ions as precipitants for nucleic acids and their use in the isolation of polynucleotides from leaves. *Biochem. J.* 73:298–304.

Tsuji, M., H. Mori, T. Yamamoto, S. Tanaka, Y. Kang, K. Sakurai, and K. Iwasaki. 2006. Manganese toxicity of melon plants growing on an isolated soil bed after steam sterilization. *J. Soil Sci. Plant Nutr.* 77:257–263.

Tyerman, S.D. 1992. Anion channels in plants. *Annu. Rev. Plant Physiol. Plant Mol. Biol.* 43:351–373.

Uchida, R. 2000. Essential nutrients for plant growth: Nutrient functions and deficiency symptoms. In *Plant Nutrient Management in Hawaii's Soils, Approaches for Tropical and Subtropical Agriculture*, eds. J.A. Silva and R. Uchida, pp. 31–54. Manoa, HI: College of Tropical Agriculture and Human Resources, University of Hawaii.

Uehara, K., S. Fujimoto, and T. Taniguchi. 1974. Studies on violet-colored acid phosphatase of sweet potato. II. Enzymatic properties and amino acid composition. *J. Biochem. (Tokyo)* 75:639–649.

Van Assche F. and H. Clijsters. 1990. Effect of metals on enzyme activity in plants. *Plant Cell Environ.* 3:195–206.

Van Der Vorm, P.D.J. and A. Van Diest. 1979. Aspect of the Fe and Mn nutrition of rice plants. I. Iron and manganese uptake by rice plants grown under aerobic and anaerobic conditions. *Plant Soil* 51:233–246.

Varotto, C., P. Pesaresi, P. Jahns et al. 2002. Single and double knockouts of the genes for photosystem I subunits G, K, and H of arabidopsis. Effects on photosystem I composition, photosynthetic electron flow, and state transitions. *Plant Physiol.* 129:1–9.

Verta, G., N. Grotzb, F. Dédaldéchampa et al. 2002. IRT1, an arabidopsis transporter essential for iron uptake from the soil and for plant growth. *Plant Cell* 14:1223–1233.

Vielemeyer, H.P., F. Fisher, and W. Bergmann. 1969. Investigations on the effect of micronutrients iron and manganese on the nitrogen metabolism of agricultural culture plants. 2nd communication: Investigations on the effect of manganese on the reduction of nitrate and the content of free amino acids in young bush bean plants. *Albercht-Thaerr-Arch.* 13:393–404.

Vlamis, J. and D.E. Williams. 1962. Ion competition in manganese uptake by barley plants. *Plant Physiol.* 37:650–655.

Vlamis, J. and D.E. Williams. 1964. Iron and manganese relations in rice and barley. *Plant Soil* 20:221–231.

Vlamis, J. and D.E. Williams. 1967. Manganese and silicon interaction in the Gramineae. *Plant Soil* 27:131–140.

Vose, P.B. 1963. The translocation and redistribution of manganese in Avena. *J. Exp. Bot.* 14:448–457.

Vose, P.B. and D.G. Jones. 1963. The interaction of Mn and Ca on nodulation and growth in varieties of *Trifolium repens*. *Plant Soil* 18:372–385.

Wada, T. and J.N.A. Lott. 1997. Light and electron microscopic and energy dispersive X-ray microanalysis studies of globoids in protein bodies of embryo tissues and aleurone layer of rice (*Oryza sativa* L.) grains. *Can. J. Bot.* 75:1137–1147.

Wall, M.J., A.J. Quinn, and G.B. D'Cunha. 2008. Manganese (Mn2+)-dependent storage stabilization of *Rhodotorula glutinis* phenylalanine ammonia-lyase activity. *J. Agric. Food Chem.* 56:894–902.

Wallace, T. 1946. Mineral deficiencies in plants. *Endeavour* 5:58–64.

Wallace, T., E.J. Hewitt, and D.J.D. Nicholas. 1945. Determination of factors injurious to plants in acid soils. *Nature* 156:778–779.

Wallace, A., R.T. Mueller, J.W. Cha, and G.V. Alexander. 1974. Soil pH, excess lime, and chelation agent on micronutrients in soybeans and Bush beans. *Agron. J.* 66:698–700.

Walton, G.H. 1978. The effect of manganese on seed yield and the split seed disorder of sweet and bitter phenotypes of *Lupinus angustifolius* and *L. cosentinii*. *Aust. J. Agric. Res.* 29:1177–1189.

Walworth, J.L. 1998. Potatoes-field crop fertilizer recommendations for Alaska. Crop Production and Soil Management Series. FGV-00246A. http://www.uaf.edu/coop-ext/ Accessed July 2004.

Wang, J., H. Raman, M. Zhou et al. 2007. High resolution mapping of Alp, the aluminum tolerance locus in barley (*Hordeum vulgare* L.), identifies a candidate gene controlling tolerance. *Theor. Appl. Genet.* 115:265–276.

Wang, F.Z., Q.B., Wang, S.Y. Kwon, S.S. Kwak, and W.A. Su. 2005. Enhanced drought tolerance of transgenic rice plants expressing a pea manganese superoxide dismutase. *J. Plant Physiol.* 162(4):465–472.

Wang, Y., Y. Ying, J. Chen, and X. Wang. 2004. Transgenic *Arabidopsis* overexpressing Mn-DOD enhanced salt-tolerance. *Plant Sci.* 167:671–677.

Wang, Y.X., P. Wu, Y.R. Wu, and X.L. Yan. 2002. Molecular marker analysis of manganese toxicity tolerance in rice under greenhouse conditions. *Plant Soil* 238:227–233.

Waraich, E.A., R. Ahmad, A. Halim, and T. Aziz. 2012. Alleviation of temperature stress by nutrient management in crop plants: A review. *J. Soil Sci. Plant Nutr.* 12:221–244.

Ward, G.M. 1977. Manganese deficiency and toxicity in greenhouse tomatoes. *Can. J. Plant Sci.* 57:107–115.

Warden, B.T. and H.M. Reisenauer. 1991. Manganese-iron interactions in the plant-soil system. *J. Plant Nutr.* 14:7–30.

Watmough, S.A., M.C. Eimersb, and P.J. Dillona. 2007. Manganese cycling in central Ontario forests: Response to soil acidification. *Appl. Geochem.* 22:1241–1247.

Webb, M.J., W.A. Norvell, R.M. Welch, and R.D. Graham. 1993. Using a chelate-buffered nutrient solution to establish the critical solution activity of Mn^{2+} required by barley (*Hordeum vulgare* L.). *Plant Soil* 153:195–205.

Weil, R.R., C.D. Foy, and C.A. Coradetti.1997. Influence of soil moisture regimes on subsequent soil manganese availability and toxicity in two cotton genotypes. *Agron. J.* 89(1):1–8.

Welch, R.M. 1986. Effects of nutrient deficiencies on seed production and quality. *Adv. Plant Nutr.* 2:205–247.

Welch, R.M., W.H. Allaway, W.A. House, and J. Kubota. 1991. Geographic distribution of problem areas. In *Micronutrients in Agriculture*, eds. J.J Mortvedt, F.R Cox, L.M. Shuman, and R.M Welch, pp. 96–99. Madison, WI: Soil Science Society of America.

Welch, R.M. and W.A. Norvell. 1993. Growth and nutrient uptake by barley (*Hordeum vulgare* L. cv. Herta): Studies using an n-(2-hydroxyethyl) ethylenedinitrilotriacetic acid-buffered nutrient solution technique (ii. role of zinc in the uptake and root leakage of mineral nutrients). *Plant Physiol.* 101:627–631.

Welch, R.M., W.A. Norvell, S.C. Schaefer, J.E. Shaft, and L.V. Kochian. 1993. Induction of iron (Ill) and copper (II) reduction in pea (*Pisum sativum* L.) roots by Fe and Cu status: Does the root-cell plasmalemma Fe(III)-chelate reductase perform a general role in regulating cation uptake? *Planta* 190:555–561.

White, R.P. 1970. Effects of lime upon soil and plant manganese levels in an acid soil. *Proc. Soil Sci Soc. Am.* 34:625–29.

White, C.L., A.D. Robson, and H.M. Fisher. 1981. Variation in nitrogen, sulfur, selenium, cobalt, manganese, copper and zinc contents of grain from wheat and two lupin species grown in a range of Mediterranean environments. *Aust. J. Agric. Res.* 32:47–59.

Wiehe, W. and G. Hoflich. 1995. Establishment of plant-growth promoting bacteria in the rhizosphere of subsequent plants after harvest of the inoculated pre-crops. *Microbiol. Res.* 150(3):331–336.

Wilhelm, N.S., J.M. Fisher, and R.D. Graham. 1985. The effect of manganese deficiency and cereal cyst nematode infection on the growth of barley. *Plant Soil* 85:23–32.

Wilhelm, N.S., R.D. Graham, and A.D. Rovira. 1988. Application of different sources of manganese sulfate decreases take-all (*Gaeumannomyces graminis* var. *tritici*) of wheat grown in a manganese deficient soil. *Aust. J. Agric. Res.* 39:1–10.

Wilhelm, N.S., R.D. Graham, and A.D. Rovira. 1990. Control of Mn status and infection rate by genotypes of both host and pathogen in the wheat take-all interaction. *Plant Soil* 123:267–275.

Wilkinson, R.E. and K. Ohki. 1988. Influence of manganese deficiency and toxicity on isoprenoid synthesis. *Plant Physiol.* 87:841–846.

Williams, C.H., J. Vlamis, H. Hall, and K.D. Gowans. 1971. Effects of urbanization, fertilization and manganese toxicity on mustard plants. *Calif. Agric.* 25(11):8–10.

Williams, D.E. and J. Vlamis. 1957a. Manganese and boron toxicities in standard culture solutions. *Proc. Soil Sci. Soc. Am.* 21:205–209.

Williams, D.E. and J. Vlamis. 1957b. The effect of silicon on the yield and manganese-54 uptake and distribution in the leaves of barley plants grown in culture solution. *Plant Physiol.* 32:404–409.

Williams, D.E. and J. Vlamis, 1957c. Manganese toxicity in standard culture solutions. *Plant Soil* 8:183–193.

Williams, D.E. and J. Vlamis. 1971. Comparative accumulation of manganese and boron in barley tissues. *Plant Soil* 33:623–628.

Wilson, D.O., F.C. Boswell, K. Ohki et al. 1982. Changes in soybean seed oil and protein as influences by manganese nutrition. *Crop Sci.* 22:948–952.

Winkler, R.G., D.G. Blevins, J.C. Polacco, and D.D. Randall. 1987. Ureide catabolism of soybeans. II. Pathway of catabolism in intact leaf tissue. *Plant Physiol.* 83:585–591.

Winkler, R.G., J.C. Polacco, D.G. Blevins, and D.D. Randall. 1985. Enzymatic degradation of allantoate in developing soybeans. *Plant Physiol.* 79:878–793.

Winterhalder, E.K. 1963. Differential resistance of two species of *Eucalyptus* to toxic soil manganese levels. *Aust. J. Sci.* 25:363–364.

Wissemeier, A.H. and W.J. Horst. 1987. Callose deposition in leaves of cowpea (*Vigna unguiculata* L. Walp.) as a sensitive response to high Mn supply. *Plant Soil* 102:283–286.

Wissemeier, A.H. and W.J. Horst. 1991. Simplifies methods for screening cowpea cultivars for manganese leaf-tissue tolerance. *Crop Sci.* 31:435–439.

Wissemeier, A.H. and W.J. Horst. 1992. Effect of light intensity on manganese toxicity symptoms and callose formation in cowpea (*Vigna unguiculata* (L.) Walp.). *Plant Soil* 143:299–309.

Wu, S. 1994. Effect of manganese excess on the soybean plant cultivated under various growth conditions. *J. Plant Nutr.* 17:991–1003.

Wu, Z.Y., F. Liang, B.M. Hong et al. 2002. An endoplasmic reticulum-bound Ca^{2+}/Mn^{2+} pump, ECA1, supports plant growth and confers tolerance to Mn^{2+} stress. *Plant Physiol.* 130:128–137.

Yang, X.E., W.R. Chen, and Y. Feng. 2007. Improving human micronutrient nutrition through biofortification in the soil-plant system: China as a case study. *Environ. Geochem. Health* 29:413–428.

Yang, J.L., Y.Y. Li, and Y.J. Zhang et al. 2008. Cell wall polysaccharides are specifically involved in the exclusion of aluminum from the rice root apex. *Plant Physiol.* 146:602–611.

Yano, J. 2010. X-ray spectroscopy of the biological water splitting catalysis. In LCLS/SSRL Annual Users' Meeting and Workshops, Menlo Park, CA. http://www-conf.slac.stanford.edu/ssrl-lcls/2010/ Accessed July 2014.

Yu, Q., L.D. Osborne, and Z. Rengel. 1999. Increased tolerance to Mn deficiency in transgenic tobacco over-producing superoxide dismutase. *Ann. Bot.* 84:543–547.

Zhang, H.M., B.R. Wang, M.G. Xu, and T.L. Fan. 2009. Crop yield and soil responses to long-term Fertilization on a red soil in southern China. *Pedosphere* 19:199–207.

Zouni, A., H.T. Witt, J. Kern et al. 2001. Crystal structure of photosystem II from *Synechococcus elongatus* at 3.8 Å resolution. *Nature* 409:739–743.

13 Molybdenum

Dean A. Kopsell, David E. Kopsell, and Russell L. Hamlin

CONTENTS

13.1 HISTORICAL BACKGROUND

13.1.1 DETERMINATION OF MOLYBDENUM ESSENTIALITY

Prior to the eighteenth century, the principal ore of molybdenum, molybdenite, was confused with graphite and lead. Back then, lead referred to any black mineral that left a mark on paper. It was not until 1778 that the Swedish chemist Carl Wilhelm Scheele confirmed that molybdenite was not the element lead (known commonly as galena) or graphite. Scheele identified the new element as molybdenum, named after the ore in which it was found. Peter Jacob Hjelm successfully extracted and purified molybdenum from molybdenite in 1781. The biological significance of molybdenum was established in 1930 when Bortels discovered that it was required for the assimilation of gaseous nitrogen by the free-living nitrogen-fixing bacterium *Azotobacter chroococcum* (Bortels, 1930; Mulder, 1948).

The essentiality of molybdenum for plant growth was demonstrated by Arnon and Stout in 1939 (Arnon and Stout, 1939a; Kaiser et al., 2005). Tomato (*Lycopersicon esculentum* Mill.) grown in nutrient solutions devoid of molybdenum developed mottled lesions on leaves and a morphology, commonly referred to as *whiptail*. The only element that reversed these conditions was molybdenum. Previously, Arnon reported that small amounts of molybdenum added to nutrient solutions improved growth of plants such as lettuce (*Lactuca sativa* L.) and asparagus (*Asparagus officinalis* L.) (Arnon, 1937, 1938). This work led to further investigations to see if molybdenum could fit the newly established criteria for plant nutrients (Arnon and Stout, 1939b).

The first report of molybdenum deficiency existing in an agricultural production system occurred in a mixed pasture stand in Australia (Anderson, 1942). Crop failures were reported for mixed stands of bulbous canary grass (*Phalaris aquatica* L.), perennial ryegrass (*Lolium perenne* L.), and subterranean clover (*Trifolium subterraneum* L.) growing on well-irrigated pastures. Further investigation found that the forages were sown on slightly acidic, sandy loam soils that were low in available nitrogen and high in iron oxides. Growth was improved through additions of lime or wood ash, materials later found to contain trace amounts of soluble and insoluble molybdenum. Moreover, alfalfa (*Medicago sativa* L.) yields improved threefold following applications of 2.24 kg molybdate ha^{-1} (Anderson, 1942). Reversal of characteristic deficiency symptoms through additions of molybdenum to soils came shortly thereafter for such crops as cauliflower (*Brassica oleracea* var. *botrytis* L.) and tomato (Davies, 1945; Mitchell, 1945; Walker, 1948). Molybdenum toxicity in plants is rare in most field agricultural systems. However, chlorosis or anthocyanin accumulations in leaf tissues for plants grown under high molybdenum concentrations have been reported (Bergmann, 1992; Gupta, 1997a). The main concern with excessive tissue molybdenum in plants is related to molybdenosis in ruminant mammals, a disorder caused by the depressing effect of molybdenum on the physiological availability of copper (Picco et al., 2012).

13.1.2 Roles of Molybdenum in Plant Physiology

Molybdenum is an essential microelement for most organisms and is involved in more than 60 different enzymes. Similar to other metals required for plant growth, molybdenum is needed by specific plant enzymes to participate in oxidative and reduction reactions (Mendel and Schwarz, 1999; Zimmer and Mendel, 1999). Molybdenum exists in oxidation states from 0 to VI; however, only the higher order oxidation states of IV–VI are considered biologically important. The functions of molybdenum in plants and other organisms are related to the valence charge that the element undergoes as a metallic component of enzymes (Marschner, 1995).

Biosynthesis of the molybdenum cofactor (Moco) is the beginning step in many of the enzymatic pathways that require molybdenum (Mendel, 2013). A recent review of molybdenum metabolism in plants is presented by Bittner (2014). With the exception of bacterial nitrogenase enzymes, molybdenum-containing enzymes in most organisms share a similar molybdopterin compound at their catalytic sites (Mendel and Schwarz, 1999; Zimmer and Mendel, 1999). The pterin is a Moco that is responsible for the correct anchoring and positioning of the element within the enzyme. This structure allows molybdenum to interact with the other components of the electron-transport chain in which the enzyme participates (Mendel and Schwarz, 1999). The biosynthesis of Moco in the mitochondria and cytosol requires molybdenum or molybdate (Bittner, 2014), and mutations in Moco biosynthesis result in loss of all molybdenum-dependent cellular processes (Mendel, 2013). Molybdenum itself is thought to be biologically inactive until the element is complexed with the Moco cofactor. Total molybdenum content in plant tissues correlates to Moco levels in arabidopsis (*Arabidopsis thaliana* Heynh.) and may indicate that molybdate levels in tissues regulate Moco synthesis (Gasber et al., 2011).

Because of its involvement in nitrogen fixation, nitrate reduction, and nitrogen transport, molybdenum plays crucial roles in nitrogen metabolism in plants (Srivastava, 1997). Several important molybdo-enzymes are significant in plant physiology and include nitrogenase, nitrate reductase, xanthine dehydrogenase, aldehyde oxidase, and possibly sulfite oxidase. Even though the physiological functions of some molybdo-enzymes are not understood fully, it is evident that any factor that influences Moco biosynthesis and molybdate supply will result in restrictions in the activity of the molybdo-enzymes and ultimately plant survival (Bittner, 2014).

13.1.2.1 Nitrogenase

The discovery that molybdenum was necessary for the growth of *Azotobacter* sp. was the first indication of the essential role that molybdenum plays in biological processes (Bortels, 1930). It is now well established that molybdenum is required for biological N_2 fixation, an activity that is facilitated by the molybdenum-containing enzyme nitrogenase. Several types of asymbiotic bacteria, such as *Azotobacter* sp., *Rhodospirillum* sp., and *Klebsiella* sp., are able to fix atmospheric N_2 into plant-available forms. The most important symbiotic relationship for agricultural production is between *Rhizobium* sp. and leguminous crops (Srivastavac, 1997). Nitrogenase enzymes are conserved among a wide range of different organisms, and these enzymes catalyze the reduction of N_2 to ammonia (NH_3) in the following reaction (Mishra et al., 1999):

$$N_2 + 8H^+ + 8e^- + 16ATP \rightarrow 2NH_3 + H_2 + 16ADP + 16Pi$$

One of the most impact-producing processes in nature is N_2 fixation, which takes place biologically at normal temperatures and atmospheric pressure (Gupta and Lipsett, 1981). In contrast, the same chemical reaction performed synthetically in the Haber–Bosch process requires temperatures of 300°C–500°C and pressures of >300 atm (Voet and Voet, 1995). According to Mishra et al. (1999), nearly all nitrogenase enzymes contain the same two proteins, both of which are inactivated irreversibly in the presence of oxygen and include a Mo–Fe protein (MW 200,000) and an Fe protein (MW 50,000–65,000). The Mo–Fe protein contains two atoms of molybdenum and has oxidation–reduction centers of two distinct types, two iron–molybdenum cofactors called Fe–Mo cofactor (FeMoco) and four Fe–S (4Fe–4S) centers. The FeMoco of nitrogenase constitutes the active site of the molybdenum-containing nitrogenase protein (Allen et al., 1999).

The effect of biological N_2 fixation on the global nitrogen cycle is substantial, with terrestrial nitrogen inputs in the range of 125–150 × 10^6 Mg nitrogen per year (Peoples and Craswell, 1992). Despite the importance of molybdenum to N_2-fixing organisms and the nitrogen cycle, the essential nature of molybdenum for plants is not based entirely on its role in N_2 fixation. Plants that lack the ability to fix atmospheric N_2 have no nitrogenase and have no need for molybdenum for this action. In addition, the process of N_2 fixation is not essential for the growth of legumes if sufficient levels of nitrogen fertilizers are supplied (Mishra, et al., 1999; George et al., 1992).

13.1.2.2 Nitrate Reductase

The essential nature of molybdenum as a plant nutrient is based solely on its role in the reduction of nitrate (NO_3^-) via nitrate reductase. This enzyme converts nitrate to nitrite (NO_2^-) and occurs in most plant species as well as in fungi and bacteria (Gupta and Lipsett, 1981). It is the principal molybdenum protein of vegetative plant tissues (Hewitt, 1983). However, in leguminous plants, the requirement of molybdenum for nitrogenase activity in root nodules is greater than the requirement of molybdenum for the activity of nitrate reductase in the vegetative tissues (Parker and Harris, 1977). Because nitrate is the major form of soil nitrogen absorbed by plant roots (Mengel and Kirkby, 1987), the role of molybdenum as a functional component of nitrate reductase is of greater importance in plant nutrition than its role in N_2 fixation.

As a molybdo-enzyme, nitrate reductase requires Moco, heme, FAD, and NADH or NADPH for enzymatic activity (Bittner, 2014). Like other molybdenum enzymes in plants, nitrate reductase is a homodimeric protein. Each identical subunit can function independently in nitrate reduction (Marschner, 1995), and each consists of three functional domains, the N-terminal domain associated with a Moco, the central heme domain (cytochrome b_{557}), and the C-terminal FAD domain (Mendel and Schwarz, 1999; Notton, 1983). The enzyme is located in the cytoplasm and catalyzes the following reaction (Mengel and Kirkby, 1987):

$$NO_3^- + 2H^+ + 2e^- \rightarrow NO_2^- + 2H_2O$$

Nitrate and molybdenum are required for the induction of nitrate reductase in plants, and if either nutrient is deficient, enzymatic activity is limited severely (Randall, 1969). In nitrogen-deficient plants, the induction of nitrate reductase activity by nitrate is a slow process, whereas the induction of enzyme activity by molybdenum is much faster (Srivastava, 1997). The molybdenum requirement of plants is higher if they are supplied NO_3^- rather than ammonium (NH_4^+) (Giordano et al., 1966), a need that can be accounted for almost completely by the molybdenum in nitrate reductase (Gupta and Lipsett, 1981).

13.1.2.3   Xanthine Dehydrogenase

In addition to nitrogenase and nitrate reductase, molybdenum is also a functional component of xanthine dehydrogenase, which is involved in purine degradation by oxidizing hypoxanthine to xanthine and xanthine to uric acid in the cytosol (Bittner, 2014; Zimmer and Mendel, 1999). This enzyme is a homodimeric protein of identical subunits, each of which contains one molecule of FAD, four Fe–S groups, and a molybdenum complex that cycles between Mo(VI) and Mo(IV) oxidation states (Marschner, 1995). Xanthine dehydrogenase carries out the catabolism of purines to uric acid via a xanthine intermediate (Mendel and Schwarz, 1999). Xanthine dehydrogenase requires Moco, FAD, and two iron–sulfur clusters (Bittner, 2014). In some legumes, the transport of symbiotically fixed N_2 from root to shoot occurs in the form of ureides, allantoin, and allantoic acid, which are synthesized from uric acid (Srivastava, 1997). Although xanthine dehydrogenase is apparently not essential for plants (Srivastava, 1997), it can play a key role in nitrogen metabolism for certain legumes for which ureides are the most prevalent nitrogen compounds formed in root nodules (Marschner, 1995). The poor growth of molybdenum-deficient legumes can be attributed in part to poor upward transport of nitrogen because of disturbed xanthine catabolism (Srivastava, 1997).

13.1.2.4   Aldehyde Oxidase

Aldehyde oxidases in animals have been well characterized, but only recently has this molybdo-enzyme been purified from plant tissue and described (Koshiba et al., 1996). The enzyme shares catalytic and structural characteristics with xanthine dehydrogenase. In contrast to xanthine dehydrogenase, aldehyde oxidase preferably oxidizes aldehydes to the respective carboxylic acid (Bittner, 2014). In plants, aldehyde oxidase is considered to be located in the cytoplasm where it catalyzes the final step in the biosynthesis of the phytohormones indole-3-acetic acid (IAA) and abscisic acid (ABA) (Zimmer and Mendel, 1999). These hormones control diverse processes and plant responses, such as stomatal aperture, germination, seed development, apical dominance, and the regulation of phototropic and gravitropic behavior (Merlot and Giraudat, 1997; Normanly et al., 1995). Mutations in aldehyde oxidases, Moco biosynthesis, or Moco activation disrupt ABA synthesis (Kaiser et al., 2005). Molybdenum therefore may play an important role in plant development and adaptation to environmental stresses through its effect on the activity of aldehyde oxidase, although other minor pathways exist for the formation of IAA and ABA in plants (Mendel and Schwarz, 1999).

13.1.2.5   Sulfite Oxidase

Molybdenum may play a role in sulfur metabolism in plants. In biological systems, the oxidation of sulfite (SO_3^{2-}) to SO_4^{2-} is mediated by the molybdo-enzyme, sulfite oxidase (Srivastava, 1997).

Although this enzyme has been well studied in animals (Kisker et al., 1997), details of the action of sulfite oxidase in plants are not well established. Plant sulfite oxidase is a peroxisomal enzyme consisting exclusively of the Moco-binding domain required for oxidizing sulfite to sulfate (SO_4^{2-}) (Bittner, 2014; Nowak et al., 2004). Marschner (1995) explains that the oxidation of sulfite can be brought about by other enzymes, such as peroxidases and cytochrome oxidase, as well as a number of metals and superoxide radicals. It is therefore not clear whether a specific sulfite oxidase is involved in the oxidation of sulfite in plants (DeKok, 1990) and, consequently, also whether molybdenum is essential in plants for sulfite oxidation.

13.2 MOLYBDENUM UPTAKE BY PLANTS

Molybdenum is absorbed by plant roots as the molybdate ion (MoO_4^{2-}), and its uptake is controlled metabolically (Mengel and Kirkby, 1987). In long-distance ion transport, molybdenum is highly mobile in plant xylem and phloem (Bittner, 2014; Kannan and Ramani, 1978). The exact form of molybdenum transported in conductive tissues is unknown, but it is highly likely that it is transported as MoO_4^{2-} rather than in a complexed form (Marschner, 1995). The proportion of various molybdenum constituents in plants naturally depends on the quantity of molybdenum absorbed and accumulated in the tissue. Molybdenum-containing enzymes, such as nitrogenase and nitrate reductase, constitute a major pool for absorbed molybdenum, but under conditions of luxury consumption, excessive molybdenum also can be stored as MoO_4^{2-} in the vacuoles of peripheral cell layers of the plant (Bittner, 2014; Hale et al., 2001).

At concentrations >0.04 μM Mo in the soil solution, molybdenum is transported to plant roots by mass flow. At levels <0.04 μM Mo, molybdenum moves by diffusion (Barber, 1995). These actions translate to a greater severity of molybdenum deficiency under dry soil conditions due to limited mass flow or diffusion under low soil water content. The MoO_4^{2-} anion can be leached readily through the soil in considerable quantities, but movement and availability depend on the pH of the soil solution (Jones and Belling, 1967). For a given soil, molybdenum uptake by plants increases as the soil pH increases. The molybdenum content of crop plants grown in a soil of pH 5.0 will, on average, double if the soil is limed to pH 6.0 and increase sixfold if the soil is further limed to pH 7.0 (Barber, 1995).

Sulfate can inhibit uptake of MoO_4^{2-} (Kannan and Ramani, 1978), and low soil sulfate concentration facilitates higher MoO_4^{2-} uptake by plants (Shinmachi et al., 2010). The competitive nature of these two ions comes from the chemical similarities that they share as both possess a double negative charge (SO_4^{2-}, MoO_4^{2-}), share similar molecular size, and have tetrahedral structures (Bittner, 2014). Recently, Tomatsu et al. (2007) identified a MoO_4^{2-} transporter (MOT1) in arabidopsis (*A. thaliana* Heyn.); MOT1 is a high-affinity transporter essential for plants to take up MoO_4^{2-} from the soil in times of limited supply. Expression of the transporter was detected in roots and shoots, with the highest expression levels in endodermis and stele cells. A second molybdate transporter, MOT2, was identified to have similarities to sulfate transporters (Gasber et al., 2011). The MOT2 transcripts accumulate in senescing leaves and may facilitate export of stored MoO_4^{2-} from the vacuole into maturing seed tissues. This action is confirmed by the fact that MOT-deficient arabidopsis show high molybdenum accumulations in the leaves and low molybdenum concentrations in seeds.

13.3 PHYSIOLOGICAL RESPONSES OF PLANTS TO SUPPLY OF MOLYBDENUM

The effect of molybdenum fertilization on increasing plant yield often is related to an increased ability of the plant to utilize nitrogen. The activities of nitrogenase and nitrate reductase are affected by the molybdenum status of plants, and their activities often are suppressed in plants suffering from molybdenum deficiency (Randall, 1969; Vieira et al., 1998a). Foliar application of molybdenum at 40 g ha^{-1} at 25 days after plant emergence greatly enhanced nitrogenase and nitrate reductase activities of common bean (*Phaseolus vulgaris* L.), resulting in an increase in total nitrogen accumulation in shoots (Vieira et al., 1998a). In addition, foliar fertilization of common bean with 40 g Mo ha^{-1} increased nodule size but not the number of root nodules (Vieira et al., 1998b). Therefore, the main effect of molybdenum on nodulation was suggested to be the avoidance of nodule senescence, thus maintaining a longer period of effective N_2 fixation.

Crop plants vary in their requirement for molybdenum and thus require different levels of molybdenum fertilization to achieve maximum growth. The application of molybdenum to soils with low amounts of available molybdenum can improve crop yield dramatically, particularly for legumes, which have a high molybdenum requirement (Gupta and Lipsett, 1981). Large-seeded legumes often do not require molybdenum fertilization as their seeds contain enough molybdenum to meet the requirements of the plant (Hewitt, 1956). However, for plants suffering from molybdenum deficiency, the response to molybdenum fertility often varies. The lack of response to molybdenum can be related to other nutritional problems, such as the toxic effects of aluminum and manganese in acid soils, which mask the effects of molybdenum nutrition (Adams, 1997). In addition, molybdenum can be rendered unavailable to plants in acid soils if molybdenum is fixed by iron and manganese oxides (Gupta and MacKay, 1968).

In acid soils, the availability of applied molybdenum can be limited due to the fixation of MoO_4^{2-} by iron, manganese, and aluminum oxides; however, the concentration of molybdenum in the soil solution increases with increasing soil pH (Adriano, 1986). Liming materials can be used in conjunction with molybdenum fertilization to increase molybdenum uptake by plants, but the effect on plant growth is limited to soil pH levels <7.0 (Mortvedt, 1981). Liming alone may liberate enough soil-bound molybdenum to sustain plant growth (Mortvedt, 1997). However, the effect of lime depends on the total molybdenum content of soils. On acid soils where aluminum toxicity can limit plant growth, adding both lime and molybdenum is often more beneficial than adding only one of them (Adriano, 1986; Burmester et al., 1988).

Soybean (*Glycine max* Merr.) yields on acid soils in southeastern United States increased by 30%–80% following molybdenum fertilization (Hagstrom and Berger, 1965a; Parker and Harris, 1962). Similar results occurred for soybean fertilized with molybdenum in Brazil (Neto et al., 2012) and for peanut (*Arachis hypogaea* L.) grown on acid soils in western Africa (Hafner et al., 1992). However, Rhoades and Nangju (1979) reported that at soil pH 4.5, soybeans did not respond to molybdenum. Differences in the response of legumes to molybdenum may be related to the timing of fertilizer applications. During the lag phase between infection and active N_2 fixation (between 10 and 21 days) (Marschner, 1995), the addition of molybdenum fertilizers may be ineffective because the growth response to added molybdenum is related primarily to the molybdenum requirements of the N_2-fixing bacteria (Parker and Harris, 1977). In other studies where molybdenum was seed-applied, cowpea (*Vigna unguiculata* Walp.) yields increased by 25% (Rhoades and Nangju, 1979), and oat (*Avena sativa* L.) yields increased by 48% (Fitzgerald, 1954). Molybdenum fertilization also increased the production of melon (*Cucumis melo* L.) (Gubler et al., 1982).

Other soil amendments, such as fertilizers containing phosphorus or sulfur, may influence the efficiency of molybdenum fertilizers by affecting the fixation of molybdenum in soils or uptake by plant roots. The use of phosphate ($H_2PO_4^-$), which has a high affinity with iron oxides, can lead to the release of adsorbed molybdenum and to an increase in the water-soluble MoO_4^{2-} concentration of the soil (Zimmer and Mendel, 1999). As a result, phosphorus fertilization often increases the molybdenum absorption by roots and its accumulation in plant tissues (Gupta, 1997b; Gupta and Lipsett, 1981). In contrast, SO_4^{2-} and MoO_4^{2-} are strongly competitive during root absorption, and sulfur fertilization can decrease molybdenum uptake by plants (MacLeod et al., 1997). Studies with peanut have shown that providing phosphorus in the form of triple superphosphate ($Ca(H_2PO_4)_2$) is superior to single superphosphate ($Ca(H_2PO_4)_2$ and $CaSO_4$) for plants grown in molybdenum-deficient soils. This difference was attributed to the sulfur component of single superphosphate and its effect on inhibiting molybdenum uptake and suppressing plant growth (Rebafka et al., 1993). It is possible that the many chemical similarities shared between SO_4^{2-} and MoO_4^{2-} result in competitive inhibition during uptake and translocation in plants (Shinmachi et al., 2010).

13.4 GENETICS OF ACQUISITION OF MOLYBDENUM BY PLANTS

The requirement for molybdenum in plants is lower than any other mineral element with the exception of nickel (Marschner, 1995). Even with the narrow sufficiency range of 0.1–1.0 µg molybdenum per g leaf dry mass, plant species will differ in the ability to accumulate molybdenum from the environment. Because of the essentiality of molybdenum in structural and catalytic function of enzymes involved in N_2 fixation, the functions of molybdenum are related closely to nitrogen metabolism. Therefore, leguminous and nonleguminous plant species dependent on N_2 fixation have high requirements for molybdenum (Marschner, 1995). Molybdo-enzymes require a Moco prosthetic group for proper function, and it is nearly identical for many of these enzymes. Mutations in genes controlling Moco synthesis and assembly will therefore have impacts on the function of several different molybdo-enzymes (Warner and Kleinhofs, 1992). In barley (*Hordeum vulgare* L.), genetic mutants with defective Moco genes lacked nitrate reductase, xanthine dehydrogenase, and aldehyde oxidase activities (Walker-Simmons et al., 1989). Species of Brassicaceae are sensitive to molybdenum deficiency and usually have higher concentrations of the element than other plants. Grasses, cereals, and large-seeded legumes contain low levels of tissue molybdenum (Adriano, 1986). Cultivar variation within species for molybdenum accumulation has been established for alfalfa (*M. sativa* L.) (Younge and Takahashi, 1953), cauliflower (*B. oleracea* var. *botrytis* L.) (Chipman et al., 1970; Gammon et al., 1954), corn (*Zea mays* L.) (Brown and Clark, 1974), and cotton (*Gossypium hirsutum* L.) (Bhatt and Appukuttan, 1982).

13.5 CONCENTRATIONS OF MOLYBDENUM IN PLANTS

The sufficiency range for molybdenum in plant tissues varies widely as plants differ in ability to absorb molybdenum from the root medium (Mortvedt, 1981) (Table 13.1). Most plants contain a sufficient level of molybdenum (0.2–2.0 mg Mo kg^{-1} dry weight); however, the difference between critical deficiency and toxicity levels can vary up to a factor of 10^4 (e.g., 0.1–1000 mg Mo kg^{-1} dry mass) (Marschner, 1995).

The allocation of molybdenum to the various plant organs varies considerably among plant species, but generally the concentration of molybdenum is highest in seeds (Gupta and Lipsett, 1981) and in the nodules of N_2-fixing plants (Marschner, 1995). Recently identified molybdenum transporter genes (*MOT1*, *MOT2*) control molybdenum partitioning in plants, with activity influenced by molybdenum concentration in plant tissues and plant ontogeny (Gasber et al., 2011). When molybdenum is limiting, preferential accumulation in root nodules may lead to considerably lower molybdenum content in the shoots and seeds of nodulated legumes (Ishizuka, 1982). Molybdenum concentrations in leaves have been reported to exceed concentrations in the stems of several crop species such as tomato, alfalfa, and soybean (Gupta and Lipsett, 1981).

13.6 RATIOS OF MOLYBDENUM AND INTERACTIONS WITH OTHER ELEMENTS

The source of nitrogen supplied to plants influences their requirements for molybdenum. Nitrate-fed plants generally have a high requirement for molybdenum (Agarwala and Hewitt, 1954), but reports conflict as to whether plants supplied with reduced nitrogen have a molybdenum requirement. Cauliflower developed symptoms of molybdenum deficiency when grown with ammonium salts, urea, glutamate, or nitrate, in the absence of molybdenum (Notton, 1983). However, Hewitt (1963) suggested that the molybdenum requirement, in the presence of reduced nitrogen, may result from the effects of traces of nitrate derived from bacterial nitrification. When cauliflower plants were supplied ammonium sulfate and no molybdenum under sterile conditions, plants showed no abnormalities and apparently had no molybdenum requirement (Hewitt and Gundry, 1970). On transfer to nonsterile conditions, whiptail symptoms appeared as a characteristic symptom of molybdenum deficiency.

TABLE 13.1

Plant Species Variation for Deficient and Sufficient Molybdenum Concentrations

Plant Species by Crop Type	Plant Part Samples	Mo Concentration (mg kg⁻¹ Dry Mass) Sufficient	References
Agronomic Crops			
Alfalfa (*Medicago sativa* L.)	Upper portion of tops; prior to bloom	0.5–5.0	Jones (1967)
	Leaves	0.10–0.72	Adriano (1986)
	Stems	0.08–0.28	Adriano (1986)
Barley (*Hordeum vulgare* L.)	Whole tops; boot stage	0.09–0.18	Gupta (1971)
Canola (*Brassica napus* L.)	Mature leaves without petioles	0.25–0.60	Bryson et al. (2014)
Corn (*Zea mays* L.)	Stems	1.4–7.0	Dios & Broyer (1965)
	Ear leaves; silk stage		Vos (1993)
Cotton (*Gossypium hirsutum* L.)	Fully mature leaves; after bloom	0.6–2.0	Bergmann (1992)
Oats (*Avena sativa* L.)	Whole tops	0.2–0.3	Bryson et al. (2014)
Peanut (*Arachis hypogaea* L.)	Upper fully developed leaves	0.5–1.0	Bergmann (1992); Smith et al. (1993)
Red clover (*Trifolium pratense* L.)	Total aboveground mass; bloom	0.3–1.59	Gupta (1970); Gupta et al. (1978)
	Whole plants; bud stage	0.46–1.08	Gupta (1970); Gupta et al. (1978)
	Leaves	0.06–0.23	Adriano (1986)
	Stems	0.08–0.35	Adriano (1986)
Rice (*Oryza sativa* L.)	Upper fully developed leaves; prior to flowering	0.4–1.0	Bergmann (1992)
Soybean (*Glycine max* Merr.)	Upper fully developed leaves; end of bloom	0.5–1.0	Bergmann (1992)
Sugar beet (*Beta vulgaris* L.)	Leaf blades	0.2–20.0	Ulrich and Hills (1973)
	Fully developed leaf without stem	0.2–20.0	Ulrich and Hills (1973)
Sunflower (*Helianthus annuus* L.)	Mature leaves from new growth	0.25–0.75	Bryson et al. (2014)
Tobacco (*Nicotiana tabacum* L.)	Mature leaves from new growth	0.1–0.6	Bryson et al. (2014)
Wheat (*Triticum aestivum* L.)	Whole tops; boot stage	0.09–0.18	Gupta (1971)
Vegetable Crops			
Bean (*Phaseolus vulgaris* L.)	Youngest fully expanded leaf; flowering	0.2–5.0	Weir and Creswell (1993)
Beet (*Beta vulgaris* L.)	Young mature leaves	0.15–0.6	Weir and Creswell (1993)
Broccoli (*Brassica oleracea* var. *italica* Plenck)	Mature leaves from new growth	0.3–0.5	Bryson et al. (2014)
	Leaves	2.16–11.16	Adriano (1986)
	Leaf petioles	0.37–4.28	Adriano (1986)
Brussels sprouts (*Brassica oleracea* var. *gemmifera* Zenker)	Leaves	0.11–2.98	Adriano (1986)
	Leaf petioles	0.12–0.62	Adriano (1986)
Cabbage (*Brassica oleracea* var. *capitata* L.)	Wrapper leaves	0.3–3.0	Bryson et al. (2014); Weir and Creswell (1993)

(Continued)

TABLE 13.1 (*Continued*)
Plant Species Variation for Deficient and Sufficient Molybdenum Concentrations

Plant Species by Crop Type	Plant Part Samples	Mo Concentration (mg kg⁻¹ Dry Mass) Sufficient	References
Carrot (*Daucus carota* L.)	Mature leaves from new growth	0.5–1.5	Bryson et al. (2014)
Cauliflower (*Brassica oleracea* var. *botrytis* L.)	Mature leaves from new growth	0.5–0.8	Bryson et al. (2014)
	Aboveground portion of plant; appearance of curd	0.68–1.49	Chipman et al. (1970)
	Leaves	0.54–3.72	Adriano (1986)
	Leaf petioles	0.37–1.10	Adriano (1986)
Cucumber (*Cucumis sativa* L.)	Youngest fully mature leaf	0.2–2.0	Weir and Creswell (1993)
Lettuce (*Lactuca sativa* L.)	Leaves	0.08–0.14	Gupta (1997a); Plant (1952)
Onion (*Allium cepa* L.)	Whole tops; mature	0.1	Gupta and LeBlanc (1990)
Pea (*Pisum sativum* L.)	Recently fully developed leaves; early bloom	0.4–1.0	Bergmann (1992)
Potato (*Solanum tuberosum* L.)	Fully developed leaves; early bloom	0.2–0.5	Bergmann (1992)
Rutabaga (*Brassica napobrassica* Mill.)	Leaves	0.20–1.59	Adriano (1986)
	Leaf petioles	0.17–0.78	Adriano (1986)
Tomato (*Lycopersicon esculentum* Mill.)	Recently matured leaf and petiole	0.2–0.6	Byson et al. (2014)
Fruit Crops			
Apple (*Malus sylvestris* Mill.)	Mature leaves from new growth	0.10–2.00	Bryson et al. (2014)
Avocado (*Persea americana* Mill.)	Mature leaves from new flush	0.05–1.00	Bryson et al. (2014)
Orange (*Citrus* × *sinensis* Osbeck)	Mature leaves from nonfruiting plants	0.1–0.9	Bryson et al. (2014)
Pear (*Pyrus communis* L.)	Mid-shoot leaves from new growth	0.10–2.00	Bryson et al. (2014)
Peach (*Prunus persica* Stokes)	Mid-shoot leaves	1.6–2.8	Bryson et al. (2014)
Strawberry (*Fragaria* × *ananassa* Duchesne)	Mature leaves from new growth	0.25–0.50	Bryson et al. (2014)
Ornamental Plants			
New Guinea impatiens (*Impatiens hawkeri* W. Bull)	Mature leaves from new growth	0.15–1.00	Bryson et al. (2014)
Poinsettia (*Euphorbia pulcherrima* Willd.)	Mature leaves from new growth	0.12–0.50	Bryson et al. (2014); Cox (1992)
Rose, hybrid tea (*Rosa* sp. L.)	Upper leaflets from mature growth	0.1–0.9	Bryson et al. (2014)
Scarlet sage (*Salvia splendens* Sello)	Mature leaves from new growth	0.2–1.08	Bryson et al. (2014)
Snapdragon (*Antirrhinum majus* L.)	Mature leaves from new growth	0.12–2.00	Bryson et al. (2014)
Verbena (*Verbena hybrida* Voss)	Mature leaves from new growth	0.14–0.80	Bryson et al. (2014)

(Continued)

TABLE 13.1 (*Continued*)
Plant Species Variation for Deficient and Sufficient Molybdenum Concentrations

Plant Species by Crop Type	Plant Part Samples	Mo Concentration (mg kg⁻¹ Dry Mass) Sufficient	References
Trees and Shrubs			
Common lilac (*Syringa vulgaris* L.)	Mature leaves from new growth	0.12–4.00	Bryson et al. (2014)
Douglas fir (*Pseudotsuga menziesii* Franco)	Terminal cuttings	0.02–0.25	Bryson et al. (2014)
Loblolly pine (*Pinus taeda* L.)	Needle leaves from terminal cuttings	0.12–0.56	Bryson et al. (2014)
White pine (*Pinus strobus* L.)	Needle leaves from terminal cuttings	0.05–0.20	Bryson et al. (2014)

Source: Adapted from Gupta, U.C., *Molybdenum in Agriculture*, Cambridge University Press, New York, 1997.

Hewitt (1983) later stated that molybdenum is of very little importance for some plants if nitrate reduction is not necessary for nitrogen assimilation but that it is impossible to say that an element is not required by plants given the limits of current analytical techniques. Evidence shows that acidification of soils by nitrogen fertilization and cation removal by crops may be a causal agent in the decreased molybdenum concentrations in crop plants (Phillips and Meyer, 1993). There are examples of crop plants where tissue molybdenum concentrations decrease with higher nitrogen application rates (Gupta, 1997b).

The molybdenum nutrition of plants is affected by the interaction of molybdenum with other nutrients in the soil, most importantly phosphorus and sulfur. It is well established that plant uptake of molybdenum is enhanced by the presence of soluble phosphorus (Reed et al., 2013) and decreased by the presence of available sulfur (Gupta, 1997b). Molybdenum and phosphorus applications significantly increased molybdenum availability in soil and molybdenum plant uptake in white clover (*Trifolium repens* L.) on either unlimed or limed andisols than with molybdenum fertilizers applied alone (Vistoso et al., 2012). In comparison to MoO_4^{2-}, phosphate has a greater affinity for sorption sites in soils, such as on sesquioxides (Smith et al., 1997). Phosphorus fertilization often liberates soil-bound molybdenum into the soil solution and increases molybdenum accumulation by plants (Srivastava and Gupta, 1996; Xie and MacKenzie, 1991). Phosphorus also may stimulate molybdenum absorption through the formation of a phosphomolybdate complex in soils, which may be absorbed readily by plants (Barshad, 1951a). In a recent study, Reed et al. (2013) showed that N_2 fixation rates of soils in tropical ecosystems correlated highly with phosphorus and not molybdenum fertilization. Stout and Meagher (1948) showed that the addition of SO_4^{2-} to the culture medium reduced absorption of radioactive molybdenum by tomatoes and decreased molybdenum absorption by tomato and pea (*Pisum sativum* L.) in soil (Stout et al., 1951). As mentioned previously, the effect of sulfur on molybdenum uptake by plants appears to be related to the direct competition between SO_4^{2-} and MoO_4^{2-} during root absorption.

13.7 DIAGNOSIS OF MOLYBDENUM STATUS IN PLANTS

13.7.1 DEFICIENCY CONDITIONS

The discovery of molybdenum as a plant nutrient led to the diagnosis of the deficiency in various crop species. Molybdate deficiency affects expression of genes involved in ion transport, stress responses, signal transduction, and metabolism of nitrogen, sulfur, and phosphate and subsequently levels of amino acids, sugars, organic acids and purine metabolites (Bittner, 2014).

The first report of molybdenum deficiency in the field was in subterranean clover pasture production where the element was required for N_2 fixation by the rhizobium–legume complex (Anderson, 1942). The critical deficiency concentration for most crop plants is very low, ranging from 0.1 to 1.0 mg Mo kg^{-1} dry weight (Gupta and Lipsett, 1981). Symptoms of molybdenum deficiency are common in plants grown on acid or strongly acidic mineral soils, partly due to the presence of hydrous iron and aluminum oxides (Adriano, 1986). Plants may also become molybdenum deficient in organic peat soils where molybdenum can complex with humic acids (Mengel and Kirkby, 1987; Szalay and Szilagyi, 1968). It is also common for molybdenum deficiency to appear in conditions of low soil molybdenum, low soil temperatures, and high soil nitrogen, and sulfur fertility (Wang et al., 1999).

Because of the low molybdenum requirement, there is only a narrow tissue concentration range between deficiency and sufficiency levels for most plants (Adriano, 1986). Molybdenum is highly mobile in xylem and phloem tissues; therefore, deficiency symptoms often appear on the entire plant (Kannan and Ramani, 1978). This characteristic is unique to molybdenum as the deficiency symptoms of the other micronutrients primarily occur in the younger plant tissues due to limited mobility. Visual diagnosis of molybdenum deficiency is complicated further by the fact that it often manifests itself as nitrogen deficiency, most notably in legumes. These symptoms are related to the function of molybdenum in nitrogen metabolism, such as N_2 fixation and nitrate reductase. Under extreme deficiency, unique molybdenum deficiency symptoms are exhibited.

Legumes often require higher concentrations of molybdenum than other crop plants. This requirement is especially true for legumes dependent on N_2 fixation as a source of nitrogen (Marschner, 1995). Molybdenum deficiency symptoms in legumes are chlorosis, stunted growth, and a restriction in the weight and quantity of root nodules (Hagstrom and Berger, 1965b). In dicotyledonous species, a drastic suppression in leaf size and irregularities in leaf blade formation (known as whiptail) are the most common visual symptoms of molybdenum deficiency (Figure 13.1). Molybdenum deficiency also can cause localized tissue necrosis and insufficient differentiation of vascular bundles in the early stages of leaf development (Bussler, 1970).

FIGURE 13.1 Molybdenum deficiency whiptail of cauliflower (*B. oleracea* var. *botrytis* L.). (Photograph courtesy of Dr. Douglas A. Cox, University of Massachusetts, Amherst, MA.)

FIGURE 13.2 (**See color insert.**) Molybdenum deficiency of poinsettia (*Euphorbia pulcherrima* Willd.). (Photograph courtesy of Dr. Douglas A. Cox, University of Massachusetts, Amherst, MA.)

Marginal and interveinal necrosis in leaf tissues is a symptom of extreme molybdenum deficiency (Figure 13.2). Leaf necrosis often is associated with elevated nitrate nitrogen concentrations, indicating a lack of nitrate reductase activity under molybdenum deficiency (Gupta and Lipsett, 1981). Molybdenum-deficient beans (*P. vulgaris* L.) often develop scald, where leaves appear pale with interveinal and marginal chlorosis that is quickly followed by burning of the leaf margin (Weir and Creswell, 1993; Wilson, 1949).

Whiptail describes symptoms of yellowing of the leaf tissue on the outer leaves followed by necrosis at the leaf edges. The absence of tissue on leaf edges results in the formation of narrow, distorted leaves that are usually slightly thickened causing leaf edges to curl upward. The whiptail disorder is observed on molybdenum-deficient cruciferous crops, with cauliflower being the most sensitive among the family to molybdenum deficiency (Weir and Creswell, 1993). In molybdenum-deficient tomato, lower leaves appear mottled and eventually cup upward and develop marginal chlorosis (Arnon and Stout, 1939a). Molybdenum deficiency in corn (*Z. mays* L.) causes suppressed tasseling, which inhibits anthesis and pollen formation (Römheld and Marschner, 1991). Molybdenum deficiency in corn shortens internodes, decreases overall leaf area, and causes leaf tissue chlorosis (Agarwala et al., 1978). In grapevines (*Vitis* sp. L.), molybdenum deficiency recently has been suggested as the primary cause of millerandage, a disorder where fruiting bunches develop unevenly (Williams et al., 2004). The inhibition of pollen formation by molybdenum deficiency may explain the lack of fruit formation common on molybdenum-deficient watermelon (*Citrullus vulgaris* Schrad.) (Gubler et al., 1982; Marschner, 1995).

13.7.2 EXCESSIVE CONDITIONS

Critical molybdenum toxicity concentrations vary widely among plant species. Molybdenum toxicity is rare under field conditions; however, toxicity has been induced under experimental conditions (Adriano, 1986). Toxic levels of molybdenum in barley occurred at 135 mg Mo kg^{-1} in leaf tissues in sand culture (Davis et al., 1978). Using similar growing techniques, Hunter and Vergnano (1953) reported the critical toxic level of molybdenum to be 200 mg kg^{-1} for oats. However, *Brassica* and *Allium* species can tolerate levels of >600 mg Mo kg^{-1} without exhibiting toxic symptoms (Gupta et al., 1978). Lyon and Beeson (1948) reported turnips (*Brassica rapa* L.) to accumulate 1800 mg Mo kg^{-1} in leaf tissues without any deleterious impacts on plant growth when cultured in nutrient solutions containing 20 μg Mo mL^{-1}. Normal growth also was reported for soybean with leaf concentrations of 80 mg Mo kg^{-1} (Golov and Kazakhko, 1973) and for cotton with leaf concentrations of >1500 mg Mo kg^{-1} (Joham, 1953). However, most crop species exhibit toxic symptomology when tissue levels reach concentrations of >500 mg Mo kg^{-1} (Gupta, 1997a). Molybdenum toxicity in plants is characterized by a malformation of leaf tissues, a golden-yellow discoloration of shoots, and restricted root and shoot growth (Kevresan et al., 2001; Marschner, 1995).

Most plants are not particularly sensitive to excessive molybdenum in the root environment. Even though molybdenum toxicity under field conditions is rare, the main concern with elevated tissue molybdenum concentrations is potential toxicity to animals grazing on the plants. Molybdenum concentrations >10 mg Mo kg^{-1} (dry mass) in forage crops can cause a nutritional disorder in grazing ruminants called molybdenosis (Marschner, 1995). Molybdenosis is a result of a copper deficiency induced by high molybdenum concentrations. Molybdate reacts with sulfur in the rumen to form thiomolybdate complexes that inhibit copper metabolism (Stark and Redente, 1990). Agricultural practices used to decrease ruminant susceptibility to molybdenosis include field applications of copper and sulfur fertilizers. The strong effects that SO_4^{2-} has on MoO_4^{2-} uptake can lower molybdenum concentration in plants below toxic levels (Pasricha et al., 1977). Increasing the copper content in forage crops through copper fertilization may also help to reduce copper deficiency in animals as a result of elevated molybdenum concentrations (Stark and Redente, 1990).

13.8 FORMS AND CONCENTRATIONS OF MOLYBDENUM IN SOILS AND AVAILABILITY TO PLANTS

13.8.1 FORMS AND CONCENTRATIONS OF MOLYBDENUM IN SOILS

Molybdenum is present in many different complexes in agricultural systems, influenced mainly by chemical properties in the soil environment. Mineral forms of molybdenum commonly present in soils include molybdenite (MoS_2), powellite ($CaMoO_4$), wulfenite ($PbMoO_4$), ferrimolybdite ($Fe_2(MoO_4)$), ilsemanite (molybdenum oxysulfate), and jordisite (amorphous molybdenum disulphide) (Adriano, 1986; Reddy et al., 1997). Of these minerals, only molybdenite and ferrimolybdite are mined commercially (Reddy et al., 1997). Soil parent materials with the highest concentrations of molybdenum include phosphorite, shale, limestone, and sandstone (Adriano, 1986). The release of molybdenum from primary minerals is through weathering. The distribution of molybdenum in the soil profile is determined by soil pH and the degree of leaching. Under alkaline soil conditions, molybdenum is more soluble and thus more susceptible to leaching.

The amount of naturally occurring molybdenum in soils depends on the molybdenum concentrations in the parent materials. Igneous rock makes up some 95% of the earth's crust (Mitchell, 1964) and contains ~2 mg Mo kg^{-1}. Similar amounts of molybdenum are present in sedimentary rock (Norrish, 1975). The total molybdenum content of soils differs by soil type and sometimes by geographical region (Table 13.2). Molybdenum concentrations in normal soils average between 0.013 and 17.0 mg Mo kg^{-1} (44), but molybdenum concentrations can exceed 300 mg Mo kg^{-1} in soils derived from organic-rich shale (Reddy et al., 1997). Large quantities of molybdenum also occur in soils polluted by mining activities (Stark and Redente, 1990). Most agricultural soils contain a relatively low amount of molybdenum by comparison, with an average of 2.0 mg kg^{-1} total molybdenum, with only about 0.2 mg kg^{-1} of available molybdenum (Mengel and Kirkby, 1987). Therefore, the amount of molybdenum present in the soil solution is relatively small (<10 µg Mo L^{-1}) and is highly dependent on soil temperature with concentrations being highest at higher soil temperatures (Barber, 1995).

Soils derived from granite, organic-rich shale, or limestone, and those high in organic matter usually contain the highest concentrations of molybdenum (Srivastava and Gupta, 1996; Welch et al., 1991), and the available molybdenum content generally increases with alkalinity or fineness of the soil texture (Srivastava and Gupta, 1996). Concentrations of extracted molybdenum in granitic and basaltic igneous-derived soils can range from 0.9 to 7.0 mg Mo kg^{-1}; soils derived from

TABLE 13.2
Molybdenum Concentrations on Surface Soils in Different Countries

Soil	Country	Range (mg kg⁻¹ Dry Weight)
Podzol and sandy soils	Australia	2.6–3.7
	Canada	0.40–2.46
	New Zealand	1–2[a]
	Poland	0.2–3.0
	Yugoslavia	0.17–0.51[b]
	Russia	0.3–2.9
Loess and silty soils	New Zealand	2.2–3.1[a]
	China	0.4–1.1
	Poland	0.6–3.0
	United States	0.75–6.40
	Russia	1.8–3.3
Loamy and clayey soils	Great Britain	0.7–4.5
	Canada	0.93–4.74
	Mali Republic	0.5–0.75
	New Zealand	2.1–4.2[a]
	Poland	0.1–6.0
	United States	1.2–7.2
	United States[c]	1.5–17.8
	Russia	1.8–3.0
Fluvisols	India	0.4–3.1[b]
	Czech Republic	2.8–3.5
	Mali Republic	0.44–0.65
	Yugoslavia	0.35–0.53[b]
	Russia	1.8–3.0
Gleysols	Australia	2.5–3.5
	India	1.1–1.8[b]
	Ivory Coast	0.18–0.60
	Yugoslavia	0.52–0.74
	Russia	0.6–2.0
Histosols and other organic soils	Canada	0.69–3.2
	Russia	0.3–1.9
Forest soils	Bulgaria	0.3–4.6
	Former Soviet Union	0.2–8.3
Various other soils	Great Britain	1–5
	India	0.013–2.5
	Italy	0.4–2.2
	Japan	0.2–11.3
	United States	0.8–3.3
	Russia	0.8–3.6

Source: Adapted from Kabata-Pendias, A. and Pendias, H. *Trace Elements in Soils and Plants,* 4th edn. Boca Raton, FL: CRC Press, 2011.

[a] Soils derived from basalts and andesites.

[b] Data from whole soil profiles.

[c] Soils from area of the western states with incidences of molybdenum toxicity to grazing animals.

shales can range from 1 to 300 mg Mo kg^{-1}, and those derived from limestones and sandstones can range from 3 to 30 mg Mo kg^{-1} (Adriano, 1986). In contrast, molybdenum is often deficient in well-drained, coarse-textured soils or in soils that are highly weathered or acidic (Gupta, 1997b; Reddy et al., 1997). In soils, molybdenum can occur in four fractions: (1) dissolved molybdenum in the soil solution, (2) molybdenum occluded with oxides, (3) molybdenum as a constituent in minerals, and (4) molybdenum associated with organic matter (Srivastava and Gupta, 1996). The accumulation of molybdenum varies with depth in the soil, but molybdenum is normally highest in the A horizons of well-drained soils and is highest in the subsoil of poorly drained mineral soils (Reddy et al., 1997).

13.8.2 FACTORS AFFECTING MOLYBDENUM AVAILABILITY TO PLANTS

The availability of molybdenum for plant growth is highly dependent on soil pH conditions, concentrations of adsorbing oxides, soil drainage characteristics, and interactions with organic compounds present in the soil colloids (Kaiser et al., 2005). Under neutral or slightly alkaline soil conditions, molybdenum is relatively soluble and is taken up by plants as MoO_4^{2-}. However, when soil pH becomes strongly acidic (about pH 5.0–5.5), molybdenum availability decreases sharply as the anion form adsorbs to soil oxides (Reddy et al., 1997). At water-measured pH >5.0, molybdenum exists primarily as MoO_4^{2-} (Sims, 1996), but at lower pH levels, the $HMoO_4^-$ and $H_2MoO_4^-$ forms dominate (Kabata-Pendias and Pendias, 2011). For each unit increase in soil pH above pH 5.0, the soluble molybdenum concentration increases 100-fold (Lindsay, 1972). Plants preferentially absorb MoO_4^{2-}, and therefore, the molybdenum nutrition of plants can be manipulated by altering soil acidity (Barber, 1995). Soil liming is used commonly to alleviate molybdenum deficiencies in plants by increasing the quantity of plant-available molybdenum in the soil solution (Mortvedt, 1997), but the effect of liming on molybdenum nutrition varies by soil and plant type. Excessive lime use may decrease the solubility of molybdenum through the formation of $CaMoO_4$ (Kabata-Pendias and Pendias, 2011), but Lindsay (1979) suggests that this complex is too soluble to persist in soils. Using lime to change the acidity of a clay loam from pH 5.0 to 6.5 resulted in greater molybdenum accumulation in cauliflower, alfalfa, and bromegrass (*Bromus inermis* Leyss.), but molybdenum accumulation was relatively unaffected if plants were grown in a sandy loam (Gupta, 1997b). For plants grown in sandy loam, lime and molybdenum were both required to significantly increase the molybdenum content of the plant tissue.

Significant amounts of molybdenum can be bound, or fixed, in soils by iron and aluminum oxides, particularly under acidic conditions (Mengel and Kirby, 1987). These sesquioxides have a pH-dependent surface charge that becomes more electrically positive as soil pH decreases and more negative as soil pH increases. Changes in the surface charge are due to the protonation and deprotonation of surface functional groups (Sparks, 1995). Under acidic soil conditions, MoO_4^{2-} is adsorbed strongly to the surface of iron and aluminum oxides by a ligand-exchange mechanism (Smith et al., 1997), and adsorption is greatest at pH 4.0 (Sims, 1996). In acid soils, the molybdenum concentration in the soil solution can be reduced greatly, but because molybdenum is adsorbed weakly to soils and hydrous oxides at alkaline pH, these soils have a relatively large proportion of molybdenum in the solution phase (Reisenauer et al., 1962). Compared with adsorption on hydrous iron oxides, the strength of molybdenum adsorption to aluminum oxide is much weaker (Jones, 1957). Despite this difference, aluminum oxides play an important role in the sorption of molybdenum in soils. For instance, the adsorption capacity of montmorillonite increases in the presence of interlayered aluminum hydroxide polymers (Srivastava and Gupta, 1996).

In water-saturated soils, the availability of molybdenum is influenced by its reaction with other redox-active elements, such as sulfur. Under strongly reducing conditions, molybdenum forms sparingly soluble thiomolybdate complexes, with MoS_2 being the most important mineral controlling molybdenum solubility (Kabata-Pendias and Pendias, 2011). Other minerals whose ions are also affected by oxidation–reduction state, such as $MnMoO_4$ or $FeMoO_4$, are too soluble to precipitate in soils (Smith et al., 1997). Soil pH greatly influences the availability of molybdenum from these mineral sources; even $PbMoO_4$, the least soluble of the possible soil compounds, becomes more soluble as pH increases (Gupta, 1997b).

Soil organic matter can complex or fix molybdenum in soils, but the mechanisms of sorption are not well understood. Molybdenum binds strongly to humic and fulvic acids (Smith et al., 1997). Owing to the great affinity of molybdenum to be fixed by organic matter, its concentration in forest litter can reach 50 mg Mo kg^{-1} (Kabata-Pendias and Pendias, 2011). The accumulation of molybdenum in organic matter can be particularly high if soil drainage is impeded (Sharma and Chatterjee, 1997). Organic-matter-rich soils can supply adequate amounts of molybdenum for plant growth due to a slow release of molybdenum from the organic complex (Kabata-Pendias and Pendias, 2011). However, there are conflicting reports concerning the effect of soil organic matter on the availability of molybdenum in the soil solution. Plant-available molybdenum can be low in soils having high quantities of organic matter (Mulder, 1954), particularly on peat soils due to the strong fixation of molybdenum by humic acid (Kabata-Pendias and Pendias, 2011). In contrast, Srivastava and Gupta (1996) suggested that soil organic matter increases the available molybdenum content of acid soils by inhibiting the fixation of MoO_4^{2-} by sesquioxides.

13.9 ASSESSMENT OF MOLYBDENUM STATUS IN SOILS

The use of soil testing to predict the capacity of soils to supply molybdenum for plant growth can be difficult because of the relatively small amounts of molybdenum in soil, the differences in plant requirement for molybdenum, and the importance of molybdenum reserves in seeds in supplying crop needs (Sims and Eivazi, 1997). In addition, the total molybdenum content of soils can differ considerably from the plant-available molybdenum fraction (Eivazi and Sims, 1997). The total molybdenum content in soils usually ranges between 0.013 and 17.0 mg Mo kg^{-1} (Kabata-Pendias and Pendias, 2011) and is dependent on the molybdenum content of the parent material (Adriano, 1986). However, the quantity of molybdenum available for plant uptake can be substantially less than the total and is dependent on soil pH and other chemical and biological factors. Detailed information on the determination of molybdenum in soils is provided by Sims (1996), Eivazi and Sims (1997), and Sims and Eivazi (1997).

13.9.1 DETERMINATION OF TOTAL MOLYBDENUM IN SOIL

Several extraction methods have been developed for the determination of molybdenum in soils. The most common method of soil extraction is by perchloric acid digestion (Reisenauer, 1965). Dry ashing followed by acid extraction is another common analytical method (Grigg, 1953). Purvis and Peterson (1956) proposed the sodium carbonate fusion method for determination of total molybdenum. The thiocyanate–stannous chloride spectrophotometric procedure revised by Johnson and Arkley (1954) and modified by Sims (1996) is used extensively for the determination of total molybdenum in soils. Molybdenum in the soil extract reacts with thiocyanate and excess iron in the presence of stannous chloride to form the colored complex $Fe(MoO(SCN)_5$. The complex is extracted from the aqueous phase with isoamyl alcohol that has been dissolved in carbon tetrachloride (CCl_4). The amount of molybdenum present is determined in a light spectrophotometer by comparison of the absorbance of the sample with appropriate standards. Difficulties associated with the thiocyanate method include interference from iron and the use of stannous chloride, which can vary in purity and consistency (Eivazi and Sims, 1997). Graphite furnace atomic absorption spectrometry has also been used for the analysis of extracts having a low concentration of molybdenum (<1.0 mg kg^{-1}) (Curtis and Grusovin, 1985). Inductively coupled plasma (ICP) spectrometry has become a popular method for molybdenum analysis in plant and soil samples. The use of ICP mass spectrometry permits analysis with greater sensitivity and with lower limits of detection than

ICP atomic emission spectrometry. Recently, Phan-Thien et al. (2012) reported a limit of detection for molybdenum of 3.0×10^{-3} μg g^{-1} with ICP mass spectrometry for plant-based sample analyses. For extracts high in molybdenum, atomic absorption spectrometry or ICP atomic emission spectrometry have been used, but Sims (1996) indicates that owing to low detection limits, interferences from other elements, or the enhancement of molybdenum readings, the usefulness of these methods is limited.

13.9.2 DETERMINATION OF AVAILABLE MOLYBDENUM IN SOIL

As with most elements, determination of soil concentrations available for plant uptake is most useful for predicting crop responses. According to Gupta and Lipsett (1981), the first report on the available molybdenum in soils was given by Grigg (1953) wherein soils were extracted with acid oxalate buffered at pH 3.0. Other extraction methods have been used with varying degrees of success for the determination of available molybdenum in soils, including ammonium oxalate, hot water, anion-exchange resin, and ammonium bicarbonate diethylenetriamine-pentaacetic acid (AB-DTPA) (Sims, 1996). The most common method for the determination of molybdenum in soil extracts is the thiocyanate method as described previously. Although the ammonium oxalate procedure is the method most commonly used to determine available molybdenum in soils, the findings have not been consistent (Eivazi and Sims, 1997). Grigg (1960) decided that the method was unreliable for diagnosis of molybdenum deficiencies, because oxalate extracts a portion of iron-bound molybdenum that is unavailable to plants. Water extraction has been shown to be well correlated with available molybdenum in some studies (Barshad, 1951b) but also has failed to give good results (Gupta and MacKay, 1965). Difficulties are encountered with water extraction because the quantities extracted are very low (Gupta and Lipsett, 1981). Sims (1996) indicates that anion-exchange resins have been used with success to extract molybdenum but that the method has not been tested widely. According to Sims and Eivazi (1997), the AB-DTPA method was developed for the simultaneous soil extraction of macronutrients and micronutrients such as phosphorus, potassium, iron, manganese, copper, and zinc and has been extended to include molybdenum. Molybdenum extracted with AB-DTPA increases with increasing soil pH (Sims, 1996), and the method has been used most often for soils or sediments high in molybdenum, such as calcareous or polluted soils (Soltanpour and Schwab, 1977; Wang et al., 1994). Because the extractant can be used in conjunction with ICP atomic emission spectrometry, it offers the added potential for measuring molybdenum during routine analysis of multiple nutrients (Sims and Eivazi, 1997).

13.10 MOLYBDENUM FERTILIZERS

There are relatively few sources of molybdenum fertilizers for use by producers, fewer than for any of the other micronutrients or beneficial elements. The sources of molybdenum available as fertilizer vary greatly in solubility and content with their effectiveness to correct deficiency symptoms depending on methods of application, plant uptake and translocation, and physical and chemical soil factors impacting availability (Gupta, 1997b). The relative solubility of various molybdenum fertilizers is as follows: sodium molybdate > ammonium molybdate > molybdic acid > molybdenum trioxide > molybdenum sulfide (Lide, 1996). Sources, chemical formulas, and concentrations for common molybdenum fertilizers are ammonium molybdate, $(NH_4)_6Mo_7O_{24} \cdot 4H_2O$, 54%; molybdic acid, $H_2MoO_4 \cdot H_2O$, 53%; molybdenum frits, fritted glass, 1%–30%; molybdenum trioxide, MoO_3, 66%; molybdenum sulfide, MoS_2, 60%; and sodium molybdate, $Na_2MoO_4 \cdot 2H_2O$, 39% (Adriano, 1986; Gupta and Lipsett, 1981; Martens and Westermann, 1991). Finely ground molybdenum frits also have been used as a fertilizer, but there are issues with limited solubility (Mortvedt, 1997). Methods of molybdenum fertilization include soil application, foliar application, and seed treatment with various types of molybdenum sources.

13.10.1 METHODS OF MOLYBDENUM APPLICATION

13.10.1.1 Soil Application

Quantities of molybdenum fertilizers needed for proper plant growth are less than most other nutrients. Molybdenum fertilizers can be incorporated in the soil profile by banding or broadcast applications. A less common practice involves spraying soluble forms of molybdenum onto the soil surface before incorporation for a uniform application (Mortvedt, 1997). Amounts of 50–100 g Mo ha^{-1} are generally sufficient to provide correct amounts for most agronomic crops. However, soil application rates as high as 400 g Mo ha^{-1} may be needed for speciality vegetable crops such as cauliflower (Gupta and Lipsett, 1981). Because of the small amount of molybdenum needed for most situations, soluble forms of molybdenum are sprayed onto the granules of phosphorus fertilizers or complete (N–P–K) fertilizers or added to the acid used in the fertilizer manufacturing process (Mortvedt, 1997).

13.10.1.2 Foliar Application

Due to the high mobility in plant tissues and the small amounts required for proper plant nutrition, foliar applications of molybdenum can be very effective at providing required concentrations. The effectiveness of foliar applications depends on molybdenum penetration through the cuticle, uptake into cells, and transport via conductive tissues (Zoz et al., 2012). Sodium molybdate and ammonium molybdate are the two most common carriers for molybdenum for foliar fertilization because of their high solubility in water. Foliar applications of molybdenum are most effective when applied at early stages of plant development at up to 200 g Mo ha^{-1} (Škarpa et al., 2013; Srivastava and Gupta, 1996; Zoz et al., 2012). Sodium molybdate applications of 125 g Mo ha^{-1} increased yield of sunflower (*Helianthus annuus* L.) when applied at the beginning of vegetation (V – 4) (Škarpa et al., 2013). Yield in wheat (*Triticum aestivum* L.) reached a maximum at foliar rates of 35 g Mo ha^{-1} (Zoz et al., 2012). Wetting agents also may be required in the spray solution to ensure uniform coverage on the foliage of crops such as onion and cauliflower (Gupta and Lipset, 1981). Foliar applications of molybdenum are much more effective than soil applications in acid soils (Marschner, 1995) or under dry conditions (Gupta, 1979).

13.10.1.3 Seed Treatments

The most common method of applying molybdenum to crops is through the seed treatments of pelleting or coating (Mortvedt, 1997). This method allows for uniform applications in field conditions through seed coatings of molybdenum to provide the required concentrations when soil molybdenum levels are too low for adequate growth (Marschner, 1995). Recommended rates of molybdenum seed treatments are 7–100 g Mo ha^{-1} (Marschner, 1995; Srivastava and Gupta, 1996), and higher rates (>115 g Mo ha^{-1}) can cause toxic effects in plants such as cauliflower (Gupta, 1979; Mortvedt, 1997). Seed application of sodium molybdate increased pea yields when applied at 55 g Mo ha^{-1}; moreover, soaking red clover (*Trifolium pratense* L.) in a 1% solution of sodium molybdate was also successful at improving yields. Applying molybdenum in dusts directly to seed proved ineffective (Hagstrom and Berger, 1965b).

13.11 CONCLUSIONS

Molybdenum is an essential microelement for plant growth and development. It is required for specific plant enzymes to participate in oxidative and reduction reactions. Molybdate is the form taken up by plants, and soil availability is most influenced by soil type and soil pH conditions. Biosynthesis of Moco is the beginning steps in many of the essential enzymatic pathways that require molybdenum. Because of its involvement in nitrogen fixation, nitrate reduction, and nitrogen transport, molybdenum plays a crucial role in nitrogen metabolism in plants. Any conditions that result in decreased biosynthesis and activation of Moco will impact functions of molybdo-enzymes and, indirectly, plant nitrogen metabolism.

Molybdenum toxicity is rare in field conditions; however, the narrow sufficiency range of molybdenum in plant nutrition makes deficiency of the element commonplace. Plant uptake of molybdenum is enhanced by the presence of soluble phosphorus and decreased by the presence of available sulfur in soils. Drastic reductions in leaf size and irregularities in leaf blade formation (whiptail) are the most common visual symptoms of molybdenum deficiency. Due to the high mobility in plant conductive tissues, foliar applications or seed treatments with molybdenum are the most effective means to supply the element to crop plants. Many aspects of plant molybdenum metabolism and transport remain unknown; therefore, more research is needed to shed light on effective use of this minor element and how crop production can be made viable in parts of the world where soil molybdenum status limits crop growth.

REFERENCES

Adams, J.F. 1997. Yield responses to molybdenum by field and horticultural crops. In *Molybdenum in Agriculture*, ed. U.C. Gupta, pp. 182–201. New York: Cambridge University Press.

Adriano, D.C. 1986. *Trace Elements in the Terrestrial Environment*, pp. 329–361. New York: Springer-Verlag.

Agarwala, S.C. and E.J. Hewitt. 1954. Molybdenum as a plant nutrient. III. The interrelationships of molybdenum and nitrate supply in the growth and molybdenum content of cauliflower plants grown in sand culture. *J. Hortic. Sci.* 29:278–290.

Agarwala, S.C., C.P. Sharma, S. Farooq, and C. Chatterjee. 1978. Effects of molybdenum deficiency on the growth and metabolism of corn plants raised in sand culture. *Can. J. Bot.* 56:1095–1909.

Allen, R.M., J.T. Roll, P. Rangaraj, V.K. Shah, G.P. Roberts, and P.W. Ludden. 1999. Incorporation of molybdenum into the iron-molybdenum cofactor of nitrogenase. *J. Biol. Chem.* 274:15869–15874.

Anderson, A.J. 1942. Molybdenum deficiency on a South Australian ironstone soil. *J. Aust. Inst. Agric. Sci.* 8:73–75.

Arnon, D.I. 1937. Ammonium and nitrate nutrition of barley at different seasons in relation to hydrogen-ion concentration, manganese, copper, and oxygen supply. *Soil Sci.* 44:91–114.

Arnon, D.I. 1938. Microelements in culture-solution experiments with higher plants. *Am. J. Bot.* 25:322–325.

Arnon, D.I. and P.R. Stout. 1939a. Molybdenum as an essential element for higher plants. *Plant Physiol.* 14:599–602.

Arnon, D.I. and P.R. Stout. 1939b. The essentiality of certain elements in minute quantity for plants with special reference to copper. *Plant Physiol.* 14:371–375.

Barber, S.A. 1995. *Soil Nutrient Bioavailability: A Mechanistic Approach.* New York: Wiley.

Barshad, I. 1951a. Factors affecting the molybdenum content of pasture plants. II Effect of soluble phosphates, available nitrogen and soluble sulfates. *Soil Sci.* 71:387–398.

Barshad, I. 1951b. Factors affecting the molybdenum content of pasture plants. I. Nature of soil molybdenum, growth of plants, and soil pH. *Soil Sci.* 71:297–313.

Bergmann, W. 1992. *Nutritional Disorders of Plants. Visual and Analytical Diagnosis.* New York: Gustav Fischer.

Bhatt, J.G. and E. Appukuttan. 1982. Manganese and molybdenum contents of the cotton plant at different stages of growth. *Commun. Soil Sci. Plant Anal.* 13:463–471.

Bittner, F. 2014. Molybdenum metabolism in plants and crosstalk to iron. *Front. Plant Sci.* 5:1–6.

Bortels, H.1930. Molybdenum as a catalysis for biological nitrogen fixation (In German). *Arch. Mikrobiol.* 1:333–342.

Brown J.C. and R.B. Clark. 1974. Differential responses of 2 maize inbreds to molybdenum stress. *Soil Sci. Soc. Am. J.* 38:331–333.

Bryson, G.M., H.A. Mills, D.N. Sasseville, J.B. Jones, Jr., and A.V. Barker. 2014. *Plant Analysis Handbook III.* Athens, GA: Micro-Macro.

Burmester, C.H., J.F. Adams, and J.W. Odom. 1988. Response of soybean to lime and molybdenum on ultisols in northern Alabama. *Soil Sci. Soc. Am. J.* 52:1391–1394.

Bussler, W. 1970. The development of Mo deficiency symptoms in cauliflower (In German). *Z. Pflanzenernähr. Bodenk.* 125:36–50.

Chipman, E.W., D.C. MacKay, U.C. Gupta, and H.B. Cannon. 1970. Response of cauliflower cultivars to molybdenum deficiency. *Can. J. Plant Sci.* 50:163–167.

Cox, D.A. 1992. Foliar-applied molybdenum for preventing or correcting molybdenum deficiency of poinsettia. *HortScience* 8:89–895.

Curtis, P.R. and J. Grusovin. 1985. Determination of molybdenum in plant tissue by graphite furnace atomic absorption spectrophotometry (GFAAS). *Commun. Soil Sci. Plant Anal.* 16:1279–1291.

Davies, E.B. 1945. A case of molybdenum deficiency in New Zealand. *Nature* 156:392.

Davis, R.D., P.H.T. Beckett, and E. Wollan. 1978. Critical levels of twenty potentially toxic elements in young spring barley. *Plant Soil* 49:395–408.

DeKok, L.J. 1990. Sulfur metabolism in plants exposed to atmospheric sulfur. In *Sulfur Nutrition and Sulfur Assimilation in Higher Plants*, eds. H. Rennenberg, C. Brunold, L.J. Dekok, and I. Stulen, pp. 111–130. The Hague, the Netherlands: SPB Academic Publishing.

Dios, R.V. and T.V.I. Broyer. 1965. Deficiency symptoms and essentiality of molybdenum in crop hybrids. *Agrochimica* 9:273–284.

Eivazi, F. and J.L. Sims. 1997. Analytical techniques for molybdenum determination in plants and soils. In *Molybdenum in Agriculture*, ed. U.C. Gupta, pp. 92–110. New York: Cambridge University Press.

Fitzgerald, J.N. 1954. Molybdenum in oats. *N. Z. J. Agric.* 89:619.

Gammon, N., G.M. Volk, E.N. McCubbin, and A.H. Eddins. 1954. Soil factors affecting molybdenum uptake by cauliflower. *Soil Sci. Soc. Am. J.* 18:302–305.

Gasber, A., S. Klaumann, O. Trentmann et al. 2011. Identification of an *Arabidopsis* solute carrier critical for intercellular transport and inter-organ allocation of molybdate. *Plant Biol.* 13:710–718.

George, T., J.K. Ladha, R.J. Buresh, and D.P. Garrity. 1992. Managing native and legume-fixed nitrogen in lowland rice-based cropping systems. *Plant Soil* 141:69–91.

Giordano, P.M., H.V. Koontz, and J.W. Rubins. 1966. C^{14} distribution in photosynthate of tomato as influenced by substrate copper and molybdenum level and nitrogen source. *Plant Soil* 24:437–446.

Golov, V.I. and Y.N. Kazakhko. 1973. Uptake of molybdenum by soybean and its residual effect when applied to soils in far-east. *Soviet Soil Sci.* 5:551–558.

Grigg, J.L. 1953. Determination of the available molybdenum in soils. *N. Z. J. Sci. Technol.* 34:405–414.

Grigg, J.L. 1960. The distribution of molybdenum in the soils of New Zealand. I. Soils of the North Island. *N. Z. J. Agric. Res.* 3:69–86.

Gubler, W.D., R.G. Gorgan, and P.P. Osterli. 1982. Yellows of melons caused by molybdenum deficiency in acid soil. *Plant Dis.* 66:449–451.

Gupta, U.C. 1970. Molybdenum requirement of crops grown on a sandy clay loam soil in the greenhouse. *Soil Sci.* 110:280–282.

Gupta, U.C. 1971. Boron and molybdenum nutrition of wheat, barley, and oats grown in Prince Edward Island soils. *Can. J. Soil Sci.* 5:41–422.

Gupta, U.C. 1979. Effect of methods of application and residual effect of molybdenum on the molybdenum concentration and yield of forages on podzol soils. *Can. J. Soil Sci.* 59:183–189.

Gupta, U.C. 1997a. Deficient, sufficient, and toxic concentrations of molybdenum in crops. In *Molybdenum in Agriculture*, ed. U.C. Gupta, pp. 150–159. New York: Cambridge University Press.

Gupta, U.C. 1997b. Soil and plant factors affecting molybdenum uptake by plants. In *Molybdenum in Agriculture*, ed. U.C. Gupta, pp. 71–91. New York: Cambridge University Press.

Gupta, U.C. and D.C. MacKay. 1965. Extraction of water soluble copper and molybdenum in podzol soils. *Soil Sci. Soc. Am. Proc.* 29:323.

Gupta, U.C. and D.C. MacKay. 1968. Crop responses to applied molybdenum and copper on podzol soils. *Can. J. Soil Sci.* 48:235–242.

Gupta, U.C. and J. Lipsett. 1981. Molybdenum in soils, plants, and animals. *Adv. Agron.* 34:73–15.

Gupta, U.C. and P.V. LeBlanc. 1990. Effect of molybdenum application on plant molybdenum concentration and crop yields on sphagnum peat soils. *Can. J. Plant Sci.* 70:717–721.

Gupta, U.C., E.W. Chipman, and D.C. MacKay. 1978. Effects of molybdenum and lime on the yield and molybdenum concentration of crops grown on acid sphagnum peat soil. *Can. J. Plant Sci.* 58:983–992.

Hafner, H., B.J. Ndunguru, A. Bationo, and H. Marschner. 1992. Effect of nitrogen, phosphorus and molybdenum application on growth and symbiotic N_2-fixation of groundnut in an acid sandy soil in Niger. *Fert. Res.* 31:69–77.

Hagstrom, G.R. and H.C. Berger. 1965a. Molybdenum status of three Wisconsin soils and its effect on four legume crops. *Agron. J.* 55:399–401.

Hagstrom, G.R. and K.C. Berger. 1965b. Molybdenum deficiencies of Wisconsin soils. *Soil Sci.* 100:52–26.

Hale, K.L., S.P. McGrath, E. Lombi et al. 2001. Molybdenum sequestration in brassica species. A role for anthocyanins? *Plant Physiol.* 126:1391–1402.

Hewitt, E.J. 1956. Symptoms of molybdenum deficiency in plants. *Soil Sci.* 81:159–171.

Hewitt, E.J. 1963. Essential nutrient elements for plants. In *Plant Physiology. Vol. III. Inorganic Nutrition of Plants*, ed. F.C. Stewart, pp. 137–360. New York: Academic Press.

Hewitt, E.J. 1983. A perspective of mineral nutrition: Essential and functional metals in plants. In *Metals and Micronutrients: Uptake and Utilization by Plants*, eds. D.A. Robb and W.S. Pierpoint, pp. 227–326. New York: Academic Press.

Hewitt, E.J. and C.S. Gundry. 1970. The molybdenum requirement of plants in relation to nitrogen supply. *J. Hortic. Sci.* 45:351–358.

Hunter, J.G. and O. Vergnano. 1953. Trace-element toxicities in oat plants. *Ann. Appl. Bot.* 40:761–777.

Ishizuka, J. 1982. Characterization of molybdenum absorption and translocation in soybean plants. *Soil Sci. Plant Nutr.* 28:63–78.

Joham, H.E. 1953. Accumulation and distribution of molybdenum in the cotton plant. *Plant Physiol.* 28:275–280.

Johnson, C.M. and T.H. Arkley. 1954. Determination of molybdenum in plant tissue. *Anal. Chem.* 26:572–574.

Jones, L.H.P. 1957. Solubility of molybdenum in simplified systems and aqueous soil suspensions. *J. Soil Sci.* 8:313–327.

Jones, J.B., Jr. 1967. Interpretation of plant analysis for several agronomic crops. In *Soil Testing and Plant Analysis*, ed. G.W. Hardy, pp. 49–58. Madison, WI: Soil Science Society of America.

Jones, G.B. and G.B. Belling. 1967. The movement of copper, molybdenum, and selenium in soils as indicated by radioactive isotopes. *Aust. J. Agric. Res.* 18:733–740.

Kabata-Pendias, A. and H. Pendias. 2011. *Trace Elements in Soils and Plants*, 4th edn. Boca Raton, FL: CRC Press.

Kaiser, B.N., K.L. Gridley, J.N. Brady, T. Phillips, and S.D. Tyerman. 2005. The role of molybdenum in agricultural plant production. *Ann. Bot.* 96:745–754.

Kannan, S. and S. Ramani. 1978. Studies on molybdenum absorption and transport in bean and rice. *Plant Physiol.* 62:179–181.

Kevresan, S., N. Petrovic, M. Popovic, and J. Kandrac. 2001. Nitrogen and protein metabolism in a young pea plants as affected by different concentrations of nickel, cadmium, lead, and molybdenum. *J. Plant Nutr.* 24:1633–164.

Kisker, C., H. Schindelin, A. Pacheco et al. 1997. Molecular basis of sulfite oxidase deficiency from the structure of sulfite oxidase. *Cell* 91:973–983.

Koshiba, T., E. Saito, N. Ono, N. Yamamoto, and M. Sato. 1996. Purification and properties of flavin- and molybdenum-containing aldehyde oxidase from coleoptiles of maize. *Plant Physiol.* 110:781–789.

Lide, D.R. 1996. *Handbook for Chemistry and Physics*, 77th edn. Boca Raton, FL: CRC Press.

Lindsay, W.L. 1972. Inorganic phase equilibria of micronutrients in soil. In *Micronutrients in Agriculture*, eds. J.J. Mortvedt, F.R. Cox, L.M. Shuman, and R.M. Welch, pp. 41–57. Madison, WI: Soil Science Society of America.

Lindsay, W.L. 1979. *Chemical Equilibria in Soils*. New York: Wiley.

Lyon, C.B. and K.C. Beeson. 1948. Influence of toxic concentrations of micro-nutrient elements in the nutrient medium on vitamin content of turnip and tomatoes. *Bot. Gazette* 109:506–520.

MacLeod, J.A., U.C. Gupta, and B. Stanfield. 1997. Molybdenum and sulfur relationships in plants. In *Molybdenum in Agriculture*, ed. U.C. Gupta, pp. 229–244. New York: Cambridge University Press.

Marschner, H. 1995. *Mineral Nutrition of Higher Plants*, 2nd edn. New York: Academic Press.

Martens, D.C. and D.T. Westermann. 1991. Fertilizer applications for correcting micronutrient deficiencies. In *Micronutrients in Agriculture*, eds. J.J. Mortvedt, F.R. Cox, L.M. Shuman, and R.M. Welch, pp. 549–582. Madison, WI: Soil Science Society of America.

Mendel, R.R. 2013. The molybdenum cofactor. *J. Biol. Chem.* 288:13165–13172.

Mendel, R.R. and G. Schwarz. 1999. Molybdoenzymes and molybdenum cofactor in plants. *Crit. Rev. Plant Sci.* 18:33–69.

Mengel, K. and E.A. Kirkby. 1987. *Principles of Plant Nutrition*, 4th edn. Bern, Switzerland: International Potash Institute.

Merlot, S. and J. Giraudat. 1997. Genetic analysis of abscisic acid signal transduction. *Plant Physiol.* 14:751–757.

Mishra, S.N., P.K. Jaiwal, R.P. Singh, and H.S. Srivastiva. 1999. Rhizobium-legume association. In *Nitrogen Nutrition and Plant Growth*, eds. H.S. Srivastava and R.P. Singh, pp. 45–102. Boca Raton, FL: CRC Press.

Mitchell, K.J. 1945. Preliminary note on the use of ammonium molybdate to control whiptail in cauliflower and broccoli crops. *N. Z. J. Sci. Technol.* A27:287–293.

Mitchell, R.L. 1964. Trace elements in soils. In *Chemistry of the Soil*, 2nd edn., ed. F.E. Bear, pp. 320–368. New York: Reinhold.

Mortvedt, J.J. 1981. Nitrogen and molybdenum uptake and dry matter relationships of soybeans and forage legumes in response to applied molybdenum on acid soil. *J. Plant Nutr.* 3:245–256.

Mortvedt, J.J. 1997. Sources and methods for molybdenum fertilization in crops. In *Molybdenum in Agriculture*, ed. U.C. Gupta, pp. 171–181. New York: Cambridge University Press.

Mulder, E.G. 1948. Importance of molybdenum in the nitrogen metabolism of microorganisms and higher plants. *Plant Soil* 1:94–119.

Mulder, E.G. 1954. Molybdenum in relation to growth of higher plants and microorganisms. *Plant Soil* 5:368–415.

Neto, D.D., G.J.A. Dario, T.N. Martin, M. Rodrigues de Silva, P.S. Pavinato, and T.L. Habitzreiter. 2012. Mineral fertilization with cobalt and molybdenum in soybean. *Semina Ciências Agrárias* 33:2741–2751.

Normanly, J., J.P. Slovin, and J.D. Cohen. 1995. Rethinking auxin biosynthesis and metabolism. *Plant Physiol.* 107:323–329.

Norrish, K. 1975. The geochemistry and mineralogy of trace elements. In *Trace Elements in Soil-Plant-Animal Systems*, eds. D.J.D. Nicholas and A.R. Egan, pp. 55–81. New York: Academic Press.

Notton, B.A. 1983. Micronutrients and nitrate reductase. In *Metals and Micronutrients: Uptake and Utilization by Plants*, eds. D.A. Robb and W.S. Pierpoint, pp. 219–240. New York: Academic Press.

Nowak, K., N. Luniak, C. Witt et al. 2004. Peroxisomal localization of sulfite oxidase separates it from chloroplast-based sulfur assimilation. *Plant Cell Physiol.* 45:1889–1894.

Parker, M.B. and H.B. Harris. 1962. Soybean response to molybdenum and lime and the relationship between yield and chemical composition. *Agron. J.* 54:480–483.

Parker, M.B. and H.B. Harris. 1977. Yield and leaf nitrogen of nodulating and non-nodulating soybean as affected by nitrogen and molybdenum. *Agron. J.* 69:551–554.

Pasricha, N.S., V.K. Nayyar, N.S. Randhawa, and M.K. Sinha. 1977. Influence of sulphur fertilization on suppression of molybdenum uptake by berseem (*Trifolium alexandrinum*) and oats (*Avena sativa*) grown on a molybdenum-toxic soil. *Plant Soil* 4:245–250.

Peoples, M.B. and E.T. Craswell. 1992. Biological nitrogen fixation: Investments, expectations, and actual contributions to agriculture. *Plant Soil* 141:13–39.

Phan-Thien, K.-Y., G.C. Wright, and N.A. Lee. 2012. Inductively coupled plasma-mass spectrometry (ICP-MS) and–optical emission spectroscopy (ICP-OES) for determination of essential mineral in closed acid digestates of peanut (*Arachis hypogaea* L.). *Food Chem.* 134:453–460.

Phillips, R.L. and R.D. Meyer. 1993. Molybdenum concentration of alfalfa in Kern County, California—1950 versus 1985. *Commun. Soil Sci. Plant Anal.* 24:19–20.

Picco, S., M.V. Ponzzinibio, G. Mattioli et al. 2012. Physiological and genotoxic effects of molybdenum-induced copper deficiency in cattle. *Agrociencia* 46:107–117.

Plant, W. 1952. The control of molybdenum deficiency in lettuce under field conditions. Annual Report Long Ashton Research Station. Bristol, UK: University of Bristol.

Purvis, E.R. and N.K. Peterson. 1956. Methods of soil and plant analysis for molybdenum. *Soil Sci.* 81:223–228.

Randall, P.J. 1969. Changes in nitrate and nitrate reductase levels on restoration of molybdenum to molybdenum-deficient plants. *Aust. J. Agric. Res.* 20:635–642.

Rebafka, F.P., B.J. Ndunguru, and H. Marschner. 1993. Single superphosphate depresses molybdenum uptake and limits yield response to phosphorus in groundnut (*Arachis hypogaea* L.) grown on an acid sandy soil in Niger, West Africa. *Fert. Res.* 34:233–242.

Reddy, K.J., L.C. Munn, and L. Wang. 1997. Chemistry and mineralogy of molybdenum in soils. In *Molybdenum in Agriculture*, ed. U.C. Gupta, pp. 4–22. New York: Cambridge University Press.

Reed, S.C., C.C. Cleveland, and A.R. Townsend. 2013. Relationships among phosphorus, molybdenum, and free-living nitrogen fixation in tropical rain forest: Results from observational and experimental analyses. *Biogeochemistry* 114:135–147.

Reisenauer, H.M. 1965. Molybdenum. In *Methods of Soil Analysis. Part 2. Chemical and Microbiological Properties*, ed. C.A. Black, pp. 1050–1058. Madison, WI: American Society of Agronomy.

Reisenauer, H.M., A.A. Tabikh, and P.R. Stout. 1962. Molybdenum reactions with soils and hydrous oxides of iron, aluminium and titanium. *Soil Sci. Soc. Am. Proc.* 26:23–27.

Rhoades, E.R. and D. Nangju. 1979. Effects of pelleting cowpea and soybean seed with fertilizer dusts. *Expt. Agric.* 15:27–32.

Römheld, V. and H. Marschner. 1991. Function of micronutrients in plants. In *Micronutrients in Agriculture*, eds. J.J. Mortvedt, F.R. Cox, L.M. Shuman, and R.M. Welch, pp. 297–328. Madison, WI: Soil Science Society America.

Sharma, C.P. and C. Chatterjee. 1997. Molybdenum availability in alkaline soils. In *Molybdenum in Agriculture*, ed. U.C. Gupta, pp. 131–149. New York: Cambridge University Press.

Shinmachi, F., P. Buchner, J.L. Stroud et al. 2010. Influence of sulfur deficiency on the expression of specific sulfate transporters and the distribution of sulfur, selenium, and molybdenum in wheat. *Plant Physiol.* 153:327–336.

Sims, J.L. 1996. Molybdenum and cobalt. In *Methods of Soil Analysis Part 3. Chemical Methods*, eds. J.M. Bartels and J.M. Bingham, pp. 723–737. Madison, WI: Soil Science Society of America.

Sims, J.L. and F. Eivazi. 1997. Testing for molybdenum availability in soils. In *Molybdenum in Agriculture*, ed. U.C. Gupta, pp. 111–130. New York: Cambridge University Press.

Škarpa, P., E. Kunzova, and H. Zukalová. 2013. Foliar fertilization with molybdenum in sunflower (*Helianthus annuus* L.). *Plant Soil Environ.* 59:156–161.

Smith, K.S., L.S. Alisterieri, S.M. Smith, and R.C. Severson. 1997. Distribution and mobility of molybdenum in the terrestrial environment. In *Molybdenum in Agriculture*, ed. U.C. Gupta, pp. 23–46. New York: Cambridge University Press.

Smith, D.H., M.A. Wells, D.M. Porter, and F.R. Cox. 1993. Peanuts. In *Nutrient Deficiencies and Toxicities in Crop Plants*, ed. W.F. Bennet, pp. 105–110. St. Paul, MN: APS Press.

Soltanpour, P.N. and A.P. Schwab. 1977. A new soil test for simultaneous extraction of macro- and micronutrients in alkaline soils. *Commun. Soil Sci. Plant Anal.* 8:195–207.

Sparks, D.L. 1995. *Environmental Soil Chemistry*. New York: Academic Press.

Srivastava, P.C. 1997. Biochemical significance of molybdenum in crop plants. In *Molybdenum in Agriculture*, ed. U.C. Gupta, pp. 47–70. New York: Cambridge University Press.

Srivastava, P.C. and U.C. Gupta. 1996. *Trace Elements in Crop Production*. Boca Raton, FL: CRC Press.

Stark, J.M. and E.F. Redente. 1990. Copper fertilization to prevent molybdenosis on retorted oil shale disposal piles. *J. Environ. Qual.* 19:502–504.

Stout, P.R. and W.R. Meagher. 1948. Studies of the molybdenum nutrition of plants with radioactive molybdenum. *Science* 108:471–473.

Stout, P.R., W.R. Meagher, G.A. Pearson, and C.M. Johnson. 1951. Molybdenum nutrition of crop plants. I Influence of phosphate and sulfate on the absorption of molybdenum from soils and solution cultures. *Plant Soil* 3:51–87.

Szalay, A. and M. Szilagyi. 1968. Laboratory experiments on the retention of micronutrients by peat humic acids. *Plant Soil* 29:219–224.

Tomatsu, H., J. Takano, H. Takahashi et al. 2007. An *Arabidopsis thaliana* high-affinity molybdate transporter required for efficient uptake of molybdate from soil. *Proc. Natl. Acad. Sci. U. S. A.* 104(47):18807–18812.

Ulrich, A. and F.J. Hills. 1973. Plant analysis as an aid in fertilizing sugar crops. Part I. Sugarbeets. pp. 271–278. In *Soil Testing and Plant Analysis*, 2nd edn., eds. L.M. Walsh and J.D. Beaton. Madison, WI: Soil Science Society of America.

Vieira, R.F., E.J.B.B. Cardoso, C. Vieira, and S.T.A. Cassini. 1998b. Foliar application of molybdenum in common bean III. Effect on nodulation. *J. Plant Nutr.* 21:2153–2161.

Vieira, R.F., C. Vieira, E.J.B.N. Cardoso, and P.R. Mosquim. 1998a. Foliar application of molybdenum in common bean. II. Nitrogenase and nitrate reductase activities in a soil of low fertility. *J. Plant Nutr.* 21:2141–2151.

Vistoso, E.M., M. Alfaro, and M.L. Mora. 2012. Role of molybdenum on yield, quality, and photosynthetic efficiency of white clover as a result of the interaction with liming and different phosphorus rates in andisols. *Commun. Soil Sci. Plant Anal.* 43:2342–2357.

Voet, D. and J.G. Voet. 1995. *Biochemistry*, 2nd edn. New York: Wiley.

Vos, R.D. 1993. Corn. In *Nutrient Deficiencies and Toxicities in Plants*, ed. W.F. Bennett, pp. 1–14. St. Paul, MN: APS Press.

Walker, R.B. 1948. Molybdenum deficiency in serpentine barren soils. *Science* 108:473–475.

Walker-Simmons, M., D.A. Kudrna, and R.L. Warner. 1989. Reduced accumulation of ABA during water stress in a molybdenum cofactor mutant of barley. *Plant Physiol.* 90:728–733.

Wang, L., K.J. Reddy, and L.C. Munn. 1994. Comparison of ammonium bicarbonate-DTPA, ammonium carbonate and ammonium oxalate to assess the availability of molybdenum in mine spoils and soils. *Commun. Soil Sci. Plant Anal.* 25:523–536.

Wang, Z.Y., Y.L. Tang, and F.S. Zhang. 1999. Effect of molybdenum on growth and nitrate reductase activity of winter wheat seedlings as influenced by temperature and nitrogen treatments. *J. Plant Nutr.* 22:387–395.

Warner, R.L. and A. Kleinhofs. 1992. Genetics and molecular biology of nitrate metabolism in higher plants. *Physiol. Plant.* 85:245–252.

Weir, R.G. and G.C. Creswell. 1993. *Plant Nutrient Disorders 3—Vegetable Crops*. Melbourne, Australia: Inkata Press.

Welch, R.M., W.H. Allaway, W.A. House, and J. Kubota. 1991. Geographic distribution of trace element problems. In *Micronutrients in Agriculture*, eds. J.J Mortvedt, F.R. Cox, L.M. Shuman, and R.M. Welch, pp. 31–57. Madison, WI: Soil Science Society of America.

Williams, C.M.J., N.A. Maier, and L. Bartlett. 2004. Effects of molybdenum foliar sprays on yield, berry size, seed formation, and petiolar nutrient composition of 'Merlot' grapevines. *J. Plant Nutr.* 27:1891–1916.

Wilson, R.D. 1949. Molybdenum in relation to the scald disease of beans. *Aust. J. Sci.* 11:209–211.

Xie, R.J. and A.F. MacKenzie. 1991. Molybdate sorption-desorption in soils treated with phosphate. *Geoderma* 48:321:333.

Younge, O.R. and M. Takahashi. 1953. Response of alfalfa to molybdenum in Hawaii. *Agron. J.* 9:420–428.

Zimmer, W. and R. Mendel. 1999. Molybdenum metabolism in plants. *Plant Biol.* 1:160–168.

Zoz, T., F. Steiner, J.V.P. Testa et al. 2012. Foliar fertilization with molybdenum in wheat. *Semina Ciências Agrárias* 33:633–638.

14 Nickel

Bruce W. Wood

CONTENTS

14.1 HISTORICAL BACKGROUND

14.1.1 INTRODUCTION

Life is an emergent property of our universe and the journeywork of stars from which the essential chemical elements for life are produced (Russell, 2007). The most nuclear stable of these elements is nickel (Ni), with ^{62}Ni having the highest binding energy per nucleon of any known element and therefore is the end product of many nuclear reactions associated with stars (Fewell, 1995). Nickel, represented by five stable isotopes, is most likely indispensable for life as known on Earth and appears to be the only element enabling certain life requisite tasks at maximum entropy, thus enabling and facilitating the organic chemistry necessary for the emergence and continuation of life (de Silva and Williams, 2006). It is divalent Ni (Ni^{2+}) that is so important to life's chemistry.

Nickel is a member of the 'iron (Fe)–cobalt (Co)–nickel (Ni)' family of transition metals; thus, Ni's physiochemistry is similar to that of Fe and Co. This physiochemical similarity, and relative rarity of Co, led to a long and evolving Ni–Fe-related "duet" of biological chemistry. As the biosphere's

'Fe–Ni ratio' changed over evolutionary time, change has also occurred in the relative dependency of evolving organisms on Ni and Fe for key life processes (Konhauser et al., 2009). This relative scarcity of Co ensures that this biologically critical dance is a duet rather than a trio.

Nickel played a far greater role in the biological chemistry of life during the Earth's early history, when reducing gases (e.g., CH_4, NH_3, H_2, CO, and H_2S) were prominent in the environmental niches where life likely emerged. The relative importance of Ni in life processes began to diminish at least by the time of 'The Great Oxidation Event', which occurred about 2.4 billion years ago and produced a dominating oxic atmosphere of O_2, N_2, and CO_2. This oxygenated biosphere greatly increased the availability of Fe for life processes at a time concomitant with an explosion in plant diversity. Life adapted to shifting metal availability to reduce Ni-dependent processes, such as oxidation-reduction transformations in enzymes and for structural integrity of peptides and proteins (Konhauser et al., 2009). With the relatively high Fe–Ni ratio present today in surface soils and waters, Ni now plays a relatively minor role in plant metabolism and physiology, with current roles likely being "metabolic fossils" conserved over evolutionary time dating from the initial emergence of life on Earth. Most ancient key Ni-associated catalytic roles are now replaced by Fe and other transition metals (i.e., Zn for enzymes requiring an electrophile, and Mo, Cu, and Mn as brokers of oxidation-reduction transformations [de Silva and Williams, 2006; Merchant, 2010]) as they became increasingly available for organic chemistry occurring within Earth's oxic biospheres. Life-supporting enzymes using Ni are now, in a sense, metal relics from the Archaean Eon (Konhauser et al., 2009; Waldron et al., 2009); yet, evolutionary processes appear to have conserved the essentiality of Ni in lower and higher plants, as well as for animals (Hertel et al., 1991; Nieboer et al., 1988; Nielsen, 1984a, 1987; Phipps et al., 2002).

Agricultural practitioners have heretofore largely ignored the need for Ni nutrition management of crops, because Ni is required at only low microconcentration or high nanoconcentration and is relatively abundant in most soils (Holmgren et al., 1993). The existence of Ni deficiency in crops is increasingly apparent as agricultural practitioners become more familiar with crop species prone to Ni deficiency, as well as with recognition of deficiency symptoms and the plant growth stage and soil conditions in which they are most likely to appear. Because Ni is potentially a trace contaminant of essentially all agricultural formulations of micronutrient fertilizers (e.g., Mn, Fe, Cu, Zn, Co, Mo), it is probable that in certain cases nutrient deficiencies heretofore attributed to these micronutrients in specific cropping situations are at least partially due to insufficient Ni at key stages of crop growth. Such situations are identifiable when plant maladies attributed to one of these other trace elements are not corrected as easily by contemporary fertilizers as they were in the past, thus reflecting either an exclusion or reduction of Ni as an otherwise inadvertent contaminating nutrient element as fertilizer processing has become more refined. As a phloem-mobile plant micronutrient, Ni transports to edible plant products, such as young foliage and seed, where upon ingestion, it contributes to animal and human health (Nielsen, 1998; Welch, 2008). The mobility of Ni in the food chain is such that it possesses a "moderate bioaccumulation index," being lower than that of Cu, Zn, Co, and B, but higher than that of Mn and Fe (Pais and Jones, 1997).

14.1.2 DETERMINATION OF NICKEL ESSENTIALITY

A quarter of a millennium passed from the initial discovery of elemental Ni (Saint Nicholas's copper, or "Devil's copper") in 1751 until discovery in 1987 that it functions as an "essential" nutrient element in higher plants (Brown et al., 1987a) based on its role in urease (Polacco et al., 2013). There is also mounting evidence that Ni likely possesses other essential roles yet to be demonstrated. It was apparent as far back as the 1940s–1950s that foliar Ni application was observed to improve the nutrient element health of certain crops, such as wheat (*Triticum aestivum* L.), potatoes (*Solanum tuberosum* L.), and broad beans (*Vicia faba* L.) (Roach and Barclay, 1946), and trigger certain seemingly essential positive effects (Eskew et al., 1983, 1984; Welch 1981). Nickel functions as a *trace element* (≥ 1 mg kg^{-1} and ≤ 1000 mg kg^{-1} DW) in many species and as an *ultratrace element* (<1 mg kg^{-1} DW) in others, depending on the environment in which the organism evolved and as mitigated by natural-selection processes (Nielsen, 2000). Dixon et al. (1975) discovered that urease (EC 3.5.1.5), an enzyme thought to be both present and essential in all lower and higher plants, possesses Ni as its cofactor. Discovery that Ni is required for normal growth of oats (*Avena sativa* L.) and wheat (*Triticum sp.*) and that Ni is essential for life cycle completion of barley (*Hordeum vulgare* L) is additional evidence of essentiality (Brown et al., 1987a,b). Once an element is proven essential for one higher plant species, it is assumed to be essential for others, because of the expense and tedium involved in such research, especially when it is a micronutrient (Bohn et al., 2001). Additional evidence for essentiality was discovery by Wood et al. (2004b), in which severely Ni-deficient pecan [*Carya illinoinensis* K. Koch] trees, a long-lived woody dicot, were found to be unable to complete their life cycle. Nickel deficiency severe enough to alter greatly growth and morphology of vegetative organs occurs in several woody perennial angiosperm genera (e.g., *Carya*, *Juglans*, *Betula*, *Pyracantha*, *Rosa*, *Coffea*, and *Prunus*). Additional evidence of essentiality is that excessive foliar concentration of a Ni inhibitory cation, such as Fe, Zn, or Cu, triggers sufficient Ni deficiency to disrupt greatly growth and development and to cause death (Wood, 2010, 2013). Nickel appears to satisfy the criteria identified by Epstein and Bloom (2005) and by Brown (2007) for an essential element. Additionally, the great diversity of adverse effects of Ni deficiency on plant metabolism, physiology, growth, and development is strong evidence that plants depend on Ni for more than a single urease-related life-critical function.

14.2 UPTAKE OF NICKEL BY PLANTS

14.2.1 FOLIAR UPTAKE

Nickel uptake can be via foliage or roots. The small amount of Ni absorbed by leaves from foliar sprays easily can correct deficiency symptoms of growing organs (Wood et al., 2004b). Most plant species probably absorb a small amount of Ni through foliar surfaces because of deposition of atmospheric Ni sources of anthropogenic origin via rain, fog, or dew from Ni-containing dust and particulate matter that has deposited on plant surfaces. Nickel binds tightly to the leaf cuticle, so it is difficult to remove from plant surfaces by washing (Kozlav et al., 2000; Nieminen et al., 2004), thus potentially contaminating and biasing the interpretation of plant Ni nutritional physiology status as assessed by standard tissue analysis. As with other micronutrients, foliar Ni absorption is greatest when at low concentration; foliage is fully turgid (usually in early morning); air is windless, humid, and cool; and a surfactant is used (Jones, 2012).

14.2.2 ROOT UPTAKE

Plant roots encounter Ni via mass flow or diffusion of Ni^{2+} released into the soil solution from minerals, ions held on exchange sites of soil colloids, or from organic matter. Soil water content, pH, temperature, physiochemical characteristics of the colloidal substances, solubility characteristics of the solid-phase Ni-containing components, and biological activity influence the Ni-uptake process. The rhizosphere environment, as influenced by other cations and protons, root exudates (such as siderophores and organic acids), bacteria, and arbuscular mycorrhizal fungi affect Ni uptake (He et al., 2012; Verbruggen et al., 2009). Plant roots easily take up Ni^{2+} directly; however, there is evidence that uptake is less when bound with large chelating agents (Cataldo et al., 1978a). Uptake is also via ion pairs $[Ni(SO_4)_2 \cdot (H_2O)_5]$ and Ni complexes (e.g., $NiOH^+$), with Ni in the soil solution being increased due to its propensity to complex with organic ligands and desorb from soil particles (He et al., 2012). As noted by He et al. (2012), several bacterial species can lead to either an increase or decrease in Ni uptake by different plant species. High concentrations of divalent cations in the soil solution can inhibit uptake of Ni ions. The most effective inhibitors are Fe^{2+}, Cu^{2+}, Co^{2+}, and Zn^{2+}.

Our understanding of the pathway of Ni uptake by roots is poor. Uptake may involve passive and active processes (Kochian, 1991), and there is likely to be more than one active pathway. Because relatively small amounts of Ni are potentially toxic to most plants and the hydrated Ni ion is very similar to that of Mg, there is strong likelihood that Ni uptake by roots is largely via an energized system (de Silva and Williams, 2006). Nickel enters the symplastic system by absorption through a selectively permeable plasma membrane of the root endodermal cells regulating xylem vessel access via ion channels/carriers. Studies with soybean (*Glycine max* Merr.) (Cataldo et al., 1978a) found a high affinity uptake phase operating between 0.075 and 0.250 μM ($K_m \approx 0.5$ μM), which appears to approximate Ni concentrations typical of most agricultural soils (Marschner, 1995). A low affinity Ni uptake mechanism appears to exist in oats (Aschmann and Zasoski, 1987). Evidence indicates that there is noncompetitive inhibition of Ni uptake by Ca^{2+}, Mg^{2+}, Cd^{2+}, and Pb^{2+} and competitive inhibition by Zn^{2+}, Cu^{2+}, Co^{2+} (Korner et al., 1986, 1987), and probably Fe^{2+}. Nickel is competitively taken up by the Fe transporter, IRT1 (Cataldo et al., 1978a; Nishida et al., 2011, 2012; Pandey and Sharma, 2002; Yang et al., 1996), which possesses relatively broad divalent cation specificity (Vert et al., 2009). Based on research in arabidopsis (*Arabidopsis thaliana* Heynh.) (Nishida et al., 2011), it is likely that the same Fe transporter is very much involved in Ni uptake in other species. Nickel also appears to be absorbed via one or more of the ZIP-family transporters (Cataldo et al., 1978a; Verbruggen et al., 2009) that enable Zn and Cu uptake (Grotz et al., 1998; Ishimaru et al., 2005; Wintz et al., 2003). While the apoplast contributes to the Ni content of roots via adsorption to cellular surfaces outside of the Casparian barrier, the contribution of this Ni to that proportion reaching aerial organs remains to be determined. There is evidence that the apoplast may contribute substantially in some way to enable Ni to bypass the root symplast xylem loading of Ni to enable direct translocation to aerial organs (Redjala et al., 2012).

14.2.3 Nickel Translocation

Nickel readily transports in xylem (Cataldo et al., 1978b, 1988; Tiffin, 1971) and phloem (Cataldo et al., 1978b) as a stable anionic organic complex. Similar to other cations, Ni complexes with organic carrier molecules via covalent bonding. The binding of Ni with either positively or negatively charged organic chemicals appears to influence Ni mobility within the vascular system. The most prominent organic molecules complexing with vascular system Ni are organic acids and amino acids. Nickel is especially likely to be bound to citrate and histidine, depending on sap pH (Grendas et al., 1999) and plant species. In certain species, xylem loading of Ni is linked to root exportation of histidine (Kerkeb and Kramer, 2003) and nicotianamine (Curie et al., 2009), and probably citrate and malate. Translocation of Ni in the xylem sap of *Alyssum* species (i.e., Ni hyperaccumulators) is as a free ion and as a carboxylic acid chelant (Alves et al., 2011). However, in Gramineae (Poaceae, the grass family), nicotianamine and the yellow stripe-like (YSL) transporter oligopeptides are important in xylem and phloem translocation of Ni (Curie et al., 2009) and also may operate in other vascular plants in regard to loading into the xylem in roots as well as in apoplastic transport. As with Mn, Fe, Co, Zn, and Cu, Ni binds to fixed negatively charged ions lining the lumen wall (e.g., carboxyl groups, Ni–pectic acid complexes, Ni-histidine, and $NiSO_4$) during transport within the newly laid down xylem vessels (Punshon et al., 2005), which possess a high cation exchange capacity (CEC). The relevance of these Ni pools to physiological processes remains unknown, but binding to lumen walls may explain partially why newly formed expanding tissues and organs are most prone to exhibit Ni deficiency.

There is considerable retranslocation of Ni from vegetative organs to developing fruit (Zeller and Feller, 1999), especially the cotyledons (Cataldo et al., 1978b). For example, senescing soybean plants mobilized 70% of their Ni to seeds (Cataldo et al., 1978b), with YSL transporters likely facilitating remobilization from senescing foliage (Curie et al., 2009). In wheat, the phloem mobility of Ni to developing seeds was equivalent to Zn and greater than that of Co or Cd, with most of the Ni supplied to seed being via the phloem rather than the xylem (Riesen and Feller, 2005). Nickel is sufficiently phloem mobile in white lupine (*Lupinus albus* L.) to be translocated substantially from expanding leaves to younger leaves (Page et al., 2006). At the level of the leaf cell membrane and vacuole, it appears that there might be at least three families of Ni transporters (i.e., ZIP [ZRT/IRT-like protein], NRAMP [natural resistance–associated macrophage protein], and YSL) operating to transport and regulate Ni within cells (Tejada-Jimenex et al., 2009).

14.3 PHYSIOLOGICAL RESPONSES OF PLANTS TO SUPPLY OF NICKEL

14.3.1 OVERVIEW

The scope of the biological chemistry of Ni has been disappearing gradually over time. This action is partially because (1) the oxidation-reduction chemistry that Ni promotes is vulnerable to the preponderance of O_2 found in our present day oxic atmosphere and (2) its ability to compete with Mg^{2+} in metabolic and physiological processes (de Silva and Williams, 2006). Visual symptoms of Ni deficiency are therefore relatively rare in agriculture due to relatively low Ni requirements and relatively high Ni content of most soils. Visual deficiency and *hidden hunger* symptoms are most likely to occur when plants are grown in high organic matter soilless artificial potting mixes, in solution culture, or in field setting involving excessive application of fertilizers containing transition metals (e.g., Fe, Zn, and Cu), in high pH soils, or with roots damaged by nematodes. The likelihood of deficiency is greatest during periods of rapid vegetative growth, such as after bud break in early spring or under high nitrate nutrition. Plants are almost always Ni sufficient within a few weeks after cessation of growth; thus, tissue analysis results often do not reflect tissue concentrations during the time when Ni was most needed by the crop, thereby being potentially misleading regarding the actual Ni status of the crop.

While approximately 70% of all enzymes are metalloenzymes, requiring a metal ion for activity, at most only a few require Ni, and only one has been identified for higher plants. There are many unknowns regarding the physiological and metabolic roles of Ni; however, the metal is likely essential to all organisms in all three phylogenetic domains of life (i.e., Archaea, bacteria, and Eukarya). In the simplest of organisms, the Archaea and bacteria (which make up the majority of the biomass of the Earth), Ni is critical in enabling carbon assimilation and energy extraction from gases and minerals (e.g., CO, CO_2, S, SO_4^{2-}, CH_3COO^-) (Ragsdale, 2004), thus highlighting the potential importance of Ni in systems where these organisms are used for biofuels or waste treatment. While the relative prominence of Ni in life has diminished as life evolved to higher forms and complexity, the plants of Eukarya still rely substantially on Ni, with known uses being linked to gas-associated metabolism; however, in the case of flowering plants (i.e., Magnoliophyta), the focus of this chapter, relatively little is known about the metabolic and physiological roles of Ni.

14.3.1.1 Metabolism

Nickel appears to play important roles in primary metabolism by altering the nitrogen (N) and carbon (C) metabolite pools in Ni-deprived annuals (Brown et al., 1987a,b; Polacco et al., 2013) and perennials (Bai et al., 2006, 2007). Nickel is also likely to play important roles in secondary metabolism, where metalloenzymes tend to have lower metal specificity than do enzymes of primary metabolism (Milo and Last, 2012; Polacco et al., 2013), thus possibly enabling Ni to replace other transition metals as cofactors for enzymes and for other metals to replace Ni (Polacco et al., 2013).

14.3.1.1.1 Urease

The essential function of Ni presently links primarily to its role in the vegetative and embryonic forms of urease, and a cytoplasmic ureolytic enzyme is ubiquitous in plant tissue (Polacco and Holland, 1993; Polacco et al., 2013); however, it is possible that Ni-dependent urease is not required for life cycle completion in all plant species (Kloth et al., 1987; Polacco, 1981).

Urease (EC 3.5.1.5; urea amidohydrolase), the first discovered metalloenzyme, requires Ni (two atoms per subunit) for activity, and other metal ions appear unable to substitute for Ni (Dixon et al., 1980b; Gerendás et al., 1998). Although primary function of urease appears to be catabolism of the linear amide urea to ammonia and carbonic acid (bicarbonate) at physiological pH, it also potentially possesses hydrolytic activity involving several other substrates (e.g., acetamide, N,N'-dihydroxyurea, N-hydroxyurea, formamide, N-methylurea, semicarbazide, phosphoric acid amides) (Dixon et al., 1980a; Zerner, 1991), although at activities of 0.001–0.01 of that for urea (Dixon et al., 1980a; Zerner, 1991). Plants' recycling of N appears to be predominately through urease (Polacco and Holland, 1993), with low Ni also reducing several amino acids and amides (Gerendas and Sattelmacher, 1997). This role in urea catabolism may be of greater practical importance to agriculture than commonly realized in that a dedicated urea transporter appears to be involved in root urea uptake and urea can be a primary source of N taken up from the soil solution when urea fertilizers are used (Witte, 2011). Because world agriculture now utilizes urea as a major N fertilizer source to crops, the ability of crops to metabolize urea rapidly is critical to crop productivity, and Ni is especially critical to ureolytic crops. These actions imply that proper Ni nutritional physiology may be more important today than in former decades of the recent agricultural era.

Nickel also interacts with RNase-A (a nonmetallonuclease) to enable urease activity. The 3D structure of this RNase–Ni complex shows Ni ions binding at amino acid 'His 105' of the RNase-A molecule (Balakrishnan, 1997), and when exposed to Ni, RNase-A catabolizes urea (Bai and Wood, 2013) but loses ribonuclease activity (Maheshwari and Dubey, 2008). This action means that exposure of RNase-A to Ni enables transformation to a different type of hydrolase, whose biological relevance is unknown. RNase-A is involved intimately in the N-cycling of plants, and two of the many N-cycling hydrolases are the ureases or ribonucleases (RNase), with the activity of both being influenced by Ni. There is also the possibility that Ni binding to nucleotides, especially adenine—and to the nucleotide ATP—is of physiological significance, especially since Ni can influence gene activity (Nielsen, 1984b).

14.3.1.1.2 Hydrogenase

Nickel is involved also in the activity of the hydrogenase within the nodules of certain N-fixing *Rhizobium*-legume systems (Cammack, 1995) where the enzyme enables oxidation of H^+ to provide the ATP needed to reduce N to ammonia (Brito et al., 2000). Strains of certain genera of Rhizobiaceae (e.g., *Rhizobium* and *Bradyrhizobium*) possess a hydrogenase system enabling either partial or full recycling of H_2 evolved by nitrogenase, thus substantially increasing the energy efficiency of N fixation (Palacios et al., 2005). The evolution of H_2 as a consequence of nitrogenase activity is a major inefficiency of the rhizobium-legume symbioses, so strains using Ni in hydrogenase exhibit much greater energy efficiency, with the host legume exhibiting greater growth and productivity (Evans et al., 1988). This result means that the physiological availability of Ni to certain rhizobium-legume symbioses is potentially a limiting factor for hydrogenase activity (Brito et al., 1994) and that there is potential for improving the productivity of legumes by bioengineering this Ni-activated hydrogenase into the various bacteroids in agricultural systems.

14.3.1.1.3 Other Enzymes

Many proteins either bind or contain Ni (Thomas, 1982); however, studies about Ni nutrition of flowering plants and its physiological significance, especially to woody perennials, are limited. There are several Ni-activated enzyme systems (NiFe-hydrogenase, carbon monoxide dehydrogenase, acetyl-CoA decarboxylase synthase, methyl-coenzyme M reductase, superoxide dismutase, Ni-dependent glyoxylase, acireductone dioxygenase, and methyleneurease) in Archaea, bacteria, and the more primitive species of Eukarya (Mulrooney and Hausinger, 2003; Walsh and Orme-Johnson, 1987). Nickel can replace Zn or Fe, and probably Cu and Co, in many metalloenzymes of primitive plants and still enable a certain level of activity (Mulrooney and Hausinger, 2003). This apparent ability to substitute, in part, for other cations raises the possibility that Ni contributes to the ability of higher plants to adapt to different environments and to the ability of species to radiate into vacant niches. Because Ni is among the "best of the best" metals for hydrogenations and organic syntheses (Lancaster, 1988), it is highly likely that there are key unidentified roles for Ni in life-essential catalytic activities. Some of these undiscovered roles are likely to involve oxidation–reduction reactions associated with secondary metabolism. Because plants must survive in a changing environment, often over many years, it is likely that there are many yet-to-be-discovered enzyme forms associated with specialized metabolism that produce compounds serving a variety of critical physiological and ecological roles (Milo and Last, 2012), and there is circumstantial evidence that Ni is associated with one or more of these processes. For example, there is evidence in rice (*Oryza sativa*) that Ni increases plasma membrane–associated Mg^{2+}-ATPase activity (Ros et al., 1990) because hydrated Ni ions are essentially the same size as the Mg^{2+} ion.

Nickel-deficient plants often exhibit such a great disruption in growth, morphology, pathogen resistance, and metabolism that the disruption is likely unexplainable based solely on impairment of urease activity, especially when such plants are not receiving urea as a fertilizer source. For example, improving Ni nutrition in species as diverse as pecan (Wood et al., 2012) and daylily (Reilly et al., 2005) helps to protect these crops from certain fungal pests. The following are a few examples of Ni deficiency–associated disruptions.

14.3.1.1.4   Nitrogen Metabolism

Ureide-transporting species, such as pecan and many temperate legumes, appear to possess a higher Ni requirement than do amide-transporting species (Wood et al., 2006b). This phenomenon raises the possibility that certain ureide-transporting species possess metalloenzymes, or enzyme isoforms, that function best with a Ni cofactor. Likely candidates are certain enzymes affecting ureide catabolism (i.e., allantoate amidohydrolase, ureidoglycolate amidinohydrolase, and urodoglycolate amidohydrolase) and the urea cycle (i.e., argininosuccinate synthase and asparagine synthase), as Ni-deficient pecan trees exhibit either direct or indirect impairment of the activity of these enzymes (Bai et al., 2006). Nickel deficiency also can reduce endogenous concentration of several amino acids. For example, in pecan trees, glycine, valine, isoleucine, tyrosine, tryptophan, arginine, and total free amino acids decline if there is Ni deficiency. Some examples of ureide-transporting genera are *Carya*, *Juglans*, *Diospyros*, *Vitis*, and *Annona*. There is increasing evidence that Ni plays an important role in N use efficiency under various N regimes by recycling urea-N generated from arginase action on arginine (Polacco et al., 2013). This occurrence is suggestive that Ni can substantially influence yield and quality of crops where protein content is important. For example, in soybean, the timely application of Ni increases the N concentration of plant organs (Kutman et al., 2013).

Based partially on the fact that severely Ni-deficient pecan (also coffee, river birch, and several woody ornamentals) trees exhibit markedly reduced lignification and auxin production, both being dependent on certain aromatic amino acids as precursors, it appears that Ni nutritional status affects functionality of the shikimate pathway—in at least a few species of woody plants, if not for higher plants in general. This life-essential pathway for plants processes ≥30% of an organism's photosynthetically fixed carbon and also produces key aromatic amino acids (L-tryptophan, L-phenylalanine, and L-tyrosine), with the amino acids serving as precursors of numerous natural products, such as pigments, alkaloids, hormones, and cell wall components (Maeda and Dudareva, 2012). Depending on the organism and the specific shikimate pathway enzyme, or isoform, Mn^{2+}, Mg^{2+}, Cu^{2+}, and Co^{2+} potentially act as cofactors, and it is likely that Ni^{2+} does also; however, Ni might alternatively affect the biochemical feedstocks supplying the shikimate pathway.

14.3.1.1.5   Respiratory Carbon Flow

There is evidence that low Ni bioavailability can alter primary metabolism via one or more glycolysis pathway and tricarboxylic acid Cycle (TCA cycle) products and, therefore, generation of chemical energy. Lactate and several amino acids (e.g., serine, glycine, leucine, valine, and tryptophan) are downstream metabolites potentially derived from intermediates of glycolysis, especially if the normal flow of carbon from glucose to citrate is blocked or hindered due to diminished enzyme activity. Such a Ni-associated disruption leads to a deficiency of TCA organic acids, such as citrate (Bai et al., 2006) and malate, as reported by Brown et al. (1990) and Bai et al. (2006). Thus, Ni deficiency somehow diminishes conversion of pyruvate to citrate via acetyl-CoA (Bai et al., 2006). In flowering plants, such a disruption links directly to the activity of one of the four enzymes of the pyruvate dehydrogenase complex [i.e., pyruvate dehydrogenase (EC 1.2.1.51), lipoamide acetyltransferase (EC 2.3.1.12), lipoamide dehydrogenase (EC 1.8.1.4), and citrate synthase (EC 4.1.3.7)] (Hill, 1998) and indirectly to a possible acetyl-CoA synthase. In prokaryotes, acetyl-CoA synthase possesses three N-containing components (adenine, pantothenic acid, and β-mercaptoethylamine) that could be affected by effects of Ni on N metabolism; however, because of the reliance of the previously mentioned enzymes of the pyruvate dehydrogenase complex on available N, any one or combination of these potentially is affected by Ni deficiency. It is noteworthy that impairment of the TCA cycle reduces energy (i.e., NADH) available for the electron transfer (oxidative phosphorylation) stage of respiration and may, in turn, contribute to the low growth vigor exhibited by Ni-deficient plants.

14.3.1.1.6 Other Functions

The wide range of morphological (e.g., blunted and thickened foliage) and growth symptoms (e.g., dwarfing of canopy organs, bonsai appearance, shorter internodes, delayed bud break, loss of apical dominance, brittle shoots, weaker shell of seed) exhibited by Ni-deficient woody perennials indicates that Ni affects primary plant metabolism by inhibiting biosynthesis of one or more key metabolic intermediates feeding into several metabolic pathways. Acetyl CoA, perhaps the most central intermediate in cellular metabolism of all life and also a biochemical fossil from the Archaean Era, feeds into many metabolic pathways (e.g., glycolysis, fatty acid synthesis, amino acids, flavonoids, sterols, isoprenoid derivatives, and TCA cycle) and does not appear to cross membranes and, as noted earlier, is affiliated with carbon flow from pyruvate to citrate. Two Ni atoms at the active site of the form of acetyl CoA synthase in acetogenic, methanogenic, and sulfate-reducing bacteria enable the use and integration of single-carbon units as a feedstock (Ito et al., 2009). To date, an identified role for Ni in a higher plant acetyl CoA synthase isoform has not occurred, and the origin of many sources of acetyl CoA in higher plant metabolism remains enigmatic, although it is likely that the ancient role of Ni in acetyl CoA metabolism has been conserved in flowering plants, in one form or another, as a life-essential relic from life's nursery.

Because Ni deficiency also can increase greatly the concentration of non-TCA cycle organic acids (e.g., lactic acid [3.2-fold] and oxalic acid [2.4-fold]) (Bai et al., 2006), insufficient Ni is a potential threat to cellular metabolism and physiology via the disruption of cytoplasmic pH and enzymatic activity. Instead of a toxic accumulation of urea due to impaired urease activity (Walker et al., 1985), it is possible that it is the accumulation of these acids, rather than urea, that is the primary cause for the poisoning (or at least a contributing factor in addition to possibly urea toxicity) of young cells and tissues of expanding foliage.

There is evidence that sufficient Ni deficiency hinders lignification to the point that shoots of woody perennials are potentially very weak and brittle (Wood et al., 2004a,b), and pericarps of seeds or nuts become relatively weak during the early stages of kernel or cotyledon filling (Wells and Wood, 2008). Trees that have long been Ni deficient exhibit a greater likelihood of major limb breakage due to winds or excessive cropping. Nickel deficiency can be confused easily with Cu deficiency, as insufficient Cu also impairs lignification (Broadley et al., 2012) to produce brittle shoots. Because many commercial Cu fertilizer sources contain Ni as a trace contaminant, the correction of lignification issues using Cu simultaneously can correct the Ni aspect of impaired lignification-associated metabolism and sporadically correct or reduce Ni deficiency symptoms, thereby misleading practitioners as to the actual nutrient element causing certain disorders (Wood et al., 2004a,b). Lignin, a product of secondary metabolism, is the second most abundant organic polymer in vascular plants and is important to plant structure and efficient water transport.

These effects indicate that Ni deficiency likely can affect plant water relations, especially if growing under substantially limiting water availability to certain species. Because lignin biosynthesis and polymerization is complex, involving many enzymes, there are many potential sites where insufficient Ni might influence directly or indirectly either the timely biosynthesis or polymerization of lignin. One path to lignin biosynthesis begins with phenylalanine, an important essential amino acid; thus, it may be that the influence of Ni on plant N metabolism is such that there is an adverse effect on the timely availability of this precursor. Because lignin is also an important component of cell walls in vascular plants and contributes to disease resistance, Ni deficiency–induced impairment of lignification is associated with greater disease susceptibility to certain pathogens. For example, improving Ni nutritional physiology reduces susceptibility to the scab fungus in pecan and rusts in daylily (Wood et al., 2012). Although it is unknown how Ni affects resistance mechanisms, it is clear that effects on lignification, respiratory carbon flow, and N metabolism potentially affect a host of resistance processes linked to secondary plant metabolism.

Severely Ni-deficient trees exhibit extremely dwarfed internodes and loss of apical dominance, thus implicating an effect of Ni on phytohormone metabolism. Both growth distortions indicate the likely direct or indirect disruption of auxin metabolism in respect to loss of apical dominance and of auxin or gibberellin metabolism in regard to internode dwarfing.

Nickel ions potentially act as cofactors in the cupin class of plant proteins, which subsequently imparts a certain unique type of chemistry within the tertiary protein structures (Dunwall et al., 2004), thus potentially enabling the use of Ni either in metalloenzymes or as metallochaperones that aid in movement and insertion of transition metals into metalloenzymes.

The cellular milieu of plants is comprised of approximately 90% water, with this water serving as solvent and transport medium for metal ions, and also as a reactant in most metabolic pathways (Farmer and Browse, 2013). Water is also a feedstock for many chemicals involved in metabolic and physiological processes; therefore, the superior catalytic ability of Ni likely plays a critical *in planta* role in chemical reactions by producing H_2, e^-, H^-, $H\cdot$, OH^-, H_3O^+, H^+, H_2O_2, O_2, and O_3 from H_2O; thus, enabling plants to use water as a feedstock, rather than simply a carrier, solvent, or hydrolyzing agent (Russell, 2007). These water-derived products are likely involved in a myriad of basic metabolic and physiological processes, thereby potentially bestowing upon Ni a physiological or metabolic role independent of a role as a cofactor for one or more metalloenzymes, and therefore conferring on Ni a long and underappreciated evolutionary pedigree.

14.3.1.2 Physiology

Nickel influences several physiological processes. For example, the germination rates of many legumes (e.g., peas, beans, castor beans [*Ricinus communis* L.], lupine, and soybeans) and grasses (e.g., wheat, timothy [*Phleum pratense* L.], and rice [*O. sativa* L.]) can be increased by elevating Ni (Das et al., 1978; Mishra and Kar, 1974), and the viability of grains (e.g., wheat, barley, and oats) is reduced by poor Ni nutrition (Aller et al., 1990; Brown et al., 1990). The Ni concentration of soybean seed influences urease activity and sensitivity and N-use efficiency of urea sprays to seedlings, and increases plant chlorophyll content and promotes shoot growth (Kutman et al., 2013). Good seed Ni nutrition also appears to protect seedlings from certain fungal pathogens. Nickel is often used to improve seed quality (Mazzafera et al., 2013). Nickel appears to be involved in phytoalexin synthesis, therefore influencing plant defense (Aller et al., 1990; Graham et al., 1985). For example, improving Ni nutrition, often even via soil application, increases resistance of daylily to rust (Reilly et al., 2005). Spraying foliage of crops can help control certain species of rust (Graham et al., 1985). Nickel was once patented by a large agrichemicals company for use against crop pathogens. There is evidence that chlorophyll destruction in association with leaf senescence is retarded by Ni (Aller et al., 1990), as Ni inhibits ethylene synthesis (Lau and Yang, 1976; Pennazio and Roggero, 1992; Polacco et al., 2013), and senescence is potentially influenced by ethylene. It is possible that the effect of Ni on ethylene biosynthesis is due to replacement of Fe in aminocyclopropane carboxylase oxidase (Polacco et al., 2013).

Nickel-deficient pecan trees exhibit restricted photosynthesis and stomatal conductance and are under greater water stress than nondeficient trees, plus foliage with Ni within the hidden hunger range is not as green as normal foliage (Wood et al., 2004a). Nickel-deficiency-induced chlorosis occurs in plants as diverse as trees (Wood et al., 2004b) and tomatoes (*Lycopersicon esculentum* Mill.) (Checkai et al., 1986). Nickel also may play a role in the protection of certain plants, especially Ni hyperaccumulators, against drought in that Ni can potentially be involved in osmotic adjustment (Bhatia et al., 2005). It also may influence the ability of a crop to withstand stress through production of oxygen scavengers (e.g., polyamides) and osmoprotectants (e.g., proline) (Polocco et al., 2013; Szabados and Savoure, 2010). Because Ni anciently played a major role in the functionality of membranes of primitive organisms, it is likely that Ni still functions to influence certain aspects of cell or organelle membrane structure and function (Welch, 1981).

From the aforementioned details, it is apparent that plant Ni nutritional physiology likely, directly or indirectly, affects several physiological and metabolic processes in flowering plants. Because Ni is a trace contaminant of essentially all metal salts used in enzyme assays, it is possible that enzymes requiring certain metallic cations for activation are cosupplied inadvertently with Ni and therefore might mislead regarding their actual metal requirements, especially in the case of one or more of isoforms. The greater role of Ni in the primary metabolism–associated enzymology of prokaryotes is suggestive of additional yet to be discovered beneficial or essential roles for Ni in higher plants.

14.4 GENETICS OF NICKEL ACQUISITION BY PLANTS

There appears to be nothing known of the genetics of Ni acquisition by plants.

14.5 CONCENTRATION OF NICKEL IN PLANTS

14.5.1 Nonhyperaccumulators

The foliar concentration of Ni in most nonhyperaccumulating plants growing on non-Ni contaminated soil, or nonserpentine soils, ranges from about 0.05 to 10 µg g^{-1} DW, depending on species (see review by Brown, 2006). These typically exhibit a foliar "lower critical" concentration (based on appearance of morphological symptoms) of roughly 0.3–0.9 µg g^{-1} for ureide-N transporters (Nyczepir et al., 2006) and perhaps as much as 10–100-fold less for amide-N transporters; however, these concentrations are likely to vary greatly depending on concentration of other divalent cations (see discussion later in this chapter). In the case of pecan, a ureide-N transporting perennial, with relatively normal concentrations of transition metal divalent cations, the Ni threshold for triggering rosetting and loss of apical dominance, both symptoms of severe deficiency, is between 0.007 and 0.640 µg g^{-1} DW (dry weight) (Nyczepir et al., 2006). The associated "lower critical" concentration appears to be ≃1 µg g^{-1} DW, with the "upper critical" level being >100 µg g^{-1}, and the "adequate" range estimated to be between 2.5 and ≥ 30 µg g^{-1} DW (Pond et al., 2006; Smith et al., 2012; personal observation by the author). All of these thresholds depend on endogenous concentrations of competing cations (e.g., Zn, Cu and Fe). An assimilation of comparable Ni concentrations in amide-N transporting species (e.g., barley, wheat, tomato, rice, zucchini [*Cucurbita pepo* L.]) (Cottenie et al., 1976; Eschew et al., 1984; Macnicol and Beckett, 1985; Singh et al., 1990; Walker et al., 1985) appears to be such that the "lower critical" concentration is ≃0.10 µg g^{-1} DW, with the adequate concentration being ≃0.08–10 µg g^{-1} DW and the "upper critical" concentration ≃10–113 µg g^{-1} DW. This concentration, of course, depends on species (e.g., tomato has a relatively low *upper critical concentration*) (Brown et al., 1987b; Gerendás et al., 1999). The *lower critical concentration* for ureide-N transporters appears to be ≃10-fold greater than that of amide-N transporting species. It is likely that crops experience a substantial hidden hunger form of physiological Ni deficiency at tissue concentrations between their lower critical and adequate concentrations, especially during phases of rapid vegetative or reproductive growth. In the case of pecan, this hidden hunger zone appears to span a range of ≈ 2 μg g^{-1} under normal leaf concentrations of competing metals such as Zn and Cu. While Ni accumulation of ≥10–100 μg g^{-1} DW, depending on species sensitivity, can limit plant growth without exhibiting obvious signs of toxicity, higher Ni concentrations can produce necrotic lesions on the lamina of expanding foliage. Table 14.1 provides examples of Ni concentration reported for a variety of vegetable, agronomic, and tree crops.

14.5.2 HYPERACCUMULATORS

The concentration of Ni in aerial organs of certain flowering plant species can be as high as 50,000 μg g^{-1} (5% DW) of leaf biomass (Baker et al., 1994; Prasad, 2005; Reeves, 1992), with Ni reaching concentrations in plants higher than any other heavy metal (He et al., 2012). Plants accumulating Ni at ≥1000 μg g^{-1} DW (i.e., 0.1%) are hyperaccumulators (Brooks et al., 1977), and most of these are endemic to metalliferous soils (Rascio and Navari-Izzo, 2011). Of roughly 450 plant species (representing 45 families) known to hyperaccumulate heavy metals, the vast majority (about 350) are Ni hyperaccumulators, with a large number of these being *Alyssum* and *Thlaspi* species (Brooks, 1998; Rascio and Navari-Izzo, 2011) of the Brassicaceae family (Prasad, 2005). Such plants are able to accumulate without phytotoxicity 100–1000-fold more Ni than nonhyperaccumulating species, thus making these plants useful for phytoextraction and phytomining.

14.6 RATIOS OF NICKEL WITH OTHER ELEMENTS AND INTERACTIONS

Nickel interacts with most essential elements, but it is those with Fe, Cu, and Zn that are likely to possess greatest biological significance. Endogenous Zn, Cu, Fe, and probably Co and Cr, influence the endogenous *in planta* availability of Ni for physiology and metabolism. Based on potted studies of pecan seedlings treated with different concentrations of soil applied Zn, Cu, or Fe, there is increasing severity of Ni deficiency as the foliar Zn–Ni ratio increases from ≈ 25 to 1000, as the Cu–Ni ratio increases from ≈ 12 to 58, and as the Fe–Ni ratio increases from ≈ 150 to 1200 (Wood, 2010, 2013); yet, in all cases, foliar Ni concentration was well within the "adequate" or "sufficiency" concentration range. Thus, in at least certain cases, Ni deficiency is not so much an issue of low Ni concentration, but rather that of low metabolic availability in cells due to competition with other transition metals, especially during early or rapid growth phases. Additionally, foliar sprays of Cu or Fe can induce Ni deficiency in dicots, especially when the Ni concentration is already within or near the "hidden hunger" range. For example, Cataldo et al. (1978b) found that Fe^{2+} suppresses Ni^{2+} absorption and translocation in soybean. Excessive Ni affects endogenous Fe concentration or bioavailability (Chen et al., 2009; de Koch, 1956; Ghasemi et al., 2009; Hewitt, 1953; Khalid and Tinsley, 1980; Kovacik et al., 2009; Misra and Ivedi, 1977; Nicholas and Thomas, 1954; Nishida et al., 2012). Khalid and Tinsley (1980) concluded that in annual ryegrass (*Lolium multiflorum* Lam.), the Ni–Fe ratio, rather than absolute concentration of either in plant tissues and organs, is most tightly associated with reduced Fe bioavailability and utilization under high Ni conditions. It appears that the interaction with Ni and Fe can be either synergistic or antagonistic, with synergism being with ferric iron (Fe^{3+}), and antagonism being with ferrous iron (Fe^{2+}) (Nielsen et al., 1982).

14.7 DIAGNOSIS OF NICKEL STATUS IN PLANTS

Most cases of Ni deficiency are in the form of hidden hunger, in which there are no visual stress symptoms, but with adverse effects on physiological and metabolic processes influencing pest resistance, plant growth, yield, quality, and resource use efficiency. Diagnosis of hidden hunger is difficult for most crops, because Ni status has not been studied, but in the case of pecan, hidden hunger in foliage

TABLE 14.1
Nickel Concentration in Selected Crop Plants

Crop Species	Plant Part	Ni Concentration (mg kg⁻¹ Dry Matter)	References
Alfalfa (*Medicago sativa* L.)	Foliage	5.3–11.9	Taylor and Allinson (1981)
Alyssum (*Alyssum corsicum* L.)	Shoot	21–15,101	Centofanti et al. (2012)
Barley (*Hordeum vulgare* L.)	Grain	0.007–0.129 −Ni/+Ni supply	Brown et al. (1987a)
Coffee (*Coffea arabica* L.)	Leaves	4–45	Tezotto et al. (2012)
	Beans	1–5	
	Branches	3–18	
Cowpea (*Vigna unguiculata* Walp.)	Leaf	0.1–3.7 −Ni/+Ni supply	Walker et al. (1985)
Maize (*Zea mays* L.)	Shoot	0.71–1.84 −Ni/+Ni supply	Sabir et al. (2011)
	Root	2.0–8.52 −Ni/+Ni supply	Sabir et al. (2011)
Tomato (*Lycopersicon esculentum* Mill.)	Fruit	3.1	Demirezen and Aksoy (2006)
Onion (*Allium cepa* L.)	Bulb	4.6	Demirezen and Aksoy (2006)
Lettuce (*Lactuca sativa* L.)	Foliage	6.3	Demirezen and Aksoy (2006)
Oats (*Avena sativa* L.)	Grain	1–4	Brown et al. (1989)
Oilseed rape (*Brassica napus* L.)	Shoot	0.021–0.173 −Ni/+Ni supply	Gerendás et al. (1999)
Pecan (*Carya illinoinensis* K. Koch)	Spring xylem sap	0.01–0.05	Bai et al. (2013)
Pecan (*Carya illinoinensis* K. Koch)	Young foliage	0.862 normal, class I 0.265 slight deficiency, class 2 0.064 deficiency, class 7 0.007 strong deficiency, class 9	Nyczepir et al. (2006)
Pecan (*Carya illinoinensis* K. Koch)	Mature foliage	1–30	Pond et al. (2006)
Rice (*Oryza sativa* L.)	Shoot	0.039–0.975 −Ni/+Ni supply	Gerendás et al. (1999)
Ryegrass (*Lolium multiflorum* L.)	Mature shoot	1–4.5	Brown et al. (1989)
Soybean (*Glycine max* Merr.)	Shoot	<0.010–0.637 −Ni/+Ni supply	Gerendás et al. (1999)
Spinach (*Spinacia oleracea* L.)	Foliage	1.83	Khan et al. (1999)
Tobacco (*Nicotiana tabacum* L.)	Shoot	0.056–0.217 −Ni/+Ni supply	Gerendás et al. (1999)
Zucchini (*Cucurbita pepo* L.)	Shoot	0.025–0.155 −Ni/+Ni supply	Gerendás et al. (1999)

appears to occur between $\simeq 1.0$ and 2.5 µg g^{-1}, with the relative concentration of other divalent ions, especially Zn, Cu, and Fe, affecting both the hidden hunger and lower critical Ni concentrations.

14.7.1 Visual Symptoms in Woody Perennials: Foliage

Visual symptoms in woody perennials are likely to appear when Ni is less than $\simeq 1.0$ µg g^{-1} in foliar tissue. The most prominent visual symptom of Ni deficiency is blunting of the apical tip (Figure 14.1a and b) and dwarfing of foliage to produce a blunted leaf or leaflet apex shape resembling the "ear of a mouse," thus producing maladies termed as "mouse-ear." Figure 14.1 illustrates several distinct symptoms of the severe form of Ni deficiency (Wood and Reilly, 2007; Wood et al., 2004a,b). Nickel-deficient foliar chlorosis, resembling mild Fe or S deficiency, is

(a)   (b)

(c)   (d)

FIGURE 14.1 (**See color insert.**) Nickel deficiency symptoms for pecan (*Carya illinoinensis* K. Koch). (a) Reduced leaf and leaflet size as severity of deficiency increases; (b) altered shape of leaflet blade to produce mouse-ear-like or little-leaf foliage; (c) blunting, crinkling, dark green zone, and necrosis of the apical portion of the leaf blade; and (d) absence of leaf lamina development in severe cases, plus deformed and pointed buds, and rosetting (loss of apical dominance).

also an early symptom of Ni deficiency in that leaves are abnormally pale or yellowish-green during the leaf, or canopy, expansion growth. This faint chlorosis tends to be uniform throughout the leaf or leaflet but often becomes the normal shade of green by midseason. This blunting can be localized to certain leaves on a shoot (usually the physiologically oldest leaves and leaflets), to certain branches, to large limbs, or can be more or less uniform throughout the canopy. In clear cases of severe Ni deficiency, the mouse-ear-like deformation is generally most prominent at the top of the tree canopy. The surface area of leaves and leaflets diminishes in proportion to the degree and onset of Ni deficiency, with the most typical reduction in area being 10%–75% compared to normal growth. This reduction in leaf or leaflet area led to the term "little-leaf" to describe the malady in certain crops, such as river birch (*Betula nigra* L.) (Reuter, 2005). As with foliage shape, the degree of reduction in area of individual leaves tends to increase with canopy height.

Affected foliage often exhibits a small dark-green zone just below the apical tip (Figure 14.1c) and disappears as the leaf or leaflet ages. This dark-green zone is most noticeable during the first few weeks after bud break and appears to possess great diagnostic power as a clear indicator of Ni deficiency in woody perennials. As the severity of Ni deficiency increases, the tips of affected leaves or leaflets exhibit cell death (Figure 14.1c). This dead zone expands with aging and degree of severity. Death is apparently due to toxic accumulation of either urea, or certain organic acids, or both, arising from reduction of the activity of one or more enzymes in affected cells. Ni deficiency is most visible in the most rapidly growing portions of the leaf. Cell expansion in the margins of leaves or leaflets, especially toward the apex, is reduced to the point that there is cupping and wrinkling of the lamina (Figure 14.1c). Affected foliage also feels thicker and less pliable and tends to be brittle. Foliage during this stage exhibits a bonsai-like appearance—that is, miniaturized. Severely deficient foliage can be such that leaf lamina development is arrested completely, displaying a leaf with a miniature vascular array but without lamina (Figure 14.1d); thus, in extreme cases, leaf area can be reduced to nothing and is most likely to occur on young transplant trees about 2–5 years after planting. This degree of deficiency generally is accompanied by an abnormal elongation and exaggerated point, or sharpening, of buds. Severe Ni deficiency also can produce protrusion of petiole wings, with each of the two wings becoming as wide as the petiole itself.

14.7.2 VISUAL SYMPTOMS IN WOODY PERENNIALS: SHOOTS, ROOTS, AND TREE CHARACTERISTICS

Severe Ni deficiency results in a loss of apical dominance, producing rosetting (Figure 14.2a) similar to that typified by severe Zn or Cu deficiency. The resulting shoot morphology varies from that of a witch's broom-like pattern to that of a dwarfed shoot possessing tightly spaced buds or shoots possessing miniature foliage. This symptom usually is associated with dwarfed, blunted, necrotic-tipped foliage or foliage with little or no leaf lamina. An increasing severity of Ni deficiency increasingly distorts bud shape (Figures 14.1d and 14.2c), with buds becoming increasingly elongated and pointed. Nickel-deficient shoots can exhibit a delay in bud break for as much as a week (Figure 14.2b). However, in the case of greenhouse-grown young seedlings, Ni deficiency can instead advance bud break. Buds on severely affected trees often die during the dormant season, giving the appearance of death from winter cold. Thus, it is easy to misdiagnose mild Ni deficiency as winter injury. Shoot length also diminishes as severity of Ni deficiency increases. Reduction is due to reduced length of internodes (Figure 14.2c) rather than node number. Shoot length can be reduced by up to 90%, but it is most usually by 10%–50%. Shoots and limbs of trees with moderate-to-severe Ni deficiency are noticeably brittle; thus, these structures are broken easily when bent. Tree size is diminished increasingly as severity and duration of Ni deficiency increases (Figure 14.2d), resulting in young orchard transplants being stunted, with little or no canopy or caliper growth. Extreme deficiency produces dwarfed trees that are less than 10% the height of their unaffected peers. The root systems of severely Ni-deficient trees are smaller than normal,

FIGURE 14.2 **(See color insert.)** Influence of Ni deficiency on pecan (*Carya illinoinensis* K. Koch). (a) Loss of apical dominance and rosetting; (b) delayed bud break (degree of Ni deficiency decreasing from left to right); (c) short internodes, dwarfed and weakened shoots possessing weak pointed buds (Ni deficient on left and Ni sufficient on right); and (d) dwarfed tree (tree is 1 m tall at age 10 years) typical of the replant form of Ni deficiency.

FIGURE 14.3 Influence of Ni deficiency on English walnut (*Juglans regia* L.) seedlings derived from tissue culture explants. Foliage and shoots are dwarfed, with foliage exhibiting marginal curling of leaflets and necrosis of leaflet tips and margins. Note normal leaflets in the upper left of the image.

with fibrous roots being dead or decayed, and trees being very slow to fruit. Severe deficiency increases shoot death during the dormant season, giving the impression of death from winter cold. Severely deficient trees usually die within a few years of transplanting in orchards, thus giving the impression of a replant-disease problem. Nickel deficiency is prone to appear as a replant malady in second-generation orchards, especially when replacement trees are at locations previously occupied crop trees.

Figure 14.3 illustrates Ni deficiency symptoms in English walnut (*Juglans regia* L.) seedlings derived from tissue culture explants. Symptoms are identical to those typically exhibited by pecan, with dwarfed shoots, and foliage exhibiting blunted shoot tips in combination with leaflet curling, wrinkling, and necrosis being associated with relatively moderate level of deficiency during organ development.

14.7.3 VISUAL SYMPTOMS IN NONWOODY PLANTS

Brown et al. (1987a,b) observed that Ni deficiency symptoms in wheat, barley, and oats include chlorosis of young leaves, reduced leaf area, and less upright growth of leaves than Ni-sufficient crops. The interveinal region of leaves first became chlorotic and then developed white lesions initiating at the midvein half way down the blade, though often appearing simultaneously at the leaf margin and leaf tip. Patchy zones of necrotic spots often developed from chlorotic regions. Leaf tips exhibit dieback, and leaves can die prematurely. The preponderance of symptoms in the youngest tissues implies that phloem Ni mobility during leaf expansion is insufficient for tissue needs of the rapidly expanding foliage. It is likely that these symptoms are typical for Gramineae (Poaceae) species. They also observed premature senescence in Ni-deficient oat plants. Incomplete unfolding of leaf tips appears to be rather typical when Ni deficiency surpasses a certain deficiency threshold. Necrosis of young leaves can transition into necrosis of the meristem. Nickel deficiency also depresses growth, inhibits grain development, causes grain inviability, and can reduce iron levels of tissues. The necrosis of leaf tips due to urea and organic acid toxicity appears to be a key defining characteristic of Ni deficiency that transcends plant families and is a reliable diagnostic trait for the recognition of Ni deficiency.

14.7.4 TOXICITY

The toxicity of elements to organisms tends to be related to oceanic abundance, with the most toxic being those elements that are not essential and occur at the lowest concentration in sea water (Ni concentration in ocean water is $\simeq 0.228$–3.00 μg L^{-1}, depending on location [Cempel and Nikel, 2006; Nielsen, 2000]); however, this relationship also is influenced greatly by natural selection processes (Nielsen, 2000). The toxicity of Ni to plants occurs when Ni overwhelms homeostatic mechanisms, with the efficiency of those mechanisms usually being dependent of the crop species exposure during its evolution (Nielsen, 2000). Nickel is readily available to plants and can therefore cause toxicity in extreme cases of very high soil Ni availability, especially in plants that evolved under conditions of relatively low Ni exposure. Nickel toxicity occurs when homeostatic processes fail to sufficiently regulate cytoplasmic Ni, at physiological pH, to prevent disruption of the functionality of proteins, nucleic acids, and peptides—especially those containing histidine (e.g., the histidine imidazole-N group) and cysteine (e.g., the cysteine sulfhydryl group)—and to induce oxidative stress (Kasprzak and Salnikow, 2007). The cytoplasmic Ni concentration appears to be held to about 10^{-10} M, with most Ni being accumulated in vacuoles. Excessive accumulation of Ni^{2+}, or any of the divalent transition metals, in cellular cytoplasm can cause toxicity by either inducing phosphate deficiency (i.e., precipitation of insoluble Ni–phosphate or metal–phosphate complexes) or by competitive displacement of a key transition metal ion within an enzyme by another metal of lower affinity and in much greater abundance. This displacement therefore limits or eliminates enzyme functionality and subsequently alters organism physiology in unpredictable ways. Excessive Ni can potentially suppress electron transport, chlorophyll content, and stomatal conductance sufficiently to reduce photosynthesis (Wood et al., 2004a). Nickel toxicity in nonhyperaccumulators can occur if plants are growing in soils that are high in available Ni, especially near mining or industrial sites or when nonadapted species are treated with sewage sludge containing excessive high Ni (Broadley et al., 2012) or are grown on serpentine soils (from hydrous magnesium iron phyllosilicate minerals).

The foliar Ni concentration threshold for phytotoxicity depends on whether the species possesses "low tolerance," "moderate tolerance," or "high tolerance" for Ni and concentration of competing cations. Relatively little is known regarding which species fall into these thresholds, but two examples of relatively "low tolerance" species are tomato and peach (Nyczepir and Wood, 2012), where foliar concentrations >10 μg g^{-1} DW are usually detrimental (personal observation). Conversely, certain "moderate tolerance" plants are the ureide-transporting genera (e.g., Carya, Juglans, Diospyros, Vitis, Annona [Schubert and Boland, 1990] and probably most legumes) and appear to tolerate Ni without adverse side effects at least as high as $\simeq 80$–120 μg g^{-1} DW. Examples of "high tolerance" species are at least 350 species of Ni hyperaccumulators that can tolerate Ni concentrations >1000 μg g^{-1} DW. Chlorosis is a common trait of Ni toxicity in monocots and dicots, with chlorotic stripes along leaves in case of monocots and interveinal chlorotic blotching in leaves of dicots. Extreme toxicity can cause necrotic spots between leaf veins and of leaf tips and margins.

Toxicity symptoms vary depending on species, but in most cases, plants appear to be mildly Fe deficient, apparently because Ni is competing with Fe in metabolic processes and the associated Fe-dependent enzymes. Recent reviews by Yusuf et al. (2011) and Ahmad and Ashraf (2011), regarding the impact of Ni toxicity on plants, note that a rather general symptom is limited growth of all plant organs (i.e., roots, stems, leaves, fruit) and impairment of a large variety of physiological processes or conditions (e.g., chlorophyll content, protein content, photosynthesis, respiration, water relations, oxidative stress, metal homeostasis) and metabolism (e.g., nitrate reductase, H+-ATPase, glutathione reductase). Nickel also potentially competes with uptake and utilization of many other divalent cations (e.g., Mg^{2+}, Ca^{2+}, Mn^{2+}, Fe^{2+}, Cu^{2+}, Co^{2+}, and Zn^{2+}) and therefore adversely affects hundreds of enzymatic processes and their downstream products and probably signaling processes within and among cells. Toxicity can be diminished by increased fertilization with Fe, Ca, Zn, Mg, or K (Crooke and Inkson 1955; Genrich et al., 1998; Goncalves et al., 2007), Cu and Zn (Parida et al., 2003), and probably by P.

Nickel is a relatively light transition metal, often considered a toxic environmental pollutant, and often classified as a "toxic heavy metal." This classification is largely because NiO and $NiSO_4$ are atmospheric contaminants produced from oil and coal burning; although, other toxic forms are also produced from certain industrial processes (Pacyna and Pacyna, 2001). As an essential ultramicronutrient (as is Mo, and probably Co, and likely certain other yet to be demonstrated metals), Ni is indeed potentially toxic, as are all essential metals (e.g., Fe, Mn, Zn, Cu) when at excessive concentration; however, its toxic potential is checked by tight homeostatic control. This property implies that the Ni requirement and sensitivity of a particular crop provides insight into that crops evolutionary history.

14.8   FORMS AND CONCENTRATION OF NICKEL IN SOILS, AVAILABILITY TO PLANTS

The average abundance of Ni in Earth's crust is $\simeq$80–100 $\mu g\ g^{-1}$, with the amount varying greatly among the various rock forms. Nickel concentration is typically highest in igneous and metaigneous rocks (i.e., ultramafic or ultrabasic rocks). Such rocks (e.g., peridotite, dunite, norite, essexite, and komatiite) originate from Earth's relatively Ni-enriched mantle and are the parent material for serpentine soils (i.e., soils high in Mg, Fe, Ni, Co, and Cd, but low in Ca, K, P, and with Ni up to 2000–6000 $\mu g\ g^{-1}$) (Barcelous, 1999; Niemineh et al., 2006). Soil Ni concentration decreases if derived from mafic (e.g., basalt, diabase, gabbro at $\simeq$150 $\mu g\ g^{-1}$) or felsic (rhyolite, aplite, granite at $\simeq\leq$15 $\mu g\ g^{-1}$) type igneous rocks, with Ni in sedimentary rock soils typically being $\simeq\leq$70 $\mu g\ g^{-1}$ (Niemineh et al., 2006). Within the United States, geometric means of Ni in mineral soils range from 6.2 $\mu g\ g^{-1}$ in North Carolina to 50.5 $\mu g\ g^{-1}$ in California, with a U.S. average of 17.1 $\mu g\ g^{-1}$ (Holmgren et al., 1993).

Certain anthropomorphic activity, such as Ni mining and metal processing, or dumping of sewage sludge from certain metal plating plants, can increase soil Ni concentration up to several thousand $\mu g\ g^{-1}$ (Barceloux, 1999). Common mineral forms of Ni are nickeline (NiAs), millerite (NiS), monenozite ($NiSO_4 \cdot 7H_2O$), and genthite [$Ni_4(Mg)Si_3O_{10}$] (Pais and Jones, 1997). Nickel availability to root uptake diminishes substantially at high pH where fixation by Ca and Mg occurs.

Although Ni content of most agricultural soils is a gross indicator of the Ni concentration of crops growing on the soil, it is not necessarily a good indicator of the Ni concentration or nutritional physiology status of plants. Availability is especially dependent on soil pH and conditions that limit root uptake, such as soil water, temperature, and competing divalent cations. Acidic soils (pH < 6.5) typically exhibit good Ni bioavailability, whereas Ni is much less available in relatively alkaline soils (pH > 6.7) (Brown et al., 1989). Cool and dry soils often are associated with Ni-deficient plants, especially if the conditions are present during rapid growth, such as spring canopy expansion. Nickel also binds strongly to soil organic matter where it can become substantially unavailable for uptake (Hamner et al., 2013). Additionally, excessive soil bioavailability of Mg, Ca, Fe, Co, Cu, and Zn can trigger Ni deficiency, especially that of Zn and Cu (Kochian, 1991; Wood et al., 2004b). The apparent competitive inhibition of Ni uptake by Zn, Co, Cu, and Fe indicates the likelihood of a shared uptake system. A field situation heavily favoring a transitory Ni deficiency occurs during early spring when soils are abnormally cool and dry, and there has been excessive long-term fertilization with P, Mg, Ca, Fe, Co, Cu, or Zn, or a combination thereof. Odds increase further if the soil possesses a low CEC and the crop receives a substantial spring application of synthetic nitrogen fertilizer (especially urea).

Depending on geological or anthropogenic source, soil Ni potentially exists in many forms. Examples include $Ni_3(PO_4)_2$, $NiSO_4$, $NiCO_3$, $Ni(OH)_2$, $NiNO_3$, NiO, NiS, and various Ni-organic ligands. Under aerobic conditions, Ni reacts strongly with O^{2-} and OH^- and tends to precipitate as insoluble hydroxyoxides or as a minor component of aluminosilicates, whereas under anaerobic conditions, Ni tends to precipitate as a carbonate or sulfide (Bohn et al., 2001). Dominant cationic forms in soil solution are Ni^{2+}, $NiOH^+$, and $NiHCO_3^+$ and exist in anionic form as $HNiO_2^-$ and $Ni(OH)_3^-$ (Kabata-Pendias and Mukherjee, 2007). Hydrated Ni (e.g., $Ni(H_2O)_6^{2+}$) appears to be the most typical form in soil solutions, with free Ni^{2+} concentration being substantially controlled by precipitation reactions with the hydrous oxides of Mn and Fe (Yusuf et al., 2011). In general, plants absorb water-soluble Ni, with uptake being a function of solubility in an aqueous solution and with the mobility of Ni in the soil being related inversely to soil pH (Kabata-Pendias and Mukherjee, 2007). However, the soil availability of Ni involves more than solubility (Centofanti et al., 2012). For example in *Alyssum corsicum*, a Ni-hyperaccumulating species, the relatively low water solubility of $Ni_3(PO_4)_2$, Ni phyllosilicate, and Ni-acid birnessite can still result in relatively high Ni plant uptake (Centofanti et al., 2012). Under relatively alkaline soil conditions, Ni increasingly occurs as $NiCO_3^0$, $NiHCO_3^+$, and $Ni(OH)_2^0$ (Nieminen et al., 2006). In high phosphate soils or those receiving high levels of phosphate fertilizer, $Ni_3(PO_4)_2$ is a common form. In acidic soils, especially if high in organic matter, Ni is likely complexed with organic ligands and therefore tends to accumulate on or near the soil surface.

14.9 SOIL TESTING

Nickel concentration in plants is a rough function of the Ni content of soils. Agricultural soils rarely are tested for Ni on a routine basis. This is because of the general presumption that all soils contain sufficient Ni to meet plant needs. The "Mehlich No.-3" (suitable for most acidic soils) or the "ammonium bicarbonate-diethylenetriaminepentaacetic acid (DPTA)" (suitable for most neutral, calcareous, or alkaline soils) procedures produce a filtrate that roughly reflects plant available Ni^{2+} where it is important to determine the amount of Ni available from the plant available Ni-pool of a particular soil (Jones, 2012). Factors such as soil type, CEC, $CaCO_3$, organic matter content, texture, and crop species potentially influence the utility of soil tests for Ni, with extraction methods undergoing slight modification in order to optimize extraction of available Ni and better assess its relationship to plant Ni nutritional physiology. To date, these soil testing procedures for Ni possess somewhat limited reliability as there has not been sufficient correlation established between the quantity of Ni extracted and plant growth for most crops. The predictive value of such tests is likely to be dependent on construction of appropriate multiple regression equations that take into account various soil properties as well as that of the extractable Ni and crop-specific field calibration. In cases where it is important to determine the amount of "environmentally available" Ni, the "US EPA Method 3050B" is used (using HNO_3 and H_2O_2), which excludes Ni bound in silicates (which are not normally environmentally mobile). While the aforementioned soil testing methods can identify Ni-enriched soils and enable moderate success at predicting plant Ni uptake, further research is needed to accurately predict "soil-Ni crop linkage."

14.10 FERTILIZERS

Soil application of Ni fertilizer is almost never necessary, as almost all soils contain sufficient Ni to satisfy plant needs. Most agricultural soils inadvertently receive Ni as a trace contaminant of most P fertilizers. Deficiency typically occurs during the early phase of either vegetative or reproductive growth, with correction being best by timely foliar application just after the inception of growth. Foliar applications of Ni salts (e.g., NO_3^-, SO_4^-) are highly effective for correction of Ni deficiency. Additionally, organic Ni ligands (e.g., lignosulfonates, heptogluconates) are also highly effective Ni foliar fertilizers, with the Ni-lignosulfonate being a preferred Ni formulation for field use due to potential worker safety concerns with the Ni-nitrate and Ni-sulfate alternatives. The aqueous extract of *Alyssum*, a Ni hyperaccumulator, biomass is a highly effective natural source of Ni fertilizer

(Wood et al., 2006a). Certain liquid or tissue culture operations may need to add Ni supplements (Ni-salts), especially if using highly purified metals for the culture solution, and elements such as Fe, Cu, and Zn are high. Additionally, sewage sludge and diluted solutions of industrial products also potentially serve as sources of fertilizer Ni.

REFERENCES

Ahmad, M.S.A. and M. Ashraf. 2011. Essential roles and hazardous effects of nickel in plants. *Rev. Environ. Contam. Toxicol.* 214:125–167.

Aller, A.J., J.L. Bernal, M.J. del Nozal, and L. Deban. 1990. Effects of selected trace elements on plant growth. *J. Sci. Food Agric.* 51:447–479.

Alves, S., C. Nabais, M. de Lurdes Simoes-Goncalves, and M.M.C. dos Santos. 2011. Nickel speciation in the xylem sap of the hyperaccumulator *Alyssum serpyllifolium* ssp. *lusitanicum* growing on serpentine soils of northeast Portugal. *J. Plant Physiol.* 168:1715–1722.

Aschmann, S.G. and R.J. Zasoski. 1987. Nickel and rubidium uptake by whole oat plants in solution culture. *Physiol. Plant* 71:191–196.

Bai, C., L. Liu, and B.W. Wood. 2013. Nickel affects xylem sap RNase A and converts RNase A to a urease. *BMC Plant Biol.* 13(1):207.

Bai, C., B.W. Wood, and C.C. Reilly. 2007. Nickel deficiency affects nitrogenous forms and urease activity in spring xylem sap of pecan. *J. Am. Soc. Hortic. Sci.* 132(3):302–309.

Bai, C.C., C. Reilly, and B.W. Wood. 2006. Nickel deficiency disrupts metabolism of ureides, amino acids, and organic acids of young pecan. *Plant Physiol.* 140:433–443.

Baker, A.J.M., S.P. McGrath, C.M.D. Sidoli, and R.D. Reeves. 1994. The possibility of in situ heavy metal decontamination of polluted soils using crops of metal-accumulating plants. *Resour. Conserv. Recy.* 11(1–4):41–49.

Balakrishnan, R., N. Ramasubbu, K.I. Varughese, and R. Parthasarathy. 1997. Crystal structures of the copper and nickel complexes of RNase A: Metal-induced interprotein interactions and identification of a novel copper binding motif. *Proc. Natl. Acad. Sci. U. S. A.* 94:9620–9625.

Barceloux, D.G. 1999. Nickel. *J. Toxicol. Clin. Toxicol.* 37:239–258.

Bhatia, N.P., A.J.M. Baker, K.B. Walsh, and D.J. Midmore. 2005. A role for nickel in osmotic adjustment in drought-stressed plants of the nickel hyperaccumulator *Stackhousia tryonii* Bailey. *Planta* 223:134–139.

Bohn, H.L., B.L. McNeal, and G.A. O'Connor. 2001. *Soil Chemistry*, 3rd edn., New York: Wiley.

Brito, B., J. Monza, J. Imperial, T. Ruiz-Argueso, and J.M. Palacios. 2000. Nickel availability and hupSL activation by heterologous regulators limit symbiotic expression of the *Rhizobium leguminosarum* bv. Viciae hydrogenase system in Hup Rhizobia. *Appl. Environ. Microbiol.* 66:937–942.

Brito, B., J.M. Palacios, E. Hidalgo, J. Imperial, and T. Ruiz-Arueso. 1994. Nickel availability to pea (*Pisum sativum* L.) plants limits hydrogenase activity in *Rhizobium leguminosarum* bv. Viciae bacteroids by affecting the processing of the hydrogenase structural subunits. *J. Bacteriol.* 176:5297–5303.

Broadley, M., P. Brown, I. Cakmak, Z. Rengel, and F. Zhao. 2012. Function of nutrients: Micronutrients. In *Mineral Nutrition of Higher Plants*, 3rd edn., ed. P. Marschner, pp. 191–248. San Diego, CA: Academic Press.

Brooks, R.R., M.F. Chambers, L.J. Nicks., and B.H. Robinson. 1998. Phytomining. *Plant Sci.* 3:359–362.

Brooks, R.R., J. Lee, R.D. Reeves, and T. Jaffre. 1977. Detection of nickeliferous rocks by analysis of herbarium species and indicator plants. *J. Geochem. Explor.* 7:49–57.

Brown, P.H., L. Dunemann, R. Schultz, and H. Marschner. 1989. Influence of redox potential and plant species on the uptake of nickel and cadmium from soils. *Z. Pflanzenernähr. Bodenkd.* 152:85–91.

Brown, P.H., R.M. Welch, E.E. Cary, and R.T. Checkai. 1987a. Nickel: A micronutrient essential for plant growth. *Plant Physiol.* 85:801–803.

Brown, P.H., R.M. Welch, E.E. Cary, and R.T. Checkai. 1987b. Beneficial effects of nickel on plant growth. *J. Plant Nutr.* 10:2125–2135.

Brown, P.H., R.M. Welch, and J.T. Madison. 1990. Effect of nickel deficiency on soluble anion, amino acid, and nitrogen levels in barley. *Plant Soil* 125:19–27.

Brown, P.H. 2007. Nickel. In *Handbook of Plant Nutrition*, eds. A.V. Barker and D.J. Pilbeam, pp. 395–409. Boca Raton, FL: Taylor and Francis Group.

Cammack, R. 1995. Splitting molecular hydrogen. *Nature* 373:556–557.

Cataldo, D.A., T.R. Garland, and R.E. Wildung. 1978a. Nickel in plants: I. Uptake kinetics using intact soybean seedlings. *Plant Physiol.* 62:563–565.

Cataldo, D.A., T.R. Garland, and R.E. Wildung. 1978b. Nickel in plants: II. Distribution and chemical form in soybean plants. *Plant Physiol.* 62:566–570.

Cataldo, K.M., T.R. McFadden, R.E. Garland, and R.E. Wildung. 1988. Organic constituents and complexation of nickel(II0, iron(III), cadmium(II) and plutonium(IV) in soybean xylem exudates. *Plant Physiol.* 86:734–739.

Cempel, M, and G. Nikel. 2006. Nickel: A review of its sources and environmental toxicology. *Pol. J. Environ. Stud.* 15(3):375–382.

Centofanti, T., M. Siebecker, R.L. Chaney, A.P. Davis, and D.L. Sparks. 2012. Hyperaccumulation of nickel by *Alyssum corsicum* is related to solubility of Ni mineral species. *Plant Soil* 359:71–83.

Checkai, R.T., W.A. Norvell, and R.M. Welch. 1986. Investigation of nickel essentiality in higher plants using a recirculating resin-buffered hydroponic system. *Agron. Abstr.* 195.

Chen, Z., T.T. Watanabe, T. Shinano et al. 2009. Element interconnections in *Lotus japonicas*: A systematic study of the effects of element additions on different natural variants. *Soil Sci. Plant Nutr.* 55:91–101.

Cottenie, A., R. Dhaese, and R. Camerlynck. 1976. Plant quality response to uptake of polluting elements. *Qual. Plant* 26:293–319.

Crooke, W.M. and H.E. Inkson. 1955. The relationship between nickel toxicity and major nutrient supply. *Plant Soil* 6:1–15.

Curie, C., G. Cassin, D. Couch et al. 2009. Metal movement within the plant: Contribution of nicotianamine and yellow stripe 1-like transporters. *Ann. Bot.* 103:1–11.

Das, P.K., M. Kar, and D. Mishra. 1978. Nickel nutrition of plants: Effect of nickel on some oxidase activities during rice (*Orzya sative* L.) seed germination. *Z. Pflanzenphysiol.* 90:225–233.

de Koch, P.C. 1956. Heavy metal toxicity and iron chlorosis. *Ann. Bot. London* 20:133–141.

de Silva, J.J.R.R. and R.J.P. Williams. 2006. *The Biological Chemistry of the Elements: The Inorganic Chemistry of Life*, 2nd edn. New York: Oxford University Press.

Demirezen, D. and A. Aksoy. 2006. Heavy metal levels in vegetables in Turkey are within safe limits for Cu, Zn, Ni and exceeded for Cd and Pb. *J. Food Qual.* 29(3):252–265.

Dixon, N.E., C. Gazzola, C.J. Asher, D.S. Lee, R.L. Blakeley, and B. Zerner. 1980a. Jack bean urease (EC 3.5.1.5). II. The relationship between nickel, enzymatic activity, and the "abnormal" ultraviolet spectrum. The nickel content of jack beans. *Can. J. Biochem.* 58:474–480.

Dixon, N.E., C. Gazzola, R.L. Blakeley, and B. Zerner. 1975. Jack Bean urease (EC 3.5.1.5), a metalloenzyme, a simple biological role for nickel? *J. Am. Chem. Soc.* 97:4131–4133.

Dixon, N.E., P.W. Riddles, C. Gazzola, R.L. Blakely, and B. Zerner. 1980b. Jack bean urease (EC 3.5.1.5). V. On the mechanism of action of urease on urea, formamide, acetamide, N-methylurea, and related compounds. *Can. J. Biochem.* 58:1335–1355.

Dunwall, J.M., A. Purvis, and S. Khuri. 2004. Cupins: The most functionally diverse protein superfamily? *Phytochemistry* 65:7–17.

Epstein, E. and A.J. Bloom. 2005. *Mineral Nutrition of Plants: Principles and Perspectives*, 2nd edn. Sunderland, MA: Sinauer.

Eskew, D.L., R.M. Welch, and W.A. Norel. 1983. Nickel: An essential micronutrient for legumes and possibly all higher plants. *Science* 222:621–623.

Eskew, D.L., R. M. Welch, and W.A. Norell. 1984. Nickel in higher plants: Further evidence for an essential role. *Plant Physiol.* 76:691–693.

Evans, H.J., S.A. Russell, F.J. Hanus, and T. Ruiz-Argueso. 1988. The importance of hydrogen recycling in nitrogen fixation by legumes. In *World Crops: Cool Season Food Legumes*, ed. R.J. Summerfield, pp. 777–791. Boston, FL: Kluwer.

Farmer, E. E. and J. Browse. 2013. Physiology and metabolism: Water for thought. *Curr. Opin. Plant Biol.* 16:271–273.

Fewell, M.P. 1995. The atomic nuclide with the highest mean binding energy. *Am. J. Phys.* 63:653–658.

Genrich, I., G.I. Burd, D.G. Dixon, and B.R. Glick. 1998. A plant growth promoting bacterium that decreases nickel toxicity in seedlings. *Appl. Environ. Microbiol.* 64:3663–3668.

Gerendás, J., J.C. Polacco, S.K. Freyermuth, and B. Sattelmacher. 1998. Co does not replace Ni with respect to urease activity in zucchini (*Cucurbita pepo* convar. *giromontiina*) and soybean (*Glycine max*). *Plant Soil* 203:127–135.

Gerendás, J., J.C. Polacco, S.K. Freyermuth, and B. Sattelmacher. 1999. Significance of nickel for plant growth and metabolism. *J. Plant Nutr. Soil Sci.* 162:241–256.

Gerendás, J. and B. Sattelmacher. 1997. Significance of Ni supply for growth, urease activity and concentrations of urea, amino acids and mineral nutrients of urea-grown plants. *Plant Soil* 190:153–162.

Ghasemi, R., S.M. Ghaderian, and U. Kramer. 2009. Interference of nickel with copper and iron homeostasis contributes to metal toxicity symptoms in the nickel hyperaccumulator plant *Alyssum inflatum*. *New Phytol.* 184:566–580.

Goncalves, S.C., A. Portugal, M.T. Goncalves, R. Vierira, M.A. Martins-Loucao, and H. Freitas. 2007. Genetic diversity and differential in vitro responses to Ni in *Cenococcum geophilum* isolates from serpentine soils in Portugal. *Mycorrhiza* 17:677–687.

Graham, R.D., R.M. Welch, and C.D. Walker. 1985. A role for nickel in the resistance of plants to rust. *Aust. Agron. Soc. Proc.* 3:159.

Grotz, N., T. Fox, E. Colnnolly, W. Park, M.L. Guerinot, and D. Eide. 1998. Identification of a family of zinc transporter genes from *Arabidopsis* that respond to zinc deficiency. *Proc. Natl. Acad. Sci. U. S. A.* 95:7220–7224.

Hamner, K., J. Eriksson, and H. Kirchmann. 2013. Nickel in Swedish soils and cereal grain in relation to soil properties, fertilization and seed germination. *Acta Agric. Scand. B—Soil Plant Sci.* 63:712–722.

He, S., Z. He, X. Yang, and V. Baligar. 2012. Mechanisms of nickel uptake and hyperaccumulatioin by plants and implications for soil remediation. *Adv. Agron.* 117:117–189.

Hertel, R.F., T. Maass, and V.R. Muller. 1991. Nickel. In *Environmental Health Criteria*. 108:12–24. Geneva, Switzerland: World Health Organization.

Hewett, E.J. 1953. Metal interrelationship in plants. *J. Exp. Bot.* 4:59–64.

Hill, S.A. 1998. Carbon metabolism in mitochondria. In *Plant Metabolism*, eds. D.T. Dennis, D.H. Turpin, D.D. Lefebvre, and D.B. Layzell, pp. 181–199. Singapore: Longman.

Holmgren, G.G.S., M.W. Meyer, R.L. Chaney, and R.B. Daniels. 1993. Cadmium, lead, zinc, copper, and nickel in agricultural soils of the United States of America. *J. Environ. Qual.* 22:335–348.

Ishimaru, Y., M. Suzuki, T. Kobayashi et al. 2005. OsZIP4, a novel zinc-regulated transporter in rice. *J. Exp. Bot.* 56:3207–3214.

Ito, M., M. Kotera, T. Matsumoto, and K. Tatsumi. 2009. Dinuclear nickel complexes modeling the structure and function of the acetyl CoA synthase active site. *Proc. Natl. Acad. Sci. U. S. A.* 106(29):11862–11866.

Jones, J.B. 2012. *Plant Nutrition and Soil Fertility Manual*. Boca Raton, FL: CRC Press.

Kabata-Pendias, A. and A.B. Mukherjee. 2007. Trace elements in group 10. In Trace Elements from Soil to Human, pp. 237–248. New York, USA: Springer.

Kasprzak, S.K. and K. Salnikow. 2007. Nickel toxicity and carcinogenesis. In *Nickel and Its Surprising Impact in Nature*, eds. A. Sigel, H. Sigel, and R.K.O. Sigel, pp. 619–647. Chichester, U.K.: Wiley.

Kerkeb, L. and U. Kramer. 2003. The role of free histidine in xylem loading of nickel *in Alyssum lesbiacum* and *Brassica juncea*. *Plant Physiol.* 131:716–724.

Khalid, B.Y. and J. Tinsley. 1980. Some effects of nickel toxicity on rye grass. *Plant Soil* 55:139–144.

Khan, N.K., M. Watanabe, and Y. Watanabe. 1999. Effect of different concentrations of urea with or without nickel addition on spinach (*Spinacia oleracea* L.) growth under hydroponic culture. *Soil Sci. Plant Nutr.* 43(3):569–575.

Kloth, R.H., J.C. Polacco, and T. Hymowitz. 1987. The inheritance of a urease-null trait in soybeans. *Theoret. Appl. Gen.* 73(3):410–418.

Kochian, L.B. 1991. Mechanisms of micronutrient uptake and translocation in plants. In *Micronutrients in Agriculture*, 2nd edn., eds. J.J. Mortvedt, F.R. Cox, L.M. Shuman, and R.M. Welch, pp. 229–296. Madison, WI: Soil Science Society of America.

Konhauser, K.O., E. Pecoits, S.V. Lalonde et al. 2009. Oceanic nickel depletion and a methanogen famine before the Great Oxidation Event. *Nature* 458:750–754.

Korner, L.E., L.M. Moller, and P. Jensen. 1986. Free space uptake and influx of Ni^{2+} in excised barley roots. *Physiol. Plant.* 68:583–289.

Korner, L.E., L.M. Moller, and P. Jensen. 1987. Effects of Ca^{2+} and other divalent cations on uptake of Ni^{2+} by excised barley roots. *Physiol. Plant.* 71:49–54.

Kovacik, J., B. Klejdus, and J. Hedbavny. 2009. Nickel uptake and its effect on some nutrient levels, amino acid contents and oxidative status in *Matricaria chamomilla* plants. *Water Air Soil Pollut.* 202:199–209.

Kozlov, M.V., E. Haukioja, A.V. Bakhtiarov, D.N. Stronganov, and S.N. Zimina. 2000. Root versus canopy uptake of heavy metals by birch in an industrially polluted area: Contrasting behavior of nickel and copper. *Environ. Pollut.* 107:413–420.

Kutman, B.Y., U.B. Kutman, and I. Cakmak. 2013. Nickel-enriched seed and externally supplied nickel improve growth and alleviate foliar urea damage in soybean. *Plant Soil* 363:61–75.

Lancaster, J.R. 1988. *The Bioinorganic Chemistry of Nickel*. Weinheim, Germany: VCH.

Lau, O.L. and S.F. Yang. 1976. Inhibition of ethylene production by cobaltous ions. *Plant Physiol.* 58:114–117.

Macnicol, R.D. and P.H.T. Beckett. 1985. Critical tissue concentrations of potentially toxic elements. *Plant Soil* 85:107–129.

Maeda, H. and N. Dudareva. 2012. The shikimate pathway and aromatic amino acid biosynthesis in plants. *Annu. Rev. Plant Biol.* 63:73–105.

Maheshwari, R. and R.S. Dubey. 2008. Inhibition of ribonuclease and protease activities in germinating rice seeds exposed to nickel. *Acta Physiol. Plant* 30:863–872.

Marschner, H. 1995. Functions of mineral nutrients. In *Mineral Nutrition of Higher Plants*, pp. 364–369. New York, USA: Academic Press.

Mazzafera, P., T. Texotto, and J.C. Polacco. 2013. Nickel in plants. In *Encylopedia of Metalloproteins*, eds. V. Uversdy, R. Kretsinger and E. Permyakov, pp. 1496–1501. New York: Springer.

Merchant, S.S. 2010. The elements of plant micronutrients. *Plant Physiol.* 154 (2):512–515.

Milo, R. and R.L.L. Last. 2012. Achieving diversity in the face of constraints: Lessons from metabolism. *Science* 336:1667–1670.

Mishra, D. and M. Kar. 1974. Nickel in plant growth and metabolism. *Bot. Rev.* 40:395–452.

Misra, S.G. and R.S. Ivedi. 1977. Residual effect of iron-nickel interactions on the availability of nutrients. *Plant Soil* 48:705–708.

Mulrooney, S.B. and R.P. Hausinger. 2003. Nickel uptake and utilization by microorganisms. *FEMS Microbiol. Rev.* 27:239–261.

Nicholas, D.J.D. and W.D.E. Thomas. 1954. Some effects of heavy metals on plants grown in the soil culture, part III. The effect of nickel on fertilizer and soil phosphate uptake and Fe and Ni status of tomato. *Plant Soil* 5:182–193.

Nieboer, E.R., T. Tom, and W.E. Sanford. 1988. Nickel metabolism in man and animals. In *Metal Ions in Biological Systems*, Vol. 23, eds. H. Sigel and A. Sigel, pp. 91–121. New York: Marcel Dekker.

Nielsen, F.H. 1984a. Ultratrace elements in nutrition. *Annu. Rev. Nutr.* 4:21–42.

Nielsen, F.H. 1984b. Nickel. In *Biochemistry of the Essential Ultratrace Elements*, ed. E. Frieden, pp. 293–308. New York: Plenum.

Nielsen, F.H. 1987. Nickel. In *Trace Elements in Human and Animal Nutrition*, 5th edn., Vol. 1, ed. W. Mertz, pp. 245–273. New York: Academic Press.

Nielsen, F.H. 1998. Ultratrace elements in nutrition: Current knowledge and speculation. *J. Trace Elem. Exp. Med.* 11:251–274.

Nielsen, F.H. 2000. Evolutionary events culminating in specific minerals becoming essential to life. *Eur. J. Nutr.* 39:62–66.

Nieminen, T.M., J. Derome, and A. Saarsalmi. 2004. The applicability of needle chemistry for diagnosing heavy metal toxicity to trees. *Water Air Soil Pollut.* 157(1–4):269–279.

Nieminen, T.M., L. Ukonmaanaho, N. Rausch, and W. Shotyk. 2006. Biogeochemistry of nickel and its release into the environment. In *Nickel and Its Surprising Impact in Nature*, eds. A. Sigel, H. Sigel, and R.K.O. Sigel, pp. 1–30. Chichester, U.K.: Wiley.

Nishida, S., A. Aisu, and T. Mizuno. 2012. Induction of IRT1 by the nickel-induced iron-dependent response in *Arabidopsis*. *Plant Signal. Behav.* 7:13–19.

Nishida, S., C. Tsuzuki, A. Kato, A. Aisu, J. Yoshida, and T. Mizuno. 2011. AtIRT1, the primary iron uptake transporter in the root, mediates excess nickel accumulation in *Arabidopsis thaliana*. *Plant Cell Physiol.* 52:1433–1442.

Nyczepir, A.P. and B.W. Wood. 2012. Foliar nickel application can increase the incidence of peach tree short life and consequent peach tree mortality. *HortScience* 47(2):224–227.

Nyczepir, A.P., B.W. Wood, and C.C. Reilly. 2006. Association of *Meloidogyne partityla* with nickel deficiency and mouse-ear of pecan. *HortScience* 41:402–404.

Pacyna, J.M. and E.G. Pacyna. 2001. An assessment of global and regional emissions of trace metals to the atmosphere from anthropogenic sources worldwide. *Environ. Rev.* 9:269–298.

Page, V., L. Weisskopf, and U. Feller. 2006. Heavy metals in white lupine: Uptake, root-to-shoot transfer and redistribution within the plant. *New Phytol.* 171:329–341.

Pais, I. and J.B. Jones. 1997. *The Handbook of Trace Elements*. Boca Raton, FL: St. Lucie.

Palacios, J.M., H. Manyani, M. Martinez et al. 2005. Genetics and biotechnology of the H_2-uptake [NiFe] hydrogenase from *Rhizobium leguminosarum* bv. Viciae, a legume endosymbiotic bacterium. *Biochem. Soc. Trans.* 33:94–96.

Pandey, N. and C.P. Sharma. 2002. Effect of heavy metals Co^{2+}, Ni^{2+} and Cd^{2+} on growth and metabolism of cabbage. *Plant Sci.* 163:753–758.

Parida, B.K., J.M. Chhibba, and V.K. Nayyar. 2003. Effect of nickel contaminated soil on fenugreek (*Trigonella corniculata* L.) growth and mineral composition. *Sci. Hortic.* 98:113–119.

Pennazio, S. and P. Roggero. 1992. Effect of cadmium and nickel on ethylene biosynthesis in soybean. *Biol. Plant.* 34:345–349.

Phipps T., S.L. Tank, J. Wirtz et al. 2002. Essentiality of nickel and homeostatic mechanisms for its regulation in terrestrial organisms. *Environ. Rev.* 10:209–261.

Polacco, J.C. 1981. A soybean seed urease-null produces urease in cell culture Glycine max. *Plant Physiol.* 66:1233–1240.

Polacco, J.C. and M.A. Holland. 1993. Roles of urease in plant cells. *Int. Rev. Cytol.* 145:65–103.

Polacco, J., P. Maxxafera, and T. Tezotto. 2013. Opinion—Nickel and urease in plants: Sill many knowledge gaps. *Plant Sci.* 199–200:79–90.

Pond, A.P., J.L. Walworth, M.W. Kilby, R.D. Gibson, R.E. Cal, and H. Nunez. 2006. Leaf nutrient levels for pecans. *HortScience* 41:1339–1341.

Prasad, M.N.V. 2005. Nickelophilous plants and their significance in phytotechnologies. *Braz. J. Plant Physiol.* 17:113–128.

Punshon, T., A. Lanzirotti, S. Harper, P.M. Bertsch, and J. Burger. 2005. Distribution and speciation of metals in annual rings of black willow. *J. Environ. Qual.* 34:1165–1173.

Ragsdale, S. W. 2004. Life with carbon monoxide. *Crit. Rev. Biochem. Mol. Biol.* 39:165–195.

Rascio, N. and F. Navari-Izzo. 2011. Heavy metal hyperaccumulationg plants: How and why do they do it, and what makes them so interesting? *Plant Sci.* 180:169–181.

Redjala, T., R. Sterckeman, S. Skiker, and G. Echevarria. 2012. Contribution of apoplast and symplast to short-term nickel uptake by maize and *Leptoplax emarginata* roots. *Environ. Exp. Bot.* 68:99–106.

Reeves, R.D. 1992. The hyperaccumulation of nickel by serpentine plants. In *The Vegetation of Ultramafic (Serpentine) Soils. Proceedings of the First International Conference on Serpentine Ecology, University of California, Davis, 1991*, eds. A.J.M. Baker, J. Proctor, and R.D. Reeves, pp. 253–277. Andover, U.K.: Intercept.

Reilly, C., M. Crawford, and J. Buck. 2005. Nickel suppresses daylily rust, *Puccinia hemerocallidis* on susceptible daylilys, *Hemerocallis* in greenhouse and field trials. (Abstr.) *Phytopathology* 95:S88.

Riesen, O. and U. Feller. 2005. Redistribution of nickel, cobalt, manganese, zinc, and cadmium via the phloem in young and maturing wheat. *J. Plant Nutr.* 28:421–430.

Roach, W.A. and C. Barclay. 1946. Nickel and multiple trace element deficiencies in agricultural crops. *Nature* 157(3995):696.

Ros, R., D.T. Cooke, R.S. Burden, and C.S. James. 1990. Effects of herbicide MCPA, and the heavy metals, cadmium and nickel on lipid composition, Mg^{2+}-ATPase activity and fluidity of plasma membranes from rice, *Oryza sativa* (cv. Bahia) shoots. *J. Exp. Bot.* 41:457–462.

Russell, M.J. 2007. The alkaline solution to the emergence of life: Energy, entropy and early evolution. *Acta Biotheor.* 55:133–179.

Ruter, J.M. 2005. Effect of nickel applications for the control of mouse ear disorder on river birch. *J. Environ. Hortic.* 23(1):17–20.

Sabir, M., A. Ghafoor, M. Saifullah, M. Zia-Ur-Rhman, H. Ahmad, and T. Aziz. 2011. Growth and metal ionic composition of *Zea mays* as affected by nickel supplementation in the nutrient solution. *Int. J. Agric. Biol.* 13(2):186–190.

Schubert, K.R. and M.J. Boland. 1990. The Ureides. In *The Biochemistry of Plants*, eds. B.J. Miflin, and P.J. Lea, Vol. 16, pp. 197–283. San Diego, CA: Academic Press.

Singh, B., Y.P. Dang, and S.C. Mehta. 1990. Influence of nitrogen on the behavior of nickel in wheat. *Plant Soil* 127:213–218.

Smith, M.W., C.T. Rohla, and W.D. Goff. 2012. Pecan leaf elemental sufficiency ranges and fertilizer recommendations. *HortTechnology* 22:594–599.

Szabodos, L. and A. Savoure. 2010. Proline: A multifunctional amino acid. *Trends Plant Sci.* 15:89–97.

Taylor, R.W. and D.W. Allinson. 1981. Influence of lead, cadmium, and nickel on the growth of *Medicago sativa* L. *Plant Soil* 60:223–236.

Tejada-Jimenez, M., A. Galvan, E. Fernandez, and A. Llamas. 2009. Homeostasis of the micronutrients Ni, Mo and Cl with specific biochemical functions. *Curr. Opin. Plant Biol.* 12:358–363.

Tezotto, T., J.L. Favarin, R.A. Azevedo, L.R.F. Alleoni, and P. Mazzafera. 2012. Coffee is highly tolerant to cadmium, nickel and zinc: Plant and soil nutritional status, metal distribution and bean yield. *Field Crops Res.* 125:25–34.

Thomson, A.J. 1982. Proteins containing nickel. *Nature* 298:602–603.

Triffin, L.O. 1971. Translocation of nickel in xylem exudates of plants. *Plant Physiol.* 48:273–277.

Verbruggen, N., C. Hermans, and H. Schat. 2009. Molecular mechanisms of metal hyperaccumulation in plants. *New Phytol.* 181:759–776.

Vert, G., M. Barberon, E. Zelazny, M. Seguela, J.F. Briat, and C. Curie. 2009. *Arabidopsis* IRT2 cooperates with the high affinity iron uptake system to maintain iron homeostasis in root epidermal cells. *Planta* 229:1171–1179.

Waldon, J.J., J.C. Rutherford, D. Ford, and N.J. Robinson. 2009. Metalloproteins and metal sensing. *Nature* 460:823–829.

Walker, C.D., R.D. Graham, J.T. Madison, E.E. Cary, and R.M. Welch. 1985. Effects of Ni deficiency on some nitrogen metabolites in cowpeas (V*igna unguiculata* L. Walp). *Plant Physiol.* 79:474–479.

Walsh, C.T. and W.H. Orme-Johnson. 1987. Nickel enzymes. *Biochemistry* 26: 4901–4906.

Welch, R.M. 1981. The biological significance of nickel. *J. Plant Physiol.* 3:345–356.

Welch, R.M. 2008. Linkages between trace elements in food crops and human health. In *Micronutrient Deficiencies in Global Crop Production*, ed. B.J. Alloway, pp. 287–309. New York: Springer.

Wells, M.L. and B.W. Wood. 2008. Foliar boron and nickel applications reduce water-stage fruit-split of pecan. *HortScience* 43:1437–1440.

Wintz, H., T. Fox, Y.Y. Wu et al. 2003. Expression profiles of *Arabidopsis thaliana* in mineral deficiencies reveal novel transporters involved in metal homeostasis. *J. Biol. Chem.* 278:47644–47653.

Witte, C.P. 2011. Urea metabolism in plants. *Plant Sci.* 180:431–438.

Wood, B.W. 2012. Nickel deficiency symptoms are influenced by foliar Zn:Ni or Cu:Ni concentration ratio. *Acta Horticulturae* 868:163–169.

Wood, B.W. 2013. Iron-induced nickel deficiency in pecan. *HortScience* 48:1145–1153.

Wood, B.W., R. Chaney, and M. Crawford. 2006a. Correcting micronutrient deficiency using metal hyper accu-mulators: *Alyssum* biomass as a natural product. *HortScience* 41:1231–1234.

Wood, B.W. and C.C. Reilly. 2007. Nickel and plant disease. In *Mineral Nutrition and Plant Disease*, eds. L.E. Datnoff, W.H. Elmer, and D.M. Huber, pp. 215–232. St. Paul, MN: The American Phytopathological Society.

Wood, B.W., C.C. Reilly, C.H. Bock, and M.W. Hotchkiss. 2012. Suppression of pecan scab by nickel. *HortScience* 47:503–508.

Wood, B.W., C.C. Reilly, and A.P. Nyczepir. 2004a. Mouse-ear of pecan: I. Symptomology and occurrence. *HortScience* 39:87–94.

Wood, B.W., C.C. Reilly, and A.P. Nyczepir. 2004b. Mouse-ear of pecan: A nickel deficiency. *HortScience* 39:1238–1242.

Wood, B.W., C.C. Reilly, and A.P. Nyczepir. 2006b. Field deficiency of nickel in trees: Symptoms and causes. *Acta Hortic.* 721:83–98.

Yang, X., V.C. Balligar, D.C. Martens, and R.B. Clark. 1996. Plant tolerance to nickel toxicity. II. Nickel effects on influx and transport of mineral nutrients of four plant species. *J. Plant Nutr.* 19:265–279.

Yusuf, M., Q. Fariduddin, S. Hayat, and A. Ahmad. 2011. Nickel: An overview of uptake, essentiality and toxicity. *Bull. Environ. Contam. Toxicol.* 86:1–17.

Zeller, S. and U. Feller. 1999. Long-distance transport of cobalt and nickel in maturing wheat. *Eur. J. Agron.* 10:91–98.

Zerner, B. 1991. Recent advances in the chemistry of an old enzyme, urease. *Bioorg. Chem.* 19:116–131.

15 Zinc

Allen V. Barker and Touria E. Eaton

CONTENTS

15.1 HISTORICAL BACKGROUND

Chapman (1965) reviewed the early research on the essentiality of zinc (Zn) and noted the work of pioneers in the field of basic plant nutrition and the work of investigators who studied crop nutrition. Recognition of the discovery of Zn as an essential element is credited to Sommer and Lipman (1926). The modern history of Zn as a nutrient started in 1918, in a pecan (*Carya illinoinensis* K. Koch) orchard in Georgia, United States (Skinner and Demaree, 1926; Storey, 2007). In this orchard, trees had increased in trunk diameter, but their tops had dieback or rosetting. Researchers and growers tried fertilizers and cover crops without success to overcome the dieback, but none of these means were successful. The common assumption among pecan growers and researchers at the time was that rosetting was due to iron deficiency (Storey, 2007). In 1932, researchers discovered that Zn and not iron was the corrective element for rosetting (Alben et al., 1932). Alben et al. (1932) used solutions of $FeCl_2$ or $FeSO_4$ in their treatments for rosetting and obtained conflicting results (Storey, 2007). Other treatments included injections into dormant trees, soil applications while the trees were dormant and after the foliage was well developed, and foliar applications. The only favorable results were obtained if the iron solutions were mixed in Zn-galvanized containers (Alben et al., 1932; Storey, 2007). Analysis of the successful solutions indicated that they contained

537

Zn dissolved from the galvanized coatings of the containers. These experiments led to the use of $ZnSO_4$ and $ZnCl_2$ solutions to correct rosetting on trees in alkaline or acid soils (Storey, 2007). The most satisfactory results were obtained with a foliar spray of 0.18% $ZnSO_4$ and 0.012% $ZnCl_2$ (Alben et al., 1932).

15.2   ZINC AS AN ESSENTIAL ELEMENT IN PLANT METABOLISM

In plants as well as in other biological systems, Zn exists only as Zn^{2+} and does not take part in oxidation–reduction reactions. The metabolic functions of Zn are based on its strong tendency to form tetrahedral complexes with ligands of nitrogen, oxygen, and particularly sulfur. These tetra-hedral complexes allow Zn to play a catalytic and structural role in enzyme reactions (Vallee and Auld, 1990).

15.2.1   ZINC INVOLVEMENT IN ENZYMATIC REACTIONS

Zinc is an integral component of enzyme structures and has catalytic or structural functions (Valee and Auld, 1990; Vallee and Falchuk, 1993). The Zn atom is coordinated to four ligands in enzymes with catalytic functions. Three of the ligands are amino acids, with histidine being the most frequent, which accounts for 28% of all of the Zn-binding ligands. Cysteine, aspartic acid, and glutamic acid are also important Zn ligands. A water molecule is the fourth ligand at all catalytic sites. The structural Zn atoms are coordinated to the S-groups of cysteine residues forming a tertiary structure of high stability.

15.2.1.1   Zinc-Containing Enzymes

Zinc is required for the structure and function of a wide range of macromolecules including hundreds of enzymes (Alloway, 2009; Broadley et al., 2012; Coleman, 1992, 1998). In biological systems, Zn is the only metal that is present in six enzyme classes, which are oxidoreductases, transferases, hydrolases, lyases, isomerases, and ligases (Barak and Helmke, 1993; Broadley et al., 2012; Sousa et al., 2009). In these enzymes, four types of Zn-binding sites have been identi-fied as catalytic, structural, cocatalytic, and protein interface sites. In enzymes with catalytic Zn sites, such as carbonic anhydrase, Zn ions are coordinated to protein ligands and water. Structural Zn sites contribute to maintenance of the configuration of enzymes, such as alcohol dehydroge-nase and ones involved in DNA replication. In these enzymes, Zn ions are coordinated to cysteine residues. Cocatalytic Zn sites are present in enzymes containing two or more Zn atoms with binding to aspartic acid and histidine as the most common ligands. At the protein interface, Zn bridges proteins or subunits and affects protein–protein interactions (Auld and Bergman, 2008; Auld, 2009).

Enzymes contain different amount of Zn. For example, alkaline phosphatase and phospholipase contain three Zn atoms each, of which at least one Zn atom has catalytic functions (Broadley et al., 2012; Coleman, 1992, 1998). Carboxypeptidase contains one Zn atom with catalytic functions. RNA polymerase contains two Zn atoms per molecule and is inactive if Zn is removed (Falchuk et al., 1977; Prask and Plocke, 1971).

15.2.1.1.1   Alcohol Dehydrogenase

This enzyme contains two Zn atoms per molecule, one with a catalytic function and one with a structural function (Auld and Bergman, 2008; Broadley et al., 2012; Coleman, 1992). This enzyme has been studied more widely in animals than in plants. Alcohol dehydrogenase catalyzes the reversible oxidation of ethanol to acetaldehyde:

$$Ethanol + NAD^+ \leftrightarrow Acetaldehyde + NADH + H^+$$

In plants, ethanol formation takes place mainly in meristematic tissues, such as root apices, under anaerobic conditions. Under aerobic conditions, the consequences of Zn deficiency on the activity of alcohol dehydrogenase are not known (Broadley et al., 2012). Under anaerobic conditions, Zn deficiency decreases the activity of alcohol dehydrogenase. In lowland rice (*Oryza sativa* L), flooding stimulates root alcohol dehydrogenase activity twice as much in Zn-sufficient as in Zn-deficient plants. The lower activity of this key enzyme with Zn deficiency may impair root functions of lowland rice (Moore and Patrick, 1988). The activities of copper–zinc superoxide dismutase (CuZnSOD) and carbonic anhydrase (see the following sections addressing these enzymes) correlate with differences in Zn efficiency in plants. It has been suggested that for economy of utilization of Zn, Zn is allocated for these essential Zn proteins during Zn deficiency. Time-based analysis after Zn resupply to previously depleted plants revealed that the activity of carbonic anhydrase, which functions in photosynthesis, was the first to recover after Zn addition, whereas the recoveries of the activities of superoxide dismutase and alcohol dehydrogenase were delayed, suggesting that carbonic anhydrase receives priority in Zn delivery (Li et al., 2013).

15.2.1.1.2 Carbonic Anhydrase

Carbonic anhydrase is a zinc-containing enzyme involved in photosynthesis. Carbonic anhydrase catalyzes the reversible hydration of CO_2:

$$CO_2 + H_2O \leftrightarrow HCO_3^- + H^+$$

Carbon dioxide (CO_2) and bicarbonate (HCO_3^-) are substrates for photosynthesis. Carbonic anhydrase is localized in chloroplasts and cytoplasm and maintains an equilibrium between CO_2 and HCO_3^- (Sandmann and Boger, 1983). Thus, Zn can have effects on photosynthesis in plants. Qiao et al. (2014) reported that foliar application of Zn (0.007 mM) benefitted rice plants at the tillering stage and increased chloroplastic carbonic anhydrase activity and thereby increased photosynthesis and chlorophyll contents of leaves. Excessive Zn can limit photosynthesis (Sagardoy et al., 2010). In this case, the limitation to photosynthesis was attributed to restricted stomatal conductance with stomata being unable to respond to physiological and chemical stimuli. The effects of excessive Zn on photochemistry were minor.

Carbonic anhydrase serves in carbon dioxide concentrating mechanisms in plants and algae. The means of supplying CO_2 to the C_3 cycle in cyanobacteria is the result of coordinated functions of high-affinity systems for the uptake of CO_2 and HCO_3^-, as well as intracellular interconversion of CO_2 and HCO_3^- by carbonic anhydrase (Kupriyanova et al., 2013). These biochemical events are under genetic control and serve to maintain cellular homeostasis and adaptation to CO_2 limitation. Zarzycki et al. (2013) noted that cyanobacteria have a carbon dioxide concentrating mechanism that consists of transporter systems for carbon dioxide and a specialized organelle, the carboxysome. The carboxysome encapsulates ribulose-1,5-bisphosphate carboxylase/oxygenase (Rubisco) together with carbonic anhydrase in a protein shell, resulting in an elevated CO_2 concentration around Rubisco and potentially enhancing photosynthesis. Sinetova et al. (2012) suggested that CO_2 might be generated from HCO_3^- by carbonic anhydrase in the thylakoid lumen with subsequent CO_2 diffusion into the pyrenoid (subcellular microcompartments in chloroplasts of many algae). Pyrenoids are associated with the operation of a CO_2 concentrating mechanism and act as centers of photosynthesis. Their main function is to act as centers of CO_2 fixation, by generating and maintaining a CO_2-rich environment around Rubisco. Pyrenoids seem to have a role analogous to that of carboxysomes in cyanobacteria.

The role of carbonic anhydrase, particularly in chloroplasts, differs between C_3 and C_4 plants and in C_4 plants differs between mesophyll and bundle-sheath chloroplasts (Broadley et al., 2012). The biochemistry and leaf anatomy of plants using C_4 photosynthesis promote the concentration of atmospheric CO_2 in leaf tissue that leads to improvements in growth and yield of C_4 plants over C_3 species in hot, dry, high-light, or saline environments (Ludwig, 2013).

In C_3 plants, carbonic anhydrase facilitates the diffusion of CO_2 to the sites of carboxylation by Rubisco. In C_4 plants, it hydrates atmospheric CO_2 to HCO_3^- in the cytosol of mesophyll cells (Ludwig, 2012), before addition of the carbon to phosphoenolpyruvate (PEP) in a reaction catalyzed by the enzyme PEP carboxylase (Badger and Price, 1994). Sasaki et al. (1998) reported that in rice, Zn deficiency resulted in a decrease in the expression of mRNAs for carbonic anhydrase.

In C_4 plants, a high carbonic anhydrase activity is required in the mesophyll chloroplasts to shift the equilibrium toward HCO_3^-, the substrate for PEP carboxylase, which forms C_4 compounds, such as malate, which is moved to the bundle-sheath chloroplasts (Burnell and Hatch, 1988; Hatch and Burnell, 1990). In the bundle-sheath chloroplasts, CO_2 is released from malate and serves as a substrate for Rubisco. Rubisco operates efficiently under the high CO_2 concentrations generated by decarboxylation of malate in the bundle-sheath cells (Gowik and Westhoff, 2011). Despite similar total activities in leaves of C_3 and C_4 plants, only 1% of the total carbonic anhydrase activity in C_4 plants is located in the bundle-sheath chloroplasts (Broadley et al., 2012; Burnell and Hatch, 1988; Utsunomiya and Muto, 1993).

Analysis of Zn-requiring enzymes in bean (*Phaseolus vulgaris* L.) leaves revealed that a Zn-efficient genotype maintained significantly higher levels of carbonic anhydrase and CuZnSOD activity under Zn deficiency than an inefficient genotype (Hacisalihoglu et al., 2004). Rengel (1995) suggested that the ability of Zn-efficient wheat genotypes to maintain greater carbonic anhydrase activity under Zn deficiency may be beneficial in maintaining the photosynthetic rate and dry matter production at a high level. Restoration of Zn supply to deficient cauliflower (*Brassica oleracea* var. *botrytis* L.) increased stomatal aperture, transpiration, and carbonic anhydrase activity within 2 h. The guard cells in epidermal peels from the Zn-deficient leaves had less K^+ than those from the Zn-sufficient plants, suggesting that Zn functions in K^+ transport into guard cells (Sharma et al., 1995). The activities of antioxidative enzymes helped to maintain carbonic anhydrase activity and photosynthesis and plant dry matter production under low Cd toxicity and stress low Zn supply (Khan et al., 2008).

15.2.1.1.3 Copper–Zinc Superoxide Dismutase

Aerobic organisms form superoxide (O_2^-) (Apel and Hirt, 2004; Day, 2014; Foyer and Noctor, 2000; Foyer and Shigeoka, 2011; Miller et al., 2012). Superoxides form during photosynthesis and in plants under biotic and abiotic stresses (Baniulis et al., 2013; Lee et al., 2010b; Mukhopadhyay et al., 2013; Perltreves and Galun, 1991; Pilon et al., 2011; Yu and Rengel, 1999). Stresses have been shown to interact with Zn deficiency, and addition of Zn to plant-growing systems sometimes increases superoxide dismutase activity and helps to overcome the growth-limiting effects of stress (Bonnet et al., 2000; Burman et al., 2013; Cakmak et al., 1997; Daneshbakhsh et al., 2013; Hajiboland and Amirazad, 2010; Lopez-Millan et al., 2005; Tavallali et al., 2010; Thi et al., 2012). Zinc-deficient plants have been reported to show signs of oxidative stress such as accumulation of H_2O_2 and other compounds and declines in activities of superoxide dismutase (Sharma et al., 2004). Daneshbakhsh et al. (2013) suggested that differential tolerance to salt stress among wheat (*Triticum aestivum* L.) genotypes was related to their tolerance to Zn deficiency, sulfhydryl content, and root activity of catalase and superoxide dismutase.

Superoxide dismutases catalyze the conversion of O_2^- to oxygen (O_2) and hydrogen peroxide (H_2O_2). A dismutation reaction, sometimes called disproportionation, is an oxidation–reduction reaction in which a molecular species is reduced and oxidized simultaneously to form two different products, as is the case with superoxide dismutase (Abreu and Cabelli, 2010):

$$O_2^- \rightarrow O_2 + H_2O_2$$

The H_2O_2 is converted to oxygen and water by catalase, which is an iron-containing enzyme.

The CuZnSOD is the most abundant superoxide dismutase in plant cells (Broadley et al., 2012). The copper atom may represent the catalytic metal component with Zn being the structural metal

(Abreu and Cabelli, 2010). In a number of plant species, Zn deficiency limits CuZnSOD activity, and a resupply of Zn rapidly restores enzyme activity (Cakmak, 2000). Activity is suppressed strongly by a low supply of Zn and is a good indicator of Zn deficiency (Broadley et al., 2012; Cakmak et al., 1997; Hacisalihoglu et al., 2004; Yu et al., 1999). The decrease in superoxide dismutase activity under Zn deficiency conditions is critical because of the increase in the rate of generation of oxygen radical O_2^-, which can lead, along with other reactive oxygen species, to peroxidation of membrane lipids and to an increase in membrane permeability (Cakmak and Marschner, 1988a,b; Wang and Jin, 2005). Overexpression of CuZnSOD in plants increases tolerance to various abiotic stress factors (Cakmak, 2000).

15.2.1.1.4 Other Zinc-Containing Enzymes

Zinc is the metal component in a number of other enzymes, for example, alkaline phosphatase, phospholipase, and carboxypeptidase (Broadley et al., 2012; Coleman, 1992, 1998; Falchuk et al., 1977; Prask and Plocke, 1971).

15.2.1.2 Zinc-Activated Enzymes

In plants, Zn is required for or affects the activities of various enzymes, including phosphatases, dehydrogenases, aldolases, isomerases, transphosphorylases, and RNA and DNA polymerases (Broadley et al., 2012). Inorganic pyrophosphatases are important components of the proton-pumping activity in cell membranes, such as the tonoplast. In addition to the Mg^{2+}-dependent enzyme, a pyrophosphatase isoenzyme in leaves is Zn^{2+} dependent (Lin and Kao, 1990). Zinc-dependent proteins (Zn metalloproteins) have been identified as being involved in DNA replication and transcription and, thus, regulation of gene expression (Andreini et al., 2009; Coleman, 1992; Vallee and Falchuk, 1993).

15.2.2 ZINC INVOLVEMENT IN PROTEIN SYNTHESIS

In Zn-deficient plants, the rate of protein synthesis and the protein concentration are suppressed, and free amino acids accumulate (Broadley et al., 2012). With resupply of Zn to deficient plants, protein synthesis resumes. In a role in protein synthesis, Zn is a structural component of ribosomes. In Zn-sufficient cells of *Euglena*, the concentration of Zn in ribosomes is from 650 to 1280 µg g^{-1} ribosomal RNA, whereas in Zn-deficient cells, it is 300–380 µg g^{-1} (Prask and Plocke, 1971). In the absence of Zn, ribosomes disintegrate but reform after resumption of Zn supply. In shoot meristems of rice, disintegration of ribosomes occurs if Zn concentration is below 100 µg g^{-1} dry weight (Broadley et al., 2012). Considerably, lower Zn concentrations of about 50 µg g^{-1} decrease protein concentration. In tobacco (*Nicotiana tabacum* L.) tissue culture cells, the corresponding concentrations were 70 µg g^{-1} for a decrease in ribosomes and 50 µg g^{-1} for a decrease in protein concentration (Obata and Umebayashi, 1988).

The Zn concentration in the growing tips of pollen tubes was about 150 µg g^{-1} dry weight compared with about 50 µg g^{-1} in more basal regions, suggesting a high requirement for protein synthesis (Ender et al., 1983). In the new root tips of wheat (*T. aestivum*), Zn concentrations were about 220 µg g^{-1} dry weight (Ozturk et al., 2006). In the shoot meristems, a Zn concentration of at least 100 µg g^{-1} dry weight was required for maintenance of protein synthesis (Broadley et al., 2012). To meet the high Zn demand in the shoot meristems, Zn is preferentially translocated there (Kitagishi and Obata, 1986).

15.2.3 ZINC INVOLVEMENT IN CARBOHYDRATE METABOLISM

In Zn-deficient leaves, a rapid decrease in carbonic anhydrase activity is the most sensitive and obvious change in activity of enzymes of carbohydrate metabolism (Broadley et al., 2012). The activity of fructose-1,6-bisphosphatase also declines fairly rapidly, whereas the activities of other enzymes, such as PEP carboxylase, Rubisco, and the malic enzyme, are affected to a much lesser extent, particularly with short durations of Zn deficiency (Broadley et al., 2012). Despite restrictions in enzyme activities and in the rate of photosynthesis, sugars and starch often accumulate in Zn-deficient plants. Soon after the Zn supply is restored, the sugar concentration and Hill reaction activity are comparable to those of the adequately supplied plants continuously receiving 1.0 μM Zn (Broadley et al., 2012). The accumulation of carbohydrates in Zn-deficient leaves increases with light intensity (Marschner and Cakmak, 1989) and is an expression of impaired new growth, particularly of the shoot apices. However, experimental evidence indicates that Zn deficiency–induced changes in carbohydrate metabolism are not primarily responsible for either growth retardation or the symptoms of Zn deficiency (Broadley et al., 2012).

15.2.4 ZINC INVOLVEMENT IN INDOLEACETIC ACID SYNTHESIS

Stunted growth, rosetting, and little leaf are distinct symptoms of zinc deficiency and are related to disturbances in the metabolism of auxin, for example, indoleacetic acid (IAA) (Storey, 2007). However, the action of Zn in auxin metabolism is unclear, and several pathways of synthesis of IAA are proposed (Mano and Nemoto, 2012; Tivendale et al., 2014). With Zn-deficient tomato (*Lycopersicon esculentum* Mill.), retarded stem elongation is correlated with a decrease in IAA synthesis (Broadley et al., 2012). With resupply of Zn, stem elongation and IAA concentration increase (Tsui, 1948). Low concentrations of IAA in Zn-deficient plants may result from inhibited synthesis or accelerated degradation of IAA (Cakmak et al., 1989), so the role of Zn may be in the maintenance of IAA.

Tryptophan is the likely precursor for the biosynthesis of IAA, and low IAA concentrations of IAA in Zn-deficient leaves and tryptophan accumulation suggest a role for Zn in biosynthesis of IAA from the amino acid (Salami and Kenefick, 1970). In leaves of Zn-deficient plants, tryptophan concentrations increase along with other amino acids (Cakmak et al., 1989; Domingo et al., 1992), likely as a result of impaired protein synthesis. Adequate Zn nutrition increases the concentrations of endogenous gibberellins (Sekimoto et al., 1997). The low concentrations of IAA and gibberellins may relate to the stunted growth, rosetting, and little leaf with Zn deficiency (Broadley et al., 2012).

On the other hand, Zn toxicity may suppress auxin activity (Kosesakal et al., 2009). Zinc toxicity with 3 or 5 mM $ZnCl_2$ in Hoagland solution caused negative geotropic behavior of tomato roots. The addition of triiodobenzoic acid (TIBA) at 100 mg L^{-1} with seed imbibition or to growth medium with 3 or 5 mM $ZnCl_2$ eliminated negative gravitropic behavior of primary roots and induced positive tropism of roots.

15.2.5 ZINC INVOLVEMENT IN MEMBRANE INTEGRITY

Zinc binds to phospholipid and sulfhydryl groups of membranes or forms complexes with cysteine in polypeptides and protects membrane lipids and proteins from oxidative damage (Broadley et al., 2012). In its function as a metal component in CuZnSOD, Zn may also control the generation of reactive oxygen species such as O_2^- (Cakmak and Marschner, 1988a,b). The permeability of the plasma membrane increases during zinc deficiency. With roots under Zn deficiency, Welch et al. (1982) reported leakage of solutes and decreases in phospholipid concentration and amount of unsaturation of fatty acids in membrane lipids. After resupplying Zn, some restoration of membrane integrity was observed. Plasma membrane vesicles isolated from Zn-deficient roots had higher permeability than vesicles from Zn-sufficient roots (Pinton et al., 1993). With Zn-deficient plants, the activities of H_2O_2 scavenging enzymes, such as catalase and ascorbate peroxidase, were restricted, probably resulting from inhibited protein synthesis (Broadley et al., 2014; Cakmak, 2000; Chen et al., 2008; Yu et al., 1999).

15.3 AVAILABILITY OF ZINC TO PLANTS

15.3.1 Zinc Content, Forms, and Dynamics in Soils

Zinc deficiency is the most widely occurring micronutrient deficiency in world crops, and crop production is limited by Zn deficiency on millions of hectares of cropland. Consequently, about one-third of humans suffer from inadequate Zn in diets. The main soil factors affecting the availability of Zn to plants are low total Zn contents, high pH, high calcite and organic matter contents, and high concentrations of Na, Ca, Mg, bicarbonate, and phosphate in the soil solution or in labile forms (Alloway, 2009). Zinc is taken up by plants predominantly as the divalent cation (Zn^{2+}); however, at high pH, it may be absorbed as a monovalent cation ($ZnOH^+$). In long-distance transport in the xylem, Zn is either bound to organic acids or occurs as the free divalent cation. In the phloem sap, Zn concentrations are fairly high, with Zn possibly complexed by low-molecular-weight organic solutes (Kochian, 1991).

The average concentrations of Zn in the crust of the earth and granitic and basaltic igneous rock are approximately 70, 40, and 100 mg kg^{-1}, respectively (Bryson et al., 2014; Taylor, 1964), whereas sedimentary rocks such as limestone, sandstone, and shale contain 20, 16, and 95 mg kg^{-1}, respectively (Turekian and Wedepohl, 1961). The total Zn content in soils varies from 3 to 770 mg kg^{-1} with the world average being 64 mg kg^{-1} (Alloway, 2009). Soils with concentrations below 10 mg kg^{-1} are low and are likely to be Zn deficient. Generally, soils with above 200 mg Zn kg^{-1} have been contaminated (Alloway, 2009).

Five major pools of Zn exist in the soil, (1) soil solution, (2) surface-adsorbed and exchangeable Zn, (3) organic matter, (4) oxides and carbonates, and (5) primary minerals and secondary aluminosilicate materials (Shuman, 1991). There is evidence that Zn^{2+} activities in the soil solution may be controlled by the oxide mineral franklinite ($ZnFe_2O_4$), the solubility of which is similar to that of soil-held Zn over pH values of 6–9 (Lindsay, 1991; Ma and Lindsay, 1993). The mineral precipitates if Zn concentration in the soil solution exceeds the equilibrium solubility of the mineral and will dissolve if the opposite condition occurs, thereby providing a Zn-buffering system. Zinc may be associated with soil organic matter, which includes water-soluble or water-insoluble compounds. Zinc is bound by incorporation into organic molecules or held by adsorption (Shuman, 1991). Zinc is associated with hydrous oxides and carbonates by adsorption, surface complex formations, ion exchange, incorporation into the crystal lattice, and coprecipitation (Shuman, 1991). Some of these reactions fix Zn strongly and are believed to control the amount of Zn in the soil solution (Jenny, 1968).

15.3.2 Uptake and Transport of Zinc by Plants

15.3.2.1 Genetic Control

As mentioned earlier, Zn is absorbed predominantly as Zn^{2+}, but at high pH, it is absorbed as a monovalent cation ($ZnOH^+$) (Marschner, 1995). Based on effects of metabolic inhibitors on Zn accumulation, early research on absorption of Zn by roots suggested that the process is not metabolic. Chandel and Saxena (1980) wrote that the initial absorption into roots was not metabolic but that translocation from roots to shoots was metabolic. Joseph et al. (1971) reported that the kinetics and effects of inhibitors indicated that Zn absorption occurred mainly by a nonmetabolic process. Recent research, however, shows that Zn absorption is controlled metabolically and regulated by genetic factors. Absorption of Zn varies with species and cultivars (Storey, 2007). For example, Khadr and Wallace (1964) reported that rough lemon (*Citrus jambhiri* Lush.) in solution culture absorbed from the solution and translocated to leaves more Zn and Fe than trifoliate orange (*Poncirus trifoliata* Raf.). Rough lemon was reported as being very resistant to Zn and iron deficiencies, and trifoliate orange was noted as being more susceptible to Zn and iron deficiency than other citrus species (Khadr and Wallace, 1964). Higher concentration of Zn occurred in pecan seedlings originating from west Texas populations than in populations from east Texas, regardless of whether they were grown in central Texas or in Georgia (Storey, 2007). Variation in rate of Zn absorption had been noted previously among seedlings of open-pollinated pecans (Storey, 2007). Storey (2007) suggested that selecting hardwood cuttings from the best of the west Texas populations would be an ideal way to use clonal materials as a means of establishing pecan orchards with uniformity of efficient Zn absorption.

With mustard (*Thlaspi* spp. L.), Lasat et al. (1996) noted that Zn absorption had a linear kinetic component that accumulated as cell-wall-bound Zn^{2+}, representing a nonmetabolic activity, and saturable component that was due to Zn^{2+} influx across the root-cell plasma membrane and that followed Michaelis–Menten kinetics of active transport. Chen et al. (2009) showed that under moderate Zn-deficient conditions, increased root length, root surface, and root tips occurred in rice genotypes but to a greater extent in a Zn-efficient type than in an inefficient type. Ishimaru et al. (2011), in a review, noted that to deal with situations of deficient or toxic amounts of available Zn, Zn uptake and transport must be regulated by plants.

Plants vary widely in susceptibility to Zn deficiency, and this response is related to their capacities to absorb and translocate Zn. Lee et al. (2010a) demonstrated that OsZIP8 is a zinc transporter that functions in Zn uptake and distribution in rice. Ishimaru et al. (2005) isolated from rice four distinct genes, *OsZIP4*, *OsZIP5*, *OsZIP6*, and *OsZIP7*, which exhibit sequence similarity to the gene for the rice ferrous ion transporter, OsIRT1. Takahashi et al. (2012) suggested that a transporter OsHMA2 plays a role in Zn and Cd loading to the xylem and in root-to-shoot translocation in rice. Progress in research on the genetic and molecular basis for Zn efficiency could facilitate the engineering of Zn-efficient plant varieties and could help the production on marginal soils as well as possibly improve the micronutrient density of foods derived from plants (Cakmak, 2008; Gao et al., 2005; Hacisalihoglu and Kochian, 2003). With barrel medic (*Medicago truncatula* Gaertn.), Stephens et al. (2011) characterized the kinetics of three Zn-transporting proteins (MtZIP1, MtZIP5, and MtZIP6) that were noted as having low or high affinity for Zn, with possible functions within different compartments of the plant.

Arnold et al. (2010) suggested that tolerance of rice to Zn deficiency was explained by complexation of Zn by an agent released from the roots and uptake of the complexed Zn by specific root transporters. They further showed by modeling that secretion of the phytosiderophore, deoxymugineic acid, accounted for differences in Zn uptake among genotypes. In studies with castor bean (*Ricinus communis* L.), Stephan et al. (1996) noted that metal complexes with the nonproteinogenic amino acid, nicotianamine, could occur in the apoplasm and in the xylem sap and in another function could control micronutrient transport in phloem.

Singh et al. (2005) noted in a review that cereal species greatly differ in their Zn efficiency, meaning the ability of a plant to grow and yield well under Zn deficiency. Efficiency was attributed to the ability to acquire Zn under conditions of low soil Zn availability rather than to its utilization or translocation within a plant. High Zn acquisition efficiency may be due to an efficient ionic Zn uptake system, good root architecture (long and fine roots) to exploit Zn from a large soil volume, or high synthesis and release of Zn-mobilizing phytosiderophores by the roots.

Rice cultivars differ widely in their susceptibility to Zn deficiency. A Zn-efficient (IR-26) and a Zn-inefficient (IR-36) cultivar were compared for Zn uptake following Zn starvation or abundance (Meng et al., 2014). At low Zn supply after pretreatment, significantly higher Zn^{2+} influx rates occurred for the Zn-efficient genotype than for the Zn-inefficient genotype. At high Zn supply levels, however, a difference (2.5-fold) in Zn^{2+} influx rate between the two genotypes was noted only with the Zn-deficient pretreatments. There appeared to be two separate Zn transport systems mediating the low- and high-affinity Zn influx in the Zn-efficient genotype. Ramesh et al. (2003) also identified Zn transporters in gene databases of rice. It has been suggested that enhanced organic acid release from the roots of Zn-efficient rice plays a strong role in plant tolerance to bicarbonate excess and Zn deficiency. Rose et al. (2011) assessed the tolerance of six rice genotypes to bicarbonate stress. In hydroponic experiments, increased root malate accumulation and efflux were responses to high solution bicarbonate in the short term (12 h) in all genotypes. Citrate and malate accumulation and efflux increased after long-term exposure (10 days) to high bicarbonate and Zn deficiency.

Pecan cultivars grafted to a common rootstock in a replicated test orchard manifested different expressions of Zn deficiency, and concentrations of Zn in leaves varied widely (Storey, 2007). The most severe deficiency symptoms occurred in cultivars with the lowest leaf Zn concentration, but generally, a poor correlation occurred between symptoms and leaf concentrations of Zn. Storey (2007) noted that to develop a molecular understanding for these nutritional observations, efforts have been initiated to identify Zn transporter genes in pecan. Zinc transport across cellular and intracellular membranes is facilitated by several types of membrane-localized proteins, in the ZIP transporter family (Storey, 2007). ZIP stands for ZRT-like, IRT-like protein, with ZRT (Zn-regulated transporter) and IRT (Fe-regulated transporter) referring to metal transporter genes identified in yeast and various plants including pecan (Eide, 1998; Guerinot, 2000; Storey, 2007). Storey (2007) suggested that characterization of pecan ZIP genes, including analysis of possible polymorphisms between genotypes of diverse geographic origin, should enhance understanding of Zn nutrition and provide tools for breeding new Zn-efficient cultivars of pecan.

15.3.2.2 Zinc Uptake and Accumulation as Affected by Other Nutrients

15.3.2.2.1 Interactions with Phosphorus

High concentrations of phosphorus have been detected in Zn-deficient plants (Cakmak and Marschner, 1986). High phosphorus content in Zn-deficient plants fertilized with phosphorus can be attributed partially to a concentration effect due to restricted growth under Zn deficiency (Lynch et al., 2012; Webb and Loneragan, 1988). High applications of phosphorus fertilizers to soils low in available Zn induced Zn deficiency, likely by imparting low solubility of Zn (Loneragan et al., 1979; Lynch et al., 2012; Robson and Pitman, 1983). Also, a high supply of phosphorus sometimes suppresses root growth and colonization of roots with arbuscular mycorrhiza and thereby restricts acquisition of Zn by limiting root mass (Ryan et al., 2008). A P–Zn antagonism on Zn accumulation in corn (*Zea mays* L.) was observed with soils that tested high for phosphorus (Izsaki, 2014). With wheat (*T. aestivum*), phosphorus fertilization reduced grain Zn concentration by 33%–39% and colonization by 33%–75% relative to the unfertilized treatment (Ryan et al., 2008). On the other hand, an increase in root length and mass with the addition of phosphorus can improve nutrient uptake from the soil and perhaps increase Zn accumulation in grains or leaves (Fageria et al., 2014). A decrease in grain and shoot Zn concentration by phosphorus fertilization was related to dilution of Zn by increased grain yield (Loneragan et al., 1979). Li et al. (2003) reported that leaf Zn concentrations decreased with an increase in phosphorus supply to barley (*Hordeum vulgare* L.) cultivars. Addition of Zn compensated for the reduction, and the role of P–Zn interactions in the context of nutritional quality of plant food was discussed. Gianquinto et al. (2000) also reported that for bean, leaf Zn concentration was restricted by the addition of phosphorus to plants grown at low Zn supply. Foliar fertilization may be needed to supply Zn to crops growing in soils in which phosphorus has accumulated (Zhang et al., 2012). A similar effect of phosphorus on manganese uptake in barley roots was reported, and it was suggested that this negative interaction can induce Mn deficiency in the shoot (Pedas et al., 2011).

For plants grown in nutrient solutions with 0.001 M phosphorus as monobasic phosphate, high or toxic concentrations of phosphorus in plant shoots occur, perhaps exceeding 10 g kg^{-1} (1%) dry weight compared with 2 g kg^{-1} (0.2%) in soil-grown crops (Barker, unpublished data). Parker (1993) reported that concentrations of phosphorus in the old leaves of several crop species were high (up to 4.6 g kg^{-1}) when grown in a chelate-buffered solution with 10 µM phosphorus and with low Zn^{2+}. Parker (1993) tested a hydroponics system with phosphorus supplied in a pouch of sparingly soluble hydroxyapatite with chelate buffering and noted that adequate phosphorus was supplied for corn or wheat, although two dicots could not obtain adequate phosphorus from this system. Zinc-deficient corn or wheat did not hyperaccumulate phosphorus.

Commonly, in corn production, an application of a high-analysis phosphorus fertilizer is applied in bands alongside the seeds at planting. The inclusion of Zn chelates in the band is a frequent practice to ensure adequate Zn nutrition for young corn plants grown in soils with low Zn status and in the presence of the high phosphorus concentration in the bands (Bryson et al., 2014).

With hydroponics, a strong effect of solution Zn, ranging from deficient to excessive supply, occurred on phosphorus concentration in potato (*Solanum tuberosum* L.) leaves, but almost no impacts of solution phosphorus on Zn concentration were observed, whereas manganese concentrations in leaves were affected by phosphorus (Barben et al., 2010). With various wheat cultivars, at Zn-deficient level of fertilization in soil, Zn-efficient cultivars had higher Zn concentrations in the shoots than Zn-inefficient cultivars (Imtiaz et al., 2006). Zinc concentrations in all cultivars but one increased if the phosphorus level in the soil was increased moderately. However, the Zn–P interaction showed that at a Zn-deficient level, Zn concentrations in the shoot decreased with an increment of phosphorus. Akhtar (2010) reported that Zn supply had little effect on tissue phosphorus concentration and uptake per unit of root dry matter of oilseed rape (*Brassica napus* L.), but an increase in phosphorus supply caused a significant reduction in Zn uptake per unit of root dry matter and tissue Zn concentration. The reduction in tissue Zn concentration was ascribed entirely to a dilution effect. Zinc concentrations and uptake by a P-efficient cultivar were significantly lower and more sensitive to phosphorus uptake than a P-inefficient cultivar, suggesting that high phosphorus-use efficiency may depress plant Zn uptake. Knight et al. (2004) reported that Zn-efficient bean genotypes had the ability to regulate phosphorus uptake and to avoid excessive accumulation of phosphorus. An increase in P availability caused a significant suppression in Zn uptake per unit of root weight and tissue concentration of Zn in two wheat cultivars (Zhu et al., 2001). The suppression in tissue Zn concentration was not explained fully by a dilution effect. Zinc uptake and Zn concentrations in a P-efficient cultivar were lower than in an inefficient cultivar. The investigators suggested that high P uptake efficiency may depress plant uptake of Zn and, therefore, limit the concentration of Zn in grains of wheat grown in low-P or low-Zn soils. With increasing phosphorus concentration in leaves, Zn deficiency symptoms become more severe, although the Zn concentration was not decreased (Broadley et al., 2012). This response is attributed to lower physiological availability of Zn in leaves (Cakmak and Marschner, 1987). On the other hand, Zhao et al. (1998) suggested that alpine pennycress (*Thlaspi caerulescens* J. and C. Presl), a hyperaccumulator of heavy metals, could accumulate and sequester huge amounts of Zn in the shoots without causing phosphorus deficiency.

Phosphorus amendments to soil have been suggested as a way to immobilize metals in soils and to restrict metal accumulation in plants in contaminated soils; however, the efficacy of treatments varies with the type and quantity of phosphorus amendments and with type of plants and metals (Shahid et al., 2014).

15.3.2.2.2 Interaction with Iron

Zinc application had an adverse effect on iron concentration in plants (Shahriaripour and Tajabadipour, 2010). Zinc-deficient plants had significantly greater concentrations of iron than Zn-sufficient plants, and as the Zn concentration in the substrate was increased, the iron, manganese, and copper concentrations in pistachio (*Pistacia vera* L.) seedlings decreased.

Soil application of Fe-EDDHA or foliar application of Fe sulfate increased iron accumulation but decreased that of Zn, manganese, and copper (Ghasemi-Fasaei and Ronaghi, 2008). Conditions that lead to Zn accumulation in plants may also favor iron accumulation. With spring or winter wheat, a strong positive correlation occurred between iron and Zn in the grain (Morgounov et al., 2007).

15.3.2.2.3 Interactions with Other Elements

Several cases of interaction of Zn with other micronutrients have been reported. The interactions of Zn and boron on plant nutrition and growth are synergistic or antagonistic depending on the concentrations of the elements supplied and on soil conditions. In greenhouse research with corn grown in a calcareous soil, applications of Zn at 5 or 10 mg kg^{-1} depressed boron accumulation in shoots (Hosseini et al., 2007). Applications of boron at 20–80 mg kg^{-1} depressed Zn accumulation, whereas applications of 2.5, 5, or 10 mg kg^{-1} did not affect Zn accumulation. On the other hand, Aref (2012) reported that neither Zn nor boron fertilization affected the accumulation of either element in leaves of field-grown corn. Symptoms of boron toxicity in Zn-deficient sour orange (*Citrus aurantium* L.) in solution culture were mitigated with Zn applications and was noted as being of practical importance since B toxicity and Zn deficiency are encountered jointly in some soils of semiarid zones (Swietlik, 1995). Ziaeyan and Rajaie (2009), based on yield increases with corn, recommended that Zn and boron be applied to corn produced in calcareous soils. Foliar application of boric acid to mandarin orange (*Citrus reticulata* Blanco.) increased Zn in leaves (Ullah et al., 2012). The application of Zn or boron increased the plant height, grain yield, total dry matter yield, and tissue Zn and boron content in wheat in soil in pot culture, and a maximum increase was obtained with the combined application of the nutrients (Singh et al., 2014). With two cultivars of oilseed rape, an increased supply of Zn enhanced B uptake under high boron supply only (Grewal et al., 1998). Also, Zn uptake in a Zn-efficient cultivar was enhanced significantly with increased rate of B supply under high Zn supply, whereas the effect was not significant in an inefficient cultivar.

Biomass production of wheat was depressed by the application of nickel alone, whereas nickel and Zn applied together produced maximum biomass (Ahmad et al., 2010). Bernal et al. (2007) reported that Zn content of soybean leaves decreased with copper treatment of leaves, whereas an enhancement in Zn accumulation was noted with copper treatment through roots. Shahriaripour and Tajabadipour (2010) reported that Zn application had an adverse effect on iron concentration in plants, whereas Zn-deficient plants had greater concentrations of iron than Zn-sufficient pistachio. As the Zn concentration in the substrate was increased, the Fe concentration in plants decreased. Zinc also antagonized the uptake of manganese and copper in the plants. Quartin et al. (2001) studied effects of excessive manganese in nutrient solutions on nutrition of triticale and wheat and reported that plants that avoided accumulation of high manganese concentrations in leaves had high Zn concentrations in leaves.

Aref (2012) reported an increase in potassium concentrations in corn leaves with increased soil application of Zn from 0 to 24 kg ha^{-1} as zinc sulfate or with a single level of foliar spraying (0.5% solution Zn_2SO_4). In this study, neither Zn nor boron had an effect on nitrogen or phosphorus accumulation in corn leaves. Ziaeyan and Rajaie (2009) observed no effect of boron or Zn on potassium concentrations in corn leaves. Moinuddin and Imas (2010) reported that with four rates of potassium application to soil-grown sorghum (*Sorghum bicolor* Moench), application of 10 kg $ZnSO_4$ per hectare surpassed no Zn application by 30%–35% in N accumulation, by 8%–15% in P accumulation, by 33%–36% in K accumulation, by 120%–140% in Zn accumulation, by 19%–21% in fresh matter yield, by 29%–31% in dry matter yield, and by 30%–34% in protein yield.

Application of silicon did not appear to have an effect on cucumber (*Cucumis sativus* L.) nutrition in hydroponics except for increased iron uptake; however, the application coincided with a lessening of leaf damage from excessive Zn. (Dominy and Bertling, 2004). This same report showed that addition of either manganese or Zn to the nutrient solution increased the uptake of all elements investigated (Ca, Mg, Fe, K, Mn, Zn, and Si). Zinc and manganese application further enhanced nutrient accumulation but also increased fruit abortion.

15.3.3 Zinc Accumulation in Plants

In seeds, most of the Zn is localized in phytate (inositol polyphosphate), which is a phosphorus-rich compound (Broadley et al., 2012). Phytate is a negatively charged molecule due to the phosphate groups and has a high affinity for Zn to form sparingly soluble complexes (Broadley et al., 2012; Lönnerdal, 2002; Schlemmer et al., 2009). These complexes are important factors in human diets as they reduce the bioavailability of Zn (Lönnerdal, 2000; Meyer et al., 2013). Zhou et al. (1992) reported that a reduction of phytate in soybean increase the bioavailability of Zn to rats (*Rattus* spp. Fischer de Waldheim). Genetic phytate reduction in sorghum by 80%–86% significantly increased

Zn availability to rats (Kruger et al., 2013). The availability of Zn from plant-based foods that are low in phytate such as sweet potato (*Ipomoea batatas* L.) could be better, and they could also be better sources of calcium and iron, as dietary absorption would be better from them than from seed-based food such as corn (Amagloh et al., 2012). As noted earlier in this chapter, Zn is a component of many proteins that are metabolic enzymes in plants. Zinc is a component also of thousands of proteins in plants (Broadley et al., 2007). Zinc being bound to proteins in seeds or leaves probably does not limit its bioavailability as protein in diets has been shown to enhance Zn absorption modestly (Miller et al., 2013).

15.3.4 BIOFORTIFICATION OF FOODS WITH ZINC

Zinc is vital for many biological functions and has roles in more than 300 enzymes in the human body (Coleman, 1992). Micronutrient deficiency leads to compromised health and economic losses and is prevalent in populations that depend on nondiversified, plant-based diets. Fruits, vegetables, and grains are perceived as principal sources for these microelements; however, fruits, vegetables, and grains are high in vitamins but are often low in essential minerals (Ashmead, 1982). More than 60% of the world population reportedly suffers from iron deficiency, and over 30% has zinc deficiency (Rawat et al., 2013). According to a nutritional survey, approximately 24% of all Chinese children suffered from a serious deficiency of iron, and over 50% showed signs of Zn deficiency (Yang et al., 2007). Imtiaz et al. (2010) reported that 70% of farmland in Pakistan is Zn deficient, yet micronutrient fertilization is negligible. Pirzadeh et al. (2010) reported that grain Zn concentration of more than 54% of rice samples from Iran was less than 20 mg kg^{-1}. This concentration indicates a low level of sufficiency for rice productivity. Also, about 55% of the rice samples were deficient in manganese, and 49% were copper deficient, whereas the iron concentration in rice grains was sufficient for crop production. Increasing mineral content of staple food crops through biofortification is a feasible strategy of combating malnutrition from micronutrient deficiency. Additionally, it will also enhance the agronomic efficiency of crops on mineral-poor soils. A multidisciplinary strategy toward enhancing mineral content of cereal grains should involve fertilization with micronutrients, increased uptake of minerals from soil, and enhanced partitioning toward grain and other edible plant parts. At the same time, it is essential to improve mineral bioavailability from grain-based diets by addressing the limitations imposed by phytates in seeds. Improvements in plant nutrition and soil fertility, application of conventional and modern breeding approaches and genetic engineering, and other actions have been suggested as means for biofortification of crop plants (see Section 15.3.2.1; Cakmak, 2008; Combs et al., 1997; Fageria et al., 2012; Genc et al., 2005; Gregorio, 2002; Hacisalihoglu and Kochian, 2003; King, 2002; Lee and An, 2009; Masuda et al., 2008, 2009; Mayer et al., 2008; McBeath and McLaughlin, 2014; Ruel and Bouis, 1998; Solomons, 2000; White and Brown, 2010; Zimmermann and Hurrell, 2002). Much of this research addressed identification of Zn-efficient cultivars of crops and how this efficiency allows the crops to grow under Zn-deficient conditions in soils as well as enhances the accumulation of Zn in edible plant parts. Some details on specific strategies to enhance Zn in plant-derived foods follow.

A lot of attention has been given to increasing the Zn concentration in rice grains. Phattarakul et al. (2012) evaluated the effect of soil, foliar, or combined soil and foliar Zn fertilization on grain yield and Zn concentration of rice grown in several countries in Asia. As the average of all trials, Zn application increased grain yield by about 5%. However, Zn concentration in brown rice (whole caryopsis with husk removed) was increased by 25% by foliar application, 32% by foliar and soil Zn applications, but only by 2.4% by soil Zn application over concentrations in unfertilized grain. The Zn concentration of unhusked rice (whole grain with husk) was increased by 66% by foliarly applying Zn. The increase in grain Zn concentration by foliar Zn spray was affected by timing of the foliar application with more distinct increases occurring in grain Zn by foliar application being achieved if Zn was applied after flowering. Hussain et al. (2012) reported that Zn application to soil increased grain yield of wheat by 29%, whole-grain Zn concentration by 95%, and whole-grain estimated Zn bioavailability by 74%. Foliar Zn application at heading also increased Zn bioavailability but with diminishing effects as soil application was increased.

Xue et al. (2012) studied the effects of nitrogen nutrition on Zn accumulation and allocation into grain of wheat. Applying at optimal rate (about 200 kg N ha^{-1}) maintained or gave higher grain Zn concentration and total accumulation than no or lower nitrogen additions. Increasing nitrogen supply from optimal to excessive gave no further increases in grain Zn concentration and content. Most of the shoot Zn requirement had been accumulated by anthesis, and accordingly, two-thirds of all of the grain Zn content was provided by Zn remobilization from preanthesis accumulation. Optimal nitrogen management ensured better shoot Zn nutrition to contribute to Zn remobilization from vegetative tissues and to maintain relatively higher grain Zn for human nutrition. Kutman et al. (2011) noted that nitrogen nutrition is a critical factor in the acquisition and grain allocation of Zn and Fe in wheat. They showed that the Zn and Fe uptake per plant of durum wheat (*Triticum durum* Desf.) was enhanced up to fourfold by high nitrogen supply and that the increases in plant growth were much less. Kutman et al. (2012) reported that by promoting tillering and grain yield, extending the grain-filling period, and prolonging the uptake of Zn, increased nitrogen supply promoted Zn accumulation in grain. Gogos et al. (2013) noted that fertilization of wheat with hydrolyzed wool increased concentrations of Zn in grain. The concentrations of Zn were 38 mg kg^{-1} with unfertilized wheat, 46 mg kg^{-1} with mineral-N fertilization, and 54 mg kg^{-1} with hydrolyzed wool. Hydrolyzed wool application increased grain yield 2-fold and grain protein 1.5-fold, compared with a 1.4-fold increase in yield and a 1.3-fold increase in protein by the mineral fertilizer.

Poletti and Sautter (2005) reviewed recent developments of biofortification with vitamin A, vitamin E, and folate and with the minerals iron, Zn, and selenium and suggested that gene technology as a supplementary technique to breeding in combination with functional genomics gene technology could contribute in the future significantly to improving nutritional quality of plant-derived foods. Lönnerdal (2003) in a review noted that genetic engineering might be used to increase the micronutrient content of staple foods from cereals and legumes. Zimmermann and Hurrell (2002) wrote that recent understanding of plant metabolism has made it possible to increase the iron, Zn, and beta-carotene content in staple foods by conventional plant breeding and genetic engineering. Peleg et al. (2008) reported that several accessions of wild emmer wheat (*Triticum turgidum* ssp. *dicoccoides* Thell.), the progenitor of domesticated durum wheat and bread wheat, had a combination of high Zn efficiency and drought stress resistance and offered a valuable source of genetic diversity to improve mineral concentrations in grains.

Zuo and Zhang (2009) proposed that intercropping of legumes and grasses can offer a pathway to iron and Zn biofortification in grains as they noted that iron and Zn concentrations of wheat and chickpea (*Cicer arietinum* L.) seeds were increased by intercropping compared with the concentrations in monocropping. They previously noted (Zuo and Zhang, 2008) that with peanut (*Arachis hypogaea* L.) intercropped with corn, wheat, barley, or oats, the chlorophyll and active iron concentrations in the young leaves of peanut in the intercropping system were much higher than those of peanut in monoculture. Also, copper and Zn in shoots of peanut increased with cropping in field and greenhouse studies with a calcareous soil. It was suggested that phytosiderophores from the grasses improved the nutrition of peanut.

15.4 DIAGNOSIS OF ZINC STATUS IN PLANTS

15.4.1 ZINC DEFICIENCY

Zinc is poorly available in calcareous or alkaline soils because of formation of sparingly soluble $Zn(OH)_2$ or $ZnCO_3$, thereby creating conditions that lead to insufficiently available Zn for crop production. As an example of sufficiency, Ahmad et al. (2012) recommended soil test extract (diethyleneaminepentaacetic acid [DTPA] or Mehlich 1) levels of 0.5–3 mg kg^{-1} and soil fertilization of 5–11 kg Zn ha^{-1} for three or four crop seasons to sustain crop production (Section 15.5). Early work by Xang et al. (1993) reported that Zn accumulation in rice cultivars was affected by bicarbonate. A Zn-inefficient cultivar was affected more by bicarbonate and by cold temperature (15°C vs. 25°C) than a Zn-efficient cultivar. The prevalence of Zn deficiency for crops may be attributed to the Zn fixation in calcareous soils rather than to low total Zn content (Rafique et al., 2012). High concentrations of bicarbonate in irrigation water also can create Zn deficiency in soils by increasing soil pH (Cartmill et al., 2007; Shahabi et al., 2005; Valdez-Aguilar and Reed, 2010). The restriction in root growth of a Zn-inefficient rice genotype was greater for plants grown with bicarbonate than with high pH treatment (Xang et al., 2003). The greater inhibitory effect of bicarbonate than high pH was suggested to result from an excessive accumulation and inefficient compartmentation of organic acids, particularly citrate and malate, in the root cells.

In calcareous soils, Zn deficiency occurs with iron deficiency (lime-induced chlorosis) with low availability of Zn being due to adsorption of Zn to clay or $CaCO_3$, rather than from the formation of sparingly soluble $Zn(OH)_2$ or $ZnCO_3$ (Broadley et al., 2012; Trehan and Sekhon, 1977). Zinc uptake and translocation are inhibited by high concentrations of bicarbonate (Broadley et al., 2012; Dogar and van Hai, 1980; Forno et al., 1975). This effect is very similar to the effect of bicarbonate on iron (Rose et al., 2012). Rose et al. (2011) noted that enhanced organic acid release from the roots of Zn-efficient rice genotypes plays a strong role in plant tolerance to bicarbonate excess and Zn deficiency. The root leakage of Zn-inefficient genotypes under bicarbonate and Zn deficiency was higher than in Zn-efficient genotypes and coincided with high root peroxide concentrations, suggesting that bicarbonate tolerance is related to the ability of Zn-efficient genotypes to overcome oxidative stress, maintain root membrane integrity, and minimize root leakage. Chen et al. (2009) suggested also that Zn efficiency is associated with maintaining an efficient antioxidative system and intact root tip cells.

High alkalinity (high pH) in nutrient solutions restricts accumulation of micronutrients. Magnesium concentration in lettuce (*Lactuca sativa* L.) leaves increased with elevating pH from 5 to 7 and decreased at pH 8; however, iron, manganese, and zinc concentrations decreased as pH rose (Roosta, 2011). The author recommended that the pH of nutrient solutions should be 5 to ensure adequate micronutrient solubility and supply. An improvement in bean plant growth under alkaline conditions in hydroponics occurred if NH_4^+, K^+, and Na^+ were blended together (Valdez-Aguilar and Reed, 2010). Optimum NH_4^+ was associated with a decrease in solution pH and an increase in shoot Fe and Zn concentration, but NH_4^+ supply greater than one-third of the total N nutrition was toxic to the plants.

Increasing bicarbonate concentration and associated high pH restricted growth and nutrient uptake of multiflora rose (*Rosa multiflora* Thunb. ex Murr.); however, inoculation with arbuscular mycorrhizal fungi enhanced plant tolerance to bicarbonate as indicated by leaf, stem, and total plant dry weight and enhanced leaf elemental concentration, including Zn (Cartmill et al., 2007).

Zinc deficiency also is prevalent in soils that are inherently low or high in organic carbon, heavily limed, light textured, salt prone, or waterlogged due to various factors that affect the availability of Zn (Ahmad et al., 2012). The effects of organic matter vary among soils, but the general principle is for organic matter to decrease plant-available Zn by adsorption. Gurpreet-Kaur et al. (2013) reported that Zn adsorption in soil was greater with organic matter added than in the absence of organic matter addition. Peat soils are inherently deficient in micronutrients, particularly copper and Zn, because of adsorption, and liming of these soils decreased plant availability of the nutrients (Abat et al., 2012). Overliming of acid soils may lead to Zn deficiency as an increase in alkalinity lowers solubility of Zn (Martens and Westermann, 1991). Charred organic matter (biochar or pyrochar) has been proposed as a sorbent for metals to restrict their availability in contaminated soils. Pyrochar increased biomass production of oats and reduced plant Zn concentrations but increased metal concentrations in soil leachates

(Wagner and Kaupenjohann, 2014). On the other hand, addition of organic matter may increase mobility of Zn. Addition of cow manure to land increased the mobility of Zn and other metals in some soils (Akpa and Agbenin, 2012). However, Zn desorption in a calcareous soil increased as Zn was applied but decreased with cow manure applied (Zahedifar et al., 2012). Long-term applications of beef manure, swine effluent, or biosolids increased micronutrient availability of the micronutrients in soils as these materials acted as fertilizers for Zn and other micronutrients (Richards et al., 2011).

Hafeez et al. (2013) reported that flooding and submergence bring about a decline in available Zn due to pH changes and the formation of insoluble Zn compounds and that rice cultivars differ in efficiency of grain yield under these Zn-limiting conditions. Zinc is more likely to be deficient in sandy soils than in silty or clayey soils. Leaching may deplete sands of Zn. Low amounts of Zn in the parent materials (e.g., quartz) that form sandy soils also lead to low Zn concentrations in these soils (Krauskopf, 1972).

15.4.2 Concentrations of Zinc in Plants

Zinc deficiency initially appears in all plants as interveinal chlorosis (mottling) in which lighter green to pale yellow color appears between the midrib and secondary veins (Figures 15.1 through 15.3). Developing leaves are smaller than normal, and the internodes are short. Popular names describe these conditions as *little leaf* and *rosetting* (Storey, 2007). Pecan trees in particular suffer from shortened internodes (rosetting) and from dieback (Storey, 2007). Shoots are more inhibited by Zn deficiency than roots (Cumbus, 1985; Zhang et al., 1991).

The concentration of Zn in plant parts varies from 15–20 to over 300 mg kg^{-1} dry weight, commonly (Table 15.1; Bryson et al., 2014). Storey (2007) reported that in most plants, the critical concentration to avoid growth limitations due to Zn deficiency ranges from 15 to 20 mg kg^{-1}. Toxicity is uncommon but may occur in soils contaminated by additions of mining or smelting by-products or sewage sludge (Broadley et al., 2012; Reed and Martens, 1996). Toxicity symptoms may appear if concentrations are 100 to more than 300 mg Zn kg^{-1}, the latter values being more typical (Storey, 2007). However, the toxic value varies with species as even concentrations exceeding 350 mg kg^{-1} commonly are in the sufficiency range for plants (Table 15.1;

FIGURE 15.1   Zinc-deficient corn (*Z. mays* L.).

FIGURE 15.2   **(See color insert.)** Zinc-deficient cucumber (*C. sativus* L.).

FIGURE 15.3   Zinc-deficient citrus (*Citrus* sp.).

Bryson et al., 2014). Increasing soil pH by liming is the most effective strategy for decreasing Zn toxicity to plants (White et al., 1979). Toxicity may be due to inhibition of several enzymes, particularly in photosynthesis. With bean, suppressed photosynthesis may be caused by competition with magnesium in Rubisco or replacing of magnesium in thylakoid membranes (Van Assche and Clijsters, 1986a,b). High concentrations of Zn in plants may induce manganese or iron deficiency by suppressing manganese or iron concentrations in leaves (Ruano et al., 1987; Sagardoy et al., 2009).

TABLE 15.1
Concentrations of Zinc in Leaves or Shoots of Various Crops

Crop Class	Selected Genera[a]	Plant Part[b]	Common Range,[c] mg kg^{-1} Dry Weight
Agronomic			
Grains			
Legumes	*Arachis, Glycine, Phaseolus*	Mature leaves, whole shoots	20–75
Grasses	*Avena, Oryza, Hordeum, Secale, Sorghum, Triticum, Zea*	Mature leaves, whole shoots	15–70
Broadleaf, nonlegume	*Beta, Brassica, Gossypium, Helianthus, Nicotiana*	Mature leaves	10–80
Hay and forage crops	*Bromus, Cynodon, Dactylis, Lotus, Medicago, Phleum, Setaria, Sorghum, Trifolium*	Mature leaves or whole shoots	30–350
Bedding plants	*Calibrachoa, Celosia, Cleome, Dianthus, Impatiens, Pelargonium, Primula, Salvia, Tagetes, Verbena, Viola, Zinnia*	Mature leaves	50–400
Forest and Landscape Trees			
Softwoods	*Abies, Cedrus,* Gingko, *Juniperus, Larix, Picea, Pinus, Taxodium, Taxus, Thuja, Tsuga*	Terminal cuttings or mature leaves	15–70
Hardwoods	*Acer, Aesculus, Alnus, Betula, Carpinus, Carya, Cercis, Cornus, Eucalyptus, Fagus, Fraxinus, Lagerstroemia, Liquidambar, Liriodendron, Magnolia, Populus, Prunus, Quercus, Salix, Ulmus*	Mature leaves	10–200
Plantation crops	*Camellia, Coffea, Theobroma*	Mature leaves	20–200
Ferns and related plants	*Adiantum, Athyrium, Dennstaedtia, Equisetum, Nephrolepis, Polystichum, Osmunda, Rumohra*	Mature fronds	20–75
Fruit and nuts	*Anacardium, Ananas, Carya, Carica, Citrus, Corylus, Diospyros, Fragaria, Juglans, Malus, Musa, Olea, Persea, Pistacia, Prunus, Rubus, Vaccinium, Vitis*	Mature leaves	20–100
Herbaceous perennials	More than 100 species	Mature leaves	20–50
Herbs	*Allium, Anethum, Mentha, Origanum, Petroselinum, Rosmarinus, Thymus*	Mature leaves, terminal cuttings, whole shoots	25–100
Ornamental monocots	*Acorus, Carex, Cyperus, Miscanthus, Pennisetum, Phalaris, Phyllostachys*	Mature leaves	25–50
Ornamental vines and groundcovers	*Ajuga, Campsis, Clematis, Euonymus, Ficus, Hedera, Liriope, Lonicera, Pachysandra, Parthenocissus, Trachelospermum, Vinca*	Mature leaves, terminal cuttings	30–100
Woody ornamental shrubs	*Buddleia, Buxus, Camellia, Cornus, Forsythia, Hydrangea, Ilex, Mahonia, Pieris, Rhododendron, Spiraea, Viburnum*	Mature leaves, terminal cuttings	10–400
Turfgrass	*Agrostis, Cynodon, Eremochloa, Festuca, Lolium, Poa, Stenotaphrum, Zoysia*	Clippings	25–200
Vegetables			
Leafy	*Amaranthus, Apium, Brassica, Lactuca, Rheum, Spinacia*	Mature leaves or wrapper leaves	20–200
Fruit	*Abelmoschus, Capsicum, Citrullus, Cucurbita, Cucumis, Lycopersicon, Solanum, Zea*	Mature leaves	20–200

(Continued)

TABLE 15.1 (*Continued*)
Concentrations of Zinc in Leaves or Shoots of Various Crops

Crop Class	Selected Genera[a]	Plant Part[b]	Common Range,[c] mg kg^{-1} Dry Weight
Bulb	*Allium*	Whole tops or mature leaves	25–75
Root	*Brassica, Daucus, Ipomoea, Raphanus*	Mature leaves	50–350
Tuber	*Solanum*	Mature leaves	25–200
Stems	*Asparagus, Brassica*	Mature leaves	20–100
Woody ornamentals	*Abelia, Aesculus, Buddleja, Buxus, Camellia, Cornus, Cotoneaster, Forsythia, Gardenia, Ilex, Kalmia, Leucothoe, Mahonia, Pieris, Rhododendron, Rhus, Spiraea, Viburnum*	Mature leaves from new growth	35–200

[a] Genera of crops that are commonly known or commonly tested for Zn accumulation from Bryson et al. (2014).
[b] Parts tested for Zn accumulation.
[c] Commonly determined sufficiency range or survey range from Bryson et al. (2014).

15.5 SOIL TESTING

As mentioned earlier, major pools of Zn in soil are (1) soil solution; (2) surface-adsorbed and exchangeable Zn; (3) organic matter, including organisms; (4) oxides and carbonates; and (5) primary minerals and secondary aluminosilicate materials (Reed and Martens, 1996; Shuman, 1991). Uncontaminated soils have from 10 to 160 mg Zn kg^{-1} dry weight, averaging about 80 mg kg^{-1} (Bryson et al., 2014; Reed and Martens, 1996). The soil solution is dilute (0.03–0.17 mg L^{-1}) and thus contains only a small portion of the total Zn (Reed and Martens, 1996). Labile forms of Zn, that in the soil solution and loosely adsorbed Zn, are readily available to plants, with the solution being the immediate source and the adsorbed Zn replenishing the soil solution. Zinc that is in organic matter or in the lattices of minerals is considered as being in nonlabile forms that have insufficient availability to nourish a crop in a growing season as these forms have to be mineralized or weathered to be in plant-available forms.

Total Zn includes all of the Zn in soil and is not a good index of plant-available forms. Total Zn is determined by heat-fusion methods or treatment with hydrofluoric and other strong acids to destroy the organic and inorganic structures totally or partially (Baker and Amacher, 1982; Reed and Martens, 1996). Determination of total Zn is of interest to scientists to compare the geochemistry of soils. Available nutrients are determined by extraction with various solvents. Zinc can be extracted by 0.1 M HCl, but other reagents such as DTPA-triethanolamine (DTPA-TEA), DTPA-ammonium bicarbonate (DTPA-AB), and ethylenediaminetetraacetic acid (EDTA) are used commonly for simultaneous extraction of Zn, iron, copper, and manganese (Almendros et al., 2013; Phattarakul et al., 2012; Pirzadeh et al., 2010; Ray et al., 2013; Reed and Martens, 1996; Richards et al., 2011; Srivastava et al., 2000). Universal extracting solutions, Mehlich 3 and Morgan, are used often by soil testing laboratories and by investigators for simultaneous determination of macronutrients and micronutrients because of the convenience and cost savings imparted by these methods (Elrashidi et al., 2003; Mehlich, 1984; Morgan, 1941; Ray et al., 2013; Sequeira et al., 2011; Sobral et al., 2013; Srivastava and Srivastava, 2008).

15.6 FERTILIZERS

Crops on average remove about 5 kg Zn ha^{-1} year^{-1} (Bryson et al., 2014). Zinc fertilizers consist of inorganic salts, chelates, and organic matter. Zinc sulfate (ZnSO$_4$), because of its solubility, is used commonly in soil or foliar application. Zinc in zinc sulfate may precipitate rapidly in calcareous soils. Zinc chelates include DPTA, EDTA, and hydroxyethyl ethylene diamine triacetic acid (HEDTA), among other organic compounds (Bryson et al., 2014). Chelates are used commonly in foliar sprays and in applications to calcareous soils. Almendros et al. (2013) reported that Zn chelates had some residual effect in the year after application and that the chelates were more effective in a weakly acid soil than in a calcareous soil. Zinc is a component (perhaps 0.1–0.5 kg Mg^{-1}) of farm manures of various types depending on the diet of the poultry or livestock, and long-term applications will lead to accumulation of Zn in soils (Richards et al., 2011). Short-term applications may not be adequate for nutrition of crops in Zn-deficient soils. Composts and farm manures performed best in potato production if their application was accompanied by phosphorus and inorganic Zn fertilization (Taheri et al., 2012).

Sparingly soluble compounds such as zinc oxide may be used as fertilizers. McBeath and McLaughlin (2014) evaluated zinc oxide with $ZnSO_4$. Either source was satisfactory for Zn nutrition of crops if the fertilizers were mixed thoroughly in the soil, but zinc oxide was not effective if banded near the seeds.

Application of soil-incorporated $ZnSO_4$ to pecan trees did not bring the Zn content of the soils to an adequate level for tree nutrition in calcareous soils because the Zn was transferred from the sulfate form to sparingly soluble $ZnCO_3$ (Lott, 1938). Storey (2007) reviewed research that showed that leaflet Zn of pecan trees in calcareous soils can be increased by soil applications of $ZnSO_4$ but that a larger increase will occur if S is applied with $ZnSO_4$.

Foliar application has proved difficult because of cuticular barriers that form as leaves become mature, and poor absorption through cuticles makes foliar application of chelates inefficient (Worley et al., 1972). Frequent Zn foliar applications are more successful than occasional treatments (Storey, 2007). Foliar $ZnSO_4$ treatments are toxic to peach leaves and to many other species, probably because the salt accumulates on leaves and results in desiccation (Storey, 2007). Storey (2007) developed a Zn nitrate–ammonium nitrate–urea fertilizer mixture (NZN) that did not burn peach (*Prunus persica* Stokes) leaves. Apparently, NZN-treated peach leaves do not suffer from salt burn because the nitrate in NZN is absorbed readily, thereby not accumulating on the surface to cause leaf burn. The inclusion of NH_4NO_3 and urea to either $Zn(NO_3)_2$ or $ZnSO_4$ gave a substantial increase in translocation of leaf-absorbed Zn. Foliar application of $Zn(NO_3)_2$ increased Zn absorption more than $ZnSO_4$.

Plants absorbed more foliarly applied Zn from NZN, $Zn(NO_3)_2$, $ZnSO_4$, or $ZnCl_2$ at high relative humidity than at low relative humidity due to the retention of surface water at high humidity (Rossi and Beauchamp, 1971; Storey, 2007). Stein and Storey (1986) evaluated adjuvants in a variety of classes, including alcohols, amines, carbohydrates, esters, ethoxylated hydrocarbons, phosphates, polyethylene glycols, proteins, silicones, sulfates, sulfonates, and alcohol alkoxylates. Glycerol was the only adjuvant that increased the percentage of nitrogen and phosphorus in leaves over the foliar fertilizer alone. Storey (2007) noted that injection of soluble Zn compounds into trunks was not generally a successful method of feeding trees.

REFERENCES

Abat, M., M.J. McLaughlin, J.K. Kirby, and S.P. Stacey. 2012. Adsorption and desorption of copper and zinc in tropical peat soils of Sarawak, Malaysia. *Geoderma* 175:58–63.

Abreu, I.A. and D.E. Cabelli. 2010. Superoxide dismutases—A review of the metal-associated mechanistic variations. *Biochim. Biophys. Acta* 1804:263–274.

Ahmad, H.R., M.A. Aziz, A. Ghafoor, M.Z. Rehman, M. Sabir, and Saifullah. 2010. Wheat assimilation of nickel and zinc added in irrigation water as affected by organic matter. *J. Plant Nutr.* 34:27–33.

Ahmad, W., M.J. Watts, M. Imtiaz, I. Ahmed, and M.H. Zia. 2012. Zinc deficiency in soils, crops and humans: A review. *Agrochimica* 56:65–97.

Akhtar, M.S., Y. Oki, and T. Adachi. 2010. Growth behavior, nitrogen-form effects on phosphorus acquisition, and phosphorus–zinc interactions in brassica cultivars under phosphorus-stress environment. *Commun. Soil Sci. Plant Anal.* 41:2011–2045.

Akpa, S.I. and J. Agbenin. 2012. Impact of cow dung manure on the solubility of copper, lead, and zinc in urban garden soils from northern Nigeria. *Commun. Soil Sci. Plant Anal.* 43:2789–2800.

Alben, A.O, J.R. Cole, and R.D. Lewis. 1932. Chemical treatment of pecan rosette. *Phytopathology* 22:595–601.

Alloway, B.J. 2009. Soil factors associated with zinc deficiency in crops and humans. *Environ. Geochem. Health* 31:537–548.

Almendros, P., D. Gonzales, and J.M. Alvarez. 2013. Residual effects of organic Zn fertilizers applied before the previous crop on Zn availability and Zn uptake by flax (*Linum usitatissimum*). *J. Plant Nutr. Soil Sci.* 176:603–615.

Amagloh, F.K., L. Brough, J.L. Weber, A.N. Mutukumira, A. Hardacre, and J. Coad. 2012. Sweet potato-based complementary food would be less inhibitory on mineral absorption than a maize-based infant food assessed by compositional analysis. *Int. J. Food Sci. Nutr.* 63:957–963.

Andreini, C., I. Bertini, and A. Rosato. 2009. Metalloproteomics: A bioinformatic approach. *Acc Chem. Res.* 42:1471–1479.

Apel, K. and H. Hirt. 2004. Reactive oxygen species: Metabolism, oxidative stress, and signal transduction *Annu. Rev. Plant Biol.* 55:373–399.

Aref, F. 2012. Evaluation of application methods and rates of zinc and boron on nitrogen, phosphorus and potassium contents of maize leaf. *J. Plant Nutr.* 35:1210–1224.

Arnold, T., G.J.D. Kirk, M. Wissuwa et al. 2010. Evidence for the mechanisms of zinc uptake by rice using isotope fractionation. *Plant Cell Environ.* 33:370–381.

Ashmead, H. 1982. *Chelated Mineral Nutrition in Plants, Animals and Man.* Springfield, IL: Charles C. Thomas.

Auld, D.S. 2009. The ins and outs of biological zinc sites. *Biometals* 2:141–148.

Auld, D.S. and T. Bergman. 2008. The role of zinc for alcohol dehydrogenase structure and function. *Cell. Mol. Life Sci.* 65:3961–3970.

Badger, M.R. and G.D. Price. 1994. The role of carbonic anhydrase in photosynthesis. *Annu Rev. Plant Physiol. Plant Mol. Biol.* 45:369–392.

Baker, D.E. and M.C. Amacher. 1982. Nickel, copper, zinc, and cadmium. In *Methods of Soil Analysis, Part 2, Chemical and Microbiological Properties*, 2nd edn., eds. A.L. Page, R.H. Miller, and D.R. Keeney, pp. 323–336. Madison, WI: Soil Science Society of America.

Baniulis, D., S.S. Hasan, J.T. Stofleth, and W.A. Cramer. 2013. Mechanism of enhanced superoxide production in the cytochrome b(6)f complex of oxygenic photosynthesis. *Biochemistry* 52:8975–8983.

Barak, P., and P.A. Helmke. 1993. The chemistry of zinc. In *Zinc in Soils and Plants*, ed. A.D. Robson, pp. 90–106. Dordrecht, the Netherlands: Kluwer Academic Publishers.

Barben, S.A., B.G. Hopkins, V.D. Jolley, et al. 2010. Optimizing phosphorus and zinc concentrations in hydroponic chelator-buffered nutrient solution for Russet Burbank potato. *J. Plant Nutr.* 33:557–570.

Bernal, M., R. Cases, R. Picorel, and I. Yruela. 2007. Foliar and root Cu supply affect differently Fe- and Zn-uptake and photosynthetic activity in soybean plants. *Environ. Exp. Bot.* 60:145–150.

Bonnet, M., O. Camares, and P. Veisseire. 2000. Effects of zinc and influence of *Acremonium lolii* on growth parameters, chlorophyll a fluorescence and antioxidant enzyme activities of ryegrass (*Lolium perenne* L. cv Apollo). *J. Exp. Bot.* 51:945–953.

Broadley, M., P. Brown, I. Cakmak, Z. Rengel, and F. Zhao. 2012. Function of nutrients: Micronutrients. In *Mineral Nutrition of Higher Plants*, 3rd edn., ed. P. Marschner, pp. 191–248. London, U.K.: Academic Press.

Broadley, M.R., P.J. White, J.P. Hammond, I. Zelko, and A. Lux. 2007. Zinc in plants. *New Phytol.* 173:677–702.

Bryson, G.M., H.A. Mills, D.N. Sasseville, J. Benton Jones, Jr., and A.V. Barker. 2014. *Plant Analysis Handbook III.* Athens, GA: Micro-Macro Publishing.

Burman, U., M. Saini, and Praveen-Kumar. 2013. Effect of zinc oxide nanoparticles on growth and antioxidant system of chickpea seedlings. *Tox. Environ. Chem.* 95:605–612.

Burnell, J.N. and M.D. Hatch. 1988. Low bundle sheath carbonic anhydrase is apparent by essential for effective C4 pathway operation. *Plant Physiol.* 86:1252–1256.

Cakmak, I. 2000. Possible roles of zinc in protecting plant cells from damage by reactive oxygen species. *New Phytol.* 146:185–205.

Cakmak, I. 2008. Enrichment of cereal grains with zinc: Agronomic or genetic biofortification? *Plant Soil* 302:1–17.

Cakmak, I., R. Derici, B. Torun, I. Tolay, H.J. Braun, and R. Schlegel. 1997. Role of rye chromosomes in improvement of zinc efficiency in wheat and triticale. *Plant Soil* 196:249–253.

Cakmak, I. and H. Marschner. 1986. Mechanism of phosphorus-induced zinc deficiency in cotton. I. Zinc deficiency-enhanced uptake rate of phosphorus. *Physiol. Plant.* 68:483–490.

Cakmak, I. and H. Marschner. 1987. Mechanism of phosphorus-induced zinc deficiency in cotton. II. Changes in physiological availability of zinc in plants. *Physiol. Plant.* 70:13–20.

Cakmak, I. and H. Marschner. 1988a. Enhanced superoxide radical production in roots of zinc-deficient plants. *J. Exp. Bot.* 39:1449–1460.

Cakmak, I. and H. Marschner. 1988b. Zinc-dependent changes in ESR signals, NADPH oxidase and plasma membrane permeability in cotton roots. *Physiol. Plant.* 73:182–186.

Cakmak, I., H. Marschner, and F. Bangerth. 1989. Effect of zinc nutritional status on growth, protein metabolism and levels of indole-3-acetic avid and other phytohormones in bean (*Phaseolus vulgaris* L.). *J. Exp. Bot.* 40:405–412.

Cakmak, I., L. Ozturk, S. Eker, B. Torun, H.I. Kalfa, and A. Yilmaz. 1997. Concentration of zinc and activity of copper/zinc-superoxide dismutase in leaves in rye and wheat cultivars differing in sensitivity to zinc deficiency. *J. Plant Physiol.* 151:91–95.

Cartmill, A.D., A. Alarcon, and L.A. Valdez-Aguilar. 2007. Arbuscular mycorrhizal fungi enhance tolerance of *Rosa multiflora* cv. burr to bicarbonate in irrigation water. *J. Plant Nutr.* 30:1517–1540.

Chandel, A.S. and M.C. Saxena. 1980. Mechanism of uptake and translocation of zinc by pea plants (*Pisum sativum* L.). *Plant Soil* 56:343–353.

Chapman, H.D. 1965. Zinc. In *Diagnostic Criteria for Plants and Soils*, ed. H.D. Chapman, pp. 484–499. Riverside, CA: Homer D. Chapman.

Chen, W., X. Yang, Z. He, Y. Feng, and F. Hu. 2008. Differential changes in photosynthetic capacity, 77 K chlorophyll fluorescence and chloroplast ultrastructure between Zn-efficient and Zn-inefficient rice genotypes (*Oryza sativa*) under low zinc stress. *Physiol. Plant.* 132:89–101.

Chen, W.R., Z.L. He, X.E. Yang, and Y. Feng. 2009. Tolerance to zinc deficiency in rice correlates with zinc uptake and translocation. *J. Plant Nutr.* 32:287–305.

Coleman, J.E. 1992. Zinc proteins: Enzymes, storage proteins, transcription factors, and replication proteins. *Annu. Rev. Biochem.* 61:897–946.

Coleman, J.E. 1998. Zinc enzymes. *Curr. Opin. Chem. Biol.* 2:222–234.

Combs, G.F., J.M. Duxbury, and R.M. Welch. 1997. Food systems for improved health: Linking agricultural production and human nutrition. *Eur. J. Clin. Nutr.* 51:S32–S33.

Cumbus, I.P. 1985. Development of wheat roots under zinc deficiency. *Plant Soil* 83:313–316.

Daneshbakhsh, B., A.H. Khoshgoftarmanesh, H. Hossein, and I. Cakmak. 2013. Effect of zinc nutrition on salinity-induced oxidative damages in wheat genotypes differing in zinc deficiency tolerance. *Acta Physiol. Plant.* 35:881–889.

Day, B.J. 2014. Antioxidant therapeutics: Pandora's box. *Free Radic. Biol. Med.* 66:58–64.

Dogar, M.A. and T. van Hai. 1980. Effect of P, N, and HCO_3^- in nutrient solution on rate of Zn absorption by rice roots and Zn content in plants. *Z. Pflanzenphysiol.* 98:203–212.

Domingo, A.L., Y. Nagatomo, M. Tamai, and H. Takaki. 1992. Free tryptophan and indoleacetic acid in zinc-deficient radish shoots. *Soil Sci. Plant Nutr.* 38:261–267.

Dominy, A. and I. Bertling. 2004. Manganese, zinc and silicon studies of cucumber (*Cucumis sativus*) using a miniature hydroponic system. In *Proceedings of the International Symposium on Growing Media & Hydroponics*, eds. B. Alsanium, P. Jensen, and H. Asp. *Acta Hort.* (Book Series) Vol. 644, pp. 393–398.

Eide, D.R. 1998. The molecular biology of metal ion transport in *Saccharomyces cerevisiae*. *Annu. Rev. Nutr.* 18:441–469.

Elrashidi, M.A., M.D. Mays, and C.W. Lee. 2003. Assessment of Mehlich3 and ammonium bicarbonate-DTPA extraction for simultaneous measurement of fifteen elements in soils. *Commun. Soil Sci. Plant Anal.* 34:2817–2838.

Ender, C., M.Q. Li, B. Martin, B. Povh, R. Nobiling, H.D. Reiss, and K. Traxel. 1983. Demonstration of polar zinc distribution in pollen tubes of *Lilium longiflorum* with the Heidelberg proton microprobe. *Protoplasma* 116:201–203.

Fageria, N.K., M. Moraes, E.P.B. Ferreira, and A.M. Knupp. 2012. Biofortification of trace elements in food crops for human health. *Commun. Soil Sci. Plant Anal.* 43:556–570.

Fageria, N.K., A. Moreira, L.A.C. Moraes, and M.F. Moraes. 2014. Root growth, nutrient uptake, and nutrient-use efficiency by roots of tropical legume cover crops as influenced by phosphorus fertilization. *Commun. Soil Sci. Plant Anal.* 45:555–569.

Falchuk, K.H., L. Ulpino, B. Mazus, and B.L. Vallee. 1977. *E. gracilis* RNA polymerase. I. A zinc metalloenzyme. *Biochem. Biophys. Res. Commun.* 74:1206–1212.

Forno, D.A., S. Yoshida, and C.J. Asher. 1975. Zinc deficiency in rice. I. Soil factors associated with the deficiency. *Plant Soil* 42:537–550.

Foyer, C.H. and G. Noctor. 2000. Oxygen processing in photosynthesis: Regulation and signaling. *New Phytol.* 146:359–388.

Foyer, C.H. and S. Shigeoka. 2011. Understanding oxidative stress and antioxidant functions to enhance photosynthesis. *Plant Physiol.* 155:93–100.

Gao, X.P., C.Q. Zou, F.S. Zhang, S.E.A.T.M. van der Zee, and E. Hoffland. 2005. Tolerance to zinc deficiency in rice correlates with zinc uptake and translocation. *Plant Soil* 278:253–261.

Genc, Y., J.M. Humphries, G.H. Lyons, and R.D. Graham. 2005. Exploiting genotypic variation in plant nutrient accumulation to alleviate micronutrient deficiency in populations. *J. Trace Elem. Med. Biol.* 18:319–324.

Ghasemi-Fasaei, R. and A. Ronaghi. 2008. Interaction of iron with copper, zinc, and manganese in wheat as affected by iron and manganese in a calcareous soil. *J. Plant Nutr.* 31:839–848.

Gianquinto, G., A. Abu-Rayyan, L. DiTola, D. Piccotino, and B. Pezzarossa. 2000. Interaction effects of phosphorus and zinc on photosynthesis, growth and yield of dwarf bean grown in two environments. *Plant Soil* 220:219–228.

Gogos, A., M.W.H. Evangelou, A. Schaffer, and R. Schulin. 2013. Hydrolysed wool: A novel soil amendment for zinc and iron biofortification of wheat. *New Zeal. J. Agric. Res.* 56:130–141.

Gowik, U. and P. Westhoff. 2011. The path from C-3 to C-4 photosynthesis. *Plant Physiol.* 155:56–63.

Gregorio, G.B. 2002. Progress in breeding for trace minerals in staple crops. *J. Nutr.* 132:500S–502S.

Grewal, H.S., R.D. Graham, and J. Stangoulis. 1998. Zinc–boron interaction effects in oilseed rape. *J. Plant Nutr.* 21:2231–2243.

Guerinot, M.L. 2000. The Zip family of metal transporters. *Biochim. Biophys. Acta* 1465:190–198.

Gurpreet-Kaur, B., D. Sharma, and S. Sharma. 2013. Effects of organic matter and ionic strength of supporting electrolyte on zinc adsorption in benchmark soils of Punjab in northwest India. *Commun. Soil Sci. Plant Anal.* 44:922–938.

Hacisalihoglu, G., J.J. Hart, C.E. Vallejos, and L.V. Kochian. 2004. The role of shoot-localized processes in the mechanism of Zn efficiency in common bean. *Planta* 218:704–711.

Hacisalihoglu, G. and L.V. Kochian. 2003. How do some plants tolerate low levels of soil zinc? Mechanisms of zinc efficiency in crop plants. *New Phytol.* 159:341–350.

Hafeez, B., Y.M. Khanif, A.W. Samsuri, O. Radziah, W. Zakaria, and M. Saleem. 2013. Direct and residual effects of zinc on zinc-efficient and zinc-inefficient rice genotypes grown under low-zinc-content submerged acidic conditions. *Commun. Soil Sci. Plant Anal.* 44:2233–2252.

Hajiboland, R. and H. Amirazad. 2010. Drought tolerance in Zn-deficient red cabbage (*Brassica oleracea* L. var. *capitata* f. *rubra*) plants. *Hort. Sci.* 37:88–98.

Hatch, M.D. and J.N. Burnell. 1990. Carbonic anhydrase activity in leaves and its role in the first stop of C4 photosynthesis. *Plant Physiol.* 93:825–828.

Hosseini, S.M., M. Maftoun, N. Karimian, A. Ronaghi, and Y. Emam. 2007. Effect of zinc × boron interaction on plant growth and tissue nutrient concentration of corn. *J. Plant Nutr.* 30:773–781.

Hussain, S., M.A. Maqsood, Z. Renge, and T. Aziz. 2012. Biofortification and estimated human bioavailability of zinc in wheat grains as influenced by methods of zinc application. *Plant Soil* 361:279–290.

Imtiaz, M., B.J. Alloway, M.Y. Memon et al. 2006. Zinc tolerance in wheat cultivars as affected by varying levels of phosphorus. *Commun. Soil Sci. Plant Anal.* 37:1689–1702.

Ishimaru, Y., K. Bashir, and N.K. Nishizawa. 2011. Zn uptake and translocation in rice plants. *Rice* 4(1):21–27.

Ishimaru, Y., M. Suzuki, T. Kobayashi et al. 2005. OsZIP4, a novel zinc-regulated zinc transporter in rice. *J. Exp. Bot.* 56:3207–3214.

Izsaki, Z. 2014. Effects of phosphorus supplies on the nutritional status of maize (*Zea mays* L.). *Commun. Soil Sci. Plant Anal.* 45:516–529.

Jenny, E.A. 1968. Controls on Mn, Fe, Co, Ni, Cu, and Zn concentrations in soils and water: The significant role of hydrous Mn and Fe oxides. *Adv. Chem.* 73:377–387.

Joseph, C., V.S. Rathore, Y.P.S. Bajaj, and S.H. Wittwer. 1971. Mechanism of zinc accumulation by intact bean plants. *Ann. Bot.* 35:683–685.

Khadr, A. and A. Wallace. 1964. Uptake and translocation of radioactive iron and zinc by trifoliate orange and rough lemon. *Proc. Am. Soc. Hort. Sci.* 85:189–200.

Khan, N.A., S. Singh, N.A. Anjum, and R. Nazar. 2008. Cadmium effects on carbonic anhydrase, photosynthesis, dry mass and antioxidative enzymes in wheat (*Triticum aestivum*) under low and sufficient zinc. *J. Plant Interact.* 3:31–37.

King, J.C. 2002. Biotechnology: A solution for improving nutrient bioavailability. *Int. J. Vitam. Nutr. Res.* 72:7–12.

Kitagishi, K. and H. Obata. 1986. Effects of zinc deficiency on the nitrogen metabolism of meristematic tissues of rice plants with reference to protein synthesis. *Soil Sci. Plant Nutr.* 32:397–405.

Knight, J.D., D. Gangotena, D.L. Allan, and C.J. Rosen. 2004. Screening common bean genotypes for tolerance to low zinc availability using a chelate-buffered hydroponics system. *J. Plant Nutr.* 27:275–293.

Kochian, L.V. 1991. Mechanism of micronutrient uptake and translocation in plants. In *Micronutrients in agriculture*, ed. J.J. Mortvedt, pp. 229–296. Madison, WI: Soil Science Society of America.

Kosesakal, T., M. Unal, and G.C. Oz. 2009. Influence of zinc toxicity on gravitropic response of tomato (*Lycopersicon esculentum* Mill.) roots. *Fresenius Environ. Bull.* 18:2401–2407.

Krauskopf, K.B. 1972. Geochemistry of micronutrients. In *Micronutrients in Agriculture*, ed. J.J. Mortvedt, pp. 7–40. Madison, WI: Soil Science Society of America.

Kruger, J., J.R.N. Taylor, X.G. Du, F.F. De Moura, B. Lönnerdal, and A. Oelofse. 2013. Effect of phytate reduction of sorghum, through genetic modification, on iron and zinc availability as assessed by an in vitro dialysability bioaccessibility assay, Caco-2 cell uptake assay, and suckling rat pup absorption model. *Food Chem.* 141:1019–1025.

Kupriyanova, E., M.A. Sinetova, S.M. Cho, Y.I. Park, D. Los, and N.A. Pronina. 2013. CO_2-concentrating mechanism in cyanobacterial photosynthesis: Organization, physiological role, and evolutionary origin. *Photosynth. Res.* 117:133–146.

Kutman, U.B., B.Y. Kutman, Y. Ceylan, E.A. Ova, and I. Cakmak. 2012. Contributions of root uptake and remobilization to grain zinc accumulation in wheat depending on post-anthesis zinc availability and nitrogen nutrition. *Plant Soil* 361:177–187.

Kutman, U.B., B. Yildiz, and I. Cakmak. 2011. Effect of nitrogen on uptake, remobilization and partitioning of zinc and iron throughout the development of durum wheat. *Plant Soil* 342:149–164.

Lasat, M.M., A.J.M. Baker, and L.V. Kochian. 1996. Physiological characterization of root Zn^{2+} absorption and translocation to shoots in Zn hyperaccumulator and nonaccumulator species of *Thlaspi*. *Plant Physiol.* 112:1715–1722.

Lee, S. and G. An. 2009. Over-expression of *OsIRT1* leads to increased iron and zinc accumulations in rice. *Plant Cell Environ.* 32:408–416.

Lee, S., S.A. Kim, J. Lee, M.L. Guerinot, and G. An. 2010a. Zinc deficiency-inducible *OsZIP8* encodes a plasma membrane-localized zinc transporter in rice. *Mol. Cells* 29:551–558.

Lee, Y.P., K.H. Baek, H.S. Lee, S.S. Kwak, J.W. Bang, and S.Y. Kwon. 2010b. Tobacco seeds simultaneously over-expressing Cu/Zn-superoxide dismutase and ascorbate peroxidase display enhanced seed longevity and germination rates under stress conditions. *J. Exp. Bot.* 61:2499–2506.

Li, H.Y., Y.G. Zhu, S.E. Smith, and F.A. Smith. 2003. Phosphorus–zinc interactions in two barley cultivars differing in phosphorus and zinc efficiencies. *J. Plant Nutr.* 26:1085–1099.

Li, Y.T., Y. Zhang, D.Q. Shi et al. 2013. Spatial-temporal analysis of zinc homeostasis reveals the response mechanisms to acute zinc deficiency in *Sorghum bicolor*. *New Phytol.* 200:1102–1115.

Lin, M.S. and C.H. Kao. 1990. Senescence of rice leaves. 23. Changes of Zn^{2+}-dependent acid inorganic pyrophosphatase. *J. Plant Physiol.* 137:41–45.

Lindsay, W.L. 1991. Iron oxide solutes solubilization by organic matter and its effects on iron availability. In *Iron Nutrition and Interaction in Plants*, ed. Y. Hadar, pp. 29–36. Dordrecht, the Netherlands: Kluwer Academic Publishers.

Loneragan, J.F., T.S. Grove, A.D. Robson, and K. Snowball. 1979. Phosphorus toxicity as a factor in zinc-phosphorus interactions in plants. *Soil Sci. Soc. Am. J.* 43:966–972.

Lönnerdal, B. 2000. Dietary factors influencing zinc absorption. *J. Nutr.* 130:1378–1383.

Lönnerdal, B. 2002. Phytic acid-trace element (Zn, Cu, Mn) interactions. *J. Food Sci. Technol.* 37:749–758.

Lönnerdal, B. 2003. Genetically modified plants for improved trace element nutrition. *J. Nutr.* 133:1490S–1493S.

Lopez-Millan, A.F., D.R. Ellis, and M.A. Grusak. 2005. Effect of zinc and manganese supply on the activities of superoxide dismutase and carbonic anhydrase in *Medicago truncatula* wild type and *raz* mutant plants. *Plant Sci.* 168:1015–1022.

Lott, W.L. 1938. The relation of hydrogen ion concentration to the availability of zinc in soil. *Proc. Soil Sci. Soc. Am.* 3:115–121.

Ludwig, M. 2012. Carbonic anhydrase and the molecular evolution of C-4 photosynthesis. *Plant Cell Environ.* 35:22–37.

Ludwig, M. 2013. Evolution of the C-4 photosynthetic pathway: Events at the cellular and molecular levels. *Photosynth. Res.* 117:147–161.

Lynch, J., P. Marschner, and Z. Rengel. 2012. Effect of internal and external factors on root growth and development. In *Mineral Nutrition of Higher Plants*, 3rd edn., ed. P. Marschner, pp. 331–346. London, U.K.: Academic Press.

Ma, Q.Y. and W.L. Lindsay. 1993. Measurements of free Zn2 activity in uncontaminated and contaminated soils using chelation. *Soil Sci. Soc. Am. J.* 57:963–967.

Mano, Y. and K. Nemoto. 2012. The pathway of auxin biosynthesis in plants. *J. Exp. Bot.* 63:2853–2872.

Marschner, H. 1995. *Mineral Nutrition of Higher Plants*, 2nd edn. New York: Academic Press.

Marschner, H. and I. Cakmak. 1989. High light intensity enhances chlorosis and necrosis in leaves of zinc, potassium, and magnesium deficient bean (*Phaseolus vulgaris*) plants. *J. Plant Physiol.* 134:308–315.

Martens, D.C. and D.T. Westermann. 1991. Fertilizer applications for correcting micronutrient deficiencies. In *Micronutrients in Agriculture*, 2nd edn., ed. J.J. Mortvedt, pp. 549–592. Madison, WI: Soil Science Society of America.

Masuda, H., M. Suzuki, K.C. Morikawa et al. 2008. Increase in iron and zinc concentrations in rice grains via the introduction of barley genes involved in phytosiderophore synthesis. *Rice* 1:100–108.

Masuda, H., K. Usuda, T. Kobayashi et al. 2009. Overexpression of the barley nicotianamine synthase gene *HvNAS1* increases iron and zinc concentrations in rice grains. *Rice* 2:155–166.

Mayer, J.E., W.H. Pfeiffer, and P. Beyer. 2008. Biofortified crops to alleviate micronutrient malnutrition. *Curr. Opin. Plant Biol.* 11:166–170.

McBeath, T.M. and M.J. McLaughlin. 2014. Efficacy of zinc oxides as fertilizers. *Plant Soil* 374:843–855.

Mehlich, A. 1984. Mehlich 3 soil test extractant: A modification of Mehlich 2 extractant. *Commun. Soil Sci. Plant Anal.* 15(12):1409–1416.

Meng, F.H., D. Liu, X.E. Yang et al. 2014. Zinc uptake kinetics in the low and high-affinity systems of two contrasting rice genotypes. *J. Plant Nutr. Soil Sci.* 177:412–420.

Meyer, M.R.M., A. Rojas, A. Santanen, and F.L. Stoddard. 2013. Content of zinc, iron and their absorption inhibitors in Nicaraguan common beans (*Phaseolus vulgaris* L.). *Food Chem.* 136:87–93.

Miller, L.V., N.F. Krebs, and K.M. Hambidge. 2012. Mathematical model of zinc absorption: Effects of dietary calcium, protein and iron on zinc absorption. *Br. J. Nutr.* 109:695–700.

Moinuddin and P. Imas. 2010. Effects of zinc nutrition on growth, yield, and quality of forage sorghum in respect with increasing potassium application rates. *J. Plant Nutr.* 33:2062–2081.

Moore, P.A., Jr. and W.H. Patrick, Jr. 1988. Effect of zinc deficiency on alcohol dehydrogenase activity and nutrient uptake in rice. *Agron. J.* 80:883–885.

Morgan, M.F. 1941. Chemical soil diagnosis by the universal soil testing system. *Conn. Agric. Exp. Stn. Bull.* No. 450.

Morgounov, A., H.F. Gomez-Becerra, A. Abugalieva et al. 2007. Iron and zinc grain density in common wheat grown in Central Asia. *Euphytica* 155:193–201.

Mukhopadhyay, M., A. Das, P. Subba et al. 2013. Structural, physiological, and biochemical profiling of tea plantlets under zinc stress. *Biol. Plant.* 57:474–480.

Obata, H. and M. Umebayashi. 1988. Effect of zinc deficiency of protein synthesis in cultured tobacco plant cells. *Soil Sci. Plant Nutr.* 34:351–357.

Ozturk, L., M.A. Yazici, C. Yucel, et al. 2006. Concentration and localization of zinc during seed development and germination in wheat. *Physiol. Plant.* 128:144–152.

Parker, D.R. 1993. Novel nutrient solutions for zinc nutrition research—Buffering free zinc(2+) with synthetic chelators and P with hydroxyapatite. *Plant Soil* 155:461–464.

Pedas, P., S. Husted, K. Skyette, and J.K. Schjoerring. 2011. Elevated phosphorus impedes manganese acquisition by barley plants. *Front. Plant Sci.* 2:Article 37.

Peleg, Z., Y. Saranga, A. Yazici, T. Fahima, L. Ozturk, and I. Cakmak. 2008. Grain zinc, iron and protein concentrations and zinc-efficiency in wild emmer wheat under contrasting irrigation regimes. *Plant Soil* 306:57–67.

Perltreves, R. and E. Galun. 1991. The tomato Cu, Zn superoxide-dismutase genes are developmentally regulated and respond to light and stress. *Plant Mol. Biol.* 17:745–760.

Phattarakul, N., B. Rerkasem, L.J. Li et al. 2012. Biofortification of rice grain with zinc through zinc fertilization in different countries. *Plant Soil* 361:131–141.

Pilon, M., K. Ravet, and W. Tapken. 2011. The biogenesis and physiological function of chloroplast superoxide dismutases. *Biochim. Biophys. Acta* 1807:989–998.

Pinton, R., I. Cakmak, and H. Marschner. 1993. Effect of zinc deficiency on proton fluxes in plasma membrane-enriched vesicles isolated from bean roots. *J. Exp. Bot.* 44:623–630.

Pirzadeh, M., M. Afyuni, A. Khoshgoftarmanesh, and R. Schulin. 2010. Micronutrient status of calcareous paddy soils and rice products: Implication for human health. *Biol. Fert. Soils* 46:317–322.

Poletti, S. and C. Sautter. 2005. We review recent developments of biofortification with vitamin A, vitamin E and folate, and with the minerals iron, zinc and selenium. *Minerva Biotecnol.* 17:1–11.

Prask, J.A. and D.J. Plocke. 1971. A role for zinc in the structural integrity of the cytoplasmic ribosomes of *Euglena gracilis. Plant Physiol.* 48:150–155.

Qiao, X., Y. He, Z.M. Wang, X. Li, K.Y. Zhang, and H.L. Zeng. 2014. Effect of foliar spray of zinc on chloroplast beta-carbonic anhydrase expression and enzyme activity in rice (*Oryza sativa* L.) leaves. *Acta Physiol. Plant.* 36:263–272.

Quartin, V.M.L., M.L. Antunes, M.C. Muralha, M.M. Sousa, and M.A. Nunes. 2001. Mineral imbalance due to manganese excess in triticales. *J. Plant Nutr.* 24:175–179.

Rafique, E., A. Rashid, and M. Mahmood-ul-Hassan. 2012. Value of soil zinc balances in predicting fertilizer zinc requirement for cotton-wheat cropping system in irrigated Aridisols. *Plant Soil* 361:43–55.

Ramesh, S.A., R. Shin, D.J. Eide, and D.P. Schachtman. 2003. Differential metal selectivity and gene expression of two zinc transporters from rice. *Plant Physiol.* 133:126–134.

Rawat, N., K. Neelam, V.K. Tiwari, and H.S. Dhaliwal. 2013. Biofortification of cereals to overcome hidden hunger. *Plant Breed.* 132:347–445.

Ray, P., S.K. Singhal, S.P. Datta, and R.K Rattan. 2013. Evaluation of suitability of chemical extractants for assessing available zinc in acid and alkaline soils amended with farmyard manure and sludge. *Agrochimica* 57:347–360.

Reed, S.T. and D.C. Martens. 1996. Copper and zinc. In *Methods of Soil Analysis, Part 3, Chemical Methods*, eds. D.L. Sparks, A.L. Page, P.A. Helmke, and R.H. Loeppert, pp. 703–722. Madison, WI: Soil Science Society of America.

Rengel, Z. 1995. Carbonic anhydrase activity in leaves of wheat genotypes differing in Zn efficiency. *J. Plant Physiol.* 147:251–256.

Richards, J.R., H.L. Zhang, J.L. Schroder, J.A. Hattey, W.R. Raun, and M.E. Payton. 2011. Micronutrient availability as affected by the long-term application of phosphorus fertilizer and organic amendments. *Soil Sci. Soc. Am. J.* 75:927–929.

Robson, A.D. and M.G. Pitman. 1983. Interactions between nutrients in higher plants. In *Encyclopedia of Plant Physiology, New Series*, Vol. 15A, eds. A. Läuchli and R.L. Bieleski, pp. 147–180. Berlin, Germany: Springer-Verlag.

Roosta, H.R. 2011. Interaction between water alkalinity and nutrient solution pH on the vegetative growth, chlorophyll fluorescence and leaf magnesium, iron, manganese, and zinc concentrations in lettuce. *J. Plant Nutr.* 34:717–731.

Rose, M.T., T.J. Rose, J. Pariasca-Tanaka, Widodo, and M. Matthias. 2011. Revisiting the role of organic acids in the bicarbonate tolerance of zinc-efficient rice genotypes. *Funct. Plant Biol.* 6:493–504.

Rose, M.T., T.J. Rose, J. Pariasca-Tanaka et al. 2012. Root metabolic response of rice (*Oryza sativa* L.) genotypes with contrasting tolerance to zinc deficiency and bicarbonate excess. *Planta* 236:959–973.

Rossi, N. and E.G. Beauchamp. 1971. Influence of relative humidity and associated anion on the absorption of Mn and Zn by soybean leaves. *Agron. J.* 63:860–863.

Ruano, A., J. Barcelo, and C. Poschenrieder. 1987. Zinc toxicity-induced variation of mineral element composition in hydroponically grown bush bean plants. *J. Plant Nutr.* 10:373–384.

Ruel, M.T. and H.E. Bouis. 1998. Plant breeding: A long-term strategy for the control of zinc deficiency in vulnerable populations. *Am. J. Clin. Nutr.* 68:488S–494S.

Ryan, M.H., J.K. McInerney, J.K. Record, and J.F. Angus. 2008. Zinc bioavailability in wheat grain in relation to phosphorus fertilization, crop sequence and mycorrhizal fungi. *J. Sci. Food Agric.* 88:208–1216.

Sagardoy, R., F. Morales, A.F. Lopez-Milan, A. Abadia, and J. Abadia. 2009. Effects of zinc toxicity on sugar beet (*Beta vulgaris* L.) plants grown in hydroponics. *Plant Biol.* 11:339–350.

Sagardoy, R., S. Vazquez, I.D. Florez-Sarasa et al. 2010. Stomatal and mesophyll conductances to CO_2 are the main limitations to photosynthesis in sugar beet (*Beta vulgaris*) plants grown with excess zinc. *New Phytol.* 187:145–158.

Salami, A.U. and D.G. Kenefick. 1970. Stimulation of growth in zinc-deficient corn seedlings by the addition of tryptophan. *Crop Sci.* 10:291–294.

Sandmann, G. and P. Boger. 1983. The enzymatological function of heavy metals and their role in electron transfer processes of plants. In *Encyclopedia of Plant Physiology, New Series*, Vol. 15A, eds. A. Lauchli and R.L. Bieleski, pp. 563–593. Berlin, Germany: Springer-Verlag.

Sasaki, H., T. Hirose, Y. Watanabe, and Y. Ohsugi. 1998. Carbonic anhydrase activity and CO_2-transfer resistance in An-deficient rice leaves. *Plant Physiol.* 118:929–934.

Schlemmer, U., W. Frolich, R.M. Prieto, and F. Grases. 2009. Phytate in foods and significance for humans: Food sources, intake, processing, bioavailability, protective role and analysis. *Mol. Nutr. Food Res.* 53:S330–S375.

Sekimoto, H., M. Hoshi, T. Nomura, and T. Yokota. 1997. Zinc deficiency affects the levels of endogenous gibberellins in *Zea mays* L. *Plant Cell Physiol.* 38:1087–1090.

Sequeira, C.H., N.F. Baros, J.C.L. Neves, R.F. Novais, I.R. Silva, and M. Alley. 2011. Micronutrient soil-test levels and eucalyptus foliar contents. *Commun. Soil Sci. Plant Anal.* 42:475–488.

Shahabi, A., M.J. Malakouti, and E. Fallahi. 2005. Effects of bicarbonate content of irrigation water on nutritional disorders of some apple varieties. *J. Plant Nutr.* 28:1663–1678.

Shahid, M., T.T. Xiong, N. Masood et al. 2014. Influence of plant species and phosphorus amendments on metal speciation and bioavailability in a smelter impacted soil: A case study of food-chain contamination. *J. Soils Sediments* 14:655–665.

Shahriaripour, R. and A. Tajabadipour. 2010. Zinc nutrition of pistachio: Interaction of zinc with other trace elements. *Commun. Soil Sci. Plant Anal.* 41:1885–1888.

Sharma, P.N., P. Kumar, and R.K. Tewari. 2004. Early signs of oxidative stress in wheat plants subjected to zinc deficiency. *J. Plant Nutr.* 27:451–463.

Sharma, P.N., A. Tripathi, and S.S. Alka. 1995. Zinc requirement for stomatal opening in cauliflower. *Plant Physiol.* 107:751–756.

Shuman, L.M. 1991. Chemical forms of micronutrients in soils. In *Micronutrients in Agriculture*, 2nd edn., eds. J.J. Mortvedt, F.R. Cox, L.M. Shuman, and R.M. Welch, pp. 113–114. Madison, WI: Soil Science Society of America.

Sinetova, M.A., E.V. Kupriyanova, A.G. Markelova, S. Allakhverdiev, and N.A. Pronina. 2012. Identification and functional role of the carbonic anhydrase Cah3 in thylakoid membranes of pyrenoid of *Chlamydomonas reinhardtii*. *Biochim. Biophys. Acta* 1817:1248–1255.

Singh, B., S.K.A. Natesan, B.K. Singh, and K. Usha. 2005. Improving zinc efficiency of cereals under zinc deficiency. *Curr. Sci.* 88:36–44.

Singh, D.,S. Yadav, and N. Nautiyal. 2014. Evaluation of growth responses in wheat as affected by application of zinc and boron to a soil deficient in available zinc and boron. *Commun. Soil Sci. Plant Anal.* 45:765–76.

Skinner, J.J. and J.B. Demaree. 1926. Relation of soil conditions and orchard management to the rosette of pecan trees. *US Dept. Agric. Bull.* 1378. 16 pages.

Sobral, L.F., J.T. Smyth, N.K. Fageria, and L.F. Stone. 2013. Mehlich 1, Mehlich 3, and DTPA solutions for soils of the Brazilian Coastal Tablelands. *Commun. Soil Sci. Plant Anal.* 44:2507–2513.

Solomons, N.W. 2000. Plant-based diets are traditional in developing countries: 21st century challenges for better nutrition and health. *Asia Pac. J. Clin. Nutr.* 9:S41–S54.

Sommer, A. and C.B. Lipman. 1926. Evidence on the indispensable nature of zinc and boron for higher green plants. *Plant Physiol.* 1:231–249.

Sousa, S.F., A.B. Lopes, P.A. Fernandes, and M.J. Ramos. 2009. The zinc proteome: A tale of stability and functionality. *Dalton Trans.* 38:7946–7956.

Srivastava, P.C., A. Dobermann, and D. Ghosh. 2000. Assessment of zinc availability to rice in Mollisols of North India. *Commun. Soil Sci. Plant Anal.* 31:2457–2471.

Srivastava, P.C. and P. Srivastava. 2008. Integration of soil pH with soil-test values of zinc for prediction of yield response in rice grown in mollisols. *Commun. Soil Sci. Plant Anal.* 39:15–16.

Stein, L.A. and J.B. Storey. 1986. Influence of adjuvants on foliar absorption of nitrogen and phosphorus by soy-beans. *J. Am. Soc. Hort. Sci.* 111:829–832.

Stephan, U.W., I. Schmidke, V.W. Stephan, and G. Scholz. 1996. The nicotianamine molecule is made-to-measure for complexation of metal micronutrients in plants. *Biometals* 9(1):84–90.

Stephens, B.W., D.R. Cook, and M.A. Grusak. 2011. Characterization of zinc transport by divalent metal transporters of the ZIP family from the model legume *Medicago truncatula*. *Biometals* 24:51–58.

Storey, J.B. 2007. Zinc. In *Handbook of Plant Nutrition*, ed. A.V. Barker and D.J. Pilbeam, pp. 411–435. Boca Raton, FL: CRC Press.

Swietlik, D. 1995. Interaction between zinc deficiency and boron toxicity on growth and mineral nutrition of sour orange. *J. Plant Nutr.* 18:1191–1207.

Taheri, N., H.H. Sharif-Abad, K. Yousefi, and S. Roholla-Mousavi. 2012. Effect of compost and animal manure with phosphorus and zinc fertilizer on yield of seed potatoes. *J. Soil Sci. Plant Nutr.* 12:705–714.

Takahashi, R., Y. Ishimaru, H. Shimo et al. 2012. The OsHMA2 transporter is involved in root-to-shoot translocation of Zn and Cd in rice. *Plant Cell Environ.* 35:1948–1957.

Tavallali, V., M. Rahemi, S. Eshghi, B. Kholdebarin, and A. Ramezanian. 2010. Zinc alleviates salt stress and increases antioxidant enzyme activity in the leaves of pistachio (*Pistacia vera* L. 'Badami') seedlings. *Turk. J. Agr. For.* 34:349–359.

Taylor, S.R. 1964. Abundance of chemical elements in the continental crust: A new table. *Cosmochim. Acta* 28:1273–1286.

Thi, L.N.N.D., A. Caruso, T.T. Li et al. 2012. Zinc affects differently growth, photosynthesis, antioxidant enzyme activities and phytochelatin synthase expression of four marine diatoms. *Sci. World J.* 2012:Article Number 982957. http://www.hindawi.com/journals/tswj/, accessed January 13, 2015.

Tivendale, N.D., J.J. Ross, and J.D. Cohen. 2014. The shifting paradigms of auxin biosynthesis. *Trends Plant Sci.* 19(1):44–51.

Trehan, S.P. and G.S.S. Sekhon. 1977. Effect of clay, organic matter and CaCO$_3$ content on zinc absorption by soils. *Plant Soil* 46:329–336.

Tsui, C. 1948. The role of zinc in auxin synthesis in the tomato plant. *Am. J. Bot.* 35:172–179.

Turekian, K.K. and K.H. Wedepohl. 1961. Distribution of the elements in some major units of the earth's crust. *Geol. Soc. Am. Bull.* 72:175–192.

Ullah, S., A.S. Khan, A.U. Malik, I. Afzal, M. Shahid, and K. Razzaq. 2012. Foliar application of boron influences the leaf mineral status, vegetative and reproductive growth, yield and fruit quality of 'Kinnow' mandarin (*Citrus reticulata* Blanco.). *J. Plant Nutr.* 35:2067–2069.

Utsunomiya, E. and S. Muto. 1993. Carbonic anhydrase in the plasma membranes from leaves of C3 and C4 plants. *Physiol. Plant.* 88:413–419.

Valdez-Aguilar, L.A. and D.W. Reed. 2010. Growth and nutrition of young bean plants under high alkalinity as affected by mixtures of ammonium, potassium, and sodium. *J. Plant Nutr.* 33:1472–1488.

Vallee, B.L. and D.S. Auld. 1990. Zinc coordination, function, and structure of zinc enzymes and other proteins. *Biochemistry* 29:5647–5659.

Vallee, B.L. and K.H. Falchuk. 1993. The biochemical basis of zinc physiology. *Physiol. Rev.* 73:79–118.

Van Asshe, F. and H. Clijsters. 1986a. Inhibition of photosynthesis in *Phaseolus vulgaris* by treatment with toxic concentrations of zinc: Effect of ribulose-1,5-bisphosphate carboxylase/oxygenase. *J. Plant Physiol.* 125:355–360.

Van Asshe, F. and H. Clijsters. 1986b. Inhibition of photosynthesis in *Phaseolus vulgaris* by treatment with toxic concentrations of zinc: Effect on electron transport and photophosphorylation. *Physiol. Plant.* 66:717–721.

Wagner, A. and M. Kaupenjohann. 2014. Suitability of biochars (pyro- and hydrochars) for metal immobilization on former sewage-field soils. *Eur. J. Soil Sci.* 65:139–148.

Wang, H. and J.Y. Jin. 2005. Photosynthetic rate, chlorophyll fluorescence parameters, and lipid peroxidation of maize leaves as affected by zinc deficiency. *Photosynthetica* 43:591–596.

Webb, M.J. and J.F. Loneragan. 1988. Effect of zinc deficiency on growth, phosphorus concentration and phosphorus toxicity of wheat plants. *Soil Sci. Soc. Am. J.* 52:1676–1680.

Welch, R.N., M.J. Webb, and J.F. Loneragan. 1982. Zinc in membrane function and its role in phosphorus toxicity. In *Proceedings of the Ninth Plant Nutrition Colloquium, Warwick, England*, ed. A. Scaife, pp. 710–715. Farnham Royal, Bucks, U.K.: Commonwealth Agricultural Bureau.

White, P.J. and P.H. Brown. 2010. Plant nutrition for sustainable development and global health. *Ann. Bot.* 105:1073–1080.

White, P.J., A.M. Decker, and R.L. Cheney. 1979. Differential cultivar tolerance of soybean to phytotoxic levels of soil Zn. I. Range of cultivar response. *Agron. J.* 71:121–126.

Worley, R.E., S.A. Hammon, and R.L. Carter. 1972. Effect of zinc sources and methods of application on yield and mineral concentration of pecan, *Carya illinoensis*, Koch. *J. Am. Soc. Hort. Sci.* 97:364–359.

Xang, X., R. Hajiboland, and V. Römheld. 2003. Bicarbonate had greater effects than high pH on inhibiting root growth of zinc-inefficient rice genotype. *J. Plant Nutr.* 26:399–415.

Xang, X., V. Romheld, and H. Marschner. 1993. Effect of bicarbonate and root-zone temperature on uptake of Zn, Fe, Mn and Cu by different rice cultivars (*Oryza sativa* L) grown in calcareous soil. *Plant Soil* 155:441–444.

Xue, Y.F., S.C. Yue, Y.Q. Zhang et al. 2012. Grain and shoot zinc accumulation in winter wheat affected by nitrogen management. *Plant Soil* 361:153–163.

Yang, X.E., W.R. Chen, and Y. Feng. 2007. Improving human micronutrient nutrition through biofortification in the soil-plant system: China as a case study. *Environ. Geochem. Health* 29:413–428.

Yu, Q. and Z. Rengel. 1999. Micronutrient deficiency influences plant growth and activities of superoxide dismutases in narrow-leafed lupins. *Ann. Bot.* 83:175–182.

Yu, Q., C. Wortha, and Z. Rengel. 1999. Using capillary electrophoresis to measure Cu/Zn superoxide dismutase concentration in leaves of wheat genotypes differing in tolerance to zinc deficiency. *Plant Sci.* 143:231–239.

Zahedifar, M., N. Karimian, and J. Yasrebi. 2012. Influence of applied zinc and organic matter on zinc desorption kinetics in calcareous soils. *Arch. Agron. Soil Sci.* 58:169–178.

Zarzycki, J., S.D. Axen, J.N. Kinney, and C.A. Kerfield. 2013. Cyanobacterial-based approaches to improving photosynthesis in plants. *J. Exp. Bot.* 64:787–798.

Zhang, F., V. Römheld, and H. Marshner. 1991. Release of zinc mobilizing root exudates in different plant species as affected by zinc nutritional status. *J. Plant Nutr.* 14:675–686.

Zhang, Y.Q., Y. Deng, R.Y. Chen et al. 2012. The reduction in zinc concentration of wheat grain upon increased phosphorus-fertilization and its mitigation by foliar zinc application. *Plant Soil* 361:143–152.

Zhao, F.J., Z.G. Shen, and S.P. McGrath. 1998. Solubility of zinc and interactions between zinc and phosphorus in the hyperaccumulator *Thlaspi caerulescens. Plant Cell Environ.* 21:108–114.

Zhou, J.R., E.J. Fordyce, V. Raboy et al. 1992. Reduction of phytic acid in soybean products improves zinc bioavailability in rats. *J. Nutr.* 122:2466–2473.

Zhu, Y.G., S.E. Smith, and F.A. Smith. 2001. Zinc (Zn)-phosphorus (P) interactions in two cultivars of spring wheat (*Triticum aestivum* L.) differing in P uptake efficiency. *Ann. Bot.* 88:941–945.

Ziaeyan, A.H. and M. Rajaie. 2009. Combined effect of zinc and boron on yield and nutrients accumulation in corn. *Int. J. Plant Prod.* 3:35–44.

Zimmermann, M.B. and R.F. Hurrell. 2002. Improving iron, zinc and vitamin A nutrition through plant biotechnology. *Curr. Opin. Biotechnol.* 13:142–145.

Zuo, Y. and F. Zhang. 2009. Iron and zinc biofortification strategies in dicot plants by intercropping with gramineous species. A review. *Agron. Sustain. Dev.* 29:63–71.

Zuo, Y.M. and F.S. Zhang. 2008. Effect of peanut mixed cropping with gramineous species on micronutrient concentrations and iron chlorosis of peanut plants grown in a calcareous soil. *Plant Soil* 306:23–36.

Section IV

Beneficial Elements

16 Aluminum

F. Pax C. Blamey, Peter M. Kopittke, J. Bernhard Wehr, and Neal W. Menzies

CONTENTS

16.1 INTRODUCTION

Aluminum (Al) is the most common metal in the Earth's crust (8.2% by weight), the main Al-containing primary minerals being feldspars and micas, with a median Al concentration in soils of 7.2% (0.07%–30%) (Sparks, 2003). Only two elements, both nonmetals, have higher median concentrations in soils, O (46%) and Si (28%). By way of contrast, the median concentrations of the major basic metals are much lower—with 4.2% Ca, 2.3% Mg, 2.1% K, and 2.4% Na (Lide, 2007). The four major basic cations are essential for plant growth, and some studies have shown Al to promote plant growth. However, no beneficial actions of Al in plant metabolism have been demonstrated.

It has been known for nearly a century that Al suppresses root growth (Miyake, 1916) and that Al restricts the uptake of nutrients by plants. Adverse effects of Al extend to many other biota, including microbes, soil flora and fauna, freshwater aquatic organisms, and terrestrial animals. There are two instances where limited plant root growth occurs in response to Al, acid and highly alkaline conditions in soils and mine spoils. This response occurs because Al minerals dissolve at acid or alkaline pH, releasing soluble Al into solution. The adverse effects of soluble Al have been studied mostly in acid soils; only sand and peat soils are exceptions because they are low in aluminosilicate (Al_2SiO_5) clays. In the alkaline range, it is only at pH > 8.5 that sufficient Al enters into solution potentially to cause problems other than those associated with alkalinity, including deficiencies of Fe, Mn, Cu, and Zn. In acid soils, however, soluble Al may restrict plant growth before the occurrence of other limitations such as Mn toxicity and deficiencies of P, Ca, Mg, K, Mo, or B. As was pointed out by Foy (1984), poor plant growth in acid soils is due to a complex of factors acting through different biochemical pathways. Aluminum toxicity, however, is probably the most important growth-limiting factor for plants in strongly acid soils and mine spoils worldwide.

The problems of acid soils for plant growth have long been known (Section 16.2) as have the ways in which the problems may be overcome (Sections 16.2 and 16.9). History has shown that some plants (calcifuges) are more tolerant than others of limitations present in acid soils. The application of certain compounds, known generally as agricultural lime (materials with Ca and Mg that neutralize soil acidity), overcomes the detrimental effects of Al or other limitations to plant growth.

At its simplest, acid soils may be defined as those with pH < 7. This definition is not useful for biological processes, however, because many plants grow better at slightly acidic pH (Farina et al., 1980). Rather, specific limitations to biological functions that occur in acid soils need to be identified so that appropriate remedial action can be taken.

Acid soils occur naturally on parent materials low in basic cations and where these have been removed by leaching, erosion, or in harvested materials. The rate of soil acidification is affected by the C, N, and S cycles (Helyar and Porter, 1989). Dissolved CO_2 and low levels of oxides of N and S in precipitation contributed to these cycles up until the Industrial Revolution. Over the last two centuries, however, the C, N, and S cycles have been modified greatly by human activities through atmospheric pollution and agricultural practices, thereby exacerbating soil acidification.

Besides effects of acid soil infertility on biological processes in individual acid soils, their global extent poses major issues, as they are often in regions otherwise well suited to agricultural production. Acid soils occupy ~40 billion hectares, or ~30% of the world's ice-free land area (von Uexküll and Mutert, 1995). More recently, the Food and Agriculture Organization (FAO) has classified the world's topsoils and subsoils based on pH and has produced maps (available at http://www.fao. org/fileadmin/templates/nr/images/resources/images/Maps/geonetwork/ph_t.png). Acid soils are mostly in two broad regions, the northern cold and temperate regions and the humid tropics. Many soils outside these regions, however, are naturally acid or have been acidified in recent times.

Recently, Miyasaka et al. (2007) provided a comprehensive overview of Al in soils and its effects on plants and other biota. This chapter does not intend to duplicate this work nor that of other recent reviews (e.g., Kochian et al., 2005; Ma, 2005; Poschenrieder et al., 2008; Horst et al., 2010; Inostroza-Blancheteau et al., 2012; Ryan and Delhaize, 2012). Rather, the adverse effects of Al on root growth, where Al has its initial effects, are addressed from physiological and practical viewpoints. A history is provided of the increasing realization that Al toxicity is responsible for much of the poor plant growth in acid soils (Section 16.2) before addressing the complex chemistry of Al (Section 16.3) and the presence of soluble Al in acid upland soils, acid sulfate soils, and highly alkaline soils and mine spoils (Section 16.4). Thereafter, the focus is on the use of solution culture (Section 16.5), Al uptake and transport by plants (Section 16.6), their physiological responses to Al in the soil solution (Section 16.7), and the genetic components of these responses (Section 16.8). The practical ways in which the detrimental effects of soluble Al on plant growth may be overcome are addressed in Section 16.9 before a final summary and conclusions (Section 16.10).

16.2 HISTORY

Liming of acid soils, though probably not recognized as such, is an ancient practice. Barber (1984) referred to Cato and Varro in ancient Rome reporting ~200 BC on the use of lime. Virgil (70–19 BC) published his extended Latin poem Georgics in ~29 BC. The Greek title means "agriculture" or "to farm," and in Book I, Virgil exhorts:

tantum ne pudeat saturare arida sola pingui fimo, neve jactare immumdum cinerem per effetos agros

which translates as

do not be mortified by improving poor soil with rich manure, nor by scattering charred ashes on worn-out fields

Gardner and Garner (1953) noted that lime, in the form of chalk or marl, was used in Britain more than two millennia ago and that Pliny the Elder, in the first century AD, reported that applying marl "greatly enriched the soils of the Gaelic Provinces and the British Islands." However, they found scant further records of use until the thirteenth century at which time there was mention of marl pits and tenants being required to marl their land. In the eighteenth century in Britain, there was increasing attention to the use of marl and quicklime (CaO) with a first reference in 1758 to Thomas Hale who encouraged the correct identification of marl, not to apply it to the "wrong" soil, and the need to grind seashells if used as a liming material. By the early twentieth century, liming in Britain had fallen into disuse, resulting in a lime subsidy being introduced in 1937 (Gardner and Garner, 1953).

In the nineteenth century, Ruffin (1852) promoted the use of lime in the United States. Edmund Ruffin was, besides other activities, a farmer and agronomist who published "*An Essay on Calcareous Manures.*" The fifth edition of his comprehensive book (available at http://archive.org/ details/essayoncalcareou03ruff) provides fascinating insights into acid soil fertility problems, defining various soils, including those acidified by cultivation, with many chapters on lime products, the value of liming, and "the damage caused by too heavy dressings of calcareous manures."

It was only in the early twentieth century that the importance of Al in acid soils was first identified. Hopkins et al. (1903) noted that "acids of the soils" are poorly soluble, "and it is practically impossible to completely extract them from the soil with distilled water". Extraction with a neutral salt by Veitch (1904) showed that filtrates of acid soils leached with NaCl did not have "appreciable quantities of free strong acids... but that the acidity of the filtrate... is due to the solution of alumina or some other acid-salt yielding base." Daikuhara (1914) also noted that leaching with KNO_3 displaced Al and Fe compounds that are "injurious to plant growth." This action was confirmed by Miyake (1916) noting that "the apparent acidity of the extract from the soil ... is found to be proportional to the amount of aluminum salts present in the solution, and evidently represents the amount of alkali required to precipitate the aluminum rather than actual free acidity." Miyake (1916) referred also to previous research on the toxicity of Al to plants: In 1905, Micheels and De Heen found that soluble Al salts reduce germination of wheat (*Triticum aestivum* L.) seeds; a pot experiment by Yamano showed that moderate amounts of Al salts improve the growth of barley (*Hordeum vulgare* L.) and flax (*Linum usitatissimum* L.) plants but that 0.2% alum ($KAl(SO_4)_2 \cdot 12H_2O$) (4 mM Al) in solution is injurious with 0.8% (16 mM Al) lethal to plants within a few days; in 1907, Hébert noted the highly toxic nature of $Al_2(SO_4)_3$; in 1911, Duggar found that $AlCl_3$ and $Al_2(SO_4)_3$ in solution are toxic to pea (*Pisum sativum* L.); and, in 1915, Ruprecht showed >1.5 mM Al is toxic to clover (*Trifolium* sp.). It is fascinating to realize that about one century ago, Miyake (1916) showed that ≥ 0.04 mM Al decreased rice (*Oryza sativa* L.) root length (Figure 16.1) more than root number or leaf length. Miyake (1916) concluded also that toxicity of $AlCl_3$ at low concentration "is not due to the hydrogen ion formed by hydrolysis of the salt in solution." Strangely, however, no comparison was made between the brittle, distorted roots caused by soluble Al and the flaccid roots resulting from H^+ injury. Later work by Magistad (1925), however, showed that added Al at pH 4.0 and 4.2 in sand culture "caused a stubby, unbranched root system" particularly in corn (*Zea mays* L.), oat (*Avena sativa* L.), and soybean (*Glycine max* Merr.).

Hartwell and Pember (1918) noted that liming an acid soil improved the growth of barley but not of rye (*Secale cereale* L.); low pH nutrient solutions, however, had similar effects on both crop plants. Solution culture experiments subsequently showed that "aluminum and the accompanying acidity to be much less deleterious to rye than to barley seedlings" with Al present at about the same concentration as that in the soil extract. They discussed the possibility that Al might either protect rye more

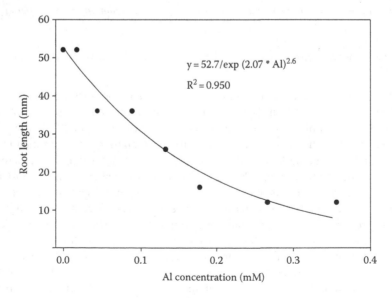

$$y = 52.7/\exp(2.07 * Al)^{2.6}$$

$$R^2 = 0.950$$

FIGURE 16.1 The effect of soluble Al added as $AlCl_3$ on the primary root length of rice seedlings plotted from data presented by Miyake (1916).

than barley from the deleterious effect of acidity or be less toxic to rye than to barley. Their research results, in which barley growth was decreased by 75% ± 2% and that of rye by 55% ± 2% ($n = 14$), and those of Miyake (1916) led Hartwell and Pember (1918) to favor the latter hypothesis.

Truog (1918) referred to research by Abbott and co-workers in 1913 on an acid, peaty soil in Indiana, United States, indicating that $Al(NO_3)_3$ "was present in sufficient amounts in the soil solution to be decidedly toxic to plants." Truog (1918) further raised issues of low Ca availability, low N_2 fixation, differences among plant species grown on acid soils, H^+ toxicity, and the increased toxic effects evident on the addition of K_2SO_4 fertilizer to the soil. It is noteworthy that Truog (1918) was much in favor of "the lime requirement of plants"—not of soils. Mirasol (1920) confirmed the findings of Veitch (1904) that the acidity of a KNO_3 soil extract is due largely to Al in solution, Al precipitating on the addition of alkali. Further studies were conducted by Mirasol (1920) on the role of nitrification in soil acidification and the toxicity to plants of Al salts in solution culture and in soils.

McLean and Gilbert (1927), referring also to the work of Hoffer and Carr in 1923, concluded that Al causes damage to the roots first, then injury to entire corn plants, with the greatest accumulation of Al in the roots. They thought that Al accumulated in the cytoplasm in living root cells, a finding not confirmed by recent research (Section 16.6). McLean and Gilbert (1927) classified crop plants into three groups based on Al added to solution (Section 16.8). There was no evidence that Al tolerance is based on the ability of plants to increase solution pH. In this work, root injury was evident first as discoloration, followed by stunting of branch roots, their failure to develop, and root death. Corn plants exposed to Al had decreased NO_3^- uptake and a 40% decrease in water loss indicating that Al "had seriously decreased the permeability of the roots." Magistad (1925) studied the complex effects of P and pH on Al solubility in solutions and their effects on plant growth.

By the mid-1920s, therefore, the toxicity of Al to plants was well established (Hardy, 1926). However, during this period, research in Europe and the United States shifted to the concept of H-clays (Jenny, 1961). Sharp and Hoagland (1916) and Hoagland (1918) addressed the roles of H^+ in soils and in solutions, the latter concluding that pH ~ 5.2 is favorable to the growth of barley seedlings but that "decided injury was caused" at pH ~ 3.5. Fred and Loomis (1917) showed that low pH decreases the growth of N_2-fixing bacteria of alfalfa (*Medicago sativa* L.) in culture medium. Mirasol (1920) was of the opinion that the discovery of Veitch (1904) "has never been appreciated, and the nature of soil acidity has generally been ascribed to the presence of free organic and mineral acids in the soil." Much later, Jenny (1961) concluded that by 1936, "H-clays and pH reigned supreme," providing an intriguing assessment of the views on acid soils at that time. Sparks (2006) has provided additional information on the history of soil acidity and other aspects of soil chemistry.

Thomas and Hargrove (1984) acknowledged that the toxic effects of Al had been neglected since the 1920s, concluding that the rediscovery of exchangeable (exch.) Al^{3+} is an important contribution to understanding the behavior of acid soils. This statement echoed the assessment by Jenny (1961) who stressed the importance of the observation that H-clays become Al-clays when stored moist. The H^+ ions react with the clay crystals releasing Al^{3+} (McLean and Brown, 1984).

It was only in the 1960s and 1970s that a major focus was redirected to soluble Al in acid soils with seminal papers on Al in soils (Jackson, 1963) and on the toxic effects of Al^{3+} on root growth (Adams and Lund, 1966). By the 1970s, research in South Africa (Reeve and Sumner, 1970) and in the United States (Kamprath, 1970) focused on using exchangeable Al as a basis for determining the lime requirement for improved crop growth. As detailed by Jenny (1961), researchers had rediscovered the results of Veitch (1904) on Al in soil solutions and the toxic effects of low Al concentration on root growth as shown by Miyake (1916).

Research over the past 40 years has focused on the complex chemistry of Al, especially Al speciation (Section 16.3) and the toxicity of various Al species to roots (Section 16.7.2). Current research appears largely focused on three areas: the physiology of Al toxicity, how some plants are better able to tolerate soluble Al in the rhizosphere, and the genetics of Al tolerance. There is still no consensus on the primary mechanism of Al toxicity, with proposed effects on the cell wall, cell plasma membrane (PM), or symplasmic components (Kochian, 1995). Greater progress has been made, however, on the physiology of Al tolerance (Kochian et al., 2004, 2005) and its genetic basis (Ryan et al., 2010, 2011). But all is not well, with many studies being conducted with apparently little cognizance of Al chemistry in solution (Sections 16.3 and 16.5).

16.3 COMPLEXITY OF ALUMINUM CHEMISTRY IN SOLUTION

16.3.1 ALUMINUM CHEMISTRY

Aluminum is a posttransition metal with atomic number 13 in Group 13 of the periodic table, preceded in this group by B (atomic number 5) and followed by Ga (atomic number 31). The electron configuration of Al, [Ne]$3s^2 2p^1$, plus its small ionic radius (54 pm) (Kinraide and Yermiyahu, 2007), makes Al highly reactive so that it is almost never found free in nature. By far most commonly found in the trivalent oxidation state, Al(III), Al reacts strongly with hard ligands (i.e., those containing O, P, and F) forming octahedral and tetrahedral coordination compounds. Aluminum also forms stable complexes with di- and multidentate ligands (chelate effect), especially important with organic ligands of low and high relative molecular mass (RMM) (*syn.* molecular weight) in soils and plants. The high reactivity of Al means that most terrestrial Al is found in numerous minerals, most of which have few direct biological effects.

Many Al salts, such as $Al(NO_3)_3$ and $AlCl_3$, are highly soluble in water, dissociating into the anions and Al^{3+}, or more correctly $Al(H_2O)_6^{3+}$, in which Al^{3+} is coordinated octahedrally with six water molecules. Under slightly acid conditions, Al^{3+} undergoes hydrolysis, resulting in monomeric hydroxy Al species, $Al(OH)^{2+}$, $Al(OH)_2^+$, and $Al(OH)_3^0$ (Jackson, 1963; Lindsay, 1979), and complex formation with many other inorganic and organic ligands present in natural systems (Lindsay and Walthall, 1996).

Adams (1971) applied theoretical considerations to displaced soil solutions so as to convert measured ionic concentrations to activities of those ions. Calculation of ionic strength (I) as $I = (1/2)\sum_{i=1}^{n} c_i z_i^2$, where c is the molar concentration of the ion and z is the valency of that ion, allows estimation of the single ion activity coefficient (f_i) of each ion (i) by means of the extended Debye–Hückel equation. This equation is defined as $-\log f_i = A z_i^2 \sqrt{I}/(1 + B a_i \sqrt{I})$, in which A and B are temperature-dependent constants, z is the valency of the ion, and a_i is an ion size parameter for each ion. Therefore, ionic strength has a much greater effect on the activity of Al^{3+}, given its trivalent nature and small ionic size, than on that of Ca^{2+}, for example, with its divalent state, Ca(II), and large ionic radius (99 pm). Ion pairing must be considered also, which is more important with multivalent cations and anions than with univalent ones (Adams, 1971).

In contrast to the minimal effects on plants and animals of Al minerals, except possibly particulate dusts, Al^{3+} at low concentration in solution is highly toxic to most biota (Section 16.7). The solubility of Al minerals increases greatly at acid and highly alkaline pH as can be shown theoretically using speciation programs such as PhreeqcI (Parkhurst, 2013) or GEOCHEM and GEOCHEM-PC (Parker et al., 1987, 1995). GEOCHEM-PC was recently modified to GEOCHEM-EZ by adding a graphical user interface (Shaff et al., 2010).

GEOCHEM-EZ was used to calculate soluble and solid Al in a hypothetical soil solution based on that reported by Wheeler et al. (1993) with 50 μM Al. The solubility of Al is high at pH ≤ 4.0, decreasing markedly with increase in pH such that virtually all Al is precipitated as $Al(OH)_3$ at pH > 5.0 until Al solubility increases at pH > 8.5 (Figure 16.2a). The trivalent Al^{3+} cation dominates at low pH with the monomeric hydroxy Al species present at much lower activities (Figure 16.2b). Their proportion, however, increases up to pH 5.0 (Miyasaka et al., 2007). Soluble Al is low up to pH ~ 8.5 above which the aluminate ion ($Al(OH)_4^-$) becomes important (Lindsay and Walthall, 1996), but its activity is <2.5 μM even at pH 9.

Besides precipitation of gibbsite (λ-$Al(OH)_3$) and variscite ($Al(PO_4)\cdot 2H_2O$) and the formation of monomeric Al species, polycationic Al species form in solution under some conditions

(Bertsch, 1987; Bertsch and Parker, 1996). Of particular interest is the tridecameric polycation, $AlO_4Al_{12}(OH)_{24}(H_2O)^{7+}$, generally referred to as Al_{13}, which is highly toxic to plant roots (Parker et al., 1989; Bertsch and Parker, 1996). Bertsch (1987) showed that the formation of Al_{13} occurs in partially neutralized Al solutions and depends on synthesis conditions including inhomogeneous pH and the initial formation of $Al(OH)_4^-$. The presence of Al_{13} in the organic horizons of two forest soils was reported by Hunter and Ross (1991) but has not been confirmed by subsequent research (Bertsch and Parker, 1996). The presence of SO_4^{2-} (Kerven et al., 1995) or silicic acid (SiH_4O_4) (Larsen et al., 1995) is detrimental to Al_{13} formation in more complex solutions. The ubiquity of these anions in the soil solution and in aquatic systems makes it unlikely that the highly charged free Al_{13} is present in natural systems. However, Al_{13} may form in cell walls even though solution conditions may not be favorable (Masion and Bertsch, 1997). Furthermore, Al_{13} may form in some solution culture systems studying Al effects in which pH has been raised to prevent H^+ toxicity (Parker et al., 1989).

16.3.2 ALUMINUM AND ORGANIC LIGANDS IN SOLUTION

Chelation of Al with many organic ligands of low RMM greatly decreases the activity of Al^{3+} in solution, thereby decreasing its biological effects. The strength of binding of Al by low RMM organic ligands is in the order: citrate > oxalate > tartrate > malate > acetate (Miyasaka et al., 2007). Importantly, L-malate is required to bind Al (Kerven et al., 1991). These ligands are commonly present in topsoil solutions and are excreted by the roots of some plant species (Section 16.7). It has been claimed that up to 80% of Al in topsoil solution at pH < 5.5 is complexed with dissolved organic C (Alleoni et al., 2010). The lower biological toxicity of Al chelates (Bartlett and Reigo, 1972; Hue et al., 1986) led to considerable research to resolve their relative effects on root growth.

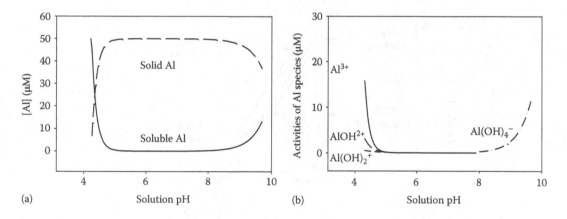

FIGURE 16.2 The effects of solution pH 4.0–9.5 on (a) the estimated total Al concentration in solution and on (b) the estimated activities of Al^{3+}, $AlOH^{2+}$, $Al(OH)_2^+$, and $Al(OH)_4^-$ as determined by GEOCHEM-EZ (Shaff et al., 2010). Solution composition was simplified from soil solution data presented by Wheeler et al. (1993) to contain (μM): 240 KCl, 450 CaCl$_2$, 150 Mg(NO$_3$)$_2 \cdot$ 6H$_2$O, 510 NaCl, and 50 AlCl$_3$. Aluminum hydrolysis reactions and pK values are based on those of Nordstrom and May (1996):

$$Al^{3+} + H_2O \leftrightarrows Al(OH)^{2+} + H^+ \qquad -5.00$$

$$Al^{3+} + 2H_2O \leftrightarrows Al(OH)_2^+ + H^+ \qquad -10.10$$

$$Al^{3+} + 3H_2O \leftrightarrows Al(OH)_3^0 + H^+ \qquad -16.80$$

$$Al^{3+} + 4H_2O \leftrightarrows Al(OH)_4^- + H^+ \qquad -22.70$$

$$Al^{3+} + 3H_2O \leftrightarrows Al(OH)_{3\,(solid)} + 3H^+ \qquad -8.10$$

16.4 ALUMINUM IN SOILS

16.4.1 SOIL ACIDIFICATION

As indicated in Section 16.1, acid soils develop from parent materials low in basic cations, with granite, for example, containing 3.8% Al and only 1.3% Ca, 0.4% Mg, 3.4% K, and 2.7% Na (Blatt et al., 2006). By comparison, calcium carbonate ($CaCO_3$) makes up 95% of carbonate rocks that contain 38% Ca. The primary Al-containing minerals (Section 16.1) weather initially to 2:1 layer aluminosilicate clay minerals (e.g., vermiculite, smectite). Further weathering leads to the formation of 1:1 layer aluminosilicates (e.g., kaolinite) and then to the leaching of Si and basic cations. A combination of good drainage and more intense weathering leaves only Al and Fe hydrous oxides behind. Increased weathering (e.g., aridisols → mollisols → ultisols) results in lower base saturation as explained by Jobbágy and Jackson (2001). Mirasol (1920) raised the question as to which compound breaks up readily in the soil to form soluble salts, concluding that gibbsite is the answer. Lindsay (1979) came to the same conclusion, gibbsite being a common mineral in many soils (Figure 16.3).

Additional loss of basic cations occurs through erosion, leaching, and removal in plant material. Addition of dissolved oxides of C, N, and S in precipitation and from the excretion of organic acids associated with the C and N cycles adds further to the acidification of soils (Helyar and Porter, 1989). Older soils, and those in regions of high precipitation and high temperature, are generally more acidic than younger soils and those in arid regions—past climate, of course, must be considered also.

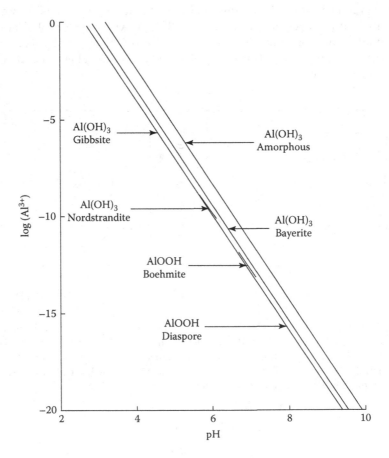

FIGURE 16.3 The solubility of various oxides and hydroxides present in soils as affected by pH. (From Lindsay, W.L. and Walthall, P.M., The solubility of aluminum in soils, in: *The Environmental Chemistry of Aluminum*, Sposito, G. (ed.), Lewis Publishers, Boca Raton, FL, 1996, pp. 333–361.)

Rates of acidification in sandy soils used for agriculture in Western Australia have been estimated at 0.15–0.23 kmol H^+ ha^{-1} $year^{-1}$, requiring 8–11 kg $CaCO_3$ ha^{-1} $year^{-1}$ to neutralize (Dolling and Porter, 1994; Dolling et al., 1994). In these instances, soil acidification was ascribed to the C cycle (removal of grain and animal products and waste to sheep camps) (Dolling and Porter, 1994) and to NO_3–N leaching, removal of alkaline products and buildup of organic matter (Dolling et al., 1994). In northern South Africa, acidification to a depth of 0.25 m has been estimated at 0.21–10.31 (mean 3.70 kmol H^+ ha^{-1} $year^{-1}$) under crop production, requiring ~190 kg lime ha^{-1} $year^{-1}$ to maintain soil pH (Jansen van Rensburg et al., 2009). Rates of acidification differ with land use, elevated rates being evident with high amounts of N fertilizers applied to sugarcane (*Saccharum* spp. L.) (Hartemink, 2008). Global fertilizer N consumption increased between 2000 and 2009 from 82 to 102 Tg, mostly in China (11 Tg) and India (5 Tg) (Cui et al., 2013). Farmers in the North China Plain grow two crops per year, applying between 80 and 800 kg N ha^{-1} $year^{-1}$ to wheat and between 80 and 730 kg N ha^{-1} $year^{-1}$ to corn; corresponding N removal averaged only 140 and 130 kg N ha^{-1} in these crops (Cui et al., 2013). Not surprisingly, Huang et al. (2013) reported 360 mg NO_3^- L^{-1} (5.7 mM) in the vadose zone to 5 m. Great effort is being expended to match N inputs and outputs in China to reduce NO_3^- losses and pollution of groundwater (Miao et al., 2011; Cui et al., 2013). From the 1980s to 2000s, high rates of N fertilization of four major Chinese cropping areas had released 20–220 kmol H^+ ha^{-1} $year^{-1}$ to corn (Guo et al., 2011), decreasing soil pH by up to 2.2 units in other than highly alkaline soils (Miao et al., 2011). This would theoretically require up to 2.5 t $CaCO_3$ ha^{-1} $year^{-1}$ to counteract the acidity generated.

Acid precipitation results from the dissolution of atmospheric gases such as CO_2, NO_2, and SO_2. As little as 30 years ago, Thomas and Hargrove (1984) thought that the effects of acid precipitation (pH ~ 4) would be insignificant on most agricultural soils where liming and heavy fertilization practices are common but that problems could occur in shallow soils and those with low organic matter or low cation-exchange capacity (CEC). More recent data, however, have shown that annual atmospheric acid inputs range from 0.2 to 4 kmol ha^{-1} $year^{-1}$ (Lawrence, 2002) and higher. Lövblad et al. (1995) estimated annual depositions in southwest Sweden were 10–15 kg N ha^{-1} and 15–25 kg S ha^{-1}. These acid inputs defoliate trees, dissolve Al minerals in soils, increase nutrient levels, and release Al from organic complexes in surface waters. Changes in soils can be surprisingly rapid; Ulrich et al. (1980) found that in 1966, Al comprised <10% of cations in the soil solution of a loess soil in Germany. This amount increased to 20% and 30% by 1973 and 1979, respectively, having marked detrimental effects on fine roots of beech (*Fagus sylvatica* L.). There have been increasing concerns regarding N depositions near Beijing, China, with average values of 30 (Liu et al., 2006) and 80 (He et al., 2007) kg N ha^{-1} $year^{-1}$ recorded from 1998 to 2004. Improved gaseous emission standards in many industrialized countries have lowered acid precipitation to the consequent benefit of the biota in soils and surface waters. Hrkal et al. (2006), for example, noted that annual S deposition in the Czech Republic has decreased from 108 to 17 kg ha^{-1} over the past half century, concluding that S deposition should not have a substantial impact on the environment by 2010. However, Lawrence (2002) noted continued acidification of streams in the Catskill region, United States, from chronic and episodic events despite a 40% decrease in S emissions in recent years. This acidification occurred because of previous N and S accumulation on the forest floor and decreased Ca in soils.

Besides acid precipitation, high salt concentration in the atmosphere increases Al in drainage waters through desorption of Al^{3+} from sorption sites. Soil solutions from 21 sites in Norway contained on average <30 μM Al, with episodic events of sea salt incursions greatly increasing maximum concentrations (Lange et al., 2006). It seems, therefore, that episodic leaching events are important in landscape acidification, possibly accounting for acidification of soils in subhumid environments such as the cropping lands of Southeastern Australia.

Agricultural activities also result in greater acidification through perturbations of the C and N cycles (Helyar and Porter, 1989). For example, Williams (1980) determined that N_2 fixation by subterranean clover (*Trifolium subterraneum* L.) and the consequent changes in soil N and organic C over 50 years in Southeastern Australia acidified the topsoil from pH 6.0 to pH 5.0. The consequent increase in soluble Al was detrimental to plant growth. Adams (1984) concluded that a major factor in decreased fertility of soils in the Southern United States was the widespread use of high rates of N fertilizer since the 1950s. The removal of increased corn grain, and especially of silage, would remove considerable quantities of Ca, Mg, and K and increase the leaching of these basic cations through increased NO_3^- in the soil solution.

16.4.2 MEASUREMENT OF TOXIC ALUMINUM IN SOILS

Mineral soils contain on average ~70 kg total Al in 1 m^3 on a dry mass (DM) basis, but total soluble Al is usually <350 μM, with concentrations >1000 μM Al occurring only in acid sulfate soils (Ritchie, 1989) (Section 16.4.3). Root length of many plants is decreased by 50% at an activity of <25 μM Al^{3+} (Poschenrieder et al., 2008). Clearly, any method to determine the potential toxicity of Al needs to discriminate between total and soluble Al in the soil—and between forms of Al in the soil solution that are not equally toxic to root growth (Asher et al., 1992) (Table 16.1). The forms of Al in solution can be discriminated by colorimetric procedures using ligands such as aluminon, pyrocatechol violet (Kerven et al., 1989a), and ferron (Parker and Bertsch, 1992). It is worth noting that inductively coupled plasma optical emission spectroscopy (ICPOES) has simplified the determination of Al in solution but does not discriminate between the forms of Al present in solution. Therefore, extracted soil solutions need ultrafiltration to 0.05 or 0.025 μm prior to analysis; filtration to 0.22 μm is not adequate (Menzies et al., 1991). It is worth noting that Jenny (1961) found that "clear" supernatant solution "swarmed with tiny colloidal particles." Ultrafiltration does not, however, discriminate between Al^{3+} and Al bound to organic ligands; a technique like that of Kerven et al. (1989b) is necessary to do so.

In subsoils and in topsoils low in organic matter, biologically active Al in the soil solution is determined by soil pH, soil mineralogy, and solution ionic strength. Complexation of Al^{3+} by organic ligands in many topsoils, however, is the governing mechanism in many situations. Van Hees et al. (2001), for example, found that no minerals, including gibbsite, adequately explained soil solution Al^{3+} equilibrium conditions in podzolic forest soils in Sweden. Complexation of Al with organic acids was the major factor in the E horizon though exchange and complexation reactions with minerals governed Al^{3+} in B and C horizons.

TABLE 16.1
Forms of Aluminum Potentially Present in Solution Extracted from Acid Soils

Form	Examples
Microcrystals	Kaolinite, Gibbsite
Amorphous precipitates	$AlSiO_5$, $Al(OH)_3$, $AlPO_4$
Inorganic polycations	$Al_2(OH)_2^{4+}$, $Al_3(OH)_4^{5+}$, $AlO_4Al_{12}(OH)_{24}(H_2O)^{7+}$
Organic complexes	
Low RMM[a]	Al citrate, Al oxalate, Al malate
High RMM	Al fulvate, Al humate
Inorganic complexes	AlF^{2+}, AlF_2^+, $AlSO_4^+$
Inorganic monomers	Al^{3+}, $AlOH^{2+}$, $Al(OH)_2^+$

Source: After Asher, C.J. et al., Recent research on soil acidity: Identifying toxic aluminium in soils, in: *La Investgacion Edafologica en Mexico 1991–1992. Memorias del 15 Congreso Nacional de la Sociedad Mexicana de la Ciencia del Suelo,* Tovar, S.J.L. and Quintero, L. (eds.), Society of Mexicana de la Ciencia del Suelo, Acapulco, Mexico, 1992, pp. 61–68.

[a] Relative molecular mass (*syn.* molecular weight).

TABLE 16.2

Mean Quantities and Mole Fractions of Ca^{2+}, Mg^{2+}, Na^+, K^+, and Al^{3+} Present in Exchangeable Forms and in Soil Solutions Filtered to 0.025 μm of 40 Unamended, Highly Weathered Acid Soils of Tropical and Subtropical Origin

Cation	Exchangeable (cmol$_c$ kg^{-1})	Soil Solution (mM)	Saturation of Exchange Sites (Mole Fraction)	Presence in Soil Solution (Mole Fraction)
Ca^{2+}	1.24	0.11	0.27	0.08
Mg^{2+}	1.25	0.21	0.28	0.15
Na^+	0.12	0.74	0.05	0.52
K^+	0.19	0.32	0.08	0.22
Al^{3+}	2.13	0.04	0.31	0.03

Source: After Menzies, N.W., Solution phase aluminium in highly weathered soils: Evaluaton and relationship to aluminium toxicity to plants, PhD thesis, The University of Queensland, Brisbane, Queensland, Australia, 1991.

Plotting solubility data on a log scale provides a useful way of inspecting Al speciation (Lindsay, 1979; Lindsay and Walthall, 1996) (Figure 16.3). The common occurrence of gibbsite in many mineral soils is used as reference mineral that limits the activity of Al^{3+} (Lindsay, 1979). Indeed, nearly a century ago, Mirasol (1920) recognized gibbsite to be the mineral that "breaks up" to form soluble salts that are injurious to root growth.

Soluble Al in acid soils adds further complexity to the chemistry of Al in solution (Section 16.3) because of reactions with the solid phase. Sposito (1996), in a definitive overview of Al in soils, pointed out there is great complexity to the story of Al biogeochemistry. Nonspecific binding (exch. Al) occurs through electrostatic forces at negatively charged sites on clays and hydrous oxides of Fe, Al, or Mn. Specific binding of Al ions occurs at variably charged sites on hydrous oxides and at the edges of clay minerals (Ritchie, 1989). The mole fraction of exch. Al^{3+} greatly exceeds that of Al^{3+} in the soil solution as shown by Menzies (1991) (Table 16.2); the opposite is evident with K^+ and Na^+.

16.4.3 Acid Sulfate Soils

Acid sulfate soils may be regarded as the extreme of soil acidification that occurs when soils and sediments containing iron sulfides, especially iron pyrite (FeS_2), are exposed to the atmosphere. Sulfidic materials in soils and sediments under anaerobic conditions, generally at pH 6.5–7.5, indicate potential acid sulfate soil problems. For these soils and sediments to form requires S (usually from seawater) and Fe (from terrestrial sources), anaerobic conditions, and organic matter as an energy source for bacteria (Fitzpatrick et al., 1996). Upon exposure to the atmosphere through mining sulfidic ores or through drainage of potentially acid soils for agriculture and the building of housing or other infrastructure, S-oxidizing bacteria produce jarosite ($KFe_3(SO_4)_2(OH)_6$) and sulfuric acid via complex reactions (Willett et al., 1992). The dissolution of soil particles at as low as pH 2 results in the release of large quantities of Al, Fe, and many other trace metals (Cravotta, 2008) into solution that may remain in situ in dry periods but move off site into waterways during wet periods. The dissolution of soil on site causes damage to roadways and housing, for example. Off-site movement may have the same effects and adversely affect aquatic organisms, including massive fish kills on occasion (Nystrand and Österholm, 2013).

Soils and sediments containing iron sulfides are found commonly in relic and current marine estuarine environments, especially of the tropics and subtropics, but can occur far from the sea also. For example, FeS_2 may be in quartz veins, metamorphic rocks, and in association with coal seams. Acid sulfate soils may, therefore, develop in coal mine spoils causing postmining problems. For example, the U.S. Environmental Protection Agency estimated that there are 7600 km of streams with low pH that have been impacted by extraction of resources, primarily coal, in the mid-Atlantic region (EPA, 2013). West Virginia and Pennsylvania each have an estimated 3500 km of streams impacted by acid mine drainage. In Northern Australia, drainage water from disturbed acid sulfate soils of estuarine areas of Trinity Inlet contains high levels of Al. A further example is the construction of an airport near Washington, DC, United States, in which the scalped surface had pH 1.8–3.5 causing major problems of revegetation (Fanning et al., 2004).

The concentration of soluble Al in solutions of acid sulfate soils can be 1000 μM (Ritchie, 1989) or higher, mostly as Al^{3+} and $AlSO_4^+$ given the very low pH and high SO_4^{2-} concentration. With 3000 μM Al in groundwater of acid sulfate soils, flooding in Eastern Australia in 1994 resulted in >270 μM Al in the Richmond River with 90 km affected in this one event (Sammut et al., 1996). These levels of Al (Section 16.7.2) cause major damage to the biota of these regions.

16.5 TESTING PLANT RESPONSES TO ALUMINUM IN SOLUTION CULTURE

16.5.1 Solution Composition

The direct study of Al effects in soils is difficult because of (1) the many interrelated components of the acid soil infertility complex, (2) the many factors that are changed when pH of an acid soil is raised or lowered, and (3) the poor direct access to roots in the soil. Because of these difficulties, solution culture studies have been used over the decades to elucidate the physiological effects of Al on root growth (Section 16.2). Interestingly, Miyake (1916) showed that 80% of maximum rice root length occurs at 40 μM Al (Figure 16.1), a result not dissimilar to those of modern solution culture studies.

The complexity of Al chemistry, however, poses many traps for the unwary, as is evidenced by many scientific publications purporting to show either an absence of Al effects or Al tolerance at Al concentrations much higher than those in the soil solution as indicated by Ritchie (1989) and Menzies et al. (1994a,b). Thus, the first step in preparing solutions for plant growth studies is an approximation of soil solution conditions including the concentration of soluble Al present. Importantly, solution pH (Figure 16.2) and concentrations of nutrients, especially of basic cations and P, need to be similar to those found in the soil solution. This requirement raises further important issues, most notably the presence of organic ligands that may or may not form stable complexes with Al and that are either nontoxic or considerably less toxic than Al^{3+} (Bartlett and Reigo, 1972; Hue et al., 1986; Ma, 2000). The possible microbial degradation of low RMM organic ligands poses a further problem (Kerven et al., 1991). Additionally, adding NaOH or KOH in the presence of Al has marked effects on Al in solution because of localized regions of high pH (Bertsch, 1987; Bertsch and Parker, 1996). Finally, solution culture must be regarded as an approximation of soil conditions with regard to its buffer capacity in supplying nutrients and regulation of soluble Al.

Once these potential problems are accepted and overcome, solution culture may be used with great benefit to study the physiological effects of Al on plant root growth. Solutions approximating the soil solution were shown to produce good plant growth provided sufficiently large solution volumes are used (Loneragan, 1968; Asher and Edwards, 1983), the solution is renewed regularly, or the nutrients replenished (Asher and Blamey, 1987; Taylor et al., 1998). Good growth of many plant species is evident at as low as 1 μM P (Bell et al., 1989) as occurs in many soil solutions (Gillman and Bell, 1978; Wheeler and Edmeades, 1995). Importantly, low P in solution limits ion pairing and the precipitation of Al as variscite. Adams and Lund (1966) provided excellent evidence for the similar effects of Al in the soil solution and in solution culture on root length of cotton (*Gossypium hirsutum* L.) (Figure 16.4).

In the 1920s, Crone's solution (Hoagland and Arnon, 1950) was used widely in the study of plant growth in solution culture, and it is noteworthy that Magistad (1925) measured only small amounts of Al present at pH > 3.6. Magistad (1925) concluded that this occurred due to the presence of 650 μM P,

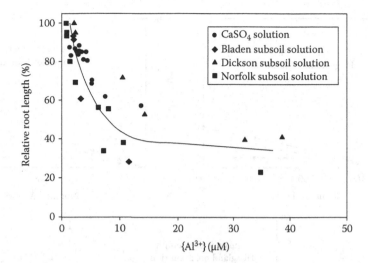

FIGURE 16.4 Relationship between the activity of Al^{3+} $\{Al^{3+}\}$ and the relative length of cotton primary roots grown in $CaSO_4$ solution and in subsoil solution of four acid soils. (After Adams, F. and Lund, Z.F., *Soil Sci.*, 101, 193, 1966.)

leading to Crone's solution not being used in subsequent research. Decades later, Shaff et al. (2010) provided a salutary lesson in that they had taken the common approach of using a nutrient solution that had appeared previously in the scientific literature but found that the majority of the Al, P, and Fe precipitated. Much research, however, is conducted at high Al, at pH $\geq$ 5, and with high P. This activity is despite findings that Al is toxic at low concentration in many early (e.g., Miyake, 1916; Adams and Lund, 1966; Lance and Pearson, 1969) and recent (e.g., Yamamoto et al., 2002; Kopittke et al., 2008) studies.

As part of work to simplify the use of GEOCHEM and GEOCHEM-PC (Parker et al., 1987, 1995), Shaff et al. (2010) questioned why the program should be used to design experimental solutions. In short, the answer provided is that it can help predict potential problems in experimental media, presenting as an example calculation of the composition of a solution recommended by Yoshida et al. (1971) and used to select rice lines for tolerance to soluble Al (Wu et al., 2000). Addition of 1.3 mM Al at pH 4.0 resulted in a solution in which only 10% of the original Al was present as Al^{3+} and the predicted concentrations between the control and Al treatment differed with regard to S, P, and Fe.

Hoagland No. 2 solution (Hoagland and Arnon, 1950) has been used successfully for many years to grow plants. This solution has the advantage of supplying high concentrations of nutrients (mM): 14 NO_3^--N, 1 NH_4^+-N, 6 K, 4 Ca, 2 Mg, 2 $SO_4^{2-}-S$, 1 P and (µM) 46 B, 25 Fe, 18 Cl, 9 Mn, 0.8 Zn, and 0.3 Cu. These concentrations, however, are far in excess of those in soil solutions of acid soils (e.g., Table 16.2) and of fertile temperate soils (Edmeades et al., 1985), resulting in Al^{3+} present in solution that is markedly lower than that intended (Figure 16.5). Using full-strength Hoagland No. 2 solution, the calculated activity of Al^{3+} was only ~4% of the nominal value at pH 4.1. The corresponding values increased to ~7 and 15% in one-half and one-quarter strength solutions because of decreased solution ionic strength. At pH 4.3 and 4.5, there was a further decrease in soluble Al through precipitation especially at higher nominal Al levels. It is important to realize that these results relate to equilibrium conditions, with Al^{3+} activity likely to be lower with inhomogeneous conditions in solution or alkalization by plant roots as often occurs with high NO_3^- in solution.

16.5.2 Measurement of Solution Culture Conditions

Finally, unexpected problems may arise when preparing nutrient solutions for the study of the biotic effects of Al. These phenomena often relate to differences between thermodynamics and kinetics of solution composition. Thermodynamic calculations, as used in GEOCHEM-EZ (Shaff et al., 2010)

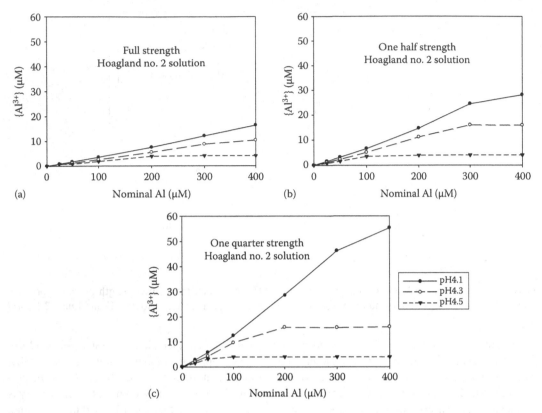

FIGURE 16.5 The activity of Al^{3+} {Al^{3+}} calculated using GEOCHEM-EZ (Shaff et al., 2010) in Hoagland No. 2 solution that contains (mM) (a) $14NO_3^--N$, $1NH_4^+-N$, 6K, 4Ca, 2Mg, $2SO_4^{2-}-S$, and 1P and (µM) 46B, 25Fe, 18Cl, 9Mn, 0.8Zn, and 0.3Cu and in (b) one-half and (c) one-quarter Hoagland No. 2 solutions is markedly lower than the nominal concentration through the effects of high solution ionic strength and precipitation of Al.

or PhreeqcI (Parkhurst, 2013), provide the theoretical equilibrium according to input solution components. The addition of base or phosphate to solutions containing Al, however, may precipitate Al hydroxides or phosphates that only slowly redissolve. Illustrative of this action is the conclusion of Blamey et al. (1983) and Alva et al. (1986) that polycationic Al is not toxic to roots. Solution preparation in these studies, however, probably resulted not in soluble polycationic Al species but rather in microparticulate forms of Al remaining in suspension (Kerven et al., 1995). This conclusion precluded a valid assessment of polycationic Al effects; Al_{13} is considerably more toxic than Al^{3+} (Parker et al., 1989; Bertsch and Parker, 1996).

To prevent problems arising, researchers contemplating solution culture should (1) use solutions comparable to those of soil solutions, (2) estimate theoretical equilibrium conditions prior to conducting an experiment (Shaff et al., 2010), (3) microfilter solutions to ≤0.05 µm (Menzies et al., 1991), and (4) measure soluble Al actually present in solution. In some instances, it would be necessary to determine the forms of Al in solution so as to discriminate between Al^{3+} and Al bound to organic ligands. Neglecting to follow these procedures leaves researchers open to incorrect conclusions regarding the effects of Al on plants.

16.6 UPTAKE AND TRANSPORT OF ALUMINUM BY PLANTS

Most plant species retain Al in their roots (Horst et al., 2010), with little translocation to the shoots. Some plants, however, are classified as Al accumulator species (Chenery, 1948; Webb, 1954; Jansen et al., 2002), presenting intriguing research opportunities related to the physiology of Al uptake and transport.

16.6.1 ALUMINUM ACCUMULATOR PLANT SPECIES

The shoots of most plants contain 0.02%–0.08% Al on a DM basis, but Gardner and Garner (1953) noted that several tropical plants may accumulate up to 4% Al. Initially, a few species with this trait were identified based on their use as a mordant in the dyeing of fabrics (Chenery, 1948), findings later extended to other plant species (e.g., Webb, 1954). Recent research has shown an increasing number of Al accumulator species (Jansen et al., 2002). The mostly perennial, woody Al accumulator species of the tropics have >1000 mg Al kg^{-1} compared to <200 mg Al kg^{-1} in leaves of most plants (Jansen et al., 2002). Interestingly, plantain (*Plantago almogravensis* Franco) is a rare, endemic Al accumulator plant of Portugal (Branquinho et al., 2007). Accumulation of Al in shoots is considered a primitive trait that may be used to characterize many families (Jansen et al., 2002).

Tea (*Camellia sinensis* Kuntze) is an Al accumulator, with Matsumoto et al. (1976) reporting 3% Al in old leaves. Despite this accumulation, tea infusions probably do not contribute greatly to increased Al exposure in humans (Flaten, 2002; Yokel and Florence, 2008) with young leaves used for infusion containing only 600 mg Al kg^{-1}. This concentration is <2% of that in the old leaves, where accumulation probably results from continued transpiration (Matsumoto et al., 1976). This occurs also in camellia (*Camellia oleifera* Abel), though phloem mobility may occur also (Zeng et al., 2013). Old tea leaves are characterized by thick epidermal cell walls, with Al accumulating especially in the upper epidermis. Synchrotron-based low-energy x-ray fluorescence (LEXRF) spectromicroscopy by Tolrà et al. (2011) has confirmed that Al is localized in the cell walls of the leaf epidermal cells. Young, fully expanded leaves of plants exposed to 200 µM Al for 2 weeks contained 1000 mg Al kg^{-1}, with almost none in the symplast. This determination suggests that localization of Al in the epidermal leaf apoplast is the major Al tolerance mechanism. Similarly, the accumulation of Al in the epidermis was evident in one species of Vochysiaceae from the Brazilian Cerrado, but two other species accumulated Al in palisade and spongy parenchyma cells (de Andrade et al., 2011).

Buckwheat (*Fagopyrum esculentum* Moench) is interesting in that it is an annual crop plant that accumulates a high concentration of Al in its leaves (Shen and Ma, 2001). As with tea, buckwheat accumulates Al mostly in the old leaves; the old leaves of 28-day-old plants contained 1680 mg Al kg^{-1} compared to only 200 mg Al kg^{-1} in the young leaves. It is noteworthy that Shen and Ma (2001) determined that Al accumulation in the old leaves is dependent on continued transpiration.

Some mistletoe species (e.g., *Phthirusa ovata* Eichler or *Phoradendron crassifolium* Eichler) are hemiparasites of Al-accumulating or non-Al-accumulating plants of the Brazilian Cerrado (Luttge and Clarkson, 1992). The mistletoe species on Al-accumulating hosts amass more Al than those on Al-non-accumulating hosts, but there were no differences in their photosynthesis or stomatal regulation on the two hosts. This result indicates good adaptation at high tissue Al. The mechanisms involved have not been identified but may be due to Al complexation within the host species. Recent research by Scanlon et al. (2013) found that a mistletoe species that parasitizes Al accumulator hosts accumulated Al in leaves, branches, and seeds; another species that parasitizes a wider range of plants had Al only in the leaves.

There are a number of possible mechanisms involved in the ability of Al accumulator plant species to overcome levels of soluble Al that are toxic to most species (Watanabe and Osaki, 2001; Jansen et al., 2002). Absorption of Al by the roots and translocation to the shoots implies that initial detoxification of Al occurs in the rhizosphere or in the root itself, possibly through chelation. Ma (2000) and Ma et al. (2001) suggested that root tip organic acids that chelate Al play an important role in this regard. Once absorbed, Al must remain detoxified in the transpiration stream and in the leaves. Accumulation of Al occurs preferentially in the cell wall of epidermal cells of leaves (Matsumoto et al., 1976; Jansen et al., 2002) and the vacuole (Jansen et al., 2002). However, Al needs to be transported prior to sequestration possibly through continued chelation by organic acids, proteins, or other organic ligands, through the synthesis of Al-tolerant proteins, or elevated enzyme activity.

It is interesting that Watanabe and Asaki (2001) concluded that there are often different chelates for Al translocation and for Al storage in the leaves. Given the differences among species in Al accumulation sites (de Andrade et al., 2011), it is possible that a range of mechanisms is involved; more research is needed to elucidate this intriguing matter.

16.7 PHYSIOLOGICAL RESPONSES OF PLANTS TO ALUMINUM

16.7.1 POSITIVE RESPONSES TO ALUMINUM

Positive responses to Al may be divided into two classes: those reported for Al accumulator species and those for the majority of plants that translocate only a small quantity of Al from the roots to the shoots. High levels of Al in the shoots of Al accumulators would imply some benefit to the plant that may include, for example, protection from herbivory or physiological roles for improved growth. Konishi et al. (1985) reported that the growth of tea plants was stimulated by the addition of Al in solution culture, an effect that was especially evident at high P supply. However, the 0.1 and 0.8 mM P and Al ($\leq$0.8 mM at low P and $\leq$6.4 mM at high P) were far in excess of those in soil solutions of acid soils on which tea is grown. Maximum growth occurred at 0.4 mM Al at low P and at 1.6 mM Al at high P, leading Konishi et al. (1985) to conclude that Al enhanced the absorption of P but also alleviated the toxicity of excess P. It is not clear if these results apply to tea plants grown in soil given the very high concentrations of Al and P used.

Positive responses to Al have been reported in species that do not accumulate Al although there is no clear evidence of a beneficial physiological role (Foy, 1984; Miyasaka et al., 2007; Poschenrieder et al., 2013). This result raises the question as to why plants respond positively to added Al. To answer this question, it is necessary to assess the experimental conditions under which positive growth responses were demonstrated. As far as we are aware, improved plant growth has not been shown by decreasing soil pH—an effect that would increase soluble Al. Rather, beneficial Al effects have been reported in solution culture. In these instances, the solutions contained concentrations of P in excess, sometimes far in excess, of those present in soil solutions. Campbell et al. (1990), for example, studied red clover (*Trifolium pratense* L.) grown in an acid soil and in solutions with 100 μM P and 111 μM Al. Aluminum depressed growth in soil but not in nutrient solution; indeed, some genotypes responded positively to Al in nutrient solution. Furthermore, the way in which solution pH has been increased by additions of NaOH or KOH has not been well documented in many studies (Section 16.3.1).

One situation in solution culture described by Kinraide (1993) may occur in which Al^{3+} enhances growth by alleviating the toxic effects of H^+ (see also Kobayashi et al., 2013; Poschenrieder et al., 2013). These cations have opposing effects on cell wall tension, at least in the short term, with H^+ relaxing the cell wall (Winch and Pritchard, 1999; Blamey et al., 2004) and Al^{3+} increasing its rigidity (Horst et al., 2010; Kopittke et al., 2011).

Overall, we conclude that positive responses to the addition of Al in many nutrient solution culture studies most likely result from ion pairing and the precipitation of high P.

16.7.2 TOXIC EFFECTS OF ALUMINUM

Under nonlimiting conditions, growth of an individual root is a function of cell division in the meristematic zone and anisotropic cell growth in the elongation zone (Baluška et al., 1996). This latter zone is responsible for root growth in the short term; in the longer term, cells that have ceased expanding need to be replaced by other cells generated in the meristem and that have passed through the transition zone.

Rediscovery of Al as a rhizotoxin in the 1960s (Jenny 1961) led plant physiologists to examine the possible mechanisms involved. This research also involved studies on the toxicity of the various Al species present in solution (Section 16.3). The free Al^{3+} and hydroxy-Al species are in equilibrium even if the data on which speciation is dependent may be uncertain in some instances (Nordstrom and May, 1996). Kinraide and Parker (1989) found no evidence for the toxicity of hydroxy-Al species, despite some studies (e.g., Alva et al., 1986) suggesting that $Al(OH)^{2+}$ is more toxic than Al^{3+}. Research has shown that $AlSO_4^+$ is nontoxic or only 1/10th as toxic as Al^{3+} (Kinraide and Parker, 1987). Both Al^{3+} and F^- are toxic to plants, but complexes of Al^{3+} and F^- are either nontoxic (MacLean et al., 1992) or at least less toxic than Al^{3+} (Kinraide, 1997) to wheat. Complexes of Al and some organic ligands also reduce Al toxicity, effects evident in topsoils high in organic C. Overall, therefore, there seems to be a consensus that Al^{3+} is the form generally most toxic to roots, with the possibility of Al_{13} being present under some conditions.

16.7.3 PERCEPTION OF SOLUBLE ALUMINUM BY PLANT ROOTS

The ways in which Al is perceived by plant roots and translated into biochemical and visible responses has been the focus of much research, as has been the case with Ca (McLaughlin and Wimmer, 1999). Bennet et al. (1987) proposed a mechanism whereby Al has its initial effect on the root cap by affecting the differentiation of peripheral cap cells and cell loss. Importantly, exposure to 19 μM Al for 29 h decreased amyloplast numbers and decreased Golgi apparatus activity; these changes preceded a reduction in mitotic activity. (Clarkson (1965) had noted that Al causes mitotic abnormalities.) Further investigations, however, led Ryan et al. (1993) to conclude that removal of the root cap has no effect on inhibition of root growth by Al but that the terminal 2–3 mm of root must be exposed to Al for inhibition to occur. Application of Al to the elongation zone damaged the rhizodermis but did not inhibit growth. Elegant experiments by Sivaguru and Horst (1998) showed that Al must be applied to the distal transition zone (Baluška et al., 1996) for root growth to be decreased. These studies indicate, therefore, that a signal is transmitted from the root cap, meristem, or transition zone to the elongation zone where visible ruptures occur (Yamamoto et al., 2001; Kopittke et al., 2008). Ma et al. (2004), however, found that 10 μM Al decreased the viscous and elastic extensibility of walls of cells in the elongation zone in root apices of wheat cv. Scout 66 (Al-sensitive) but not in cv. Atlas (Al-resistant). These effects were evident in ~3 h, leading to the conclusion that Al reacts with the walls of actively elongating cells. These conflicting results indicate that more research is needed to elucidate the region where Al is perceived by plant roots.

16.7.4 LOCATION OF ALUMINUM IN ROOTS

In contrast to the findings of McLean and Gilbert (1927) that Al accumulates in the cytoplasm, recent research has shown that Al is present mostly in the walls of cells, especially those of the rhizodermis and outer cortex (Ishikawa et al., 1996; Marienfeld et al., 2000; Rangel et al., 2009; Horst et al., 2010). Bennet et al. (1985a) noted that the severe cellular disorganization of the outer root did not result in appreciable quantities of Al being found in the inner cortical regions or the stele. Marienfeld et al. (2000) found that the Al concentration was fourfold higher in the rhizodermis and outer cortex than in the inner cortex of corn roots exposed to 50 μM Al for 3 h. The corresponding value was 30-fold for bean (*Phaseolus vulgaris* L.). It was concluded by Marienfeld et al. (2000), therefore, that the reduced elongation of inner cortical cells is mediated indirectly by Al injury to the outer root cells. This conclusion was echoed by Kopittke et al. (2008, 2009) as related to the effects of Al^{3+} and a number of cations that also bind strongly to hard ligands (e.g., those containing O, P, or F) (Kinraide, 2009). Furthermore, increased Al sensitivity in bean has been related to the higher content of unmethylated pectin with higher negative charge of the cell walls (Rangel et al., 2009). In contrast, Kinraide et al. (2005) concluded that much higher Al is tolerated by the rhizodermis than the cortex. High Al in the apoplast does not rule out effects on other cellular components (Section 16.7.6). Rather, the question that needs to be addressed is whether or not apoplastic Al plays any role in this ubiquitous element's toxic effects.

Although the majority of Al is present in the apoplast, research has shown that Al is present in the cytoplasm, albeit at low levels. Taylor et al. (2000) used a precise novel technique with ^{26}Al to quantify Al movement in the cell wall and cell PM over 30 min. There was a 1000-fold higher rate of Al accumulation in the cell wall (71–318 µg m^{-2} min^{-1}) than Al crossing the PM (71–540 ng m^{-2} min^{-1}). Not surprisingly, therefore, the cell PM has also been implicated in Al toxicity (see Kobayashi et al., 2013 and references therein). The interference of Al with PM H$^+$-ATPase (Kim et al., 2010) may be implicated because low pH promotes cell expansion through the action of auxin (Rayle and Cleland, 1970)—H$^+$-ATPases are essential for the movement of ions and metabolites across the PM (Murphy et al., 2011). It is likely that Al interferes initially with rhizodermal cell wall relaxation and the cell PM given their location but does not rule out other effects occurring in the longer term (Yamamoto et al., 2001).

Many difficulties exist in measuring cytoplasmic Al (Rengel, 1996), but it appears that little Al crosses the cell PM, initially suggesting that there may be limited effects of Al on cytoplasmic components. This action may not be so, however, as shown by studies with Ca: only 300 µM Ca$^{2+}_{free}$ is required in the apoplast for structural integrity of the cell wall and PM; only 0.1 µM Ca$^{2+}_{free}$ is present in the cytoplasm, playing a crucial role in signaling (Stael et al., 2012).

16.7.5 Visible Symptoms of Aluminum Toxicity in Plants

16.7.5.1 Timing

A nutritional disorder may be diagnosed by the symptoms caused, including their timing, appearance, location, and development. In most plant species, the first toxic effect of soluble Al is evident as a suppression of root growth. This action was shown to occur 1 h after exposure to 40 µM Al in wheat (Ownby and Popham, 1989) with a similar timeframe in soybean (Horst et al., 1991) and corn (Llugany et al., 1995). Blamey et al. (2004) found a 56%–75% reduction in root elongation rate 18–52 min after exposing mung bean (*Vigna radiata* (L.) R. Wilczek) roots to 50 µM Al.

16.7.5.2 Appearance and Location of Symptoms

Besides detrimental effects on root growth, soluble Al has striking effects on the morphology of root tips. Field observations have shown that many crop plants develop short, stubby roots in soils high in exchangeable Al. Miyake (1916) identified roots as the initial site of Al injury by suppressing root growth (Figure 16.1), and Magistad (1925) noted that the roots are short and stubby. Many roots develop ruptures near their tips soon after an Al-induced suppression of root growth becomes apparent (e.g., Ryan et al., 1993; Yamamoto et al., 2001). In solution culture, Blamey et al. (2004) reported the development of a kink and transverse ruptures ~4 h after exposing mung bean roots to 50 µM Al; these initial ruptures were located ~2.4 mm from the tip. Similar findings were evident in cowpea (*Vigna unguiculata* L.) roots at 54–600 µM Al, with transverse ruptures in the elongation zone (Kopittke et al., 2008) as illustrated in Figure 16.6.

In solution culture, Al injury is evident as ruptures to the outer cellular layers (viz., the rhizodermis and outer cortex) ~2–6 mm from the apex, which break and separate from inner layers (Yamamoto et al., 2001; Jones et al., 2006; Kopittke et al., 2008). This damage, in turn, results in the loss of apical dominance and development of lateral roots (which, in turn, become damaged) close to the apex—the short, stubby roots initially described by Magistad (1925). Wagatsuma et al. (1987), Marienfeld et al. (2000), and Kopittke et al. (2008) all concluded that the ruptures result from the differential expansion of the outer and inner root layers—the former slowing while the latter continue to grow.

Besides the ruptures evident in many species, Bennet et al. (1985b) found that Al causes swelling of cortical root cells and distortion of cell walls in corn. Isotropic enlargement of cortical cells 1–2 mm from the tip was evident 6.5 h after exposure to 300 µM Al, continuing for 48 h after exposure at which time there was complete disruption of root tip cells. In similar vein, Kinraide (1988) found that wheat roots had swollen cells in the epidermis and outer cortex of the elongation

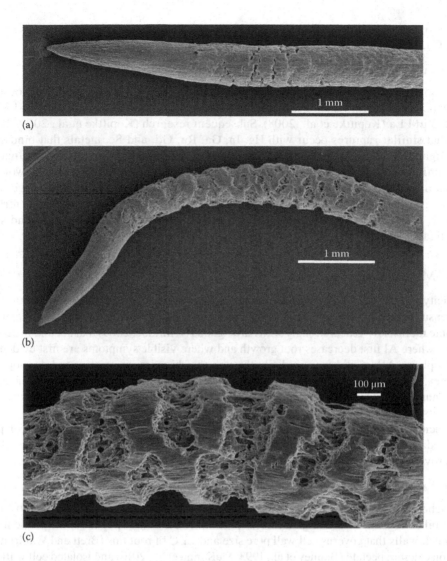

FIGURE 16.6 Scanning electron micrographs (SEM) of cowpea roots exposed to 40 μM Al for (a) 4 and (b) 12 h illustrating the location of the ruptures ~2 mm from the root tip and their expansion over time. (c) An SEM at greater magnification illustrates the smooth rhizodermis separated by expanded ruptures in cowpea roots exposed to 40 μM Al for 12 h.

zone after exposure to <3 μM Al for 2 days. This finding is similar to that of Wheeler et al. (1992) with the swollen cells 4–6 mm from the wheat root tip after 2–3 day exposure to Al. It is possible that the swollen rhizodermal cells may reflect disorders with root hair formation (J.B. Wehr, unpublished). These symptoms differ markedly in appearance from the transverse ruptures (Figure 16.6) evident in Al intoxicated roots of many species. Reasons for these differences, however, have not been elucidated.

Many root injuries result in the formation of callose (β-1,3 glucan), and this occurs with Al exposure also (Wissemeier et al., 1987; Yamamoto et al., 2001). Indeed, Horst et al. (1997) used callose as a sensitive marker for differences in Al sensitivity among corn genotypes.

Symptoms of Al toxicity are not easily identified in the shoots (Foy, 1984) because injury to roots causes problems with the uptake of essential nutrients and exacerbates drought effects. Lance and Pearson (1969) showed that cotton roots exposed to 6–11 μM Al had decreased uptake of water, Ca, Mg, P, and NO_3–N from solution. Indeed, wilting of leaves is a common symptom of root damage due to Al, even with adequate water in the soil profile, as are symptoms of Ca, Mg, P, and N deficiencies.

16.7.5.3   Possible Confusion with Other Disorders

Similar ruptures to those caused by Al were noted in onion (*Allium cepa* L.) roots exposed to Ga, In, or La (Clarkson, 1965) and in cowpea roots exposed to a restricted range of 0.85–1.8 μM Cu or 2.0–5.5 μM La (Kopittke et al., 2008). Subsequent research (Kopittke et al., 2009, 2011) has shown that similar ruptures occur with Hg, In, Ga, Ru, Gd, and Sc, metals that bind strongly to hard ligands as does Al. The only exception is Ag, which binds strongly to soft ligands but causes similar ruptures in the root elongation zone (Blamey et al., 2010). It is noteworthy that these metals are toxic to cowpea roots at concentrations considerably lower than that of Al. The large global extent of Al toxicity in acid soils, however, arises from the high concentration of Al in soils (~7%) relative to those of the other trace metals (Sparks, 2003) that bind strongly to hard ligands.

16.7.6   Mechanisms of Aluminum Toxicity

The toxicity of Al to plant roots has been known for about 100 years (Section 16.2), but there is still no consensus regarding the biochemical mechanism or mechanisms involved. As a trivalent ion of small ionic radius, Al^{3+} reacts strongly with hard ligands (Kinraide, 2009) that abound in cells of the root apex where Al first decreases root growth and where visible symptoms are first evident. From the rhizosphere, Al^{3+} would move radially through the rhizodermis, cortex, and developing stele of the apex. On a subcellular basis, Al^{3+} would move sequentially, having impact on the apoplast, cell PM, numerous cytoplasmic components, nucleus, tonoplast, and vacuole. All these subcellular components have been implicated in the toxic effects of Al.

As concluded by Zheng and Yang (2005), it still remains difficult to distinguish between primary and secondary effects of toxic Al levels in the root environment. The review by Miyasaka et al. (2007) provides insight into the complexity of this issue.

16.7.6.1   Apoplast

Sattelmacher et al. (1998) pointed out that the root apoplast is the first plant component to encounter adverse soil conditions such as high Al^{3+}. Of special interest in the apoplast is the pectic matrix of primary cell walls that governs cell wall pore size and CEC of root tips (Brett and Waldron, 1996). Studies on calcium pectate (Blamey et al., 1993; McKenna et al., 2010) and isolated cell walls (Wehr et al., 2010) have provided evidence of rapid, strong binding of Al and other metals that bind to hard ligands. The recent review by Horst et al. (2010) illustrates ways in which the root apoplast may be involved in Al-induced inhibition of root elongation and in Al resistance of plants. While most Al accumulates in the cell wall, with possible adverse effects on root growth, Al movement across the PM (Taylor et al., 2000) allows for effects on other root components.

16.7.6.2   Cell Plasma Membrane

Calcium is required for the structural integrity of cell wall and PM, and the toxic effect of Al has been implicated in disruption of Ca homeostasis (Rengel, 1992). It is not surprising, therefore, that soluble Al has been regarded as interfering with PM integrity. Indeed, the phospholipids of the PM are hard ligands to which Al^{3+} would bind strongly, decreasing the fluidity of the PM and thereby triggering the production of callose (Ishikawa et al., 1996). Wagatsuma et al. (1995) considered the PM of younger and outer cells of roots as the primary site for Al toxicity in roots. The accumulation of Al and leakage of K was in the order pea > corn > rice, the same order as their Al sensitivity. Protoplasts were damaged by Al with callose being formed; 1,3-β-glucan synthase is embedded in the PM. More recent results by Khan et al. (2009) have shown that exposure to 20 μM Al for 1 h increased PM permeability and Al uptake by an Al-sensitive rice cultivar.

Although Horst et al. (2010) addressed the effects of Al on the apoplast, the PM was considered important also. Indeed, Ishikawa et al. (1996) concluded that the PM of rhizodermal and outer cortical cells of the root tip are the primary sites for Al^{3+} accumulation. Further research is necessary, however, to determine the specific effects of Al^{3+} on the PM's structure and function.

16.7.6.3 Cytoplasmic Components

Aluminum has been shown to cross the cell PM (Taylor et al., 2000), but entry into the cytoplasm at pH 7.2–7.5 (Felle, 2001) may result in precipitation (Figure 16.2) or change in speciation by complexation with organic ligands (Miyasaka et al., 2007). Huang et al. (2012) found that more accumulation of Al in the cytoplasm and nucleus of three rice mutants increased their sensitivity to soluble Al.

Many studies have shown that Al produces reactive oxygen species (ROS) in plant tissues that result in oxidative damage to plant membranes (Cakmak and Horst, 1991; Yamamoto et al., 2001, 2002), but ROS formation may be the result, not the cause, of Al toxic effects (Yamamoto et al., 2001; Navascues et al., 2012). Changes to the activities of antioxidant enzymes occur also (Darkó et al., 2004; Simonovicová et al., 2004; Cartes et al., 2012). Cartes et al. (2012) found that adverse effects of Al were reduced by superoxide dismutase in Al-tolerant ryegrass (*Lolium perenne* L.). Nitric oxide (NO) has been implicated in Al toxicity, with some studies suggesting that Al-induced NO formation increases sensitivity to Al by affecting cell wall or PM properties (Zhang et al., 2011; Zhou et al., 2012) or plant hormones (He et al., 2012).

Despite uncertainty about the speciation and reactions of Al in the cytoplasm, Horst et al. (1999) found that Al caused disintegration of microtubules and changes in the actin cytoskeleton within 1 h of exposure. Therefore, the effects of Al may be integrated across the cell wall, PM, and cytoplasm.

16.7.6.4 Nucleus

Of great interest to plant physiologists is the possible effect of soluble Al on cell division in the root meristem as suggested by Silva et al. (2000) based on Al in nuclei of cells in the meristematic region of the root tip in an Al-sensitive soybean cultivar within 30 min. Indeed, exposure of the roots of Al-sensitive corn plants to 50 µM Al for 5 min was sufficient to inhibit cell division 250–800 µm from the tip (Doncheva et al., 2005) and exposure of bean roots for 12 h to ≤10 µM Al decreased mitosis and increased mitotic abnormalities in root tips (Yi et al., 2010). Interestingly, continued lateral root development indicates that the latent meristematic activity of the pericycle is not affected.

16.7.6.5 Tonoplast and Vacuole

In contrast to the cytoplasm, the vacuole has an acidic pH that acts as a cytoplasmic pH buffer along with the apoplast (Felle, 2001). Sequestration of Al in the vacuole occurs as an oxalate complex in buckwheat, an Al accumulator species (Shen and Ma, 2001). Sequestration of Al in the vacuole occurs also in rice, which is Al tolerant but not an Al accumulator (Huang et al., 2012). Interestingly, sequestration occurred via the *OsALS1* gene present in all root cells and localized in the tonoplast; knockout of this gene in three rice lines increased their sensitivity to Al.

16.7.7 RELATIVE RHIZOTOXICITY OF METAL CATIONS

Recent theoretical and experimental studies by Kinraide (2006), Kinraide and Yermiyahu (2007), Wang et al. (2008), Kinraide (2009), and Kinraide and Wang (2010) have put the toxic effects of Al^{3+} in a general framework of other metal cation effects. There are two components to this research: (1) the physicochemical properties of cations and (2) the electrical potential at the outer surface of the PM. Pearson (1963) had classified cations into three groups: hard (e.g., Al^{3+}), borderline (e.g., Co^{2+}), and soft (e.g., Ag^+). Kinraide and Yermiyahu (2007) and Kinraide (2009) quantified this concept, with cations allocated values on normalized scales based on the strength with which cations bind to hard and soft ligands (i.e., hard ligand scale [HLScale] and SLScale). The second component

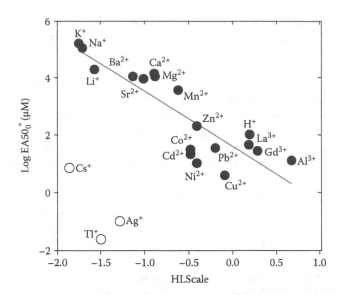

FIGURE 16.7 Relationship between the HLScale determined by Kinraide (2009) of 21 cations and the activity of these cations at the outer surface of the root cell plasma membrane that causes a 50% reduction in cowpea root elongation rate (EA50$_0$°) (Kopittke et al., 2011 Supplemental Data). The plotted linear regression does not include data for three cations (Cs+, Ag+, and Tl+) that are probably toxic through strong binding to soft ligands.

permits calculation of the activity of an ion at the outer surface of the PM as a function of the ion's activity in the bulk solution and the electrical potential at the outer surface of the cell PM (Kinraide, 2006; Kinraide and Wang, 2010). This approach has been detailed in the supporting information by Wang et al. (2013). Kinraide and Wang (2010) concluded that PM surface activities are better than bulk-phase activities for interpreting physiological responses to ions.

Kopittke et al. (2011) used this approach to assess the rhizotoxic effects of metal cations, including Al^{3+}, using the Speciation Gouy Chapman Stern program (Kopittke et al., 2014). Cowpea seedlings were grown for 48 h in solutions containing 1 mM Ca, 5 μM B, and 1 of 21 cations to evaluate their toxic effects on root growth. Cations that bind strongly to hard ligands (e.g., COO⁻ of the cell wall pectic matrix or P(OH)$_3$O⁻ of the PM) were more toxic than those that bind less strongly (Figure 16.7). It was concluded that the strength with which cations bind to hard ligands is important, generally, but nonspecific mechanism of toxicity—the only exceptions were the highly toxic effects of Tl+, Ag+, and Cs+. This concept is in keeping with the conclusion of Kinraide and Yermiyahu (2007) that a cation will be at least as toxic as the strength with which it binds to hard ligands.

16.7.8 ALUMINUM IN ALKALINE SOILS

High pH increases the solubility of Al with the aluminate ion, Al(OH)$_4^-$, predominating at pH > 8.5 (Figure 16.2). Whether or not Al(OH)$_4^-$ is toxic to plant roots has been assessed in many studies since first investigated by Magistad (1925). Conflicting results have arisen because of the many factors that limit root growth at high pH plus the difficulties in maintaining high solution pH and measuring soluble Al in such conditions. Confirming some previous studies, Fuller and Richardson (1986) demonstrated decreased growth of desert salt grass (*Distichlis spicata* Green) on bauxite residue, which is highly saline and has a pH 9–12, and in solution culture with 1.0 mM added Al(OH)$_4^-$ at pH 9.5 and 10.5. This led to the conclusion that Al(OH)$_4^-$ toxicity is a major factor reducing plant growth in bauxite residue. A more recent study by Brautigan et al. (2012) came to a similar conclusion.

However, Kinraide (1990) found poorer root growth of wheat and red clover in solutions at pH 8.0 than at pH 9.0 despite the theoretical increase in $Al(OH)_4^-$ with increase in pH (Figure 16.2b). This result was attributed to the formation of Al_{13} in the acidic free space within the roots. Kopittke et al. (2004) found that mung bean root growth was reduced and that ruptures to roots formed at a calculated activity of 16 μM $Al(OH)_4^-$ at pH 9.5. This finding was attributed to the formation of Al_{13} in solution at concentrations below the detection limit of the ferron procedure. Both studies concluded, therefore, that $Al(OH)_4^-$ is not rhizotoxic.

The complex effects of high pH on plant growth and difficulties in the control of rhizosphere conditions, however, limit the confidence with which firm conclusions can be made. Research is still needed to clearly elucidate the biotic effects of $Al(OH)_4^-$.

16.7.9 EFFECTS OF ALUMINUM ON N₂ FIXATION

Acid soils have detrimental effects on many soil microbes, with bacteria considered to be more sensitive than fungi (Shirokikh et al., 2004). Of particular relevance to the growth of legumes are the effects of Al on rhizobia, nodulation, and N_2 fixation. Fred and Loomis (1917) showed that addition of sulfuric acid (H_2SO_4) to culture media decreased the growth of *Bacillus radicicola*, which infects alfalfa roots, subsequent research continuing to focus on pH effects. Kinraide and Sweeney (2003), however, showed that 1 μM $AlCl_3$ in the medium was highly toxic to rhizobia. Interestingly, a decrease in solution pH alleviated the toxic effects of Al, Cu, and La (Section 16.7.1).

Hecht-Buchholz et al. (1990) and Brady et al. (1990, 1994) addressed the effects of Al on root structures as they affect infection by rhizobia. In peanut (*Arachis hypogaea* L.), infection is via root hairs in rhizodermal ruptures associated with lateral roots. In soybean and many other legumes, infection occurs at the tips root hairs and subsequent development of an infection thread. In both situations, inhibition of nodulation occurred at a lower Al concentration than that which decreased root growth. In peanut, Al decreased the rosette of multicellular root hairs found in normal plants (Brady et al., 1994). In soybean, Al reduced root hair proliferation, thereby decreasing nodulation (Hecht-Buchholz et al., 1990). Jones et al. (1995) also noted that as little as 2 μM Al stopped root hair growth and caused swelling of the tip as seen with microtubule inhibitors.

The Ca status of topsoils to increase Ca uptake by peanut pods is often achieved by applying gypsum (Section 16.9.1.2), but field studies on an Al-toxic ultisol showed little effect (Blamey and Chapman, 1982). Indeed, a general chlorosis of the foliage became evident ~90 days after planting. Liming, however, increased nodulation, improved crop appearance, and had a consequent benefit on hay and kernel yields. More recently, Murata et al. (2002) found that 4 t calcitic lime ha^{-1} increased peanut kernel yield from 1860 to 2960 kg ha^{-1} and from 960 to 2020 kg ha^{-1} on two acid granitic sandy soils in Zimbabwe.

Overall, therefore, Al^{3+} has detrimental effects on rhizobia and disrupts the infection of legume roots via alterations to root structure, with consequent decreases in N_2 fixation.

16.7.10 ALUMINUM EFFECTS ON MYCORRHIZAE

Numerous studies have shown the beneficial effects of mycorrhizal symbioses with plants, especially under low fertility conditions, with a particular benefit to plants through uptake of sparingly soluble P. This condition is common in acid soils given the precipitation of aluminum phosphates. Seguel et al. (2013) proposed various mechanisms, including organic ligand excretion, whereby arbuscular mycorrhizae may decrease Al phytotoxicity. Recently, Aguilera et al. (2011) suggested that glomalin-related proteins and fungal cell walls have a high capacity to immobilize Al. It seems, however, that information on mycorrhizal effects in Al-toxic soils remains inconclusive.

16.7.11 Effects of Aluminum in Aquatic Systems

Soluble Al in soils may move off site through leaching or erosion, with Al in solution being highly toxic to freshwater aquatic organisms, including invertebrates and fish. Many invertebrates, such as insects, snails, and leeches, are sensitive to the acidification of freshwaters (Raddum and Fjellheim, 2003). Freshwater fish populations decline in acidified streams, rivers, and lakes; indeed, their populations can be used as a measure of the health of aquatic systems. Raddum and Fjellheim (2003) reported that Atlantic salmon (*Salmo salar* L.) populations in the Audna River in Southern Norway declined from the 1920s and that they were extinct in the early 1970s. Brown trout (*Salmo trutta* L.) populations declined also but were not yet extinct.

As with plants, mechanisms of Al toxicity to aquatic biota are not well understood. Soluble Al was found to impair gill function in freshwater crayfish (*Pacifastacus leniusculus* Dana) causing O_2 stress (Ward et al., 2006). Furthermore, the toxicity of metals in freshwater systems is related to their binding to oxygen-containing ligands (i.e., hard ligands) as has been reported in Atlantic salmon (Wilkinson and Campbell, 1993) and in macroinvertebrates (Stockdale et al., 2010; Iwasaki et al., 2013). It would be intriguing to determine if Al has similar reactions with the mucus of aquatic organisms and the pectic matrix of the root cell wall or the phospholipids of the cell PM.

16.7.12 Physiology of Aluminum Tolerance in Plants

It may be expected that the physiology of Al tolerance is uncertain in view of the many possible detrimental effects of Al on plant root growth. This is not the case, however, because many responses that reduce the toxic effects of Al have been clarified (Kochian et al., 2002, 2004, 2005). Miyasaka et al. (2007) listed various mechanisms whereby Al may be excluded from entering the symplasm (Al avoidance) and of Al tolerance in the symplasm.

Aluminum resistance is often associated with less Al accumulation in the tips of roots (Horst et al., 2010). (This occurrence was not found by Cartes et al., 2012, however.) Reduced Al in the root tip may occur through less binding of Al to components of the cell wall where most Al is found (Marienfeld et al., 2000; Taylor et al., 2000; Rangel et al., 2009) or due to changes to cell wall components that result in decreased Al binding (Horst et al., 2010). Decreased binding of Al to phospholipids of the cell PM may have similar effects.

Many studies have shown that the toxic effects of Al are decreased through excretion of organic ligands on exposure of roots to Al (Ryan and Delhaize, 2012). This mechanism has received the most attention from plant physiologists and geneticists. Indeed, Ma et al. (2001) considered that organic ligands play a "central role" in Al tolerance mechanisms. Many plant species excrete low RMM organic ligands as first shown by Miyasaka et al. (1991) in snap bean (*Phaseolus vulgaris* L.). In this study, Al exposure resulted in an Al-tolerant cultivar excreting 10 times more citrate than an Al-sensitive one. Malate excretion was shown to occur on Al exposure in wheat (Delhaize et al., 1993); again Al-tolerant genotypes excreted 5–10 times more malate than did Al-sensitive genotypes. Malate excretion was evident within 15 min and continued for 24 h. Importantly, malate excretion was shown to occur at the root tip (Ryan et al., 1993).

Variable responses have been shown with respect to organic ligand excretion on exposure of roots to other metals. In wheat, for example, Delhaize et al. (1993) found that La did not elicit malate excretion, but Nian et al. (2002) showed that both citrate and malate were excreted on exposure to Cu.

16.8 GENETICS OF PLANT RESPONSES TO ALUMINUM

16.8.1 Genetic Differences in Toxic Effects of Aluminum

As detailed in Section 16.2, McLean and Gilbert (1927) classified crop plants into three groups based on Al added to solution: sensitive (74 µM), medium sensitive (260 µM), and resistant (520 µM), values somewhat higher than recent findings. This research was followed some 40 years later in what turned out to be pioneering work at the Plant Stress Laboratory at Beltsville, MD, United States (see Foy, 1984, and references therein), at a time when there was still controversy as to whether soil acidity problems were due to H^+ or Al^{3+} (Jenny, 1961). Research was conducted on a range of plants both in soils high in Al (e.g., Foy and Brown, 1964; Armiger et al., 1968) and in solution culture (e.g., Fleming and Foy, 1968; Foy et al., 1969). Of particular interest was the greater Al sensitivity of wheat cultivars from Indiana than those from Ohio that were indirectly selected for the more acid soils of Ohio (Foy et al., 1974).

Numerous studies have since shown that plant species and genotypes within species differ in sensitivity to soluble Al. The mechanisms involved may be classified as (1) Al exclusion through the excretion of organic ligands, (2) internal tolerance through chelation or sequestration, and (3) differences in cell wall binding of Al. These actions would occur in situations of organic ligand excretion but may not account for all instances of lower Al binding in the apoplast (Horst et al., 2010).

16.8.2 Genetic Differences in Aluminum Tolerance

Modern molecular techniques have been used over the past decade to make major advances in understanding Al tolerance with increasing evidence that excretion of organic acids plays an important role in detoxifying Al (Ma, 2000, 2005; Ma et al., 2001; Kochian et al., 2004). Sasaki et al. (2004) identified an Al^{3+}-activated *malate transporter* (ALMT) gene that encodes anion channels in the cell PM in wheat. This was followed by the identification of *multidrug and toxic compound extrusion* (MATE) genes that encode a co-transporter protein for the release of citrate in sorghum (*Sorghum bicolor* Moench) (Magalhaes et al., 2007), barley (Furukawa et al., 2007; Wang et al., 2007), corn (Maron et al., 2010), and rice (Yokosho et al., 2011). As concluded by Ryan et al. (2010), these gene families control Al^{3+} resistance by the same general mechanism, with different proteins achieving the same phenotype.

The progenitors of wheat are sensitive to Al, but malate excretion "has multiple independent origins" (Ryan et al., 2010)—an example of convergent evolution allowing greater distribution of this important crop (Ryan and Delhaize, 2010). Both ALMT and MATE genes have been shown to operate in arabidopsis (*Arabidopsis thaliana* Heynh.) (Liu et al., 2009). Two genes specific for transport of Al^{3+} in rice suggest that Al detoxification may occur through sequestration of Al into vacuoles (Xia et al., 2010; Huang et al., 2012).

Overall, genetics and molecular biology have identified 14 genes from 7 plant species that enable roots to tolerate high soluble Al based on the efflux of citrate or malate (Ryan et al., 2011).

16.9 SOIL TESTING, AMELIORATION OF ACID SOILS, AND PLANT BREEDING

There are many gaps in knowledge regarding the physiology of Al toxicity and tolerance (Section 16.7), but practical ways of overcoming problems associated with the acid soil infertility complex have been known for millennia (Section 16.2). These include the use of liming materials, the application of nutrients in short supply, and the cultivation of plants that grow well on acid soils.

16.9.1 Soil Amelioration

16.9.1.1 Liming

Materials used in agricultural liming contain Ca and Mg compounds that are capable of neutralizing soil acidity (Barber, 1984). The materials commonly include calcite ($CaCO_3$), dolomite ($CaCO_3 \cdot MgCO_3$), quicklime (CaO), and slaked or hydrated lime ($Ca(OH)_2$) and less commonly marl, seashells, ash, cement, and various industry by-products such as basic slag. Besides raising soil pH, cement and basic slag apply Si that may be beneficial to some crops. Organic materials of high ash alkalinity (Noble et al., 1996) may be used also.

Liming in the United Kingdom goes back more than two millennia, indicating that its benefi-cial result must have been easily observable (Gardner and Garner, 1953). Besides improved crop growth, the stated benefits of liming included physical improvement of soils that crust or have poor friability, suppression of certain weeds (e.g., sheep's sorrel *Rumex acetosella*, corn spurrey *Spergula arvensis*, knawels *Scleranthus* spp.), prevention of poor growth of root crops (e.g., turnip *Brassica rapa* L.), improved feeding value of pastures, improved beneficial bacterial processes (e.g., nitrifica-tion, N_2 fixation), increased P availability to plants, and decreased deficiencies of Ca, Mg, Mo, or B. In 1942, however, staff of Rothamsted Experimental Station showed that >25% of arable land in the United Kingdom was moderately to severely acid (pH ≤ 6) (Gardner and Garner, 1953). This situ-ation is, most likely, present to a greater or lesser extent throughout much agriculturally productive land worldwide (von Uexküll and Mutert, 1995).

Leaving the economics of liming aside, the question remains as to how much lime should be applied for optimum crop growth. Gardner and Garner (1953) pointed to the problem of relating laboratory tests, in which finely ground lime is mixed thoroughly with soil, to the situation of a famer needing to apply and incorporate several tonnes of lime to the land. They also pointed out that the term "requirement" assumes "an ideal or desirable lime status for all soils in general and for all crops" and that only one depth of soil need be considered. Overall, liming is often based on soil pH and on classifying soils into light, medium, and heavy soils.

Liming materials differ greatly in quality including such characteristics as purity (i.e., concentra-tions of Ca and Mg), crystallinity, and fineness, all factors that compound the conversion of labora-tory test results to the field. Liming materials are most often applied to the surface of the land with subsequent incorporation by plowing—with varying degrees of mixing. Because the mobility of lime through the soil profile is very low, rapid alleviation of subsoil acidity can only be achieved by deep placement, a practice that is economically not possible for many crops. Therefore, it is impor-tant to limit subsoil acidification.

16.9.1.2 Gypsum

Gypsum ($CaSO_4 \cdot 2H_2O$), which supplies readily available Ca, does not increase soil pH as liming materials do. Indeed, gypsum may decrease soil pH measured in water, through the salt effect given the solubility of gypsum, but not when measured in $CaCl_2$ or KCl (Takahashi et al., 2006). Increased subsoil pH, however, has been reported following the application of 15 t gypsum ha^{-1} to a strongly acid ultisol (Farina et al., 2000a,b). The surface application of gypsum alleviates the adverse effects of soluble Al in some subsoils through increased Ca in the rhizosphere and the self-liming effect through SO_4^{2-}–sorption (Reeve and Sumner, 1972; Sumner, 1993). Phosphogypsum, a by-product of the wet acid process for producing phosphoric acid from phosphate rock, may be used also.

Gypsum applications may be especially beneficial in alleviating toxic Al effects in subsoils because of its greater solubility than liming materials (Farina and Channon, 1988). This study found that an application of 10 t gypsum ha^{-1} over 4 years to an ultisol increased subsoil root development, with cumulative corn grain yield increasing by >3 t ha^{-1}. Importantly, Toma et al. (1999) found that beneficial gypsum effects were evident 16 years after application, increasing corn and alfalfa yields by up to 50%. Improved plant growth appears to be due to better root growth through increased Ca and SO_4^{2-}–S in the subsoil.

The application of both lime and gypsum proved beneficial in overcoming the adverse effects of subsoil acidity on the yield of corn on a strongly acidic ultisol (Farina et al., 2000a,b). Beneficial effects of gypsum were especially evident in drought years due to improved subsoil conditions increasing deeper root growth. Lime and phosphogypsum may be applied together, Blum et al. (2013) reporting marked improvements in soil characteristics (e.g., increases in pH, SO_4^{2-}–S, and exch. Ca; decrease in exch. Al) but decreased exch. Mg and K. Applications of lime and phosphogypsum increased yields of corn, wheat, and triticale (x *Triticosecale* Wittm. ex A. Camus.).

16.9.1.3 Organic Matter

Besides improving soil structure, organic materials, especially those of high ash alkalinity (Noble et al., 1996), are effective in overcoming acid soil problems. The application of organic matter as farmyard manure, for example, can lower Al toxicity by adding basic cations, especially Ca and Mg, and by complexing Al. However, high rates (often >10 t ha⁻¹) are needed. This often involves removal of organic materials from one site to another, with basic cations and other nutrients being depleted at the former site.

Butterly et al. (2013) determined that crop residues increase the pH of acid soils with temporal changes dependent on residue and soil type. Interestingly, alkalinity from canola (*Brassica napus* L.) and chickpea (*Cicer arietinum* L.) residues moved to the subsoil, an effect attributed to immobilization of NO_3–N and organic ligand decomposition by microbes. As found by Noble et al. (1996), Butterly et al. (2013) noted that the increase in soil pH was positively related to excess cations in residues.

16.9.2 Determining Rates of Soil Ameliorants

Estimates have been made over many centuries to determine liming rates to overcome the adverse effects of soil acidity. The major advance over the past century has been the use of data from soil tests to improve the accuracy of these estimates. Besides soil tests, cognizance must be taken of a plant's tolerances to Al as initially noted by Truog (1918) (Section 16.2). Fageria and Baligar (2008) have provided a summary of crop's tolerances to soil acidity constraints in oxisols and ultisols.

16.9.2.1 pH

The pH of a soil is probably the single, most useful measure that can be made since this provides information that can be used to infer the fertility of a soil. However, the method used to measure pH needs to be stated, common methods varying, among others, in the ratio of soil/solution (e.g., 1:5 or 1:10) and in the solution used. Most commonly, pH is measured in H_2O, 0.01 M $CaCl_2$ (to mimic the soil solution), or 1 M KCl (to desorb H^+ and Al^{3+} from exchange sites). In acid soils, measurements in H_2O are generally 0.1–1.0 pH unit higher than those in salt solutions. Numerous sources indicate the acceptable pH range for good crop growth. McLean and Brown (1984), for example, provided comprehensive tables of permissible soil pH ranges for various crops and of the average minimum pH values for crops of the Midwestern United States.

16.9.2.2 Buffer Methods

While useful, soil pH alone is not able to accurately determine how much lime should be applied to a soil for optimum plant growth because it does not take into account soil buffer capacity. This shortcoming led to the development of a number of methods in which soil pH is measured prior to and after adding a buffered solution. Examples include the Shoemaker–McLean–Pratt (SMP) (Shoemaker et al., 1961) and Mehlich (Mehlich, 1976) methods. As with soil pH alone, numerous sources are available for the interpretation of results.

16.9.2.3 Exchangeable Aluminum and Exchangeable Acidity

Both Veitch (1904) and Daikuhara (1914) proposed using the quantity of Al displaced by a neutral salt as a basis for determining lime requirement. Reeve and Sumner (1970) and Kamprath (1970) showed that the lime rate needed to reduce exch. Al to a level tolerated by crops was considerably lower than that required to increase soil pH to 6.5 in the ultisols and oxisols studied.

Fageria and Baligar (2008) summarized a range of recommendations based on using exch. Al to determine the amount of lime to be applied. With low exch. Ca and Mg levels in oxisols, Fageria and Baligar (2008) recommended that lime rate be based on the following equation:

$$\text{Lime (t ha}^{-1}) = (2 \times \text{exch. Al}^{3+}) + (2 - (\text{exch. Ca}^{2+} + \text{exch. Mg}^{2+}))$$

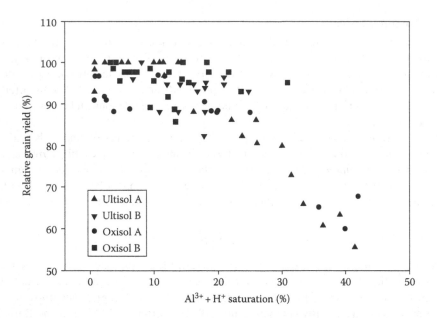

FIGURE 16.8 Relationship between acid (Al^{3+} and H^+) saturation and relative yield of corn grain conducted over 22 site-years on two ultisols and two oxisols in KwaZulu-Natal, South Africa. (Redrawn from Farina, M.P.W. and Channon, P., *Soil Sci. Soc. Am. J.*, 52, 175, 1991.)

with exchangeable Al^{3+}, Ca^{2+}, and Mg^{2+} values in $cmol_c$ kg^{-1}. However, multiplication of the exch. Al^{3+} value (×2 in the aforementioned equation) needs to be modified for soil texture: namely, ≤1 is for sandy soils (0%–15% clay), 1–2 for soils with 15%–35% clay, 2–3 for soils with 35%–60%, and 3–4 for soils with 60%–100% clay.

Farina and Channon (1991) determined the response of corn to the liming of oxisols and ultisols with a wide range in physical and chemical characteristics. In this study over 22 site-years, grain yield was related to soil pH, exch. Al^{3+}, or exch. acidity ($Al^{3+} + H^+$). Soil pH was inferior to Al-based indices, with the readily obtained exchangeable acid saturation of the effective cation-exchange capacity providing the best fit to the yield data (Figure 16.8). Corn grain yield decreased at an acid saturation level >20%, in keeping with findings from pot experiments with a wider range of soils (Farina et al., 1980). Kariuki et al. (2007) found that exchangeable Al saturation, which is related closely to exchangeable acidity (Farina and Channon, 1991), was a good measure to select red winter wheat cultivars for Al tolerance.

16.9.3 PLANTS ADAPTED TO ACID SOIL CONDITIONS

Gardner and Garner (1953) classified crops and native plants of Britain into four categories from most to least sensitive to acidity, pointing out the incorrect assumption that a crop suddenly fails at a critical point. Rather, a range of pH over which a crop can grow successfully needs to be considered.

As detailed in Section 16.2, McLean and Gilbert (1927) appear to have been the first to classify plants on their ability to grow in solution containing Al, work later extended by Foy and coworkers (see Foy, 1984, and references therein). In practical terms, it has long been known that some plants grow better than others on acid soils and vice versa. This knowledge has led to many plants being selected or bred for better performance when grown on soils high in soluble Al.

With improved knowledge of the physiology of Al tolerance and of the underlying genetics, there has been increased interest in selecting and breeding crops able to grow well on acid soils.

16.9.4   Using a Combination of Liming and Growing Aluminum-Tolerant Plants

A major problem of using Al-tolerant crops alone is continued soil acidification—a situation that will eventually result in the poor growth of many plants, even those deemed Al tolerant under present conditions. Acidification and loss of essential nutrients from the soil would continue to such an extent as to limit plant growth. Overcoming the problem, therefore, requires soil amelioration and replacement of essential nutrients to meet a plant's needs.

A rational approach by farmers would be to continually limit soil acidification through limiting the rate of fertilizer N (especially NH_4^+ and urea) and using practices that limit leaching of NO_3^-. Overall, however, only an increase in soil pH and the replacement of lost basic cations and nutrients will overcome soil acidification. While it is straightforward to list the routes to be taken, it should be recognized that these is often difficult to implement because of a farmer's financial constraints.

16.9.5   Liming of Lakes

Since the 1970s, many industrialized countries have carried out extensive programs of liming lakes to counteract the deleterious effects of acid precipitation and acidification of adjacent land (Andersen, 2006). Clearly, the rate of lime to be applied varies with acid input. Applying 3 t lime ha^{-1} to the terrestrial area surrounding two small lakes in Southern Norway rapidly increased their pH from 4.5 to 7.0 and decreased reactive Al from 10 to 3 μM (Traaen et al., 1997). Over the following 11 years, however, the lakes gradually reverted to pretreatment conditions. The application of high rates of lime to the lakes and adjacent wetlands near Stockholm, Sweden, had similar effects, initially increasing pH and decreasing soluble Al; cessation of liming resulted in reacidification (Wallstedt et al., 2009).

With decreased N and S emissions, many European countries are reconsidering the need to continue applying lime to lakes, but Sjöstedt et al. (2013) modeled such a scenario and concluded that many of the ~3000 lakes treated in Sweden would decrease by 1 pH unit with inorganic soluble Al increasing from 17 to 32 μg L^{-1} (63–118 μM). Lakes in southwestern Sweden would be most affected. Lake Store Hovvatn and the adjacent Lake Lille Hovvatn in Norway were acidified to pH 4.6 by 1974 through 150 years of increasing depositions of N and S (Hindar and Wright, 2005). Liming improved water characteristics up until 2003. The good quality of data on the lime applied and of water characteristics has allowed modeling to predict that liming will be required for many decades to ensure a healthy brown trout population. These findings suggest that industrial pollution has created an ongoing problem despite improvements in recent decades.

16.10   SUMMARY AND CONCLUSIONS

Being the most common metal in the Earth's crust, Al is ubiquitous in nature, but the Al in most minerals has few or no biological effects. Aluminum has a complex chemistry in solution, with further complexity arising by the many forms that Al present in soil solutions (Table 16.1) and the reactions of Al with the soil solid phase.

There is no conclusive evidence of Al benefits to living organisms, but Al^{3+} in solution is toxic at concentrations <25 μM to the roots of many plants (Poschenrieder et al., 2008). Despite this toxicity being known for about one century, there is still no consensus as to the primary toxic mechanism involved. As a small trivalent ion, Al^{3+} binds strongly to the many hard ligands in plant roots. This action has led to many hypotheses as to the initial site of Al toxicity, resulting from evidence that Al^{3+} and other Al ions impact on the cell wall, cell PM, cytoplasm, and nucleus.

Many plant species and lines within species show greater tolerance than others to soluble Al. The greatest difference is between plants that accumulate Al and those that do not (Section 16.6.1). Plants that accumulate Al do so via chelation in the rhizosphere or root with continued chelation during transport and sequestration in the leaf epidermis or other subcellular locations. Plants that do not accumulate Al but tolerate soluble Al in the rhizosphere do so by binding less Al in the root tip (Horst et al., 2010) through excretion of organic ligands or other means not yet identified.

Knowledge of the genetics of Al tolerance has increased greatly in recent years, with a focus on organic ligand excretion by the root tip (Section 16.8.2). The organic ligands identified to be most commonly excreted are citrate and malate (Ryan and Delhaize, 2012). Great progress has been made in identifying the genes responsible for this means of resisting the adverse effects of soluble Al. It needs to be established, however, if organic ligand excretion is the only Al tolerance mechanism used by plants.

Although uncertainty exists on the physiological basis for Al toxicity, it has been known for over two millennia that the application of liming materials or organic matter overcomes problems of acid soils. Problems remain, however, regarding the economics of liming, given the extensive areas of acid soils, continued acidification, and the large quantities of lime that are often required. Furthermore, uncertainty remains as the quantity of lime that should be applied although it would seem that liming rates based on exchangeable Al or exchangeable acidity provide the best estimate.

The rate of soil acidification has been reduced in many countries through improved air pollution standards to decrease emissions of atmospheric oxides of C, N, and S. However, the increased intensity of agricultural practices, especially high rates of N fertilizer and of N_2 fixation by leguminous plants, has increased rates of soil acidification, thereby exacerbating Al toxicity. This increased intensity consequently has increased the rates of soil ameliorants that need to be applied. Furthermore, it is difficult to identify the actual cause of poor plant growth in many instances. This conclusion was that of Ritchie (1989) who stated: "The challenge facing researchers in the field of soil acidity today is no different from the situation 30 years ago. The major problem is still our inability to identify accurately which of the many effects of soil acidity is reducing plant growth". It is now more than 20 years since this statement was made—and the question may be asked as to whether any further progress has been made in this regard.

Finally, it is apposite to urge soil scientists, agronomists, plant physiologists, and geneticists working on Al toxicity to heed the words of Jackson (1963): "It seems abundantly clear (1), that (we) will have to be ever-increasingly familiar with the findings in basic chemistry, physics, geology, biochemistry, biology, and meteorology; (2), the necessary great specialization in research must be combined with great breadth of knowledge; and (3), individual research effort must be combined with team research effort to encompass the basic knowledge, the specialization, and the breadth required."

REFERENCES

Adams, F. 1971. Ionic concentrations and activities in soil solutions. *Soil Sci. Soc. Am. Proc.* 35:420–426.

Adams, F. 1984. Crop response to lime in the southern United States. In *Soil Acidity and Liming*, 2nd edn. Agronomy No. 12, ed. F. Adams, pp. 211–265. Madison, WI: American Society of Agronomy.

Adams, F. and Z.F. Lund. 1966. Effect of chemical activity of soil solution aluminum on cotton root penetration of acid subsoils. *Soil Sci.* 101:193–198.

Aguilera, P., F. Borie, A. Seguel, and P. Cornejo. 2011. Fluorescence detection of aluminum in arbuscular mycorrhizal fungal structures and glomalin using confocal laser scanning microscopy. *Soil Biol. Biochem.* 43:2427–2431.

Alleoni, L.R.F., M.A. Cambri, E.F. Caires, and F.J. Garbuio. 2010. Acidity and aluminum speciation as affected by surface liming in tropical no-till soils. *Soil Sci. Soc. Am. J.* 74:1010–1017.

Alva, A.K., F.P.C. Blamey, D.G. Edwards, and C.J. Asher. 1986. An evaluation of aluminum indices to predict aluminum toxicity in plants grown in nutrient solutions. *Commun. Soil Sci. Plant Anal.* 17:1271–1280.

Andersen, D.O. 2006. Labile aluminium chemistry downstream of a limestone treated lake and an acid tributary: Effects of warm winters and extreme rainstorms. *Sci. Tot. Environ.* 366:739–748.

Armiger, W.H., C.D. Foy, A.L. Fleming, and B.E. Caldwell. 1968. Different tolerance of soybean varieties to an acid soil high in exchangeable aluminum. *Agron. J.* 60:67–70.

Asher, C.J. and F.P.C. Blamey. 1987. Experimental control of plant nutrient status using programmed nutrient addition. *J. Plant Nutr.* 10:1371–1380.

Asher, C.J., F.P.C. Blamey, and G.L. Kerven. 1992. Recent research on soil acidity: Identifying toxic aluminium in soils. In *La Investgacion Edafologica en Mexico 1991–1992. Memorias del 15 Congreso Nacional de la Sociedad Mexicana de la Ciencia del Suelo*, eds. S.J.L. Tovar and L. Quintero, pp. 61–68. Acapulco, Mexico: Society of Mexicana de la Ciencia del Suelo.

Asher, C.J. and D.G. Edwards. 1983. Modern solution culture techniques. In *Encyclopedia of Plant Physiology*, New Series 15, eds. A. Lauchli and R.L. Bieleski, pp. 94–119. Berlin, Germany: Springer-Verlag.

Baluška, F., D. Volkman, and P.W. Barlow. 1996. Specialized zones of development in roots: View from the cellular level. *Plant Physiol.* 112:3–4.

Barber, S.A. 1984. Liming materials and practices. In *Soil Acidity and Liming*, 2nd edn. Agronomy No. 12, ed. F. Adams, pp. 171–209. Madison, WI: American Society of Agronomy.

Bartlett, R.J. and D.C. Reigo. 1972. Effect of chelation on the toxicity of aluminium. *Plant Soil* 37:419–423.

Bell, R.W., D.G. Edwards, and C.J. Asher. 1989. External calcium requirements for growth and nodulation of six tropical food legumes grown in flowing solution culture. *Aust. J. Agric. Res.* 40:85–96.

Bennet, R.J., C.M. Breen, and V. Bandu. 1985a. Aluminium toxicity and regeneration of the root cap: Preliminary evidence for a Golgi apparatus derived morphogen in the primary root of *Zea mays*. *S. Afr. J. Bot.* 5:363–370.

Bennet, R.J., C.M. Breen, and M.V. Fey. 1985b. The primary site of aluminium injury in the root of *Zea mays* L. *S. Afr. J. Plant Soil* 2:8–17.

Bennet, R.J., C.M. Breen, and M.V. Fey. 1987. The effects of aluminium on root cap function and root development in *Zea mays* L. *Environ. Exp. Bot.* 27:91–104.

Bertsch, P.M. 1987. Conditions for Al_{13} polymer formation in partially neutralized aluminum solutions. *Soil Sci. Soc. Am. J.* 51:825–828.

Bertsch, P.M. and D.R. Parker. 1996. Aqueous polynuclear aluminum species. In *The Environmental Chemistry of Aluminum*, ed. G. Sposito, pp. 117–168. Boca Raton, FL: CRC Lewis Publishers.

Blamey, F.P.C., C.J. Asher, G.L. Kerven, and D.G. Edwards. 1993. Factors affecting aluminium sorption by calcium pectate. *Plant Soil* 149:87–94.

Blamey, F.P.C. and J. Chapman. 1982. Soil amelioration effects on peanut growth, yield, and quality. *Plant Soil* 65:319–334.

Blamey, F.P.C., D.G. Edwards, and C.J. Asher. 1983. Effect of aluminum, OH:Al and P:Al molar ratios, and ionic strength on soybean root elongation in solution culture. *Soil Sci.* 136:197–207.

Blamey, F.P.C., P.M. Kopittke, J.B. Wehr, T.B. Kinraide, and N.W. Menzies. 2010. Rhizotoxic effects of silver in cowpea seedlings. *Environ. Toxicicol. Chem.* 29:2071–2078.

Blamey, F.P.C., N.K. Nishizawa, and E. Yoshimura. 2004. Timing, magnitude, and location of initial soluble aluminum injuries to mungbean roots. *Soil Sci. Plant Nutr.* 50:67–76.

Blatt, H., R.J. Tracy, and B.E. Owens. 2006. *Petrology: Igneous, Sedimentary, and Metamorphic*, 3rd edn. New York: W.H. Freeman.

Blum, S.C., E.F. Caires, and L.R.F. Alleoni. 2013. Lime and phosphogypsum application and sulfate retention in subtropical soils under no-till system. *J. Soil Sci. Plant Nutr.* 13:279–300.

Brady, D.J., D.G. Edwards, and C.J. Asher. 1994. Effects of aluminium on the peanut (*Arachis hypogaea* L.)/ *Bradyrhizobium* symbiosis. *Plant Soil* 159:265–276.

Brady, D.J., C. Hecht-Buchholz, C.J. Asher, and D.G. Edwards. 1990. Effects of low activities of aluminium on soybean (*Glycine max*). I. Early growth and nodulation. In *Plant Nutrition—Physiology and Applications*, ed. M.L. van Beusichem, pp. 329–334. Haren, the Netherlands: Kluwer Academic Publishers.

Branquinho, C., H.C. Serrano, M.J. Pinto, and M.A. Martins-Loucao. 2007. Revisiting the plant hyperaccumulation criteria to rare plants and earth abundant elements. *Environ. Pollut.* 146:437–443.

Brautigan, D.J., P. Rengasamy, and D.J. Chittleborough. 2012. Aluminium speciation and phytotoxicity in alkaline soils. *Plant Soil* 360:187–196.

Brett, C.T. and K.W. Waldron. 1996. *Physiology and Biochemistry of Plant Cell Walls*, 2nd edn. London, U.K.: Chapman & Hall.

Butterly, C.R., J.A. Baldock, and C. Tang. 2013. The contribution of crop residues to changes in soil pH under field conditions. *Plant Soil* 366:185–198.

Cakmak, I. and W.J. Horst. 1991. Effect of aluminum on net efflux of nitrate and potassium from root-tips of soybean (*Glycine max* L.). *J. Plant Physiol.* 138:400–403.

Campbell, T.A., N.J. Nuernberg, and C.D. Foy. 1990. Differential responses of red clover germplasms to aluminum stress. *J. Plant Nutr.* 13:1463–1474.

Cartes, P., M. McManus, C. Wulff-Zottele, S. Leung, A. Gutiérrez-Moraga, and M.D. Mora. 2012. Differential superoxide dismutase expression in ryegrass cultivars in response to short term aluminium stress. *Plant Soil* 350:353–363.

Chenery, E.M. 1948. Aluminium in the plant world. *Kew Bull.* 3:173–183.

Clarkson, D.T. 1965. The effect of aluminium and some other trivalent metal cations on cell division in root apices of *Allium cepa. Ann. Bot.* 29:309–315.

Cravotta, C.A. 2008. Dissolved metals and associated constituents in abandoned coal-mine discharges, Pennsylvania, USA. Part 1: Constituent quantities and correlations. *Appl. Geochem.* 23:166–202.

Cui, Z.L., X.P. Chen, and F.S. Zhang. 2013. Development of regional nitrogen rate guidelines for intensive cropping systems in China. *Agron. J.* 105:1411–1416.

Daikuhara, G. 1914. Ueber saure mineral boden. *Bulletin of the Imperial Central Agricultural Experiment Stations Japan*, Tokyo, Japan, vol. 2, p. 18.

Darkó, É., H. Ambrus, É. Stefanovits-Bányai, J. Fodor, F. Bakos, and B. Barnabás. 2004. Aluminium toxicity, Al tolerance and oxidative stress in an Al-sensitive wheat genotype and in Al-tolerant lines developed by in vitro microspore selection. *Plant Sci.* 166:583–591.

de Andrade, L.R.M., L.M.G. Barros, G.F. Echevarria et al. 2011. Al-hyperaccumulator Vochysiaceae from the Brazilian Cerrado store aluminum in their chloroplasts without apparent damage. *Environ. Exp. Bot.* 70:37–42.

Delhaize, E., P.R. Ryan, and P.J. Randall. 1993. Aluminum tolerance in wheat (*Triticum aestivum* L.) II. Aluminum-stimulated excretion of malic acid from root apices. *Plant Physiol.* 103:695–702.

Dolling, P.J. and W.M. Porter. 1994. Acidification rates in the central wheat belt of Western Australia 1. On a deep yellow sand. *Aust. J. Exp. Agric.* 34:1155–1164.

Dolling, P.J., W.M. Porter, and I.C. Rowland. 1994. Acidification rates in the central wheat-belt of Western Australia. 2. On a sandy duplex soil. *Aust. J. Exp. Agric.* 34:1165–1172.

Doncheva, S., M. Amenós, C. Poschenrieder, and J. Barceló. 2005. Root cell patterning: A primary target for aluminium toxicity in maize. *J. Exp. Bot.* 56:1213–1220.

Edmeades, D.C., D.M. Wheeler, and O.E. Clinton. 1985. The chemical composition and ionic strength of soil solutions from New Zealand topsoils. *Aust. J. Soil Res.* 23:151–165.

EPA. 2013. Mid-Atlantic Water. Mining operations as nonpoint source pollution. US Environmental Protection Agency: Washington, DC, United States. http//www.epa.gov/reg3.wapd/nps/mining.html, accessed on October 25, 2013.

Fageria, N.K. and V.C. Baligar. 2008. Ameliorating soil acidity of tropical oxisols by liming for sustainable production. *Adv. Agron.* 99:345–399.

Fanning, D.S., C. Coppock, Z.W. Orndorff, W.L. Daniels, and M.C. Rabenhorst. 2004. Upland active acid sulfate soils from construction of new Stafford County, Virginia, USA, Airport. *Aust. J. Soil Res.* 42:527–536.

Farina, M.P.W. and P. Channon. 1988. Acid-subsoil amelioration: II. Gypsum effects on growth and subsoil chemical properties. *Soil Sci. Soc. Am. J.* 52:175–180.

Farina, M.P.W. and P. Channon. 1991. A field comparison of lime requirement indices for maize. *Plant Soil* 134:127–135.

Farina, M.P.W., P. Channon, and G.R. Thibaud. 2000a. A comparison of strategies for ameliorating subsoil acidity: I. Long-term growth effects. *Soil Sci. Soc. Am. J.* 64:646–651.

Farina, M.P.W., P. Channon, and G.R. Thibaud. 2000b. A comparison of strategies for ameliorating subsoil acidity: II. Long-term soil effects. *Soil Sci. Soc. Am. J.* 64:652–658.

Farina, M.P.W., M.E. Sumner, C.O. Plank, and W.S. Letzsch. 1980. Exchangeable aluminum and pH as indicators of lime requirement for corn. *Soil Sci. Soc. Am. J.* 44:1036–1041.

Felle, H.H. 2001. pH: Signal and messenger in plant cells. *Plant Biol.* 3:577–591.

Fitzpatrick, R.W., E. Fritsch, and P.G. Self. 1996. Interpretation of soil features produced by ancient and modern processes in degraded landscapes. 5. Development of saline sulfidic features in non-tidal seepage areas. *Geoderma* 69:1–29.

Flaten, T.P. 2002. Aluminium in tea—Concentrations, speciation and bioavailability. *Coord. Chem. Rev.* 228:385–395.

Fleming, A.L. and C.D. Foy. 1968. Root structure reflects differential aluminum tolerance in wheat varieties. *Agron. J.* 60:172–176.

Foy, C.D. 1984. Physiological effects of hydrogen, aluminum, and manganese toxicities in acid soil. In *Soil Acidity and Liming*, 2nd edn. Agronomy No. 12, ed. F. Adams, pp. 57–97. Madison, WI: American Society of Agronomy.

Foy, C.D. and J.C. Brown. 1964. Toxic factors in acid soils. II. Differential aluminum tolerance of plant species. *Soil Sci. Soc. Am. Proc.* 28:27–32.

Foy, C.D., A.L. Fleming, and W.H. Armiger. 1969. Aluminum tolerance of soybean varieties in relation to calcium nutrition. *Agron. J.* 61:505–511.

Foy, C.D., H.N. Lafever, J.W. Schwartz, and A.L. Fleming. 1974. Aluminum tolerance of wheat cultivars related to region of origin. *Agron. J.* 66:751–758.

Fred, E.B. and N.E. Loomis. 1917. Influence of hydrogen-ion concentration of medium on the reproduction of alfalfa bacteria. *J. Bacteriol.* 2:629–633.

Fuller, R.D. and C.J. Richardson. 1986. Aluminate toxicity as a factor controlling plant growth in bauxite residue. *Environ. Toxicicol. Chem.* 5:905–916.

Furukawa, J., N. Yamaji, H. Wang et al. 2007. An aluminum-activated citrate transporter in barley. *Plant Cell Physiol.* 48:1081–1091.

Gardner, H.W. and H.V. Garner. 1953. *The Use of Lime in British Agriculture*. London, U.K.: Farmer & Stockbreeder Publications Ltd.

Gillman, G.P. and L.C. Bell. 1978. Soil solution studies on weathered soils from tropical north Queensland. *Aust. J. Soil Res.* 16:67–77.

Guo, J.H., X.J. Liu, Y. Zhang et al. 2011. Significant acidification in major Chinese croplands. *Science* 327:1008–1010.

Hardy, F. 1926. The role of aluminium in soil infertility and toxicity. *J. Agric. Sci.* 16:616–631.

Hartemink, A.E. 2008. Sugarcane for bioethanol: Soil and environmental issues. *Adv. Agron.* 99:125–182.

Hartwell, B.L. and F.R. Pember. 1918. The presence of aluminum as a reason for the difference in the effect of so-called acid soil on barley and rye. *Soil Sci.* 6:259–279.

He, C.E., X.J. Liu, A. Fangmeier, and F.S. Zhang. 2007. Quantifying the total airborne nitrogen input into agroecosystems in the North China Plain. *Agric. Ecosyst. Environ.* 121:395–400.

He, H.Y., L.F. He, M.H. Gu, and X.F. Li. 2012. Nitric oxide improves aluminum tolerance by regulating hormonal equilibrium in the root apices of rye and wheat. *Plant Sci.* 183:123–130.

Hecht-Buchholz, C., D.J. Brady, C.J. Asher, and D.G. Edwards. 1990. Effects of low activities of aluminium on soybean (*Glycine max*) II. Root cell structure and root hair development. In *Plant Nutrition—Physiology and Applications*, ed. M.L. van Beusichem, pp. 335–343. Haren, the Netherlands: Kluwer Academic Publishers.

Helyar, K.R. and W.M. Porter. 1989. Soil acidification, its measurement and the processes involved. In *Soil Acidity and Plant Growth*, ed. A.D. Robson, pp. 61–101. Sydney, New South Wales, Australia: Academic Press.

Hindar, A. and R.F. Wright. 2005. Long-term records and modelling of acidification, recovery, and liming at Lake Hovvatn, Norway. *Canad. J. Fish. Aquat. Sci.* 62:2620–2631.

Hoagland, D.R. 1918. The effect of hydrogen and hydroxyl ion on the growth of barley seedlings. *Soil Sci.* 3:547–560.

Hoagland, D.R. and D.I. Arnon. 1950. *The Water-Culture Method for Growing Plants without Soil*. Berkley, CA: The College of Agriculture University of California, Circular, vol. 347, pp. 1–32.

Hopkins, E.G., V.H. Knox, and J.H. Pettit. 1903. A quantitative method for determining the acidity of soils. *USDA Bureau of Chemistry Bulletin* 73:114–119.

Horst, W.J., C.J. Asher, J. Cakmak, P. Szulkiewicz, and A.H. Wissemeier. 1991. Short-term responses of soybean roots to aluminium. In *Plant-Soil Interactions at Low pH. Developments in Plant and Soil*, eds. R.J. Wright, V.C. Baligar, and R.P. Murrmann, pp. 733–739. Dordrecht, the Netherlands: Kluwer Academic Publishers.

Horst, W.J., A.K. Puschel, and N. Schmohl. 1997. Induction of callose formation is a sensitive marker for genotypic aluminium sensitivity in maize. *Plant Soil* 192:23–30.

Horst, W.J., N. Schmohl, M. Kollmeier, F. Baluška, and M. Sivaguru. 1999. Does aluminium affect root growth of maize through interaction with the cell wall—Plasma membrane—Cytoskeleton continuum? *Plant Soil* 215:163–174.

Horst, W.J., Y. Wang, and D. Eticha. 2010. The role of the root apoplast in aluminium-induced inhibition of root elongation and in aluminium resistance of plants: A review. *Ann. Bot.* 106:185–197.

Hrkal, Z., H. Prchalová, and D. Fottová. 2006. Trends in impact of acidification on groundwater bodies in the Czech Republic: An estimation of atmospheric deposition at the horizon 2015. *J. Atmosph. Chem.* 53:1–12.

Huang, C.F., N. Yamaji, Z.C. Chen, and J.F. Ma. 2012. A tonoplast-localized half-size ABC transporter is required for internal detoxification of aluminum in rice. *Plant J.* 69:857–867.

Huang, T.M., Z.H. Pang, and L.J. Yuan. 2013. Nitrate in groundwater and the unsaturated zone in (semi)arid northern China: Baseline and factors controlling its transport and fate. *Environ. Earth Sci.* 70:145–156.

Hue, N.V., G.R. Craddock, and F. Adams. 1986. Effect of organic acids on aluminum toxicity in subsoils. *Soil Sci. Soc. Am. J.* 50:28–34.

Hunter, D. and D.S. Ross. 1991. Evidence for a phytotoxic hydroxy-aluminum polymer in organic soil horizons. *Science* 251:1056–1058.

Inostroza-Blancheteau, C., Z. Rengel, M. Alberdi et al. 2012. Molecular and physiological strategies to increase aluminum resistance in plants. *Mol. Biol. Rep.* 39:2069–2079.

Ishikawa, S., T. Wagatsuma, and T. Ikarashi. 1996. Comparative toxicity of Al^{3+}, Yb^{3+}, and La^{3+} to root-tip cells differing in tolerance to high Al^{3+} in terms of ionic potentials of dehydrated trivalent cations. *Soil Sci. Plant Nutr.* 42:613–625.

Iwasaki, Y., P. Cadmus, and W.H. Clements. 2013. Comparison of different predictors of exposure for modeling impacts of metal mixtures on macroinvertebrates in stream microcosms. *Aquat. Toxicol.* 132–133:151–156.

Jackson, M.L. 1963. Aluminum bonding in soils: A unifying principle in soil science. *Soil Sci. Soc. Am. Proc.* 27:1–10.

Jansen, S., T. Watanabe, and E. Smets. 2002. Aluminium accumulation in leaves of 127 species in Melastomataceae with comments on the order Myrtales. *Ann. Bot.* 90:53–64.

Jansen van Rensburg, H.G., A.S. Claassens, D.J. Beukes, H.L. Weepener, and P.J. Beukes. 2009. Assessing the potential soil acidification risk under dryland agriculture in the Mlondozi district in the Mpumalanga Province of South Africa. *South Afr. J. Plant Soil* 26:244–253.

Jenny, H. 1961. Reflections on the soil acidity merry-go-round. *Soil Sci. Soc. Am. Proc.* 25:248.

Jobbágy, E.G. and R.B. Jackson. 2001. The distribution of soil nutrients with depth: Global patterns and the imprint of plants. *Biogeochemistry* 53:51–77.

Jones, D.L., E.B. Blancaflor, L.V. Kochian, and S. Gilroy. 2006. Spatial coordination of aluminium uptake, production of reactive oxygen species, callose production and wall rigidification in maize roots. *Plant Cell Environ.* 29:1309–1318.

Jones, D.L., J.E. Shaff, and L.V. Kochian. 1995. Role of calcium and other ions in directing root hair tip growth in *Limnobium stoloniferum. Planta* 197:672–680.

Kamprath, E.J. 1970. Exchangeable aluminum as a criterion for liming leached mineral soils. *Soil Sci. Soc. Am. J.* 34:252–254.

Kariuki, S.K., H. Zhang, J.L. Schroder et al. 2007. Hard red winter wheat cultivar responses to a pH and aluminum concentration gradient. *Agron. J.* 99:88–98.

Kerven, G.L., C.J. Asher, D.G. Edwards, and Z. Ostatek-Boczynski. 1991. Sterile solution culture techniques for aluminum toxicity studies involving organic acids. *J. Plant Nutr.* 14:975–985.

Kerven, G.L., D.G. Edwards, C.J. Asher, P.S. Hallman, and S. Kokot. 1989a. Aluminium determination in soil solution. I. Evaluation of existing colorimetric and separation methods for the determination of inorganic monomeric aluminium in the presence of organic acid ligands. *Aust. J. Soil Res.* 27:79–90.

Kerven, G.L., D.G. Edwards, C.J. Asher, P.S. Hallman, and S. Kokot. 1989b. Aluminium determination in soil solution. II. Short term colorimetric procedures for the measurement of inorganic monomeric aluminium in the presence of organic acid ligands. *Aust. J. Soil Res.* 27:91–102.

Kerven, G.L., P.L. Larsen, and F.P.C. Blamey. 1995. Detrimental sulfate effects on formation of Al-13 tridecameric polycation in synthetic soil solutions. *Soil Sci. Soc. Am. J.* 59:765–771.

Khan, M.S.H., K. Tawaraya, H. Sekimoto et al. 2009. Relative abundance of Δ^5-sterols in plasma membrane lipids of root-tip cells correlates with aluminum tolerance of rice. *Physiol. Plant.* 135:73–83.

Kim, Y.S., W. Park, H. Nian et al. 2010. Aluminum tolerance associated with enhancement of plasma membrane H^+-ATPase in the root apex of soybean. *Soil Sci. Plant Nutr.* 56:140–149.

Kinraide, T.B. 1988. Proton extrusion by wheat roots exhibiting severe Al toxicity symptoms. *Plant Physiol.* 88:418–423.

Kinraide, T.B. 1990. Assessing the rhizotoxicity of the aluminate ion, $Al(OH)_4^-$. *Plant Physiol.* 93:1620–1625.

Kinraide, T.B. 1993. Aluminum enhancement of plant growth in acid rooting media. A case of reciprocal alleviation of toxicity by two toxic cations. *Physiol. Plant.* 88:619–625.

Kinraide, T.B. 1997. Reconsidering the rhizotoxicity of hydroxyl, sulphate, and fluoride complexes of aluminium. *J. Exp. Bot.* 48:1115–1124.

Kinraide, T.B. 2006. Plasma membrane surface potential (Ψ_{PM}) as a determinant of ion bioavailability: A critical analysis of new and published toxicological studies and a simplified method for the computation of plant Ψ_{PM}. *Environ. Toxicol. Chem.* 25:3188–3198.

Kinraide, T.B. 2009. Improved scales for metal ion softness and toxicity. *Environ. Toxicol. Chem.* 28:525–533.

Kinraide, T.B. and D.R. Parker. 1987. Nonphytotoxicity of the aluminum sulphate ion, $AlSO_4^+$. *Physiol. Plant.* 71:207–212.

Kinraide, T.B. and D.R. Parker. 1989. Assessing the phytotoxicity of mononuclear hydroxy-aluminum. *Plant Cell Environ.* 12:479–487.

Kinraide, T.B., D.R. Parker, and R.W. Zobel. 2005. Organic acid secretion as a mechanism of aluminium resistance: A model incorporating the root cortex, epidermis, and the external unstirred layer. *J. Exp. Bot.* 56:1853–1865.

Kinraide, T.B. and B.K. Sweeney. 2003. Proton alleviation of growth inhibition by toxic metals (Al, La, and Cu) in rhizobia. *Soil Biol. Biochem.* 35:199–205.

Kinraide, T.B. and P. Wang. 2010. The surface charge density of plant cell membranes (σ): An attempt to resolve conflicting values for intrinsic σ. *J. Exp. Bot.* 61:2057–2518.

Kinraide, T.B. and U. Yermiyahu. 2007. A scale of metal ion binding strengths correlating with ionic charge, Pauling electronegativity, toxicity, and other physiological effects. *J. Inorg. Biochem.* 101:1201–1213.

Kobayashi, Y., T. Watanabe, J.E. Shaff et al. 2013. Molecular and physiological analysis of Al^{3+} and H^+ rhizotoxicities at moderately acidic conditions. *Plant Physiol.* 163:180–192.

Kochian, L.V. 1995. Cellular mechanisms of aluminum toxicity and resistance in plants. *Annu. Rev. Plant Physiol. Plant Mol. Biol.* 46:237–260.

Kochian, L.V., O.A. Hoekenga, and M.A. Piñeros. 2004. How do crop plants tolerate acid soils? Mechanisms of aluminum tolerance and phosphorus efficiency. *Annu. Rev. Plant Biol.* 55:459–493.

Kochian, L.V., N.S. Pence, D.L.D. Letham et al. 2002. Mechanisms of metal resistance in plants: Aluminum and heavy metals. *Plant Soil* 247:109–122.

Kochian, L.V., M.A. Piñeros, and O.A. Hoekenga. 2005. The physiology, genetics and molecular biology of plant aluminum resistance and toxicity. *Plant Soil* 274:175–195.

Konishi, S., S. Miyamoto, and T. Taki. 1985. Stimulatory effects of aluminum on tea plants grown under low and high phosphorus supply. *Soil Sci. Plant Nutr.* 31:361–368.

Kopittke, P.M., F.P.C. Blamey, B.A. McKenna, P. Wang, and N.W. Menzies. 2011. Toxicity of metals to roots of cowpea in relation to their binding strength. *Environ. Toxicol. Chem.* 30:1827–1833.

Kopittke, P.M., F.P.C. Blamey, and N.W. Menzies. 2008. Toxicities of soluble Al, Cu, and La include ruptures to rhizodermal and root cortical cells of cowpea. *Plant Soil* 303:217–227.

Kopittke, P.M., B.A. McKenna, F.P.C. Blamey, J.B. Wehr, and N.W. Menzies. 2009. Metal-induced cell rupture in elongating roots is associated with metal ion binding strengths. *Plant Soil* 322:303–315.

Kopittke, P.M., N.W. Menzies, and F.P.C. Blamey. 2004. Rhizotoxicity of aluminate and polycationic aluminium at high pH. *Plant Soil* 266:177–186.

Kopittke, P.M., P. Wang, N.W. Menzies, R. Naidu, and T.B. Kinraide. 2014. A web-accessible computer program for calculating electrical potentials and ion activities at cell-membrane surfaces. *Plant Soil* 375:35–46.

Lance, J.C. and R.W. Pearson. 1969. Effect of low concentrations of aluminum on growth and water and nutrient uptake by cotton roots. *Soil Sci. Soc. Am. Proc.* 33:95–98.

Lange, H., S. Solberg, and N. Clarke. 2006. Aluminum dynamics in forest soil waters in Norway. *Sci. Total Environ.* 367:942–957.

Larsen, P.L., G.L. Kerven, L.C. Bell, and D.G. Edwards. 1995. Effects of silicic acid on the chemistry of monomeric and polymeric (Al_{13}) aluminium species in solutions. In *Plant–Soil Interactions at Low pH: Principles and Management*, eds. R.A. Date, N.J. Grundon, G.E. Rayment, and M.E. Probert, pp. 617–621. Dordrecht, the Netherlands: Kluwer Academic Publishers.

Lawrence, G.B. 2002. Persistent episodic acidification of streams linked to acid rain effects on soil. *Atmos. Environ.* 36:1589–1598.

Lide, D.R. 2007. *CRC Handbook of Chemistry and Physics*. Boca Raton, FL: CRC Press.

Lindsay, W.L. 1979. *Chemical Equilibria in Soils*. New York: John Wiley & Sons.

Lindsay, W.L. and P.M. Walthall. 1996. The solubility of aluminum in soils. In *The Environmental Chemistry of Aluminum*, ed. G. Sposito, pp. 333–361. Boca Raton, FL: Lewis Publishers.

Liu, J.P., J.V. Magalhaes, J. Shaff, and L.V. Kochian. 2009. Aluminum-activated citrate and malate transporters from the MATE and ALMT families function independently to confer *Arabidopsis* aluminum tolerance. *Plant J.* 57:389–399.

Liu, X.J., X.T. Ju, Y. Zhang, C. He, J. Kopsch, and Z. Fusuo. 2006. Nitrogen deposition in agroecosystems in the Beijing area. *Agric. Ecosyst. Environ.* 113:370–377.

Llugany, M., C. Poschenrieder, and J. Barceló. 1995. Monitoring the aluminium-induced inhibition of root elongation in four maize cultivars differing in tolerance to aluminium and proton toxicity. *Physiol. Plant.* 93:265–271.

Loneragan, J. 1968. Nutrient requirements of plants. *Nature* 220 (5174):1307–1308.

Lövblad, G., K. Kindbom, P. Grennfelt, H. Hultberg, and O. Westling. 1995. Deposition of acidifying substances in Sweden. *Ecol. Bull.* 44:17–34.

Luttge, U. and D.T. Clarkson. 1992. Mineral nutrition: Aluminium. *Progr. Bot.* 53: 63–77.

Ma, J.F. 2000. Role of organic acids in detoxification of aluminum in higher plants. *Plant Cell Physiol.* 41:383–390.

Ma, J.F. 2005. Plant root responses to three abundant soil minerals: Silicon, aluminum and iron. *Crit. Rev. Plant Sci.* 24:267–281.

Ma, J.F., P.R. Ryan, and E. Delhaize. 2001. Aluminium tolerance in plants and the complexing role of organic acids. *Trends Plant Sci.* 6:273–278.

Ma, J.F., R. Shen, S. Nagao, and E. Tanimoto. 2004. Aluminum targets elongating cells by reducing cell wall extensibility in wheat roots. *Plant Cell Physiol.* 45:583–589.

MacLean, D.C., K.S. Hansen, and R.E. Schneider. 1992. Amelioration of aluminum toxicity in wheat by fluoride. *New Phytol.* 121:81–88.

Magalhaes, J.V., J. Liu, C.T. Guimaraes et al. 2007. A gene in the multidrug and toxic compound extrusion (MATE) family confers aluminum tolerance in sorghum. *Nat. Genet.* 39:1156–1161.

Magistad, O.C. 1925. The aluminum content of the soil solution and its relation to soil reaction and plant content. *Soil Sci.* 20:183–225.

Marienfeld, S., N. Schmohl, M. Klein, W.H. Schroder, A.J. Kuhn, and W.J. Horst. 2000. Localisation of aluminium in root tips of *Zea mays* and *Vicia faba*. *J. Plant Physiol.* 156:666–671.

Maron, L.G., M.A. Piñeros, C.T. Guimaraes et al. 2010. Two functionally distinct members of the MATE (multi-drug and toxic compound extrusion) family of transporters potentially underlie two major aluminum tolerance QTLs in maize. *Plant J.* 61:728–740.

Masion, A. and P.M. Bertsch. 1997. Aluminium speciation in the presence of wheat root cell walls: A wet chemical study. *Plant Cell Environ.* 20:504–512.

Matsumoto, H., E. Hirasawa, S. Morimura, and E. Takahashi. 1976. Localization of aluminium in tea leaves. *Plant Cell Physiol.* 17:627–631.

McKenna, B.A., T.M. Nicholson, J.B. Wehr, and N.W. Menzies. 2010. Effects of Ca, Cu, Al, and La on pectin gel strength: Implications for plant cell walls. *Carbohydr. Res.* 345:1174–1179.

McLaughlin, S.B. and R. Wimmer. 1999. Calcium physiology and terrestrial ecosystem processes. *New Phytol.* 142:373–417.

McLean, E.O. and J.R. Brown. 1984. Crop response to lime in the midwestern United States. In *Soil Acidity and Liming*, 2nd edn. Agronomy No. 12, ed. F. Adams, pp. 267–303. Madison, WI: American Society of Agronomy.

McLean, F.T. and B.E. Gilbert. 1927. The relative aluminum tolerance of crop plants. *Soil Sci.* 24:163–175.

Mehlich, A. 1976. New buffer pH method for rapid estimation off exchangeable acidity. *Commun. Soil Sci. Plant Anal.* 7:637–652.

Menzies, N.W. 1991. Solution phase aluminium in highly weathered soils: Evaluaton and relationship to aluminium toxicity to plants. PhD thesis, The University of Queensland, Brisbane, Queensland, Australia.

Menzies, N.W., L.C. Bell, and D.G. Edwards. 1991. Characteristics of membrane filters in relation to aluminium studies in soil solutions and natural waters. *J. Soil Sci.* 42:585–597.

Menzies, N.W., L.C. Bell, and D.G. Edwards. 1994a. Exchange and solution phase chemistry of acid, highly weathered soils. I. Characteristics of soils and effects of lime and gypsum amendments. *Aust. J. Soil Res.* 32:251–267.

Menzies, N.W., L.C. Bell, and D.G. Edwards. 1994b. Exchange and solution phase chemistry of acid, highly weathered soils. II. Investigation of mechanisms controlling Al release into solution. *Aust. J. Soil Res.* 32:269–283.

Miao, Y.X., B.A. Stewart, and F.S. Zhang. 2011. Long-term experiments for sustainable nutrient management in China. A review. *Agron. Sust. Dev.* 31:397–414.

Mirasol, J.J. 1920. Aluminum as a factor in soil acidity. *Soil Sci.* 10:153–217.

Miyake, K. 1916. The toxic action of soluble aluminium salts upon the growth of the rice plant. *J. Biol. Chem.* 25:23–28.

Miyasaka, S.C., J.G. Buta, R.K. Howell, and C.D. Foy. 1991. Mechanism of aluminum tolerance in snapbeans: Root exudation of citric acid. *Plant Physiol.* 96:737–743.

Miyasaka, S.C., N.V. Hue, and M.A. Dunn. 2007. Aluminum. In *Handbook of Plant Nutrition*, eds. A.V. Barker and D.J. Pilbeam, pp. 439–497. Boca Raton, FL: CRC Press.

Murata, M.R., P.S. Hammes, and G.E. Zharare. 2002. Soil amelioration effects on nutrient availability and productivity of groundnut on acid sandy soils of Zimbabwe. *Exp. Agric.* 38:317–331.

Murphy, A.S., W. Peer, and B. Schulz. 2011. *The Plant Plasma Membrane*. Heidelberg, Germany: Springer.

Navascués, J., C. Perez-Rontomé, D.H. Sánchez et al. 2012. Oxidative stress is a consequence, not a cause, of aluminum toxicity in the forage legume *Lotus corniculatus*. *New Phytol.* 193:625–636.

Nian, H., Z.M. Yang, S.J. Ahn, Z.J. Cheng, and H. Matsumoto. 2002. A comparative study on the aluminium- and copper-induced organic acid exudation from wheat roots. *Physiol. Plant.* 116:328–335.

Noble, A.D., I. Zenneck, and P.J. Randall. 1996. Leaf litter alkalinity and neutralisation of soil acidity. *Plant Soil* 179:293–302.

Nordstrom, D.K. and H.M. May. 1996. Aquous equilibrium data for mononuclear amuminum species. In *The Environmental Chemistry of Aluminum*, ed. G. Sposito, pp. 39–80. Boca Raton, FL: Lewis Publishers.

Nystrand, M.I. and P. Österholm. 2013. Metal species in a Boreal river system affected by acid sulfate soils. *Appl. Geochem.* 31:133–141.

Ownby, J.D. and H.R. Popham. 1989. Citrate reverses the inhibition of wheat root growth caused by aluminum. *J. Plant Physiol.* 135:588–591.

Parker, D.R. and P. Bertsch. 1992. Formation of the "Al$_{13}$" tridecameric polycation under diverse synthesis conditions. *Environ. Sci. Technol.* 26:914–920.

Parker, D.R., T.B. Kinraide, and L.W. Zelazny. 1989. On the phytotoxicity of polynuclear hydroxy-aluminum complexes. *Soil Sci. Soc. Am. J.* 53:789–796.

Parker, D.R., W.A. Norvell, and R.L. Chaney. 1995. GEOCHEM-PC: A chemical speciation program for IBM and compatible personal computers. In *Soil Chemical Equilibrium and Reaction Models*, eds. R.H. Loeppert, A.P. Schwab, and S. Goldberg, pp. 253–269. Madison, WI: Soil Science Society of America, Inc.

Parker, D.R., L.W. Zelazny, and T.B. Kinraide. 1987. Improvements to the program GEOCHEM. *Soil Sci. Soc. Am. J.* 51:488–491.

Parkhurst, D. 2013. PhreeqcI v3.0.6.7757. United States Geological Survey. www.cr.usgs.gov/projects/GWC_coupled/phreeqc/, accessed October 25, 2013.

Pearson, R.G. 1963. Hard and soft acids and bases. *J. Am. Chem. Soc.* 85:3533–3539.

Poschenrieder, C., C. Cabot, S. Martos, B. Gallego, and J. Barceló. 2013. Do toxic ions induce hormesis in plants? *Plant Sci.* 212:15–25.

Poschenrieder, C., B. Gunse, I. Corrales, and J. Barceló. 2008. A glance into aluminum toxicity and resistance in plants. *Sci. Total Environ.* 400:356–368.

Raddum, G.G. and A. Fjellheim. 2003. Liming of River Audna, southern Norway: A large-scale experiment of benthic invertebrate recovery. *Ambio* 32:230–234.

Rangel, A.F., I.M. Rao, and W.J. Horst. 2009. Intracellular distribution and binding state of aluminum in root apices of two common bean (*Phaseolus vulgaris*) genotypes in relation to Al toxicity. *Physiol. Plant.* 135:162–173.

Rayle, D. and R.E. Cleland. 1970. Enhancement of wall loosening and elongation by acid solutions. *Plant Physiol.* 46:250–253.

Reeve, N.G. and M.E. Sumner. 1970. Lime requirements of Natal oxisols based on exchangeable aluminum. *Soil Sci. Soc. Am. Proc.* 34:595–598.

Reeve, N.G. and M.E. Sumner. 1972. Amelioration of subsoil acidity in Natal oxisols by leaching of surface-applied amendments. *Agrochemophysica* 4:1–6.

Rengel, Z. 1992. Role of calcium in aluminium toxicity. *New Phytol.* 121:499–513.

Rengel, Z. 1996. Uptake of aluminium by plant cells. *New Phytol.* 134:389–406.

Ritchie, G.S.P. 1989. The chemical behaviour of aluminium, hydrogen and manganese in acid soils. In *Soil Acidity and Plant Growth*, ed. A.D. Robson, pp. 1–60. Sydney, New South Wales, Australia: Academic Press.

Ruffin, E. 1852. *An Essay on Calcareous Manures*, 5 edn. Richmond, VA: J.W. Randolph.

Ryan, P.R. and E. Delhaize. 2012. Adaptations to aluminium toxicity. In *Plant Stress Physiology*, ed. S. Shabala, pp. 171–193. Cambridge, MA: CABI.

Ryan, P.R., J.M. Ditomaso, and L.V. Kochian. 1993. Aluminium toxicity in roots: An investigation of spatial sensitivity and the role of the root cap. *J. Exp. Bot.* 44:437–446.

Ryan, P.R., H. Raman, S. Gupta, T. Sasaki, Y. Yamamoto, and E. Delhaize. 2010. The multiple origins of aluminium resistance in hexaploid wheat include *Aegilops tauschii* and more recent cis mutations to TaALMT1. *Plant J.* 64:446–455.

Ryan, P.R., S.D. Tyerman, T. Sasaki et al. 2011. The identification of aluminium-resistance genes provides opportunities for enhancing crop production on acid soils. *J. Exp. Bot.* 62:9–20.

Ryan, P.R. and E.A. Delhaize. 2010. The convergent evolution of aluminium resistance in plants exploits a convenient currency. *Funct. Plant Biol.* 37:275–284.

Sammut, J., I. White, and M.D. Melville. 1996. Acidification of an estuarine tributary in eastern Australia due to drainage of acid sulfate soils. *Mar. Freshwater Res.* 47:669–684.

Sasaki, T., Y. Yamamoto, B. Ezaki et al. 2004. A wheat gene encoding an aluminum-activated malate transporter. *Plant J.* 37:645–653.

Sattelmacher, B., K.-H. Mühling, and K. Pennewiss. 1998. The apoplast—Its significance for the nutrition of higher plants. *Zeit. Pflanzen. Boden.* 161:485–498.

Scalon, M.C., M. Haridasan, and A.C. Franco. 2013. A comparative study of aluminium and nutrient concentrations in mistletoes on aluminium-accumulating and non-accumulating hosts. *Plant Biol.* 15:851–857.

Seguel, A., J.R. Cumming, K. Klugh-Stewart, P. Cornejo, and F. Borie. 2013. The role of arbuscular mycorrhizas in decreasing aluminium phytotoxicity in acidic soils: A review. *Mycorrhiza* 23:167–183.

Shaff, J.E., B.A. Schultz, E.J. Craft, R.T. Clark, and L.V. Kochian. 2010. GEOCHEM-EZ: A chemical speciation program with greater power and flexibility. *Plant Soil* 330:207–214.

Sharp, L.T. and D.R. Hoagland. 1916. Acidity and adsorption in soils as measured by the hydrogen electrode. *J. Agric. Res.* 7:123–145.

Shen, R.F. and J.F. Ma. 2001. Distribution and mobility of aluminium in an Al-accumulating plant, *Fagopyrum esculentum* Moench. *J. Exp. Bot.* 52:1683–1687.

Shirokikh, I.G., A.A. Shirokikh, N.A. Rodina, L.M. Polyanskaya, and O.A. Burkanova. 2004. Effects of soil acidity and aluminum on the structure of microbial biomass in the rhizosphere of barley. *Eurasian Soil Sci.* 37:839–843.

Shoemaker, H.E., E.O. McLean, and P.F. Pratt. 1961. Buffer methods for determining lime requirement of soils with appreciable amounts of extractable aluminum. *Soil Sci. Soc. Am. Proc.* 25:274–277.

Silva, I.R., T.J. Smyth, D.F. Moxley, T.E. Carter, N.S. Allen, and T.W. Rufty. 2000. Aluminum accumulation at nuclei of cells in the root tip. Fluorescence detection using lumogallion and confocal laser scanning microscopy. *Plant Physiol.* 123:543–552.

Simonovicová, M., J. Huttová, I. Mistrik, B. Siroká, and L. Tamás. 2004. Root growth inhibition by aluminum is probably caused by cell death due to peroxidase-mediated hydrogen peroxide production. *Protoplasma* 224:91–98.

Sivaguru, M. and W.J. Horst. 1998. The distal part of the transition zone is the most aluminum-sensitive apical root zone of maize. *Plant Physiol.* 116:155–163.

Sjöstedt, C., C. Andrén, J. Fölster, and J.P. Gustafsson. 2013. Modelling of pH and inorganic aluminium after termination of liming in 3000 Swedish lakes. *Appl. Geochem.* 35:221–229.

Sparks, D.L. 2003. *Environmental Soil Chemistry*. San Diego, CA: Academic Press.

Sparks, D.L. 2006. Historical aspects of soil chemistry. In *Footprints in the Soil: People and Ideas in Soil History*, ed. B.P. Warkentin, pp. 307–337. Amsterdam, the Netherlands: Elsevier.

Sposito, G. 1996. *The Environmental Chemistry of Aluminum*, 2nd edn. Boca Raton, FL: CRC Lewis Publishers.

Stael, S., B. Wurzinger, A. Mair, N. Mehlmer, U.C. Vothknecht, and M. Teige. 2012. Plant organellar calcium signalling: An emerging field. *J. Exp. Bot.* 63:1525–1542.

Stockdale, A., E. Tipping, S. Lofts, S.J. Ormerod, W.H. Clements, and R. Blust. 2010. Toxicity of proton–metal mixtures in the field: Linking stream macroinvertebrate species diversity to chemical speciation and bio-availability. *Aquat. Toxicol.* 100:112–119.

Sumner, M.E. 1993. Gypsum and acid soils—The world scene. *Adv. Agron.* 51:1–32.

Takahashi, T., Y. Ikeda, H. Nakamura, and M. Nanzyo. 2006. Efficiency of gypsum application to acid Andosols estimated using aluminum release rates and plant root growth. *Soil Sci. Plant Nutr.* 52:584–592.

Taylor, G.J., F.P.C. Blamey, and D.G. Edwards. 1998. Antagonistic and synergistic interactions between aluminium and manganese on growth of *Vigna unguiculata* at low ionic strength. *Physiol. Plant.* 104:183–194.

Taylor, G.J., J.L. McDonald-Stephens, D.B. Hunter et al. 2000. Direct measurement of aluminum uptake and distribution in single cells of *Chara corallina*. *Plant Physiol.* 123:987–996.

Thomas, G.W. and W.L. Hargrove. 1984. The chemistry of soil acidity. In *Soil Acidity and Liming*, 2nd edn. Agronomy No. 12, ed. F. Adams, pp. 3–56. Madison, WI: American Society of Agronomy.

Tolrà, R., K. Vogel-Miks, R. Hajiboland et al. 2011. Localization of aluminium in tea (*Camellia sinensis*) leaves using low energy x-ray fluorescence spectro-microscopy. *J. Plant Res.* 124:165–172.

Toma, M., M.E. Sumner, G. Weeks, and M. Saigusa. 1999. Long-term effects of gypsum on crop yield and subsoil chemical properties. *Soil Sci. Soc. Am. J.* 39:891–895.

Traaen, T.S., T. Frogner, A. Hindar, E. Kleiven, A. Lande, and R.F. Wright. 1997. Whole-catchment liming at Tjonnstrond, Norway: An 11-year record. *Water Air Soil Pollut.* 94:163–180.

Truog, E. 1918. Soil acidity: I. Its relation to the growth of plants. *Soil Sci.* 5:169–195.

Ulrich, B., R. Mayer, and P.K. Khanna. 1980. Chemical changes due to acid precipitation in a loess-derived soil in central Europe. *Soil Sci.* 130:193–199.

van Hees, P., U. Lundström, R. Danielsson, and L. Nyberg. 2001. Controlling mechanisms of aluminium in soil solution—An evaluation of 180 podzolic forest soils. *Chemosphere* 45:1091–1101.

Veitch, F.P. 1904. Comparison of methods for the estimation of soil acidity. *J. Am. Chem. Soc.* 26:637–662.

von Uexküll, H.R. and E. Mutert. 1995. Global extent, development and economic impact of acid soils. *Plant Soil* 171:1–15.

Wagatsuma, T., S. Ishikawa, H. Obata, K. Tawaraya, and S. Katohda. 1995. Plasma membrane of younger and outer cells is the primary specific site for aluminium toxicity in roots. *Plant Soil* 171:105–112.

Wagatsuma, T., M. Kaneko, and Y. Haysaka. 1987. Destruction process in plant root cells by aluminum. *Soil Sci. Plant Nutr.* 33:161–175.

Wällstedt, T., F. Edberg, and H. Borg. 2009. Long-term water chemical trends in two Swedish lakes after termination of liming. *Sci. Total Environ.* 407:3554–3562.

Wang, J.P., H. Raman, M.X. Zhou et al. 2007. High-resolution mapping of the Alp locus and identification of a candidate gene HvMATE controlling aluminium tolerance in barley (*Hordeum vulgare* L.). *Theor. Appl. Genet.* 115:265–276.

Wang, P., N.W. Menzies, E. Lombi et al. 2013. Quantitative determination of metal and metalloid spatial distribution in hydrated and fresh roots of cowpea using synchrotron-based x-ray fluorescence microscopy. *Sci. Total Environ.* 463–464:131–139.

Wang, P., D.M. Zhou, T.B. Kinraide et al. 2008. Cell membrane surface potential (ψ_0) plays a dominant role in the phytotoxicity of copper and arsenate. *Plant Physiol.* 148:2134–2143.

Ward, R.J.S., C.R. McCrohan, and K.N. White. 2006. Influence of aqueous aluminium on the immune system of the freshwater crayfish *Pacifasticus leniusculus*. *Aquat. Toxicol.* 77:222–228.

Watanabe, T. and M. Osaki. 2001. Influence of aluminum and phosphorus on growth and xylem sap composition in *Melastoma malabathricum* L. *Plant Soil* 237:63–70.

Webb, L.J. 1954. Aluminium accumulation in the Australian—New Guinea flora. *Aust. J. Bot.* 2:176–198.

Wehr, J.B., F.P.C. Blamey, P.M. Kopittke, and N.W. Menzies. 2010. Comparative hydrolysis and sorption of Al and La onto plant cell wall material and pectic materials. *Plant Soil* 332:319–330.

Wheeler, D.M. and D.C. Edmeades. 1995. Effect of depth and lime or phosphorus fertilizer applications on the soil solution chemistry of some New Zealand pastoral soils. *Aust. J. Soil Res.* 33:461–476.

Wheeler, D.M., I.L. Power, and D.C. Edmeades. 1993. Effect of various metal ions on growth of two wheat lines known to differ in aluminium tolerance. *Plant Soil* 155/156:489–492.

Wheeler, D.M., D.J.C. Wild, and D.C. Edmeades. 1992. Preliminary results from a microscopic examination on the effects of aluminum on the root tips of wheat. *Plant Soil* 146:83–87.

Wilkinson, K.J. and P.G.C. Campbell. 1993. Aluminum bioconcentration at the gill surface of juvenile Atlantic salmon in acidic media. *Environ. Toxicol. Chem.* 12:2083–2095.

Willett, I.R., R.H. Crockford, and A.R. Milnes. 1992. Transformations of iron, manganese and aluminium during oxidation of a sulfidic material from an acid sulfate soil. In *Biomineralization Processes of Iron and Manganese—Modern and Ancient Environments*, eds. H.C.W. Skinner and R.W. Fitzpatrick, pp. 287–302. Catena Supplement, Cremlingen, Germany: Catena Verlag.

Williams, C.H. 1980. Soil acidification under clover pasture. *Aust. J. Exp. Agric. Anim. Husb.* 20:561–567.

Winch, S. and J. Pritchard. 1999. Acid-induced wall loosening is confined to the accelerating region of the root growing zone. *J. Exp. Bot.* 50:1481–1487.

Wissemeier, A.H., F. Klotz, and W.J. Horst. 1987. Aluminium induced callose synthesis in roots of soybean (*Glycine max* L.). *J. Plant Physiol.* 129:487–492.

Wu, P., C.Y. Liao, B. Hu et al. 2000. QTLs and epistasis for aluminum tolerance in rice (*Oryza sativa* L.) at different seedling stages. *Theor. Appl. Genet.* 100:1295–1303.

Xia, J.X., N. Yamaji, T. Kasai, and J.F. Ma. 2010. Plasma membrane-localized transporter for aluminum in rice. *Proc. Nat. Acad. Sci. USA* 107:18381–18385.

Yamamoto, Y., Y. Kobayashi, S.R. Devi, S. Rikiishi, and H. Matsumoto. 2002. Aluminum toxicity is associated with mitochondrial dysfunction and the production of reactive oxygen species in plant cells. *Plant Physiol.* 128:63–72.

Yamamoto, Y., Y. Kobayashi, and H. Matsumoto. 2001. Lipid peroxidation is an early symptom triggered by aluminum, but not the primary cause of elongation inhibition in pea roots. *Plant Physiol.* 125:199–208.

Yi, M., H.L. Yi, H.H. Li, and L.H. Wu. 2010. Aluminum induces chromosome aberrations, micronuclei, and cell cycle dysfunction in root cells of *Vicia faba*. *Environ. Toxicol.* 25:124–129.

Yokel, R.A. and R.L. Florence. 2008. Aluminum bioavailability from tea infusion. *Food Chem. Toxicol.* 46:3659–3663.

Yokosho, K., N. Yamaji, and J.F. Ma. 2011. An Al-inducible MATE gene is involved in external detoxification of Al in rice. *Plant J.* 68:1061–1069.

Yoshida, S., D.A. Forno, and K.A. Cook. 1971. *Laboratory Manual for Physiological Studies of Rice*, 2nd edn. Los Baños, CA: International Rice Research Institute, pp. 53–57.

Zeng, Q.L., R.F. Chen, X.Q. Zhao et al. 2013. Aluminum could be transported via phloem in *Camellia oleifera* Abel. *Tree Physiol.* 33:96–105.

Zhang, Z.Y., H.H. Wang, X.M. Wang, and Y.R. Bi. 2011. Nitric oxide enhances aluminum tolerance by affecting cell wall polysaccharides in rice roots. *Plant Cell Rep.* 30:1701–1711.

Zheng, S.J. and J.L. Yang. 2005. Target sites of aluminum phytotoxicity. *Biol. Plant.* 49:321–331.

Zhou, Y., X.Y. Xu, L.Q. Chen, J.L. Yang, and S.J. Zheng. 2012. Nitric oxide exacerbates Al-induced inhibition of root elongation in rice bean by affecting cell wall and plasma membrane properties. *Phytochemistry* 76:46–51.

17 Cobalt

Aurelia Pérez-Espinosa, Raúl Moral Herrero,
María Dolores Pérez-Murcia, Concepción Paredes Gil,
and María De Los Ángeles Bustamante Muñoz

CONTENTS

17.1   HISTORICAL BACKGROUND

Cobalt is considered an essential element in trace amounts for humans and other mammals as it is an integral component of the vitamin B_{12} complex (Smith and Carson, 1981). Cobalt in this form is obtained from microorganisms or from animal sources (USEPA, 2005). It is a naturally occurring element, widely distributed in rocks, soils, vegetation, and water (Gál et al., 2008). This element usually is found in association with nickel and appears in two oxidation states (Co^{2+} and Co^{3+}), Co^{3+} being, except in certain complexes, thermodynamically unstable under typical oxidation–reduction and pH conditions (Palit et al., 1994; Nagpal, 2004).

However, in nonleguminous plants, a physiological function for this element has not been clearly identified, since Co-containing vitamin B_{12} does not occur in plants (Bakkaus et al., 2005). The only physiological role as an essential element has been reported for the growth of many marine algal species, including diatoms, chrysophytes, and dinoflagellates (Gál et al., 2008) and for some blue-green algae (Nagpal, 2004). Furthermore, cobalt is required by microorganisms for the fixation of molecular nitrogen in root nodules of leguminous plants that possess vitamin B_{12} and cobamide coenzymes, and in the nonlegumes alder (*Alnus glutinosa* L.) and river she-oak or Australian pine (*Casuarina cunninghamiana* Miq.) (Bond and Hewitt, 1962; Gál et al., 2008). As has been reported in several studies (Reisenauer, 1960; Delwiche et al., 1961; Bakken et al., 2004), cobalt may be involved in symbiotic nitrogen fixation, since it constitutes an important component of the soil bacterium *Rhizobium* involved in the legume symbiosis (Gál et al., 2008). Also, different authors (Marschner, 1995; O'Hara, 2001) have shown a positive effect on the vigor, nodule development, and N content of some legumes of additional supply of Co in soils with low concentrations of the element.

However, although Co is a nutrient for humans and animals, in high concentrations, it is toxic to humans and to terrestrial and aquatic animals and plants (Nagpal, 2004). According to different studies (Dekock, 1956; Chatterjee and Chatterjee, 2000, 2003; Osman et al., 2004), cobalt is relatively toxic to plants when given in high doses, affecting the growth and metabolism, to different degrees, depending on the concentration and status of Co in the rhizosphere and soil (Palit et al., 1994).

This element interacts with other elements, forming complexes, and thus, the phytotoxic and cytotoxic activities of Co and its compounds depend on the physicochemical properties of the complexes formed (Smith and Carson, 1981; Gál et al., 2008). Therefore, the competitive absorption and mutual activation of related metals influence the role of cobalt in several phytochemical reactions (Palit et al., 1994).

17.2 UPTAKE OF COBALT BY PLANTS

The plant uptake of specific metal ions is based on a balance between regulation of the metal uptake and detoxification and storage mechanisms (Krämer et al., 2007). This section includes the relationship between uptake and regulation, showing a relatively narrow range of Co concentrations between toxicity and deficiency.

The transport processes at the solution–plant interface imply transport by organic complexes and sorption to nonspecific sites on the plant cell wall (Gál et al., 2008). Primary transport in the xylem, retranslocation in the phloem, and transfer from the xylem to the phloem constitute important processes for the redistribution of an element within a plant (Marschner, 1995). Transport in the xylem is directed from the roots to above-ground plant parts in the transpiration stream, whereas phloem transport is more selective and occurs from sources to sinks (Marschner, 1995).

The uptake and distribution of Co in plants are species dependent and controlled by different mechanisms (Li et al., 2009). In addition, soil properties also strongly influence Co availability and thus uptake by plants, since it depends on the concentration of the element in the ionic form in the soil solution and the concentration present on the exchange sites of the cation exchangeable soil surfaces (Alloway, 1997; Mollah and Begum, 2001; Kukier et al., 2004; Li et al., 2004). The absorption of Co^{2+} by roots implies active transport across cell membranes, although the molecular mechanisms are still unknown. Then, subsequent transport through the cortical cells implies passive diffusion and an active uptake process. A comparison of concentration of Co in cytoplasm and vacuoles indicates that if passive diffusion alone is operative, active transport is produced in an outward direction at the plasmalemma and into the vacuoles at the tonoplast (Palit et al., 1994). The distribution of Co may involve organic complexes, although the low mobility of Co^{2+} in plants restricts its transport from roots to shoots (Palit et al., 1994; Barysas et al., 2002; Bakkaus et al., 2005). The absorption of Co^{2+} into the roots of plants involves active transport; Co in plants is accumulated mainly in the roots, although other parts of the plant also can accumulate it to varying degrees (Palit et al., 1994; EURAS, 2005; Sasmaz and Yaman, 2006).

A few studies describe the interactions of Co^{2+} within the plant, reporting that the interaction of Co with other metals mainly depends on the concentration of the metals considered (Gál et al., 2008). Palit et al. (1994) observed that high levels of Co^{2+} induce Fe deficiency in plants and suppress uptake of Cd by roots; Co interacts synergistically with Zn, Cr, and Sn, and also it seems to show an antagonistic relationship with Ni.

In the xylem, Co is transported mainly by the transpirational flow, whereas the distribution through the sieve tubes is accomplished by complexing with organic compounds (Meharg, 2005; Page and Feller, 2005). If transport is not affected by transpirational water flux through the plant, about 10% of the Co absorbed from roots is moved to the shoots. Most of this amount is stored in the cortical cell vacuoles and removed from the transport pathway (Macklon and Sim, 1987; Palit et al., 1994). Also, Zeller and Feller (1998) reported that Co in wheat showed an intermediate mobility in the phloem, lower than Ni, as was also reported by Riesen and Feller (2005) in an experiment to study the redistribution of different heavy metals via the phloem in young and maturing wheat.

The specific organ distribution of heavy metals in the plants determines a concentration gradient from roots to the stems and from leaves to the fruits. However, the lower mobility of this element in relation to other divalent cations restricts the transport to leaves from stems, being transported in small amounts (Palit et al., 1994). Page and Feller (2005) reported in a study of mobility of different heavy metals in young wheat plants that cobalt was retained in the root system, being not (or only very slowly) released to the shoot, suggesting a poor loading into the xylem vessels.

In addition, Collins et al. (2010) studied the uptake, localization, and speciation of Co in a monocotyledon (wheat) and a dicotyledon (*Solanum lycopersicum* L. syn. *Lycopersicon esculentum* Mill., tomato) species grown in nutrient solutions containing Co. These authors found a greater transport from roots to shoots for tomato plants, with cobalt being located mainly in the vascular system of both plant species and maintaining the oxidation state of Co^{2+}, complexed with organic acids.

17.3 EFFECT OF COBALT IN PLANTS: PHYSIOLOGICAL RESPONSES AND VISUAL SYMPTOMS

The doses of cobalt affect the growth and metabolism of plants to different degrees, depending on the concentration and status of Co in rhizosphere and soil (Palit et al., 1994) (Table 17.1). Low concentrations of Co^{2+} stimulate growth (Gál et al., 2008), as was reported by Atta-Aly et al. (1991) in an experiment with tomato plants, observing an improvement in growth, flowering, and fruiting when the nutrient solution was supplemented with a low level of cobalt (0.25 ppm). These levels of Co can also increase drought resistance in melon (Atta-Aly, 2003). Boureto et al. (2001) also found that 2.5 mg Co/L in sand culture enhanced the absorption of N, P, and K by sugar beet (*Beta vulgaris* L.). Liu et al. (1995) showed that growth of onion (*Allium cepa* L.) roots increased with increasing Co addition, more than shoot growth. Gad and Kandil (2010) also reported that the addition of low concentrations of Co enhanced tomato growth and yield.

However, in supranormal doses, Co is relatively toxic to plants (Dekock, 1956). The toxic effects on plant morphology associated with high doses of Co include a stimulation of leaf fall, an inhibition of greening, discoloration of the veins, chlorotic mottling (e.g., in cauliflower) (Chatterjee and Chatterjee, 2000), premature leaf closure, as well as restricted shoot weight (Palit et al., 1994). In addition, another potential effect of cobalt is the alteration of the sex of plants such as hemp (*Cannabis sativa* L.), duckweed (*Lemna aequinoctialis* Welw.), and melon

TABLE 17.1
Main Effects Depending on the Concentration and Status of Cobalt

Toxic Effects	Beneficial Effects	Other Effects
Morphological changes: leaf fall, inhibition of greening, discolored veins, premature leaf closure, and reduced shoot weight, decrease germination.	Retardation of senescence of leaf	Sex alteration
	Increase in drought resistance in seeds	Effect on photoperiodism
		Interference on heme biosynthesis
Inhibit active ion transport	Regulation of alkaloid accumulation in medicinal plants	Enhancement of N fixation of leguminous plants
Changes in chlorophyll and carotenoid contents	Biocidal and antifungal activity	Effect on cell wall and cell membrane
Mutagenic action	Inhibition of ethylene biosynthesis	
	Stimulation of growth (low concentrations)	

Sources: Gal, J. et al., *Environ. Int.*, 34, 821, 2008; Palit, S. et al., *Bot. Rev.* 60(2), 151, 1994; Chatterjee, J. and Chatterjee, C., *Environ. Pollut.*, 109, 69, 2000; Liu, J. et al., *Physiol. Plant.*, 110, 104, 2000; Moreno-Caselles, J. et al., *J. Plant Nutr.*, 20(9), 1231, 1997b; Moreno-Caselles, J. et al., *Acta Hort.*, 458, 239, 1998a.

(a)

(b)

(c)

FIGURE 17.1 Toxic effect of increasing concentrations of cobalt in plants of (a) tomato (*Solanum lycopersicum* L.) and (b and c) lettuce (*Lactuca sativa* L.) grown in nutrient solutions. (From Pérez-Espinosa, A., Dinámica y efectos del Co en el sistema suelo–planta, Doctoral thesis, University of Alicante, Alicante, Spain, 1997.)

(*Cucumis melo* L.) cultivars (Palit et al., 1994). In Figure 17.1, some of these toxic effects on tomato and lettuce (*Lactuca sativa* L.) plants are shown.

In this sense, different studies on Co toxicity have been developed, these studies being often based on experiments in solution culture (Hogan and Rauser, 1979; Chatterjee and Chatterjee, 2000; Liu et al., 2000; Bakkaus et al., 2005), which try to estimate or establish phytotoxic concentrations of Co in plants. In general, toxic metals impose a significant stress on plants, the severity of which may be indicated by changes in chlorophyll and carotenoid contents in plant leaves. Soudek et al. (2011) reported a suppression in chlorophyll and carotenoid contents in the leaves of garlic hydroponically cultivated in the presence of Co at different concentrations, especially at 0.25 mM metal concentration. Also, Liu et al. (2000) reported a significant depression of growth at 5 μM in mung beans, whereas Bakkaus et al. (2005) observed toxicity symptoms in wheat and tomato after exposure to 50 μM of Co in solution. Moreno-Caselles et al. (1997b) investigated the effects of increasing concentrations of Co in tomato plants grown in nutrient solutions, observing a drastic reduction of the yield and of the total contents of chlorophylls. Also, in an experiment developed in soilless culture to study the effect of several levels of Co^{2+} (0, 5, 10, and 20 mg L^{-1}) on lettuce, Moreno-Caselles et al. (1998a) reported that the increasing concentration of Co drastically limited the yield of the plants and affected negatively other morphological parameters, such as root weight and root length.

However, there is little information about toxicity threshold values of cobalt in soil. Kapustka et al. (2006) established different toxicity threshold values for Co in an experiment to derive soil screening levels for Co using two soils, three plant species (alfalfa, *Medicago sativa* L.;

barley, *Hordeum vulgare* L.; and radish, *Raphanus sativus* L.), and different endpoints such as emergence, mortality, shoot height, or root length.

Micó et al. (2008) also designed an experiment to derive Co toxicity threshold values for barley on a range of 10 European and North American soils, with very different properties, observing that soil effective cation exchange capacity (eCEC) and exchangeable Ca were the best predictors of the toxicity threshold values for Co. Li et al. (2009) also studied the phytotoxicity and bioavailability of cobalt to barley, oilseed rape (*Brassica napus* L.), and tomato grown in ten soils with very different chemical and physical properties, and the results indicated that solubility of Co is a key factor influencing its toxicity to plants.

17.4 GENETICS OF ACQUISITION OF COBALT BY PLANTS

Many studies in the literature show that varieties or species differ in the concentrations of metals in their shoots when grown under standard conditions (Macnair et al., 2000). These differences must reflect genetic differences, but in general, little is known about the nature of the genes causing this variation and especially in relation to the uptake of Co. Many genes are involved in metal uptake, translocation, and sequestration and reflect many different ways that the uptake or translocation of metals could be affected by genetic variation (Macnair et al., 2000; Eapen and D'Souza, 2005). There are a number of major mutants of plants in which metal uptake and translocation are affected, for example, the *brz* mutant of peas (Cary et al., 1994), the *man-1* mutant of arabidopsis (*Arabidopsis thaliana* Heynh.) (Delhaize, 1996), although it is not clear if these loci are involved in natural variation in accumulation.

Macnair (1993) reviewed the genetic basis of plant tolerance (the ability to survive in a soil that is toxic to other plants of the same or different species) in wild and agricultural species and showed that metal tolerance was governed by a small number of major genes. Also, classic genetic studies have shown that only a few genes (one to three) are responsible for metal tolerance (Macnair et al., 2000). Additionally, other authors have suggested that there are additional minor "modifier" genes contributing to the phenomenon (Macnair et al., 2000).

One of the bases of metal tolerance, and even accumulation, is the storage of metals, once they are translocated to the shoot cells, in cellular locations where the metal will not damage the vital cellular processes (Eapen and D'Souza, 2005). In this process of sequestration, metal transporters and metal-binding proteins play an essential role. Thio-reactive metals may be sequestered by cysteine-rich peptides—the metallothioneins (MTs) and phytochelatins (PCs). Metals, such as Co^{2+}, are sequestered by bonding with organic sulfur (R-SH) on the cysteine residues of these peptides. In this sense, plants have a family of MT genes encoding peptides that generally are composed of 60–80 amino acids and contain 9–16 cysteine residues (Zhou and Goldsbrough, 1995; Chatthai et al., 1997), with the MTs protecting plants from effects of toxic metal ions (Eapen and D'Souza, 2005). Brouwer et al. (1993) found that MT–metal complexes can be glutathionated, suggesting that they may be transported into vacuoles for long-term sequestration. The other group of metal-binding proteins, PCs, composed of only three amino acids, glutamate, cysteine, and glycine, are also involved in heavy metal sequestration, but these proteins probably are not directly encoded by genes, being products of a biosynthetic pathway—presumably using glutathione (GSH) as a substrate (Cobbett and Goldsbrough, 2000).

17.5 COBALT CONCENTRATIONS IN PLANTS

The distribution of cobalt in plants is entirely species dependent (Palit et al., 1994). Table 17.2 shows the average normal concentrations of cobalt in different plant species. Most of the terrestrial plant species do not have the capacity of bioconcentrating Co (Robinson et al., 1999a,b; Dzantor, 2004; Kukier et al., 2004). Plants accumulating above 50–100 mg Co kg^{-1} dry weight from soil are exposed to severe phytotoxicity (Chaney, 1983). Only certain plant species (e.g., thistle,

TABLE 17.2
Average Normal Cobalt Concentrations in Different Plant Species (Dry Material)

Plant Species	Range/Mean Value (µg kg⁻¹)	References
Clover (*Trifolium* spp.)	0.10–0.57	Mitchell (1972); Berrow and Burridge (1979); Curylo (1981)
Grasses (*Echinochloa crus-galli* (L.) Beauv.)	0.03–0.27	Johansen and Steinnes (1972); Mitchell (1972); Berrow and Burridge (1979)
Barley grain (*Hordeum vulgare* L.)	27.3	Houba and Uittenbogaard (1997)
Pea seed (*Pisum sativum* L.)	42.2	Houba and Uittenbogaard (1997)
Maize plant (*Zea mays* L.)	108	Houba and Uittenbogaard (1997)
Rye grain (*Secale cereale* L.)	36.8	Houba and Uittenbogaard (1997)
Cucumber fruit (*Cucumis sativus* L.)	667	Houba and Uittenbogaard (1997)
Orange fruits (*Citrus sinensis* (L.) Osbeck)	19–45	Shacklette et al. (1978); Shacklette (1980)
Banana fruits (*Musa paradisiaca* L.)	13.3	Houba and Uittenbogaard (1997)
Tomato fruits (*Solanum lycopersicum* L.)	62–200	Shacklette et al. (1978); Shacklette(1980)
Tomato leaf (*Solanum lycopersicum* L.)	235	Houba and Uittenbogaard (1997)
Lettuce (*Lactuca sativa* L.)	177	Houba and Uittenbogaard (1997)
Spinach (*Spinacia oleracea* L.)	272	Houba and Uittenbogaard (1997)
Melon plant (*Cucumis melo* L.)	187	Houba and Uittenbogaard (1997)
Broad bean seeds (*Vicia faba* L.)	188	Houba and Uittenbogaard (1997)
Broccoli (*Brassica oleracea* L. var. *botrytis*)	100	Houba and Uittenbogaard (1997)
Wheat straw (*Triticum* L.)	139	Houba and Uittenbogaard (1997)
Potato tubers (*Solanum tuberosum* L.)	37–160	Shacklette et al. (1978); Shacklette (1980)

Berkheya coddii Ehrh.; yellow tuft, *Alyssum corsicum* Duby) and other species in the Lamiaceae, Scrophulariaceae, Asteraceae, and Cyperaceae (Gál et al., 2008) have the ability to extract metals (such as Co) from soils, showing potential in phytoremediation technologies. As defined by Baker and Brookes (1989), a plant containing >1000 mg kg⁻¹ of metal in its tissues is considered to be a metal hyperaccumulator (Chaney, 1983; Baker and Brooks, 1989; Malaisse et al., 1999).

Different studies have shown that the greatest accumulation of Co in plants occurs in the roots, although other parts of the plant also accumulate it to varying degrees (EURAS, 2005; Sasmaz and Yaman, 2006). Bakkaus et al. (2005) reported in an experiment to study the patterns of Co accumulation and storage in tomato and wheat that the concentration of Co in the roots for both species was much higher than in the shoots.

Plants can accumulate small amounts of Co from the soil, especially in the parts of the plant that are more routinely consumed, such as the fruit, grain, and seeds (ATSDR, 2004). Bakken et al. (2004) found that legumes contain larger amounts of Co than grasses and cereals. In spiking experiments, it has been observed that bioconcentration factors (plant/soil) for ⁶⁰Co and stable Co fertilizer vary from crop to crop (Gál et al., 2008). In the case of tomato, uptake from irrigated experiments exhibited strongest accumulation in the aboveground components, with leaves showing higher levels than fruits (Sabbarese et al., 2002). This fact also was found in studies where tomato was stressed by increasing Co concentrations and associated P and Fe (Chatterjee and Chatterjee, 2005). Similarly, Pérez-Espinosa et al. (2005) observed in an experiment to study Co phytoavailability for a tomato cultivar grown on an agricultural soil amended with sewage sludge that the gradient of Co accumulation in tomato plants was root > leaf > stem and branches > fruit, with a concentration in the edible parts ranging between 4 and 25 mg kg⁻¹, whereas Woodard et al. (2003) found that the Co concentration in tomato followed the following order: root > stem > leaf > fruit and flower. Moreno-Caselles et al. (1997a) reported levels around 250 µg Co g⁻¹ in fruit in an experiment designed to study the effect of different concentrations of Co²⁺ (0, 5, 15, and 30 mg L⁻¹) on yield and nutrient composition of tomato fruits, whereas, in a study of cultivation of lettuce in soilless culture with different concentrations of Co, Moreno-Caselles et al. (1998a) found that the internal concentration of Co in some cases rises to values around levels of 600 µg Co g^{-1} in the edible part. Chatterjee and Chatterjee (2000) also carried out a similar experiment using cauliflower (*Brassica oleracea* var. *botrytis* L.), subjecting the plants to an excess supply of Co in refined sand. The leaves of untreated plants showed higher accumulation of Co (10.1 mg kg^{-1}) than the roots (5.1 mg kg^{-1}). The plants treated with Co showed more accumulation of the metal in the roots (1274 mg kg^{-1}) than the leaves (507 mg kg^{-1}). Excessive supply of Co caused an increase in the P concentration (Chatterjee and Chatterjee, 2000). Chatterjee and Dube (2005) reported that the Co concentrations in leaves of potato, cauliflower, and cabbage grown in soils amended with sewage sludge for long periods were 4.97, 7.12, and 8.60 mg kg^{-1}, respectively, and the concentration of Co in potato (*Solanum tuberosum* L.), cabbage (*Brassica oleracea* var. *capitata* L.), and cauliflower all decreased in the following order: roots > leaves > stem. Also, Murtaza et al. (2008) found that the concentrations of Co in the leaves and edible parts of vegetables irrigated with city effluent were in the order of okra (*Abelmoschus esculentus* Moench) < bitter gourd (*Momordica charantia* L.) < spinach (*Spinacia oleracea* L.) < spearmint (*Mentha spicata* L.), whereas Bhattacharyya et al. (2008) reported that rice (*Oryza sativa* L.) straw contained much more Co than rice grain.

In addition, Luo et al. (2010), in an experiment to study the bioavailability of cobalt and its transfer from soil to 20 commonly consumed vegetables and rice, reported average Co concentrations in the edible parts of vegetables and rice of 15.4 and 15.5 mg kg^{-1}, respectively. These authors also found that the Co concentrations of the vegetables studied generally follow the order of fruit vegetables > leafy vegetables > root vegetables.

17.6 INTERACTION OF COBALT WITH UPTAKE OF OTHER ELEMENTS

There are few studies regarding the interaction of Co^{2+} with other metals within the plant, and these show that the interaction of Co with other metals principally depends on the concentration of the metals used. High concentrations of Co^{2+} induce iron deficiency in plants, since the excess of metals, such as Co, accumulated in shoots and especially in roots, reduces iron absorption and distribution in these organs, followed by other effects (e.g., chlorosis, decrease in catalase activity, and increase in nonreducing sugar concentration in barley) (Palit et al., 1994). In the case of Co, its lower mobility compared with other divalent cations restricts its transport to leaves from stems, and thus, active Co localization in plant roots (Macklon and Sim, 1987; Liu et al., 2000) suggests that toxicity may relate to competitive reactions disrupting nutrient transport (e.g., of Fe and S in mung beans). Also, Bakkaus et al. (2005) investigated by micro-proton-induced x-ray emission (PIXE) the patterns of Co accumulation and storage in tomato and wheat grown in nutrient solutions, with different Co treatments. These authors reported that the spatial distributions of K, Ca, Fe, and Co in whole plants, and in leaf and stem sections examined, revealed that in wheat leaves (monocotyledon), Co was distributed in exactly the same pattern as K and Ca, following the network of parallel veins. By contrast, in tomato leaves (dicotyledon), elemental maps showed that Co is not distributed in the same way as K, Ca, and Fe.

In addition, a competitive absorption and mutual activation of the transport of cobalt and zinc toward the parts aboveground was reported in pea (*Pisum sativum* L.) (Babalakova et al., 1986) and wheat seedlings (Chaudhry and Loneragan, 1972). In excess, cobalt tends to interact synergistically with Zn, this synergism being sometimes related to induced nutrient deficiency (Wallace and Abou-Zamzam, 1989). However, in bush beans (*Phaseolus vulgaris* L.), supply of supplemental Co at high concentration with supplemental Zn at low concentration gave bigger leaf yields (although not significantly so) than for either element supplied alone at the same concentration. This protective effect was accompanied by a nonsignificantly lower leaf Zn concentration than in the sole Zn treatment plants and much lower leaf Co concentration than the sole Co treatment plants (Wallace and Abou-Zamzam, 1989). Cobalt also interacts synergistically with other elements, such as Mn, Cr, and Sn, and suppresses uptake of Cd by roots. Nickel overcomes the inhibitory effect of Co on protonemal growth of moss, thus indicating an antagonistic relationship (Palit et al., 1994; Gál et al., 2008).

Micó et al. (2008) found in an experiment to study the influence of soil properties on the bioavailability and toxicity of Co to barley that the toxicity of Co^{2+} ions in soil solution appeared to be modified by the presence of other competitive ions, demonstrating a tendency of decreasing Co toxicity with decreasing pH or increasing solution Mg concentration, at the same free Co^{2+} activity. In this case, Co probably displaces H^+ and Mg^{2+} from binding sites on the soil solid phase (e.g., organic matter). Therefore, the competition between Co^{2+} and H^+ and Mg^{2+} for binding sites on soil particles to some extent cancels out the competition for the biotic ligands (Lock et al., 2006). Similar effects of protons have been reported also for Cd, Cu, Pb, and Zn (Lofts et al., 2004; Zhao et al., 2006). Lock et al. (2007) reported competition between Co^{2+} and some cations in the development of a terrestrial biotic ligand model to predict the effect of Co on barley root growth in nutrient solutions.

Pérez-Espinosa et al. (2002) studied the effect of increasing cobalt additions (0, 50, 100, and 200 mg kg^{-1}) on the nitrogen forms in soils amended with sewage sludge, as well as the transference to tomato plants. The results obtained in this experiment showed an increase in the ammonium concentration in the soils with the presence of Co, suggesting that the Co soluble pool in soil could induce a diminution, even a partial blocking, of the nitrification processes, probably due to a negative incidence of Co on the activity of the microorganisms involved in organic matter mineralization (Sauerbeck, 1986). Mandal and Parsons (1989) also reported a progressive suppression of the nitrification with the increment of Co in peat, whereas Maliszewska et al. (1985) found a negative effect of Co on the number of nitrificant bacteria.

In addition, Moreno-Caselles et al. (1998b) investigated the incidence of several concentrations of Co^{2+} (0, 5, 10, and 20 mg L^{-1}) on the nutrient content of lettuce plants (cv. Malta) hydroponically cultivated. These authors observed a negative effect of Co treatments on N, P, K, Fe, Cu, Mn, Cu, and Zn, as well as an increase in the Ca and Mg contents, also reporting a synergistic Co–B effect.

17.7 METHODS FOR DIAGNOSING COBALT STATUS IN PLANTS

The determination of the distribution of the element in plant tissues has been carried out using different methods (Wójcik et al., 2005). Histochemical detection (Vázquez et al., 1992b; Seregin and Ivanov, 1997), for example, by using silver sulfide staining, is a highly sensitive method and allows detection of very small metal concentrations in light and electron microscopy. However, the reaction is not specific, and positive results are obtained with not only Cd sulfide but also with Fe, Zn, or Cu sulfides, which makes it impossible to distinguish between the deposits of these metals (Wójcik et al., 2005). Another method, x-ray microanalysis (Khan et al., 1984; Vázquez et al., 1992a, 1992b), although it is not always useful at very low metal concentrations, provides more specific signals (Wójcik et al., 2005). Other methods used for the identification of the distribution of Co are cell fractionation (Lozano-Rodriguez et al., 1997), micro-PIXE (Ager et al., 2002), or the nuclear micro-probe technique (NMP) (Ager et al., 2003), the latter methods being more often used recently. Micro-PIXE with the use of focused ion beams constitutes a useful tool with imaging capabilities for the localization and quantification of toxic elements in plants (Ager et al., 2003). The technique has sufficient spatial resolution and analytical sensitivity to provide information as far down as the cellular level, and it can be performed simultaneously with other nuclear techniques, such as Rutherford backscattering spectrometry (RBS) and electron imaging (SEM), which give complementary information about sample composition and structure (Ager et al., 2003). Also, the nuclear microprobe technique allows us to obtain quantitative or semi-quantitative elemental distribution maps of major and trace elements with high resolution and sensitivity and is becoming a useful tool for localizing the sites of element accumulation in a wide variety of studies (Ager et al., 2002).

17.8   FORMS AND CONCENTRATIONS OF COBALT IN SOILS: PLANT AVAILABILITY

Cobalt normally occurs in the environment in association with other metals, such as Cu, Ni, Mn, and As, and usually is found in soils as Co^{2+} (USEPA, 2005). The total content in the soil varies depending on the parent material; also, there are differences with depth in the soil profile and between different soil types derived from a common parent material due to natural pedological processes (Gál et al., 2008). Therefore, the abundance of this element is variable, from 0.05 to 300 mg kg^{-1}, with an average content in the range 0.1–15.0 mg kg^{-1} from which the available part (the proportion of Co that is taken up by vegetation) is between 0.1 and 2.0 mg kg^{-1} (Hamilton, 1994; Gál et al., 2008). In worldwide soils, the average value of the concentration of Co is between 1 and 40 mg kg^{-1} (Alloway, 1997; He et al., 2005). More specifically, Yang and Yang (2000) reported average values of cobalt in Chinese soils of 5–40 mg kg^{-1}, whereas Kabata-Pendias and Pendias (2001) found average Co concentrations in soils from the United States and from Japan of <3–70 mg/kg^{-1} and 1.3–116 mg kg^{-1}, respectively. In Europe, Díez (2006) reported cobalt average concentrations in Spanish soils between 0 and 86.2 mg kg^{-1}, whereas Ferreira et al. (2001) reported values between 1 and 84 mg kg^{-1} in Portuguese soils, and Baize (1997) found average Co concentrations of 0–148 mg kg^{-1} in soils of France. In Table 17.3, the ranges of concentration and the average values of cobalt in soils of other European countries are shown (Kabata-Pendias and Pendias, 2001).

In most soils, trace metals such as Co are present in the form of salt carbonates, oxides, and sulfides, and the dominant minerals of each trace metal vary among different soils (He et al., 2005). Cobalt is found typically at higher concentrations in ultrabasic rocks, where it is associated with olivine minerals (Nagpal, 2004). Cobalt-containing minerals in soil include cobaltite (CoAsS), skutterudite (CoAs$_{2-3}$), and erythrite (Co$_3$(AsO$_4$)$_2 \cdot$8H$_2$O), these minerals being confined mainly in igneous rocks (He et al., 2005).

Cobalt is mobilized relatively easily during weathering processes (Gál et al., 2008), and it is one of the metals that are susceptible to oxidation–reduction reactions, being also involved in dissolution–precipitation processes in the soil. Therefore, cobalt can be either oxidized or reduced under variable soil conditions and can influence the mobility and bioavailability of associated trace elements (Kabata-Pendias, 2004). Cobalt bioavailability is dependent on the physicochemical and chemical properties of the soil, such as organic matter (quantity and quality), clay mineral content, texture, presence of complexing ions, and water content of the soil, as well as on other aspects, such as adsorption, temperature, oxidation–reduction potential, and pH (Kabata-Pendias, 2004; Gál et al., 2008). However, some of the main factors that control the availability of Co to plants are the concentrations of MnO_2 in soil, since Co is sorbed strongly and coprecipitated with MnO_2 (Palit et al., 1994; Alloway, 1997; Gál et al., 2008). The interactions with other chemical elements have a strong influence on the phytoavailability of some trace elements. Most cobalt (up to 79%, on average)

TABLE 17.3
Average Values and Range of Concentrations of Cobalt in European Soils

Country	Average Concentration (mg kg^{-1})	Range of Concentrations (mg kg^{-1})
Denmark	2.3	—
Germany	14.5	—
Italy	14	4–275
Sweden	17	2–58
Bulgaria	21.5	3.8–65
Romania	3.1	1–6.9

Source: Kabata-Pendias, A. and Pendias, H., *Trace Elements in Soils and Plants*, 3rd edn., CRC Press, Boca Raton, FL, 2001.

in soils is contained in, or associated with, Mn in various mineral forms (Nagpal, 2004). Thus, the presence of Mn and Fe is very likely to have the greatest influence on cobalt coprecipitation or adsorption, and therefore, on its bioavailability, these elements bind with Co forming less biologically active species (Gál et al., 2008). Therefore, in soil–plant systems, the behavior of Co follows that of Fe–Mn (Alloway, 1997; Li et al., 2009) and is modified by the presence of organic matter and clay minerals.

Despite the importance of the factors previously commented on, pH constitutes the dominant parameter in the terrestrial environment that regulates Co availability. Simple divalent Co species dominate at low pH, in most situations. Adsorption of Co^{2+} on soil colloids is high between pH 6 and 7, whereas leaching and plant uptake of Co are enhanced by lower pH (Nagpal, 2004; Watmough et al., 2005), since the more acidic is the soil, the less sorbed is Co (Gál et al., 2008).

In addition, Co is taken up as the divalent cation that is the only oxidation state commonly found in soil minerals. Therefore, Ca, P, and Mg are often the main antagonistic elements in the absorption of some microcations (Kabata-Pendias, 2004). In a study to evaluate the influence of soil properties on the bioavailability and toxicity of Co to barley, Micó et al. (2008) reported that soil eCEC and exchangeable Ca were the best predictors of the toxicity threshold values for Co, also showing that there was a competition of Co^{2+} with H^+ and Mg^{2+} from binding sites on the soil phase. Kukier et al. (2004) also found differences in the uptake pattern of Co by alyssum (*Alyssum* spp.), a Ni and Co hyperaccumulator plant species, from different soils and pH, probably due to differences in the contents of organic matter and iron in the studied soils. Soil water status also plays an important role in the availability of Co for plant uptake (Hamilton, 1994; Robinson et al., 1999a,b). In poorly drained soils, the amount of extractable Co is greater than in areas that are well drained (Alloway, 1997), resulting in significantly increased plant uptake.

The solid–liquid distribution of trace metals such as Co strongly affects their mobility and bioavailability in soils. Therefore, partition between solid and solution phases often is quantified by a distribution coefficient (Kd), which gives the ratio of the concentration in the solid phase to that in the solution phase. This coefficient is used usually as an indicator of potential bioavailability, since the sensitivity of Kd to soil properties has been shown to be a good predictor of plant uptake (Watmough et al., 2005; Degryse et al., 2009). In this sense, Bakkaus et al. (2008) studied the potential phytoavailability of Co in eight soils subjected to the atmospheric deposition of anthropogenic Co using techniques of isotope dilution. For this investigation, these authors evaluated the distribution coefficient and the potential phytoavailability (E value) of Co, reporting that the Kd of isotopically exchangeable Co in these soils was modeled best using two parameters, soil pH and organic C, which showed the most significant influence on labile Co concentrations. However, as also was reported by Degryse et al. (2009), although soil pH is the most important soil property determining the retention of a free metal ion, the Kd values based on total dissolved metal in solution may show little pH dependence for metal ions that have strong affinity for dissolved organic matter.

17.9 METHODS FOR DIAGNOSING COBALT PHYTOAVAILABILITY IN SOILS

The availability of Co for biological uptake is determined by the detailed chemical speciation, defined as the process and quantification of the different defined species, forms, and phases of a trace element (Kabata-Pendias, 2004), and usually is approached through sequential chemical extraction. In addition, several authors suggest that the available soil metal concentration, assessed using a chemical extraction, predicts metal transfer from soil to crops better than the total metal concentration (Brun et al., 1998). In this sense, several studies have focused on the selection of suitable extractants and the development of extraction procedures aimed at the better prediction of their soil-to-plant transfer (McGrath, 1996; Kabata-Pendias, 2004; Pueyo et al., 2004; Meers et al., 2007). Among the numerous extractants used for assessing the availability of Co, there are various solutions most commonly used (Aery and Jagetiya, 2000; Kabata-Pendias, 2004; Pérez-Espinosa et al., 2005; Bhattacharyya et al., 2008; Li et al., 2009): (1) acids (mineral acids at various concentrations); (2) chelating agents (DTPA, EDTA); (3) buffered salts (NH_4OAc); and (4) neutral salts ($CaCl_2$, $MgCl_2$, $Sr(NO_3)_2$, NH_4NO_3). Some other methods like electrodialysis, diffusion through membrane, diffusive gradient in thin film (DGT), and bioindicators have been proposed also to estimate the pool of the soluble (available) trace metals in soils. However, since several soil parameters and climatic factors have a significant impact on the absorption of trace elements such as Co by roots, any of these methods has to be related to given soil and plant conditions (Kabata-Pendias, 2004).

On the other hand, isotope dilution is considered a useful technique and conceptually the most attractive method to determine the potential phytoavailability of labile metals in soil, as it represents the fraction of metals that is in dynamic equilibrium with metals in the solution phase within a certain time-frame (Bakkaus et al., 2008; Degryse et al., 2009). Isotope dilution techniques now are accepted widely as methodologies to evaluate the potential phytoavailable contents of nutritive and toxic elements in soils (Hamon et al., 2002). However, currently, there is some limited application of isotopic dilution to Co determinations (although all of the associated publications are focused solely on agricultural soils, most of which are considered as Co deficient) (Bakkaus et al., 2008). In addition, the use of radioisotopes is not suitable for routine measurements. Therefore, it is still necessary to carry out more research about the methodologies to determine the potential phytoavailability of Co and to evaluate the analytical and sampling errors associated (Kabata-Pendias, 2004; Degryse et al., 2009).

17.10 AGRICULTURAL INPUTS OF COBALT

Cobalt, like the other trace elements, can enter the soil by a number of pathways, the most important anthropogenic sources of trace elements for soils being associated with atmospheric deposition from industrial, urban, and road emissions and especially with agricultural activities, through the use of commercial fertilizers, liming materials, agrochemicals, and irrigation waters, especially wastewaters, as well as organic wastes and composts used as soil amendments (Senesi et al., 1999; Micó et al., 2006).

Application to soil of commercial fertilizers frequently implies the incorporation of small amounts of trace elements contained in them as impurities originating from the parent rock, the corrosion of equipment, catalysts and reagents used in their manufacture, and materials added to commercial preparation as coatings, fillers, and conditioners, for example, gypsum, kaolin, and limestone (Senesi et al., 1999). In Table 17.4, the contents of Co in several mineral and synthetic commercial fertilizers of different nature and origin and limestones collected from different manufacturers and distributors in Italy are shown (Senesi et al., 1999).

On the other hand, organic materials such as farm manures, sewage sludges (biosolids), anaerobic digestates, or composts, which are increasingly and beneficially recycled in agriculture as soil amendments, also contain variable and, usually, higher concentrations of various trace elements than most agricultural soils. There are numerous studies about the incorporation of trace elements into agricultural soils (Chaney et al., 2001; McBride, 2004); however, there is little information about the concentration of Co in these materials.

Regarding the Co contents in sewage sludges, Fytili sand Zabaniotou (2008) reported typical concentration ranges in dry sludges of about 11.3–2490 mg kg^{-1}, whereas Moreno-Caselles et al. (1997c) found values of Co concentration between 9 and 14 mg kg^{-1} in sewage sludges from different wastewater treatment plants of southeast Spain.

In relation to the concentrations of Co in manures and slurries, Moral et al. (2008) studied several hazard-associated parameters (salinity, organic content, and heavy metal contents) in samples of pig slurries of 36 pig farms in southeastern Spain, reporting levels of Co between 0.06 and 0.23 g m^{-3}, depending on the animal production stage, observing the highest contents for the finisher stage.

Additionally, there are other materials that can be susceptible to be used as organic amendments in agricultural soils, such as digestates, the by-product of the process of anaerobic digestion of organic wastes. In this regard, Bustamante et al. (2013) reported Co concentrations of 1.88 mg kg^{-1}

TABLE 17.4

Concentrations of Cobalt (mg kg⁻¹) in Different Mineral and Synthetic Commercial Fertilizers and in Limestones

Product	Average	Range of Co
Nitrogen-based fertilizers		
Ammonium sulfate	8.3	7.7–12.0
Ammonium nitrate	7.8	5.4–11.5
Calcium nitrate	9.7	8.6–12.8
Urea	1.2	1.0–1.4
Calcium cyanamide	—	22.0
Phosphorus-based fertilizers		
Superphosphate	13.3	8.8–21.0
Triple superphosphate	—	16.7
Potassium sulfate	6.1	5.8–7.0
Other fertilizers		
NP compounds	13.9	10.0–17.0
NPK compounds	12.1	10.5–15.7
Limestones	2.7	0.4–6.0

Sources: Senesi, N. and Polemio, M., *Fertil. Res.*, 2, 289, 1981; Senesi, G.S. et al., *Chemosphere*, 30, 343, 1999.

(dry matter) in the solid fraction of anaerobic digestates from pig slurry and of 15.5 µg kg⁻¹ (fresh matter) in the liquid fraction of pig slurry digestate.

Also, several studies reported information concerning the concentration of trace elements, including cobalt, in composts from different origins. Bustamante et al. (2008, 2009) reported concentrations of cobalt in the range of 19–33 mg kg⁻¹ in composts mainly derived from winery and distillery wastes, while Baykov et al. (2005) reported a concentration of Co of 1.64 mg kg⁻¹ in composts from litter from broiler production. Ciba et al. (1999) reported a concentration of 12.0 mg kg⁻¹ in compost from municipal solid wastes, and Trubetskaya et al. (2001) found concentrations of 30 mg kg⁻¹ in a compost derived from sewage sludge and yard trimmings. Also, Ogundare and Lajide (2013) found in composts from waste materials of cassava (*Manihot esculenta* Crantz), vegetable, banana (*Musa* spp. L.), orange (*Citrus* spp. L.), and cow dung fortified each with 100 g of NPK, 100 g of kaolin, and 100 g of ammonium chloride, concentrations of Co between 0.55 and 4.49 mg kg⁻¹, depending on the raw materials used and the particle size of the compost sample. In addition, the concentrations of cobalt in composts from anaerobic digestates ranged between 1.42 and 1.96 mg kg⁻¹ for anaerobic digestates from cattle manure (Bustamante et al., 2012) and between 1.22 and 1.93 mg kg⁻¹ for composts from anaerobic digestates from pig slurry (Bustamante et al., 2013). Therefore, all these studies show that the concentrations of cobalt in an organic material present a high variability and are strongly influenced by the nature and origin of the organic material considered.

REFERENCES

Aery, N.C. and B.L. Jagetiya. 2000. Effect of cobalt treatments on dry matter production of wheat and DTPA extractable cobalt content in soils. *Commun. Soil Sci. Plant Anal.* 3:1275–1286.

Ager, F.J., M.D. Ynsa, J.R. Domınguez-Sol, C. Gotor, M.A. Respaldiza, and L.C. Romero. 2002. Cadmium localization and quantification in the plant *Arabidopsis thaliana* using Micro-PIXE. *Nucl. Instrum. Methods B* 189:494–498.

Ager, F.J., M.D. Ynsa, J.R. Domínguez-Sol, M.C. Lopez-Martinez, C. Gotor, and L.C. Romero. 2003. Nuclear micro-probe analysis of *Arabidopsis thaliana* leaves. *Nucl. Instrum. Methods B* 210:401–406.

Alloway, B.J. (ed.). 1997. *Heavy Metals in Soils*. London, U.K.: Blackie Academic and Professional.

ATSDR. 2004. Toxicological profile for cobalt. Agency for toxic substances and disease registry. Atlanta, GA: U.S. Department of Health and Human Services, Public Health Service.

Atta-Aly, M.A. 2003. Effect of galia melon seed soaking in cobalt solution on plant growth, fruit yield, quality and sudden wilt disease infection. http://www.aaaid.org/pdf/magazine/2003/Cobalt%2067-72.pdf, accessed May 20, 2008.

Atta-Aly, M.A., N.G. Shehata, and T.M. Kobbia. 1991. Effect of cobalt on tomato plant growth and mineral content. *Ann. Agric. Sci.* 36(2):617–624.

Babalakova, N., T. Kudrev, and I. Petrov. 1986. Cu, Cd, Zn and Co interactions in their absorption by pea plants. *Fiziol. Rast.* 12:67–73.

Baize, D. 1997. *Teneur Totales en Elements Traces Metalliques Dans Les Sols (France)*. Paris, France: INRA Editions.

Baker, A.J.M. and R.R. Brooks. 1989. Terrestrial plants which hyperaccumulate metallic elements: A review of their distribution, ecology and phytochemistry. *Biorecovery* 1:81–126.

Bakkaus E., R.N. Collins, J.L. Morel, and B. Gouget. 2008. Potential phytoavailability of anthropogenic cobalt in soils as measured by isotope dilution techniques. *Sci. Total Environ.* 406:108–111.

Bakkaus, E., B. Gouget, J.P. Gallien et al. 2005. Concentration and distribution of cobalt in higher plants: The use of micro-PIXE spectroscopy. *Nucl. Instrum. Methods B* 231:350–356.

Bakken, A.K., O.M. Synnes, and S. Hansen. 2004. Nitrogen fixation by red clover as related to the supply of cobalt and molybdenum from some Norwegian soils. *Acta Agric. Scand. B* 54(2):97–101(5).

Barysas, D., T. Cesniene, L. Balciuniene, V. Vaitkuniene, and V. Rancelis. 2002. Genotoxicity of Co^{2+} in plants and other organisms. *Biologija* 2:58–63.

Baykov, B., I. Popova, B. Zaharinov, K. Kirov, and I. Simeonov. 2005. Assessment of the compost from the methane fermentation of litter from broiler production with a view to its utilization in organic plant production. *Animals and Environment*, Volume 2: *Proceedings of the XIIth ASAH Congress on Animal Hygiene*, pp. 69–72. Warsaw, Poland: ISAH.

Berrow, M.L. and J.C. Burridge. 1979. Sources and distribution of trace element in soils and related crops. In: *Proceedings International Conference on Management and Control of Heavy Metals in the Environment*, pp. 304–311. Edinburgh, Scotland: CEP Consultants Ltd.

Bhattacharyya, P., K. Chakrabarti, A. Chakraborty, S. Tripathy, K. Kima, and M.A. Powell. 2008. Cobalt and nickel uptake by rice and accumulation in soil amended with municipal solid waste compost. *Ecotoxicol. Environ. Saf.* 69:506–512.

Bond, G. and E.J. Hewitt. 1962. Cobalt and the fixation of nitrogen by root nodules of *Alnus* and *Casuarina*. *Nature* 195:94–95.

Boureto, A.E., M.C. Castro, and J.N. Kagawa. 2001. Effect of cobalt on sugar beet growth and mineral content. *Rev. Bras. Sementes* 18:63–68.

Brouwer, M., T. Hoexum-Brouwer, and R.E. Cashon. 1993. A putative glutathione-binding site in Cd Zn-metallothionein identified by equilibrium binding and molecular modeling studies. *Biochem. J.* 294:219–225.

Brun, L.A., J. Maillet, J. Richarte, P. Herrmann, and J.C. Remy. 1998. Relationships between extractable copper, soil properties and copper uptake by wild plants in vineyard soils. *Environ. Pollut.* 102:151–161.

Bustamante, M.A., J.A. Alburquerque, A.P. Restrepo et al. 2012. Co-composting of the solid fraction of anaerobic digestates, to obtain added-value materials for use in agriculture. *Biomass Bioenerg.* 43:26–35.

Bustamante, M.A., C. Paredes, F.C. Marhuenda-Egea, A. Pérez-Espinosa, M.P. Bernal, and R. Moral. 2008. Co-composting distillery wastes with animal manure: carbon and nitrogen transformations and evaluation of compost stability. *Chemosphere* 72(4):551–557.

Bustamante, M.A., C. Paredes, J. Morales, A.M. Mayoral, and R. Moral. 2009. Study of the composting process of winery and distillery wastes using multivariate techniques. *Bioresour. Technol.* 100:4766–4772.

Bustamante, M.A., A.P. Restrepo, J.A. Alburquerque et al. 2013. Recycling of anaerobic digestates by composting: effect of the bulking agent used. *J. Cleaner Prod.* 47:61–69.

Cary, E.E., W.A. Norvell, D.L. Grunes, R.M. Welch, and W.S. Reid. 1994. Iron and manganese accumulation by the *brz* pea mutant grown in soils. *Agron. J.* 86:938–941.

Chaney, R.L. 1983. *Potential Effects of Waste Constituents on the Food Chain*. Park Ridge, NJ: Noyes Data Corporation.

Chaney, R.L., J.A. Ryan, U. Kukier et al. 2001. Heavy metal aspects of compost use. In: *Compost Utilization in Horticultural Cropping*, eds. P.J. Stoffella and B.A. Khan, pp. 324–359. Boca Raton, FL: CRC Press.

Chatterjee, J. and C. Chatterjee. 2000. Phytotoxicity of cobalt, chromium and copper in cauliflower. *Environ. Pollut.* 109:69–74.

Chatterjee, J. and C. Chatterjee. 2003. Management of phytotoxicity of cobalt in tomato by chemical measures. *Plant Sci.* 164:793–801.

Chatterjee, J. and C. Chatterjee. 2005. Deterioration of fruit quality for tomato by excess cobalt and its amelioration. *Commun. Soil Sci. Plant Anal.* 36:1931–1945.

Chatterjee, C. and B.K. Dube. 2005. Impact of pollutant elements on vegetables growing in sewage-sludge-treated soils. *J. Plant Nutr.* 28:1811–1820.

Chatthai, M., K.H. Kaukinen, T.J. Tranbarger, P.K. Gupta, and S. Misra. 1997. The isolation of a novel metallo-thionein related cDNA expressed in somatic and zygotic embryos of Douglas-fir: Regulation of ABA, osmoticum and metal ions. *Plant Mol. Biol.* 34:243–254.

Chaudhry, F.M. and J.F. Loneragan. 1972. Zinc absorption by wheat seedling: II inhibition by hydrogen ions and by micronutrient cations. *Proc. Soil Sci. Soc. Am.* 36:327–331.

Ciba, J., T. Korolewicz, and M. Turek. 1999. The occurrence of metals in composted municipal wastes and their removal. *Water Air Soil Pollut.* 111:159–170.

Cobbett, C.S. and P.B. Goldsbrough. 2000. Mechanisms of metal resistance phytochelatins and metallothioneins. In: *Phytoremediation of Toxic Metals: Using Plants to Clean Up the Environment*, eds. I. Raskin and B.D. Ensley, pp. 247–268. New York/Chichester, U.K.: John Wiley & Sons.

Collins, R.N., E. Bakkaus, H.K. Carrière, O. Proux, J.L. Morel, and B. Gouget. 2010. Uptake, localization, and speciation of cobalt in *Triticum aestivum* L. (wheat) and *Lycopersicon esculentum* M. (tomato). *Environ. Sci. Technol.* 44:2904–2910.

Curylo, T. 1981. The effect of some soil properties and of NPK fertilization levels on the uptake of cobalt by plants. *Acta Agrar. Silvestria* 20:57–60.

Degryse, F., E. Smolders, and D.R. Parker. 2009. Partitioning of metals (Cd, Co, Cu, Ni, Pb, Zn) in soils: Concepts, methodologies, prediction and applications—A review. *Eur. J. Soil Sci.* 60:590–612.

Dekock, P.C. 1956. Heavy metal toxicity and iron chlorosis. *Ann. Bot.* 20:133–141.

Delhaize, E. 1996. A metal-accumulator mutant of *Arabidopsis thaliana*. *Plant Physiol.* 111:849–855.

Delwiche, C.C., C.M. Johnson, and H.M. Reisenauer. 1961. Influence of cobalt on nitrogen fixation by Medicago. *Plant Physiol.* 36(1):73–78.

Díez, M. 2006. Background levels of trace elements in soils of Granada (Southern Spain). Doctoral thesis, Departamento de Edafología y Química Agrícola, University of Granada, Granada, Spain.

Dzantor, E.K. 2004. Phytoremediation-plant-based strategies for cleaning up contaminated soils, vol. 768. Maryland Extension Services Fact Sheet. College Park, MD: University of Maryland, pp. 1–4.

Eapen, S. and S.F. D'Souza. 2005. Prospects of genetic engineering of plants for phytoremediation of toxic metals. *Biotechnol. Adv.* 23:97–114.

EURAS. 2005. Fact sheet 8-MERAG program-building block: Bioconcentration, bioaccumulation and biomagnification of metals. Ghent, Belgium: EURAS.

Ferreira, A., M.M. Inacio, P. Morgado et al. 2001. Low-density geochemical mapping in Portugal. *Appl. Geochem.* 16:1323–1331.

Fytili, D. and A. Zabaniotou. 2008. Utilization of sewage sludge in EU application of old and new methods—A review. *Renew. Sust. Energy Rev.* 12:116–140.

Gad, N. and H. Kandil. 2010. Influence of cobalt on phosphorus uptake, growth and yield of tomato. *Agric. Biol. J. North Am.* 1(5):1069–1075.

Gál, J., A. Hursthouse, P. Tatner, F. Stewart, and R. Welton. 2008. Cobalt and secondary poisoning in the terrestrial food chain: Data review and research gaps to support risk assessment. *Environ. Int.* 34:821–838.

Hamilton, E. 1994. The geobiochemistry of cobalt. *Sci. Total Environ.* 150(1–3):7–39.

Hamon, R.E., I. Bertrand, and M.J. McLaughlin. 2002. Use and abuse of isotopic exchange data in soil chemistry. *Aust. J. Soil Res.* 40:1371–1381.

He, Z.L., X.E. Yang, and P.J. Stofella. 2005. Trace elements in agroecosystems and impacts on the environment. *J. Trace Elem. Med. Biol.* 19:125–140.

Hogan, G.D. and W.E. Rauser. 1979. Tolerance and toxicity of cobalt, copper, nickel and zinc in clones of *Agrostis gigantea*. *New Phytol.* 83:665–670.

Houba, V.J.G. and J. Uittenbogaard. 1997. International plant-analytical exchange, Report 1997. International Plant-Analytical Exchange (IPE). Wageningen, the Netherlands: Department of Soil Science and Plant Nutrition, Wageningen Agricultural University.

Johansen, O. and E. Steinnes. 1972. Routine determination of traces of cobalt in soils and plant tissue by instrumental neutron activation analysis. *Acta Agric. Scand.* 22:103.

Kabata-Pendias, A. 2004. Soil-plant transfer of trace elements-an environmental issue. *Geoderma* 122:143–149.

Kabata-Pendias, A. and H. Pendias. 2001. *Trace Elements in Soils and Plants*, 3rd edn. Boca Raton, FL: CRC Press.

Kapustka, L., D. Eskew, and J.M. Yocum. 2006. Plant toxicity testing to derive ecological soil screening levels for cobalt and nickel. *Environ. Toxicol. Chem.* 25:865–874.

Khan, D.H., J.G. Duckett, B. Frankland, and J.B. Kirkham. 1984. An X-ray microanalytical study of the distribution of cadmium in roots of *Zea mays* L. *J. Plant Physiol.* 115:19–28.

Krämer, U., I.N. Talke, and M. Hanikenne. 2007. Transition metal transport. *FEBS Lett.* 581:2263–2272.

Kukier, U., C.A. Peters, R.L. Chaney, J.S. Angle, and R.J. Roseberg. 2004. The effect of pH on metal accumulation in two *Alyssum* species. *J. Environ. Qual.* 33:2090–2102.

Li, H.-F., C. Gray, C. Mico, F.-J. Zhao, and S.P. McGrath. 2009. Phytotoxicity and bioavailability of cobalt to plants in a range of soils. *Chemosphere* 75:979–986.

Li, Z., R.G. McLaren, and A.K. Metherell. 2004. The availability of native and applied soil cobalt to ryegrass in relation to soil cobalt and manganese status and other soil properties. *N. Z. J. Agric. Res.* 47:33–43.

Liu, D., L. Zhai, W. Jiag, and W. Wang. 1995. Effect of Mg^{+2}, Co^{+2} and Hg^{+2} on the nucleus in root tip cells of *Allium cepa*. *Bull. Environ. Contam. Toxicol.* 55: 779–787.

Liu, J., R.J. Reid, and F.A. Smith. 2000. The mechanism of cobalt toxicity in mung beans. *Physiol. Plant.* 110:104–110.

Lock, K., K.A.C. De Schamphelaere, S. Becaus, P. Criel, H. Van Eeckhout, and C.R. Janssen. 2006. Development and validation of an acute biotic ligand model (BLM) predicting cobalt toxicity in soil to the potworm *Enchytraeus albidus*. *Soil Biol. Biochem.* 38:1924–1932.

Lock, K., K.A.C. De Schamphelaere, S. Becaus, P. Criel, H. Van Eeckhout, and C.R. Janssen. 2007. Development and validation of a terrestrial biotic ligand model predicting the effect of cobalt on root growth of barley (*Hordeum vulgare*). *Environ. Pollut.* 147:626–633.

Lofts, S., D.J. Spurgeon, C. Svendsen, and E. Tipping. 2004. Deriving soil critical limits for Cu, Zn, Cd, and Pb: a method based on free ion concentrations. *Environ. Sci. Technol.* 38:3623–3631.

Lozano-Rodriguez, E., L.E. Hernández, P. Bonay, and R.O. Carpena-Ruiz. 1997. Distribution of cadmium in shoot and root tissues of maize and pea plants: Physiological disturbances. *J. Exp. Bot.* 48:123–128.

Luo, D., H. Zheng, Y. Chen, G. Wang, and D. Fenghua. 2010. Transfer characteristics of cobalt from soil to crops in the suburban areas of Fujian Province, southeast China. *J. Environ. Manage.* 91:2248–2253.

Macklon, A.E.S. and A. Sim. 1987. Cellular cobalt fluxes in roots and transport of the shoots of wheat seedling. *J. Exp. Bot.* 38:1663–1677.

Macnair, M.R. 1993. The genetics of metal tolerance in vascular plants. *New Phytol.* 124:541–559.

Macnair, M.R., G.H. Tilstone, and S.E. Smith. 2000. The genetics of metal tolerance and accumulation in higher plants. In *Phytoremediation of Contaminated Soil and Water*, eds. N. Terry and G. Bañuelos, pp. 235–251. Boca Raton, FL: Lewis Publishers.

Malaisse, F., A.J.M. Baker, and S. Ruelle. 1999. Diversity of plant communities and leaf heavy metal content at Luiswishi copper/cobalt mineralization, Upper Katanga, Dem. Rep. Congo. *Biotechnol. Agron. Soc. Environ.* 3(2):104–114.

Maliszewska, W., S. Dec, H. Wierzbicka, and A. Wozniakowska. 1985. The influence of various heavy metal compounds on the development and activity of soil micro-organisms. *Environ. Pollut.* (Series A) 37:195–215.

Mandal, R. and J.W. Parsons. 1989. Effects of chlorides of cobalt, nickel and copper on nitrification in peat. *Pak. J. Sci. Ind. Res.* 32:584–586.

Marschner, H. 1995. *Mineral Nutrition of Higher Plants*, 2nd edn. London, U.K.: Academic Press.

McBride, M.B. 2004. Molybdenum, sulfur, and other trace elements in farm soils and forages after sewage sludge application. *Commun. Soil Sci. Plant Anal.* 35:517–535.

McGrath, D. 1996. Application of single and sequential procedures to polluted and unpolluted soils. *Sci. Total Environ.* 178:37–44.

Meers, E., R. Samson, F.M.G. Tack, A. Ruttens, J. Vangronsveld, and M.G. Verloo. 2007. Phytoavailability assessment of heavy metals in soils by single extractions and accumulation in *Phaseolus vulgaris*. *Environ. Exp. Bot.* 60:385–396.

Meharg, A.A. 2005. Mechanisms of plant resistance to metal and metalloid ions and potential biotechnical applications. *Plant Soil* 274:163–174.

Micó, C., H.F. Li, F.J. Zhao, and S.P. McGrath. 2008. Use of Co speciation and soil properties to explain variation in Co toxicity to root growth of barley (*Hordeum vulgare* L.) in different soils. *Environ. Pollut.* 156:883–890.

Micó, C., L. Recatalá, M. Peris, and J. Sánchez. 2006. Assessing heavy metal sources in agricultural soils of an European Mediterranean area by multivariate analysis. *Chemosphere* 65:863–872.

Mitchell, R.L. 1972. Cobalt in soil and its uptake by plants. *Agrochimica* 16:521–532.

Mollah, A.S. and A. Begum. 2001. A study on transfer factors of Co-60 and Zn-65 from soil to plants in the tropical environment of Bangladesh. *Environ. Mon. Assess.* 68:91–97.

Moral, R., M.D. Perez-Murcia, A. Perez-Espinosa, J. Moreno-Caselles, C. Paredes, and B. Rufete. 2008. Salinity, organic content, micronutrients and heavy metals in pig slurries from South-eastern Spain. *Waste Manage.* 28:367–371.

Moreno-Caselles, J., A. Pérez-Espinosa, R. Moral, M.D. Pérez-Murcia, and I. Gómez. 1998a. Cobalt-Induced stress in soilless lettuce cultivation: I. Effect on yield and pollutant accumulation. *Acta Hort.* 458:239–241.

Moreno-Caselles, J., A. Pérez-Espinosa, M.D. Pérez-Murcia, R. Moral, and I. Gómez. 1997a. Effect of increased cobalt treatments on cobalt concentration and growth of tomato plants. *J. Plant Nutr.* 20(7–8):805–811.

Moreno-Caselles, J., A. Pérez-Espinosa, M.D. Pérez-Murcia, R. Moral, and I. Gómez. 1997b. Cobalt-Induced stress in tomato plants. Effect on yield, chlorophyll content and nutrient evolution. *J. Plant Nutr.* 20(9):1231–1237.

Moreno-Caselles, J., M.D. Pérez, A. Pérez, and R. Moral. 1997c. Heavy metal pollution in sewage sludge and agricultural impact. *Fresenius Environ. Bull.* 6:519–524.

Moreno-Caselles, J., A. Pérez-Espinosa, M.D. Pérez-Murcia, R. Moral, and I. Gómez. 1998b. Cobalt-Induced stress in soilless lettuce cultivation: II. Effect on nutrient evolution. *Acta Hort.* 458:243–246.

Murtaza, G., A. Ghafoor, and M.J. Qadir. 2008. Accumulation and implications of cadmium, cobalt and manganese in soils and vegetables irrigated with city effluent. *J. Sci. Food Agric.* 88(1):100–107.

Nagpal, N.K. 2004. *Water Quality Guidelines for Cobalt.* Victoria, British Columbia, Canada: Ministry of Water, Land and Air Protection, Water Protection Section; Water, Air and Climate Change Branch.

O'Hara, G.W. 2001. Nutritional constraints on root nodule bacteria affecting symbiotic nitrogen fixation: A review. *Aust. J. Exp. Agric.* 41(3):417–433.

Ogundare, M.O. and L. Lajide. 2013. Physico-chemical and mineral analysis of composts fortified with NPK fertilizer, ammonium chloride and kaolin. *J. Agric. Chem. Environ.* 2:27–33.

Osman, M.E.H., A.H. El-Naggar, M.M. El-Sheekh, and E.E. El-Mazally. 2004. Differential effects of Co^{2+} and Ni^{2+} on protein metabolism in *Scenedesmus obliquus* and *Nitzschia perminuta*. *Environ. Toxicol. Pharmacol.* 16:169–178.

Page, V. and U. Feller. 2005. Selective transport of zinc, manganese, nickel, cobalt and cadmium in the root system and transfer to the leaves in young wheat plants. *Ann. Bot.* 96:425–434.

Palit, S., A. Sharma, and G. Talukder. 1994. Effects of cobalt on plants. *Bot. Rev.* 60(2):151–181.

Pérez Espinosa, A. 1997. Dinámica y efectos del Co en el sistema suelo—Planta. Doctoral thesis, University of Alicante, Alicante, Spain.

Pérez-Espinosa, A., R. Moral, J. Moreno-Caselles, A. Cortés, M.D. Perez-Murcia, and I. Gómez. 2005. Co phytoavailability for tomato in amended calcareous soils. *Bioresour. Technol.* 96:649–655.

Pérez-Espinosa, A., J. Moreno-Caselles, R. Moral, M.D. Pérez-Murcia, and I. Gomez. 2002. Effect of increased cobalt treatments on sewage sludge amended soil: Nitrogen species in soil and transference to tomato plants. *Arch. Acker-Pfl. Boden.* 48:273–278.

Pueyo M., J.F. Lopez-Sanchez, and G. Rauret. 2004. Assessment of $CaCl_2$, $NaNO_3$ and NH_4NO_3 extraction procedures for the study of Cd, Cu, Pb and Zn extractability in contaminated soils. *Anal. Chim. Acta* 504:217–226.

Reisenauer, H.M. 1960. Cobalt in nitrogen fixation by a legume. *Nature* 186:375–376.

Riesen, O. and U. Feller. 2005. Redistribution of nickel, cobalt, manganese, zinc and cadmium via the phloem in young and in maturing wheat. *J. Plant Nutr.* 28:421–430.

Robinson, B.H., R.R. Brooks, and B.E. Clothier. 1999a. Soil amendments affecting nickel and cobalt uptake by *Berkheya coddii*: Potential use for phytomining and phytoremediation. *Ann. Bot.* 84:689–694.

Robinson, B.H., R.R. Brooks, and M.J. Hedley. 1999b. Cobalt and nickel accumulation in *Nyssa* (tupelo) species and its significance for New Zealand agriculture. *N. Z. J. Agric. Res.* 42:235–240.

Sabbarese, C., L. Stellato, M.F. Cotrufo et al. 2002. Transfer of Cs-137 and Co-60 from irrigation water to a soil–tomato plant system. *J. Environ. Radioact.* 61:21–31.

Sasmaz, A. and M. Yaman. 2006. Distribution of chromium, nickel and cobalt in different parts of plant species and soil in mining area of Keban, Turkey. *Commun. Soil Sci. Plant Anal.* 37:1845–1857.

Sauerbeck, D.R. (ed.). 1986. Effects of agricultural practices on the physical, chemical and biological properties of soils: Part II. Use of sewage sludge and agricultural wastes. *Scientific Basis for Soil Protection in the European Community*, pp. 181–210. Essex, U.K.: Elsevier Applied Science Publishers Ltd.

Senesi, G.S., G. Baldassarre, N. Senesi, and B. Radina. 1999. Trace element inputs into soils by anthropogenic activities and implications for human health. *Chemosphere* 30:343–377.

Senesi, N. and M. Polemio. 1981. Trace element addition to soil by application of NPK fertilizers. *Fertil. Res.* 2:289–302.

Seregin, I.V. and V.B. Ivanov. 1997. Histochemical investigation of cadmium and lead distribution in plants. *Russ. J. Plant Physiol.* 44:791–796.

Shacklette, H.T. 1980. Elements in fruits and vegetables from areas of commercial production in the contermi-nous United States. U.S. Geological Survey of Professional Paper 1178. Washington, DC: United States Government Printing Office.

Shacklette, H.T., J.A. Erdman, T.F. Harms, and C.S.E. Papp. 1978. Trace elements in plant foodstuffs. In: *Toxicity of Heavy Metals in the Environment*, ed. F.W. Oehme, pp. 25–68. New York: Marcel Dekker.

Smith, I.C. and B.L. Carson. 1981. *Trace Metals in the Environment*, Vol. 6: *Cobalt*. Ann Arbor, MI: Ann Arbor Science Publication.

Soudek, P., S. Petrova, and T. Vanek. 2011. Heavy metal uptake and stress responses of hydroponically culti-vated garlic (*Allium sativum* L.). *Environ. Exp. Bot.* 74:289–295.

Trubetskaya, O.E., O.A. Trubetskoj, and C. Ciavatta. 2001. Evaluation of the transformation of organic matter to humic substances in compost by coupling sec-page. *Bioresour. Technol.* 77:51–56.

U.S. EPA. 2005. *Ecological Soil Screening Levels for Cobalt*. Washington, DC: U.S. EPA.

Vázquez, M.D., J. Barceló, Ch. Poschenrieder et al. 1992a. Localization of zinc and cadmium in *Thlaspi caer-ulescens* (Brassicaceae), a metallophyte that can accumulate both metals. *J. Plant Physiol.* 140:350–355.

Vázquez, M.D., Ch. Poschenrieder, and J. Barceló. 1992b. Ultrastructural effects and localization of low cad-mium concentrations in bean roots. *New Phytol.* 120:215–226.

Wallace, A. and A.M. Abou-Zamzam. 1989. Cobalt-zinc interactions in bush beans grown in solution culture. *Soil Sci.* 147(6):436–438.

Watmough, S.A., P.J. Dillon, and E.N. Epova. 2005. Metal partitioning and uptake in central Ontario forests. *Environ. Pollut.* 134(3):493–502.

Wójcik, M., J. Vangronsveld, J. D'Haen, and A. Tukiendorf. 2005. Cadmium tolerance in *Thlaspi caerulescens* II. Localization of cadmium in *Thlaspi caerulescens*. *Environ. Exp. Bot.* 53:163–171.

Woodard, T.L., R.J. Thomas, and B. Xing. 2003. Potential for phytoextraction of cobalt by tomato. *Commun. Soil Sci. Plant Anal.* 34:645–654.

Yang, Y.A. and X.E. Yang. 2000. Micronutrients in sustainable agriculture. In: *Micronutrients and Biohealth*, ed. Q.L. Wu, pp. 120–134. Guiyan, China: Guizhou Science & Technology Press.

Zeller, S. and U. Feller. 1998. Redistribution of cobalt and nickel in detached wheat shoots: Effects of steam-girdling and of cobalt and nickel supply. *Biol. Plant.* 41:427–434.

Zhao, F.J., C.P. Rooney, H. Zhang, and S.P. McGrath. 2006. Comparison of soil solution speciation and dif-fusive gradients in thin-films measurement as an indicator of copper bioavailability to plants. *Environ. Toxicol. Chem.* 25:733–742.

Zhou, J. and P.B. Goldsbrough. 1995. Structure, organization and expression of the metallothionein gene family in *Arabidopsis*. *Mol. Gen. Genet.* 248:318–328.

18 Lanthanides

Silvia H. Haneklaus, Ewald Schnug,
Bernd G. Lottermoser, and Zhengyi Hu

CONTENTS

18.1 INTRODUCTION

The rare earth elements (REEs) comprise the elements scandium (Sc, Z = 21) and yttrium (Y, Z = 39) and 14 lanthanides with successive atomic numbers (Z) from 57 to 71. There is a demarcation between light (lanthanum, La; cerium, Ce; praseodymium, Pr; neodymium, Nd; samarium, Sm; and europium, Eu) and heavy (gadolinium, Gd; terbium, Tb; dysprosium, Dy; holmium, Ho; erbium, Er; thulium, Tm; ytterbium, Yb; and lutetium, Lu) REEs. Promethium belongs chemically to the group of lanthanides but is an artificial radioactive element that is not found in nature. REEs are often used synonymously in terminology except for Y and Sc, which are treated separately. The reason is that Sc shows a distinctly different geochemical behavior, and Y mimics heavy REEs (Nozaki, 2001).

Lanthanides show similar physicochemical characteristics. The reason is the identical configuration of valence electrons in their outermost shell, and 4f orbitals are filled with increasing atomic number, while the ionic radius decreases (Zawisza et al., 2011). The latter phenomenon is called lanthanide contraction. This occurrence comes along with decreasing basicity and increasing solubility (Haley, 1991). The chemistry, toxicity, and behavior of lanthanides in environmental compartments are discussed regularly as a group because of their analogy. The ionic radius of lanthanides is comparable to that of Ca so that it is assumed widely that their functions and behavior are comparable (Evans, 1990).

The light REEs La, Ce, Pr, Nd, Sm, and Eu account for 90% of the total REE content in soils (Hu et al., 2003). The term "rare earth elements" suggests their scarcity in parent materials and soils, but it was rather the fact that these elements had been extracted from rare minerals that coined their name. As a result, the reserves of REEs that can be mined economically are limited compared with the current and expected demand, although the global resources of REE oxides are not critical (De Boer and Lammertsma, 2012; Erdmann and Graedel, 2011). De Boer and Lammertsma (2012) estimate global mineable rare earth oxide (REO) reserves to be 114 million tons, of which 50% are in China.

"Vorsprung durch Technik" (progress through technology) in the field of green and high technologies only became reality at the beginning of the twenty-first century by taking advantage of the physicochemical properties of REEs, such as superconductivity. Technical innovations will depend in the future highly on a sufficient and reliable supply of REEs (Erdmann and Graedel, 2011). However, the risk of supply disruption exists on a short- to medium-term basis (De Boer and Lammertsma, 2012; Erdmann and Graedel, 2011). The fields of applications where REEs are indispensable are manifold and comprise catalysts, ceramics, nuclear technology, water treatment, fertilizers, medical examination, and treatment (Erdmann and Graedel, 2011; Fricker, 2006; Huang and Tsourkas, 2013; Zhang et al., 2011). Before the start of the computer era, the demand of industry for REEs was clear, so that profitable alternative fields of applications were explored. Agricultural research in China focused on experimentation with these elements in order to increase not only crop productivity and quality but also animal performance (Redling, 2006). Usually, mixtures of lanthanides have been applied rather than individual elements. However, differences in plant uptake and translocation exist and have been assessed, for instance, by Xu et al. (2002) and El-Ramady (2008).

Lanthanides are rated regularly as beneficial for plants, but increasing stock market prices make it improbable that putative positive effects will outweigh costs for REE fertilizer products. Nevertheless, research data on the influence of graded rates of lanthanides on crop growth are required as these elements are applied to soils with common mineral fertilizer products and contaminations of industrial sludge. In the first case, low-dose effects prevail, while in the latter case, critical loads may be applied.

It is the objective of this chapter to assess the effect of lanthanides on crop productivity and quality. Hu et al. (2004, 2006) summarized comprehensive experimentation in China on the topic, thus providing accessibility to underlying research. Other recommended review articles are provided by Andrès et al. (2003), Redling (2006), and Tyler (2004). In addition, an appraisal was performed to assess possible consequences when lanthanides are discharged in significant amounts to the environment and how an increased intake of lanthanides by humans might affect health.

18.2 LANTHANIDES IN AGRICULTURAL SOILS

The total content of REEs in soils is related to the parent material and decreases in the order granite > quaternary > basalt > purple sandstone > red sandstone (Zhu and Liu, 1988). REE concentrations in these soils vary between 174 and 219 µg/g, whereas soils from loess and calcareous rock show lower REE concentrations between 137 and 174 µg/g (Hu et al., 2006). The elements La, Ce, Nd, Sm, and Eu constitute 90% of the total lanthanide content in soils with a quantity basis of 41, 74, 7, 28, and 6 mg/kg in soils (Hu et al., 2006). In general, the ratio of lanthanides decreases in the following order: Ce > La > Nd > Pr > Sm > Gd > Dy > Er > Tb > Ho > Tm > Lu (Hu et al., 2006).

In Table 18.1, mean concentrations of lanthanides in soils are listed for different countries (Hu et al., 2006). The mean REE content proved to be highest in China at 177 µg/g and lowest in Germany at 15 µg/g (Hu et al., 2006). Reimann et al. (2003) found close relationships between individual lanthanides in agricultural soils of Northern Europe. Median values for total Ce and La contents ranged countrywise between 24 and 53 mg/kg. In the same study, between <0.6% and 3.0% of the La content was available to plants.

Basically, five to eight different binding forms of lanthanides can be distinguished (Hu et al., 2006). These are water soluble, exchangeable, adsorbed, carbonate bound, loosely and tightly organic matter bound, colloidal bound to Fe/Mn oxides, and residual lanthanide configurations.

TABLE 18.1
Total Lanthanide Concentrations (µg/g) in Soils of Different Countries

Lanthanide	Australia	Poland	Switzerland	Germany	Sweden	Japan	Malaysia	United States	China
La	15.4	13.0	17.8	3.5	17.7	18.2	30.5	13.6	37.5
Ce	60.5	25.7	36.1	5.9	29.0	39.8	52.8	25.7	77.3
Pr	4.1	2.4	—	0.9	7.2	4.5	—	2.4	7.8
Nd	14.6	9.9	15.0	2.5	13.5	17.6	28.7	9.9	29.3
Sm	2.8	1.4	2.8	0.5	3.0	3.6	4.8	1.4	5.7
Eu	0.8	0.3	0.5	0.1	0.7	0.9	0.9	0.3	1.1
Gd	2.6	2.8	—	0.5	2.5	3.7	21.1	2.8	5.1
Tb	0.4	0.1	0.3	0.1	0.6	0.5	1.3	0.1	0.8
Dy	2.1	0.7	—	0.5	2.5	3.2	4.9	0.7	4.6
Ho	0.2	0.2	—	0.1	0.5	0.6	—	0.2	0.9
Er	0.8	<0.1	—	0.2	0.8	1.9	5.5	<0.1	2.6
Tm	0.1	<0.1	—	<0.1	0.3	0.3	—	<0.1	0.4
Yb	0.6	<0.1	1.4	0.2	1.4	2.0	2.9	<0.1	2.5
Lu	0.1	<0.1	—	—	—	0.2	0.9	<0.1	0.4
n	9	52	6	5	2	77	12	30	279
Sum	105	57.3	74.0	15.4	80.2	97.5	155	57.4	177

Source: Adapted from Hu, Z. et al., *Commun. Soil Sci. Plant Anal.*, 37, 1381, 2006.

In the latter fraction, about 107 mg/kg of lanthanides was fixed in lattices of minerals by isomorphic exchange, which accounts for roughly 70% of the total inventory (Zhu and Xing, 1992a).

Pedogenesis exerts the strongest influence on the lanthanide concentration in soils, and it is the physicochemical properties of soils that govern translocation, mobilization, and immobilization processes and thus their bioavailability (Shan et al., 2004). Commonly lighter lanthanides show a higher mobility in soils, are translocated faster by runoff, and are taken up preferentially by plants (Brioschi et al. 2013). The speciation of lanthanides in the studies of Zhu and Xing (1992b) showed that only 0.05%–0.17% were water soluble, 0.02%–6.5% were exchangeable, and 60%–89% were bound in residual forms that are not available for plant uptake. In general, a higher content of water-soluble lanthanides was found for elements with odd atomic number (Hu et al., 2006). Water-soluble lanthanides can be taken up directly by plant roots and soil microorganisms or pass through the soil porous system. In 34 soils from China, the average water-soluble REE content was 0.27 µg/g (Hu et al., 2006). The plant-available lanthanide concentration may be lower than the practical detection limit but may well exceed 200 µg/g (Xiong et al., 2000). On average, the plant-available lanthanide content in Chinese soils was 11.8 µg/g (n = 1790) (Xiong et al., 2000). Preferential binding of heavy REESs to Fe oxyhydroxides increases mobility of light REEs, which are consequently more easily taken up by plants and prone to runoff (Brioschi et al., 2013). Concentration and speciation of lanthanides, with heavy REESs being found most abundantly as dissolved complexes, are major factors influencing plant uptake and explain preferential uptake of free light REE ions (Brioschi et al., 2013).

Low values for soil pH (Cao et al., 2001; Diatloff et al., 1996), cation-exchange capacity, organic matter content (Jones 1997; Shan et al., 2002), and redox potential (Cao et al., 2001) all increase the solubility of REEs in soils. In the pH range from 4.1 to 6.3, desorption decreases from 90% to 29.5% (Ran and Liu, 1992). Correspondingly, translocation is negligible on alkaline soils with high adsorption capacity, whereas on acid soils with low adsorption capacity, a value of 0.5 cm per year was measured (Zhu et al., 1996). The discharge of humic substances by surface runoff will result in concomitant losses of lanthanides (Hu et al., 2006).

The plant-available lanthanide content increases with increasing clay and organic matter content of soils (Xiong et al., 2000). In the case of acetic acid-extractable lanthanides, soil pH had the strongest impact on translocation, and for HCl and HNO_3-extractable lanthanides, it was the content of Fe–Mn oxides in soils (Hu et al., 2006). The adsorption of lanthanides becomes stronger with increasing soil pH (Ran and Liu, 1992, 1993). As a result, the plant-available lanthanide content in acid soils is usually higher than in calcareous soils. Fluorides, carbonates, phosphates, and hydroxides form neutral complexes with lanthanides that show a low solubility (Weltje, 1997).

For the determination of the plant-available lanthanide pool in Chinese soils, 1 M acetic acid–1 M sodium acetate (pH 4.8) is used (Liu, 1988a). In comparison, El-Ramady (2008) extracted soil samples by acid ammonium acetate EDTA extraction (0.5 M CH_3COONH_4, 0.5 M CH_3COOH, and 0.02 M Na_2-EDTA; pH 4.65) according to Lakanen and Erviö (1971). The latter method has been applied universally and calibrated to determine the plant-available content of nutrients and heavy metals in soils.

18.2.1 Soil Microorganisms and Enzymes

The numbers of soil microorganisms and activity of soil enzymes are indicators for soil health on heavy metal–contaminated soils (Mathre and Pankhurst, 1997). Redling (2006) summarized data concerning the impact of lanthanides on soil fauna. The influence of lanthanides on soil microorganisms is dose dependent in such a way that low doses promote and high doses inhibit growth and activity (Chen and Zhao, 2007; Chu et al., 2001a; Zhu et al., 2002). Data provided by Andrès et al. (2003) showed that accumulation of lanthanides varies between 5 and 140 µg/g for La, 1–126 µg/g for Eu, 18–56 µg/g for Yb, and 0.8–54 µg/g for Gd in different species.

Enzyme activities in soils have been used as indicators for evaluating the degree of heavy metal pollution in soil (Banerjee et al. 1997; Chu et al., 2002). Soil enzymes catalyze reactions in soils that are important in cycling of nutrients such as C, N, P, and S. The dehydrogenase activity indicates the overall microbiological activity of soils as it is part of each viable cell. Other enzymes, such as acid and alkaline phosphatase, may occur in the extracellular state, too (Eichler et al., 2004). Interestingly, Brookes et al. (1984) reported that dehydrogenase activity was lower in metal-contaminated soils, but phosphatase activity remained unaffected. It is the physical and chemical properties of soil samples that are related closely to the strength of effects that lanthanides exert on acid phosphatase and dehydrogenase activities (Redling, 2006). El-Ramady (2008) determined microbial counts of actinomycetes, bacteria, and fungi and dehydrogenase and alkaline phosphatase activity on a Cambisol with a loamy sand texture after application of graded doses of REE fertilizer and single lanthanides and compared this effect to that of copper (Cu). The main results of this experiment are summarized in Table 18.2.

With respect to dose-dependent effects of La, Ce, and REE fertilizer on microbial counts, the results proved to be not consistent. Graded rates of REE fertilizer decreased the number of fungi significantly at the highest rate that corresponded with an amendment that was 100 times higher than the REE content of the soil used for experimentation (El-Ramady, 2008; Table 18.2). The highest Cu rate reduced all microbial counts effectively; fungi showed a significant reduction already when the plant-available Cu content of the soil doubled.

The dehydrogenase activity in the vegetated soil was about six to eight times higher than in the nonvegetated soil (El-Ramady, 2008; Table 18.2). Although the highest Ce rate increased the dehydrogenase activity, differences to other treatments were only small. In contrast, Cu reduced the dehydrogenase activity explicitly. Regarding the alkaline phosphatase, activities were consistently higher in the nonvegetated soils except at the highest Cu rate. La and REE fertilizer application gave rise to detrimental effects, which were not observed in the nonvegetated soil, while for Ce inverse results were found (Table 18.2). Based on the results of El-Ramady (2008) with maize and oilseed rape in experiments that have been duplicated, it seems difficult to use enzymatic activities as an indicator for the size of the lanthanide pool in soils.

TABLE 18.2

Influence of Graded Rates of Single Lanthanides, REE Fertilizer, and Copper on Soil Microbial Counts (CFU) and Enzymatic Activities in Vegetated (Oilseed Rape) and Nonvegetated Soil

	Vegetated			Vegetated	Nonvegetated	Vegetated	Nonvegetated
Rate	Heterotrophic Bacteria	Actinomycetes	Fungi	Dehydrogenase Activity		Alkaline Phosphatase Activity	
(µg/g)	(CFU[a])			(TPF)		(p-NP)	
La							
0[b]	5.3×10^6 a	3.7×10^6 b	1.9×10^6 a	19.5 a	3.1 a	133.8 b	185.3 a
1.0	5.1×10^6 a	1.8×10^6 a	1.6×10^6 a	19.2 a	3.4 a	52.4 a	189.8 a
10	5.1×10^7 b	3.0×10^6 ab	1.3×10^6 a	19.5 a	3.8 a	101.1 ab	204.7 a
50	4.9×10^6 a	3.3×10^6 b	1.2×10^6 a	21.7 a	2.9 a	116.2 b	210.9 a
100	4.8×10^6 a	3.6×10^6 b	4.7×10^6 a	19.1 a	2.2 a	91.5 ab	197.6 a
Ce							
0	5.3×10^6 a	3.7×10^6 a	1.9×10^6 a	19.5 ab	3.1 a	133.8 a	185.3 ab
0.8	4.7×10^6 a	3.2×10^6 a	1.1×10^6 a	20.0 ab	3.2 a	92.1 a	152.4 ab
8.0	4.8×10^6 a	2.9×10^6 a	1.2×10^6 a	17.9 a	3.4 a	127.3 a	141.5 a
40	6.8×10^8 b	3.7×10^6 a	1.3×10^6 a	21.0 ab	2.9 a	105.7 a	192.6 b
80	1.7×10^8 a	3.4×10^6 a	1.7×10^6 a	23.2 b	3.0 a	108.8 a	201.3 b
REEs							
0	5.3×10^6 a	3.7×10^6 a	1.9×10^6 b	19.5 a	3.1 a	133.8 b	185.3 a
2.7	5.3×10^6 a	2.9×10^6 a	1.5×10^6 ab	21.8 a	3.4 a	92.3 ab	136.5 a
27	2.8×10^7 b	3.5×10^6 a	1.7×10^6 ab	21.0 a	3.8 a	99.9 ab	126.4 a
135	3.4×10^7 b	3.5×10^6 a	1.6×10^6 ab	22.0 a	2.9 a	130.6 b	184.0 a
270	1.1×10^7 a	2.6×10^6 a	6.9×10^5 a	14.2 a	2.2 a	86.2 a	131.9 a
Cu							
0	5.3×10^6 c	3.7×10^6 b	1.9×10^6 b	19.5 c	3.1 bc	133.8 a	185.3 c
4.3	5.3×10^6 c	3.6×10^6 b	8.6×10^5 a	18.1 c	3.8 c	178.3 a	152.4 c
43	4.1×10^6 bc	2.9×10^6 b	3.9×10^5 a	9.6 b	2.6 bc	61.4 a	95.7 b
215	2.9×10^6 ab	2.4×10^6 b	2.6×10^5 a	1.5 a	0.8 ab	77.1 a	71.5 ab
430	1.3×10^6 a	1.9×10^5 a	6.9×10^5 a	0.4 a	0.3 a	124.3 a	25.8 a

Source: Extracted from El-Ramady, H., A contribution on the bio-actions of rare earth elements in the soil/plant environment, PhD thesis, Technical University Braunschweig, Braunschweig, Germany, 2008, p. 173.

Note: Values for graded rates of each element followed by the same letter denote no significant difference from the control by Tukey's test at the 0.05 level.

[a] CFU, colony-forming unit/g dry soil.

[b] Concentration in soil (control): 1.0 µg/g La, 0.8 µg/g Ce, and 2.7 µg/g total lanthanides.

18.3 RATIONALE FOR BENEFICIAL POTENCY

An element needs to comply with the criteria defined by Arnon (1950) to be declared as being essential for plant growth. In comparison, minerals are rated as beneficial for plants if they promote growth and yield, replace the function of essential nutrients in plant metabolism, or strengthen the natural resistance of plants against abiotic and biotic stress. Prominent examples are, for instance, Na and Si. Sodium may replace K under conditions of an insufficient K supply. Yield of salt-tolerant plants such as sugar beet is increased significantly by Na on lighter soils under drought stress, and the sufficiency range for Na varies between 2–7 mg Na/g foliar dry matter Na for root yield and 4–6.5 mg/g Na for sugar yield if K is sufficiently (36–60 mg/g foliar dry matter) supplied

(Haneklaus et al., 1998). Silicon was shown to promote growth of crop plants and improves resistance against salt and biotic stress (Datnoff et al., 2001). It is difficult to come to an ultimate decision whether lanthanides are beneficial for plant growth as contradictory findings exist. On all accounts, hormesis seems to play a key role for inducing positive effects.

18.3.1 SUPPORTING EVIDENCE FOR HORMESIS INDUCED BY ENHANCED UPTAKE OF LANTHANIDES

Hormesis describes the stimulatory effect of low doses of toxins, including radiation, xenobiotics, and heavy metals, which expose significant adverse effects at higher rates (Stebbing, 1982). Such protective feedback mechanisms imply the stimulation of metabolic detoxification and repair processes and result in beneficial effects from the cell to the species level (Macklis and Bereford, 1991). In contrast, higher doses of heavy metals will prompt inhibitory effects that culminate finally in yield losses.

Sterner and Elser (2002) provide a comprehensive overview of homeostasis in the context of stoichiometry in the animate and inanimate world. Homeostasis assumes that complex physiological processes maintain a steady state in the plant unless any severe imbalance of ions taken up by the plant will impair, for instance, the functionality of chloroplasts, and reduce biomass production (Solymosi and Bertrand, 2010). Homeostasis is maintained efficiently by balanced nutrient ratios in plant tissues to foster crop productivity, quality, and plant health, whereas excessive loads of minerals will reduce yield. An applicable example is provided by Haneklaus et al. (2006), who determined threshold values for the S supply of different crops including symptomatological values when macroscopic symptoms of deficiency become visible and upper critical values for total S in shoots of cereals, oilseed rape, and sugar beet reduce yield by 10%. Uptake and assimilation of S and N by plants are strongly interrelated and dependent upon each other. Up to 70% of the total S is present in the reduced form in cysteine and methionine so that protein biosynthesis is disturbed if S is deficient. In comparison, cross talk between Ca and S metabolic pathways under excessive S stress could trigger dysfunction of homeostasis and thus unfold detrimental effects on crop growth (Haneklaus et al., 2006). Interactions between lanthanides, in particular La and Ce, in plant metabolism might contribute to positive or negative growth effects (see Sections 18.4.1 and 18.4.2).

Hormesis, together with the principle of homeostasis, may give some deeper insight into the relationship between heavy metal stress and biomass production. Scientific proof for hormetic effects will be difficult to establish at a plant level as it is more or less impossible to replicate any treatment so that it will fully comply with *ceteris paribus* conditions. Yet, upper boundary lines that describe the "pure effect of a nutrient" on crop yield under *ceteris paribus* conditions may substantiate the phenomenon (Haneklaus et al., 2007a; Schnug and Haneklaus, 2008). The relationship between an element in plant tissue and yield reflects physiological patterns in the internal mineral utilization, which is specific for different nutrients, minerals, and plant species. In comparison, the relationship between fertilizer dose and elemental concentration in plant tissue is much less dependent on physiological factors but is strongly influenced by factors affecting the physical mobility and losses from soils.

In general, boundary lines show a steep increase at the beginning, which reflects the response of the photosynthetic system to nutrient deficiency. Hereafter, the boundary line continues over a long range asymptotically toward the value above which no further yield increase is to be expected from increasing nutrient concentrations. In the case of major plant nutrients such as S, this part of the boundary line reflects most likely the proportion of S that is bound, for example, in the protein fraction of the cereal grain. In plants with a lower S demand such as sugar beet, the boundary line reaches the no effect value much faster after a steep increase as sugar beet roots take up only small amounts of S (Haneklaus et al., 2007a). A comparison of the course of upper boundary lines for essential major and minor plant nutrients that are provided by Schnug and Haneklaus (2008) for cereals, oilseed rape, and sugar beet reveals constitutive differences between macro- and micronutrients. These can be summarized as follows: in general, ranges for severe deficiency, optimum, and

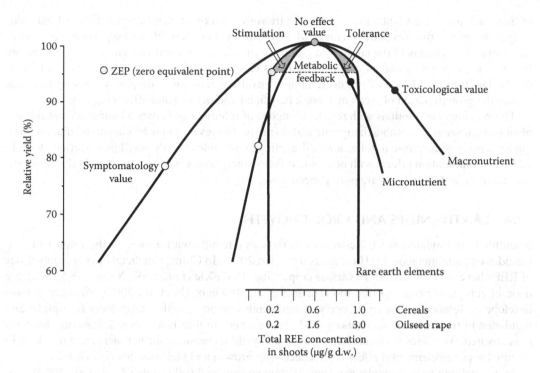

FIGURE 18.1 Common, abstract boundary line progression for essential macro- and micronutrients and predicted progression of boundary lines for REEs. (Based on data provided by El-Ramady, 2008; Haneklaus et al., 1998; Haneklaus et al., 2006, 2007a; Lukey, 1990; Schnug and Haneklaus, 2008; Hu et al., 2004, 2006.)

noxious supply are significantly smaller for minor nutrients. Hidden deficiency, where no symptoms are macroscopically visible yet yield response can be verified, is not pronounced clearly.

In Figure 18.1, upper boundary lines are shown schematically for macro- and micronutrients, and important threshold values are denoted next to the predicted growth curve for lanthanides, and the putative range of concentration where a hormetic effect can be expected has been deduced.

Busby and Schnug (2007) observed that living systems generally employ few elements of atomic numbers above calcium (Z = 20) for accomplishing their life cycle. The highest atomic number elements existing in any quantity in plants are Zn (Z = 30) and Mo (Z = 42). Plotting the logarithm of the minimum concentration of each essential macro- and micronutrient required for the optimum growth (Schnug and Haneklaus, 1992) of plants (Y) against the logarithm of the fourth power of Z reveals that the nutrient demand decreases linearly with increasing Z to values that are slightly higher than the fourth power of the atomic number (Busby and Schnug, 2007). Following the approach of Busby and Schnug (2007), diagnostic reference values of 0.220 μg/g for light REE and 0.186 μg/g for heavy REE concentrations comply with optimum plant growth, well aware that these elements are rated as beneficial and not essential for living organisms. Lukey defined minimum concentrations that trigger biostimulation of plant metabolism as the zero equivalent point (ZEP). Following this basic rationale, hormesis will be conceivable only if the REE contents are higher than these minimum concentrations in the plant tissue. The metadata analysis of REE concentrations in plant shoots and biomass production delivered the following results: the bioactivity range of REEs where stimulating, positive effects can be expected in the lower concentrations and tolerance toward higher concentrations in the upper spectrum differs between monocots and dicots. Indicative optimum REE concentrations vary between 0.2 and 1 μg/g for cereals and 0.2 and 3.0 μg/g REEs for oilseed rape (Figure 18.1).

In principle, the progression of the cause–effect relationship between lanthanide dose and plant growth is suggested to proceed as follows: plants show no traceable physiological response to uptake of marginal amounts of lanthanides, while the transgression of an element-specific threshold value triggers hormesis; the concentration range where hormesis is conceivable will supposedly be similar for individual elements of the lanthanide group, as will be detrimental concentrations, as there are limits to what combinations of chemical elements can function in a plant cell (Sterner and Elser, 2002). The typical dilution effect, decreasing mineral concentrations despite increasing biomass, during the growth period of crops may mask beneficial and detrimental effects equally.

The previous assumptions with respect to modes of interactions between lanthanide uptake and plant growth seem admissible though difficult to prove. However, it may be concluded that the margin for a positive response to foliar and soil-applied lanthanides is only small (see Section 18.4). In addition, combination effects with other stress factors may well contribute to undesired physiological reactions, although these are purely speculative.

18.4   LANTHANIDES AND CROP GROWTH

Scientific experimentation has delivered contradictory results with a view to the impact of lanthanides on plant metabolism (Hu et al., 2004; Tyler, 2004). In China, beneficial effects of low doses of REEs have been published for various crops (Guo, 1987; Xie et al., 2002; Xiong, 1995), whereas toxic effects have been reported too (Diatloff et al., 1995a,b; Hu et al., 2002). Plants may have developed tolerance to low-level supply of lanthanides as no specific transporters for uptake and regulation of translocation in plants evolved. Above a certain threshold value, defense mechanisms are activated, which result in enhanced growth and yield increase. A further increasing uptake will eventually yield detrimental effects and reduce crop growth and yield (see Section 18.3).

In general, lanthanide uptake is related to plant species and soil supply (Zeng et al., 2006). Yet, knowledge about plant uptake mechanisms of lanthanides is restricted. Possibly, lanthanides are reduced first to a divalent state (Tian et al., 2005), although Ce(III) was shown to be taken up by horseradish (Guo et al., 2007).

The total content of lanthanides in plants is highly variable and may range from <0.2 µg/g in corn (*Zea mays* L.) leaves up to >200 µg/g in fern fronds (El-Ramady, 2008). In the study by Ozaki et al. (2000), 96 fern species were analyzed; La and Ce concentrations in the genera *Dryopteris*, *Asplenium*, *Adiantum*, and *Dicranopteris* were as high as 10–40 µg/g for La and 3–30 µg/g dry weight for Ce, whereas in all other ferns, these varied between 0.003–2.7 µg/g La and 0.076–3.6 µg/g Ce. The concentration of lanthanides is highest in roots as the Casparian strip prevents uncontrolled translocation into shoots. The lanthanide content in stems and leaves of vascular plants is higher than that of fruits and seeds, which regularly show the lowest concentration (Xu et al., 2003). About 20% of mobile lanthanides in soils are taken up by plants. After soil-applied fertilization, this value increases to 55%–60% (Hu et al., 2004). The highest concentrations of lanthanides can be found regularly in roots if this is the main pathway into the plant; only if lanthanides are foliarly applied will leaves show the highest concentrations (Hu et al., 2004). Only 1% of Nd in roots was translocated to leaves and stems, whereas 3% was taken up directly from the leaf surface (Hu et al., 2004). Inhibitory effects of La and Nd on root development were reported for oats (*Avena sativa* L.) and barley (*Hordeum vulgare* L.), whereby this effect seems to be dose dependent (Hu et al., 2004).

Synergistic and antagonistic effects were reported for uptake, translocation, and function among individual lanthanides and between lanthanides and essential major plant nutrients (Hu et al., 2004). Beneficial effects of lanthanides have been attributed to an improved bioavailability of calcium and manganese in soils (Chang, 1991), the stimulation of chlorophyll synthesis (Guo, 1988), enhancement of seedling development (Chang, 1991; Wu et al., 1983), and promotion of root and shoot growth in wheat (*Triticum aestivum* L.), cucumber (*Cucumis sativus* L.), soybean (*Glycine max* Merr.), and corn (Wu et al., 1983, 1984). However, Wheeler and Power (1995) showed that La exhibited plant toxicity, which proved to be stronger than that of Mn, Zn, and Fe, but lower than that of Cu. Negative impacts of lanthanides on plant metabolism may result from interference with enzyme functions, replacement of essential metals in pigments, and synthesis of reactive oxygen species (Babula et al., 2008).

In plants, lanthanides were shown to bind not only to proteins, nucleic acids, and amino acids but also to pigments and cellulose (Hu et al., 2004). So lanthanides have been suggested to compete with Ca for binding sites in proteins and replace the nutrient, thus stabilizing cell membranes (Brown et al., 1990; Dong et al., 1993; Hu and Ye, 1996; Mikkelson, 1976; Qiao et al., 1993). REEs changed the membrane fluidity so that their leakiness was restrained (Redling, 2006; Shen and Yan, 2002; Tian, 1990). This effect is dose dependent as high concentrations of lanthanides may increase cell permeability and consequently destroy membrane stability (Chang, 1991). Yet another implication might be that higher membrane stability is effective against oxidative stress (Wang et al., 2003). Changes in membrane physiological characteristics (Redling, 2006) also could result from binding to hormones and interactions with their receptors (Brown et al., 1990; Enyeart et al., 2002).

An increased net photosynthetic rate of 11%–31% was attributed to the replacement of Mg by La and Ce if the Mg supply was insufficient (Hong et al., 2002; Xiong et al., 2000). REE fertilizer and Ce and Nd applications were shown to increase the chlorophyll content by 5%–30% and 9%–40%, respectively (Hu et al., 2004).

With respect to putative hormetic effects of lanthanides (see Section 18.3), the findings of Hong et al. (2000) and Liu et al. (2004) seem relevant as they found increased activities of superoxide dismutase, catalase, and peroxidase after lanthanide application. Superoxide dismutase activity increased, for instance, up to 3.7- and 4.3-fold after La and Nd application, respectively (Liu et al., 2004). These enzymes are involved in defense reactions against oxidative stress (Foyer et al., 1994). For a detailed summary of physiological effects of lanthanides on plant metabolism, see Hu et al. (2004).

Lanthanides were reported to increase proline content and as a consequence to increase the share of bound water (Hu et al., 2004; Yu and Liu, 1992). Lanthanum was shown to increase the resistance of barley against drought and crop yield by 18% in a field experiment if the water content was not higher than 50% of the field capacity (Meehan et al., 2001).

18.4.1 Lanthanides and Crop Yield

Experimentation in China showed that the highest yield response can be expected on less fertile soils with a plant-available lanthanide content of <10 mg/kg, whereas it is not likely that there will be a positive response on fertile soils with a lanthanide content of >20 mg/kg (Liu, 1988b).

Increase of biomass production of up to 50% has been reported, although on average the promotion varied between 5% and 15% (Brown et al., 1990; Hu et al., 2004; Xiong, 1995). In Table 18.3, positive yield effects of lanthanides are summarized for different crops. Even higher yield increases

TABLE 18.3
Monitored Yield Increases after Application of Lanthanides to Different Crops in Field Experimentation

| Crop | Field Experiments | | |
	Total No.	No. of Responsive Sites	Yield Increase (%)
Corn	No data	No data	8.5–103
Cotton	19	15	5–12
Potato	No data	43	13.8
Rice	65	61	7
Soybean	No data	40	8.1–8.9
Sugar beet	104	124	7.1 (tuber)
			5 (sugar)
Wheat	39	34	7.5

Source: Extracted from Hu, Z. et al., *J. Plant Nutr.*, 27, 183, 2004.

than those summarized in Table 18.3 were found for corn, cotton (*Gossypium hirsutum* L.), sugar beet (*Beta vulgaris* L.), and wheat with 103%, 12%, 24%, and 17%, respectively, being seen (Hu et al., 2004). Other, extraordinarily high, yield increases were determined for oilseed rape (*Brassica napus* L.), cabbage (*Brassica oleracea* var. *capitata* L.), and sugarcane (*Saccharum officinarum* L.) with an increase of 48%, 10%–20%, and 10%–15%, respectively (Hu et al., 2004). Hong et al. (1996) suggested that it is possible to bring about such positive yield effect each year as the authors found yield increases of 4%–10% consistently over a period of 10 years if 600 g/ha REEs were applied. While the majority of field experiments in China showed positive yield responses, field trials in other countries reported no, or negative, effects for barley and corn (Diatloff et al., 1999; Meehan et al., 2001).

Xu et al. (2002) determined in field experiments that a soil-applied dosage of <10 kg/ha REEs did not result in REE accumulation of maize grains. Uptake of La was two to three times higher than that of Ce.

A regular shortcoming of field experimentation has been that it operated with single doses rather than graded rates (Hu et al., 2004). In addition, commercial REE fertilizer products have been employed, so that it is not possible to attribute effects to an individual element. In the greenhouse experiments of El-Ramady (2008), graded doses of individual lanthanides or REE mixtures have been applied to oilseed rape, whereby rates were based on manifolds of the plant-available lanthanide content in the soil; effects have been compared with that of Cu (see also Table 18.2, Section 18.2). The results from these experiments are summarized in Table 18.4. The treatments had no significant effect on plant height, but in both years of experimentation, El-Ramady (2008) found the shortest plants at the highest La and REE fertilizer rates. Soil-applied lanthanide caused yield effects in the experiments of El-Ramady (2008), however, not consistently in two separate experiments and in relation to doses and form of minerals (Table 18.4). Vaguely, a tendency toward higher yields seems to show at rates of La, Ce, and REE fertilizers that are 10 times higher than the plant-available concentration of these elements in the untreated soil. Cerium applications yielded a stronger positive response than La in oilseed rape or corn, whereas a REE fertilizer rate of 135 µg/g caused distinct yield reductions in corn in two experiments (El-Ramady, 2008; Table 18.5). El-Ramady (2008) suggested that a 100-fold increase of the plant-available lanthanide content in soils boosted the translocation into shoots overproportionally. In separate experiments with corn, the concentration factors for lanthanides ($CF_{soil/plant} = C_{Plant}/C_{Soil}$, with C_{Plant} = total concentration of lanthanides in plants; C_{Soil} = plant-available background concentration + REE rate) decreased with increasing supply up to 50-fold of the natural supply. Hereafter, the concentration factor increased again to a level comparable to that of a 10-fold increase.

Expectedly, roots showed higher La, Ce, Pr, and Nd concentrations than shoots (Table 18.4). Striking is the fact that the La and Ce concentrations in roots are more than three times higher in oilseed rape than in corn. Differences in the shoot concentrations between both crops were even more pronounced for Ca and Ce and Pr and Nd. Obviously, differences between monocots and dicots in accumulation of Ca and Mg also apply for lanthanides.

Treatment effects expressed by significantly higher concentrations of these elements in shoots existed only for La, Ce, and Nd at the highest REE fertilizer application rate in two independent experiments (El-Ramady, 2008; Table 18.4). The concentration factors for roots and shoots of corn and oilseed rape decreased in the order Ce > La ≥ Nd ≥ Pr.

El-Ramady (2008) determined close linear relationships for corn and oilseed rape between Ca and La, Ce, and Pr in shoots. In addition La, Ce, and Pr concentrations in roots and shoots showed positive linear regressions. In general, correlations were distinctly tighter for corn than for oilseed rape. The studies of El-Ramady (2008) reveal that crop and element-specific differences in root uptake of REEs exist and that translocation of individual REEs within the plant seems to be controlled by different transporter systems in oilseed rape and corn. Another important result of the same studies is that even 100 times higher concentrations of La, Ce, and REEs in soils cause no

TABLE 18.4

Influence of Graded Rates of Single Lanthanides and REE Fertilizer on Uptake of Lanthanides and Biomass Production of Oilseed Rape 60 Days after Sowing

Rate (µg/g)	Biomass Production (g/Pot)			Mean Accumulation by Shoots (µg/Pot)			
	Roots	Shoots	Total	La	Ce	Pr	Nd
Control[b]	1.0 a	9.0 a	10.0 a	0.5 a	0.6 a	<LLD[a]	0.3 a
La							
1.0	1.9 a	16.9 a	18.9 a	1.1 a	1.1 a	<LLD	0.6 a
10	2.1 a	15.5 a	17.6 a	2.6 a	1.2 a	<LLD	0.5 a
50	2.4 a	18.2 a	20.5 a	10.5 ab	1.0 a	<LLD	0.4 a
100	1.0 a	11.6 a	12.7 a	19.1 b	0.9 a	<LLD	0.4 a
Ce							
0.8	2.5 a	17.1 ab	19.5 a	1.4 a	1.6 a	<LLD	0.6 a
8.0	1.3 a	32.4 b	33.8 a	2.2 a	4.4 a	<LLD	1.5 b
40	1.6 a	13.4 ab	15.0 a	0.9 a	4.2 a	<LLD	0.5 a
80	2.0 a	18.3 ab	20.4 a	1.2 a	19.8 b	<LLD	0.5 a
REEs							
2.7	1.8 ab	18.5 b	20.3 b	1.4 a	2.0 a	<LLD	0.6 a
27	2.8 b	18.2 b	20.9 b	3.2 a	3.7 a	0.4 a	1.1 a
135	2.4 ab	17.3 b	19.7 b	11.2 b	12.4 b	1.2 a	3.9 b
270	0.4 a	3.6 a	4.0 a	8.0 a	7.2 ab	0.7 a	2.4 ab

	Mean Concentration in Shoots (µg/g)				Mean Concentration in Roots (µg/g)			
	La	Ce	Pr	Nd	La	Ce	Pr	Nd
Control	0.057 a	0.067 a	<LLD	0.030 a	1.64 a	3.12 a	0.306 a	1.067 a
La								
1.0	0.067 a	0.063 a	<LLD	0.033 a	2.10 a	3.32 a	0.323 a	1.129 a
10	0.158 a	0.077 a	<LLD	0.034 a	4.94 a	2.87 a	0.282 a	0.983 a
50	0.779 ab	0.079 a	<LLD	0.034 a	16.84 b	2.83 a	0.271 a	0.952 a
100	1.589 a	0.076 a	<LLD	0.031 a	54.60 c	3.74 a	0.359 a	1.253 a
Ce								
0.8	0.084 a	0.098 a	<LLD	0.036 a	2.49 a	3.79 a	0.349 a	1.205 a
8.0	0.062 a	0.120 a	<LLD	0.035 a	1.99 a	5.59 a	0.374 a	1.315 a
40	0.071 a	0.333 b	<LLD	0.035 a	1.56 a	9.87 a	0.291 a	1.049 a
80	0.067 a	1.112 b	<LLD	0.028 a	2.01 a	38.69 b	0.365 a	1.294 a
REEs								
2.7	0.072 a	0.107 a	<LLD	0.047 a	2.83 a	5.23 a	0.514 a	1.764 a
27	0.181 a	0.209 a	0.025	0.032 a	4.69 a	7.95 a	0.787 a	2.584 a
135	0.608 a	0.674 ab	0.067	0.060 ab	19.95 a	30.25 a	3.019 a	9.761 a
270	1.586 b	1.486 b	0.146	0.211 b	110.66 b	136.54 b	12.927 b	41.457 b

Source: Extracted from El-Ramady, H., A contribution on the bio-actions of rare earth elements in the soil/plant environment, PhD thesis, Technical University Braunschweig, Braunschweig, Germany, 2008, p. 173.

Note: Values for graded rates of each element followed by the same letter denote no significant difference from the control by Tukey's test at the 0.05 level.

[a] <LLD – below lower limit of detection.

[b] Concentration in soil (control): 1.0 µg/g La, 0.8 µg/g Ce, and 2.7 µg/g total lanthanides.

TABLE 18.5
Monitored Increases of Yield and Quality Parameters in Different Crop Plants after Application of Lanthanides

Crop	Yield Increase (%)	Quality Parameter	Increase/Improvement
Maize	7–14	TGW[a]	+0.2 to 0.35 g
Oilseed rape	14–24	Oil content	+2%
Potato	10–14	Starch content	+1%
Alfalfa	5–33	Crude protein content	+3% to 10%
Apple tree	10–22	Sugar content	+0.5% to 10%
		Ascorbic acid content	+20%
Chinese gooseberry[b]	10–25	Sugar content	+1.3% to 2.9%
		Ascorbic acid content	+4.6%
Banana	8–14	Sugar content	+3% to 4%
		Ascorbic acid content	+4.6%
Flax		Fiber content	+10% to 15%
Stems	8–12		
Seeds	10–10		
Rubber tree	6–20	Rubber	First grade quality
Tobacco	7–16	Leaves	+10% higher grades
Cotton	5–12	Boll weight	+0.1 g
		Fiber length	+0.1% to 0.4%

Source: Adapted from Xiong, B.K., Application of rare earths in Chinese agriculture and their perspective of development, in *Proceedings of the Rare Earths in Agriculture Seminar*, Australian Academy of Technological Sciences and Engineering, Victoria, Australia, 1995, pp. 5–9.

[a] TGW, thousand grain weight.
[b] Rate/plant.

acute stress in plants as the α-tocopherol and chlorophyll content in shoots of corn and oilseed rape showed no significant changes, although levels were elevated (El-Ramady, 2008).

18.4.2 LANTHANIDES AND CROP QUALITY

Lanthanides were reported to improve the nutritional quality of crop plants, particularly under stress conditions, but such beneficial effects may be restricted to certain growth stages (von Tucher et al., 2001). Lanthanide application improved crop quality in terms of sugar, vitamin C, starch, ascorbic acid, fiber, fat, and protein content in different species (Table 18.5; Brown et al., 1990; Chen and Zheng, 1990; Wan et al., 1998; Xiong, 1995). Tested doses deviated distinctly from the mean annual lanthanide application rate of <230 g/ha REEs in China (Xu et al., 2002). Wang (1998) reported that 750–1000 g/ha REEs were applied on more than three million hectares of agricultural land.

18.4.3 LANTHANIDE APPLICATIONS BY FERTILIZATION

Due to an increasing global demand, the annual REE production has been expanding from several thousand tons in the 1950s to almost 100,000 t in 2000 (Haxel et al., 2004). In the late 1990s, the annual consumption of REOs in agriculture was about 1100 t (Yan, 1999). In addition, significant amounts of lanthanides have been applied with mineral phosphate fertilizers. Meanwhile, the interest in purchasing REEs for fertilization diminished drastically because of high market prices. So, the price for lanthanides increased from 2002 to 2011 about 60-fold for La, Ce, Nd, and Pr

(Anonymous, 2011). Actually, the prices for REEs as nitrates and La and Ce as chlorides (99.99% purity) vary between 2450 and 3310 €/t (exchange rate RMB to € of 0.1225 on November 11, 2013; Hu, Z., unpublished data). Thus, soil-applied applications of REEs will be profitable only if a rate of <15–21 kg/ha La/CeCl$_3$ will increase yield of wheat from 5 t/ha by >5% (price of 203.5 €/t wheat on November 11, 2013).

Soil-amended lanthanides were shown to be present in the soils in water-soluble, exchangeable, carbonate-bound, Fe–Mn-oxide-bound, organic matter and sulfide-bound form (Wen et al., 2001). Over time, the share of soluble and exchangeable forms decreases, while the percentage of lanthanides that binds to organic matter increases (Liu et al., 1999).

Basically, four types of REE fertilizers are commercially available in China: Changel-Yizhisu with lanthanides bound as nitrates; Nongle with lanthanides bound as chlorides; MAR (rare earth complex of mixed amino acids) with La, Ce, Pr, and Nd and 17 amino acids (Pang et al., 2002); and REE ammonium bicarbonate (Hu et al., 2004). All products contain lanthanides in water-soluble, instantly plant-available form. For the efficient application of lanthanides, wet and dry coating of seeds and foliar sprays has been employed (El-Ramady, 2008). Light REEs and heavy REEs were translocated at similar rates if foliar applied to pepper (*Capsicum annuum* L.), whereas the rate was one to two orders of magnitude higher for light REEs than for heavy REEs in mustard (*Brassica juncea* L.) (Cao et al., 2000).

Besides a direct application, lanthanides enter the soil/plant system by P fertilization, either with mineral fertilizer products or in the form of phosphogypsum and sewage sludges (Hu et al., 2006). About 30–170 g/ha REEs are applied with common rates of P fertilization (Todorovsky et al., 1997). Similar amounts of about 160 g/ha REEs are applied with phosphogypsum if used as a Ca source at a rate of 1 t/ha for peanut (*Arachis hypogaea* L.) (Traxler, 1996). The lanthanide content of phosphogypsum decreased in the following order (concentration in mg/kg in brackets): La (48) > Nd (32) > Ce (20) > Gd, Dy (9) > Sm, Er (6) > Yb (5) >Ho, Eu (2) > Tb, Lu (1) (Elbaz-Pulichet and Dupuy, 1999).

The total REE concentration varied between 16 and 1600 mg/kg in compound fertilizers for basal and top dressings, whereas that in straight and compound fertilizers for foliar application and in irrigation water, it was below the detection limit (Otero et al., 2005). The same authors used the REE pattern (logarithmic scale of single REE concentration in sample divided by single REE concentration in reference material), and the La/Lu ratio normalized to post-Archean Australian average shale successfully to distinguish between compound fertilizers derived from carbonatite and phosphorite. The REE pattern has also been used to trace illegal dumping of manure (Hu et al., 1998).

With a view to the accumulation of lanthanides in soils, a simple calculation reveals that annual foliar fertilization or seed coating with about 60 g/ha lanthanides will statistically increase the plant-available lanthanide content in soil of 10 mg/kg by 1 mg/kg in approximately 50 years. As removal of lanthanides with harvest products is only marginal, runoff and leaching will account for losses from agroecosystems so that the impact of lanthanides on biological systems in water bodies requires special attention (see Section 18.5). Xu et al. (2002) determined in field trials that a soil-applied dosage of <10 kg/ha REEs did not result in REE accumulation in maize grains. In these experiments, REE uptake decreased in the order roots ≫ leaves > stems > grain. Highest concentrations of lanthanides can be found regularly in roots if this is the main site of uptake (Hu et al., 2004). Uptake of La was two to three times higher than that of Ce. Only if lanthanides were foliarly applied did leaves show the highest concentrations (Hu et al., 2004).

When it comes to recommended doses of lanthanides on production fields, it will be impossible to provide reliable and safe advice as the influence on yield and quality cannot be foreseen. A simple comparison of recommended and tested rates of micronutrient and lanthanide fertilization rates may underline extreme practices in the case of lanthanides (Table 18.6). El-Ramady (2008) tested in greenhouse experiments graded doses of REE fertilizers and individual lanthanides. His results showed that low doses exhibited positive growth effects, whereas noxious rates reduced biomass up

TABLE 18.6
Recommended and Tested Rates of Foliar and Soil-Applied Micronutrient and Lanthanide Fertilizers and Dose-Dependent Yield Effects

| Element | No Effect Value in Shoots (µg/g) | Fertilization (kg/ha) | | Nutrient Demand (g/ha) |
		Foliar	Soil[a]	
Mn	40[a]	0.2/1[b]	—[b]	300–500[c]
Zn	30[a]	0.1/0.4[b]	10–20[a]	200–400[c]
Cu	5[a]	0.5[a]	5–10[a]	30–90[d]

	Peak Hormesis (µg/g)			Yield Increase[e] (%)	Yield Decrease (%)
Lanthanides	0.6–1.6[f]	0.75–1.0[g] (>0.1% toxic)[e]	2–100[h]	5–15[i]	47–52[j]

[a] Finck (1976).
[b] Not recommended as deficiency being related to immobility of Mn in soils, if applied in chelate/sulfate form (Schnug, 1991).
[c] For 6–10 t/ha wheat yield (Pissarek and Schnug, 1982).
[d] Kratz et al. (2009).
[e] If insufficiently supplied.
[f] See Figure 18.1.
[g] Wang (1998).
[h] Xu et al. (2002); Hu et al. (2004).
[i] Diatloff et al. (1995a–c).
[j] El-Ramady (2008): if soil-applied rates were 100 times that of plant-available content in soils.

to 47%. The latter corresponded to a plant-available lanthanide content that was 100 times higher than the natural supply.

With a view to the applicability of experimental data to the field level, two variables need to be considered. These are scale of experimentation (see Section 18.6) and experimental design, as both of them tend to overestimate the validity of treatment effects. Quantum and type (natural and spiked, fresh or aged, soils) of lanthanide contamination will govern the physicochemical and biological behavior of lanthanides in soils and their carryover into soil fauna and crop plants. This factor has been addressed previously in studies with other environmentally relevant heavy metals (Ali et al., 2004; Amorim et al., 2005; De Brouwere et al., 2007). In the case of lanthanides, hardly any data exist that reflect changes over time in the bioavailable content in soils. Commonly, experimentation assumes steady state rather than aging and sequestration. Sijm et al. (2000) proposed the principle of soil availability ratio (SARA) to determine the bioavailable fraction of organic compounds and metal substances. SARA uses the ratio of the concentration in the organism (for instance, soil macrofauna and plants) divided by the concentration in soil of the aged chemical to that of the spiked soil. This approach seems applicable for inorganic materials such as lanthanides, too.

In Tables 18.7 and 18.8, lanthanide concentrations in different waste materials are summarized. Comparing the studies of Kawasaki et al. (1998) and Zhang et al. (2001), it is striking that sewage sludge from China displayed significantly higher concentrations of lanthanides than that in Japan. La, Ce, Pr, and Nd concentrations were about 2–3 times higher, while that of Gd was 5 and that of Sm even 11 times higher.

It can be expected that contamination of manures will increase proportionally if lanthanides are used as feed supplements (Redling, 2006).

TABLE 18.7
Mean REE Concentration in Compost, Sewage Sludge, Food Industry Sludge, and Chemical Sludge in Japan

| Element | Mean REE Concentration (µg/g) | | | |
	Compost[a]	Sewage Sludge[b]	Food Industry Sludge[c]	Chemical Industry Sludge[d]
La	2.2	6.7	0.9	2.5
Ce	3.8	14.1	1.8	2.7
Pr	0.5	1.5	0.2	0.5
Nd	2.2	6.0	0.9	2.0
Sm	0.5	1.0	0.2	0.4
Gd	0.5	1.2	0.2	0.5
Tb	0.1	0.2	<0.1	0.1
Dy	0.4	0.9	0.1	0.3
Ho	0.1	0.2	<0.1	0.1
Er	0.3	0.6	0.1	0.3
Tm	0.1	0.1	<0.1	<0.1
Yb	0.4	0.5	0.1	0.1
Lu	0.1	0.1	<0.1	<0.1
Sum	11.0	33.0	4.6	9.5

Source: Adapted from Kawasaki, A. et al., *Soil Sci. Plant Nutr.*, 44, 443, 1998.

[a] Compost based on pork manure, animals housed with sawdust.
[b] Municipal sewage sludge.
[c] Sludge from wastewater treatment plants in the food industry.
[d] Sludge from wastewater treatment plants in the chemical industry.

TABLE 18.8
Mean REE Concentration (µg/g) in Ashes of Waste Materials from Different Sources

| Element | Food Scrap | | Animals | | Horticulture | | Sewage Sludge | | Incinerators[a] | |
	Mean	Range	Mean	Range	Mean	Range	Mean	Range	Mean	Range
Sc	5.1	3.2–13.8	8.1	3.3–15.9	12.7	7.1–22.5	19.2	7.1–32.3	6.4	3.8–9.9
Y	8.7	5.1–13.0	10.2	6.3–13.8	17.1	12.0–24.7	16.5	11.6–24.1	15.9	8.6–19.7
La	8.5	7.2–9.8	11.8	9.3–14.5	14.3	11.3–16.8	19.3	14.5–26.3	14.7	6.8–24.4
Ce	15.5	12.5–20	23.5	21.1–29.1	27.3	20.0–33.1	35.4	26.6–43.8	24.6	11.2–41
Pr	1.6	1.1–2.2	3.1	2.3–3.6	3.3	2.6–3.8	3.5	2.2–4.89	2.5	1.1–4.2
Nd	5.9	3.9–8.4	12.2	8.9–15.1	12.9	10.1–14.8	13.7	9.4–19.0	9.3	4.1–16.1
Sm	3.5	1.7–5.1	2.4	1.7–2.8	2.8	2.2–3.6	10.7	8.1–17.9	2.3	0.9–3.8
Eu	0.7	0.2–1.2	0.7	0.5–1.1	0.6	0.5–1.4	1.6	1.1–2.5	1.4	0.3–2.5
Gd	1.3	0.9–1.9	2.9	1.8–4.4	3.1	2.4–3.6	4.1	2.2–6.1	1.9	0.8–3.3
Tb	0.3	0.2–0.4	0.5	0.2–0.7	0.5	0.4–0.6	0.8	0.4–1.2	0.6	0.3–0.9
Dy	0.8	0.5–1.4	2.1	1.2–2.6	2.6	1.9–3.3	2.1	1.3–3.4	1.2	0.5–2.2
Ho	0.1	0.1–0.2	0.4	0.2–0.5	0.5	0.4–0.6	0.4	0.2–0.6	0.2	0.1–0.4
Er	0.4	0.3–0.8	1.1	0.6–1.4	1.5	1.2–2.0	1.1	0.7–1.5	0.7	0.3–1.3
Tm	0.1	0.01–0.1	0.1	0.1–0.2	0.2	0.1–0.3	0.1	0.1–0.2	0.1	0.1–0.2
Yb	0.4	0.3–0.8	1.1	0.6–1.4	1.5	1.2–1.9	1.1	0.7–1.6	0.7	0.3–1.2
Lu	0.1	0.05–0.1	0.1	0.1–0.2	0.2	0.1–0.3	0.1	0.1–0.3	0.1	0.1–0.2
Total REEs	53.7		80.9		101		130		83.1	

Source: Adapted from Zhang, F. et al., *Environ. Int.*, 27, 393, 2001.
[a] Bottom ashes.

18.5 IMPACTS OF ANTHROPOGENIC DISPERSAL OF LANTHANIDES

The continuously increasing dispersal of lanthanides in the environment will inevitably influence biological processes in different ecosystems. Consequently, maximum permissible concentrations of some of REEs in surface water, sediments, and soils have been adopted in the Netherlands (Kučera et al., 2007; Sneller et al., 2000). In Figure 18.2, pathways of lanthanides into environmental compartments and typical concentrations are depicted.

Atmospheric deposition

Rainwater
0.06 µg/L

Particulate matter
0.22–33 ng/m^3

Riverine and marine water

Sea, 1.22 (Eu)–66.3 (Ce) pmol/kg
River, 120–3300 (Nd) pmol/kg

Ld$_{50}$ for zebra fish 14–25 µg/L
NOEC for zebra fish 1.2–3.8 µg/L

Industrial activities

Content in tailings
U mining 2%–4%

Geological deposits

Carbonates, 0.009%–6.3%
Alkalic igneous, 0.05%–8%

Fertilization

Straight P fertilizer
24 kg/ha P, 30–170 g/ha*year
Compound P fertilizer
24 kg/ha P, up to 734 g/ha
Phosphogypsum
1 t/ha, 160 g/ha
Recommended
maximum Ce load, 44 mg/kg soil

Translocation

Soil water, 1–4 µg/L
Surface run-off, 5–7 µg/L

Leaching, 0.5 cm/year on acid soils

Drinking water

Surface water, 1µg/L
Groundwater up to 30 µg/L

Recommended value 2 µg/L

Crops

Concentration in shoots
Monocts, 0.16–1.84 µg/g (d.w.)
Dicots, 0.12–53.4 µg/g (d.w.)

Meat and meat products

Meat 17 µg/kg
Eggs 15 µg/kg
Fish 13 µg/kg

Herbal food

Cereals   17µg/kg
(safety limit 2 mg/kg)
Vegetables   15 µg/kg
(safety limit 0.7 mg/kg)

Human nutrition

Daily intake, 133 µg

ADI 0.6 mg/kg body weight (REOs)

Health impacts

Stringent doses may cause
Pulmonary fibrosis
Sulmonary heptic lesions
Peritonitis
Inflammation
Neoplasm

FIGURE 18.2 Benchmark data on lanthanides in environmental compartments. (*Sources:* Casado-Martinez, 2011; El-Ramady, 2008; Elderfield et al., 1990; Elderfield and Grieves 1982; Goldstein und Jacobsen, 1988; Haley, 1991; Ji and Cui, 1988; Jiang et al., 2012; Keasler and Loveland, 1982; Orris and Grauch, 2002; Otero et al., 2005; Sneller et al., 2000; Todorovsky et al., 1997; Traxler, 1996; Wang et al., 2000; Wang et al., 2001; Zhu et al., 1996.)

18.5.1 LANTHANIDES IN WATER BODIES

Complexation of lanthanides with organic and inorganic ligands influences their mobility, effective solubility, reactivity, and chemical fractionation in the environment (Ding et al., 2005; Shan et al., 2002). For example, carbonate complexes seem to dominate lanthanide complexations in neutral to alkaline water bodies, while free lanthanide ions and lanthanide-sulfate complexes occur in acidic water (Tang and Johannesson, 2003). There are a few studies that claim that toxicity and bioavailability of lanthanides can be attributed to their free-ionic forms (Cacheris et al., 1990; Stanley and Byrne, 1990; Wang et al., 2004; Weltje et al., 2004; Wen et al., 2006). Lanthanides are bound to humic substances in many natural waters (Ingri et al., 2000; Tanizaki et al., 1992; Viers et al., 1997). Humic substances such as humic acid and fulvic acid are the main organic ligands interacting with metal ions in natural aquifers.

Steinmann (2008) showed that for assessing transport and fractionation of REEs in water samples, colloidal particles and vegetation rather than pH, E_h, complexation, and sorption are highly relevant. The mobility of REEs was directly related to the presence and stability of accessory minerals such as apatite and Fe/Mn oxyhydroxides (Steinmann, 2008).

Nozaki (2001) showed that the REE content of seawater is in the pg range and significantly lower than in river waters. The REE content was roughly twice as high in the Northwest Pacific than in the West Indian Ocean and North Atlantic. The dominant element in these marine bodies was Ce. Riverine input is the main source of REEs (Nozaki, 2001), and REEs are discharged by mining, refining industry, fertilization, and medical diagnosis (Casado-Martinez, 2011). Actually, an increasing discharge of Gd that is used as a contrast medium in magnetic resonance imaging with hospital effluents led to a positive Gd anomaly of surface waters in Europe (Brioschi et al., 2013). Kulakisiz and Bau (2013) calculated that annually up to 5700 kg of La, 584 kg of Sm, and 730 kg Gd of anthropogenic origin is dispersed into the North Sea. La, Sm, and Gd concentrations in the river Rhine were more than ten times higher compared to the natural background (Kulaksiz and Bau, 2011, 2013). As a result REEs can be traced in rivers, coastal seawaters, groundwater, and tap water (Casado-Martinez, 2011). The lethal concentration (LD_{50}) to freshwater zebra fish ranged between 14 and 25 µg/L in short-term exposure, and the no observed effect concentration in chronic exposures varied between 1.2 and 3.8 µg/L (Casado-Martinez, 2011). The recommended maximum value for drinking water is 2 µg/L in the Netherlands (de Boer et al., 1996).

18.5.2 LANTHANIDES IN SOILS, PLANTS, AND ANIMALS

The impact of lanthanides on soil life, plant metabolism, and crop performance was outlined in Sections 18.2 and 18.3. Lanthanides may well give rise to toxic effects on soil microorganisms (Chu et al., 2001b, 2003a,b) and negatively affect their activities (Chu et al., 2003b; Xu et al., 2004; Xu and Wang, 2001; Zhu et al., 2002). Sneller et al. (2000) determined a maximum permissible addition of Ce of 44 µg/g to soils in order to exclude negative effects.

A negative influence on plant metabolism can be attributed to the replacement of Ca, high affinity to phosphate groups, and blocking of enzymes (Casado-Martinez, 2011). Zeng et al. (2006) calculated that La concentrations of up to 42 µg/g in red soils and 83 µg/g in paddy soils will have no negative impact on crop productivity. As these values are based on a single experiment, it is difficult to assess their validity on production fields (see Section 18.6).

Jiang et al. (2012) determined the lanthanide content in almost 20,000 samples of various food products in China. The median REE values for cereals, fresh vegetables, fresh meat, and eggs were similar at 0.017, 0.015, 0.016, and 0.018 mg/kg, whereas that of fresh aquatic products was somewhat lower at 0.013 mg/kg. The Chinese food safety standards denote a limit of 2.0 mg/kg REOs for cereals and 0.7 mg/kg REOs for fresh vegetables (Jiang et al., 2012). Though the median values proved to be lower, the 95 and 97.5 percentile exceeded the standard limit (Jiang et al., 2012). Fertilizer practices, here the use of REE fertilizers, mineral phosphates, and manures from livestock that have been supplemented with REEs and the use of surface water for irrigation, may contribute to an accumulation of lanthanides in food products.

Lanthanide supplements have been used to promote animal performance in livestock husbandry. Results suggest that lanthanides administered in organic form yielded better results than salts and mixtures of lanthanides may foster growth better than single lanthanides (Redling, 2006). Increased body gain weight of 15%–25% and feed conversion rates of up to 24% at rates of 100–200 mg/kg feed were reported for different animal species in China (Redling, 2006). Experimentation in Europe delivered inconsistent results in such a way that body weight gain of pigs was reduced by up to 10%, but in other experiments, this parameter was increased 20%; the feed conversion rate was increased by at maximum 11% (Redling, 2006). Redling (2006) concluded from the existing litera-ture that intake of lanthanides by meat and fish will not be significantly influenced by lanthanide supplements in livestock husbandry. The reason for this according to Krafka (1999) is the poor gas-trointestinal absorption of lanthanides. As a consequence, supplemented lanthanides are excreted by animals and enter the soil/plant system in manure applications.

18.5.3 LANTHANIDES AND HUMAN HEALTH

Fricker (2006), Redling (2006), and Thompson and Orvig (2006) provide a comprehensive sum-mary of relevant literature on the use for medical applications and health effects of lanthanides, so that at this point the main findings are only briefly addressed. The phosphate-binding capacity and substitution of Ca make them predestined for use in renal diseases such as hyperphospha-temia in order to regulate the phosphate level; Ce nitrate has been used successfully for treating burns (Redling, 2006). The efficacy has been attributed to the direct interaction with the burn toxin and immunosuppressive factors or alternatively direct binding of Ce to lipid protein complexes (Redling, 2006).

When lanthanides were first used as anticancer agents, high doses needed to be administered and gave rise to toxic effects (Kostova, 2005). Kostova (2005) describes that the challenge of devel-oping effective lanthanide-based anticarcinogenics is to modulate duality and diversity of single lanthanides. The same author postulated that beneficial hormetic effects are restricted to low con-centrations and any transgression of critical concentrations will trigger inhibitory effects. The anti-carcinogenic features of lanthanides have been described, and various modes of action have been identified (Redling, 2006). An innovative approach is the use of radioactive lanthanide isotopes (^{177}Lu, [153]Sm, [171]Eu, [157]Dy) in nuclear medicine against metastases and isotopic diagnostics for bone imaging (Redling, 2006). The paramagnetic properties of lanthanides, in particular Gd, make them ideal contrasting agents in magnetic resonance imaging (Redling, 2006). The anti-inflammatory properties of lanthanides make them a promising aid for the treatment of arthritis by intra-articular injection of [169]Er (Redling, 2006).

The mean dietary intake of REEs is about 133 µg/day (Jiang et al., 2012). A no-observed-adverse-effect level of 60 mg/kg and an acceptable daily intake of 0.6 mg/kg body weight for REOs were calculated by Ji and Cui (1988). Excessive intake of lanthanides by humans may negatively impede functions of the immune, circulatory, digestive, and nervous systems, effects that are expressed in negative impacts on IQ, physical growth, and development of children, but may also cause cancer (Yuan et al., 2003).

The consequences of lanthanides on humans are recapitulated by Haley (1991). The impact of lanthanides on the metabolism of animals and humans depends on the modality of administration, chemical form (soluble, insoluble, chelated), dose, and treatment duration (Haley, 1991). In general, the toxicity of lanthanides decreases with increasing atomic number (Haley, 1991).

The inhalation of stable lanthanides may induce progressive pulmonary fibrosis in a dose- and duration-dependent way (Haley, 1991). Up to 2 g/kg Gd and Sm can be administered orally without causing lethality. After oral intake, lanthanides are found preferentially in liver, kidneys, bone, and teeth (Redling, 2006). Lanthanide contraction (see Section 18.1), here affecting basicity and solubility, is reflected in tissue distribution in such a way that elements with large ionic radius tend to be deposited in the liver and those with smaller radius accumulate in the bone (Haley, 1991). Symptoms of acute toxicity by relatively high doses gave rise to, for instance, pulmonary hepatic lesions, peritonitis, inflammation, and neoplasm (Haley, 1991).

In short-term animal experiments with high doses of La, Y, and Eu, there was reduced body weight, with females proving to be more sensitive (TERA, 1999). In chronic studies, body weights of female mice were 17% lower after 18 months when 0.95 mg/kg-day Y had been administered (TERA, 1999). In comparison, a dose of up to 500 mg/kg-day of individual lanthanides did not affect weight gain of the tested animals in subchronic tests (TERA, 1999). In addition, the treatments had no effect on hematology and histopathology of exposed animals.

18.6 CONCLUSIONS

Several spectroscopic techniques exist to determine the lanthanide content in biological samples. The suitability of different methods for the determination of lanthanides in different matrices has been assessed by Zawisza et al. (2011). Though it is analytically possible to determine individual lanthanides precisely in the pg range, two other confounding variables need to be kept in mind, which may obscure treatment effects. Firstly, in agricultural research, the biggest source of error is posed by sampling (Buczko et al., 2012), and secondly, increasing the scale of experimentation from laboratory to the production field will multiply responses (Haneklaus, 1989). In general, the effect of treatments can be studied unbiased at a molecular or cell level. Thus, a basic understanding of metabolic processes can be acquired. In greenhouse and field experimentation, interactions with any other factor affecting plant growth might modify metabolic response and finally plant growth so that beneficial effects might be obscured and detrimental effects mitigated or compensated (Haneklaus et al., 2007b). For an agricultural use of such effects, it is, however, vital that positive response can be triggered consistently.

Lanthanides are heavy metals and graded applications can be expected to exhibit typical dose/response curves. Obviously, plants are able to metabolize significant amounts of lanthanides before negative impacts on crop performance are manifested (see Section 18.4). Interesting from the viewpoint of agricultural production, however, is the range of supply that triggers hormesis and gives rise to positive effects on plant growth. In the case of heavy metals, hormesis is expressed by increased metabolic activity, for instance, enhanced synthesis of stress proteins and metallothioneins (Lukey, 1990; Damelin et al., 2000). The phenomenon that mild stress induces yield gains and quality improvements is well known from the cultivation and processing of herbal plants (Bloem et al., 2007; Selmar and Kleinwaechter, 2013). In both cases, the effect is explained by a stimulation of metabolic processes in order to defy oxidative stress. Yet, it is a demanding challenge to establish crop-specific threshold concentrations of lanthanides in shoots (see Section 18.3). Such investigations are of scientific rather than practical interest as the world economy for REEs changed drastically within the last decade (see Section 18.1). Consequently, data about point and nonpoint losses of lanthanides to the environment and their impact on the animate and inanimate world are of prime interest (see Section 18.5).

REFERENCES

Ali, N.A., M. Ater, G.I. Sunahara, and P.Y. Robidoux. 2004. Phytotoxicity and bioaccumulation of copper and chromium using barley (*Hordeum vulgare* L.) in spiked artificial and natural forest soils. *Ecotoxicol. Environ. Saf.* 57:363–374.

Amorim, M.J., J. Römbke, H.J. Schallnass, and A.M. Soares. 2005. Effect of soil properties and aging on the toxicity of copper for *Enchytraeus albidus*, *Enchytraeus luxuriosus*, and *Folsomia candida*. *Environ. Toxicol. Chem.* 24:1875–1885.

Andrès, Y.A., C. Texier, and P. Le Cloire. 2003. Rare earth elements removal by microbial biosorption: A review. *Environ. Technol.* 24:1367–1375.

Anonymous. 2011. Rare earth elements. Nottingham, U.K.: British Geological Survey. www.mineralsUK.com (accessed on July 8, 2013).

Arnon, D.L. 1950. Criteria for essentiality of inorganic micronutrients for plants. *Loytsia* 3:31–38.

Babula, P., A. Vojtech, R. Opatrilova, J. Zehnalek, L. Havel, and R. Kizek. 2008. Uncommon heavy metals, metalloids and their plant toxicity: A review. *Environ. Chem. Lett.* 6:189–213.

Banerjee, M.R., D.L. Burton, and S. Depoe. 1997. Impact of sewage sludge application on soil biological characteristics. *J. Agric. Ecosyst. Environ.* 66:241–249.

Bloem, E., S. Haneklaus, and E. Schnug. 2007. Comparative effects of sulfur and nitrogen fertilization and post-harvest processing parameters on the glucotropaeolin content of *Tropaeolum majus* L. *J. Sci. Food Agric.* 87:1576–1585.

Brioschi, L., M. Steinmann, E. Lucot et al. 2013. Transfer of rare earth elements (REE) from natural soil to plant systems: Implications for the environmental availability of anthropogenic REE. *Plant Soil* 366:143–163.

Brookes, P.C., S.P. McGrath, D.A. Klein, and E.T. Elliott. 1984. Effects of heavy metals on microbial activity and biomass in field soils treated with sewage sludge. In *International Conference on Environmental Contamination*, pp. 573–583. London, U.K.: CEP Consultants Ltd.

Brown, P.H., A.H. Rathjien, R.D. Graham, and D.E. Tribe. 1990. Rare earth elements in biological systems. In *Handbook on the Physics and Chemistry of Rare Earths*, eds. K.A. Geschneider and L. Eyring, vol. 13, pp. 423–453. New York: Elsevier Science Publisher B.V.

Buczko, U., R.O. Kuchenbuch, W. Ubelhor, and L. Natscher. 2012. Assessment of sampling and analytical uncertainty of trace element contents in arable field soils. *Environ. Monit. Assess.* 184:4517–4538.

Busby, C. and E. Schnug. 2007. Advanced biochemical and biophysical aspects of uranium contamination. In *Loads and Fate of Fertilizer Derived Uranium*, eds. L.J. De Kok and E. Schnug, pp. 11–22. Leiden, the Netherlands: Backhys Academic Publishers.

Cacheris, W.P., S.C. Quay, and S.M. Rocklage. 1990. The relationship between thermodynamics and the toxicity of gadolinium complexes. *Magn. Reson. Imaging* 8:467–481.

Cao, X., Y. Chen, X. Wang, and X. Deng. 2001. Effects of redox potential and pH value on the release of rare earth elements from soil. *Chemosphere* 44:655–661.

Cao, X.Y., Z.M. Gu, and X.R. Wang. 2000. Determination of trace rare elements in plant and soil samples by inductively coupled plasma-mass spectrometry. *Int. J. Environ. Anal. Chem.* 76:295–309.

Casado-Martinez, C. 2011. Ecotoxicology of rare earth elements. Dübendorf, Switzerland: Eawag. http://www.oekotoxzentrum.ch/dokumentation/info/doc/rareearth.pdf (accessed on July 9, 2013).

Chang, J. 1991. Effects of lanthanum on the permeability of root plasmalemma and the absorption and accumulation of nutrients in rice and wheat. *Plant Physiol. Commun.* 27:17–21.

Chen, F.M. and T.J. Zheng. 1990. Study application of REEs in potatoes. *J. Potatoes* 4:224–227.

Chen, X.H. and B. Zhao. 2007. Arbuscular mycorrhizal fungi mediated uptake of lanthanum in Chinese milk vetch (*Astragalus sinicus* L.). *Chemosphere* 68:1548–1555.

Chu, H.Y., J.G. Zhu, Z.B. Xie et al. 2002. Effect of lanthanum on hydrolytic enzyme activities in red soil. *J. Rare Earth.* 20:158–160.

Chu, H.Y., J.G. Zhu, Z.B. Xie, Z. Cao, Z. Li, and Q. Zeng. 2001b. Effects of lanthanum on microbial biomass carbon and nitrogen in red soil. *J. Rare Earth.* 19:63–66.

Chu, H.Y., J.G. Zhu, Z.B. Xie, H.Y. Zhang, Z.H. Cao, and Z.G. Li. 2003a. Effects of lanthanum on dehydrogenase activity and carbon dioxide evolution in a Haplic Acrisol. *Aust. J. Soil Res.* 41:731–739.

Chu, H.Y., J.G. Zhu, Z.B. Xie, Q. Zheng, Z.G. Li, and Z.H. Cao. 2003b. Availability and toxicity of exogenous lanthanum in a Haplic Acrisols. *Geoderma* 115:121–128.

Chu, H.Y., W. Zunhua, Z.B. Xie, Z. Jianguo, Z.L. Zhengao, and Z.H. Cao. 2001a. Effect of lanthanum on major microbial populations in red soil. *Pedosphere* 11:73–76.

Damelin, L.H., S. Vokes, J.M. Whittcut, S.B. Damelin, and J.J. Alexander. 2000. Hormesis: A stress response in cells exposed to low levels of heavy metals. *Hum. Exp. Toxicol.* 19:420–430.

Datnoff, L.E., G.H. Snyder, and G.H. Korndörfer (eds.). 2001. *Silicon in Agriculture*, vol. 8: *Studies in Plant Science*, pp. 1–403. New York: Elsevier.

De Boer, J.L.M., W. Verweij, T. van der Velde-Koerts, and W. Mennes. 1996. Levels of rare earth elements in Dutch drinking water and its sources. Determination by inductively coupled plasma mass spectrometry and toxicological implications. A pilot study. *Water Res.* 30:190–198.

De Boer, M.A. and K. Lammertsma. 2012. Scarcity of rare earth elements. *R. Neth. Chem. Soc. (KNCV)* 2:1–25.

De Brouwere, K., S. Hertigers, and E. Smolders. 2007. Zinc toxicity on N₂O reduction declines with time in laboratory spiked soils and is undetectable in field contaminated soils. *Soil Biol. Biochem.* 39:3167–3176.

Diatloff, E., C.J. Asher, and F.W. Smith. 1995c. Effects of rare earth elements (REE) on the growth and mineral nutrition of plants. In *Proceedings of the Rare Earth in Agriculture Seminar*, Canberra, Australian Capital Territory, Australia, pp. 11–20.

Diatloff, E., C.J. Asher, and F.W. Smith. 1996. Concentration of rare earth elements in some Australian soils. *Aust. J. Soil Res.* 34:735–747.

Diatloff, E., C.J. Asher, and F.W. Smith. 1999. Foliar application of rare earth elements to maize and mungbean. *Aust. J. Exp. Agric.* 39:189–194.

Diatloff, E., F.W. Smith, and C.J. Asher. 1995a. Rare earth elements and plant growth. I. Effects of lanthanum and cerium on root elongation of corn and mungbean. *J. Plant Nutr.* 18:1963–1976.

Diatloff, E., F.W. Smith, and C.J. Asher. 1995b. Rare earth elements and plant growth. II. Response of corn and mungbean to low concentrations of lanthanum in dilute, continuously flowing nutrient solutions. *J. Plant Nutr.* 18:1977–1989.

Ding, S., T. Liang, C. Zhang, J. Yan, and Z. Zhang. 2005. Accumulation and fractionation of rare earth elements (REEs) in wheat: Controlled by phosphate precipitation, cell wall absorption and solution complexation. *J. Exp. Bot.* 56:2765–2775.

Dong, B., Z.M. Wu, and X.K. Tan. 1993. Effects of $LaCl_3$ on physiology of cucumber with Ca deficiency. *J. Chin. Rare Earth Soc.* 11:65–68.

Eichler, B., M. Caus, E. Schnug, and D. Köppen. 2004. Soil acid and alkaline phosphatase activities in relation to crop species and fungal treatment. *Landbauforschung Völkenrode* 1(54):1–5.

Elbaz-Poulichet, F. and C. Dupuy. 1999. Behaviour of rare earth elements at the freshwater-seawater interface of two acid mine rivers: The Tinto and Odiel (Andalucia, Spain). *Appl. Geochem.* 14:1063–1072.

Elderfield, H. and M.J. Greaves. 1982. The rare earth elements in seawater. *Nature* 296:214–219.

Elderfield, H., R. Upstill-Goddard, and E.R. Sholkovitz. 1990. The rare earth elements in rivers, estuaries, and coastal seas and their significance to the composition of ocean waters. *Geochim. Cosmochim. Acta* 54:971–991.

El-Ramady, H. 2008. A contribution on the bio-actions of rare earth elements in the soil/plant environment. PhD thesis, Technical University Braunschweig, Braunschweig, Germany, p. 173.

Enyeart, J., J.L. Xu, and J.A. Enyeart. 2002. Dual actions of lanthanides on ACTH-inhibited leak K^+ channels. *Am. J. Physiol. Endoc. Metab.* 282:1255–1266.

Erdmann, L. and T.E. Graedel. 2011. Criticality of non-fuel minerals: A review of major approaches and analyses. *Environ. Sci. Technol.* 45:7620–7630.

Evans, C.H. 1990. *Biochemistry of the Lanthanides*. New York: Plenum Press.

Finck, A. 1976. *Pflanzenernährung in Stichworten*. Kiel, Germany: Hirt Verlag, p. 130.

Foyer, C.H., M. Leiandais, and K.J. Kunert. 1994. Photooxidative stress in plants. *Physiol. Plant.* 92:696–717.

Fricker, S.P. 2006. The therapeutic application of lanthanides. *Chem. Soc. Rev.* 35:524–533.

Goldstein, S.J. and S.B. Jacobsen. 1988. Rare earth elements in river waters. *Earth Planet. Sci. Lett.* 89:35–47.

Guo, B.S. 1987. A new application of rare earth-agriculture. In *Rare Earth Horizons*, pp. 237–246. Canberra, Australian Capital Territory, Australia: Australia Department of Industry and Commerce.

Guo, B.S. 1988. *Rare Earth Elements in Agriculture*. Beijing, China: Agriculture Science and Technology Press.

Guo, X.S., Q. Zhou, X.D. Zhu, T.H. Lu, and X.H. Huang. 2007. Migration of a rare earth element cerium(III) in horseradish. *Acta Chim. Sin.* 65:1922–1924.

Haley, P.J. 1991. Pulmonary toxicity of stable and radioactive lanthanides. *Health Phys.* 61:809–820.

Haneklaus, S. 1989. Strontiumgehalte in Pflanzen und Böden Schleswig-Holsteins und Bewertung von Düngungsmaßnahmen zur Verminderung der Strontiumaufnahme von Kulturpflanzen. PhD thesis, University of Kiel, Kiel, Germany, p. 169.

Haneklaus, S., E. Bloem, and E. Schnug. 2006. Sulphur interactions in crop ecosystems. In *Sulfur in Plants: An Ecological Perspective*, eds. M.J. Hawkesford and L.J. De Kok, pp. 17–58. Dordrecht, the Netherlands: Springer.

Haneklaus, S., E. Bloem, and E. Schnug. 2007b. Sulfur and plant disease. In *Mineral Nutrition and Plant Diseases*, eds. L. Datnoff, W. Elmer, and D. Huber, pp. 101–118. St. Paul, MN: APS Press.

Haneklaus, S., E. Bloem, E. Schnug, L.V. De Kok, and I. Stulen. 2007a. Sulfur. In *Handbook of Plant Nutrition*, eds. A.V. Barker and D.J. Pilbeam, pp. 183–238. Boca Raton, FL: CRC Press.

Haneklaus, S., L. Knudsen, and E. Schnug. 1998. Minimum factors in the mineral nutrition of field grown sugar beets in northern Germany and eastern Denmark. *Aspects Appl. Biol.* 52:57–64.

Haxel, G.B., B. James, and G.J. Orris. 2004. Rare earth elements—Critical resources for high technology. Science for a changing world. Washington, DC: U.S. Geological Survey. www.minerals.usgs.gov/minerals/pubs/commodity/rare_earths/ (accessed on October 10, 2004).

Hong, W.M., X.B. Duan, Z.S. Cao, C.P. Hu, W. Zheng, and H.J. Qu. 1996. Long-term location test of REEs on agriculture and REE residual analysis in wheat seeds. In *Proceedings of the First Sino–Dutch Workshop on the Environmental Behavior and Ecotoxicology of Rare Earth Elements*, Beijing, China, October 15–16, 1996, pp. 83–87.

Hong, F.H., Z.G. Wei, and G.W. Zhao. 2000. Effect of lanthanum on aged seed germination of rice. *Biol. Trace Elem. Res.* 75:205–213.

Hong, F.S., Z.G. Wei, and G.W. Zhao. 2002. Mechanism of lanthanum effect on the chlorophyll of spinach. *J. Sci. China C* 45:166.

Hu, Q.H. and Z.J. Ye. 1996. Physiological effects of rare earth elements on plants. *Chin. Plant Physiol. Commun.* 32:296–300.

Hu, X., Z.H. Ding, Y. Chen, X. Wang, and L. Dai. 2002. Bioaccumulation of lanthanum and cerium and their effects on the grown of wheat (*Triticum aestivum* L.) seedlings. *Chemosphere* 48:621–629.

Hu, Y., F. Vanhaecke, L. Moens, R. Dams, P. del Castilho, and J. Japenga. 1998. Determination of the aqua regia soluble content of rare earth elements in fertilizer, animal fodder phosphate and manure samples using inductively coupled plasma spectrometry. *Anal. Chim. Acta* 373:95–105.

Hu, Z., S. Haneklaus, G. Sparovek, and E. Schnug. 2006. Rare earth elements in soils. *Commun. Soil Sci. Plant Anal.* 37:1381–1420.

Hu, Z., H. Richter, G. Sparovek, and E. Schnug. 2004. Physiological and biochemical effects of rare earth elements on plants and their agricultural significance: A review. *J. Plant Nutr.* 27:183–220.

Hu, Z., G. Sparovek, S. Haneklaus, and E. Schnug. 2003. Rare earth elements. In *Encyclopedia of Soil Science*, ed. R. Lal. New York: Marcel Dekker Inc. http://www.dekker.com/servlet/product/productid/E-ESS.

Huang, C.-H. and A. Tsourkas. 2013. Gd-based macromolecules and nanoparticles as magnetic resonance contrast agents for molecular imaging. *Curr. Top. Med. Chem.* 13:411–421.

Ingri, J., A. Winderlund, M. Land, Ö. Gustafsson, P. Andersson, and B. Ölander. 2000. Temporal variation in the fractionation of the rare earth elements in a boreal river; the role of colloidal particles. *Chem. Geol.* 166:23–45.

Ji, Y.J. and M.Z. Cui. 1988. Toxicological studies on safety of rare earths used in agriculture. *Biomed. Environ. Sci.* 1:270–276.

Jiang, D.G., J. Yang, S. Zhang, and D.J. Yang. 2012. A survey of 16 rare earth elements in the major foods in China. *Biomed. Environ. Sci.* 25:267–271.

Jones, D.L. 1997. Trivalent metal (Cr, Y, Rh, La, Pr, Gd) sorption in two acid soils and its consequences for bioremediation. *Eur. J. Soil Sci.* 48:697–702.

Kawasaki, A., R. Kimura, and S. Arai. 1998. Rare earth elements and other trace elements in wastewater treatment sludges. *Soil Sci. Plant Nutr.* 44:443–441.

Keasler, K.M. and W.D. Loveland. 1982. Rare earth elemental concentrations in some Pacific northwest rivers. *Earth Planet. Sci. Lett.* 61:68–72.

Kostova, I. 2005. Lanthanides as anticancer agents. *Curr. Med. Chem.* 5:591–602.

Krafka, B. 1999. Neutronenaktivierungsanalyse an Boden und Pfanzenproben: Untersuchungen zum Gehalt an Lanthanoiden sowie Vergleich der Multielementanalytik mit aufschlussabhängigen Analysemethoden. Dissertation, Technische Universität, München, Germany.

Kratz, S., S. Haneklaus, and E. Schnug. 2009. Kupfergehalte in Acker- und Grünlandböden und das Verhältnis dieser Gehalte zu den durch Pflanzenschutz ausgebrachten Kupfermengen. *J. Kulturpflanzen* 61:112–116.

Kučera, J., J. Mizera, Z. Řanda, and M. Vávrová. 2007. Pollution of agricultural crops with lanthanides, thorium and uranium studied by instrumental and radiochemical neutron activation analysis. *J. Radioanal. Nucl. Chem.* 271:581–587.

Kulaksiz, S. and M. Bau. 2011. Rare earth elements in the Rhine River, Germany: First case of anthropogenic lanthanum as a dissolved microcontaminant in the hydrosphere. *Environ. Int.* 37:973–979.

Kulaksiz, S. and M. Bau. 2013. Anthropogenic dissolved and colloid/nanoparticle-bound samarium, lanthanum and gadolinium in the Rhine River and the impending destruction of the natural rare earth element distribution in rivers. *Earth Planet. Sci. Lett.* 362:43–50.

Lakanen, E. and R. Erviö. 1971. A comparison of eight extractants for the determination of plant available micronutrients in soils. *Acta Agric. Fenn.* 122:223–232.

Liu, C., F. Hong, L. Zheng, P. Tang, and Z. Wang. 2004. Effects of rare earth elements on vigor enhancement of aged spinach seeds. *J. Rare Earth.* 22:547–551.

Liu, Y.L., J.H. Liu, Z.J. Wang, and A. Peng. 1999. Speciation transformation of added rare earth elements in soil. *Environ. Chem.* 18:393–397.

Liu, Z. 1988a. Rare earth elements in soil. In *Rare Earth Elements in Agriculture* (in Chinese), eds. B.S. Guo, W.M. Zhu, B.K. Xiong, Y.J. Ji, Z. Liu, and Z.M. Wu, pp. 23–44. Beijing, China: China Agricultural Science and Technology Press.

Liu, Z. 1988b. Effects of rare earth elements on growth on crops. In *Proceedings of the Fourth International Symposium of New Results in the Research of Hardly Known Trace Elements*, Budapest, Hungary, July 15–18, 1988, pp. 150–161.

Lukey, T.D. 1990. Radiation hormesis overview. *RSO Mag.* 8(4):22–41.

Macklis, R.M. and B. Beresford. 1991. Radiation hormesis. *J. Nucl. Med.* 32:250–260.

Mathre, G.N. and C.E. Pankhurst. 1997. Bioindicators to detect contamination of soils with special reference to heavy metals. In *Biological Indicators of Soil Health*, eds. C.E. Pankhurst, B.M. Doube, and V.V.S.R. Gupta, pp. 349–369. New York: CAB International.

Meehan, B., K. Peverill, J. Maheswaran, and S. Bidwell. 2001. The application of rare earth elements in enhancement of crop production in Australia, Part 1. In *Proceedings of the Fourth International Conference on Rare Earth Development and Application*, eds. Z.S. Yu, C.H. Yan, G.Y. Xu, J.K. Niu, and Z.H. Chen, Beijing, China: Metallurgical Industry Press, June 16–18, 2001, pp. 244–250.

Mikkelson, R.B. 1976. Lanthanides as calcium probes. In *Biological Membranes*, eds. D. Chapman and D.F. Wallach, vol. III, pp. 153–190. New York: Academic Press.

Nozaki, Y. 2001. Rare earth elements and their isotopes. In *Encyclopedia of Ocean Sciences*, eds. J.H. Steele, S.A. Thorpe, and K.K. Turekian, vol. 4, pp. 2354–2366. London, U.K.: Academic Press.

Orris, G.J. and R.I. Grauch. 2002. Rare earth element mines, deposits and occurrences. Open file report 02-189, 2002. Reston, VA: U.S. Geological Survey.

Otero, N., L. Vitoria, A. Soler, and A. Canals. 2005. Fertiliser characterization: Major, trace and rare earth elements. *Appl. Geochem.* 20:1473–1488.

Ozaki, T., S. Enomoto, Y. Minai, S. Ambe, and Y. Makide. 2000. A survey of trace elements in pteridophytes. *Biol. Trace Elem. Res.* 74:259–273.

Pang, X., D. Li, and A. Peng. 2002. Application of rare earth elements in the agriculture of China and its environmental behavior in soil. *Environ. Sci. Pollut. Res. Int.* 9:143–148.

Pissarek, H.-P. and E. Schnug. 1982. Wege zur besseren Spurennährstoffversorgung von Getreide. *DLG-Mitteilungen* 97:75–77.

Quiao, G.L., X.K. Tang, and Z.M. Wu. 1993. Study dynamics of NdCl3 on K+ exosmosis form corn roots. *Acta Bot. Sin.* 35:286–290.

Ran, Y. and Z. Liu. 1992. Characteristics of adsorption and desorption of REEs on main type soils of China. *J. Chin. Rare Earth Soc.* 10:377–380.

Ran, Y. and Z. Liu. 1993. Adsorption and desorption of rare earth elements on soils and synthetic oxides. *Acta Sci. Circumst.* 3:288–293.

Redling, K. 2006. Rare earth elements in agriculture with emphasis on animal husbandry. PhD thesis, University of Munich, Munich, Germany.

Reimann, C., U. Siewers, T. Tarvainen et al. 2003. Agricultural soils in Northern Europe: A geochemical atlas. Hannover/Stuttgart, Germany: Bundesanstalt für Geowissenschaften und Rohstoffe and Staatliche Geologische Dienste in der Bundesrepublik Deutschland.

Schnug, E. 1991. *Das Raps-Handbuch*, 5. Auflage. Bad-Homburg, Germany: Elanco, pp. 72–87.

Schnug, E. and S. Haneklaus. 1992. PIPPA: Un programme d'interprétation des analyses de plantes pour le colza et les céréales. *Suppl. Perspect. Agric.* 171:30–33.

Schnug, E. and S. Haneklaus. 2008. Evaluation of the relative significance of sulfur and other essential mineral elements in oilseed rape, cereals, and sugar beet production. In *Sulfur: A Missing Link between Soils, Crops, and Nutrition*, ed. J. Jez, pp. 219–233. Madison, WI: Crop Science Society of America, American Society of Agronomy, Soil Science Society of America.

Selmar, D. and M. Kleinwaechter. 2013. Influencing the product quality by deliberately applying drought stress during the cultivation of medicinal plants. *Ind. Crop. Prod.* 42:558–566.

Shan, X., J. Lian, and B. Wen. 2002. Effect of organic acids on adsorption and desorption of rare earth elements. *Chemosphere* 47:701–710.

Shan, X.Q., S.Z. Zhang, and B. Wen. 2004. Fractionation and bioavailability of rare earth elements in soils. In *Trace and Ultratrace Elements in Plants and Soil*, ed. I. Shtangeeva, pp. 249–285. Southampton, U.K.: WIT Press.

Shen, H. and X.L. Yan. 2002. Membrane permeability in roots of *Crotalaria* seedlings as affected by low temperature and low phosphorus stress. *J. Plant Nutr.* 25:1033–1047.

Sijm, D., R. Kraaij, and A. Belfroid. 2000. Bioavailability in soil or sediment: Exposure of different organisms and approaches to study it. *Environ. Pollut.* 108:113–119.

Sneller, F.E.C., D.F. Kalf, L. Weltje, and A.P. van Wezel. 2000. Maximum permissible concentrations for negligible concentrations for rare earth elements (REEs). RIVM report no. 601501011. Bilthoven, the Netherlands: National Institute of Public Health and the Environment.

Solymosi, K. and M. Bertrand. 2010. Soil metals, chloroplast, and secure crop production: A review. *Agron. Sustain. Dev.* 32:245–272.

Stanley, J.K. and R.H. Byrne. 1990. The influence of solution chemistry on REE uptake by *Ulva lactuca* L. in seawater. *Geochim. Cosmochim. Acta* 54:1587–1595.

Stebbing, A.R.D. 1982. Hormesis—The stimulation of growth by low levels of inhibitors. *Sci. Total Environ.* 22:213–234.

Steinmann, M. 2008. Les terres rares et les isotopes radiogéniques comme traceurs d'échange et de transfert dans les cycles géochimiques externs. DSc thesis, UFC Université, Besançon, France.

Sterner, R.W. and J.J. Elser. 2002. *Ecological Stoichiometry: The Biology of Elements from Molecules to the Biosphere*. Princeton, NJ: Princeton University Press.

Tang, J. and K.H. Johannesson. 2003. Rare earth element speciation along groundwater flow paths in two different aquifer types (i.e., sand vs. carbonate). *American Geophysical Union, Fall Meeting 2003*, San Francisco, CA, Abstract H11G-0955.

Tanizaki, Y., T. Shimokawa, and M. Nakamura. 1992. Physico-chemical speciation of trace elements in river waters by size fractionation. *Environ. Sci. Technol.* 26:1433–1444.

TERA (Toxicology Excellence for Risk Assessment). 1999. Development of reference doses and reference concentrations for lanthanides, p. 52. http://www.tera.org/Publications/Lanthanides.pdf (accessed on July 22, 2013).

Thompson, K.H. and C. Orvig. 2006. Lanthanide compounds for therapeutic and diagnostic applications. *Chem. Soc. Rev.* 35:499–502.

Tian, H.E., F.Y. Gao, F.L. Zeng, F.M. Li, and L. Shan. 2005. Effects of Eu^{3+} on the metabolism of amino acid and protein in xerophytic *Lathyrus sativus* L. *Biol. Trace Elem. Res.* 105:257–267.

Tian, W.X. 1990. Effect of REEs on seed germination, growth of seedling and yield of tuber of sugar beet. *Bot. Commun.* 7:37–40.

Todorovsky, D.S., N.L. Minkova, and D.P. Bakalova. 1997. Effect of the application of superphosphate on rare earth content in the soils. *Sci. Total Environ.* 203:13–16.

Traxler, G. 1996. The economic benefit of phosphogypsum use in agriculture in the Southeastern Unites States. Bartow, FL: Florida Institute of Phosphate Research, FIPR report no. 01-124-119.

Tyler, G. 2004. Rare earth elements in soil and plant systems—A review. *Plant Soil* 267:191–206.

Viers, J., B. Dupré, M. Polvé, J. Schott, J.L. Dandurand, and J.J. Braun. 1997. Chemical weathering in the drainage basin of a tropical watershed (Nsimi-Zoetele site, Cameroon): Comparison between organic-poor and organic-rich waters. *Chem. Geol.* 140:181–206.

von Tucher, S., C. Goy, and U. Schmidhalter. 2001. Effect of lanthanum on growth and composition of mineral nutrients of *Phaseolus vulgaris* L var. nanus and *Zea mays* L. conv. Saccharata. In *Plant Nutrition. Food Security and Sustainability of Agro-Ecosystems*, eds. W.J. Horst, M. Schenk, A. Bürkert et al., pp. 524–525. Dordrecht, the Netherlands: Kluwer Academic Publishers.

Wan, Q., J. Tian, H. Peng et al. 1998. The effects of rare earth on increasing yield, improving quality and reducing agricultural chemical remained in crop production. In *Proceedings of Second International Symposium on Trace Elements and Food Chain*, Wuhan, Hubei, China, November 12–15, 1998, p. 25.

Wang, C.X., W. Zhu, Z.J. Wang, R. Guicherit, C.X. Wang, and Z.J. Wang. 2000. Rare earth elements and other metals in atmospheric particle matter in the western part of the Netherlands. *Water Air Soil Pollut.* 121:109–118.

Wang, H.Y. 1998. Special topic in agricultural use of rare earths in China. *Chin. Rare Earth.* 19:68–72.

Wang, J., N. Chunji, Y. Kuiyue, and N. Jiazuan. 2004. Formation equilibria of ternary metal complexes with citric acid and glutamine (alanine) in aqueous solution. *J. Rare Earth.* 22:183–186.

Wang, X.-P., X.-Q. Shan, and S.-Z. Zhang. 2003. Distribution of rare earth elements among chloroplast components of hyperaccumulator *Dicranopteris dichotoma*. *Anal. Bioanal. Chem.* 376:913–917.

Wang, Z., C. Wang, P. Lu, and W. Zhu. 2001. Concentrations and flux of rare earth elements in a semifield plot as influenced by their agricultural application. *Biol. Trace Elem. Res.* 84:213–226.

Weltje, L. 1997. Uptake and bioconcentration of lanthanides in higher plants: linking terrestrial and aquatic studies. In *Proceedings of the Second Sino-Dutch Workshop on the Environmental Behaviour and Ecotoxicology of REEs and Heavy Metals*, Delft, the Netherlands, pp. 1–12.

Weltje, L., L.R. Verhoof, W. Verweij, and T. Hamers. 2004. Lutetium speciation and toxicity in a microbial bioassay: Testing the free-ion model for lanthanides. *Environ. Sci. Technol.* 38:6597–6604.

Wen, B., Y. Liu, X. Hu, and X. Shan. 2006. Effect of earthworms (*Eisenia fetida*) on the fractionation and bio-availability of rare earth elements in nine Chinese soils. *Chemosphere* 63:1179–1186.

Wen, B., D.A. Yuan, X.Q. Shan, F.L. Li, and S.Z. Zhang. 2001. The influence of rare earth element fertilizer application on the distribution and bioaccumulation of rare earth elements in plants under field conditions. *Chem. Spec. Bioavailab.* 13:39–48.

Wheeler, D.M. and I.L. Power. 1995. Comparison of plant uptake and plant toxicity of various ions in wheat. *Plant Soil* 172:167–173.

Wu, Z., X. Tang, and C. Tsui. 1983. Studies on the effect of rare earth elements on the increasement of yield in agriculture. *J. Chin. Rare Earth Soc.* 1:70–75.

Wu, Z.M., X.K. Tang, Z.W. Jia, and X.X. Cao. 1984. Studies effect of application of REEs on the increment of yield in agriculture II. Effect of REEs on physiological metabolism in crops. *J. Chin. Rare Earth Soc.* 2:75–79.

Xie, Z., J.G. Zhu, H.Y. Chu, Y.L. Zhang, H.L. Ma, and Z.H. Cao. 2002. Effect of lanthanum on rice production, nutrient uptake and distribution. *J. Plant Nutr.* 25:2315–2331.

Xiong, B.K. 1995. Application of rare earths in Chinese agriculture and their perspective of development. In *Proceedings of the Rare Earths in Agriculture Seminar*, Australian Academy of Technological Sciences and Engineering, Victoria, Australia, pp. 5–9.

Xiong, B.K., P. Chen, B.S. Guo, and W. Zheng. 2000. *Rare Earth Element Research and Application in Chinese Agriculture and Forest* (in Chinese). Beijing, China: Metallurgical Industry Press, pp. 1–151.

Xu, D., L. Guangshen, X. Jie, and L. Weiping. 2004. Effects of lanthanum and cerium on acid phosphatase activities in two soils. *J. Rare Earth.* 22:725–728.

Xu, X. and Z. Wang. 2001. Effects of lanthanum and mixture of rare earths on ammonium oxidation and mineralization of nitrogen in soil. *Eur. J. Soil Sci.* 52:323–329.

Xu, X., W. Zhu, Z. Wang, and G.-J. Witkamp. 2002. Distribution of rare earths and heavy metals in field-grown maize after application of rare earth containing fertilizers. *Sci. Total Environ.* 293:97–105.

Xu, X.K., W.Z. Zhu, Z.J. Wang, and G.-J. Witkamp. 2003. Accumulation of rare earth elements in maize plants (*Zea mays* L.) after application of mixtures of rare earth elements and lanthanum. *Plant Soil* 252:267–277.

Yan, J.C. 1999. China rare earth: The brilliant fifty years. In *China Rare Earth Information*, 5. Baotou, Inner Mongolia, China: China Rare-Earth Information Centre, October 5, 1999.

Yu, Y.H. and Q.Y. Liu. 1992. Distribution cumulative of REEs in sugarcane plant and effect of REEs on adversity tolerance in sugarcane. *J. South China Agric. Univ.* 13:47–54.

Yuan, Z.K., Y. Liu, H.Q. Yu et al. 2003. Study on relationship between rare earth level in blood and health condition of residents. *China Public Health* 19:133–135.

Zawisza, B., K. Pytlakowska, B. Feist, M. Polowniak, A. Kita, and R. Sitko. 2011. Determination of rare earth elements by spectroscopic techniques: A review. *J. Anal. Atom. Spectrom.* 26:2373–2390.

Zeng, Q., J.G. Zhu, H.L. Cheng, Z.B. Xie, and H.Y. Chu. 2006. Phytotoxicity of lanthanum in rice in haplic acrisols and cambisols. *Ecotoxicol. Environ. Saf.* 64:226–233.

Zhang, F., S. Yamasaki, and K. Kimura. 2001. Rare earth element content in various waste ashes and potential risk to Japanese soils. *Environ. Int.* 27:393–398.

Zhang, J., Y. Li, X. Hao et al. 2011. Recent progress in therapeutic and diagnostic applications of lanthanides. *Mini-Rev. Med. Chem.* 11:678–694.

Zhu, J.G., H.Y. Chu, Z.B. Xie, and K. Yagi. 2002. Effects of lanthanum on nitrification and ammonification in three Chinese soils. *Nutr. Cycl. Agroecosyst.* 63:309–314.

Zhu, J.G. and G.X. Xing. 1992a. Forms of rare earth elements in soils: I. Distribution. *Pedosphere* 2:125–134.

Zhu, J.G. and G.X. Xing. 1992b. Sequential extraction procedure for the speciation of REEs in soils. *Soils* (in Chinese) 24:215–218.

Zhu, Q.Q. and Z. Liu. 1988. REEs in soils of Eastern China. *J. Chin. Rare Earth Soc.* 6:59–65.

Zhu, W.M., J.Z. Zhang, and L.G. Zhang. 1996. Numerical stimulation of REE migration in soils. *J. Chin. Rare Earth Soc.* 14:341–346.

19 Selenium

David J. Pilbeam, Henry M.R. Greathead, and Khaled Drihem

CONTENTS

19.1 HISTORICAL BACKGROUND

Selenium (Se) is a nonmetallic element in Group VIa of the periodic table, between sulfur and tellurium, and with properties like both. It is an essential element for most animals in low concentrations, although it is toxic at higher concentrations. It was claimed to be beneficial for the growth of some plants in the 1920s and 1930s and even to be possibly essential (Hewitt, 1966). Levine (1925) showed that selenium inhibited the germination and growth of white lupin (*Lupinus alba* L.) and timothy grass (*Phleum pratense* L.), but at low concentration (0.0001% and 0.001%), selenous acid and selenic acid stimulated the growth of the lupin plants. Some indication of beneficial effects of Se on plant growth had been published by Stoklasa (1922). Trelease and Trelease (1938) knew it could accumulate in high enough concentrations in plants to be toxic to animals and that some plants (including species of *Astragalus*, *Stanleya*, *Xylorhiza*, and *Oonopsis* in three different plant families) accumulated Se at concentrations of up to several thousand ppm if grown in seleniferous soils. Some of these species could be considered indicator plants of high soil Se concentrations.

Trelease and Trelease (1938) showed that sodium selenite stimulated the growth of cream milkvetch (*Astragalus racemosus* Pursh) at low concentration and that sodium selenite and potassium selenate stimulated the growth of Patterson's milkvetch (*Astragalus pattersonii* A. Gray ex Brandegee). Their work also showed that there tended to be an inverse relationship between the amount of selenium and sulfur supplied and their effects on plant growth. They suggested that Se may be an essential element for indicator species. Selenium has been shown to be essential for the green alga *Chlamydomonas reinhardtii* P.A. Dang. (Novoselov et al., 2002). Plants suffering from selenium toxicity show symptoms including stunting, chlorosis, withering of leaves, and decreased protein synthesis (Terry et al., 2000).

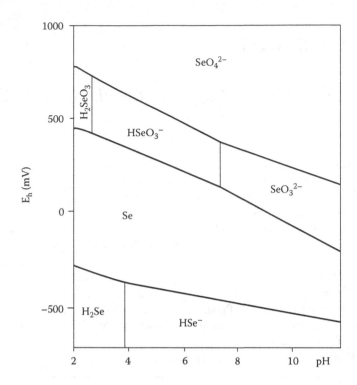

FIGURE 19.1 Occurrence of selenium species at different pHs and oxidation–reduction potentials E_h. (From Neal, R.H. et al., *Soil Sci. Am. J.*, 51, 1161, 1987a.)

Selenium can exist in its two most abundant forms as elemental selenium (Se^0) or in the -2 oxidation state as selenide (Se^{2-}), but in oxidizing conditions it can occur in the $+4$ oxidation state as selenite (SeO_3^{2-}) or in the $+6$ oxidation state as selenate (SeO_4^{2-}) (Kopsell and Kopsell, 2007; Munier-Lamy et al., 2007) (Figure 19.1). Selenite is, in effect, selenium dioxide (SeO_2) dissolved in water, giving selenous acid (H_2SeO_3). This acid is weak and occurs primarily as the biselenite ($HSeO_3^-$) ion between pH 3 and 8 and as selenite above pH 8 (Elrashidi et al., 1987; Missana et al., 2009). The selenite ion forms $Fe_2(SeO_3)_3$ and $Fe_2(OH)_4SeO_3$, which are sparingly soluble in soil water (Wang and Gao, 2001). Selenate (SeO_4^{2-}) is the anion of selenic acid (H_2SeO_4), which is dissociated fully at normal soil pH values, and because it has a low binding constant selenate does not sorb strongly to soil components (Barrow and Whelan, 1989). In natural environments, the selenate form occurs at pH values above 6, but only in more oxidizing conditions than those under which biselenite and selenite occur (Missana et al., 2009). The selenate minerals in well-aerated soils are generally more soluble than the selenite minerals present (Elrashidi et al., 1987).

19.2 UPTAKE OF SELENIUM BY PLANTS

Plants take up Se as selenate or selenite ions or as organic selenium compounds such as selenomethionine (Sors et al., 2005b; Kopsell and Kopsell, 2007), so it is the availability of these Se forms in soils that affects uptake of the element.

In four different plant species supplied Se as selenate, selenite, or selenomethionine at 2 μM concentration in hydroponic culture, with broccoli (*Brassica oleracea* var. *botrytis* L.) and Indian mustard (*Brassica juncea* Czern.), there was more uptake per plant from selenate than selenomethionine and considerably more uptake from either of these forms than from selenite. With sugar beet (*Beta vulgaris* L.), there was more uptake from selenomethionine than from selenate, with selenite again giving the lowest uptake per plant. With rice (*Oryza sativa* L.), uptake followed the order selenomethionine > selenite > selenate (Zayed et al., 1998). Another study on Indian mustard showed a rate of uptake of selenate from 20 µM concentration in hydroponic culture that was double the rate of selenite uptake (de Souza et al., 1998). Black cabbage (*Brassica campestris* L.) accumulated more Se when it was supplied as selenate than as selenite (Zhao et al., 2005). Accumulation of Se in roots and shoots of cowpea (*Vigna unguiculata* Walp.) in hydroponic culture was 18 and 61 times lower, respectively, 24 h after 20 µM selenite was supplied than when selenate was the Se source (Wang et al., 2013).

With common bean (*Phaseolus vulgaris* L.) grown in hydroponic culture, selenate and selenite supplied at 5 µM concentration at pH 5.1 were taken up in hydroponic culture, in approximately equal amounts (Arvy, 1993). Similarly, wheat (*Triticum aestivum* L.) plants took up selenate and selenite at approximately equal rates where the Se forms were supplied individually in hydroponic culture, although when they were supplied together selenite uptake was slightly faster than selenate uptake (Li et al., 2008). In contrast, in corn (*Zea mays* L.) grown in hydroponic culture the concentration of Se in the whole plants after 2 weeks was much higher when selenite rather than selenate was supplied (Longchamp et al., 2013).

The uptake of selenate has been known for some time to be an active process. The SeO_4^{2-} ion shares at least one high-affinity carrier with the sulfate (SO_4^{2-}) ion (Takahashi et al., 2000; Li et al., 2008), and as a consequence high availability of sulfate in the rooting medium lowers its uptake (Hopper and Parker, 1999). Se-hyperaccumulator *Astragalus* species seem to have a higher uptake capacity for sulfur than nonaccumulator *Astragalus* species, a further indication that uptake of selenate occurs by means of sulfate transporters (Cabannes et al., 2011). The SULTR1;1 and SULTR1;2 high-affinity sulfate transporters in the lateral root cap, root hairs, epidermis, and cortex of roots have been shown to transport selenate into arabidopsis (*Arabidopsis thaliana* Heynh.) (Takahashi et al., 2000; Shibagaki et al., 2002; El Kassis et al., 2007; Barberon et al., 2008; Van Hœwyk et al., 2008).

Although it was long thought that uptake of selenite is mainly passive (Arvy, 1993), it is now known that it is an active process (Li et al., 2008). There is evidence that high concentration of phosphate in the rooting medium lowers uptake of the selenite ion (Hopper and Parker, 1999), and it is thought that selenite is taken up at least in part through a phosphate transporter as the presence of phosphate markedly lowers the affinity of the uptake system for selenite (Li et al., 2008). Phosphate is transported into plants mainly as the $H_2PO_4^-$ ion (Ullrich-Eberius et al., 1981), and as in the mildly acid conditions used in many plant nutrition experiments selenite would actually be in the $HSeO_3^-$ form, it is apparent that similar charge and shape properties could allow the two ions to compete for uptake. In rice, the NIP2;1 silicon transporter (an aquaporin) is able to take up selenite (Zhao et al., 2010).

Inoculation with the mycorrhizal fungus *Glomus mosseae* had no effect on the biomass of roots or shoots of ryegrass (*Lolium perenne* L.) in soil artificially contaminated with 12 mg Se kg^{-1} as sodium selenate, but the Se content of the plants was much smaller in the mycorrhizal plants as the concentration of Se per unit shoot dry mass was only about one-third of the concentration in the nonmycorrhizal plants (Munier-Lamy et al., 2007). It is not at all clear why being mycorrhizal should have this effect, which could be caused by changes in soil properties, changes in other microorganisms that affect soil properties, or another unidentified reason.

When selenate or selenite is the form of Se taken up, the inorganic ion must be converted to organic forms. When selenate is the Se source, it is translocated rapidly to the shoots and remains as selenate in the plant for some time (Asher et al., 1967; Arvy, 1993; de Souza et al., 1998; Zayed et al., 1998; Li et al., 2008). Of the Se taken up as selenate by cowpea seedlings, 29% had been translocated to the shoots after 24 h, compared with 15% of the smaller amounts of Se taken up as selenite (Wang et al., 2013). When selenite is the Se source, it can be reduced nonenzymatically to selenide (S^{2-}) and remain in the roots, or it can be converted to organic forms and transported to the shoots as Se-amino acid (Gissel-Nielsen, 1979; de Souza et al., 1998; Li et al., 2008). In an experiment on corn, although the plants took up selenite more than selenate, there was a higher transfer factor for Se with selenate supply, and the leaves of both groups of plants accumulated similar Se concentrations (Longchamp et al., 2013). In durum wheat (*Triticum turgidum* subsp. *durum* Husn.) and spring canola (*Brassica napus* L.), a much higher proportion of the Se was translocated to the shoots within 24 h of uptake when selenate was the source than when the Se was supplied as selenite (Renkema et al., 2012).

The SULTR2;1 transporter, present in the xylem parenchyma cells of roots and leaves and in root pericycle and leaf phloem cells, which loads sulfate into the vascular tissue in roots, may be involved in moving selenate from roots to shoots (Takahashi et al., 2000; Van Hœwyk et al., 2008). The overall process of uptake and loading of Se into the xylem is certainly active, as in canola the transpiration stream concentration factor (the mass of Se in the shoots/mass of water transpired) was up to a value of 8.1 within 8 h of supplying selenate (Renkema et al., 2012). However, in durum wheat it took 24 h to get to values approaching 1.0, indicating that either overall uptake and translocation had been a passive process during that time or transporters moved Se out of the xylem in the roots as well as into it. Supply of selenite to both plant species gave values well below 1.0, indicating that the uptake and translocation of selenite over the 24 h was not an active process overall (Renkema et al., 2012).

Selenium can be mobilized from leaves to developing fruits and seeds, and hyperaccumulator plants contain high concentrations of Se in their seeds, as well as in vegetative organs (Galeas et al., 2007). Uptake of Se by wheat continues throughout the growth of the plants, but Se concentration in the leaves and stems decreases during growth as it increases in the grains, so some at least of the grain Se comes from remobilization within the plants (Keskinen et al., 2010; Shinmachi et al., 2010).

19.3 PHYSIOLOGICAL RESPONSES OF PLANTS TO SUPPLY OF SELENIUM

Selenate is similar to sulfate, so not only is it taken up by sulfate transporters but the selenium in it is incorporated into the amino acids cysteine and methionine. Selenate taken up by plants is translocated to the shoot, and in the chloroplasts it is reduced to selenite in reactions catalyzed by the enzymes ATP sulfurylase and adenosine phosphosulfate reductase (Van Hœwyk, 2013). This reduction seems to be the rate-limiting step of its assimilation, as selenate ions accumulate in plants. Selenite arising from selenate in the shoots, or accumulated in the roots after uptake in that form, is converted to selenide and then to the amino acids selenocysteine (catalyzed by the enzyme cysteine synthase) and selenomethionine. Cysteine synthases from both accumulator and nonaccumulator plants may have higher affinities for Se^{2-} than for sulfide (S^{2-}) (Ng and Anderson, 1978). Selenocysteine can be converted further to methylselenocysteine and ultimately to volatile dimethyl selenide and dimethyl diselenide. Selenomethionine can go through a similar route to form these two volatile Se compounds (Van Hœwyk, 2013). The rate of volatilization in plants is lower when Se is supplied as selenite rather than selenomethionine and lower still with selenate supply (Zayed et al., 1998). Selenocysteine also can be broken down to the amino acid alanine and elemental Se^0 by the enzyme selenocysteine lyase (Bañuelos et al., 2007).

Selenocysteine may be incorporated into proteins, and as it is not quite the same shape as normal cysteine protein folding can be affected (Van Hœwyk, 2013). This could affect the action of key enzymes and be dangerous to plants. The growth of arabidopsis has been shown to be related inversely to the ratio of selenium/sulfur in the shoots, indicating a competition between Se and S for incorporation into cysteine and methionine so that plants are most at risk of Se toxicity when S supply is low (White et al., 2004). Direct evidence of misfolding of selenoproteins has not been found, but in the hyperaccumulator prince's plume (*Stanleya pinnata* Britton), selenate seems to promote the breakdown of selenoproteins, with the possibility that these are degraded because of their misfolding (Sabbagh and Van Hœwyk, 2012).

The toxic effect may be due also to selenium accumulation causing formation of reactive oxygen species (ROS), which in turn degrade lipids in membranes (Gomes-Junior et al., 2007; Tamaoki et al., 2008). In white clover (*Trifolium repens* L.), lipid peroxidation decreased with increasing shoot concentration up to approximately 0.20 mg Se kg^{-1} dry mass and then increased above this concentration. Radical scavenging activity increased up to 0.20 mg Se kg^{-1} shoot dry mass and decreased above this concentration (de la Luz Mora et al., 2008). Accumulation of ROS seems to be an effect caused by both selenate and selenite (Van Hœwyk, 2013).

Selenium induces antioxidant defenses in plants, and this is possibly why low doses of Se are beneficial to plant growth even though higher concentrations are inhibitory (Hartikainen, 2005). However, the concentration of the reduced form of the antioxidant glutathione (which is a tripeptide that has cysteine as one of the amino acids present) has been shown to be fourfold lower in shoots of arabidopsis after treatment with selenate (Van Hœwyk et al., 2008). A decrease in the ratio of the reduced form of the antioxidant ascorbic acid to its oxidized form also was seen.

Glutathione and ascorbic acid are important in maintaining oxidation–reduction potential in cells, and the fact that the ratio of reduced forms to oxidized forms decreased implies that oxidizing conditions had arisen following the uptake of selenate. Increased concentrations of Se in shoots of ryegrass arising from either sodium selenate or sodium selenite gave rise to increased activity of the antioxidant enzyme glutathione peroxidase [EC 1.11.1.9] (Cartes et al., 2005, 2011). This enzyme, which catalyzes the reduction of peroxides by reduced glutathione, is a selenoprotein, and an increase in its activity could be one of the reasons why levels of reduced glutathione decrease with selenate toxicity.

Supply of selenate to arabidopsis was shown to affect the transcription of 893 genes in the roots and 385 genes in the shoots (Van Hœwyk et al., 2008). Nearly 50% of the upregulated genes in both parts of the plants were involved in responses to abiotic and biotic stress, including genes for calcium signaling and synthesis of jasmonic acid and ethylene and genes for the synthesis of stress proteins such as members of the heat shock protein family. Signaling by jasmonic acid and ethylene seems to be strongly involved in resistance to selenite as in mutants of a selenite-resistant accession of arabidopsis in which the synthesis of these plant growth regulators was lowered some of the resistance was lost (Tamaoki et al., 2008).

As different plant species accumulate Se to different extents, they must also show different tolerances to Se as well. In a comparison of Indian mustard, corn, rice, and wheat, significant reductions in plant yield were seen above 105, 77, 41.5, and 19 mg Se kg^{-1} shoot mass, respectively, but rice could tolerate a higher rate of Se supply to the soil before it accumulated its own critical tissue Se concentration (Rani et al., 2005). The most extreme tolerance is seen in hyperaccumulators, which can accumulate so much selenium that they are toxic to animals (Van Hœwyk, 2013).

In a comparison of the hyperaccumulator prince's plume (*S. pinnata*) with the secondary accumulator white prince's plume (*Stanleya albescens* M. E. Jones), the hyperaccumulator had a 50% higher capacity to scavenge radicals than the secondary accumulator in the absence of Se, with a 40% higher concentration of glutathione in the young leaves and no decrease in the ratio of reduced glutathione to oxidized glutathione on supply of Se, unlike the secondary accumulator (Freeman et al., 2010). Prince's plume is able to convert much of its selenocysteine into methylselenocysteine, which minimizes the risk of selenocysteine being incorporated into proteins. It was shown that 80% of the Se in young leaves was present as methylselenocysteine and 20% as selenocysteine (Freeman et al., 2010). Three hyperaccumulator *Astragalus* species were shown to accumulate fivefold higher concentrations of methylselenocysteine in their shoots than five secondary accumulator *Astragalus* species (Sors et al., 2005a). Both nonaccumulator and accumulator *Astragalus* species contain a protein that has potential homocysteine methyltransferase enzyme activity, but a mutation in the ancestors of the hyperaccumulator species appears to have given them selenocysteine methyltransferase activity (Sors et al., 2009).

19.4 GENETICS OF ACQUISITION OF SELENIUM BY PLANTS

As both the transporters of selenium and the enzymes involved in its assimilation are coded for by different genes, it is only to be expected that there should be genotypic differences between plants in their acquisition of the element. The study of mutants that have deletions or overexpression of any of these genes is a valuable technique in trying to ascertain how uptake of the element occurs. For example, the *sultr1;2* mutant of arabidopsis that has no SULTR1;2 activity contained lower concentrations of selenate than the *sultr1;1* mutant (deficient in SULTR1;1), which, however, did not contain lower concentrations than the wild type (Barberon et al., 2008). The *sultr1;1–sultr1;2* double mutant had lower selenate concentrations than either of the single mutants. It therefore seems as if SULTR1;2 is more important for selenate uptake in arabidopsis than SULTR1;1, although absence of both transporters was required to lower uptake of selenate substantially and give increased tolerance to Se phytotoxicity (Barberon et al., 2008). This emphasizes the genetic basis of selenium acquisition and gives the possibility of selecting crop plants that accumulate more Se to benefit humans and livestock (genetic biofortification), plants that are able to grow on seleniferous soils without taking up Se at levels that would bring about phytotoxic effects, and plants that can grow on seleniferous soils and take up Se to the extent that they can be used in phytoremediation.

Various studies have investigated crop species, to see if there are genotypic differences in rates of Se uptake. In rice, a study of 10 genotypes showed differences of 1.72 times the rate of uptake of selenite between the cultivar with the fastest rate and the one with the slowest rate (Zhang et al., 2006). In a study comparing wheat grown in different areas of South Dakota, within the main wheat-growing region of North America, in which 8 winter wheat and 10 spring wheat cultivars were grown in soils naturally rich in Se and soils with a lower Se concentration, there were found to be significant differences in grain Se concentration between the different field sites, but not between the cultivars (Lee et al., 2011c). In the 2006 experimental season, the winter wheat cultivars had a grain concentration that ranged from 0.40 to 1.18 mg Se kg^{-1} across four field sites, and the spring wheat cultivars had mean values of 0.38 and 1.07 mg Se kg^{-1} in two field sites. There was a positive correlation between wheat grain Se concentration and the concentration of plant-available Se in the soil (Lee et al., 2011c). Field trials of 100 bread wheat cultivars and landraces from a range of countries showed large variation in grain Se concentration between replicates (Lyons et al., 2005), with no consistent pattern of one genotype accumulating more Se in its grains than others. For wheat, at least, it appears that increasing the supply of Se to the crop (agronomic biofortification) would be a more effective way of increasing grain Se content than genetic biofortification.

There are obviously big differences between the concentrations of selenium in hyperaccumulator plants and nonaccumulators (see Section 19.5), and many hyperaccumulators are in the families Fabaceae (e.g., *Astragalus* species) and Brassicaceae (e.g., *S. pinnata*). However, even closely related species may be nonaccumulators, and silky milkvetch (*Astragalus sericoleucus* A. Gray) was found to accumulate 100 times lower Se concentrations in its leaves than the hyperaccumulator two-grooved milkvetch (*Astragalus bisulcatus* A. Gray) (Galeas et al., 2007). Perhaps the fact that hyperaccumulators take up more Se than other species when grown under the same conditions could be used to breed enhanced Se uptake into crop species, but they tend to grow on seleniferous soils, so they may not be more effective at acquiring Se than crop plants at normal soil Se concentrations but are merely more tolerant of Se in high-Se soils (Zhu et al., 2009).

Attempts to increase uptake of Se by increasing the tolerance of plants to its potential phytotoxic effects have been made by overexpressing genes for the enzymes selenocysteine lyase or selenocysteine methyltransferase in Indian mustard (Bañuelos et al., 2007). Overexpression of either of these enzymes gave greater accumulation of selenium in the shoots of plants grown in seleniferous drainage sediment in the field than in wild-type plants, and it was suggested that this would be a useful mechanism in breeding for phytoremediation of selenium-contaminated land (Bañuelos et al., 2007). Other approaches for genetic manipulation of plants for phytoremediation have been suggested (Zhu et al., 2009).

19.5 CONCENTRATIONS OF SELENIUM IN PLANTS

Most plant species start to suffer from Se toxicity at comparatively low tissue concentrations, below 10–100 mg Se kg^{-1} dry mass, and these nonaccumulator species typically contain less than 25 mg Se kg^{-1} dry matter (White et al., 2004). Other species, grown in both nonseleniferous and seleniferous soils, can accumulate concentrations of up to 1000 mg Se kg^{-1} safely. These are sometimes referred to as Se-indicator plants (White et al., 2004) or secondary accumulators (Brown and Shrift, 1982; Terry et al., 2000; Freeman et al., 2010). A small number of species, rarely found in nonseleniferous soils, tolerate higher tissue concentrations and are Se accumulators (White et al., 2004). This group is usually referred to as hyperaccumulators (e.g., Sors et al., 2005a), and concentrations of over 12,000 mg Se kg^{-1} dry matter were seen in spring in young leaves of the hyperaccumulator two-grooved milkvetch grown in Se-rich soil in its native habitat (Table 19.1) (Galeas et al., 2007).

Within the nonaccumulators that are used as crops, total uptake of Se is variable according to species (Table 19.1). Pasture grasses tend to contain two times higher Se concentrations than cereals and plants in the family Brassicaceae even higher concentrations (Gissel-Nielsen et al., 1984). High concentrations of Se in brassicas may occur because these plants contain high concentrations of sulfur-containing compounds, but it has also been suggested that since Se occurs as selenoamino acids in proteins, those plants with higher protein concentrations could have higher Se concentrations (Gupta and Gupta, 2000). In a study of durum wheat and spring canola, the Se concentration (and also the rate of uptake of Se over 24 h) was much higher in the canola than in the wheat (Renkema et al., 2012). Within the cereals, rye (*Secale cereale* L.) appears to have a higher grain Se concentration than bread wheat, although ancestral diploid wheats may have higher grain Se concentration than modern hexaploids (Lyons et al., 2005).

In a series of crops grown on the low-sulfur soils of Prince Edward Island, Canada, peas (*Pisum sativum* L.) contained nearly four times higher concentrations of Se than maize (Gupta and Gupta, 2000). Plants in the family Fabaceae certainly seem to take up more Se after selenium fertilization than plants in the Gramineae, and soybeans (*Glycine max* Merr.) contained 0.93 mg Se kg^{-1} in their leaves after fertilization with sodium selenate compared with 0.17 mg kg^{-1} in barley at the boot stage and 0.56 mg kg^{-1} in oats at the boot stage (Gupta and Macleod, 1994). In that study, only ryegrass (*L. perenne*) of the species tested had concentrations of Se close to those seen in soybeans after fertilization (0.73 mg kg^{-1}). In a study on the effects of adding fly ash to low-Se soils, the leguminous species bird's-foot trefoil (*Lotus corniculatus* L.), alfalfa (*Medicago sativa* L.), and bean (*P. vulgaris*) contained higher concentrations of Se than maize when the fly ash supplemented the soils (Mbagwu, 1983).

Vegetables in the *Allium* genus (in the subfamily Allioideae of the Amaryllidaceae) contain high amounts of selenium. Green onions (spring onions, scallions, *Allium fistulosum* L.) watered for two weeks with selenate, selenite, or selenomethionine contained 66, 12, and 11 mg Se kg^{-1} dry mass, respectively; chives (*Allium schoenoprasum* L.) contained 613, 222, and 265 mg Se kg^{-1} dry mass, respectively (Kápolna and Fodor, 2007). As can be seen in Table 19.1, most species contain higher tissue Se concentrations when selenate is the source of Se than with selenite. Concentrations of Se mostly tend to be higher in leaves than in roots of both accumulator and nonaccumulator species (Table 19.1; Kopsell and Kopsell, 2007), probably indicative of selenate being the Se source most commonly taken up, and being translocated to the shoots.

The concentration of Se in plants is affected strongly by the availability of Se in the soil, as well as which Se form is available. For example, concentrations in kenaf (*Hibiscus cannabinus* L.) shoots were raised from less than 1.0 mg Se kg^{-1} dry matter with no Se fertilizer to 8 mg Se kg^{-1}

TABLE 19.1
Total Selenium Concentrations in Plants

Plant	Plant Part	Se Concentration (mg kg⁻¹ Dry Mass)	References
Crop species			
Alfalfa (*Medicago sativa* L.)	Hay (grown in Oregon, United States)	0.07, 0.95, 1.55, and 3.26 in a second cut after the application of 0, 22.5, 45.0, and 89.9 g Se ha⁻¹ as selenate 0.16, 0.28, and 0.60 in a third cut after the application of 22.5, 45.0, and 89.9 g Se ha⁻¹ as selenate	Hall et al. (2013)
Barley (*Hordeum vulgare* L.), one accession grown in comparison with rye, triticale, and wheat	Grain	0.064	Lyons et al. (2005)
Rapid-cycling brassica (*B. oleracea* L., model for *B. oleracea* crops)	Leaves	0–373 in hydroponics with 0–1.5 mg Se L⁻¹ sodium selenate supplied	Toler et al. (2007)
Broccoli (*B. oleracea* L. var. *italica* Plenck)	Leaves	1,200 (with 40 μM selenate supplied) 130 (with 40 μM selenite supplied)	Lyi et al. (2005)
	Florets	Approx. 1,200 (with 20 μM selenate supplied) Less than 100 (with 20 μM selenite supplied)	
Broccoli (*B. oleracea* var. *italica*)	Shoots	801–1,798 (range in 38 accessions, 20 μM selenate in hydroponic culture)	Ramos et al. (2011)
Carrot (*Daucus carota* L. subsp. *sativus*)	Leaves	25.0	Kápolna et al. (2012)
	Roots	2.8 (50 μM ⁷⁷Se selenite supplied regularly during growth)	
Chicory (*Cichorium intybus* L.)	Leaves	480, 460, 456, 167 (four cvs. supplied 7 mg Se L⁻¹ as selenate)	Mazej et al. (2008)
Chives (*Allium schoenoprasum* L.)	Leaves	613 (selenate) 222 (selenite) 265 (selenomethionine)	Kápolna and Fodor (2007); Kápolna et al. (2007)
	Roots	722 (selenate) 494 (selenite) 422 (selenomethionine) Watered with solutions containing 10 mg L⁻¹ of Se	
Cowpea (*Vigna unguiculata* Walp.)	Roots	97.4/39.5/(mean at 0.2 mm from root apex, selenate/selenite supplied) 151/91.9 (mean at 0.6 mm from root apex, selenate/selenite supplied) 307/55.9 (mean at 2 mm from root apex, selenate/selenite supplied)	Wang et al. (2013)
Dandelion (*Taraxacum officinale* F. H. Wigg.)	Leaves	49 (7 mg Se L⁻¹ supplied as selenate)	Mazej et al. (2008)
Lamb's lettuce (*Valerianella locusta* Laterr.)	Leaves	455 (7 mg Se L⁻¹ supplied as selenate)	Mazej et al. (2008)

(Continued)

TABLE 19.1 (*Continued*)
Total Selenium Concentrations in Plants

Plant	Plant Part	Se Concentration (mg kg⁻¹ Dry Mass)	References
Leek (*Allium ampeloprasum* L. subsp. *ampeloprasum*, syn. *A. ampeloprasum* var. *porrum*)	Plant	102–982 (0.2–3.8 mg Se kg⁻¹ soil supplied as selenate) 24.1–103 (0.2–3.8 mg Se kg⁻¹ soil supplied as selenite)	Srikanth Lavu et al. (2012)
Lettuce (butterhead lettuce, *Lactuca sativa* L. var *capitata*)	Shoots	0.7 (no added Se in hydroponics) 14.6/11.3 (6 μM Se as selenate/selenite) 43.3/30.6 (15 μM Se as selenate/selenite)	Hawrylak-Nowak (2013)
	Roots	1.1 (no added Se in hydroponics) 21.1/127.9 (6 μM Se as selenate/selenite) 44.2/201.4 (15 μM Se as selenate/selenite)	
Maize, corn (*Zea mays* L.)	Leaves	*ca* 0.9 (10 μg Se L⁻¹ supplied as either selenate or selenite)	Longchamp et al. (2013)
6-week old plants	Roots	*ca* 3.3 (50 μg Se L⁻¹ supplied as either selenate or selenite) *ca* 2 (10 μg Se L⁻¹ as selenate) *ca* 17 (10 μg Se L⁻¹ as selenite) *ca* 5 (50 μg Se L⁻¹ as selenate) *ca* 53 (50 μg Se L⁻¹ as selenite)	
Onion (*Allium cepa* L.)	Leaves	601 (5 mg Se L⁻¹ supplied as selenate) 154 (5 mg Se L⁻¹ supplied as selenite)	Wróbel et al. (2004)
	Bulb	51.3 (5 mg Se L⁻¹ supplied as selenate) 15.6 (5 mg Se L⁻¹ supplied as selenite)	
Onion (*Allium cepa*)	Leaves	6.3	Kápolna et al. (2012)
	Bulb	6.8 (50 μM ⁷⁷Se selenite supplied regularly during growth)	
	Leaves	<0.045 (control plants in soil/quartz) 5.2/4.0 (late foliar application of 10 mg Se L⁻¹ as selenate/selenite) 9.6/7.8 (late foliar application of 100 mg L⁻¹ selenate/selenite)	
	Bulb	<0.045, 0.2 (control plants in soil/quartz) 2.1/1.1 (late foliar application of 10 mg Se L⁻¹ as selenate/selenite) 9.1/4.3 (late foliar application of 100 mg Se L⁻¹ as selenate/selenite)	
Parsley (*Petroselinum crispum* Nyman)	Leaves	290 (7 mg Se L⁻¹ supplied as selenate)	Mazej et al. (2008)
Rape (*Brassica napus* L.) rosette stage	Shoots	0.75/160/690/679 (0/1/2/4 mg Se kg⁻¹ soil supplied as selenate)	Sharma et al. (2010)
	Roots	0.14/27.4/23.6/19.4	
	Shoots	0.75/1.9/4.3/9.0 (0/1/2/4 mg Se kg⁻¹ soil supplied as selenite)	
	Roots	0.14/1.5/5.1/9.0	

(Continued)

TABLE 19.1 (*Continued*)
Total Selenium Concentrations in Plants

Plant	Plant Part	Se Concentration (mg kg⁻¹ Dry Mass)	References
Rye (*Secale cereale* L.)	Grain	0.093	Lyons et al. (2005)
One accession grown in comparison with barley, triticale, and wheat seedlings	Shoot	19, 25 (two accessions in hydroponics, Se supplied at 80 µg L⁻¹ as selenate)	
Ryegrass (*Lolium multiflorum* L.)	Shoots	*Sandy soil* 0.018 and 0.019 (first and second cuts, 0 Se added in the previous year) 0.030 and 0.026 (first and second cuts, 0.0025 mg Se added L⁻¹ soil in the previous year) 0.066 and 0.053 (first and second cuts, 0.005 mg Se added L⁻¹ soil in the previous year) *Silty clay* 0.029 and 0.020 (1st and 2nd cuts, 0.0025 mg Se added L⁻¹ soil in the previous year) 0.057 and 0.037 (1st and 2nd cuts, 0.005 mg Se added L⁻¹ soil in the previous year)	Keskinen et al. (2010)
	Roots	*Sandy soil* 0.291 (0 Se added in the previous year) 0.365 (0.005 mg Se added L⁻¹ soil in the previous year) *Silty clay* 0.322 (0 Se added in the previous year) 0.337 (0.005 mg Se added L⁻¹ soil in the previous year)	
Scallions, spring onions, green onions, Welsh onions (*Allium fistulosum* L.)	Leaves	66.4 (selenate) 11.8 (selenite) 11.2 (selenomethionine) Watered with solutions containing 10 mg L⁻¹ of Se	Kápolna and Fodor (2007)
Soybean (*Glycine max* (L.) Merr.)	Seed	0.051 (0 added Se, Andosol[a] soil 1) 0.047 (20 µg Se as selenite added to 2 kg Andosol soil 1) 0.024 (0 added Se, Andosol soil 2) 0.037 (20 µg Se as selenite added to 2 kg Andosol soil 2) 0.008 (0 added Se, Andosol soil 3)	Nakamaru and Sekine (2008)
Tobacco (*Nicotiana tabacum* L.)	Leaves Stems Roots	2.6–37.3 (2.2–22 mg Se kg⁻¹ soil as selenite) 2.2–10.3 (2.2–22 mg Se kg⁻¹ soil as selenite) 16.7–58.6 (2.2–22 mg Se kg⁻¹ soil as selenite)	Han et al. (2013)
Triticale (× *Triticosecale* Wittm. ex A. Camus), one accession grown in comparison with barley, rye, and wheat	Grain	0.071	Lyons et al. (2005)
Wheat (*Triticum aestivum* L.)	Grain	*Winter wheat in the United Kingdom* 0.010–0.115 (2003 field experiments) 0.011–0.054 (2004 field experiments)	Zhao et al. (2007)

(Continued)

TABLE 19.1 (*Continued*)

Total Selenium Concentrations in Plants

Plant	Plant Part	Se Concentration (mg kg⁻¹ Dry Mass)	References
Wheat (*Triticum aestivum*)	Grain	Mean values for two separate field experiments in the United Kingdom 0.039/0.099 (0 Se, 0 S) 0.016/0.039 (0 Se, 20 kg S ha⁻¹) 0.266/0.373 (20 g Se ha⁻¹, 0 S)	Stroud et al. (2010b)
	Straw	0.432/0.322 (20 g Se ha⁻¹, 20 kg S ha⁻¹) 0.014/0.023 (0 Se, 0 S) 0.012/0.051 (0 Se, 20 kg S ha⁻¹) 0.104/0.151 (20 g Se ha⁻¹, 0 S) 0.216/0.179 (20 g Se ha⁻¹, 20 kg S ha⁻¹)	
Wheat (*Triticum aestivum*)	Grain	*Winter wheat* 0.36 (site 1) 0.43 (site 2) 0.44 (site 3) *Spring wheat* 0.58 (site 1) 1.05 (site 4) Value means over two seasons, all field sites in S. Dakota, United States	Lee et al. (2011c)
Wheat (and ancestors)			Lyons et al. (2005)
Triticum aestivum 100 genotypes, some durum wheat genotypes included	Grain	0.056 (range 0.009–0.244)	
90 landraces, some durum wheat genotypes included		0.164 (range 0.015–0.510)	
10 commercial bread wheats in South Australia		0.155 (range 0.005–0.720)	
1 cv. bread wheat in comparison with barley, triticale, and rye		0.071 0.135	
Triticum dicoccum Schrank (BBAA genome)	Shoot		
Aegilops tauschii Coss. (DD genome)		0.179	
		21, 27 (2 accessions in hydroponics, Se supplied at 80 μg L⁻¹ as selenate)	
T. aestivum seedlings			
Wheat (*T. aestivum*) grown in seleniferous soil in Punjab, India	Shoot Grain	11–146 (lowest–highest values) 29–185 (lowest–highest values)	Cubbada et al. (2010)
White clover (*Trifolium repens* L.)	Shoots	Nonlimed/limed Andisol soil in Chile 0.062/0.023 (0 P, 0 Se) 0.359/0.857 (0 P, 40 g Se ha⁻¹ as selenite) 0.065/0.033 (400 mg P kg⁻¹ soil, 0 Se) 0.311/0.625 (400 mg P kg⁻¹ soil, 40 g Se ha⁻¹ as selenite)	de la Luz Mora et al. (2008)

(Continued)

TABLE 19.1 (*Continued*)
Total Selenium Concentrations in Plants

Plant	Plant Part	Se Concentration (mg kg⁻¹ Dry Mass)	References
Other species			
Astragalus bisulcatus (Hook.)	Young leaf	Up to 6,000 (2004) Up to 12,700 (2005)	Galeas et al. (2007)
Gray (hyperaccumulator)	Roots	Up to 1,500	
	Seeds	6,500	
Stanleya pinnata (Pursh)	Young leaf	Up to 1,900 (2004) Up to 4,400 (2005)	
Britton (hyperaccumulator)	Roots	Up to 2,300 (2005)	
	Seeds	3,300 in field at Fort Collins, Colorado, United States	
Stanleya albescens M. E.	Leaf	818	Freeman et al. (2010)
Jones (secondary accumulator)	Root	201	
Stanleya pinnata	Leaf	2,973	
(hyperaccumulator)	Root	721	
		(after 16 weeks on agar with 20 μM selenate)	

[a] In the USDA classification, Andosol soils are known as Andisols.

with sodium selenite applied at 4 mg Se kg⁻¹ soil and 1000 mg Se kg⁻¹ dry matter at the same rate of application of sodium selenate (Srikanth Lavu et al., 2013). Se concentrations in shoots of white clover increased linearly up to 0.6 mg Se kg⁻¹ dry mass with increase in supply of sodium selenite up to 60 g Se ha⁻¹, and Se concentrations in the roots were increased up to 6.0 mg kg⁻¹ by 20–25 g Se ha⁻¹ (de la Luz Mora et al., 2008).

In plants, Se occurs as selenoamino acids in proteins (Section 19.3), and in a study on ryegrass, over 73% of the Se in the plants was incorporated into the organic fraction (Cartes et al., 2006). Most of this accumulation was in insoluble form (in membrane-bound proteins or insoluble proteins), with less than 10% of total shoot selenium in soluble proteins and amino acids.

Selenoamino acids in, or bound to, proteins are liberated by extracting plant materials in the presence of protease. In protease extracts of kenaf plants supplied selenate, nearly 50% of the Se in the shoots was as selenate, with selenomethionine and selenocystine accounting for a further 10% and unidentified compounds the rest, but in plants supplied selenite approximately 30%, 20%, and 18% of the Se in the plants was present as selenomethionine, selenocystine, and selenate, respectively (Srikanth Lavu et al., 2013). Selenite did not accumulate under either treatment.

Although the green onion and chives plants of Kápolna and Fodor (2007) contained three to six times more Se with selenate supply, a high proportion of it was as selenate, whereas there was a bigger proportion as organic Se compounds with supply of selenite or selenomethionine. Approximately 55% of the total Se in leek (*Allium ampeloprasum* subsp. *ampeloprasum* L., syn. *A. ampeloprasum* var. *porrum* L.) plants supplied selenate was in inorganic form, but only 21% when the plants were supplied selenite. Methylselenocysteine and selenomethionine were the major Se forms with supply of both selenate and selenite (Srikanth Lavu et al., 2012). Concentrations of Se forms in selected crop species are given in Table 19.2, and more detailed presentation of similar data is given by Pyrzynska (2009) and by Srikanth Lavu et al. (2012). It is clear from Table 19.2 that with selenate supply, a high proportion of the total plant Se remains as selenate, whereas with supply of selenite a much higher proportion (although not necessarily a higher amount) is present as organic Se forms.

TABLE 19.2
Concentrations of Selenium Compounds in Selected Plant Species (Values in mg Se kg⁻¹ Dry Mass)

Plant	Se Source	Se(VI)	Se(IV)	Se-Met	Se-Cys$_2$	MeSeCys	γ-glu-MeSeCys	References
Carrot (*Daucus carota* L. subsp. *sativus*) roots	Selenite (50 mg Se L⁻¹ supplied throughout growth)	0.071	0.20	1.47		1.23	<0.09	Kápolna et al. (2012)
Chicory (*Cichorium intybus* L.) leaves, 4 cvs.	Selenate (7 mg Se L⁻¹, aeroponics)	313	0.9	40	—	3.4		Mazej et al. (2008)
		390	1.0	28	0.9	2.3		
		264	1.8	68	3.7	7.4		
		113	0.2	20	ND	2.7		
Chives (*Allium schoenoprasum* L.) leaves	Selenate	313	—	—	135	127		Kápolna et al. (2007)
	Selenite	11.1	6.7	11.1	93	80		
	Selenomethionine (pots watered with 10 mg L⁻¹)*	—	2.7	8.0	98	127		
Dandelion (*Taraxacum officinale* F. H. Wigg.) leaves	Selenate (7 mg Se L⁻¹, aeroponics)	27	0.1	10	—	1.0		Mazej et al. (2008)
Lamb's lettuce (*Valerianella locusta* Later.) leaves	Selenate (7 mg Se L⁻¹, aeroponics)	182	3.1	32	3.5	13		Mazej et al. (2008)
Leek (*Allium ampeloprasum* L. subsp. *ampeloprasum*) leaves	Selenate (3 mg Se kg⁻¹ soil)	466	5.2	101.9	6.2	35.2	5.3	Srikanth Lavu et al. (2012)
	Selenate (3 mg Se kg⁻¹ soil)	26.5	—	17.3	1.1	5.6	—	
Onion (*Allium cepa* L.) bulb	Selenite (50 mg Se L⁻¹ supplied throughout growth)	<0.06	<0.06	0.23		2.41	1.83	Kápolna et al. (2012)
	Late foliar application of selenate	0.68		0.17		0.12	3.1	
	and selenite at 100 mg Se L⁻¹	0.05		0.28		0.17	3.9	
Parsley (*Petroselinum crispum* Nyman) leaves	Selenate (7 mg Se L⁻¹, aeroponics)	175	2.0	25	2.3	12.8		Mazej et al. (2008)
Wheat (*Triticum aestivum* L.) grains	Seleniferous soil	0.35 (low total Se sample)	ND	17.1	Trace	0.04	—	Cubadda et al. (2010)
		5.1 (high total Se sample)	ND	122	Trace	0.74	—	

Note: All extracts had protease added to release selenoamino acids in proteins.

Abbreviations: Se(VI), selenate; Se(IV), selenite; Se-Met, selenomethionine; Se-Cys$_2$, Se-selenocystine or Se-selenocystine (indistinguishable); MeSeCys, Se-methylselenocysteine; γ-glu-MeSeCys, γ-glutamyl-Se-methylselenocysteine; Ref., reference; ND, not determined; —, indicates investigated, but not detectable; no entry denotes no information; * 10 mg L⁻¹ presumably means "10 mg Se L⁻¹" in watering solution, but not made clear by authors.

19.6 INTERACTION OF SELENIUM WITH UPTAKE OF OTHER ELEMENTS

19.6.1 SULFUR

Selenate is taken up by means of sulfate transporters (Section 19.2), and it has been known for a long time that sulfate strongly inhibits uptake of selenate (e.g., Zayed et al., 1998). These authors showed that uptake of selenite and selenomethionine were not affected by sulfate supply, although in a more recent study uptake of Se from sodium selenite was lowered by supply of sulfate to ryegrass, and the decreased concentration of Se in the shoots gave lowered activity of glutathione peroxidase and increased peroxidation of lipids (Cartes et al., 2006). In rice, there was a slight inhibition of uptake of selenite by high concentration of sulfate in a cultivar with a fast rate of selenite uptake when the selenite was supplied at low concentration. However, there was a strong inhibition of selenite uptake by sulfite in both that cultivar and one with a much slower rate of selenite uptake (Zhang et al., 2006).

SULTR1;1 and SULTR2;1 sulfate transporters are induced by sulfur deficiency, but in roots of S-sufficient plants the amount of mRNA of the *SULTR1;1* and *SULTR2;1* genes is also increased by supply of selenate (Takahashi et al., 2000; Van Hœwyk et al., 2008). In contrast, the arabidopsis *SULTR1;2* gene seems to be expressed irrespective of sulfate supply (El Kassis et al., 2007). The increased expression of the *SULTR1;1* and *SULTR2;1* genes (and of a gene for another possible sulfate transporter) with supply of selenate gives rise to fivefold higher concentrations of sulfate in the shoots of arabidopsis than in plants not receiving selenate (Van Hœwyk et al., 2008). These authors also showed induction of enzymes responsible for mobilizing sulfur, such as ATP sulfurylase and enzymes for the degradation of glucosinolates with supply of selenate. As the *SULTR1;1* gene is expressed less under sulfur sufficiency and its expression is increased under sulfur deficiency as well as with supply of selenate, supply of high levels of sulfate would lower uptake of selenate into roots, whereas a low rate of sulfate supply would enhance it (El Kassis et al., 2007). In rapid-cycling brassica (*B. oleracea*), supply of Se gave increased concentration of sulfur, as well as selenium, with increasing concentration of sodium selenate given (Toler et al., 2007).

The *SULTR3;1* gene for a sulfate transporter was induced by supply of selenite in a selenite-tolerant accession of *Arabidopsis*, but not in a selenite-susceptible accession, and the *SULTR2;2* and *SULTR3;5* genes were induced more in the selenite-resistant than in the selenite-sensitive accession, as were genes coding for several enzymes of sulfur assimilation, indicating that enhanced uptake and assimilation of sulfur may be one of the mechanisms that gives selenite tolerance (Tamaoki et al., 2008). The selenite-sensitive accession contained lower concentrations of sulfur in its shoots after treatment with selenite than without selenite, and also lower concentrations than the selenite-tolerant accession after selenite treatment.

Durum wheat and canola took up much more selenate from hydroponic culture without sulfate than from a plus sulfate treatment (Renkema et al., 2012). Sulfur deficiency increased accumulation of naturally available selenium in field-grown wheat, caused at least in part by increased expression of *SULTR1;1* and other sulfate transporter genes (Shinmachi et al., 2010). S deficiency in wheat specifically gives increased uptake of Se as selenate, not selenite (Li et al., 2008).

Sulfate lowered retranslocation of ^{75}Se supplied as selenate from stems of wheat to maturing leaves and from stems and leaves to grains (Govasmark and Salbu, 2011a,b). In the wheat plants of Shinmachi et al. (2010), most of the increased Se taken up during S deficiency accumulated in the grains, indicating that there was efficient remobilization from vegetative tissues. There was no enhancement of accumulation of Se in the roots under S shortage. In wheat, SeO_4^{2-} and SO_4^{2-} actively compete for both uptake and translocation to the shoots, but in canola supply of SO_4^{2-} seems to stimulate the movement of Se to the shoot. Canola is a brassica crop, and brassicas have a high S requirement. Elevated sulfate supply did not lower the accumulation of Se. Low sulfate supply gives increased volatilization of Se taken up by broccoli plants when selenate is the Se form supplied (Zayed et al., 1998), indicating that there is competition between sulfate and selenate for metabolic enzymes. In plants with high rates of sulfur supply, the critical concentration of Se for phytotoxicity appears to be higher, not only because sulfate competes with selenate for uptake but because a high internal S/Se ratio probably lowers the incorporation of Se into selenoamino acids (White et al., 2004).

Shortage of sulfur has also been shown to give increased accumulation of Se supplied as selenate in both Se-hyperaccumulator and nonaccumulator *Astragalus* species, whereas increased selenate supply increased sulfate accumulation in shoots (Cabannes et al., 2011). The increased accumulation of Se was due, at least in part, to lowered competition between SeO_4^{2-} and SO_4^{2-} uptake as the expression of genes for sulfate transporters in the roots was only increased slightly by the withdrawal of sulfate. Hyperaccumulators have high ratios of tissue Se/S, so their uptake systems seem to have evolved to take up Se preferentially, whereas nonaccumulator species appear to be unable to discriminate so much between the two elements (Galeas et al., 2007; White et al., 2007).

19.6.2 PHOSPHORUS

As selenite seems to be taken up by phosphate transporters, interaction between selenium and phosphorus nutrition is to be expected. Uptake of selenite is inhibited by phosphate, but this inhibition is not as strong as inhibition of uptake of selenate by sulfate (Hopper and Parker, 1999). The presence of phosphate causes a big decrease in affinity of the uptake system for selenite, and phosphorus starvation gave an increase in selenite uptake (Li et al., 2008).

19.6.3 NITROGEN

Nitrate inhibits the uptake of selenite in rice roots in a dose-dependent manner when supplied to the part of the root system where the uptake of selenite is occurring, but may stimulate uptake of selenite when supplied to other parts of the root system (Xu et al., 2010). Under the waterlogged conditions of a paddy soil, selenite is likely to be the major Se form available, but most of the N will be present as ammonium, so the effect of N may be of limited significance to acquisition of selenium by rice plants.

19.7 DIAGNOSIS OF SELENIUM STATUS IN PLANTS

Routine testing of crops for selenium content is not carried out, but there are various methods for measuring Se concentration in plant parts. One of the common methods is inductively coupled plasma spectrometry, frequently combined with mass spectrometry (ICP-MS) (e.g. Zhao et al., 2007; Li et al., 2008; Broadley et al., 2010; Shinmachi et al., 2010; Srikanth Lavu et al., 2012; Longchamp et al., 2013), but a technique that uses less costly equipment is atomic absorption spectrometry (AAS). In order to obtain adequate sensitivity, selenium forms must be converted to volatile hydrides by the addition of sodium borohydride to the plant extract first. This step works more efficiently on selenite than selenate, so typically plant material is digested in $HNO_3/HClO_4/H_2SO_4$ under reducing conditions that reduce selenate to selenite and is then subjected to AAS with hydride generation (e.g., Zayed et al., 1998; Hopper and Parker, 1999; Laser, 2007; Cartes et al., 2011; Lee et al., 2011b). An alternative to these techniques is hydride-inductively coupled plasma optical emission spectrometry (e.g., Lyons et al., 2005).

There are also fluorimetric methods for measuring total Se concentrations in plant materials. Selenate in extracts is reduced to selenite, which is then reacted with diaminonaphthalene, and the precipitate is separated into an organic solvent and the fluorescence of this fraction is measured (e.g., Rani et al., 2005; Zhao et al., 2005). Alternatively, the fluorescence of hydrides, formed similarly as for AAS, can be measured (e.g., Zhang et al., 2006; Xu et al., 2010; Zhao et al., 2010; Han et al., 2013).

If there is a requirement to identify individual Se forms in plant materials, other methods are required. Commonly used methods include separation by HPLC before measuring the Se concentration by ICP-MS or using HPLC followed by electrospray mass spectrometry (HPLC-EIS-MS/MS) (e.g., Kápolna et al., 2012). Another useful technique is x-ray absorbance spectroscopy, where a synchrotron source is used to generate x-rays, and their absorption by different Se forms is then quantified (Zayed et al., 1998; Sors et al., 2005a; El Kassis et al., 2007). Variants include x-ray absorption near-edge structure analysis (e.g., Freeman et al., 2010; Wang et al., 2013). The cellular location of selenium can be visualized by x-ray fluorescence (XRF) microscopy (e.g., Freeman et al., 2010; Wang et al., 2013).

19.8 FORMS AND CONCENTRATIONS OF SELENIUM IN SOILS AND ITS AVAILABILITY TO PLANTS

Selenium is found in major ore deposits as selenide, substituting for iron sulfides, and is weathered into elemental selenium or ferric selenite ($Fe_2(SeO_3)_3$) (Lakin, 1972). It is in relatively low concentration in igneous rocks, although in higher concentration in areas of more recent volcanic activity, and it is in higher concentration in some shales (Lakin, 1972). This gives a range of soil concentrations worldwide dependent on distribution of underlying rocks, typically ranging from 0.1 to 2 mg Se kg^{-1}. Soils in the wheat-growing areas of the United States and Canada, such as Nebraska and the Dakotas, Saskatchewan, and Manitoba, are high in selenium, while soils in New Zealand, parts of China, parts of Russia, Finland, and other Scandinavian countries are low in selenium (Gissel-Nielsen et al., 1984; Combs, 2001; Kopsell and Kopsell, 2007). In oxidizing conditions above pH 6, selenite is oxidized to selenate, and both ions are available for plant uptake. Within a soil, microorganisms reduce these ions to elemental Se^0, which is not soluble, or to volatile selenides, which are lost from the soil. Neither of those forms are available to plants (Munier-Lamy et al., 2007).

Precise values below which a soil is selenium deficient are difficult to arrive at as the availability of Se in a soil varies according to chemical and physical conditions. However, an Inceptisol from Yunnan province, China (where the Se-deficiency diseases Keshan and Kashin–Beck occur in the human population), had a concentration of 0.0318 mg Se kg^{-1}, a concentration that can be seen to be low because the plants grown in it accumulated low concentrations of Se (He et al., 1994). Soils regarded as being Se deficient tend to have less than 0.4 mg Se kg^{-1}, but seleniferous soils may contain up to 100 mg Se kg^{-1} (Srikanth Lavu et al., 2013). Where these soils occur, agriculture faces the problem of selenium toxicity rather than selenium deficiency.

It is the soluble Se that is probably a better indicator as to whether or not a soil is deficient in plant-available Se, and in low-Se soils this soluble Se can be as low as 2 µg kg^{-1} soil, representing about 1%–2% of the total soil Se. In a Se-sufficient soil, it can constitute a higher proportion of the total soil Se and will be at a concentration as high as 18 µg kg^{-1} soil (Wang and Gao, 2001).

Both selenite and selenate ions bind to iron oxyhydroxides, such as the mineral goethite, so the clay content and composition of soils affects the extent to which Se is fixed. In a study of 50 soils in South Carolina there was a negative correlation between clay content and Se concentration (Franklin et al., 2003), and in a study of Se uptake from 6 Danish soils there was a negative correlation between plant Se concentration and soil clay content when Se was supplied as either selenate or selenite (Bisbjerg and Gissel-Nielsen, 1969). In moist and semimoist acidic soils, the Se present binds to Fe(III), Mn, and Al oxides, whereas in semidry, alkaline soils, it binds to Ca, Mg, and K oxides. Therefore, concentrations of soluble Se tend to decrease in the order Ferralisols > Ferrisols > Luviols > Isohumisols > Aridisols (Wang and Gao, 2001).

In a study of soils containing a representative range of Se concentrations from low to adequate, an Inceptisol derived from purplish sandstone had a low concentration of 0.0318 mg Se kg^{-1}, an Ultisol derived from a quaternary red earth contained 0.1517 mg Se kg^{-1}, and an Alfisol derived from loess had an Se concentration of 0.2559 mg kg^{-1} (He et al., 1994). The Inceptisol adsorbed selenite more strongly than the Ultisol and the Alfisol, and this, as well as the lower total Se concentration, contributed to giving it a lower concentration of plant-available Se (8.1 µg kg^{-1}, compared with 22.0 and 28.1 µg kg^{-1} in the Ultisol and Alfisol, respectively). In five different alluvial soils, in the acidic pH range, sorption of selenite correlated positively with the amounts of solubilized Al, Fe, and Mn (Neal et al., 1987a). In a study of 58 agricultural soils from Japan, the adsorption of selenite was strongly correlated with aluminum and iron oxides, particularly the former, and 80% of added selenite was recovered in Al-bound and Fe-bound pools. Andosols showed more adsorption of selenite than other soil types investigated (Nakamura et al., 2005).

The pH of a soil is particularly important in adsorption of different Se forms. Selenite sorbs more strongly than selenate, with sorption of both ions decreasing with increased pH (Barrow and Whelan, 1989; Zhang and Sparks, 1990). Sorption of selenite to four soils from South Dakota decreased to zero from pH 5 to pH 9, an effect that was probably caused by higher concentrations of OH$^-$ in the soils at high pH (Lee et al., 2011a). In a study in which SeO$_3{}^{2-}$ was added to two soils, one at pH 4.5 and one at pH 7, there was a higher concentration of selenate than selenite present initially in soil solutions from both soils, although the ions were present in the solution from the pH 7 soil at much higher concentration than in the solution from the pH 4.5 soil (van Dorst and Peterson, 1984). This result indicates that selenate was sorbed less strongly than selenite in both soils and that both ions were sorbed less strongly at pH 7 than at pH 4.5. Goldberg and Glaubig (1988) observed no adsorption of selenate in a calcareous montmorillonitic soil, but there was adsorption of selenite, which decreased with increase in pH from 3 to 6. The clays kaolinite and montmorillonite both sorbed selenite, with a pH optimum close to 5. There was a small amount of sorption of selenite above pH 7 in this soil, which was shown to be due to the presence of calcite.

Soil pH also affects what forms of selenium are present in the rooting environment. Under aerobic conditions and high pH or under very highly oxidizing conditions at neutral pH, selenate predominates, and at slightly less oxidizing (although still aerobic) conditions the Se is present as selenite or biselenite (Elrashidi et al., 1987; Missana et al., 2009). In the aerobic conditions of a normal agricultural soil, SeO$_3{}^{2-}$ predominates in alkaline conditions, and HSeO$_3{}^-$ predominates when such soils are acidic (Elrashidi et al., 1987; Missana et al., 2009). The different solubilities and mobilities of these ions in soil account for the different extents of their sorption to soil components, and the ions are taken up by plants to differing extents. In soils with acid pH and reducing conditions, there is likely to be selenium deficiency in the plant biomass, due to the interacting effects of most of the Se being present as selenite rather than selenate and the strong adsorption of this ion under these conditions.

There is competition between selenite and phosphate for adsorption sites, and the presence of phosphate affects surface charge and the electrical potential of soil surfaces (Barrow et al., 2005). The soil/solution distribution coefficients for Se in Japanese Andosol and Fluvisol soils decreased with increasing concentration of H$_2$PO$_4{}^-$ supplied (Nakamaru and Sekine, 2008). Sorption of selenite to four South Dakota soils was decreased progressively by increasing concentrations of phosphate (Lee et al., 2011a), and Neal et al. (1987b) reported that with phosphate present there is less adsorption of selenite in soils. In a soil containing 6% clay (dominated by kaolinite, but with gibbsite, goethite, and hematite also present) and with a pH of 4.5, the selenite was held less strongly than phosphate (Barrow et al., 2005). In the experiments by He et al. (1994) on Inceptisol, Ultisol, and Alfisol soils, phosphate fertilization increased the Se concentration in the plants grown in the Inceptisol that absorbed Se most strongly, but not in the other two soils. This indicates that supply of phosphate makes Se available from soils in which it is tightly adsorbed. Soybean plants were shown to take up more Se from soils with high P content or low P retention capacity than from soils with low P content or a high P retention capacity (Nakamaru and Sekine, 2008). The soils with a higher capacity to sorb phosphate would also sorb selenite strongly. Singh et al. (1981) found that the addition of phosphate to soil leads to desorption of selenate as well as selenite.

Ions other than phosphate also affect Se adsorption. Addition of the sulfate ion to soil desorbs selenate, but less effectively than addition of phosphate desorbs selenite or selenate (Singh et al., 1981).

However, the competition between SO_4^{2-} and SeO_4^{2-} for uptake means that any increased availability of selenate brought about by desorption would not necessarily be apparent in terms of plant selenate uptake. In experiments on typically low-Se soils of the United Kingdom at 10 sites, there was a negative effect of soil sulfur content on the grain Se concentration in wheat, with extractable sulfur and total Se concentration accounting for over 70% of the variance in a model of Se uptake and adding in extractable Se raising this figure to over 86% (Stroud et al., 2010a). Up to 70% of the extractable soil Se was in the selenite form, even though soil pH was high and even when selenate fertilizer had been supplied, and no selenate was detected, so there is an effect of sulfur on Se uptake that is not just competition between sulfate and selenate for adsorption sites and for plant transporters.

High salt content of soil in general lowers sorption of selenate and selenite (Singh et al., 1981), and the competitive ability of anions for selenite adsorption sites seems to follow the order phosphate > silicate > citrate > molybdate > bicarbonate/carbonate > oxalate > fluoride > sulfate (Balistrieri and Chao, 1987). Selenite also binds to the organic fractions of soil, particularly the fraction containing hydrophobic fulvates (Gustafsson and Johnsson, 1992). Singh et al. (1981) showed that adsorption of both selenate and selenite is affected positively by soil organic carbon content, and in a study of 54 Canadian soils, Lévesque (1974) showed that the distribution of Se was related strongly to the organic matter present. The selenium in that study was mainly in the upper soil horizons, where most of the organic matter occurred. Although this occurrence could be because Se was entering the system from atmospheric pollution, it is undoubtedly the case that the organic matter helped fix it in the soil. In a study of 66 sites of low-input grazed grassland in central Europe carried out over 2 years, there was a close relationship between the concentration of soil Se and soil organic matter content (Laser, 2004). However, there was no relationship between the soil organic matter content and plant Se concentration, and indeed there was no relationship between soil Se and plant Se concentrations.

There is obviously a temporal effect here, as conditions in which Se is fixed less strongly to soil particles make it more available to plants but also make it more likely to be prone to leaching. The conditions that change the form of Se between selenate and selenite also change the extent of adsorption (selenite binds more than selenate), uptake (in many plant species, selenate is taken up preferentially), and leaching (selenate is more water soluble than selenite). With the addition of a Se source to a soil, the physical conditions will influence the form in which the element occurs. After the addition of selenate to a sandy soil in Finland, 84% of it was present in a salt-soluble pool at the first sampling time a week later, while 13% was present in a pool associated with aluminum oxides and a trivial amount in a pool associated with iron oxides (Keskinen et al., 2010). The plant uptake of Se was largely restricted to the salt-soluble pool. Little selenate was reduced to selenite over the 10 weeks of the experiment. In a study on two Finnish soils approximately 1% of the Se present was in the soluble fraction, 15%–20% was adsorbed on to oxides, 50% was bound to organic matter, 10% was in elemental form, and 20% was in recalcitrant forms (Keskinen et al., 2009).

19.9 SOIL TESTING FOR SELENIUM

Various methods have been proposed for extraction of selenium from soils in order to measure the concentrations of different Se forms and to evaluate the size of different Se pools.

In order to determine total Se concentration in a soil, the traditional method of extraction with aqua regia (a mixture or HCl and HNO_3) has been used. Keskinen et al. (2009) boiled soil samples with aqua regia under reflux for two hours and found higher concentrations in one soil but slightly lower (although not significantly so) concentrations of Se in another soil than with a sequential extraction procedure. However, they pointed out that in other soils, the aqua regia method may underestimate total Se content of a soil as the soil matrix may not be totally degraded.

This underestimate of Se concentration in soils is unimportant as much of the Se present in a soil is in a form that is not available for plant uptake anyway, and what is really required is a measure of the Se in different pools that may be plant available under different circumstances. One of the techniques for estimating the sizes of different pools of elements in soils is to carry out sequential extraction with different extractants, but for each pool the calculated size varies according to the concentration of the extractant, the pH, and also what extractant is actually used.

One of the classic sequential procedures, developed by Chang and Jackson (1957) for soil phosphorus, gives a salt-soluble Se pool (extracted with 1 M NH_4Cl), a pool associated with aluminum oxides (extract the residue with 0.5 M NH_4F), a pool associated with iron oxides (extract the residue from the aluminum-associated pool with 0.1 M NaOH), and an acid-soluble Se pool (carry out a final extraction with 0.25 M H_2SO_4) (Keskinen et al., 2009). However, a second sequential procedure that separated soil Se into a soluble pool (extracted with 0.25 M KCl), an adsorbed pool (extracted with 0.1 M KH_2PO_4 at pH 8.0), an organically associated pool (extracted with 0.1 M NaOH), a pool of elemental Se (extracted with 0.25 M Na_2SO_3 at pH 7.0), and a recalcitrant pool of organic Se (extracted with 5% NaOCl at pH 9.5) was found to give higher concentrations of total Se when the values from all the pools were combined than the Chang and Jackson procedure (Keskinen et al., 2009). The Al-bound and Fe-bound fractions of Se in the clay soil of this study together measured nearly 0.14 mg Se kg^{-1} soil dry mass in the first sequential procedure, yet in the second procedure, the adsorbed Se pool measured less than 0.07 mg Se kg^{-1}, with the organically associated pool measuring approximately 0.18 mg Se kg^{-1}, so the sizes of pools measured depend very much on the extractants used. A slightly different version of this procedure involved extracting in water, then 0.25 M KCl, then 0.1 M KH_2PO_4, then 4 M HCl, and finally concentrated HNO_3. The water and KCl fractions were deemed to comprise the soluble Se; the KH_2PO_4 fraction exchangeable Se; the HCl fraction Se that was bound to Fe, Al, and Mn oxides and hydrolyzable organic matter; and the HNO_3 fraction the residual Se (Munier-Lamy et al., 2007).

Irrespective of the difficulty of accurately measuring the size of a particular pool, it is not always certain which pools plants acquire their Se from. One such fraction could be the Se that is dissolved in hot water extracts, used in Finland (Keskinen et al., 2009, 2010). Selenate readily dissolves in hot water, which probably also solubilizes some weakly adsorbed selenite. Keskinen et al. (2010) showed that the amount of Se taken up into wheat (*T. aestivum*) and ryegrass (*Lolium multiflorum* Lam.) ranged from 12% to 47% of the NH_4Cl-extractable soil Se, so it seems likely that plant Se uptake is restricted to this most easily soluble Se pool. Other workers have used 0.01 M $CaCl_2$ to extract plant-available Se (e.g., Zhao et al., 2007). Another possible extractant is phosphate buffer, which dissolves soluble Se forms but also releases some adsorbed selenate and selenite (Singh et al., 1981). Increased concentration of the buffer increases the amount of Se extracted (Zhao et al., 2005; Keskinen et al., 2009), as does increased pH (Keskinen et al., 2009), so it is unclear to what extent this pool represents a pool of Se that can be acquired by plants in its entirety.

In the experiments of Keskinen et al. (2010), phosphate buffer was more powerful than the ryegrass at removing Se. The phosphate-extractable Se fraction from 24 Japanese soils was more than 10× higher than the water-soluble fraction and comprised 0.9%–12% of the total Se in the soils (Nakamaru and Sekine, 2008). 1 M phosphate buffer at pH 7.0 extracted considerably more Se than hot water from two Finnish soils, although the latter was a more efficient extractant than the salt solutions used in the two sequential procedures (Keskinen et al., 2009). Extraction of soil with 0.1 M KH_2PO_4 or 0.25 M KCl both gave a pool of Se that correlated strongly with the concentration of Se in leaves of tea plants (*Camellia sinensis* L.), and this fraction also correlated strongly with the total Se content of rye seedlings (Zhao et al., 2005). Munier-Lamy et al. (2007) deemed that their soluble and exchangeable fractions comprised Se that was directly available to plants. Lee and coworkers extracted soil with 0.1 M KH_2PO_4 to extract plant-available Se and then 4 M HCl to extract *conditionally* available Se (Lee et al., 2011b,c).

Once the Se has been extracted, it has to be quantified in the extract. This is carried out by many of the same methods used for plant analysis (Section 19.7).

19.10 SELENIUM FERTILIZERS

Selenium is an essential trace mineral in human and animal nutrition, and some selenoproteins are known to be essential in animal metabolism (Papp et al., 2007). In fact, there is a specific codon that gives the incorporation of selenocysteine into proteins, a codon that is otherwise involved in the termination of translation (Papp et al., 2007). The main biochemical function of Se in animals is as a component (as selenocysteine) in glutathione peroxidases, enzymes that degrade peroxides and hydroperoxides formed from polyunsaturated fatty acids (Rayman, 2000; Papp et al., 2007). Selenium is also a component of enzymes that deiodinate thyroid hormones (Papp et al., 2007). In humans, Se deficiency has been implicated in myopathy, cardiomyopathy, and immune system dysfunction and in livestock in white muscle disease (Gupta and Gupta, 2000). Low selenium in the diet gives rise to low sperm motility in men (Rayman, 2000). Levels of intake of *ca* 120 µg day^{-1} have been implicated in having chemopreventative action against chronic diseases such as cancer, cardiovascular disease, and asthma (Thomson, 2004). Supply of Se also helps maintain levels of tocopherols in plants, enhancing supply of vitamin E to animals that consume them (Hartikainen, 2005).

Despite its importance in human health, the amount of selenium in the western diet is decreasing. Dietary Se intake of humans in the United Kingdom, for example, declined from *ca* 60 µg day^{-1} in 1975 to *ca* 34 µg day^{-1} in 1995 (Givens et al., 2004), below the reference nutrient intake of 75 µg Se day^{-1} for an adult male and 60 µg Se day^{-1} for an adult female (MAFF, 1999). Levels of intake of selenium below 20 µg day^{-1} give rise to acute deficiency symptoms, which manifest themselves as endemic cardiomyopathy (Keshan disease) and endemic osteoarthritis (Kashin–Beck disease) in countries such as China. Concentrations below 0.05–0.10 mg Se kg^{-1} dry matter in crops are likely to be inadequate for human nutrition (Gissel-Nielsen et al., 1984), but Se-deficiency diseases are uncommon because in most areas with low Se availability in the soil, the human population imports some of its food from outside. Subacute deficiencies of Se in the diet are more common.

One reason for the decline in Se in a western diet is the trend for consuming European rather than North American wheat, which has a higher Se content, together with an overall decline in the consumption of bread (Barclay and MacPherson, 1992). U.K. soils are comparatively low in selenium, with concentrations in soils and stream sediments ranging from 0.1 to 4 mg Se kg^{-1} (Broadley et al., 2006). Barclay and MacPherson showed that Scottish wheat grown for bread making in 1989 contained 0.028 mg Se kg^{-1} dry mass of grains, compared with 0.518 mg kg^{-1} in Canadian wheat. Approximately 80% of wheat grain samples from 160 years of the Broadbalk experiment at Rothamsted Experimental Station had concentrations of less than 0.05 mg Se kg^{-1} (Fan et al., 2008). This trend of declining use of North American wheat occurs in many countries worldwide (Combs, 2001). Another problem is that wheat grain yield is negatively related to grain Se concentration (Zhao et al., 2007), so any agricultural activity that increases grain yield can give lower concentrations of Se in the grain by a dilution effect.

Meat and dairy products are other good sources of Se in the human diet. However, pasture in countries with low soil Se concentrations is likely to be Se poor. Normal concentrations of Se in pasture herbage range from 0.10 to 0.30 mg kg^{-1} dry mass. Deficiency symptoms arise when livestock diets contain less than 0.030 mg kg^{-1} (Agricultural Research Council, 1980), and many pastures in Europe are Se deficient. In a study of long-term grassland at Rothamsted, there was shown to be a decrease in herbage Se concentration from the 1960s (Haygarth et al., 1993), and in many parts of the United Kingdom the soil Se concentration is now so low that the concentration of Se in grass is less than 0.10 mg kg^{-1}. Unsupplemented cattle at pasture, such as late lactation or dry cows, and cycling heifers, are likely to show selenium deficiency under such conditions. In a study on feeding grass, grass silage, and maize silage to dairy cows in Germany, all three diets were regarded as being Se deficient (Gierus et al., 2002). In the past, some Se was added to farmland in N-P-K fertilizers, but with changes in their production and use this input has decreased. This is partially offset by the fact that some decrease in Se uptake was due to atmospheric sulfur dioxide pollution (Fan et al., 2008), which is now decreasing in intensity. Removal of Se in the heavy yields of crops that the use of NPK fertilizers has given (nutrient mining) also has the potential to give a shortfall in Se supplies for subsequent crops.

In Finland, a country where soil Se concentrations are naturally low, there has been a policy to improve human Se intake by giving selenium fertilizers to crops as well as giving selenium supplements to livestock. From 1984, fertilizers for food or forage crops in Finland have been enriched with sodium selenate (Hartikainen, 2005). Typically, 16 mg Se was added per kg of compound fertilizer used for grain production, and 6 mg per kg of compound fertilizer was used for forage crops, to take into account the relative lack of mobility of Se in the plant and the need to allow the element to accumulate in cereal grains (Eurola et al., 1990). Mean Se concentration in samples of Finnish wheat grains from 1982 to 1984 was 0.01 mg kg^{-1}, whereas from 1985 onward the mean Se concentration was more than 0.03 mg kg^{-1} in winter wheat and more than 0.2 mg kg^{-1} in spring wheat (Eurola et al., 1990). The added selenium worked through the food chain very quickly, as samples of unskimmed milk from Finnish shops were found to contain 0.06 mg Se kg^{-1} dry matter in 1984, and this concentration had doubled to 0.12 mg kg^{-1} by as quickly as 1985 (Varo et al., 1988). Prior to 1984, human Se intake was only 30–50 µg day^{-1}, but nowadays, Finland is not noted for Se deficiency in its population (Hartikainen, 2005). In fact, the fertilization policy has been so successful that the Se concentration in compound fertilizers was cut in 1991 and then increased again in 1998 to take into account changes in selenium concentration in human diets arising from this alteration (Hartikainen, 2005).

Supplying 10 or 20 g Se ha^{-1} in experiments on Se-deficient pasture in the United Kingdom rapidly raised herbage Se concentrations above the critical value of 0.030 mg kg^{-1}, and although the concentration declined after the initial cut the critical value was exceeded for over nine months (Rimmer et al., 1990). Hay made from alfalfa fertilized with sodium selenate at 22.5, 45.0, and 89.9 g Se ha^{-1} showed increased Se concentration in linear proportion to the rate of Se applied, and calves fed on the hay for 7 weeks had higher whole-blood Se concentrations, again in linear proportion to the rate of Se supply (Hall et al., 2013). The calves fed hay from the two highest Se application rates had higher body weight. Experiments on a subterranean clover (*Trifolium subterraneum* L.) pasture, in which quick release and slow release forms of selenate fertilizer were compared, showed that the former gave concentrations in the pasture of just over 4.5 mg Se kg^{-1} within 43 days and the latter gave concentrations of nearly 0.4 mg Se kg^{-1} within 73 days (Whelan and Barrow, 1994). For both forms of the fertilizer, the Se concentrations declined from these initial values.

One of the problems with a fertilization policy is that there is only a narrow range of Se concentrations in plant tissues between deficiency and toxicity. Butterhead lettuce (*Lactuca sativa* L. var. *capitata*) grown in hydroponics showed increased fresh weight of shoots and roots with supply of selenite up to 10 µM, but with 15 µM toxic effects lowered plant mass (Hawrylak-Nowak, 2013). In soil-grown white clover, supply of selenite at up to 20 g Se ha^{-1} increased shoot Se concentration from less than 0.062 mg kg^{-1} shoot dry mass with no Se supply to over 0.20 mg kg^{-1}, but higher rates of selenite application depressed plant growth (de la Luz Mora et al., 2008). In a pot experiment on perennial ryegrass in an Inceptisol soil, concentrations of Se in the shoots were significantly higher in two cuts of the grass with application of 1.0 mg Se kg^{-1} soil than with 0.1 mg Se kg^{-1} or no Se supply, but when the ryegrass received 10 mg Se kg^{-1} it had died by the second cut (Hartikainen et al., 2000). However, Se fertilization still seems to offer a good prospect for agronomic biofortification if an optimum rate of Se supply is given.

The most effective source of selenium appears to be selenate rather than selenite. In experiments on perennial ryegrass grown in a soil of pH 4.5, Se concentrations of up to nearly 250 mg kg^{-1} shoot matter were obtained with selenate fertilizer at the highest rate of supply, but only just over 4 mg Se kg^{-1} with selenite (Cartes et al., 2005). Concentrations of available Se in the soil were always more than 100 times higher with supply of selenate rather than selenite at equivalent Se concentrations. However, although the selenate was a better source of Se for the ryegrass, the balance between alleviating Se deficiency and bringing about Se toxicity was more difficult to achieve, and with selenate the plants could exceed concentrations of 150 mg Se kg^{-1}, above which their dry matter yield was depressed, whereas the Se concentrations in plants given selenite never came close to this value irrespective of the amount supplied. However, they were well above the critical concentration of 0.030 mg Se kg^{-1}, and as selenite is more rapidly converted to organic Se forms in plants than selenate the Se could have been more bioavailable to livestock. Concentrations of Se in soil-grown plants of kenaf, another fodder crop, were increased in a dose-dependent manner by the application of either sodium selenate or sodium selenite, although the selenate had an effect that was much greater (Srikanth Lavu et al., 2013). The tissue Se concentrations reached with selenate (up to 1000 mg Se kg^{-1} dry matter) would be toxic to livestock, so kenaf fertilized in this manner would be suitable only as a feed supplement, although when fertilized with sodium selenite at 2 mg Se kg^{-1} soil might produce forage with enhanced, but safe, levels of Se (Srikanth Lavu et al., 2013).

Some workers have suggested that agronomic biofortification of vegetable crops would be a good strategy for increasing Se in the human diet. Watering green onions and chives with selenate gave much higher concentrations of Se in the plants than watering with selenite or selenomethionine, although as selenate tends to accumulate in plants unchanged and inorganic forms of selenium have lower bioavailability in the human diet than organic Se (Thomson, 2004), the Se arising from selenate treatment may be less beneficial (Kápolna and Fodor, 2007). Leeks were shown to accumulate Se with supply of either selenate or selenite, with the highest concentrations accumulating with selenate (Srikanth Lavu et al., 2012). Although the organic forms of Se that are presumed to give the greatest health benefits formed a greater proportion of the Se in the plants with selenite, their concentrations were still not as high as with selenate. Supply of selenate to rapid-cycling brassica (*B. oleracea*) led to lower concentrations of some glucosinolates in the leaves, so if this were to occur in *B. oleracea* cultivars used as crops (such as broccoli, Brussels sprouts, cabbage, cauliflower, kale, and kohlrabi), increasing the concentration of Se for human consumption might come at the expense of another group of compounds perceived to be beneficial to human health (Toler et al., 2007; Barickman et al., 2013). However, an experiment in which sodium selenate was supplied to broccoli seedlings gave increased Se concentrations in the shoots without total glucosinolate concentrations being any lower than in control plants (Hsu et al., 2011). With supply of sodium selenate to 38 accessions of broccoli in hydroponic culture, only 13 had reduced concentrations of glucosinolates in their shoots (Ramos et al., 2011).

Supply of 10 or 20 g Se ha^{-1} significantly increased the Se concentration in wheat grains in a field experiment in the United Kingdom (Stroud et al., 2010b). However, the beneficial effect on grain Se concentration did not carry over until the following season. Supply of sodium selenate in spring (but not winter) increased the Se concentration in grain of winter wheat from 0.024 mg kg^{-1} grain fresh weight, close to the U.K. average, by 16–26 µg Se kg^{-1} fresh weight for every g Se ha^{-1} supplied (Broadley et al., 2010). No yield reductions were seen at application rates of up to 100 g Se ha^{-1}, and the authors recommended that an application rate of 10 g Se ha^{-1} would raise the U.K. wheat grain Se concentration to approximately half of the Se concentration in U.S. or Canadian wheat without incurring the risk of Se accumulating in soils and harming subsequent crops. The application of Se as a sodium selenate drench, by incorporation of selenate with compound NPK fertilizer, or the enrichment of calcium ammonium nitrate with selenate gave increased grain Se concentrations in maize in Malawi (Chilimba et al., 2012). The increases were 20, 21, and 15 µg Se kg^{-1} dry weight of grain for each gram of Se supplied in the three treatments, respectively.

It was noted in a study on wheat that when Se was supplied as selenate there was more accumulation of Se in the grains when the NPK fertilizer used was chloride-based rather than sulfate-based, but when selenite was applied accumulation of Se in the grains was greater when the NPK fertilizer was sulfate-based (Singh, 1991). This effect can be explained by competition between anions for uptake.

Sulfate fertilizers could be applied to desorb selenate, making it available for plant uptake. In field-grown wheat that was deficient in S to the extent that grain yield was lowered, there was no significant effect of supply of supplemental sulfate on grain Se concentration (Stroud et al., 2010b).

However, in a season in which the S supply was not limiting to crop growth supply of supplemental sulfate gave higher grain Se concentrations, at least when sodium selenate fertilizer was also applied. Additional fertilization with sulfate seemed to aid the recovery of added selenate by the crop by suppressing its conversion to unavailable Se forms in the soil. However, in a study on bread-making wheat in the United Kingdom supply of sulfur decreased the Se concentrations in the grains (Adams et al., 2002). In ryegrass, shoot Se concentration was decreased by application of sulfur fertilizers (Cartes et al., 2006). High concentrations of sulfate would competitively inhibit uptake of selenate, so even if the desorbed selenate was not leached from the soil quickly it might not be taken up by the plants. Furthermore, in Europe there are not many soils that are not already well loaded with sulfur due to the deposition of atmospheric sulfur dioxide from pollution.

In field experiments on wheat in South Dakota, supply of sulfate at increasing rates gave an increasing ratio of S/Se in the grains, and this gave rise to a significant decrease in grain Se concentration where plant-available Se in the soil was comparatively high (Lee et al., 2011b). Competition between S and Se for uptake gave lowered grain Se concentration in plants grown on that soil. Nitrogen aids the remobilization of Se in wheat plants, so it has been suggested that supply of N and Se together at heading may increase grain Se content (Govasmark and Salbu, 2011a,b). These authors suggested that sulfate-free fertilizers should be used to prevent competition for uptake and translocation between sulfate and selenate, although this would not be suitable if wheat grain yield was depressed by shortage of sulfur. Sulfate is commonly applied in South Dakota (Lee et al., 2011a), so sulfate-free fertilizers would not be suitable in this major wheat-producing area.

Given that phosphate can desorb selenite and selenate from soils, it may be the case that phosphate fertilization could make Se available for plant uptake. It at least seems likely that supply of phosphate with Se fertilizer may prevent the Se from being adsorbed (He et al., 1994). However, supply of triple superphosphate fertilizer at recommended rate gave decreased wheat grain Se concentration in a site in South Dakota, but had no significant effect at another site (Lee et al., 2011b). The soil in the first site had a concentration of plant-available Se three to four times higher than the second site, and at the first site the P fertilizer increased grain yield, lowering the grain Se concentration due to a dilution effect.

Rock phosphate is a source of exogenous Se, and superphosphate derived from it also contains the element, so amendment of soils with these P sources may be beneficial for Se availability to crops by actually supplying Se. An analysis of commercial triple superphosphate, monoammonium phosphate, and diammonium phosphate fertilizers and rock phosphate samples from around the world gave median values of 1.1, 0.3, 0.1, and 1.0 mg Se kg^{-1}, respectively, although one of the rock phosphate sources gave a noticeably higher value (Charter et al., 1995). Robbins and Carter (1970) found a rock phosphate supply with a concentration of 178 mg Se kg^{-1}. Supply of 50 kg of either triple superphosphate or rock phosphate per hectare would supplement the soil by 55–50 mg Se ha^{-1} at the values of Se concentration seen by Charter et al. (1995), although values of 200 times higher would be required to give 10 g Se ha^{-1}. Selenium is present in single superphosphate at higher concentration than in triple superphosphate (Robbins and Carter, 1970); more single superphosphate is required to give the same rate of P application, and single superphosphate is also a source of sulfate that could desorb soil selenate, so supply of single superphosphate would give more Se to crops than triple superphosphate, and the increased use of triple superphosphate may have been one of the causes of decreasing Se concentrations in European crops (Broadley et al., 2006).

The concentration of plant-available selenium is increased by soil $CaCO_3$ (Zhao et al., 2005), probably because the high pH lowers the amount of sorption of selenate and selenite. In the experiments of van Dorst and Peterson (1984), in which $^{75}SeO_3^{2-}$ was supplied to a soil of pH 4.5 and a soil of pH 7, perennial ryegrass plants took up much more of the labelled selenium from the pH 7 soil than from the pH 4.5 soil. It has already been pointed out that the pH 7 soil contained much more of both selenite and selenate in its soil solution, particularly more selenate, so the lowered adsorption of these ions in this soil apparently made them more available for uptake by the plants.

Could liming be carried out to increase availability of Se? Because of the decreased adsorption of selenite at high pH, it might be expected that liming would make soil Se more available to crops. In a study on white clover there was higher Se concentration of plants grown in limed soil than in unlimed soil when sodium selenite was supplied, although not in plants without supplemental Se (de la Luz Mora et al., 2008). However, in a study on three pasturelands in Germany that gave crops with concentrations below 0.10 mg Se kg^{-1} plant dry matter, there was no clear effect of liming on plant Se concentrations (Laser, 2007). The author indicated that the increase in soil pH brought about by the liming may not have been sufficient to improve Se availability. He pointed out that even if it does, the higher soil pH would make essential nutrients such as copper and zinc less available, and this would not be desirable for livestock production (Laser, 2007). It has also been reported that liming pastureland results in lower uptake of Se by plants (Haygarth et al., 1993); perhaps in that study, the decreased sorption led to more Se being leached from the soil. In a study of 66 sites of low-input grazed grassland in central Europe carried out over 2 years, there was no relationship between plant Se uptake and soil pH (Laser, 2004).

Selenium fertilizers can be supplied as solids or as foliar sprays. Another possible source is with pelleted seeds, and pelletization of ryegrass seeds with sodium selenite has been shown to give rise to increases in shoot Se concentrations in proportion to the rate of Se supply, increases that were less apparent in successive cuts of the herbage but which were still obvious for higher rates of Se supply at the fourth cut (Cartes et al., 2011).

If selenium fertilization is carried out, its timing needs to be considered carefully. In the experiments of Rimmer et al. (1990), the control plots of the pasture showed higher Se concentrations in winter and spring than in summer. In some situations overwintered livestock or animals put out to pasture in spring may obtain sufficient Se, but concentrations in the herbage may become inadequate in summer due to dilution effects of the enhanced plant growth in the warmer months. Under these conditions, herbage conserved as silage or hay is also likely to be deficient. Spring application of nitrogen with Se fertilizer can give enhanced Se concentration of pasture plants, whereas in some locations summer application of N with Se fertilizer can give decreased plant Se concentration compared with plants supplied Se without N (Laser, 2007). This again may be a dilution effect brought about by the stimulation of plant growth by the nitrogen.

Selenium for commercial use is currently produced as a by-product in electrolysis of copper (and also zinc and nickel), but if Se fertilization in agriculture is to become prevalent, steps should be taken to safeguard selenium supplies through recycling and development of other sources (Broadley et al., 2010).

REFERENCES

Adams, M.L., E. Lombi, F.-J. Zhao, and S.P. McGrath. 2002. Evidence of low selenium concentrations in UK bread-making wheat grain. *J. Sci. Food Agric.* 82:1160–1165.

Agricultural Research Council. 1980. *The Nutrient Requirements of Ruminant Livestock.* Commonwealth Agricultural Bureau: Farnham, U.K.

Arvy, M.P. 1993. Selenate and selenite uptake and translocation in bean plants (*Phaseolus vulgaris*). *J. Exp. Bot.* 44:1083–1087.

Asher, C.J., C.S. Evans, and C.M. Johnson. 1967. Collection and partial characterization of volatile selenium compounds from *Medicago sativa* L. *Aust. J. Biol. Sci.* 20:737–748.

Balistrieri, L.S. and T.T. Chao. 1987. Selenium adsorption by goethite. *Soil Sci. Soc. Am. J.* 51:1145–1151.

Bañuelos, G., D.L. Leduc, E.A.H. Pilon-Smits, and N. Terry. 2007. Transgenic Indian mustard overexpressing selenocysteine lyase or selenocysteine methyltransferase exhibit enhanced potential for phytoremediation under field conditions. *Environ. Sci. Technol.* 41:599–605.

Barberon, M., P. Berthomieu, M. Clairotte, N. Shibagaki, J.-C. Davidian, and F. Gosti. 2008. Unequal functional redundancy between the two *Arabidopsis thaliana* high-affinity sulphate transporters SULTR1;1 and SULTR1;2. *New Phytol.* 180:608–619.

Barclay, M.N.I. and A. MacPherson. 1992. Selenium content of wheat for bread making in Scotland and the relationship between glutathione peroxidase (*EC* 1.11.1.9) levels in whole blood and bread consumption. *Br. J. Nutr.* 68:261–270.

Barickman, T.C., D.A. Kopsell, and C.E. Sams. 2013. Selenium influences glucosinolate and isothiocyanates and increases sulfur uptake in *Arabidopsis thaliana* and rapid-cycling *Brassica oleracea. J. Agric. Food Chem.* 61:202–209.

Barrow, N.J., P. Cartes, and M.L. Mora. 2005. Modifications to the Freundlich equation to describe anion sorption over a large range and to describe competition between pairs of ions. *Eur. J. Soil Sci.* 56:601–606.

Barrow, N.J. and B.R. Whelan. 1989. Testing a mechanistic model. VII. The effects of pH and of electrolyte on the reaction of selenite and selenite with a soil. *J. Soil Sci.* 40:17–28.

Bisbjerg, B. and G. Gissel-Nielsen. 1969. The uptake of applied selenium by agricultural plants. 1. The influence of soil type and plant species. *Plant Soil* 31:287–298.

Broadley, M.R., J. Alcock, J. Alford et al. 2010. Selenium biofortification of high-yielding winter wheat (*Triticum aestivum* L.) by liquid or granular Se fertilisation. *Plant Soil* 332:5–18.

Broadley M.R., P.J. White, R.J. Bryson et al. 2006. Biofortification of UK food crops with selenium. *Proc. Nutr. Soc.* 65:169–181.

Brown, T.A. and A. Shrift. 1982. Selenium: Toxicity and tolerance in higher plants. *Biol. Rev.* 57:59–84.

Cabannes, E., P. Buchner, M.R. Broadley, and M.J. Hawkesford. 2011. A comparison of sulfate and selenium accumulation in relation to the expression of sulfate transporter genes in *Astragalus* species. *Plant Physiol.* 157:2227–2239.

Cartes, P., L. Gianfreda, and M.L. Mora. 2005. Uptake of selenium and its antioxidant activity in ryegrass when applied as selenate and selenite forms. *Plant Soil* 276:359–367.

Cartes, P., L. Gianfreda, C. Paredes, and M.L. Mora. 2011. Selenium uptake and its antioxidant role in ryegrass cultivars as affected by selenite seed pelletization. *J. Soil Sci. Plant Nutr.* 11:1–14.

Cartes, P., C. Shene, and M.L. Mora. 2006. Selenium distribution in ryegrass and its antioxidant role as affected by sulfur fertilization. *Plant Soil* 285:187–195.

Chang, S.C. and M.L. Jackson. 1957. Fractionation of soil phosphorus. *Soil Sci.* 84:133–144.

Charter, R.A., M.A. Tabatabai, and J.W. Schafer. 1995. Arsenic, molybdenum, selenium, and tungsten contents of fertilizers and phosphate rocks. *Commun. Soil Sci. Plant Anal.* 26:3051–3062.

Chilimba, A.D.C., S.D. Young, C.R. Black, M.C. Meacham, J. Lammel, and M.R. Broadley. 2012. Agronomic biofortification of maize with selenium (Se) in Malawi. *Field Crops Res.* 125:118–128.

Combs, G.F. 2001. Selenium in global food systems. *Br. J. Nutr.* 85:517–547.

Cubadda, F., F. Aureli, S. Ciardullo et al. 2010. Changes in selenium speciation associated with increasing tissue concentrations of selenium in wheat grain. *J. Agric. Food Chem.* 58:2295–2301.

de la Luz Mora, M.L., L. Pinilla, A. Rosas, and P. Cartes. 2008. Selenium uptake and its influence on the antioxidative system of white clover as affected by lime and phosphorus fertilization. *Plant Soil* 303:139–149.

de Souza, M.P., E.A.H. Pilon-Smits, C.M. Lytle et al. 1998. Rate-limiting steps in selenium assimilation and volatilization by Indian mustard. *Plant Physiol.* 117:1487–1494.

El Kassis, E., N. Cathala, H. Rouached et al. 2007. Characterization of a selenate-resistant *Arabidopsis* mutant. Root growth as a potential target for selenate toxicity. *Plant Physiol.* 143:1231–1241.

Elrashidi, M.A., D.C. Adriano, S.M. Workman, and W.L. Lindsay. 1987. Chemical equilibria of selenium in soils: A theoretical development. *Soil Sci.* 144:141–152.

Eurola, M., P. Ekhlom, M. Ylinen, P. Koivistoinen, and P. Varo. 1990. Effects of selenium fertilization on the selenium content of cereal grains, flour, and bread produced in Finland. *Cereal Chem.* 67:334–337.

Fan, M.-S., F.-J. Zhao, P.R. Poulton, and S.P. McGrath. 2008. Historical changes in the concentrations of selenium in soil and wheat grain from the Broadbalk experiment over the last 160 years. *Sci. Total Environ.* 389:532–538.

Franklin, R.E., L. Duis, B.R. Smith, R. Brown, and J.I. Toler. 2003. Elemental concentrations in soils of South Carolina. *Soil Sci.* 168:280–291.

Freeman, J.L., M. Tamaoki, C. Stushnoff et al. 2010. Molecular mechanisms of selenium tolerance and hyperaccumulation in *Stanleya pinnata. Plant Physiol.* 153:1630–1652.

Galeas, M.L., L.H. Zhang, J.L. Freeman, M. Wegner, and E.A.H. Pilon-Smits. 2007. Seasonal fluctuations of selenium and sulfur accumulation in selenium hyperaccumulators and related nonaccumulators. *New Phytol.* 173:517–525.

Gierus, M., F.J. Schwarz, and M. Kirchgessner. 2002. Selenium supplementation and selenium status of dairy cows fed diets based on grass, grass silage or maize silage. *J. Anim. Physiol. Anim. Nutr.* 86:74–82.

Gissel-Nielsen, G. 1979. Uptake and translocation of selenium-75 in *Zea mays*. In *Isotopes and Radiation in Research on Soil-Plant Relationships*, pp. 427–436. International Atomic Energy Agency: Vienna, Austria.

Gissel-Nielsen, G., U.C. Gupta, M. Lamand, and T. Westermarck. 1984. Selenium in soils and plants and its importance in livestock and human nutrition. *Adv. Agron.* 37:397–460.

Givens, D.I., R. Allison, B. Cottrill, and J.S. Blake. 2004. Enhancing the selenium content of bovine milk through alteration of the form and concentration of selenium in the diet of the dairy cow. *J. Sci. Food Agric.* 84:811–817.

Goldberg, S. and R.A. Glaubig. 1988. Anion sorption on a calcareous, montmorillonitic soil—Selenium. *Soil Sci. Soc. Am. J.* 52:954–958.

Gomes-Junior, R.A., P.L. Gratão, S.A. Gaziola, P. Mazzafera, P.J. Lea, and R.A. Azevedo. 2007. Selenium-induced oxidative stress in coffee cell suspension cultures. *Funct. Plant Biol.* 34:449–456.

Govasmark, E. and B. Salbu. 2011a. Translocation and re-translocation of selenium taken up from nutrient solution during vegetative growth in spring wheat. *J. Sci. Food Agric.* 91:1367–1372.

Govasmark, E. and B. Salbu. 2011b. Re-translocation of selenium during generative growth in wheat. *J. Plant Nutr.* 34:1919–1929.

Gupta, U.C. and S.C. Gupta. 2000. Selenium in soils and crops, its deficiencies in livestock and humans: Implications for management. *Commun. Soil Sci. Plant Anal.* 31:1791–1807.

Gupta, U.C. and J.A. Macleod. 1994. Effect of various sources of selenium fertilization on the selenium concentration of feed crops. *Can. J. Soil Sci.* 74:285–290.

Gustafsson, J.P. and L. Johnsson. 1992. Selenium retention in the organic matter of Swedish forest soils. *Eur. J. Soil Sci.* 43:461–472.

Hall, J.A., G. Bobe, J.K. Hunter et al. 2013. Effect of feeding selenium-fertilized alfalfa hay on performance of weaned beef calves. *PLoS ONE* 8:e58188.

Han, D., X.-H. Li, S.-L. Xiong et al. 2013. Selenium uptake, speciation and stressed response of *Nicotiana tabacum* L. *Environ. Exp. Bot.* 95:6–14.

Hartikainen, H. 2005. Biogeochemistry of selenium and its impact on food chain quality and human health. *J. Trace Elem. Med. Biol.* 18:309–318.

Hartikainen H, T. Hu, and V. Piironen. 2000. Selenium as an anti-oxidant and pro-oxidant in ryegrass. *Plant Soil* 225:193–200.

Hawrylak-Nowak, B. 2013. Comparative effects of selenite and selenate on growth and selenium accumulation in lettuce plants under hydroponic conditions. *Plant Growth Regul.* 70:149–157.

Haygarth, P.M., A.I. Cooke, K.C. Jones, A.F. Harrison, and A.E. Johnston. 1993. Long-term change in the biogeochemical cycling of atmospheric selenium: Deposition to plants and soil. *J. Geophys. Res.* 98:16769–16776.

He Z, X. Yang, Z. Zhu, Q. Zhang, W. Xia, and J. Tan. 1994. Effect of phosphate on the sorption, desorption and plant-availability of selenium in the soil. *Fert. Res.* (currently *Nutr. Cycl. Agroecosyst.*) 3:189–197.

Hewitt, E.J. 1966. *Sand and Water Culture Methods Used in the Study of Plant Nutrition*, 2nd edn. Commonwealth Agricultural Bureau: Farnham, U.K.

Hopper, J.L. and D.R. Parker. 1999. Plant availability of selenite and selenate as influenced by the competing ions phosphate and sulfate. *Plant Soil* 210:199–207.

Hsu, F.-C., M. Wirtz, S.C. Heppel et al. 2011. Generation of Se-fortified broccoli as functional food: Impact of Se fertilization on S metabolism. *Plant Cell Environ.* 34:192–207.

Kápolna, E. and P. Fodor. 2007. Bioavailability of selenium from selenium-enriched green onions (*Allium fistulosum*) and chives (*Allium schoenoprasum*) after 'in vitro' gastrointestinal digestion. *Int. J. Food Sci. Nutr.* 58:282–296.

Kápolna, E., K.H. Laursen, S. Husted, and E.H. Larsen. 2012. Bio-fortification and isotopic labelling of Se metabolites in onions and carrots following foliar application of Se and [77]Se. *Food Chem.* 133:650–657.

Kápolna, E., M. Shah, J.A. Caruso, and P. Fodor. 2007. Selenium speciation studies in Se-enriched chives (*Allium schoenoprasum*) by HPLC-ICP-MS. *Food Chem.* 101:1398–1406.

Keskinen, R., P. Ekholm, M. Yli-Halla, and H. Hartikainen. 2009. Efficiency of different methods in extracting selenium from agricultural soils of Finland. *Geoderma* 153:87–93.

Keskinen, R., M. Turakainen, and H. Hartikainen. 2010. Plant availability of soil selenate additions and selenium distribution within wheat and ryegrass. *Plant Soil* 333:301–313.

Kopsell, D.A. and D.E. Kopsell 2007. Selenium. In *Handbook of Plant Nutrition*, eds. A.V. Barker and D.J. Pilbeam, pp. 515–549. CRC Press: Boca Raton, FL.

Lakin, H.W. 1972. Selenium accumulation in soils and its absorption by plants and animals. *Geol. Soc. Am. Bull.* 83:181–190.

Laser, H. 2004. Selenium concentrations in herbage from various swards in low-input grassland systems. *Grassl. Sci. Eur.* 9:1008–1010.

Laser, H. 2007. Effects of liming and nitrogen application on the trace element concentrations of pastures in low mountain range. *Plant Soil Environ.* 53:258–266.

Lee, S., J.J. Doolittle, and H.J. Woodard. 2011a. Selenite adsorption and desorption in selected South Dakota soils as a function of pH and other oxyanions. *Soil Sci.* 176:73–79.

Lee, S., H.J. Woodard, and J.J. Doolittle. 2011b. Effect of phosphate and sulfate fertilizers on selenium uptake by wheat (*Triticum aestivum*). *Soil Sci. Plant Nutr.* 57:696–704.

Lee, S., H.J. Woodard, and J.J. Doolittle. 2011c. Selenium uptake response among selected wheat (*Triticum aestivum*) varieties and relationship with soil selenium fractions. *Soil Sci. Plant Nutr.* 57:823–832.

Lévesque, M. 1974. Selenium distribution in Canadian soil profiles. *Can. J. Soil Sci.* 54:63–68.

Levine, V.E. 1925. The effect of selenium compounds upon growth and germination in plants. *Am. J. Bot.* 12:82–90.

Li, H.-F., S.P. McGrath, and F.-J. Zhao. 2008. Selenium uptake, translocation and speciation in wheat supplied with selenate or selenite. *New Phytol.* 178:92–102.

Longchamp, M., N. Angeli, and M. Castrec-Rouelle. 2013. Selenium uptake in *Zea mays* supplied with selenate or selenite under hydroponic conditions. *Plant Soil* 362:107–117.

Lyi, S.M., L.I. Heller, M. Rutzke, R.M. Welch, L.V. Kochian, and L. Li. 2005. Molecular and biochemical characterization of the selenocysteine Se-methyltransferase gene and Se-methylselenocysteine synthesis in broccoli. *Plant Physiol.* 138:409–420.

Lyons, G., I. Ortiz-Monasterio, J. Stangoulis, and R. Graham. 2005. Selenium concentration in wheat grain: Is there sufficient genotypic variation to use in breeding? *Plant Soil* 269:369–380.

MAFF. 1999. *MAFF UK - 1997 Total Diet Study – Aluminium, Arsenic, Cadmium, Chromium, Copper, Lead, Mercury, Nickel, Selenium, Tin and Zinc*. Food Surveillance Information Sheet No. 191. MAFF: London.

Mazej, D., J. Osvald, and V. Stibilj. 2008. Selenium species in leaves of chicory, dandelion, lamb's lettuce and parsley. *Food Chem.* 107:75–83.

Mbagwu, J.S.C. 1983. Selenium concentrations in crops grown on low-selenium soils as affected by fly-ash amendment. *Plant Soil* 74:75–81.

Missana, T., U. Alonso, and M. García-Guttiérez. 2009. Experimental study and modelling of selenite onto illite and smectite clays. *J. Colloid Interface Sci.* 334:132–138.

Munier-Lamy, G., S. Deneux-Mustin, C. Mustin, D. Merlet, J. Berthelin, and C. Leyval. 2007. Selenium bioavailability and uptake as affected by four different plants in a loamy clay soil with particular attention to mycorrhizae inoculated ryegrass. *J. Environ. Radioact.* 97:148–158.

Nakamaru, Y., K. Tagami, and S. Uchida. 2005. Distribution coefficient of selenium in Japanese agricultural soils. *Chemosphere* 58:1347–1354.

Nakamaru, Y.M. and K. Sekine. 2008. Sorption behavior of selenium and antimony in soils as a function of phosphate ion concentration. *Soil Sci. Plant Nutr.* 54:332–341.

Neal, R.H., G. Sposito, K.M. Holtzclaw, and S.J. Traina. 1987a. Selenite adsorption on alluvial soils. I. Soil composition and pH effects. *Soil Sci. Soc. Am. J.* 51:1161–1165.

Neal, R.H., G. Sposito, K.M. Holtzclaw, and S.J. Traina. 1987b. Selenite adsorption on alluvial soils. II. Solution composition effects. *Soil Sci. Soc. Am. J.* 51:1165–1169.

Ng, B.H. and J.W. Anderson. 1978. Synthesis of selenocysteine by cysteine synthases from selenium accumulator and non-accumulator plants. *Phytochemistry* 17:2069–2074.

Novoselov, S.V., M. Rao, N.V. Onoshko et al. 2002. Selenoproteins and selenocysteine insertion system in the model plant cell system, *Chlamydomonas reinhardtii*. *EMBO J.* 21:3681–3693.

Papp, L.V., J. Lu, A. Holmgren, and K.K. Khanna. 2007. From selenium to selenoproteins: Synthesis, identity, and their role in human health. *Antioxid. Redox Signal.* 9:775–806.

Pyrzynska, K. 2009. Selenium speciation in enriched vegetables. *Food Chem.* 114:1183–1191.

Ramos, S.J., Y. Yuan, V. Faquin, L.R.G. Guilherme, and L. Li. 2011. Evaluation of genotypic variation of broccoli (*Brassica oleracea* var. *italic*) in response to selenium treatment. *J. Food Agric. Chem.* 59:3657–3665.

Rani, N., K.S. Dhillon, and S.K. Dhillon. 2005. Critical levels of selenium in different crops grown in an alkaline silty loam soil treated with selenite-Se. *Plant Soil* 277:367–374.

Rayman, M.P. 2000. The importance of selenium to human health. *Lancet* 356:233–241.

Renkema, H., A. Koopmans, L. Kersbergen, J. Kikkert, B. Hale, and E. Berkelaar. 2012. The effect of transpiration on selenium uptake and mobility in durum wheat and spring canola. *Plant Soil* 354:239–250.

Rimmer, D.L., D.S. Shiel, J.K. Syers, and M. Wilkinson. 1990. Effect of soil application of selenium on pasture composition. *J. Sci. Food Agric.* 51:407–410.

Robbins, C.W. and D.L. Carter. 1970. Selenium concentrations in phosphorus fertilizer materials and associated uptake by plants. *Soil Sci. Soc. Am. J.* 34:506–509.

Sabbagh, M. and D. Van Hœwyk. 2012. Malformed selenoproteins are removed by the ubiquitin-proteasome pathway in *Stanleya pinnata*. *Plant Cell Physiol.* 53:555–564.

Sharma, S., A. Bansal, S.K. Dhillon, and K.S. Dhillon. 2010. Comparative effects of selenate and selenite on growth and biochemical composition of rapeseed (*Brassica napus* L.). *Plant Soil* 329:339–348.

Shibagaki, N., A. Rose, J.P. McDermott et al. 2002. Selenate-resistant mutants of *Arabidopsis thaliana* identify *Sultr1;2*, a sulfate transporter required for efficient transport of sulfate into roots. *Plant J.* 29:475–486.

Shinmachi, F., P. Buchner, J.L. Stroud et al. 2010. Influence of sulfur deficiency on the expression of specific sulfate transporters and the distribution of sulfur, selenium, and molybdenum in wheat. *Plant Physiol.* 153:327–336.

Singh, B.R. 1991. Selenium content of wheat as affected by selenate and selenite contained in a Cl- or SO$_4$-based NPK fertilizer. *Fert. Res.* (currently *Nutr. Cycl. Agroeco.*) 30:1–7.

Singh, M., N. Singh, and P.S. Relan. 1981. Adsorption and desorption of selenite and selenate selenium on different soils. *Soil Sci.* 132:134–141.

Sors, T.G., D.R. Ellis, G.N. Na et al. 2005a. Analysis of sulfur and selenium assimilation in *Astragalus* plants with varying capacities to accumulate selenium. *Plant J.* 42:785–797.

Sors, T.G., D.R. Ellis, and S.E. Salt. 2005b. Selenium uptake, translocation, assimilation and metabolic fate in plants. *Photosynth. Res.* 86:373–389.

Sors, T.G., C.P. Martin, and D.E. Salt. 2009. Characterization of selenocysteine methyltransferases from *Astragalus* species with contrasting selenium accumulation capacity. *Plant J.* 59:110–122.

Srikanth Lavu, R.V., V. De Schepper, K. Steppe, P.N.V. Majeti, F. Tack, and G. Du Laing. 2013. Use of selenium fertilizers for production of Se-enriched Kenaf (*Hibiscus cannabinus*): Effect of Se concentration and plant productivity. *J. Plant Nutr. Soil Sci.* 176:634–639.

Srikanth Lavu, R.V., G. Du Laing, T. Van De Wiele et al. 2012. Fertilizing soil with selenium fertilizers: Impact on concentration, speciation, and bioaccessibility of selenium in leek (*Allium ampeloprasum*). *J. Agric. Food Chem.* 60:10930–10935.

Stoklasa, J. 1922. Über die Einwirkung des Selens auf den Bau- und Betriebsstoffwechsel der Pflanze bei Anwesenheit der Radioactivität der Luft und des Bodens. *Biochem. Zeitschrift* 130:604–643 (cited in Trelease and Trelease, 1938).

Stroud, J.L., M.R. Broadley, I. Foot et al. 2010a. Soil factors affecting selenium concentration in wheat grain and the fate and speciation of Se fertilisers applied to soil. *Plant Soil* 332:19–30.

Stroud, J.L., H.F. Li, F.J. Lopez-Bellido et al. 2010b. Impact of sulphur fertilisation on crop response to selenium fertilisation. *Plant Soil* 332:31–40.

Takahashi, H., A. Watanabe-Takahashi, F.W. Smith, M. Blake-Kalff, M.J. Hawkesford, and K. Saito. 2000. The roles of three functional sulphate transporters involved in uptake and translocation of sulphate in *Arabidopsis thaliana*. *Plant J.* 23:171–182.

Tamaoki, M., J.L. Freeman, and E.A.H. Pilon-Smits. 2008. Cooperative ethylene and jasmonic acid signalling regulates selenite resistance in *Arabidopsis*. *Plant Physiol.* 146:1219–1230.

Terry, N., A.M. Zayed, M.P. de Souza, and A.S. Tarun. 2000. Selenium in higher plants. *Annu. Rev. Plant Physiol. Mol. Biol.* 51:401–432.

Thomson, C.D. 2004. Assessment of requirements for selenium and adequacy of selenium status: A review. *Eur. J. Clin. Nutr.* 58:391–402.

Toler, H.D., C.S. Charron, and C.E. Sams. 2007. Selenium increases sulfur uptake and regulates glucosinolate metabolism in rapid-cycling *Brassica oleracea*. *J. Am. Soc. Hort. Sci.* 132:14–19.

Trelease, S.F. and H.M. Trelease. 1938. Selenium as a stimulating and possibly essential element for indicator plants. *Am. J. Bot.* 25:372–380.

Ullrich-Eberius, C.I., A. Novacky, E. Fischer, and U. Lüttge. 1981. Relationship between energy-dependent phosphate uptake and the electrical membrane potential in *Lemna gibba* G1. *Plant Physiol.* 67:797–801.

van Dorst, S.H. and P.J. Peterson. 1984. Selenium speciation in the soil solution and its relevance to plant uptake. *J. Sci. Food Agric.* 35:601–605.

Van Hœwyk, D. 2013. A tale of two toxicities: malformed selenoproteins and oxidative stress both contribute to selenium stress in plants. *Ann. Bot.* 112:965–972.

Van Hœwyk, D., H. Takahashi, E. Inoue, A. Hess, M. Tamaoki, and E.A.H. Pilon-Smits. 2008. Transcriptome analyses give insight into selenium-stress responses and selenium tolerance mechanisms in *Arabidopsis*. *Physiol. Plant.* 132:236–253.

Varo, P., G. Alfthan, P. Ekholm, and A. Aro. 1988. Selenium intake and serum selenium in Finland: effects of soil fertilization with selenium. *Am. J. Clin. Nutr.* 48:324–329.

Wang, P., N.W. Menzies, E. Lombi et al. 2013. In situ speciation and distribution of toxic selenium in hydrated roots of cowpea. *Plant Physiol.* 163:407–418.

Wang, Z. and Y. Gao. 2001. Biogeochemical cycling of selenium in Chinese environments. *Appl. Geochem.* 16:1345–1351.

Whelan, B.R. and N.J. Barrow. 1994. Slow-release selenium fertilizers to correct selenium deficiency in grazing sheep in Western Australia. *Fert. Res.* (currently *Nutr. Cycl. Agroeco.*) 38:183–188.

White, P.J., H.C. Bowen, B. Marshall, and M.R. Broadley. 2007. Extraordinarily high leaf selenium to sulfur ratios define 'Se-accumulator' plants. *Ann. Bot.* 100:111–118.

White, P.J., H.C. Bowen, P. Parmaguru et al. 2004. Interactions between selenium and sulphur nutrition in *Arabidopsis thaliana*. *J. Exp. Bot.* 55:1927–1937.

Wróbel, K., K. Wróbel, S.S. Kannamkumarath et al. 2004. HPLC-ICP-MS speciation of selenium in enriched onion leaves—A potential dietary source of Se-methylselenocysteine. *Food Chem.* 86: 617–623.

Xu, W.F., Q.X. Chen, and W.M. Shi. 2010. Effects of nitrate supply site on selenite uptake by rice roots. *J. Agric. Food Chem.* 58:11075–11080.

Zayed, A, C.M. Lytle, and N. Terry. 1998. Accumulation and volatilization of different chemical species of selenium by plants. *Planta* 206:284–292.

Zhang, L., W. Shi, and X. Wang. 2006. Difference in selenite absorption between high- and low-selenium rice cultivars and its mechanism. *Plant Soil* 282:183–193.

Zhang, P. and D.L. Sparks. 1990. Kinetics of selenate and selenite adsorption/desorption at the goethite/water interface. *Environ. Sci. Technol.* 24:1848–1856.

Zhao, C., J. Ren, C. Xue, and E. Lin. 2005. Study on the relationship between soil selenium and plant selenium uptake. *Plant Soil* 277:197–206.

Zhao, F.-J., F.J. Lopez-Bellido, C.W. Gray, W.R. Whalley, L.J. Clark, and S.P. McGrath. 2007. Effects of soil compaction and irrigation on the concentrations of selenium and arsenic in wheat grains. *Sci. Total Environ.* 372:433–439.

Zhao, X.Q., N. Mitani, N. Yamaji, R.F. Shen, and J.F. Ma. 2010. Involvement of silicon influx transporter OsNIP2;1 in selenite uptake in rice. *Plant Physiol.* 153:1871–1877.

Zhu, Y.-G., E.A.H. Pilon-Smits, F.-J. Zhao, P.N. Williams, and A.A. Meharg. 2009. Selenium in higher plants: Understanding mechanisms for biofortification and phytoremediation. *Trends Plant Sci.* 14:436–442.

20 Silicon

Jian Feng Ma

CONTENTS

20.1 INTRODUCTION

Silicon (Si) is the second most abundant element, both in terms of weight and number of atoms, in the earth's crust. As silicon dioxides comprise more than 50% of soil, all plants rooting in soil contain Si in their tissues. However, the impact of Si on plant growth was hardly paid attention to for a long time. This inattention is because, unlike other nutrients, the symptoms of Si deficiency are not easily observed visibly in plants. In 1862, Sachs, one of the early pioneers of plant nutrition, first asked the following questions in his article on Si nutrition "...whether silicic acid is an indispensable substance for those plants that contain silica, whether it takes part in nutritional processes, and what is the relationship that exists between silicic acid and the life of the plant?" (Sachs, 1862). However, he could not find beneficial effects of Si on corn (*Zea mays* L.) growth. Nowadays, although Si has not been recognized as an essential element for plant growth, the beneficial effects of Si have been observed in a wide variety of plant species.

20.2 SILICON IN SOIL

Although Si is a major component of soil, the solubility of Si compounds is usually extremely low. This is due to strong affinity of Si with oxygen, so that Si in soil always exists as silica or silicate, which is combined with various metals. Most soils contain Si in solution at concentrations between 100 and 500 μM.

Soil Si can be divided into two parts: one is Si solubilized without change in redox potential toward soil reduction, while the other is Si solubilized with progress of soil reduction. Soluble Si concentration in paddy soil is affected by fall of floodwater, reduction, pH, temperature of soil,

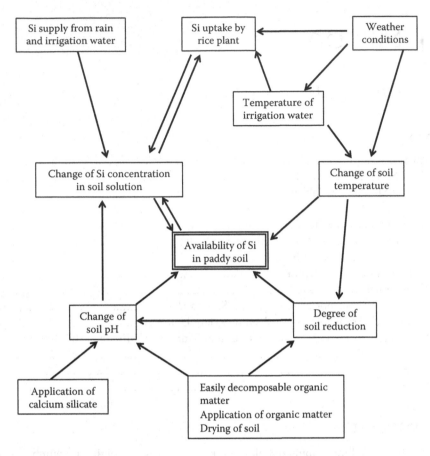

FIGURE 20.1 Dynamics of soluble silicon in paddy soil. (Based on Sumida, H., *Jpn. J. Soil Sci. Plant Nutr.*, 67, 435, 1996. With kind permission of the Japanese Society of Soil Science and Plant Nutrition.)

and other factors (Sumida, 1996) (Figure 20.1). The effect of soil reduction on availability of Si is less than that on P. Soil temperature also affects the availability of soil Si; more is available under high temperatures. Moreover, soil has a capacity for keeping Si concentration in soil solution at a constant level even though in the paddy soil of Figure 20.1, it would be expected to decrease with Si uptake by the rice plants. This capacity is controlled by characteristics of Si dissolution–adsorption of soil.

The form of Si in soil solution differs depending on the pH. At pH below 8.0, Si is present as a nondissociated silicic acid molecule, H_4SiO_4 (Figure 20.2). At higher pH, silicic acid is dissociated to negatively charged silicates. In most soils, therefore, Si is considered to be present in the form of an uncharged molecule, which is the form of Si taken up by plant roots.

20.3 TEST OF SOIL SILICON AVAILABILITY

Many methods for evaluating available Si in soil have been developed. Using 1 N acetate buffer (pH 4.0) is a simple and convenient method for estimating available soil Si in natural soils (Imaizumi and Yoshida, 1958). However, in soils with application of silicate fertilizers, this method cannot reflect the availability of soil Si because acetate can also extract Al-bound Si, which is not available for plants. As alternatives, "incubation under a submerged condition" (incubation at 40°C for 1 week) and "phosphate buffer" (0.04 M phosphate buffer [pH 6.2] at 40°C) methods have been developed (Takahashi and Nonaka, 1986; Kato and Sumida, 2000).

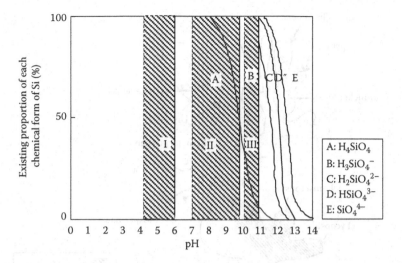

FIGURE 20.2   Silicon forms at different pH values.

20.4   SILICON IN PLANTS

20.4.1   Differences in Extent of Accumulation of Silicon in Plants

Although all plants contain Si in their tissues, there is a wide variation in the concentration of Si in the aboveground tissues, ranging from 0.1% to 10% of dry weight (Ma and Takahashi, 2002). Table 20.1 shows the concentration of Si in some common crops grown under the same conditions. Rice (*Oryza sativa* L.) contains 39. 1 mg Si g^{-1} in the shoots, but chickpea (*Cicer arietinum* L.) contains only 3.0 mg Si g^{-1}.

Based on Si content and Si/Ca ratio, three types of plants, Si accumulating, nonaccumulating, and intermediate, have been proposed (Ma and Takahashi, 2002). Si-accumulating plants have a Si concentration and Si/Ca ratio of higher than 1.0% and 1.0%, respectively, while nonaccumulating plants show less than 0.5% Si and a Si/Ca ratio of 0.5. Those between accumulating and nonaccumulating plants belong to an intermediate type. In the phylogenetic tree, high Si accumulation can be seen in Bryophyta, Lycopsida, and Equisetopsida of Pteridophyta, but this accumulation appears to have been lost from some of the Filicopsida in Pteridophyta through the Gymnospermae and from most of the Angiospermae (Figure 20.3). However, high Si accumulation appears again in the Cyperaceae and Gramineae in the monocots (Ma and Takahashi, 2002). Plants of the orders Cucurbitales and Urticales and the family Commelinaceae show an intermediate accumulation of Si, whereas most

TABLE 20.1
Silicon Concentration in Different Plant Species Grown under the Same Conditions

Plant Species	Si Concentration (mg Si g^{-1} dry wt.)
Rice (*Oryza sativa* L.)	39.1
Wheat (*Triticum aestivum* L.)	15.4
Pumpkin (*Cucurbita moschata* Duch.)	13.4
Zucchini (*Cucurbita pepo* L.)	19.8
Chickpea (*Cicer arietinum* L.)	3.0
Cucumber (*Cucumis sativus* L.)	22.9
Corn (*Zea mays* L.)	21.0

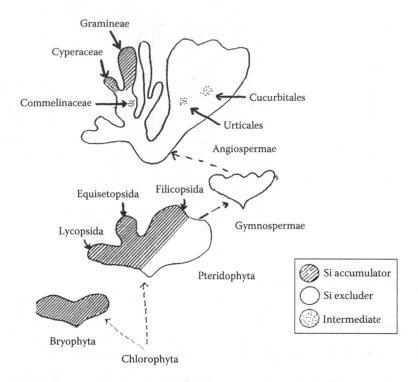

FIGURE 20.3 Distribution of Si-accumulating and nonaccumulating plants in the plant kingdom.

other species show a low accumulation of Si (Ma and Takahashi, 2002; Hodson et al., 2005). These differences have been attributed to the capacity of the roots to take up Si (Takahashi et al., 1990).

There is also a variation in Si concentration between varieties within a species. The Si concentration of leaves varied from 5.4% to 10.6% in 38 varieties of wild rice (*Oryza perennis* Moench) grown under the same conditions (Takahashi et al., 1981). The Si concentration in barley (*Hordeum vulgare* L.) grains showed a large variation among different varieties (Ma et al., 2003), ranging from 0% to 0.36%. The Si concentration is much lower in hulless barley than in covered barley. This is because most Si is localized in the hull. The Si content of the hull was between 1.53% and 2.71% in the varieties tested.

20.4.2 FORM AND LOCALIZATION OF SILICON IN PLANTS

Plant roots take up Si in the form of silicic acid [$Si(OH)_4$], an uncharged monomeric molecule (Takahashi and Hino, 1978), when the solution pH is below 9. This form is also present in the xylem sap (Casey et al., 2003; Mitani et al., 2005). However, if silicic acid is translocated to the shoots with the transpiration stream, it is concentrated through loss of water (due to transpiration) and then starts to be polymerized because silicic acid has a property of polymerization if its concentration exceeds 2 mM at 25°C. The process of Si polymerization converts silicic acid to colloidal silicic acid and finally to silica gel with increasing silicic acid concentration (Ma and Takahashi, 2002). In rice plants, more than 90% of total Si in the shoot is present in the form of silica gel, while the concentration of colloidal plus monomeric Si is kept below 140–230 mg Si L^{-1}. A similar pattern of accumulation is observed in cucumber (*Cucumis sativus* L.) leaves, although the total Si concentration of cucumber is much lower than that of rice.

Polymerized Si is deposited under the cuticle and in particular cells (Figure 20.4). There are two types of silicified cells that are termed as "silica cell" and "silica body" or "silica bulliform cell" in rice leaf blades. Silica cells are located in vascular bundles, showing a dumbbell shape, while

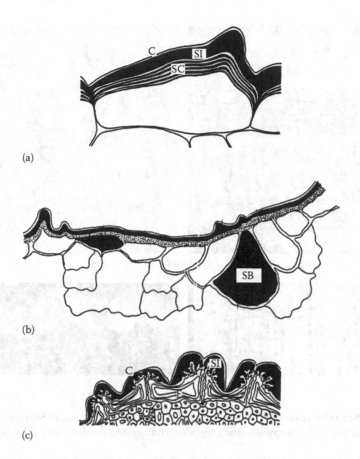

FIGURE 20.4 Diagram of silicon deposition in different tissues of rice: (a) cuticle–silica double layer in the leaf blade, (b) silicified cells, and (c) cuticle–silica double layer in hull of rice grain. Blackened area represents location of Si deposition. C, cuticle; SI, silica layer; E, epidermis; SC, silica cellulose membrane; SB, silica body. (From Yoshida, S., *Bull. Natl. Inst. Agric. Sci. Series B.* 15, 1, 1965.)

silica bodies are in bulliform cells of rice leaves (Figure 20.5). When the Si concentration of the shoot is below 5% SiO_2, only silica cells are formed. With increasing Si concentration, silica bodies increase, indicating that the silicification process of epidermal cells is from silica cells to silica bodies (Ma and Takahashi, 2002). There is a good correlation between Si content of a rice shoot and the number of Si bodies.

In addition to leaf blades, silicified cells are also observed in the epidermis and vascular tissues of stem, leaf sheath, and hull. These silicifications contribute to enhancing the strength and rigidity of the cell wall, thus increasing the resistance of rice plants to diseases and lodging, improving the form of the plants in the community for receiving light, and decreasing transpiration as described in the following section.

20.4.3 BENEFICIAL EFFECTS OF SILICON

Silicon has not been recognized as an essential element for plant growth. The major reason is that there is no evidence to show that Si is involved in the metabolism of plants, which is one of the three criteria required for essentiality established by Arnon and Stout (1939). However, the beneficial effects of Si have been observed in a wide variety of plant species. Silicon plays an important role in increasing the resistance to both biotic and abiotic stresses including diseases, pests, lodging, drought, and nutrient imbalance (Figure 20.6) (Ma and Takahashi, 2002; Ma, 2004; Ma and Yamaji, 2006). Therefore, the beneficial effect of Si is more evident under stress conditions.

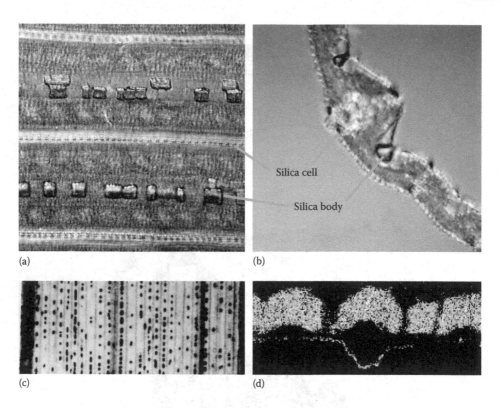

FIGURE 20.5 **(See color insert.)** Silicon accumulation in rice leaf blade. (a, b) Silica cell and silica body. (c) Silica body detected by soft x-ray. (d) Silica deposition on epidermal cells detected by SEM–EDX.

Silicon is effective in controlling diseases caused by both fungi and bacteria in different plant species. For example, Si increases rice resistance to leaf and neck blast (*Pyricularia grisea* Sacc.), sheath blight (*Thanatephorus cucumeris* Donk), brown spot (*Cochliobolus miyabeanus* Drechs.), leaf scald (*Microdochium oryzae* Samuels and I.C. Hallett), and stem rot (*Magnaporthe salvinii* R. Krause and Webster) (Figure 20.7a) (Datnoff and Rodrigues, 2005). Silicon also decreases the incidence of powdery mildew (*Erysiphe cichoracearum* DC. and *Blumeria graminis* (DC.) Speer f. sp. *tritici* [*Erysiphe graminis* DC. f. sp. *tritici* Em. Marchal]) in cucumber, barley, and wheat (*Triticum aestivum* L.), ring spot (*Leptosphaeria sacchari* Breda de Haan) in sugarcane (*Saccharum officinarum* L.), rust (*Uromyces vignae* Barc.) in cowpea (*Vigna unguiculata* Walp.), leaf spot in Bermuda grass (*Cynodon dactylon* (L.) Pers.), and gray leaf spot (*P. grisea*) in St. Augustine grass (*Stenotaphrum secundatum* (Walt.) Kuntze) and perennial ryegrass (*Lolium perenne* L.) (Bélanger et al., 2003; Fauteux et al., 2005). Silicon also suppresses insect pests such as stem borers (*Chilo suppressalis* Walker, *Scirpophaga incertulas* Walker), brown plant hopper (*Nilaparvata lugens* Stål), rice green leafhopper (*Nephotettix virescens* Distant), whitebacked plant hopper (*Sogatella furcifera* Horvath), and arachnid pests such as leaf spiders and mites (Figure 20.7b) (Savant et al., 1997). Resistance to the damage by wild rabbits is also enhanced by an increased amount of Si in wheat (Cotterill et al., 2007).

Si plays an important role in alleviating various abiotic stresses including physical stress and chemical stress (Ma, 2004; Ma and Yamaji, 2006). For example, Si can alleviate water stress and climatic stress such as from typhoons, low temperature, and insufficient sunshine during the summer season (Ma and Takahashi, 2002; Ma, 2004; Ma and Yamaji, 2006). Excess transpiration from the husk causes whitehead, resulting in low fertility in rice (Figure 20.7c and d); however, Si is effective in protecting the panicles from excess transpiration due to heavy deposition of Si on the husk. The beneficial effects of Si under chemical stresses, including P deficiency, P excess, and Mn

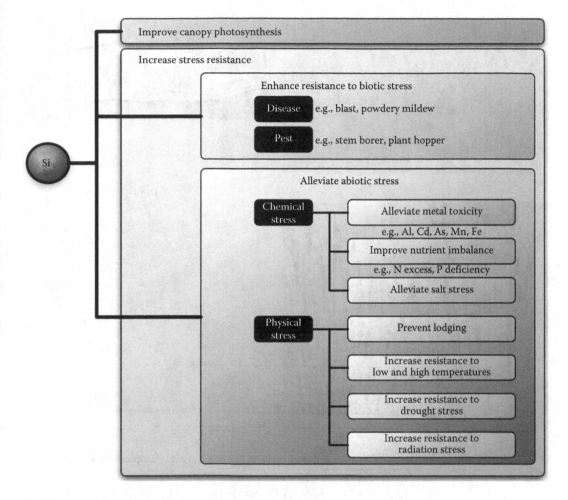

FIGURE 20.6 Beneficial effects of silicon on plant growth.

and salt toxicities, have been observed in many plants (Ma and Takahashi, 2002). Manganese toxicity in pumpkin (*Cucurbita moschata* Duch.) is significantly reduced by Si application (Iwasaki and Matsumura, 1999).

In addition to the role of Si in alleviating various stresses, Si improves light interception by keeping leaves erect, thereby stimulating canopy photosynthesis in rice (Ma and Takahashi, 2002). This is particularly important in dense plant stands and when nitrogen fertilizers are heavily applied, to minimize mutual shading. Indeed, silicon is especially important for healthy growth and high production of rice (Tamai and Ma, 2008), which is able to accumulate Si to over 10% of its dry weight in the shoots. If Si accumulation is not enough, the yield is greatly reduced mainly due to decreased fertility (Tamai and Ma, 2008) (Figure 20.8).

Most of these beneficial effects are attributed to Si deposition in cell walls of roots, leaves, stems, and hulls, a deposition that functions as a physiological barrier. Heavy Si deposition in the plant tissues can impede penetration by fungi mechanically, and thereby avoid the infection process. In addition, soluble Si acts as a modulator of host resistance to pathogens (Rodrigues et al., 2004). Several studies in monocotyledons (rice and wheat) and dicotyledons (cucumber) have shown that plants supplied with Si can produce phenolics and phytoalexins in response to fungal infection such as those causing rice blast and powdery mildew (Chérif et al., 1994; Fawe et al., 1998). Silicon deposition also provides a mechanical barrier against probing and chewing by insects. Deposition of Si in the roots reduces apoplastic bypass flow and provides binding

FIGURE 20.7 **(See color insert.)** Effect of silicon on (a) blast, (b) pest, (c) overtranspiration, and (d) fertility in rice.

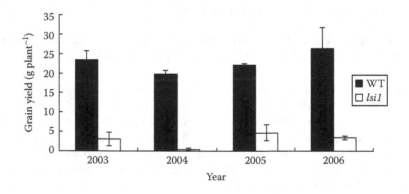

FIGURE 20.8 Rice grain yield in wild-type rice (WT) and a mutant defective in silicon uptake (*lsi1*) grown in a field in different years. (From Tamai, K. and Ma, J.F., *Plant Soil*, 307, 21, 2008.)

sites for metals, resulting in decreased uptake and translocation of toxic metals and salts from the roots to the shoots. Deposition of Si in the culms, leaves, and hulls also enhances the strength and rigidity of cell walls and decreases transpiration from the cuticle and, thus, increases the resistance to lodging, low and high temperatures, radiation, UV, and drought stresses (Ma and Takahashi, 2002).

20.4.4 MODES OF SILICON UPTAKE

Three different modes of Si uptake have been proposed for plants having different degrees of Si accumulation, that is, active, passive, and rejective uptake (Ma and Takahashi, 2002; Mitani and Ma, 2005). Plants with an active mode of uptake such as rice take up Si faster than water, resulting in a depletion of Si in the solution of the medium. Plants with a passive mode of uptake such as cucumber take up Si at a rate that is similar to water uptake; thus, no significant changes in the concentration of Si in the solution are observed. In contrast, plants with a rejective mode of uptake such as tomato tend to exclude Si, which is demonstrated by the increasing concentration of Si in the solution. These differences are attributed to different expression of transporters involved in Si uptake.

20.4.5 SILICON TRANSPORTERS

20.4.5.1 Transporters Involved in Silicon Uptake

The first Si transporter Lsi1 was identified in rice by using a mutant defective in Si uptake (Ma et al., 2006). Lsi1 belongs to a Nod26-like major intrinsic protein (NIP) subfamily of aquaporin-like proteins and shows influx activity for silicic acid in *Xenopus* oocytes. The predicted amino acid sequence has six transmembrane domains and two Asn-Pro-Ala (NPA) motifs, which are well conserved in typical aquaporins. *Lsi1* is expressed constitutively in the roots, but its expression is decreased to one-fourth by Si supply (Ma et al., 2006). Within a root, the expression of *Lsi1* is much lower in the root tip region between 0 and 10 mm than in the basal regions of the root (>10 mm) (Yamaji and Ma, 2007). Silicon uptake in the root tip region (0–10 mm) comprising the apical meristem and the elongation zone is also much lower than that in the basal regions (>10 mm from the root tips). Therefore, the site of Si uptake is located in the mature regions of the roots rather than in the root tips. Lsi1 is localized in the main and lateral roots, but not in root hairs (Figure 20.9a) (Ma et al., 2001); therefore, root hairs do not play a role in Si uptake. In the roots including seminal, lateral, and crown roots, the Lsi1 protein is localized on the plasma membranes of the exodermis and the endodermis (Ma et al., 2006; Yamaji and Ma, 2007), where the Casparian strip prevents

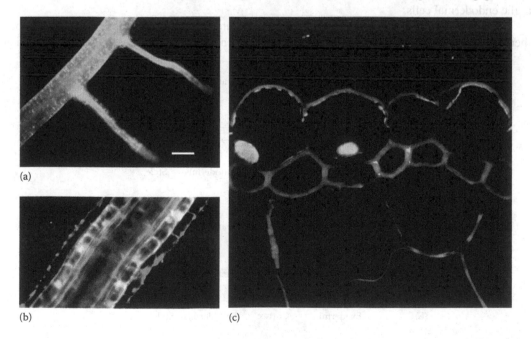

(a)

(b)　　　　　(c)

FIGURE 20.9 **(See color insert.)** Tissue localization of (a) Lsi1 (green), (b) Lsi2 (red), and (c) polar localization of Lsi1 (green) and Lsi2 (red) in rice.

apoplastic transport into the root stele. Furthermore, Lsi1 shows polar localization at the distal side of both the exodermis and endodermis cells (Figure 20.9c).

Homologs of rice Lsi1 (OsLsi1) have been identified in barley (HvLsi1), corn (ZmLsi1), pumpkin (CmLsi1), wheat (TaLsi1), and horsetail (*Equisetum arvense* L.) (EaLsi1) (Chiba et al., 2009; Mitani et al., 2009, 2011; Grégoire et al., 2012; Montpetit et al., 2012). However, unlike OsLsi1, HvLsi1, and ZmLsi1 are localized at epidermal, hypodermal, and cortical cells (Chiba et al., 2009; Mitani et al., 2009). CmLsi1 from the pumpkin is localized at all root cells (Mitani et al., 2011). Interestingly, mutation of the 242 residue from proline to leucine in CmLsi1 resulted in significant decrease in Si uptake in some pumpkin cultivars due to failure of the protein to be located at the plasma membrane (Mitani et al., 2011). In contrast to *OsLsi1*, the expression levels of *HvLsi1* and *ZmLsi1* are unaffected by Si (Chiba et al., 2009; Mitani et al., 2009).

The second Si transporter Lsi2 in rice also was identified by a similar approach of investigating mutants as used in the identification of Lsi1. However, unlike Lsi1, Lsi2 is an efflux transporter of Si in rice (Ma et al., 2007). Lsi2 belongs to a putative anion transporter without any similarity to the silicon influx transporter Lsi1. The expression pattern and tissue and cellular localizations of Lsi2 are the same as that of Lsi1 (Ma et al., 2007), but in contrast to Lsi1, Lsi2 is localized at the proximal side of the exodermis and the endodermis cells (Figure 20.9b,c). Transport of Si by Lsi2 is driven by a proton gradient.

Similar transporters of Lsi2 also have been identified in barley and corn (Mitani et al., 2009). However, ZmLsi2 and HvLsi2 are localized only to the endodermis of roots in corn and barley, without polarity. These differences result in a different pathway of Si from external solution to the xylem between upland crops (barley and maize) and a paddy crop (rice). In barley and corn, Si can be taken up from external solution (soil solution) by HvLsi1/ZmLsi1 at different cells including epidermal, hypodermal, and cortical cells (Figure 20.10a). In contrast, in rice, Si is taken up only at the exodermal cells by OsLsi1 (Figure 20.10b). After being taken up into the root cells, Si is transported to the endodermis by the symplastic pathway and then released to the stele by ZmLsi2 in corn and HvLsi2 barley. By contrast, in rice, Si taken up by OsLsi1 at the exodermal cells is released by OsLsi2 to the apoplast and is then transported into the stele by both OsLsi1 and OsLsi2, again at the endodermal cells.

The difference in the uptake system may be attributed to the root structures. In rice roots, there are two Casparian strips, at the exodermis and endodermis, whereas only one Casparian

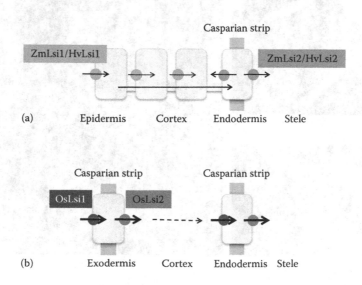

FIGURE 20.10 (**See color insert.**) Lsi1- and Lsi2-mediated silicon uptake system in (a) upland crop (barley/maize) and (b) rice.

strip usually is present at the endodermis of corn and barley roots under nonstressed conditions. Moreover, mature roots in rice have a distinct structure, a highly developed aerenchyma, wherein almost all cortex cells between exodermis and endodermis are destroyed. Therefore, Si transported into the exodermis cells by the influx transporter, OsLsi1, has to be released by the efflux transporter, OsLsi2, into the apoplast of a spoke-like structure across the aerenchyma. However, in corn and barley roots, there is no such structure or, if any, it is developed poorly. These differences in the localization of transporters and polarity may be one of the reasons for the different Si uptake capacities among species.

20.4.5.2 Transporters Involved in Xylem Unloading

Following uptake by the roots through Lsi1 and Lsi2, Si is translocated to the shoot by transpirational volume flow through the xylem. More than 90% of Si taken up by the roots is translocated to the shoots in rice (Ma and Takahashi, 2002). Silicic acid in the xylem sap must be exported to other tissues. Lsi6, a homolog of Lsi1, is responsible for this xylem unloading (Yamaji et al., 2008). Lsi6 also shows transport activity for silicic acid. However, unlike Lsi1 and Lsi2, Lsi6 also is expressed in the leaf sheaths and leaf blades in addition to the root tips. Lsi6 is localized in the adaxial side of the xylem parenchyma cells in the leaf sheaths and leaf blades (Figure 20.11a). Knockout of *Lsi6* does not affect the uptake of Si by the roots, but affects the silica deposition pattern in the leaf blades and sheaths (Yamaji et al., 2008). When *Lsi6* was knocked out, some abaxial epidermal cells were observed to be silicified, whereas in the wild-type plants, silica bodies were arranged above the veins on the adaxial surface. Also in the leaf sheaths, silicified epidermal cells were observed frequently in the knockout line, but they are not present in the wild-type rice. Furthermore, knockout of *Lsi6* also caused increased excretion of Si in the guttation fluid. These changes in the knockout lines are caused by an altered pathway of Si transport due to loss of function of Lsi6.

A homolog of rice Lsi6 also was identified in barley (Yamaji et al., 2012). HvLsi6 also is localized to the plasma membrane. At the vegetative growth stage, *HvLsi6* was expressed in the roots and shoots. The expression level was unaffected by Si supply. In the roots, HvLsi6 is localized in epidermis and cortex cells of the tips, whereas in the leaf blades and sheaths, HvLsi6 is localized only at parenchyma cells of vascular bundles. HvLsi6 probably is involved in Si uptake in the root tips and xylem unloading of Si in leaf blade and sheath.

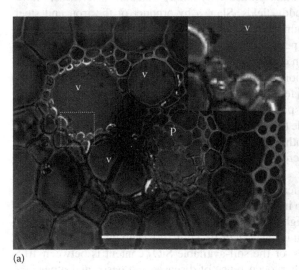 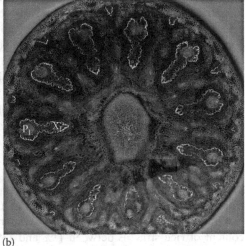

(a)   (b)

FIGURE 20.11 **(See color insert.)** Localization of Lsi6 protein in (a) leaf blade and (b) node I in rice. Red color shows signal of Lsi6 protein. v, vessel; p, phloem; xylem (X_L) and phloem regions (P_L) of large vascular bundles.

20.4.5.3 Transporters for Intervascular Transfer of Silicon at the Reproductive Stage

In gramineous plants, minerals taken up by the roots are not transported directly to the grains but are redirected at plant nodes. Therefore, the distribution of minerals at the nodes is considered to be a key step in the selective control of mineral accumulation in the panicles. At node I beneath the panicles of rice, there are enlarged large and small vascular bundles and diffuse vascular bundles. Large and small vascular bundles come from the lower nodes and connect to the flag leaf and are enlarged markedly at the node. Diffuse vascular bundles are parallel to and surround the enlarged vascular bundles. They also are assembled in the upper internode (peduncle) to form regular bundles, which connect toward the panicle tissues. Therefore, intervascular transfer of minerals from the enlarged vascular bundles to diffuse vascular bundles is required to deliver minerals taken up by the roots to developing seeds.

Lsi6 is involved in the intervascular transfer of Si in rice (Yamaji and Ma, 2009). At the reproductive stage, Lsi6 is highly expressed at node I below the panicles. Lsi6 is localized mainly at the xylem transfer cells with polarity facing toward the xylem vessel (Figure 20.11b). These cells are located at the outer boundary region of the enlarged vascular bundles and are characterized by large surface area due to cell wall ingrowth. Knockout of *Lsi6* decreased Si accumulation in the panicles but increased Si accumulation in the flag leaf (Yamaji and Ma, 2009). Therefore, Lsi6 is required for transfer of Si from the enlarged vascular bundles coming from the roots to the diffuse vascular bundles connected to the panicles.

HvLsi6 in barley also shows high expression in the nodes (Yamaji et al., 2012). HvLsi6 in node I also is located polarly at the transfer cells surrounding the enlarged vascular bundles toward the numerous xylem vessels, like OsLsi6. Furthermore, HvLsi2 is localized at the parenchyma cell layer adjacent to the transfer cells with opposite polarity to HvLsi6, suggesting that the coupling of HvLsi6 and HvLsi2 is involved in the intervascular transfer of Si at the nodes. Silicon translocated via the enlarged vascular bundles is unloaded to the transfer cells by HvLsi6, followed by HvLsi2 to reload Si to the diffuse vascular bundles, which are connected to the upper part of the plant, especially the panicles, the ultimate Si sink.

20.5 SILICATE FERTILIZERS

Due to beneficial effects of Si on crop growth and productivity, many kinds of Si fertilizers have been developed and applied, especially in paddy fields. Slag, a by-product of the iron and steel industries, is used as a Si fertilizer in many countries. The major component of slag is calcium silicate. Usually, 1.5–2.0 tons of calcium silicate per hectare is applied. A trial in Florida showed that application of slag is effective as a fungicide in controlling rice blast (Figure 20.12) (Datnoff et al., 1997). In addition to rice, an effect of application of calcium silicate also is observed in other crops such as sugar cane. In addition to slag, other Si fertilizers such as fused magnesium phosphate made by melting phosphate rock with serpentinite, potassium silicate, porous hydrate calcium silicate, and silica gel have been commercially available (Ma and Takahashi, 2002). They differ in availability of Si and composition, particularly of the other minerals present. In addition to soil application, foliar application is also sometimes useful for crops that are not able to actively take up Si (Menzies et al., 1992).

In Japan, the criterion for application of Si fertilizers has been established (Table 20.2) (Ma and Takahashi, 2002). This criterion divides a crop into three classes. A profitable increase in rice yield from fertilization can be expected with the highest probability if the SiO_2 content of rice straw is less than 11%, or the soil-available SiO_2 content is lower than 10.5 mg/100 g soil (class I). If the SiO_2 content of rice straw is between 11% and 13% or the soil-available SiO_2 content is between 10.5 and 13 mg/100 g soil, it is dependent on weather, occurrence of diseases and pests, and other factors whether the application of silicate fertilizers is effective in increasing rice yield (class II). No increase in rice yield can be expected if the SiO_2 content of rice straw or soil-available SiO_2 content is higher than 13% and 13 mg/100 g soil (class III).

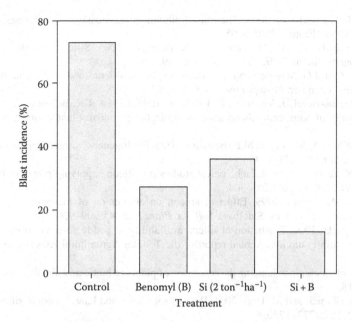

FIGURE 20.12 Effect of slag and fungicide on incidence of blast in rice. (From Datnoff, L.E. et al., *Crop Prot.,* 16, 525, 1997.)

TABLE 20.2
Criteria for Predication of Silicon Fertilizer Application in Rice

Class	SiO$_2$ Content in Rice Straw at Harvest (%)	Available Si Content in 100 g Soil (mg)	Response to Slag Application
I	<11	<10.5	A profitable increase in rice yield is expected with the highest probability.
II	11–13	10.5–13	A profitable increase in rice yield is expected in many cases, but not in some cases.
III	>13	>13	No profitable increase in rice yield is expected.

Classes of rice crops depending on their likelihood to benefit from application of slag.

REFERENCES

Arnon, D.I. and P.R. Stout. 1939. The essentiality of certain elements in minute quantity for plants with special reference to copper. *Plant Physiol.* 14:371–375.

Bélanger, R..R., N. Benhamou, and J.G. Menzies. 2003. Cytological evidence of an active role of silicon in wheat resistance to powdery mildew (*Blumeria graminis* f. sp. *tritici*). *Phytopathology* 93:402–412.

Casey, W.H., S.D. Kinrade, C.T.G. Knight, D.W. Rains, and E. Epstein. 2003. Aqueous silicate complexes in wheat, *Triticum aestivum* L. *Plant Cell Environ.* 27:51–54.

Chérif, M., A. Asselin, and R.R. Bélanger. 1994. Defense responses induced by soluble silicon in cucumber roots infected by *Pythium* spp. *Phytopathology* 84:236–242.

Chiba, Y., N. Mitani, N. Yamaji, and J.F. Ma. 2009. HvLsi1 is a silicon influx transporter in barley. *Plant J.* 57:810–818.

Cotterill, J.V., R.W. Watkins, C.B. Brennon, and D.P. Cowan. 2007. Boosting silica levels in wheat leaves reduces grazing by rabbits. *Pest Manage. Sci.* 63:247–253.

Datnoff, L.E., C.W. Deren, and G.H. Snyder. 1997. Silicon fertilization for disease management of rice in Florida. *Crop Prot.* 16:525–531.

Datnoff, L.E. and F.A. Rodrigues. 2005. The role of silicon in suppressing rice diseases. APSnet features. doi 10.1094/APSnetFeature-2005-0205

Fauteux, F., W. Remus-Borel, J.G. Menzies, and R.R. Bélanger. 2005. Silicon and plant disease resistance against pathogenic fungi. *FEMS Microbiol. Lett.* 249:1–6.

Fawe, A., M. Abou-Zaid, J.G. Menzies, and R.R. Bélanger. 1998. Silicon-mediated accumulation of flavonoid phytoalexins in cucumber. *Phytopathology* 88:396–401.

Grégoire, C., W. Rémus-Borel, J. Vivancos, C. Labbé, F. Belzile, and R.R. Bélanger. 2012. Discovery of a multigene family of aquaporin silicon transporters in the primitive plant *Equisetum arvense. Plant J.* 72:320–330.

Hodson, M.J., P.J. White, A. Mead, and M.R. Broadley. 2005. Phylogenetic variation in the silicon composition of plants. *Ann. Bot.* 96:1027–1046.

Imaizumi, K. and S. Yoshida. 1958. Edaphological studies on silicon supplying power of paddy fields. *Bull. Natl. Inst. Agric. Sci. Ser. B* 8:261–304.

Iwasaki, K. and A. Matsumura. 1999. Effect of silicon on alleviation of manganese toxicity in pumpkin (*Cucurbita moschata* Duch cv. Shintosa). *Soil Sci. Plant Nutr.* 45:909–920.

Kato, N. and H. Sumida. 2000. Evaluation of silicon availability in paddy soils by an extraction using a pH 6.2 phosphate buffer solution. Annual report of the Tohoku Agricultural Experiment Station for 1999, pp. 73–75.

Ma, J.F. 2004. Role of silicon in enhancing the resistance of plants to biotic and abiotic stresses. *Soil Sci. Plant Nutr.* 50:11–18.

Ma, J.F., S. Goto, K. Tamai, and M. Ichii. 2001. Role of root hairs and lateral roots in silicon uptake by rice. *Plant Physiol.* 127:1773–1780.

Ma, J.F., A. Higashitani, H. Sato, and K. Takeda. 2003. Genotypic variation in silicon concentration of barley grain. *Plant Soil* 249:383–387.

Ma, J.F. and E. Takahashi. 2002. *Soil, Fertilizer, and Plant Silicon Research in Japan.* Amsterdam, the Netherlands: Elsevier Science.

Ma, J.F., K. Tamai, N. Yamaji et al. 2006. A silicon transporter in rice. *Nature* 440:688–691.

Ma, J.F. and N. Yamaji. 2006. Silicon uptake and accumulation in higher plants. *Trends Plant Sci.* 11:392–397.

Ma, J.F., N. Yamaji, N. Mitani et al. 2007. An efflux transporter of silicon in rice. *Nature* 448:209–212.

Menzies, J., P. Bowen, and D. Ehret. 1992. Foliar application of potassium silicate reduces severity of powdery mildew on cucumber, muskmelon, and zucchini squash. *J. Am. Soc. Hort. Sci.* 117:902–905.

Mitani, N., Y. Chiba, N. Yamaji, and J.F. Ma. 2009. Identification and characterization of maize and barley Lsi2-like silicon efflux transporters reveals a distinct silicon uptake system from that in rice. *Plant Cell* 21:2133–2142.

Mitani, N. and J.F. Ma. 2005. Uptake system of silicon in different plant species. *J. Exp. Bot.* 56:1255–1261.

Mitani, N., J.F. Ma, and T. Iwashita. 2005. Identification of the silicon form in xylem sap of rice (*Oryza sativa* L.). *Plant Cell Physiol.* 46:279–283.

Mitani, N., N. Yamaji, Y. Ago, K. Iwasaki, and J.F. Ma. 2011. Isolation and functional characterization of an influx silicon transporter in two pumpkin cultivars contrasting in silicon accumulation. *Plant J.* 66:231–240.

Mitani, N., N. Yamaji, and J.F. Ma. 2009. Identification of maize silicon influx transporters. *Plant Cell Physiol.* 50:5–12.

Montpetit, J., J. Vivancos, N. Mitani-Ueno et al. 2012. Cloning, functional characterization and heterologous expression of TaLsi1, a wheat silicon transporter gene. *Plant Mol. Biol.* 79:35–46.

Rodrigues, F.A., D.J. McNally, L.E. Datnoff et al. 2004. Silicon enhances the accumulation of diterpenoid phytoalexins in rice: A potential mechanism for blast resistance. *Phytopathology* 94:177–183.

Sachs, J. 1862. Ergebnisse einiger neuerer untersuchungen uber die in Pflanzen enthaltene Kieselsaure. *Flora* 33:65–71.

Savant, N.K., G.H. Snyder, and L.E. Datnoff. 1997. Silicon management and sustainable rice production. *Adv. Agron.* 58:151–199.

Sumida, H. 1996. Progress and prospect of the research on paddy soil management under various rice growing system. 1 Progress in nutrient behavior and management research on paddy soil (3) silicic acid. *Jpn. J. Soil Sci. Plant Nutr.* 67:435–439.

Takahashi, E. and K. Hino. 1978. Silica uptake by plant with special reference to the forms of dissolved silica. *J. Sci. Soil Manure Jpn.* 49:357–360.

Takahashi, E., J.F. Ma, and Y. Miyake. 1990. The possibility of silicon as an essential element for higher plants. *Comments Agric. Food Chem.* 2:99–122.

Takahashi, E., T. Tanaka, and Y. Miyake. 1981. Distribution of silica accumulator plants in the plant kingdom (5) Gramineae. *Jpn. J. Soil Sci. Plant Nutr.* 52:503–510.

Takahashi, K. and K. Nonaka. 1986. Method of measurement of available silicate in paddy field. *Jpn. J. Soil Sci. Plant Nutr.* 57:515–517.

Tamai, K. and J.F. Ma. 2008. Reexamination of silicon effects on rice growth and production under field conditions using a low silicon mutant. *Plant Soil* 307:21–27.

Yamaji, N., Y. Chiba, N. Mitani-Ueno, and J.F. Ma. 2012. Functional characterization of a silicon transporter gene implicated in silicon distribution in barley. *Plant Physiol.* 160:1491–1497.

Yamaji, N. and J.F. Ma. 2007. Spatial distribution and temporal variation of the rice silicon transporter Lsi1. *Plant Physiol.* 143:1306–1313.

Yamaji, N. and J.F. Ma. 2009. Silicon transporter Lsi6 at the node is responsible for inter-vascular transfer of silicon in rice. *Plant Cell* 21:2878–2883.

Yamaji, N., N. Mitani, and J.F. Ma. 2008. A transporter regulating silicon distribution in rice shoots. *Plant Cell* 20:1381–1389.

Yoshida, S. 1965. Chemical aspects of the role of silicon in physiology of the rice plant. *Bull. Natl. Inst. Agric. Sci. Series B.* 15:1–58.

21 Sodium

Sven Schubert

CONTENTS

21.1 BACKGROUND

Sodium (Na), like potassium (K), is an alkali metal belonging to the first group in the periodic table of the chemical elements. Both of the elements are abundant in the earth's crust, which is 2.6% Na and 2.4% K. They occur in various salts, and after dissolution of the salts form monovalent cations. Despite these striking similarities, K and Na differ considerably in their significance for the biosphere. Whereas K is considered a plant nutrient (see Chapter 4), Na, depending on the plant species, may act as a beneficial or toxic element (Kronzucker et al., 2013). The difference between the two monovalent cations is based on the ionic diameter, which in turn determines the surface charge. Because the ionic diameter of the nonhydrated Na^+ is smaller (0.20 nm) than that of the nonhydrated K^+ (0.27 nm), the surface charge of Na^+ is higher, resulting in a larger diameter of the hydration shell. This size difference is the reason why K^+ in most cases cannot be substituted by Na^+ in the three major physiological functions, namely,

1. Charge balance
2. Enzyme activation
3. Osmotic effects

Charge balance requires high permeability of cell membranes to allow free movement of the ion following an electrochemical gradient. In most of the plant cell membranes, this action is facilitated by cation channels with a high specificity for K^+. The larger hydration shell that is not abandoned during channel-mediated transport hinders the passage of Na^+. The same applies to enzyme activation.

Many enzymes depend on the sorption of K^+. This sorption is defined as an unspecific electrostatic binding with retention of the hydration shell. The sorption energy is determined by Coulomb's law:

$$E = \frac{Q_1 \times Q_2}{\varepsilon \times r^2} \tag{21.1}$$

where
  E is the binding energy
  Q_1 is the charge of the enzyme
  Q_2 is the charge of the ion
  ε is the dielectric coefficient
  r is the distance between enzyme and ion

Binding of K^+ results in a conformational change of the protein and arrangement of the tertiary structure allowing optimal enzyme action. Since the distance r is determined by the hydration shell, the distance between cation and enzyme, and particularly r^2, is much larger for Na^+ than for K^+. The resulting difference in binding energy has consequences for enzyme structure and is responsible for the difference in enzyme activation by Na^+ and K^+. It is assumed that during evolution, pumping Na^+ out of cells helped to regulate turgor and at the same time resulted in the formation of enzymes with a high preference for K^+. As a consequence, in plant cells as in other cells, K^+ accumulates in intracellular (particularly cytoplasmic) and Na^+ in extracellular compartments (Leigh and Wyn Jones, 1986). Although unspecific osmotic functions of K^+ can be taken over by Na^+, it is also the low permeability of biological membranes that prevents an exclusive role of Na^+ in the water economy of many plants. This function applies particularly to the efficient regulation of stomata (Slabu et al., 2009).

From all these restraints, it may be inferred that plants exclude Na^+ to avoid detrimental effects on metabolism and water economy, which is particularly important under saline soil conditions (Koyro and Stelzer, 1988). The exclusion strategy is not only in glycophytes, but Na^+ exclusion is a quantitatively important strategy of the halophyte gray mangrove (*Avicennia marina* Vierh.) to cope with saline water (Burchett et al., 1984). In this regard, there is no fundamental difference between halophytic and glycophytic plants; rather, the two plant groups differ in quantitative and strategic aspects (Cheeseman, 1988; Flowers and Yeo, 1989; Gorham, 1994).

21.2 UPTAKE OF SODIUM AND DISTRIBUTION WITHIN PLANTS

21.2.1 UPTAKE OF SODIUM

Uptake of Na^+ is determined by passive influx and active efflux processes (Figure 21.1). Sodium enters root cells passively, that is, along an electrochemical gradient ($\Delta\mu_{Na+}$). The plasma membrane H^+-ATPase establishes an electrochemical proton gradient ($\Delta\mu_{H+}$) with an electrical membrane potential of about 100–200 mV that is the main driving force for Na^+. In addition, passive exclusion, sequestration in cell vacuoles, and transversal transport of Na^+ in the root tissue are responsible for relatively low cytosolic Na^+ concentrations that may contribute to the driving force for passive Na^+ uptake, particularly under saline conditions when rhizosphere Na^+ concentrations are high. The lipid membrane bilayer forms an effective barrier for Na^+ influx. However, an apoplastic bypass of Na^+ in undifferentiated root tissue, circumventing the Casparian strip, may allow uncontrolled Na^+ uptake of root tips as shown by Yeo et al. (1977) and Yadav et al. (1996) for rice (*Oryza sativa* L.).

The passive exclusion of Na^+ at the plasma membranes of epidermal and cortex cells is abolished partially by transport proteins (Figure 21.1). These transporters are required for nutrient cation uptake but are not sufficiently selective. There are three major groups of ion transporters that are responsible for Na^+ influx: low-affinity cation transporters (LCTs; Schachtman et al., 1997;

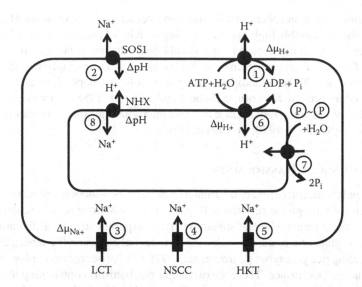

FIGURE 21.1   General model illustrating cellular Na$^+$ transport processes. (1) Plasma membrane H$^+$-ATPase, generating an electrochemical proton gradient; (2) SOS1 Na$^+$/H$^+$-antiporter, mediating electroneutral sodium/proton exchange; (3) LCT, mediating sodium influx; (4) NSCC, mediating sodium influx; (5) HKT, mediating sodium influx; (6) tonoplast H$^+$-ATPase, generating an electrochemical proton gradient; (7) tonoplast H$^+$-pyrophosphatase, generating an electrochemical proton gradient; (8) NHX Na$^+$/H$^+$-antiporter, mediating electroneutral sodium/proton exchange, ΔpH = pH gradient, $\Delta\mu_{H+}$ = electrochemical proton gradient, $\Delta\mu_{Na+}$ = electrochemical sodium gradient.

Amtmann et al., 2001), nonselective cation channels (NSCCs; Tyerman et al., 1997; Amtmann and Sanders, 1999; Davenport and Tester, 2000; Horie and Schroeder, 2004), and high-affinity K$^+$ transporters (HKTs; Schachtman and Schroeder, 1994; Rus et al., 2001; Laurie et al., 2002; Yao et al., 2010; Platten et al., 2013). The occurrence of channels with high selectivity of K$^+$ over Na$^+$ allows efficient Na$^+$ exclusion at the root surface and is responsible for cultivar differences in Na$^+$ uptake by maize (Schubert and Läuchli, 1990). There is evidence for both the LCT (Schachtman et al., 1997) and the NSCC (Tyerman et al., 1997; Davenport and Tester, 2000) that increasing the calcium concentration reduces the Na$^+$ current and thus reduces Na$^+$ uptake by roots. In addition, Na$^+$ influx can be restricted by polyamines that are known to alleviate abiotic stress (Zhao et al., 2007). Since Na$^+$ transport in the soil solution to the root surface is mediated by mass flow, Na$^+$ exclusion at the root surface enhances rhizosphere Na$^+$ concentrations and may aggravate osmotic stress under saline conditions (Vetterlein et al., 2004).

Sodium uptake is not only determined by passive influx but may be reduced by active efflux that allows active exclusion (Kronzucker and Britto, 2011). Uptake is mediated by an electroneutral Na$^+$/H$^+$-antiporter (salt overly sensitive [SOS1]; Shi et al., 2000, 2003; Qiu et al., 2003; Zhu, 2003; Martínez et al., 2007; Oh et al., 2009) that is driven by the pH gradient (ΔpH; Figure 21.1). Although there is evidence for in vitro Na$^+$/H$^+$-antiport in plasma membrane vesicles not only in glycophytes but also in the halophyte blue-green saltbush (*Atriplex nummularia* Lindl.) (Braun et al., 1988), the contribution of the SOS1 Na$^+$/H$^+$-antiporter to Na$^+$ exclusion at the root surface under saline conditions is uncertain because of the high pH values prevailing in saline soils. In addition, dominating nitrate over ammonium nutrition under these environmental conditions makes an apoplastic acidification by root epidermal and cortex cells unlikely (Schubert and Yan, 1997). Therefore, a pH gradient to drive outward-directed active Na$^+$ transport is probably lacking under the relevant environmental conditions. The question remains why a Na$^+$/H$^+$-antiporter is maintained in the plasma membrane of root cells. At first glance, a contribution to cytoplasmic pH regulation, as in animal cells, seems unlikely because in plant cells cytosolic pH regulation is achieved by plasma membrane and tonoplast proton pumps and because this would counteract the Na^+ exclusion. However, it cannot be excluded that the presumably high proton-pumping activity of the plasma membrane H^+-ATPase at high rhizosphere pH leads to increased cytosolic pH that is downregulated by the Na^+/H^+-antiport.

An ortholog of the *SOS1* gene exists in corn (*Zea mays* L.) and transcription of the gene was demonstrated (Sekhon et al., 2011). In addition, strong metabolically dependent and active Na^+ efflux from roots was demonstrated for corn (Schubert and Läuchli, 1988). However, only weak Na^+/H^+-antiporter activity was at the root surface of intact plants and in isolated membrane vesicles from roots of this species (Schubert, 1990; Fortmeier, 2000).

21.2.2 ROOT-TO-SHOOT TRANSLOCATION

After uptake by epidermal, cortex, or endodermal root cells, root-to-shoot translocation of Na^+ requires transversal movement through the root tissue (Figure 21.2). This movement is achieved by diffusion and may be accelerated by cytoplasmic streaming. The Casparian strip in differentiated root zones forces the Na^+ on a symplastic route, thus allowing its potential exclusion (Schubert, 2011). Except for inefficiently excluding rice genotypes (Garcia et al., 1997), this barrier seems to play an important role in Na^+ exclusion at the root surface. There are two major mechanisms contributing to exclusion of Na^+ from the shoot. First, on its way through cortex cells, vacuoles absorb Na^+ from the cytosol via NHX Na^+/H^+-antiporters (NHX: sodium proton exchanger) (Xia et al., 2002; Figure 21.1), thus hindering its further movement into the stele. Because of the large volume of cortical cells, these may play a dominant role in this Na^+-exclusion strategy (Neubert et al., 2005, Pitann et al., 2013), although the contribution of stelar cells should not be neglected (Yeo et al., 1977). The significance of the NHX Na^+/H^+-antiporter for Na^+ exclusion from shoot was demonstrated in transgenic wheat by Xue et al. (2004).

Second, Na^+ delivered into xylem vessels may be reabsorbed by xylem parenchyma cells (Figure 21.2; Kramer et al., 1977; Läuchli, 1984). Since the pH gradient is directed into the cells (acid xylem sap vs. neutral cytosolic pH of xylem parenchyma cells), the SOS1 Na^+/H^+-antiporter cannot contribute to this process. Instead, the Na^+ transporter HKT (Byrt et al., 2007; Davenport et al., 2007; Hauser and Horie, 2010) may allow passive influx of Na^+ from xylem to xylem parenchyma cells, which may be enforced by active Na^+ sequestration into vacuoles driven by NHX Na^+/H^+-antiporters. It recently has been demonstrated that the *HKT* gene improved the salt resistance of wheat by improving the Na^+ exclusion from the shoot (Munns et al., 2012).

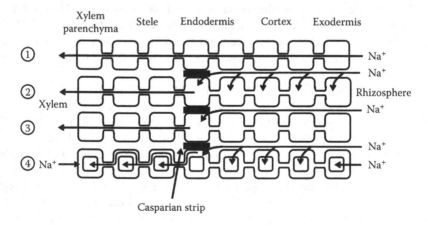

FIGURE 21.2 General model illustrating transversal Na^+ transport in differentiated root zones. (1) Cytoplasmic Na^+ movement from rhizosphere to xylem; (2) apoplastic Na^+ transport in the cortex and Na^+ uptake by epidermal, cortex, and endodermal cells; (3) apoplastic Na^+ transport in the cortex and enforcement of the symplastic pathway by the Casparian strip and at the endodermal cells; (4) Na^+ exclusion from xylem by sequestration into vacuoles and by reabsorption into xylem parenchyma cells.

Sodium exclusion from the shoot also is mediated by retranslocation via the phloem (Jacoby, 1979; Marschner and Ossenberg-Neuhaus, 1976). However, since Na^+ tends to be excluded from phloem, the quantitative contribution of this third exclusion strategy is uncertain.

21.2.3 INTRACELLULAR COMPARTMENTALIZATION

Tonoplast-located NHX Na^+/H^+-antiporters are responsible for Na^+ sequestration in cell vacuoles (Blumwald and Poole, 1985; Apse et al., 1999; Blumwald, 2000; Xia et al., 2002). The electroneutral sodium/proton exchange is driven by a pH gradient established by the tonoplast H^+-ATPase and the H^+-pyrophosphatase (Figure 21.1). In roots, vacuolar Na^+ compartmentalization is presumably less needed for stress avoidance than in shoots because root growth under saline conditions is less sensitive to Na^+ than shoot growth (Schubert and Läuchli, 1986). Therefore, the primary function of vacuolar sequestration in root cells may be Na^+ exclusion from the extremely sensitive shoot tissues (Saqib et al., 2005).

In shoots, vacuolar sequestration serves a dual role. First, toxic cytoplasmic Na^+ concentrations are avoided (Carden et al., 2003), and, second, vacuolar Na^+ serves as an important solute that contributes to osmotic adjustment under saline environmental conditions. In contrast to Na^+ exclusion that has been regarded as a typical glycophytic strategy of sodium resistance, Na^+ inclusion has been considered a typical halophytic trait that allows halophytes to cope with low soil water potentials and high soil Na^+ concentrations. However, research of the past few years has revealed that both strategies are used by glycophytes as well as halophytes, though to varying degree. The combination of exclusion and inclusion strategies has led to significant improvement of the salt resistance of newly developed maize hybrids (Schubert et al., 2009; Pitann et al., 2013).

21.3 PHYSIOLOGICAL RESPONSES OF PLANTS TO SODIUM

21.3.1 SODIUM TOXICITY

Due to its large hydration shell, it is generally accepted that Na^+ cannot take over the specific functions of K^+. Although high tissue concentrations of Na^+ have been reported to be toxic (Greenway and Munns, 1980; Moftah and Michel, 1987; Fortmeier and Schubert, 1995), there are many reports that found no correlation between Na^+ exclusion and salt resistance (Rush and Epstein, 1976; Lessani and Marschner, 1978; Läuchli, 1984; Cerda et al., 1982; Lauter and Munns, 1986). The reason for this paradox is the disregard of the osmotic salt-stress component (Phase 1 according to the biphasic model of Munns [1993] and Munns and Tester [2008]) on the one hand and vacuolar sequestration of Na^+ on the other hand. Relatively low Na^+ concentrations in the root medium (25 mM), which do not affect apparent corn growth, result in a drastic unspecific change of the proteome (Zörb et al., 2004). In contrast, in sugar beet (*Beta vulgaris* L.), most of the detected proteins showed only slight changes under salt stress, and the proteome showed high stability despite significant growth reduction (Wakeel et al., 2011a). Cytoplasmic Na^+ accumulation interferes with K^+, which is required by numerous enzymes (Evans and Sorger, 1966).

Although there is agreement that compartmentalization is the reason that halophytes can tolerate high tissue concentrations of Na^+, and although there is evidence that glycophytic and halophytic enzymes do not differ in K^+ requirement (Flowers, 1972; Greenway and Osmond, 1972; Pollard and Wyn Jones, 1979; Matoh et al., 1989), there is also evidence that specific K^+-requiring key enzymes may have adapted to high Na^+ concentration. Wakeel et al. (2011b), investigating the effect of increasing Na^+/K^+ ratios on the in vitro activity of the plasma membrane H^+-ATPase, found that the hydrolytic activity of H^+-ATPase in vesicles from sugar beet shoots was maintained even at 100 mM Na^+ and 0 mM K^+, whereas the H^+-ATPase activity in vesicles from corn shoots was decreased significantly when 25 mM Na^+ or 75 mM K^+ was applied. This result suggests that not only compartmentalization but also specific adaptation of key enzymes may contribute to tolerance of high Na^+ concentrations in sugar beet. When substituting K^+ by Na^+, it was shown that physiological Ca^{2+} deficiency rather than K^+ deficiency was the primary

FIGURE 21.3   **(See color insert.)** Induced Ca^{2+} deficiency by substitution of K^+ by Na^+.

problem in sugar beet nutrition (Figure 21.3; Wakeel et al., 2009a). This problem was ascribed to cation antagonism between Na^+ and Ca^{2+} during transport processes mediated by NSCCs.

The need for efficient Na^+ compartmentalization also is demonstrated by the sensitivity of cell organelles to Na^+. In corn root tips, NaCl treatment inhibited respiration, which was compensated by a higher number of mitochondria (Hecht-Buchholz et al., 1971). In contrast to salt-resistant sugar beet, chloroplasts of salt-sensitive bean (*Phaseolus vulgaris* L.) plants were deformed by salt treatment (Marschner and Mix, 1973). The proteome of corn chloroplasts showed extreme sensitivity to short-term salt stress, which was ascribed to Na^+ toxicity (Zörb et al., 2009). There are large species (and varietal) differences in Na^+ sensitivity that may be primarily (but probably not exclusively) attributed to efficient Na^+ compartmentalization.

Grasses are particularly sensitive to Na^+, and, under NaCl treatment, chloride toxicity can be neglected in this plant group (Fortmeier and Schubert, 1995). The symptom of Na^+ toxicity is very similar to that of K^+ deficiency and hard to distinguish. With the cumulative increase of Na^+ concentration, symptoms first appear on old leaves (Figure 21.4). The leaf tissue becomes necrotic, mostly starting at the leaf margins. The symptoms may be related to disturbance of the water economy and the triggered oxidative stress. For faba bean (*Vicia faba* L.), Slabu et al. (2009) have shown that the necrosis starts in the vicinity of stomata that apparently lack sufficient regulation due to induced K^+ deficiency. Accumulation of Na^+ in the guard cells is irreversible and results in permanent stomatal aperture and unproductive water loss. Also, Oertli's hypothesis that apoplastic salt accumulation in leaves due to exclusion may disturb water economy recently has been supported for faba bean that accumulated Na^+ in the apoplast (Shahzad et al., 2013). Sometimes, chlorotic symptoms also can be observed on old leaves before necrosis prevails and the leaves die.

21.3.2   SODIUM AS A PLANT NUTRIENT

Currently, Na is not considered a nutrient for most plant species. In halophytes, growth stimulation and avoidance of deficiency symptoms by Na^+ have been demonstrated (Redondo-Gómez et al., 2010), and Na^+ has been established as a micronutrient in saltbush (*Atriplex* sp. L.) (Brownell, 1965; Brownell and Crossland, 1972). The main function of Na^+ seems to be to maintain osmolality in the plant tissues of the halophyte seepweed (*Suaeda salsa* Pall.) under high NaCl conditions

FIGURE 21.4 **(See color insert.)** Corn (*Z. mays* L.) plant showing typical Na+ toxicity symptoms.

(Mori et al., 2010). Sodium can substitute for potassium in *Beta* sp. to a large extent (Subbarao et al., 1999). In addition, Na+ has been accepted as a micronutrient in some C_4 species that require Na+ for cotransport with pyruvate into mesophyll chloroplasts that serves to regenerate phosphoenolpyruvate (PEP). These species deviate from other C_4 species that use NADP+-dependent malic enzyme for pyruvate generation and that depend on the plant characteristic pyruvate/proton cotransport. In addition, there is evidence that in sorghum (*Sorghum* sp. Moench.) and CAM plants, PEP carboxylase kinase that regulates the activity of PEP carboxylase depends on Na+. The question of Na+ essentiality for plants has been discussed in detail by Marschner (1995) and Gorham (2007).

21.4 INTERACTIONS OF SODIUM WITH OTHER ELEMENTS

There are interactions of potassium, calcium, and silicon (Si) with Na, which play an important role for the salt resistance of plants. The strong interaction of Na+ with K+ has been described in the previous parts of this chapter and has been reviewed by Wakeel et al. (2011c) and Wakeel (2013).

It has been known for a long time that there are strong interactions of Na+ with Ca^{2+} (La Haye and Epstein, 1969; Cramer, 1994). Displacement of Ca^{2+} by Na+ from the plasma membrane of root cells changes the membrane potential (Läuchli and Schubert, 1989) and the ion selectivity of the membrane (Lynch et al., 1987). As a consequence, the Na+/K+ ratio in the plant tissue increases (Cramer et al., 1985, 1986). There is evidence that Ca^{2+} reduces Na+ influx via HKT (Essah et al., 2003), LCT (Schachtman et al., 1997), and particularly NSCC (Tyerman et al., 1997; Davenport and Tester, 2000; Anil et al., 2007). Sodium–calcium interactions have been reviewed by Cramer (2002).

The resistance to salt stress is also improved by silicon. Saqib et al. (2008) have shown that Na^+ uptake and Na^+ root-to-shoot translocation in wheat (*Triticum aestivum* L.) were decreased by Si. Although the precise mechanisms are not clear, it is evident that Si can improve the salt resistance via these avoidance strategies.

21.5 SODIUM IN SOILS

Due to its minor role for plants, even under humid conditions, available Na^+ concentrations in soils are usually sufficient to sustain crop development (for exceptions see succeeding text). In contrast, Na^+ accumulation under arid conditions contributes substantially to the worldwide problem of soil salinization. If evapotranspiration exceeds precipitation, there is a capillary upward movement of water in the soil profile. Upon evaporation at the soil surface, salt enrichment gradually results in salinization. It is estimated that worldwide, 10^9 ha are affected by soil salinity (Qadir et al., 2001). Soils with an electrical conductivity (EC) of ≥ 4 dS m^{-1} in the saturation extract are defined as saline (Qadir and Schubert, 2002). If the exchangeable Na^+ percentage (ESP) exceeds 15%, soils are referred to as sodic soils. A combination of excess salts (EC ≥ 4 dS m^{-1}) and a high degree of sorbed Na^+ (ESP $\geq 15\%$) characterizes saline–sodic soils. The occurrence of excess Na^+ results in osmotic problems, ion toxicity, and ion imbalances in crop plants. In addition, excess Na^+ induces soil structural problems such as slaking, swelling, dispersion of clay minerals, and surface crusting. As in the case of the physiological processes described earlier, the problems are related to the large hydration shell of Na^+. As a consequence, poor soil structure limits root penetration into the soil and affects water and air movements, with consequences for nutrient acquisition by the plants and for microbial activity that has a bearing on the carbon, nitrogen, and sulfur cycles.

Amelioration strategies for saline and saline–sodic soils usually comprise the leaching of sodium salts by excess irrigation and the exchange of Na^+ by Ca^{2+} at the exchange sites. The limited availability of high-quality water for leaching and the high costs of the chemical amelioration methods usually restrict the applicability in the affected areas. Therefore, phytoremediation as an alternative approach has been suggested for calcareous saline–sodic soils. The principle is not based on phytoextraction of Na^+ as is the case for phytoremediation of heavy metal–polluted soils. Instead, the release of respiratory CO_2 and protons by roots is used to dissolve indigenous soil $CaCO_3$ and exchange Na^+ by Ca^{2+} from the exchange sites. Excess Na^+ is then leached out of the root zone. It has been shown that phytoremediation of calcareous saline–sodic soils can be as effective as the chemical amendment strategies (Qadir et al., 2003). A comprehensive review of phytoremediation of sodic and saline–sodic soils has been given by Qadir et al. (2007).

21.6 SOIL TESTING

Usually, because Na is not generally regarded as a plant nutrient, it is not analyzed in routine soil testing. However, the determination of Na^+ availability in grassland soils or in K^+-fixing soils (see succeeding text) may be necessary to estimate fertilizer requirements. Soil testing to determine Na^+ availability can rely on extraction methods that are used to determine exchangeable K^+ (see Chapter 4). Potential Na^+ toxicity is estimated with the following parameters (Qadir and Schubert, 2002):

1. Soil salinity is measured in soil saturation extracts. The EC (dS m^{-1}) is a measure for the total ion concentration in the soil solution. Usually, Na^+ is the cation that contributes significantly to this value.
2. A more specific parameter to characterize sodicity is ESP (%). It describes the percentage of Na^+ relative to all exchangeable cations sorbed by soil particles. Soils with an ESP ≥ 15 are defined as sodic soils, and those that combine an EC ≥ 4 dS m^{-1} and an ESP ≥ 15 are regarded as saline–sodic soils.

3. An alternative parameter of sodicity is sodium adsorption ratio (SAR):

$$SAR = \frac{\left[Na^+\right]}{\left[Ca^{2+} + Mg^{2+}\right]^{1/2}}$$

(21.2)

A threshold value of 10 is defined for sodic soils.

21.7 FERTILIZER APPLICATION

Since Na is a micronutrient for only some C_4 species (see preceding text) and is beneficial for halophytes to improve osmotic adjustment under saline conditions (Marschner, 1995), it is not surprising that Na fertilizers are generally not applied to arable crops. However, K fertilizers with relatively high Na concentrations may be an economical alternative to high-purity K fertilizers if applied to natrophilic crops such as sugar beet under humid conditions. Critical K concentrations of sugar beet vary in a wide range as long as Na is available for substitution (Bergmann, 1992). This variability indicates that in sugar beet, a high proportion of K can be substituted by Na. Provided that typical fertilizer rates are not exceeded, the negative effects of Na^+ on soil structure described earlier can be neglected. However, two important limitations in the crop rotation must be considered, namely, Cl-sensitive and Na-sensitive plant species.

The application of K fertilizers with high Na concentration recently has attracted interest for K-fixing soils. These soils with a high content of 2:1 clay minerals may specifically bind up to 10×10^4 kg K ha^{-1} without a fertilizer response of the crops. Sodium fertilizer application on these soils avoids fixation because the Na^+ with its smaller ion diameter in the nonhydrated state is not fixed by these clay minerals (Figure 21.5). Therefore, Na fertilizer application at a rate needed for nonfixing soils improved white-sugar yield of beets under these conditions in a similar way as the very high K rate needed for K-fixing soils (Wakeel et al., 2009b). The use of a fertilizer such as Magnesia-Kainit® (K+S KALI GmbH, Kassel, Germany), a natural ore with potassium, magnesium, sodium, and sulfur and with a more than twofold higher Na concentration than K concentration, may represent an efficient strategy to supply sugar beet with sufficient Na (as a substitute for K) and, later in the rotation, to rely on the efficient K acquisition of grasses (i.e., corn, wheat, and barley, *Hordeum vulgare* L.) from interlayer K (Schubert and Mengel, 1989).

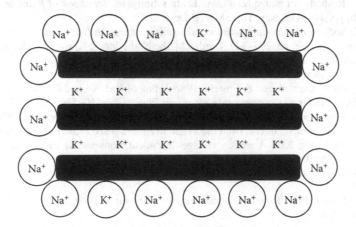

FIGURE 21.5 Different binding behaviors of Na^+ and K^+ in soils rich in 2:1 clay minerals. Whereas Na^+ is only unspecifically sorbed at the surface of the 2:1 clay minerals (hydration shell is maintained), K^+ is additionally fixed by specific binding in the clay interlayers (lost hydration shell). The fixation of K^+ decreases K^+ availability.

In contrast to arable crops, the application of Na-containing K fertilizers is a long-established measure for meadows and pastures (Marschner, 1995), particularly to meet the demands of animal nutrition. Because of permanent vegetation, possible negative effects of Na^+ on soil structure can be excluded safely in the relevant concentration range. However, the Na resistance of forage plants that contribute to the plant community determines the amount of K that can be substituted by Na.

REFERENCES

Amtmann, A., M. Fischer, E.L. Marsh, A. Stefanovich, D. Sanders, and D.P. Schachtman. 2001. The wheat cDNA *LCT1* generates hypersensitivity to sodium in a salt-sensitive yeast strain. *Plant Physiol.* 126:1061–1071.

Amtmann, A. and D. Sanders. 1999. Mechanisms of Na^+ uptake by plant cells. *Adv. Bot. Res.* 29:75–112.

Anil, V.S., H. Krishnamurthy, and M.K. Mathew. 2007. Limiting cytosolic Na^+ confers salt tolerance to rice cells in culture: A two photon microscopy study of SBFI-loaded cells. *Physiol. Plant.* 129:607–621.

Apse, M.P., G.S. Aharon, W.A. Snedden, and E. Blumwald. 1999. Salt tolerance conferred by overexpression of a vacuolar Na^+/H^+ antiport in *Arabidopsis*. *Science* 285:1256–1258.

Bergmann, W. 1992. *Nutritional Disorders of Plants: Development, Visual and Analytical Diagnosis*. Jena, Germany: Gustav Fischer.

Blumwald, E. 2000. Sodium transport and salt tolerance in plants. *Curr. Opin. Cell Biol.* 12:431–434.

Blumwald, E. and R.J. Poole. 1985. Na^+/H^+ antiport in isolated tonoplast vesicles from storage tissue of *Beta vulgaris*. *Plant Physiol.* 78:163–167.

Braun, Y., M. Hassidim, H.R. Lerner, and L. Reinhold. 1988. Evidence for a Na^+/H^+ antiporter in membrane vesicles isolated from roots of the halophyte *Atriplex nummularia*. *Plant Physiol.* 87:104–108.

Brownell, P.F. 1965. Sodium as an essential micronutrient element for a higher plant (*Atriplex vesicaria*). *Plant Physiol.* 40:460–468.

Brownell, P.F. and C.J. Crossland. 1972. The requirement for sodium as a micronutrient by species having the C_4 dicarboxylic photosynthetic pathway. *Plant Physiol.* 49:794–797.

Burchett, M.D., C.D. Field, and A. Pulkownik. 1984. Salinity, growth and root respiration in the grey mangrove, *Avicennia marina*. *Physiol. Plant.* 60:113–118.

Byrt, C.S., J.D. Platten, W. Spielmeyer et al. 2007. HKT1;5-like cation transporters linked to Na^+ exclusion loci in wheat, *Nax* and *Kna1*. *Plant Physiol.* 143:1918–1928.

Carden, D.E., D.J. Walker, T.J. Flowers, and A.J. Miller. 2003. Single-cell measurements of the contributions of cytosolic Na^+ and K^+ to salt tolerance. *Plant Physiol.* 131:676–683.

Cerda, A., M. Caro, and F.G. Fernández. 1982. Salt tolerance of two pea cultivars. *Agron. J.* 74:796–798.

Cheeseman, J.M. 1988. Mechanisms of salinity tolerance in plants. *Plant Physiol.* 87:547–550.

Cramer, G.R. 1994. Response of maize (*Zea mays* L.) to salinity. In *Handbook of Plant and Crop Stress*, ed. M. Pessarakli, pp. 449–459. New York: Marcel Dekker.

Cramer, G.R. 2002. Sodium-calcium interactions under salinity stress. In *Environment-Plants-Molecules*, eds. A. Läuchli and U. Lüttge, pp. 205–227. Dordrecht, the Netherlands: Kluwer Academic Publishers.

Cramer, G.R., A. Läuchli, and V.S. Polito. 1985. Displacement of Ca^{2+} by Na^+ from the plasmalemma of root cells: A primary response to salt stress? *Plant Physiol.* 79:207–211.

Cramer, G.R., J. Lynch, A. Läuchli, and E. Epstein. 1986. Influx of Na^+, K^+, and Ca^{2+} into roots of salt-stressed cotton seedlings: Effects of supplemental Ca^{2+}. *Plant Physiol.* 83:510–516.

Davenport, R.J., A. Muñoz-Mayor, D. Jha, D., P.A. Essah, A. Rus, and M. Tester. 2007. The Na^+ transporter AtHKT1;1 controls retrieval of Na^+ from the xylem in *Arabidopsis*. *Plant Cell Environ.* 30:497–507.

Davenport, R.J. and M. Tester. 2000. A weakly voltage-dependent, nonselective cation channel mediates toxic sodium influx in wheat. *Plant Physiol.* 122:823–834.

Essah, P.A., R. Davenport, and M. Tester. 2003. Sodium influx and accumulation in *Arabidopsis*. *Plant Physiol.* 133:307–318.

Evans, H.J. and G.J. Sorger. 1966. Role of mineral elements with emphasis on the univalent cations. *Annu. Rev. Plant Physiol.* 17:47–77.

Flowers, T.J. 1972. The effect of sodium chloride on enzyme activities from four halophyte species of Chenopodiaceae. *Phytochemistry* 11:1881–1886.

Flowers, T.J. and A.R. Yeo. 1989. Effects of salinity on plant growth and crop yield. In *Environmental Stress in Plants: Biochemical and Physiological Mechanisms*, ed. J.H. Cherry, pp. 101–119. Berlin, Germany: Springer.

Fortmeier, H. 2000. Na⁺/H⁺-antiport in Maiswurzeln? In vitro-Untersuchungen zum Mechanismus des aktiven Na⁺-Transports am Plasmalemma von Maiswurzelzellen (*Zea mays* L.). PhD thesis, Faculty of Agriculture, Nutrition and Environmental Management, Justus Liebig University Giessen, Giessen, Germany.

Fortmeier, R. and S. Schubert. 1995. Salt tolerance of maize (*Zea mays* L.): The role of sodium exclusion. *Plant Cell Environ.* 18:1041–1047.

Garcia, A., C.A. Rizzo, J. Ud-Din et al. 1997. Sodium and potassium transport to the xylem are inherited independently in rice, and the mechanism of sodium:potassium selectivity differs between rice and wheat. *Plant Cell Environ.* 20:1167–1174.

Gorham, J. 1994. K/Na discrimination and its significance for selection and breeding for salt tolerance. In *Genetic Techniques for Adapting Crop Plants to Lowered Fertilizer Inputs and Nutrition Stresses*, ed. D.T. Clarkson, pp. 133–147. *Proceedings of EC Workshop*, Brussels: Commission of the European Community.

Gorham, J. 2007. Sodium. In *Handbook of Plant Nutrition*, eds. A.V. Barker and D.J. Pilbeam, pp. 569–583. Boca Raton, FL: CRC Press.

Greenway, H. and R. Munns. 1980. Mechanisms of salt tolerance in nonhalophytes. *Annu. Rev. Plant Physiol.* 31:149–190.

Greenway, H. and C.B. Osmond. 1972. Salt responses of enzymes from species differing in salt tolerance. *Plant Physiol.* 49:256–259.

Hauser, F. and T. Horie. 2010. A conserved primary salt tolerance mechanism mediated by HKT transporters: A mechanism for sodium exclusion and maintenance of high K⁺/Na⁺ ratio in leaves during salinity stress. *Plant Cell Environ.* 33:552–565.

Hecht-Buchholz, C., R. Pflüger, and H. Marschner. 1971. Einfluss von Natriumchlorid auf die Mitochondrienzahl und Atmung von Maiswurzelspitzen. *Z. Pflanzenphysiol.* 65:410–417.

Horie, T. and J.I. Schroeder. 2004. Sodium transporters in plants. Diverse genes and physiological functions. *Plant Physiol.* 136:2457–2462.

Jacoby, B. 1979. Sodium recirculation and loss from *Phaseolus vulgaris* L. *Ann. Bot.* 43:741–744.

Koyro, H.-W. and R. Stelzer. 1988. Ion concentrations in the cytoplasm and vacuoles of rhizodermis cells from NaCl treated *Sorghum*, *Spartina* and *Pucinella* plants. *J. Plant Physiol.* 133:441–446.

Kramer, D., A. Läuchli, A.R. Yeo, and J. Gullasch. 1977. Transfer cells in roots of *Phaseolus coccineus*: Ultrastructure and possible function in exclusion of sodium from the shoot. *Ann. Bot.* 41:1031–1040.

Kronzucker, H.J. and D.T. Britto. 2011. Sodium transport in plants: A critical review. *New Phytol.* 189:54–81.

Kronzucker, H.J., D. Coskun, L.M. Schulze, J.R. Wong, and D.T. Britto. 2013. Sodium as nutrient and toxicant. *Plant Soil* 369:1–23.

La Haye, P.A. and E. Epstein. 1969. Salt toleration by plants: Enhancement with calcium. *Science* 166:395–396.

Läuchli, A. 1984. Salt exclusion: An adaptation of legumes for crops and pastures under saline conditions. In *Salinity Tolerance in Plants: Strategies for Crop Improvement*, eds. R.C. Staples and G.H. Toennissen, pp. 171–187. New York: John Wiley.

Läuchli, A. and S. Schubert. 1989. The role of calcium in the regulation of membrane and cellular growth processes under salt stress. In *Environmental Stress in Plants*, ed. J.H. Cherry, *Nato ASJ Series*. Berlin, Germany: Springer.

Laurie, S., K.A. Feeney, F.J.M. Maathuis, P.J. Heard, S.J. Brown, and R.A. Leigh. 2002. A role for HKT1 in sodium uptake by wheat roots. *Plant J.* 32:139–149.

Lauter, D.J. and D.N. Munns. 1986. Salt resistance of chickpea genotypes in solutions salinized with NaCl and Na₂SO₄. *Plant Soil* 95:271–279.

Leigh, R.A. and R.G. Wyn Jones. 1986. Cellular compartmentation in plant nutrition: The selective cytoplasm and the promiscuous vacuole. In *Advances in Plant Nutrition*, Vol. 2, eds. B. Tinker and A. Läuchli, pp. 249–279. New York: Praeger.

Lessani, H. and H. Marschner. 1978. Relation between salt tolerance and long-distance transport of sodium and chloride in various crop species. *Aust. J. Plant Physiol.* 5:27–37.

Lynch, J., G.R. Cramer, and A. Läuchli. 1987. Salinity reduces membrane-associated calcium in corn root protoplasts. *Plant Physiol.* 83:390–394.

Marschner, H. 1995. *Mineral Nutrition of Higher Plants*, 2nd edn. London, U.K.: Academic Press.

Marschner, H. and G. Mix. 1973. Einfluss von Natriumchlorid und Mycostatin auf den Mineralstoffgehalt im Blattgewebe und die Feinstruktur der Chloroplasten. *Z. Pflanzenernähr. Bodenk.* 136:203–219.

Marschner, H. and H. Ossenberg-Neuhaus. 1976. Langstreckentransport von Natrium in Bohnenpflanzen. *Z. Pflanzenernähr. Bodenk.* 139:129–142.

Martínez-Atienza, J., X. Jiang, B. Garciadeblas et al. 2007. Conservation of the salt overly sensitive pathway in rice. *Plant Physiol.* 143:1001–1012.

Matoh, T., T. Ishikawa, and E. Takahashi. 1989. Collapse of ATP-induced pH gradient by sodium ions in micro-somal membrane vesicles prepared from *Atriplex gmelini* leaves. Possibility of Na^+/H^+ antiport. *Plant Physiol.* 89:180–183.

Moftah, A.E. and B.E. Michel. 1987. The effect of sodium chloride on solute potential and proline accumulation in soybean leaves. *Plant Physiol.* 83:238–240.

Munns, R. 1993. Physiological processes limiting plant growth in saline soils: Some dogmas and hypotheses. *Plant Cell Environ.* 16:15–24.

Munns, R., R.A. James, B. Xu et al. 2012. Wheat grain yield on saline soils is improved by an ancestral Na^+ transporter gene. *Nat. Biotechnol.* 30:360–364.

Munns, R. and M. Tester. 2008. Mechanisms of salinity tolerance. *Annu. Rev. Plant Biol.* 59:651–681.

Mori, S., M. Akiya, K. Yamamura et al. 2010. Physiological role of sodium in the growth of the halophyte *Suaeda salsa* (L.) Pall. under high-sodium conditions. *Crop Sci.* 50:2492–2498.

Neubert, A.B., C. Zörb, and S. Schubert. 2005. Expression of vacuolar Na^+/H^+ antiporters (*ZMNHX*) and Na^+ exclusion in roots of maize (*Zea mays* L.) genotypes with improved salt resistance. In *Plant Nutrition for Food Security, Human Health and Environmental Protection*, eds. C.J. Li, F.S. Zhang, A. Dobermann et al., pp. 544–545. Beijing, China: Tsinghua University Press.

Oh, D.-H., E. Leidi, Q. Zhang et al. 2009. Loss of halophytism by interference with SOS1 expression. *Plant Physiol.* 151:210–222.

Pitann, B., A.-K. Mohamed, A.B. Neubert, and S. Schubert. 2013. Tonoplast Na^+/H^+ antiporters of newly developed maize (*Zea mays*) hybrids contribute to salt resistance during the second phase of salt stress. *J. Plant Nutr. Soil Sci.* 176:148–156.

Platten, J.D., J.A. Egdane, and A.M. Ismail. 2013. Salinity tolerance, Na^+ exclusion and allele mining of *HKT1;5* in *Oryza sativa* and *O. glaberrima*: Many sources, many genes, one mechanism? *BMC Plant Biol.* 13:32–48.

Pollard, A. and R.G. Wyn Jones. 1979. Enzyme activities in concentrated solutions of glycinebetaine and other solutes. *Planta* 144:291–298.

Qadir, M., J.D. Oster, S. Schubert, A.D. Noble, and K.L. Sahrawat. 2007. Phytoremediation of sodic and saline-sodic soils. *Adv. Agron.* 96:197–247.

Qadir, M. and S. Schubert. 2002. Degradation processes and nutrient constraints in sodic soils. *Land Degrad. Dev.* 13:275–294.

Qadir, M., S. Schubert, and A. Ghafoor. 2001. Amelioration strategies for sodic soils: A review. *Land Degrad. Dev.* 12:357–386.

Qadir, M., D. Steffens, F. Yan, and S. Schubert. 2003. Proton release by N_2-fixing plant roots: A possible contribution to phytoremediation of calcareous sodic soils. *J. Plant Nutr. Soil Sci.* 166:14–22.

Qiu, Q.-S., B.J. Barkla, R. Vera-Estrella, J.-K. Zhu, and K.S. Shumaker. 2003. Na^+/H^+ exchange activity in the plasma membrane of *Arabidopsis*. *Plant Physiol.* 132:1041–1052.

Redondo-Gómez, S., E. Mateos-Naranjo, M.E. Figueroa, and A.J. Davy. 2010. Salt stimulation of growth and photosynthesis in an extreme halophyte, *Arthrocnemum macrostachyum*. *Plant Biol.* 12:79–87.

Rus, A., S. Yokoi, A. Sharkhuu et al. 2001. AtHKT1 is a salt tolerance determinant that controls Na^+ entry into plant roots. *Proc. Natl. Acad. Sci. U. S. A.* 98:14150–14155.

Rush, D.W. and E. Epstein. 1976. Genotypic responses to salinity. Differences between salt-sensitive and salt-tolerant genotypes of the tomato. *Plant Physiol.* 57:162–166.

Saqib, M., C. Zörb, Z. Rengel et al. 2005. The expression of the endogenous vacuolar Na^+/H^+ antiporters in roots and shoots correlates positively with the salt resistance of wheat (*Triticum aestivum* L.). *Plant Sci.* 169:959–965.

Saqib, M., C. Zörb, and S. Schubert. 2008. Silicon-mediated improvement in the salt resistance of wheat (*Triticum aestivum*) results from increased sodium exclusion and resistance to oxidative stress. *Funct. Plant Biol.* 35:633–639.

Schachtman, D.P., R. Kumar, J.I. Schroeder, and E.L. Marsh. 1997. Molecular and functional characterization of a novel low-affinity cation transporter (LCT1) in higher plants. *Proc. Natl. Acad. Sci. USA* 94:11079–11084.

Schachtman, D.P. and J.I. Schroeder. 1994. Structure and transport mechanism of a high-affinity potassium uptake transporter from higher plants. *Nature* 370:655–658.

Schubert, S. 1990. *Natriumexklusion von Maiswurzeln und ihre Bedeutung für die Salzresistenz der Pflanze.* Habilitationsschrift at the Faculty of Nutrition and Household Economics, Justus Liebig University, Giessen, Germany.

Schubert, S. 2011. Salt resistance of crop plants—Physiological characterization of a multigenic trait. In *The Molecular Basis of Nutrient Use Efficiency in Crops*, eds. M. Hawkesford and P. Barraclough, pp. 443–455. Chichester, U.K.: Wiley Blackwell.

Schubert, S. and A. Läuchli. 1986. Na^+ exclusion, H^+ release, and growth of two different maize cultivars under NaCl salinity. *J. Plant Physiol.* 126:145–154.

Schubert, S. and A. Läuchli. 1988. Metabolic dependence of Na^+ efflux from roots of intact maize seedlings. *J. Plant Physiol.* 133:193–198.

Schubert, S. and A. Läuchli. 1990. Sodium exclusion mechanisms at the root surface of two maize cultivars. *Plant Soil* 123:205–209.

Schubert, S. and K. Mengel. 1989. Important factors in nutrient availability: Root morphology and physiology. *Z. Pflanzenernähr. Bodenk.* 152:169–174.

Schubert, S., A. Neubert, A. Schierholt, A. Sümer, and C. Zörb. 2009. Development of salt-resistant maize hybrids: The combination of physiological strategies using conventional breeding methods. *Plant Sci.* 177:196–202.

Schubert, S. and F. Yan. 1997. Nitrate and ammonium nutrition of plants: Effects on acid/base balance and adaptation of root cell plasmalemma H^+ ATPase. *Z. Pflanzenernähr. Bodenk.* 160:275–281.

Sekhon, R.S., H. Lin, K.L. Childs et al. 2011. Genome-wide atlas of transcription during maize development. *Plant J.* 66:553–563.

Shahzad, M., C. Zörb, C.-M. Geilfus, and K.H. Mühling. 2013. Apoplastic Na^+ in *Vicia faba* leaves rises after short-term salt stress and is remediated by silicon. *J. Agron. Crop Sci.* 199:161–170.

Shi, H., M. Ishitani, C. Kim, and J.-K. Zhu. 2000. The *Arabidopsis thaliana* salt tolerance gene *SOS1* encodes a putative Na^+/H^+ antiporter. *Proc. Natl. Acad. Sci.* 97:6896–6901.

Shi, H., B.-H. Lee, S.-J. Wu, and J.-K. Zhu. 2003. Overexpression of a plasma membrane Na^+/H^+ antiporter gene improves salt tolerance in *Arabidopsis thaliana*. *Nat. Biotechnol.* 21:81–85.

Slabu, C., C. Zörb, D. Steffens, and S. Schubert. 2009. Is salt stress of faba bean (*Vicia faba* L.) caused by Na^+ or Cl^- toxicity? *J. Plant Nutr. Soil Sci.* 172:644–650.

Subbarao, G.V., R.M. Wheeler, G.W. Stutte, and L.H. Levine. 1999. How far can sodium substitute for potassium in red beet? *J. Plant Nutr.* 22:1745–1761.

Tyerman, S.D., M. Skerrett, A. Garrill, G.P. Findlay, and R.A. Leigh. 1997. Pathways for the permeation of Na^+ and Cl^- into protoplasts derived from the cortex of wheat roots. *J. Exp. Bot.* 48:459–480.

Vetterlein, D., K. Kuhn, S. Schubert, and R. Jahn. 2004. Consequences of sodium exclusion for the osmotic potential in the rhizosphere. Comparison of two maize cultivars differing in Na^+ uptake. *J. Plant Nutr. Soil Sci.* 167:337–344.

Wakeel, A. 2013. Potassium-sodium interactions in soil and plant under saline-sodic conditions. *J. Plant Nutr. Soil Sci.* 176:344–354.

Wakeel, A., F. Abd-El-Motagally, D. Steffens, and S. Schubert. 2009a. Sodium-induced calcium deficiency in sugar beet during substitution of potassium by sodium. *J. Plant Nutr. Soil Sci.* 172:254–260.

Wakeel, A., A.R. Asif, B. Pitann, and S. Schubert. 2011a. Proteome analysis of sugar beet (*Beta vulgaris* L.) elucidates constitutive adaptation during the first phase of salt stress. *J. Plant Physiol.* 168:519–526.

Wakeel, A., M. Farooq, M. Qadir, and S. Schubert. 2011b. Potassium substitution by sodium in plants. *Crit. Rev. Plant Sci.* 30:401–413.

Wakeel, A., D. Steffens, and S. Schubert. 2009b. Potassium substitution by sodium in sugar beet (*Beta vulgaris*) nutrition on K-fixing soils. *J. Plant Nutr. Soil Sci.* 173:127–134.

Wakeel, A., A. Sümer, S. Hanstein, F. Yan, and S. Schubert. 2011c. In vitro effect of different Na^+/K^+ ratios on plasma membrane H^+-ATPase activity in maize and sugar beet shoot. *Plant Physiol. Biochem.* 49:341–345.

Xia, T., M.P. Apse, G.S. Aharon, and E. Blumwald. 2002. Identification and characterization of a NaCl-inducible vacuolar Na^+/H^+ antiporter in *Beta vulgaris*. *Physiol. Plant.* 116:206–212.

Xue, Z.-Y., D.-Y. Zhi, G.-P. Xue, H. Zhang, Y.-X. Zhao, and G.-M. Xia. 2004. Enhanced salt tolerance of transgenic wheat (*Triticum aestivum* L.) expressing a vacuolar Na^+/H^+ antiporter gene with improved grain yields in saline soils in the field and a reduced level of Na^+. *Plant Sci.* 167:849–859.

Yadav, R., T.J. Flowers, and A.R. Yeo. 1996. The involvement of the transpirational bypass flow in sodium uptake by high- and low-sodium-transporting lines of rice developed through intravarietal selection. *Plant Cell Environ.* 19:329–336.

Yao, X., T. Horie, S. Xue et al. 2010. Differential sodium and potassium transport selectivities of the rice OsHKT2;1 and OsHKT2;2 transporters in plant cells. *Plant Physiol.* 152:341–355.

Yeo, A.R., M.E. Yeo, and T.J. Flowers. 1977. The contribution of an apoplastic pathway to sodium uptake by rice roots in saline conditions. *J. Exp. Bot.* 38:1141–1153.

Zhao, F., C.-P. Song, J. He, and H. Zhu. 2007. Polyamines improve K^+/Na^+ homeostasis in barley seedlings by regulating root ion channel activities. *Plant Physiol.* 145:1061–1072.

Zhu, J.K. 2003. Regulation of ion homeostasis under salt stress. *Curr. Opin. Plant Biol.* 6:441–445.

Zörb, C., R. Herbst, C. Forreiter, and S. Schubert. 2009. Short-term effects of salt exposure on the maize chloroplast pattern: A proteomic study. *Proteomics* 9:4209–4220.

Zörb, C., S. Schmitt, A. Neeb, S. Karl, M. Linder, and S. Schubert. 2004. The biochemical reaction of maize (*Zea mays* L.) to salt stress is characterized by a mitigation of symptoms and not by a specific adaptation. *Plant Sci.* 167:91–100.

22 Vanadium

David J. Pilbeam

CONTENTS

22.1 HISTORICAL BACKGROUND

As shown in the first edition of *Handbook of Plant Nutrition*, vanadium at low concentrations appears to stimulate the growth of plants, whereas at higher concentrations it is toxic (Pilbeam and Drihem, 2007). Although some vanadium comes from rocks, particularly igneous rocks (Cappuyns and Slabbinck, 2012), its origin in the soil is largely from anthropogenic activity. This includes as a pollutant from the use of vanadium as a catalyst in various industries, from the smelting of iron and other metals, from the addition of ash after burning of coal, and also from deposition from the atmosphere where it arises from the burning of oil (Pilbeam and Drihem, 2007; Cappuyns and Slabbinck, 2012). Another source of the element is fertilizers. Rock phosphate contains variable amounts of vanadium, depending on its place of origin, with a sample of South African rock phosphate being shown to contain 19 mg V kg^{-1}, rock phosphate from Morocco 91 mg V kg^{-1}, and a Jordanian sample as much as 123 mg V kg^{-1} (Vachirapatama et al., 2002). Some vanadium is present in commercial fertilizers, with a sample of NPK fertilizer from Thailand being shown to contain 181 mg V kg^{-1}. A sample from Norway had 37 mg V kg^{-1}, whereas an Australian sample was much purer and contained only 7.5 mg V kg^{-1} (Vachirapatama et al., 2002). The concentration of vanadium in the environment has been increasing since the Industrial Revolution, including in plant material.

Little work has been carried out on the physiology of vanadium in plants. Furthermore, although concentrations of the element in soils and plants frequently are measured as one potential pollutant of many in studies on heavy metal pollution, there have not been large numbers of studies on vanadium pollution as a single factor. Studies on the uptake of vanadium by plants, physiological effects of vanadium in plants, and the availability of vanadium in soils carried out up to 2006 were reviewed in the first edition (Pilbeam and Drihem, 2007). A comprehensive table of vanadium concentrations in plants was also presented. Work on the importance of the element in plant nutrition that has been published since 2006 is limited and is reviewed below.

22.2 UPTAKE OF VANADIUM BY PLANTS

At pH values below 5, vanadium occurs in solution predominantly in +3 (V^{3+} ions) and +4 oxidation states (as VO^{2+} and $VOOH^+$ vanadyl ions), but at alkaline pH values there is a bigger diversity of forms (Morrell et al., 1986). At pH values between 5 and 8, the +5 state predominates. In soil, vanadium probably is fully oxidized to this +5 state, as vanadate (VO_3^-, $VO_2(OH)_2^-$, or VO_3OH^{2-}) is soluble across a range of pH values and probably represents the mobile form of the element in soils (Morrell et al., 1986; Peacock and Sherman, 2004). In experiments with excised barley (*Hordeum vulgare* L.) roots, the rate of uptake of V was highest at pH 4, but then decreased with increase in pH to a very low level by pH 10 (Welch, 1973). Rate of uptake was not affected by metabolic inhibitors, and the kinetics of uptake were linear with external V concentration, so the mechanism of uptake appears to be passive (Welch, 1973). In experiments on maize, vanadium was taken up if supplied as either vanadyl sulfate (VO^{2+}) or ammonium metavanadate, again showing a linear relationship between rate of uptake and rate of supply (Morrell et al., 1986). Experiments on barley seedlings indicated that vanadate is reduced to vanadyl inside the root (Morrell et al., 1986).

22.3 PHYSIOLOGICAL RESPONSES OF PLANTS TO SUPPLY OF VANADIUM

22.3.1 BENEFICIAL

Studies on the role of vanadium in plant nutrition are hampered by the fact that if it is an essential or beneficial element, it is required in such low concentrations that it is difficult to grow plants in control treatments without inadvertently supplying adequate vanadium as an impurity in the chemicals and water used. Indeed, in a study on the accumulation of vanadium by Chinese green mustard (*Brassica campestris* L. ssp. *chinensis* var. *parachinensis*, syn. *B. rapa* L. Parachinensis Group), roots of the control plants grown without supply of vanadium contained 58 mg V kg^{-1} dry matter at the end of the study (Vachirapatama et al., 2011).

However, despite this technical difficulty, it seems to be the case that vanadium is an essential element in green algae and a beneficial element in higher plants (Pilbeam and Drihem, 2007). Its beneficial role appears to be related to chlorophyll biosynthesis, possibly through an effect on iron nutrition (Pilbeam and Drihem, 2007). No further evidence of a beneficial role for vanadium has been reported recently.

22.3.2 TOXIC

The use of vanadate in countless biochemical experiments to inhibit ATPases gives a good indication that vanadium could be toxic to plants at quite low concentrations. Morrell et al. (1986) suggest that the reduction of the +5 oxidation state to the +4 state in the plant may prevent this inhibition from occurring, but even so it is known that supply of vanadium at comparatively low concentrations can reduce plant growth. Vachirapatama et al. (2011) observed a noticeable depression in growth of Chinese green mustard and tomato (*Solanum lycopersicum* L., syn. *Lycopersicon esculentum* Mill.) with supply of 40 mg ammonium vanadate per liter (0.34 mM). One obvious symptom was that the growth of lateral roots of the Chinese green mustard was inhibited. Vanadium as 1 mM sodium orthovanadate gave a significant reduction in rice (*Oryza sativa* L.) root growth (Lin et al., 2009, 2013).

This depression in rice root growth is accompanied by accumulation of reactive oxygen species (ROS) (Lin et al., 2009, 2013), as is commonly found with heavy metal toxicity and other abiotic and biotic stresses. Concentration of malondialdehyde increased within 1 h of supply of 5 mM vanadate, indicative of lipid peroxidation of membranes occurring due to the presence of ROS (Lin et al., 2009).

The onset of vanadium toxicity in rice roots is accompanied by upregulation of some genes associated with breakdown of ROS within 3 h of supply of 1 mM vanadate. Genes for biosynthesis and signaling of auxins, ABA, and jasmonic acid, and also for some protein kinases and transporters such as ATP-binding cassette (ABC) transporters, were upregulated within 1 h, and by a concentration of vanadate as low as 0.25 mM (Lin et al., 2013). Treatment with Cu or Cd gave a different pattern of gene expression, so these were not merely generic responses to heavy metal toxicity.

22.4 GENETICS OF ACQUISITION OF VANADIUM BY PLANTS

A study on the accumulation of vanadium in pea (*Pisum sativum* L.) cultivars showed that the bioaccumulation factor (concentration in plant biomass/concentration in substrate) was 11.275, 11.770, and 13.153 in the roots and 0.809, 0.467, and 0.749 in shoots of the cultivars Opal, Laser, and Ramir, respectively (Nowakowski, 1993). This result indicates genetic differences in the accumulation of vanadium by plant species, but this topic has been studied very little.

22.5 CONCENTRATIONS OF VANADIUM IN PLANTS

As can be seen in the earlier text, vanadium accumulates in the roots of plants much more than in the shoots. Furthermore, despite its uptake being by passive mechanisms, it appears to be able to accumulate in the roots to concentrations considerably greater than those present in the external medium. Recent research has shown that in *Brassica juncea* L. and *B. campestris*, vanadium accumulates predominantly in the roots, and in both roots and leaves a large proportion of the vanadium taken up is associated with the cell walls (Hou et al., 2013). In a study of six dominant plant species growing in soil polluted with vanadium, there were much higher concentrations of vanadium in the roots than in the leaves or stems, and although there was a positive relationship between the concentration of available vanadium in the soil and root vanadium concentration there was no relationship between soil vanadium concentration and leaf vanadium concentration (Qian et al., 2014). This indicates exclusion of vanadium from the aerial parts in these species. For all but three of the 22 measurements on the six species, concentration of vanadium in the roots relative to concentration of available vanadium in the soil was greater than 1.0 and was particularly high at low soil vanadium concentrations (Qian et al., 2014).

This accumulation could be facilitated by the conversion of vanadate to vanadyl in the roots (Morrell et al., 1986), thereby allowing diffusion of vanadate into the roots to continue as long as this ion continues to be reduced from the +5 to the +4 oxidation state. Both +5 and +4 forms can form stable chelates with ketones, aldehydes, catechols, amino compounds, and phenols (Morrell et al., 1986), and formation of complexes would enable the diffusion of free vanadate or vanadyl into the roots to continue.

Concentrations of vanadium presented in the first edition ranged from 0.25 to 5680 mg kg^{-1} dry matter in roots (in 9 plant species), from 0.115 to 530 mg kg^{-1} in vegetative aerial parts (in 29 plant species), and from 0.018 to 0.184 mg kg^{-1} in fruits or seeds (in 11 plant species) (Pilbeam and Drihem, 2007). An earlier review, in which data for a number of samples from different plant groups were considered, showed mean concentrations of 1.4 mg V kg^{-1} dry matter in grasses, 0.84 in legumes, 1.2 in nonleguminous forbs, 2.7 in deciduous shrubs, 1.65 in deciduous trees, and 0.69 mg kg^{-1} in conifers. For the smaller number of lichen and moss samples reviewed, the mean values were 8.6 and 108 mg V kg^{-1} dry matter, respectively (Cannon, 1963). Some plants growing in areas of the United States with high vanadium content of the soil accumulate vanadium at concentrations 10–100 times higher than those shown earlier (Cannon, 1963).

Vanadium concentrations in plant parts determined more recently are shown in Table 22.1. From those plant samples included, it seems as if crop products typically contain below 1 mg V kg^{-1} dry mass, with aerial parts of herbaceous plants in general containing 0.1–4 mg V kg^{-1}. This is

TABLE 22.1
Vanadium Concentrations in Plants

Plant	Plant Part and Treatment	V Concentration (mg kg⁻¹ Dry Mass)	Reference
Crop species			
Alfalfa, lucerne (*Medicago sativa* L.)	Near power station/control soil	9.57/3.43	Khan et al. (2011)
Beet		0.50	Madrakian et al. (2011)
Beet	Leaves		Teng et al. (2011)
	Agricultural site	7.6–42.8 (n = 3)	
	Mining site	6.5–27.2 (n = 4)	
	Both in area with high-V soil		
Berseem clover (*Trifolium alexandrinum* L.)	Near power station/control soil	13.65/4.86	Khan et al. (2011)
Bitter gourd (*Momordica charantia* L.)	Near power station/control soil	9.78/3.12	Khan et al. (2011)
Cabbage		0.43	Madrakian et al. (2011)
Carrot (*Daucus carota* L.)	Near power station/control soil	14.9/4.98	Khan et al. (2011)
Carrot		0.74	Madrakian et al. (2011)
Carrot (*Daucus carota*)	Storage root	0.53×10^{-1}–0.93×10^{-1}	Smoleń et al. (2012)
Celery		0.47	Madrakian et al. (2011)
Chinese cabbage	Leaves	20.8, 40.6	Teng et al. (2011)
	Mining site in area with high-V soil		
Chinese green mustard (*Brassica campestris* L. ssp. *chinensis* var. *parachinensis*)	Leaf (no supplemental V)	4.73	Vachirapatama et al. (2011)
	(1 mg L⁻¹ NH₄VO₃)	16.7	
	(40 mg L⁻¹ NH₄VO₃)	328	
	(80 mg L⁻¹ NH₄VO₃)	477	
	Stem (no supplemental V)	2.06	
	(1 mg L⁻¹ NH₄VO₃)	25.7	
	(40 mg L⁻¹ NH₄VO₃)	340	
	(80 mg L⁻¹ NH₄VO₃)	823	
	Root (no supplemental V)	58.4	
	(1 mg L⁻¹ NH₄VO₃)	1.95×10^3	
	(40 mg L⁻¹ NH₄VO₃)	9.66×10^3	
	(80 mg L⁻¹ NH₄VO₃)	1.35×10^4	
Cluster bean (*Cyamopsis tetragonoloba* (L.) Taub.)	Near power station/control soil	8.74/2.95	Khan et al. (2011)
Coriander (*Coriandrum sativum* L.)	Near power station/control soil	10.12/3.45	Khan et al. (2011)
Cucumber		0.45	Madrakian et al. (2011)
Lettuce		0.52	Madrakian et al. (2011)
Mango	Leaves		Teng et al. (2011)
	Agricultural site	10.3–22.7 (n = 3)	
	Urban park	3.0–11.5 (n = 9)	
	Mining site	15.9	
	Smelting site	15.1	
	All in area with high-V soil		
Okra (*Abelmoschus esculentus* (L.) Moench)	Near power station/control soil	12.53/4.15	Khan et al. (2011)
Onion (*Allium cepa* L.)	Near power station/control soil	13.98/4.85	Khan et al. (2011)
Pea (*Pisum sativum* L.)	Near power station/control soil	9.82/4.13	Khan et al. (2011)

(Continued)

TABLE 22.1 (*Continued*)
Vanadium Concentrations in Plants

Plant	Plant Part and Treatment	V Concentration (mg kg^{-1} Dry Mass)	Reference
Peppermint (*Mentha × piperita* L.)	Near power station/control soil	10.85/4.47	Khan et al. (2011)
Potato (*Solanum tuberosum* L.)	Near power station/control soil	13.54/4.55	Khan et al. (2011)
Potato		0.40	Madrakian et al. (2011)
Radish	Root	2.6	Teng et al. (2011)
	Urban park site in area with high-V soil		
Rape	Leaves	26.6	Teng et al. (2011)
	Root	12.6	
	Seed	18.8	
	3 separate samples from urban park in area with high-V soil		
Soybean	Leaves	27.3	Teng et al. (2011)
	Urban park site in area with high-V soil		
Spinach (*Spinacia oleracea* L.)	Near power station/control soil	9.15/3.83	Khan et al. (2011)
Spinach		0.49	Madrakian et al. (2011)
Sweet corn	Leaves	33.2	Teng et al. (2011)
	Urban park site in area with high-V soil		
Tomato		0.75	Madrakian et al. (2011)
Tomato (*Solanum lycopersicum* L.)	Fruit (no supplemental V)	0.21	Vachirapatama et al. (2011)
	(1 mg L^{-1} NH$_4$VO$_3$)	0.49	
	(40 mg L^{-1} NH$_4$VO$_3$)	4.04	
	(80 mg L^{-1} NH$_4$VO$_3$)	8.15	
	Root (no supplemental V)	4.92	
	(1 mg L^{-1} NH$_4$VO$_3$)	273	
	(40 mg L^{-1} NH$_4$VO$_3$)	4.01×10^3	
	(80 mg L^{-1} NH$_4$VO$_3$)	5.24×10^3	
Other species			
Aescinate grass	Leaves	7.1–14.0 (n = 3)	Teng et al. (2011)
	Smelting site in area with high-V soil		
Artemisia vulgaris L.	Leaves	6.98–11.6	Qian et al. (2014)
	Roots	25.7–113	
	Range of mean values from four contaminated sites		
Banyan	Leaves	26.3	Teng et al. (2011)
	Smelting site in area with high-V soil		
Betula populifolia Marshall	Leaves	5.96–12.1	Qian et al. (2014)
	Roots	33.1–280	
	Range of mean values from eight contaminated sites		
Buxus microphylla	Leaves	21.0	Teng et al. (2011)
	Mining site in area with high-V soil		

(Continued)

TABLE 22.1 (*Continued*)
Vanadium Concentrations in Plants

Plant	Plant Part and Treatment	V Concentration (mg kg⁻¹ Dry Mass)	Reference
Blue cohosh (*Caulophyllum thalictroides* (L.) Michx)	Roots	2.17	Avula et al. (2010)
Centella (*Centella asiatica* (L.) Urb)	Aerial parts	0.06	Avula et al. (2010)
Erect centella (*Centella erecta* (L. f.) Fernald)	Aerial parts	0.26	Avula et al. (2010)
Veldt grape (*Cissus quadrangularis* L.)	Aerial parts	0.15	Avula et al. (2010)
Turmeric (*Curcuma longa* L.)	Rhizomes	0.99	Avula et al. (2010)
Cycas	Leaves Mining site in area with high-V soil	14.2	Teng et al. (2011)
Acai palm (*Euterpe oleracea* Mart.)	Aerial parts	0.006	Avula et al. (2010)
Hoodia (*Hoodia gordonii* (Masson) Sweet ex Decne.)	Aerial parts	0.78	Avula et al. (2010)
Phragmites australis (Cav.) Trin. ex Steud.	Leaves Roots Mean value from one contaminated site	2.06 218	Qian et al. (2014)
Plantain	Leaves Urban park in area with high-V soil	15.2	Teng et al. (2011)
Polygonum cuspidatum Siebold and Zucc.	Leaves Roots Mean value from one contaminated site	9.40 225	Qian et al. (2014)
Eastern cottonwood (*Populus deltoides* W. Bartram ex Marshall)	Leaves Roots Range of mean values from three contaminated sites	4.86–8.24 67.3–119	Qian et al. (2014)
Populus tremula L. × *tremuloides* Michx.	Leaves Branches Highest value in polluted site/ control site	1.0/0.7 0.6/0.3	Hassinen et al. (2009)
Kudzu (*Pueraria lobata* var. *lobata* (Willd.) Ohwi)	Leaves Roots	nd 1.32	Avula et al. (2010)
Rhus copallinum L.	Leaves Roots Range of mean values from five contaminated sites	4.28–8.71 32.5–118	Qian et al. (2014)
Cancer bush (*Sutherlandia frutescens* (L.) R. Br.)	Leaves and stems	0.94	Avula et al. (2010)
Damiana (*Turnera diffusa* Willd. ex Schult.)	Leaves	2.33	Avula et al. (2010)

Notes: nd, not detected. Crop species in Madrakian et al. (2011) are commercial foodstuffs from Iran; binomial names not given in original publication, so are not shown; plants sampled by Teng et al. (2011) mostly did not have binomial names given; plant part sampled was not listed in Khan et al. (2011) or Madrakian et al. (2011), but as their work comprised studies of food products, it is probably the part normally eaten by livestock or humans.

not universal, and in a comparison of crops grown in a soil from an agricultural area in Pakistan with crops grown in a soil from near a power plant, the vegetables investigated contained up to 5 mg V kg^{-1} in the agricultural area, and more than this near to the power station (Khan et al., 2011). It was apparent from the data in the first edition (Pilbeam and Drihem, 2007) and more recent studies (e.g., Qian et al., 2014) that many plants accumulate higher concentrations of vanadium when its supply is increased. This is true for crop species, as well as species adapted to contaminated soils, and supply of high concentrations of vanadium to Chinese green mustard gave concentrations of up to 477 mg V kg^{-1} dry matter (Vachirapatama et al., 2011). However, in the high-vanadium soils of Panzhihua Province, China (see Section 22.8), crops growing in the locality only accumulated vanadium at levels up to one-tenth of this concentration (Teng et al., 2011).

According to the U.K. government's Expert Group on Vitamins and Minerals (2003), vanadium has not been proven to be an essential element for mammals. The group warned that supplements giving doses of 7.5–10 mg V per day may not be safe to humans, and the University of Maryland Medical Center (2013) gives a safe upper level of 1.8 mg for vanadium. An adult would only have to eat approximately 36 g fresh weight of leaves containing 500 mg V kg^{-1} dry mass to exceed this level. However, Chinese green mustard plants were noticeably smaller than normal with leaf concentrations of 328 mg V kg^{-1} dry matter, and even with 50.9 mg V kg^{-1} leaf dry matter there was some inhibition of growth (Vachirapatama et al., 2011), so it seems unlikely that crops would be harvested from land that was sufficiently polluted for very high concentrations of vanadium to be present in the edible parts. This is reinforced by the fact that polluted land often contains other heavy metals, which could cause additional phytotoxicity and prevent a crop from being harvested.

22.6 INTERACTION OF VANADIUM WITH UPTAKE OF OTHER ELEMENTS

As vanadate is a well-known metabolic inhibitor of ATPases in cells, it is only to be expected that uptake of vanadium by plants could interfere with active transporters, and hence the uptake of other elements. Concentrations of various ions have been shown to decrease or increase in plants with supply of vanadium, as reported in the first edition (Pilbeam and Drihem, 2007). One such effect is that there is an interaction between uptake of vanadium and that of other heavy metals by plants, with uptake of V^{5+} into mustard (*Sinapis alba* L.) seedlings being inhibited by Ni^{2+}, Mn^{2+}, and Cu^{2+}, while the uptake of Mn^{2+} and Cu^{2+} was stimulated by the presence of V^{5+} (Fargašová and Beinrohr, 1998). The uptake of Mo^{6+} was inhibited by the presence of V^{5+}. Manganese concentrations in primary leaves of a manganese-sensitive cultivar of bush bean (*Phaseolus vulgaris* L.) increased with increasing supply of vanadium (Kohno, 1986). The increased manganese concentration was matched by a decreased Fe concentration, and it seems as if the vanadium caused phytotoxicity by increasing manganese uptake and lowering Fe uptake.

Some plant species that grow on seleniferous soils, and accumulate selenium, also accumulate vanadium (Cannon, 1963).

22.7 DIAGNOSIS OF VANADIUM STATUS IN PLANTS

Analysis of vanadium concentrations in plant extracts is carried out by techniques commonly used for other heavy metals. These include inductively coupled plasma atomic emission spectrometry (ICP-AES) (Teng et al., 2011), inductively coupled plasma mass spectroscopy (ICP-MS) (Avula et al., 2010; Qian et al., 2014), ICP-MS after ion interaction reversed phase high-performance liquid chromatography (HPLC) of vanadium/4-(2-pyridylazo)resorcinol-hydrogen peroxide ternary complexes to clean the samples (Vachirapatama et al., 2005, 2011), and visible spectrophotometry of vanadium/bromopyrogallol red complexes cleaned by micelle-mediated extraction (Madrakian et al., 2011).

22.8   FORMS AND CONCENTRATIONS OF VANADIUM IN SOILS, AVAILABILITY TO PLANTS

The form of vanadium present in soil and soil solution depends on redox conditions and pH, and as discussed, the element is present in +3, +4, and +5 oxidation states in aqueous solution, with the +4 and +5 states predominating. As the +4 form (vanadyl) is a cation and the +5 form (vanadate) is an anion (except in the VO_2^+ state that occurs at low pH and under oxidizing conditions), this can influence the binding of vanadium to soil constituents and hence its availability to plants. Vanadate adsorbs to iron hydroxides and oxides in soils such as goethite (α-FeOOH) (Peacock and Sherman, 2004), and also to manganese and aluminum hydroxides and oxides (Gäbler et al., 2009). Vanadyl seems to coat the organic matter in a soil, which is itself adsorbed on ferric oxide particles (Flogeac et al., 2005). As VO^{2+} vanadyl ions and the VO_2^+ cationic vanadate form predominate at low pH, and anionic vanadate ions at high pH, there can be different amounts of adsorption not only according to composition of inorganic and organic components of the soil to which the ions bind but also according to the pH and redox conditions that influence the proportions of these two vanadium forms. In a study of three soils from Finland, the maximum adsorption of vanadium was seen at approximately pH 4, although there was still significant adsorption at pH 6 (Mikkonen and Tummavuori, 1994).

The amount of vanadium adsorption varies considerably between different soils at the same pH value, and in a study of five European soils a 10-fold difference among soils occurred in the amount of vanadium that was adsorbed at a concentration of 2.5 mg V L^{-1} in the soil solution (Larsson et al., 2013). Gäbler et al. (2009) were able to break down 30 different German soils into four groups, with sandy soils showing the lowest adsorption of vanadium, subsoils with a pH of less than 5.5 having the highest adsorption, and topsoils and subsoils with a pH greater than 5.5 being intermediate. The sandy soils had low concentrations of oxalate-extractable Fe, Mn, and Al, indicating that adsorption was low as there were low concentrations of minerals for vanadate to sorb to. The higher adsorption of the low pH subsoils than the high pH subsoils was also indicative of adsorption by vanadate, which tends to sorb more strongly at pH values below 7 due to increased adsorption due to the $H_2VO_4^-$ ion, then undissociated H_3VO_4 and then VO_2^+ as pH is lowered toward a value of 2 (Mikkonen and Tummavuori, 1994).

Because of the deposition of vanadium as a pollutant, it would be expected to be in higher concentrations in soil in industrial areas rather than in more remote rural areas. In a soil from a site in Italy previously used for producing polyvinyl chloride, a process where vanadium is used as a catalyst, a total V concentration of 1688 mg kg^{-1} soil was found (Terzano et al., 2007). In a soil from an industrial area of Poland, total V concentration was 25 mg kg^{-1} soil, compared with only 5.1 mg kg^{-1} in a soil from a more agricultural area (Połedniok and Buhl, 2003). In a study of crops grown in soils from an agricultural area in Pakistan with crops grown near to a power station, the soil in the vicinity of the power station had a concentration of 110 mg V kg^{-1}, whereas the agricultural soil contained 18.7 mg V kg^{-1} (Khan et al., 2011). The soil from the agricultural area, while having a much lower concentration of vanadium than the soil near to the power station, contained more than the agricultural soil in the Polish study, and almost as high a concentration as in the industrial area of Poland.

A more extreme example of high vanadium concentrations in agricultural soil was seen in work carried out in Panzhihua city in Sichuan Province, SW China, an area with mining of extensive vanadium titanium magnetite deposits (Teng et al., 2011). Here, the agricultural soil contained a mean concentration of 113.0 mg V kg^{-1}. This is compared with mean values of 119.8, 312.7, and 532.1 mg V kg^{-1} in an urban park, an area associated with mining, and an area associated with smelting, respectively, in the same locality (Teng et al., 2011).

In the study on soil from the industrial area in Poland, more than 40% of the soil vanadium was associated with organic matter, but in the agricultural area, this fraction only accounted for 15% of the total vanadium (Połedniok and Buhl, 2003). This could indicate that vanadium arising from atmospheric pollution may be found in the organic horizon, the uppermost layer of soils.

In unpolluted soil in the Puszcza Borecka forest in Poland left after the retreat of the glaciers from the last European Ice Age, mean total vanadium concentrations were 8.23 mg kg^{-1} dry mass of soil in the organic horizons and 34.04 mg kg^{-1} in the Bk enrichment horizon (Agnieszka and Barbara, 2012), so where vanadium does not have an exogenous origin it can be seen to be more of a component of the mineral fraction of the soil.

Other soils at risk of pollution are those in floodplains. In a comparison of V concentrations in soils from 3 Belgian floodplains and other floodplains across Europe, total V was found to be 459, 71, and 87 mg V kg^{-1} in the 3 Belgian soils and a mean of 59 mg kg^{-1} in 749 soils from 26 European countries (Cappuyns and Slabbinck, 2012). Vanadium extracted by 10 mM CaCl$_2$ (*pore water vanadium*) was present at 2 and 4 µg L^{-1} in the Belgian soils with the lowest and the highest total V concentrations, and vanadium extracted by 50 mM ammonium EDTA (*potentially available vanadium*) was present at 2.4 and 11.2 µg L^{-1} in these two soils. This implies that much of the vanadium that may arise from pollution, as with vanadium present in soil from the underlying rock, may be unavailable to plants. In none of the horizons of the soils from the Puszcza Borecka forest did available vanadium (i.e., V in the water-soluble and exchangeable fractions plus V bound to carbonates or weakly specifically adsorbed) exceed 4% of the total vanadium (Agnieszka and Barbara, 2012).

In the two soils studied by Poledniok and Buhl (2003), there was no detectable vanadium in the fraction adsorbed to carbonates, and the concentrations of vanadium that were in the water-soluble and exchangeable fractions were 3.39 and 2.07 mg kg^{-1} in the industrial and agricultural regions, respectively (Poledniok and Buhl, 2003). Most of the vanadium present in the Puszcza Borecka soils and the soil from the industrial region was strongly associated with iron and manganese oxides or was in the residual fraction (Poledniok and Buhl, 2003; Agnieszka and Barbara, 2012). In their contaminated soil from Italy, Terzano et al. (2007) found 40% of the vanadium was in a reducible fraction, which was associated with iron oxides, and 43% was in the residual fraction. They showed that vanadium in the +5 oxidation state was associated with the iron (3) oxide hematite and the iron (2)/iron (3) oxide magnetite, and it was suggested that the 40% in the reducible fraction was adsorbed to the iron minerals, and the 43% residual fraction was deep within the structure of these minerals and volborthite (Cu$_3$(OH)$_2$V$_2$O$_7$2H$_2$O) (Terzano et al., 2007). As seen in other studies, very little of the vanadium was in a mobile fraction. In a study of a soil in Italy polluted in the past by a stream containing wastewater from local industries, a mean total vanadium concentration of 26.8 mg kg^{-1} dry mass of soil was seen, with most of the vanadium being in the residual fraction and 10%–20% being split between the fraction adsorbed to iron and manganese oxides and the fraction adsorbed to organic matter (Malandrino et al., 2011). In the high-vanadium soil of Panzhihua city, 75%–93% of the total vanadium in all of the soil samples was in a residual fraction, and hence unavailable to plants, and the concentrations of mobile vanadium ranged from 9.8–26.4 mg V kg^{-1} soil in the agricultural area, 9.9–25.2 mg kg^{-1} in the urban park, 18.8–83.6 mg kg^{-1} in the mining area, and 41.7–132.1 mg kg^{-1} in the smelting area (Teng et al., 2011). Much of the vanadium in the soils was still presumably in vanadium titanium magnetite particles in the soil, although the concentrations of mobile vanadium were still high enough for the plants growing in the locality to contain high concentrations of the element compared with plants in many other studies (Table 22.1).

In a bioassay of root elongation of barley and shoot growth of barley and tomato on the five European soils with a 10-fold variation in vanadium adsorption, there was a strong correlation between the strength of adsorption in each soil and elongation or growth, showing that the amount of vanadium in a soil that would give a deleterious effect on plant growth would have to be greater in a soil with a strong capacity to adsorb vanadate or vanadyl ions than in a soil with a lower adsorption capacity (Larsson et al., 2013). Sodium metavanadate freshly added to soils has been found to give a greater danger of phytotoxicity than when it has been present for some time, as the vanadium became more strongly adsorbed with time (Baken et al., 2012). Vanadium concentrations in CaCl$_2$ extracts of soil had diminished to less than half their initial values within 25 days.

22.9 SOIL TESTING FOR VANADIUM

For measurement of total vanadium content of soils, samples can be extracted with aqua regia (a mixture of concentrated nitric and hydrochloric acids) (Gäbler et al., 2009; Baken et al., 2012; Cappuyns and Slabbinck, 2012). Some workers digest soil samples in an HNO_3–$HClO_4$–HF mixture at high temperature to release the most recalcitrant of the soil vanadium (Qian et al., 2014).

However, given that much of the vanadium in soils is adsorbed, and therefore is not in soil solution and is not available for plant uptake, Larsson et al. (2013) suggest that measuring concentrations of vanadium in soil solution gives a better indication of potentially phytotoxic levels of the element in a soil. This soil solution vanadium can be extracted in $CaCl_2$ and exchangeable vanadium with ammonium EDTA (Cappuyns and Slabbinck, 2012), referred to by the authors as being *pore water* vanadium and *potentially available* vanadium and developed as a standardized European extraction method.

A common extraction technique to subdivide soil vanadium into different pools is to use a sequence of extractants, based on the method developed for heavy metals in soils by Tessier et al. (1979). This gives rise to five fractions, namely, a fraction of water-soluble and exchangeable vanadium, a fraction bound to carbonates, a fraction bound to iron and manganese oxides, a fraction bound to organic matter, and a residual fraction. Alternatively, some workers use a three-step fractionation procedure developed by the European Community Bureau of Reference for extraction of trace metals where vanadium is separated into an exchangeable and weak acid-soluble fraction, a reducible fraction, an oxidizable fraction, and a residual fraction (Teng et al., 2011). These fractions correspond to water-extractable vanadium plus weakly adsorbed vanadium, vanadium sorbed to iron and manganese oxides, vanadium sorbed to or within organic matter, and vanadium in the crystalline structure of the soil (Bakircioglu et al., 2011).

Analysis of vanadium concentrations in soil extracts has been carried out by techniques such as ICP-MS (Gäbler et al., 2009; Cappuyns and Slabbinck, 2012; Qian et al., 2014), inductively coupled plasma optical emission spectrometry (Baken et al., 2012; Larsson et al., 2013), and ICP-AES (Połedniok and Buhl, 2003; Terzano et al., 2007; Malandrino et al., 2011; Teng et al., 2011). There are also spectrophotometric methods available (Połedniok and Buhl, 2003).

An alternative to making extracts is to measure vanadium in a soil paste by energy dispersive polarized x-ray fluorescence spectrophotometry, a technique used in the Forum of European Geological Surveys Directors' project to measure mineral components of European soils (Sandström et al., 2005). Visualization of position of vanadium forms has been carried out by extended x-ray absorption fine structure spectroscopy (Peacock and Sherman, 2004) and micro-x-ray fluorescence spectroscopy (Terzano et al., 2007). Oxidation states in soil samples have been determined by micro-x-ray absorption near edge structure spectroscopy (Terzano et al., 2007). Vanadium as a contaminant in fertilizers has been measured by ion interaction reversed phase HPLC (Vachirapatama et al., 2002).

Further details of methods used for the study of vanadium in environmental samples are given by Chen and Owens (2008).

REFERENCES

Agnieszka, J. and G. Barbara. 2012. Chromium, nickel and vanadium mobility in soils derived from fluvioglacial sands. *J. Hazard. Mater.* 237–238:315–322.
Avula, B., Y.-H. Wang, T.J. Smillie, N.S. Duzgoren-Aydin, and I.A. Khan. 2010. Quantitative determination of multiple elements in botanicals and dietary supplements using ICP-MS. *J. Agric. Food Chem.* 58:8887–8894.
Baken, S., M.A. Larsson, J.P. Gustafsson, F. Cubadda, and E. Smolders. 2012. Ageing of vanadium in soils and consequences for bioavailability. *Eur. J. Soil Sci.* 63:839–847.
Bakircioglu, D., Y.B. Kurtulus, and H. Ibar. 2011. Investigation of trace elements in agricultural soils by BCR sequential extraction method and its transfer to wheat plants. *Environ. Monit. Assess.* 175:303–314.

Cannon, H.L. 1963. The biogeochemistry of vanadium. *Soil Sci.* 96:196–204.

Cappuyns, V. and E. Slabbinck. 2012. Occurrence of vanadium in Belgian and European alluvial soils. *App. Environ. Soil Sci.* 2012:1–12.

Chen, Z.L. and G. Owens. 2008. Trends in speciation analysis of vanadium in environmental samples and biological fluids—A review. *Anal. Chim. Acta* 607:1–14.

Fargašová, A. and E. Beinrohr. 1998. Metal-metal interactions in accumulation of V^{5+}, Ni^{2+}, Mo^{6+}, Mn^{2+} and Cu^{2+} in under- and above-ground parts of *Sinapis alba*. *Chemosphere* 36:1305–1317.

Flogeac, K., E. Guillon, M. Aplincourt et al. 2005. Characterization of soil particles by X-ray diffraction (XRD), X-ray photoelectron spectroscopy (XPS), electron paramagnetic resonance (EPR) and transmission electron microscopy (TEM). *Agron. Sustain. Dev.* 25:345–353.

Gäbler, H.-E., K. Glüh, A. Bahr, and J. Utermann. 2009. Quantification of vanadium adsorption by German soils. *J. Geochem. Explor.* 103:37–44.

Hassinen, V., V.-M. Vallinkoski, S. Issakainen, A. Tervahauta, S. Kärenlampi, and K. Servomaa. 2009. Correlation of foliar *MT2b* expression with Cd and Zn concentrations in hybrid aspen (*Populus tremula* × *tremuloides*) grown in contaminated soil. *Environ. Pollut.* 157:922–930.

Hou, M., C. Hu, L. Xiong, and C. Lu. 2013. Tissue accumulation and subcellular distribution of vanadium in *Brassica juncea* and *Brassica chinensis*. *Microchem. J.* 110:575–578.

Khan, S., T.G. Kazi, N.F. Kolachi et al. 2011. Hazardous impact and translocation of vanadium (V) species from soil to different vegetables and grasses grown in the vicinity of thermal power plant. *J. Hazard. Mater.* 190:738–743.

Kohno, Y. 1986. Vanadium induced manganese toxicity in bush bean plants grown in solution culture. *J. Plant Nutr.* 9:1261–1272.

Larsson, M.A., S. Baken, J.P. Gustafsson, G. Hadialhejazi, and E. Smolders. 2013. Vanadium bioavailability and toxicity to soil microorganisms and plants. *Environ. Toxicol. Chem.* 32:2266–2273.

Lin, C.-W., C.-Y. Lin, C.-C. Chang, R.-H. Lee, T.-M. Tsai, P.-Y. Chen, W.-C. Chi, and H.-J. Huang. 2009. Early signalling pathways in rice roots under vanadate stress. *Plant Physiol. Biochem.* 47:369–376.

Lin, C.-Y., N.N. Trinh, C.-W. Lin, and H.-J. Huang. 2013. Transcriptome analysis of phytohormones, transporters and signaling pathways in response to vanadium stress in rice roots. *Plant Physiol. Biochem.* 66:98–104.

Madrakian, T., A. Afkhami, R. Siri, and M. Mohammadnejad. 2011. Micelle mediated extraction and simultaneous spectrophotometric determination of vanadium (V) and molybdenum (VI) in plant foodstuff samples. *Food Chem.* 127:769–773.

Malandrino, M., O. Abollino, S. Buoso, A. Giacomino, C. La Gioia, and E. Mentasti. 2011. Accumulation of heavy metals from contaminated soil to plants and evaluation of soil remediation by vermiculite. *Chemosphere* 82:169–178.

Mikkonen, A. and J. Tummavuori. 1994. Retention of vanadium (V) by three Finnish mineral soils. *Eur. J. Soil Sci.* 45:361–368.

Morrell, B.G., N.W. Lepp, and D.A. Phipps. 1986. Vanadium uptake by higher plants: Some recent developments. *Environ. Geochem. Health* 8:14–18.

Nowakowski, W. 1993. Vanadium bioaccumulation in *Pisum sativum* seedlings. *Biol. Plant.* 35:461–465.

Peacock, C.L. and D.M. Sherman. 2004. Vanadium (V) adsorption onto goethite (α-FeOOH) at pH 1.5 to 12: A surface complexation model based on ab initio molecular geometries and EXAFS spectroscopy. *Geochim. Cosmochim. Acta* 68:1723–1733.

Pilbeam, D.J. and K. Drihem. 2007. Vanadium. In *Handbook of Plant Nutrition*, eds. A.V. Barker and D.J. Pilbeam, pp. 585–596. Boca Raton, FL: CRC Press.

Połedniok, J. and F. Buhl. 2003. Speciation of vanadium in soil. *Talanta* 59:1–8.

Qian, Y., F.J. Gallagher, H. Feng, M. Wu, and Q. Zhu. 2014. Vanadium uptake and translocation in dominant plant species on an urban coastal brownfield site. *Sci. Tot. Environ.* 476/7:696–704.

Sandström, H., S. Reeder, A. Bartha et al. 2005. Sample preparation and analysis. In *Geochemical Atlas of Europe. Part 1. Background Information, Methodology and Maps*, ed. R. Salminen. Espoo, Finland: Geological Survey of Finland.

Smoleń, S., W. Sady, and J. Wierzbińska. 2012. The influence of nitrogen fertilization with ENTEC-26 and ammonium nitrate on the concentration of thirty-one elements in carrot (*Daucus carota* L.) storage roots. *J. Elementol.* 17:115–137.

Teng, Y., J. Yang, Z. Sun, J. Wang, R. Zuo, and J. Zheng. 2011. Environmental vanadium distribution, mobility and bioaccumulation in different land-use Districts in Panzhihua Region, SW China. *Environ. Monit. Assess.* 176:605–620.

Terzano, R., M. Spagnuolo, B. Vekemans et al. 2007. Assessing the origin and fate of Cr, Ni, Cu, Zn, Pb, and V in industrial polluted soil by combined microspectroscopic techniques and bulk extraction methods. *Environ. Sci. Technol.* 41:6762–6769.

Tessier, A., P.G.C. Campbell, and M. Bisson. 1979. Sequential extraction procedure for the speciation of particulate trace metals. *Anal. Chem.* 51:844–851.

UK Government Expert Group on Vitamins and Minerals. 2003. Safe upper levels for vitamins and minerals. http://multimedia.food.gov.uk/multimedia/pdfs/vitmin2003.pdf, Accessed on February 13, 2014.

University of Maryland Medical Center. 2013. Vanadium. http://umm.edu/health/medical/altmed/supplement/vanadium, Accessed on February 13, 2014.

Vachirapatama, N., G. Dicinoski, A.T. Townsend, and P.R. Haddad. 2002. Determination of vanadium as PAR-hydrogen peroxide complex in fertilisers by ion-interaction PR-HPLC. *J. Chromatogr. A* 956:221–227.

Vachirapatama, N., Y. Jirakiattikul, G. Dicinoski, A.T. Townsend, and P.R. Haddad. 2005. On-line preconcentration and sample clean-up system for the determination of vanadium as a 4-(2-pyridylazo)resorcinol-hydrogen peroxide ternary complex in plant tissues by ion-interaction high performance liquid chromatography. *Anal. Chim. Acta* 543:70–76.

Vachirapatama, N., Y. Jirakiattikul, G. Dicinoski, A.T. Townsend, and P.R. Haddad. 2011. Effect of vanadium on plant growth and its accumulation in plant tissues. *Songklanakarin J. Sci. Technol.* 33:255–261.

Welch, R.M. 1973. Vanadium uptake by plants. Absorption kinetics and the effects of pH, metabolic inhibitors, and other anions and cations. *Plant Physiol.* 51:828–832.

Section V

Conclusion

23 Conclusion

Allen V. Barker and David J. Pilbeam

CONTENTS

Detailed information is given in the individual chapters about current knowledge on each of the nutrient and beneficial elements used by plants. The chapters were written between June 2013 and April 2014, so "current" in this context refers to that time period. Comparison of the chapters with the equivalent ones in the first edition gives some indication of how research in plant nutrition has moved on since 2007, and although there are mostly different contributors of the chapters in this second edition, there are also changes in the way that each element is covered due to personal choice. It is also possible to draw out from the chapters common themes that link work on the different elements.

One of the main changes from the first edition is that of our knowledge of the genetics of plant nutrition. Many of the chapters give details of not only protein transporters of ions taken up by plants but also of the genes that code for these proteins and how the expression of those genes is switched on or off by environmental and physiological cues. Some of these cues are the individual nutrients themselves, which not only switch on genes for different transporter proteins but which can modulate root system architecture (Giehl et al., 2014). From the number of papers that appear on this subject every month, it is obvious that our knowledge of genetic control of nutrient acquisition and use will increase rapidly, and in fact will have already done so between the writing of the chapters and publication of this book. That information in turn will lead to rational methods in breeding crops for nutrient-use efficiency.

Many of the transporters are highly specific for particular ions or molecules, but others are more general and are featured in more than one chapter in this book. For example, aquaporins (water channels) seem to have a role in the uptake of nitrate, boron, and silicon, as well as water, and possibly also selenite. Transporters in the NIP subfamily of aquaporins also can take up toxic arsenite (As(III)) (Ma et al., 2008) and antimonite (Sb(III)) (Kamiya and Fujiwara, 2009). In general, elements that have forms with similar sizes and charges are likely to be transported and metabolized by the same pathway in plants (Baxter and Dilkes, 2012).

Other transporters that are important for the uptake of more than one element include ZIP transporters, which are involved in the movement of transition metals such as copper, iron, manganese, nickel, and zinc. P-type heavy metal ATPases transport copper and zinc, among other heavy metals, across membranes. One group moves monovalent copper and silver, another zinc, cobalt, cadmium and lead, and divalent copper. Transporters in the NRAMP family seem to be involved in transport of iron, manganese, and zinc (Merlot et al., 2014). As nickel and iron are transition metals, it is hardly surprising that nickel is taken up by the iron transporter IRT1, as is manganese. The high-affinity sulfate transporter SULTR1;2 also takes up selenate, so there is interaction between uptake of selenium and sulfur by plants, and then further interaction during their assimilation. Sulfur also interacts with molybdenum, as the divalent sulfate and molybdate anions have similar properties and compete for uptake. Cation-chloride cotransporters allow the translocation of Cl^-, K^+, and Na^+ ions in plants (Colmenero-Flores et al., 2007). This effect may be why supply of KCl to crops gives a higher K^+ content of the plants than supply of K_2SO_4.

Potassium and sodium ions have similar uptake mechanisms, with both being taken up by nonselective cation channels (NSCCs), low-affinity cation transporters, and high-affinity K^+ transporters. Voltage-independent NSCCs contain a group of channels that may allow uptake of NH_4^+, K^+, Na^+, Ca^{2+}, Mg^{2+}, Mn^{2+}, and Zn^{2+} (Demidchik and Maathuis, 2007), so they could be important for uptake of a large range of plant nutrients. However, they could also allow uptake of harmful Cs^+, Pb^{2+}, Hg^+, and Cd^{2+} ions (Demidchik and Maathuis, 2007). Further interaction between elements in their uptake comes from the fact that phytosiderophores released by strategy 2 plants as a response to iron deficiency can also bind to metals other than iron. The elements are taken up by plants as complexes with the phytosiderophore mugineic acid by transporters in the yellow stripe-like (YSL) family, and may include Co^{2+}, Cu^{2+}, Mn^{2+}, Ni^{2+}, and Zn^{2+} ions as well as Fe^{3+} (Haydon and Cobbett, 2007).

It is shown in many of the chapters that large numbers of the transporters involved in movement of nutrient ions are not involved in uptake of ions into roots, but in movement of the ions within the plant once they have been taken up. The transition elements cobalt, copper, and vanadium have restricted movement to the shoots, and so accumulate in the roots. Some transition elements, for example, Cu, Fe, Mn, and Zn, are not very mobile in plants, so deficiency symptoms do not tend to show on old leaves but only on younger leaves. Heavy metal ions taken up by the plants end up being sequestered in the vacuole, presumably as a defense against their potential interference in metabolic processes, and similar mechanisms for them entering the vacuoles occur for a range of ions. For example, the P_{1B}-ATPase transporter, heavy metal associated 3 (HMA3), is able to transport Zn^{2+}, Co^{2+}, Cd^{2+}, and Pb^{2+} into vacuoles in *Arabidopsis thaliana* (Morel et al., 2009). HMA2 and HMA4 are plasmalemma-bound transporters involved in translocation of Zn^{2+} and Cd^{2+} from roots to shoots, and HMA3 and HMA4 seem to have a major role in the sequestration of heavy metals in the shoots of hyperaccumulator species (Park and Ahn, 2014).

Metal ions taken up by plants are also rendered metabolically inactive by immobilization with various ligands, including phytochelatins, metallothioneins, nicotianamine, and also organic ions such as citrate and malate and histidine (either as the free amino acid or histidine residues in proteins) (Haydon and Cobbett, 2007). Phytochelatins, which are oligomers of glutathione, make complexes with metals, and these complexes are transported from the cytoplasm into the vacuole, where they give less interference in metabolic processes. Exposure of plants to cadmium causes enhancement of sulfur uptake and more production of phytochelatins, with phytochelatin synthase being upregulated by a number of trace elements. Metallothioneins are low-MW proteins with primary structures rich in highly conserved motifs that give an ability to bind mono- and divalent metal ions, in particular ions of Cu, Zn, and Cd (Hassinen et al., 2011). Metal ions such as Co^{2+} bind to the sulfur (R–SH) of cysteine residues of these proteins (Leszczyszyn et al., 2013). Fe^{2+} and Zn^{2+} are transported by an iron transporter, and accumulation of iron and zinc in shoot vacuoles is affected by the chelating compound nicotianamine (Baxter and Dilkes, 2012). This formation of complexes is important in transport, and metals complexed with siderophores and nicotianamine may be transported over long distances through plants by YSL transporters (Yruela, Chapter 10, Wood, Chapter 14, this book; Haydon and Cobbett, 2007). These authors point out that only small proportions of the cellular ion content of copper and other metals are likely to be present as free ions.

Recent analysis of the patterns of expression of eight metallothionein genes in barley indicates that these proteins should not be thought of as merely having a protective function against heavy metal toxicity, but rather that they are important in regulating tissue concentrations of micronutrients during growth and development (Schiller et al., 2014). Metallothioneins seem to be important in the redistribution of copper from senescing leaves into seeds (Benatti et al., 2014), something that may be important in attempts to carry out genetic biofortification of crop species.

Another function of metallothioneins is that they can act as powerful antioxidants, and they may help protect plants against reactive oxygen species (ROS) that the presence of transition elements in the tissues causes to form (Hassinen et al., 2011). Such formation of ROS can be seen following exposure of plants to vanadium and also is seen following exposure to selenium (although Se is not a transition element). In fact, selenium seems to cause reactions in plants similar to those caused by transition elements, and supply of selenite induces an isozyme of glutathione reductase that is also induced by the supply of Ni or Cd (Gomes-Junior et al., 2007). Selenite and vanadate toxicities can cause an increase in jasmonic acid and other stress-signaling plant growth regulators (PGRs) (e.g., for vanadate, Lin et al., 2013). The formation of ROS due to deficiency of magnesium and the functioning of antioxidant systems is discussed in detail in Chapter 6.

Common use of transporters by different ions can give rise to competition between nutrients for uptake. Other interactions between nutrient elements occur where different ions can be alternative substrates for assimilatory enzymes. Examples of such physiological interactions are given in the individual chapters. Other interactions between different elements may occur in the soil, for example, as competition for adsorption or ion exchange sites. Selenate and sulfate ions compete for adsorption sites in soils, although to a lesser extent than selenite and phosphate ions. Supply of Al^{3+} and Mn^{2+} ions displaces other cations from cation exchange sites in the soil. Other soil interactions between essential and beneficial elements can occur. Liming of soils supplies calcium and also increases soil pH. This amendment lowers the availability of potentially harmful aluminum but increases the availability of molybdenum to plants. Supply of lime also increases silicon concentration in the rooting environment.

Uptake and translocation of individual elements can give rise to different concentrations of each element in plant tissues. This accumulation forms the basis of the tables of nutrient concentrations in plants in many of the chapters. As can be seen in Chapter 2, there is a direct relationship between internal nitrogen concentration and rate of plant growth, and a plant given low rates of N supply remains as a small plant. In an experiment on *Salix* genotypes, fertilization and watering gave increased concentrations of N (along with P, S, Mn, and Cu) in the leaves and greater leaf biomass (Ågren and Weih, 2012). The P, S, and Mn were in relatively constant ratio to the N in the leaves. However, K, Ca, and Mg were taken up in direct proportion to leaf growth, so their concentrations were not affected by the watering and fertilization regime and were the same in the larger dry biomass of these plants as in the untreated plants. This result means that these three elements each had a decreased concentration relative to nitrogen (i.e., K/N, Ca/N, and Mg/N ratios were lower in the watered and fertilized plants). Concentrations of Fe, B, Zn, and Al decreased in the fertilized, watered plants (Ågren and Weih, 2012). It is therefore apparent that environmental factors can cause noticeable differences in proportions of different elements in plants, something that is shown in the chapters in this book.

Although increased uptake of N, P, S, and some other elements can have a direct, positive impact on relative growth rates across a big range of concentrations, many micronutrients appear to give rise to hormesis. This response is the effect whereby low doses of toxins (or radiation) may give beneficial effects to the recipients, even though higher doses would be dangerous. Whether or not potential toxins can exert a beneficial effect is controversial in human medicine and toxicology, but it is a common response in plant nutrition that many micronutrients and beneficial elements are toxic to plants in concentrations above a threshold level. This action for beneficial elements is discussed in detail for lanthanides in Chapter 18 and is an addition in this book. As discussed earlier, exposure to transition metals commonly results in plants producing enhanced levels of ROS, and it seems plausible that exposure to low concentrations of these potential toxins switches on the antioxidant defenses of plants, so improving their growth, even though higher concentrations of the elements would overwhelm these defenses and interfere with growth processes (Poschenrieder et al., 2013).

Some of the transition metals that are essential or beneficial to plant growth at low rates of supply and are potentially toxic at higher rates of supply are essential elements in human nutrition yet are in short supply in the diet. Provision of these elements in the human diet has been affected adversely by changes in food preferences and through advances in agricultural practice. A crop receiving a high rate of supply of N-P-K fertilizers has the potential to achieve a high yield, and this response can dilute micronutrient concentrations in crops. Additionally, even if micronutrient concentrations in a crop remain high after application of NPK fertilizers, these micronutrients are removed at harvest, and this nutrient mining can lower their availability in the soil for subsequent crops (Moran, 2011).

Furthermore, it seems as if growing C3 crops in the high concentrations of atmospheric CO_2 predicted for the middle of this century will give lower concentrations of zinc and iron in the harvested products by what is known as the "dilution effect" of dry matter accumulation on nutrient concentrations (Myers et al., 2014). As discussed in Chapter 1, nutrient mining may be less of a problem in newer soils than in soils that are old and weathered, but it is obvious from many of the chapters that there is increasing interest in growing crops with enhanced levels of micronutrients for livestock and human consumption, either by increasing the efficiency of uptake and transport of the ions into edible tissues or by increasing the amount of bioavailable mineral accumulation in the plant (Hirschi, 2009). This increased accumulation can be brought about by plant breeding (genetic biofortification) or through changes to agricultural practices (agronomic biofortification). Both are discussed in many of the chapters in this book, and it is also apparent that not only micronutrient concentrations are affected by dilution in crops but also steps need to be taken to increase concentrations of some macronutrients, such as magnesium.

Genetic biofortification may involve the manipulation of transporters of the elements in question, particularly in terms of loading them into the fruits and seeds that form major parts of human diets. Enhancing the movement of Fe^{2+} into rice grains by altering expression of the iron–nicotianamine *Oryza sativa* yellow stripe-like 2 transporter, or increasing production of phytosiderophores and nicotianamine in the plants, has given experimental plants with a three- to fourfold higher iron concentration in polished grains (Schroeder et al., 2013). Because chelators such as nicotianamine are involved in the uptake and translocation of a range of metal ions, manipulation of their synthesis can improve uptake of more than one element that is in short supply in the human diet, and increased expression of the enzyme that synthesizes nicotianamine can increase concentrations of zinc in rice grains as well as iron (Schroeder et al., 2013). The HarvestPlus program, which aims to increase concentrations of iron, zinc, and provitamin A in staple food crops, relies on conventional breeding for this, thus avoiding public opposition that genetic modification technology attracts in some countries (Hirschi, 2009). It is important to note that to date, there have not been detailed studies investigating the effects of biofortified crop products on human populations (Hirschi, 2009). Supply of lanthanides to plants has been shown to increase their concentrations of ascorbic acid, vitamin C, sugars, proteins, and other compounds deemed desirable in the human diet. Supply of selenium maintains concentrations of tocopherol in plants, making crops well supplied with selenium a more valuable vitamin E source for animals and man. This effect gives a potential health benefit, as does the role of selenium in removing peroxides through its presence as a component of glutathione peroxidases.

There is ongoing interest in the importance of macronutrients in agriculture, as well as in agronomic biofortification with micronutrients. This knowledge could lead to more efficient use of fertilizers. More efficient fertilizer use is important to save resources used in their manufacture and to avoid pollution problems from their overuse, issues that are covered in detail in relation to phosphorus in Chapter 3. It seems to be the case that agronomic N-use efficiency (crop yield per unit of N available) is more affected by physiological N-use efficiency (yield per unit of N acquired) than by N-acquisition efficiency (N acquired per unit of N available) (White et al., 2013). For example, in a study of modern high-yielding rice cultivars in Japan, their grain yield was 1.2–1.7 times higher than the grain yield of conventional cultivars, and this increase was matched by a 1.3–1.5 times higher efficiency for sink capacity per unit of total N in the shoots at maturity (Mae, 2011). By contrast, agronomic P-use efficiency generally relies on P-acquisition efficiency, and efficiency in use of K to produce crops owes more to the efficiency in K uptake rather than the efficiency with which it is utilized in the plants (White et al., 2013). Many nutrients are associated with the organic matter and negatively charged particles in the upper layers of the soil, so breeding for shallow root systems may be a good idea for low-input agriculture, but nitrate and sulfate can be leached down through the soil profile, so a deeper rooting plant ought to be able to utilize these ions more efficiently (Postma et al., 2014). It has been suggested that better use of P can be obtained by breeding crops with roots that proliferate in the top soil, whereas N use efficiency would be improved by having roots that go down steeply to deeper positions in the soil and that are metabolically cheap to produce

(White et al., 2013). Potassium-use efficiency would be improved by breeding for root systems intermediate between these two extremes (White et al., 2013).

Improvements in the efficiency of use of macronutrient fertilizers, not only to increase crop yields but also to minimize the oversupply of nutrients that are not taken up by a crop and are lost into the environment, are still possible. In a long-term study of a wheat–sugar beet rotation in northern France, 45%–50% of ^{15}N added in N fertilizers was taken up by the crop in the first year, with 32%–37% being incorporated into soil organic matter within the first 3 years (Sebilo et al., 2013). The N in the organic matter decreased with time, so that by 27 years after application, 61%–65% of the ^{15}N had been taken up by the crops, and 8%–12% had been leached out after nitrification of the organic matter. Agricultural systems in northern France are efficiently managed, and fertilizer use has developed after years of research, but in many other countries, crop yields have been increased rapidly through overuse of fertilizers, and there is scope for more efficient supply of nutrients without compromising crop yields (see Chapter 1). In other countries, yields are still held back by low application of fertilizers and the resulting poor soil fertility.

Some of the chapters cover diagnosis of nutrient deficiency in crops and the modeling of such deficiency as a tool to aid in fertilizer recommendations. Increased access to remote imaging by systems, such as satellites and aircraft or land-based vehicles, is making it more routine for farmers to evaluate the nutrient requirement of their crops, and the availability of Global Positioning Systems is enabling them to use remote sensing data in order to apply nutrients differentially within fields. At the same time, handheld sensors are being developed for small-scale farmers. However, work still is required to formulate indices of growth of different crops in relation to availability of different nutrients in the soil and how these can be generated by measurement of reflectance of different wavelengths of radiation from the crops (e.g., Kawamura et al., 2011). Future advances in sensing of nutrient status and modeling its impact on plant growth will enable farmers not only to supply more nutrients to crops in areas where crop yields have been depressed by poor soil fertility but also to avoid waste of nutrients in parts of the world where yields are not limited by availability of nutrients but fertilizers are oversupplied and cause environmental problems due to their surplus.

REFERENCES

Ågren, G.I. and M. Weih. 2012. Åeferenceslus.ause environmental problems due re crop yields have been dElement concentration patterns reflect environment more than genotype. *New Phytol.* 194:944–952.

Baxter, I. and B.P. Dilkes. 2012. Elemental profiles reflect plant adaptations to the environment. *Science* 336:1661–1663.

Benatti, M.R., N. Yookongkaew, M. Meetam et al. 2014. Metallothionein deficiency impacts copper accumulation and redistribution in leaves and seeds of *Arabidopsis*. *New Phytol.* 202:940–951.

Colmenero-Flores, J.M., G. Martínez, G. Gamba et al. 2007. Identification and functional characterization of cation-chloride cotransporters in plants. *Plant J.* 50:278–292.

Demidchik, V. and F.J.M. Maathuis. 2007. Physiological roles of nonselective cation channels in plants: From salt stress to signalling and development. *New Phytol.* 175:387–404.

Giehl, R.F.H., B.D. Gruber, and N. von Wirén. 2014. It's time to make changes: Modulation of root system architecture by nutrient signals. *J. Exp. Bot.* 65:769–778.

Gomes-Junior, R.A., P.L. Gratão, S.A. Gaziola, P. Mazzafera, P.J. Lea, and R.A. Azevedo. 2007. Selenium-induced oxidative stress in coffee cell suspension cultures. *Funct. Plant Biol.* 34:449–456.

Hassinen, V.H., A.I. Tervahauta, H. Schat, and S.O. Kärenlampi. 2011. Plant metallothioneins—metal chelators with ROS scavenging activity? *Plant Biol.* 13:225–232.

Haydon, M.J. and C.S. Cobbett. 2007. Transporters of ligands for essential metal ions in plants. *New Phytol.* 174:499–506.

Hirschi, K.D. 2009. Nutrient biofortification of food crops. *Annu. Rev. Nutr.* 29:401–421.

Kamiya, T. and T. Fujiwara. 2009. *Arabidopsis* NIP1;1 transports antimonite and determines antimonite sensitivity. *Plant Cell Physiol.* 50:1977–1981.

Kawamura, K., A.D. Mackay, M.P. Tuohy, K. Betteridge, I.D. Sanches, and Y. Inoue. 2011. Potential for spectral indices to remotely sense phosphorus and potassium content of legume-based pasture as a means of assessing soil phosphorus and potassium fertility status. *Int. J. Remote Sens.* 32:103–124.

Leszczyszyn, O.I., H.T. Imam, and C.A. Blindauer. 2013. Diversity and distribution of plant metallothioneins: A review of structure, properties and functions. *Metallomics* 5:1146–1169.

Lin, C.-Y., N.N. Trinh, C.-W. Lin, and H.-J. Huang. 2013. Transcriptome analysis of phytohormones, transporters and signaling pathways in response to vanadium stress in rice roots. *Plant Physiol. Biochem.* 66:98–104.

Ma, J.F., N. Yamaji, N. Mitani et al. 2008. Transporters of arsenite in rice and their role in arsenic accumulation in rice grain. *Proc. Natl. Acad. Sci. U.S.A.* 105:9931–9935.

Mae, T. 2011. Nitrogen acquisition and its relation to growth and yield in recent high-yielding cultivars of rice (*Oryza sativa* L.) in Japan. *Soil Sci. Plant Nutr.* 57:625–635.

Merlot, S., L. Hannibal, S. Martins et al. 2014. The metal transporter PgIREG1 from the hyperaccumulator *Psychotria gabriellae* is a candidate gene for nickel tolerance and accumulation. *J. Exp. Bot.* 65:1551–1564.

Moran, K. 2011. Role of micronutrients in maximising yields and in bio-fortification of food crops. *Proceedings International Fertiliser Society 702*, Leek, U.K.

Morel, M., J. Crouzet, A. Gravot et al. 2009. AtHMA3, a P_{1B}-ATPase allowing Cd/Zn/Co/Pb vacuolar storage in *Arabidopsis*. *Plant Physiol.* 149:894–904.

Myers, S.S., A. Zanobetti, I. Kloog et al. 2014. Increasing CO_2 threatens human nutrition. *Nature* 510:139–142.

Park, W. and S.-J. Ahn. 2014. How do heavy metal ATPases contribute to hyperaccumulation? *J. Plant Nutr. Soil Sci.* 177:121–127.

Poschenrieder, C., C. Cabot, S. Martos, B. Gallego, and J. Barceló. 2013. Do toxic ions induce hormesis in plants? *Plant Sci.* 212:15–25.

Postma, J.A., U. Schurr, and F. Fiorani. 2014. Dynamic root growth and architecture responses to limiting nutrient availability: Linking physiological models and experimentation. *Biotech. Adv.* 32:53–65.

Schiller, M., Hegelund, J.N., Pedas, P. et al. 2014. Barley metallothioneins differ in ontogenetic pattern and response to metals. *Plant Cell Environ.* 37:353–367.

Schroeder, J.I., E. Delhaize, W.B. Frommer et al. 2013. Using membrane transporters to improve crops for sustainable food production. *Nature* 497:60–66.

Sebilo, M., B. Mayer, B. Nicolardot, G. Pinay, and A. Mariotti. 2013. Long-term fate of nitrogen fertilizer in agricultural soils. *Proc. Natl. Acad. Sci. U.S.A.* 110:18185–18189.

White, P.J., T.S. George, P.J. Gregory, A.G. Bengough, P.D. Hallett, and B.M. McKenzie. 2013. Matching roots to their environment. *Ann. Bot.* 112:207–222.

Index

Printed in the United States
by Baker & Taylor Publisher Services